KIRK-OTHMER

ENCYCLOPEDIA OF CHEMICAL TECHNOLOGY

FOURTH EDITION

VOLUME **1**

A

TO

ALKALOIDS

EXECUTIVE EDITOR
Jacqueline I. Kroschwitz

Editor
Mary Howe-Grant

KIRK-OTHMER

ENCYCLOPEDIA OF CHEMICAL TECHNOLOGY

FOURTH EDITION

VOLUME **1**

A
TO
ALKALOIDS

A Wiley-Interscience Publication
JOHN WILEY & SONS
New York • Chichester • Brisbane • Toronto • Singapore

Library of Congress Cataloging-in-Publication Data

Encyclopedia of chemical technology / executive editor, Jacqueline
   I. Kroschwitz; editor, Mary Howe-Grant.—4th ed.
      p. cm.
    At head of title: Kirk-Othmer.
    "A Wiley-Interscience publication."
    Includes index.
    Contents: v. 1. A to alkaloids.
    ISBN 0-471-52669-X (v. 1)
    1. Chemistry, Technical—Encyclopedias. I. Kirk, Raymond E.
(Raymond Eller), 1890–1957. II. Othmer, Donald F. (Donald
Frederick), 1904– . III. Kroschwitz, Jacqueline I., 1942– . IV. Howe-Grant,
Mary, 1943– . V. Title: Kirk-Othmer encyclopedia of chemical
technology.
  Ref TP9.E685 1991 v.1   91-16789
  660'.03—dc20   CIP

Printed in the United States of America

CONTENTS

EDITORIAL STAFF
FOR VOLUME 1

Executive Editor: **Jacqueline I. Kroschwitz**
Editor: **Mary Howe-Grant**
Editorial Supervisor: **Michalina Bickford**
Copy Editor: **Lee Gray**

CONTRIBUTORS
TO VOLUME 1

Lawrence E. Ball, *BP Research, Cleveland, Ohio,* Survey and SAN (under Acrylonitrile polymers)

William Bauer, Jr., *Rohm and Haas Company, Spring House, Pennsylvania,* Acrylic acid and derivatives

Tilak V. Bommaraju, *Occidental Chemical Corporation, Niagara Falls, New York,* Chlorine and sodium hydroxide (under Alkali and chlorine products)

John C. Bost, *The Dow Chemical Company, Midland, Michigan,* Halogenated derivatives (under Acetic acid and derivatives)

James F. Brazdil, *BP Research, Cleveland, Ohio,* Acrylonitrile

Angelo Brisimitzakis, *GE Plastics, Pittsfield, Massachusetts,* ABS resins (under Acrylonitrile polymers)

Kenneth W. Cooper, *Poolpak, Inc., Seven Valleys, Pennsylvania,* Air conditioning

Burton B. Crocker, *Consultant, Chesterfield, Missouri,* Air pollution control methods

Benedict S. Curatolo, *BP Research, Cleveland, Ohio,* Survey and SAN (under Acrylonitrile polymers)

L. Calvert Curlin, *OxyTech Systems, Inc., Chardon, Ohio,* Chlorine and sodium hydroxide (under Alkali and chlorine products)

David R. Dalton, *Temple University, Philadelphia, Pennsylvania,* Alkaloids

Darwin D. Davis, *E. I. du Pont de Nemours & Co., Inc., Victoria, Texas,* Adipic acid

W. G. Etzkorn, *Union Carbide Chemicals & Plastics Company Inc., South Charleston, West Virginia,* Acrolein and derivatives

Michael Favaloro, *Textron Defense Systems, Gloucester, Massachusetts,* Ablative materials

Leroy W. Fritch, Jr., *GE Plastics, Pittsfield, Massachusetts,* ABS resins (under Acrylonitrile polymers)

Stanley A. Gembicki, *UOP, Des Plaines, Illinois,* Adsorption, liquid separation

C. E. Habermann, *Dow Chemical, USA, Midland, Michigan,* Acrylamide

H. J. Hagemeyer, *Texas Eastman Company, Longview, Texas,* Acetaldehyde

Constance B. Hansson, *Occidental Chemical Corporation, Dallas, Texas,* Chlorine and sodium hydroxide (under Alkali and chlorine products)

Eugene V. Hort, *GAF Corporation, Wayne, New Jersey,* Acetylene-derived chemicals

William L. Howard, *The Dow Chemical Company, Freeport, Texas,* Acetone

William N. Hunter, *Celanese Canada Inc., Edmonton, Alberta, Canada,* Alcohols, polyhydric

Michael D. Jackson, *Acurex Corporation, Mountain View, California,* Alcohol fuels

James A. Johnson, *UOP, Des Plaines, Illinois,* Adsorption, liquid separation

Donald R. Kemp, *E. I. du Pont de Nemours & Co., Inc., Victoria, Texas,* Adipic acid

Joseph Kozakiewicz, *American Cyanamid Company, Stamford, Connecticut,* Acrylamide polymers

Donald M. Kulich, *GE Plastics, Pittsfield, Massachusetts,* ABS resins (under Acrylonitrile polymers)

J. J. Kurland, *Union Carbide Chemicals & Plastics Company Inc., South Charleston, West Virginia,* Acrolein and derivatives

George R. Lappin, *Ethyl Corporation, Baton Rouge, Louisiana,* Synthetic processes (under Alcohols, higher aliphatic)

P. A. Larson, *E. I. du Pont de Nemours & Co., Inc., Wilmington, Delaware,* Dimethylacetamide (under Acetic acid and derivatives)

David Lipp, *American Cyanamid Company, Stamford, Connecticut,* Acrylamide polymers

R. P. Lukens, *ASTM Committee E-43 on SI Practice, Woodbury, New Jersey,* Conversion Factors, Abbreviations, and Unit Symbols

David J. Miller, *Union Carbide Chemicals & Plastics Corporation, South Charleston, West Virginia,* Aldehydes

Earl D. Morris, *The Dow Chemical Company, Midland, Michigan,* Halogenated derivatives (under Acetic acid and derivatives)

Carl B. Moyer, *Acurex Corporation, Mountain View, California,* Alcohol fuels

Samuel Natansohn, *GTE Laboratories, Inc., Waltham, Massachusetts,* Structural ceramics (under Advanced ceramics)

Jeffrey T. Neil, *GTE Laboratories, Inc., Waltham, Massachusetts,* Structural ceramics (under Advanced ceramics)

W. D. Neilsen, *Union Carbide Chemicals & Plastics Company Inc., South Charleston, West Virginia,* Acrolein and derivatives

Robert E. Newnham, *Pennsylvania State University, University Park, Pennsylvania,* Electronic ceramics (under Advanced ceramics)

Alvin W. Nienow, *University of Birmingham, Birmingham, United Kingdom,* Biotechnology (under Aeration)

Ronald W. Novak, *Rohm and Haas Company, Spring House, Pennsylvania,* Survey (under Acrylic ester polymers)

J. T. O'Connor, *Loctite Corporation, Newington, Connecticut,* 2-Cyanoacrylic ester polymers (under Acrylic ester polymers)

Anil R. Oroskar, *UOP, Des Plaines, Illinois,* Adsorption, liquid separation

John E. Pace, *GE Plastics, Pittsfield, Massachusetts,* ABS resins (under Acrylonitrile polymers)

Marina R. Pascucci, *GTE Laboratories, Inc., Waltham, Massachusetts,* Structural ceramics (under Advanced ceramics)

Richard A. Peters, *The Procter & Gamble Company, Cincinnati, Ohio,* Survey and natural alcohols manufacture (under Alcohols, higher aliphatic)

Alphonsus V. Pocius, *The 3M Company, St. Paul, Minnesota,* Adhesives

Richard B. Rajendaen, *Aeromix Systems, Inc., Minneapolis, Minnesota,* Water treatment (under Aeration)

Frank Rauh, *FMC Corporation, Princeton, New Jersey,* Sodium carbonate (under Alkali and chlorine products)

Charles V. Rue, *Norton Company, Worcester, Massachusetts,* Abrasives

Douglas M. Ruthven, *University of New Brunswick, Fredricton, New Brunswick, Canada,* Adsorption

John J. Sciarra, *Sciarra Aeromed Development Corporation, Brooklyn, New York,* Aerosols

Glenn T. Seaborg, *University of California, Berkeley, Berkeley, California,* Actinides and transactinides

John D. Sherman, *UOP, Tarrytown, New York,* Adsorption, gas separation

Thomas R. Shrout, *Pennsylvania State University, University Park, Pennsylvania,* Electronic ceramics (under Advanced ceramics)

John B. Starr, *Hoechst Celanese Corporation, Summit, New Jersey,* Acetal resins

Paul Taylor, *GAF Corporation, Wayne, New Jersey,* Acetylene-derived chemicals

Urs von Stockar, *École Polytechnique Fédérale, Lausanne, Switzerland,* Absorption

Frank S. Wagner, Jr., *Nandina Corporation, Corpus Christi, Texas,* Acetamide; Acetic acid; Acetic anhydride; Acetyl chloride (all under Acetic acid and derivatives)

John D. Wagner, *Ethyl Corporation, Baton Rouge, Louisiana,* Synthetic processes (under Alcohols, higher aliphatic)

J. C. Watts, *E. I. du Pont de Nemours & Co., Inc., Wilmington, Delaware,* Dimethyl-acetamide (under Acetic acid and derivatives)

Charles R. Wilke, *University of California, Berkeley, Berkeley, California,* Absorption

George T. Wolff, *General Motors, Warren, Michigan,* Air pollution

Carmen M. Yon, *UOP, Tarrytown, New York,* Adsorption, gas separation

J. Richard Zietz, *Ethyl Corporation, Baton Rouge, Louisiana,* Synthetic processes (under Alcohols, higher aliphatic)

PREFACE

The Fourth Edition of the *Encyclopedia of Chemical Technology* is built on the solid foundation of the previous editions and also looks forward into the 21st century. The First Edition, published between 1949 and 1956, demonstrated the enormous progress the American chemical industry made during World War II and the postwar period. The Second Edition, published between 1963 and 1972, reflected the chemical industry as an international enterprise with the interchange of experience and know-how on a global scale. The Third Edition, published between 1978 and 1984, continued the presentation of the best worldwide practices in the process technologies and emphasized topics of great concern to all scientists and engineers, ie, energy, safety, and the environment.

The Fourth Edition is in many ways an entirely new encyclopedia in a format familiar to those acquainted with the earlier editions. All of the articles in this new edition have been rewritten and updated and many new subjects have been added, reflecting changes in chemical technology since the Third Edition. The results, however, will be familiar to the users of the earlier editions: comprehensive, authoritative, accessible, lucid.

The use of SI units as well as common units, Chemical Abstracts Services Registry Numbers, and complete indexing based on automated retrieval from a machine-readable composition system continue. The *Encyclopedia* is an indispensible information source for all producers and users of chemical products and materials.

New subjects have been added, especially in biotechnology, computer topics, analytical techniques and instrumentation, environmental concerns, fuels and energy, inorganic and solid state chemistry, composite materials and materials science in general, and pharmaceuticals.

The Fourth Edition could not be published without the assistance of the many advisors, reviewers, chemists, and engineers who have given their advice and guidance. To all of them our thanks is due.

INTRODUCTION

The main subject of the *Encyclopedia* is chemical technology, and about one-half of all the articles deal with chemical substances, either single substances, such as Sulfuric acid, or groups of substances, such as Aluminum compounds. There are also articles on industrial processes, such as Alkylation; on uses, such as Abrasives; Adhesives; on pharmaceuticals, dyes, fibers; on foods and other human uses, such as Cosmetics. There are articles on the unit operations and unit processes of chemical engineering; on fundamentals, such as Absorption; Mass transfer; and on scientific and technological subjects, such as Catalysis, Color, Electrochemical processing, Magnetic materials, and Ultrasonics. Still other articles deal with such general subjects as Computer technology, Information retrieval, Patents, Regulatory agencies, Technical service, and Transportation.

In general, the properties and manufacture of any substance are given in one article, which makes cross reference to one or more articles where the uses of that substance are described. Thus the manufacture of fused alumina is described under Aluminum compounds, but for its uses the reader will be directed to such articles as Abrasives and Refractories.

For inorganic compounds, in some cases it is the anion, in others the cation that has the greater industrial significance. Thus calcium phosphate, sodium phosphate, and ammonium phosphate are important primarily as phosphates and are discussed under Phosphoric acid and phosphates. Similarly, chromates and borates are under Chromium compounds and Boron compounds, respectively, and salts of organic acids (except acetates and formates) are discussed with the acids. On the other hand, barium chloride, barium nitrate, and barium sulfate would be thought of together and are therefore described in Barium compounds. In general, compounds of the following anions are dealt with in articles such as Aluminum compounds and Calcium compounds: acetates; carbonates; formates; chlorides, bromides, and iodides (under halides); nitrates; nitrites; oxides (including hydroxides and oxygen acids and their salts, but excluding true peroxides); sulfates; sulfites; and sulfides. The organic compounds of a metal, containing a metal-to-carbon bond, are also discussed with the compounds of that metal or under

Organometallics. However, fluorine, in its industrial applications, is so different from the other halogens that the metallic fluorides are usually grouped together under Fluorine compounds, inorganic.

Organic compounds containing fluorine (with or without other halogens) are discussed under Fluorine compounds, organic. There are also articles on Bromine compounds and Iodine compounds. Chlorine is treated somewhat differently. The article Chlorocarbons and chlorohydrocarbons covers a large number of industrially important compounds; compounds containing other elements as well as carbon, hydrogen, and chlorine are sometimes grouped together (as, Chlorophenols; Chlorohydrins), sometimes treated as derivatives under a parent compound (thus chloroanilines appear as derivatives under Amines, aromatic, aniline).

In general, the treatment of a compound will be found either under its own name, or under a group of substances (for example, ethyl acetate under Esters, organic), or as a derivative under a parent compound (for example, ethyl acrylate under Acrylic acid and derivatives). The cross references provided will, it is hoped, in almost all cases direct the reader to the appropriate part of the *Encyclopedia*.

NOTE ON CHEMICAL ABSTRACTS SERVICE REGISTRY NUMBERS AND NOMENCLATURE

Chemical Abstracts Service (CAS) Registry Numbers are unique numerical identifiers assigned to substances recorded in the CAS Registry System. They appear in brackets in the *Chemical Abstracts* (CA) substance and formula indexes following the names of compounds. A single compound may have synonyms in the chemical literature. A simple compound like phenethylamine can be named β-phenylethylamine or, as in *Chemical Abstracts,* benzeneethanamine. The usefulness of the *Encyclopedia* depends on accessibility through the most common correct name of a substance. Because of this diversity in nomenclature careful attention has been given to the problem in order to assist the reader as much as possible, especially in locating the systematic CA index name by means of the Registry Number. For this purpose, the reader may refer to the CAS Registry Handbook—Number Section which lists in numerical order the Registry Number with the *Chemical Abstracts* index name and the molecular formula; eg, **458-88-8,** Piperidine, 2-propyl-, (S)-, $C_8H_{17}N$; in the *Encyclopedia* this compound would be found under its common name, coniine [*458-88-8*]. Alternatively, this information can be retrieved electronically from CAS Online. In many cases molecular formulas have also been provided in the *Encyclopedia* text to facilitate electronic searching. The Registry Number is a valuable link for the reader in retrieving additional published information on substances and also as a point of access for on-line data bases.

In all cases, the CAS Registry Numbers have been given for title compounds in articles and for all compounds in the index. All specific substances indexed in *Chemical Abstracts* since 1965 are included in the CAS Registry System as are a large number of substances derived from a variety of reference works. The CAS Registry System identifies a substance on the basis of an unambiguous computer-language description of its molecular structure including stereochemical detail. The Registry Number is a machine-checkable number (like a Social Security number) assigned in sequential order to each substance as it enters the registry system. The value of the number lies in the fact that it is a concise and unique means of substance identification, which is independent of, and therefore bridges, many systems of chemical nomenclature. For polymers, one Registry Number may be used for the entire family; eg, polyoxyethylene (20) sorbitan monolaurate has the same number as all of its polyoxyethylene homologues.

Cross-references are inserted in the index for many common names and for some systematic names. Trademark names appear in the index. Names that are incorrect, misleading, or ambiguous are avoided. Formulas are given very frequently in the text to help in identifying compounds. The spelling and form used, even for industrial names, follow American chemical usage, but not always the usage of *Chemical Abstracts* (eg, *coniine* is used instead of *(S)-2-propylpiperidine, aniline* instead of *benzenamine,* and *acrylic acid* instead of *2-propenoic acid*).

There are variations in representation of rings in different disciplines. The dye industry does not designate aromaticity or double bonds in rings. All double bonds and aromaticity are shown in the *Encyclopedia* as a matter of course. For example, tetralin has an aromatic ring and a saturated ring and its structure

appears in the *Encyclopedia* with its common name, Registry Number enclosed in brackets, and parenthetical CA index name, ie, tetralin [*119-64-2*] (1,2,3,4-tetrahydronaphthalene). With names and structural formulas, and especially with CAS Registry Numbers, the aim is to help the reader have a concise means of substance identification.

CONVERSION FACTORS, ABBREVIATIONS, AND UNIT SYMBOLS

SI Units (Adopted 1960)

The International System of Units (abbreviated SI), is being implemented throughout the world. This measurement system is a modernized version of the MKSA (meter, kilogram, second, ampere) system, and its details are published and controlled by an international treaty organization (The International Bureau of Weights and Measures) (1).

SI units are divided into three classes:

BASE UNITS

length	meter[†] (m)
mass	kilogram (kg)
time	second (s)
electric current	ampere (A)
thermodynamic temperature[‡]	kelvin (K)
amount of substance	mole (mol)
luminous intensity	candela (cd)

SUPPLEMENTARY UNITS

plane angle	radian (rad)
solid angle	steradian (sr)

[†]The spellings "metre" and "litre" are preferred by ASTM; however "-er" is used in the *Encyclopedia*.

[‡]Wide use is made of Celsius temperature (*t*) defined by

$$t = T - T_0$$

where T is the thermodynamic temperature, expressed in kelvin, and $T_0 = 273.15$ K by definition. A temperature interval may be expressed in degrees Celsius as well as in kelvin.

DERIVED UNITS AND OTHER ACCEPTABLE UNITS

These units are formed by combining base units, supplementary units, and other derived units (2–4). Those derived units having special names and symbols are marked with an asterisk in the list below.

Quantity	Unit	Symbol	Acceptable equivalent
*absorbed dose	gray	Gy	J/kg
acceleration	meter per second squared	m/s^2	
*activity (of a radionuclide)	becquerel	Bq	1/s
area	square kilometer	km^2	
	square hectometer	hm^2	ha (hectare)
	square meter	m^2	
concentration (of amount of substance)	mole per cubic meter	mol/m^3	
current density	ampere per square meter	A/m^2	
density, mass density	kilogram per cubic meter	kg/m^3	g/L; mg/cm^3
dipole moment (quantity)	coulomb meter	C·m	
*dose equivalent	sievert	Sv	J/kg
*electric capacitance	farad	F	C/V
*electric charge, quantity of electricity	coulomb	C	A·s
electric charge density	coulomb per cubic meter	C/m^3	
*electric conductance	siemens	S	A/V
electric field strength	volt per meter	V/m	
electric flux density	coulomb per square meter	C/m^2	
*electric potential, potential difference, electromotive force	volt	V	W/A
*electric resistance	ohm	Ω	V/A
*energy, work, quantity of heat	megajoule	MJ	
	kilojoule	kJ	
	joule	J	N·m
	electronvolt[†]	eV[†]	
	kilowatt-hour[†]	kW·h[†]	
energy density	joule per cubic meter	J/m^3	
*force	kilonewton	kN	
	newton	N	$kg·m/s^2$

[†]This non-SI unit is recognized by the CIPM as having to be retained because of practical importance or use in specialized fields (1).

Quantity	Unit	Symbol	Acceptable equivalent
*frequency	megahertz	MHz	
	hertz	Hz	1/s
heat capacity, entropy	joule per kelvin	J/K	
heat capacity (specific), specific entropy	joule per kilogram kelvin	J/(kg·K)	
heat transfer coefficient	watt per square meter kelvin	W/(m²·K)	
*illuminance	lux	lx	lm/m²
*inductance	henry	H	Wb/A
linear density	kilogram per meter	kg/m	
luminance	candela per square meter	cd/m²	
*luminous flux	lumen	lm	cd·sr
magnetic field strength	ampere per meter	A/m	
*magnetic flux	weber	Wb	V·s
*magnetic flux density	tesla	T	Wb/m²
molar energy	joule per mole	J/mol	
molar entropy, molar heat capacity	joule per mole kelvin	J/(mol·K)	
moment of force, torque	newton meter	N·m	
momentum	kilogram meter per second	kg·m/s	
permeability	henry per meter	H/m	
permittivity	farad per meter	F/m	
*power, heat flow rate, radiant flux	kilowatt	kW	
	watt	W	J/s
power density, heat flux density, irradiance	watt per square meter	W/m²	
*pressure, stress	megapascal	MPa	
	kilopascal	kPa	
	pascal	Pa	N/m²
sound level	decibel	dB	
specific energy	joule per kilogram	J/kg	
specific volume	cubic meter per kilogram	m³/kg	
surface tension	newton per meter	N/m	
thermal conductivity	watt per meter kelvin	W/(m·K)	
velocity	meter per second	m/s	
	kilometer per hour	km/h	
viscosity, dynamic	pascal second	Pa·s	
	millipascal second	mPa·s	
viscosity, kinematic	square meter per second	m²/s	
	square millimeter per second	mm²/s	

Quantity	Unit	Symbol	Acceptable equivalent
volume	cubic meter	m^3	
	cubic decimeter	dm^3	L (liter) (5)
	cubic centimeter	cm^3	mL
wave number	1 per meter	m^{-1}	
	1 per centimeter	cm^{-1}	

In addition, there are 16 prefixes used to indicate order of magnitude, as follows:

Multiplication factor	Prefix	Symbol	Note
10^{18}	exa	E	
10^{15}	peta	P	
10^{12}	tera	T	
10^9	giga	G	
10^6	mega	M	
10^3	kilo	k	
10^2	hecto	h^a	[a]Although hecto, deka, deci, and
10	deka	da^a	centi are SI prefixes, their use
10^{-1}	deci	d^a	should be avoided except for SI
10^{-2}	centi	c^a	unit-multiples for area and volume
10^{-3}	milli	m	and nontechnical use of
10^{-6}	micro	μ	centimeter, as for body and
10^{-9}	nano	n	clothing measurement.
10^{-12}	pico	p	
10^{-15}	femto	f	
10^{-18}	atto	a	

For a complete description of SI and its use the reader is referred to ASTM E 380 (4) and the article UNITS AND CONVERSION FACTORS which appears in Vol. 24.

A representative list of conversion factors from non-SI to SI units is presented herewith. Factors are given to four significant figures. Exact relationships are followed by a dagger. A more complete list is given in the latest editions of ASTM E 380 (4) and ANSI Z210.1 (6).

Conversion Factors to SI Units

To convert from	To	Multiply by
acre	square meter (m^2)	4.047×10^3
angstrom	meter (m)	$1.0 \times 10^{-10\dagger}$
are	square meter (m^2)	$1.0 \times 10^{2\dagger}$
astronomical unit	meter (m)	1.496×10^{11}

[†]Exact.

To convert from	To	Multiply by
atmosphere, standard	pascal (Pa)	1.013×10^5
bar	pascal (Pa)	$1.0 \times 10^{5\dagger}$
barn	square meter (m^2)	$1.0 \times 10^{-28\dagger}$
barrel (42 U.S. liquid gallons)	cubic meter (m^3)	0.1590
Bohr magneton (μ_B)	J/T	9.274×10^{-24}
Btu (International Table)	joule (J)	1.055×10^3
Btu (mean)	joule (J)	1.056×10^3
Btu (thermochemical)	joule (J)	1.054×10^3
bushel	cubic meter (m^3)	3.524×10^{-2}
calorie (International Table)	joule (J)	4.187
calorie (mean)	joule (J)	4.190
calorie (thermochemical)	joule (J)	$4.184^\dagger$
centipoise	pascal second (Pa·s)	$1.0 \times 10^{-3\dagger}$
centistokes	square millimeter per second (mm^2/s)	$1.0^\dagger$
cfm (cubic foot per minute)	cubic meter per second (m^3/s)	4.72×10^{-4}
cubic inch	cubic meter (m^3)	1.639×10^{-5}
cubic foot	cubic meter (m^3)	2.832×10^{-2}
cubic yard	cubic meter (m^3)	0.7646
curie	becquerel (Bq)	$3.70 \times 10^{10\dagger}$
debye	coulomb meter (C·m)	3.336×10^{-30}
degree (angle)	radian (rad)	1.745×10^{-2}
denier (international)	kilogram per meter (kg/m)	1.111×10^{-7}
	tex‡	0.1111
dram (apothecaries')	kilogram (kg)	3.888×10^{-3}
dram (avoirdupois)	kilogram (kg)	1.772×10^{-3}
dram (U.S. fluid)	cubic meter (m^3)	3.697×10^{-6}
dyne	newton (N)	$1.0 \times 10^{-5\dagger}$
dyne/cm	newton per meter (N/m)	$1.0 \times 10^{-3\dagger}$
electronvolt	joule (J)	1.602×10^{-19}
erg	joule (J)	$1.0 \times 10^{-7\dagger}$
fathom	meter (m)	1.829
fluid ounce (U.S.)	cubic meter (m^3)	2.957×10^{-5}
foot	meter (m)	$0.3048^\dagger$
footcandle	lux (lx)	10.76
furlong	meter (m)	2.012×10^{-2}
gal	meter per second squared (m/s^2)	$1.0 \times 10^{-2\dagger}$
gallon (U.S. dry)	cubic meter (m^3)	4.405×10^{-3}
gallon (U.S. liquid)	cubic meter (m^3)	3.785×10^{-3}
gallon per minute (gpm)	cubic meter per second (m^3/s)	6.309×10^{-5}
	cubic meter per hour (m^3/h)	0.2271

†Exact.
‡See footnote on p. xviii.

To convert from	To	Multiply by
gauss	tesla (T)	1.0×10^{-4}
gilbert	ampere (A)	0.7958
gill (U.S.)	cubic meter (m^3)	1.183×10^{-4}
grade	radian	1.571×10^{-2}
grain	kilogram (kg)	6.480×10^{-5}
gram force per denier	newton per tex (N/tex)	8.826×10^{-2}
hectare	square meter (m^2)	$1.0 \times 10^{4\dagger}$
horsepower (550 ft·lbf/s)	watt (W)	7.457×10^2
horsepower (boiler)	watt (W)	9.810×10^3
horsepower (electric)	watt (W)	$7.46 \times 10^{2\dagger}$
hundredweight (long)	kilogram (kg)	50.80
hundredweight (short)	kilogram (kg)	45.36
inch	meter (m)	$2.54 \times 10^{-2\dagger}$
inch of mercury (32°F)	pascal (Pa)	3.386×10^3
inch of water (39.2°F)	pascal (Pa)	2.491×10^2
kilogram-force	newton (N)	9.807
kilowatt hour	megajoule (MJ)	$3.6^\dagger$
kip	newton(N)	4.448×10^3
knot (international)	meter per second (m/s)	0.5144
lambert	candela per square meter (cd/m^3)	3.183×10^3
league (British nautical)	meter (m)	5.559×10^3
league (statute)	meter (m)	4.828×10^3
light year	meter (m)	9.461×10^{15}
liter (for fluids only)	cubic meter (m^3)	$1.0 \times 10^{-3\dagger}$
maxwell	weber (Wb)	$1.0 \times 10^{-8\dagger}$
micron	meter (m)	$1.0 \times 10^{-6\dagger}$
mil	meter (m)	$2.54 \times 10^{-5\dagger}$
mile (statute)	meter (m)	1.609×10^3
mile (U.S. nautical)	meter (m)	$1.852 \times 10^{3\dagger}$
mile per hour	meter per second (m/s)	0.4470
millibar	pascal (Pa)	1.0×10^2
millimeter of mercury (0°C)	pascal (Pa)	$1.333 \times 10^{2\dagger}$
minute (angular)	radian	2.909×10^{-4}
myriagram	kilogram (kg)	10
myriameter	kilometer (km)	10
oersted	ampere per meter (A/m)	79.58
ounce (avoirdupois)	kilogram (kg)	2.835×10^{-2}
ounce (troy)	kilogram (kg)	3.110×10^{-2}
ounce (U.S. fluid)	cubic meter (m^3)	2.957×10^{-5}
ounce-force	newton (N)	0.2780
peck (U.S.)	cubic meter (m^3)	8.810×10^{-3}
pennyweight	kilogram (kg)	1.555×10^{-3}
pint (U.S. dry)	cubic meter (m^3)	5.506×10^{-4}

†Exact.

To convert from	To	Multiply by
pint (U.S. liquid)	cubic meter (m^3)	4.732×10^{-4}
poise (absolute viscosity)	pascal second (Pa·s)	$0.10^\dagger$
pound (avoirdupois)	kilogram (kg)	0.4536
pound (troy)	kilogram (kg)	0.3732
poundal	newton (N)	0.1383
pound-force	newton (N)	4.448
pound force per square inch (psi)	pascal (Pa)	6.895×10^3
quart (U.S. dry)	cubic meter (m^3)	1.101×10^{-3}
quart (U.S. liquid)	cubic meter (m^3)	9.464×10^{-4}
quintal	kilogram (kg)	$1.0 \times 10^{2\dagger}$
rad	gray (Gy)	$1.0 \times 10^{-2\dagger}$
rod	meter (m)	5.029
roentgen	coulomb per kilogram (C/kg)	2.58×10^{-4}
second (angle)	radian (rad)	$4.848 \times 10^{-6\dagger}$
section	square meter (m^2)	2.590×10^6
slug	kilogram (kg)	14.59
spherical candle power	lumen (lm)	12.57
square inch	square meter (m^2)	6.452×10^{-4}
square foot	square meter (m^2)	9.290×10^{-2}
square mile	square meter (m^2)	2.590×10^6
square yard	square meter (m^2)	0.8361
stere	cubic meter (m^3)	$1.0^\dagger$
stokes (kinematic viscosity)	square meter per second (m^2/s)	$1.0 \times 10^{-4\dagger}$
tex	kilogram per meter (kg/m)	$1.0 \times 10^{-6\dagger}$
ton (long, 2240 pounds)	kilogram (kg)	1.016×10^3
ton (metric) (tonne)	kilogram (kg)	$1.0 \times 10^{3\dagger}$
ton (short, 2000 pounds)	kilogram (kg)	9.072×10^2
torr	pascal (Pa)	1.333×10^2
unit pole	weber (Wb)	1.257×10^{-7}
yard	meter (m)	$0.9144^\dagger$

Abbreviations and Unit Symbols

Following is a list of common abbreviations and unit symbols used in the *Encyclopedia*. In general they agree with those listed in *American National Standard Abbreviations for Use on Drawings and in Text (ANSI Y1.1)* (6) and *American National Standard Letter Symbols for Units in Science and Technology (ANSI Y10)* (6). Also included is a list of acronyms for a number of private and

†Exact.

government organizations as well as common industrial solvents, polymers, and other chemicals.

Rules for Writing Unit Symbols (4):

1. Unit symbols are printed in upright letters (roman) regardless of the type style used in the surrounding text.
2. Unit symbols are unaltered in the plural.
3. Unit symbols are not followed by a period except when used at the end of a sentence.
4. Letter unit symbols are generally printed lower-case (for example, cd for candela) unless the unit name has been derived from a proper name, in which case the first letter of the symbol is capitalized (W, Pa). Prefixes and unit symbols retain their prescribed form regardless of the surrounding typography.
5. In the complete expression for a quantity, a space should be left between the numerical value and the unit symbol. For example, write 2.37 lm, *not* 2.37lm, and 35 mm, *not* 35mm. When the quantity is used in an adjectival sense, a hyphen is often used, for example, 35-mm film. *Exception:* No space is left between the numerical value and the symbols for degree, minute, and second of plane angle, degree Celsius, and the percent sign.
6. No space is used between the prefix and unit symbol (for example, kg).
7. Symbols, not abbreviations, should be used for units. For example, use "A," not "amp," for ampere.
8. When multiplying unit symbols, use a raised dot:

$$N \cdot m \quad \text{for} \quad \text{newton meter}$$

In the case of W·h, the dot may be omitted, thus:

$$Wh$$

An exception to this practice is made for computer printouts, automatic typewriter work, etc, where the raised dot is not possible, and a dot on the line may be used.
9. When dividing unit symbols, use one of the following forms:

$$m/s \quad or \quad m \cdot s^{-1} \quad or \quad \frac{m}{s}$$

In no case should more than one slash be used in the same expression unless parentheses are inserted to avoid ambiguity. For example, write:

$$J/(mol \cdot K) \quad or \quad J \cdot mol^{-1} \cdot K^{-1} \quad or \quad (J/mol)/K$$

but *not*

$$J/mol/K$$

10. Do not mix symbols and unit names in the same expression. Write:

$$\text{joules per kilogram} \quad or \quad \text{J/kg} \quad or \quad \text{J·kg}^{-1}$$

but *not*

$$\text{joules/kilogram} \quad nor \quad \text{joules/kg} \quad nor \quad \text{joules·kg}^{-1}$$

ABBREVIATIONS AND UNITS

A	ampere		AOAC	Association of Official Analytical Chemists
A	anion (eg, HA)			
A	mass number		AOCS	American Oil Chemists' Society
a	atto (prefix for 10^{-18})			
AATCC	American Association of Textile Chemists and Colorists		APHA	American Public Health Association
			API	American Petroleum Institute
ABS	acrylonitrile–butadiene–styrene			
			aq	aqueous
abs	absolute		Ar	aryl
ac	alternating current, *n.*		*ar-*	aromatic
a-c	alternating current, *adj.*		*as-*	asymmetric(al)
ac-	alicyclic		ASHRAE	American Society of Heating, Refrigerating, and Air Conditioning Engineers
acac	acetylacetonate			
ACGIH	American Conference of Governmental Industrial Hygienists			
			ASM	American Society for Metals
ACS	American Chemical Society		ASME	American Society of Mechanical Engineers
AGA	American Gas Association			
Ah	ampere hour		ASTM	American Society for Testing and Materials
AIChE	American Institute of Chemical Engineers			
			at no.	atomic number
AIME	American Institute of Mining, Metallurgical, and Petroleum Engineers		at wt	atomic weight
			av(g)	average
			AWS	American Welding Society
			b	bonding orbital
AIP	American Institute of Physics		bbl	barrel
			bcc	body-centered cubic
AISI	American Iron and Steel Institute		BCT	body-centered tetragonal
			Bé	Baumé
alc	alcohol(ic)		BET	Brunauer-Emmett-Teller (adsorption equation)
Alk	alkyl			
alk	alkaline (not alkali)		bid	twice daily
amt	amount		Boc	*t*-butyloxycarbonyl
amu	atomic mass unit		BOD	biochemical (biological) oxygen demand
ANSI	American National Standards Institute			
			bp	boiling point
AO	atomic orbital		Bq	becquerel

C	coulomb	DIN	Deutsche Industrie Normen
°C	degree Celsius		
C-	denoting attachment to carbon	*dl-*; DL-	racemic
		DMA	dimethylacetamide
c	centi (prefix for 10^{-2})	DMF	dimethylformamide
c	critical	DMG	dimethyl glyoxime
ca	circa (approximately)	DMSO	dimethyl sulfoxide
cd	candela; current density; circular dichroism	DOD	Department of Defense
		DOE	Department of Energy
CFR	Code of Federal Regulations	DOT	Department of Transportation
cgs	centimeter-gram-second	DP	degree of polymerization
CI	Color Index	dp	dew point
cis-	isomer in which substituted groups are on same side of double bond between C atoms	DPH	diamond pyramid hardness
		dstl(d)	distill(ed)
		dta	differential thermal analysis
cl	carload		
cm	centimeter	*(E)-*	entgegen; opposed
cmil	circular mil	ϵ	dielectric constant (unitless number)
cmpd	compound		
CNS	central nervous system	e	electron
CoA	coenzyme A	ECU	electrochemical unit
COD	chemical oxygen demand	ed.	edited, edition, editor
coml	commercial(ly)	ED	effective dose
cp	chemically pure	EDTA	ethylenediaminetetraacetic acid
cph	close-packed hexagonal		
CPSC	Consumer Product Safety Commission	emf	electromotive force
		emu	electromagnetic unit
cryst	crystalline	en	ethylene diamine
cub	cubic	eng	engineering
D	Debye	EPA	Environmental Protection Agency
D-	denoting configurational relationship		
		epr	electron paramagnetic resonance
d	differential operator		
d	day; deci (prefix for 10^{-1})	eq.	equation
d-	*dextro-*, dextrorotatory	esca	electron spectroscopy for chemical analysis
da	deka (prefix for 10^1)		
dB	decibel	esp	especially
dc	direct current, *n.*	esr	electron-spin resonance
d-c	direct current, *adj.*	est(d)	estimate(d)
dec	decompose	estn	estimation
detd	determined	esu	electrostatic unit
detn	determination	exp	experiment, experimental
Di	didymium, a mixture of all lanthanons	ext(d)	extract(ed)
		F	farad (capacitance)
dia	diameter	*F*	faraday (96,487 C)
dil	dilute	f	femto (prefix for 10^{-15})

FAO	Food and Agriculture Organization (United Nations)
fcc	face-centered cubic
FDA	Food and Drug Administration
FEA	Federal Energy Administration
FHSA	Federal Hazardous Substances Act
fob	free on board
fp	freezing point
FPC	Federal Power Commission
FRB	Federal Reserve Board
frz	freezing
G	giga (prefix for 10^9)
G	gravitational constant = 6.67×10^{11} N·m^2/kg^2
g	gram
(g)	gas, only as in $H_2O(g)$
g	gravitational acceleration
gc	gas chromatography
gem-	geminal
glc	gas–liquid chromatography
g-mol wt; gmw	gram-molecular weight
GNP	gross national product
gpc	gel-permeation chromatography
GRAS	Generally Recognized as Safe
grd	ground
Gy	gray
H	henry
h	hour; hecto (prefix for 10^2)
ha	hectare
HB	Brinell hardness number
Hb	hemoglobin
hcp	hexagonal close-packed
hex	hexagonal
HK	Knoop hardness number
hplc	high performance liquid chromatography
HRC	Rockwell hardness (C scale)
HV	Vickers hardness number

hyd	hydrated, hydrous
hyg	hygroscopic
Hz	hertz
^{i} (eg, Pr^i)	iso (eg, isopropyl)
i-	inactive (eg, i-methionine)
IACS	International Annealed Copper Standard
ibp	initial boiling point
IC	integrated circuit
ICC	Interstate Commerce Commission
ICT	International Critical Table
ID	inside diameter; infective dose
ip	intraperitoneal
IPS	iron pipe size
ir	infrared
IRLG	Interagency Regulatory Liaison Group
ISO	International Organization Standardization
ITS-90	International Temperature Scale (NIST)
IU	International Unit
IUPAC	International Union of Pure and Applied Chemistry
IV	iodine value
iv	intravenous
J	joule
K	kelvin
k	kilo (prefix for 10^3)
kg	kilogram
L	denoting configurational relationship
L	liter (for fluids only) (5)
l-	levo-, levorotatory
(l)	liquid, only as in $NH_3(l)$
LC$_{50}$	conc lethal to 50% of the animals tests
LCAO	linear combination of atomic orbitals
lc	liquid chromatography
LCD	liquid crystal display
lcl	less than carload lots

LD_{50}	dose lethal to 50% of the animals tested	$N\text{-}$	denoting attachment to nitrogen
LED	light-emitting diode	n (as n_D^{20})	index of refraction (for 20°C and sodium light)
liq	liquid		
lm	lumen	^{n} (as Bu^n), $n\text{-}$	normal (straight-chain structure)
ln	logarithm (natural)		
LNG	liquefied natural gas	n	neutron
log	logarithm (common)	n	nano (prefix for 10^9)
LPG	liquefied petroleum gas	na	not available
ltl	less than truckload lots	NAS	National Academy of Sciences
lx	lux		
M	mega (prefix for 10^6); metal (as in MA)	NASA	National Aeronautics and Space Administration
M	molar; actual mass	nat	natural
$\overline{M}_w$	weight-average mol wt	ndt	nondestructive testing
$\overline{M}_n$	number-average mol wt	neg	negative
m	meter; milli (prefix for 10^{-3})	NF	*National Formulary*
		NIH	National Institutes of Health
m	molal		
$m\text{-}$	meta	NIOSH	National Institute of Occupational Safety and Health
max	maximum		
MCA	Chemical Manufacturers' Association (was Manufacturing Chemists Association)	NIST	National Institute of Standards and Technology (formerly National Bureau of Standards)
MEK	methyl ethyl ketone		
meq	milliequivalent		
mfd	manufactured	nmr	nuclear magnetic resonance
mfg	manufacturing		
mfr	manufacturer	NND	New and Nonofficial Drugs (AMA)
MIBC	methyl isobutyl carbinol		
MIBK	methyl isobutyl ketone	no.	number
MIC	minimum inhibiting concentration	NOI-(BN)	not otherwise indexed (by name)
min	minute; minimum	NOS	not otherwise specified
mL	milliliter	nqr	nuclear quadruple resonance
MLD	minimum lethal dose		
MO	molecular orbital	NRC	Nuclear Regulatory Commission; National Research Council
mo	month		
mol	mole		
mol wt	molecular weight	NRI	New Ring Index
mp	melting point	NSF	National Science Foundation
MR	molar refraction		
ms	mass spectrum	NTA	nitrilotriacetic acid
mxt	mixture	NTP	normal temperature and pressure (25°C and 101.3 kPa or 1 atm)
μ	micro (prefix for 10^{-6})		
N	newton (force)		
N	normal (concentration); neutron number	NTSB	National Transportation Safety Board

O-	denoting attachment to oxygen	qv	quod vide (which see)
o-	ortho	R	univalent hydrocarbon radical
OD	outside diameter	*(R)-*	rectus (clockwise configuration)
OPEC	Organization of Petroleum Exporting Countries	*r*	precision of data
o-phen	*o*-phenanthridine	rad	radian; radius
OSHA	Occupational Safety and Health Administration	RCRA	Resource Conservation and Recovery Act
owf	on weight of fiber	rds	rate-determining step
Ω	ohm	ref.	reference
P	peta (prefix for 10^{15})	rf	radio frequency, *n.*
p	pico (prefix for 10^{-12})	r-f	radio frequency, *adj.*
p-	para	rh	relative humidity
p	proton	RI	Ring Index
p.	page	rms	root-mean square
Pa	pascal (pressure)	rpm	rotations per minute
PEL	personal exposure limit based on an 8-h exposure	rps	revolutions per second
		RT	room temperature
pd	potential difference	^{s} (eg, Bu^{s}); *sec-*	secondary (eg, secondary butyl)
pH	negative logarithm of the effective hydrogen ion concentration	S	siemens
		(S)-	sinister (counterclockwise configuration)
phr	parts per hundred of resin (rubber)	*S-*	denoting attachment to sulfur
p-i-n	positive-intrinsic-negative	*s-*	symmetric(al)
pmr	proton magnetic resonance	s	second
		(s)	solid, only as in $H_2O(s)$
p-n	positive-negative	SAE	Society of Automotive Engineers
po	per os (oral)		
POP	polyoxypropylene	SAN	styrene–acrylonitrile
pos	positive	sat(d)	saturate(d)
pp.	pages	satn	saturation
ppb	parts per billion (10^9)	SBS	styrene–butadiene–styrene
ppm	parts per million (10^6)	sc	subcutaneous
ppmv	parts per million by volume	SCF	self-consistent field; standard cubic feet
ppmwt	parts per million by weight	Sch	Schultz number
		SFs	Saybolt Furol seconds
PPO	poly(phenyl oxide)	SI	Le Système International d'Unités (International System of Units)
ppt(d)	precipitate(d)		
pptn	precipitation		
Pr (no.)	foreign prototype (number)	sl sol	slightly soluble
		sol	soluble
pt	point; part	soln	solution
PVC	poly(vinyl chloride)	soly	solubility
pwd	powder	sp	specific; species
py	pyridine	sp gr	specific gravity

sr	steradian
std	standard
STP	standard temperature and pressure (0°C and 101.3 kPa)
sub	sublime(s)
SUs	Saybolt Universal seconds
syn	synthetic
^{t} (eg, Bu^t), t-, tert-	tertiary (eg, tertiary butyl)
T	tera (prefix for 10^{12}); tesla (magnetic flux density)
t	metric ton (tonne)
t	temperature
TAPPI	Technical Association of the Pulp and Paper Industry
TCC	Tagliabue closed up
tex	tex (linear density)
T_g	glass-transition temperature
tga	thermogravimetric analysis
THF	tetrahydrofuran
tlc	thin layer chromatography
TLV	threshold limit value
trans-	isomer in which substituted groups are

	on opposite sides of double bond between C atoms
TSCA	Toxic Substances Control Act
TWA	time-weighted average
Twad	Twaddell
UL	Underwriters' Laboratory
USDA	United States Department of Agriculture
USP	*United States Pharmacopeia*
uv	ultraviolet
V	volt (emf)
var	variable
vic-	vicinal
vol	volume (not volatile)
vs	versus
v sol	very soluble
W	watt
Wb	weber
Wh	watt hour
WHO	World Health Organization (United Nations)
wk	week
yr	year
(Z)-	zusammen; together; atomic number

Non-SI (Unacceptable and Obsolete) Units		Use
Å	angstrom	nm
at	atmosphere, technical	Pa
atm	atmosphere, standard	Pa
b	barn	cm^2
bar†	bar	Pa
bbl	barrel	m^3
bhp	brake horsepower	W
Btu	British thermal unit	J
bu	bushel	m^3; L
cal	calorie	J
cfm	cubic foot per minute	m^3/s
Ci	curie	Bq
cSt	centistokes	mm^2/s
c/s	cycle per second	Hz
cu	cubic	exponential form
D	debye	C·m

†Do not use bar (10^5Pa) or millibar (10^2Pa) because they are not SI units, and are accepted internationally only for a limited time in special fields because of existing usage.

Non-SI (Unacceptable and Obsolete) Units		Use
den	denier	tex
dr	dram	kg
dyn	dyne	N
dyn/cm	dyne per centimeter	mN/m
erg	erg	J
eu	entropy unit	J/K
°F	degree Fahrenheit	°C; K
fc	footcandle	lx
fl	footlambert	lx
fl oz	fluid ounce	m^3; L
ft	foot	m
ft·lbf	foot pound-force	J
gf den	gram-force per denier	N/tex
G	gauss	T
Gal	gal	m/s^2
gal	gallon	m^3; L
Gb	gilbert	A
gpm	gallon per minute	(m^3/s); (m^3/h)
gr	grain	kg
hp	horsepower	W
ihp	indicated horsepower	W
in.	inch	m
in. Hg	inch of mercury	Pa
in. H_2O	inch of water	Pa
in.-lbf	inch pound-force	J
kcal	kilo-calorie	J
kgf	kilogram-force	N
kilo	for kilogram	kg
L	lambert	lx
lb	pound	kg
lbf	pound-force	N
mho	mho	S
mi	mile	m
MM	million	M
mm Hg	millimeter of mercury	Pa
mμ	millimicron	nm
mph	miles per hour	km/h
μ	micron	μm
Oe	oersted	A/m
oz	ounce	kg
ozf	ounce-force	N
η	poise	Pa·s
P	poise	Pa·s
ph	phot	lx
psi	pounds-force per square inch	Pa
psia	pounds-force per square inch absolute	Pa
psig	pounds-force per square inch gage	Pa
qt	quart	m^3; L
°R	degree Rankine	K
rd	rad	Gy
sb	stilb	lx
SCF	standard cubic foot	m^3
sq	square	exponential form
thm	therm	J
yd	yard	m

BIBLIOGRAPHY

1. The International Bureau of Weights and Measures, BIPM (Parc de Saint-Cloud, France) is described in Appendix X2 of Ref. 4. This bureau operates under the exclusive supervision of the International Committee for Weights and Measures (CIPM).
2. *Metric Editorial Guide (ANMC-78-1)*, latest ed., American National Metric Council, 5410 Grosvenor Lane, Bethesda, Md. 20814, 1981.
3. *SI Units and Recommendations for the Use of Their Multiples and of Certain Other Units (ISO 1000-1981)*, American National Standards Institute, 1430 Broadway, New York, N.Y. 10018, 1981.
4. Based on *ASTM E 380-89a (Standard Practice for Use of the International System of Units (SI)*, American Society for Testing and Materials, 1916 Race Street, Philadelphia, Pa. 19103, 1989.
5. *Fed. Regist.*, Dec. 10, 1976 (41 FR 36414).
6. For ANSI address, see Ref. 3.

R. P. LUKENS
ASTM Committee E-43 on SI Practice

ABACA FIBER. See Fibers, vegetable.

ABHERENTS. See Release agents.

ABIETIC ACID. See Terpenoids.

ABLATIVE MATERIALS

The word ablation is derived from the suppletive past particle of the Latin *auferre,* which means to remove. It was originally used in the geologic sense to describe the combined, predominantly thermal, processes by which a glacier wastes. The present use of the word maintains the thermal aspect and describes the absorption, dissipation, and blockage of heat associated with high speed entry into the atmosphere. Thus ablative, thermal protection materials are used to protect vehicles from damage during atmospheric reentry. The need for these materials was first realized during the development of operational ballistic missiles in Pennemunde, Germany, when a large percentage of V-2's failed to reach their targets because of missile skin disintegration caused by aerodynamic heating (1). Ablative materials are also used to protect rocket nozzles and ship hulls from propellant gas erosion, as protection from laser beams, and to protect land-based structures from high heat environments.

The Ablation Environment

The functional requirements of the ablative heatshield must be well understood before selection of the proper material can occur. Ablative heatshield materials not only protect a vehicle from excessive heating, they also act as an aerodynamic body and sometimes as a structural component (2,3). Intensity and duration of heating, thermostructural requirements and shape stability (4,5), potential for particle erosion (6), weight limitations (7–10), and reusability (11) are some of the factors which must be considered in selection of an ablative material.

Some typical ablative environments are shown in Figure 1. Each of the altitude–velocity profiles results in a specific heating rate and radiation equilibrium temperature for a given material. When a vehicle decelerates at high altitudes under low pressure conditions, and the flight angle with respect to the horizon is low, the heating rate is low but the heating time period is long, eg, the Apollo trajectory. In this situation, material insulating ability becomes important. Conversely, a sharp atmospheric entry angle results in severe heating rates but for a shorter duration, eg, the ballistic missile trajectory, which requires less emphasis on the insulating capability of the heatshield material.

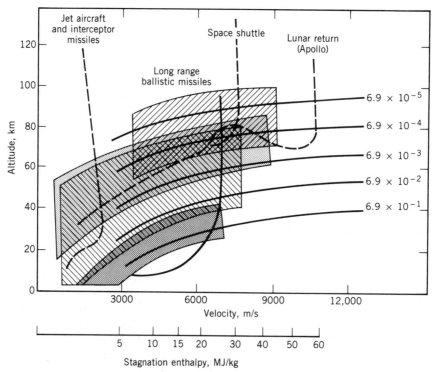

Fig. 1.   Simulation facility capabilities and mission requirements: ▨ ROVERS arc and ▨ 10-MW arc, Textron Defense Systems (TDS); ▨ Interaction Heating Facility and ▨ Aero Heating Facility, NASA/Ames. Numbers on the curves indicate stagnation pressure P_s in MPa; ——, ballistic entry; – – –, lifting entry. To convert MPa to psi multiply by 145. To convert MJ to kcal divide by 4.18×10^{-3}.

Several other factors must also be considered with respect to heating conditions. At the front end of a vehicle, ie, at the nosetip, the heating rate is most severe, generally decreasing toward the aft end of the vehicle in instances of laminar flow. Because of this variation in heating conditions, the nosetip material is usually different from the heatshield or aft end materials. Vehicle design is, of course, influential. Sharp, heavy ballistic vehicles having a high mass-to-drag ratio drop in altitude at higher velocities than blunt, lightweight Apollo-type vehicles, resulting in much higher heating rates for the former. Then also, efficient aerodynamic vehicles, such as long range glide vehicles, utilize sharp leading edges on the nose and wings at the expense of high local heating. Moreover, the rate of heat transfer for a turbulent gas boundary results in higher heating conditions than a laminar gas stream. Thus to reduce or delay the tendency for turbulent flow, smooth, uniform vehicle contours are preferred and, whenever possible, high density materials are avoided to minimize the weight-to-drag ratio.

Heatshield thickness and weight requirements are determined using a thermal prediction model based on measured thermophysical properties. The models typically include transient heat conduction, surface ablation, and charring in a heatshield having multiple sublayers such as bond, insulation, and substructure. These models can then be employed for any specific heating environment to determine material thickness requirements and to identify the lightest heatshield materials.

In a very simplified first-order analysis the ablative heatshield is considered to be of two components: the ablated thickness and the remaining thickness, or the insulation. The ablated weight is determined by the total aerodynamic heat load divided by the heat of ablation, that is, the heat absorbed per unit weight. The insulation weight is determined by the heat conduction parameter, $\rho k/C_p$, the product of density and conductivity, divided by the specific heat, and the ratio of temperature rise at the back surface to that at the front. As shown in Figure 2, the ablated weight increases as the total heat load increases. However, the insulation weight, which initially increases with increased heating, exhibits a maximum and

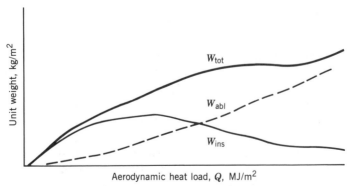

Fig. 2. A simplified material thermal performance analysis for a reentry vehicle thermal protection system where W_{abl} = density × surface recession thickness = total aerodynamic heat/heat of ablation; W_{ins} = density × insulative layer thickness = $f(\rho k/C_p, T_{structure}, T_{surface})$; and W_{tot} = unit weight of heatshield = density × thickness = $W_{abl} + W_{ins}$.

then decreases. Thus at some level ablation begins to dominate, temperature gradients become very steep, and the need for insulation decreases. As a result, the total heatshield weight requirements may not monotonically reflect increases in heat load.

The selection of a material having the right balance of ablation and insulation properties is needed to produce optimum heatshield performance. This material selection is complicated because the higher density materials that usually offer better ablation performance also have higher thermal conductivities and are therefore poor insulators. Properties of known materials are given in Table 1, whereas the desired trends in properties and characteristics of thermal protection (TP) materials are summarized in Table 2 (12). In the high flux region most of the aerodynamic heating is absorbed by ablation. The parameter $\rho k/C_p$ gives some indication of material performance when the rate at which heat is conducted into the shield is very small compared to that at which the heat is radiated away from the surface; ρk alone is somewhat less indicative. Q_{eff} indicates material (and system) performance in all regions because it is essentially a weight parameter including the interaction of all other variables. It must be calculated based on a knowledge of all material properties and mission environment.

Table 1. Heat of Ablation and Relative Thermal Conductivity for Reentry Vehicle Materials Assuming Laminar Flow

	Cold wall heat of ablation[a], J/g[b]		Relative thermal conductivity
Material	$H_s = 12{,}000$ J/g[b] $V_\infty = 4800$ m/s	$H_s = 24{,}000$ J/g[b] $V_\infty = 6{,}800$ m/s	
carbon–carbon	32,000	39,500	high
carbon–phenolic	24,000	29,600	↑
silica–phenolic	13,000	19,000	↓
Teflon	6,650	10,350	low

[a]H_s is the stagnation enthalpy at the surface of the leading edge; V_∞ is the velocity of the airstream at the leading edge.
[b]To convert J to cal divide by 4.184.

The thermostructural requirements of the heatshield are important to material selection and both aerodynamic and attachment load requirements must be met. In the case of a charring ablator, surface char must be of sufficient strength to survive aerodynamic shear. Changes in the ablative material's mechanical and thermal properties occur as a result of the thermal gradient through the depth of the material. Backface surface temperatures, the temperatures on the inside surface of the heatshield, dictate attachment methods and materials. An excessive backface surface temperature caused by inadequate insulation characteristics may weaken an adhesive bond to a substructure or even weaken a load-carrying substructure. In the event that the heatshield also serves as the load-carrying structure, sufficient thickness must be provided for both ablation and insulation so that enough material remains cold (uncharred). A mismatch in axial vs radial thermal expansion can result in severe thermal stresses and

Table 2. Summary of Criteria for Material Selection and Performance Evaluation[a]

Parameter	Definition	Desired trend	Best region of application as a figure of merit
effective heat of ablation[b], q^*	sensible heat + heat of decomposition + mass transfer shielding + shear effects	$\longrightarrow \infty$	High-flux region.
$\dfrac{\rho k}{C_p}$	$\dfrac{\text{density} \times \text{conductivity}}{\text{specific heat}}$	$\longrightarrow 0$	Predominant parameter when no ablation occurs.
ρk	density × conductivity	$\longrightarrow 0$	
T_A	ablation temperature	No general trend. Depends on interaction with other parameters and design criteria.	None except as an indicator of whether or not ablation will occur.
T_r	backface temperature	Low or as prescribed.	A design and comparative criterion for testing.
ϵT_E^4	emissivity × (radiation equilibrium temperature)[4]	$\epsilon \rightarrow 1$ depends on interaction with other parameters and design criteria.	Short duration, or when heat leakage into system is minimized.
Q_{eff}[b]	$\dfrac{\text{total cold wall heat input}}{\text{total required weight of TP system for a given backface temperature}}$	$\longrightarrow \infty$	All regions.
W_t or ρL_t	total required weight of TP system	$\longrightarrow 0$	All regions.

[a]Ref. 12.
[b]Also referred to by many other symbols with various heat fluxes as reference.

subsequent failure, as has been noted with the use of thick sections of pyrolitic graphite [7782-42-5] (13,14). In reentry, erosion from rain or ice particles is also a consideration, particularly at the tip. In addition, in rocket nozzles and on surfaces exposed to propellant gases, erosion resistance from solid particulates must also be considered.

The practice of employing reusable thermal protection systems for reentry is becoming more common. These are essentially ablative materials exposed to environments where very little ablation actually occurs. Examples include the space shuttle tiles and leading edges, exhaust nozzle flaps for advanced engines, and the proposed structural surface skin for the National Aerospace plane.

Another environmental issue important to low earth orbit materials is atomic erosion. At an altitude of 300 km, absorption of solar radiation produces atmospheric temperatures of 1150°C, and at these temperatures gas molecules decompose. Erosion of surface materials by oxygen atoms or nitrogen–oxygen radicals is a serious issue for low altitude orbiting satellites. Experiments conducted on early shuttle flights determined that organic materials that would normally be found on a heatshield erode more rapidly than metallic ones (15). Thus, the effects of atomic erosion must be considered for any vehicle that is subject to long term exposure at low earth altitudes.

A variety of test methods and facilities have been developed to address the process of ablation. These utilize lasers, chemical flames, plasma arcs, electric arc heaters, and other heat sources and sometimes include high velocity wind tunnel facilities that introduce particles to simulate high speed erosion. Examples of ablation facilities used to simulate a variety of reentry conditions are shown in Figure 1. The TDS 10-MW arc facility simulates high speed, high pressure (up to 2.5 MPa) ablation conditions for ballistic reentry. It is mainly used for examining the ablative performance of high velocity nosetip and heatshield materials. The TDS ROVERS (radiation orbital vehicle reentry simulator) arc is a combined convection–radiative heating arc used to simulate high altitude, low pressure reentry conditions. The Interaction Heating Facility and Aero Heating Facility at NASA/Ames are used to simulate a wide range of pressures and enthalpies and have the capacity for much larger specimen sizes than the Textron facilities. A summary of simulation capabilities is given in Table 3 and Figure 3. A listing of nationwide arc facilities and corresponding test capabilities is also available (16).

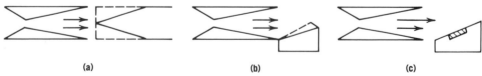

(a)   (b)   (c)

Fig. 3. Simulation parameters: (**a**) splash onto cylindrical nosetip; (**b**) attached wedge (——, fixed; – – –, varied); and (**c**) detached wedge. See Table 3.

The Ablation Process

Thermophysically, the ablation process can be described as the elimination of a large amount of thermal energy by sacrifice of surface material. Principles operating during this highly efficient and orderly heat and mass transfer process are (1)

Table 3. Summary of Arc Simulation Capabilities

Parameter	TDS 10-MW arc[a]			TDS ROVERS arc[a]		NASA/Ames Facilities[a,b]	
simulation parameter	splash	detached wedge	attached wedge	splash	detached wedge	Aero Heating	Interaction Heating
cylindrical diameter, cm	to 7.6			to 7.6		20.3	45.7
wedge cutout, cm[c]		5.1 × 17.8	2.5 × 7.6		7.6 × 7.6	66 × 60	61 × 61
enthalpy, MJ/kg	0.7–21.2	0.7–21.2	0.7–21.2	0.7–40.2	0.7–40.2	1–31.2	7–44.6
convective heat flux, MW/m^2	1.1–45.4	0.1–7.9	0.6–45.4	0.02–7.9	0.02–7.9	0.006–3.4	0.006–1.5
test time, max, s	20	25	25	continuous	continuous		
gas	air	air	air	air, N_2, others	air, N_2, others	air	air
jet mach no.	1–2	0.5–2.0	1–2.5	2–3.5	2–3.5	2.5–12	5.5–7.5
model surface shear, kg/m^2		4.9–97.6	24.4–488.0	0.2–2.4	0.2–2.4		
particle erosion	yes	yes	yes	yes	yes		
programmed heating: enthalpy variation	yes	yes	yes	yes	yes		
heat flux variation	yes	yes	yes	yes	yes		
model pressure, MPa[d]	0.1–2.43	0.10–0.51	0.10–0.30	to 0.01	to 0.01	0.0005–0.2	10^{-5}–0.0015
jet diameter, cm	to 5.1	3.2–5.7	2.5–6.4	to 7.6	to 7.6	7.6–106.7	to 104.1

[a]See Figure 3 for illustration of simulation parameters.
[b]These facilities can test larger specimens over a wide range of enthalpies and pressures.
[c]To convert MJ to kcal divide by 4.184 × 10^{-3}.
[d]To convert MPa to psi multiply by 145.

phase changes such as melting, vaporization, and sublimation, (2) conduction and storage of heat in the material substrate, (3) absorption of heat by gases as they are forced to the surface, (4) heat convection in a liquid layer, (5) transpiration of gases and liquids and subsequent heat absorption from the surface into the boundary layer, (6) exothermic and endothermic chemical reactions, and (7) radiation on the surface and in bulk (17).

The relationship between heat transfer and the boundary layer species distribution should be emphasized. As vaporization occurs, chemical species are transported to the boundary layer and act to cool by transpiration. These gaseous products may undergo additional thermochemical reactions with the boundary-layer gas, further impacting heat transfer. Thus species concentrations are needed for accurate calculation of transport properties, as well as for calculations of convective heating and radiative transport.

Ablative Materials

Ablative materials are classified according to dominant ablation mechanism. There are three groups: subliming or melting ablators, charring ablators, and intumescent ablators. Figure 4 shows the physical zones of each. Because of the basic thermal and physical differences, the classes of ablative materials are used in different types of applications.

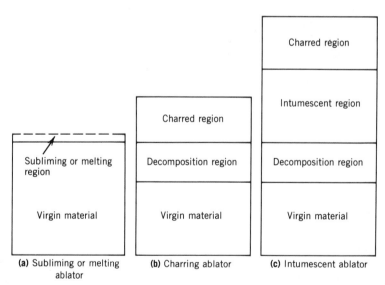

Fig. 4. Physical zones of ablators. Typical time-integrated heat flux, J/m^2, (a) 500, (b) 5000, (c) <50; maximum instantaneous heat flux, MW/m^2, (a) 0.5, (b) >1, (c) 0.1. To convert J to cal divide by 4.184.

Subliming and Melting Ablators. Subliming ablators act as heat sinks to the incident heat flux until the temperature on the surface reaches the sublimation or melting temperature, also known as the reaction temperature in these cases. At this time the sublimation or melting action removes heat from the insulation material. In the sublimation process the convective transfer of heat from the boundary layer to the material surface is also blocked by the gas evolving from the ablative material, concurrently thickening the boundary layer. This blocking action can reduce the net heating of the ablative material by more than 50%.

Some of the early reentry vehicles utilized metallic heat sinks of copper [7440-50-8] or beryllium [7440-41-7] to absorb reentry heat. Other metallic materials that have been evaluated for nosetip applications include tungsten [7440-33-7] and molybdenum [7439-98-7]. The melt layers of these materials are believed to be very thin because of the high rate at which volatile oxide species are formed.

One of the first subliming ablative materials to be identified was polytetrafluoroethylene [9002-84-0], Teflon, which offers light weight, good insulating properties as a result of its decomposition temperature (about 500°C), and a high endothermic value for the depolymerization or ablative heat of reaction. An added advantage is that Teflon ablates to form a volatile monomer without forming a conductive char, thereby maintaining the low dielectric properties of the virgin material. A dielectric material without a conductive char is very useful for transmitting and receiving radiofrequency signals during reentry. For higher heat loads, Teflon and a high temperature dielectric fiber such as quartz can be mixed to reduce the necessary wall thickness and improve overall thermostructural performance. Teflon was proposed as a heatshield for a Venus probe, using a reflective coating on the back end for additional insulation (18). Reinforced Teflon has been suggested for use in high speed ablative missile radome applications (19).

Subliming ablators are used for vehicles subject to long term, low altitude exposure and subsequent atomic erosion. In experiments conducted on early shuttle flights, metallic materials exhibited a significantly greater atomic erosion resistance than organic-based materials with the exception of Teflon, which does not react strongly with atomic oxygen. Very thin coatings of the erosion-resistant materials were found to protect the substrate from atomic erosion. In addition, these coatings sublime cleanly upon reentry (20,21).

Graphite. Carbon [7440-44-0] has been identified as having the highest heat of ablation. This high ablation efficiency often identifies carbon or carbon composites for use in high heating environments where a minimum of shape change is important, such as in missile nosetips and small radius leading edges. Graphite [7782-42-5] sublimes at temperatures as high as 3900 K (22) (see CARBON, CARBON AND ARTIFICIAL GRAPHITE; CARBON, NATURAL GRAPHITE).

When monolithic graphite is used for ablation, the critical factors affecting performance include high uniform density and small uniform pore size. Surface roughening of the graphite, however, caused by ablation down to subsurface porosity, can affect the surface heating as the flow is changed from laminar to turbulent (23). Pyrolitic graphite, which is free of open porosity and very high in density compared to other forms (Table 4), has been shown to be superior in resistance to laser penetration (24). However, pyrolitic graphite exhibits high thermal expansion anisotropy and is therefore subject to thermal fracture. The other commercial forms of graphite are less susceptible to thermal fracture and have also been evaluated for reentry applications. ATJ graphite was found to be

Table 4. Bulk Graphite Properties

Material	Specific gravity	% Thermal expansion from 300 to 2500 K	
		Radial, AB, direction	Axial, C, direction
pyrolytic graphite	2.2	0.5	6
Union Carbide ATJ-S	1.83	0.9	1.2
Unocal Poco AXF-5Q	1.81	1.9	1.9

an order of magnitude greater in resistance to laser penetration than reinforced charring ablators (25). However, monolithic graphite has been used less in recent years because of the increased variety in forms of reinforced carbon or carbon–carbon composites that are available.

Carbon–Carbon Composites. Carbon–carbon composites are simply described as a carbon fiber reinforcement in one or many directions using a carbon or graphite matrix material (see CARBON FIBERS; COMPOSITE MATERIALS).

Techniques available for densifying woven carbon fiber preforms into carbon–carbon composites include (1) high pressure impregnation of the preform using molten coal tar or petroleum pitch, followed by pyrolysis and high temperature graphitization for multiple cycles, (2) low pressure impregnation using high char yield resin matrices, followed by pyrolysis and graphitization for multiple cycles, and (3) carbon vapor deposition–infiltration into the preform using a high strength graphitic structure. Processes are often combined to yield a high density composite (26). These materials exhibit improved thermal stress performance over monolithic graphite.

For nosetip materials 3-directional-reinforced (3D) carbon preforms are formed using small cell sizes for uniform ablation and small pore size. Figure 5

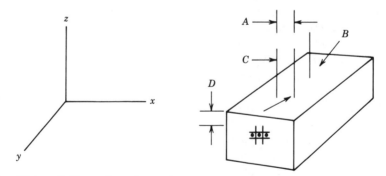

Fig. 5.   Unit cell dimensions for carbon–carbon nosetip materials. Parameters are:

	fine-weave pierced fabric	3D orthogonal weave
A, z: fiber spacing, mm	1.32	0.76
B, z: fiber spacing, mm	1.32	0.76
C, z: fiber size, mm	0.635 diameter	0.38×0.38
D, x, y: cell spacing, mm	0.25	0.86

shows typical unit cell dimensions for two of the most common 3D nosetip materials. Carbon–carbon woven preforms have been made with a variety of cell dimensions for different applications (27–33). Fibers common to these composites include rayon, polyacrylonitrile, and pitch precursor carbon fibers. Strength of these fibers ranges from 1 to 5 GPa (145,000–725,000 psi) and modulus ranges from 300 to 800 GPa.

Carbon–carbon composites for rocket nozzles or exit cones are usually made by weaving a 3D preform composed of radial, axial, and circumferential carbon or graphite fibers to near net shape, followed by densification to high densities. Because of the high relative volume cost of the process, looms have been designed for semiautomatic fabrication of parts, taking advantage of selective reinforcement placement for optimum thermal performance.

Other forms of carbon–carbon composites have been or are being developed for space shuttle leading edges, nuclear fuel containers for satellites, aircraft engine adjustable exhaust nozzles, and the main structure for the proposed National Aerospace plane (34). For reusable applications, a silicon carbide [409-21-2] based coating is added to retard oxidation (35,36), with a boron [7440-42-8] based sublayer to seal any cracks that may form in the coating.

Ceramic Ablators.   Several types of subliming or melting ceramic ablators have been used or considered for use in dielectric applications; particularly with quartz or boron nitride [10043-11-5] fiber reinforcements to form a nonconductive char. Fused silica is available in both nonporous (optically transparent) and porous (slip cast) forms. Ford Aerospace manufactures a 3D silica-fiber-reinforced composite densified with colloidal silica (37). The material, designated AS-3DX, demonstrates improved mechanical toughness compared to monolithic ceramics. Other dielectric ceramic composites have been used with performance improvements over monolithic ceramics (see COMPOSITE MATERIALS, CERAMIC MATRIX).

Melting ablators such as nylon and quartz perform essentially as subliming ablators do, except that they melt. In general, melting ablators have heats of reaction similar to subliming ablators but have much higher thermal conductivities. When compared to other types of ablative materials, there are very few advantages to using melting ablators. However, they are often combined with charring ablative materials in a reinforcing fiber form to improve ablation performance by transpirational cooling as the endothermic melt is forced to the surface (38). Some melting ablators have also found application as dielectric ablators, when no electrically conductive residue is formed. Silicon carbide and silicon nitride [12033-89-5] have also been considered as effective ablators for specific thermal protection applications (39).

Subliming ablators are being used in a variety of manufacturing applications. The exposure of some organic polymers to pulsed uv-laser radiation results in spontaneous ablation by the sublimation of a controlled thickness of the material. This photoetching technique is utilized in the patterning of polymer films (40,41) (see PHOTOCHEMICAL TECHNOLOGY).

The thermal protection system of the space shuttle is composed mainly of subliming or melting ablators that are used below their fusion or vaporization reaction temperatures (42). In addition to the carbon–carbon systems discussed above, a flexible reusable surface insulation composed of Nomex felt substrate, a Du Pont polyamide fiber material, is used on a large portion of the upper surface.

High and low temperature reusable surface insulation composed of silica-based low density tiles are used on the bottom surface of the vehicle, which sees a more severe reentry heating environment than does the upper surface of the vehicle (43).

Charring Ablators.   Charring ablators are used in a greater variety of thermal environments than either subliming or intumescent ablators because of their ability to withstand a much higher heat flux. In the charring ablator, the ablative material acts as a heat sink, absorbing all of the incident heat flux and causing the surface temperature to increase quickly. At reaction temperature, endothermic chemical decomposition occurs: the organic matrix pyrolizes into carbonized material and gaseous products. The passage of heat-absorbing gases through the charred surface provides further insulative performance and thickens the boundary layer, reducing the convective heat transfer. The charring is a continuous process: as the charred surface is eroded by the severe surface environment, more char forms to take its place.

Charring ablators are often used in combination with subliming or melting reinforcement materials. Melting reinforcements such as silica or nylon provide transpirational cooling. Carbon-fiber-reinforced phenolic composites are commonly used as heatshields for high load reentry vehicles (44,45). This material was also evaluated for survival of the severe heating environment anticipated with the Jovian probe Galileo (46). High strength, high temperature subliming reinforcements such as carbon fibers provide substantial strength, both to withstand high shear environments and to act as a structural heatshield material. A laminated carbon–phenolic composite is typically made by using an 8-harness satin-weave fabric prepreg having fibers at 45° to the wrapping direction (bias orientation) and then laying up on a cylinder or frustum so that the plane of tape makes an angle of 20° to the surface of the shell. This type of configuration improves resistance to delamination, which can occur in a simple cylindrical or scroll wrap. Also, the fibers are at a low angle to the ablating surface, thereby minimizing thermal conduction. In some instances the types of reinforcements are varied along the thickness of the material to improve insulation (47) (see COMPOSITE MATERIALS, POLYMER MATRIX).

High density charring ablators such as carbon–phenolic contain high density reinforcements to improve shear resistance. In contrast, lower density charring ablators as a rule are used for low shear environments. The Apollo mission reentry conditions are typical of a relatively low shear environment, so low density ablators consisting of epoxy–novolac resin containing phenolic microballoons and silica fiber reinforcement have been used. In order to improve the shear resistance and safety factor of the material for this mission, the ablator was injected into the cavities of a fiberglass-reinforced phenolic honeycomb that was bonded to the substructure of the craft (48).

Elastomeric shield materials (ESM) have been developed as low density flexible ablators for low shear applications (49). General Electric's RTV 560 is a foamed silicone elastomer loaded with silicon dioxide [7631-86-9] and iron oxide [1317-61-9] particles, which decomposes to a similar foam of SiO_2, SiC, and $FeSiO_3$. Silicone resins are relatively resistant to thermal decomposition and the silicon dioxide forms a viscous liquid when molten (50) (see SILICON COMPOUNDS, SILICONES).

One indication of the performance of a charring ablator resin is the ability of the organic material to form a high density char. As shown in Figure 6, silicone is quite resistant to decomposition, even after exposure to high temperatures. Phenolic is also shown to be a relatively high char yield material (50). However, epoxy which has a higher decomposition rate, is commonly used because of ease of handling and processing. In addition, the char structure of epoxy-based ablators can be improved by the addition of a variety of reinforcements (51). For example, a graphite-fiber-reinforced epoxy composite has been found to be a cost-effective substitute for typical low density ablators in a low shear lifting environment (52).

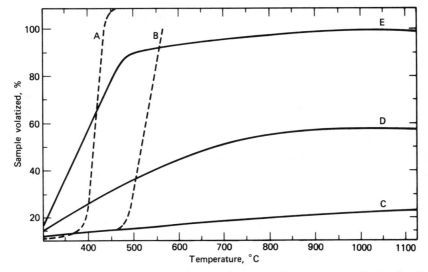

Fig. 6. Decomposition of polymers as a function of temperature during heating. A, polymethylene; B, polytetrafluoroethylene; C, silicone; D, phenolic resin; E, epoxy resin.

Cork [61789-98-8] is an effective low cost charring ablator. In order to reduce moisture absorption and related poor performance, cork particles are often blended in a silicone or phenolic resin. The result is a uniform ablative material in a sheet form that is easy to apply.

It should be noted that a number of low density ablators contain either glass or phenolic microballoons. Advantages include reduction of the total unit weight of the heatshield and lower thermal conductivity of the base, resulting in improved insulative properties. A very light weight ablative material, composed of glass and phenolic microballoons and cork particles in a silicone resin, has been shown to protect the fuel tank on the space shuttle. A unique gas-injection method was developed to mold the material to the proper configuration (53). Polyurethane foams have also been considered as fuel tank ablative material (54).

Wood has been used as an effective low cost charring ablator. The Chinese successfully used white oak as the heatshield for their RRS FSW vehicle. Wood was recommended as an alternative to more expensive heatshield materials for commercial reusable satellites. However, the safety factors for this type of material should be very high, since there is no easy way to guarantee uniformity.

Standard NDT techniques cannot distinguish between cracks and naturally oc-curring growth rings of various sizes in thick wooden parts.

Intumescent Ablators

Additives in an intumescent ablator form a foamlike region on exposure to heat. This process causes the material thickness to increase significantly, resulting in improved insulation performance. Intumescent ablators are sometimes classified as charring ablators because they form a surface char. However, there are several basic differences between intumescent and charring ablators. The basic intu-mescent decomposition reaction is exothermic, whereas the charring decomposi-tion is endothermic. However, inorganic fillers are usually added to intumescent materials to produce a net endothermic reaction (55). In addition, an intumescent reaction results in decreased thermal conductivity and increased specific heat as the material temperature increases. Conversely, a charring reaction produces a net increase in thermal conductivity and a decrease in specific heat as the material temperature increases. Thus as the net material temperature increases, the thermal diffusivity of an intumescent ablator decreases and that of a charring ablator increases.

The low thermal diffusivity and high depth of penetration make intumescent ablators useful as insulators in transient heat conduction systems. A typical application is as a protective coating for munitions stored on naval ships. In the event of a fire, the insulative properties of the intumescent ablator should provide for more escape time before the munitions detonate from the heat. Intumescent ablators are also used to coat load-carrying beams for bridges, oil rigs, and other structures. The insulative properties of the intumescent ablator should keep the temperatures of the load-carrying beams low, thereby maintaining the high strength of the beam material and delaying the collapse of the structure.

Intumescent ablators are not generally used in severe thermal environments such as reentry, because these materials usually have higher densities than char-ring ablators and the drastic shape changes they undergo would be detrimental to aerodynamic performance. Intumescent ablators usually possess good mechani-cal strength, however, and the ablative coating is capable of adding strength to a structure at high temperatures. In some cases a metallic mesh is incorporated to improve resistance of the char to erosive forces.

Several forms of intumescent materials are available from a number of suppliers (56). Some are available as rubberized sheets that can be bonded to simple shaped structures. Others are supplied as a tape, paste, or spray-on coating. A typical example of an intumescent coating material is CHARTEK 59, a high performance, lightweight, epoxy-based material from Textron that can be applied by either spray coating or troweling (57). It adheres well to steel and is used heavily in the hydrocarbon processing industry where protection from high temperature fires is a serious design consideration for support structures. An-other example is Interam, a rubber-based intumescent material manufactured by the 3M Co. In one test, Interam was used as a wrap to protect bags of howitzer propellant from heat generated by a nearby explosion. The material successfully prevented ignition of the howitzer propellant, thereby enhancing survival of personnel on military vehicles in the event of a hit in the munitions area (58).

BIBLIOGRAPHY

"Ablation" in *ECT* 2nd ed., Vol. 1, pp. 11–21, by I. J. Gruntfest, General Electric Company; "Ablative Materials" in *ECT* 3rd ed., Vol. 1, pp. 10–26, by E. R. Stover, P. W. Juneau, Jr., and J. P. Brazel, General Electric Company.

1. D. K. Huzel, *Pennemunde to Canaveral,* Prentice-Hall, Inc., Englewood Cliffs, N.J., 1962.
2. H. J. Allen, *J. Aeronaut. Sci.* **25,** 217 (1958).
3. M. C. Adams, *Jet Propul.* **29,** 625 (1959).
4. A. M. Morrison, *J. Spacecr. Rockets* **12,** 633 (1975).
5. P. J. Schneider and co-workers, *J. Spacecr. Rockets* **10,** 592 (1973).
6. H. L. Moody and co-workers, *J. Spacecr. Rockets* **13,** 746 (1976).
7. D. L. Schmidt in D. V. Rosato and R. T. Schwartz, eds., *Environmental Effects on Polymeric Materials,* Wiley-Interscience, New York, 1968, pp. 487–587.
8. D. L. Schmidt in G. F. D'Alelio and J. A. Parker, eds., *Ablative Plastics,* Marcel Dekker, Inc., New York, 1971, pp. 1–39 (a good reference book); *J. Macromol. Sci. Chem.* **3,** 327 (1969).
9. M. L. Minges in G. F. D'Alelio and J. A. Parker, eds., *Ablative Plastics,* Marcel Dekker, Inc., New York, 1971, pp. 287–313.
10. H. K. Hurwicz, T. Munson, R. E. Mascola, and J. Cordero, *Astronaut. Aerosp. Eng.* **1**(7), 64–73 (Aug. 1963).
11. H. N. Kelley and G. L. Webb, *Assessment of Alternate Thermal Protection Systems for the Space Shuttle Orbiter* (AIAA/ASME 3rd Joint Thermophysics, Fluids, Plasma and Heat Transfer Conference, June 7–11, 1982, St. Louis, Mo., AIAA-82-0899, 1982.
12. Ref. 10, p. 65.
13. C. D. Pears, *Characterization of Several Typical Polygraphites with Some Convergence on Solid Billet ATJ-S* (Proceeding of the Conference on Continuum Aspects on Graphite Design, 1970), CONF-7001105, NTIS, Springfield, Va., 1972, pp. 115–136.
14. J. G. Baetz, *Characterization of Advanced Solid Rocket Nozzle Materials* (SAMSO-TR-75-301), Air Force Rocket Propulsion Laboratories, Edwards AFB, Calif., Dec. 1975.
15. D. E. Hunton, *Sci. Am.* **261,** 92–98 (Nov. 1989).
16. Arnold Engineering Development Center, *Test Facility Data Base, Vol. 3: Aerothermal Test Facilities, Aeroballistic and Impact Ranges, and Space Environmental Chambers,* Arnold Air Force Base, Tenn., Oct. 1988.
17. D. L. Schmidt in ref. 9, p. 5.
18. D. L. Peterson and W. E. Nicolet, *J. Spacecr. Rockets* **11,** 382 (1974).
19. M. R. McHenry and B. Laub, *Ablative Radome Materials Thermal-Ablation and Erosion Modelling* (13th Intersociety Conf. on Environmental Systems, San Francisco, Calif., July 11–13), 1983.
20. S. L. Koontz, K. Albyn, and L. Leger, *Inst. Environ. Sci.* 50–59 (March/April 1990).
21. R. R. Laher and L. R. Megill, *Planet. Space Sci.* **36,** 1497–1508 (1988).
22. H. Hurwicz and J. E. Rogan, "Ablation," in W. M. Rohsenow and J. P. Hartlett, eds., *Handbook of Heat Transfer,* McGraw-Hill Book Co., Inc., New York, 1973, Sect. 16.
23. K. M. Kratsch and co-workers, AFML-TR-70-307, Vol. IV, ASD, Wright-Patterson AFB, Ohio, May, 1973.
24. J. H. Lundell, R. R. Dickey, and J. T. Howe, *Simulation of Planetary Entry Radiative Heating With a CO_2 Gasdynamic Laser* (ASME Conference on Environmental Systems, San Francisco, Calif., July 1975), American Society of Mechanical Engineers, New York, 1975.
25. P. D. Zavitsanos, J. A. Golden, and W. G. Browne, *Study of Laser Effects on Heatshield Materials* (final report), General Electric Co., Philadelphia, Pa., Jan. 1979.

26. J. Delmonte, *Technology of Carbon and Graphite Fiber Composites,* Van Nostrand Reinhold Company, New York, 1981, p. 398.
27. A. R. Taverna and L. E. Mcallister in J. Buckley, ed., *Advanced Materials, Composites and Carbon,* American Ceramic Society, Columbus, Ohio, 1972, pp. 203–211.
28. K. M. Kratsch, J. C. Schutzler, and D. A. Eitman, *Carbon–Carbon 3D Orthogonal Material Behavior* (AIAA Paper No. 72365, AIAA-ASME-SAE 13th Structural Dynamics and Materials Conference, 1972), American Institute of Aeronautics and Astronautics, New York, 1972.
29. E. R. Stover and co-workers, *11th Biennial Conference on Carbon,* CONF-730601, NTIS, Springfield, Va., 1973, pp. 277, 335–336.
30. J. L. Perry and D. F. Adams, *Carbon* **14,** 61 (1976).
31. Product data, Textron Specialty Materials Division, Lowell, Mass., 1990.
32. Product data, Fiber Materials, Inc., Biddeford, Maine, Jan. 1975.
33. J. J. Gebhardt and co-workers, in M. L. Deviney and T. M. O'Grady, eds., *Petroleum Derived Carbons* (ACS Symposium Series No. 21), American Chemical Society, Washington, D.C., 1976, pp. 212–217.
34. J. Brahney, *Aerosp. Eng.* **7**(6) (June 1987).
35. J. M. Williams and R. J. Imprescia, *J. Spacecr. Rockets* **12,** 151 (1975).
36. *Materials,* Office National d'Etudes et de Recherches Aerospatiales, France, 1985, pp. 18–19.
37. T. M. Place, *Proceedings of the 12th Symposium on Electromagnetic Windows,* Georgia Institute of Technology, Atlanta, Ga., 1974, pp. 47–51.
38. H. E. Goldstein in ref. 9, pp. 12, 323.
39. H. Shirai, K. Tabei, and S. Akiba, *JSME Int. J.* **30**(264) 945–949 (1987).
40. R. Srinivasan, *Polym. Degrad. Stab.* **17**(3), 193–203 (1987); *Chem. Abstr.* **107,** 67818u (1987).
41. D. Dijkkamp and A. S. Gozdz, *Phys. Rev. Lett.* **58**(20), 2142–2145 (1987); *Chem. Abstr.* **107,** 40493x (1987).
42. C. D. Lutes, *Nonlinear Modeling and Initial Condition Estimation for Identifying the Aerothermodynamic Environment of the Space Shuttle Orbiter,* Masters thesis, Air Force Institute of Technology, WPAFB, Ohio, Jan. 1984.
43. *A NASA Spinoff Cools the Fire,* NASA Tech Briefs, March, 1988, p. 12.
44. R. W. Farmer, *Extended Heating Ablation of Carbon Phenolic and Silica Phenolic,* AFML-TR-74-75, ASD, WPAFB, Ohio, Sept. 1974.
45. *Ablative Materials Handbook,* U.S. Polymeric, Inc., Santa Ana, Calif., 1964.
46. A. Balkrishnan, W. Nicolet, S. Sandhu, and J. Dodson, *Galileo Probe Thermal Protection: Entry Heating Environments and Spallation Experiment Design,* Acurex Corp./ Aerotherm, Mountain View, Calif., Nov. 1979.
47. J. I. Yuck and S. Y. Mo, *Han'guk Somyu Konghakhoechi* **24**(5), 444–452 (1987) (Korean); *Chem. Abstr.* **108,** 132936t (1988).
48. E. P. Bartlett and L. W. Anderson, *J. Spacecr. Rockets* **8,** 463 (1971).
49. T. F. McKeon in ref. 9, pp. 259–286.
50. P. W. Juneau, Jr., *Third Annual Polymer Conference Series, Program VIII, 1972,* University of Detroit, Detroit, Mich., 1972.
51. U.S. Pat. 4,772,495 (Sept. 20, 1988), S. E. Headrick and R. L. Hill (to United Technologies Co.).
52. W. A. Sigur, *SAMPE Q.* **17**(2), 25–33 (1986).
53. K. A. Seeler and L. Erwin, *SAMPE Q.* **17**(3), 40–48 (1986).
54. C. Williams and L. Ronquillo, *Thermal Protection System for the Space Shuttle External Tanks,* 6th SPI Intl., Tech./Mark. Conf., 1983, pp. 90–100.
55. U.S. Pat. 4,088,806 (May 9, 1978), P. W. Sawko and S. R. Riccitiello (to NASA).
56. J. M. Leary, *Characteristics of Various Types of Ablative Materials with Associated*

Naval Applications, Thesis, Massachusetts Institute of Technology, Cambridge, Mass., 1983 (a very good reference paper).
57. Product Brochure, Textron Specialty Materials Division, Lowell, Mass., 1990.
58. C. Paone, *Preventing Cook-Off with Intumescent Materials,* Army RD&A Bulletin, Jan.–Feb. 1990.

MICHAEL FAVALORO
Textron Defense Systems

ABRASIVES

An abrasive is a substance used to abrade, smooth, or polish an object. If the object is soft, such as wood, then relatively soft abrasive materials may be used. Usually, however, abrasive connotes very hard substances ranging from naturally occuring sands to the hardest material known, diamond.

The use of abrasive materials to shape, scour, and polish implements or weapons has roots deep in antiquity. Much of the history is lost but we know abrasives were used over two million years ago to sharpen and shape weapons, other implements, and ornamental items. An archaeological find in Egypt of an iron dagger accompanied by a sharpening sandstone is probably the earliest dated use of an abrasive for sharpening metal (1). These items were dated to between 1550 and 1100 BC. Three thousand years later Leonardo da Vinci's notebooks (ca AD 1500) show three intricate grinding machines utilizing either grinding wheels or grinding belts, including one designed to polish and sharpen needles (2). The use of diamonds as engraving tools dates back to biblical times; bonded diamond abrasive wheels were in use as early as 1824 (3).

Abrasives have evolved into an essential component of modern industry. Sandstone, emery, and corundum were the abrasives of choice until the late 1800s when artificial materials were developed. Today synthetic abrasives offer such improved performance that the natural ones have been largely replaced except for jobs where cost is paramount. In 1987 U.S. statistics (4) showed natural abrasive production to be about $7 million while that of crude manufactured abrasives was over $182 million. Total value of abrasives and abrasive products worldwide is estimated to be over 6 billion dollars.

There are three basic forms of abrasives: grit (loose, granular, or powdered particles); bonded materials (particles are bonded into wheels, segments, or stick shapes); and coated materials (particles are bonded to paper, plastic, cloth, or metal).

Advances in grinding wheels, abrasive belts, and the grinding process have been controlled primarily by the development of abrasives and to a lesser extent by advances in bonding and manufacturing methods. Without abrasives, modern industrial production would be impossible. The U.S. Government alone has over 300,000 tons of abrasives in its strategic National Defense Stockpile (4).

Properties of Abrasive Materials

Hardness.   Table 1 lists the various scales of hardness (qv) used for abrasives. The earliest scale was developed by the German mineralogist Friedrich Mohs in 1820. It is based on the relative scratch hardness of one mineral compared to another ranging from talc, assigned a value of one, to diamond, assigned a value of ten. Mohs' scale has two limitations; it is not linear and, because most modern abrasives fall between 9 and 10, there is insufficient delineation. Ridgeway (5) modified Mohs' scale by giving garnet a hardness value of 10 ($H = 10$) and making diamond 15. Woodell (6) extended the scale even further by using resistance to abrasion, where diamond equals 42.5. This method is dynamic and less affected by surface hardness variations than the other methods which involve indentation.

Table 1. Scales of Hardness

Material	CAS Registry Number	Mohs' scale	Ridgeway's[a] scale	Woodell's[b] scale	Knoop hardness[c], $kN/m^{2\ d}$
talc	[14807-96-6]	1			
gypsum	[13397-24-5]	2			
calcite	[13397-26-7]	3			
fluorite	[7789-75-5]	4			
apatite	[1306-05-4]	5			
orthoclase	[12251-44-4]	6	6		
vitreous silica	[60676-86-0]		7		
quartz	[14808-60-7]	7	8	7	8
topaz	[1302-59-6]	8	9		13
garnet	[12178-41-5]		10		13
corundum	[1302-74-5]	9		9	20
fused ZrO_2	[1314-23-4]		11		11
fused ZrO_2/Al_2O_3[e]					16
fused Al_2O_3	[1344-28-1]		12		21[f]
SiC	[409-21-2]		13	14	24[f]
boron carbide	[12069-32-8]		14		27
cubic boron nitride	[10043-11-5]				46[f]
diamond	[7782-40-3]	10	15	42.5	78

[a]Ref. 5.
[b]Ref. 6.
[c]At a 100-g load (K-100) average. Ref. 7.
[d]To convert kN/m^2 to kgf/mm^2 divide by 0.00981.
[e]39% ZrO_2 (NZ Alundum).
[f]Ref. 8.

Knoop developed an accepted method of measuring abrasive hardness using a diamond indenter of pyramidal shape and forcing it into the material to be evaluated with a fixed, often 100-g, load. The depth of penetration is then determined from the length and width of the indentation produced. Unlike Woodell's method, Knoop values are static and primarily measure resistance to plastic flow and surface deformation. Variables such as load, temperature, and environment, which affect determination of hardness by the Knoop procedure, have been examined in detail (9).

A linear relationship exists between the cohesive energy density of an abrasive (10) and the Woodell wear resistance values occurring between corundum ($H = 9$) and diamond ($H = 42.5$). The cohesive energy density is a measure of the lattice energy per unit volume.

Toughness. An abrasive's toughness is often measured and expressed as the degree of friability, the ability of an abrasive grit to withstand impact without cracking, spalling, or shattering. Toughness is often considered a measure of resistance to fracture and given the symbol K_c. This value is directly related to the load on an indenter required to initiate cracking and leads to a brittleness index defined as hardness/K_c (11).

A practical industry friability test (12) for abrasives involves careful sizing of subject grains to pass a given sieve size while being retained on the next finer screen. A unit weight of this grain is then ball-milled using a standard steel ball load for a given time. The percentage of milled grain retained on the original screen is a measure of toughness or lack of friability. Other methods of evaluating this property involve centrifugally impacting sized grits and then evaluating the debris (13).

Refractoriness (Melting Temperature). Instantaneous grinding temperatures may exceed 3500°C at the interface between an abrasive and the workpiece being ground (14). Hence melting temperature is an important property. Additionally, for alumina, silicon carbide, B_4C, and many other materials, hardness decreases rapidly with increasing temperature (7). Fortunately, ferrous metals also soften with increasing temperatures and do so even more rapidly than abrasives (15).

Chemical Reactivity. Any chemical interaction between abrasive grains and the material being abraded affects the abrasion. Endurance scratch tests made on polished glass and iron rolls using conical grains of aluminum oxide and silicon carbide (16) showed that silicon carbide produced a long scratch path on the glass roll and a short path on the steel roll. Exactly the opposite was true for aluminum oxide. These effects are explained by the reactivities of the two abrasives toward glass and steel. Silicon carbide resists attack by glass but readily dissolves in steel, whereas aluminum oxide is attacked by glass and is inert to steel. The advent of boron carbide, harder than either fused aluminum oxide or silicon carbide, brought grand hopes for its use in grinding wheels and belts. Boron carbide's ease of oxidation and its reactivity toward both metals and ceramics prevented these developments.

Thermal Conductivity. Abrasive materials may transfer heat from the cutting tip of the grain to the bond posts, retaining the heat in a bonded wheel or coated belt. The cooler the cutting point, the harder it is. Fused zirconium oxide has a relatively low thermal conductivity compared to other abrasive materials. It also has a lower hardness than aluminum oxide, yet it performs quite well on hard-to-grind materials. This is attributed to the decreased heat flow from the grinding interface into the grain (whose hardness decreases as its temperature rises) and to the bond (subject to heat degradation).

Fracture. Fracture characteristics of abrasive materials are important, as well as the resulting grain shapes. Equiaxed grains are generally preferred for bonded abrasive products and sharp, acicular grains are preferred for coated ones. How the grains fracture in the grinding process determines the wear resistance and self-sharpening characteristics of the wheel or belt.

Microstructure.   Crystal size, porosity, and impurity phases play a major role in fixing the fracture characteristics and toughness of an abrasive grain. As an example, rapidly cooled fused aluminum oxide has a microcrystalline structure promoting toughness for heavy-duty grinding applications, whereas the same composition cooled slowly has a macrocrystalline structure more suitable for medium-duty grinding.

Natural Abrasives

Naturally occurring abrasives are still an important item of commerce, although synthetic abrasives now fill many of their former uses. In 1987 about 156 million metric tons of natural abrasives were produced in the United States. Production was up from 1986 because of increased nonabrasive uses and increased use of garnet in sandblasting (4).

Diamond.   Diamond [7782-40-3] is the hardest substance known (see CARBON, DIAMOND). It has a Knoop hardness of 78–80 kN/m^2 (8000–8200 kgf/m^2). The next hardest substance is cubic boron nitride with a Knoop value of 46 kN/m^2, and its inventor, Wentorf, believes that no manufactured material will ever exceed diamond's hardness (17). In 1987 the world production of natural industrial diamonds (4) was about 110 t (1 g = 5 carats). It should be noted that whereas the United States was the leading consumer of industrial diamonds in 1987 (140 t) only 260 kg of natural industrial diamonds were consumed; this is the lowest figure in 48 years (4), illustrating the impact that synthetic diamonds have made on the natural diamond abrasive market.

Although all diamonds are carbon in a crystalline cubic structure, industrial diamonds occur as three types. Bort is diamond in single crystal fragments which are off-color or otherwise unsatisfactory for gems. Ballas consists of spherical masses of minute, intergrown diamond crystals arranged more or less concentrically. Carbonado, the toughest form of natural industrial diamond, is an impure form of diamond consisting of diamond, graphite, and amorphous carbon. The latter two types are found mainly in Brazil, but bort, found mostly in Africa, makes up more than 90% of all natural industrial diamond production.

Abrasive applications for industrial diamonds include their use in rock drilling, as tools for dressing and trueing abrasive wheels, in polishing and cutting operations (as a loose powder), and as abrasive grits in bonded wheels and coated abrasive products.

Corundum.   Corundum [1302-75-5] (see ALUMINUM COMPOUNDS) is a naturally occurring massive crystalline mineral composed of aluminum oxide. It is an impure form of the gems ruby and sapphire. Prior to 1900 corundum was an important abrasive for the production of grinding wheels. Today it is mainly employed as a loose abrasive for grinding and polishing optical lenses. Almost all the world's supply of corundum now comes from Africa, primarily from Zimbabwe.

Emery.   Emery [57407-26-8] is a dark granular rock consisting of an intimate mixture of corundum (Al_2O_3) and magnetite (Fe_3O_4) or hematite (Fe_2O_3)

together with impurities of titania, magnesia, and silica. The best grades of emery for abrasive use are mined in Turkey and Greece. A small quantity of emery is used in coated abrasive products, but its principal use in the United States is in nonskid, wear-resistant floors and pavements. High tonnages are reportedly shipped to the Far East to be used in millstones for grinding rice (18). Turkey and Greece, the major world suppliers of emery, produced about 17,500 t in 1987 (4).

Garnet. Garnet [12178-41-5] is the name given to a group of silicate minerals possessing similar physical properties and crystal forms but differing in chemical composition. Seven species exist but the two most important are pyrope, a magnesium aluminum silicate, and almandine, an iron aluminum silicate. The formula for garnet is $(MO)_3M_2'O_3(SiO_2)_3$: The divalent element, M, can be calcium, magnesium, iron, or manganese; the trivalent, M', is aluminum, iron, chromium, or titanium.

The United States produces 63% of the world's annual crop of garnet, followed by India, Australia, China, and the USSR (4). The largest U.S. producer is Barton Mines Corporation in upper New York State. This mine has been in continuous operation since the 1880s and produces a high grade abrasive quality garnet (18). U.S. production and use of garnet in 1987 totaled almost 38,350 t, valued at a record $4.35 million (4). Garnet usage has increased in recent years because its use as a sandblast medium reduces the risk of silicosis. It is also increasingly used in water filtration.

Silica. Silica (qv) comes in various forms including quartz [14808-60-7]. It has found wide use as an abrasive in the past, particularly as an inexpensive coated abrasive for woodworking. The term sandpaper is still used as a generic term for coated abrasives in many quarters although the use of sand in coated abrasives has been almost entirely eliminated because of the hazard of silicosis to the user and its inferior grinding properties (especially for metals).

Sandstone. Sandstone wheels were once quarried extensively for farm and industrial use, and special grades of stone for precision honing, sharpening, and lapping are a small but important portion of today's abrasive industry. Production of honing and sharpening stones from deposits of dense, fine grain sandstone in Arkansas account for 76% of the value (about $2 million in 1987) and 88% of the total quantity of such stones in the United States (4).

Tripoli. Tripoli [1317-95-9] is a fine grained, porous, decomposed siliceous rock produced mainly in Arkansas, Illinois, and Oklahoma. It is widely used for polishing and buffing metals, lacquer finishing, and plated products. Since tripoli particles are rounded, not sharp, it has a mild abrasive action particularly suited for polishing. Tripoli is also used in toothpastes, in jewelry polishing, and as filler in paints, plastics, and rubber. Rottenstone and amorphous silica are similar to tripoli and find the same uses. In 1987 the abrasive use of tripoli in the United States totaled 26.6 million tons and was valued at about $3.1 million; however, the portion used as a filler totaled 71.1 million tons and had a value of almost $10 million (4).

Pumice and Pumicite. Pumice and pumicite are porous, glassy forms of lava, rich in silica. Both pumice, the massive form, and pumicite, the powder or dust form, have been widely used as a mild abrasive for polishing operations. This use, however, has continued to decline. Currently, only about 1% of their production is for abrasives (18).

Miscellaneous Natural Abrasives.   Powdered feldspar [68476-25-5] is used as a mild abrasive in cleansing powders, and clays are sometimes used in polishing powders. Staurolite [12182-56-8] is a complex hydrated aluminosilicate of iron, of high density (3.74–3.83 g/mL) and a hardness of 7 to 8 on Mohs' scale. It is primarily used as a sandblasting grit, but silicosis hazards had cut production in 1987 about 25% compared to that of 1986 (4).

Manufactured Abrasives

The use of automatic machine tools, often computer controlled and programmable, requires abrasive elements which have performance and lifetime reproducibility. Natural abrasives suffer on both counts: performance is limited and inconsistent quality leads to unpredictable lifetimes. Manufactured abrasives have both superior performance and consistency; consequently, they have largely replaced the natural ones.

Silicon Carbide.   The first artificial abrasive, silicon carbide [409-21-2], SiC, was produced by Edward Acheson in 1891 (19). This invention led to the formation of the Carborundum Company. The registered trademark, Carborundum, is essentially synonymous with silicon carbide.

Silicon carbide is produced from quartz sand and carbon in large electric furnaces in which the charge acts as the refractory container and thermal insulator for the ingot being formed. Reaction temperatures range from 1800 to 2200°C, melting the quartz sand which then reacts with the solid carbon to form crystalline silicon carbide. Both green SiC and black SiC are produced. Woodell's (6) lapping tests indicate that they have the same hardness, although others (18) state that the green is harder than the black. Green SiC is more expensive and is often preferred for tool grinding wheels because the tool is more visible against a green background. Black SiC is preferred for grinding low tensile strength materials such as cast iron,. chilled iron rolls, marble, ceramics, and aluminum. It also is widely used as a wire-sawing abrasive in quarrying building stone.

In 1989 U.S. and Canadian plants produced 116,600 t of crude silicon carbide valued at $56.4 million (20). Abrasive uses accounted for 46% of the total tonnage and 48% of the total value.

Fused Aluminum Oxide.   Fused aluminum oxide [1344-28-1] was the next manufactured abrasive to appear (19). By 1900 the first commercially successful fused alumina of controlled friability was produced from bauxite [1318-16-7]. The Norton Company obtained the rights and patents in 1901 and constructed the first plant to produce this material on a commercial basis. In 1904 the Higgins arc furnace, which used a water-cooled steel shell instead of a refractory lining, was first used (21). A typical Higgins furnace is about 2 m in diameter by 2 m in height, producing an ingot of about 5½ t. The fusion and slow cooling of a mixture of bauxite, coke, and iron turnings gives a coarse crystalline product of about 95% alumina and 0.7% titania [13403-67-7], designated regular aluminum oxide. By adding more coke to the charge, greater reduction is obtained and the percentage of residual titania is reduced, producing semifriable alumina (about 97% alumina). As the name implies, this material is less tough than the regular alumina. After crushing, further heat treatment in rotary furnaces is used to alter the valence state and solubility of titania, producing an exsolved disperse phase which affects the impact strength of the resulting product. Some Higgins furnaces are still in use, but most aluminum oxide is now fused in tilting furnaces and poured into ingots of sizes suitable for the desired rate of cooling and resulting crystal size. Small ingots cool rapidly and produce microcrystalline alumina, tougher and stronger than regular.

Bayer alumina (see ALUMINUM COMPOUNDS), containing about 0.5% soda as its only significant impurity, is also the starting material for the production of fused white aluminum oxide abrasive. During the fusion process much of the soda is volatilized, producing small bubbles and fissures in the final product and giving a slightly less dense, much more friable abrasive than regular or semifriable aluminum oxide. This white abrasive is widely used in tool grinding as well as in other applications requiring cool cutting, self-sharpening, or a damage-free workpiece.

Special pink or ruby variations of the white abrasive are produced by adding small amounts of chromium compounds to the melt. The color is dependent on the amount of chromium added. A green alumina, developed by Simonds Abrasives, results from small additions of vanadia [11099-11-9]. Each was developed to improve on the suitability of white abrasive for tool and precision grinding.

A "single-crystal" fused alumina abrasive, produced by fusing bauxite in a sulfide matrix and then controlling the cooling and solidification rate to grow the required sizes of single crystals, has also been developed (22). This abrasive is even better than white abrasive for tool grinding and is particularly effective for large area vertical spindle surface grinding. The process is not environment-friendly and it is difficult to change the output range of grit sizes quickly to match the desires of the marketplace. An abrasive of this type formerly produced in Sweden was abandoned when customers would not purchase the complete range of grit sizes produced.

In 1989, 164,200 t of regular fused alumina abrasive and 30,630 t of high purity fused alumina were produced in the United States and Canada, valued at U.S.$62.3 and $19.4 million, respectively (20).

Sintered Aluminum Oxide. The first commercially successful sintered alumina abrasive was patented by Ueltz in 1963 (23). Powdered, calcined bauxite was pressed into blocks, granulated to size, and fired at high temperature until sintered to maximum density. The impurities in the bauxite act as sintering aids, producing a very strong product in which the crystals are quite small (less than 10 μm). This abrasive, designated 75A, found immediate use in heavy-duty snagging wheels used to condition the surfaces of stainless steel slabs and billets. It could condition from 5 to 10 times the tonnage of steel as the fast-cooled microcrystalline fused alumina previously employed.

One advantage of sintering is the close control of size and shape of the abrasive particle. Extruded, cylindrically shaped, sintered abrasives of circular cross section were produced from bauxite (24) and from calcined alumina (25). The Ueltz sintered bauxite was also later produced in extruded cylinder form and designated as 76A. Extruded sintered abrasives of a wide variety of cross-sectional configurations, eg, square and triangular, were later patented (26).

Sintered abrasives made from bauxite and calcined alumina are heavy-duty abrasives; they are much too strong and tough for precision grinding.

Sol–Gel Sintered Aluminum Oxide. A new and much more versatile sintered alumina abrasive is now produced from aluminum monohydrate, with or without small additions of modifiers such as magnesia, by the sol–gel process (see SOL–GEL TECHNOLOGY). The first modified sol–gel abrasive on the market, Cubitron, was patented (27) and produced by the 3M Corporation for products such as coated belts and disks. The success of this material promoted intensive research into sol–gel abrasives.

A higher density sol–gel abrasive, produced by the introduction of seed crystallites formed by wet-milling with high alumina media or by introduction of submicrometer α-alumina particles, was patented (28) and designated Norton SG. The microstructure of this abrasive consists of submicrometer α-alumina crystals (Fig. 1) and its bulk density approaches that of fused alumina. Norton SG has proven to be an exceptional performer in coated and bonded abrasive products; it was awarded the 1989 ASM Engineering Materials Achievement Award (29).

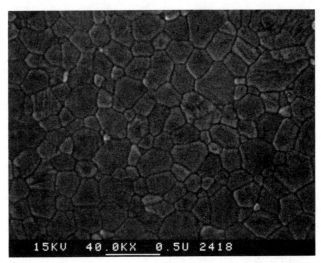

Fig. 1.   SEM photomicrograph of polished and thermally etched section of Norton SG sol–gel alumina abrasive grain.

Another sol–gel abrasive, produced by seeding with α-ferric oxide or its precursors, has been patented (30). A magnesium-modified version of this abrasive, also called Cubitron, is being produced as a replacement for the earlier type. Yttria [1314-36-9]-modified sol–gel abrasives have also been patented (31), as well as rare earth oxide modified materials (32). These abrasives are all produced by 3M Corporation; they have performed very well in various applications such as in coated abrasives for grinding stainless steel and exotic alloys.

Sol–gel abrasives proved to be exceptional performers in coated applications and in resin-bonded wheels for cut-off, flute-grinding, or other operations. Initial attempts to form vitreous bonded wheels using these materials failed because the standard high temperature (1250°C) bonds reacted with the abrasive to produce a soft, punky wheel. Firing at lower temperatures or using more viscous, less reactive, glass bonds (33) produces vitrified wheels that are now finding wide application in precision grinding, particularly for difficult-to-grind alloys. They are also used where a minimum of grinding heat is needed, such as in tool grinding and the shaping of aerospace parts. The quantity of metal removed per unit volume of grinding wheel (*G*-ratio) in dry grinding applications can be increased by a factor of ten compared to the best available fused alumina wheel.

Fused Zirconia–Alumina.   Fused zirconia [*1314-23-4*] would be an excellent abrasive for heavy snagging of steel slabs and billets except for its cost, density, and lack of toughness. Quick-cooled fusions of zirconia and alumina, however, are widely used in large resin-bonded snagging wheels for heavy-duty conditioning of steel slabs and billets and for weld-bead removal in pipeline construction. These materials, called AZ abrasives, are also used extensively in coated abrasive applications (34), but they are not used in vitrified bonded wheels because of thermal instability. There are two principal varieties of zirconia–alumina abrasives: a near eutectic combination of 40% ZrO_2 and 60% Al_2O_3 (Fig. 2), and a less costly 25% ZrO_2 and 75% Al_2O_3. Extremely rapid cooling of the melt achieved by casting into a bed of steel balls, between steel plates, or into thin sheets is required. Yearly production figures are not available but production increased in both tonnage and value in 1987 (4).

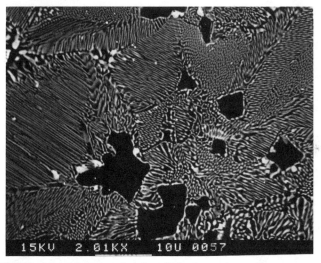

Fig. 2.   SEM photomicrograph of polished section of near eutectic alumina–zirconia abrasive grain showing white zirconia in dark alumina matrix.

Synthetic Diamond.   In 1955 the General Electric Company announced the successful production of diamonds (see CARBON, DIAMOND) from graphite under very high pressure and temperature in the presence of a metal catalyst. It was later reported that a Swedish company, Allmana Svenska Electriska AB (ASEA), had succeeded in producing diamond in 1953 (35).

Today there are four producers of diamond in the United States, the largest consumer of industrial diamonds, in addition to producers in Ireland, Sweden, USSR, and Japan. Manufactured diamond grit and powder have almost completely replaced natural diamond for abrasive work except for sizes larger than 20 grit (4). Sintered diamond, or diamond compacts, while successfully competing with carbonado and finding new uses in tool bits and wear resistant parts, is not an important diamond abrasive.

Cubic Boron Nitride. Cubic boron nitride [10043-11-5] (see BORON COMPOUNDS), or CBN, is a synthetic mineral not found in nature. It was first produced by the General Electric Company using the same equipment used for the production of synthetic diamond and is designated as Borazon. CBN is nearly as hard as diamond, yet it does not perform as well in the usual diamond grinding applications such as abrading ceramics, rock, or cemented tungsten carbide. However, CBN is an extremely efficient abrasive for grinding steel (17). Although its cost is comparable to that of synthetic diamond, it successfully competes with the inexpensive fused aluminum oxide in grinding steel tools. CBN improves grinding wheel life by as much as 100 times over that of alumina, thus increasing productivity, reducing downtime for wheel changes and dressing, and improving the quality of parts (36).

In 1985 world sales of CBN abrasive products totaled $41.4 million, 37% of which was in the Pacific area (primarily Japan), 34% in Europe, and 29% in the Americas (37).

Boron Carbide. Boron carbide [12069-32-8], B_4C, is produced by the reaction of boron oxide and coke in an electric arc furnace (70% B_4C) or by that of carbon and boric anhydride in a carbon resistance furnace (80% B_4C) (see BORON COMPOUNDS, REFRACTORY BORON COMPOUNDS). It is primarily used as a loose abrasive for grinding and lapping hard metals, gems, and optics (18). Although B_4C is oxidation-prone, the slow speed of lapping does not generate enough heat to oxidize the abrasive.

Slags. Slags from metals smelting and from coal-fired power plants find considerable use as blasting media for ships, offshore oil rigs, and other ironwork. The silica content of most slags is quite high, but in this form is not a health hazard (18).

Steel Shot and Grit. Steel shot and grit are also widely used in grit blasting and abrasive finishing. In 1989, 220,196 metric tons of metallic abrasives were produced in the United States with a combined value of $89.55 million (20).

Miscellaneous Manufactured Abrasives. Steel wool is made from a variety of steels by shaving or scraping a continuously moving wire with a fixed serrated cutting tool. It is used for finishing wood and soft metals and for scouring and cleaning. Brass wool and copper wool are used primarily for domestic cleaning.

Many metal oxides and other compounds are precipitated chemically as very fine particles and used as polishing and lapping agents. Iron oxide (as rouge) [8011-97-0], cerium oxide [1306-38-3], and zirconia are widely used in polishing glass. Crocus [5124-00-1], another form of iron oxide, is used for finishing cutlery and brass. Chromium oxide [12018-01-8], manganese dioxide [1313-13-9], and tin oxide [18282-10-5] are also used for special polishing operations. Softer abrasives such as magnesia [1309-48-4] and calcium carbonate [471-34-1] are widely used in scouring and buffing compounds.

Sizing, Shaping, and Testing of Abrasive Grains

Sizing. Manufactured abrasives are produced in a variety of sizes that range from a pea-sized grit of 4 (5.2 mm) to submicrometer diameters. It is almost impossible to produce an abrasive grit which will just pass through one sieve size yet be 100% retained on the next smaller sieve. Thus a standard range was adopted in the United States which specifies a screen size through which 99.9% of the grit must pass, maximum oversize, minimum on-size, maximum through-size, and fines. The original Bureau of Commerce size standards, although nonmetric, have been internationally recognized; even the Russian GOST standards are an exact duplicate (38). The Bureau of Commerce standards have been updated by *ANSI Standard B74.12-1982* in the United States (39) and by *FEPA Standard 42-GB-1984* in Europe (40). Table 2 shows the average diameter of grit sizes ranging from 4 to 1200. Designations for the finest grit sizes vary in the United States, Europe, and Japan. Mizuho, a Japanese company lists 10,000 grit (0.5 μm) as available.

Table 2. Bonded Abrasive Grit Sizes

Grit	Diameter, μm	Grit	Diameter, μm	Grit	Diameter, μm
4	5200	24	775	120	115
5	4500	30	650	150	95
6	3650	36	550	180	80
7	3050	40	460	220	69
8	2550	46	388	240	58
10	2150	54	328	280	48
12	1850	60	275	320	35
14	1550	70	230	400	23
16	1300	80	195	500	16
18	1100	90	165	600	8
20	925	100	138	1200	3

The permissible variation in the openings of U.S. standard test sieves varies from 15% for the coarser sizes to 60% for the range of 200 to 400 grit. To reduce this built-in error, the diamond industry has developed precision, electroformed test screens that are produced by a combination of photoengraving and electroplating. These screens, in which the accuracy and uniformity of aperture size can be controlled to 2 μm, are too expensive for routine testing; they are used instead to calibrate standard wire sieves (41). The sizing of diamond abrasive grains is much tighter than that of other abrasives; details can be found in reference 42.

Shaping. Screening is a two-dimensional process and cannot give information about the shape of the abrasive particle. Desired shapes are obtained by controlling the method of crushing and by impacting or mulling. Shape determinations are made optically and by measuring the loose-packed density of the abrasive particles: cubical particles pack better than acicular. In general, cubical particles are preferred for grinding wheels, whereas high aspect-ratio acicular particles are preferred for coated abrasive belts and disks.

Testing. Chemical analyses are done on all manufactured abrasives, as well as physical tests such as sieve analyses, specific gravity, impact strength, and loose poured density (a rough measure of particle shape). Special abrasives such as sintered sol–gel aluminas require more sophisticated tests such as electron microscope measurement of α-alumina crystal size, and indentation microhardness.

Coated Abrasives

Coated abrasive products, once limited to sandpaper in woodworking shops, are versatile and efficient industrial tools. Machines ranging from portable sanders to giant slab conditioners and roll grinders utilize coated abrasives. Abrasive belt machines now perform many of the operations that were once the exclusive province of grinding wheels.

Wearable coated abrasive wheels of the radial-flap type can grind and polish contours that are almost impossible for bonded abrasive wheels. Coated bands, sleeves, and cartridge rolls (coated abrasives wound around a spindle into a multilayered straight or tapered cartridge) add to the utility of these wheels. Additionally, coated abrasive disks are widely used for a variety of grinders, including disk grinders employed in auto body finishing and refinishing.

Coated abrasives consist of a flexible backing on which films of adhesive hold a coating of abrasive grains. The backing may be paper, cloth, open-mesh cloth, vulcanized fiber (a specially treated cotton rag base paper), or any combination of these materials. The abrasives most generally used are fused aluminum oxide, sol–gel alumina, alumina–zirconia, silicon carbide, garnet, emery, and flint.

The wear rates of alumina and alumnia–zirconia abrasive belts have been compared in reference 43. Alumina–zirconia was found to be much freer-cutting and requires less energy to remove a given volume of metal. Under moderate grinding conditions the alumina–zirconia abrasive removed ten times more metal than the alumina abrasive and under severe conditions about 30% more. The newly developed sintered sol–gel alumina abrasives are more effective in some applications than alumina–zirconia in increasing cut-rate, reducing power requirements, and increasing belt or disk life.

A new form of coated abrasive has been developed that consists of tiny aggregates of abrasive material in the form of hollow spheres. As these spheres break down in use, fresh cutting grains are exposed; this maintains cut-rate and keeps power low (44).

Manufacture of Coated Abrasives. In the manufacture of coated abrasives, the first step is called the making process. A smooth film of adhesive, either hide glue, phenolic resin, or synthetic varnish, or a combination of one of these and a mineral filler, is applied to the backing in a thickness dependent on the size of grit to be applied. Each adhesive has advantages and disadvantages (see ADHESIVES): hide glue has high initial strength but becomes soft in water and gummy in high humidity; phenolic resin is good for severe operations and where water resistance is required; varnish is limited to paper production for wet sanding operations.

In the second step, the "make coat" of abrasive grit is deposited on the coated backing either mechanically or electrostatically. Electrostatic deposition is particularly useful for coarse grit paper and cloth products because it gives a faster-cutting product. Each grain is drawn electrostatically to the backing surface, the flat side to the backing and peak side up. Grit sizes range from 12 to 600, and spacing may be close (close-packed) or open (sparsely populated). Although close coating is most often employed, open coating is used in applications on gummy materials that would clog close-spaced grits. Open coatings cover only 50 to 70% of the backing with abrasive grit.

After drying, a second coating called the "size coat," usually the same composition as the first coat, is applied; this serves to anchor the abrasive grits. Drying completes the operation and the product is then wound into large rolls and stored for later conversion. Resin-bonded products may require additional curing at elevated temperatures in roll form. If loading or clogging is anticipated, a coating of antistick stearate can be applied over the abrasive coating.

Further finishing treatment is required to convert the large rolls to consumer products such as belts, disks, and sheets. The first such operation is usually "flexing," a controlled cracking of the adhesive layers to promote flexibility in the final product. Items such as garnet and SiC finishing papers do not require flexing, but the majority of paper and cloth products are at least single-flexed by being pulled at a 90° angle over a steel bar. If more flexibility is desired, the product may be flexed a second or third time and at different angles.

Coated abrasives are supplied in widths ranging from 3.175 mm to 2.2 m in standard 45.7-m rolls. They are also formed into sheets, disks, and molded coils or rolls. Belts, regularly supplied in the widths mentioned, have been made up to 3 m wide.

Coated Abrasive Sales.   There has been a steady rise in the sale of coated abrasives in the United States: in 1938 sales were about $12 million; in 1988 sales exceeded $700 million. World sales of coated abrasives for 1988 were estimated at over $2 billion.

Bonded Abrasives

Grinding wheels are by far the most important bonded abrasive product both in production volume and utility. They are produced in grit sizes ranging from 4, for steel mill snagging wheels, to 1200, for polishing the surface of rotogravure rolls. Wheel sizes vary in diameter from tiny mounted wheels for internal bore grinding to 1.83-m wheels for cutting off steel billets. Bonded shapes other than wheels, such as segments, cylinders, blocks, and honing stones, are also widely used.

Marking System.   Grinding wheels and other bonded abrasive products are specified by a standard marking system which is used throughout most of the world. This system allows the user to recognize the type of abrasive, the size and shaping of the abrasive grit, and the relative amount and type of bonding material. The individual symbols chosen by each manufacturer may vary, but the relative position for each item in the marking system is standard, as shown in Figure 3.

Grain Size.   The surface finish, or degree of roughness, produced by a grinding wheel on the workpiece being ground is roughly proportional to the size of abrasive grains in the wheel.

Grade.   The grade of a bonded abrasive product is represented alphabetically with A being the softest acting (least bond) and Z being the hardest (most bond).

Structure.   The structure designation may or may not be given. It is numeric: 0 represents the closest possible packing of abrasive grits, and ascending numbers correspond to incrementally less abrasive per unit volume of wheel (wider spaced grits).

Bond Type.   Most bonded abrasive products are produced with either a vitreous (glass or ceramic) or a resinoid (usually phenolic resin) bond. Bonding

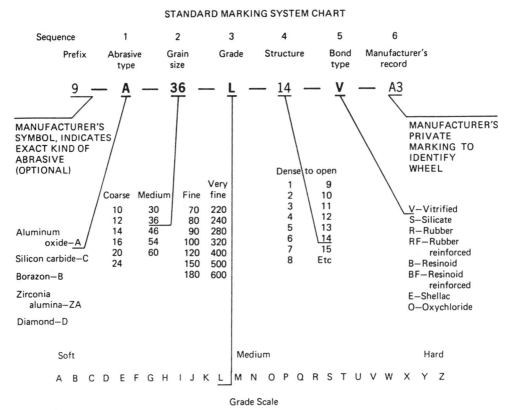

STANDARD MARKING SYSTEM CHART

Sequence	1	2	3	4	5	6
Prefix	Abrasive type	Grain size	Grade	Structure	Bond type	Manufacturer's record

9 — A — 36 — L — 14 — V — A3

MANUFACTURER'S
SYMBOL, INDICATES
EXACT KIND OF
ABRASIVE
(OPTIONAL)

MANUFACTURER'S
PRIVATE
MARKING TO
IDENTIFY
WHEEL

Dense to open

			Very	1	9
Coarse	Medium	Fine	fine	2	10
10	30	70	220	3	11
12	36	80	240	4	12
14	46	90	280	5	13
16	54	100	320	6	14
20	60	120	400	7	15
24		150	500	8	Etc
		180	600		

Aluminum oxide—A
Silicon carbide—C
Borazon—B
Zirconia alumina—ZA
Diamond—D

V—Vitrified
S—Silicate
R—Rubber
RF—Rubber reinforced
B—Resinoid
BF—Resinoid reinforced
E—Shellac
O—Oxychloride

Soft   Medium   Hard

A B C D E F G H I J K L M N O P Q R S T U V W X Y Z

Grade Scale

Fig. 3. Standard wheel markings recommended and approved by the Standards Committee of The Grinding Wheel Institute.

agents such as rubber, shellac, sodium silicate, magnesium oxychloride, or metal are used for special applications.

Ceramic (Vitrified). The first successful vitrified-bonded grinding wheels were produced in 1869 by Sven Pulson in the Worcester, Massachusetts pottery shop of Franklin P. Norton. In 1884 the business was sold to a group founding the Norton Company, which today is the world's largest producer of abrasive products. For many years a low melting slip clay was the leading vitreous bond. Natural variations in slip clay led to the development of bonds of more controllable composition: glass for aluminum oxide wheels and porcelain for grinding wheels containing the more reactive silicon carbide abrasive. Firing temperatures for vitreous bonded wheels vary widely, but they are commonly in the pyrometric cone-12 range (about 1250°C).

Resin (Resinoid). The resinoid bond, originally called Bakelite, was named for its inventor, Leo Baekeland (see PHENOLIC RESINS). Baekeland's original patent was issued in 1909, but it was not until the 1920s that this type of bonding was perfected for use in abrasives. The resin consists basically of phenol and formaldehyde. It is thermosetting in nature, making it particularly suitable for tough grinding and cutting-off operations at high speeds. Resin bonds cure at temperatures of 150 to 200°C. Thus there are no problems with grain reactivity and fiberglass or metal reinforcements may be molded in to make a much safer product for high speed grinding. Additionally, this low processing temperature allows the use of inert fillers to strengthen the bonded product, or of "active" fillers to increase the efficiency of grinding. Active fillers include such materials as cryolite, pyrites, potassium fluoroborate, sodium and potassium chloride, zinc sulfide, antimony sulfide, and tin powder. Such materials, alone or in combination, aid grinding by acting as extreme pressure lubricants or as reactants for the metal being ground; this prevents rewelding of the chips being removed. Lead and lead compound fillers are still used in Europe for some cut-off wheels (special marking required) but are not used in the United States for health reasons.

Resinoid-bonded wheels find wide use in heavy-duty snagging operations where large amounts of metal must be removed quickly, in cutting-off operations, portable disk grinding (as for weld-beads), roll grinding, and vertical spindle disk grinding.

Rubber.   Both natural and synthetic rubber are used as bonding agents for abrasive wheels. Rubber-bond wheels are ideal for thin cut-off and slicing wheels and centerless grinding feed wheels. They are more flexible and more water-resistant than resinoid wheels.

In manufacture, the abrasive grain is mixed with crude rubber, sulfur, and other ingredients for curing, then passed through calender rolls to produce a sheet of desired thickness. The wheels are stamped from this sheet and heated under pressure to vulcanize the rubber.

Shellac.   Shellac wheels are limited to a few applications where extreme coolness of cut is required and wheel life is immaterial. They are produced by mixing shellac [9000-59-3] and abrasive grain in a heated mixer, then rolling or shaping to the desired configuration.

Silicate.   Once important, silicate-bonded wheels have faded almost to oblivion; increasing wheel speeds have made this inherently weak bond obsolete. Some wheels are produced, however, for old grinding equipment still in existence.

Magnesium Oxychloride.   A mixture of abrasive grains, MgO, water, and $MgCl_2$ placed in an appropriate mold will cold-set to form a grinding wheel that is then cured for a long period of time in a moist atmosphere. This type of bond finds some use in disk grinding applications and cutlery grinding. Like silicate wheels, however, these have been largely replaced by soft-acting resinoid bonds.

Methods of Manufacture

Bonded abrasive products are made as wheels, disks, cylinders, sticks, blocks, and segments, all of which are defined in *ASA-B74.2-1974,* "USA Standard Specifications for Shapes and Sizes of Grinding Wheels." This bulletin is sponsored by the Grinding Wheel Institute; it is obtainable from most wheel manufacturers.

The first step in the manufacture of both resin- and ceramic-bonded abrasives is the wetting of a weighed quantity of abrasive grain with a liquid pick-up agent. This agent is usually water or a glue solution for ceramics and liquid resin or furfural for resins. A weighed quantity of the bonding material is then mixed with the wetted grain until the bond is uniformly coated onto the abrasive. A weighed portion of this mixture is leveled into a mold and pressed to a predetermined bulk density. All ceramic-bonded wheels are pressed cold, but some resin-bonded wheels are pressed hot to achieve almost zero porosity. After pressing, the wheel has the calculated volume of abrasive, bond, and pores corresponding to the grade and structure desired.

Pressed wheels are heat-treated to fuse the ceramic bond or cure the resin bond. Ceramic-bonded wheels are fired to about 1250°C in continuous tunnel or periodic kilns, and resin-bonded ones are baked in periodic ovens at 175 to 200°C. After firing or baking, the wheels are checked for proper dimension, density, and modulus of elasticity, and sometimes for sandblast resistance to assure proper grade. These slightly oversize wheels are then finished, ie, machined to final size. To achieve desired thickness, the wheel sides are rubbed under heavy pressure on large rotating steel tables charged with steel shot. Diameter reduction is accomplished using rotating conical steel cutters mounted in a lathe post holder. Holes may be trued using conical cutters or diamond tools but many times they are simply bushed with a plastic, lead, or sulfur compound by pouring or injection molding.

After finishing, the wheels are again checked for dimension and carefully balanced. All wheels over 15 cm in diameter are tested at rotation speeds higher than their final operating speeds. Test and operating speeds, specification, and balance points for large wheels are marked on the wheel itself or on wheel blotters. Blotters are thin disks of paper or fiber placed on the sides of wheels to equalize the clamping pressure of mounting flanges on a grinding machine. Testing speeds and maximum safe operating speeds for the various sizes and types of wheels are given in "ANSI Safety Code for the Use, Care, and Protection of Abrasive Wheels" (*ANSI B7.1-1978*). This also may be obtained from most wheel manufacturers. Both users and manufacturers are required to adhere to this code to assure the personal safety of the grinding operator.

Special Forms of Bonded Abrasives

There are many specialized forms and uses of bonded abrasives, a detailed discussion of which is found in reference 45.

Honing and Superfinishing. Honing and superfinish stones are produced from large vitrified-bonded abrasive blocks that are diamond sawed to smaller rectangular pieces suitable for mounting in metal or plastic holders. Honing stones, used to true engine and hydraulic cylinders, can vary in grit size from 36 (0.55 mm) to 600 (8 μm); superfinish stones, used to polish the external diameters of machine and automotive parts, vary in grit size from 600 to 1200 (3 μm). Both types of stones are quite soft; steel ball indentation hardness and density are often used as a quality control to measure grade.

Pulpstone Wheels. Grinding wheels play an important role in the production of paper pulp (qv). Massive pulpstone wheels are made from vitrified abrasive segments, bolted and cemented together around a reinforced concrete central body. They may be up to 1.80 m in diameter and have a breadth of 1.70 m. In operation, debarked wood logs are fed into a machine and forced against the rotating pulpstone, which shreds the wood into fibers under a torrent of water.

The ground fibers are then screened and passed through subsequent operations to produce various types of paper.

Crush-Form Grinding Wheels. In crush-form grinding, a rotating, contoured crushing wheel is forced into the face of a revolving vitrified wheel, crushing the face to the exact contour needed on the metal object to be ground. The contoured wheel is then placed in production and when wear or dulling occurs, the face is again crushed to regain proper contour. Many parts formerly turned with metal-cutting tools and then surface ground are now shaped and surface finished in one pass of a crush-formed wheel.

Creep Feed Wheels. Abrasive machining in which the grinding wheel does not merely finish the surface of a machine part but actually forms it by removing a significant amount of metal is a relatively new grinding procedure called creep feed grinding. In this process, a shape is generated in the face of an open-structure vitrified grinding wheel by diamond tooling or crush-form roller. The profiled wheel is then fed into the metal to be ground in a deep cut, ranging from 1 to 10 mm, at a very slow rate under a flood of coolant. In normal grinding, cuts are shallow, from 0.025 to 0.13 mm, and formed at a fast travel rate. Creep feed grinding finds its greatest use in the aerospace industry where hard-to-grind, heat-resistant alloy parts must be ground without surface damage. The intricate "Christmas tree" shape at the base of a turbine blade can be ground in one pass using this process and there is less wheel wear, better retention of form, and higher overall productivity than with other shaping methods. Reference 46 provides a complete treatment of the creep feed grinding process.

Superabrasive Wheels

Diamond Wheels. Synthetic diamond has almost entirely replaced natural diamond in diamond grinding wheels except for the coarsest grit sizes (4). Diamond grits, ranging from 16 to 1200, are embedded in resin, vitreous, or metal bonds in a thin rim on the periphery of the wheel body, which may be of metal-filled plastic, steel, ceramic, or bronze. Both resin- and ceramic-bonded diamond wheels are widely used to shape and sharpen carbide tooling and all kinds of ceramics. New heat-resistant, expensive resins, such as polyimides, are widely used in superabrasive bonded products to give exceptional life and productivity. The majority of resin-bonded diamond wheels are produced using diamond grits that have been plated or coated with a refractory metal such as nickel to provide increased resistance to pull-out and to serve as a heat sink in grinding.

Diamond wheel specifications show diamond concentration in the grinding rim: a concentration of 100 equals 25 vol % diamond. Most wheels have diamond concentrations in the range of 50 to 200 and selection depends on use. Lower concentrations work best on wide contact surfaces; higher concentrations work best on narrow edge widths (47).

The construction and quarrying industries make wide use of metal-bonded diamond cut-off wheels for road work, including expansion joints, surface planing, and antiskid slotting, and for production of dimension stone of all kinds. Diamond wheels are not recommended for grinding or cutting steel except for certain wear-resistant, high carbon, high vanadium content steels that contain hard vanadium carbide particles. Even for these steels, a cubic boron nitride wheel is usually a better choice (48).

Cubic Boron Nitride (CBN) Wheels.   Bonded CBN wheels were introduced to the world market in 1969 and initial reception was poor: these wheels were even more expensive than diamond wheels, they were extremely difficult to true and dress, and they had not found suitable grinding applications. Initially CBN was tried on such typical diamond applications as carbides and ceramics with very limited success. When it was tried on steel, however, it was phenomenally successful. CBN's extremely low wheel wear and its ability to stay sharp lends itself to computer-aided manufacturing (CAM) (see COMPUTER-AIDED DESIGN AND MANUFACTURING). Japanese engineers, especially, were quick to realize the production advantages in tool grinding, bearings, and auto engine parts. By 1986 Japan consumed over 5 million carats (about one-third of the world production) of CBN (49). Like diamond grit, most CBN intended for resin-bonded applications is coated with metal for increased grain retention.

Grinding Fluids

Grinding fluids or coolants are fluids employed in grinding to cool the work being ground, to act as a lubricant, and to act as a grinding aid. Water is still the best liquid for removing heat from a workpiece but it has no lubricating properties. Soluble oil coolants in which petroleum oils are emulsified in water have been developed to impart some lubricity along with rust-preventive properties. These coolants must be carefully controlled to prevent health and dermatitis problems from bacterial growth and to avoid contamination with leaking "tramp" hydraulic oil. It should be noted that the wear rate of aluminum oxide wheels is much higher when water or water-containing coolants are used than when the same wheel is used dry. Although this has been attributed to the formation of iron spinel during the grinding process (50), the same effect has been observed in grinding iron-free refractory alloys and may be attributed to the formation of aluminum hydrate at the cutting tip of the abrasive grains, where temperatures and pressures are extremely high. The soft hydrate layer is removed on each pass through the work. This hypothesis is supported by the fact that dry and wet grinding G-ratios (volume of steel removed/volume of wheel wear) are more nearly equal in vertical spindle surface grinding, where each grit enters the work and remains buried for a long distance (protected from reaction with water) as opposed to horizontal spindle grinding, where cut-path is very short and reaction can occur at each wheel revolution.

The highest G-ratios are obtained when grinding with straight oil coolants. Such oils reduce power, increase maximum depth of cut, and produce smoother finishes. Disadvantages include inability to remove heat from the work, oil mist in the work area, fire hazard, and tendency to hold grinding swarf (fine metal chips and abrasive particles produced in the grinding process) in suspension. Reference 51 is an excellent survey article for grinding fluids.

Loose Abrasives

In addition to their use in bonded and coated products, both natural and manufactured abrasive grains are used loose in such operations as polishing, buffing, lapping, pressure blasting, and barrel finishing. All of these operations are characterized by very low metal removal rates and are used to improve the surface quality of the workpiece.

Jet Cutting. High pressure jet cutting with abrasive grit can be used on metals to produce burn-free cuts with no thermal or mechanical distortion. It is also effective on ceramics and metal–ceramic composites (52). The speed with which jet cuts can be made is illustrated by a report (53) that a suspension of 60-grit garnet in water forced through a tungsten carbide orifice pierced an armor plate 50 mm thick and cut a slot 254 mm long in 10 min total time. Jet cutting will be even more effective when improved nozzle materials which permit the economical use of harder abrasives such as aluminum oxide are developed (54).

Health and Safety

Except for silica and natural abrasives containing free silica, the abrasive materials used today are classified by NIOSH as nuisance dust materials and have relatively high permissable dust levels (55). The OSHA TWA allowable total dust level for aluminum oxide, silicon carbide, boron carbide, ceria, and other nuisance dusts is 10 mg/m^3. Silica, in contrast, is quite toxic as a respirable dust: for cristobalite [14464-46-1] and tridymite [15468-32-3] the allowable TWA level drops to 0.05 mg/m^3 and the TWA for quartz [14808-60-7] is set at 0.1 mg/m^3. Any abrasive that contains free silica in excess of 1% should be treated as a potential health hazard if it is in the form of respirable dust. Dust masks are required for those exposed to such materials (see INDUSTRIAL HYGIENE AND PLANT SAFETY).

Economic Aspects

The world's leading manufacturers of abrasives and their locations are presented by product category and order of sales volume:

Abrasives as a Whole: Norton Company, U.S.; Washington Mills, U.S.; Treibacher, Austria; Pechiney, France; ESK, Germany; and Exolon-ESK, U.S.;

Superabrasives (Synthetic Diamond and CBN): General Electric, U.S.; De Beers, UK; and Tomei, Japan;

Coated Abrasives: 3M Corporation, U.S.; Norton Company, U.S.; Hermes, Germany; SIA, Switzerland; and VSM, Germany;

Bonded Abrasives: Norton Company, U.S.; Tyrolit, Austria; Noritake, Japan; American Industries, U.S.; and Naxos Union, Germany.

The abrasive industry is highly competitive and many small companies worldwide successfully compete by specializing in a particular segment of the business, eg, disk wheels, mounted points, and rubber wheels. Costs in the fused abrasive industry are primarily in materials and electric power. Thus manufacturers seek out plant sites having the lowest power costs. Costs for coated abrasive manufacturers are capital and labor intensive and they seek out sources of low cost labor.

In the long term, the bonded abrasive industry for precision grinding will be impacted by the superior performance of CBN and sol–gel abrasives. In Japan, MITI records show that precision-bonded abrasives increased only 0.47% from 1985 to 1989 but CBN usage in the same time period increased 18.24%.

BIBLIOGRAPHY

"Abrasives" in *ECT* 2nd ed., Vol. 1, pp. 22–43 by J. R. Gregor; in *ECT* 3rd ed., Vol. 1, pp. 26–52 by W. G. Pinkstone.

1. M. F. Collie, *The Saga of the Abrasive Industry,* Grinding Wheel Institute and Abrasive Grain Association, Cleveland, Ohio, 1951, p. 2.
2. R. S. Woodbury, *History of the Grinding Machine,* The Technology Press, MIT, Cambridge, Mass., 1959, pp. 18–22.
3. S. Tolansky in J. Burls, ed., *Science and Technology of Industrial Diamonds, Proceedings of the 1966 International Industrial Diamond Conference,* Industrial Diamonds Information Bureau, London, 1967, pp. 341–349.
4. G. T. Austin, "Abrasive Materials," in *Minerals Yearbook 1987,* Vol. 1, *Metals and Minerals,* U.S. Department of the Interior, 1987, pp. 71–84.
5. R. R. Ridgeway, A. H. Ballard, and B. L. Bailey, *Trans. Electrochem. Soc.* **63,** 369 (1933).
6. C. E. Woodell, *Trans. Electrochem. Soc.* **68,** 111–130 (1935).
7. L. Coes, Jr., *Abrasives,* Springer-Verlag, New York, Vienna, 1971, p. 55.
8. P. D. St. Pierre, *Conference on Ultrahard Tool Materials,* Carnegie-Mellon University, Pittsburgh, Pa., May 26, 1970.
9. J. T. Czernuska and T. F. Page, *Proc. Br. Ceram. Soc.* **34,** 145–156 (1984).
10. J. N. Plendl and P. J. Gielisse, *Phys. Rev.* **125,** 828–832 (1962).
11. B. Lamy, *Trib. Int.* **17**(1), 36–38 (1984).
12. "Procedure for Friability of Abrasive Grain," *ANSI Standard B74.8-1987,* American National Standards Institute, New York, 1987.
13. Ref. 7, pp. 152–153.
14. W. J. Sauer in M. C. Shaw, ed., *New Developments in Grinding, Proceedings of the International Grinding Conference 1972,* Carnegie Press, Carnegie-Mellon University, Pittsburgh, Pa., 1972, pp. 391–411.
15. Ref. 7, pp. 155–156.
16. L. Coes, Jr., *Ind. & Eng. Chem.* **47,** 2493 (1955).
17. R. H. Wentorf, Jr., *1986 Proceedings of the 24th Abrasive Engineering Society Conference,* Abrasive Engineering Society, Pittsburgh, Pa., 1986, pp. 27–31.
18. P. Harbin, *Ind. Min.* 49–73 (Nov. 1978).
19. H. F. G. Ueltz in M. C. Shaw, ed., *New Developments in Grinding, Proceedings of the International Grinding Conference 1972,* Carnegie Press, Carnegie-Mellon University, Pittsburgh, Pa., 1972, pp. 1–52.
20. *Mineral Industry Surveys* (Fourth Quarter 1989), U.S. Department of Interior, Bureau of Mines, Washington, D.C., May 5, 1990.
21. U.S. Pat. 775,654 (Nov. 22, 1904), A. C. Higgins (to Norton Company).
22. U.S. Pat. 2,003,867 (June 4, 1935), R. R. Ridgeway (to Norton Company).

23. U.S. Pat. 3,079,243 (Feb. 26, 1963), H. F. G. Ueltz (to Norton Company).
24. U.S. Pat. 3,387,957 (June 11, 1963), E. E. Howard (to Carborundum Company).
25. U.S. Pat. 3,183,071 (May 11, 1965), C. V. Rue, E. Fisher, and F. Heischman (to Wakefield Corporation and Electric Autolite Company).
26. U.S. Pat. 3,481,723 (Dec. 2, 1969), S. S. Kistler and C. V. Rue (to ITT Corporation).
27. U.S. Pat. 4,314,827 (Feb. 1982), M. A. Leitheiser and co-workers (to 3M Corporation).
28. U.S. Pat. 4,623,364 (Nov. 18, 1986), T. E. Cottringer, R. H. van de Merwe, and R. Bauer (to Norton Company).
29. E. J. Kubel, Jr., *ASM News* **4,** 14 (Oct. 1989).
30. U.S. Pat. 4,744,802 (May 17, 1988), M. G. Schwable (to 3M Corporation).
31. U.S. Pat. 4,770,671 (Sep. 13, 1988), L. D. Monroe and W. P. Wood (to 3M Corporation).
32. U.S. Pat. 4,881,951 (Nov. 21, 1989), W. P. Wood and co-workers (to 3M Corporation).
33. U.S. Pat. 4,543,107 (Sept. 24, 1985), C. V. Rue (to Norton Company).
34. U.S. Pat. 3,893,826 (July 8, 1975), J. R. Quinan and J. E. Patchett (to Norton Company).
35. *Ind. Min.,* **45,** 23 (1971).
36. *Norton CBN Wheels, Form 4800 LPBXM 8-83,* Company Bulletin, Norton Company, Worcester, Mass., p. 13.
37. H. Meyer, F. Klocke, and J. Sauren, *1986 Proceedings of the 24th Abrasive Engineering Society Conference,* Abrasive Engineering Society, Pittsburgh, Pa., 1986, pp. 62–92.
38. E. N. Masslow, *Grundlagen der Theorie des Metallschleifens,* Verlag Technik GmbH, Berlin, 1952, p. 42.
39. *ANSI Standard B74.12-1982* (macro sizes) and *ANSI Standard B74.10-1977 (R1983)* (micro sizes), American National Standards Institute, New York, 1982 and 1977.
40. *FEPA Standard 42-GB-1984,* British Abrasive Federation, London, 1984.
41. B. T. G. O'Carroll, *Ind. Diamond Rev.* **4** 129–131 (1973).
42. *A Review of Diamond Sizing and Standards,* IDA Bulletin, Industrial Diamond Association of America, Columbia, S.C., 1985.
43. R. G. Visser and R. C. Lokken, *Wear* **65,** 325–350 (1981).
44. U.S. Pat. 4,132,533 (Jan. 2, 1979), W. Lohmer and J. Schotten (to the Carborundum Company).
45. W. F. Schleicher, *The Grinding Wheel,* 3rd ed., The Grinding Wheel Institute, Cleveland, Ohio, 1976.
46. C. Andrew, T. D. Howes, and T. R. A. Pearce, *Creep Feed Grinding,* Industrial Press Inc., New York, 1985.
47. H. Reinhart in J. Burls, ed., *Science and Technology of Industrial Diamonds, Proceedings of the 1966 International Industrial Diamond Conference,* Industrial Diamonds Information Bureau, London, 1967, pp. 341–349.
48. J. D. Birle in M. C. Shaw, ed., *New Developments in Grinding, Proceedings of the International Grinding Conference 1972,* Carnegie Press, Carnegie-Mellon University, Pittsburgh, Pa., 1972, pp. 887–905.
49. T. Ishikawa, *1986 Proceedings of the 24th Abrasive Engineering Society Conference,* Abrasive Engineering Society, Pittsburgh, Pa., 1986, pp. 32–51.
50. L. Coes, Jr., ref. 7, p. 158.
51. C. A. Sluhan, *Lub. Eng.,* 352–374 (Oct. 1970).
52. G. Hamatani and M. Ramulu, *PED (ASME)* **35,** 49–62 (1988).
53. J. G. Sylvia and T. J. Kim, *1985 Proceedings of the 23rd Abrasive Engineering Society Conference,* Abrasive Engineering Society, Pittsburgh, Pa., 1985, pp. 83–90.
54. M. Hashish, *Manuf. Rev.* **2**(2), 142–150 (June 1989).
55. *Federal Register* **52**(12), 2418–2983 (Jan. 19, 1989).

CHARLES V. RUE
Norton Company

ABSORPTION

Absorption, or gas absorption, is a unit operation used in the chemical industry to separate gases by washing or scrubbing a gas mixture with a suitable liquid. One or more of the constituents of the gas mixture dissolves or is absorbed in the liquid and can thus be removed from the mixture. In some systems, this gaseous constituent forms a physical solution with the liquid or the solvent, and in other cases, it reacts with the liquid chemically.

The purpose of such scrubbing operations may be any of the following: gas purification (eg, removal of air pollutants from exhaust gases or contaminants from gases that will be further processed), product recovery, or production of solutions of gases for various purposes. Several examples of applied absorption processes are shown in Table 1.

Gas absorption is usually carried out in vertical countercurrent columns as shown in Figure 1. The solvent is fed at the top of the absorber, whereas the gas mixture enters from the bottom. The absorbed substance is washed out by the solvent and leaves the absorber at the bottom as a liquid solution. The solvent is often recovered in a subsequent *stripping* or desorption operation. This second step is essentially the reverse of absorption and involves countercurrent contacting of the liquid loaded with solute using an inert gas or water vapor. The absorber may be a packed column, plate tower, or simple spray column, or a bubble column. The packed column is a shell either filled with randomly packed elements or having a regular solid structure designed to disperse the liquid and bring it and the rising gas into close contact. Dumped-type packing elements come in a great variety of shapes (Fig. 2a–f) and construction materials, which are intended to create a large internal surface but a small pressure drop. Struc-

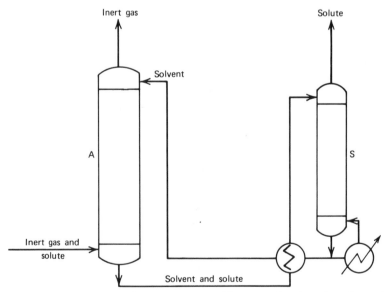

Fig. 1.   Absorption column arrangement with a gas absorber A and a stripper S to recover solvent.

Table 1. Typical Commercial Gas Absorption Processes

Treated gas	Absorbed gas, solute	Solvent	Function
coke oven gas	ammonia	water	by-product recovery
coke oven gas	benzene and toluene	straw oil	by-product recovery
reactor gases in manufacture of formaldehyde from methanol	formaldehyde	water	product recovery
drying gases in cellulose acetate fiber production	acetone	water	solvent recovery
refinery gases	hydrogen sulfide	alkaline solutions	pollutant removal
natural and refinery gases	hydrogen sulfide	solution of sodium 2,6- (and 2,7-) anthraquinonedisulfonate	pollutant removal
products of combustion	sulfur dioxide	water	pollutant removal
	carbon dioxide	ethanolamines	by-product recovery
wet well gas	propane and butane	kerosene	gas separation
ammonia synthesis gas	carbon monoxide	ammoniacal cuprous chloride solution	contaminant removal
roast gases	sulfur dioxide	water	production of calcium sulfite solution for pulping

39

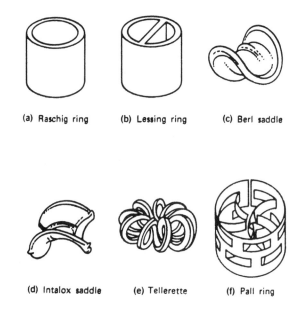

(a) Raschig ring   (b) Lessing ring   (c) Berl saddle

(d) Intalox saddle   (e) Tellerette   (f) Pall ring

(g)

Fig. 2. Packing materials for packed columns. (**a**)–(**f**) Typical packing elements generally used for random packing; (**g**) example of structured packing. (**g**) Courtesy of Sulzer Bros. S.A. Winterthur, Switzerland.

tured, or arranged packings may be made of corrugated metal or plastic sheets providing a large number of regularly arranged channels (Fig. 2**g**), but a variety of other geometries exists. In plate towers, liquid flows from plate to plate in cascade fashion and gases bubble through the flowing liquid at each plate through a multitude of dispersers (eg, holes in a sieve tray, slits in a bubble-cap tray) or through a cascade of liquid as in a shower deck tray (see DISTILLATION).

The advantages of packed columns include simple and, as long as the tower diameter is not too large, usually relatively cheaper construction. These columns are preferred for corrosive gases because packing, but not plates, can be made from ceramic or plastic materials. Packed columns are also used in vacuum applications because the pressure drop, especially for regularly structured packings, is usually less than through plate columns. Tray absorbers are used in applications where tall columns are required, because tall, random-type packed towers are subject to channeling and maldistribution of the liquid streams. Plate towers can be more easily cleaned. Plates are also preferred in applications having large heat effects since cooling coils are more easily installed in plate towers and liquid can be withdrawn more easily from plates than from packings for external cooling. Bubble trays can also be designed for large liquid holdup.

The fundamental physical principles underlying the process of gas absorption are the solubility of the absorbed gas and the rate of mass transfer. Information on both must be available when sizing equipment for a given application. In addition to the fundamental design concepts based on solubility and mass transfer, many practical details have to be considered during actual plant design and construction which may affect the performance of the absorber significantly. These details have been described in reviews (1) and in some of the more comprehensive treatments of gas absorption and absorbers (2–5) (see also DISTILLATION; HEAT EXCHANGE TECHNOLOGY; MASS TRANSFER).

Gas Solubility

At equilibrium, a component of a gas in contact with a liquid has identical fugacities in both the gas and liquid phase. For ideal solutions Raoult's law applies:

$$y_A = \frac{P_s}{P} x_A \tag{1}$$

where y_A is the mole fraction of A in the gas phase, P is the total pressure, P_s is the vapor pressure of pure A, and x_A is the mole fraction of A in the liquid. For moderately soluble gases with relatively little interaction between the gas and liquid molecules Henry's law is often applicable:

$$y_A = \frac{H}{P} x_A \tag{2}$$

where H is Henry's constant. Usually H is dependent upon temperature, but relatively independent of pressure at moderate levels. In solutions containing inorganic salts, H is also a function of the ionic strength. Henry's constants are tabulated for many of the common gases in water (6).

A more general way of expressing solubilities is through the vapor–liquid equilibrium constant m defined by

$$y_A = mx_A \tag{3}$$

The value of m, also known as equilibrium K value, is widely employed to represent hydrocarbon vapor–liquid equilibria in absorption and distillation calculations. When equation 1 or 2 is applicable at constant pressure and temperature (equivalent to constant m in eq. 3) a plot of y vs x for a given solute is linear from the origin. In other cases, the y–x plot may be approximated by a linear relationship over limited regions. Generally, for nonideal solutions or for nonisothermal conditions, y is a curving function of x and must be determined from experimental data or more rigorous theoretical relationships. The y–x plot, when applied to absorber design, is commonly called the equilibrium line.

Gas solubility has been treated extensively (7). Methods for the prediction of phase equilibria and actual solubility data have been given (8,9) and correlations of the equilibrium K values of hydrocarbons have been developed and compiled (10). Several good sources for experimental information on gas– and vapor–liquid equilibrium data of nonideal systems are also available (6,11,12).

Mass Transfer Concepts

Mass Transfer Coefficients and Driving Forces. In order to determine the size of the equipment necessary to absorb a given amount of solvent per unit time, one must know not only the equilibrium solubility of the solute in the solvent, but also the rate at which the equilibrium is established; ie, the rate at which the solute is transferred from the gas to the liquid phase must be determined. One of the first theoretical models describing the process proposed an essentially stable gas–liquid interface (13). Large fluid motions are presumed to exist at a certain distance from this interface distributing all material rapidly and equally in the bulk of the fluid so that no concentration gradients are developed. Closer to this interface, however, the fluid motions are impaired and the slow process of molecular diffusion becomes more important as a mechanism of mass transfer. The rate-governing step in gas absorption is therefore the transfer of solute through two thin gas and liquid films adjacent to the phase interface. Transfer of materials through the interface itself is normally presumed to take place instantaneously so that equilibrium exists between these two films precisely at the interface. Although this assumption has been confirmed in experiments utilizing many systems and different types of phase interface (14–18), interfacial resistances can develop in some situations (19–24).

The resulting concentration profile is shown in Figure 3. With the passage of time in a nonflowing closed system, the profiles would become straight horizontal lines as the bulk gas and bulk liquid reached equilibrium. In a flowing system, Figure 3 represents conditions at some countercurrent flow point, eg, at a certain height in an absorption tower where, as gas and liquid pass each other, the bulk materials do not have sufficient contact time to attain equilibrium. Solute is continuously transferred from the gas to the liquid and concentration gradients develop when this transfer proceeds at only a finite rate.

The experimentally observed rates of mass transfer are often proportional to the displacement from equilibrium and the rate equations for the gas and liquid

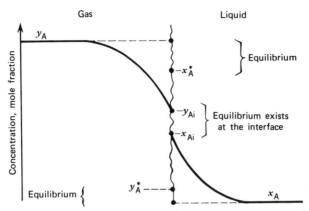

Fig. 3. The two-film concept: y_A and x_A are the concentrations in the bulk of the phases; y_{Ai} and x_{Ai} are the actual interfacial concentrations at equilibrium; y_A^* and x_A^* are the hypothetical equilibrium concentrations which would be in equilibrium with the bulk concentration of the other phase.

films are

$$N_A = k_G(p_A - p_{Ai}) = k_G P(y_A - y_{Ai}) \tag{4}$$

$$N_A = k_L(c_{Ai} - c_A) = k_L \bar{\rho}(x_{Ai} - x_A) \tag{5}$$

where $y_A - y_{Ai}$ and $x_{Ai} - x_A$ are concentration driving forces, k_G is the gas-phase mass transfer coefficient, and k_L is the liquid-phase mass transfer coefficient.

Mass transfer rates may also be expressed in terms of an overall gas-phase driving force by defining a hypothetical equilibrium mole fraction y_A^* as the concentration which would be in equilibrium with the bulk liquid concentration ($y_A^* = mx_A$):

$$N_A = K_{OG} P(y_A - y_A^*) \tag{6}$$

The relationship of the overall gas-phase mass transfer coefficient K_{OG} to the individual film coefficients may be found from equations 4 and 5, assuming a straight equilibrium line:

$$N_A = k_G P(y_A - y_{Ai}) = k_L \bar{\rho} \frac{1}{m} (y_{Ai} - y_A^*)$$

and by comparison with equation 6,

$$\frac{1}{K_{OG}} = \frac{1}{k_G} + \frac{mP}{k_L \bar{\rho}} \tag{7}$$

Expressions similar to equations 6 and 7 may be derived in terms of an overall liquid-phase driving force. Equation 7 represents an addition of the resistances to mass transfer in the gas and liquid films. The analogy of this process to the flow of electrical current through two resistances in series has been analyzed (25).

A representation of the various concentrations and driving forces in a y–x diagram is shown in Figure 4. The point representing the interfacial concentrations (y_{Ai}, x_{Ai}) must lie on the equilibrium curve since these concentrations are at equilibrium. The point representing the bulk concentrations (y_A, x_A) may be anywhere above the equilibrium line for absorption or below it for desorption. The slope of the tie line connecting the two points is given by equations 4 and 5:

$$\frac{y_A - y_{Ai}}{x_A - x_{Ai}} = -\frac{k_L \bar{\rho}}{k_G P} \qquad (8)$$

In situations where the gas film resistance is predominant (gas film-controlled situation), $k_G P$ is much smaller than $k_L \bar{\rho}$ and the tie line is very steep. y_{Ai} approaches y^* so that the overall gas-phase driving force and the gas-film driving force become approximately equal, whereas the liquid-film driving force becomes negligible. From equation 7 it also follows that in such cases $K_{OG} \approx k_G$. The reverse is true if the liquid film resistance is controlling. Since the example depicted in Figure 4 involves a strongly curved equilibrium line, equation 7 is only valid if the slope of the dashed line between x_A and x_{Ai} is substituted for m. Overall mass transfer coefficients may vary considerably over a certain concentration range as a result of variations in m even if the individual film constants stay essentially constant.

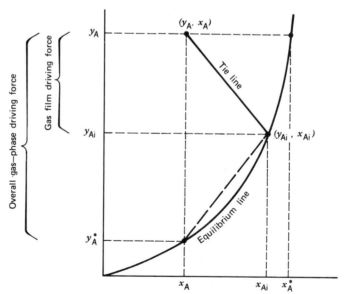

Fig. 4. The driving forces in the y–x diagram.

Film Theory. Many theories have been put forth to explain and correlate experimentally measured mass transfer coefficients. The classical model has been the film theory (13,26) that proposes to approximate the real situation at the interface by hypothetical "effective" gas and liquid films. The fluid is assumed to be essentially stagnant within these effective films making a sharp change to totally turbulent flow where the film is in contact with the bulk of the fluid. As a result, mass is transferred through the effective films only by steady-state molecular diffusion and it is possible to compute the concentration profile through the films by integrating Fick's law:

$$J_A = -D_{AB} \frac{dc_A}{dz} \tag{9}$$

where J_A is the flux of component A relative to the average molar flow of the whole mixture, D_{AB} is the diffusion coefficient of A in B, z is the distance of diffusion, and c_A is the molar concentration of A at a given point in the film. The bulk concentrations are denoted y_{Ab} and x_{Ab} in the following three sections whereas y_A and x_A stand for the concentration at a particular point within the films.

Equimolar Counterdiffusion in Binary Cases. If the flux of A is balanced by an equal flux of B in the opposite direction (frequently encountered in binary distillation columns), there is no net flow through the film and N_A, like J_A, is directly given by Fick's law. In an ideal gas, where the diffusivity can be shown to be independent of concentration, integration of Fick's law leads to a linear concentration profile through the film and to the following expression where $(P/RT)y_A$ is substituted for c_A:

$$N_A = \frac{D_{AB}P}{z_0 RT} (y_{Ab} - y_{Ai}) \tag{10}$$

thus

$$k_G^0 = \frac{D_{AB}}{z_0 RT} \tag{11}$$

where k_G is labeled with a zero to indicate equimolar counterdiffusion and z_0 is the effective film thickness. This same treatment is usually adopted for liquids, although the diffusion coefficients are customarily not completely independent of concentration. Substituting $\bar{\rho} x_A$ for c_A, the result is

$$k_L^0 = \frac{D_A}{z_0} \tag{12}$$

Equations 11 and 12 cannot be used to predict the mass transfer coefficients directly, because z_0 is usually not known. The theory, however, predicts a linear dependence of the mass transfer coefficient on diffusivity.

Unidirectional Diffusion through a Stagnant Medium. An ideally simple gas absorption process involves diffusion of only one component, either the inert gas or the solvent, through a nondiffusing medium. There exists a net flux of material through the film in this case, and therefore the mixture as a whole is not at rest. The total flux of A is now the sum of the flux with respect to the average flow of the mixture, that is still given by Fick's law of diffusion, plus the flux of A caused by the average bulk flow of the mixture itself:

$$N_A = J_A + y_A \sum_j N_j = -\frac{D_{AB}P}{RT}\frac{dy_A}{dz} + y_A N_A \tag{13}$$

The derivation of this result may be found in various texts (27). Rearranging and integrating equation 13 yields

$$N_A = \frac{D_{AB}P}{z_0 RT}\frac{1}{y_{BM}}(y_{Ab} - y_{Ai}) \tag{14}$$

where y_{BM} is the logarithmic mean of the stagnant gas concentration through the film:

$$y_{BM} = \frac{(1 - y_{Ab}) - (1 - y_{Ai})}{\ln[(1 - y_{Ab})/(1 - y_{Ai})]} \tag{15}$$

Therefore

$$k'_G = \frac{D_{AB}}{z_0 RT y_{BM}} \tag{16}$$

For liquids,

$$k'_L = \frac{D_A}{z_0 x_{BM}} \tag{17}$$

where

$$x_{BM} = \frac{(1 - x_{Ab}) - (1 - x_{Ai})}{\ln[(1 - x_{Ab})/(1 - x_{Ai})]} \tag{18}$$

The values of y_{BM} and x_{BM} are near unity for dilute gases.

   Multicomponent Diffusion. In multicomponent systems, the binary diffusion coefficient D_{AB} has to be replaced by an effective or mean diffusivity D_{Am}. Although its rigorous computation from the binary coefficients is difficult, it may be estimated by one of several methods (27–29). Any degree of counterdiffusion, including the two special cases "equimolar counterdiffusion" and "no counterdiffusion" treated above, may arise in multicomponent gas absorption. The influence of bulk flow of material through the films is corrected for by the film factor concept (28). It is based on a slightly different form of equation 13:

$$N_A = J_A + y_A t_A N_A \tag{19}$$

where

$$t_A \equiv \frac{\Sigma_j \, N_j}{N_A} \tag{20}$$

Applying the same derivation as for unidirectional diffusion through a stagnant medium, the results turn out to be

$$k_G = \frac{D_{AB}}{z_0 R T Y_f} \tag{21}$$

$$k_L = \frac{D_{AB}}{z_0 X_f} \tag{22}$$

where

$$Y_f \equiv \frac{(1 - t_A y_{Ab}) - (1 - t_A y_{Ai})}{\ln(1 - t_A y_{Ab})/(1 - t_A y_{Ai})} \tag{23}$$

$$X_f \equiv \frac{(1 - t_A x_{Ab}) - (1 - t_A x_{Ai})}{\ln(1 - t_A x_{Ab})/(1 - t_A x_{Ai})} \tag{24}$$

Y_f and X_f are called the film factors. They are generalized y_{BM} and x_{BM} factors, respectively, and are reduced to them in the case of unidirectional diffusion through a stagnant medium because $t_A = 1$ in this case. The film factors Y_f and X_f correct the mass transfer coefficients for the effect of net flux through the films. In situations having strong counterdiffusion giving rise to a net flow opposed to the diffusion of A, the film factor becomes larger than one and therefore decreases the mass transfer coefficient and the flux of A. For weak or negative counterdiffusion producing a bulk flux parallel to the diffusion of A, the film factor is smaller than unity and thus increases N_A. In extreme situations, counterdiffusion may become large enough to reverse the direction of transport of a given component and force it to diffuse against its own driving force. These situations are characterized by a negative film factor and hence a negative k_G. If equimolar counterdiffusion prevails, t_A becomes zero, the film factor is unity, irrespective of the concentration, and $k_G = k_G^0$.

Rate Equations with Concentration-Independent Mass Transfer Coefficients. Except for equimolar counterdiffusion, the mass transfer coefficients applicable to the various situations apparently depend on concentration through the y_{BM} and Y_f factors. Instead of the classical rate equations 4 and 5, containing variable mass transfer coefficients, the rate of mass transfer can be expressed in terms of the constant coefficients for equimolar counterdiffusion using the relationships

$$k_G \, Y_f = k_G' y_{BM} = k_G^0 \tag{25}$$

$$k_L X_f = k_L' x_{BM} = k_L^0 \tag{26}$$

This leads to rate equations with constant mass transfer coefficients, whereas the effect of net transport through the film is reflected separately in the y_{BM} and Y_f factors. For unidirectional mass transfer through a stagnant gas the rate equation becomes

$$N_A = k_G^0 P(y_{Ab} - y_{Ai}) \frac{1}{y_{BM}} \tag{27}$$

For multicomponent diffusion,

$$N_A = k_G^0 P(y_{Ab} - y_{Ai}) \frac{1}{Y_f} \tag{28}$$

Equation 28 and its liquid-phase equivalent are very general and valid in all situations. Similarly, the overall mass transfer coefficients may be made independent of the effect of bulk flux through the films and thus nearly concentration independent for straight equilibrium lines:

$$N_A = K_{OGP}^0(y_{Ab} - y_A^*) \frac{1}{y_{BM}^*} \tag{29}$$

$$= K_{OGP}^0(y_{Ab} - y_A^*) \frac{1}{Y_f^*} \tag{30}$$

where the logarithmic means in y_{BM}^* and Y_f^* must be taken between y_{Ab} and y_A^*.

Rate equations 28 and 30 combine the advantages of concentration-independent mass transfer coefficients, even in situations of multicomponent diffusion, and a familiar mathematical form involving concentration driving forces. The main inconvenience is the use of an effective diffusivity which may itself depend somewhat on the mixture composition and in certain cases even on the diffusion rates. This advantage can be eliminated by working with a different form of the Maxwell-Stefan equation (30–32). One thus obtains a set of rate equations of an unconventional form having concentration-independent mass transfer coefficients that are defined for each binary pair directly based on the Maxwell-Stefan diffusivities.

Other Models for Mass Transfer. In contrast to the film theory, other approaches assume that transfer of material does not occur by steady-state diffusion. Rather there are large fluid motions which constantly bring fresh masses of bulk material into direct contact with the interface. According to the penetration theory (33), diffusion proceeds from the interface into the particular element of fluid in contact with the interface. This is an unsteady state, transient process where the rate decreases with time. After a while, the element is replaced by a fresh one brought to the interface by the relative movements of gas and liquid, and the process is repeated. In order to evaluate N_A, a constant average contact time τ for the individual fluid elements is assumed (33). This leads to relations such as

$$k_L = 2\sqrt{\frac{D}{\pi\tau}} \tag{31}$$

If, on the other hand, it is assumed that contact times for the individual fluid elements vary at random, an exponential surface age distribution characterized by a fractional rate of renewal s may be used (34). This approach is called surface renewal theory and results in

$$k_{\mathrm{L}} = \sqrt{Ds} \tag{32}$$

Neither the penetration nor the surface renewal theory can be used to predict mass transfer coefficients directly because τ and s are not normally known. Each suggests, however, that mass transfer coefficients should vary as the square root of the molecular diffusivity, as opposed to the first power suggested by the film theory.

Another concept sometimes used as a basis for comparison and correlation of mass transfer data in columns is the Chilton-Colburn analogy (35). This semi-empirical relationship was developed for correlating mass- and heat-transfer data in pipes and is based on the turbulent boundary layer model and the close analogy between momentum and mass transfer. It must be considerably modified for gas-absorption columns, but it predicts that the mass transfer coefficient varies with D raised to the $\frac{2}{3}$ power (4,36). A modern theory for surface rejuvenation has also been published and compared with earlier models (37).

Absorption and Chemical Reaction. In instances where the solute gas is absorbed into a liquid or a solution where it is able to undergo chemical reaction, the driving forces of absorption become far more complex. The solute not only diffuses through the liquid film at a rate determined by the gradient of the concentration, but at the same time also reacts with the liquid at a rate determined by the concentrations of both the solute and the solvent at the point of interest. Inclusion of a term for the chemical reaction in Fick's second law of diffusion, followed by integration of the expression, allows the concentration profiles through the liquid film to be computed based on a particular mass transfer model. The calculations show that these profiles are steeper and the rate of mass transfer higher than without chemical reaction. Thus the results are often expressed as an enhancement factor ϕ, defined as the fractional increase of the liquid film mass transfer coefficient resulting from the chemical reaction $(k_{\mathrm{L}}^{\mathrm{r}}/k_{\mathrm{L}}^{0})$. The solutions that have been developed in this manner based on the film, penetration, and surface renewal theories are quite similar for a given type of reaction (38,39). Solutions and estimations of enhancement factors may be found in the literature (4,37–40).

Design of Packed Absorption Columns

Discussion of the concepts and procedures involved in designing packed gas absorption systems shall first be confined to simple gas absorption processes without complications: isothermal absorption of a solute from a mixture containing an inert gas into a nonvolatile solvent without chemical reaction. Gas and liquid are assumed to move through the packing in a plug-flow fashion. Deviations such as nonisothermal operation, multicomponent mass transfer effects, and departure from plug flow are treated in later sections.

Standard Absorber Design Methods. *Operating Line.*   As a gas mixture travels up through a gas absorption tower, as shown in Figure 5, the solute A is transferred to the liquid phase and thus gradually removed from the gas. The liquid accumulates solute on its way down through the column so x increases from the top to the bottom of the column. The steady-state concentrations y and x at any given point in the column are interrelated through a mass balance around either the upper or lower part of the column (eq. 34), whereas the four concentrations in the streams entering and leaving the system are interrelated by the overall material balance.

Since the total gas and liquid flow rates per unit cross-sectional area vary throughout the tower (Fig. 5) rigorous material balances should be based on the constant inert gas and solvent flow rates G'_M and L'_M, respectively, and expressed in terms of mole ratios Y' and X'. A balance around the upper part of the tower yields

$$G'_M Y' + L'_M X'_2 = G'_M Y'_2 + L'_M X' \tag{33}$$

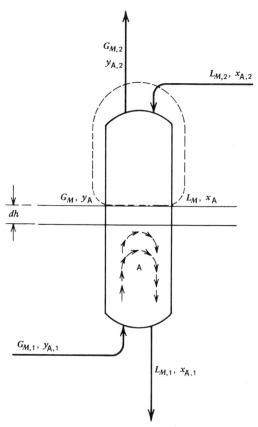

Fig. 5.   Mass balance in gas absorption columns. The curved arrows indicate the travel path of the solute A. The upper broken curve delineates the envelope for the material balance of equation 33.

which may be rearranged to give

$$Y' = \frac{L'_M}{G'_M}(X' - X'_{A,2}) + Y'_{A,2} \tag{34}$$

where G'_M, L'_M are in kg·mol/(h·m²) [lb·mol/(h·ft²)] and $Y' \equiv y/(1 - y)$ and $X' \equiv x/(1 - x)$. The overall material balance is obtained by substituting $Y' = Y'_1$ and $X' = X'_1$. For dilute gases the total molar gas and liquid flows may be assumed constant and a similar mass balance yields

$$y = \frac{L_M}{G_M}(x - x_2) + y_2 \tag{35}$$

A plot of either equation 34 or 35 is called the operating line of the process as shown in Figure 6. As indicated by equation 35, the line for dilute gases is straight, having a slope given by L_M/G_M. (This line is always straight when plotted in Y'–X' coordinates.) Together with the equilibrium line, the operating line permits the evaluation of the driving forces for gas absorption along the column (Fig. 4). The farther apart the equilibrium and operating lines, the larger the driving forces become and the faster absorption occurs, resulting in the need for a shorter column (Fig. 6).

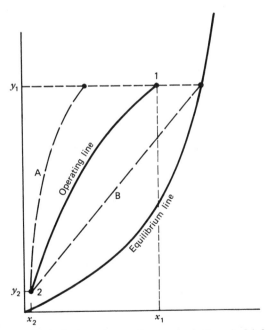

Fig. 6. Operating lines for an absorption system: line A, high L_M/G_M ratio; solid line, medium L_M/G_M ratio; line B, L_M/G_M ratio at theoretical minimum necessary for the removal of the specified quantity of solute. Subscript 1 represents the bottom of tower, 2, the top of tower.

To place the operating line, the flows, composition of the entering gas y_1, entering liquid x_2, and desired degree of absorption y_2, are usually specified. The specification of the actual liquid rate used for a given gas flow (the L_M/G_M ratio) usually depends on an economic optimization because the slope of the operating line may be seen to have a drastic effect on the driving force. For example, use of a very high liquid rate (line A in Fig. 6) results in a short column and a low absorber cost, but at the expense of a high cost for solvent circulation and subsequent recovery of the solute from a relatively dilute solution. On the other hand, a liquid rate near the theoretical minimum, which is the rate at which the operating line just touches the equilibrium line (line B in Fig. 6), requires a very tall tower because the driving force becomes very small at its bottom. Use of a liquid rate on the order of one and one-half times the theoretical minimum is not unusual. In the absence of a detailed cost analysis, the L_M/G_M ratio is often specified at 1.4 times the slope of the equilibrium line (41).

Design Procedure. The packed height of the tower required to reduce the concentration of the solute in the gas stream from $y_{A,1}$ to an acceptable residual level of $y_{A,2}$ may be calculated by combining point values of the mass transfer rate and a differential material balance for the absorbed component. Referring to a slice dh of the absorber (Fig. 5),

$$N_A a \, dh = -d(G_M y) = -G_M \, dy - y \, dG_M \tag{36}$$

and

$$dG_M = -N_A a \, dh \tag{37}$$

where a is the interfacial area present per unit volume of packing. Combining equations 36 and 37,

$$-dh = \frac{G_M \, dy}{N_A a (1 - y)} \tag{38}$$

Substituting for N_A from equation 27 and integrating over the tower,

$$h = \int_{y_2}^{y_1} \frac{G_M}{k_G^0 a P} \frac{y_{BM} \, dy}{(1 - y)(y - y_i)} \tag{39}$$

Equation 39 may be integrated numerically or graphically and its component terms evaluated at a series of points on the operating line. y_i is found by placing tie lines from each of these points; the slopes are given by equation 8. Thus equation 39 is a general expression determining the column height required to effect a given reduction in y_A.

Equation 39 can often be simplified by adopting the concept of a mass transfer unit. As explained in the film theory discussion earlier, the purpose of selecting equation 27 as a rate equation is that k_G^0 is independent of concentration. This is also true for the $G_M/k_G^0 a P$ term in equation 39. In many practical instances, this expression is fairly independent of both pressure and G_M: as G_M increases through the tower, k_G^0 increases also, nearly compensating for the variations in G_M. Thus this term is often effectively constant and can be removed from the integral:

$$h = \left(\frac{G_M}{k_G^0 a P}\right) \int_{y_2}^{y_1} \frac{y_{BM}\, dy}{(1-y)(y-y_i)} \tag{40}$$

$G_M/k_G^0 a P$ has the dimension of length or height and is thus designated the gas-phase height of one transfer unit, H_G. The integral is dimensionless and indicates how many of these transfer units it takes to make up the whole tower. Consequently, it is called the number of gas-phase transfer units, N_G. Equation 40 may therefore be written as

$$h = (H_G)(N_G) \tag{41}$$

where

$$H_G = \frac{G_M}{k_G^0 a P} \tag{42}$$

and

$$N_G = \int_{y_2}^{y_1} \frac{y_{BM}\, dy}{(1-y)(y-y_i)} \tag{43}$$

The same treatment for the liquid side yields

$$h = (H_L)(N_L) \tag{44}$$

where

$$H_L = \frac{L_M}{k_L^0 a \overline{\rho}} \tag{45}$$

and

$$N_L = \int_{x_2}^{x_1} \frac{x_{BM}\, dx}{(1-x)(x_i-x)} \tag{46}$$

A similar treatment is possible in terms of an overall gas-phase driving force by substituting equation 29 into equation 38:

$$h = (H_{OG})(N_{OG}) \tag{47}$$

where

$$H_{OG} = \frac{G_M}{K_{OG}^0 a P}. \tag{48}$$

and

$$N_{OG} = \int_{y_2}^{y_1} \frac{y_{BM}^* \, dy}{(1 - y)(y - y^*)} \tag{49}$$

H_{OG} and N_{OG} are called the overall gas-phase height of a transfer unit and the number of overall gas-phase transfer units, respectively. In the case of a straight equilibrium line, K_{OG}^0 is often nearly concentration-independent as explained earlier. In such cases, use of equation 47 is especially convenient because N_{OG}, as opposed to N_G, can be evaluated without solving for the interfacial concentrations. In all other cases, H_{OG} must be retained under the integral and its value calculated from H_G and H_L at different points of the equilibrium line as

$$H_{OG} = \frac{y_{BM}}{y_{BM}^*} H_G + \frac{m G_M}{L_M} \frac{x_{BM}}{y_{BM}^*} H_L \tag{50}$$

To use all of these equations, the heights of the transfer units or the mass transfer coefficients $k_G^0 a$ and $k_L^0 a$ must be known. Transfer data for packed columns are often measured and reported directly in terms of H_G and H_L and correlated in this form against G_M and L_M.

Sometimes the height equivalent to a theoretical plate (HETP) is employed rather than H_G and H_L to characterize the performance of packed towers. The number of heights equivalent to one theoretical plate required for a specified absorption job is equal to the number of theoretical plates, N_{TP}. It follows that

$$h = (\text{HETP})(N_{TP}) \tag{51}$$

which is similar in form to equation 47. HETP is a less fundamental variable than the heights of the transfer units, and it is more difficult to translate HETPs from one situation to another. Only for linear operating and equilibrium lines can they be related analytically to H_{OG} as shown later by equation 81.

Simplified Design Procedures for Linear Operating and Equilibrium Lines. *Logarithmic-Mean Driving Force.* As noted earlier, linear operating lines occur if all concentrations involved stay low. Where it is possible to assume that the equilibrium line is linear, it can be shown that use of the logarithmic mean of the terminal driving forces is theoretically correct. When the overall gas-film coefficient is used to express the rate of absorption, the calculation reduces to solution of the equation

$$L_M(x_1 - y_2) = G_M(y_1 - y_2) = K_{OG}aPh(y - y^*)_{av} \tag{52}$$

where

$$(y - y^*)_{av} = \frac{(y_1 - y_1^*) - (y_2 - y_2^*)}{\ln\left[(y_1 - y_1^*)/(y_2 - y_2^*)\right]} \tag{53}$$

In these cases, a quantitative significance can be given to the concept of a transfer unit. Because $H_{OG} = G_M/K_{OG}aP$, it follows from equations 52 and 47 that

$$N_{OG} = \frac{y_i - y_2}{(y - y^*)_{av}}$$

Therefore, in this case, one transfer unit corresponds to the height of packing required to effect a composition change just equal to the average driving force.

Number of Transfer Units. For relatively dilute systems the ratios involving y_{BM}, y_{BM}^*, and $1 - y$ approach unity so that the computation of H_{OG} from equation 50 and N_{OG} from equation 49 may be simplified to

$$H_{OG} = H_G + \left(\frac{mG_M}{L_M}\right) H_L \tag{54}$$

$$N_{OG} \approx N_T = \int_{y_2}^{y_1} \frac{dy}{y - y^*} \tag{55}$$

Equation 55 is a rigorous expression for the number of overall transfer units for equimolar counterdiffusion, in distillation columns, for instance.

For cases in which the equilibrium and operating lines may be assumed linear, having slopes L_M/G_M and m, respectively, an algebraic expression for the integral of equation 55 has been developed (41):

$$N_{OG} \approx N_T = \frac{\ln\left[\left(1 - \frac{mG_M}{L_M}\right)\left(\frac{y_1 - mx_2}{y_2 - mx_2}\right) + \frac{mG_M}{L_M}\right]}{1 - \frac{mG_M}{L_M}} \tag{56}$$

The required tower height may thus be easily calculated using equation 47, where H_{OG} is given by equation 54 and N_{OG} by equation 56.

Rapid Approximate Design Procedure for Curved Operating and Equilibrium Lines. If the operating or the equilibrium line is nonlinear, equation 56 is of little use because mG_M/L_M will assume a range of values over the tower. The substitution of effective average values for m and for L_M/G_M into equations 50 and 56 obviates lengthy graphical or numerical integrations and leads to a quick, approximate solution for the required tower height (4).

The effective average values of m and L_M/G_M were determined in a computational study covering hundreds of hypothetical absorber designs for gas streams containing up to 80 mol % of solute for recoveries from 81 to 99.9%. By numerical integration, precise values were obtained for N_{OG} and N_T. By solving equation 56 numerically for each of the design cases, average values of the slope of the equilibrium line $\overline{m}$ and average flow ratios $L_M/G_M = R_{av}$ were found which gave the same N_T when substituted into equation 56 as the graphical or numerical integration.

It was found that the effective average L_M/G_M ratio, R_{av}, could be correlated satisfactorily as a function of the terminal values R_1 and R_2, of the change in the mole fraction of the absorbed component over the tower, and of the fractional approach to equilibrium y_1^*/y_1 between the concentrated gas entering the tower and the liquid leaving. Figure 7 shows the resulting correlation for cases with $L_M/G_M > 1$. No correlation was obtained when this ratio was less than unity. The effective average slope of the equilibrium line, $\overline{m}$, was correlated as a function of the initial slope m_2, of the slope m_c of the cord connecting y_1^*/x_1 and y_1^*/x_2, and of various other parameters as shown in Figure 8. Figure 8a applies when the equilibrium line is concave upward, ie, $m_c > m_2$; and Figure 8b applies when the curvature is concave downward, $m_c < m_2$.

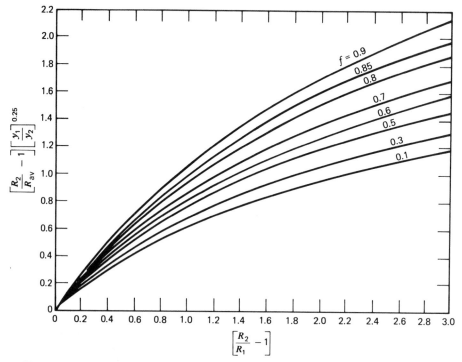

Fig. 7.   Design chart for estimation of average-flow ratio in absorption (4). $R_2 = L_M/G_M$ at gas outlet; $R_1 = G_M/L_M$ at gas inlet; y_2 = mole fraction in outlet gas; y_1 = mole fraction in inlet gas; R_{av} = effective average L_M/G_M; $f = y_1^*/y_1$ = fractional approach to equilibrium.

The recommended design procedure uses the values of $(L_M/G_M)_{av}$ and $\overline{m}$ from Figures 7 and 8 in equation 56 and yields a very good estimation of N_T despite the curvature of the operating and the equilibrium lines. This value differs from N_{OG} obtained by equation 49 because of the $y_{BM}^*/(1 - y)$ term in the latter equation. A convenient approach for purposes of approximate design is to define a correction term ΔN_{OG} which can be added to equation 55:

$$N_{OG} = N_T + \Delta N_{OG} \qquad (57)$$

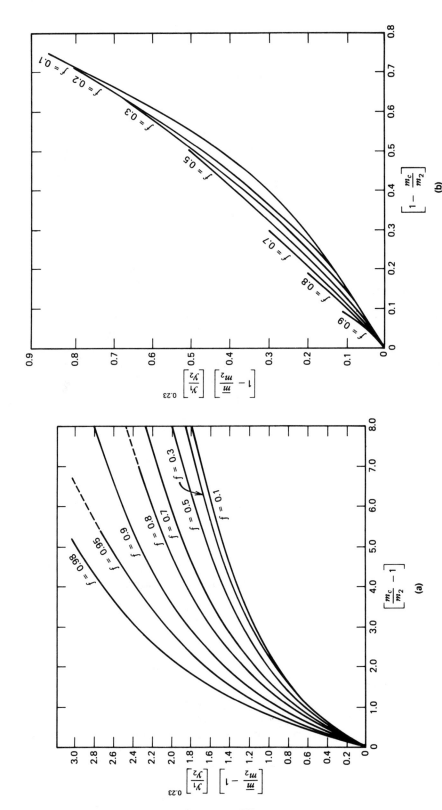

Fig. 8. Correlation of the effective average slopes $\overline{m}$ of the equilibrium line (4). **(a)** Equilibrium line curved concave upward; **(b)** equilibrium line curved concave downward.

For cases in which y_{BM}^* may be represented by an arithmetic mean (42),

$$\Delta N_{OG} = \frac{1}{2} \ln \frac{1 - y_2}{1 - y_1} \tag{58}$$

Equation 58 is sufficiently accurate for most situations.

The average slopes R_{av} and $\overline{m}$ from Figures 7 and 8 may also be used in equation 54 to compute H_{OG} although equation 50, with some suitable averages of y_{BM}^* and x_{BM}, should be preferred. Use of point values at an effective average liquid concentration given by equation 59 is suggested.

$$\overline{x} = \left(\frac{R_2}{R_{av}} - 1 \right) \bigg/ (R_2 - 1) \tag{59}$$

In many situations, however, especially when $m \geq 1$, the results using the simpler equation 54 are virtually the same. The required tower height is finally calculated by means of equation 47.

Multicomponent Mass Transfer Effects. *Equimolar Counterdiffusion.* Just as unidirectional diffusion through stagnant films represents the situation in an ideally simple gas absorption process, equimolar counterdiffusion prevails as another special case in ideal distillation columns. In this case, the total molar flows L_M and G_M are constant, and the mass balance is given by equation 35. As shown earlier, no y_{BM} factors have to be included in the derivation and the height of the packing is

$$h_T = \int_{y_2}^{y_1} H_G \frac{dy}{y - y_i} \tag{60}$$

N_{OG} is given by N_T:

$$N_{OG} = \int_{y_2}^{y_1} \frac{dy}{y - y^*} \tag{61}$$

and H_{OG} is rigorously defined by equation 54. It must, however, be retained under the integral because m usually changes over the tower:

$$h = \int_{y_2}^{y_1} H_{OG} \frac{dy}{y - y^*} \tag{62}$$

General Situation. Both unidirectional diffusion through stagnant media and equimolar diffusion are idealizations that are usually violated in real processes. In gas absorption, slight solvent evaporation may provide some counterdiffusion, and in distillation counterdiffusion may not be equimolar for a number of reasons. This is especially true for multicomponent operation.

A simple treatment is still possible if it may be assumed that the flux of the component of interest A through the interface stays in a constant proportion to the total molar transfer through the interface over the entire tower:

$$\frac{\Sigma N_j}{N_A} = t_A = \text{constant} = \frac{\Sigma \Delta g_j}{\Delta g_A} \tag{63}$$

where Δg_j = total moles of component j absorbed over the tower. It will generally suffice to compute t_A from preliminary estimates of Δg_j and Δg_A, the total mass transfer of each component over the tower.

The mass balance for A is best represented as a straight line in hypothetical coordinates Y^0 and X^0:

$$Y_A^0 = \frac{L_M^0}{G_M^0}(X_A^0 - X_{A,2}^0) + Y_{A,2}^0 \tag{64}$$

where $Y_A^0 = y_A/(1 - t_A y_A)$, $X_A^0 = x_A/(1 - t_A x_A)$, $G_M^0 = G_M(1 - t_A y_A)$, and $L_M^0 = L_M(1 - t_A x_A)$. G_M^0 and L_M^0 are always constant, whereas G_M and L_M are not. For unimolecular diffusion through stagnant gas ($t_A = 1$), Y^0 and X^0 reduce to Y' and X' and G_M^0 and L_M^0 reduce to G_M' and L_M'; equation 64 then becomes equation 34. For equimolar counterdiffusion $t_A = 0$, and the variables reduce to y, x, G_M, and L_M, respectively, and equation 64 becomes equation 35. Using the film factor concept and rate equation 28, the tower height may be computed by

$$h_T = \int_{y_2}^{y_1} H_G \frac{Y_f}{(1 - t_A y)} \frac{dy}{(y_A - y_{Ai})} \tag{65}$$

y_{Ai} is found as usual through tie lines of the slope

$$\frac{y_A - y_{Ai}}{x_A - x_{Ai}} = -\frac{L_M}{G_M} \frac{H_G}{H_L} \frac{Y_f}{X_f} \tag{66}$$

where

$$\frac{L_M}{G_M} = \frac{L_M^0(1 - t_A y_A)}{G_M^0(1 - t_A x_A)} \tag{67}$$

It may be noted that the above system of equations is very general and encompasses both the usual equations given for gas absorption and distillation as well as situations with any degree of counterdiffusion. The exact derivations may be found elsewhere (43).

Nonisothermal Gas Absorption.   *Nonvolatile Solvents.*   In practice, some gases tend to liberate such large amounts of heat when they are absorbed into a solvent that the operation cannot be assumed to be isothermal, as has been done thus far. The resulting temperature variations over the tower will displace the equilibrium line on a y–x diagram considerably because the solubility usually depends strongly on temperature. Thus nonisothermal operation affects column performance drastically.

The principles outlined so far may be used to calculate the tower height as long as it is possible to estimate the temperature as a function of liquid concentration. The classical basis for such an estimate is the assumption that the heat of solution manifests itself entirely in the liquid stream. It is possible to relate the temperature increase experienced by the liquid flowing down through the tower to the concentration increase through a simple enthalpy balance, equation 68, and thus correct the equilibrium line in a y–x diagram for the heat of solution as shown in Figure 9.

$$T_{\mathrm{L}} \approx T_{\mathrm{L2}} + \frac{(x_{\mathrm{A}} - x_{\mathrm{A2}})H_{\mathrm{OS}}}{x_{\mathrm{A}}C_{\mathrm{qA}} + (1 - x_{\mathrm{A}})C_{\mathrm{qB}}} \tag{68}$$

where T_{L} is the liquid temperature, °C; T_{L2} is the temperature of the entering liquid, °C; $C_{\mathrm{q}j}$ is the molar heat capacity of component j; H_{OS} is the integral mean heat of solution of solute. For each pair of values for x_{A} and T_{L} obtained from equation 68 it is possible to evaluate $y^*(x, T_{\mathrm{L}})$, the concentration in equilibrium with the bulk of the liquid phase, and to place the equilibrium line for the overall driving force (line A in Fig. 9). The line connecting the actual interfacial concentrations $(y_{\mathrm{i}}, x_{\mathrm{i}})$, line B on Figure 9, does not coincide with line A unless there is no liquid mass transfer resistance. However, because the interfacial temperature T_{i} and the bulk liquid temperature T_{L} usually are virtually equal, the equilibrium concentration y^* and the actual interfacial concentration y_{i} are connected by an isotherm. Line B on Figure 9 may therefore be constructed as shown on the basis

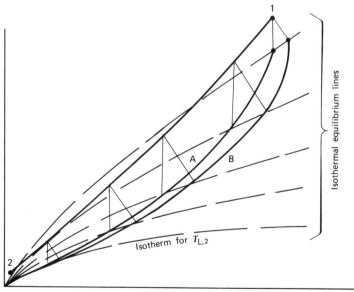

Fig. 9.   Simple model of adiabatic gas absorption. A, nonisothermal equilibrium line for overall gas-phase driving force: $y^* = f(x, T_{\mathrm{L}})$; B, nonisothermal equilibrium line for individual gas-film driving force: $y_{\mathrm{i}} = f(x_{\mathrm{i}})$.

of line A, tie lines, and isothermal equilibrium lines. Line B may be used in conjunction with equation 39 to compute the required depth of packing.

General Case.   The simple adiabatic model just discussed often represents an oversimplification, since the real situation implies a multitude of heat effects: (*1*) The heat of solution tends to increase the temperature and thus to reduce the solubility. (*2*) In the case of a volatile solvent, partial solvent evaporation absorbs some of the heat. (This effect is particularly important when using water, the cheapest solvent.) (*3*) Heat is transferred from the liquid to the gas phase and vice versa. (*4*) Heat is transferred from both phase streams to the shell of the column and from the shell to the outside or to cooling coils.

In the general case, the temperature profile is determined simultaneously by all of the four heat effects. The temperature influences the transfer of mass and heat to a large extent by changing the solubilities. This turns the simple gas absorption process into a very complex one and all factors exhibit a high degree of interaction. Computer algorithms for solving the problem rigorously have been developed (43,44). Figure 10 depicts typical profiles through an adiabatic packed gas absorber from one of these algorithms (43,45). The calculations were carried out to solve a design example calling for the removal of 90% of the acetone vapors present in an air stream by absorption into water at an L_M/G_M ratio of 2.5. The air stream contained 6 mol % acetone and was saturated with water; the ambient temperature was 15°C.

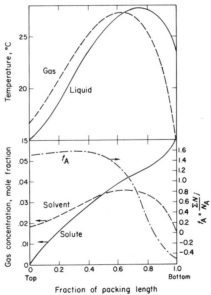

Fig. 10.   Computed rigorous profiles through an adiabatic packed absorber during the absorption of acetone into water (43).

It is a typical feature of such calculations that the shapes of the liquid temperature profiles are highly irregular and often exhibit maxima within the column. Such internal temperature maxima have been observed experimentally in plate and packed absorbers (44,46,47), and the measured temperature profiles can be shown to agree closely with rigorous computations. The temperature maximum occurs in part because the heat of solution causes the entering liquid stream to be heated. In the lower part of the tower, however, the heat of absorption is smaller than the opposite heat effects of solvent evaporation and heat transfer to the cold entering gas, so that the net effect is a cooling of the liquid phase. These transfers are reversed in the upper part of the column, as is obvious from Figure 10: the gas gives up heat to the liquid, is cooled, and some of the solvent condenses from the gas stream into the liquid stream, which is heated much faster in this part of the column than would be the case with the absorption alone.

Figure 11 shows the rigorously computed y–x diagram for the same example. The temperature maximum within the column produces a region of reduced solubility reflecting itself in the typical bulge in the middle of the rigorous equilibrium line D. Since less acetone is absorbed in this part of the equipment, the gas concentration curve exhibits a slight plateau (Fig. 10). This example may also serve as a demonstration of the difficulty in estimating the required depth of packing using simplifying assumptions (Table 2). The isothermal approximation failed completely in this case and yielded 1.95 m of required packing as opposed to the rigorously determined value of 3.63 m. Neglecting the temperature increase completely, this model assumes a solubility which is much too large, reflected by equilibrium line A, and thus underestimates the rigorous result by 90%.

The standard way to correct for the heat of solution approximately is the

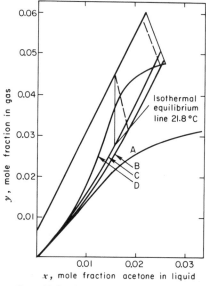

Fig. 11. y–x diagram for adiabatic absorption of acetone into water. A, isothermal equilibrium line at T_{L2}; B, equilibrium line for simple model of adiabatic gas absorption, gas-film driving force; C, equilibrium line for simple model of adiabatic gas absorption, overall gas-phase driving force; D, rigorously computed equilibrium line, gas-film driving force (43).

Table 2. Comparison of Results of Different Design Calculations[a]

Method used	N_{OG}	Required depth of packing, m
rigorous calculation	5.56	3.63
isothermal approximation	3.30	1.96
simple adiabatic model	4.01	2.38
suggested short-cut design procedure		
graphical integration	5.51	3.60
approximate analytical integration	5.56	3.73

[a]Ref. 45.

simple adiabatic model described on the preceding pages, which yields equilibrium line B if the gas-phase driving force is used and line C on the basis of the overall driving force. This model, however, is a poor representation of the conditions prevailing in the absorber, as demonstrated by the deviation of its equilibrium line B from the rigorous line D. The approximation underestimates the true packing depth value of 3.63 m by more than one-third yielding 2.4 m (Table 2).

Rapid Approximate Design Procedure. For purposes of quick, approximate design of adiabatic packed gas absorbers an empirical correlation of liquid temperature profiles has been developed (45) which may be used in a way similar to equation 68 to compute the necessary tower height. The liquid temperature profiles were rigorously computed for over 90 hypothetical design cases covering wide variations in system properties and operating conditions. The mG_M/L_M factor was varied from maximum values given by pinchpoints between the equilibrium and the operating lines down to values where heat effects were subdued by the heat capacity of the high solvent flow rate. The apparent Henry's law constants were varied from 0.0 to 186.6 kPa (1400 mm Hg) and heats of solution up to 58.6 kJ/mol (14.0 kcal/mol) were taken into account. The investigation was limited to gas concentrations below 15 mol % and to recovery fractions ranging from 90 to 99%. Water was considered to be the most important solvent, but variations in solvent properties have been included to a certain extent.

It was found that the rigorously computed liquid temperature profiles could be satisfactorily represented as a function of liquid concentration by the empirical equation

$$T_L = T_{L,2} + (T_{L,1} - T_{L,2})X_N + 74.34(X_N^{1.074} - X_N^{1.114}) \Delta T_{max} \qquad (69)$$

where

$$X_N = \frac{x_A - x_{A,2}}{x_{A,1} - x_{A,2}} \qquad (70)$$

$T_{L,1}$ is the temperature of the liquid leaving the tower and may be calculated from an enthalpy balance around the whole tower based on an estimate of the temperature of the leaving gas, $T_{G,2}$. It was found that the rigorous values of $T_{L,1}$ could be predicted rapidly within a standard deviation of 1.2°C if the following semiempirical equations were used for estimating $T_{G,2}$:

$$T_{G,2} = T_{L,2} + \left(\frac{dT_L}{dx_A}\right)_2 \left(\frac{G_M}{L_M}\right)_2 \left(\frac{H_{G,Q}}{H_{OG,A}}\right) (y_{A,2} - y_{A,2}^*) \qquad (71)$$

$$\left(\frac{dT_L}{dx_A}\right)_2 = \frac{L_{M,2}H_{OS} - G_{M,2}H_v m_{B,2}}{L_{M,2}C_{q,2} - G_{M,2}C_{P,2} - G_{M,2}H_v(1 - x_{A,2})(dm_B/dT_L)_2} \qquad (72)$$

where $H_{G,Q} = G_M C_P / h_G a$, the gas-phase height of a heat transfer unit. This quantity may be estimated from H_G for the solute using the Chilton-Colburn analogy; $H_{OG,A} = H_{OG}$ for the solute; H_{OS} is the integral heat of solution for solute; H_v is the latent heat of pure solvent; $m_{B,2}$ is the "solubility" of solvent, y_B^*/x_B at the top of the tower; $(dm_B/dT_L)_2$ is the temperature coefficient of m_B at the top of the tower; and $C_{P,2}$ and $C_{q,2}$ are mean molar heat capacities of gas and liquid at the top of the towers, respectively.

ΔT_{max} in equation 69 is related to the internal temperature maximum and may be estimated from specified variables using the correlations shown in Figure 12. All other quantities in equations 69 and 70 are usually known from design specifications. If applied to values of the liquid concentration, equation 69 yields corresponding temperature values which may serve to evaluate the equilibrium gas concentration y^* as a function of the liquid concentration. This procedure yields a good approximation of the rigorous equilibrium line accounting for all heat effects. The required number of transfer units and the tower height may then be computed on the basis of a conventional x–y diagram by graphical integration.

To obviate the tedious graphical integration, a simplified design procedure was developed on the basis of Colburn's analytical solution, equation 56. Substitution of the L_M/G_M ratio presents no problem because this ratio stays fairly constant in the tower at the low concentrations for which Figure 12 is valid. The difficulty arises with m because of the unexpected shapes of nonisothermal equilibrium lines.

The equilibrium line, as approximated by the synthetically correlated temperature profile, is cut into two parts at its inflection point as shown in Figure 13. In each part the equilibrium curve is replaced by a straight line with a slope $\overline{m}$ which, upon substitution into equation 56, yields the same number of transfer units as the graphical integration based on the curved equilibrium line. These effective average slopes $\overline{m}$ were calculated for an expanded set of well over 100 hypothetical design cases by performing the integration numerically and solving for m. They were correlated using a multidimensional, nonlinear regression analysis and the initial slopes and the slopes of the cord of the equilibrium line in the respective section (Fig. 13). The resulting correlation for the dilute section of the y–x diagram, where the equilibrium is curved concave upward, is shown in Figure 14, whereas Figure 15 represents the correlation for the concentrated section where the curvature is concave downward.

The recommended rapid approximate design procedure for adiabatic packed gas absorbers consists of the following steps: (1) After the problem is fully specified, the temperature of the leaving product is estimated (eqs. 71, 72) using an

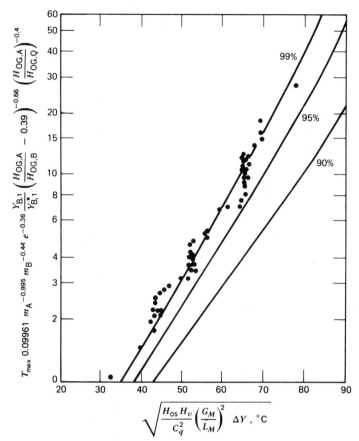

Fig. 12. Correlation of $\Delta T_{\max}$. The three lines represent the best fit of a mathematical expression obtained by multidimensional nonlinear regression techniques for 99, 95, and 90% recovery; the points are for 99% recovery. C_q = mean molar heat capacity of liquid mixture, averaged over tower; $\Delta Y = y_{A,1} - y_{A,2}$; m_A = slope of equilibrium line for solute, to be taken at liquid feed temperature; m_B = slope of equilibrium line for solvent, $y_B^*/(1 - x_A)$, to be taken at liquid feed temperature; y_B/y_B^* = saturation of solvent in feed gas; $H_{OG,Q} \approx H_{G,Q}$, the height of a gas phase heat transfer unit. (The H_{OG}s are computed from equation 54; the individual heights of a transfer unit for the solvent and for heat may be estimated from $H_{G,A}$ and $H_{L,A}$ using the Chilton-Colburn analogy. The m in equation 54 is to be evaluated at the temperature of the liquid feed.) Reprinted by permission (45).

enthalpy balance around the tower. (2) The maximum temperature $\Delta T_{\max}$ of the convex portion of the liquid temperature profile is estimated on the basis of Figure 12. (Steps 1 and 2 determine the estimated temperature profile completely (eq. 69).) (3) A special correlation (Fig. 16) may then be used to find the locus of the inflection point on the equilibrium curve on the y–x diagram without trial and error. (4) The equation for the estimated temperature profile (eq. 69) is then used to find the temperature as well as the slope of the equilibrium line at the inflection point. Based on this information, the effective average slopes for the two sections may be read from Figures 14 and 15, and used in equation 56 to determine the

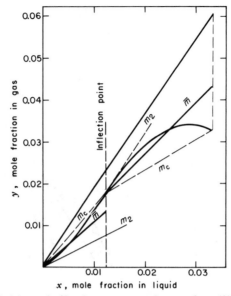

Fig. 13. Definition of effective average slopes of equilibrium line (45).

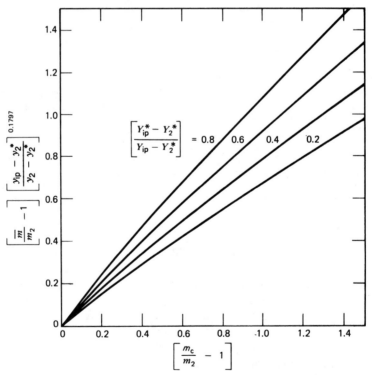

Fig. 14. Correlation of effective average slope $\overline{m}$ of equilibrium line (dilute part of absorber); equilibrium line concave upward. Y_{ip}: gas mole fraction at inflection point (45).

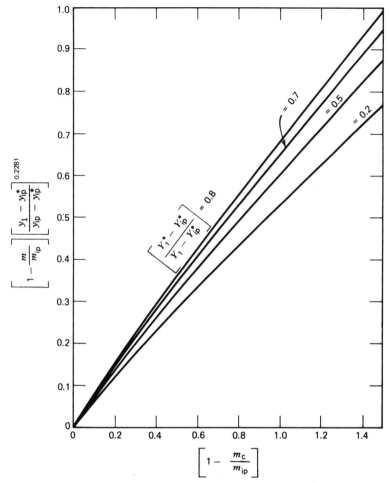

Fig. 15. Correlation of effective average slope $\overline{m}$ of nonisothermal equilibrium line (concentrated part of absorber); equilibrium line concave downward. The subscript ip indicates the inflection point (45).

necessary number of transfer units. The overall height of a transfer unit is evaluated for each section substituting the values for $\overline{m}$ from Figures 14 and 15 into equation 54.

The total number of transfer units and the total required depth of packing are obtained by adding the respective values for the two sections. Table 2 compares this design procedure, the isothermal and the simple adiabatic approximations, and the rigorous result with respect to the design example, discussed previously, and clearly demonstrates the superior accuracy of the procedure. No serious temperature bulge will develop if $(T_{L,2} - T_{L,1})/\Delta T_{max} \geq 4.3$. The simple model of adiabatic gas absorption may then be used to determine the required tower height. Figure 14 may, of course, be used also in this case together with equations 47, 54, and 56 to circumvent the graphical integration and obtain the result rapidly without significant loss of accuracy.

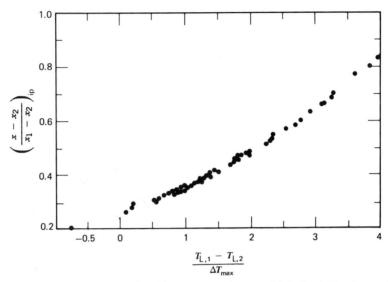

Fig. 16. Correlation of the liquid concentration at which the inflection point of the nonisothermal equilibrium occurs (45).

Axial Dispersion Effects. *Effect of Axial Dispersion on Column Performance.* Another assumption underlying standard design methods is that the gas and the liquid phases move in plug-flow fashion through the column. In reality, considerable departure from this ideal flow assumption exists (4) and different fluid particles travel through the packing at varying velocities. The impact of this effect, which is usually referred to as axial dispersion, on the concentration profiles is demonstrated in Figure 17. The effect counteracts the countercurrent contacting scheme for which the column is designed and thus lowers the driving forces throughout the packed bed. Neglect of axial dispersion results in an overestimation of the driving forces and in an underestimation of the number of transfer units needed. It may therefore lead to an unsafe design.

Determination of separation efficiencies from pilot-plant data is also affected by axial dispersion. Neglecting it yields high H_G or H_L values. Literature data for this parameter have usually not been corrected for this effect.

The extent of axial dispersion occurring in a gas absorber can be determined by measuring the residence time distribution of both gas and liquid. Based on classical flow models of axial dispersion, the result is usually expressed in terms of two Peclet numbers (Pe), one for each phase. Peclet numbers tending towards infinity indicate near-ideal plug flow, whereas vanishing values of Pe indicate axial dispersion to such an extent that the phase begins to become well backmixed. When designing packed columns, the Peclet numbers are usually estimated from literature correlations (48–50). Correlations for predicting Peclet numbers in large scale gas–liquid contactors are quite scarce, but contributions have been made (51–55). Some of the available data have been described (4) and a review of published correlations for liquid-phase Peclet numbers is also available (55). Figure 18 reproduces some data (51).

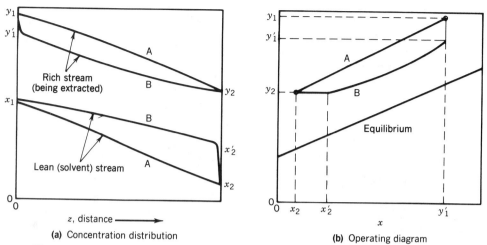

Fig. 17. Effect of axial dispersion in both phases on solute distribution through countercurrent mass transfer equipment. A, piston or plug flow; B, axial dispersion in both streams (diagrammatic). Reprinted with permission (4).

When designing packed towers, axial dispersion can be accounted for by incorporating terms for axial dispersion of the solute into the differential mass balance equation 36. The integration of the resulting differential equations is best effected by computer. Analytical solutions for cases having linear equilibrium and operating lines have been developed (56,57). They are, however, not explicit for the design case and are of such enormous complexity that application for design also requires a computer.

Rapid Approximate Design Procedure. Several simplified approximations to the rigorous solutions have been developed over the years (57–60), but they all remain too complicated for practical use. A simple method proposed in 1989 (61,62) uses a correction factor accounting for the effect of axial dispersion, which is defined as (57)

$$\text{correction factor} = \frac{\text{NTU}_{ap}}{\text{NTU}} \tag{73}$$

NTU_{ap} is the "exterior apparent" overall gas-phase number of transfer units calculated neglecting axial dispersion simply on the basis of equation 56, whereas NTU stands for the higher real number of transfer units (N_{OG}) which is actually required under the influence of axial dispersion. The correction factor ratio can be represented as a function of those parameters that are actually known at the outset of the calculation

$$\frac{\text{NTU}_{ap}}{\text{NTU}} = f\left(\text{NTU}_{ap}, \left(\frac{mG}{L}\right), \text{Pe}_x, \text{Pe}_y\right) \tag{74}$$

Equation 74 is shown graphically in Figure 19a for a given set of conditions.

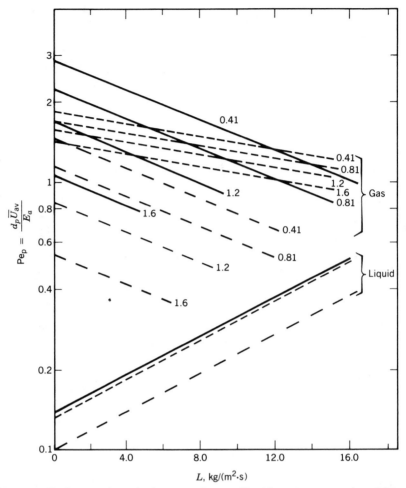

Fig. 18. Peclet numbers in large scale gas–liquid contactors using 2.54-cm Berl saddles (——) or 2.54 cm (— —) or 5.08 cm (- - -) Raschig rings (51). Numbers on lines represent G values = gas flow in kg/(m²·s); d_p = nominal packing size; U_{av} = superficial velocity. To convert kg/(m²·s) to lb/(h·ft²) multiply by 737.5. Reprinted with permission (4).

Curves such as these cannot be directly used for design, however, because the Peclet number contains the tower height h as a characteristic dimension. Therefore, new Peclet numbers are defined containing H_{OG} as the characteristic length. These relate to the conventional Pe as

$$\text{Pe}_{\text{HTU}} = \frac{u H_{\text{OG}}}{D_{\text{ax}}} = \frac{uh}{D_{\text{ax}}} \frac{H_{\text{OG}}}{h} \tag{75a}$$

$$= \text{Pe} \, \frac{1}{N_{\text{OG}}} \tag{75b}$$

The correction factor $(\text{NTU})_{\text{ap}}/\text{NTU}$ as a function of Pe_{HTU} rather than Pe is

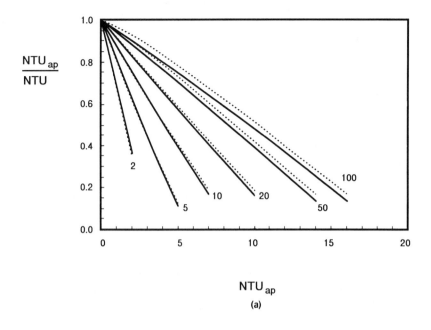

$$NTU_{ap}$$

(a)

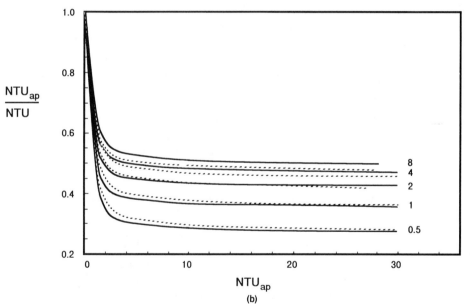

$$NTU_{ap}$$

(b)

Fig. 19. Correction factor for axial dispersion as a function of NTU_{ap}. Solid lines are rigorous calculations; broken lines, approximate formulas according to literature (61). (a) Numbers on lines represent Pe_x values; $Pe_y = 20$; $mG_M/L_M = 0.8$. (b) For design calculations. Numbers on lines represent $Pe_{HTU,x}$ values; $Pe_{HTU,y} = 1$; $mG_M/L_M = 0.8$.

shown in Figure 19**b**. The correction factors given in Figures 19**a** and 19**b** can roughly be estimated as

$$\frac{NTU_{ap}}{NTU} \approx 1 - \frac{NTU_{ap}}{\dfrac{\ln S}{S-1} + \dfrac{Pe_x Pe_y}{Pe_y + SPe_x}} \tag{76}$$

$$\frac{NTU_{ap}}{NTU} \approx \frac{Pe_{HTU,y} Pe_{HTU,x}}{Pe_{HTU,y}\, Pe_{HTU,x} + Pe_{HTU,y} + SPe_{HTU,x}} \tag{77}$$

In these equations S denotes the stripping factor, mG_M/L_M. Equation 77 is only valid for a sufficiently high number of transfer units so that the correction factor becomes independent of NTU_{ap}.

In the original study (61), NTU_{ap}/NTU was calculated for thousands of hypothetical design cases as a function of both Pe and Pe_{HTU}. The results were correlated and empirical expressions were given that can be evaluated on a handheld calculator, just as equations 76 and 77, but which approximate the computer calculation much better, to within about $\pm 5\%$.

The recommended rapid design procedure consists of the following steps: (*1*) The apparent N_{OG} is calculated using equation 56. (*2*) The extent of axial dispersion is estimated from literature correlations for each phase in terms of Pe numbers and transformed into Pe_{HTU} values. (*3*) The correction factor NTU_{ap}/NTU is estimated on the basis of the correlation given in the literature (61). A reasonable, conservative estimate may also be obtained using equation 77, provided $NTU_{ap} > 5$. When the apparent number of transfer units is divided by this correction factor, the value of N_{OG} actually required under the influence of axial dispersion is obtained. (*4*) The packed tower height is found by multiplying N_{OG} by the true H_{OG}. In order to obtain values for the latter, pilot-plant data has to be corrected for the influence of axial dispersion. This correction may be made in a manner similar to that described above, but equation 76 would be used to estimate the correction factor rather than equation 77.

Experimental Mass Transfer Coefficients.   Hundreds of papers have been published reporting mass transfer coefficients in packed columns. For some simple systems which have been studied quite extensively, mass transfer data may be obtained directly from the literature (6). The situation with respect to the prediction of mass transfer coefficients for new systems is still poor. Despite the wealth of experimental and theoretical studies, no comprehensive theory has been developed, and most generalizations are based on empirical or semiempirical equations.

Liquid-Phase Transfer.   It is difficult to measure transfer coefficients separately from the effective interfacial area; thus data is usually correlated in a lumped form, eg, as $k_L a$ or as H_L. These parameters are measured for the liquid film by absorption or desorption of sparingly soluble gases such as O_2 or CO_2 in water. The liquid film resistance is completely controlling in such cases, and $k_L a$ may be estimated as $K_{OL}a$ since $x_i \approx x^*$ (Fig. 4). This is a prerequisite because the interfacial concentrations would not be known otherwise and hence the driving force through the liquid film could not be evaluated.

The resulting correlations fall into several categories. Some are essentially empirical in nature. Examples include (63)

$$H_L = \frac{1}{\alpha}\left(\frac{L}{\mu}\right)^n\left(\frac{\mu_L}{\rho_L D_L}\right)^{0.5} \tag{78}$$

The values of α and n are given in Table 3; typical values for D_L can be found in Table 4. The exponent of 0.5 on the Schmidt number ($\mu_L/\rho_L D_L$) supports the penetration theory. Further examples of empirical correlations provide partial experimental confirmation of equation 78 (3,64–68). The correlation reflecting what is probably the most comprehensive experimental basis, the Monsanto Model, also falls in this category (68,69). It is based on 545 observations from 13 different sources and may be summarized as

$$H_L = \phi\, C_{fl}\left(\frac{h}{3.05}\right)^{0.15} Sc_L^{0.5} \tag{79}$$

The packing parameter $\phi\,(m)$ reflects the influence of the liquid flow rate as shown in Figure 20. C_{fl} reflects the influence of the gas flow rate, staying at unity below 50% of the flooding rate but beginning to decrease above this point. At 75% of the flooding velocity, $C_{fl} = 0.6$. Sc_L is the Schmidt number of the liquid.

Table 3. Values of Constants for Equation 78

Packing	Size, cm	α^a	α^b	n
Raschig rings	0.95	3120	550	0.46
	1.3	1390	280	0.35
	2.5	430	100	0.22
	3.8	380	90	0.22
	5.1	340	80	0.22
Berl saddles	1.3	685	150	0.28
	2.5	780	170	0.28
	3.8	730	160	0.28

[a]Valid for units kg, s, m.
[b]Valid for units lb, h, ft.

Table 4. Diffusion Coefficients for Dilute Solutions of Gases in Liquids at 20°C

Gas	Liquid	D_L, m^2/s $\times\ 10^{-9}$
CO_2	water	1.78
Cl_2	water	1.61
H_2	water	5.22
HCl	water	0.61
H_2S	water	1.64
N_2	water	1.92
N_2O	water	1.75
NH_3	water	1.83
O_2	water	2.08
acetone	water	1.61
benzene	kerosene	1.41

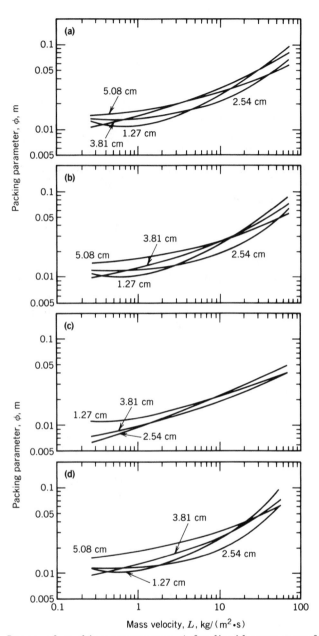

Fig. 20. Improved packing parameters ϕ for liquid mass transfer: (**a**) ceramic Raschig rings; (**b**) metal Raschig rings; (**c**) ceramic Berl saddles; (**d**) metal Pall rings (69). Reprinted with permission (70).

Other correlations based partially on theoretical considerations but made to fit existing data also exist (71–75). A number of researchers have also attempted to separate k_L from a by measuring the latter, sometimes in terms of the wetted area (76–78). Finally, a number of correlations for the mass transfer coefficient k_L itself exist. These are based on a more fundamental theory of mass transfer in packed columns (79–82). Although certain predictions were verified by experimental evidence, these models often cannot serve as design basis because the equations contain the interfacial area a as an independent variable.

   Gas-Phase Transfer.   The height of a gas-phase mass transfer unit, or $k_G a$, is normally measured either by vaporization experiments of pure liquids, in which no liquid mass transfer resistance exists, or using extremely soluble gases. In the latter case, m is so small that the liquid-film resistance in equation 7 is negligible and the gas-film mass transfer coefficient can be observed as $k_G a \approx K_{OG} a$ and $y_i \approx y^*$. The experiments are difficult because they have to be carried out in very shallow beds. Otherwise, all of the highly soluble gas is absorbed and the driving force cannot be evaluated. The resulting end effects are probably the main reason for the substantial disagreement of the published data, which have been reported to vary some threefold for the same packing and flow rates (83). Furthermore, the effective interfacial areas seem to differ for absorption and vaporization (84). During the absorption experiments, the many stagnant or semistagnant pockets of liquid which exist in a packing tend to become saturated and thus ineffective. This is not the case in vaporization experiments where the total effective interfacial area consists of the surface area of moving liquid plus the semistagnant liquid pockets.

   The correlation of H_G based on the most extensive experimental basis is again the Monsanto Model (69):

$$H_G = \psi \frac{(3.28\ d'_c)^m (h/3.05)^{1.3}}{(737 L f_\mu f_\rho f_\sigma)^n}\ \mathrm{Sc}_G^{0.5} \qquad (80)$$

where d'_c is the lesser of 0.61 or column diameter, m; $f_\mu = (\mu_L/\mu_w)^{0.16}$; $f_\rho = (\rho_L/\rho_w)^{-1.25}$; $f_\sigma = (\sigma_L/\sigma_w)^{-0.8}$; Sc_G = Schmidt number of the gas. The packing parameter ψ, in m, depends on the gas flow rate as shown in Figure 21. Values of the diffusivity of various solutes in air and typical Schmidt numbers for use in equation 80 are found in Table 5. The exponents m and n adopt the values of 1.24 and 0.6, respectively, for rings as packing materials, and 1.11 and 0.5, for saddles.

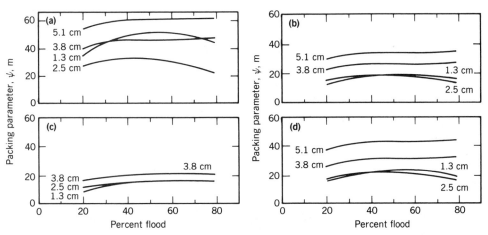

Fig. 21.   Improved packing parameters ψ for gas mass transfer: (**a**) ceramic Raschig rings; (**b**) metal Rashig rings; (**c**) ceramic Berl saddles; (**d**) metal Pall rings (69). Reprinted with permission.

Table 5. Values of the Diffusion Coefficient D_A and of $\mu_G/D_A\rho_G$ for Various Gases in Air at 0°C and at Atmospheric Pressure[a]

Gas	CAS Registry Number	D_A, m^2/s $\times$ 10^{-5}		$\dfrac{\mu_G}{\rho_G D_A}$
		Calculated	Experiment	
acetic acid	[64-19-7]		1.05	1.26
acetone	[67-64-1]	0.83		1.60
ammonia	[7664-41-7]	1.62	2.17	0.61
benzene	[71-43-2]	0.72	0.78	1.71
bromobenzene	[108-86-1]	0.67		1.71
butane	[106-97-8]	0.75		1.77
n-butyl alcohol	[71-36-3]		0.69	1.88
carbon dioxide	[124-38-9]	1.19	1.39	0.96
carbon disulfide	[75-15-0]		278.	1.48
carbon tetrachloride	[56-23-5]	0.61		2.13
chlorine	[7782-50-5]	0.92		1.42
chlorobenzene	[108-90-7]	0.61		2.13
chloropicrin	[76-06-2]	0.61		2.13
2,2'-dichloroethyl sulfide (mustard gas)	[505-60-2]	0.56		2.44
ethane	[74-84-0]	1.08		1.22
ethyl acetate	[141-78-6]	0.67	0.72	1.84
ethyl alcohol	[64-17-5]	0.94	1.03	1.30
ethyl ether	[60-29-7]	0.69	0.78	1.70
ethylene dibromide	[106-93-4]	0.67		1.97
hydrogen	[1333-74-0]	5.61		0.22
methane	[74-82-8]	1.58		0.84
methyl acetate	[79-20-9]		0.94	1.57
methyl alcohol	[67-56-1]	1.22	1.33	1.00
naphthalene	[91-20-3]		0.50	2.57
nitrogen	[7727-37-9]	1.33		0.98
n-octane	[111-65-9]		0.50	2.57
oxygen	[7782-44-7]	1.64	1.78	0.74
pentane	[109-66-0]	0.67		1.97
phosgene	[75-44-5]	0.81		1.65
propane	[74-98-6]	0.89		1.51
n-propyl acetate	[109-60-4]	0.67	0.67	1.97
n-propyl alcohol	[71-23-8]	0.81	0.86	1.55
sulfur dioxide	[7446-09-5]	1.03		1.28
toluene	[108-88-3]	0.64	0.72	1.86
water	[7732-18-5]	1.89	2.19	0.60

[a]The value of μ_G/ρ_G is that for pure air, 1.33 $\times$ 10^{-5} m^2/s. Diffusion coefficients may be corrected for other conditions by assuming them proportional to $T^{2/3}$ and inversely proportional to P. The Schmidt numbers depend only weakly on temperature (113).

Another type of experiment to measure k_G separately from other factors consists of saturating packings made from porous materials using a volatile liquid and subsequently drying it by passing a stream of inert gas through the packing (85–88). Since the surface of the packing is normally known in these experiments,

k_G can be computed. Application of this kind of data to gas absorption design is difficult, however, because of the different, unknown effective interfacial areas when two phases are flowing through the packing. A similar approach was used by evaporating naphthalene from a packing made from this material (78,84). The mass transfer coefficient k_G measured in this manner was then combined with $k_G a$ data (89) to determine the effective area, which was found to be fairly independent of gas rate up to the loading point. The data bank underlying the Monsanto Model and literature correlations for k_G and k_L (78,84) have also been used (90) to develop a new correlation for packed distillation columns.

Height Equivalent to a Theoretical Plate. Provided both the equilibrium and operating lines are straight, HETP values may be estimated by combining the H_G and H_L values predicted by the above correlations and by translating the resulting H_{OG} into HETP by combining equations 47, 51, and 56 with equation 85, which is discussed under bubble tray absorption columns:

$$\text{HETP} = \frac{\ln(mG_M/L_M)}{(mG_M/L_M) - 1} H_{OG} \tag{81}$$

HETP values obtained in this way have been compared to measured values in data banks (69) and statistical analysis reveals that the agreement is better when equations 79 and 80 are used to predict H_G and H_L than with the other models tested. Even so, a design at 95% confidence level would require a safety factor of 1.7 to account for scatter.

The situation is very much poorer for structured rather than random packings, in that hardly any data on H_G and H_L have been published. Based on a mechanistic model for mass transfer, a way to estimate HETP values for structured packings in distillation columns has been proposed (91), yet there is a clear need for more experimental data in this area.

Capacity Limitations. Thus far the discussion has been confined to factors affecting the tower height required to perform a specific absorption job. The necessary tower diameter, on the other hand, depends primarily on the total amount of gas and liquid that must be handled. At a given set of flow rates, the diameter of the packing can only be decreased at the expense of a large pressure drop, which in turn generates higher operating costs because more power is needed to blow the gas through the packing. The reason for this is the fact that handling a given total gas flow rate in a smaller tower diameter increases the superficial velocity at which the gas has to be pushed through the packing.

The relationship between the pressure drop per unit of packed height and the superficial gas velocity given in terms of the gas flow rate is shown schematically in Figure 22. In a dry packing, ΔP increases almost as the square of the gas velocity, which is in accord with the turbulent nature of the flow. At low liquid flow rates, the curves are somewhat shifted upwards because the presence of liquid films restricts the free section available for gas flow and thus increases the linear gas velocity somewhat. Because the liquid hold-up remains independent of G, the slope of the curve in this log–log plot remains close to 2. At higher pressure drops, however, the upflowing gas impairs the downflow of liquid and excess liquid starts to accumulate in the packing, thereby increasing the hold-up. In this operating region, called the loading zone, the increasing liquid hold-up

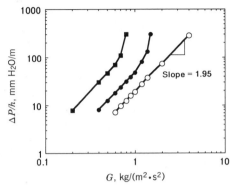

Fig. 22. Schematic representation of typical pressure drop as a function of superficial gas velocity, expressed in terms of $G = \rho_G u_G$, in packed columns. $\bigcirc$, Dry packing; $\bullet$, low liquid flow rate; $\blacksquare$, higher liquid flow rate. The points do not correspond to actual experimental data, but represent examples.

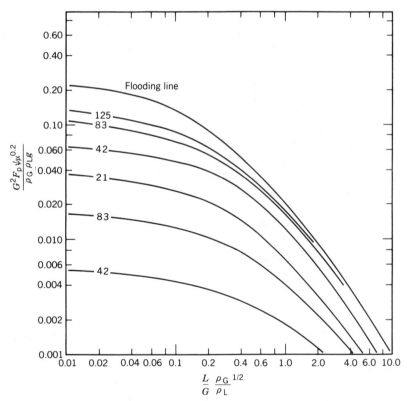

Fig. 23. Pressure drop and flooding correlation for various random packings (95). $\psi = \rho_{H_2O}/\rho_L$, g (standard acceleration of free fall) $= 9.81$ m/s^2, $\mu =$ liquid viscosity in mPa·s; numbers on lines represent pressure drop, mm H$_2$O/m of packed height; to convert to in. H$_2$O/ft multiply by 0.012. Packing factors for various packings have been published (96) and are reproduced in part in Table 6.

restricts the free section available for the gas flow further as G becomes larger. Hence the linear gas velocities increase faster than G and the power dependence of ΔP on G starts to rise above 2. The G value at which the curve begins to deviate from the straight line has been defined as the loading point.

If the tower diameter is made too small for a given total gas flow rate, that is, if u_G and G are increased above a certain critical value, ΔP becomes so great that the liquid cannot flow downward anymore over the packing, but is blown out the top of the packing. The vertical asymptotes on Figure 22 indicate the gas rates at which this condition, called flooding, occurs. This gas flow rate at flooding, G_F, determines the theoretical minimum diameter at which the tower is operable, and knowledge of it is therefore very important. Flooding rates have been correlated (92–94).

Both the pressure drop per unit length of packed tower and the gas flooding rate have been correlated for random packings as shown in Figure 23 (95). Such correlations enable predicting the gas flow rate G that will flood the packing at a given L/G ratio. In practice, the tower has to be operated at flow rates considerably less than the flooding rates for safety reasons. It is generally accepted that 50 to 80% of the flooding flow rates can be permitted. The curves plotted on Figure 23 thus also enable prediction of the pressure drop at any chosen operating value of G. The diameter of the column must then be evaluated by comparing this value to the total quantity of gas that the tower is supposed to handle. The correlation shown in Figure 23 can be applied to predict the hydraulic performance of many different packings owing to an adjustable parameter known as the packing factor F_P. Values for F_P have been compiled (96); a few examples are listed in Table 6. Similar flooding rate correlations are available for arranged packings. Figure 24 reports results for four examples of Mellapak types (97). Comparison of Figures 24 and 23 reveals higher capacity limits in structured than in many of the dumped packings. At similar loads, structured packings tend generally to give rise to less pressure drop.

Table 6. Characteristics of Dumped Tower Packings[a]

Packing type	Nominal size, mm	Surface area, m^2/m^3	Packing factor, F_P, m^{-1}
Raschig rings, ceramic	6	710	5250
	13	370	2000
	25	190	510
	50	92	215
Raschig rings, steel	25	185	450
	50	95	187
Berl saddles, ceramic	6	900	2950
	13	465	790
	25	250	360
	50	105	150
Pall rings, metal	25	205	157
	50	115	66
Pall rings, polypropylene	25	205	170
	50	100	82

[a]Ref. 96.

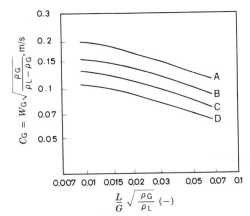

Fig. 24. Souders load diagram for capacity limit determination for four structured packings of the Sulzer-Mellapak type. The solid lines represent the capacity limits of the respective packings as defined by a pressure drop of 1.2 kPa/m: (A) 125 Y; (B) 250 Y; (C) 350 Y; (D) 500 Y. Flooding rates are about 5% higher. Reprinted with permission (97).

Bubble Tray Absorption Columns

General Design Procedure. Bubble tray absorbers may be designed graphically based on a so-called McCabe-Thiele diagram. An operating line and an equilibrium line are plotted in y–x, Y'–X', or Y^0–X^0 coordinates using the principles for packed adsorbers outlined above (see Fig. 25). The minimum number of plates required for a specified recovery may be computed by assuming that equilibrium is reached between the two phases on each bubble tray. Thus the gas and the liquid leaving a tray are at equilibrium and a hypothetical tray capable of equilibrating the phase streams is termed a theoretical plate. Starting the calculation at the bottom of the tower, where the concentrations are y_{N+1} and x_N (see Fig. 26), the concentration leaving the lowest theoretical plate y_N may be found on the design diagram (Fig. 25a) by moving from the operating line vertically to the equilibrium line, because y_N is at equilibrium with x_N. Since the concentrations between two plates are always related by the operating line, x_{N-1} may be found from y_N by moving horizontally to the operating line. By repeating this sequence of steps until the desired residual gas concentration y_1 is reached, the number of theoretical plates can be counted.

The required number of actual plates, N_P, is larger than the number of theoretical plates, N_{TP}, because it would take an infinite contacting time at each stage to establish equilibrium. The ratio N_{TP}:N_P is called the overall column efficiency. This parameter is difficult to predict from theoretical considerations, however, or to correct for new systems and operating conditions. It is therefore customary to characterize the single plate by the so-called Murphree vapor plate efficiency, E_{MV} (98):

$$E_{MV} \equiv \frac{y_n - y_{n+1}}{y_n^* - y_{n+1}} \tag{82}$$

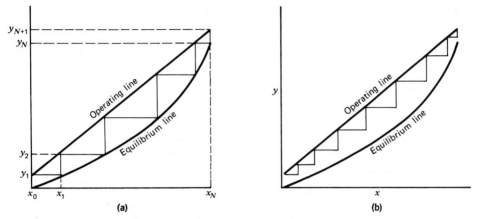

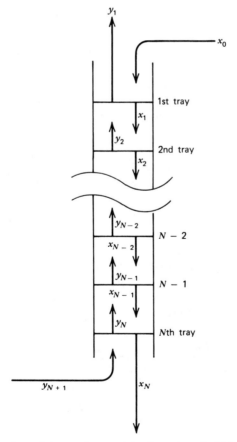

Fig. 25. McCabe-Thiele diagram. (**a**) Number of theoretical plates, 5; (**b**) number of actual plates, 8.

Fig. 26. Numbering plates and concentrations in a bubble tray column.

which indicates the fractional approach to equilibrium achieved by the plate. An efficiency of 80% means that the reduction in solute gas concentration effected by the plate is 80% of the reduction obtained from a theoretical plate. Corresponding actual plates may therefore be stepped off by moving from the operating line vertically only 80% of the distance between operating and equilibrium line (Fig. 25b). In some special cases having negligible resistance in the gas phase, E_{MV} values may become unreasonably small. It is then more logical to define a Murphree liquid plate efficiency, E_{ML}, simply by reversing the role of liquid and gas and by focusing on the change in liquid composition across the plate with respect to an equilibrium given by the leaving vapor.

Simplified Design Procedure for Linear Equilibrium and Operating Lines.   A straight operating line occurs when the concentrations are low such that L_M and G_M remain essentially constant. (The material balance is obtained from equation 35.) In cases where the equilibrium K value does not depend too much on concentration, the use of absorption and stripping factors (99–101) allows rapid calculations for absorption design. One of the simplifying assumptions made in the development of this so-called Kremser-Brown method involves the use of the absorption factor A. The following algebraic expression describes the liquid and vapor flows from the plate (102):

$$A_n = \frac{L_n}{K_n G_n} \quad \text{or} \quad A_{\mathrm{av}} = \frac{L_{\mathrm{av}}}{K_{\mathrm{av}} G_{\mathrm{av}}} \tag{83}$$

where A_n is the absorption factor for each plate n and L_n and G_n are the liquid and vapor flows from the plate. The fractional absorption of any component by an absorber of N plates is expressed in a form similar to the Kremser equation (100,101,103).

$$\frac{Y_{N+1} - Y_1}{Y_{N+1} - Kx_0} = \frac{A^{N+1} - A}{A^{N+1} - 1} \tag{84}$$

where Y_{N+1} = moles of absorbed component entering the column per mole of entering vapor and Y_1 = moles of absorbed component leaving the column per mole of entering vapor. The calculation of plate efficiency (100) is quite sensitive to the choice of equilibrium constants (4,104).

For linear equilibrium and operating lines, an explicit expression for the number of theoretical plates required for reducing the solute mole fraction from y_{N+1} to y_1 has been derived (41):

$$N_{\mathrm{TP}} = \frac{\ln\left[\left(1 - \frac{mG_M}{L_M}\right)\left(\frac{y_{N+1} - mx_0}{y_1 - mx_0}\right) + \frac{mG_M}{L_M}\right]}{\ln \frac{L_M}{mG_M}} \tag{85}$$

This is the one case where the overall column efficiency can be related analytically to the Murphree plate efficiency, so that the actual number of plates is calculable by dividing the number of theoretical plates through equation 86:

$$\frac{N_{\text{TP}}}{N_{\text{P}}} = \frac{\ln\left[1 + E_{\text{MV}}\left(\frac{mG_M}{L_M} - 1\right)\right]}{\ln\frac{mG_M}{L_M}} \tag{86}$$

Nonisothermal Gas Absorption. The computation of nonisothermal gas absorption processes is difficult because of all the interactions involved as described for packed columns. A computer is normally required for the enormous number of plate calculations necessary to establish the correct concentration and temperature profiles through the tower. Suitable algorithms have been developed (46,105) and nonisothermal gas absorption in plate columns has been studied experimentally and the measured profiles compared to the calculated results (47,106). Figure 27 shows a typical liquid temperature profile observed in an adiabatic bubble plate absorber (107). The close agreement between the calculated and observed profiles was obtained without adjusting parameters. The plate efficiencies required for the calculations were measured independently on a single exact copy of the bubble cap plates installed in the five-tray absorber.

A general, approximate, short-cut design procedure for adiabatic bubble tray absorbers has not been developed, although work has been done in the field of nonisothermal and multicomponent hydrocarbon absorbers. An analytical

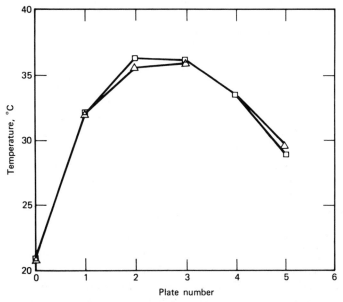

Fig. 27. Computed and experimental liquid temperature profiles in an ammonia absorber with 5 bubble cap trays (107). Water was used as a solvent. $y_{N+1} = 0.123$; $y_1 = 0.0242$; $L'_M/G'_M = 1.757$; $\triangle$, measured; $\square$, calculated.

expression which will predict the recovery of each component provided the stripping factor, ie, the group mG_M/L_M, is known for each component on each tray of the column has been developed (102). This requires knowledge of the temperature and total flow (G_M and L_M) profiles through the tower. There are many suggestions about how to estimate these profiles (102,103,108). A realistic estimate of the temperature profile for theoretical plates can probably be obtained by the short-cut method developed on the basis of rigorous computer solutions for about 40 different hypothetical designs (108) which closely resemble those of Figure 27.

Plate Efficiency Estimation.   *Rate of Mass Transfer in Bubble Plates.* The Murphree vapor efficiency, much like the height of a transfer unit in packed absorbers, characterizes the rate of mass transfer in the equipment. The value of the efficiency depends on a large number of parameters not normally known, and its prediction is therefore difficult and involved. Correlations have led to widely used empirical relationships, which can be used for rough estimates (109,110). The most fundamental approach for tray efficiency estimation, however, summarizing intensive research on this topic, may be found in reference 111.

In large plates 0.61 m or more in diameter, the efficiency of the tray as a whole may differ from the efficiency observed at some particular point of the tray because the liquid is not uniformly mixed in the direction of the flow on the whole tray. The point value of the efficiency called E_{OG} is thus more closely related to interphase diffusion than E_{MV}. As the gas passes upward through the liquid covering a small area of the plate, mass transfer from gas to liquid occurs in a manner similar to a packed tower of height h_B, the depth of the bubbling area. Under the assumption that the liquid is completely mixed in the vertical direction, and that the gas travels through that minicolumn in a plug-flow-like fashion, the number of transfer units of the bubbling area may be calculated in terms of the gas concentrations above and below the area under consideration by applying the definition of N_{OG}, equation 55. This equation may be integrated by taking y^* as constant and equal to y_n^* because of the well-mixed nature of the liquid phase. By comparing the result with the definition of the plate efficiency, equation 82, formulated for a single point on the plate, the following relationship between the point efficiency and the number of transfer units arises:

$$E_{OG} = 1 - e^{-N_{OG}} \tag{87}$$

If resistance to transfer is present in both phases, N_{OG} may be expressed as an addition of resistances using equations 54, 47, 41, and 44:

$$\frac{1}{N_{OG}} = \frac{1}{N_G} + \frac{mG_M}{L_M}\frac{1}{N_L} \tag{88}$$

Hence the point efficiency E_{OG} may be computed if both N_G and N_L in the bubbling area are known. These parameters are determined by the prevailing transfer coefficients, the interfacial area, and by h_B: $N_G = k_G a P h_B/G_M$ and $N_L = k_L a \rho h_B/L_M$.

To estimate the number of transfer units for design, the following empirical correlations which were derived from efficiency measurements employing a variety of trays and operating conditions under the aforementioned assumptions are recommended (111):

$$N_G = (0.776 + 4.63h_w - 0.238F + 0.0712L')Sc^{-0.5} \qquad (89)$$

where h_w is the weir height in m; $F = u_G\sqrt{\rho_G}$ in m/s [(kg/m^3)$^{1/2}$]; L' is the liquid flow rate per average width of stream in m^3/(s·m); and

$$N_L = 3050 \sqrt{D_L}(68\,h_w + 1)t_L \qquad (90)$$

where $t_L = h_L z_L/L'$ is the liquid residence time in s. The recommended correlation for h_L is

$$h_L = 0.0419 + 0.19h_w - 0.0135F + 2.46L' \qquad (91)$$

Effect of Different Degrees of Mixing. Once E_{OG} is evaluated on the basis of N_G and N_L using equation 87, it has to be translated into E_{MV} by considering the degree of mixing on the tray. It is obvious that for a small plate with completely backmixed liquid,

$$E_{MV} = E_{OG} \qquad (92)$$

If, on the other hand, the liquid flows in a plug-flow-like manner over the tray, but the vapor may be assumed to mix between the trays so that it enters each tray in uniform composition, the result may be calculated according to (112).

$$E_{MV} = \frac{L_M}{mG_M} [e^{(mG_M/L_M)E_{OG}} - 1] \qquad (93)$$

In the case of unmixed vapors between the plates, the equations, being implicit in E_{MV}, have also been solved numerically (112). The results depend on the arrangement of the downcomers and are not too different numerically from equation 93. In reality, however, the liquid is neither completely backmixed nor can the tray be considered as a plug-flow device.

Many theories have been put forth to handle partial liquid backmixing on plates. Early calculations (113–115) used the so-called tanks-in-series model. Recycle models have been derived by assuming partial or complete backmixing of the liquid, or the vapor, or both (116–118). The suggested procedure (111) is based on the eddy diffusion model when the parameter characterizing the degree of liquid backmixing is the dimensionless group known as the Peclet number:

$$Pe = \frac{z_L^2}{D_E t_L} \qquad (94)$$

The eddy dispersion coefficient D_E has been measured and correlated empirically as

$$\sqrt{D_E} = 0.00378 + 0.0179u_g + 3.69L' + 0.18h_w \qquad (95)$$

The nomenclature is the same as that used in equations 89 and 90.

The relationship between E_{MV}, E_{OG}, and Pe has been calculated according to the recommended formulas (110) and presented in tabular form (4). A plot of this data is shown in Figure 28. Complete backmixing is characterized by Pe = 0 ($D_E = \infty$) where $E_{MV} = E_{OG}$. In larger columns, in which the liquid is not completely mixed horizontally, E_{MV} can be seen from Figure 28 to become larger than the point efficiency and may thus even exceed 100% (119). Entrainment, as well as by-passing, are always detrimental to the plate efficiency. Analytical expressions to correct for entrainment and by-passing have been developed (120,121).

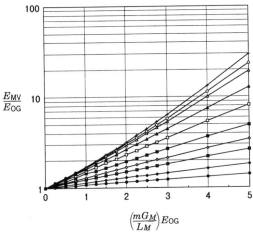

Fig. 28. Relationship between E_{OG} and E_{MV} at different degrees of liquid backmixing (4). Curves represent different Peclet numbers. From top to bottom Pe = ∞, +; Pe = 100, $\bigcirc$; Pe = 50, $\triangle$; Pe = 20, $\blacktriangle$; Pe = 10, $\square$; Pe = 5, $\blacksquare$; Pe = 3, $\Diamond$; Pe = 2, $\blacklozenge$; Pe = 1, $\blacklozenge$; Pe = 0.5, $\bullet$.

Capacity Limitations.   The fluid flow capacity of a bubble tray may be limited by any of three principal factors.

1. Flooding, often the most restrictive of the limitations, occurs when the clear liquid height in the downcomer, H_{dc}, exceeds a certain fraction of the tray spacing. During operation, the liquid level in the downcomer builds up as a result of the head necessary to overcome the various resistances to liquid flow, including the friction in the downcomer itself and the hydraulic gradient across the plate. A significant portion in the liquid backup is caused by the need for the liquid to overcome the difference in pressure between the inside and the outside of the downcomer, which in turn is caused by the pressure drop of the vapors through the next higher plate. If the diameter of the column is made too small for a given

flow, the vapor pressure drop and thus the liquid backup in the downcomers will increase to the point where some liquid spills onto the next higher tray, and flooding sets in. In principle, the condition may be corrected by increasing the diameter or the tray spacing. A conservative design requires a clear liquid head in the downcomer of no more than half the tray spacing to allow for froth entrapped in the downcomer. The maximum allowable superficial velocity based on the column cross section u_G may be roughly estimated for different tray spacings (6,122,123). A more reliable design would consider each pressure drop contributing to H_{dc} separately.

2. Entrainment occurs when spray or froth formed on one tray enters the gas passages in the tray above. In moderate amounts, entrainment will impair the countercurrent action and hence drastically decrease the efficiency. If it happens in excessive amounts, the condition is called priming and will eventually flood the downcomers.

3. At high liquid flow, the hydraulic gradient may become so large that the caps near the liquid feed point will stop bubbling, and the efficiency will suffer.

NOMENCLATURE

Symbol	Definition	Units
A	L_M/mG_M, absorption factor	
a	effective interfacial area per packed volume	m^2/m^3
c_{Ai}	concentration of solute at interface	mol/m^3 or mol/L
C_P	specific heat of gas mixture	$kJ/(mol\cdot K)$
C_q	specific heat of liquid mixture	$kJ/(mol\cdot K)$
D_{ax}	axial dispersion coefficient	m^2/s
D_{AB}	diffusion coefficient of A in B	m^2/s
D_{Am}	effective diffusion coefficient of A in a mixture	m^2/s
D_E	eddy dispersion (or diffusion) coefficient on a tray	m^2/s
D_j	diffusion coefficient of component j	m^2/s
D_L	liquid diffusion coefficient	m^2/s
E_{MV}	Murphree vapor plate efficiency, see equation 82	
E_{OG}	point value of plate efficiency	
F	$u_G\sqrt{\rho_G}$, F-factor for bubble tray gas load	$kg^{1/2}/(m^{1/2}\cdot s)$
g_j	moles of component j	mol
G_M	gas flow rate	$kmol/(m^2\cdot s)$
G'_M	flow rate of inert gas	$kmol/(m^2\cdot s)$
G_M^0	generalized molar gas flow rate, see equation 64	
H	Henry's constant	Pa
HETP	height equivalent to a theoretical plate	m
H_G	height of a gas-phase transfer unit, see equation 42	m
H_L	height of a liquid-phase transfer unit, see equation 45	m
H_{OG}	overall gas-phase height of a transfer unit, see equation 48	m
H_{OS}	integral heat of solution for the solute	kJ/mol
H_v	latent heat of pure solvent	kJ/mol
h	total height of packing	m
h_L	clear height of liquid on a tray	m

Symbol	Definition	Units
h_{w}	weir height	m
h_{B}	height of bubble layer on a tray	m
J_{A}	flux of component A due to diffusion	$\mathrm{mol/(m^2 \cdot s)}$
K	partition coefficient for gas–liquid equilibrium	
K_{OG}	overall gas-phase mass transfer coefficient	$\mathrm{mol/(s \cdot m^2 \cdot Pa)}$
k_{G}	gas-phase mass transfer coefficient	$\mathrm{mol/(s \cdot m^2 \cdot Pa)}$
k_{L}	liquid-phase mass transfer coefficient	m/s
$k_{\mathrm{L}}^{\mathrm{r}}$	mass transfer coefficient including effect of chemical reaction	m/s
k^0	mass transfer coefficient for equimolar counterdiffusion	m/s
k'	mass transfer coefficient for unidirectional molecular diffusion through a stagnant gas	m/s
k	mass transfer coefficient for general multicomponent diffusion situation	m/s
L	mass flow rate of liquid	$\mathrm{kg/(s \cdot m^2)}$
L_M	liquid flow rate	$\mathrm{kmol/(s \cdot m^2)}$
L_M^0	generalized liquid flow rate, see equation 64	$\mathrm{kmol/(s \cdot m^2)}$
L_M'	flow rate of solute-free solvent	$\mathrm{kmol/(s \cdot m^2)}$
L'	liquid flow rate on a plate per width of stream	$\mathrm{m^3/(s \cdot m)}$
m	slope of equilibrium line	
$m_{\mathrm{B},2}$	"solubility" of solvent, $y_{\mathrm{B}}^*/x_{\mathrm{B}}$ at the top of the tower	
N_{A}	flux of solute A through phase interface	$\mathrm{mol/(s \cdot m^2)}$
N_{G}	number of gas phase transfer units, see equation 43	
N_j	flux of component j through phase interface	$\mathrm{mol/(s \cdot m^2)}$
N_{L}	number of liquid-phase transfer units, see equation 46	
N_{OG}	number of overall gas-phase transfer units, see equation 49	
N_{P}	number of actual plates in the column	
N_{T}	N_{OG} for equimolar counterdiffusion, see equation 57	
N_{TP}	number of theoretical plates	
$\mathrm{NTU_{ap}}$	N_{OG} calculated by equation 56, see also equation 73	
NTU	actual value of N_{OG} under the influence of axial dispersion, see equation 74	
P	pressure	Pa
p_{A}	partial pressure of A	Pa
p_{Ai}	partial pressure of A at the interface	Pa
P_{s}	vapor pressure of a pure component	Pa
Pe	uh/D_{ax}, dimensionless Peclet number	
Pe_i	Peclet number specifically defined for phase i, $i = x$ or y	
$\mathrm{Pe_{HTU}}$	$uH_{\mathrm{OG}}/D_{\mathrm{ax}}$, dimensionless HTU Peclet number, see equation 75	
$\mathrm{Pe_{HTU},}_i$	Peclet number specifically defined for phase i, $i = x$ or y, see equation 77	
R	$L_M{:}G_M$, flow ratio	
s	surface renewal rate, equation 32	$\mathrm{s^{-1}}$
Sc	$\mu/\rho D$, the dimensionless Schmidt number	
T	temperature	°C or K
t_{A}	$\Sigma_j\, N_j/N_{\mathrm{A}}$, parameter indicating the degree of counterdiffusion	
t_{L}	residence time of liquid on plate, see equation 90	s

Symbol	Definition	Units
u	superficial velocity of gas or liquid phase	m/s
X_f	film factor, see equation 24	
$X^0{}_A$	$x_A/(1 - t_A x_A)$, generalized liquid concentration, see equation 64	
X'_A	$x_A/(1 - x_A)$, liquid mole ratio	
x_A	liquid mole fraction of solute	
x_{Ab}	mole fraction in bulk of the liquid phase	
x_{Ai} or x_i	liquid mole fraction of solute at the phase interface	
x_{BM}	logarithmic mean of solvent concentration between the phase interface and the bulk of the liquid, see equation 18	
Y	moles absorbed component per mole of entering vapor, equation 84	
Y'	$y/(1 - y)$, mole ratio	
Y^0_A	$Y_A/(1 - t_A y_A)$, generalized gas concentration, see equation 64	
Y_f	film factor, see equation 23	
y^*_A	$m x_A$, concentration in equilibrium with bulk liquid concentration	
y_{Ab}	mole fraction in bulk of gas phase	
y_{Ai} or y_i	gas mole fraction of solute at phase interface	
y_{BM}	logarithmic mean of inert gas concentration between the phase interface and the bulk of the gas phase, see equation 15	
Y^*_{BM}	$[(1 - y_A)(1 - y^*_A)]/\ln\,[(1 - y_A)/(1 - y^*_{BM})]$, logarithmic mean of inert gas concentration between the equilibrium concentration and the bulk of the gas phase	
Y^*_f	$[(1 - t_A y_A) - (1 - t_A y^*_A)]/\ln\,[(1 - t_A y_A)/(1 - t_A y^*_A)]$, overall gas-phase film factor, see equation 30	
z_L	length of liquid travel on tray	m
Δg_j	total amount of component j absorbed	kmol/(s·m^2)
ΔT_{max}	temperature associated with internal temperature maximum	°C or K
μ	viscosity	mPa·s (= cP)
$\bar{\rho}$	mean density of liquid phase	mol/m^3, mol/L, or g/cm^3
ρ	density of liquid	mol/m^3
σ	surface tension	N/m
ϕ	packing parameter for equation 79	m
ψ	packing parameter for equation 80	m

Subscripts

av	average
A	component A, solute
B	component B, usually nondiffusing
G	gas
L	liquid
i	at gas–liquid interface

Symbol	Definition	Units
b	in the bulk; this symbol is normally omitted	
o	liquid feed for plate columns	
x	pertaining to liquid phase	
y	pertaining to gas phase	
M	molar quantity	
n	referring to stream leaving the nth tray	
1	bottom of a packed column	
2	top of a packed column	

BIBLIOGRAPHY

"Absorption" in *ECT* 1st ed., Vol. 1, pp. 14–32, by E. G. Scheibel, Hoffmann-La Roche, Inc.; in *ECT* 2nd ed., Vol. 1, pp. 44–77, by F. A. Zenz, Squires International, Inc.; in *ECT* 3rd ed., Vol. 1, pp. 53–96, by C. W. Wilke and Urs von Stockar, University of California, Berkeley.

1. F. A. Zenz, *Chem. Eng.* **1972**, 120 (Nov. 13, 1972).
2. R. E. Treybal, *Mass Transfer Operations,* 3rd ed., McGraw-Hill Book Co., Inc., New York, 1980.
3. W. S. Norman, *Absorption, Distillation and Cooling Towers,* Longmans, Green & Co., Ltd. (Wiley) New York, 1962.
4. T. K. Sherwood, R. L. Pigford, and C. R. Wilke, *Mass Transfer,* McGraw-Hill Book Co., Inc., New York, 1975.
5. A. Mersmann, H. Hofer, and J. Stichlmair, *Ger. Chem. Eng.* **2,** 249–258 (1979).
6. R. H. Perry and D. Green, eds., *Perry's Chemical Engineer's Handbook,* 6th ed., McGraw-Hill Book Co., Inc., New York, 1984.
7. J. H. Hildebrand and R. L. Scott, *Regular Solutions,* Prentice-Hall, Inc., Englewood Cliffs, N.J., 1962.
8. R. C. Reid, J. M. Prausnitz, and B. E. Poling, *The Properties of Gases and Liquids,* 4th ed., McGraw-Hill Book Co., New York, 1988.
9. A. S. Kertes and co-workers, *Solubility Data Series,* Pergamon Press, Oxford, UK, 1979.
10. W. C. Edmister, *Applied Hydrocarbon Thermodynamics,* Gulf Publishing Co., Houston, Tex., 1961.
11. N. B. Vargaftik, *Tables on the Thermophysical Properties of Liquids and Gases,* John Wiley & Sons, Inc., New York, 1975.
12. J. Gmehling and U. Onken, *Vapor-Liquid Equilibrium Data Collection, Chemistry Data Series,* Dechema, Frankfurt, 1977 ff.
13. W. G. Whitman, *Chem. Metall. Eng.* **29,** 147 (1923).
14. A. F. Ward and L. H. Brooks, *Trans. Faraday Soc.* **48,** 1124 (1952).
15. E. J. Cullen and J. F. Davidson, *Trans. Faraday Soc.* **53,** 113 (1957).
16. W. J. Ward and J. A. Quinn, *AIChE J.* **11,** 1005 (1965).
17. J. L. Duda and J. S. Vrentas, *AIChE J.* **14,** 286 (1968).
18. S. Lynn, J. R. Straatemeier, and H. Kramers, *Chem. Eng. Sci.* **4,** 49, 58, 63 (1955).
19. R. W. Schrage, *A Theoretical Study of Interface Mass Transfer,* Columbia University, New York, 1953.
20. B. Paul, *J. Am. Rocket Soc.* 1321 (Sept. 1962).
21. L. V. Delaney and L. C. Eagleton, *AIChE J.* **8,** 418 (1962).
22. R. Cartier, D. Pindzola, and P. E. Bruins, *Ind. Eng. Chem.* **51,** 1409 (1959).
23. N. A. Clontz, R. T. Johnson, W. L. McCabe, and R. W. Rousseau, *Ind. Eng. Chem. Fundam.* **11,** 368 (1972).

24. J. T. Davies and E. K. Rideal, *Adv. Chem. Eng.* **4**, 1 (1963).
25. C. J. King, *AIChE J.* **10**, 671 (1964).
26. W. K. Lewis and W. G. Whitman, *Ind. Eng. Chem.* **16**, 1215 (1924).
27. R. B. Bird, W. E. Stewart, and E. N. Lightfoot, *Transport Phenomena,* John Wiley & Sons, Inc., New York, 1960.
28. C. R. Wilke, *Chem. Eng. Prog.* **46**, 95 (1950).
29. R. E. Treybal, *Ind. Eng. Chem.* **61**, 36 (1969).
30. R. Krishna and R. Taylor in N. P. Chemerisinoff, ed., *Handbook for Heat and Mass Transfer,* Vol. 2, Gulf Publishing Corporation, Houston, Tex., 1986.
31. J. A. Wesselingh and R. Krishna, *Elements of Mass Transfer,* Technical University Delft, Delft, the Netherlands, 1989.
32. R. Krishna and R. Taylor, *Multicomponent Mass Transfer,* John Wiley & Sons, Inc., New York, 1991.
33. R. Higbie, *Trans. Am Inst. Chem. Eng.* **31**, 365 (1935).
34. P. V. Danckwerts, *Ind. Eng. Chem.* **43**, 1460 (1951).
35. T. H. Chilton and A. P. Colburn, *Ind. Eng. Chem.* **26**, 1183 (1934).
36. G. F. Froment in *Chemical Reaction Engineering* (Advances in Chemistry Series 109), American Chemical Society, Washington, D.C., 1972, p. 19.
37. K. F. Loughlin, M. A. Abul-Hamayel, and L. C. Thomas, *AIChE* **31**, 1614 (1985).
38. P. V. Danckwerts, *Gas-Liquid Reactions,* McGraw-Hill Book Co., Inc., New York, 1970.
39. D. W. van Krevelen and P. J. Hoftijzer, *Rec. Trav. Chim.* **67**, 563 (1948).
40. G. Astarita, *Mass Transfer with Chemical Reaction,* Elsevier, Amsterdam, 1966.
41. A. P. Colburn, *Trans. Am. Inst. Chem. Eng.* **35**, 211 (1939).
42. J. H. Wiegand, *Trans. Am. Inst. Chem. Eng.* **36**, 679 (1940).
43. C. R. Wilke and U. v. Stockar, *Ind. Eng. Chem. Fundam.* **16**(2), 88 (1977).
44. J. D. Raal and M. K. Khurana, *Can. J. Chem. Eng.* **51**, 162 (1973).
45. C. R. Wilke and U. v. Stockar, *Ind. Eng. Chem. Fundam.* **16**(2), 94 (1977).
46. J. Stichlmair, *Chem. Ind. Technol.* **44**, 411 (1972).
47. J. R. Bourne, U. v. Stockar, and G. C. Coggan, *Ind. Eng. Chem. Process Des. Dev.* **13**, 124 (1974).
48. I. A. Furzer, *Ind. Eng. Chem. Fundam.* **23**, 159 (1984).
49. N. Kolev and Kr. Semkov, *Chem. Eng. Prog.* **19**, 175 (1985).
50. N. Kolev and Kr. Semkov, *Vt Verfahrenstechnik* **17**, 474 (1983).
51. W. E. Dunn, T. Vermeulen, C. R. Wilke, and T. T. Word, *Report UCRL 10394 (1962),* Univ. of California Radiation Laboratory as cited in reference 4.
52. W. E. Dunn, T. Vermeulen, C. R. Wilke, and T. T. Word, *Ind. Eng. Chem. Fundam.* **16**, 116 (1977).
53. E. T. Woodburn, *AIChE J.* **20**, 1003 (1974).
54. M. Richter, *Chem. Tech.* **30**, 294 (1978).
55. U. von Stockar and P. F. Cevey, *Ind. Eng. Chem. Process Res. Dev.* **23**, 717 (1984).
56. T. Miyauchi and T. Vermeulen, *Ind. Eng. Chem. Fundam.* **2**, 113 (1963).
57. C. A. Sleicher, *AIChE J.* **5**, 145 (1959).
58. S. Stemerding and F. J. Zuiderweg, *Chem. Engr.* CE 156–CE 159 (May 1963).
59. J. C. Mecklenburgh and S. Hartland, *The Theory of Backmixing,* Wiley-Interscience, New York, 1975, Chapt. 10.
60. J. S. Watson and H. D. Cochran, *Ind. Eng. Chem. Process Res. Dev.* **10**, 83–85 (1971).
61. U. von Stockar and Xiao-Ping Lu, *A Simple and Accurate Short-cut Procedure to Account for Axial Dispersion in Counter-current Separation Columns, Ind. Eng. Chem. Research,* 1991, in press.
62. U. von Stockar, *ACHEMASIA, 1989* (International Meeting on Chemical Engineering & Biotechnology, Beijing, China), Dechema & Ciesc, 1989, p. 43.

63. T. K. Sherwood and F. A. L. Holloway, *Trans. Inst. Am. Chem. Eng.* **36**, 39 (1940).
64. F. F. Rixon, *Trans. Instn. Chem. Engrs.* **26**, 119 (1948).
65. H. A. Koch, L. F. Stutzman, H. A. Blum, and L. E. Hutchings, *Chem. Eng. Prog.* **45**, 677 (1949).
66. E. L. Knoedler and C. F. Bonilla, *Chem. Eng. Prog.* **50**, 125 (1954).
67. J. E. Vivian and C. J. King, *AIChE J.* **120**, 221 (1964).
68. D. Cornell, W. G. Knapp, and J. R. Fair, *Chem. Eng. Prog.* **56**(7), 68 (1960).
69. W. L. Bolles and J. R. Fair, *Chem. Eng.* **89**(July 12), 109 (1982).
70. P. H. Au-Yeung and A. B. Ponter, *Can. J. Chem. Eng.* **61**, 481 (1983).
71. D. M. Mohunta, A. S. Vaidyanathan, and G. S. Laddha, *Ind. Chem. Eng.* **11**, 73 (1969).
72. J. B. Zech and A. B. Mersmann, *I. Chem. E. Symp. Ser.* **56**, 39 (1979).
73. R. Mangers and A. B. Ponter, *Ind. Eng. Chem. Process Des. Dev.* **19**, 530 (1980).
74. Mei Geng Shi and A. B. Mersmann, *Ger. Chem. Eng.* **8**, 87 (1985).
75. R. Billet and M. Schultes, *Paper given at AIChE Annual Meeting,* Washington, D.C. (1988).
76. D. W. van Krevelen and P. J. Hoftijzer, *Rec. Trav. Chim. Pays-Bas* **66**, 49 (1947).
77. K. Onda, H. Takeuchi, and Y. Okumoto, *J. Chem. Eng., Jpn.* **1**, 56 (1968).
78. H. L. Schulman, C. F. Ulrich, A. Z. Proulx, and J. O. Zimmerman, *AIChE J.* **1**, 253 (1955).
79. J. F. Davidson, *Trans. Instn. Chem. Engrs.* **37**, 131 (1959).
80. J. Bridgewater and A. M. Scott, *Trans. Instn. Chem. Engrs.* **52**, 317 (1974).
81. A. B. Ponter and P. H. Au-Yeung, *Can. J. Chem. Eng.* **60**, 94 (1982).
82. R. Echarte, H. Campana, and E. A. Brignole, *Ind. Eng. Chem. Process Des. Dev.* **23**, 349 (1984).
83. E. J. Lynch and C. R. Wilke, *AIChE J.* **1**, 9 (1955).
84. H. L. Shulman, C. F. Ullrich, and N. Wells, *AIChE J.* **1**, 247 (1955).
85. B. W. Gamson, G. Thodos, and O. A. Hougen, *Trans. Am. Inst. Chem. Eng.* **39**, 1 (1943).
86. R. G. Eckert and O. A. Hougen, *Chem. Eng. Prog.* **45**, 188 (1949).
87. C. R. Wilke and O. A. Hougen, *Trans. Am. Inst. Chem. Eng.* **41**, 445 (1945).
88. M. Hobson and G. Thodos, *Chem. Eng. Prog.* **47**, 370 (1951).
89. L. Fellinger, Dissertation, Massachusetts Institute of Technology, 1941; see also reference 113.
90. J. L. Bravo and J. R. Fair, *Ind. Eng. Chem. Process Des. Dev.* **21**, 162 (1982).
91. J. R. Hufton, J. L. Bravo, and J. R. Fair, *Ind. Eng. Chem. Res.* **27**, 2096 (1988).
92. T. K. Sherwood, G. H. Shipley, and F. A. L. Holloway, *Ind. Eng. Chem.* **30**, 765 (1938).
93. W. E. Lobo, L. Friend, F. Hashmall, and F. Zenz, *Trans. Am. Inst. Chem. Eng.* **41**, 693 (1945).
94. F. A. Zenz and R. A. Eckert, *Pet. Refiner.* **40**, 130 (1961).
95. R. A. Eckert, *Chem. Eng. Prog.* **66**(3), 39 (1970).
96. J. R. Fair, D. E. Steinmeyer, W. R. Penney, and B. B. Crocker, in ref. 6, pp. 18–23.
97. L. Spiegel and W. Meier, *Chem. Eng. Symp. Ser.* **104**, A203 (1987).
98. E. V. Murphree, *Ind. Eng. Chem.* **17**, 474 (1925).
99. A. Kremser, *Nat. Pet. News* **22**(21), 42 (1930).
100. G. G. Brown and M. Souders, *Oil Gas J.* **31**(5), 34 (1932).
101. M. Souders and G. G. Brown, *Ind. Eng. Chem.* **24**, 519 (1932).
102. G. Horton and W. B. Franklin, *Ind. Eng. Chem.* **32**, 1384 (1940).
103. W. C. Edmister, *Ind. Eng. Chem.* **35**, 837 (1943).
104. G. G. Brown and M. Souders, *The Science of Petroleum,* Vol. 2, Oxford University Press, New York, 1938, Sect. 25, p. 1557.
105. J. R. Bourne, U. v. Stockar, and G. C. Coggan, *Ind. Eng. Chem. Process Des. Dev.* **13**, 115 (1974).
106. J. Stichlmair and A. Mersmann, *Chem. Ing. Technol.* **43**, 17 (1971).

107. U. von Stockar, *Gasabsorption mit Wärmeeffekten, Diss. Nr 4917,* ETH-Zurich, 1973.
108. W. R. Owens and R. N. Maddox, *Ind. Eng. Chem.* **60**(12), 14 (1968).
109. H. G. Drickamer and J. R. Bradford, *Trans. Am. Inst. Chem. Eng.* **39,** 319 (1943).
110. A. E. O'Connell, *Trans. Am. Inst. Chem. Eng.* **42,** 741 (1946).
111. Research Committee, *Bubble Tray Design Manual,* American Institute of Chemical Engineers, New York, 1958.
112. W. K. Lewis, *Ind. Eng. Chem.* **28,** 399 (1936).
113. T. K. Sherwood and R. L. Pigford, *Absorption and Extraction,* McGraw-Hill Book Co., Inc., New York, 1952.
114. M. Nord, *Trans. Am. Inst. Chem. Eng.* **42,** 863 (1946).
115. M. F. Gautreaux and H. E. O'Connell, *Chem. Eng. Prog.* **51,** 232 (1955).
116. E. D. Oliver and C. C. Watson, *AIChE J.* **2,** 18 (1956).
117. L. A. Warzel, Dissertation, University of Michigan, Ann Arbor, Mich., 1955.
118. V. M. Ramm, *Absorption of Gases,* Israel Program for Scientific Translations, Jerusalem, 1968, Chapt. 3.
119. G. G. Brown, M. Souders, H. V. Nyland, and W. H. Hessler, *Ind. Eng. Chem.* **27,** 383 (1935).
120. A. P. Colburn, *Ind. Eng. Chem.* **28,** 526 (1936).
121. C. P. Strand, *Chem. Eng. Prog.* **59,** 58 (1963).
122. M. Souders and G. G. Brown, *Ind. Eng. Chem.* **26,** 98 (1934).
123. J. R. Fair, *Petro/Chem Eng.* **33**(10), 45 (Sept. 1961).

General references

References 2, 3, 4, and 113 are general references.
A. H. P. Skelland, *Diffusional Mass Transfer,* John Wiley & Sons, Inc., New York, 1974.

URS VON STOCKAR
École Polytechnique Fédérale, Lausanne

CHARLES R. WILKE
University of California, Berkeley

ACARICIDES.   See INSECT CONTROL TECHNOLOGY.

ACACIA.   See DIURETICS; GUMS.

ACAROID, ACCROIDES.   See RESINS, NATURAL.

ACETALDEHYDE

Acetaldehyde [75-07-0] (ethanal), CH_3CHO, was first prepared by Scheele in 1774, by the action of manganese dioxide [1313-13-9] and sulfuric acid [7664-93-9] on ethanol [64-17-5]. The structure of acetaldehyde was established in 1835 by Liebig from a pure sample prepared by oxidizing ethyl alcohol with chromic acid. Liebig named the compound "aldehyde" from the Latin words translated as al(cohol) dehyd(rogenated). The formation of acetaldehyde by the addition of water [7732-18-5] to acetylene [74-86-2] was observed by Kutscherow in 1881.

Acetaldehyde, first used extensively during World War I as a starting material for making acetone [67-64-1] from acetic acid [64-19-7], is currently an important intermediate in the production of acetic acid, acetic anhydride [108-24-7], ethyl acetate [141-78-6], peracetic acid [79-21-0], pentaerythritol [115-77-5], chloral [302-17-0], glyoxal [107-22-2], alkylamines, and pyridines. Commercial processes for acetaldehyde production include the oxidation or dehydrogenation of ethanol, the addition of water to acetylene, the partial oxidation of hydrocarbons, and the direct oxidation of ethylene [74-85-1]. In 1989, it was estimated that 28 companies having more than 98% of the world's 2.5 megaton per year plant capacity used the Wacker-Hoechst processes for the direct oxidation of ethylene.

Acetaldehyde is a product of most hydrocarbon oxidations. It is an intermediate product in the respiration of higher plants and occurs in trace amounts in all ripe fruits that have a tart taste before ripening. The aldehyde content of volatiles has been suggested as a chemical index of ripening during cold storage of apples. Acetaldehyde is also an intermediate product of fermentation (qv), but it is reduced almost immediately to ethanol. It may form in wine (qv) and other alcoholic beverages after exposure to air, imparting an unpleasant taste; the aldehyde reacts to form diethyl acetal [105-57-7] and ethyl acetate [141-78-6]. Acetaldehyde is an intermediate product in the decomposition of sugars in the body and hence occurs in trace quantities in blood.

Physical Properties

Acetaldehyde is a colorless, mobile liquid having a pungent, suffocating odor that is somewhat fruity and quite pleasant in dilute concentrations. Its physical properties are given in Table 1; the vapor pressure of acetaldehyde and its aqueous solutions appear in Tables 2 and 3, respectively; and the solubilities of acetylene, carbon dioxide [124-38-9], and nitrogen [7727-37-9] in liquid acetaldehyde are given in Table 4. Acetaldehyde is miscible in all proportions with water and most common organic solvents, eg, acetone, benzene [71-43-2], ethyl alcohol, ethyl ether [60-29-7], gasoline, paraldehyde [123-63-7], toluene [108-88-3], xylenes [1330-20-7], turpentine, and acetic acid.

The freezing points of aqueous solutions of acetaldehyde are 4.8 wt %, −2.5°C; 13.5 wt %, −7.8°C; and 31.0 wt %, −23.0°C.

Given in the literature are: vapor pressure data for acetaldehyde and its aqueous solutions (1–3); vapor–liquid equilibria data for acetaldehyde–ethylene oxide [75-21-8] (1), acetaldehyde–methanol [67-56-1] (4), sulfur dioxide [7446-09-5]–acetaldehyde–water (5), acetaldehyde–water–methanol (6); the azeotropes of

Table 1. Physical Properties of Acetaldehyde

Properties	Values
formula weight	44.053
melting point, °C	−123.5
boiling point at 101.3 kPa[a] (1 atm), °C	20.16
density, g/mL	
d_4^0	0.8045
d_4^{11}	0.7901
d_4^{15}	0.7846
d_4^{20}	0.7780
coefficient of expansion per °C (0–30°C)	0.00169
refractive index, n_D^{20}	1.33113
vapor density (air = 1)	1.52
surface tension at 20°C, mN/m (= dyn/cm)	21.2
absolute viscosity at 15°C, mPa·s (= cP)	0.02456
specific heat at 0°C, J/(g·K)[b]	
15°C	2.18
25°C	1.41
$\alpha = C_p/C_v$ at 30°C and 101.3 kPa[a] (1 atm)	1.145
latent heat of fusion, kJ/mol[b]	3.24
latent heat of vaporization, kJ/mol[b]	25.71
heat of solution in water, kJ/mol[b]	
at 0°C	−8.20
at 25°C	−6.82
heat of combustion of liquid at constant pressure, kJ/mol[b]	12867.9
heat of formation at 273 K, kJ/mol[b]	−165.48
free energy of formation at 273 K, kJ/mol[b]	−136.40
critical temperature, °C	181.5
critical pressure, MPa[c]	6.40
dipole moment, C·m[d]	8.97×10^{-30}
ionization potential, eV	10.50
dissociation constant at 0°C, K_a	0.7×10^{-14}
flash point, closed cup, °C	−38
ignition temperature in air, °C	165
explosive limits of mixtures with air, vol % acetaldehyde	4.5–60.5

[a]To convert kPa to psi, multiply by 0.14503.
[b]To convert J to cal, divide by 4.187.
[c]To convert MPa to psi, multiply by 145.
[d]To convert C·m to debyes, multiply by 2.998×10^{29}.

acetaldehyde–butane [106-97-8] and acetaldehyde–ethyl ether (7); solubility data for acetaldehyde–water–methane [74-82-8] (8), acetaldehyde–methane (9); densities and refractive indexes of acetaldehyde for temperatures 0–20°C (2); compressibility and viscosity at high pressure (10); thermodynamic data (11–13); pressure–enthalpy diagram for acetaldehyde (14); specific gravities of acetaldehyde–paraldehyde and acetaldehyde–acetaldol mixtures at 20/20°C vs composition (7); boiling point vs composition of acetaldehyde–water at 101.3 kPa (1 atm) and integral heat of solution of acetaldehyde in water at 11°C (7).

Table 2. Vapor Pressure of Acetaldehyde

Temperature, °C	Vapor pressure, kPa[a]	Temperature, °C	Vapor pressure, kPa[a]
−50	2.5	20	100.6
−20	16.4	20.16	101.3
0	44.0	30	145.2
5	54.8	50	279.4
10	67.7	70	492.6
15	82.9	100	1,014

[a]To convert kPa to mm Hg, multiply by 7.5.

Table 3. Vapor Pressure of Aqueous Solutions of Acetaldehyde

Temperature, °C	Mol %	Total vapor pressure, kPa[a]
10	4.9	9.9
10	10.5	18.6
10	46.6	48.4
20	5.4	16.7
20	12.9	39.3
20	21.8	57.7

[a]To convert kPa to mm Hg, multiply by 7.5.

Table 4. Solubility of Gases in Liquid Acetaldehyde at Atmospheric Pressure

Temperature, °C	Volume of gas (STP) dissolved in 1 volume acetaldehyde		
	Acetylene	Carbon dioxide	Nitrogen
−16	54		
−6	27	11	
0	17	6.6	
12	7.3	14.5	0.15
16	5	1.5	
20	3		

Chemical Properties

The limits and products of the various combustion zones for acetaldehyde–oxygen and acetaldehyde–air have been described (15–18); the effect of pressure on the explosive limits of acetaldehyde–air mixtures has been investigated (19). In a study of the spontaneous ignition of fuels injected into hot air streams, it was found that acetaldehyde was the least ignitable of the aldehydes examined (20,21). The influence of surfaces on the ignition and detonation of fuels containing acetaldehyde has been reported (22,23). Ignition data have been published for the systems acetaldehyde–oxygen–peroxyacetic acid–acetic acid (24), acetaldehyde–oxygen–peroxyacetic acid (25), and ethylene oxide–air–acetaldehyde (26).

Acetaldehyde is a highly reactive compound exhibiting the general reactivity of aldehydes (qv). Acetaldehyde undergoes numerous condensation, addition, and polymerization reactions; under suitable conditions, the oxygen or any of the hydrogens can be replaced.

Decomposition. Acetaldehyde decomposes at temperatures above 400°C, forming principally methane and carbon monoxide [630-08-0]. The activation energy of the pyrolysis reaction is 97.7 kJ/mol (408.8 kcal/mol) (27). There have been many investigations of the photolytic and radical-induced decomposition of acetaldehyde and deuterated acetaldehyde (28–30).

The Hydrate and Enol Form. In aqueous solutions, acetaldehyde exists in equilibrium with the acetaldehyde hydrate [4433-56-1], $(CH_3CH(OH)_2)$. The degree of hydration can be computed from an equation derived by Bell and Clunie (31). Hydration, the mean heat of which is -21.34 kJ/mol (-89.29 kcal/mol), has been attributed to hyperconjugation (32). The enol form, vinyl alcohol [557-75-5] $(CH_2{=}CHOH)$ exists in equilibrium with acetaldehyde to the extent of approximately one molecule per 30,000. Acetaldehyde enol has been acetylated with ketene [463-51-4] to form vinyl acetate [108-05-4] (33).

Oxidation. Acetaldehyde is readily oxidized with oxygen or air to acetic acid, acetic anhydride, and peracetic acid (see ACETIC ACID AND DERIVATIVES). The principal product depends on the reaction conditions. Acetic acid [64-19-7] may be produced commercially by the liquid-phase oxidation of acetaldehyde at 65°C using cobalt or manganese acetate dissolved in acetic acid as a catalyst (34). Liquid-phase oxidation in the presence of mixed acetates of copper and cobalt yields acetic anhydride [108-24-7] (35). Peroxyacetic acid or a perester is believed to be the precursor in both syntheses. There are two commercial processes for the production of peracetic acid [79-21-0]. Low temperature oxidation of acetaldehyde in the presence of metal salts, ultraviolet irradiation, or ozone yields acetaldehyde monoperacetate, which can be decomposed to peracetic acid and acetaldehyde (36). Peracetic acid can also be formed directly by liquid-phase oxidation at 5–50°C with a cobalt salt catalyst (37) (see PEROXIDES AND PEROXY COMPOUNDS). Nitric acid oxidation of acetaldehyde yields glyoxal [107-22-2] (38,39). Oxidations of p-xylene to terephthalic acid [100-21-0] and of ethanol to acetic acid are activated by acetaldehyde (40,41).

Reduction. Acetaldehyde is readily reduced to ethanol (qv). Suitable catalysts for vapor-phase hydrogenation of acetaldehyde are supported nickel (42) and copper oxide (43). The kinetics of the hydrogenation of acetaldehyde over a commercial nickel catalyst have been studied (44).

Polymerization. Paraldehyde, 2,4,6-trimethyl-1,3,5-trioxane [123-63-7], a cyclic trimer of acetaldehyde, is formed when a mineral acid, such as sulfuric, phosphoric, or hydrochloric acid, is added to acetaldehyde (45). Paraldehyde can also be formed continuously by feeding liquid acetaldehyde at 15–20°C over an acid ion-exchange resin (46). Depolymerization of paraldehyde occurs in the presence of acid catalysts (47); after neutralization with sodium acetate, acetaldehyde and paraldehyde are recovered by distillation. Paraldehyde is a colorless liquid, boiling at 125.35°C at 101 kPa (1 atm).

paraldehyde   metaldehyde

Metaldehyde [*9002-91-9*], a cyclic tetramer of acetaldehyde, is formed at temperatures below 0°C in the presence of dry hydrogen chloride or pyridine–hydrogen bromide. The metaldehyde crystallizes from solution and is separated from the paraldehyde by filtration (48). Metaldehyde melts in a sealed tube at 246.2°C and sublimes at 115°C with partial depolymerization.

Polyacetaldehyde, a rubbery polymer with an acetal structure, was first discovered in 1936 (49,50). More recently, it has been shown that a white, nontacky, and highly elastic polymer can be formed by cationic polymerization using BF_3 in liquid ethylene (51). At temperatures below -75°C using anionic initiators, such as metal alkyls in a hydrocarbon solvent, a crystalline, isotactic polymer is obtained (52). This polymer also has an acetal [poly(oxymethylene)] structure. Molecular weights in the range of 800,000–3,000,000 have been reported. Polyacetaldehyde is unstable and depolymerizes in a few days to acetaldehyde. The methods used for stabilizing polyformaldehyde have not been successful with polyacetaldehyde and the polymer has no practical significance (see ACETAL RESINS).

Reactions with Aldehydes and Ketones. The base-catalyzed self-addition of acetaldehyde leads to formation of the dimer, acetaldol [*107-89-1*], which can be hydrogenated to form 1,3-butanediol [*107-88-0*] or dehydrated to form crotonaldehyde [*4170-30-3*]. Crotonaldehyde (qv) can also be made directly by the vapor-phase condensation of acetaldehyde over a catalyst (53).

Acetaldehyde forms aldols with other carbonyl compounds containing active hydrogen atoms. Kinetic studies of the aldol condensation of acetaldehyde and deuterated acetaldehydes have shown that only the hydrogen atoms bound to the carbon adjacent to the CHO group take part in the condensation reactions and hydrogen exchange (54,55). A hexyl alcohol, 2-ethyl-1-butanol [*97-95-0*], is produced industrially by the condensation of acetaldehyde and butyraldehyde in dilute caustic solution followed by hydrogenation of the enal intermediate (see ALCOHOLS, HIGHER ALIPHATIC). Condensation of acetaldehyde in the presence of dimethylamine hydrochloride yields polyenals which can be hydrogenated to a mixture of alcohols containing from 4 to 22 carbon atoms (56).

The base-catalyzed reaction of acetaldehyde with excess formaldehyde [*50-00-0*] is the commercial route to pentaerythritol [*115-77-5*]. The aldol condensation of three moles of formaldehyde with one mole of acetaldehyde is followed by a crossed Cannizzaro reaction between pentaerythrose, the intermediate product, and formaldehyde to give pentaerythritol (57). The process proceeds to completion without isolation of the intermediate. Pentaerythrose [*3818-32-4*] has also been made by condensing acetaldehyde and formaldehyde at 45°C using magnesium oxide as a catalyst (58). The vapor-phase reaction of acetaldehyde and formaldehyde at 475°C over a catalyst composed of lanthanum oxide on silica gel gives acrolein [*107-02-8*] (59).

Ethyl acetate [*141-78-6*] is produced commercially by the Tischenko condensation of acetaldehyde using an aluminum ethoxide catalyst (60). The Tischenko reaction of acetaldehyde with isobutyraldehyde [*78-84-2*] yields a mixture of ethyl acetate, isobutyl acetate [*110-19-0*], and isobutyl isobutyrate [*97-85-8*] (61).

Reactions with Ammonia and Amines. Acetaldehyde readily adds ammonia to form acetaldehyde–ammonia. Diethylamine [*109-89-7*] is obtained when acetaldehyde is added to a saturated aqueous or alcoholic solution of ammonia and the mixture is heated to 50–75°C in the presence of a nickel catalyst and hydrogen at 1.2 MPa (12 atm). Pyridine [110-86-1] and pyridine derivatives are made from paraldehyde and aqueous ammonia in the presence of a catalyst at elevated temperatures (62); acetaldehyde may also be used but the yields of pyridine are generally lower than when paraldehyde is the starting material. The vapor-phase reaction of formaldehyde, acetaldehyde, and ammonia at 360°C over oxide catalyst was studied; a 49% yield of pyridine and picolines was obtained using an activated silica–alumina catalyst (63). Brown polymers result when acetaldehyde reacts with ammonia or amines at a pH of 6–7 and temperature of 3–25°C (64). Primary amines and acetaldehyde condense to give Schiff bases: $CH_3CH=NR$. The Schiff base reverts to the starting materials in the presence of acids.

Reactions with Alcohols, Mercaptans, and Phenols. Alcohols add readily to acetaldehyde in the presence of trace quantities of mineral acid to form acetals; eg, ethanol and acetaldehyde form diethyl acetal [105-57-7] (65). Similarly, cyclic acetals are formed by reactions with glycols and other polyhydroxy compounds; eg, ethylene glycol [107-21-1] and acetaldehyde give 2-methyl-1,3-dioxolane [497-26-7] (66):

$$
\begin{array}{c}
\overset{\displaystyle \frown}{\underset{\underset{CH_3 \qquad H}{\diagup \diagdown}}{O \qquad O}}
\end{array}
$$

Mercaptals, $CH_3CH(SR)_2$, are formed in a like manner by the addition of mercaptans. The formation of acetals by noncatalytic vapor-phase reactions of acetaldehyde and various alcohols at 35°C has been reported (67). Butadiene [106-99-0] can be made by the reaction of acetaldehyde and ethyl alcohol at temperatures above 300°C over a tantala–silica catalyst (68). Aldol and crotonaldehyde are believed to be intermediates. Butyl acetate [123-86-4] has been prepared by the catalytic reaction of acetaldehyde with 1-butanol [71-36-3] at 300°C (69).

Reaction of one mole of acetaldehyde and excess phenol in the presence of a mineral acid catalyst gives 1,1-bis(p-hydroxyphenyl)ethane [2081-08-5]; acid catalysts, acetaldehyde, and three moles or less of phenol yield soluble resins. Hardenable resins are difficult to produce by alkaline condensation of acetaldehyde and phenol because the acetaldehyde tends to undergo aldol condensation and self-resinification (see PHENOLIC RESINS).

Reactions with Halogens and Halogen Compounds. Halogens readily replace the hydrogen atoms of the acetaldehyde's methyl group: chlorine reacts with acetaldehyde or paraldehyde at room temperature to give chloroacetaldehyde [107-20-0]; increasing the temperature to 70–80°C gives dichloroacetaldehyde [79-02-7]; and at a temperature of 80–90°C chloral [302-17-0] is formed (70). Catalytic chlorination using antimony powder or aluminum chloride–ferric chloride has also been described (71). Bromal [115-17-3] is formed by an analogous series of reactions (72). It has been postulated that acetyl bromide [506-96-7] is an intermediate in the bromination of acetaldehyde in aqueous ethanol (73). Acetyl chloride [75-36-5] has been prepared by the gas-phase reaction of acetaldehyde and chlorine (74).

Acetaldehyde reacts with phosphorus pentachloride to produce 1,1-dichloroethane [75-34-3] and with hypochlorite and hypoiodite to yield chloroform [67-66-3] and iodoform [75-47-8], respectively. Phosgene [75-44-5] is produced by the reaction of carbon tetrachloride with acetaldehyde in the presence of anhydrous aluminum chloride (75). Chloroform reacts with acetaldehyde in the presence of potassium hydroxide and sodium amide to form 1,1,1-trichloro-2-propanol [7789-89-1] (76).

Miscellaneous Reactions. Sodium bisulfite adds to acetaldehyde to form a white crystalline addition compound, insoluble in ethyl alcohol and ether. This bisulfite addition compound is frequently used to isolate and purify acetaldehyde, which may be regenerated with dilute acid. Hydrocyanic acid adds to acetaldehyde in the presence of an alkali catalyst to form cyanohydrin; the cyanohydrin may also be prepared from sodium cyanide and the bisulfite addition compound. Acrylonitrile [107-13-1] (qv) can be made from acetaldehyde and hydrocyanic acid by heating the cyanohydrin that is formed to 600–700°C (77). Alanine [302-72-7] can be prepared by the reaction of an ammonium salt and an alkali metal cyanide with acetaldehyde; this is a general method for the preparation of α-amino acids called the Strecker amino acid synthesis. Grignard reagents add readily to acetaldehyde, the final product being a secondary alcohol. Thioacetaldehyde [2765-04-0] is formed by reaction of acetaldehyde with hydrogen sulfide; thioacetaldehyde polymerizes readily to the trimer.

Acetic anhydride adds to acetaldehyde in the presence of dilute acid to form ethylidene diacetate [542-10-9]; boron fluoride also catalyzes the reaction (78). Ethylidene diacetate decomposes to the anhydride and aldehyde at temperatures of 220–268°C and initial pressures of 14.6–21.3 kPa (110–160 mm Hg) (79), or upon heating to 150°C in the presence of a zinc chloride catalyst (80). Acetone (qv) [67-64-1] has been prepared in 90% yield by heating an aqueous solution of acetaldehyde to 410°C in the presence of a catalyst (81). Active methylene groups condense acetaldehyde. The reaction of isobutylene [115-11-7] and aqueous solutions of acetaldehyde in the presence of 1–2% sulfuric acid yields alkyl-m-dioxanes; 2,4,4,6-tetramethyl-m-dioxane [5182-37-6] is produced in yields up to 90% (82).

Manufacture

Since 1960, the liquid-phase oxidation of ethylene has been the process of choice for the manufacture of acetaldehyde. There is, however, still some commercial production by the partial oxidation of ethyl alcohol and hydration of acetylene. The economics of the various processes are strongly dependent on the prices of the feedstocks. Acetaldehyde is also formed as a coproduct in the high temperature oxidation of butane. A more recently developed rhodium catalyzed process produces acetaldehyde from synthesis gas as a coproduct with ethyl alcohol and acetic acid (83–94).

Oxidation of Ethylene. In 1894 F. C. Phillips observed the reaction of ethylene [74-85-1] in an aqueous palladium(II) chloride solution to form acetaldehyde.

$$C_2H_4 + PdCl_2 + H_2O \longrightarrow CH_3CHO + Pd + 2 HCl$$

The direct liquid phase oxidation of ethylene was developed in 1957–1959 by Wacker-Chemie and Farbwerke Hoechst in which the catalyst is an aqueous solution of $PdCl_2$ and $CuCl_2$ (86).

Studies of the reaction mechanism of the catalytic oxidation suggest that a *cis*-hydroxyethylene–palladium π-complex is formed initially, followed by an intramolecular exchange of hydrogen and palladium to give a *gem*-hydroxyethylpalladium species that leads to acetaldehyde and metallic palladium (88–90).

The metallic palladium is reoxidized to $PdCl_2$ by the $CuCl_2$ and the resultant cuprous chloride is then reoxidized by oxygen or air as shown.

$$Pd + 2\ CuCl_2 \longrightarrow PdCl_2 + 2\ CuCl$$

$$2\ CuCl + \tfrac{1}{2}\ O_2 + 2\ HCl \longrightarrow 2\ CuCl_2 + H_2O$$

Thus ethylene is oxidized continuously through a series of oxidation–reduction reactions (87,88). The overall reaction is

$$C_2H_4 + \tfrac{1}{2}\ O_2 \longrightarrow CH_3CHO \quad \Delta H = 427\ \text{kJ} \ (102\ \text{kcal})$$

There are two variations for this commercial production route: the two-stage process developed by Wacker-Chemie and the one-stage process developed by Farbwerke Hoechst (91–92). In the two-stage process shown in Figure 1, ethylene is almost completely oxidized by air to acetaldehyde in one pass in a tubular plug-flow reactor made of titanium (93,94). The reaction is conducted at 125–130°C and 1.13 MPa (150 psig) using the palladium and cupric chloride catalysts. Acetaldehyde produced in the first reactor is removed from the reaction loop by adiabatic flashing in a tower. The flash step also removes the heat of reaction. The catalyst solution is recycled from the flash-tower base to the second stage (or oxidation reactor) where the cuprous salt is oxidized to the cupric state with air. The high pressure off-gas from the oxidation reactor, mostly nitrogen, is

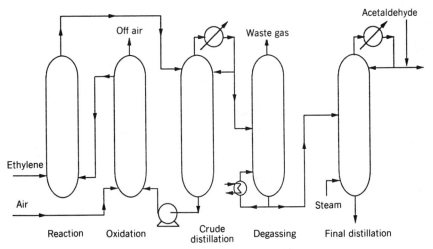

Fig. 1. The two-stage acetaldehyde process.

separated from the liquid catalyst solution and scrubbed to remove acetaldehyde before venting. A small portion of the catalyst stream is heated in the catalyst regenerator to destroy any undesirable copper oxalate. The flasher overhead is fed to a distillation system where water is removed for recycle to the reactor system and organic impurities, including chlorinated aldehydes, are separated from the purified acetaldehyde product. Synthesis techniques purported to reduce the quantity of chlorinated by-products generated have been patented (95).

In the one-stage process (Fig. 2), ethylene, oxygen, and recycle gas are directed to a vertical reactor for contact with the catalyst solution under slight pressure. The water evaporated during the reaction absorbs the heat evolved, and make-up water is fed as necessary to maintain the desired catalyst concentration. The gases are water-scrubbed and the resulting acetaldehyde solution is fed to a distillation column. The tail-gas from the scrubber is recycled to the reactor. Inert materials are eliminated from the recycle gas in a bleed-stream which flows to an auxiliary reactor for additional ethylene conversion.

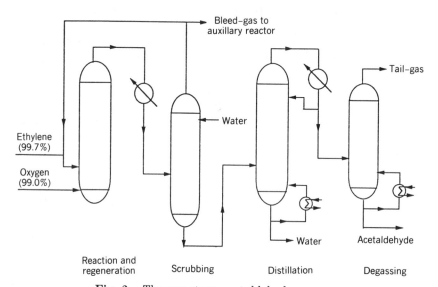

Fig. 2.   The one-stage acetaldehyde process.

This oxidation process for olefins has been exploited commercially principally for the production of acetaldehyde, but the reaction can also be applied to the production of acetone from propylene and methyl ethyl ketone [78-93-3] from butenes (87,88). Careful control of the potential of the catalyst with the oxygen stream in the regenerator minimizes the formation of chloroketones (94). Vinyl acetate can also be produced commercially by a variation of this reaction (96,97).

From Ethyl Alcohol.   Some acetaldehyde is produced commercially by the catalytic oxidation of ethyl alcohol. The oxidation is carried out by passing alcohol vapors and preheated air over a silver catalyst at 480°C (98).

$$CH_3CH_2OH + \tfrac{1}{2} O_2 \longrightarrow CH_3CHO + H_2O \quad \Delta H = -242 \text{ kJ/mol } (-57.84 \text{ kcal/mol})$$

With a multitubular reactor, conversions of 74–82% per pass can be obtained while generating steam to be used elsewhere in the process (99).

From Acetylene. Although acetaldehyde has been produced commercially by the hydration of acetylene since 1916, this procedure has been almost completely replaced by the direct oxidation of ethylene. In the hydration process, high purity acetylene under a pressure of 103.4 kPa (15 psi) is passed into a vertical reactor containing a mercury catalyst dissolved in 18–25% sulfuric acid at 70–90°C (see ACETYLENE-DERIVED CHEMICALS).

$$HC\equiv CH + H_2O \xrightarrow[\text{H}_2\text{SO}_4\ (70-90^\circ\text{C})]{\text{Hg}^{2+}} CH_3CHO$$

Fresh catalyst is fed to the reactor periodically; the catalyst may be added in the mercurous form but the catalytic species has been shown to be a mercuric ion complex (100). The excess acetylene sweeps out the dissolved acetaldehyde, which is condensed by water and refrigerated brine and then scrubbed with water; this crude acetaldehyde is purified by distillation; the unreacted acetylene is recycled. The catalytic mercuric ion is reduced to catalytically inactive mercurous sulfate and metallic mercury. Sludge, consisting of reduced catalyst and tars, is drained from the reactor at intervals and resulfated. The rate of catalyst depletion can be reduced by adding ferric or other suitable ions to the reaction solution. These ions reoxidize the mercurous ion to the mercuric ion; consequently, the quantity of sludge which must be recovered is reduced (80,101). In one variation, acetylene is completely hydrated with water in a single operation at 68–73°C using the mercuric–iron salt catalyst. The acetaldehyde is partially removed by vacuum distillation and the mother liquor recycled to the reactor. The aldehyde vapors are cooled to about 35°C, compressed to 253 kPa (2.5 atm), and condensed. It is claimed that this combination of vacuum and pressure operations substantially reduces heating and refrigeration costs (102).

From Synthesis Gas. A rhodium-catalyzed process capable of converting synthesis gas directly into acetaldehyde in a single step has been reported (83,84).

$$CO + H_2 \longrightarrow CH_3CHO + \text{other products}$$

This process comprises passing synthesis gas over 5% rhodium on SiO_2 at 300°C and 2.0 MPa (20 atm). Principal coproducts are acetaldehyde, 24%; acetic acid, 20%; and ethanol, 16%. Although interest in new routes to acetaldehyde has fallen as a result of the reduced demand for this chemical, one possible new route to both acetaldehyde and ethanol is the reductive carbonylation of methanol (85).

$$CH_3OH + CO + H_2 \longrightarrow CH_3CHO + H_2O$$

The catalyst of choice is cobalt iodide with various promotors from Group 15 elements. The process is run at 140–200°C, 28–41 MPa (4,000–6,000 psi), and gives an 88% conversion with 90% selectively to acetaldehyde. Neither of these acetaldehyde syntheses have been commercialized.

Specifications, Analytical and Test Methods

Commercial acetaldehyde has the following typical specifications: assay, 99% min; color, water-white; acidity, 0.5% max (acetic acid); specific gravity, 0.790 at 20°C; bp, 20.8°C at 101.3 kPa (1 atm). It is shipped in steel drums and tank cars bearing the ICC red label. In the liquid state, it is noncorrosive to most metals; however, acetaldehyde oxidizes readily, particularly in the vapor state, to acetic acid. Precautions to be observed in the handling of acetaldehyde have been published (103).

Analytical methods based on many of the reactions common to aldehydes have been developed for acetaldehyde determination. In the absence of other aldehydes, it can be detected by the formation of a mirror from an alkaline silver nitrate solution (Tollens' reagent) and by the reduction of Fehling's solution. It can be determined quantitatively by fuchsin–sulfur dioxide solution (Schiff's reagent), or by reaction with sodium bisulfite, the excess bisulfite being estimated iodometrically. Acetaldehyde present in mixtures with other carbonyl compounds, such as organic acids, can be determined by paper chromatography of 2,4-dinitrophenylhydrazones (104), polarographic analysis either of the untreated mixture (105) or of the semicarbazones (106), the color reaction with thymol blue on silica gel (detector tube method) (107), mercurimetric oxidation (108), argentometric titration (109), microscopic (110) and spectrophotometric methods (111), and gas–liquid chromatographic analysis (112). However, gas–liquid chromatographic techniques have superseded most chemical tests for routine analyses.

Acetaldehyde can be isolated and identified by the characteristic melting points of the crystalline compounds formed with hydrazines, semicarbazides, etc; these derivatives of aldehydes can be separated by paper and column chromatography (104,113). Acetaldehyde has been separated quantitatively from other carbonyl compounds on an ion-exchange resin in the bisulfite form; the aldehyde is then eluted from the column with a solution of sodium chloride (114). In larger quantities, acetaldehyde may be isolated by passing the vapor into ether, then saturating with dry ammonia; acetaldehyde–ammonia crystallizes from the solution. Reactions with bisulfite, hydrazines, oximes, semicarbazides, and 5,5-dimethyl-1,3-cyclohexanedione [126-81-8] (dimedone) have also been used to isolate acetaldehyde from various solutions.

Health and Safety Factors

Acetaldehyde appears to paralyze respiratory muscles, causing panic. It has a general narcotic action which prevents coughing, causes irritation of the eyes and mucous membranes, and accelerates heart action. When breathed in high concentration, it causes headache and sore throat. Carbon dioxide solutions in acetaldehyde are particularly pernicious because the acetaldehyde odor is weakened by the carbon dioxide. Prolonged exposure causes a decrease of both red and white blood cells; there is also a sustained rise in blood pressure (115–117). The threshold limit value (TLV) of acetaldehyde in air is 100 ppm (118). In normal industrial operations there is no health hazard in handling acetaldehyde provided normal precautions are taken. Mixtures of acetaldehyde vapor and air are flammable; they are explosive if the concentrations of aldehyde and oxygen rise above 4 and 9%, respectively. Reference 103 discusses handling precautions.

Economic Aspects and Uses

The production pattern for acetaldehyde has undergone significant changes since the principal industrial routes to acetaldehyde were hydration of acetylene and oxidation of ethyl alcohol. First, increasing acetylene costs made this feedstock economically unattractive. Then the two aldehyde Wacker-Hoechst GmbH processes for the liquid-phase oxidation of ethylene to acetaldehyde began commercial operation. By 1968 more acetaldehyde was produced by the direct oxidation of ethylene using the Wacker process than from ethanol. Union Carbide discontinued its annual production of 90,700 t of acetaldehyde from ethanol in late 1977 (119). The percentage of ethanol consumed for acetaldehyde in the United States dropped from 20% in 1971 to zero in 1984.

   Figure 3 shows the production of acetaldehyde in the years 1969 through 1987 as well as an estimate of 1989–1995 production. The year 1969 was a peak year for acetaldehyde with a reported production of 748,000 t. Acetaldehyde production is linked with the demand for acetic acid, acetic anhydride, cellulose acetate, vinyl acetate resins, acetate esters, pentaerythritol, synthetic pyridine derivatives, terephthalic acid, and peracetic acid. In 1976 acetic acid production represented 60% of the acetaldehyde demand. That demand has diminished as a result of the rising cost of ethylene as feedstock and methanol carbonylation as the preferred route to acetic acid (qv).

   The nameplate capacities for acetaldehyde production for the United States in 1989 are shown in Table 5 (120). Synthetic pyridine derivatives, peracetic acid,

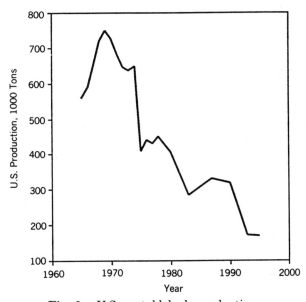

Fig. 3.   U.S. acetaldehyde production.

Table 5. United States Acetaldehyde Producers, 1989

Producer, location	Capacity, 10^3 t/yr
Hoechst-Celanese, Bay City, Tex.	217
Texas Eastman Co., Longview, Tex.	226

acetate esters by the Tischenko route, and pentaerythritol account for 40% of acetaldehyde demand. This sector may show strong growth in some products but all of these materials may be prepared from alternative processes.

The price of acetaldehyde during the period 1950 to 1973 ranged from $0.20 to 0.22/kg. Increased prices for hydrocarbon cracking feedstocks beginning in late 1973 resulted in higher costs for ethylene and concurrent higher costs for acetaldehyde. The posted prices for acetaldehyde were $0.26/kg in 1974, $0.78/kg in 1985, and $0.92/kg in 1988. The future of acetaldehyde growth appears to depend on the development of a lower cost production process based on synthesis gas and an increase in demand for processes based on acetaldehyde activation techniques and peracetic acid.

BIBLIOGRAPHY

"Acetaldehyde" in *ECT* 1st ed., Vol. 1, pp. 32–43, by M. S. W. Small, Shawinigan Chemicals Limited; in *ECT* 2nd ed., Vol. 1, pp. 77–95, by E. R. Hayes, Shawinigan Chemicals Limited (1963); in *ECT* 3rd ed., Vol. 1, pp. 97–112, by H. J. Hagemeyer, Jr., Texas Eastman Company (1978).

1. K. F. Coles and F. Popper, *Ind. Eng. Chem.* **42,** 1434 (1950).
2. T. E. Smith and R. F. Bonner, *Ind. Eng. Chem.* **43,** 1169 (1951).
3. A. A. Dobrinskaya, V. C. Markovich, and M. B. Nelman, *Izv. Akad. Nauk SSR Ser. Khim.* **434,** (1953); *Chem. Abstr.* **48,** 4378 (1955).
4. R. P. Kirsanova and S. Sh. Byk, *Zh. Prikl. Khim. Leningrad* **31,** 1610 (1958).
5. A. E. Rabe, University Microfilm, Ann Arbor, Mich., L. C. Card Mic. 58–1920.
6. D. S. Tsiklis and A. M. Kofman, *Zh. Fiz. Khim.* **31,** 100 (1957).
7. W. H. Horsley, *Adv. Chem. Ser.* **35,** 27 (1962).
8. D. S. Tsiklis, *Zh. Fiz. Khim.* **32,** 1367 (1958).
9. D. S. Tsiklis and Ya. D. Shvarts, *Zh. Fiz. Khim.* **31,** 2302 (1957).
10. P. M. Chaudhuri and co-workers, *J. Chem. Eng. Data* **13,** 9 (1968).
11. C. F. Coleman and T. DeVries, *J. Am. Chem. Soc.* **71,** 2839 (1949).
12. K. A. Kobe and H. R. Crawford, *Pet. Refiner* **37**(7), 125 (1958).
13. K. S. Pitzer and W. Weltner, Jr., *J. Am. Chem. Soc.* **71,** 2842 (1949).
14. L. D. Christensen and J. M. Smith, *Ind. Eng. Chem.* **42,** 2128 (1950).
15. D. M. Newitt, L. M. Baxt, and V. V. Kelkar, *J. Chem. Soc.* 1703, 1711 (1939).
16. J. H. Burgoyne and R. F. Neale, *Fuel* **32,** 5 (1953).
17. J. Chamboux and M. Lucquin, *Compt. Rend.* **246,** 2489 (1958).
18. A. G. White and E. Jones, *J. Soc. Chem. Ind. London* **69,** 2006, 209 (1950).
19. F. C. Mitchell and H. C. Vernon, *Chem. Metall. Eng.* **44,** 733 (1937).
20. B. P. Mullins, *Fuel* **32,** 481 (1953).
21. S. S. Penner and B. P. Mullins, *Explosions, Detonations, Flammability and Ignition,* Pergamon Press, Inc., New York, 1959, pp. 199–203.

22. R. O. King, S. Sandler, and R. Strom, *Can. J. Technol.* **32,** 103 (1954).
23. S. Ono, *Rev. Phys. Chem. Jpn.* **1946,** 42; *Chem. Abstr.* **44,** 4661 (1950).
24. N. M. Emanuel, *Dokl. Akad. Nauk SSSR* **59,** 1137 (1948); *Chem. Abstr.* **42,** 7142 (1948).
25. T. E. Pavlovskaya and N. M. Emanuel, *Dokl. Akad. Nauk SSSR* **58,** 1693 (1947); *Chem. Abstr.* **46,** 4231 (1952).
26. E. M. Wilson, *J. Am. Rocket Soc.* **23,** (1953).
27. H. Nilsen, *Tidsskr. Kjemi Bergves. Metall.* **17,** 149 (1957); *Chem. Abstr.* **53,** 10916 (1959).
28. F. P. Lossing, *Can. J. Chem.* **35,** 305 (1957).
29. R. K. Brinton and D. H. Volman, *J. Chem. Phys.* **20,** 1053, 1054 (1952).
30. P. D. Zemany and M. Burton, *J. Am. Chem. Soc.* **73,** 499, 500 (1951).
31. R. P. Bell and J. C. Clunie, *Trans. Faraday Soc.* **48,** 439 (1952).
32. R. P. Bell and M. Rand, *Bull Soc. Chim. Fr.* 115 (1955).
33. H. J. Hagemeyer, Jr., *Ind. Eng. Chem.* **41,** 766 (1949).
34. A. F. Cadenhead, *Can. Chem. Process.* **39**(7), 78 (1955).
35. G. Benson, *Chem. Metall. Eng.* **47,** 150, 151 (1940).
36. B. Phillips, F. C. Frostick, Jr., and P. S. Starcher, *J. Am. Chem. Soc.* **79,** 5982 (1957).
37. U.S. Pat. 2,8309,080 (reissued RE-25057) (Apr. 8, 1958), H. B. Stevens (to Shawinigan Chemicals Ltd.).
38. U.S. Pat. 3,290,378 (Dec. 6, 1966), K. Tsunemitsu and Y. Tsujino.
39. Ger. Pat. 573,721 (Apr. 5, 1933), M. Mugdam and J. Sixt (to Consortium für electro-chemische Industrie G.m.b.H.).
40. U.S. Pat. 3,240,803 (Mar. 15, 1966), B. Thompson and S. D. Neeley (to Eastman Kodak Co.).
41. U.S. Pat. 2,287,803 (June 30, 1943), D. C. Hull (to Eastman Kodak Co.).
42. T. Sasa, *J. Soc. Org. Syn. Chem. Jpn.* **12,** 60 (1954); *Chem. Abstr.* **51,** 2780 (1957).
43. Fr. Pat. 973,322 (Feb. 9, 1951), H. M. Guinot (to Usines de Melle).
44. C. C. Oldenburg and H. F. Rase, *Am. Inst. Chem. Eng. J.* **3,** 462 (1957).
45. U.S. Pat. 2,318,341 (May 1943), B. Thompson (to Eastman Kodak Co.).
46. U.S. Pat. 2,479,559 (Aug. 1949), A. A. Dolnick and co-workers (to Publicker Ind. Inc.).
47. T. Kawaguchi and S. Hasegawa, *Tokyo Gakugei Daigaku Kiyo Dai 4 Bu* **21,** 64 (1969).
48. U.S. Pat. 2,426,961 (Sept. 2, 1947), R. S. Wilder (to Publicker Ind. Inc.).
49. M. W. Travers, *Trans. Faraday Soc.* **32,** 246 (1936).
50. M. Letort, *Compt. Rend.* **202,** 767 (1936).
51. O. Vogl, *J. Poly. Sci. A* **2,** 4591 (1964).
52. J. T. Furukawa, *Macromol. Chem.* **37,** 149 (1969).
53. U.S. Pat. 2,810,6 (Oct. 22, 1957), J. F. Gabbet, Jr. (to Escambia Chemical Corp.).
54. Y. Pocker, *Chem. Ind.* **1959,** 599 (1959).
55. R. P. Bell and M. J. Smith, *J. Chem. Soc.* 1691 (1958).
56. W. Langenbeck, J. Alm, and K. W. Knitsch, *J. Prakt. Chem.* **8,** 112 (1959).
57. E. Berlow, R. H. Barth, and J. E. Snow, *The Pentaerythritols,* Reinhold Publishing Corporation, New York, 1959, pp. 4–24.
58. Fr. Pat. 962,381 (June 8, 1950) (to Etablissements Kuhlmann).
59. U.S. Pat. 3,701,798 (Oct. 31, 1972), T. C. Snapp, A. E. Blood, and H. J. Hagemeyer, Jr. (to Eastman Kodak Co.).
60. A. S. Hester and K. Himmler, *Ind. Eng. Chem.* **51,** 1428 (1959).
61. U.S. Pat. 3,714,236 (Jan. 30, 1973), H. N. Wright and H. J. Hagemeyer, Jr. (to Eastman Kodak Co.).
62. M. S. Astle, *Industrial Organic Nitrogen Compounds,* Reinhold Publishing Corporation, New York, 1961, pp. 134–136.
63. S. L. Levy and D. F. Othmer, *Ind. Eng. Chem.* **47,** 789 (1955).
64. J. F. Carson and H. S. Olcott, *J. Am. Chem. Soc.* **76,** 2257 (1954).
65. J. Deschamps, M. Paty, and P. Pineau, *Compt. Rend.* **238,** 911 (1954).

66. Neth. Pat. Appl. 6,510,968 (Feb. 21, 1966) (to Lummus Co.); W. G. Lloyd, *J. Org. Chem.* **34,** 3949 (1969).
67. U.S. Pat. 2,691,684 (Oct. 12, 1954), L. K. Frevel and J. W. Hedelund (to The Dow Chemical Company).
68. B. B. Corson and co-workers, *Ind. Eng. Chem.* **42,** 359 (1950).
69. S. L. Lel'chuk, *Khim. Prom.* **1946**(9), 16 (1946); *Chem. Abstr.* **41,** 3756 (1947).
70. W. T. Cave, *Ind. Eng. Chem.* **45,** 1854 (1953).
71. Jpn. Pat. 4713 (Nov. 14, 1952), J. Imamura (to Bureau of Industrial Technics); *Chem. Abstr.* **47,** 11224 (1953).
72. M. N. Shchukina, *Zh. Obshch. Khim.* **18,** 1653 (1948); *Chem. Abstr.* **43,** 2575 (1949).
73. N. N. Lichton and F. Granchelli, *J. Am. Chem. Soc.* **76,** 3729 (1954).
74. Jpn. Pat. 153,599 (Nov. 2, 1942), Y. Kato; *Chem. Abstr.* **43,** 3027 (1949).
75. G. Illari, *Gazz. Chim. Ital.* **81,** 439 (1951); *Chem. Abstr.* **46,** 5532 (1952).
76. R. Lombard and R. Boesch, *Bull. Soc. Chim. Fr.* **1953**(10), C23 (1953).
77. K. Sennewald and K. H. Steil, *Chem. Ing. Tech.* **30,** 440 (1958).
78. E. H. Man, J. J. Sanderson, and C. R. Hauser, *J. Am. Chem. Soc.* **72,** 847 (1950).
79. C. C. Coffin, *Can. J. Res.* **5,** 639 (1931).
80. P. W. Sherwood, *Pet. Refiner* **34**(3), 203 (1955).
81. St. Grzelczyk, *Przem. Chem.* **12**(35), 696 (1956); *Chem. Abstr.* **52,** 12753 (1958).
82. M. I. Farberov and K. A. Machtina, *Uch. Zap. Yarosl. Tekhnol. Inst.* **2,** 5 (1957); *Chem. Abstr.* **53,** 18041 (1959).
83. W. Ger. Offen. 2,503,204 (Jan. 28, 1974), M. M. Bhasin (to Union Carbide Co.).
84. Belg. Pat. 824,822 (July 28, 1975) (to Union Carbide Co.).
85. U.S. Pat. 4,337,365 (June 29, 1982), W. E. Wakler (to Union Carbide Co.).
86. R. Jira and W. Freiesleben, *Organomet. React.* **3,** 22 (1972).
87. J. Smidt and co-workers, *Angew. Chem.* **71,** 176 (1959).
88. J. Smidt, *Chem. Ind. London* **1962**(2), 54 (1962).
89. R. Jira, J. Sedlmeier, and J. Smidt, *Justus Liebigs Ann. Chem.* 6993, 99 (1966).
90. R. Jira, W. Blan, and D. Grimm, *Hydrocarbon Process.* **55**(3), 97 (1976).
91. Can. Pat. 625,430 (Aug. 8, 1961), J. Smidt and co-workers (to Consortium fur elektrochemische Industrie (G.m.b.H.).
92. *Pet. Refiner* **40**(11), 206 (1961).
93. R. P. Lowry, *Hydrocarbon Process.* **53**(11), 105 (1974).
94. R. Jira, in S. A. Miller, ed., *Ethylene and Its Industrial Derivatives,* Ernest Benn Ltd., London, 1969, pp. 639–553.
95. U.S. Pat. 4,720,474 (Sept. 24, 1985), J. Vasilevskis and co-workers (to Catalytica Associates).
96. Brit. Pat. 1,109,483 (Apr. 10, 1968), H. J. Hagemeyer, Jr., and co-workers (to Eastman Kodak Co.).
97. K. R. Bedell and H. A. Rainbird, *Hydrocarbon Process.* **51**(11), 141 (1972).
98. W. L. Faith, D. B. Keyes, and R. L. Clark, *Industrial Chemicals,* 2nd ed., John Wiley & Sons, Inc., New York, 1957, pp. 2–3.
99. U.S. Pat. 3,284,170 (Nov. 8, 1966), S. D. Neeley (to Eastman Kodak Co.).
100. K. Schwabe and J. Voigt, *Z. Phys. Chem. Leipzig* **203,** 383 (1954).
101. *Pet. Refiner* **40**(11), 207 (1961).
102. D. F. Othmer, K. Kon, and T. Igarashi, *Ind. Eng. Chem.* **48,** 1258 (1956).
103. Chemical Safety Data Sheet SD-43, *Properties and Essential Information for Safe Handling and Use of Acetaldehyde,* Manufacturing Chemists Association, Inc., Washington, D.C., 1952.
104. R. Ellis, A. M. Gaddis, and G. T. Currie, *Anal. Chem.* **30,** 475 (1958).
105. S. Sandler and Y. H. Chung, *Anal. Chem.* **30,** 1252 (1958).
106. D. M. Coulson, *Anal. Chim. Acta.* **19,** 284 (1958).

107. Y. Kobayashi, *Yuki Gosei Kagaku Kyokai Shi* **16,** 625 (1958); *Chem. Abstr.* **53,** 984 (1959).
108. J. E. Ruch and J. B. Johnson, *Anal. Chem.* **28,** 69 (1956).
109. H. Siegel and F. T. Weiss, *Anal. Chem.* **26,** 917 (1954).
110. R. E. Dunbar and A. E. Aaland, *Microchem. J.* **2,** 113 (1954).
111. J. H. Ross, *Anal. Chem.* **25,** 1288 (1953).
112. R. Stevens, *Anal. Chem.* **33,** 1126 (1961).
113. L. Nebbia and F. Guerrieri, *Chim. Ind. Milan* **39,** 749 (1957).
114. G. Gabrielson and O. Samuelson, *Sven. Km. Tidskr.* **64,** 150 (1952) (in English); *Chem. Abstr.* **46,** 9018 (1952).
115. E. W. Page and R. Reed, *Am. J. Physiol.* **143,** 122 (1945).
116. E. Skog, *Acta Pharmacol. Toxicol.* **6,** 299 (1950).
117. H. F. Smyth, Jr., C. P. Carpenter, and C. S. Weil, *J. Ind. Hyg. Toxicol.* **31,** 60 (1949).
118. H. I. Sax, *Dangerous Properties of Industrial Materials,* Van Nostrand-Reinhold Publishing Corp., New York, 1989, p. 6.
119. *Chem. Mark. Rep.* 3 (Dec. 27, 1976).
120. "Acetaldehyde, 601.500H," *Chemical Economics Handbook,* SRI International, Menlo Park, Calif., 1989.

H. J. HAGEMEYER
Texas Eastman Company

ACETAL RESINS

The term "acetal resins" commonly denotes the family of homopolymers and copolymers whose main chains are completely or essentially composed of repeating oxymethylene units $(-CH_2-O-)_n$. The polymers are derived chiefly from formaldehyde or methanal [*50-00-00*], either directly or through its cyclic trimer, trioxane or 1,3,5-trioxacyclohexane [*110-88-3*].

Formaldehyde polymers have been known for some time (1) and early investigations of formaldehyde polymerization contributed significantly to the development of several basic concepts of polymer science (2). Polymers of higher aliphatic homologues of formaldehyde are also well known (3) and frequently referred to as aldehyde polymers (4). Some have curious properties, but none are commercially important.

Formaldehyde homopolymer is composed exclusively of repeating oxymethylene units and is described by the term polyoxymethylene (POM) [*9002-81-7*]. Commercially significant copolymers, for example [*95327-43-8*], have a minor fraction (typically less than 5 mol %) of alkylidene or other units, derived from cyclic ethers or cyclic formals, distributed along the polymer chain. The occasional break in the oxymethylene sequences has significant ramifications for polymer stabilization.

Acetal resins were first commercialized by Du Pont in 1960 under the tradename Delrin (registered trademark of E. I. du Pont de Nemours and Co., Inc.) Introduction of the then new engineering plastics followed development of suitable processes for the preparation of high molecular weight formaldehyde homopolymer (5) and for the requisite conversion of the homopolymer's unstable end groups by a procedure known as end-capping (6). Development and commercialization of copolymers of trioxane with cyclic ethers (eg, ethylene oxide) by Celanese (7) quickly followed in 1962. The copolymers did not require end-capping. Rather, good polymer stability could be achieved by ablation of unstable polymer fractions by a suitable process (8). Today there are ten producers of acetal resins, mostly copolymers.

Throughout the remainder of this article the term homopolymer refers to Delrin acetal resin manufactured and sold by Du Pont; the term copolymer refers to Celcon acetal copolymer resins (registered trademark of Hoechst Celanese Corporation).

Structure and Properties

The many commercially attractive properties of acetal resins are due in large part to the inherent high crystallinity of the base polymers. Values reported for percentage crystallinity (x ray, density) range from 60 to 77%. The lower values are typical of copolymer. Polyoxymethylene most commonly crystallizes in a hexagonal unit cell (9) with the polymer chains in a 9/5 helix (10,11). An orthorhombic unit cell has also been reported (9). The oxyethylene units in copolymers of trioxane and ethylene oxide can be incorporated in the crystal lattice (12). The nominal value of the melting point of homopolymer is 175°C, that of the copolymer is 165°C. Other thermal properties which depend substantially on the crystallization or melting of the polymer are listed in Table 1. See also reference 13.

The high crystallinity of acetal resins contributes significantly to their excellent resistance to most chemicals, including many organic solvents. Acetal

Table 1. Thermal Characteristics of Acetal Resins

Property	ASTM test method	Homopolymer	Copolymer
heat deflection temperature, °C	D648		
at 1.82 MPa[a]		136	110
at 0.45 MPa		172	158
coefficient of linear thermal			
expansion × 10^6 per °C, −40 to 30°C	D696	75	84
thermal conductivity, W/(m·K)		0.2307	0.2307
specific heat, J/(kg·K)[b]		1465	1465
melting point, °C		175	165
flow temperature, °C	D569	184	174

[a]To convert MPa to psi, multiply by 145.
[b]To convert J to cal, divide by 4.184.

resins retain their properties after exposure to a wide range of chemicals and environments. More detailed data are available (14).

The polymers dissolve in 1,1,1,3,3,3-hexafluoro-2-propanol [920-66-1], hot phenols, and N,N-dimethylformamide [68-12-2] near its boiling point. The excellent solvent resistance notwithstanding, solvents suitable for measurement of intrinsic viscosity, useful for estimation of molecular weight, are known (13,15).

Mechanical Properties. Stiffness, resistance to deformation under constant applied load (creep resistance), resistance to damage by cyclical loading (fatigue resistance), and excellent lubricity are mechanical properties for which acetal resins are perhaps best known and which have contributed significantly to their excellent commercial success. General purpose acetal resins are substantially stiffer than general purpose polyamides (nylon-6 or -6,6 types) when the latter have reached equilibrium water content. The creep and fatigue properties are known and predictable and very valuable to the design engineer.

Typical values of important properties of general purpose acetal resins (homopolymer and copolymer) are collected in Table 2. Properties in the table were determined on specimens subjected only to the conditioning required by the ASTM procedure. In this case, values measured for homopolymer are characteristically higher than those for copolymer.

Table 2. Mechanical Properties of Acetal Resin

Property	ASTM test method	Homopolymer	Copolymer
tensile strength, yield, MPa[a], 23°C	D638	68.9	60.6
elongation, break, %	D638	25–75	40–75
tensile modulus, MPa[a], 23°C	D638	3100	2825
flexural strength, MPa[a], 23°C	D790	97.1	89.6
flexural modulus, MPa[a], 23°C	D790	2830	2584
compressive stress, MPa[a], 23°C	D695	35.8	31
1% deflection		35.8	31
10% deflection		124	110
shear strength, MPa[a], 23°C	D732	65	53
Izod impact strength, notched, 3.175 mm, J/m[b],	D256		
23°C		69–122	53–80
−40°C		53–95	43–64
water absorption, %	D570		
24-h immersion		0.25	0.22
equilibrium, 50% rh, 23°C		0.22	0.16
equilibrium, immersion		0.9	0.80
Rockwell hardness, M scale	D785	94	80
coefficient of friction, dynamic	D1894		
steel		0.1–0.3	0.15
aluminum, brass			0.15
acetal resin			0.35
specific gravity	D792	1.42	1.41

[a]To convert MPa to psi, multiply by 145.
[b]To convert J/m to ft·lb/in., divide by 53.39.

Electrical Properties. Electrical properties of acetal resin are collected in Table 3. The dielectric constant is constant over the temperature range of most interest (-40 to $50°C$).

Table 3. Electrical Properties of Acetal Resins

Property	ASTM test method	Homopolymer	Copolymer
dielectric constant, 10^2–10^6 Hz	D150	3.7	3.7
dissipation factor	D150		
10^2 Hz			0.0010
10^3 Hz			0.0010
10^4 Hz			0.0015
10^6 Hz		0.0048	0.006
dielectric strength, kV/mm, short time, 2.29-mm sheet	D149	20	20
surface resistivity, ohm	D257	1×10^{15}	1.3×10^{16}
volume resistivity, ohm·cm	D257	1×10^{15}	1×10^{14}

Chemical Structure and Properties. Homopolymer consists exclusively of repeating oxymethylene units. The copolymer contains alkylidene units (eg, ethylidene $-CH_2-CH_2-$) randomly distributed along the chain. A variety of end groups may be present in the polymers. Both homopolymer and copolymer may have alkoxy, especially methoxy (CH_3O-), or formate ($HCOO-$) end groups. Copolymer made with ethylene oxide has 2-hydroxyethoxy end groups. Homopolymer generally has acetate end groups.

The number-average molecular weight of most commercially available acetal resins is between 20,000 and 90,000. Weight-average molecular weight may be estimated from solution viscosities.

The details of the commercial preparation of acetal homo- and copolymers are discussed later. One aspect of the polymerization so pervades the chemistry of the resulting polymers that familiarity with it is a prerequisite for understanding the chemistry of the polymers, the often subtle differences between homo- and copolymers, and the difficulties which had to be overcome to make the polymers commercially useful. The ionic polymerizations of formaldehyde and trioxane are equilibrium reactions. Unless suitable measures are taken, polymer will begin to revert to monomeric formaldehyde at processing temperatures by depolymerization (called unzipping) which begins at chain ends.

The intrinsic stability of the polymer can be substantially improved by converting unstable hemiacetal end groups to more stable acetate esters in a process, referred to as endcapping, which is routinely practiced with homopolymer. In copolymers, depolymerization proceeds only until a comonomer unit (eg, $-CH_2CH_2-$) is encountered and unzipping stops. Any chemical process which leads to cleavage of the polymer chain or to formation of unstable end groups presages complete destruction of a homopolymer chain. In copolymers, the comonomer units (eg, $-CH_2CH_2-$) add further stability by acting as chain blockers. That is, once a chain is cleaved, the effective length of chain that can unzip is smaller for copolymer. In addition, the greater prevalence of ester end groups in homopolymer reduces their stability in alkaline environments compared to copolymer.

Acetal resins are generally stable in mildly alkaline environments. However, bases can catalyze hydrolysis of ester end groups, resulting in less thermally stable polymer.

Properly end-capped acetal resins, substantially free of ionic impurities, are relatively thermally stable. However, the methylene groups in the polymer backbone are sites for peroxidation or hydroperoxidation reactions which ultimately lead to scission and depolymerization. Thus antioxidants (qv), especially hindered phenols, are included in most commercially available acetal resins for optimal thermal oxidative stability.

Like most other engineering thermoplastics, acetal resins are susceptible to photooxidation by oxidative radical chain reactions. Carbon–hydrogen bonds in the methylene groups are principal sites for initial attack. Photooxidative degradation is typically first manifested as chalking on the surfaces of parts.

Other aspects of stabilization of acetal resins are briefly discussed under processing and fabrication. Reference 15 provides a more detailed discussion of the mechanism of polymer degradation.

When ignited, nonfilled acetal resins burn in air with a characteristic dull blue flame.

Testing.   Melt index or melt flow rate at 190°C, according to ASTM D1238, is the test most frequently applied to the characterization of commercial acetal resins. The materials are typically grouped or differentiated according to their melt flow rate. Several other ASTM tests are commonly used for the characterization and specification of acetal resins.

The weight average molecular weight of acetal copolymers may be estimated from their melt index (MI, expressed in g/10 min) according to the relation

$$MI = 3.30 \times 10^{18} \, M_w^{-3.55}$$

The comonomer content of copolymers may be estimated by nmr or by controlled solvolysis of the copolymer followed by quantitative chromatographic analysis of the residues.

Manufacturing

Although there is a substantial body of information in the public domain concerning the preparation of polyacetals, the details of processes for manufacturing acetal resins are kept highly confidential by the companies that practice them. Nevertheless, enough information is available that reasonably accurate overviews can be surmised. Manufacture of both homopolymer and copolymer involves critical monomer purification operations, discussion of which is outside the scope of this article (see FORMALDEHYDE). Homopolymer and copolymer are manufactured by substantially different processes for accomplishing substantially different polymerization chemistries.

Homopolymer.   Formaldehyde polymerizes by both anionic and cationic mechanisms. Strong acids are needed to initiate cationic polymerization. Anionic polymerization, which can be initiated by relatively weak bases (eg, pyridine), can be represented by the following equations:

Initiation

$$M^+B^- + H_2C{=}O \longrightarrow B{-}CH_2{-}O^-M^+$$

Propagation

$$B{-}CH_2{-}O^-M^+ + (n + 1)\, H_2C{=}O \longrightarrow B{-}CH_2{-}O(CH_2{-}O)_n{-}CH_2{-}O^-M^+$$

The exact structure of M^+ and B^- depends on solvent, the initiator used, and incidental impurities (eg, water).

The anionic polymerization of formaldehyde is free of most spontaneous termination reactions common to free-radical chain growth polymerization of vinyl monomers. Chain transfer to monomer is an intrinsic molecular-weight-regulating reaction and results in one formate and one methoxy end group. Molecular-weight-regulating agents, especially alcohols, are customarily used for optimum control of molecular weight. If an alcohol is used, an alkoxy end group results.

In production, anhydrous formaldehyde is continuously fed to a reactor containing well-agitated inert solvent, especially a hydrocarbon, in which monomer is sparingly soluble. Initiator, especially amine, and chain-transfer agent are also fed to the reactor (5,16,17). The reaction is quite exothermic and polymerization temperature is maintained below 75°C (typically near 40°C) by evaporation of the solvent. Polymer is not soluble in the solvent and precipitates early in the reaction.

Polyoxymethylene is obtained as a finely divided solid. The bulk density of the product, which is very important for ease of handling in subsequent manufacturing steps, is influenced by many reaction variables, including solvent type, polymerization temperature, and agitation.

Polymer is separated from the polymerization slurry and slurried with acetic anhydride and sodium acetate catalyst. Acetylation of polymer end groups is carried out in a series of stirred tank reactors at temperatures up to 140°C. End-capped polymer is separated by filtration and washed at least twice, once with acetone and then with water. Polymer is made ready for extrusion compounding and other finishing steps by drying in a steam-tube drier.

Copolymer.   Copolymerization of trioxane with cyclic ethers or formals is accomplished with cationic initiators. Boron trifluoride dibutyl etherate is used in one process. In this case, the actual initiating species is formed by reaction with water (18). Polymerization by ring opening of the 6-membered ring to form high molecular weight polymer does not commence immediately upon mixing monomer and initiator. Rather, an induction period is observed during which an equilibrium concentration of formaldehyde is produced (18,19).

$$I^+ + \overline{(O{-}CH_2)_3} \rightleftharpoons I{-}OCH_2{-}OCH_2{-}O{\doteq}\overset{+}{C}H_2 \rightleftharpoons I{-}OCH_2{-}O{\doteq}\overset{+}{C}H_2 + CH_2O$$

When the equilibrium formaldehyde concentration is reached, polymer begins to precipitate. Further polymerization takes place in trioxane solution and, more importantly, at the surface of precipitated polymer.

Comonomer is exhausted at relatively low conversion (20), but a random copolymer is nevertheless obtained. This is because a very facile transacetalization reaction allows for essentially random redistribution of the comonomer units (18) and also results in a polydispersity index near 2.0 (21).

Transfer reactions, analogous to those discussed for homopolymer, also occur during the copolymerization reaction and result in formate end groups.

The enthalpy of the copolymerization of trioxane is such that bulk polymerization is feasible. For production, molten trioxane, initiator, and comonomer are fed to the reactor; a chain-transfer agent is included if desired. Polymerization proceeds in bulk with precipitation of polymer and the reactor must supply enough shearing to continually break up the polymer bed, reduce particle size, and provide good heat transfer. The mixing requirements for the bulk polymerization of trioxane have been reviewed (22). Raw copolymer is obtained as fine crumb or flake containing imbibed formaldehyde and trioxane which are substantially removed in subsequent treatments which may be combined with removal of unstable end groups.

Acetal copolymer may be end-capped in a process completely analogous to that used for homopolymer. However, the presence of comonomer units (eg, $-O-CH_2-CH_2-O-$) in the backbone and the relative instability to base of hemiacetal end groups allow for another convenient route to polymer with stable end groups. The hemiacetal end groups may be subjected to base catalyzed (especially amine) hydrolysis in the melt (23,24) or in solution (25) or suspension (26) and the chain segments between the end group and the nearest comonomer unit deliberately depolymerized until the depropagating chain encounters the "zipper-jamming" comonomer unit. If ethylene oxide or dioxolane is used as comonomer, a stable hydroxyethyl ether end group results ($-O-CH_2CH_2-OH$). Some formate end groups, which are intermediate in thermal stability between hemiacetal and ether end groups, may also be removed by this process.

Product from melt or suspension treatment is obtained directly as crumb or powder. Polymer recovered from solution treatment is obtained by precipitative cooling or spray drying. Polymer with now stable end groups may be washed and dried to remove impurities, especially acids or their precursors, prior to finishing operations.

Processing and Fabrication

Finishing. All acetal resins contain various stabilizers introduced by the supplier in a finishing extrusion (compounding) step. The particular stabilizers used and the exact method of their incorporation are generally not revealed. Thermal oxidative and photooxidative stabilizers have already been mentioned. These must be carefully chosen and tested so that they do not aggravate more degradation (eg, by acidolysis) than they mitigate.

Traces of formaldehyde, present in neat end-capped polymer or produced by processing polymer under abusive conditions, detract from polymer stability. Commercial resins typically contain formaldehyde scavengers. Nitrogen compounds, especially amines and amides, epoxies, and polyhydroxy compounds, are particularly efficacious scavengers.

A variety of other additives may be incorporated during finishing extrusion to produce acetal resins especially formulated to enhance certain characteristics for specific applications.

Fabrication. Acetal resins are most commonly fabricated by injection molding. A homogeneous melt is essential for optimum appearance and for performance of injection molded parts. A screw compression ratio of no less than 3:1 is advised and the size of the injection molded shot should be 50 to 75% of the rated capacity (based on polystyrene) of the barrel.

Acetal resins may also be fabricated into rod, slab, and other shapes by profile extrusion. Extruded shapes are frequently further machined. Parts fabricated by molding or extrusion are ammenable to all typical postforming processes.

Reference 27 gives a concise overview of the processing and fabrication of acetal resins.

Scrap and Recycle. Acetal resins can be processed with very little waste. Sprues, runners, and out-of-tolerance parts can, in general, be ground and the resins reused. Up to about 25% of regrind can usually be safely recycled into virgin resin. However, the amount of regrind that can be used in a particular circumstance varies. The appropriate literature from the supplier should be consulted.

Acetal resins are one of several plastic materials specifically targeted for recycle in at least one soon-to-be-constructed recycling facility (28).

Resin Grades. Nonfilled and unmodified (except for stabilizers) grades of acetal resin are generally differentiated on the basis of melt index. Table 4 gives the nominal melt index of various grades of nonfilled Celcon acetal copolymer from Hoechst Celanese. Where possible, the grade of Delrin or Ultraform (registered trademark of the BASF Corporation) that has the most similar nominal melt index is listed. Similarity in MI does not a priori imply that particular resin grades are interchangeable. Standard grades are generally available in a wide variety of colors.

Table 4. Nonfilled Acetal Resins, General Purpose

Nominal MI,[a] g/10 min	Hoechst Celanese Celcon	Du Pont Delrin	BASF Ultraform
2.5	M25	100	
9.0[b]	M90	500	H2320
27.0	M270	900	H2330
45.0	M450	1700	

[a]MI = melt index according to ASTM.
[b]This is perhaps the most commonly encountered melt index.

Grades of acetal resins specifically designed to enhance the excellent lubricity of the material, without sacrifice in other properties, are known. These include Delrin AF, Ultraform N2311, and Celcon LW materials. Antistatic grades, particularly useful in electronic applications, have been developed. At least one recently introduced new grade, Celcon SR90, claims good scuff resistance.

Development of toughened acetals has recently flourished. One such grade, Delrin 100ST, boasts a notched-impact strength of 900 J/m (16.86 ft·lb/in.) at 23°C, substantially higher than that of nonmodified, general purpose grades (Table 2).

Glass-fiber-reinforced (increased stiffness and tensile strength) and mineral filled (reduced shrink and warp) grades also have been developed.

Economic Aspects

Supply and demand statistics for 1988 for all regions of the world as compiled by SRI International are given in Table 5. The world producers of acetal resins and their annual capacities are listed in Table 6 (29). Hoechst Celanese and Ultraform Corporation (a joint venture of Degussa and BASF) have announced capacity expansions in the United States to 77,000 t and 16,000 t, respectively; both were due in place in 1990. Part of general capacity expansion plans, announced by Du Pont for completion in 1991, are believed to apply to acetal resins.

Table 5. World Supply/Demand for Polyacetal Resins for 1988[a,b]

	United States	Western Europe	Japan	Other[c]	Total
capacity	125	116	127	21	*389*
production	118	100	110	11	*339*
imports	3	17	4	56–60	*80–84*
exports	36–41	16	24		*79–84*
apparent consumption	80–84	101	87	67–71	*335–343*

[a]Ref. 29. Courtesy of SRI International.
[b]As of October 1989 in thousands of tons.
[c]Includes Mexico, USSR, and Poland.

Specifications and Standards

ASTM D4181 calls out standard specifications for acetal molding and extrusion materials. Homopolymer and copolymer are treated separately. Within each class of resin, materials are graded according to melt flow rate. The International Standards Organization (ISO) is expected to issue a specification for acetal resins before 1992.

Many grades of acetal resins are listed in *Underwriters' Laboratories* (UL) *Recognized Component Directory*. UL assigns temperature index ratings indicating expected continuous-use retention of mechanical and electrical properties. UL also classifies materials on the basis of flammability characteristics; homopolymer and copolymer are both classified 94HB.

Many grades of acetal are accepted under the component acceptance program of the Canadian Standards Association (CSA) and are listed in the *CSA Plastics Directory*.

Table 6. World Producers of Polyacetal Resins[a]

Company and plant location	Annual capacity, 10^3 t
North America	
United States	
E.I. du Pont de Nemours & Company, Inc.	45
Hoechst Celanese Corporation	68
Ultraform Company[b]	12
Mexico	
Du Pont, S.A. de C.V.	1.4
Western Europe	
Germany, Federal Republic of	
Ticona Polymerwerke GmbH[c]	45
Ultraform GmbH	25
Netherlands	
Du Pont de Nemours (Nederland) BV	45
Asia	
Japan	
Asahi Chemical Industry Co., Ltd	30
Mitsubishi Gas Chemical Co., Ltd.	20
Polyplastics Co., Ltd.[d]	77
Korea, Republic of	
Korea Engineering Plastics[d]	10
Eastern Europe	
State Complexes[e]	>10
Total	*388*

[a]Ref. 29. Courtesy of SRI International as of October 1, 1989.
[b]Joint venture, BASF and Degussa
[c]Joint venture, Hoechst Celanese and Hoechst AG.
[d]Joint venture, Mitsubishi Gas Chemical and Tongyoung Nylon
[e]Plants are believed to be located in the USSR and Poland.

The U.S. Food and Drug Administration regulates acetal resins intended for repeated contact with food. The FDA regulation for homopolymer is *21CFR 177.2480* and that for copolymer is *21CFR 177.2470*. The U.S. Department of Agriculture regulates the use of acetal resins in contact with meat and poultry.

The National Sanitation Foundation publishes a list of acetal resins which they find acceptable for use in potable water applications.

Acetal resins are typically supplied in 25-kg multiwall bags and 500-kg Gaylords (rigid containers). Precompounded custom colors are available from manufacturers or the customer may use color concentrates.

Health and Safety

When processed and used according to manufacturer's recommendations, acetal resins present no extraordinary health risks. Before the use of any plastic material, including acetal, the *Material Safety Data Sheet* (MSDS) applicable to the grade in question should be consulted.

If acetal resins are processed at temperatures substantially above those recommended for the particular grade, minor amounts of formaldehyde may be liberated. Formaldehyde (qv) is a colorless, lacrimatory gas with a pungent odor and is intensely irritating to mucous membranes. The human nose is sensitive to concentrations in the range of 0.1 to 0.5 ppm. The current threshold limit value for formaldehyde is 1 ppm.

Uses

The mechanical strength, friction and wear characteristics, predictable long and short term properties, and generally excellent solvent resistance of acetal resins make them ideal candidates for industrial applications. Conveying devices and gears are excellent examples of this type of application. Information on the design of such devices is available from suppliers (30). Optimization of design and molding of gears from acetal resins has been discussed in the technical literature (31). Acetal resins compete effectively with nylons and thermoplastic polyester among other resins in many industrial applications.

Molding of parts for a wide variety of plumbing and irrigation applications consumes as much acetal resin as the industrial applications. Rod and slab stock can be machined into components for precision flow control devices.

Acetal resins are also used extensively in transportation, especially automotive. Handles and internal components (gears, gear racks, cables) for window lifts and other similar devices are examples. Most of the applications which do not involve painting or plating are below the window line. Many common consumer items are manufactured essentially entirely from acetal resin (eg, disposable lighters) or have critical components molded from acetal resin (eg, hubs and platforms for videocassettes). The properties that make acetal resins useful in industrial applications make them useful for internal components, especially mechanical drive systems, of many household appliances.

BIBLIOGRAPHY

"Acetal Resins" in *ECT* 2nd ed., Vol. 1, pp. 95–107, by C. E. Schweitzer, E. I. du Pont de Nemours & Co., Inc.; in *ECT* 3rd ed., Vol. 1, pp. 112–123, by K. J. Persak and L. M. Blair, E. I. du Pont de Nemours & Co., Inc.

1. A. M. Butlerov, *Ann.* **3,** 242 (1859).
2. H. Staudinger, *Die Hochmolekularen organische Verbindungen Kautschuk und Cellolose,* Springer-Verlag, Berlin, 1932.
3. I. Negulescu and O. Vogl, *J. Polym. Sci., Polym. Chem. Ed.* **14,** 2415 (1976).
4. O. Vogl, *J. Macromol. Sci. Chem.* 1(2), 203 (1967).
5. U.S. Pat. 2,768,994 (Oct. 1956), R. N. MacDonald (to E.I. du Pont de Nemours & Co., Inc.).
6. C. E. Schweitzer, R. N. MacDonald, and I. O. Pundersson, *J. Appl. Polym. Sci.* **1,** 158 (1959).
7. U.S. Pat. 3,027,352 (Mar. 1962), C. Walling, F. Brown, and K. Bartz (to Celanese Corp.).
8. U.S. Pat. 3,103,499 (Sept. 1963), T. J. Dolce and F. M. Berardinelli (to Celanese Corp.).

9. G. A. Carazzilo and M. Mammi, *J. Polym. Sci. Part A* **1**, 965 (1963).
10. M. L. Huggins, *J. Chem. Phys.* **3**, 37 (1945).
11. H. Tadokoro and co-workers, *J. Polym. Sci.* **44**, 266 (1960).
12. M. Droscher and G. Wegner, *Ind. Eng. Chem. Prod. Res. Dev.* **18**, 275 (1979).
13. J. Brandrup and E. H. Immergut, *Polymer Handbook,* 3rd ed., John Wiley & Sons, Inc. New York, 1989, pp. V-87–V-97.
14. T. Dolce and J. Grates, in J. I. Kroschwitz, ed., *Encyclopedia of Polymer Science and Engineering,* 2nd ed., Wiley-Interscience, New York, 1985, Vol. 1, p. 42.
15. D. J. Carlsson and D. M. Wiles, in J. I. Kroschwitz, ed., *Encyclopedia of Polymer Science and Engineering,* 2nd ed., Wiley-Interscience, New York, 1985, Vol. 4, p. 361.
16. U.S. Pat. 3,172,736 (Mar. 1965), R. E. Gee (to E.I. du Pont de Nemours & Co., Inc.).
17. R. N. MacDonald, *Macromolecular Synthesis,* Vol. 3, John Wiley & Sons, Inc., New York, 1968.
18. W. Kern and V. Jaacks, *J. Polym. Sci.* **48**, 399 (1960).
19. L. Leese and W. Bauber, *Polymer* **6**, 269 (1965).
20. G. L. Collins and co-workers, *J. Polym. Sci., Polym. Chem. Ed.* **19**, 1597 (1981).
21. K. Weissermel, E. Fischer, and K. Gutweiler, *Kunstoffe* **54**, 410 (1964).
22. D. B. Todd, *Polymn.–Plast. Technol. Eng.* **28**, 123 (1989).
23. U.S. Pat. 3,152,343 (Dec. 1963), T. J. Dolce and F. M. Berardinelli (to Celanese Corp.).
24. U.S. Pat. 4,301,273 (Nov. 1981), A. Sugio and co-workers (to Mitsubishi Gas Chemical Company, Ltd.).
25. U.S. Pat. 3,174,948 (Mar. 1965), J. E. Wall, E. T. Smith, and G. F. Fishcer (to Celanese Corp.).
26. U.S. Pat. 3,318,848 (May 1967), C. M. Clarke (to Celanese Corp.).
27. A. Serle in J. M. Margolis, ed., *Engineering Thermoplastics,* Marcel Dekker, New York, 1985, p. 151.
28. *J. of Comm.* 7 (Aug. 25, 1989).
29. K. Wheeler with W. Cox and Y. Sakuma, *Chemical Economics Handbook,* SRI International, Menlo Park, Calif., December 1989, 580.0930 A–G.
30. *Bulletins CS-03* and *PIE-04,* Hoechst Celanese Engineering Plastics Company, 1985.
31. H. Käufer and A. Burr, *Kunststoffe* **73**, 11 (1983).

General References

References 14 and 27 are good general references.
M. Sittig, *Polyacetal Resins,* Gulf Publishing Co., Houston, Tex., 1963.

JOHN B. STARR
Hoechst Celanese Corporation

ACETAMIDE.   See ACETIC ACID AND DERIVATIVES.

ACETANILIDE.   See AMINES—AROMATIC AMINES, ANILINE AND ITS
                                     DERIVATIVES.

ACETARSONE.   See ANTIPARASITIC AGENTS—ANTIPROTOZOALS; ARSENIC
                                   COMPOUNDS.

ACETATE AND TRIACETATE FIBERS.   See FIBERS, CELLULOSE
                                       ESTERS.

ACETIC ACID AND DERIVATIVES

ACETIC ACID

Acetic acid [*64-19-7*], CH_3COOH, is a corrosive organic acid having a sharp odor, burning taste, and pernicious blistering properties. It is found in ocean water, oilfield brines, rain, and at trace concentrations in many plant and animal liquids. It is central to all biological energy pathways. Fermentation of fruit and vegetable juices yields 2–12% acetic acid solutions, usually called vinegar (qv). Any sugar-containing sap or juice can be transformed by bacterial or fungal processes to dilute acetic acid.

Theophrastos (272–287 BC) studied the utilization of acetic acid to make white lead and verdigris [*52503-64-7*]. Acetic acid was also well-known to alchemists of the Renaissance. Andreas Libavius (AD 1540–1600) distinguished the properties of vinegar from those of icelike (glacial) acetic acid obtained by dry distillation of copper acetate or similar heavy metal acetates. Numerous attempts to prepare glacial acetic acid by distillation of vinegar proved to be in vain, however.

Lavoisier believed he could distinguish acetic acid from acetous acid, the hypothetical acid of vinegar, which he thought was converted into acetic acid by oxidation. Following Lavoisier's demise, Adet proved the essential identity of acetic acid and acetous acid, the latter being the monohydrate, and in 1847, Kolbe finally prepared acetic acid from the elements.

Most of the acetic acid is produced in the United States, Germany, Great Britain, Japan, France, Canada, and Mexico. Total annual production in these countries is close to four million tons. Uses include the manufacture of vinyl acetate [*108-05-4*] and acetic anhydride [*108-24-7*]. Vinyl acetate is used to make latex emulsion resins for paints, adhesives, paper coatings, and textile finishing agents. Acetic anhydride is used in making cellulose acetate fibers, cigarette filter tow, and cellulosic plastics.

Physical Properties

Acetic acid, fp 16.635°C (1), bp 117.87°C at 101.3 kPa (2), is a clear, colorless liquid. Water is the chief impurity in acetic acid although other materials such as acetaldehyde, acetic anhydride, formic acid, biacetyl, methyl acetate, ethyl acetoacetate, iron, and mercury are also sometimes found. Water significantly lowers the freezing point of glacial acetic acid as do acetic anhydride and methyl acetate (3). The presence of acetaldehyde [*75-07-0*] or formic acid [*64-18-6*] is commonly revealed by permanganate tests; biacetyl [431-03-8] and iron are indicated by color. Ethyl acetoacetate [141-97-9] may cause slight color in acetic acid and is often mistaken for formic acid because it reduces mercuric chloride to calomel. Traces of mercury provoke catastrophic corrosion of aluminum metal, often employed in shipping the acid.

The vapor density of acetic acid suggests a molecular weight much higher than the formula weight, 60.06. Indeed, the acid normally exists as a dimer (4), both in the vapor phase (5) and in solution (6). This vapor density anomaly has important consequences in engineering computations, particularly in distillations.

Acetic acid containing less than 1% water is called glacial. It is hygroscopic and the freezing point is a convenient way to determine purity (7). Water is nearly always present in far greater quantities than any other impurity. Table 1 shows the freezing points for acetic acid–water mixtures.

Table 1. Acetic Acid–Water Freezing Points

Acetic acid, wt %	Freezing point, °C
100	16.635
99.95	16.50
99.70	16.06
99.60	15.84
99.2	15.12
98.8	14.49
98.4	13.86
98.0	13.25
97.6	12.66
97.2	12.09
96.8	11.48
96.4	10.83
96.0	10.17

The Antoine equation for acetic acid has recently been revised (2)

$$\ln (P) = 15.19234 + (-3654.622)/T + (-45.392)$$

The pressure P is measured in kPa and the temperature T in K. The vapor pressure of pure acetic acid is tabulated in Table 2. Precise liquid density measurements are significant for determining the mass of tank car quantities of acid. Liquid density data (8) as a function of temperature are given in Table 3.

Acetic acid forms a monohydrate containing about 23% water; thus the density of acetic acid–water mixtures goes through a maximum between 77–80 wt % acid at 15°C. When water is mixed with acetic acid at 15–18°C, heat is given off. At greater acetic acid concentrations, heat is taken up. The measured heat of mixing is consistent with dimer formation in the pure acid. The monohydrate, sometimes called acetous acid, was formerly the main article of commerce. Data on solidification points of aqueous acetic acid mixtures have been tabulated, and the eutectic formation mapped (9). The aqueous eutectic temperature is about

Table 2. Acetic Acid Vapor Pressure

Temperature, °C	Pressure, kPa[a]	Temperature, °C	Pressure, kPa[a]
0	4.7	110	776.7
10	8.5	118.2	1013
20	15.7	130	1386.5
30	26.5	140	1841.1
40	45.3	150	2461.1
50	74.9	160	3160
60	117.7	170	4041
70	182.8	180	5091
80	269.4	190	6333
90	390.4	200	7813
100	555.3	210	9612

[a]To convert kPa to psi, multiply by 0.145.

Table 3. Density of Acetic Acid (Liquid)

Temperature, °C	Density, kg/m^3
20	1049.55
25	1043.92
30	1038.25
47	1019.19
67	996.46
87	973.42
107	949.90
127	925.60
147	900.27
167	873.56
187	845.04
197	829.88
207	814.07
217	797.44

−26°C. A procedure for concentrating acetic acid by freezing, hampered by eutectic formation, has been sought for some time. The eutectic can be decomposed through adding a substance to form a compound with acetic acid, eg, urea or potassium acetate. Glacial acetic acid can then be distilled. The densities of acetic acid–water mixtures at 15°C are given in Table 4.

A summary of the physical properties of glacial acetic acid is given in Table 5.

Chemical Properties

Decomposition Reactions. Minute traces of acetic anhydride are formed when very dry acetic acid is distilled. Without a catalyst, equilibrium is reached after about 7 h of boiling, but a trace of acid catalyst produces equilibrium in 20 min. At equilibrium, about 4.2 mmol of anhydride is present per liter of acetic

Table 4. Density of Aqueous Acetic Acid

Acetic acid, wt %	Density, g/cm^3
1	1.007
5	1.0067
10	1.0142
15	1.0214
20	1.0284
30	1.0412
40	1.0523
50	1.0615
60	1.0685
70	1.0733
80	1.0748
90	1.0713
95	1.0660
100	1.0550

Table 5. Properties of Glacial Acetic Acid

Property	Value	Reference
freezing point, °C	16.635	1
boiling point, °C	117.87	4
density, g/mL at 20°C	1.0495	8
refractive index, n_D^{25}	1.36965	10
heat of vaporization ΔH_v, J/g^a at bp	394.5	11
specific heat (vapor), J/(g·K)^a at 124°C	5.029	11
critical temperature, K	592.71	2
critical pressure, MPa^b	4.53	2
enthalpy of formation, kJ/mol^a at 25°C		
liquid	−484.50	12
gas	−432.25	12
normal entropy, J/(mol·K)^a at 25°C		
liquid	159.8	13
gas	282.5	13
liquid viscosity, mPa (= cP)		
20°C	11.83	14
40°C	8.18	14
surface tension, mN/m (= dyn/cm) at 20.1°C	27.57	15
flammability limits, vol % in air	4.0 to 16.0	15
autoignition temperature, °C	465	
flash point, °C		16
closed cup	43	
open cup	57	

^{a}To convert J to cal, divide by 4.184.
^{b}To convert MPa to psi, multiply by 145.

acid, even at temperatures as low as 80°C (17). Thermolysis of acetic acid occurs at 442°C and 101.3 kPa (1 atm), leading by parallel pathways to methane [72-82-8] and carbon dioxide [124-38-9], and to ketene [463-57-4] and water (18). Both reactions have great industrial significance.

Single pulse, shock tube decomposition of acetic acid in argon involves the same pair of homogeneous, molecular first-order reactions as thermolysis (19). Platinum on graphite catalyzes the decomposition at 500–800 K at low pressures (20). Ketene, methane, carbon oxides, and a variety of minor products are obtained. Photochemical decomposition yields methane and carbon dioxide and a number of free radicals, which have complicated pathways (21). Electron impact and gamma rays appear to generate these same products (22). Electron cyclotron resonance plasma made from acetic acid deposits a diamond [7782-40-3] film on suitable surfaces (23). The film, having a polycrystalline structure, is a useful electrical insulator (24) and widespread industrial exploitation of diamond films appears to be on the horizon (25).

Acid–Base Chemistry. Acetic acid dissociates in water, $pK_a = 4.76$ at 25°C. It is a mild acid which can be used for analysis of bases too weak to detect in water (26). It readily neutralizes the ordinary hydroxides of the alkali metals and the alkaline earths to form the corresponding acetates. When the crude material pyroligneous acid is neutralized with limestone or magnesia the commercial acetate of lime or acetate of magnesia is obtained (7). Acetic acid accepts protons only from the strongest acids such as nitric acid and sulfuric acid. Other acids exhibit very powerful, superacid properties in acetic acid solutions and are thus useful catalysts for esterifications of olefins and alcohols (27). Nitrations conducted in acetic acid solvent are effected because of the formation of the nitronium ion, NO_2^+. Hexamethylenetetramine [100-97-0] may be nitrated in acetic acid solvent to yield the explosive cyclotrimethylenetrinitramine [121-82-4], also known as cyclonite or RDX.

Acetylation Reactions. Alcohols may be acetylated without catalysts by using a large excess of acetic acid.

$$CH_3COOH + ROH \longrightarrow CH_3COOR + H_2O$$

The reaction rate is increased by using an entraining agent such as hexane, benzene, toluene, or cyclohexane, depending on the reactant alcohol, to remove the water formed. The concentration of water in the reaction medium can be measured, either by means of the Karl-Fischer reagent, or automatically by specific conductance and used as a control of the rate. The specific electrical conductance of acetic acid containing small amounts of water is given in Table 6.

Nearly all commercial acetylations are realized using acid catalysts. Catalytic acetylation of alcohols can be carried out using mineral acids, eg, perchloric acid [7601-90-3], phosphoric acid [7664-38-2], sulfuric acid [7664-93-9], benzenesulfonic acid [98-11-3] , or methanesulfonic acid [75-75-2], as the catalyst. Certain acid-reacting ion-exchange resins may also be used, but these tend to decompose in hot acetic acid. Mordenite [12445-20-4], a decationized Y-zeolite, is a useful acetylation catalyst (28) and aluminum chloride [7446-70-0], Al_2Cl_6, catalyzes n-butanol [71-36-3] acetylation (29).

Olefins add anhydrous acetic acid to give esters, usually of secondary or tertiary alcohols: propylene [115-07-1] yields isopropyl acetate [108-21-4]; isobutylene [115-11-7] gives tert-butyl acetate [540-88-5]. Minute amounts of water inhibit the reaction. Unsaturated esters can be prepared by a combined oxidative esterification over a platinum group metal catalyst. For example, ethylene–air–acetic acid passed over a palladium–lithium acetate catalyst yields vinyl acetate.

Table 6. Specific Conductance of Aqueous Acetic Acid

Acetic acid, wt %	Specific conductance κ, S/cm $\times$ 10^7
100	0.060
99.9515	0.065
99.746	0.103
99.320	0.261
98.84	0.531
97.66	2.19
96.68	5.45
94.82	20.1
92.50	59.9
90.75	111
82.30	688

Acetylation of acetaldehyde to ethylidene diacetate [542-10-9], a precursor of vinyl acetate, has long been known (7), but the condensation of formaldehyde [50-00-0] and acetic acid vapors to furnish acrylic acid [79-10-7] is more recent (30). These reactions consume relatively more energy than other routes for manufacturing vinyl acetate or acrylic acid, and thus are not likely to be further developed. Vapor-phase methanol–methyl acetate oxidation using simultaneous condensation to yield methyl acrylate is still being developed (28). A vanadium–titania phosphate catalyst is employed in that process.

Industrial Manufacture

Commercial production of acetic acid has been revolutionized in the decade 1978–1988. Butane–naphtha liquid-phase catalytic oxidation has declined precipitously as methanol [67-56-1] or methyl acetate [79-20-9] carbonylation has become the technology of choice in the world market. By-product acetic acid recovery in other hydrocarbon oxidations, eg, in xylene oxidation to terephthalic acid and propylene conversion to acrylic acid, has also grown. Production from synthesis gas is increasing and the development of alternative raw materials is under serious consideration following widespread dislocations in the cost of raw material (see CHEMURGY).

Ethanol fermentation is still used in vinegar production. Research on fermentative routes to glacial acetic acid is also being pursued. Thermophilic, anaerobic microbial fermentations of carbohydrates can be realized at high rates, if practical schemes can be developed for removing acetic acid as fast as it is formed. Under usual conditions, about 5% acid brings the anaerobic reactions to a halt, but continuous separation produces high yields at high production rates. Heat for the reaction is provided by the metabolic activity of the microorganisms. Fermentative condensation of CO_2 is another possible route to acetic acid.

Currently, almost all acetic acid produced commercially comes from acetaldehyde oxidation, methanol or methyl acetate carbonylation, or light hydrocarbon liquid-phase oxidation. Comparatively small amounts are generated by butane liquid-phase oxidation, direct ethanol oxidation, and synthesis gas. Large amounts of acetic acid are recycled industrially in the production of cellulose acetate, poly(vinyl alcohol), and aspirin and in a broad array of other proprietary processes. (These recycling processes are not regarded as production and are not discussed herein.)

Acetaldehyde Oxidation.   Ethanol [64-17-5] is easily dehydrogenated oxidatively to acetaldehyde (qv) using silver, brass, or bronze catalysts. Acetaldehyde can then be oxidized in the liquid phase in the presence of cobalt or manganese salts to yield acetic acid. Peracetic acid [79-21-0] formation is prevented by the transition metal catalysts (7). (Most transition metal salts decompose any peroxides that form, but manganese is uniquely effective.) Kinetic system models are useful for visualizing the industrial operation (31,32). Stirred-tank and sparger reactor rates have been compared for this reaction and both are so high that they are negligible in the reaction's mathematical description.

Figure 1 is a typical flow sheet for acetaldehyde oxidation. The reactor is an upright vessel, fitted with baffles to redistribute and redirect the air bubbles. Oxygen is fully depleted by the time a bubble reaches the first baffle and bubbles above the first baffle serve mainly for liquid agitation. Such mechanical contacting decomposes transitory intermediates and stabilizes the reactor solution. Even though the oxidizer-reactor operates under mild pressure, sufficient aldehyde boils away to require an off-gas scrubber. Oxidate is passed into a column operated under a positive nitrogen pressure, thence to an acetaldehyde recovery column where unreacted aldehyde is recycled. More importantly, many dangerous peroxides are decomposed in this column, some into acetic acid, while traces of ethanol are esterified to ethyl acetate. Crude acid is taken off at the bottom and led to a column for stripping off the low boiling constituents other than aldehyde.

Crude oxidate is passed to a still where any remaining unreacted acetaldehyde and low-boiling by-products, eg, methyl acetate and acetone [67-64-1], are removed as are CO, CO_2, and N_2. High concentration aqueous acetic acid is obtained. The main impurities are ethyl acetate, formaldehyde, and formic acid although sometimes traces of a powerful oxidizing agent, possibly diacetyl peroxide, are present. If the acetaldehyde contains ethylene oxide, then ethylene glycol diacetate is present as an impurity. Formic acid can be entrained using hexane or heptane, ethyl acetate, or a similar azeotroping agent. Often the total contaminant mass is low enough to permit destruction by chemical oxidation. The oxidizing agent, such as sodium dichromate, is fed down the finishing column as a concentrated solution. Potassium permanganate solution is also effective, but it often clogs the plates of the distillation tower.

Final purification is effected by distillation giving high purity acid. Some designs add ethyl acetate to entrain water and formic acid overhead in the finishing column. The acid product is removed as a sidestream. Potassium permanganate has been employed to oxidize formaldehyde and formic acid because the finished acid must pass a permanganate test. The quantity of water in the chemical oxidizer solution is important for regulating the corrosion rate of the finishing column: acetic acid having a purity of 99.90–100% corrodes stainless steel SS–316 or SS–320. Lowering the acid concentration to 99.75–99.80% with distilled water in the permanganate solution diminishes the corrosion rate dra-

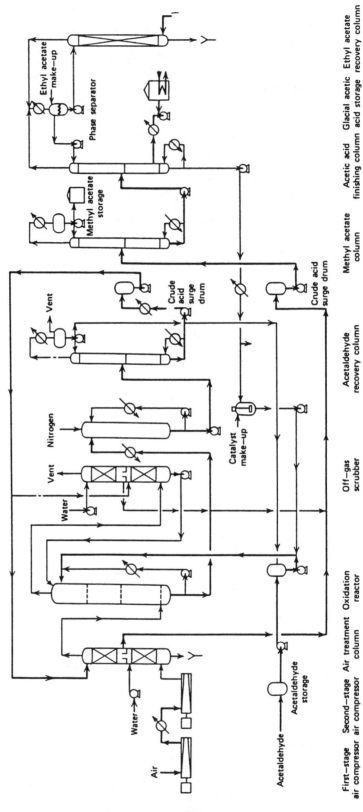

First—stage   Second—stage   Air treatment   Oxidation   Off-gas   Acetaldehyde   Methyl acetate   Acetic acid   Glacial acetic   Ethyl acetate
air compressor   air compressor   column   reactor   scrubber   recovery column   column   finishing column   acid storage   recovery column

Fig. 1. A typical acetaldehyde oxidation flow sheet.

128

matically. Residues containing manganese acetate or chromium acetate are washed with a two-phase mixture of water–butyl acetate or water–toluene. The organic solvent removes high boiling materials, tars and residual acid, and the metallic acetates remain in the water layer (33).

Alternative purification treatments for acetic acid from acetaldehyde have been explored but have no industrial application. Nitric acid or sodium nitrate causes the oxidation of formic acid and formaldehyde, but provokes serious corrosion problems. Schemes have been devised to reduce rather than oxidize the impurities; for example, injecting a current of hydrogen and passing the acid over a metallic catalyst such as nickel or copper turnings. Since reduction occurs at $110–120°C$, the reaction can be run in the final column. The risk of an explosion from the hydrogen passing through the column and venting at the top probably discourages the use of this treatment. Certain simple salts, eg, $FeSO_4$ or $MnSO_4$, may be introduced in the same way as the permanganate or dichromate discussed earlier. These serve to eliminate most of the quality-damaging impurities.

Conversion of acetaldehyde is typically more than 90% and the selectivity to acetic acid is higher than 95%. Stainless steel must be used in constructing the plant. This is an established process and most of the engineering is well-understood. The problems that exist are related to more extensively automating control of the system, notably at start-up and shutdown, although even these matters have been largely solved. This route is the most reliable of acetic acid processes.

Methanol Carbonylation. Several processes were patented in the 1920s for adding carbon monoxide to methanol to produce acetic acid (34). The earliest reaction systems used phosphoric acid at $300–400°C$ under high CO pressures. Copper phosphate, hydrated tungstic oxide, iodides, and other materials were tried as catalysts or promoters. Nickel iodide proved to be particularly valuable. At that time only gold and graphite were recognized as adequate to resist temperatures of $300–320°C$ and pressures of 20 MPa (2900 psi). In 1945–1946, when German work was disclosed by capture of the Central Research Files at Badische Anilin, a virtually complete plant design became public. Although this high pressure methanol carbonylation system suffered many of the difficulties experienced in earlier processes, eg, loss of iodine, corrosive conditions, and dangerously high pressures, new alloys such as Hastelloy C permitted the containment of nearly all the practical problems. Experimental and pilot-plant units were operated successfully and by 1963 BASF opened a large plant at Ludwigshafen, Germany providing license to Borden Chemical Company for a similar unit in Louisiana in 1966.

In 1968 a new methanol carbonylation process using rhodium promoted with iodide as catalyst was introduced by a modest letter (35). This catalyst possessed remarkable activity and selectivity for conversion to acetic acid. Nearly quantitative yields based on methanol were obtained at atmospheric pressure and a plant was built and operated in 1970 at Texas City, Tex. The effect on the world market has been exceptional (36).

Low pressure methanol carbonylation transformed the market because of lower cost raw materials, gentler, lower cost operating conditions, and higher yields. Reaction temperatures are $150–200°C$ and the reaction is conducted at 3.3–6.6 MPa (33–65 atm). The chief efficiency loss is conversion of carbon monoxide to CO_2 and H_2 through a water-gas shift as shown.

$$CO + H_2O \longrightarrow CO_2 + H_2$$

The subject has been reviewed (37,38). Water may be added to the feed to suppress methyl acetate formation, but is probably not when operating on an industrial scale. Water increases methanol conversion, but it is involved in the unavoidable loss of carbon monoxide. A typical methanol carbonylation flow sheet is given in Figure 2.

Low boiling substances are removed from the chilled reactor product by distilling up to a cut point of 80°C. These low boilers are gaseous dimethyl ether, methyl acetate, acetaldehyde, butyraldehyde, and ethyl acetate. The bottoms are flash-distilled to recover the rhodium catalyst. Flash distilled acid is azeotropically dehydrated. In the final distillation, glacial acid is obtained. Traces of iodine that may remain in the finished acid may be removed by fractional crystallization or by addition of a trace of methanol followed by distillation of the methyl iodide that forms. Somewhere in the carbonylation reaction, a minute amount of propionic acid seems to be made. It typically is found in the residues of the acetic acid finishing system and can be removed by purging the finishing column bottoms.

Vapor-phase methanol carbonylation over a supported metal catalyst has been described (39,40). Methanol itself is obtained from synthesis gas, so the possibility of making acetic acid directly without isolating methanol has been explored (41) in both vapor and liquid phases. Alcohols are generated from CO and H_2 using halide promoted ruthenium, but acetic acid can be produced by addition of $Ru_3(CO)_{12}$ (42). A complex metallic catalyst containing rhodium, manganese, iridium, and lithium supported on silica has been used to provide selective synthesis (43). Ruthenium melt catalyst has been patented (44). Catalysts can be improved by running them in by stages to the optimum operating temperature over prolonged time periods, eg, 100–1000 h (45). A rhodium–nickel–silver catalyst for this reaction has been developed (46).

Synthesis gas is obtained either from methane reforming or from coal gasification (see COAL CONVERSION PROCESSES). Telescoping the methanol carbonylation into an esterification scheme furnishes methyl acetate directly. Thermal decomposition of methyl acetate yields carbon and acetic anhydride,

$$2 CH_3COCH_3 \longrightarrow (CH_3CO)_2O + C(soot)$$

but a pyrolytic route is not attractive because of excessive energy consumption. Methyl acetate carbonylation yields both anhydride and acetic acid, controllable in part by the conditions. A plant based on this process was put in operation in October 1983 (47).

Butane–Naphtha Catalytic Liquid-Phase Oxidation. Direct liquid-phase oxidation of butane and/or naphtha [8030-30-6] was once the most favored worldwide route to acetic acid because of the low cost of these hydrocarbons. Butane [106-97-8], in the presence of metallic ions, eg, cobalt, chromium, or manganese, undergoes simple air oxidation in acetic acid solvent (48). The peroxidic intermediates are decomposed by high temperature, by mechanical agitation, and by action of the metallic catalysts, to form acetic acid and a comparatively small suite of other compounds (49). Ethyl acetate and butanone are produced, and the process can be altered to provide larger quantities of these valuable materials. Ethanol is thought to be an important intermediate (50); acetone forms through a

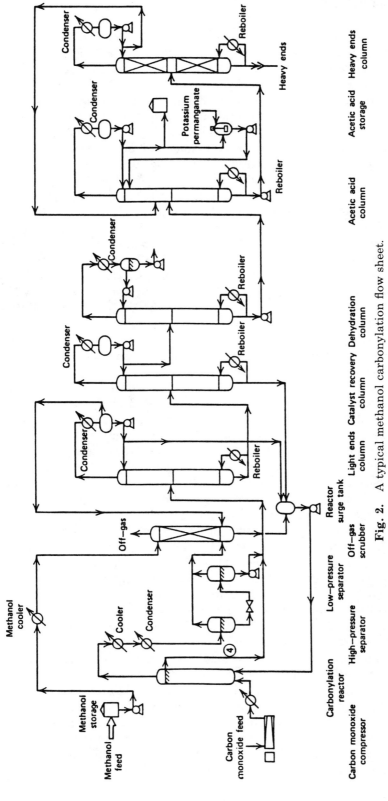

Fig. 2. A typical methanol carbonylation flow sheet.

131

minor pathway from isobutane present in the hydrocarbon feed. Formic acid, propionic acid, and minor quantities of butyric acid are also formed.

The theoretical explanation of the butane reaction mechanism is as fully developed as is that of acetaldehyde oxidation (51). The theory of the naphtha oxidation reaction is more troublesome, however, and less well understood. This is largely because of a back-biting reaction which leads to cyclic products (52).

Liquid-phase butane oxidation is realized in a sparged column, fabricated of high alloy stainless steel. Cobalt, chromium, and manganese acetate catalyst is dissolved in acetic acid and introduced with the butane–acetic acid solution. Air or O_2-enriched air may be used. The temperature is kept just below the critical temperature of butane, 152°C. Pressure is about 5.6 MPa (812 psi) (53). The reactor product is cooled and the pressure slowly lowered. In stripping the low boiling constituents away from the reactor product, the first obtained is unreacted butane, which is often led through an expansion turbine that powers air compressors for the reactor and cools the product. The butane is then recycled to the reactor. After cooling, the reactor oxidate appears as two phases: a hydrocarbon-rich phase and a denser aqueous phase. The former is decanted and led back into the reactor; the latter is distilled to obtain the boiling oxygenates. Butanone [78-93-3] and ethyl acetate [141-78-6] are the chief constituents of the aqueous layer, but there are traces of methyl vinyl ketone (an unpleasant lacrimator), aldehydes, and esters. Formic acid, also present, forms a maximum-boiling azeotrope that boils higher than either of the chief constituents.

Although acetic acid and water are not believed to form an azeotrope, acetic acid is hard to separate from aqueous mixtures. Because a number of common hydrocarbons such as heptane or isooctane form azeotropes with formic acid, one of these hydrocarbons can be added to the reactor oxidate permitting separation of formic acid. Water is decanted in a separator from the condensate. Much greater quantities of formic acid are produced from naphtha than from butane, hence formic acid recovery is more extensive in such plants. Through judicious recycling of the less desirable oxygenates, nearly all major impurities can be oxidized to acetic acid. Final acetic acid purification follows much the same treatments as are used in acetaldehyde oxidation. Acid quality equivalent to the best analytical grade can be produced in tank car quantities without difficulties.

Two explosions, on November 14, 1987, at the largest butane liquid-phase oxidation plant resulted in 3 deaths and 37 people injured (54). The plant, which had operated since December 1952 free of such disasters, was rebuilt (55).

Prospective Processes. There has been much effort invested in examining routes to acetic acid by olefin oxidation or from ethylene, butenes, or sec-butyl acetate. No product from these sources is known to have reached the world market; the cost of the raw materials is generally prohibitive.

Recovery of acid from wood distillate has long been viewed as a desirable prospect. Nearly all common lumber woods can be destructively distilled to furnish pyroligneous acid containing about 5–8% acetic acid. Coupled with azeotropic or extractive distillation, good quality glacial acetic acid could be prepared (56,57). Indeed, glacial acetic acid used to be prepared from gray acetate of lime, formed by the reaction of pyroligneous acid and limestone, by reaction of the acetate with bicarbonate followed by sulfuric acid:

$$Ca(CH_3COO)_2 + Na_2CO_3 \longrightarrow 2\ Na(CH_3COO) + CaCO_3$$

$$NaOOCCH_3 + H_2SO_4 \longrightarrow NaHSO_4 + CH_3COOH$$

The acetic acid was distilled to give soda acetic acid and upon further purification, it was often used for food or pharmaceutical applications.

Mixtures of trioctylamine and 2-ethylhexanol have been employed to extract 1–9% by volume acetic acid from its aqueous solutions. Reverse osmosis for acid separation has been patented and solvent membranes for concentrating acetic acid have been described (58,59). Decalin and trioctylphosphine were selected as solvents (60). Liquid–liquid interfacial kinetics is an especially significant factor in such extractions (61).

The fermentative fixing of CO_2 and water to acetic acid by a species of acetobacterium has been patented; acetyl coenzyme A is the primary reduction product (62). Different species of clostridia have also been used. Pseudomonads (63) have been patented for the fermentation of certain C_1 compounds and their derivatives, eg, methyl formate. These methods have been reviewed (64). The manufacture of acetic acid from CO_2 and its dewatering and refining to glacial acid has been discussed (65,66).

The autotropic pathway for acetate synthesis among the acetogenic bacteria has been examined (67). Quantitative fermentation of one mole of glucose [50-99-7], $C_6H_{12}O_6$, yields three moles of acetic acid, while two moles of xylose [58-86-6], $C_5H_{10}O_5$, yields five moles. The glucose reaction is

$$C_6H_{12}O_6 + 2 H_2O \longrightarrow 2 CH_3COOH + 2 CO_2 + 8 H^+ + 8 e^-$$

$$2 CO_2 + 8 H^+ + 8 e^- \longrightarrow CH_3COOH + 2 H_2O$$

Simply by passing gaseous H_2–CO_2 through an aqueous sugar mixture, the carbon dioxide is fixed into acetic acid:

$$2 CO_2 + 4 H_2 \longrightarrow CH_3COOH + 2 H_2O$$

Using carbon monoxide, the reaction becomes

$$4 CO + 2 H_2O \longrightarrow CH_3COOH + 2 CO_2$$

A number of C_1 compounds act as surrogates for reduction.

The possibility of using fermentation to generate a safe, noncorrosive road deicing composition has been studied (68). Calcium magnesium acetate [76123-46-1] is readily prepared from low concentration acetic acid produced from glucose or other inexpensive sugars. The U.S. Federal Highway Administration has financed development of anaerobic, thermophilic bacteria in an industrial process to manufacture calcium magnesium acetate from hydrolyzed corn starch and dolime [50933-69-2]. Economic evaluation shows the product can be made for about $0.11/kg, marginally competitive with acid from methanol carbonylation (69). It is equally possible to utilize methanol–CO_2 or H_2–CO_2 mixtures. These fermentative processes lead to dilute acid, often no more than 5 wt %, so that a concentration procedure is essential to recovery. Another new method for acetic acid production is electrodialytic conversion of dilute sodium acetate into concentrated acid (70). Electrodialytic fermentation systems have been subjected to computerized control and the energy cost for recovery is rapidly being diminished.

Economic Aspects

Acetic acid has a place in organic processes comparable to sulfuric acid in the mineral chemical industries and its movements mirror the industry. Growth of synthetic acetic acid production in the United States was greatly affected by the dislocations in fuel resources of the 1970s. The growth rate for 1988 was 1.5%.

About half of the world production comes from methanol carbonylation and about one-third from acetaldehyde oxidation. Another tenth of the world capacity can be attributed to butane–naphtha liquid-phase oxidation. Appreciable quantities of acetic acid are recovered from reactions involving peracetic acid. Precise statistics on acetic acid production are complicated by recycling of acid from cellulose acetate and poly(vinyl alcohol) production. Acetic acid that is by-product from peracetic acid [79-21-0] is normally designated as virgin acid, yet acid from hydrolysis of cellulose acetate or poly(vinyl acetate) is designated recycle acid. Indeterminate quantities of acetic acid are coproduced with acetic anhydride from coal-based carbon monoxide and unknown amounts are bartered or exchanged between corporations as a device to lessen transport costs.

Production Routes. The capacity for manufacturing acetic acid through methanol carbonylation grew from almost nothing in 1969 to 80% of actual plant operational capacity in 1988 for the United States. Almost all new plants use the low pressure carbonylation route. In 1989, Eastman Chemical's plant at Kingsport, Tenn. had the ability to coproduce acetic acid in excess of 68 t/yr. The plant produces both acid and anhydride from coal by carbonylation of methyl acetate (51). American production of acetic acid from butane liquid-phase oxidation at the Hoechst-Celanese plant at Pampa, Tex., is at a nominal 1989 capacity of 250 kg/yr. Favorable prices for the butane raw material and high efficiency of the plant design favor this production method. Acetaldehyde oxidation is extensively employed in Europe where acetaldehyde is produced from the palladium–copper-catalyzed oxidation of ethylene. Process yields for acetic acid on the order of 90% may be obtained. Sophisticated engineering and design gives the acetaldehyde route much of its competitiveness.

By-product acetic acid is obtained chiefly from partial hydrolysis of cellulose acetate [9004-35-7]. Lesser amounts are obtained through the reaction of acetic anhydride and cellulose. Acetylation of salicylic acid [69-72-7] produces one mole of acetic acid per mole of product and the oxidation of allyl alcohol using peracetic acid to yield glycerol furnishes by-product acid, but the net yield is low.

Transportation

Acetic acid, providing the concentration is greater than about 99%, may be stored and shipped in aluminum. Aluminum slowly corrodes forming a layer of basic aluminum acetate that prevents further corrosion. Some of this basic oxide coating is suspended in the acid and the heels of tank cars often have a white or gray, cloudy appearance. Water increases the corrosion rate significantly; hence every effort must be devoted to maintaining a high acid concentration. Mercury in minute quantities catalyzes corrosion of aluminum by acetic acid so that a single broken thermometer can provoke catastrophic and dangerous corrosion in 99.6–99.7% glacial acid. Mercurial thermometers are often absolutely prohibited near aluminium tank cars or barrels. Acid can also be stored and shipped in stainless steel, glass carboys, and polyethylene drums.

Because glacial acetic acid freezes at about 16°C, exceptional care must be taken for melting the product in cool weather. Electrical or steam heaters may be employed. Tank cars or tank wagons must be fitted with heating coils, which can be attached to a steam line and trap. Tank vents must be traced with electrical or steam lines to prevent crystallization. Acetic acid sublimes so that a single, large crystal can appear and completely fill an otherwise adequate vent.

According to the U.S. DOT regulations, acetic acid is a corrosive material (71). It may be shipped in metal or plastic packaging when no more than 0.45 kg is involved. Greater quantities may be shipped in boxed glass carboys, kegs or plywood drums, wooden barrels lined with asphalt or paraffin, earthenware containers in protective boxes, or plastic drums. It may not be shipped in plastic bags. Steel drums having polyethylene liners or polyethylene drums having steel overpacks are acceptable. Polyethylene drums do not appear to cause trace contamination after 12 months storage. Nonreturnable containers ought to be emptied and rinsed with fresh water. No other chemical ought to be shipped or stored in acetic acid containers.

Tank wagons are used for acetic acid deliveries in amounts intermediate between drum and tank car shipments, or to destinations not served by railways. Tank wagons are usually fabricated of stainless steel or aluminum alloy. Most shipments are carried in railway tank cars with nominal capacities of 38–76 m^3 (10,000–20,000 gal). For bookkeeping, capacities are determined by weight instead of volume measurements. Some tank cars are unloaded by suction, others by applying a positive nitrogen pressure and then siphoning the acetic acid into an eduction tube. Siphoning is frequently required by local laws.

Acetic acid is also transported in barges, sometimes in amounts of 1500 to 1750 tons. Acetic acid is not as hygroscopic as some other anhydrous organic substances, but barge shipments occasionally have specification problems because of wave splashing into the tanks or other careless handling.

Specifications and Analysis

Most specifications and analytical methods have been given (72). Most of the standards have remained unchanged for the past half-century. They were designed for acid recovered from wood tar condensates. All acid of commerce easily passes these tests.

Acetic acid made by methanol carbonylation sometimes has traces of iodine or bromine if the acid comes from the high pressure route. Qualitatively, these may be quickly detected by the Beilstein test for halogens: a copper wire is heated in a gas burner until no color can be seen and the coil plunged into the acetic acid, then brought into the gas flame again. Any trace of green or blue-green flame shows the presence of halogen. The lower identification threshold is about 0.7 ppm for chloride, about 0.65 ppm for bromide, and about 0.55 for iodide.

Super-pure acid is often specified by performance tests. Acetic acid to be used for Wijs Reagent must be very highly purified, otherwise the reagent deteriorates quickly. The dichromate–sulfuric acid test, made by dissolving potassium dichromate in concentrated H_2SO_4, and then mixing with an equal volume of acetic acid, is a sensitive test for minute quantities of certain oxidizing agents. These substances can be removed only by such special treatments as refluxing the acid with dichromate or permanganate, followed by redistillation. In commercial practice, instrumental methods are used to monitor quality but these methods are seldom given officially as standards. Gas chromatographic and mass spectrometric methods are capable of very high sensitivity.

Glacial acetic acid is considered to be 99.50 wt % or higher. A different grade has a minimum concentration of 99.70 wt %. Specialty users require water solutions of 86% and 36%. Such grades are prepared on special order. Only minor quantities of these grades are marketed, and their use is vanishing.

Health and Safety

Acetic acid has a sharp odor and the glacial acid has a fiery taste and will penetrate unbroken skin to make blisters. Prolonged exposure to air containing 5–10 mg/m^3 does not seem to be seriously harmful, but there are pronounced, undesirable effects from constant exposure to as high as 26 mg/m^3 over a 10-day period (8).

Humans exude about 90 mg/day of volatile fatty acids in exhaled breath and perspiration, 80% of which is acetic acid (73). In a confined environment, as much as 15–20 mg/m^3 can accumulate and such concentrations can become serious in submarines or space capsules.

Concentrated aqueous or organic solutions can be strongly damaging to skin. Any solution containing more than 50% acetic acid should be considered a corrosive acid. Acetic acid can irreparably scar delicate tissues of the eyes, nose, or mouth. The acid penetrates the mucosa of the tongue which is near the pK of the acid (74). The action of acetic acid is insidious. There is no quick burning sensation when applied to the unbroken skin. Blisters appear within 30 min to 4 h. Little or no pain is experienced at first but when sensory nerve receptors are attacked, severe and unremitting pain results. Once blistering occurs, washing with water or bicarbonate seldom relieves the pain. Medical care should be sought immediately.

Care ought to be taken in handling acetic acid to avoid spillage or otherwise breathing vapors. Wash any exposed areas with large amounts of water. Once the odor of acetic acid vapors is noticeable, the area should be abandoned immediately. The U.S. threshold limit value for acetic acid is 10 ppm (25 mg/m^3). Similar values prevail in Germany (75).

Glacial acetic acid is dangerous, but its precise toxic dose is not known for humans. The LD_{50} for rats is said to be 3310 mg/kg, and for rabbits 1200 mg/kg (76). Ingestion of 80–90 g must be considered extraordinarily dangerous for humans. Vinegar, on the other hand, which is dilute acetic acid, has been used in foods and beverages since the most ancient of times. Although vinegar is subject to excise taxation in many countries of the world other than the United States, acetic acid for nonfood applications is commonly exempted (77).

Industrial plants for acetic acid production sometimes have waste streams containing formic acid and acetic acid. The quantity of acids in these streams is not significant to the overall plant efficiency, but from a safety viewpoint such material must either be recycled, treated with alkaline substances, or consumed microbially. Recycling is probably the most expensive route because the cost of processing is not repaid by an increase in efficiency. In Europe these streams must be neutralized or degraded biologically. In Germany, no more than 3 kg/h may be emitted in vent gases, with a maximum of 150 mg/m^3 (76). The acid must be removed from the vent gas by scrubbing or chilling.

BIBLIOGRAPHY

"Acetic Acid" in *ECT* 1st ed., Vol. 1, pp. 56–74, by W. F. Schurig, The College of the City of New York; "Ethanoic Acid" in *ECT* 2nd ed., Vol. 8, pp. 386–404, by E. Le Monnier, Celanese Corporation of America; "Acetic Acid" in *ECT* 3rd ed., Vol. 1, pp. 124–147, by F. S. Wagner, Jr., Celanese Chemical Company.

1. K. Hess and H. Haber, *Ber. Dtsh. Chem. Ges.* **70**, 2205 (1937).
2. D. Ambrose and N. B. Ghiassee, *J. Chem. Thermodyn.* **19**, 505–519 (1987).
3. D. D. Perrin and co-workers, *Purification of Laboratory Chemicals,* Pergamon Press, New York, 1966, p. 56.
4. I. Malijevska, *Collect. Czech. Chem. Commun.* 48(8), 2147–2155 (1983); O. K. Mikhailova and N. P. Markuzin, *Zh. Obshch. Khim.* **52**(10), 2164–2166 (1982); *Zh. Obshch. Khim.* **53**(4), 713–716 (1983).
5. R. Buettner and G. Maurer, *Ber. Bunsenges. Phys. Chem.* **87**(10), 877–882 (1983).
6. Y. Fujii, H. Yamada, and M. Mizuta, *J. Phys. Chem.* **92**, 6768–6772 (1988).
7. J. F. Thorpe and M. A. Whiteley, *Thorpe's Dictionary of Applied Chemistry,* 4th ed., Longmans, Green & Co., London, 1937, Vol. 1, p. 51.
8. J. L. Hales, H. A. Gundry, and J. H. Ellender, *J. Chem. Thermodyn.* **15**, 211–215 (1983).
9. R. S. Barr and D. M. T. Newsham, *Chem. Eng. J. (Lausanne)* **33**(2), 79–86 (1986).
10. K. S. Howard, *J. Phys. Chem.* **62**, 1597 (1958).
11. W. Weltner, *J. Am. Chem. Soc.* **77**, 3941 (1955).
12. R. J. W. LeFévre, *Trans. Faraday Soc.* **34**, 1127 (1938).
13. D. D. Wagman and co-workers, *J. Phys. Chem. Ref. Data* **11**, Suppl. 2.
14. S. P. Miskidzh'yan and N. A. Trifonov, *Zh. Obshch. Khim.* **17**, 1033 (1947).
15. A. I. Vogel, *J. Chem. Soc.* **1948**, 1814 (1948).
16. N. I. Sax, *Dangerous Properties of Industrial Materials,* 5th ed., Van Nostrand-Reinhold, New York, 1979, p. 333.
17. L. W. Hessel and E. C. Kooyman, *Pharm. Weekbl.* **104**, 687 (1969).
18. P. G. Blake and G. E. Jackson, *J. Chem. Soc.* **1968B**, 1153–1155 (1968); **1969B**, 94–96 (1969).
19. J. C. Mackie and K. R. Doolan, *Int. J. Chem. Kinet.* **16**(5), 525–4 (1984).
20. J. J. Vajo, Y. K. Sun, and W. H. Winberg, *J. Phys. Chem.* **91**, 1153–1158 (1987).
21. J. G. Calvert and J. N. Pitts, *Photochemistry,* 428–431 (1966).
22. A. S. Newton, *J. Chem. Phys.* **26**, 1764–1765 (1957).
23. Jpn. Pat. JP 62-96,397 (May 2, 1987), S. Kawachi and K. Nakamura (to Ashai Chemical Industries).

24. Jpn. Pat. JP 62-113,797 (May 2, 1987), S. Kawachi and K. Katsuyuki (Asahi Chemical Industries).
25. Y. Hirose, *Hyomen,* **25**(12), 734–743 (1987); *Chem. Abstr.* **108,** 189283g. Y. Hirose, *Seimitsu Kogaku Kaishi* **53**(10), 1507–1510 (1987); *Chem. Abstr.* **108,** 61009e. N. Koshino, M. Kawarada, and K. Kurihara, *Denshi Zairyo* **27**(1), 49–54 (1988); *Chem. Abstr.* **109,** 16121c. H. Kawarada, J. Suzuki, and A. Hiraki, *Kagaku Kogyo* **39**(9), 784–790 (1988); *Chem. Abstr.* **110,** 41327v. Y. Hirose and F. Akatsuka, *Ibid.* **39**(8), 673–681, 693 (1988); *Chem. Abstr.* **110,** 10257t. Y. Namba, *Ibid.* **39**(8), 666–672, 689 (1988); *Chem. Abstr.* 110:10258u. F. Akatsuka, Y. Hirose, and K. Komaki, *Jpn. J. Appl. Phys. Pt 2* **27**(9), L1600–L1602 (1988).
26. A. Popoff, "Anhydrous Acetic Acid as a Nonaqueous Solvent", in J. J. Lagowski, ed., *Chemistry of Nonaqueous Solvents,* Vol. 3, Academic Press, 1970.
27. "Kohlenstoff," in K. von Baczko, ed., *Gmelins Handbuch der Anorganischen Chemie,* 8th ed., Teil C4, Frankfurt, 1975, pp. 141–197.
28. M. Ai, *J. Catal.* **112**(1), 194–200 (1988).
29. P. S. T. Sai, *Reg. J. Energy, Heat Mass Transfer* **10**(2), 181–189 (1988).
30. J. F. Vitcha and V. A. Sims, *Ind. Eng. Chem. Prod. Res. Dev.* **5,** 50 (1966).
31. H. Hartig, *Chem. Ing. Tech.* **45,** 467 (1973).
32. A. Y. Yau, A. Manielec, and A. I. Johnson, in E. Rhodes and D. S. Scott, eds., *Proceedings of International Symposium on Research in Cocurrent Gas-Liquid Flow,* Plenum Press, New York, 1969, pp. 607–632.
33. Ger. Pat. 2,153,767 (May 3, 1973), H. Schaum and H. Goessell (to Farbwerke Hoechst A.G.).
34. F. J. Weymouth and A. F. Millidge, *Chem. Ind. (London)* **1966,** 887–893 (May 28, 1966).
35. F. E. Paulik and J. F. Roth, *Chem. Commun.* **1968,** 1578 (1968).
36. *Chem. Eng. News,* 24 (Feb. 13, 1984); U.S. Pat. 4,690,912 (Sept. 1, 1987), F. E. Paulik, A. Hershman, W. R. Knox, and J. F. Roth (to Monsanto Company).
37. F. E. Paulik, *Catal. Rev.* **6,** 49 (1972).
38. R. S. Dickson, *Homogeneous Catalysis with Compounds of Rhodium and Iridium,* Reidel, Dordrecht, The Netherlands, 1985.
39. K. Omata and co-workers, *Stud. Surf. Sci. Catal.* **86,** 245–249 (1988); *Chem. Abstr.* **109,** 8355j (1988).
40. K. Fujimoto and co-workers, *Ind. Eng. Chem. Prod. Res. Dev.* **22,** 436–439 (1983).
41. A. F. Borowski, *Wiad. Chem.* **39**(10–12), 667–687 (1985); J. Ogonowski, *Chemik* **40**(2), 38–42; **40**(3), 67–71; **40**(7), 199–201 (1987).
42. H. Ono and co-workers, *J. Organomet. Chem.* **331,** 387–395 (1987).
43. T. Nakajo, K. Sano, S. Matsuhira, and H. Arakawa, *Chem. Commun. (London)* **1987**(9), 647–649 (1987).
44. U.S. Pats. 4,440,570 (Apr. 3, 1984), 4,442,304 (Apr. 10, 1984), 4,557,760 (Dec. 10, 1985), H. Erpenbach and co-workers (to Hoechst A.G.).
45. J. F. Knifton, *Platinum Met. Rev.* **29**(2), 63–72 (1985); Jpn. Pat. Kokai Tokkyo Koho JP 59–25340 (Feb. 9, 1984) (to Agency of Industrial Sciences and Technology).
46. U.S. Pat. 4,351,908 (Sept. 28, 1984), H. J. Schmidt and E. I. Leupold (to Hoechst A.G.).
47. V. H. Agreda, *CHEM-TECH* **18**(4), 250–253 (1988); *Chem. Eng. News* 30–32 (May 21, 1990).
48. F. Broich, *Chem. Ing. Tech.* **36**(5), 417–422 (1964).
49. H. Höfermann, *Chem. Ing. Tech.* **36**(5), 422–429 (1964).
50. C. C. Hobbs and co-workers, *Ind. Eng. Chem. Prod. Res. Dev.* **9,** 497 (1970); **11,** 220 (1972); *Ind. Eng. Chem. Process Des. Dev.* **59** (1972).
51. J. B. Saunby and B. W. Kiff, *Hydrocarbon Process.* **55**(11), 247 (1976).
52. R. K. Jensen and co-workers, *J. Am. Chem. Soc.* **101,** 7574 (1979).
53. N. M. Emanuel, E. T. Denisov, and Z. K. Maizus, *Liquid Phase Oxidation of Hydrocarbons,* Plenum Press, New York, 1967.

54. *New York Times Sect. I,* 15 (Nov. 16, 1987).
55. M. S. Reisch, *Chem. Eng. News* **65** (Nov. 23, 1987); **66,** 9–11 (Feb. 8, 1988); **66,** 5 (May 23, 1988).
56. D. F. Othmer, *DECHEMA Monogr.* **33,** 9–22 (1959).
57. D. F. Othmer, R. E. White, and E. Trueger, *Ind. Eng. Chem.* **33,** 1240–1248 (1941).
58. Jpn. Pat. Kokai Tokkyo Koho, JP 61–176,552 (Aug. 8, 1986), M. Tanaka, N. Kawada, and T. Morinaga (to Agency of Industrial Sciences and Technology).
59. Y. Kuo and H. P. Gregor, *Sepn. Sci. Technol.* **18**(5), 421–440 (1983).
60. U.S. Pat. 3,980,701 (Apr. 21, 1975), R. R. Grinstead (to The Dow Chemical Company); R. W. Hellsell, *Chem. Eng. Prog.* **73,** 55–59 (May 1977).
61. G. J. Hanna and R. D. Noble, *Chem. Rev.* **85,** 583–598 (1985).
62. T. Morinaga, *Hakko to Kogyo* **43**(11), 1015–1023 (1985); Jpn. Pat. JP 63–84,495 (Apr. 15, 1988), T. Morinaga (to Agency of Industrial and Scientific Technology).
63. Jpn. Pat. Kokai Tokkyo Koho JP 59–179,089 (Oct. 11, 1984) (to Agency of Industrial Sciences and Technology).
64. H. Wood, H. L. Drake, and S. Hu in E. E. Snell, ed., *Some Historical and Modern Aspects of Amino Acid, Fermentations, and Nucleic Acids,* (symposium: June 3, 1981, St. Louis, Mo.), Annual Reviews Inc., Palo Alto, Calif., 1982, pp. 29–56.
65. J. G. Zeikus, R. Kerby, and J. A. Krzycki, *Science* **227**(4691), 1167–1173 (1985).
66. L. G. Ljungdahl, *Ann. Rev. Microbiol.* **40,** 415–450 (1986).
67. L. G. Ljungdahl and co-workers, *Biotechnology and Bioengineering, Symp. No. 15,* pp. 207–223 (1985).
68. L. G. Ljungdahl and co-workers, *CMA Manufacture (II) Improved Bacterial Strain for Acetate Production,* Final Report, FHWA/RD-86/117, U.S. Dept. of Transportation, Washington, D.C.
69. J. Kassotis and co-workers, *J. Electrochem. Soc.* **131,** 2810–2814 (Dec. 1984).
70. Y. Nomura and co-workers, *Appl. Environ. Microbiol.* **54,** 137–142 (1988).
71. *Code of Federal Regulations,* Title 49, 173.244–173.245, U.S. Dept. of Transportation, Washington, D.C.
72. E. F. Joy and A. J. Barnard, Jr., *Encyclopedia of Industrial Chemical Analysis,* Vol. 4, John Wiley & Sons, Inc., New York, 1967, pp. 93–101.
73. H. S. Christensen and T. Luginbyl, eds., *Registry of Toxic Effects of Chemical Substances,* U.S. Dept. of Health, Education, Welfare, Rockville, Md., 1975.
74. V. P. Savina and B. V. Anisimov, *Kosm. Biol. Aviakosmicheskaya. Med.* **22**(1), 57–61 (1988).
75. I. A. Siegel, *Arch. Oral Biol.* **29**(1), 13–16 (1988).
76. American Conference Governmental Industrial Hygienists, *Threshold Limit Values,* Cincinnati, Ohio, 1982; cf. Deutsche Forschungsgemeinschaft, *Maximale Arbeitsplatzkonzentrationen, 1983,* Verlag Chemie, Weinheim, 1983, p. 15.
77. *Vinegar Tax,* Bundesrepublik Deutschland, (April 25, 1972).

FRANK S. WAGNER, JR.
Nandina Corporation

ACETAMIDE

Acetamide [*60-35-5*], C_2H_5NO, mol wt 59.07, is a white, odorless, hygroscopic solid derived from acetic acid and ammonia. The stable crystalline habit is trigonal; the metastable is orthorhombic. The melt is a solvent for organic substances; it is used in electrochemistry and organic synthesis. Pure acetamide has a bitter taste.

Unknown impurities, possibly derived from acetonitrile, cause its mousy odor (1). It is found in coal mine waste dumps (2).

Physical and Chemical Properties

Table 1 lists many of acetamide's important physical properties. Acetamide, CH_3CONH_2, dissolves easily in water, exhibiting amphoteric behavior. It is slow to hydrolyze unless an acid or base is present. The autodissociation constant is about 3.2×10^{-11} at 94°C. It combines with acids, eg, HBr, HCl, HNO_3, to form solid complexes. The chemistry of metal salts in acetamide melts has been researched with a view to developing electroplating methods. The literature of acetamide melts and complexes, their electrochemistry and spectroscopy, has been critically reviewed (9).

Table 1. Physical Properties of Acetamide

Property	Value	Reference
melting point (trigonal), °C	80.0 –80.1	3
triple point, K	353.33	4
heat of melting, ΔH_m, kJ/kg^a	264	5
dielectric constant	59	6
dipole moment, C·m^b	12.41×10^{-30}	3
density equation^c	$1.357 - 0.0012T + 0.64 \times 10^{-6}T^2$	6
melt density at 85°C, g/mL	0.9986	7
vapor pressure at T in K, kPa^d		8
272	10	
278	20	
281	30	
284	40	
285	50	
287	60	
288	70	
290	80	
291	90	
292	100	

^{a}To convert kJ to kcal, divide by 4.184.
^{b}To convert C·m to debyes, multiply by 2.998×10^{29}.
^{c}Kelvin temperature.
^{d}To convert kPa to mm Hg, multiply by 7.50.

Preparation and Manufacture

Most commercial routes for the production of acetamide involve dehydration of ammonium acetate [631-31-8]:

$$NH_4OOCCH_3 \longrightarrow H_2O + CH_3CONH_2$$

Industrial production is often based on transformation of this laboratory method into a continuous process (10). Another route is acetonitrile [75-05-8] hydration:

$$CH_3CN + H_2O \longrightarrow CH_3CONH_2$$

Because huge quantities of by-product acetonitrile are generated by ammoxidation of propylene, the nitrile may be a low cost raw material for acetamide production. Copper-catalyzed hydration gives conversions up to 83% (11), and certain bacteria can effect the same reaction at near room temperature (12).

Health and Safety

Acetamide has been used experimentally as a source of nonprotein nitrogen for sheep and dairy cattle (13). It does not appear to be toxic in amounts of about 2–3% of ration. Buffering the diet with dibasic acids serves to allow higher levels of intake because the ammonia liberated in the digestive process is then scavenged.

Economic Aspects

Heico Chemicals is the only producer of acetamide in the United States. Small amounts are imported from Europe and Asia. It is shipped in 32-L (35-gal) drums weighing about 80 kg. Acetamide appears to have a wide spectrum of applications. It suppresses acid buildup in printing inks, lacquers, explosives, and perfumes. It is a mild moisturizer and is used as a softener for leather, textiles, paper, and certain plastics. It finds some applications in the synthesis of pharmaceuticals, pesticides, and antioxidants for plastics.

BIBLIOGRAPHY

"Acetamide," in *ECT* 1st ed., Vol. 1, pp. 45–48 by S. J. Cohen, Hardesty Chemical Co., Inc.; in *ECT* 2nd ed., Vol. 1, pp. 142–145 by A. P. Lurie, Eastman Kodak Company; in *ECT* 3rd ed., Vol. 1, pp. 148–151 by T. A. Moretti, American Hoechst Corporation.

1. N. V. Sidgwick, *The Organic Chemistry of Nitrogen*, Oxford University Press, New York, 1966, p. 228.
2. B. I. Srebrodol'ski, *Zap. Vses. Mineral. O.-va*, **104**(3), 326–328 (1975); *Chem. Abstr.* **83**, 166946b (1975).
3. W. D. Kumler and C. W. Porter, *J. Am. Chem. Soc.* **56**, 2549–2554 (1934); E. C. Wagner, *J. Chem. Educ.* **7**, 1135–1137 (1930).
4. H. G. M. De Wit and co-workers, *J. Chem. Thermodyn.* **15**, 891–902 (1983).
5. H. H. Emons and co-workers, *Thermochim. Acta* **104**, 127 (1986).
6. L. Vogel and H. Schuberth, *Wiss. Z. Martin-Luther-Uni., Halle-Wittenberg, Math.-Naturwiss. Reihe* **34**(3), 79–80 (1985).
7. W. E. S. Turner and E. W. Merry, *J. Chem. Soc. London,* **97**, 2069–2083 (1910).
8. H. G. M. De Wit and coworkers, *J. Chem. Thermodyn.* **15**, 651–663 (1983).
9. D. H. Kerridge, *Chem. Soc. Rev.* **17**, 181–227 (1988).

10. G. H. Coleman and A. M. Alvarado, *Org. Synth. Coll. Vol.* **1**, 3, 4 (1941).
11. M. Ravindranatha, N. Kalyanam and S. Sivaram, *J. Org. Chem.* **47**, 4812, 4813 (1982).
12. Eur. Pat. 204,555 (Dec. 10, 1986), K. Kawakami, T. Tanabe, and O. Nagano (to Asahi Chemical Industries Co. Ltd.); *Chem. Abstr.* **106**, 212584a (1987).
13. Eur. Pat. 317,092 (May 24, 1989), J. M. Moxley and K. R. Zone (to OMI International Corp.); *Chem. Abstr.* **111**, 107394u (1989); *Can. J. Anim. Sci.* **64 suppl.**, 37, 38 (1984).

FRANK S. WAGNER, JR.
Nandina Corporation

ACETIC ANHYDRIDE

Acetic anhydride [*108-24-7*], $(CH_3CO)_2O$, is a mobile, colorless liquid that has an acrid odor and is a more piercing lacrimator than acetic acid [*64-19-7*]. It is the largest commercially produced carboxylic acid anhydride: U.S. production capacity is over 900,000 t yearly. Its chief industrial application is for acetylation reactions; it is also used in many other applications in organic synthesis, and it has some utility as a solvent in chemical analysis.

First prepared by C. F. Gerhardt from benzoyl chloride and carefully dried potassium acetate (1), acetic anhydride is a symmetrical intermolecular anhydride of acetic acid; the intramolecular anhydride is ketene [*463-51-4*]. Benzoic acetic anhydride [*2819-08-1*] undergoes exchange upon distillation to yield benzoic anhydride [*93-97-0*] and acetic anhydride.

Physical and Chemical Properties

No dimerization of acetic anhydride has been observed in either the liquid or solid state. Decomposition, accelerated by heat and catalysts such as mineral acids, leads slowly to acetic acid (2). Acetic anhydride is soluble in many common solvents, including cold water. As much as 10.7 wt % of anhydride will dissolve in water. The unbuffered hydrolysis rate constant k at 20°C is 0.107 min^{-1} and at 40°C is 0.248 min^{-1}. The corresponding activation energy is about 31.8 kJ/mol (7.6 kcal/mol) (3). Although aqueous solutions are initially neutral to litmus, they show acid properties once hydrolysis appreciably progresses. Acetic anhydride ionizes to acetylium, CH_3CO^+, and acetate, $CH_3CO_2^-$, ions in the presence of salts or acids (4). Acetate ions promote anhydride hydrolysis. A summary of acetic anhydride's physical properties is given in Table 1.

Acetic anhydride acetylates free hydroxyl groups without a catalyst, but esterification is smoother and more complete in the presence of acids. For example, in the presence of *p*-toluenesulfonic acid [*104-15-4*], the heat of reaction for ethanol and acetic anhydride is -60.17 kJ/mol (-14.38 kcal/mol) (13):

$$ROH + (CH_3CO)_2O \longrightarrow CH_3COOR + CH_3COOH$$

Amines undergo an analogous reaction to yield acetamides, the more basic amines having the greater activity:

$$RNH_2 + (CH_3CO)_2O \longrightarrow CH_3CONHR + CH_3COOH$$

Table 1. Physical Properties of Acetic Anhydride

Property	Value	Reference
freezing point, °C	−73.13	5
boiling point, °C at 101.3 kPa[a]	139.5	5
density, d_4^{20}, g/cm³	1.0820	5
refractive index, n_D^{20}	1.39038	6
vapor pressure (Antoine equation), P in kPa[a] and T in K	$\ln(P) = \dfrac{14.6497 - 3467.76}{T - 67.0}$	7
heat of vaporization, ΔH_v, at bp, J/g[b]	406.6	5
specific heat, J/kg[c] at 20°C	1817	8
surface tension, mN/m (= dyn/cm)		
25°C	32.16	9
40°C	30.20	9
viscosity, mPa·s (= cP)		
15°C	0.971	10
30°C	0.783	10
heat conductivity, mW/(m·K)[d] at 30°C	136	11
electric conductivity, S/cm	2.3×10^{-8}	12

[a]To convert kPa to mm Hg, multiply by 7.5.
[b]To convert J/g to Btu/lb, multiply by 0.4302.
[c]To convert J to cal, divide by 4.184.
[d]To convert mW/(m·K) to (Btu·ft)/(h·ft²·°F), multiply by 578.

Potassium acetate, rubidium acetate, and cesium acetate are very soluble in anhydride in contrast to the only slightly soluble sodium salt. Barium forms the only soluble alkaline earth acetate. Heavy metal acetates are poorly soluble.

Triacetylboron [4887-24-5], $C_6H_9BO_6$, is generated when boric acid is added to acetic anhydride and warmed. Although explosions have resulted from carrying out this reaction, slowly adding the boric acid to a zinc chloride solution in acetic anhydride, and maintaining a temperature below 60°C, gives a good yield (14). Acetic acid is also formed

$$H_3BO_3 + 3\ (CH_3CO)_2O \longrightarrow (CH_3CO)_3BO_3 + 3\ CH_3COOH$$

Heating triacetylboron at temperatures above its melting point, 123°C, causes a rearrangement to $B_2O(OCCH_3)_4$ (15). An explosive hazard is also generated by dissolving BF_3 in anhydride (see BORON COMPOUNDS).

Hydrogen peroxide undergoes two reactions with anhydride:

$$(CH_3CO)_2O + H_2O_2 \longrightarrow CH_3COOOH + CH_3COOH + H_2O$$

$$(CH_3CO)_2O + 2\ H_2O_2 \longrightarrow 2\ CH_3COOOH + H_2O$$

Peroxyacetic acid [79-21-0] is used in many epoxidations (16) where ion-exchange resins, eg, Amberlite IR–1180M, serve as catalysts. Pinene is epoxidized to sobrerol [498-71-5], $C_{10}H_{18}O_2$, using peroxyacetic acid at −5 to −10°C (17). Care must be taken, however, to avoid formation of the highly explosive diacetyl peroxide [110-22-5]:

$$(CH_3CO)_2O + CH_3COOOH \longrightarrow CH_3CO-OO-COCH_3 + CH_3COOH$$

Acetic anhydride can be used to synthesize methyl ketones in Friedel-Crafts reactions. For example, benzene [71-43-2] can be acetylated to furnish acetophenone [98-86-2]. Ketones can be converted to their enol acetates and aldehydes to their alkylidene diacetates. Acetaldehyde reacts with acetic anhydride to yield ethylidene diacetate [542-10-9] (18):

$$CH_3CHO + (CH_3CO)_2O \longrightarrow CH_3CH(O_2CCH_3)_2$$

Isopropenyl acetate [108-22-5], which forms upon reaction of acetone [67-64-1] with anhydride, rearranges to acetylacetone [123-54-6] in the presence of BF_3 (19):

$$CH_3COOC{=}CH_2 \xrightarrow{BF_3} \underset{\displaystyle CH_3}{CH_3C}{-}CH_2{-}\underset{\displaystyle}{C}{=}CH_3$$

(with carbonyl oxygens shown: $CH_3\overset{O}{\overset{||}{C}}-CH_2-\overset{O}{\overset{||}{C}}=CH_3$, and CH_3 substituent on the $COOC=CH_2$)

Unsaturated aldehydes undergo a similar reaction in the presence of strongly acid ion-exchange resins to produce alkenylidene diacetates. Thus acrolein [107-02-8] or methacrolein [78-85-3] react with equimolar amounts of anhydride at $-10°C$ to give high yields of the *gem*-diacetates from acetic anhydride, useful for soap fragrances.

Acids react with acetic anhydride to furnish higher anhydrides (20). An acid which has a higher boiling point than acetic acid is refluxed with acetic anhydride until an equilibrium is established. The low boiling acetic acid is distilled off and the anhydride of the higher acid is left. Adipic polyanhydride is obtained in this manner (21).

Manufacture

The Acetic Acid Process. Prior to the energy crisis of the 1970s, acetic anhydride was manufactured by thermal decomposition of acetic acid at pressures of 15–20 kPa (2.2–2.9 psi) (22), beginning with the first step:

$$CH_3COOH \longrightarrow CH_2{=}C{=}O + H_2O$$

The heat of reaction is approximately 147 kJ/mol (35.1 kcal/mol) (23). Optimum yields of ketene [463-51-4] require a temperature of about 730–750°C. Low pressure increases the yield, but not the efficiency of the process. Competitive reactions are

$$CH_3COOH \longrightarrow CH_4 + CO_2 \quad \text{(endothermic)}$$
$$CH_3COOH \longrightarrow 2\,CO + 2\,H_2 \quad \text{(exothermic)}$$
$$2\,CH_2{=}C{=}O \longrightarrow CH_4 + C + 2\,CO \quad \text{(exothermic)}$$
$$2\,CH_2{=}C{=}O \longrightarrow CH_2{=}CH_2 + 2\,CO \quad \text{(exothermic)}$$

The second step is the liquid-phase ketene and acetic acid reaction

$$CH_2{=}C{=}O + CH_3COOH \longrightarrow (CH_3CO)_2O \quad \text{(exothermic)}$$

Triethyl phosphate is commonly used as dehydration catalyst for the water formed in the first step. It is neutralized in the exit gases with ammonia. Aqueous 30% ammonia is employed as solvent in the second step because water facilitates the reaction, and the small amount of water introduced is not significant overall. Compression of ketene using the liquid-ring pump substantially improves the formation of anhydride. Nickel-free alloys, eg, ferrochrome alloy, chrome–aluminum steel, are needed for the acetic acid pyrolysis tubes, because nickel promotes the formation of soot and coke, and reacts with carbon monoxide yielding a highly toxic metal carbonyl. Coke formation is a serious efficiency loss. Conventional operating conditions furnish 85–88% conversion, selectivity to ketene 90–95 mol %. High petroleum energy costs make these routes only marginally economic.

Acetone cracks to ketene, and may then be converted to anhydride by reaction with acetic acid. This process consumes somewhat less energy and is a popular subject for chemical engineering problems (24,25). The cost of acetone works against widespread application of this process, however.

The Acetaldehyde Oxidation Process. Liquid-phase catalytic oxidation of acetaldehyde (qv) can be directed by appropriate catalysts, such as transition metal salts of cobalt or manganese, to produce anhydride (26). Either ethyl acetate or acetic acid may be used as reaction solvent. The reaction proceeds according to the sequence

$$CH_3CHO + O_2 \longrightarrow CH_3COOOH$$

$$CH_3COOOH + CH_3CHO \longrightarrow CH_3COOOCH(OH)CH_3$$

$$CH_3COOOCH(OH)CH_3 \longrightarrow (CH_3CO)_2O + H_2O$$

$$CH_3COOOCH(OH)CH_3 \longrightarrow CH_3COOH + CH_3COOH$$

Acetaldehyde oxidation generates peroxyacetic acid which then reacts with more acetaldehyde to yield acetaldehyde monoperoxyacetate [7416-48-0], the Loesch ester (26). Subsequently, parallel reactions lead to formation of acetic acid and anhydride plus water.

Under sufficient pressure to permit a liquid phase at 55–56°C, the acetaldehyde monoperoxyacetate decomposes nearly quantitatively into anhydride and water in the presence of copper. Anhydride hydrolysis is unavoidable, however, because of the presence of water. When the product is removed as a vapor, an equilibrium concentration of anhydride higher than that of acetic acid remains in the reactor. Water is normally quite low. Air entrains the acetic anhydride and water as soon as they form.

High purity acetaldehyde is desirable for oxidation. The aldehyde is diluted with solvent to moderate oxidation and to permit safer operation. In the liquid take-off process, acetaldehyde is maintained at 30–40 wt % and when a vapor product is taken, no more than 6 wt % aldehyde is in the reactor solvent. A considerable recycle stream is returned to the oxidation reactor to increase selectivity. Recycle air, chiefly nitrogen, is added to the air introduced to the reactor at 4000–4500 times the reactor volume per hour. The customary catalyst is a mixture of three parts copper acetate to one part cobalt acetate by weight. Either salt alone is less effective than the mixture. Copper acetate may be as high as 2 wt % in the reaction solvent, but cobalt acetate ought not rise above 0.5 wt %. The reaction is carried out at 45–60°C under 100–300 kPa (15–44 psi). The reaction solvent is far above the boiling point of acetaldehyde, but the reaction is so fast that little escapes unoxidized. This temperature helps oxygen absorption, reduces acetaldehyde losses, and inhibits anhydride hydrolysis.

Product refining is quite facile, following the same general pattern for acetic acid (qv) recovery from acetaldehyde liquid-phase oxidation. Low boilers are stripped off using ethyl acetate or acetic acid, then anhydride is distilled. Residues, largely ethylidene diacetate and certain higher condensation products with the catalyst salts, can be recycled (27).

Methyl Acetate Carbonylation.   Anhydride can be made by carbonylation of methyl acetate [79-20-9] (28) in a manner analogous to methanol carbonylation to acetic acid. Methanol acetylation is an essential first step in anhydride manufacture by carbonylation. See Figure 1. The reactions are

$$CH_3COOH + CH_3OH \longrightarrow CH_3COOCH_3 + H_2O \qquad -4.89 \text{ kJ/mol} \ (-1.2 \text{ kcal/mol})$$

$$CH_3COOCH_3 + CO \longrightarrow (CH_3CO)_2O \qquad -94.8 \text{ kJ/mol} \ (-22.7 \text{ kcal/mol})$$

Surprisingly, there is limited nonproprietary experimental data on methanol esterification with acetic acid (29). Studies have been confined to liquid-phase systems distant from equilibrium (30), in regions where hydrolysis is unimportant. A physical study of the ternary methanol–methyl acetate–water system is useful for design work (31). Methyl acetate and methanol form an azeotrope which boils at 53.8°C and contains 18.7% alcohol. An apparent methanol–water azeotrope exists, boiling at 64.4°C and containing about 2.9% water. These azeotropes seriously complicate methyl acetate recovery. Methyl acetate is quite soluble in water, and very soluble in water–methanol mixtures, hence two liquid phases suitable for decanting are seldom found.

The reaction mechanism and rates of methyl acetate carbonylation are not fully understood. In the nickel-catalyzed reaction, rate constants for formation of methyl acetate from methanol, formation of dimethyl ether, and carbonylation of dimethyl ether have been reported, as well as their sensitivity to partial pressure of the reactants (32). For the rhodium chloride [10049-07-7] catalyzed reaction, methyl acetate carbonylation is considered to go through formation of ethylidene diacetate (33):

$$2 \ CH_3COOCH_3 + 2 \ CO \longrightarrow 2 \ (CH_3CO)_2O$$

$$2 \ (CH_3CO)_2O + H_2 \longrightarrow CH_3CH(OOCCH_3)_2 + CH_3COOH$$

The role of iodides, especially methyl iodide, is not known. The reaction occurs scarcely at all without iodides. Impurities and co-products are poorly reported in the patent literature on the process.

The catalyst system for the modern methyl acetate carbonylation process involves rhodium chloride trihydrate [13569-65-8], methyl iodide [74-88-4], chromium metal powder, and an alumina support or a nickel carbonyl complex with triphenylphosphine, methyl iodide, and chromium hexacarbonyl (34). The use of nitrogen–heterocyclic complexes and rhodium chloride is disclosed in one Euro-

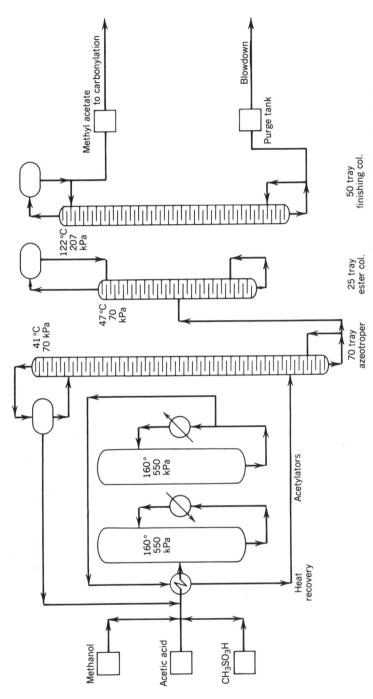

Fig. 1. Flow sheet for methyl acetate manufacture. To convert kPa to psi multiply by 0.145.

147

pean patent (35). In another, the alumina catalyst support is treated with an organosilicon compound having either a terminal organophosphine or similar ligands and rhodium or a similar noble metal (36). Such a catalyst enabled methyl acetate carbonylation at 200°C under about 20 MPa (2900 psi) carbon monoxide, with a space-time yield of 140 g anhydride per g rhodium per hour. Conversion was 42.8% with 97.5% selectivity. A homogeneous catalyst system for methyl acetate carbonylation has also been disclosed (37). A description of another synthesis is given where anhydride conversion is about 30%, with 95% selectivity. The reaction occurs at 445 K under 11 MPa partial pressure of carbon monoxide (37). A process based on a montmorillonite support with nickel chloride coordinated with imidazole has been developed (38). Other related processes for carbonylation to yield anhydride are also available (39,40).

The first anhydride plant in actual operation using methyl acetate carbonylation was at Kingsport, Tennessee (41). A general description has been given (42) indicating that about 900 tons of coal are processed daily in Texaco gasifiers. Carbon monoxide is used to make 227,000 t/yr of anhydride from 177,000 t/yr of methyl acetate; 166,000 t/yr of methanol is generated. Infrared spectroscopy has been used to follow the apparent reaction mechanism (43).

The unit has virtually the same flow sheet (see Fig. 2) as that of methanol carbonylation to acetic acid (qv). Any water present in the methyl acetate feed is destroyed by recycle anhydride. Water impairs the catalyst. Carbonylation occurs in a sparged reactor, fitted with baffles to diminish entrainment of the catalyst-rich liquid. Carbon monoxide is introduced at about 15–18 MPa from centrifugal, multistage compressors. Gaseous dimethyl ether from the reactor is recycled with the CO and occasional injections of methyl iodide and methyl acetate may be introduced. Near the end of the life of a catalyst charge, additional rhodium chloride, with or without a ligand, can be put into the system to increase anhydride production based on net noble metal introduced. The reaction is exothermic, thus no heat need be added and surplus heat can be recovered as low pressure steam.

Catalyst recovery is a major operational problem because rhodium is a costly noble metal and every trace must be recovered for an economic process. Several methods have been patented (44–46). The catalyst is often reactivated by heating in the presence of an alcohol. In another technique, water is added to the homogeneous catalyst solution so that the rhodium compounds precipitate. Another way to separate rhodium involves a two-phase liquid such as the immiscible mixture of octane or cyclohexane and aliphatic alcohols having 4–8 carbon atoms. In a typical instance, the carbonylation reactor is operated so the desired products and other low boiling materials are flash-distilled. The reacting mixture itself may be boiled, or a sidestream can be distilled, returning the heavy ends to the reactor. In either case, the heavier materials tend to accumulate. A part of these materials is separated, then concentrated to leave only the heaviest residues, and treated with the immiscible liquid pair. The rhodium precipitates and is taken up in anhydride for recycling.

By-products remain unmentioned in most patents. Possibly there are none other than methyl acetate, acetic acid, and ethylidene diacetate, which are all precursors of anhydride.

In anhydride purification, iodide removal is of considerable significance;

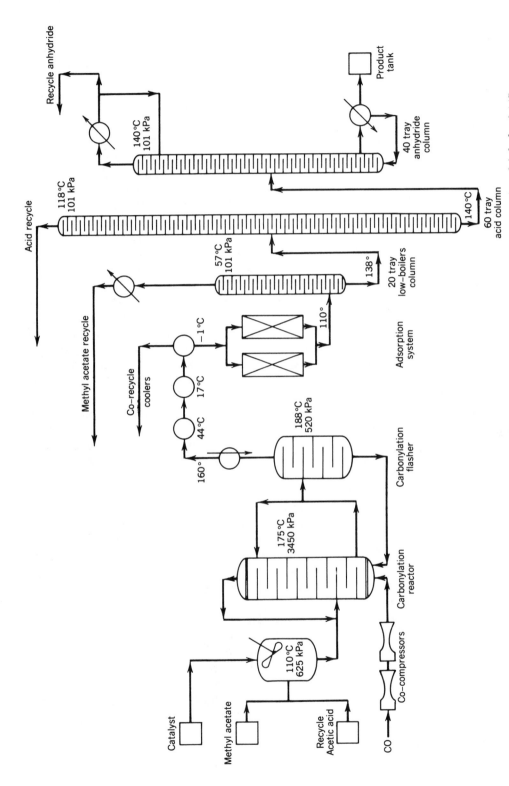

Fig. 2. Flow sheet for methyl acetate carbonylation to anhydride. To convert kPa to psi multiply by 0.145.

149

potassium acetate has been suggested for this procedure (47). Because of the presence of iodide in the reaction system, titanium is the most suitable material of construction. Conventional stainless steel, SS-316, can be used for the acetylation of methanol, but it is unsuited for any part of the plant where halogen is likely to be present. Although such materials of construction add substantially to the capital cost of the plant, savings in energy consumed are expected to compensate. No authentic data on process efficiency has been published, but a rough estimate suggests a steam consumption of 1.9–2.0 kg/kg product anhydride.

Prospective Routes to Acetic Anhydride.   Methyl acetate–dimethyl ether carbonylation seems to be the leading new route to acetic anhydride production. The high energy costs of older routes proceeding through ketene preclude their reintroduction. Thermolysis of acetone, methyl acetate, and ethylidene diacetate suffers from the same costly energy consumption. Acetic acid cracking, under vacuum and at atmospheric pressure, continues to be used for anhydride manufacture in spite of the clear obsolescence of the processes. The plants have high capital investment, and dismantling them demands a large economic advantage from any supplanting process.

Acetaldehyde oxidation to anhydride does not consume great amounts of energy. The strongly exothermic reaction actually furnishes energy and the process is widely used in Europe. Acetaldehyde must be prepared from either acetylene or ethylene. Unfortunately, use of these raw materials cancels the other advantages of this route. Further development of more efficient acetaldehyde oxidation as well as less expensive materials of construction would make that process more favorable.

Copper acetate, ferrous acetate, silver acetate [563-63-3], basic aluminum acetate, nickel acetate [373-02-4], cobalt acetate, and other acetate salts have been reported to furnish anhydride when heated. In principle, these acetates could be obtained from low concentration acetic acid. Complications of solids processing and the scarcity of knowledge about these thermolyses make industrial development of this process expensive. In the early 1930s, Soviet investigators discovered the reaction of dinitrogen tetroxide [10544-72-6] and sodium acetate [127-09-3] to form anhydride:

$$2\ NaOOCCH_3 + 2\ N_2O_4 \longrightarrow (CH_3CO)_2O + 2\ NaNO_3 + N_2O_3$$

Yields of the order of 85% were secured in the dry reaction (48). (Propionic anhydride and butyric anhydride can be obtained similarly from their sodium salts.) Inasmuch as dinitrogen tetroxide can be regenerated, the economic prospects of this novel way of making anhydride are feasible.

Sodium acetate reacts with carbon dioxide in aqueous solution to produce acetic anhydride and sodium bicarbonate (49). Under suitable conditions, the sodium bicarbonate precipitates and can be removed by centrifugal separation. Presumably, the cold water solution can be extracted with an organic solvent, eg, chloroform or ethyl acetate, to furnish acetic anhydride. The half-life of aqueous acetic anhydride at 19°C is said to be no more than 1 h (2) and some other data suggests a 6 min half-life at 20°C (50). The free energy of acetic anhydride hydrolysis is given as −65.7 kJ/mol (−15.7 kcal/mol) (51) in water. In wet chloroform, an extractant for anhydride, the free energy of hydrolysis is strangely much lower, −50.0 kJ/mol (−12.0 kcal/mol) (51). Half-life of anhydride in moist chloroform may be as much as 120 min. Ethyl acetate, chloroform, isooctane, and n-octane may have promise for extraction of acetic anhydride. Benzene extracts acetic anhydride from acetic acid–water solutions (52).

Ketene can be obtained by reaction of carbon oxides with ethylene (53). Because ketene combines readily with acetic acid, forming anhydride, this route may have practical applications. Little is known about the engineering possibilities of these reactions.

Economic Aspects

Acetic anhydride is a mature commodity chemical in the United States and its growth rate in the 1970s and 1980s was negative until 1988 when foreign demand nearly doubled the exports of 1986. This increase in exports was almost certainly attributable to the decline in the value of the U.S. dollar. Over four-fifths of all anhydride production is utilized in cellulose acetate [9004-35-7] manufacture (see CELLULOSE ESTERS). Many anhydride plants are integrated with cellulose acetate production and thus employ the acetic acid pyrolysis route. About 1.25 kg acetic acid is pyrolyzed to produce 1.0 kg anhydride. (Methyl acetate carbonylation demands only 0.59 kg acetic acid per kg anhydride manufactured.)

Anhydride has been used for the illegal manufacture of heroin [561-27-3] (acetylmorphine) and certain other addictive drugs. Regulations on acetic anhydride commerce have long been a feature of European practice. After passage in 1988 of the Chemical Diversion and Trafficking Act, there is also U.S. control. Orders for as much as 1,023 kg acetic anhydride, for either domestic sale or export, require a report to the Department of Justice, Drug Enforcement Administration (54).

Acetic anhydride is used in acetylation processes. There has been some diversification of anhydride usage in recent years. Acetylation of salicylic acid [69-72-7] using anhydride to furnish acetylsalicyclic acid [50-78-2] (aspirin) is a mature process, although N-acetyl-p-aminophenol (acetaminophen) is currently making inroads on the aspirin market. Acetic anhydride is used to acetylate various fragrance alcohols to transform them into esters having much higher unit value and vitamins are metabolically enhanced by acetylation. Anhydride is also extensively employed in metallography, etching, and polishing of metals, and in semiconductor manufacture. Starch acetylation furnishes textile sizing agents.

Acetic anhydride is a useful solvent in certain nitrations, acetylation of amines and organosulfur compounds for rubber processing, and in pesticides. Though acetic acid is unexceptional as a fungicide, small percentages of anhydride in acetic acid, or in cold water solutions are powerful fungicides and bactericides. There are no reports of this application in commerce. It is possible that anhydride may replace formaldehyde for certain mycocidal applications.

Analysis and Specifications

Analytical and control methods for acetic anhydride are fully discussed in reference 55. Performance tests are customarily used where the quality of the product is crucial, as in food or pharmaceutical products. Typical specifications are:

assay	99.0 wt % as anhydride
specific gravity$_{20}^{20}$	1.080 to 1.085
color, Pt–Co	10 max
KMnO$_4$ time	5 min needed to reduce 2 mL having no more than 0.1 mL of 0.1 N KMnO$_4$.
trace ions	less than 1 ppm Al, Cl, PO$_4^{3-}$, SO$_4^{2-}$, and Fe.
heavy metals	none
nitrates	none

A technical quality anhydride, assay about 97% maximum, often contains color bodies, heavy metals, phosphorus, and sulfur compounds. Anhydride manufactured by acetic acid pyrolysis sometimes contains ketene polymers, eg, acetylacetone, diketene, dehydroacetic acid, and particulate carbon, or soot, is occasionally encountered. Polymers of allene, or its equilibrium mixture, methylacetylene–allene, are reactive and refractory impurities, which if exposed to air, slowly autoxidize to dangerous peroxidic compounds.

Health and Safety

Acetic anhydride penetrates the skin quickly and painfully forming burns and blisters that are slow to heal. Anhydride is especially dangerous to the delicate tissues of the eyes, ears, nose, and mouth. The odor threshold is 0.49 mg/m^3, but the eyes are affected by as little as 0.36 mg/m^3 and electroencephalogram patterns are altered by only 0.18 mg/m^3. When handling acetic anhydride, rubber gloves that are free of pinholes are recommended for the hands, as well as plastic goggles for the eyes, and face-masks to cover the face and ears.

Acetic anhydride is dangerous in combination with various oxidizing substances and strong acids. Chromium trioxide [1333-82-0] and anhydride react violently to burn (56); mixtures containing nitric acid [7692-37-2] are said to be more sensitive than nitroglycerin (57). Thermal decomposition of nitric acid in acetic acid solutions is accelerated by the presence of anhydride (58). The critical detonation diameter for a nitric acid–acetic anhydride mixture has been subject to much study in the Soviet Union (59). The greatest explosion involving anhydride took place when a mixture of 568 L perchloric acid [7601-90-3] were admixed with 227 L of acetic anhydride. The mixture detonated, killing 17 people and destroying 116 buildings over several city blocks (60). The plant was an illegal metal-treating facility where the mixture was used to finish aluminum surfaces. Acetyl perchlorate is probably present in such solutions. These perchloric acid solutions are useful in metal finishing, but the risks in using them must be recognized. Such solutions are used commonly in metallography.

BIBLIOGRAPHY

"Acetic Anhydride" in *ECT* 1st ed., Vol. 1, pp. 78–86, by Gwynn Benson, Shawinigan Chemicals Limited; in *ECT* 2nd ed., Vol. 8, pp. 405–414, by Ernest Le Monnier, Celanese

Corporation of America; in *ECT* 3rd ed., Vol. 1, pp. 151–161, by Frank S. Wagner, Jr., Celanese Chemical Corporation.

1. C. F. Gerhardt, *Ann. Chim. Phys.* **37**(3), 285 (1852).
2. A. Lumière, L. Lumière, and H. Barbier, *Bull. Soc. Chim.* **33**, [iii], 783 (1905).
3. M. S. Sytilin, A. I. Morozov, and I. A. Makolkin, *Zh. Fiz. Khim.* **46**, 2266 (1972).
4. H. Schmidt, I. Wittkopf, and G. Jander, *Z. Anorg. Chem.* **256**, 113 (1948).
5. R. A. McDonald, S. A. Shrader, and D. R. Stull, *J. Chem. Eng. Data* **4**, 311 (1959).
6. Luck and co-workers, *J. Am. Chem. Soc.* **81**, 2784–2786 (1959).
7. D. Ambrose and N. B. Ghiassee, *J. Chem. Thermodyn.* **19**, 911 (1987).
8. F. J. Wright, *J. Chem. Eng. Data* **6**, 454 (1961).
9. K. N. Kovalenko and co-workers, *Zh. Obshch. Khim.* **26**, 403–407 (1956).
10. T. V. Malkova, *Zh. Obshch. Khim.* **24**, 1157 (1954).
11. N. V. Tsederberg, *The Thermal Conductivity of Gases and Liquids,* Massachusetts Institute of Technology Press, Cambridge, Mass., 1965.
12. G. Jander and H. Surawski, *Z. Elektrochem.* **65**, 469 (1961).
13. Eur. Pat. 170,173 (Feb. 12, 1986), G. W. Stockton, D. H. Chidester, and S. J. Ehrlich (to American Cyanamid Company).
14. L. M. Lerner, *Chem. Eng. News* **51**(34), 42 (1973).
15. I. G. Ryss and V. N. Plakhotnik, *Ukr. Khim. Zh. (Russ. ed.)* **36**, 423 (1970).
16. Ger. Pat. 3,447,864 (July 10, 1986), K. Eckwert and co-workers (to Henkel K.-G.a.A.).
17. Span. Pat. 548,516 (Dec. 1, 1986), J. L. Diez Amez; *Chem. Abstr.* **108**, 22107h (1988).
18. N. Rizkalla and A. Goliaszewski in D. A. Fahey, ed., *Industrial Chemicals via C₁ Processes* (ACS Symposium Series, 328), American Chemical Society, Washington, D.C., 1987, pp. 136–153.
19. Ger. Pat. 1,904,141 (Sept. 11, 1969), W. Gay, A. F. Vellturo, and D. Sheehan.
20. Ger. Offen. 3,510,035 (Sept. 25, 1986), K. Bott and co-workers (to BASF A.G.); Jpn. Pats. JP 61-151,152 and JP 61-151,153 (July 9, 1986) and JP 61-161,241 (Jan. 8, 1985), K. Inoue, H. Takeda, and M. Kobayashi (to Mitsubishi Rayon Co. Ltd.); Ger. Pat. 3,644,222 (July 30, 1987), M. Hinenoya and M. Endo (to Daicel Chemical Industries Ltd.).
21. V. Zvonar, *Coll. Czech. Chem. Communs.* **38**(8), 2187 (1973).
22. Ger. Pat. 1,076,090 (Feb. 25, 1960), corresponding to U.S. 3,111,548 (Nov. 19, 1963), T. Alternschoepfer, H. Spes, and L. Vornehm (to Wacker-Chemie); H. Spes, *Chem. Ing. Tech.* **38**, 963 (1966).
23. W. Hunter, *BIOS Field Report 1050, # 22,* Feb. 1, 1947; reissued as *PB 68,123,* U.S. Dept. of Commerce, Washington, D.C.
24. G. V. Jeffreys, *The Manufacture of Acetic Anhydride,* 2nd ed., The Institution of Chemical Engineers, London, 1964.
25. "Acetic Anhydride Design Problem" in J. J. McKetta and W. A. Cunningham, eds., *Encyclopedia of Chemical Processing and Design,* Marcel Dekker, Inc., New York, Vol. 1, p. 271.
26. Ger. Pats. 699,709 (Nov. 7, 1940) and 708,822 (June 19, 1941), J. Loesch and co-workers (to A.G. für Stickstoffdünger); *Chem. Abstr.* **35**, 5133 (1941); **36**, 1955 (1942).
27. Fr. Pat. 1,346,360 (Dec. 20, 1963), G. Sitaud and P. Biarais (to Les Usines de Melle).
28. U.S. Pat. 2,730,546 (April 16, 1957), W. Reppe (to BASF A.G.).
29. H. Goldschmidt and co-workers, *Z. Physik. Chem.* **60**, 728 (1907); *Z. Physik. Chem.* **81**, 30 (1912); *Z. Physik. Chem.* **143**, 139 (1929); *Z. Physik. Chem.* **143**, 278 (1929); A. T. Williamson and co-workers, *Trans. Faraday Soc.* **34**, 1145–1149 (1934); H. A. Smith, *J. Am. Chem. Soc.* **61**(1), 254–260 (1939).
30. A. G. Crawford and co-workers, *J. Chem. Soc.* **1949**, 1054–1058 (1949).
31. Ger. Pat. 1,070,165 (Dec. 3, 1959), K. Kummerle (to Farbwerke Hoechst); Br. Pat. 2,033,385 (May 21, 1980), C. G. Wan (to Halcon).
32. K. Fujimoto and co-workers, *Am. Chem. Soc. Div. Pet. Chem., Prepr.,* **31**(1), 85–90, 91–96 (1986).

33. E. Drent, *Am. Chem. Soc. Div. Pet. Chem. Prepr.* **31**(1), 97–103 (1986).
34. U.S. Pats. 4,002,678 and 4,002,677 (Jan. 11, 1977), A. N. Naglieri and co-workers (to Halcon International).
35. Eur. Pat. 8396 (March 5, 1980), H. Erpenbach and co-workers (to Hoechst).
36. Eur. Pat. 180,802 (Oct. 10, 1986), G. Luft and G. Ritter (to Hoechst A.G.).
37. G. Ritter and G. Luft, *Chem. Ing. Tech.* **59**(6), 485–486 (1987); **58**(8), 668–669 (1986).
38. Jpn. Pat. 62-135,445 (June 16, 1987), N. Okada and O. Takahashi (to Idemitsu Kosan K.K.).
39. U.S. Pat. 4,690,912 (Sept. 1, 1987), F. E. Paulik, A. Hershman, W. R. Knox, and J. F. Roth (to Monsanto Company).
40. Jpn. Pat. 62-226,940 (Oct. 5, 1987), H. Koyama, N. Noda, and H. Kojima (to Daicel Chemical Industry K.K.).
41. V. H. Agreda, *CHEMTECH* 250–253 (April 1988).
42. T. H. Larkins, Jr., *Am. Chem. Soc. Div. Pet. Chem. Prepr.* **31**(1), 74–78 (1986); G. G. Mayfield and V. H. Agreda, *Energy Progress* **6**(4), 214–218 (Dec. 1986); *Chem. Eng. News,* 30–32 (May 21, 1990).
43. S. W. Polichnowski, *J. Chem. Educ.* **63**(3), 206–209 (1986).
44. U.S. Pat. 4,605,541 (Aug. 12, 1986), J. Pugach (to The Halcon SD Group, Inc.).
45. U.S. Pat. 4,650,649 (Mar. 17, 1987) and U.S. Pat. 4,578,368 (Mar. 25, 1986), J. R. Zoeller (to Eastman Kodak Company).
46. U.S. Pat. 4,440,570 (April 3, 1984), U.S. Pat. 4,442,304 (April 10, 1984), U.S. Pat. 4,557,760 (Dec. 1985), H. Erpenbach and co-workers (to Hoechst A.G.).
47. Brit. Pat. 2,033,385 (May 21, 1980), P. L. Szecsi (to Halcon Research & Development).
48. V. M. Rodionov, A. I. Smarin, and T. A. Obletzove, *Chem. Abstr.* **30**, 1740, 4149 (1936); V. M. Rodionov and T. A. Oblitseva, *Chem. Zentr.*, **1938**(II), 4054; *Chem. Abstr.* **34**, 6572 (1940).
49. Fr. Pat. 47,873 addn. to Fr. Pat. 809,731 (Aug. 14, 1937), A. Consalvo; Brit. Pat. 480,953 (Feb. 28, 1938); Brit. Pat. 486,964, addn. to 480,953 (June 14, 1938).
50. W. P. Jencks and co-workers, *J. Am. Chem. Soc.* **88**, 4464–4467 (1966).
51. R. Wolfenden and R. Williams, *J. Am. Chem. Soc.* **107**, 4345–4346 (1985).
52. Jpn. Pat. 60-204,738 (Oct. 16, 1985), M. Ichino and co-workers (to Daicel Chemical Industries Ltd.).
53. Fr. Pat. 973,160 (Feb. 8, 1951), G. H. van Hoek; Fr. Pat. 1,040,934 (Oct. 20, 1953), J. Francon.
54. D. Hanson, *Chem. Eng. News,* 17 (Feb. 27, 1989).
55. E. F. Joy and A. J. Barnard, Jr. in F. D. Snell and C. L. Hilton, eds., *Encyclopedia of Industrial Chemical Analysis,* Vol. 4, John Wiley & Sons Inc., New York, 1967, pp. 102–107.
56. G. A. P. Tuey, *Chem. Ind. London,* **1948,** 766 (1948); D. A. Peak, *Chem. Ind. London* **1949,** 14 (1949).
57. J. Dubar and J. Calzia, *C.R. Acad. Sci., Ser. C.* **266,** 1114 (1968).
58. V. I. Semenikhim and co-workers, *Tr. Mosks. Khim. Tekhnol. Inst. im. D. I. Mendeleeva* **112,** 43–47 (1980); *Chem. Abstr.* **97,** 99151n (1982).
59. V. M. Raikova, *Tr. Mosks. Khim. Tekhnol. Inst.* **112,** 97–102 (1980); *Chem. Abstr.* **98,** 5958b (1983).
60. J. H. Kuney, *Chem. Eng. News* **25,** 1658 (1947).

FRANK S. WAGNER, JR.
Nandina Corporation

ACETYL CHLORIDE

Acetyl chloride [75-36-5], C_2H_3OCl, mol wt 78.50, is a colorless, corrosive, irritating liquid that fumes in air. It has a stifling odor and reacts very rapidly with water, readily hydrolyzing to acetic acid and hydrochloric acid. As little as 0.5 parts per million activate the flow of tears, and often provoke a burning sensation in the eyes, nose, and throat. Acetyl chloride is toxic. Its high reactivity with hydroxyl, sulfhydryl, and amine groups leads to modifications that block the action of many important enzymes needed by living tissue.

Physical and Chemical Properties

The common physical properties of acetyl chloride are given in Table 1. The vapor pressure has been measured (2,7), but the experimental difficulties are considerable. An equation has been worked out to represent the heat capacity (8), and the thermodynamic ideal gas properties have been conveniently organized (9).

Table 1. Physical Properties of Acetyl Chloride

Property	Value	Reference
freezing point, °C	−112.0	1
boiling point, °C at 101.3 kPa[a]	50.2	2
density, g/mL		3
4°C	1.1358	4
20°C	1.1051	3
25°C	1.0982	4
heat of formation, ΔH_f, kJ/mol[b]	−243.93	5
heat of vaporization at bp, ΔH_v, kJ/g[b]	0.36459	6
refractive index, n_D^{20}	1.38976	3

[a]To convert kPa to atm, divide by 101.3.
[b]To convert kJ to kcal, divide by 4.184.

The important chemical properties of acetyl chloride, CH_3COCl, were described in the 1850s (10). Acetyl chloride was prepared by distilling a mixture of anhydrous sodium acetate [127-09-3], $C_2H_3O_2Na$, and phosphorous oxychloride [10025-87-3], $POCl_3$, and used it to interact with acetic acid yielding acetic anhydride. Acetyl chloride's violent reaction with water has been used to model liquid-phase reactions.

$$CH_3COCl + H_2O \longrightarrow CH_3COOH + HCl$$

A fixed-bed reactor for this hydrolysis that uses feed-forward control has been described (11); the reaction, which is first order in both reactants, has also been studied kinetically (12–14). Hydrogen peroxide interacts with acetyl chloride to yield both peroxyacetic acid [79-21-0], $C_2H_4O_3$, and acetyl peroxide [110-22-5], $C_4H_6O_4$ (15). The latter is a very dangerous explosive.

Reactions of acetyl chloride that are formally analogous to hydrolysis occur with alcohols, mercaptans, and amines: primary or secondary compounds form corresponding acetates or amides; tertiary alcohols generally yield the tertiary alkyl chlorides. Acetyl chloride can split the ether linkages of many ordinary ethers and acetals. It equilibrates with fatty acids to provide measureable amounts of the mixed acetic–alkylcarboxylic anhydride or acyl chloride, either of which may be employed in esterifications. For example, lauric acid [143-07-7], $C_{12}H_{24}O_2$, and acetyl chloride undergo the reactions

$$CH_3(CH_2)_{10}COOH + CH_3COCl \longrightarrow CH_3(CH_2)_{10}COCl + CH_3COOH$$

$$CH_3(CH_2)_{10}COCl + CH_3(CH_2)_{10}COOH \longrightarrow (CH_3(CH_2)_{10}CO)_2O + HCl$$

Acetyl chloride reacts with aromatic hydrocarbons and olefins in suitably inert solvents, such as carbon disulfide or petroleum ether, to furnish ketones (16). These reactions are catalyzed by anhydrous aluminum chloride and by other inorganic chlorides (17). The order of catalytic activity increases in the order

$$ZnCl_2 < BiCl_3 < TeCl_4 < TiCl_4 < SnCl_4 < TeCl_2 < FeCl_3 < SbCl_5 < AlCl_3$$

Acetyl chloride is reduced by various organometallic compounds, eg, LiAlH$_4$ (18). *tert*-Butyl alcohol lessens the activity of LiAlH$_4$ to form lithium tri-*t*-butoxyalumium hydride [17476-04-9], $C_{12}H_{28}AlO_3Li$, which can convert acetyl chloride to acetaldehyde [75-07-0] (19). Triphenyltin hydride also reduces acetyl chloride (20). Acetyl chloride in the presence of Pt(II) or Rh(I) complexes, can cleave tetrahydrofuran [109-99-9], C_4H_8O, to form chlorobutyl acetate [13398-04-4] in about 72% yield (21). Although catalytic hydrogenation of acetyl chloride in the Rosenmund reaction is not very satisfactory, it is catalytically possible to reduce acetic anhydride to ethylidene diacetate [542-10-9] in the presence of acetyl chloride over palladium complexes (22). Rhodium trichloride, methyl iodide, and triphenylphosphine combine into a complex that is active in reducing acetyl chloride (23).

Manufacture

Acetyl chloride is manufactured commercially in Europe and the Far East. Some acetyl chloride is produced in the United States for captive applications such as acetylation of pharmaceuticals.

Acetyl chloride was formerly manufactured by the action of thionyl chloride [7719-09-7], Cl_2OS, on gray acetate of lime, but this route has been largely supplanted by the reaction of sodium acetate or acetic acid and phosphorus trichloride [7719-12-2] (24). A similar route apparently is still being used in the Soviet Union (25). Both pathways are inherently costly.

Patents on the carbonylation of methyl chloride [74-87-3] using carbon monoxide [630-08-0] in the presence of rhodium, palladium, and iridium complexes, iodo compounds, and phosphonium iodides or phosphine oxides have been obtained (26). In one example the reaction was conducted for 35 min at 453 K and 8360 kPa (82.5 atm) to furnish a 56% conversion to acetyl chloride. Reactions of this kind are now possible because of improved corrosion-resistant alloys (27,28). It is not known whether these methods are practiced industrially.

Other acetyl chloride preparations include: the reaction of acetic acid and chlorinated ethylenes in the presence of ferric chloride [7705-08-0] (29); a combination of benzyl chloride [100-44-3] and acetic acid at 85% yield (30); conversion of ethylidene dichloride, in 91% yield (31); and decomposition of ethyl acetate [141-78-6] by the action of phosgene [75-44-5], producing also ethyl chloride [75-00-3] (32). The expense of raw material and capital cost of plant probably make this last route prohibitive. Chlorination of acetic acid to monochloroacetic acid [79-11-8] also generates acetyl chloride as a by-product (33). Because acetyl chloride is costly to recover, it is usually recycled to be converted into monochloroacetic acid. A salvage method in which the mixture of HCl and acetyl chloride is scrubbed with H_2SO_4 to form acetyl sulfate has been patented (33).

Analysis and Quality Control

Acetyl chloride frequently contains 1–2% by weight of acetic acid or hydrochloric acid. Phosphorus or sulfur-containing acids may also be present in the commercial material. A simple test for purity involves addition of a few drops of Crystal Violet solution in $CHCl_3$. Pure acetyl chloride will retain the color for as long as 10 min, but hydrochloric, sulfuric, or acetic acid will cause the solution to become first green, then yellow (34).

Economic Aspects and Shipping

Little is known of the market for acetyl chloride. The production and sales are believed to be small, but may have potential for very large scale-up. The total U.S. market may amount to only 500 t annually. Acetyl chloride must be shipped in polyethylene-lined drums having capacities of only 220 L; it must be labeled as a corrosive substance. Acetyl chloride generated captively from purchased raw materials probably has a unit value of no more than $0.92–0.95/kg. Shipping costs and other factors set the price at about $3/kg for the commercial trade.

Health and Safety

Acetyl chloride has stifling fumes and an irritating odor. It is highly poisonous. Because of acetyl chloride's combustibility and its high reactivity toward water and many alkalies, extreme caution must be exercised in handling. Provisions must be made for drawing the vapors away from people, and handlers should wear impervious, protective clothing. The large containers ought to be stored in cool, dry areas that are separate from noncorrosive, flammable chemicals.

Uses

A small amount of acetyl chloride is consumed in the start-up of acetic acid chlorination to monochloroacetic acid. After initiation, the acetyl chloride by-product provides sufficient catalysis. Acetyl chloride is a powerful acetylating agent. It is used in the manufacture of aspirin, acetaminophen, acetanilide, and acetophenone. Liquid crystal compositions for optical display and memory devices frequently require acetyl chloride. Reactions which are difficult or lethargic using acetylation agents such as acetic anhydride or acetic acid become facile using acetyl chloride (35,36). Polymers made by reactions involving acetyl chloride can chelate with metal ions, such as copper, and may be prepared with superior electrical and magnetic properties (37).

Anthralin [1143-38-0] is acetylated using acetyl chloride in toluene and a pyridine catalyst to furnish 1,8-dihydroxy-10-acetylanthrone [3022-61-5], an intermediate in the preparation of medications used in treating skin disorders, such as warts, psoriasis, and acne (38). Sugar esters can be similarly prepared from acetyl chloride under anhydrous conditions (39).

Although acetyl chloride is a convenient reagent for determination of hydroxyl groups, spectroscopic methods have largely replaced this application in organic chemical analysis. Acetyl chloride does form derivatives of phenols, uncomplicated by the presence of strong acid catalysts, however, and it finds some use in acetylating primary and secondary amines.

Acetyl chloride can be used as a substitute for acetic anhydride in many reactions. Whereas the anhydride requires a mineral acid catalyst for acetylation, acetyl chloride does not. Acetyl chloride is utilized in a wide range of reactions wherein its comparatively high price is offset by convenience. Should its nominal cost be lowered, acetyl chloride would be a powerful competitor for acetic anhydride in large scale manufacturing.

BIBLIOGRAPHY

"Acetyl Chloride" in *ECT* 1st ed., Vol. 1, pp. 98–100, by M S. W. Small, Shawinigan Chemicals Limited; in *ECT* 2nd ed., Vol. 1, pp. 138–142, by A. P. Laurie, Eastman Kodak Company; in *ECT* 3rd ed., Vol. 1, pp. 162–165, by Theresa A. Moretti, American Hoechst Corporation.

1. J. Timmermans and Th. J. F. Mattaar, *Bull. Soc. Chim. Belg.* **30**, 213–219 (1921).
2. J. A. Devore and H. E. O'Neal, *J. Phys. Chem.* **73**, 2644–2648 (1969).
3. P. Walden, *Z. Phys. Chem. Leipzig* **55**, 212–222 (1909); **70**, 569–619 (1910).
4. J. A. Brühl, *Justus Liebigs Ann. Chem.* **203**, 11 (1880).
5. R. A. McDonald, S. A. Shrader, and D. R. Stull, *J. Chem. Eng. Data* **4**, 311–313 (1959).
6. J. H. Matthews and P. R. Fehlandt, *J. Am. Chem. Soc.* **53**, 3212–3217 (1931).
7. C. L. Yaws, H. M. Ni, and P. Y. Chiang, *Chem. Eng. New York* **98**, 91–98 (May 9, 1988).
8. C. L. Yaws and P. Y. Chiang, *Chem Eng. New York* **98**, 81–88 (Sept. 26, 1988).
9. D. R. Stull, E. F. Westrum, Jr., and G. C. Sinke, *The Chemical Thermodynamics of Organic Compounds,* John Wiley & Sons, Inc., New York, 1969, pp. 536–537; cf. S. G. Frankiss and W. Kynaston, *Spectrochim. Acta Part A* **31A**, 661–678 (1975).
10. C. Gerhardt, *Compt. Rend.* **34**, 755–902 (1852).
11. J. D. Tinker and D. E. Lamb, *Chem. Eng. Prog. Symp. Ser.* **61**(55), 155–167 (1965).
12. C. G. Swain and C. B. Scott, *J. Am. Chem. Soc.* **75**, 246–248 (1953).
13. V. Macho and M. Rusina, *Petrochemia* **25**(5–6), 127–136 (1985); *Chem. Abstr.* **107**, 22761d (1987).
14. V. Gold and J. Hilton, *J. Chem. Soc. London* **1955**, 838–842 (1955).

15. J. D'Ans and W. Friederich, *Z. Anorg. Chem.* **73**, 355 (1912).

16. S. Krapiwin, *Bull. Soc. Imp. Nat. Moscou* **1908**, 1–38; cf. C. Friedel and J. M. Crafts, *Ann. Chim. (Paris)* **[6]**, 1, 507.

17. O. C. Dermer, D. M. Wilson, F. M. Johnson, and V. H. Dermer, *J. Am. Chem. Soc.* **63**, 2881–2883 (1941); O. C. Dermer and R. A. Billmaier, *J. Am. Chem. Soc.* **64**, 464–465 (1942).

18. C. E. Sroog and H. M. Woodburn, *Org. Syn.* **32**, 46–48 (1952).

19. H. C. Brown and B. C. Subba Rao, *J. Am. Chem. Soc.* **80**, 5377–5380 (1958).

20. H. Patin and R. Dabard, *Bull. Soc. Chim. Fr.* **1973**(9–10, part 2), 2760–2764 (1973).

21. J. W. Fitch, W. G. Payne, and D. Westmoreland, *J. Org. Chem.* **48**, 751–753 (1983).

22. Ger. Offen. DE 3,140,214 (May 19, 1982), J. W. Brockington, W. R. Koch, and C. M. Bartish (to Air Products & Chemicals, Inc.); *Chem. Abstr.* **97**, 25526s (1982).

23. Eur. Pat. EP-58,442 (Aug. 25, 1982), B. V. Maatschappij (to Shell Internationale Research); *Chem. Abstr.* **97**, 215584g (1982).

24. Hun. Pat. 30,692 (Mar. 28, 1984), J. Besan (to Nehezvegyipari Kutato Intezet Nitrokemia Ipartelepek); *Chem. Abstr.* **101**, 72287a (1984).

25. G. Ya. Gordon, N. N. Davydova, and S. U. Kreigol'd, *Tr. IREA* **48**, 129–133 (1986); *Chem. Abstr.* **111**, 16943f (1989).

26. Ger. Offen. 3,016,900 (Nov. 5, 1981), H. Erpenbach and co-workers (to Hoechst A.G.); *Chem. Abstr.* **96**, 19680a (1982).

27. Ger. Offen. 3,248,468 (July 12, 1984), H. Erpenbach and co-workers (to Hoechst A.G.); *Chem. Abstr.* **101**, 191160p (1984).

28. Eur. Pat. 48,335 (Mar. 31, 1982), H. Erpenbach and co-workers (to Hoechst A.G.); *Chem. Abstr.* **97**, 109563q (1982).

29. Ger. Offen. 2,059,597 (June 9, 1971), J. C. Strini (to Produits Chimiques Pechiney-Saint Gobain); *Chem. Abstr.* **75**, 109879t (1971).

30. H. C. Brown, *J. Am. Chem. Soc.* **60**, 1325–1328 (1938).

31. U.S. Pat. 2,778,852 (Jan. 22, 1957) (corresponding to Brit. Pat. 743,557), K. Adam and H. G. Trieschmann (to Badische Anilin und Soda-Fabrik A.G.); *Chem. Abstr.* **51**, 5819h (1957).

32. Jpn. Pat. 60-65701 [85-65,701] (Apr. 15, 1985) (to Daicel Chemical Industries Ltd.); *Chem. Abstr.* **104**, 70666m (1986).

33. Jpn. Pat. 63-50303 [88-50303] (Mar. 3, 1988), T. Myazaki and K. Murata (to Daicel Chemical Industries Ltd.); *Chem. Abstr.* **108**, 223885s (1988).

34. J. Singh, R. C. Paul, and S. S. Sandhu, *J. Chem. Soc.* [**1959**, 845–847 (1959); R. C. Paul, S. P. Narula, P. Meyer, and S. K. Gondal, *J. Sci. Ind. Res.* **21B**, 552–554 (1962).

35. PCT Int. Appl. WO-88-05,803 (Aug. 11, 1988), I. C. Sage, S. J. Lewis, and D. Chaplin (to Merck Patent G.m.b.H); *Chem. Abstr.* **109**, 240837r (1988).

36. Jpn. Pat. 63-281,830 [87-281-830] (Dec. 7, 1987), S. Tatsuki, T. Ikeda, and K. Tachibana (to Daicel Chemical Industries Ltd.); *Chem. Abstr.* **109**, 45862t (1988).

37. Eur. Pat. 242,278 (Oct. 21, 1987), S. I. Stupp and J. S. Moore (to University of Illinois); *Chem. Abstr.* **108**, 159,114k (1988).

38. Ger. Offen. 3,523,231 (Jan. 9, 1986), B. Shroot, G. Lang, and J. Maignan (to Centre International de Recherches Dermatologiques); *Chem. Abstr.* **105**, 6323r (1986).

39. Br. Pat. 2,163,425 (Feb. 26, 1986), S. Kea and C. E. Walker (to Nebraska Department of Economic Development); *Chem. Abstr.* **105**, 224,935h (1986).

FRANK S. WAGNER, JR.
Nandina Corporation

DIMETHYLACETAMIDE

Dimethylacetamide [127-19-5], DMAC, mol wt 87.12, $CH_3CON(CH_3)_2$, is a colorless, high boiling polar solvent. DMAC is a good solvent for a wide range of organic and inorganic compounds and it is miscible with water, ethers, esters, ketones, and aromatic compounds. Unsaturated aliphatics are highly soluble, but saturated aliphatics have limited solubility in DMAC. The polar nature of DMAC enables it to act as a combined solvent and reaction catalyst, in many instances producing high yields and pure product in short time periods.

The rate of hydrolysis of DMAC is very low, but increases somewhat in the presence of acids or bases. DMAC is a stable compound, but is mildly hygroscopic and desiccation and/or dry nitrogen blanketing of storage vessels are sometimes used to reduce water pick-up. In the absence of water, acids, or bases, DMAC is stable at temperatures up to its boiling point at atmospheric pressure. Its greater stability enables more economical recovery by distillation relative to that of other similar solvents.

Physical Properties

Selected physical properties of DMAC are: boiling point, 166.1°C; melting point, −20°C; vapor pressure at 25°C, 0.27 kPa (2 mm Hg); density at 15.6°C, 0.945 g/mL; viscosity at 25°C, 0.92 mPa·s (= cP); surface tension at 30°C, 32.43 mN/m (= dyn/cm); refractive index n_D^{25}, 1.4356; heat of vaporization at 166°C, 43.1 kJ/mol (10.3 kcal/mol); heat of combustion, 2544 kJ/mol (608 kcal/mol); thermal conductivity at 22.2°C, 0.1835 W/(m·K) (0.1579 kcal/(m·h·°C)); flash point (Tag closed cup), 63°C; ignition temperature, 490°C; flammability limits in air, 1.8 vol % LEL (lower explosive limit) (100°C), 8.6 wt % UL (upper limit) (120°C); critical temperature, 385°C; critical pressure, 4.02 MPa (39.7 atm); dielectric constant at 25°C, 10 kHz, 37.8; dipole moment at 20°C, 1.64×10^{-29} C·m (4.6 Debye units); solubility parameter, 10.8; hydrogen bonding index, 6.6; DMAC–acetic acid azeotrope at 170.8°C and 101.3 kPa (1 atm), 84.1 and 15.9 wt % (78.5 and 21.5 mol %), respectively (1).

Chemical Properties

The chemical reactions of DMAC are typical of those of disubstituted amides. Under suitable conditions, DMAC will react as follows:

Hydrolysis in the Presence of Strong Acids

$$CH_3CON(CH_3)_2 + H_2O + HCl \longrightarrow CH_3COOH + (CH_3)_2NH \cdot HCl$$

Saponification in the Presence of Strong Bases

$$CH_3CON(CH_3)_2 + NaOH \longrightarrow CH_3COONa + (CH_3)_2NH$$

Alcoholysis in the Presence of Hydrogen Ions

$$CH_3CON(CH_3)_2 + ROH \longrightarrow CH_3COOR + (CH_3)_2NH$$

Manufacturing Processes

Dimethylacetamide can be produced by the reaction of acetic acid [64-19-7] and dimethylamine [124-40-3]:

$$CH_3COOH + (CH_3)_2NH \longrightarrow CH_3CON(CH_3)_2 + H_2O$$

The product of this reaction can be removed as an azeotrope (84.1% amide, 15.9% acetic acid) which boils at 170.8–170.9°C. Acid present in the azeotrope can be removed by the addition of solid caustic soda [1310-73-2] followed by distillation (2). The reaction can also take place in a solution having a DMAC–acetic acid ratio higher than the azeotropic composition, so that an azeotrope does not form. For this purpose, dimethylamine is added in excess of the stoichiometric proportion (3). If a substantial excess of dimethylamine reacts with acetic acid under conditions of elevated temperature and pressure, a reduced amount of azeotrope is formed. Optimum temperatures are between 250–325°C, and pressures in excess of 6200 kPa (900 psi) are required (4).

DMAC can also be made by the reaction of acetic anhydride [108-24-7] and dimethylamine:

$$(CH_3CO)_2O + 2\ (CH_3)_2NH \longrightarrow 2\ CH_3CON(CH_3)_2 + H_2O$$

Dimethylamine is added somewhat in excess of the stoichiometric proportion in this synthesis. Another method employs the reaction of methyl acetate [79-20-9] and dimethylamine:

$$CH_3CO_2CH_3 + (CH_3)_2NH \xrightarrow{\text{NaOCH}_3} CH_3CON(CH_3)_2 + CH_3OH$$

Dimethylamine also reacts with the azeotrope of methyl acetate and methanol to give DMAC in 45% yield (5).

Shipping and Storage

Dimethylacetamide is available in drums with a capacity of 0.208 m^3 (55 gal), 186 kg net, and in tank cars or trucks. Although the DOT classifies DMAC as a combustible liquid, no DOT label is required.

Mild steel is a suitable material of construction for storage and handling of DMAC at ambient temperatures. Aluminum or stainless steel is recommended for cases involving very stringent color or iron contamination requirements. Mild steel is not recommended for high temperature service or handling aqueous solutions of less than 50 mol % (82.86 wt %) DMAC.

DMAC is a good solvent for many resins; therefore, flange gaskets and pump and valve packing should be limited to Teflon fluorocarbon resins.

Analytical Methods

Determination of Water in DMAC.   DMAC is hygroscopic and precautions must be taken to minimize exposure to the atmosphere. Trace amounts of water can be determined by the Karl-Fischer method.

Determination of DMAC in Air.   DMAC can be measured in air by passing a known amount of sample through water in a gas-scrubbing vessel and then analyzing the solution either chemically or by gas chromatography.

Health and Safety Factors

DMAC is capable of producing systemic injury when repeatedly inhaled or absorbed through the skin. Symptoms of overexposure are nausea, headache, and weakness. The principal effect is cumulative damage to the liver and kidney. DMAC has a low order of acute toxicity when swallowed or upon brief contact of the liquid vapor with the eyes or skin. Although DMAC is not a skin sensitizer, it is irritating to the skin and eyes. The LD_{50} (oral, rats) for DMAC is greater than 2250 mg/kg (6).

In laboratory tests, application of DMAC to the skin of pregnant rats has caused fetal deaths when the dosages were close to the lethal dose level for the mother. Embryonal malformations have been observed at dose levels 20% of the lethal dose and higher. However, when male and female rats were exposed to mean DMAC concentrations of 31,101, and 291 ppm for six hours per day over several weeks, no reproductive effects were observed (6).

It is important to note that DMAC can rapidly penetrate the skin, leading to overexposure. Contact of DMAC liquid or mixtures containing DMAC with the eyes, skin, and clothing should be avoided. If contact is unavoidable, appropriate personal protective equipment, including chemical splash goggles, butyl rubber gloves, or an impervious butyl chemical suit having breathing air supply should be worn.

The U.S. Department of Labor (OSHA) has ruled that an employee's exposure to dimethylacetamide in any 8-h work shift of a 40-h work week shall not exceed a time-weighted average of 10 ppm DMAC vapor in air by volume or 35 mg/m^3 in air by weight (7). If there is significant potential for skin contact with DMAC, biological monitoring should be carried out to measure the level of DMAC metabolites in urine specimens collected at the end of the shift. One industrial limit is 40 ppm DMAC metabolites, expressed as N-methylacetamide [79-16-3], for individuals, and 20 ppm metabolite average for workers on the job (8).

Uses

The uses of dimethylacetamide are very similar to those for dimethylformamide [68-12-2] (see FORMIC ACID). DMAC is employed most often where higher temperatures are needed for solution of resins or activation of chemical reactions.

Resin and Polymer Solvent. Dimethylacetamide is an excellent solvent for synthetic and natural resins. It readily dissolves vinyl polymers, acrylates, cellulose derivatives, styrene polymers, and linear polyesters. Because of its high polarity, DMAC has been found particularly useful as a solvent for polyacrylonitrile, its copolymers, and interpolymers. Copolymers containing at least 85% acrylonitrile dissolve in DMAC to form solutions suitable for the production of films and yarns (9). DMAC is reportedly an excellent solvent for the copolymers of acrylonitrile and vinyl formate (10), vinylpyridine (11), or allyl glycidyl ether (12).

Polyimides for use in molded products and high temperature films can be produced by the reaction of pyromellitic dianhydride [89-32-7] and 4,4'-diaminodiphenyl ether [101-80-4] in DMAC to form a polyamide that can be converted into a polyimide (13). DMAC can also be used as a spinning solvent for polyimides. Additionally, polymers containing over 50% vinylidene chloride are soluble up to 20% at elevated temperatures in DMAC. Such solutions are useful in preparing fibers (14).

DMAC and nonpolar solvents form synergistic mixtures which dissolve high molecular weight vinyl chloride homopolymers. For example, a mixture of DMAC with an equal volume of carbon disulfide [75-15-0], a nonsolvent, dissolves 14 wt % of Geon 101 vinyl chloride homopolymer at room temperature, whereas the solubility of Geon 101 in DMAC alone is about 5 wt % (15).

Crystallization and Purification Solvent. Dimethylacetamide is useful in the purification by crystallization of aromatic dicarboxylic acids such as terephthalic acid [100-21-0] and p-carboxyphenylacetic acid [501-89-3]. These acids are not soluble in the more common solvents. DMAC and dibasic acids form crystalline complexes containing two moles of the solvent for each mole of acid (16). Microcrystalline hydrocortisone acetate [50-03-3] having low settling rate is prepared by crystallization from an aqueous DMAC solution (17).

Electrolytic Solvent. The use of DMAC as a nonaqueous electrolytic solvent is promising because salts are modestly soluble in DMAC and appear to be completely dissociated in dilute solutions (18).

Complexes. In common with other dialkylamides, highly polar DMAC forms numerous crystalline solvates and complexes. The HCN–DMAC complex has been cited as an advantage in using DMAC as a reaction medium for hydrocyanations. The complexes have vapor pressures lower than predicted and permit lower reaction pressures (19).

Complexes of DMAC and many inorganic halides have been reported (20). These complexes are of interest because they catalyze a number of organic reactions. Complexes of DMAC and such heavy metal salts as $NiBr_2$ exert a greater catalytic activity than the simple salts (21). The crystalline complex of SO_3 and dimethylacetamide has been suggested for moderating the reaction conditions in sulfation of leuco vat dyestuffs (22).

Chemical Reaction Medium and Catalyst. DMAC, as well as other alkyl carboxylic amides, has the ability to serve as a reaction catalyst, often increasing yields, improving product quality and reducing reaction time. DMAC is a highly polar solvent of high dielectric properties and is capable of converting many organic and inorganic molecules into reactive forms by solvation. Types of reactions improved by the use of DMAC are elimination reactions (23), halogenation (24), cyclization (25), alkylations (26), nitrile formation (27), formation of organic isocyanates (28), interesterification (29), phthaloylation (30), and formation of sulfonyl chlorides (31).

BIBLIOGRAPHY

"Acetic Acid Derivatives (Dimethylacetamide)" in *ECT* 2nd ed., Vol. 1, pp. 145–148, by D. C. Hubinger, E. I. du Pont de Nemours and Co., Inc.; in *ECT* 3rd ed., Vol., pp. 167–171, by J. C. Siegle, E. I. du Pont de Nemours and Co., Inc.

1. J. R. Ruhoff and E. E. Reid, *J. Am. Chem. Soc.* **59,** 401 (1937).
2. U.S. Pat. 2,667,511 (Jan. 26, 1954), C. Downing (to Chemstrand Corp.).
3. U.S. Pat. 3,006,956 (Oct. 31, 1961), A. W. Campbell (to Commercial Solvents Corp.).
4. S. Collis, *The Development and Evaluation of Paint Remover Used by the U.S. Air Force,* Air Force Technical Report 5714, Suppl. 1, Jan. 1955.
5. Ger. Offen. 2,437,702 (Feb. 13, 1975), W. B. Pearson (to Imperial Chemical Industries Ltd.).
6. G. L. Kennedy, *CRC Crit. Rev. Toxicol.* **17,** 129 (1986).
7. *Code of Federal Regulations,* Title 29, Section 1910.1000 (Sept. 1989).
8. *Dimethylacetamide,* Material Safety Data Sheet, E.I. du Pont de Nemours and Co., Inc., Wilmington, Del., 1988.
9. U.S. Pat. 2,649,427 (Aug. 18, 1953), C. S. Marvel (to E.I. du Pont de Nemours and Co., Inc.); U.S. Pat. 2,531,407 (Nov. 28, 1950), G. F. D'Alelio (to Industrial Rayon Corp.).
10. U.S. Pat. 2,558,793 (July 3, 1951), E. Stanin and J. B. Dickey (to Eastman Kodak Co.).
11. U.S. Pat. 2,676,952 (Apr. 27, 1954), E. Ham (to Chemstrand Corp.).
12. U.S. Pat. 2,650,151 (Aug. 25, 1953), E. Ham (to Chemstrand Corp.).
13. Belg. Pat. 627,628 (July 25, 1963), W. M. Edwards and A. L. Endrey (to E.I. du Pont de Nemours and Co., Inc.).
14. U.S. Pat. 2,531,406 (Nov. 28, 1950), G. F. D'Alelio (to Industrial Rayon Corp.).
15. R. L. Adelman and I. M. Klein, *J. Polym. Sci.* **31,** 77 (1958).
16. U.S. Pat. 2,811,548 (Oct. 29, 1957), G. E. Ham and A. B. Beindorff (to Chemstrand Corp.).
17. U.S. Pat. 2,805,232 (Sept. 3, 1957), W. H. Baade and T. J. Macek (to Merck & Co., Inc.).
18. G. R. Lester, T. A. Grover, and P. G. Sears, *J. Phys. Chem.* **60,** 1076 (1956).
19. U.S. Pat. 2,698,337 (Dec. 28, 1954), R. L. Heider and H. M. Walker (to Monsanto Chemical Co.).
20. W. Gerrard, M. F. Lappert, and J. W. Wallis, *J. Chem. Soc.* **1960,** 2141 (1960); C. L. Rollinson and R. C. White, *Inorg. Chem.* **1,** 281 (1962); W. E. Bull, S. K. Madan, and J. E. Willis, *Inorg. Chem.* **2,** 203 (1963).
21. U.S. Pat. 2,854,458 (Sept. 30, 1958), W. Reppe, H. Friederick, and H. Liantenschlager (to Badische Anilin- and Soda-Fabrik A.G.).
22. U.S. Pat. 2,506,580 (May 9, 1950), S. Coffey, G. W. Driver, D. A. Whyte, and F. Irving (to Imperial Chemical Industries, Ltd.).
23. U.S. Pat. 2,833,790 (May 6, 1958), J. M. Chemerda, E. M. Chamberlin, and E. W. Tristam (to Merck & Co., Inc.).
24. U.S. Pat. 2,823,232 (Feb. 11, 1958), H. F. Wilson (to Rohm and Haas Co.).
25. U.S. Pat. 2,872,447 (Feb. 3, 1959), H. F. Oehlschlaeger (to Emery Industries, Inc.).
26. U.S. Pat. 2,846,491 (Aug. 5, 1958), T. F. Rutledge (to Air Reduction Co., Inc.).
27. U.S. Pat. 2,715,137 (Aug. 9, 1955), H. B. Copelin (to E.I. du Pont de Nemours and Co., Inc.).
28. U.S. Pat. 2,866,801 (Dec. 30, 1958), C. M. Himel and L. M. Richards (to Ethyl Corp.).
29. U.S. Pat. 2,831,854 (Apr. 22, 1958), N. B. Tucker and J. B. Martin (to Procter & Gamble Co.).

30. U.S. Pat. 2,783,245 (Feb. 26, 1957), J. F. Weidenheimer, L. Ritter, and F. J. Richter (to American Cyanamid Co.).
31. U.S. Pat. 2,888,486 (May 26, 1959), W. A. Gregory (to E.I. du Pont de Nemours and Co., Inc.).

J. C. WATTS
P. A. LARSON
E.I. du Pont de Nemours and Co., Inc.

HALOGENATED DERIVATIVES

The most important of the halogenated derivatives of acetic acid is chloroacetic acid. Fluorine, chlorine, bromine, and iodine derivatives are all known, as are mixed halogenated acids. For a discussion of the fluorine derivatives see FLUORINE COMPOUNDS, ORGANIC.

Chloroacetic Acid

Physical Properties. Pure chloroacetic acid [79-11-8] ($ClCH_2COOH$), mol wt 94.50, $C_2H_3ClO_2$, is a colorless, white deliquescent solid. It has been isolated in three crystal modifications: α, mp 63°C, β, mp 56.2°C, and γ, mp 52.5°C. Commercial chloroacetic acid consists of the α form. Physical properties are given in Table 1.

Table 1. Physical Properties of Chloroacetic Acid

Property	Value
boiling point, °C	189.1
density, at 25°C, g/mL	1.4043
dielectric constant at 60°C	12.3
free energy of formation, at 100°C, ΔG_f, kJ/mol[a]	−368.7
heat capacity, J/(mol·k)[b] at 100°C	181.0
heat of formation, at 100°C, ΔH_f, kJ/mol[a]	−490.1
heat of sublimation, at 25°C, ΔH_s, kJ/mol[a]	88.1
vapor pressure, kPa[c],	
at 25°C	8.68×10^{-3}
at 100°C	3.24
viscosity, at 100°C, mPa·s (= cP)	1.29
refractive index at 55°C	1.435
surface tension, at 100°C, mN/m (= dyn/cm)	35.17
dissociation constant K_a	1.4×10^{-3}
solubility (g/100 g solvent)	
water	614
acetone	257
methylene chloride	51
benzene	26
carbon tetrachloride	2.75

[a]To convert kJ/mol to kcal/mol, divide by 4.184.
[b]To convert J/(mol·K) to cal/(mol·K), divide by 4.184.
[c]To convert kPa to mm Hg, multiply by 7.5.

Chloroacetic acid forms azeotropes with a number of organic compounds. It can be recrystallized from chlorinated hydrocarbons such as trichloroethylene, perchloroethylene, and carbon tetrachloride. The freezing point of aqueous chloroacetic acid is shown in Figure 1.

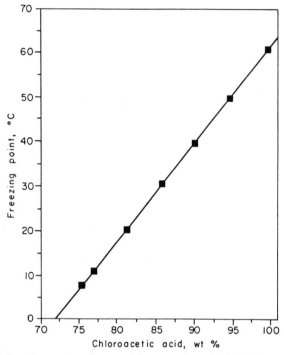

Fig. 1.   The freezing point of monochloroacetic acid (MCAA)–water mixtures. For wt % acid > 75%: fp(°C) = 2.17 × (wt % MCAA) − 156.5. Data courtesy of the Dow Chemical Company.

Chemical Properties and Industrial Uses.   Chloroacetic acid has wide applications as an industrial chemical intermediate. Both the carboxylic acid group and the α-chlorine are very reactive. It readily forms esters and amides, and can undergo a variety of α-chlorine substitutions. Electron withdrawing effects of the α-chlorine give chloroacetic acid a higher dissociation constant than that of acetic acid.

Major industrial uses for chloroacetic acid are in the manufacture of cellulose ethers (mainly carboxymethylcellulose, CMC), herbicides, and thioglycolic acid. Other industrial uses include manufacture of glycine, amphoteric surfactants, and cyanoacetic acid.

The reaction of chloroacetic acid with alkaline cellulose yields carboxymethylcellulose which is used as a thickener and viscosity control agent (1) (see CELLULOSE ETHERS). Sodium chloroacetate reacts readily with phenols and cresols. 2,4,-Dichlorophenol and 4-chloro-2-cresol react with chloroacetic acid to give the herbicides 2,4-D and MCPA (2-methyl-4-chlorophenoxy acetic acid), respectively (2,3) (see HERBICIDES). Thioglycolic acid [68-11-1], $C_2H_4O_2S$, is prepared commercially by treating chloroacetic acid with sodium hydrosulfide. Sodium and ammonium salts of thioglycolic acid (qv) have been used for cold-waving human hair; the calcium salt is used as a depilating agent (see COSMETICS). Thioglycolic acid derivatives are also used as stabilizers in poly(vinyl chloride) products.

Chloroacetic acid can be esterified and aminated to provide useful chemical intermediates. Amphoteric agents suitable as shampoos have been synthesized by reaction of sodium chloroacetate with fatty amines (4,5). Reactions with amines (6) such as ammonia, methylamine, and trimethylamine yield glycine [56-40-6], sarcosine [107-97-1], and carboxymethyltrimethylammonium chloride, respectively. Reaction with aniline forms N-phenylglycine [103-01-5], a starting point for the synthesis of indigo (7).

Reaction of chloroacetic acid with cyanide ion yields cyanoacetic acid [372-09-8], $C_3H_3NO_2$, (8) which is used in the formation of coumarin, malonic acid and esters, and barbiturates. Reaction of chloroacetic acid with hydroxide results in the formation of glycolic acid [79-14-1].

Manufacture.   Most chloroacetic acid is produced by the chlorination of acetic acid using a suitable catalyst such as acetic anhydride (9–12). The remainder is produced by the hydrolysis of trichloroethylene with sulfuric acid (13,14) or by reaction of chloroacetyl chloride with water.

A major disadvantage of the chlorination process is residual acetic acid and overchlorination to dichloroacetic acid. Although various inhibitors have been tried to reduce dichloroacetic acid formation, chloroacetic acid is usually purified by crystallization (15–17). Dichloroacetic acid can be selectively dechlorinated to chloroacetic acid with hydrogen and a catalyst such as palladium (18–20). Extractive distillation (21) and reaction with ketene (22) have also been suggested for removing dichloroacetic acid. Whereas the hydrolysis of trichloroethylene with sulfuric acid yields high purity chloroacetic acid, free of dichloracetic acid, it has the disadvantage of utilizing a relatively more expensive starting material and producing a sulfur containing waste stream.

Corrosive conditions of the chlorination process necessitate the use of glass-lined or lead-lined steel vessels in the manufacture of chloroacetic acid. Process piping and valves also are either glass-lined, or steel lined with a suitable polymer, eg, polytetrafluoroethylene (PTFE). Pumps, heat exchangers, and other process equipment can be fabricated from ceramic, graphite composite, tantalum, titanium, or certain high performance stainless steels. Chloroacetic acid can be stored as a molten liquid in glass-lined tanks for a short period of time, but develops color on aging. For long-term storage, the solid, flaked form of the acid is held in a polyethylene-lined fiber drum. The drum is constructed so that no chloroacetic acid vapors contact the fiber or metallic portions. Stainless steels are acceptable for the shipment of 80% solutions of chloroacetic acid provided the temperature is maintained below 50°C. However, long-term or continuous storage of the 80% solution results in signficant pickup of iron from stainless steel.

Gas chromatography or liquid chromatography (23) are commonly used to measure impurities such as acetic, dichloroacetic, and trichloroacetic acids. High purity 99 + % chloroacetic acid will contain less than 0.5% of either acetic acid or dichloroacetic acid. Other impurities that may be present in small amounts are water and hydrochloric acid.

Economic Aspects. Figures on U.S. production, imports, and projected demand of chloroacetic acid are listed in Table 2 (24). The majority of imported chloroacetic acid is produced in Germany. Western European capacity for chloroacetic acid is in excess of 225,000 metric tons per year. In 1990 the price was $1.25 to $1.36/kg (24,25).

Table 2. Production and Demand for Chloroacetic Acid

Year	U.S. production, 10^3 t	Imports, 10^3 t
1980	33	17.3
1981	35	16.7
1982	38	13.5
1983	43	8.6
1984	48	3.7
1985	36	9.6
1986	22	14.2
1987	15–19	15.9
	Demand[a]	
1988	38.6	
1989	39.5	
1993[b]	43.5	

[a]Ref. 25.
[b]Projected.

Toxicity and Handling. Chloroacetic acid is extremely corrosive and will cause serious chemical burns. It also is readily absorbed through the skin in toxic amounts. Contamination of 5 to 10% of the skin area is usually fatal (26). The symptoms are often delayed for several hours. Single exposure to accidental spillage on the skin has caused human fatalities. The toxic mechanism appears to be blocking of metabolic cycles. Chloroacetic acid is 30–40 times more toxic than acetic, dichloroacetic, or trichloroacetic acid (27). When handling chloroacetic acid and its derivatives, rubber gloves, boots, and protective clothing must be worn. In case of skin exposure, the area should immediately be washed with large amounts of water and medical help should be obtained at once. Oral LD_{50} for chloroacetic acid is 76 mg/kg in rats (28).

Sodium Chloroacetate

Sodium chloroacetate [3926-62-3], mol wt 116.5, $C_2H_2ClO_2Na$, is produced by reaction of chloroacetic acid with sodium hydroxide or sodium carbonate. In many applications chloroacetic acid or the sodium salt can be used interchangeably. As an industrial intermediate, sodium chloroacetate may be purchased or formed *in situ* from free acid. The sodium salt is quite stable in dry solid form, but is hydrolyzed to glycolic acid in aqueous solutions. The hydrolysis rate is a function of pH and temperature (29).

Dichloroacetic Acid

Dichloroacetic acid [79-43-6] ($Cl_2CHCOOH$), mol wt 128.94, $C_2H_2Cl_2O_2$, is a reactive intermediate in organic synthesis. Physical properties are: mp 13.9°C, bp 194°C, density 1.5634 g/mL, and refractive index 1.4658, both at 20°C. The liquid is totally miscible in water, ethyl alcohol, and ether. Dichloroacetic acid ($K_a = 5.14 \times 10^{-2}$) is a stronger acid than chloroacetic acid. Most chemical reactions are similar to those of chloroacetic acid, although both chlorine atoms are susceptible to reaction. An example is the reaction with phenol to form diphenoxyacetic acid (30). Dichloroacetic acid is much more stable to hydrolysis than chloroacetic acid.

Dichloroacetic acid is produced in the laboratory by the reaction of chloral hydrate [302-17-0] with sodium cyanide (31). It has been manufactured by the chlorination of acetic and chloroacetic acids (32), reduction of trichloroacetic acid (33), hydrolysis of pentachloroethane [76-01-7] (34), and hydrolysis of dichloroacetyl chloride. Due to similar boiling points, the separation of dichloroacetic acid from chloroacetic acid is not practical by conventional distillation. However, this separation has been accomplished by the addition of azeotrope-forming hydrocarbons such as bromobenzene (35) or by distillation of the methyl or ethyl ester.

Dichloroacetic acid is used in the synthesis of chloramphenicol [56-75-7] and allantoin [97-59-6]. Dichloroacetic acid has virucidal and fungicidal activity. It was found to be active against several staphylococci (36). The oral toxicity is low: the LD_{50} in rats is 4.48 g/kg. It can, however, cause caustic burns of the skin and eyes and the vapors are very irritating and injurious (28).

Trichloroacetic Acid

Trichloroacetic acid [76-03-9] (Cl_3CCOOH), mol wt 163.39, $C_2HCl_3O_2$, forms white deliquescent crystals and has a characteristic odor. Physical properties are given in Table 3.

Table 3. Physical Properties of Trichloroacetic Acid

Property	Value
melting point, °C	59
boiling point, °C	197.5
density, at 64°C, g/mL	1.6218
refractive index at 61°C	1.4603
heat of combustion, kJ/g[a]	3.05
solubility, at 25°C, g/100 g solvent	
water	1306
methanol	2143
ethyl ether	617
acetone	850
benzene	201
o-xylene	110

[a]To convert kJ/g to kcal/g, divide by 4.184.

Trichloroacetic acid ($K_a = 0.2159$) is as strong an acid as hydrochloric acid. Esters and amides are readily formed. Trichloroacetic acid undergoes decarboxylation when heated with caustic or amines to yield chloroform . The decomposition of trichloroacetic acid in acetone with a variety of aliphatic and aromatic amines has been studied (37). As with dichloroacetic acid, trichloroacetic acid can be converted to chloroacetic acid by the action of hydrogen and palladium on carbon (17).

Trichloroacetic acid is manufactured in the United States by the exhaustive chlorination of acetic acid (38). The patent literature suggests two alternative methods of synthesis: hydrogen peroxide oxidation of chloral (39) and hydrolytic oxidation of tetrachloroethene (40).

Sodium trichloroacetate [650-51-1], $C_2Cl_3O_2Na$, is used as a herbicide for various grasses and cattails (2). The free acid has been used as an astringent, antiseptic, and polymerization catalyst. The esters have antimicrobial activity. The oral toxicity of sodium trichloroacetate is quite low (LD_{50} rats, 5.0 g/kg). Although very corrosive to skin, trichloroacetic acid does not have the skin absorption toxicity found with chloroacetic acid (28).

Chloroacetyl Chloride

Chloroacetyl chloride [79-04-9] ($ClCH_2COCl$) is the corresponding acid chloride of chloroacetic acid (see ACETYL CHLORIDE). Physical properties include: mol wt 112.94, $C_2H_2Cl_2O$, mp $-21.8°C$, bp 106°C, vapor pressure 3.3 kPa (25 mm Hg) at 25°C, 12 kPa (90 mm Hg) at 50°C, and density 1.4202 g/mL and refractive index 1.4530, both at 20°C. Chloroacetyl chloride has a sharp, pungent, irritating odor. It is miscible with acetone and benzene and is initially insoluble in water. A slow reaction at the water–chloroactyl chloride interface, however, produces chloroacetic acid. When sufficient acid is formed to solubilize the two phases, a violent reaction forming chloroacetic acid and HCl occurs.

Since chloroacetyl chloride can react with water in the skin or eyes to form chloroacetic acid, its toxicity parallels that of the parent acid. Chloroacetyl chloride can be absorbed through the skin in lethal amounts. The oral LD_{50} for rats is between 120 and 250 mg/kg. Inhalation of 4 ppm causes respiratory distress. A TLV of 0.05 ppm is recommended (28,41).

Chloroacetyl chloride is manufactured by reaction of chloroacetic acid with chlorinating agents such as phosphorus oxychloride, phosphorus trichloride, sulfuryl chloride, or phosgene (42–44). Various catalysts have been used to promote the reaction. Chloroacetyl chloride is also produced by chlorination of acetyl chloride (45–47), the oxidation of 1,1-dichloroethene (48,49), and the addition of chlorine to ketene (50,51). Dichloroacetyl and trichloroacetyl chloride are produced by oxidation of trichloroethylene or tetrachloroethylene, respectively.

Much of the chloroacetyl chloride produced is used captively as a reactive intermediate. It is useful in many acylation reactions and in the production of adrenalin [51-43-4], diazepam [439-15-5], chloroacetophenone [532-27-4], chloroacetate esters, and chloroacetic anhydride [541-88-8]. A major use is in the production of chloroacetamide herbicides (3) such as alachlor [15972-60-8].

Chloroacetate Esters

Two chloroacetate esters of industrial importance are methyl chloroacetate [96-34-4], $C_3H_5ClO_2$, and ethyl chloroacetate [105-39-5], $C_4H_7ClO_2$. Their properties are given in Table 4.

Table 4. Physical Properties of Chloroacetate Esters

Property	Methyl ester	Ethyl ester
molecular weight	108.52	122.55
melting point, °C	−32.1	−26
boiling point, °C	129.8	143.3
density, at 20°C, g/mL	1.2337	1.159
flash point, °C	57	66
vapor pressure, kPa[a]		
at 25°C	0.96	0.61
at 50°C	4.3	2.6
structural formula	$ClCH_2COOCH_3$	$ClCH_2COOC_2H_5$

[a]To convert kPa to mm Hg, multiply by 7.5.

Both esters have a sweet pungent odor and present a vapor inhalation hazard. They are rapidly absorbed through the skin and hydrolyzed to chloroacetic acid. The oral LD_{50} for ethyl chloroacetate is between 50 and 100 mg/kg (52).

Chloroacetate esters are usually made by removing water from a mixture of chloroacetic acid and the corresponding alcohol. Reaction of alcohol with chloroacetyl chloride is an anhydrous process which liberates HCl. Chloroacetic acid will react with olefins in the presence of a catalyst to yield chloroacetate esters. Dichloroacetic and trichloroacetic acid esters are also known. These esters are useful in synthesis. They are more reactive than the parent acids. Ethyl chloroacetate can be converted to sodium fluoroacetate by reaction with potassium fluoride (see FLUORINE COMPOUNDS, ORGANIC). Both methyl and ethyl chloroacetate are used as agricultural and pharmaceutical intermediates, specialty solvents, flavors, and fragrances. Methyl chloroacetate and β-ionone undergo a Darzens reaction to form an intermediate in the synthesis of Vitamin A. Reaction of methyl chloroacetate with ammonia produces chloroacetamide [79-07-2], C_2H_4ClNO (53).

Bromoacetic Acid

Bromoacetic acid [79-08-3] ($BrCH_2COOH$), mol wt 138.96, $C_2H_3BrO_2$, occurs as hexagonal or rhomboidal hygroscopic crystals, mp 49°C, bp 208°C, d^{50} 1.9335, n_D^{50} 1.4804. It is soluble in water, methanol, and ethyl ether. Bromoacetic acid undergoes many of the same reactions as chloroacetic acid under milder conditions, but is not often used because of its greater cost. Bromoacetic acid must be protected from air and moisture, since it is readily hydrolyzed to glycolic acid. The simple derivatives such as the acid chloride, amides, and esters are well known. Esters of bromoacetic acid are the reagents of choice in the Reformatsky reaction which is used to prepare β-hydroxy acids or α,β-unsaturated acids. Similar reactions with chloroacetate esters proceed slowly or not at all (54).

Bromoacetic acid can be prepared by the bromination of acetic acid in the presence of acetic anhydride and a trace of pyridine (55), by the Hell-Volhard-Zelinsky bromination catalyzed by phosphorus, and by direct bromination of acetic acid at high temperatures or with hydrogen chloride as catalyst. Other methods of preparation include treatment of chloroacetic acid with hydrobromic acid at elevated temperatures (56), oxidation of ethylene bromide with fuming nitric acid, hydrolysis of dibromovinyl ether, and air oxidation of bromoacetylene in ethanol.

Dibromoacetic Acid

Dibromoacetic acid [631-64-1] ($Br_2CHCOOH$), mol wt 217.8, $C_2H_2Br_2O_2$, mp 48°C, bp 232–234°C (decomposition), is soluble in water and ethyl alcohol. It is prepared by adding bromine to boiling acetic acid, or by oxidizing tribromoethene [598-16-3] with peracetic acid.

Tribromoacetic Acid

Tribromoacetic acid [75-96-7] (Br_3CCOOH), mol wt 296.74, $C_2HBr_3O_2$, mp 135°C, bp 245°C (decomposition), is soluble in water, ethyl alcohol, and diethyl ether. This acid is relatively unstable to hydrolytic conditions and can be decomposed to bromoform in boiling water. Tribromoacetic acid can be prepared by the oxidation of bromal [115-17-3] or perbromoethene [79-28-7] with fuming nitric acid and by treating an aqueous solution of malonic acid with bromine.

Iodoacetic Acid

Iodoacetic acid [64-69-7] (ICH_2COOH), mol wt 185.95, $C_2H_3IO_2$, is commercially available. The colorless, white crystals (mp 83°C) are unstable upon heating. It has a K_a of 7.1×10^{-4}. Iodoacetic acid is soluble in hot water and alcohol, and slightly soluble in ethyl ether. Iodoacetic acid can be reduced with hydroiodic acid at 85°C to give acetic acid and iodine (57). Iodoacetic acid cannot be prepared by the direct iodination of acetic acid (58), but has been prepared by iodination of acetic anhydride in the presence of sulfuric or nitric acid (59). Iodoacetic acid can also be prepared by reaction of chloroacetic or bromoacetic acid with sodium or potassium iodide (60).

Diiodoacetic Acid

Diiodoacetic acid [598-89-0] ($I_2CHCOOH$), mol wt 311.85, $C_2H_2I_2O_2$, mp 110°C, occurs as white needles and is soluble in water, ethyl alcohol, and benzene. It has been prepared by heating diiodomaleic acid with water (61) and by treating malonic acid with iodic acid in a boiling water solution (62).

Triiodoacetic Acid

Triiodoacetic acid [594-68-3] (I_3CCOOH), mol wt 437.74, $C_2HO_2I_3$, mp 150°C (decomposition), is soluble in water, ethyl alcohol, and ethyl ether. It has been prepared by heating iodic acid and malonic acid in boiling water (63). Solutions of triiodoacetic acid are unstable as evidenced by the formation of iodine. Triiodoacetic acid decomposes when heated above room temperature to give iodine, iodoform, and carbon dioxide. The sodium and lead salts have been prepared.

BIBLIOGRAPHY

"Acetic Acid Derivatives" under "Acetic Acid," in *ECT* 1st ed., Vol. 1, pp. 74–78, by L. F. Berhenke, F. C. Amstutz, and U. A. Stenger, The Dow Chemical Company; "Ethanoic Acid (Halogenated)" in *ECT,* 2nd ed., Vol. 8, pp. 415–422, by A. P. Lurie, Eastman Kodak Company; "Acetic Acid Derivatives (Halogenated)" in *ECT* 3rd ed., Vol. 1, pp. 171–178, by E. R. Freiter, The Dow Chemical Company.

1. E. Ott, *Cellulose and Cellulose Derivatives,* 2nd ed., Wiley-Interscience, New York, 1955.
2. *Herbicide Handbook,* 5th ed., Weed Science Society of America, Champaign, Ill., 1983, p. 128.
3. M. Sittig, ed., *Pesticide Manufacturing and Toxic Materials Control Encyclopedia,* Noyes Data Corp., Park Ridge, N.J., 1980.
4. U.S. Pat. 2,961,451 (Nov. 22, 1960), A. Keough (to Johnson & Johnson Co., Inc.).
5. Jpn. Pat. 70/23,925 (Aug. 11, 1970), H. Marushige (to Lion Fat & Oil Co.); *Chem. Abstr.* **74,** 12622v, 1971.
6. N. D. Cheronis and K. H. Spitzmueller, *J. Org. Chem.* **6,** 349 (1941).
7. K. Venkataraman, *The Chemistry of Synthetic Dyes,* Vol. 2, Academic Press, Inc., New York, 1952, p. 1013.
8. L. Baker, *Org. Synth. Coll. Vol.* **1,** 181 (1941).
9. G. Koenig, E. Lohmar, and N. Rupprich, "Chloroacetic Acids," in *Ullmann's Encyclopedia of Industrial Chemistry,* Vol. **A6,** VCH Publishers, New York, 1986.
10. U.S. Pat. 2,503,334 (Apr. 11, 1950), A. Hammond (to Celanese Corporation); *Chem. Abstr.* **44,** 5901g (1950).
11. U.S. Pat. 2,539,238 (Jan. 23, 1951), C. Eaker (to Monsanto Corporation); *Chem. Abstr.* **45,** 4739i (1951).
12. G. Sioli, *Hydrocarbon Process.* **58** (part 2), 111–113 (1979).
13. U.S. Pat. 1,304,108 (May 20, 1919), J. Simon; *Chem. Abstr.* **13,** 2039 (1919).
14. Eur. Pat. Appl. 4,496 (Oct. 3, 1979), Y. Correia (to Rhône-Poulenc); *Chem. Abstr.* **92,** 58239a (1980).
15. U.S. Pat. 3,365,493 (Jan. 23, 1968); D. D. De Line (to The Dow Chemical Company); *Chem. Abstr.* **66,** 46104c (1967).
16. Brit. Pat. 949,393 (Feb. 12, 1964) (to Uddeholms Aktiebolag); *Chem. Abstr.* **60,** 15739h (1964).
17. Eur. Pat. Appl. EP 32,816 (July 29, 1981), R. Sugamiya (to Tsukishima Kikai); *Chem. Abstr.* **96,** 34575p (1982).
18. Neth. Pat. 109,768 (Oct. 15, 1964), G. van Messel (to N. V. Koninklijke Nederlandse Zoutindustrie); *Chem. Abstr.* **62,** 7643e (1965).

19. Ger. Offen. 2,240,466 (Feb. 21, 1974), A. Ohorodnik (to Knapsack Company); *Chem. Abstr.* **80,** 132809 (1974).
20. Can. Pat. 757,667 (Apr. 25, 1967), A. B. Foster (to Shawinigan Chemicals); *Chem. Abstr.* **67,** 53692s (1967).
21. U.S. Pat. 4,246,074 (Jan. 20, 1981), E. Fumaux (to Lonza Company); *Chem. Abstr.* **94,** 139253u (1981).
22. Ger. Offen. 2,640,658 (Feb. 9, 1978), E. Greth (to Lonza Company); *Chem. Abstr.* **88,** 152045y (1978).
23. B. De Spiegeleer, *J. Liq. Chromatogra.* **11,** 863 (1988).
24. *Chemical Economics Handbook,* SRI International, Menlo Park, Calif., Dec. 1988, p. 676.1000D.
25. "Chemical Profile," *Chem. Mark. Rep.* (May 8, 1989).
26. "Chloroacetic Acid—Toxicity Profile," The British Industrial Biological Research Association (BIBRA), Surrey, England, 1988.
27. G. Woodward, *J. Ind. Hyg. Toxicol.* **23,** 78 (1941).
28. G. Weiss, ed., *Hazardous Chemicals Data Book,* Noyes Data Corp., Park Ridge, N.J., 1980, pp. 245, 424, 628.
29. L. F. Berhenke and E. C. Britton, *Ind. Eng. Chem.* **38,** 544 (1946).
30. J. van Alphen, *Rec. Trav. Chim.* **46,** 144 (1927); *Chem. Abstr.* **21,** 1641 (1927).
31. A. C. Cope, J. R. Clark, and R. Connor, *Org. Synth. Coll. Vol.* **2,** 181–183 (1950).
32. U.S. Pat. 1,921,717 (Aug. 8, 1933), F. C. Amstutz (to The Dow Chemical Company); *Chem. Abstr.* **27,** 5084 (1933).
33. Ger. Pat. 246,661 (Apr. 27, 1911), K. Brand; *Chem. Abstr.* **6,** 2496 (1912).
34. Fr. Pat. 773,623 (Nov. 22, 1934), Alias, Froges, and Camargue (to Compagnie de Produits Chimiques et Electrometalluriques); *Chem. Abstr.* **29,** 1437 (1935).
35. U.S. Pat 3,772,157 (Nov. 13, 1973), L. H. Horsley (to The Dow Chemical Company); *Chem. Abstr.* **80,** 36713a (1974).
36. Jpn. Kokai 74/109,525 (Oct. 18, 1974), Y. Momotari (to Hardness Chemical Industries Ltd.); *Chem. Abstr.* **82,** 165885y (1975).
37. T. Jasinkski and Z. Pawlak, *Zesz. Nauk. Wyzsz. Szk. Pedagog. Gdansku. Mat. Fiz. Chem.* **8,** 131 (1968); *Chem. Abstr.* **71,** 12355s (1969).
38. U.S. Pat. 2,613,220 (Oct. 7, 1952), C. M. Eaker (to Monsanto Company); *Chem. Abstr.* **47,** 8773c (1953).
39. Jpn. Kokai 72/42,619 (Dec. 16, 1972), H. Miyamori (to Mitsubishi Edogawa Chemical Co.); *Chem. Abstr.* **78,** 83833h (1973).
40. B. G. Yasnitskii, E. B. Dolberg, and C. I. Kovelenka, *Metody Poluch. Khim. Reakt. Prep.* **21,** 106 (1970); *Chem. Abstr.* **76,** 85321x (1972).
41. *Threshold Limit Values and Biological Exposure Indicies,* 5th ed., American Conference of Government Industrial Hygienists, Cincinnati, Ohio, 1986, p. 122.
42. Ger. Offen. 2,943,433 (May 7, 1981), G. Rauchschwalbe (to Bayer A.G./FRG); *Chem. Abstr.* **95,** 61533j (1981).
43. Ger. Offen. 2,943,432 (May 21, 1981), W. Heykamp (to Bayer A.G./FRG); *Chem. Abstr.* **95,** 80511s (1981).
44. Brit. Pat. 1,361,018 (July 24, 1974), O. Hertel (to Bayer A.G./FRG); *Chem. Abstr.* **77,** 74841m (1972).
45. Eur. Pat. Appl. 22,185 (Jan. 14, 1981), A. Ohorodnik (to Hoechst Corporation); *Chem. Abstr.* **94,** 174374f (1981).
46. U.S. Pat. 3,880,923 (Apr. 29, 1975), U. Bressel (to BASF A.G./FRG); *Chem. Abstr.* **81,** 120007b (1974).
47. U.S. Pat. 4,169,847 (Oct. 2, 1979), G. Degisher (to Saeurefabrik Schweizerhall); *Chem. Abstr.* **88,** 104701n (1978).
48. U.S. Pat. 3,674,664 (July 4, 1972), E. R. Larson (to The Dow Chemical Company).

49. K. Shinoda, *Kagaku Kaishi,* 1973, p. 527; *Chem. Abstr.* **79,** 4644s, 4649x, 4650r (1973).
50. U.S. Pat. 3,883,589 (May 13,1975), V. W. Gash (to Monsanto Company). Also see related earlier patents by the same author: 3,758,569; 3,758,571; 3,794,679; 3,812,183.
51. F. B. Erickson and E. J. Prill, *J. Org. Chem.* **23,** 141 (1958).
52. G. D. Clayton and F. E. Clayton, *Patty's Industrial Hygiene and Toxicology,* 3rd ed., Wiley-Interscience, New York, 1981, p. 2391.
53. U.S. Pat. 2,321,278 (June 8, 1941), E. C. Britton (to The Dow Chemical Company); *Chem. Abstr.* **37,** 6677 (1943).
54. E. E. Royals, *Advanced Organic Chemistry,* Prentice-Hall, Englewood Cliffs, N.J., 1954, p. 706.
55. S. Natelson and C. Gottfried, *Org. Synth. Coll. Vol.* **3,** 381–384 (1955).
56. Ger. Offen. 2,151,565 (Apr. 19, 1973), H. Jenker and R. Karsten (to Chemische Fabrik Kalk); *Chem. Abstr.* **79,** 18121f (1973).
57. K. Ichikawa and E. Miura, *J. Chem. Soc. Jpn.* **74,** 798 (1953).
58. F. Fieser and M. Fieser, *Organic Chemistry,* 3rd. ed., D. C. Heath & Co., Lexington, Mass., 1958, p. 171.
59. USSR Pat. 213,014 (Mar. 12, 1968), A. N. Novikou (to Tornsk Polytechnic Institute); *Chem. Abstr.* **69,** 51528k (1968).
60. Czech. Pat. 152,947 (Apr. 15, 1974), E. Prochaszka; *Chem. Abstr.* **81,** 135465y (1974).
61. L. Clarke and E. K. Bolton, *J. Am. Chem. Soc.* **36,** 1899 (1914).
62. R. A. Fairclough, *J. Chem. Soc.* **1938,** 1186 (1938).
63. R. L. Cobb, *J. Org. Chem.* **23,** 1368 (1958).

EARL D. MORRIS
JOHN C. BOST
The Dow Chemical Company

ACETIC ACID—TRIALKYLACETIC ACID. See CARBOXYLIC ACIDS.

ACETIC ANHYDRIDE. See ACETIC ACID AND DERIVATIVES.

ACETINS (ACETATES OF GLYCEROL). See GLYCEROL.

ACETOACETIC ACID AND ESTER. See KETENES AND OTHER RELATED SUBSTANCES.

ACETONE

Acetone [67-64-1] (2-propanone, dimethyl ketone, CH_3COCH_3), molecular weight 58.08 (C_3H_6O), is the simplest and most important of the ketones. It is a colorless, mobile, flammable liquid with a mildly pungent, somewhat aromatic odor, and is miscible in all proportions with water and most organic solvents. Acetone is an excellent solvent for a wide range of gums, waxes, resins, fats, greases, oils, dyestuffs, and cellulosics. It is used as a carrier for acetylene, in the manufacture of a variety of coatings and plastics, and as a raw material for the chemical synthesis of a wide range of products such as ketene, methyl methacrylate, bisphenol A, diacetone alcohol, methyl isobutyl ketone, hexylene glycol (2-methyl-2,4-pentanediol), and isophorone. World production of acetone in 1990 was about three million metric tons per year, of which about one million are made in the United States. Most of the world's manufactured acetone is obtained as a coproduct in the process for phenol from cumene and most of the remainder from the dehydrogenation of isopropyl alcohol. Numerous natural sources of acetone make it a normal constituent of the environment. It is readily biodegradable.

Physical and Thermodynamic Properties

Selected physical properties are given in Table 1 and some thermodynamic properties in Table 2. Vapor pressure (P) and enthalpy of vaporization (H) over the temperature range 178.45 to 508.2 K can be calculated with an error of less than 3% from the following equations wherein the units are P, kPa; H, mJ/mol; T, K; and T_r = reduced temperature, T/T_c (1):

$$\log(P) = 70.72 - 5685/T - 7.351 \ln(T) + 0.0000063 T^2$$

$$\log(H) = \log(49170000) + (1.036 - 1.294 T_r + 0.672 T_r^2) \log(1 - T_r)$$

Spectral characterization data are given in Table 3 (2).

Chemical Properties

The closed cup flash point of acetone is $-18°C$ and open cup $-9°C$. The auto ignition temperature is $538°C$, and the flammability limits are 2.6 to 12.8 vol % in air at 25°C (3).

Acetone shows the typical reactions of saturated aliphatic ketones. It forms crystalline compounds such as acetone sodium bisulfite [540-92-1], $(CH_3)_2C(OH)SO_3Na$, with alkali bisulfites. The highly reactive compound ketene [463-51-4], $CH_2=C=O$, results from the pyrolysis of acetone. Reducing agents convert acetone to pinacol [76-09-5], isopropyl alcohol [67-63-0], or propane [74-98-6]. Reductive ammonolysis produces isopropyl amines. Acetone is stable to many of the usual oxidants such as Fehling's solution, silver nitrate, cold nitric acid, and neutral potassium permanganate, but it can be oxidized with some of the stronger oxidants such as alkaline permanganate, chromic acid, and hot nitric

Table 1. Physical Properties[a]

Property	Value
melting point, °C	−94.6
boiling point at 101.3 kPa[b], °C	56.29
refractive index, n_D	
at 20°C	1.3588
at 25°C	1.35596
electrical conductivity at 298.15 K, S/cm	5.5×10^{-8}
critical temperature, °C	235.05
critical pressure, kPa[b]	4701
critical volume, L/mol	0.209
critical compressibility	0.233
triple point temperature, °C	−94.7
triple point pressure, Pa[b]	2.59375
acentric factor	0.306416
solubility parameter at 298.15 K, $(J/m^3)^{1/2}$ [c]	19773.5
dipole moment, C·m[d]	9.61×10^{-30}
molar volume at 298.15 K, L/mol	0.0739
molar density, mol/L	
solid at −99°C	16.677
liquid at 298.15 K	13.506

Selected physical properties as a function of temperature

temperature, °C	0	20	40
surface tension, mN/m (= dyn/cm)	26.2	23.7	21.2
vapor pressure, kPa[b]	9.3	24.7	54.6
specific gravity at 20°C	0.807	0.783	0.759
viscosity, mPa·s (= cP)	0.40	0.32	0.27

[a] Extensive tables and equations are given in ref. 1 for viscosity, surface tension, thermal conductivity, molar density, vapor pressure, and second virial coefficient as functions of temperature.
[b] To convert kPa to mm Hg, multiply by 7.501.
[c] To convert $(J/m^3)^{1/2}$ to $(cal/m^3)^{1/2}$, divide by 2.045.
[d] To convert C·m to debyes, divide by 3.336×10^{-30}.

acid. Metal hypohalite, or halogen in the presence of a base, oxidizes acetone to the metal acetate and a haloform, eg, iodoform. Halogens alone substitute for the H atoms, yielding haloacetones. Acetone is a metabolic product in humans and some other mammals and is a normal constituent of their blood and urine. In diabetics it is present in relatively large amounts.

Compounds with active hydrogen add to the carbonyl group of acetone, often followed by the condensation of another molecule of the addend or loss of water. Hydrogen sulfide forms hexamethyl-1,3,5-trithiane probably through the transitory intermediate thioacetone which readily trimerizes. Hydrogen cyanide forms acetone cyanohydrin [75-86-5] $(CH_3)_2C(OH)CN$, which is further processed to methacrylates. Ammonia and hydrogen cyanide give $(CH_3)_2C(NH_2)CN$ [19355-69-2] from which the widely used polymerization initiator, azobisisobutyronitrile [78-67-1] is made (4).

Primary amines form Schiff bases, $(CH_3)_2C{=}NR$. Ammonia induces an

Table 2. Thermodynamic Properties[a,b,c]

Property	Value
specific heat of liquid at 20°C, J/g	2.6
specific heat of vapor at 102°C, J/(mol·K)	92.1
heat of vaporization at 56.1°C, kJ/mol	29.1
enthalpy of vaporization, kJ/mol	30.836
enthalpy of fusion at melting point, J/mol	5691.22
heat of combustion of liquid, kJ/mol	1787
enthalpy of combustion, kJ/mol	−1659.17
entropy of liquid, J/(mol·K)	200.1
entropy of ideal gas, J/(mol·K)	295.349
Gibbs energy of formation, kJ/mol	−152.716
enthalpy of formation, kJ/mol	
ideal gas	−217.15
gas	−216.5
liquid	−248

[a]Extensive tables and equations are given in ref. 1 for enthalpy of vaporization and heat capacity at constant pressure.
[b]At 298.15 K unless otherwise noted.
[c]To convert J to cal, divide by 4.184.

Table 3. Spectral Parameters for Acetone[a]

Method	Property
Absorption peaks, cm^{-1}	
infrared: SADG 77	3000, 1715, 1420, 1360, 1220, 1090, 900, 790, 530
Raman: SAD 162	3010, 2930, 2850, 2700, 1740, 1710, 1430, 1360, 1220, 1060, 900, 790, 520, 490, 390
Absorption peaks, nm	
ultraviolet: SAD 89	270 (in methanol)
Chemical shift, ppm	
^{1}H nmr: SAD 9228	2.1 ($CDCl_3$)
^{13}C nmr: JJ 28 FT	30.6, 206.0 ($CDCl_3$)
m/e (relative abundance)	
mass spec: Wiley 30	43(100), 58(42), 15(14), 42(6), 27(4), 39(3), 26(3), 29(2) molecular ion = 58.04

[a]Ref. 2.

aldol condensation followed by 1,4-addition of ammonia to produce diacetone amine (from mesityl oxide), 4-amino-4-methyl-2-pentanone [625-04-7], $(CH_3)_2C(NH_2)CH_2COCH_3$, and triacetone amine (from phorone), 2,2,6,6-tetra-methyl-4-piperidinone [826-36-8]. Hydroxylamine forms the oxime and hydrazine compounds ($RNHNH_2$) form hydrazones ($RNHN=C(CH_3)_2$). Acetone and nitrous acid give the isonitroso compound which is the monoxime of pyruvaldehyde [306-44-5], $CH_3COCH=NOH$. Mercaptans form hemimercaptols by addition and mercaptols, $(CH_3)_2C(SR)_2$, by substitution following the addition.

With aldehydes, primary alcohols readily form acetals, $RCH(OR')_2$. Acetone also forms acetals (often called ketals), $(CH_3)_2C(OR)_2$, in an exothermic reaction, but the equilibrium concentration is small at ambient temperature. However, the methyl acetal of acetone, 2,2-dimethoxypropane [77-76-9], was once made commercially by reaction with methanol at low temperature for use as a gasoline additive (5). Isopropenyl methyl ether [116-11-0], useful as a hydroxyl blocking agent in urethane and epoxy polymer chemistry (6), is obtained in good yield by thermal pyrolysis of 2,2-dimethoxypropane. With other primary, secondary, and tertiary alcohols, the equilibrium is progressively less favorable to the formation of ketals, in that order. However, acetals of acetone with other primary and secondary alcohols, and of other ketones, can be made from 2,2-dimethoxypropane by transacetalation procedures (7,8). Because they hydrolyze extensively, ketals of primary and especially secondary alcohols are effective water scavengers.

Acetone has long been used as an agent to block the reactivity of hydroxyl groups in 1,2 and 1,3-diols, especially in carbohydrate chemistry. The equilibrium for the formation of acetals with hydroxyls in these compounds is more favorable because the products are five- and six-membered ring compounds, 1,3-dioxolanes and dioxanes, respectively. With glycerol the equilibrium constant for formation of the dioxolane is about 0.50 at 23°C and 0.29 at 48°C in a mixture resulting from acidification of equal volumes of acetone and glycerol at ambient temperature. The equilibrium can be displaced toward acetal formation by the use of a water scavenger, eg, an anhydrous metal salt such as copper sulfate.

Acetone undergoes aldol additions,

$$R^1R^2CH-\overset{\overset{\displaystyle O}{\|}}{C}-R^3 + CH_3-\overset{\overset{\displaystyle O}{\|}}{C}-CH_3 \longrightarrow R^1R^2C-\overset{\overset{\displaystyle \overset{\overset{\displaystyle O}{\|}}{C}-R^3}{}}{\underset{OH}{\overset{|}{C}}}\overset{\diagup CH_3}{\diagdown CH_3}, \quad R = H \text{ or alkyl}$$

and further reacts with the products, forming aldol chemicals, diacetone alcohol (4-hydroxy-4-methyl-2-pentanone [123-42-2]), mesityl oxide (4-methyl-3-penten-2-one [141-79-7]), isophorone (3,5,5-trimethyl-2-cyclohexenone [78-59-1]), phorone (2,6-dimethyl-2,5-heptadien-4-one [504-20-1]), and mesitylene (1,3,5-trimethylbenzene [108-67-8]). From these are produced the industrial solvents methyl isobutyl ketone (MIBK, 4-methyl-2-pentanone [108-10-1]), methylisobutylcarbinol (MIBC, 4-methyl-2-pentanol [108-11-2]), hexylene glycol (2-methyl-2,4-pentanediol [107-41-5]), and others. Acetone enters the typical nucleophilic addition and condensation reactions of ketones both at its carbonyl group and at its methyl groups, with aldehydes, other ketones, and esters. The Claisen reaction with ethyl acetate gives acetylacetone (2,4-pentanedione [123-54-6]); Mannich reaction with secondary amines gives R_2NCH_2-substituted acetones; and the Reformatzky reaction gives β-hydroxy esters. Glycidic esters (esters with a 2-epoxy group) can be made by condensation of acetone and chloroacetic esters with a metal alkoxide.

The para and ortho positions of phenols condense at the carbonyl group of acetone to make bisphenols, eg, bisphenol A, 4,4'-(1-methylethylidene)bisphenol [80-05-07]). If the H atom is activated, ClCH— compounds add to the carbonyl group in the presence of strong base; chloroform gives chloretone (1,1,1-trichloro-2-methyl-2-propanol [57-15-8]).

Manufacture

Acetone was originally observed about 1595 as a product of the distillation of sugar of lead (lead acetate). In the nineteenth century it was obtained by the destructive distillation of metal acetates, wood, and carbohydrates with lime, and pyrolysis of citric acid. Its composition was determined by Liebig and Dumas in 1832.

Until World War I acetone was manufactured commercially by the dry distillation of calcium acetate from lime and pyroligneous acid (wood distillate) (9). During the war processes for acetic acid from acetylene and by fermentation supplanted the pyroligneous acid (10). In turn these methods were displaced by the process developed for the bacterial fermentation of carbohydrates (cornstarch and molasses) to acetone and alcohols (11). At one time Publicker Industries, Commercial Solvents, and National Distillers had combined biofermentation capacity of 22,700 metric tons of acetone per year. Biofermentation became noncompetitive around 1960 because of the economics of scale of the isopropyl alcohol dehydrogenation and cumene hydroperoxide processes.

Production of acetone by dehydrogenation of isopropyl alcohol began in the early 1920s and remained the dominant production method through the 1960s. In the mid-1960s virtually all United States acetone was produced from propylene. A process for direct oxidation of propylene to acetone was developed by Wacker Chemie (12), but is not believed to have been used in the United States. However, by the mid-1970s 60% of United States acetone capacity was based on cumene hydroperoxide [80-15-9], which accounted for about 65% of the acetone produced.

Acetone was a coproduct of the Shell process for glycerol [56-8-5]. Propylene was hydrated to isopropyl alcohol. Some of the alcohol was catalytically oxidized to acrolein and some was oxidized to give hydrogen peroxide and acetone. Some more of the isopropyl alcohol and the acrolein reacted to give allyl alcohol and acetone. The allyl alcohol was then treated with the peroxide to give glycerol. About 1.26 kg of acetone resulted per kilogram of glycerol. In 1974 23,000 to 32,000 t of acetone may have been produced by this method.

Direct oxidation of hydrocarbons and catalytic oxidation of isopropyl alcohol have also been used for commercial production of acetone.

Most of the world's acetone is now obtained as a coproduct of phenol by the cumene process, which is used by 21 of 31 producing companies in North America,

Western Europe, and Japan. Cumene is oxidized to the hydroperoxide and cleaved to acetone and phenol . The yield of acetone is believed to average about 94%, and about 0.60–0.62 unit weight of acetone is obtained per unit of phenol (13).

Dehydrogenation of isopropyl alcohol accounts for most of the acetone production not obtained from cumene. The vapor is passed over a brass, copper, or other catalyst at 400–500°C, and a yield of about 95% is achieved (1.09 unit weight of alcohol per unit of acetone) (13).

Almost 95% of the acetone produced in the United States in 1987 and 1988 was made from cumene and 4% from isopropyl alcohol (13).

Minor amounts of acetone are made by other processes. Until mid-1980 Shell Chemical Company obtained acetone and hydrogen peroxide as coproducts of noncatalytic oxidation of isopropyl alcohol with oxygen in the liquid phase. Yield to acetone was about 90%. Acetone is a coproduct of propylene oxide in the ARCO Chemical Company process, but is hydrogenated to isopropyl alcohol. In a process analogous to the cumene process, Eastman Chemical Products, Inc., and The Goodyear Tire & Rubber Company produce hydroquinone and acetone from diisopropylbenzene [25321-09-9] in the United States. Similarly, Sumitomo Chemical Co., Ltd., and Mitsui Petrochemical Industries, Ltd., in Japan produce coproduct acetone with cresol from cymene [25755-15-1]. Wacker process oxidation of propylene [115-07-1] is used by Kyowa Yuka Company and Mitsubishi Kasei Corporation in Japan. BP Chemicals, Ltd., in the United Kingdom recovers by-product acetone from the manufacture of acetic acid by the oxidation of light petroleum distillate. Usina Victor Sence SA in Brazil may be operating a fermentation process for acetone and butyl alcohol (13).

Producers of acetone in the United States and their capacities and feed-stocks are given in Table 4 (14). Data on world production and processes by regions are shown in Tables 5 and 6 (15).

Cumene Hydroperoxide Process for Phenol and Acetone. Benzene is alkylated to cumene, which is oxidized to cumene hydroperoxide, which in turn is cleaved to phenol and acetone.

$$C_6H_5CH(CH_3)_2 \xrightarrow[\text{oxygen}]{\text{air or}} C_6H_5\overset{\overset{\displaystyle OOH}{|}}{C}(CH_3)_2 \xrightarrow[\text{heat}]{\text{acid}} C_6H_5OH + (CH_3)_2C{=}O$$

One kilogram of phenol production results in about 0.6 kg of acetone or about 0.40–0.45 kg of acetone per kilogram of cumene used.

There are many variations of the basic process and the patent literature is extensive. Several key patents describe the technology (16). The process steps are oxidation of cumene to a concentrated hydroperoxide, cleavage of the hydroperoxide, neutralization of the cleaved products, and distillation to recover acetone.

In the first step cumene is oxidized to cumene hydroperoxide with atmospheric air or air enriched with oxygen in one or a series of oxidizers. The temperature is generally between 80 and 130°C and pressure and promoters, such as sodium hydroxide, may be used (17). A typical process involves the use of three or four oxidation reactors in series. Feed to the first reactor is fresh cumene and cumene recycled from the concentrator and other reactors. Each reactor is

Table 4. U.S. Producers of Acetone

Company	Location	Annual[a] capacity, 10^3 t
Cumene feedstock		
Allied Signal Corporation	Frankford, Pa.	221
Aristech Chemical Corporation	Haverhill, Ohio	172
BTL Specialty Resins Corporation	Blue Island, Ill.	24
Dow Chemical U.S.A.	Oyster Creek, Tex.	152
General Electric Company	Mount Vernon, Ind.	177
Georgia Gulf Corporation	Plaquemine, La.	109
Shell Oil Company	Deer Park, Tex.	166
Texaco Corporation	El Dorado, Kans.	25
Isopropyl alcohol feedstock		
Shell Oil Company[b]	Deer Park, Tex.	45
Union Carbide Corporation	Institute, W.Va.	77
Diisopropylbenzene feedstock		
Eastman Kodak Company	Kingsport, Tenn.	c
The Goodyear Tire and Rubber Company, Chemical Division	Bayport, Tex.	c

[a]As of January 1, 1990.
[b]Currently not active.
[c]Small amounts, in manufacture of hydroquinone.

Table 5. World Acetone Production Data by Regions Other Than the United States, 1987

	Canada	Mexico	Western Europe	Japan
producing companies	2	2	14	6
capacity (year end), 10^3 t	44	94	988	324
percentage of capacity from				
isopropyl alcohol	61	77	26	
cumene	39	23	69	70
other[a]			5	30
production, 10^3 t	41	47	867	260
imports, 10^3 t		0.6		8
exports, 10^3 t	18	1.4		31
net imports, 10^3 t			35	
consumption, 10^3 t	23	47	877	247
solvent applications	15	28[b]	337	
ACH/MMA[c]		11	268	88
aldol chemicals	8		78	
bisphenol A		1	74	25
other		7	120	134[d]
production/capacity ratio	0.93	0.5	0.88	0.8

[a]Other feedstocks are propylene and cymene.
[b]Includes use in the manufacture of other solvents.
[c]ACH = acetone cyanohydrin; MMA = methyl methacrylate.
[d]Includes solvent uses.

Table 6. Acetone Capacities in Western Europe, 10^3 t, 1989

Country	No. of companies	Capacity from		
		2-Propanol	Cumene	Other
Finland	1		45	
France	2	65	72	
West Germany	14	36	250	
Italy			220	
Netherlands	1	100		
Spain	2	9	52	
United Kingdom	3	45	50	50

partitioned. At the bottom there may be a layer of fresh 2–3% sodium hydroxide if a promoter (stabilizer) is used. Cumene enters the side of the reactor, overflows the partition to the other side, and then goes on to the next reactor. The air (oxygen) is bubbled in at the bottom and leaves at the top of each reactor.

The temperatures decline from a high of 115°C in the first reactor to 90°C in the last. The oxygen ratio as a function of consumable oxygen is also higher in the later reactors. In this way the rate of reaction is maintained as high as possible, while minimizing the temperature-promoted decomposition of the hydroperoxide.

This procedure may result in a concentration of cumene hydroperoxide of 9–12% in the first reactor, 15–20% in the second, 24–29% in the third, and 32–39% in the fourth. Yields of cumene hydroperoxide may be in the range of 90–95% (18). The total residence time in each reactor is likely to be in the range of 3–6 h. The product is then concentrated by evaporation to 75–85% cumene hydroperoxide. The hydroperoxide is cleaved under acid conditions with agitation in a vessel at 60–100°C. A large number of nonoxidizing inorganic acids are useful for this reaction, eg, sulfur dioxide (19).

After cleavage the reaction mass is a mixture of phenol, acetone, and a variety of other products such as cumylphenols, acetophenone, dimethyl-phenylcarbinol, α-methylstyrene, and hydroxyacetone. It may be neutralized with a sodium phenoxide solution (20) or other suitable base or ion-exchange resins. Process water may be added to facilitate removal of any inorganic salts. The product may then go through a separation and a wash stage, or go directly to a distillation tower.

A crude acetone product is recovered by distillation from the reaction mass. One or two additional distillation columns may be required to obtain the desired purity. If two columns are used, the first tower removes impurities such as acetaldehyde and propionaldehyde. The second tower removes undesired heavies, the major component being water.

The yield of acetone from the cumene/phenol process is believed to average 94%. By-products include significant amounts of α-methylstyrene [98-83-9] and acetophenone [98-86-2] as well as small amounts of hydroxyacetone [116-09-6] and mesityl oxide [141-79-7]. By-product yields vary with the producer. The α-methylstyrene may be hydrogenated to cumene for recycle or recovered for monomer use. Yields of phenol and acetone decline by 3.5–5.5% when the α-methylstyrene is not recycled (21).

Dehydrogenation of Isopropyl Alcohol.   In the United States about 4% of the acetone is made by this process, and in Western Europe about 19% (22). Isopropyl alcohol is dehydrogenated in an endothermic reaction.

$$\text{CH}_3\text{CHOHCH}_3 + 66.5 \text{ kJ/mol (at } 327°\text{C)} \longrightarrow \text{CH}_3\text{COCH}_3 + \text{H}_2$$

The equilibrium is more favorable to acetone at higher temperatures. At 325°C 97% conversion is theoretically possible. The kinetics of the reaction has been studied (23). A large number of catalysts have been investigated, including copper, silver, platinum, and palladium metals, as well as sulfides of transition metals of groups 4, 5, and 6 of the periodic table. These catalysts are made with inert supports and are used at 400–600°C (24). Lower temperature reactions (315–482°C) have been successfully conducted using zinc oxide–zirconium oxide combinations (25), and combinations of copper–chromium oxide and of copper and silicon dioxide (26).

It is usual practice to raise the temperature of the reactor as time progresses to compensate for the loss of catalyst activity. When brass spelter is used as a catalyst, the catalyst must be removed at intervals of 500–1000 h and treated with a mineral acid to regenerate catalytically active surface (27). When 6–12% zirconium oxide is added to a zinc oxide catalyst and the reaction temperatures are not excessive, the catalyst life is said to be a minimum of 3 months (25). The dehydrogenation is carried out in a tubular reactor. Conversions are in the range of 75–95 mol %. A process described by Shell International Research (24) is a useful two-stage reaction to attain high conversion, with lower energy cost and lower capital cost. The first stage uses a tubular reactor at 420–550°C to convert up to 70% of the alcohol to acetone. The second stage employs an unheated fixed-bed reactor with the same catalyst used in the tube reaction to complete the conversion at about 85%.

Although the selectivity of isopropyl alcohol to acetone via vapor-phase dehydrogenation is high, there are a number of by-products that must be removed from the acetone. The hot reactor effluent contains acetone, unconverted isopropyl alcohol, and hydrogen, and may also contain propylene, polypropylene, mesityl oxide, diisopropyl ether, acetaldehyde, propionaldehyde, and many other hydrocarbons and carbon oxides (25,28).

The mixture is cooled and noncondensable gases are scrubbed with water. Some of the resultant gas stream, mainly hydrogen, may be recycled to control catalyst fouling. The liquids are fractionally distilled, taking acetone overhead and a mixture of isopropyl alcohol and water as bottoms. A caustic treatment may be used to remove minor aldehyde contaminants prior to this distillation (29). In another fractionating column, the aqueous isopropyl alcohol is concentrated to about 88% for recycle to the reactor.

A yield of about 95% of theoretical is achieved using this process (1.09 units of isopropyl alcohol per unit of acetone produced). Depending on the process technology and catalyst system, such coproducts as methyl isobutyl ketone and diisobutyl ketone can be produced with acetone (30).

Production and Shipment

Acetone is produced in large quantities and usually shipped by producers to consumers and distributors in drums and larger containers. Distributors repackage the acetone into containers ranging in size from small bottles to drums or even tank trucks. Specialty processors make available various grades and forms of acetone such as high purity, specially analyzed, analytical reagent grade, chromatography and spectrophotometric grades, and isotopically labeled forms, and ship them in ampoules, vials, bottles, or other containers convenient for the buyers.

The Department of Transportation (DOT) hazard classification for acetone is Flammable Liquid, identification number UN1090. DOT regulations concerning the containers, packaging, marking, and transportation for overland shipment of acetone are published in the *Federal Register* (31). Regulations and information for transportation by water in the United States are published in the *Federal Register* (32) and by the U.S. Coast Guard (33). Rules and regulations for ocean shipping have been published by the International Maritime Organization (IMO), a United Nations convention of nations with shipping interests, in the IMOBCH Code (34). The IMO identification number is 3.1. Because additions and changes to the regulations appear occasionally, the latest issue of the regulations should be consulted.

Small containers up to 4–5 L (about 1 gal) are usually glass. Acetone is also shipped by suppliers of small quantities in steel pails of 18 L. Depending on the size of the container, small amounts are shipped by parcel delivery services or truck freight. Quantities that can be accepted by some carriers are limited by law and special "over-pack" outer packaging may be required. Usual materials for larger containers are carbon steel for 55-gal (0.21 m^3) drums, stainless steel or aluminum for tank trucks, and carbon steel, lined steel, or aluminum for rail tank cars. The types of tank cars and trucks that can be used are specified by law, and shippers may have particular preferences. Barges and ships are usually steel, but may have special inner or deck-mounted tanks. Increasing in use, especially for international shipments, are intermodal (IM) portable containers, tanks suspended in frameworks suitable for interchanging among truck, rail, and ship modes of transportation.

Containers less than bulk must bear the red diamond-shaped "FLAMMA-BLE LIQUID" label. Bulk containers must display the red "FLAMMABLE" placard in association with the UN1090 identification. Fire is the main hazard in emergencies resulting from spills. Some manufacturers provide transportation emergency response information. A listing of properties and hazard response information for acetone is published by the U.S. Coast Guard in its CHRIS manual (35). Two books on transportation emergencies are available (36). Immediate information can be obtained from CHEMTREC (37). Interested parties may contact their suppliers for more detailed information on transportation and transportation emergencies.

Tank cars contain up to 10, 20, or 30 thousand gal (10,000 gal = 38 m^3) of material, tank trucks 6000 gal (22.7 m^3), and barges 438,000 gal (about 1270 tons). International shipments by sea are typically about 2000 tons.

Economic Aspects

The economics of acetone production and its consequent market position are unusual. Traditional laws of supply and demand cannot be applied because supply depends on the production of phenol and demand is controlled by the uses of acetone. Therefore, coproduct acetone from the cumene to phenol process will continue to dominate market supply. Deliberate production of acetone from isopropyl alcohol accommodates demand in excess of that supplied by the phenol process. More than 75% of world and 90% of U.S. production comes from the cumene to phenol process.

World Capacity, Production, and Consumption. Current and future world capacity, based on announced new plants and expansions, and 1987 production and consumption data are shown in Table 7 (38). Consumption of acetone is expected to grow at a rate of about 2% annually until 1992, but phenol demand and consequent coproduct acetone production are expected to grow at a rate of 2.5–3%, thus resulting in excess supplies. The fastest growing outlet for acetone is for bisphenol A, mainly for growth in polycarbonate. Although bisphenol A production consumes one mole of acetone, it yields a net amount of one mole of acetone production because two moles of acetone accompany the production of the required phenol. Production of "on-purpose" acetone will probably decline as supplies of by-product acetone increase.

Table 7. World Capacity, Production, and Consumption for Acetone, 10^3 t

	Capacity[a]		Production 1987	Imports 1987	Exports 1987	Consumption[b] 1987
	1/1/1989	1/1/1994				
North America						
Canada	44	87	41		18	23
Mexico	94	96	47	0.5	1.4	47
United States	1084	1345	950	75	118	908
Western Europe	994	1062	867	35[c]		877
Japan	324	324	260	8	31	247
Other	857[d]	1046	833	429	439	777
Total	*3397*	*3960*	*2998*	*548*	*607*	*2879*

[a]Current and projected future capacity based on announced new plants and expansions.
[b]Actual consumption; the difference between actual and apparent consumption is primarily due to inventory changes. U.S. datum is apparent consumption.
[c]Net imports.
[d]Includes Africa, Asia, Eastern Europe (460,000 t), Australia, and South America.

World consumption data by end use in 1987 are shown in Table 8 (39). Solvent applications account for the largest use of acetone worldwide, followed by production of acetone cyanohydrin for conversion to methacrylates. Aldol chemicals are derivatives of acetone used mainly as solvents (40).

U.S. Capacity, Production, and Consumption. U.S. acetone capacity in 1989 was about 32% of world capacity. Planned additions that have been announced will increase U.S. capacity to 1.34×10^6 t by 1992, if all are attained (41). Historically, the percentage of U.S. capacity from cumene has steadily increased from 1970 (37%) to 1989 (87%), except for a slight dip in the early 1980s. During the same period the percentage from isopropyl alcohol steadily declined from 60 to 11%, except for a compensating slight rise in the early 1980s. Total capacity ranged from 887,000 t in 1970 to 1.66×10^6 t in 1982 to 1.09×10^6 t in 1989 (42).

Table 8. World Demand for Acetone by Use in 1987

| Use | Consumption by region, percentage | | | | |
	Canada	Mexico	United States	Western Europe	Japan
solvent applications	65	60[a]	21	38	b
acetone cyanohydrin and methyl methacrylate		23	41	31	36
bisphenol A		2	14	8	10
aldol chemicals	35	a	12	9	
other		15	12	14	54[b]

[a]Use both as a solvent and in production of aldol chemicals used as solvents.
[b]Other for Japan includes solvent applications.

Production of acetone from isopropyl alcohol ranged from about 250,000 to 300,000 t/yr from 1970 to 1981 with the exception of three of those years when it was higher. Then production dropped sharply to the 30,000–50,000 t range for 1984 to 1988. From 1970 to 1988 production of acetone from cumene had three periods of growth and two periods of decline which roughly paralleled fluctuations in the U.S. economy. The amount produced ranged from 300,000 t in 1970 to about 900,000 t in 1988, with an average growth rate of 26,800 t/yr, or about 4.4% per annum. Total annual acetone production during the period 1970–1987 has been in the range of 0.7–1 $\times$ 10^6 t except for a peak of about 1.15 $\times$ 10^6 t in 1979, with a trend slightly upward. Average total production for the period was 880,000 t (43).

U.S. list prices for acetone were around $100 per metric ton from 1970 to 1973, rose sharply in 1974, and then slowly for the next five years to around $300, rose sharply again to a peak of nearly $750 in 1982, and then declined to the $600 range. The lowest annual list prices from 1982 to 1989 were in the $500 range, but tended toward the highest prices during the decade. Unit sales prices (negotiated for very large quantities) were closely parallel to but a little lower than list prices until 1980 when they diverged sharply and remained about $200 per metric ton lower than the highest list prices for the rest of the decade (44). Current pricing is strongly tied to acetone supply and demand and only loosely to raw material costs.

The United States was a net exporter in the 1980s except for 1984–1986 when it was a net importer. This reversal resulted from a number of conditions including the global recession, foreign relationships, new overseas capacity especially in Japan and South Africa (which in 1988 was the largest exporter to the United States), and the increase in other countries of the ratio of phenol production to acetone demand (45).

Trends in acetone consumption by uses are shown in Table 9 (46). The amount of acetone going to acetone cyanohydrin has increased rather steadily at an average rate of about 10,600 t/yr (4% annual rate), with minor dips in 1975 and 1982. Acetone to bisphenol A followed a similar curve with growth of about 5,600 t/yr (9%). Total acetone consumption fluctuated a bit more but was otherwise similar, with average annual growth for 1970–1988 of about 5,600 t/yr (0.7%). Except for 1982, solvent uses were within about 10% of 200,000 t/yr. Amounts for

Table 9. U.S. Consumption of Acetone by Uses, 10^3 t

Use	1970	1980	1988
acetone cyanohydrin	176	298	406
bisphenol A	26	67	137
aldol chemicals			
methyl isobutyl ketone	113	95	63
methylisobutylcarbinol	25	20	17
other aldol chemicals	66	55	33
solvent uses		194	222
other uses	342^a	103	73

[a]Includes solvent uses.

other uses generally declined as indicated in the table, with some fluctuations. Little or no growth in the solvents market is expected because acetone is classified as a volatile organic compound. Eleven large volume consumers in the United States and their products from acetone are listed in Table 10 (47). The largest distributors are Ashland Chemical Company, Unocal Chemicals, ChemCentral, Van Waters & Rogers, and JLM Industries (47).

Table 10. Large Volume Consumers of Acetone for Derivatives in the United States

Derivative product	Consumer[a]
acetone cyanohydrin	B, Cy, Du, R
methyl isobutyl ketone	E, S, U
methylisobutylcarbinol	S, U
other aldol chemicals	E, S, U, Ce
bisphenol A	A, Do, G, S

[a] A = Aristech Chemical B = BP Chemicals Ce = Celanese Cy = CYRO Industries
Do = Dow Chemical Du = Du Pont E = Eastman Chemical G = General Electric
R = Rohm & Haas S = Shell Chemical U = Union Carbide

Specifications, Standards, and Quality Control

The ASTM "Standard Specification for ACETONE," D329, requires 99.5% grade acetone to conform to the following: apparent specific gravity 20/20°C, 0.7905 to 0.7930; 25/25°C, 0.7860 to 0.7885 (ASTM D891); color, not more than No. 5 on the platinum–cobalt scale (ASTM D1209); distillation range, 1.0°C which shall include 56.1°C (ASTM D1078); nonvolatile matter, not more than 5 mg/100 mL (D1353); odor, characteristic, nonresidual (ASTM D1296); water, not more than 0.5 wt % (ASTM D1364); acidity (as free acetic acid), not more than 0.002 wt %, equivalent to 0.019 mg of KOH per gram of sample (ASTM D1613); water miscibility, no turbidity or cloudiness at 1:10 dilution with water (ASTM D1722); alkalinity (as ammonia), not more than 0.001 wt % (ASTM D1614); and permanganate time, color of added $KMnO_4$ must be retained at least 30 min at 25°C in the dark (ASTM D1363).

Higher or lower quality at more or less cost will meet the needs of some consumers. Acetone is often produced under contract to meet customer specifications which are different from those of ASTM D329. Some specialty grades are analyzed reagent, isotopically labeled, clean room, liquid chromatography, spectroscopic, ACS reagent (48), semiconductor (low metals), and Federal Specification O-A-51G.

Specification tests are performed on plant streams once or twice per worker shift, or even more often if necessary, to assure the continuing quality of the product. The tests are also performed on a sample from an outgoing shipment, and a sample of the shipment is usually retained for checking on possible subsequent contamination. Tests on specialty types of acetone may require sophisticated instruments, eg, mass spectrometry for isotopically labeled acetone.

Analytical and Test Methods, Storage

In current industrial practice gas chromatographic analysis (glc) is used for quality control. The impurities, mainly a small amount of water (by Karl-Fischer) and some organic trace constituents (by glc), are determined quantitatively, and the balance to 100% is taken as the acetone content. Compliance to specified ranges of individual impurities can also be assured by this analysis. The gas chromatographic method is accurately correlated to any other tests specified for the assay of acetone in the product. Contract specification tests are performed on product to be shipped. Typical wet methods for the determination of acetone are acidimetry (49), titration of the liberated hydrochloric acid after treating the acetone with hydroxylamine hydrochloride; and iodimetry (50), titrating the excess of iodine after treating the acetone with iodine and base (iodoform reaction).

Carbon steel tanks of welded construction, as specified in the American Petroleum Institute Standard 650 (51), are recommended for acetone storage. Gaskets should be ethylene–propylene rubber or Viton rubber. An inert gas pad should be used. Provisions should be made to prevent static charge buildup during filling. Design considerations of the National Fire Prevention Association Code 30 and local fire codes should be followed. Tank venting systems should comply with local vapor emission standards and conform with National Fire Prevention Association recommendations. Where the purity of the acetone is to be optimized, an inorganic zinc lining is recommended (52). One such lining is Carbozinc 11, metallic zinc in an ethyl silicate binder, available from Carboline Co., St. Louis, Missouri.

Health and Safety

Acetone is among the solvents of comparatively low acute and chronic toxicity. High vapor concentrations produce anesthesia, and such levels may be irritating to the eyes, nose, and throat, and the odor may be disagreeable. Acetone does not have sufficient warning properties to prevent repeated exposures to concentrations which may cause adverse effects. In industry no injurious effects have been reported other than skin irritation resulting from its defatting action, or headache from prolonged inhalation (53). Direct contact with the eyes may produce irritation and transient corneal injury.

Material Safety Data Sheets (MSDS) issued by suppliers of acetone are required to be revised within 90 days to include new permissible exposure limits (PEL). Current OSHA PEL (54) and ACGIH threshold limit values (TLV) (55) are the same, 750 ppm TWA and 1000 ppm STEL. For comparison, the ACGIH TWA values for the common rubbing alcohols are ethyl, 1000, and isopropyl, 400 ppm. A report on human experience (56) concluded that exposure to 1000 ppm for an 8-h day produced no effects other than slight, transient irritation of the eyes, nose, and throat.

There are many natural sources of acetone including forest fires, volcanoes, and the normal metabolism of vegetation, insects, and higher animals (57). Acetone is a normal constituent of human blood, and it occurs in much higher concentrations in diabetics. Its toxicity appears to be low to most organisms. Acetone is ubiquitous in the environment, but is not environmentally persistent because it is readily biodegraded. In general acetone is an environmentally benign compound, widely detected but in concentrations that are orders of magnitude below toxicity thresholds.

Acetone can be handled safely if common sense precautions are taken. It should be used in a well-ventilated area, and because of its low flash point, ignition sources should be absent. Flame will travel from an ignition source along vapor flows on floors or bench tops to the point of use. Sinks should be rinsed with water while acetone is being used to clean glassware, to prevent the accumulation of vapors. If prolonged or repeated skin contact with acetone could occur, impermeable protective equipment such as gloves and aprons should be worn.

Compatibility of acetone with other materials should be carefully considered, especially in disposal of wastes. It reacts with chlorinating substances to form toxic chloroketones, and potentially explosively with some peroxy compounds and a number of oxidizing mixtures. Mixed with chloroform, acetone will react violently in the presence of bases. Other incompatibilities are listed in the Sax handbook (53).

Vapor flammability range in air (2.6–12.8 vol %) and low flash point ($-18°C$, $0°F$) make fire the major hazard of acetone. Quantities larger than laboratory hand bottles should be stored in closed metal containers. Gallon glass bottles should be protected against impacts. Areas where acetone is in contact with the ambient air should be free of ignition sources (flames, sparks, static charges, and hot surfaces above the autoignition temperature, about 500–600°C depending on the reference consulted). Fires may be controlled with carbon dioxide or dry chemical extinguishers. Recommended methods of handling, loading, unloading, and storage can be obtained from Material Safety Data Sheets and inquiries directed to suppliers of acetone.

Uses

Acetone is used as a solvent and as a reaction intermediate for the production of other compounds which are mainly used as solvents and/or intermediates for consumer products.

Direct Solvent Use. A large volume, direct solvent use of acetone is in formulations for surface coatings and related washes and thinners, mainly for acrylic and nitrocellulose lacquers and paints. It is used as a solvent in the manufacture of pharmaceuticals and cosmetics (about 7000 metric tons in nail polish removers), in spinning cellulose acetate fibers, in gas cylinders to store acetylene safely, in adhesives and contact cements, in various extraction processes, and in the manufacture of smokeless powder. It is a wash solvent in fiberglass boat manufacturing, a cleaning solvent in the electronics industry, and a solvent for degreasing wool and degumming silk.

Acrylics. Acetone is converted via the intermediate acetone cyanohydrin to the monomer methyl methacrylate (MMA) [80-62-6]. The MMA is polymerized to poly(methyl methacrylate) (PMMA) to make the familiar clear acrylic sheet. PMMA is also used in molding and extrusion powders. Hydrolysis of acetone cyanohydrin gives methacrylic acid (MAA), a monomer which goes directly into acrylic latexes, carboxylated styrene–butadiene polymers, or ethylene–MAA ionomers. As part of the methacrylic structure, acetone is found in the following major end use products: acrylic sheet molding resins, impact modifiers and processing aids, acrylic film, ABS and polyester resin modifiers, surface coatings, acrylic lacquers, emulsion polymers, petroleum chemicals, and various copolymers (see METHACRYLIC ACID AND DERIVATIVES; METHACRYLIC POLYMERS).

Bisphenol A. One mole of acetone condenses with two moles of phenol to form bisphenol A [80-05-07], which is used mainly in the production of polycarbonate and epoxy resins. Polycarbonates (qv) are high strength plastics used widely in automotive applications and appliances, multilayer containers, and housing applications. Epoxy resins (qv) are used in fiber-reinforced laminates, for encapsulating electronic components, and in advanced composites for aircraft–aerospace and automotive applications. Bisphenol A is also used for the production of corrosion- and chemical-resistant polyester resins, polysulfone resins, polyetherimide resins, and polyarylate resins.

Aldol Chemicals. The aldol condensation of acetone molecules leads to the group of aldol chemicals which are themselves used mainly as solvents. The initial condensation product is diacetone alcohol (DAA) which is dehydrated to mesityl oxide. Because of its toxicity effects, mesityl oxide is no longer produced for sale, but is used captively to make methyl isobutyl ketone (MIBK) and methylisobutylcarbinol (MIBC) by hydrogenation. DAA is hydrogenated to hexylene glycol. Three molecules of acetone give isophorone and phorone which is hydrogenated to diisobutyl ketone (DIBK) [108-83-8] and diisobutylcarbinol (DIBC) [108-82-7].

MIBK is a coatings solvent for nitrocellulose lacquers and vinyl and acrylic polymer coatings, an intermediate for rubber antioxidants and specialty surfactants, and a solvent for the extraction of antibiotics. MIBC is used mainly for the production of zinc dialkyl dithiophosphates which are used as lubricating oil additives. It is a flotation agent for minerals and a solvent for coatings. Besides its use as a chemical intermediate, DAA is used as a solvent for nitrocellulose, cellulose acetate, oils, resins, and waxes, and in metal cleaning compounds. Hexylene glycol is a component in brake fluids and printing inks. Isophorone is a solvent for industrial coatings and enamels. DIBK is used in coatings and leather finishes.

Other Uses. More than 70 thousand metric tons of acetone is used in small volume applications some of which are to make functional compounds such as antioxidants, herbicides, higher ketones, condensates with formaldehyde or diphenylamine, and vitamin intermediates.

Further information on uses, current volumes, and estimates of growth is available in reference 58.

BIBLIOGRAPHY

"Acetone" in *ECT* 1st ed., Vol. 1, pp. 88–95, by C. L. Gabriel and A. A. Dolnick, Publicker Industries; in *ECT* 2nd ed., Vol. 1, pp. 159–167, by R. J. Miller, California Research Corporation; in *ECT* 3rd ed., Vol. 1, pp. 179–191, by D. L. Nelson and B. P. Webb, The Dow Chemical Company.

1. American Institute of Chemical Engineers, *Design Institute for Physical Property Data,* (DIPPR File), University Park, Pa., 1989. For other listings of properties, see *Beilsteins Handbuch der Organischen Chemie,* Springer-Verlag, Berlin, Vol. 1 and supplement; and J. A. Riddick, W. B. Bunger, and T. K. Sakano, "Organic Solvents, Physical Properties, and Methods of Purification," in *Techniques of Organic Chemistry,* Vol. 2, John Wiley & Sons, Inc., New York, 1986.
2. R. C. Weast and J. G. Grasselli, eds., *Handbook of Data on Organic Compounds,* 2nd ed., Vol. 6, CRC Press, Inc., Boca Raton, Fla., Compound no. 21433, p. 3731.
3. *Fire Hazard Properties of Flammable Liquids, Gases, and Volatile Solids, Report 325M-1984, National Fire Codes,* Vol. 8, National Fire Protection Association, Batterymarch Park, Quincy, Mass.
4. R. A. Smiley, "Nitriles" in M. Grayson, ed., *Kirk-Othmer Encyclopedia of Chemical Technology,* 3rd ed., Vol. 15, Wiley-Interscience, New York, 1981, p. 901.
5. U.S. Pat. 2,827,494 (Mar. 18, 1958), J. H. Brown, Jr., and N. B. Lorette (to The Dow Chemical Company); *Chem. Abstr.* **52,** 14655i (1958). U.S. Pat. 2,827,495 (March 18, 1958), G. C. Bond and L. A. Klar (to The Dow Chemical Company); *Chem. Abstr.* **52,** 14656a (1958). N. B. Lorette, W. L. Howard, and J. H. Brown, Jr., *J. Org. Chem.* **24,** 1731 (1959); *Chem. Abstr.* **55,** 12275g (1961).
6. U.S. Pat. 3,804,795 (Apr. 16, 1974), W. O. Perry, M. W. Sorenson, and T. J. Hairston, (to The Dow Chemical Company); *Chem. Abstr.* **81,** 65384v (1974). Ger. Offen. 2,424,522 (Dec. 12, 1974) and U.S. Pat. 3,923,744 (Dec. 2, 1975), M. W. Sorenson, R. C. Whiteside, and R. A. Hickner (to The Dow Chemical Company); *Chem. Abstr.* **82,** 141725v (1975).
7. N. B. Lorette and W. L. Howard, *J. Org. Chem.* **25,** 521 (1960); *Chem. Abstr.* **54,** 19531c (1960). W. L. Howard and N. B. Lorette, *J. Org. Chem.* **25,** 525 (1960); *Chem. Abstr.* **54,** 19528f (1960).
8. U.S. Pat. 3,127,450 (March 31, 1964), W. L. Howard and N. B. Lorette (to The Dow Chemical Company); *Chem. Abstr.* **60,** 15737f (1964). U.S. Pat. 3,166,600 (January 19, 1965), N. B. Lorette and W. L. Howard (to The Dow Chemical Company); *Chem. Abstr.* **62,** 7656i (1965).
9. E. G. R. Ardah, A. D. Barbour, G. E. McClellan, and E. W. McBride, *Ind. Eng. Chem.* **16,** 1133 (1924).
10. J. M. Weiss, *Chem. Eng. News* **36,** 70 (June 9, 1958).
11. U.S. Pat. 1,329,214 (Jan. 27, 1920), C. Weizmann and A. Hamlyn; *Chem. Abstr.* **14,** 998 (1920).
12. Brit. Pat. 876,025 (Aug. 30, 1961) and Ger. Pat. 1,080,994 (to Consortium Fuer Elektrochemische Industrie G.m.b.H.). Brit. Pat. 884,962 (Dec. 20, 1961) (to Consortium Fuer Elektrochemische Industrie G.m.b.H.); *Chem. Abstr.* **59,** 5024h (1963). Brit. Pat. 892,158 (Mar. 21, 1962) (to Consortium Fuer Elektrochemische Industrie G.m.b.H.); *Chem. Abstr.* **59,** 13826d (1963).

13. C. S. Read with T. Gibson and Z. Sedaghat-Pour, "Acetone" in *Chemical Economics Handbook,* SRI International, Menlo Park, Calif., 1989, p. 604.5000 H.

14. K. D. McCracken, "The Dow Chemical Company, private communication, Feb. 1990.

15. Ref. 13, p. 604.5001 O to W.

16. Brit. Pat. 1,257,595 (Dec. 22, 1971) and Fr. Pat. 2,050,175, R. L. Feder and co-workers (to Allied Chemical); *Chem. Abstr.* **76,** 3548q (1972). U.S. Pat. 2,632,774 (Mar. 24, 1953), J. C. Conner, Jr., and A. D. Lohr (to Hercules, Inc.). U.S. Pat. 2,744,143 (Sept. 2, 1953), L. J. Filar (to Hercules, Inc.). U.S. Pat. 3,365,375 (Jan. 23, 1968) and Brit. Pat. 1,193,119, J. R. Nixon, Jr. (to Hercules, Inc.). Brit. Pat. 999,441 (July 28, 1965) (to Allied Chemical); *Chem. Abstr.* **63,** 14764g (1965).

17. Brit. Pat. 1,257,595 of ref. 16. U.S. Pat. 2,799,711 (July 16, 1957), E. Beati and F. Severini (to Montecatini). Brit. Pat. 895,622 (May 2, 1962) (to Societa Italiana Resine); *Chem. Abstr.* **57,** 11108h (1962).

18. Brit. Pat. 1,257,595 of ref. 16.

19. Brit. Pat. 970,945 (Sept. 23, 1964) (to Societa Italiana Resine); *Chem. Abstr.* **61,** 14586a (1964). U.S. Pat. 2,757,209 (July 31, 1956), G. C. Joris (to Allied Chemical).

20. U.S. Pat. 2,632,774 of ref. 16.

21. Ref. 13, p. 604.5000 F. Detailed process information is available in *Phenol, Report No. 22B,* Process Economics Program, SRI International, Menlo Park, Calif., December 1977.

22. Ref. 13, p. 604.5001 T.

23. C. Sheely, Jr., *Kinetics of Catalytic Dehydrogenation of Isopropanol,* University Microfilms, Ann Arbor, Michigan, 1953, p. 3.

24. Brit. Pat. 938,854 (Oct. 9, 1953) and Belg. Pat. 617965, J. B. Anderson, K. B. Cofer, and G. E. Coury (to Shell Chemical); *Chem. Abstr.* **59,** 13826a (1963).

25. Brit. Pat. 665,376 (Jan. 23, 1952), H. O. Mottern (to Standard Oil Development Co.); *Chem. Abstr.* **46,** 6142h (1952); this work is also U.S. Pat. 2,549,844; *Chem. Abstr.* **46,** 524c (1952).

26. Brit. Pat. 804,132 (Nov. 5, 1958) (to Knapsack-Griesheim Aktiengesellschaft); *Chem. Abstr.* **53,** 7990c (1959).

27. Brit. Pat. 817,622 (Aug. 6, 1959), W. Edyvean (to Shell Research Limited); *Chem. Abstr.* **54,** 7562e (1960).

28. Brit. Pat. 1,097,819 (Jan. 3, 1968) (to Les Usines DeMelle); *Chem. Abstr.* **68,** 63105n (1968). Brit. Pat. 610,397 (Oct. 14, 1948) (to Universal Oil Products); *Chem. Abstr.* **43,** 4287f (1949).

29. Brit. Pat. 742,496 (Dec. 30, 1955), W. G. Emerson, Jr., and J. R. Quelly (to Esso Research & Engineering Co.); *Chem. Abstr.* **50,** 8710a (1956); this work is also U.S. Pat. 2,662,848; *Chem. Abstr.* **48,** 8815a (1954).

30. *Acetone, Methyl Ethyl Ketone, and Methyl Isobutyl Ketone, Report No. 77,* May 1972, Process Economics Program, SRI CEH, SRI International, Menlo Park, Calif. p. 604.5000 F. Contains detailed process information.

31. "Code of Federal Regulations, Title 49, pt. 100 to 177," *Federal Register,* Washington, D.C., October 1, 1988, paragraphs 173.118, 173.119, 173.32C, and 172.101. Because of occasional changes, the latest issue of *Federal Register* should be consulted.

32. "Code of Federal Regulations, Titles 33 and 46," *Federal Register,* Washington, D.C. See also ref. 31.

33. *Chemical Data Guide for Bulk Shipment by Water,* United States Coast Guard, Washington, D.C.

34. *Code for Construction and Equipment of Ships Carrying Dangerous Chemicals in Bulk,* International Maritime Organization, Publications Section, London, England.

35. *Chemical Hazard Response Information System, Commandant Instruction M.16465.12A,* U.S. Coast Guard, U.S. Department of Transportation, Washington, D.C.

36. *Guidebook for Initial Response to Hazardous Materials Incidents, DOT P 5800.4,* U.S. Department of Transportation, Washington, D.C., 1987. *Emergency Handling of Hazardous Materials in Surface Transportation,* Bureau of Explosives, Association of American Railroads, Washington, D.C., 1981.
37. Chemical Transportation Emergency Center, a public service of the Chemical Manufacturers' Association, 2501 M Street, N.W., Washington, D.C. 20037-1303.
38. Ref. 13, pp. 604.5000 C,D.
39. Ref. 13, p. 604.5000 E.
40. Ref. 13, pp. 604.5000 D,E.
41. Ref. 13, p. 604.5000 H.
42. Ref. 13, pp. 604.5000 H,S.
43. Ref. 13, pp. 604.5000 U,V.
44. Ref. 13, pp. 604.5001 I,J.
45. Ref. 13, p. 604.5001 M.
46. Ref. 13, p. 604.5000 Z.
47. Ref. 13, p. 604.5000 T.
48. J. A. Riddick, W. B. Bunger, and T. K. Sakano, "Organic Solvents, Physical Properties, and Methods of Purification," in *Techniques of Organic Chemistry,* Vol. 2, John Wiley & Sons, Inc., New York, 1986, p. 954.
49. M. Morosco, *Ind. Eng. Chem.* **18,** 701 (1926).
50. L. F. Goodwin, *J. Am. Chem. Soc.* **42,** 39 (1920).
51. *American Petroleum Institute Standard 650,* 1977 ed., American Petroleum Institute, Washington, D.C., paragraphs 3.5.2e1, 3.5.2e3.
52. *Acetone,* Form No. 115-598-84, product bulletin of The Dow Chemical Company, Midland, Mich., 1984.
53. N. I. Sax and R. J. Lewis, Sr., *Dangerous Properties of Industrial Materials,* 7th ed., Vol. 2, Van Nostrand Reinhold, New York, 1989.
54. *Federal Register* **54**(12), 2332 (1989).
55. *Threshold Limit Values and Biological Exposure Indices for 1987–1988,* American Conference of Government Industrial Hygienists, Cincinnati, Ohio.
56. W. J. Krasavage, J. L. O'Donoghue, and G. D. Divincenzo, "Ketones" in G. D. Clayton and F. E. Clayton, eds., *Patty's Industrial Hygiene and Toxicology,* 3rd rev. ed., Vol. 2C, *Toxicology,* John Wiley & Sons, Inc., 1982, p. 4724. Several human experience case histories are cited.
57. T. E. Graedel, D. T. Hawkins, and L. D. Claxton, *Atmospheric Chemical Compounds,* Academic Press, Orlando, Fla., 1986, p. 263, cited in *Hazardous Substances Data Bank, Acetone* from Toxicology Data Network (TOXNET), National Library of Medicine, Bethesda, Md., Jan. 1990, NATS section in the review.
58. Ref. 13, pp. 604.5000T to 604.5001Z.

General Reference

C. S. Read with T. Gibson and Z. Sedaghat-Pour, "Acetone" in *Chemical Economics Handbook,* SRI International, Menlo Park, Calif., 1989. An excellent information source.

WILLIAM L. HOWARD
The Dow Chemical Company

ACETONITRILE.   See Nitriles.

ACETONYLACETONE.   See Ketones.

ACETOPHENETIDIN.   See Analgesics, antipyretics, and anti-
inflammatory agents.

ACETOPHENONE.   See Phenyl ketones.

ACETYL CHLORIDE.   See Acetic acid and derivatives.

ACETYLCHOLINE.   See Choline; Enzymes, inhibitors and antagonists.

ACETYLENE.   See Hydrocarbons, acetylene.

ACETYLENE BLACK.   See Carbon, carbon and artificial graphite.

ACETYLENE-DERIVED CHEMICALS

Acetylene [74-86-2], C_2H_2, is an extremely reactive hydrocarbon, principally used as a chemical intermediate (see Hydrocarbons, acetylene). Because of its thermodynamic instability, it cannot easily or economically be transported for long distances. To avoid large free volumes or high pressures, acetylene cylinders contain a porous solid packing and an organic solvent. Acetylene pipelines are severely restricted in size and must be used at relatively low pressures. Hence, for large-scale operations, the acetylene consumer must be near the place of acetylene manufacture.

Historically, the use of acetylene as raw material for chemical synthesis has depended strongly upon the availability of alternative raw materials. The United States, which until recently appeared to have limitless stocks of hydrocarbon feeds, has never depended upon acetylene to the same extent as Germany, which had more limited access to hydrocarbons (1). During World War I the first manufacture of a synthetic rubber was undertaken in Germany to replace imported natural rubber, which was no longer accessible. Acetylene derived from calcium carbide was used for preparation of 2,3-dimethyl-1,3-butadiene by the following steps:

acetylene $\longrightarrow$ acetaldehyde $\longrightarrow$ acetic acid $\longrightarrow$ acetone $\longrightarrow$
2,3-dimethyl-2,3-butanediol $\longrightarrow$ 2,3-dimethyl-1,3-butadiene

Methyl rubber, obtained by polymerization of this monomer, was expensive and had inferior properties, and its manufacture was discontinued at the end of World

War I. By the time World War II again shut off access to natural rubber, Germany had developed better synthetic rubbers based upon butadiene [106-99-0] (see ELASTOMERS, SYNTHETIC).

In the United States butadiene was prepared initially from ethanol and later by cracking four-carbon hydrocarbon streams (see BUTADIENE). In Germany butadiene was prepared from acetylene via the following steps: acetylene → acetaldehyde → 3-hydroxybutyraldehyde → 1,3-butanediol → 1,3-butadiene.

Toward the end of the war, an alternative German route to butadiene was introduced which required much less acetylene:

$$HC{\equiv}CH + \quad 2\ CH_2O \quad \longrightarrow\ HOCH_2C{\equiv}CCH_2OH \longrightarrow$$

acetylene   formaldehyde   1,4-butynediol

$$HOCH_2CH_2CH_2CH_2OH \longrightarrow CH_2{=}CHCH{=}CH_2$$

1,4-butanediol   1,3-butadiene

Because of its relatively high price, there have been continuing efforts to replace acetylene in its major applications with cheaper raw materials. Such efforts have been successful, particularly in the United States, where ethylene has displaced acetylene as raw material for acetaldehyde, acetic acid, vinyl acetate, and chlorinated solvents. Only a few percent of U.S. vinyl chloride production is still based on acetylene. Propylene has replaced acetylene as feed for acrylates and acrylonitrile. Even some recent production of traditional Reppe acetylene chemicals, such as butanediol and butyrolactone, is based on new raw materials.

Reaction Products

Acetaldehyde.   Acetaldehyde [75-07-0], C_2H_4O, (qv) was formerly manufactured principally by hydration of acetylene.

$$HC{\equiv}CH + H_2O \longrightarrow CH_3CHO$$

Many catalytic systems have been described; acidic solutions of mercuric salts are the most generally used. This process has long been superseded by more economical routes involving oxidation of ethylene or other hydrocarbons.

Acrylic Acid, Acrylates, and Acrylonitrile.   Acrylic acid [79-10-7], $C_3H_4O_2$, and acrylates were once prepared by reaction of acetylene and carbon monoxide with water or an alcohol, using nickel carbonyl as catalyst. In recent years this process has been completely superseded in the United States by newer processes involving oxidation of propylene (2). In western Europe, however, acetylene is still important in acrylate manufacture (see ACRYLIC ACID AND DERIVATIVES; ACRYLIC ESTER POLYMERS).

In the presence of such catalysts as a solution of cuprous and ammonium chlorides, hydrogen cyanide adds to acetylene to give acrylonitrile [107-13-1], C_3H_3N (qv).

$$HC{\equiv}CH + HCN \longrightarrow CH_2{=}CHCN$$

Since the early 1970s this process has been completely replaced by processes involving ammoxidation of propylene (3).

Chlorinated Solvents.   Originally, successive chlorination and dehydrochlorination of acetylene was the route to trichloroethylene [79-01-6], C_2HCl_3, and perchloroethylene [127-18-4], C_2Cl_4.

$$HC\equiv CH + 2\ Cl_2 \longrightarrow CHCl_2CHCl_2 \longrightarrow CHCl=CCl_2 + HCl$$

$$CHCl=CCl_2 + Cl_2 \longrightarrow CHCl_2CCl_3 \longrightarrow CCl_2=CCl_2 + HCl$$

This route has been completely displaced, first by chlorination and dehydrochlorination of ethylene or vinyl chloride, and more recently by oxychlorination of two-carbon raw materials (2) (see CHLOROCARBONS AND CHLOROHYDROCARBONS).

Cyclooctatetraene (COT).   Tetramerization of acetylene to cyclooctatetraene [629-20-9], C_8H_8, although interesting, does not seem to have been used commercially. Nickel salts serve as catalysts. Other catalysts give benzene. The mechanism of this cyclotetramerization has been studied (4).

Ethylene.   During World War II the Germans manufactured more than 60,000 t/yr of ethylene [74-85-1], C_2H_4, by hydrogenation of acetylene, using palladium on silica gel as catalyst. Subsequently, cracking of hydrocarbons displaced this process. However, it is still utilized for purification of ethylene containing small amounts of acetylene as contaminant (5) (see ETHYLENE).

Vinyl Acetate.   Vinyl acetate [108-05-04], $C_4H_6O_2$, used to be manufactured by addition of acetic acid to acetylene.

$$HC\equiv CH + CH_3COOH \longrightarrow CH_2=CHOOCCH_3$$

Liquid- and vapor-phase processes have been described; the latter appear to be advantageous. Supported cadmium, zinc, or mercury salts are used as catalysts. In 1963 it was estimated that 85% of U.S. vinyl acetate capacity was based on acetylene, but it has been completely replaced since about 1982 by newer technology using oxidative addition of acetic acid to ethylene (2) (see VINYL POLYMERS). In western Europe production of vinyl acetate from acetylene still remains a significant commercial route.

Vinylacetylene and Chloroprene.   In the presence of cuprous salt solutions, acetylene dimerizes to vinylacetylene [689-97-4], C_4H_4. Yields of 87% monovinylacetylene, together with 10% of divinylacetylene, have been described (6).

$$2\ HC\equiv CH \longrightarrow HC\equiv CCH=CH_2$$

Using cuprous chloride as catalyst, hydrogen chloride adds to acetylene, giving 2-chloro-1,3-butadiene [126-99-8], chloroprene, C_4H_5Cl, the monomer for neoprene rubber.

$$HC\equiv CCH=CH_2 + HCl \longrightarrow CH_2=CClCH=CH_2$$

Manufacture via this process has been completely replaced by chlorination of butadiene (3) (see CHLOROCARBONS AND CHLOROHYDROCARBONS, CHLOROPRENE; ELASTOMERS, SYNTHETIC, POLYCHLOROPRENE).

Vinyl Chloride and Vinylidene Chloride.   In the presence of mercuric salts, hydrogen chloride adds to acetylene giving vinyl chloride [75-01-4], C_2H_3Cl.

$$HC{\equiv}CH + HCl \longrightarrow CH_2{=}CHCl$$

Once the principal route to vinyl chloride, in all but a few percent of current U.S. capacity this has been replaced by dehydrochlorination of ethylene dichloride. A combined process in which hydrogen chloride cracked from ethylene dichloride was added to acetylene was advantageous but it is rarely used because processes to oxidize hydrogen chloride to chlorine with air or oxygen are cheaper (7) (see VINYL POLYMERS).

In similar fashion, vinylidene chloride [75-35-4], $C_2H_2Cl_2$, has been prepared by successive chlorination and dehydrochlorination of vinyl chloride (see VINYLIDENE CHLORIDE AND POLY(VINYLIDENE CHLORIDE)).

$$CH_2{=}CHCl + Cl_2 \longrightarrow CH_2ClCHCl_2$$
$$CH_2ClCHCl_2 \longrightarrow CH_2{=}CCl_2 + HCl$$

Vinyl Fluoride.   Vinyl fluoride [75-02-5], C_2H_3F, the monomer for poly(vinyl fluoride), is manufactured by addition of hydrogen fluoride to acetylene (see FLUORINE COMPOUNDS, ORGANIC, POLY(VINYL FLUORIDE)).

$$HC{\equiv}CH + HF \longrightarrow CH_2{=}CHF$$

Ethynylation Reaction Products

The name ethynylation was coined by Reppe to describe the addition of acetylene to carbonyl compounds (8).

$$HC{\equiv}CH + RCOR' \longrightarrow HC{\equiv}CC(OH)RR'$$

Although stoichiometric ethynylation of carbonyl compounds with metal acetylides was known as early as 1899 (9), Reppe's contribution was the development of catalytic ethynylation. Heavy metal acetylides, particularly cuprous acetylide, were found to catalyze the addition of acetylene to aldehydes. Although ethynylation of many aldehydes has been described (10), only formaldehyde has been catalytically ethynylated on a commercial scale. Copper acetylide is not effective as catalyst for ethynylation of ketones. For these, and for higher aldehydes, alkaline promoters have been used.

The following series of reactions illustrates the manufacture of the principal Reppe acetylene chemicals.

$$HC{\equiv}CH + 1 \text{ or } 2 \text{ HCHO} \longrightarrow HC{\equiv}CCH_2OH + HOCH_2C{\equiv}CCH_2OH$$

propargyl alcohol   2-butyne-1,4-diol

$$HOCH_2C{\equiv}CCH_2OH + H_2 \longrightarrow HOCH_2CH{=}CHCH_2OH$$

2-butene-1,4-diol

$$HOCH_2C{\equiv}CCH_2OH + 2\ H_2 \longrightarrow HOCH_2CH_2CH_2CH_2OH$$

1,4-butanediol

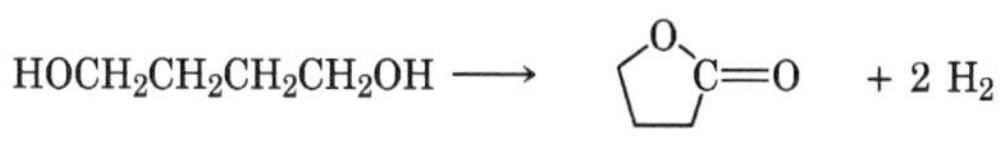

$$HOCH_2CH_2CH_2CH_2OH \longrightarrow \underset{\gamma\text{-butyrolactone}}{\text{[structure]}} C{=}O\ + 2\ H_2$$

$$\text{[structure]} C{=}O + RNH_2 \longrightarrow \underset{R}{\text{[structure]}} C{=}O\ + H_2O$$

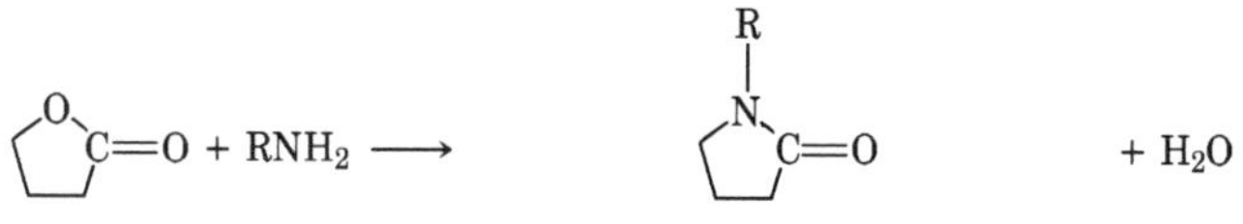

2-pyrrolidinone (R = H)
1-methyl-2-pyrrolidinone (R = CH₃)

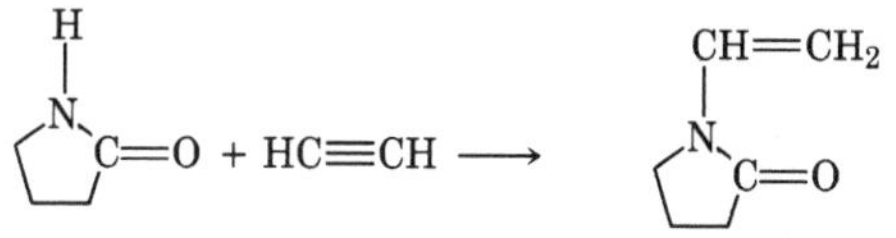

$$\underset{H}{\text{[structure]}} C{=}O + HC{\equiv}CH \longrightarrow \underset{CH{=}CH_2}{\text{[structure]}} C{=}O$$

N-vinyl-2-pyrrolidinone

Except for the pyrrolidinones (see PYRROLE AND PYRROLE DERIVATIVES), these products are discussed in the following.

Propargyl Alcohol. Propargyl alcohol [107-19-7], 2-propyn-1-ol, C_3H_4O, is the only commercially available acetylenic primary alcohol. A colorless, volatile liquid, with an unpleasant odor that has been described as "mild geranium," it was first prepared in 1872 from β-bromoallyl alcohol (11). Propargyl alcohol is miscible with water and with many organic solvents. Physical properties are listed in Table 1.

Reactions. Propargyl alcohol has three reactive sites—a primary hydroxyl group, a triple bond, and an acetylenic hydrogen—making it an extremely versatile chemical intermediate.

The hydroxyl group can be esterified with acid chlorides, anhydrides, or carboxylic acids and it reacts with aldehydes (12) or vinyl ethers (13) in the presence of an acid catalyst to form acetals.

$$RCHO + 2\ HC{\equiv}CCH_2OH \longrightarrow RCH(OCH_2C{\equiv}CH)_2 + H_2O$$

$$CH_2{=}CHOR + HC{\equiv}CCH_2OH \longrightarrow CH_3\underset{OR}{CH}OCH_2C{\equiv}CH$$

At low temperatures, oxidation with chromic acid gives propynal [624-67-9], C_3H_2O (14), or propynoic acid [471-25-0], $C_3H_2O_2$ (15), which can also be prepared in high yields by anodic oxidation (16).

$$HC{\equiv}CCH_2OH \longrightarrow HC{\equiv}CCHO$$

$$HC{\equiv}CCH_2OH \longrightarrow HC{\equiv}CCOOH$$

Table 1. Physical Properties of Propargyl Alcohol

Property	Value
melting point, °C	−52
boiling point, °C	114
specific gravity, d_4^{20}	0.948
refractive index, n_D^{20}	1.4310
viscosity at 20°C, mPa·s (= cP)	1.65
dielectric constant, ϵ	24.5
specific heat, C_p^{20}, J/(g·K)[a]	2.577
heat of combustion at constant vol, kJ/mol[b]	1731
heat of vaporization at 112°C, kJ/mol[b]	42.09
flash point, Tagliabue open cup, °C	36

Vapor pressure data

Temperature, °C	Vapor pressure, kPa[c]	Temperature, °C	Vapor pressure, kPa[c]
20	1.55	80	29.7
30	2.80	90	43.6
40	4.80	100	63.3
50	7.87	110	88.7
60	12.8	114	101.3
70	19.9		

Azeotropes

Other component	Other component, mol%	bp, °C
water	79.8	97
benzene	87.2	78

[a]To convert J/(g·K) to cal/(g·°C) divide by 4.184.
[b]To convert kJ/mol to kcal/mol divide by 4.184.
[c]To convert kPa to mm Hg (torr) multiply by 7.5.

Various halogenating agents have been used to replace hydroxyl with chlorine or bromine. Phosphorus trihalides, especially in the presence of pyridine, are particularly suitable (17,18). Propargyl iodide is easily prepared from propargyl bromide by halogen exchange (19).

Hydrogenation gives allyl alcohol [*107-18-6*], C_3H_6O, its isomer propanal [*123-38-6*] (20), or propanol, C_3H_8O [*71-23-8*] (21). With acidic mercuric salt catalysts, water adds to give acetol, hydroxyacetone, $C_3H_6O_2$ [*116-09-6*] (22).

$$HC\equiv CCH_2OH + H_2O \longrightarrow CH_3COCH_2OH$$

Using alcohols instead of water under similar conditions gives cyclic ketals (23), which can be hydrolyzed to acetol.

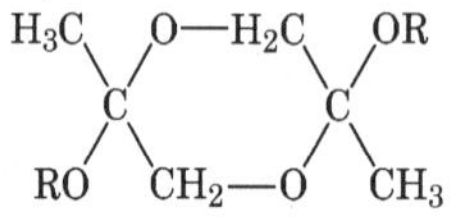

Halogens add stepwise, giving almost exclusively dihaloallyl alcohols (24,25).

$$HC{\equiv}CCH_2OH + X_2 \longrightarrow CHX{=}CXCH_2OH$$

A second mole of halogen adds with greater difficulty; oxidative side reactions can be minimized by halogenating an ester instead of the free alcohol (26).

$$(HC{\equiv}CCH_2O)_3PO + 6\ Br_2 \longrightarrow (HCBr_2CBr_2CH_2O)_3PO$$

With mercuric salt catalysts, hydrogen chloride adds to give 2-chloroallyl alcohol, 2-chloroprop-2-en-1-ol [5976-47-6] (27).

$$HC{\equiv}CCH_2OH + HCl \longrightarrow CH_2{=}CClCH_2OH$$

In the presence of suitable nickel or cobalt complexes, propargyl alcohol trimerizes to a mixture of 1,3,5-benzenetrimethanol [4464-18-0] and 1,2,4-trimethanol [25147-76-6] benzene (28).

Cyclization with various nickel complex catalysts gives up to 97% selectivity to a mixture of cyclooctatetraene derivatives, with only 3% of benzene derivatives. The principal isomer is the symmetrical 1,3,5,7-cyclooctatetraene-1,3,5,7-tetramethanol (29).

Nickel halide complexes with amines give mixtures of linear polymer and cyclic trimers (30). Nickel chelates give up to 40% of linear polymer (31).

When heated with ammonia over cadmium calcium phosphate catalysts, propargyl alcohol gives a mixture of pyridines (32).

In the presence of copper acetylide catalysts, propargyl alcohol and aldehydes give acetylenic glycols (33). When dialkylamines are also present, dialkylaminobutynols are formed (34).

$$HC{\equiv}CCH_2OH + RCHO \longrightarrow R\overset{\overset{\displaystyle OH}{|}}{C}HC{\equiv}CCH_2OH$$

$$HC{\equiv}CCH_2OH + HCHO + R_2NH \longrightarrow R_2NCH_2C{\equiv}CCH_2OH + H_2O$$

With two equivalents of an organomagnesium halide, a Grignard reagent is formed, capable of use in further syntheses (35,36). Cuprous salts catalyze oxidative dimerization of propargyl alcohol to 2,4-hexadiyne-1,6-diol [3031-68-3] (37).

$$2 \ HC{\equiv}CCH_2OH + \tfrac{1}{2} \ O_2 \longrightarrow HOCH_2C{\equiv}C-C{\equiv}CCH_2OH + H_2O$$

Manufacture. Propargyl alcohol is a by-product of butynediol manufacture. The original high pressure butynediol processes gave about 5% of the by-product; newer lower pressure processes give much less. Processes have been described that give much higher proportions of propargyl alcohol (38,39).

Shipment, Storage, and Price. Propargyl alcohol is available in tank cars, tank trailers, and drums. It is usually shipped in unlined steel containers and transferred through standard steel pipes or braided steel hoses; rubber is not recommended. Clean, rust-free steel is acceptable for short-term storage. For longer storage, stainless steel (types 304 and 316), glass lining, or phenolic linings (Lithcote LC-19 and LC-24, Unichrome B-124, and Heresite) are suitable. Aluminum, epoxies, and epoxy-phenolics should be avoided.

The 1991 U.S. bulk price for propargyl alcohol was about $5.64/kg.

Specifications and Analytical Methods. The commercial material is specified as 97% minimum purity, determined by gas chromatography or acetylation. Moisture is specified at 0.05% maximum (Karl-Fischer titration). Formaldehyde content is determined by bisulfite titration.

Health and Safety Factors. Although propargyl alcohol is stable, violent reactions can occur in the presence of contaminants, particularly at elevated temperatures. Heating in undiluted form with bases or strong acids should be avoided. Weak acids have been used to stabilize propargyl alcohol prior to distillation. Since its flash point is low, the usual precautions against ignition of vapors should be observed.

Propargyl alcohol is a primary skin irritant and a severe eye irritant and is toxic by all means of ingestion; all necessary precautions must be taken to avoid contact with liquid or vapors. The LD_{50} is 0.07 mL/kg for white rats and 0.06 mL/kg for guinea pigs.

Uses. Propargyl alcohol is a component of oil-well acidizing compositions, inhibiting the attack of mineral acids on steel (see CORROSION AND CORROSION INHIBITORS). It is also employed in the pickling and plating of metals.

It is used as an intermediate in preparation of the miticide Omite [2312-35-8], 2-(4'-tert-butylphenoxy)cyclohexyl 2-propynyl sulfite (40); sulfadiazine [68-35-9] (41); and halogenated propargyl carbonate fungicides (42).

Butynediol. Butynediol, 2-butyne-1,4-diol, [110-65-6] was first synthesized in 1906 by reaction of acetylene bis(magnesium bromide) with paraformaldehyde (43). It is available commercially as a crystalline solid or a 35% aqueous solution manufactured by ethynylation of formaldehyde. Physical properties are listed in Table 2.

Reactions. Butynediol undergoes the usual reactions of primary alcohols. Because of its rigid, linear structure, many reactions forming cyclic products from butanediol or cis-butenediol give only polymers with butynediol.

Both hydroxyl groups can be esterified normally (44). The monoesters are readily prepared as mixtures with diesters and unesterified butynediol, but care

Table 2. Physical Properties of Butynediol, Butenediol, and Butanediol

Property	Butynediol	Butenediol	Butanediol
molecular formula	$C_4H_6O_2$	$C_4H_8O_2$	$C_4H_{10}O_2$
CAS Registry Number	[110-65-6]	[110-64-5]	[110-63-4]
melting point, °C	58	11.8	20.2
boiling point, °C at kPa[a]			
0.133	101	84	86
1.33	141	122	123
13.3	194	176	171
101.3	248	234	228
specific gravity	$d_4^{20} = 1.114$	$d_{15}^{25} = 1.070$	$d_4^{20} = 1.017$
refractive index n_D^{25}	$\alpha = 1.450 \pm 0.002$	1.4770	1.4445
	$\beta = 1.528 \pm 0.002$		
heat of combustion,[b] kJ/mol[c]	2204		
flash point, Tagliabue open cup, °C	152	128	121
viscosity at 20°C, mPa·s (= cP)		22	84.9
at 38°C		10.8	
at 99°C		2.5	
surface tension at 20°C, mN/m (= dyn/cm)			44.6
dielectric constant at 20°C, ϵ			31.5
solubility, g/100 mL solvent at 25°C			
water (0°C)	2		
water	374		miscible
ethanol	83		miscible
acetone	70		miscible
ether	2.6		3.1
benzene	0.04		0.3
hexane			<0.1

[a]To convert kPa, to mm Hg, multiply by 7.5.
[b]At constant volume.
[c]To convert kJ to kcal, divide by 4.184.

must be taken in separating them because the monoesters disproportionate easily (45).

The hydroxyl groups can be alkylated with the usual alkylating agents. To obtain aryl ethers a reverse treatment is used, such as treatment of butynediol toluenesulfonate or dibromobutyne with a phenol (44). Alkylene oxides give ether alcohols (46).

In the presence of acid catalysts, butynediol and aldehydes (47) or acetals (48) give polymeric acetals, useful intermediates for acetylenic polyurethanes suitable for high energy solid propellants.

$$HOCH_2C{\equiv}CCH_2OH + CH_2O \longrightarrow HO(CH_2C{\equiv}CCH_2OCH_2O)_nH$$

Electrolytic oxidation gives acetylene dicarboxylic acid [142-45-0] (2-butynedioic acid) in good yields (49); chromic acid oxidation gives poor yields (50). Oxidation with peroxyacetic acid gives malonic acid [141-82-2] (qv) (51).

Butynediol can be hydrogenated partway to butenediol or completely to butanediol.

$$HOCH_2C\equiv CCH_2OH \longrightarrow HOCH_2CH=CHCH_2OH \longrightarrow HOCH_2CH_2CH_2CH_2OH$$

Dichlorobutyne [821-10-3] and dibromobutyne [2219-66-1] are readily prepared by treatment with thionyl or phosphorus halides. The less-stable diiodobutyne is prepared by treatment of dichloro- or dibromobutyne with an iodide salt (52).

Addition of halogens proceeds stepwise, sometimes accompanied by oxidation. Iodine forms 2,3-diiodo-2-butene-1,4-diol (53). Depending on conditions, bromine gives 2,3-dibromo-2-butene-1,4-diol, 2,2,3,3-tetrabromobutane-1,4-diol, mucobromic acid, or 2-hydroxy-3,3,4,4-tetrabromotetrahydrofuran (54). Addition of chlorine is attended by more oxidation (55–57), which can be lessened by esterification of the hydroxyl groups.

Uncatalyzed addition of hydrochloric acid is accompanied by replacement of one hydroxyl group, giving high yields of 2,4-dichloro-2-buten-1-ol (58); with mercuric or cupric salt catalysts, addition occurs without substitution (59,60).

$$HOCH_2C\equiv CCH_2OH + 2\ HCl \longrightarrow ClCH_2CH=CClCH_2OH + H_2O$$

$$HOCH_2C\equiv CCH_2OH \xrightarrow[\text{catalyst}]{\text{HCl}} HOCH_2CCl=CHCH_2OH \xrightarrow[\text{catalyst}]{\text{HCl}} HOCH_2CCl_2CH_2CH_2OH$$

When aqueous solutions of sodium bisulfite are heated with butynediol, one or two moles add to the triple bond, forming sodium salts of sulfonic acids (61).

In the presence of mercuric salts, butynediol rapidly isomerizes to 1-hydroxy-3-buten-2-one (62).

$$HOCH_2C\equiv CCH_2OH \longrightarrow CH_2=CH\overset{\overset{\displaystyle O}{\|}}{C}CH_2OH$$

This adds compounds with active hydrogen such as water, alcohols, and carboxylic acids (63), to give 1,4-dihydroxy-2-butanone or its derivatives.

$$CH_2=CH\overset{\overset{\displaystyle O}{\|}}{C}CH_2OH + H_2O \longrightarrow HOCH_2CH_2\overset{\overset{\displaystyle O}{\|}}{C}CH_2OH$$

Butynediol is more difficult to polymerize than propargyl alcohol, but it cyclotrimerizes to hexamethylolbenzene [2715-91-5] (benzenehexamethanol) with a nickel carbonyl–phosphine catalyst (64); with a rhodium chloride–arsine catalyst a yield of 70% is claimed (65).

When heated with acidic oxide catalysts, mixtures of butynediol with ammonia or amines give pyrroles (66) (see PYRROLE AND PYRROLE DERIVATIVES).

$$HOCH_2C\equiv CCH_2OH + RNH_2 \longrightarrow \underset{\underset{R}{\overset{|}{N}}}{\boxed{}}$$

Manufacture. All manufacturers of butynediol use formaldehyde ethynylation processes. The earliest entrant was BASF, which, as successor to I. G. Farben, continued operations at Ludwigshafen, FRG, after World War II. Later BASF also set up a U.S. plant at Geismar, La. The first company to manufacture in the United States was GAF in 1956 at Calvert City, Ky., and later at Texas City, Tex. and Seadrift, Tex. The most recent U.S. manufacturer is Du Pont, which went on stream at La Porte, Tex. about 1969. Joint ventures of GAF and Hüls in Marl, Germany, and of Du Pont and Idemitsu in Chiba, Japan, are the newest producers.

At the end of World War II the butynediol plant and process at Ludwigshafen were studied extensively (67,68). Variations of the original high pressure, fixed-bed process, which is described below, are still in use. However, all of the recent plants use low pressures and suspended catalysts (69–75).

The hazards of handling acetylene under pressure must be considered in plant design and construction. Although means of completely preventing acetylene decomposition have not been found, techniques have been developed that prevent acetylene decompositions from becoming explosive. The original German plant was designed for pressures up to 20.26 MPa (200 atm), considered adequate for deflagration (nonexplosive decomposition), which could increase pressures approximately tenfold. It was not practical to design for control of a detonation (explosive decomposition), which could increase pressure nearly 200-fold (76).

The reactors were thick-walled stainless steel towers packed with a catalyst containing copper and bismuth oxides on a siliceous carrier. This was activated by formaldehyde and acetylene to give the copper acetylide complex that functioned as the true catalyst. Acetylene and an aqueous solution of formaldehyde were passed together through one or more reactors at about 90–100°C and an acetylene partial pressure of about 500–600 kPa (5–6 atm) with recycling as required. Yields of butynediol were over 90%, in addition to 4–5% propargyl alcohol.

Shipment, Storage, and Price. Butynediol, 35% solution, is available in tank cars, tank trailers, and drums. Stainless steel, nickel, aluminum, glass, and various plastic and epoxy or phenolic liners have all been found satisfactory. Rubber hose is suitable for transferring. The solution is nonflammable and freezes at about −5°C.

Butynediol solid flakes are packed in polyethylene bags inside drums. The product is hygroscopic and must be protected from moisture.

In 1991, U.S. prices were about $4.63/kg for solid butynediol and about $1.40/kg for the 35% aqueous solution.

Specifications and Analytical Methods. The commercial aqueous solution is specified as 34% minimum butynediol, as determined by bromination or refractive index. Propargyl alcohol is limited to 0.2% and formaldehyde to 0.7%.

The commercial flake is specified as 96.0% minimum butynediol content, with a maximum of 2.0% moisture. Purity is calculated from the freezing point (at least 52°C).

Health and Safety Factors. Although butynediol is stable, violent reactions can take place in the presence of certain contaminants, particularly at elevated temperatures. In the presence of certain heavy metal salts, such as mercuric chloride, dry butynediol can decompose violently. Heating with strongly alkaline materials should be avoided.

Butynediol is a primary skin irritant and sensitizer, requiring appropriate precautions. Acute oral toxicity is relatively high: LD_{50} is 0.06 g/kg for white rats.

Uses.   Most butynediol produced is consumed by the manufacturers in manufacture of butanediol and butenediol. Small amounts are converted to ethers with ethylene oxide.

Butynediol is principally used in pickling and plating baths. Small amounts are used in the manufacture of brominated derivatives, useful as flame retardants. It was formerly used in a wild oat herbicide, Carbyne (Barban), 4-chloro-2-butynyl-*N*-(3-chlorophenyl)carbamate [*101-27-9*], $C_{11}H_9Cl_2NO_2$ (77).

Butenediol.   2-Butene-1,4-diol [*110-64-5*] is the only commercially available olefinic diol with primary hydroxyl groups. The commercial product consists almost entirely of the cis isomer.

trans-2-Butene-1,4-diol diacetate was prepared from 1,4-dibromo-2-butene in 1893 (78) and hydrolyzed to the diol in 1926 (79). The original preparation of the cis diol utilized the present commercial route, partial hydrogenation of butynediol.

Physical properties are listed in Table 2. Butenediol is very soluble in water, lower alcohols, and acetone. It is nearly insoluble in aliphatic or aromatic hydrocarbons.

Reactions.   In addition to the usual reactions of primary hydroxyl groups and of double bonds, *cis*-butenediol undergoes a number of cyclization reactions.

The hydroxyl groups can be esterified normally: the interesting diacrylate monomer (80) and the biologically active haloacetates (81) have been prepared in this manner. Reactions with dibasic acids have given polymers capable of being cross-linked (82) or suitable for use as soft segments in polyurethanes (83). Polycarbamic esters are obtained by treatment with a diisocyanate (84) or via the bischloroformate (85).

$$HOCH_2CH{=}CHCH_2OH + R(NCO)_2 \longrightarrow H(OC_4H_6O\overset{O}{\overset{\|}{C}}{-}NHRNH\overset{O}{\overset{\|}{C}}{-})_nOC_4H_6OH$$

$$HOCH_2CH{=}CHCH_2OH + 2\ COCl_2 \longrightarrow Cl\overset{O}{\overset{\|}{O}}CH_2CH{=}CHCH_2O\overset{O}{\overset{\|}{C}}Cl + \xrightarrow{R(NH_2)_2}$$

$$H_2NRNH(\overset{O}{\overset{\|}{C}}OC_4H_6O\overset{O}{\overset{\|}{C}}NHRNH)_nH$$

The hydroxyl groups can be alkylated in the usual manner. Hydroxyalkyl ethers may be prepared with alkylene oxides and chloromethyl ethers by reaction with formaldehyde and hydrogen chloride (86). The terminal chlorides can be easily converted to additional ether groups.

$$HOCH_2CH{=}CHCH_2OH + HCHO + HCl \longrightarrow$$

$$ClCH_2OCH_2CH{=}CHCH_2OCH_2Cl \xrightarrow{NaOR} ROCH_2OCH_2CH{=}CHCH_2OCH_2OR$$

cis-Butenediol reacts readily with aldehydes (87), vinyl ethers (88), or dialkoxyalkanes (89) in the presence of acidic catalysts to give seven-membered cyclic acetals (4,7-dihydro-1,3-dioxepins).

$$HOCH_2CH=CHCH_2OH + RCHO \longrightarrow R\overline{CHOCH_2CH=CHCH_2O} + H_2O$$

$$HOCH_2CH=CHCH_2OH + ROCH=CH_2 \longrightarrow CH_3\overline{CHOCH_2CH=CHCH_2O} + ROH$$

$$HOCH_2CH=CHCH_2OH + RR'C(OR'')_2 \longrightarrow RR'\overline{COCH_2CH=CHCH_2O} + 2\ R''OH$$

The hydroxyl groups of butenediol are replaced by halogens by treatment with thionyl chloride or phosphorus tribromide (90,91); by stopping short of total halogenation, mixtures can be obtained containing 4-halobutanols as the major constituent (92). The hydroxyl groups undergo typical allylic reactions such as being replaced by cyanide with cuprous cyanide as catalyst (93).

With a palladium chloride catalyst, butenediol is carbonylated by carbon monoxide, giving 3-hexenedioic acid [4436-74-2], $C_6H_8O_4$ (94).

$$HOCH_2CH=CHCH_2OH + 2\ CO \longrightarrow HOOCCH_2CH=CHCH_2COOH$$

An early attempt to hydroformylate butenediol using a cobalt carbonyl catalyst gave tetrahydro-2-furanmethanol (95), presumably by allylic rearrangement to 3-butene-1,2-diol before hydroformylation. Later, hydroformylation of butenediol diacetate with a rhodium complex as catalyst gave the acetate of 3-formyl-3-buten-1-ol (96). Hydrogenation in such a system gave 2-methyl-1,4-butanediol (97).

Heating with cuprous chloride in aqueous hydrochloric acid isomerizes 2-butene-1,4-diol to 3-butene-1,2-diol (98)] Various hydrogen-transfer catalysts isomerize it to 4-hydroxybutyraldehyde [25714-71-0], $C_4H_8O_2$ (99), acetals of which are found as impurities in commercial butanediol and butenediol.

$$HOCH_2CH=CHCH_2OH \longrightarrow HOCH_2CH_2CH_2CHO$$

Treatment with acidic catalysts dehydrates *cis*-butenediol to 2,5-dihydrofuran [1708-29-8], C_4H_6O (100). Cupric (101) or mercuric (102) salts give 2,5-divinyl-1,4-dioxane [21485-51-8], presumably via 3-butene-1,2-diol.

Mixtures of butenediol and ammonia or amines cyclize to pyrrolines when heated with acidic catalysts (66).

Halogens add to butenediol, giving 2,3-dihalo-1,4-butanediol (90,91). In a reaction typical of allylic alcohols, hydrogen halides cause substitution of halogen for hydroxyl (103).

When butenediol is treated with acidic dichromate solution, dehydration and oxidation combine to give a high yield of furan [110-00-9], $C_4H_4O_2$ (104) (see FURAN DERIVATIVES).

Treatment with hydrogen peroxide converts butenediol to 2,3-epoxy-1,4-butanediol (105) or gives hydroxylation to erythritol [149-32-6], $C_4H_{10}O_4$ (106). Under strongly acidic conditions, tetrahydro-3,4-furanediol is the principal product (107).

$$HOCH_2CH{=}CHCH_2OH \rightleftarrows \begin{array}{l} HOCH_2CH{-}CHCH_2OH \\ HOCH_2CHOHCHOHCH_2OH \\ \end{array}$$

Butenediol is a weak dienophile in Diels-Alder reactions. Adducts have been described with anthracene (108) and with hexachlorocyclopentadiene (109).

Butenediol does not undergo free-radical polymerization. A copolymer with vinyl acetate can be prepared with a low proportion of butenediol (110).

Manufacture. Butenediol is manufactured by partial hydrogenation of butynediol. Although suitable conditions can lead to either cis or trans isomers (111), the commercial product contains almost exclusively *cis*-2-butene-1,4-diol. Trans isomer, available at one time by hydrolysis of 1,4-dichloro-2-butene, is unsuitable for the major uses of butenediol involving Diels-Alder reactions. The liquid-phase heat of hydrogenation of butynediol to butenediol is 156 kJ/mol (37.28 kcal/mol) (112).

The original German process used either carbonyl iron or electrolytic iron as hydrogenation catalyst (113). The fixed-bed reactor was maintained at 50–100°C and 20.26 MPa (200 atm) of hydrogen pressure, giving a product containing substantial amounts of both butynediol and butanediol. Newer, more selective processes use more active catalysts at lower pressures. In particular, supported palladium, alone (49) or with promoters (114,115), has been found useful.

Shipment, Storage, and Price. Butenediol is available in unlined steel tank cars, tank trailers, and various sized drums. Because of its relatively high freezing point, tank cars are fitted with heating coils. The 1991 U.S. bulk price of technical grade butenediol was about $5.36/kg.

Specifications and Analytical Methods. Purity is determined by gas chromatography. Technical grade butenediol, specified at 95% minimum, is typically 96–98% butenediol. The cis isomer is the predominant constituent; 2–4% is trans.

Principal impurities are butynediol (specified as 2.0% maximum, typically less than 1%), butanediol, and the 4-hydroxybutyraldehyde acetal of butenediol. Moisture is specified at 0.75% maximum (Karl-Fischer titration). Typical technical grade butenediol freezes at about 8°C.

Health and Safety Factors. Butenediol is noncorrosive and stable under normal handling conditions. It is a primary skin irritant but not a sensitizer; contact with skin and eyes should be avoided. It is much less toxic than butynediol. The LD_{50} is 1.25 mL/kg for white rats and 1.25–1.5 mL/kg for guinea pigs.

Uses. Butenediol is used to manufacture the insecticide Endosulfan, other agricultural chemicals, and pyridoxine (vitamin B_6) (see VITAMINS) (116). Small amounts are consumed as a diol by the polymer industry.

Butanediol. 1,4-Butanediol [*110-63-4*], tetramethylene glycol, 1,4-butylene glycol, was first prepared in 1890 by acid hydrolysis of *N,N'*-dinitro-1,4-bu- tanediamine (117). Other early preparations were by reduction of succinaldehyde (118) or succinic esters (119) and by saponification of the diacetate prepared from 1,4-dihalobutanes (120). Catalytic hydrogenation of butynediol, now the principal commercial route, was first described in 1910 (121). Other processes used for commercial manufacture are described in the section on Manufacture. Physical properties of butanediol are listed in Table 2.

Reactions. The chemistry of butanediol is determined by the two primary hydroxyls. Esterification is normal. It is advisable to use nonacidic catalysts for esterification and transesterification (122) to avoid cyclic dehydration. When carbonate esters are prepared at high dilutions, some cyclic ester is formed; more concentrated solutions give a polymeric product (123). With excess phosgene the useful bischloroformate can be prepared (124).

$$HO(CH_2)_4OH + COCl_2 \longrightarrow \text{[cyclic carbonate]} + H[O(CH_2)_4O\overset{O}{\overset{||}{C}}]_nOH$$

$$HO(CH_2)_4OH + 2\ COCl_2 \longrightarrow Cl\overset{O}{\overset{||}{C}}O(CH_2)_4O\overset{O}{\overset{||}{C}}Cl$$

Ethers are formed in the usual way (125). The bischloromethyl ether is obtained using formaldehyde and hydrogen chloride (86).

$$HO(CH_2)_4OH + 2\ HCHO + 2\ HCl \longrightarrow ClCH_2O(CH_2)_4OCH_2Cl + 2\ H_2O$$

With aldehydes or their derivatives, butanediol forms acetals, either 7-membered rings (1,3-dioxepanes) or linear polyacetals; the rings and chains are easily intraconverted (126,127).

$$HO(CH_2)_4OH + RCHO \longrightarrow \text{[cyclic acetal]} + H[O(CH_2)_4OCHR]_nOH$$

Heating butanediol with acetylene in the presence of an acidic mercuric salt gives the cyclic acetal expected from butanediol and acetaldehyde (128).

A commercially important reaction is with diisocyanates to form poly-urethanes (129) (see URETHANE POLYMERS).

$$HO(CH_2)_4OH + R(NCO)_2 \longrightarrow H[O(CH_2)_4O\overset{O}{\overset{||}{C}}NHRNH\overset{O}{\overset{||}{C}}]_nO(CH_2)_4OH$$

Thionyl chloride readily converts butanediol to 1,4-dichlorobutane [110-56-5] (130) and hydrogen bromide gives 1,4-dibromobutane [110-52-1] (131). A procedure using 48% HBr with a Dean-Stark water trap gives good yields of 4-bromobutanol [33036-62-3], free of diol and dibromo compound (132).

With various catalysts, butanediol adds carbon monoxide to form adipic acid. Heating with acidic catalysts dehydrates butanediol to tetrahydrofuran [109-99-9], C_4H_8O (see FURAN DERIVATIVES). With dehydrogenation catalysts, such as copper chromite, butanediol forms butyrolactone (133). With certain cobalt catalysts both dehydration and dehydrogenation occur, giving 2,3-dihydrofuran (134).

$$HO(CH_2)_4OH \xrightarrow{-H_2O} \text{(structure)}$$

$$HO(CH_2)_4OH \xrightarrow{-H_2} \text{(structure)}C=O$$

$$HO(CH_2)_4OH \xrightarrow{-H_2O,\ -H_2} \text{(structure)}$$

Heating butanediol or tetrahydrofuran with ammonia or an amine in the presence of an acidic heterogeneous catalyst gives pyrrolidines (135,136). With a dehydrogenation catalyst, one or both of the hydroxyl groups are replaced by amino groups (137).

$$HO(CH_2)_4OH + R_2NH \longrightarrow \text{mixture of } R_2N(CH_2)_4OH \text{ and } R_2N(CH_2)_4NR_2$$

With an acidic catalyst, butanediol and hydrogen sulfide give tetrahydrothiophene [110-01-0], C_4H_8S (138).

$$HO(CH_2)_4OH + H_2S \longrightarrow \text{(structure)}$$

Vapor-phase oxidation over a promoted vanadium pentoxide catalyst gives a 90% yield of maleic anhydride [108-31-6] (139). Liquid-phase oxidation with a supported palladium catalyst gives 55% of succinic acid [110-15-6] (140).

Manufacture. Most butanediol is manufactured in Reppe plants via hydrogenation of butynediol. Recently an alternative route involving acetoxylation of butadiene has come on stream and, more recently, a route based upon hydroformylation of allyl alcohol. Worldwide butanediol capacity has climbed steadily for many years. In 1990 it was estimated to be 428,000 metric tons (141), as compared to a little more than 70,000 metric tons in 1975 (142,143). Table 3 lists the manufacturers of butanediol, their locations, and the processes used.

Another process, involving chlorination of butadiene, hydrolysis of the dichlorobutene, and hydrogenation of the resulting butenediol, was practiced by Toyo Soda in Japan until the mid-1980s (144).

Reppe Process

$$HC{\equiv}CH + 2\ HCHO \longrightarrow HOCH_2C{\equiv}CCH_2OH \xrightarrow{H_2} HOCH_2CH_2CH_2CH_2OH$$

Table 3. Butanediol Manufacturers

Manufacturer	Location(s)	Manufacturing process
BASF	Germany and U.S.	Reppe
Du Pont	U.S.	Reppe
GAF	U.S.	Reppe
GAF–Hüls	Germany	Reppe
Du Pont–Idemitsu	Japan	Reppe
Mitsubishi Kasei	Japan	acetoxylation
ARCO	U.S.	hydroformylation

Acetoxylation Process

$$H_2C{=}CHCH{=}CH_2 + 2\ CH_3COOH \xrightarrow{O_2}$$

$$\underset{CH_3COCH_2CH=CHCH_2OCCH_3}{\overset{O\qquad\qquad\qquad O}{||\qquad\qquad\qquad ||}} + H_2O \xrightarrow{\text{hydrogenation}}\xrightarrow{\text{hydrolysis}} HOCH_2CH_2CH_2CH_2OH$$

Hydroformylation Process

$$CH_3\overset{O}{\overset{\diagdown\diagup}{CH{-}CH_2}} \longrightarrow HC_2{=}CHCH_2OH \xrightarrow{H_2,\ CO}$$

$$HOCH_2CH_2CH_2CHO \xrightarrow{\text{hydrogenation}} HOCH_2CH_2CH_2CH_2OH$$

Shipment, Storage, and Price. Tank cars and tank trailers, selected to prevent color formation, are of aluminum or stainless steel, or lined with epoxy or phenolic resins; drums are lined with phenolic resins. Flexible stainless steel hose is used for transfer. Because of butanediol's high freezing point (about 20°C) tank car coil heaters are provided. The U.S. list price for bulk quantities in 1991 was about $2.18/kg, but heavy discounting was prevalent for large contracts.

Specifications and Analytical Methods. Butanediol is specified as 99.5% minimum pure, determined by gas chromatography (gc), solidifying at 19.6°C minimum. Moisture is 0.04% maximum, determined by Karl-Fischer analysis (directly or of a toluene azeotrope). The color is APHA 5 maximum, and the Hardy color (polyester test) is APHA 200 maximum. The carbonyl number is 0.5 mg KOH/g maximum; the acetal content can also be measured directly by gc.

Health and Safety Factors. Butanediol is much less toxic than its unsaturated analogs. It is neither a primary skin irritant nor a sensitizer. Because of its low vapor pressure, there is ordinarily no inhalation problem. As with all chemicals, unnecessary exposure should be avoided. The LD_{50} for white rats is 1.55 g/kg.

Uses. The largest uses of butanediol are internal consumption in manufacture of tetrahydrofuran and butyrolactone (145). The largest merchant uses are for poly(butylene terephthalate) resins (see POLYESTERS, THERMOPLASTIC) and in polyurethanes, both as a chain extender and as an ingredient in a hydroxyl-terminated polyester used as a macroglycol. Butanediol is also used as a solvent, as a monomer for various condensation polymers, and as an intermediate in the manufacture of other chemicals.

Butyrolactone. γ-Butyrolactone [96-48-0], dihydro-2(3H)-furanone, was first synthesized in 1884 via internal esterification of 4-hydroxybutyric acid (146).

In 1991 the principal commercial source of this material is dehydrogenation of butanediol. Manufacture by hydrogenation of maleic anhydride (147) was discontinued in the early 1980s and resumed in the late 1980s. Physical properties are listed in Table 4.

Table 4. Physical Properties of Butyrolactone

Property	Value
freezing point, °C	−44
boiling point, °C at kPa[a]	
0.133	35
1.33	77
13.3	134
101.3	204
specific gravity, d_4^{20}	1.129
d_4^{25}	1.125
refractive index, n_D^{20}	1.4362
n_D^{25}	1.4348
viscosity at 25°C, mPa·s (= cP)	1.75
dielectric constant at 20°C, ϵ	39.1
heat capacity at 20°C, J/(g·K)[b]	1.60
critical temperature, °C	436
critical pressure, MPa[c]	3.43
flash point, Tagliabue open cup, °C	98

[a]To convert kPa to mm Hg multiply by 7.5.
[b]To convert J to cal divide by 4.184.
[c]To convert MPa to atm divide by 0.1013.

Butyrolactone is completely miscible with water and most organic solvents. It is only slightly soluble in aliphatic hydrocarbons. It is a good solvent for many gases, for most organic compounds, and for a wide variety of polymers.

Reactions. Butyrolactone undergoes the reactions typical of γ-lactones. Particularly characteristic are ring openings and reactions in which ring oxygen is replaced by another heteroatom. There is also marked reactivity of the hydrogen atoms alpha to the carbonyl group.

Hydrolysis in neutral aqueous solutions proceeds slowly at room temperature and more rapidly at acidic conditions and elevated temperatures. The hydrolysis–esterification reaction is reversible. Under alkaline conditions hydrolysis is rapid and irreversible. Heating the alkaline hydrolysis product at 200–250°C gives 4,4′-oxydibutyric acid [7423-25-8] after acidification (148).

With acid catalysts, butyrolactone reacts with alcohols rapidly even at room temperature, giving equilibrium mixtures consisting of esters of 4-hydroxybutyric acid [591-81-1] with unchanged butyrolactone as the main component. Attempts to distill such mixtures ordinarily result in complete reversal to butyrolactone and alcohol. The esters can be separated by a quick flash distillation at high vacuum (149).

$$\text{(butyrolactone)} + \text{ROH} \rightleftharpoons \text{HO(CH}_2)_4\text{COOR}$$

When butyrolactone and alcohols are heated for long times and at high temperatures in the presence of acidic catalysts, 4-alkoxybutyric esters are formed. With sodium alkoxides, sodium 4-alkoxybutyrates are formed (150).

$$\text{(butyrolactone)} + 2\ \text{ROH} \longrightarrow \text{RO(CH}_2)_3\text{COOR} + \text{H}_2\text{O}$$

$$\text{(butyrolactone)} + \text{NaOR} \longrightarrow \text{RO(CH}_2)_3\text{COONa}$$

Butyrolactone and hydrogen sulfide heated over an alumina catalyst result in replacement of ring oxygen by sulfur (151).

$$\text{(butyrolactone)} + \text{H}_2\text{S} \longrightarrow \text{(thiobutyrolactone)}$$

Heating butyrolactone with bromine at 160–170°C gives a 70% yield of α-bromobutyrolactone (152). With phosphorus tribromide as catalyst, bromination is accelerated, giving 2,4-dibromobutyric acid, which dehydrobrominates to α-bromobutyrolactone when distilled (153). Chlorination gives α-position mono-chlorination at 110–130°C and α-dichlorination at 190–200°C (154).

$$\text{(butyrolactone)} + \text{Br}_2 \xrightarrow{\text{PBr}_3} \text{BrCH}_2\text{CH}_2\text{CHBrCOOH} \xrightarrow[\text{dist.}]{-\text{HBr}} \text{Br}-\text{(butyrolactone)}$$

The α-methylene group of butyrolactone condenses easily with a number of different types of carbonyl compounds; eg, sodium alkoxides catalyze self-condensation to α-dibutyrolactone (155), benzaldehyde gives α-benzylidenebutyrolactone (156), and ethyl acetate gives α-acetobutyrolactone (157).

$$2\ \text{(butyrolactone)} \longrightarrow \text{(α-dibutyrolactone)}$$

The α-acetobutyrolactone, with or without isolation, can be used in the preparation of various 5-substituted 2-butanone derivatives, presumably by decarboxylation of the acetoacetic acid obtained by ring hydrolysis. Simple hydrolysis gives 5-hydroxybutan-2-one (158) and acidolysis with hydrochloric acid gives 5-chlorobutan-2-one in good yields (159).

The α-methylene groups also add to double bonds; eg, 1-decene at 160°C gives up to 80% of α-decylbutyrolactone (160). With photochemical initiation similar additions take place at room temperature (161).

With Friedel-Crafts catalysts, butyrolactone reacts with aromatic hydrocarbons. With benzene, depending on experimental conditions, either phenylbutyric acid or 1-tetralone can be prepared (162).

Carbonylation of butyrolactone using nickel or cobalt catalysts gives high yields of glutaric acid [110-94-1] (163).

$$\text{(butyrolactone)} + CO + H_2O \longrightarrow HOOC(CH_2)_3COOH$$

Frequently unique and synthetically useful are a series of ring-opening reactions. Butyrolactone and anhydrous hydrogen halides give high yields of 4-halobutyric acids (164). In the presence of alcohols, esters are formed.

$$\text{(butyrolactone)} + HX \longrightarrow X(CH_2)_3COOH, \quad X = Cl, Br, \text{ or } I$$

Phosgene (165) or thionyl chloride in the presence of an acid catalyst (166) gives good yields of 4-chlorobutyryl chloride. Heating butyrolactone and thionyl chloride in an alcohol gives good yields of 4-chlorobutyric esters (167).

Butyrolactone with sodium sulfide or hydrosulfide forms 4,4'-thiodibutyric acid (168); with sodium disulfide, the product is 4,4'-dithiodibutyric acid (169).

$$\text{(butyrolactone)} + Na_2S_x \longrightarrow S_x(CH_2CH_2CH_2COONa), \quad x = 1 \text{ or } 2$$

Salts of thiols (170) or of sulfinic acids (171) react like the alkoxides, giving 4-alkylthio- or 4-alkylsulfono-substituted butyrates. Alkali cyanides give 4-cyanobutyrates (172), hydroxylamine gives a hydroxamic acid (173), and hydrazine a hydrazide (174).

$$\text{(butyrolactone)} + NaCN \longrightarrow NC(CH_2)_3COONa$$

$$\text{(butyrolactone)} + HONH_2 \longrightarrow HO(CH_2)_3CONHOH$$

$$\text{(butyrolactone)} + NH_2NH_2 \longrightarrow HO(CH_2)_3CONHNH_2$$

Butyrolactone reacts rapidly and reversibly with ammonia or an amine forming 4-hydroxybutyramides (175), which dissociate to the starting materials when heated. At high temperatures and pressures the hydroxybutyramides slowly and irreversibly dehydrate to pyrrolidinones (176). A copper-exchanged Y-zeolite (177) or magnesium silicate (178) is said to accelerate this dehydration.

$$\text{(butyrolactone)} + RNH_2 \rightleftharpoons HO(CH_2)_3CONHR \longrightarrow \text{(pyrrolidinone)}\,N\text{—}R + H_2O$$

Manufacture. Butyrolactone is manufactured by dehydrogenation of butanediol. The butyrolactone plant and process in Germany, as described after World War II (179), approximates the processes presently used. The dehydrogenation was carried out with preheated butanediol vapor in a hydrogen carrier over a supported copper catalyst at 230–250°C. The yield of butyrolactone after purification by distillation was about 90%.

Shipment, Storage, and Price. Butyrolactone is shipped in unlined steel tank cars and plain steel drums. Plain steel, stainless steel, aluminum, and nickel are suitable for storage and handling; rubber, phenolics, and epoxy resins are not suitable. Butyrolactone is hygroscopic and should be protected from moisture. Because of its low freezing point (−44°C), no provision for heating storage vessels is needed.

The U.S. bulk price in 1991 was about $3.31/kg.

Specifications and Analytical Methods.   Purity is specified as 99.5% minimum, by gc area percentage, with a maxium of 0.1% moisture by Karl-Fischer titration. Color, as delivered, is 40 APHA maximum; samples may darken on long storage.

Health and Safety Factors.   Butyrolactone is neither a skin irritant nor a sensitizer; however, it is judged to be a severe eye irritant in white rabbits. The acute oral LD_{50} is 1.5 mL/kg for white rats or guinea pigs. Subacute oral feeding studies were carried out with rats and with dogs. At levels up to 0.8% of butyrolactone in the diet there were no toxicologic or pathologic effects in the three months of the test.

Because of its high boiling point (204°C), it does not ordinarily represent a vapor hazard.

Uses.   Butyrolactone is principally consumed by the manufacturers by reaction with methylamine or ammonia to produce *N*-methyl-2-pyrrolidinone [872-50-4] and 2-pyrrolidinone [616-45-5], C_4H_7NO, respectively. Considerable amounts are used as a solvent for agricultural chemicals and polymers, in dyeing and printing, and as an intermediate for various chemical syntheses.

Other Alcohols and Diols

Secondary acetylenic alcohols are prepared by ethynylation of aldehydes higher than formaldehyde. Although copper acetylide complexes will catalyze this reaction, the rates are slow and the equilibria unfavorable. The commercial products are prepared with alkaline catalysts, usually used in stoichiometric amounts.

Ethynylation of ketones is not catalyzed by copper acetylide, but potassium hydroxide has been found to be effective (180). In general, alcohols are obtained at lower temperatures and glycols at higher temperatures. Most processes use stoichiometric amounts of alkali, but true catalytic processes for manufacture of the alcohols have been described; the glycols appear to be products of stoichiometric ethynylation only.

Table 5 lists the principal commercially available acetylenic alcohols and glycols; Tables 6 and 7 list the physical properties of acetylenic alcohols and glycols, respectively.

Table 5. Commercial Secondary and Tertiary Acetylenic Alcohols and Glycols

Alcohol or glycol	Molecular formula	CAS Registry No.	Starting material
1-hexyn-3-ol	$C_6H_{10}O$	[105-31-7]	butyraldehyde
4-ethyl-1-octyn-3-ol	$C_{10}H_{18}O$	[5877-42-9]	2-ethylhexanal
2-methyl-3-butyn-2-ol	C_5H_8O	[115-19-5]	acetone
3-methyl-1-pentyn-3-ol	$C_6H_{10}O$	[77-75-8]	methyl ethyl ketone
2,5-dimethyl-3-hexyne-2,5-diol	$C_8H_{14}O_2$	[142-30-3]	acetone
3,6-dimethyl-4-octyne-3,6-diol[a]	$C_{10}H_{18}O_2$	[78-66-0]	methyl ethyl ketone
2,4,7,9-tetramethyl-5-decyne-4,7-diol[a]	$C_{14}H_{26}O_2$	[126-86-3]	methyl isobutyl ketone

[a]These glycols are commercially available as mixtures of diastereoisomers.

Table 6. Physical Properties of Acetylenic Alcohols

Property	Hexynol	Ethyl-octynol	Methyl-butynol	Methyl-pentynol
molecular weight	98	154	84	98
freezing point, °C	−80	−45	2.6	−30.6
boiling point, °C	142	197.2	103.6	121.4
specific gravity, d_{20}^{20}	0.882	0.873	0.8672	0.8721
refractive index, n_D^{20}	1.4350	1.4502	1.4211	1.4318
viscosity at 20°C, mPa·s (= cP)			3.79	2.65^a
flash point, Tagliabue open cup, °C		83	25	38
water solubility (20°C), wt %	3.8	<0.1	miscible	9.9

^{a}At 31°C.

Table 7. Physical Properties of Acetylenic Glycols

Property	Dimethyl-hexynediol	Dimethyl-octynediol	Tetramethyl-decynediol
molecular weight	142	170	226
melting point, °C	96–97	49–51	37–38
boiling point, °C	206	222	260
surface tension, mN/m (= dyn/cm) 0.1% in water at 25°C	60.9	55.3	31.6
water solubility (20°C), wt %	27.0	10.5	0.12

Methylbutynol. 2-Methyl-3-butyn-2-ol [115-19-5], prepared by ethynylation of acetone, is the simplest of the tertiary ethynols, and serves as a prototype to illustrate their versatile reactions. There are three reactive sites, ie, hydroxyl group, triple bond, and acetylenic hydrogen. Although the triple bonds and acetylenic hydrogens behave similarly in methylbutynol and in propargyl alcohol, the reactivity of the hydroxyl groups is very different.

Reactions. As with other tertiary alcohols, esterification with carboxylic acids is difficult and esters are prepared with anhydrides (181), acid chlorides (182), or ketene (183). Carbamic esters may be prepared by treatment with an isocyanate (184) or with phosgene followed by ammonia or an amine (185).

The labile hydroxyl group is easily replaced by treatment with thionyl chloride, phosphorous chlorides, or even aqueous hydrogen halides. At low temperatures aqueous hydrochloric (186) or hydrobromic (187) acids give good yields of 3-halo-3-methyl-1-butynes. At higher temperatures these rearrange, first to 1-halo-3-methyl-1,2-butadienes, then to the corresponding 1,3-butadienes (188,189).

$$
\underset{\text{HC}\equiv\text{CC(CH}_3)_2}{\overset{\text{OH}}{|}} \xrightarrow{\text{HX}} \underset{\text{HC}\equiv\text{CC(CH}_3)_2}{\overset{\text{X}}{|}} \longrightarrow \text{XCH}=\text{C}=\text{C(CH}_3)_2 \longrightarrow \underset{\qquad\qquad\quad\text{CH}_3}{\text{XCH}=\text{CH}-\overset{|}{\text{C}}=\text{CH}_2}
$$

With acid catalysts in the liquid (190) or vapor (191) phase, methylbutynol is dehydrated to isopropenylacetylene.

$$\underset{\text{HC}\equiv\overset{\overset{\displaystyle OH}{|}}{C}C(CH_3)_2}{} \longrightarrow \underset{\text{HC}\equiv\overset{\overset{\displaystyle CH_3}{|}}{C}C=CH_2}{}$$

Hydrogenation of methylbutynol gives 2-methyl-3-buten-2-ol and then 2-methylbutan-2-ol in stepwise fashion (192).

$$HC\equiv CC(OH)(CH_3)_2 \longrightarrow CH_2=CHC(OH)(CH_3)_2 \longrightarrow CH_3CH_2C(OH)(CH_3)_2$$

Acidic mercury salts catalyze hydration to form a ketone (193).

$$\underset{HC\equiv\overset{\overset{\displaystyle OH}{|}}{C}C(CH_3)_2}{} + H_2O \longrightarrow CH_3\overset{\overset{\displaystyle O}{\|}}{C}-\underset{\underset{\displaystyle OH}{|}}{C}(CH_3)_2$$

Bromination in polar solvents usually gives *trans*-3,4-dibromo-2-methyl-3-buten-2-ol; in nonpolar solvents, with incandescent light, the cis isomer is the principal product (194). Chlorine adds readily up to the tetrachloro stage, but yields are low because of side reactions (195).

$$HC\equiv\overset{\overset{\displaystyle OH}{|}}{C}C(CH_3)_2 \longrightarrow CHX=\overset{\overset{\displaystyle OH}{|}}{C}XC(CH_3)_2 \longrightarrow CHX_2CX_2\overset{\overset{\displaystyle OH}{|}}{C}(CH_3)_2$$

Upon treatment with suitable cobalt complexes, methylbutynol cyclizes to a 1,2,4-substituted benzene. Nickel complexes give the 1,3,5-isomer (196), sometimes accompanied by linear polymer (25) or a mixture of tetrasubstituted cyclooctatetraenes (26).

When bis(π-allyl)nickel is used, only small amounts of cyclic product are obtained and the principal product is formed by addition of one triple bond to another (197).

$$2\ HC\equiv\overset{\overset{\displaystyle OH}{|}}{C}C(CH_3)_2 \longrightarrow (CH_3)_2\overset{\overset{\displaystyle OH}{|}}{C}CH=CHC\equiv\overset{\overset{\displaystyle OH}{|}}{C}C(CH_3)_2$$

With a nickel carbonyl catalyst, hydrochloric acid, and an alcohol the initially formed allenic ester cyclizes on distillation (198).

$$\underset{\underset{\displaystyle (CH_3)_2\overset{\textstyle |}{C}C\equiv CH}{|}}{OH} \longrightarrow (CH_3)_2C{=}C{=}CHCOOR \longrightarrow$$

With palladium chloride catalyst, carbon monoxide, and an alcohol the labile hydroxyl is alkylated during carbonylation (199).

$$\underset{\underset{\displaystyle (CH_3)_2\overset{\textstyle |}{C}C\equiv CH}{|}}{OH} \longrightarrow \underset{\underset{\displaystyle (CH_3)_2\overset{\textstyle |}{C}CH{=}CHCOOR}{|}}{OR}$$

Copper salts catalyze oxidative dimerization to conjugated diynediols in high yields (200).

$$2\,\underset{\underset{\displaystyle (CH_3)_2\overset{\textstyle |}{C}C\equiv CH}{|}}{OH} \longrightarrow \underset{}{OH \qquad OH} (CH_3)_2CC\equiv CC\equiv CC(CH_3)_2$$

Glycols are obtained by treatment with a ketone using alkali as catalyst or with an aldehyde using alkali or copper acetylide as catalyst (201,202).

$$\underset{\underset{\displaystyle (CH_3)_2\overset{\textstyle |}{C}C\equiv CH}{|}}{OH} + RCOR' \longrightarrow \underset{}{OH \quad OH}(CH_3)_2CC\equiv CCRR'$$

Hypohalites replace the acetylenic hydrogen with chlorine, bromine, or iodine (203).

$$\underset{\underset{\displaystyle (CH_3)_2\overset{\textstyle |}{C}C\equiv CH}{|}}{OH} + NaOX \longrightarrow \underset{\underset{\displaystyle (CH_3)_2\overset{\textstyle |}{C}C\equiv CX}{|}}{OH}$$

Ethynyl carbinols rearrange to conjugated unsaturated aldehydes. Copper or silver salts catalyze isomerization of the acetate to an allenic acetate, which can be hydrolyzed to an unsaturated aldehyde (204).

$$\underset{\underset{\displaystyle (CH_3)_2\overset{\textstyle |}{C}C\equiv CH}{|}}{OOCCH_3} \xrightarrow{Ag_2CO_3} (CH_3)_2C{=}C{=}CHOOCCH_3$$

Manufacture.   In general, manufacture is carried out in batch reactors at close to atmospheric pressure. A moderate excess of finely divided potassium hydroxide is suspended in a solvent such as 1,2-dimethoxyethane. The carbonyl compound is added, followed by acetylene. The reaction is rapid and exothermic. At temperatures below 5°C the product is almost exclusively the alcohol. At 25–30°C the glycol predominates. Such synthesis also proceeds well with non-complexing solvents such as aromatic hydrocarbons, although the conversion is usually lower (205).

Continuous processes have been developed for the alcohols, operating under pressure with liquid ammonia as solvent. Potassium hydroxide (206) or anion exchange resins (207) are suitable catalysts. However, the relatively small manufacturing volumes militate against continuous production. For a while a continuous catalytic plant operated in Ravenna, Italy, designed to produce about 40,000 t/yr of methylbutynol for conversion to isoprene (208,209).

Economic Aspects.   A number of secondary and tertiary acetylenic alcohols and glycols are manufactured by Air Products and Chemicals Co. In 1990, U.S. bulk prices of ethyloctynol, methylbutynol, and tetramethyldecynediol were about \$8.35/kg, \$3.95/kg, and \$8.00/kg, respectively.

Health and Safety Factors.   Under normal conditions acetylenic alcohols are stable and free of decomposition hazard. The more volatile alcohols present a fire hazard.

The alcohols are toxic orally, through skin absorption, and through inhalation. The secondary alcohols are more toxic than the tertiary. The glycols are relatively low in toxicity.

Compound	LD_{50}, mL/kg (mice)
hexynol	0.175
ethyloctynol	2.1
methylbutynol	2.2
methylpentynol	0.7
tetramethyldecynediol	4.6

Uses.   The secondary acetylenic alcohols hexynol and ethyloctynol are used as corrosion inhibitors in oil-well acidizing compositions (see CORROSION AND CORROSION INHIBITORS). The tertiary alcohols methylbutynol and methylpentynol are used as chemical intermediates, for manufacture of Vitamin A and other products, and in metal plating and pickling operations. Dimethylhexynediol can be used in manufacture of fragrance chemicals and peroxide catalysts. Higher acetylenic glycols and ethoxylated acetylene glycols are useful as surfactants and electroplating additives.

Vinylation Reaction Products

Unlike ethynylation, in which acetylene adds across a carbonyl group and the triple bond is retained, in vinylation a labile hydrogen compound adds to acetylene, forming a double bond.

In early work, vinyl chloride had been heated with stoichiometric amounts of alkali alkoxides in excess alcohol as solvent, giving vinyl ethers as products (210). Supposedly this involved a Williamson ether synthesis, where alkali alkoxide and organic halide gave an ether and alkali halide. However, it was observed that small amounts of acetylene were formed by dehydrohalogenation of vinyl chloride, and that this acetylene was consumed as the reaction proceeded. Hence acetylene was substituted for vinyl chloride and only catalytic amounts of alkali were used. Vinylation proceeded readily with high yields (211).

Catalytic vinylation has been applied to a wide range of alcohols, phenols, thiols, carboxylic acids, and certain amines and amides. Vinyl acetate is no longer prepared this way in the United States, although some minor vinyl esters such as stearates may still be prepared this way. However, the manufacture of vinylpyrrolidinone and vinyl ethers still depends on acetylene.

N-Vinylcarbazole.   Vinylation of carbazole proceeds in high yields with alkaline catalysts (212,213). The product, 9-ethenylcarbazole, $C_{14}H_{11}N$ [1484-13-5], forms rigid high-melting polymers with outstanding electrical properties.

Neurine.   Neurine is trimethylvinylammonium hydroxide, $C_5H_{13}NO$ [463-88-7]. Tertiary amines and their salts vinylate readily at low temperatures with catalysis by free tertiary amines.

$$(CH_3)_3N + HC \equiv CH + H_2O \longrightarrow [(CH_3)_3NCH=CH_2]^+[OH]^-$$

Above about 50°C tetramethylammonium hydroxide is formed as a by-product; it is the sole product above 100°C (214).

N-Vinyl-2-pyrrolidinone.   1-Ethenyl-2-pyrrolidinone [88-12-0], C_6H_9NO, N-vinylpyrrolidinone, was developed by Reppe's laboratory in Germany at the beginning of World War II and patented in 1940 (215).

The major use of vinylpyrrolidinone is as a monomer in manufacture of poly(vinylpyrrolidinone) (PVP) homopolymer and in various copolymers, where it frequently imparts hydrophilic properties. When PVP was first produced, its principal use was as a blood plasma substitute and extender, a use no longer sanctioned. These polymers are used in pharmaceutical and cosmetic applications, soft contact lenses, and viscosity index improvers. The monomer serves as a component in radiation-cured polymer compositions, serving as a reactive diluent that reduces viscosity and increases cross-linking rates (see VINYL POLYMERS, N-VINYL MONOMERS AND POLYMERS).

Vinyl Ethers.   The principal commercial vinyl ethers are methyl vinyl ether (methoxyethene, C_3H_6O) [107-25-5]; ethyl vinyl ether (ethoxyethene, C_4H_8O) [104-92-2]; and butyl vinyl ether (1-ethenyloxybutane, $C_6H_{12}O$) [111-34-2]. (See Table 8 for physical properties.) Others such as the isopropyl, isobutyl, hydroxybutyl, decyl, hexadecyl, and octadecyl ethers, as well as the divinyl ethers of butanediol

Table 8. Physical Properties of Vinyl Ethers[a]

Property	Methyl	Ethyl	Butyl
molecular weight	58	72	100
freezing point, °C	−122.8	−115.4	−91.9
boiling point, °C	5.5	35.7	93.5
vapor pressure at 20°C, kPa[b]	156.7	57	5.6
specific gravity, d_4^{20}	0.7511	0.7541	0.7792
refractive index, n_D^{20}	1.3730	1.3767	1.4020
	(0°C)		
flash point, °C	−56	< −18	−1
water solubility at 20°C, wt %	1.5	0.9	0.2

[a]Lower vinyl ethers are miscible with nearly all organic solvents.
[b]To convert kPa to mm Hg, multiply by 7.5.

and of triethylene glycol, have been offered as development chemicals (see ETHERS).

Ethyl vinyl ether was the first to be prepared, in 1878, by treatment of diethyl chloroacetal with sodium (216). Methyl vinyl ether was first listed in Reppe patents on vinylation in 1929 and 1930 (210,211).

Reactions. Vinyl ethers undergo all of the expected reactions of olefinic compounds plus a number of reactions that are both useful and unusual.

With a suitable catalyst, usually a Lewis acid, many labile hydrogen compounds add across the vinyl ether double bond in the Markovnikov direction.

Alcohols give acetals. This reaction has been frequently used to provide blocking groups in organic synthesis. The acetals are stable under neutral or alkaline conditions and are easily hydrolyzed with dilute acid after other desired reactions have occurred (217,218). Water gives acetaldehyde and the corresponding alcohol, presumably via disproportionation of the hemiacetal (219). Carboxylic acids give 1-alkoxyethyl esters (220). Thiols give thioacetals (221).

$$CH_2{=}CHOR + R'OH \longrightarrow CH_3\overset{\displaystyle OR'}{\overset{|}{C}}HOR$$

$$CH_2{=}CHOR + H_2O \longrightarrow [CH_3\overset{\displaystyle OH}{\overset{|}{C}}HOR] \longrightarrow CH_3CHO + ROH$$

$$CH_2{=}CHOR + R'COOH \longrightarrow CH_3\overset{\displaystyle OOCR'}{\overset{|}{C}}HOR$$

Hydrogen halides react vigorously to give 1-haloethyl ethers, which are reactive intermediates for further synthesis (222). Conditions must be carefully selected to avoid polymerization of the vinyl ether. Hydrogen cyanide adds at high temperature to give a 2-alkoxypropionitrile (223).

$$CH_2{=}CHOR + HX \longrightarrow CH_3\overset{\displaystyle X}{\overset{|}{C}}HOR \quad (X = Cl\ or\ Br)$$

$$CH_2{=}CHOR + HCN \longrightarrow CH_3\overset{\displaystyle CN}{\overset{|}{C}}HOR$$

Chlorine and bromine add vigorously, giving, with proper control, high yields of 1,2-dihaloethyl ethers (224). In the presence of an alcohol, halogens add as hypohalites, which give 2-haloacetals (225,226). With methanol and iodine this is used as a method of quantitative analysis, titrating unconsumed iodine with standard thiosulfate solution (227).

$$CH_2{=}CHOR + X_2 \longrightarrow XCH_2\overset{\displaystyle X}{\overset{|}{C}}HOR$$

$$CH_2{=}CHOR + X_2 + R'OH \longrightarrow XCH_2\underset{\displaystyle OR'}{\overset{|}{C}}HOR$$

With Lewis acids as catalysts, compounds containing more than one alkoxy group on a carbon atom add across vinyl ether double bonds. Acetals give

3-alkoxyacetals; since the products are also acetals, they can react further with excess vinyl ether to give oligomers (228–230). Orthoformic esters give diacetals of malonaldehyde (231). With Lewis acids and mercuric salts as catalysts, vinyl ethers add in similar fashion to give acetals of 3-butenal (232,233).

$$\begin{array}{ccc} & & \overset{\displaystyle OR''}{\underset{|}{}} \quad \overset{\displaystyle OR''}{\underset{|}{}} \\ CH_2{=}CHOR + R'CH(OR'')_2 & \longrightarrow & R'CHCH_2CHOR \end{array}$$

$$\begin{array}{ccc} & & \overset{\displaystyle OR'}{\underset{|}{}} \\ CH_2{=}CHOR + CH(OR')_3 & \longrightarrow & CH(OR')_2CH_2CHOR \end{array}$$

$$2\ CH_2{=}CHOR \longrightarrow CH_2{=}CHCH_2CH(OR)_2$$

Vinyl ethers and α,β-unsaturated carbonyl compounds cyclize in a hetero-Diels-Alder reaction when heated together in an autoclave with small amounts of hydroquinone added to inhibit polymerization. Acrolein gives 3,4-dihydro-2-methoxy-2H-pyran (234,235), which can easily be hydrolyzed to glutaraldehyde (236) or hydrogenated to 1,5-pentanediol (237). With 2-methylene-1,3-dicarbonyl compounds the reaction is nearly quantitative (238).

Vinyl ethers cyclize with ketenes to cyclobutanones (239).

Vinyl ethers serve as a source of vinyl groups for transvinylation of such compounds as 2-pyrrolidinone or caprolactam (240,241).

Compounds such as carbon tetrachloride (242) or trinitromethane (243) can add across the double bond.

$$CH_2{=}CHOR + CCl_4 \longrightarrow ROCHClCH_2CCl_3$$

$$CH_2{=}CHOR + CH(NO_2)_3 \longrightarrow ROCH_2CH_2C(NO_2)_3$$

With thionyl chloride as catalyst, hydrogen peroxide adds to vinyl ethers in anti-Markovnikov fashion, as do monothioglycols with amine catalysts (244).

$$2\ CH_2\!=\!CHOR + HOOH \longrightarrow ROCH_2CH_2OOCH_2CH_2OR$$

$$CH_2\!=\!CHOR + HSCH_2CH_2OH \longrightarrow ROCH_2CH_2SCH_2CH_2OH$$

Substances that form carbanions, such as nitro compounds, hydrocyanic acid, malonic acid, or acetylacetone, react with vinyl ethers in the presence of water, replacing the alkyl group under mild conditions (245).

$$CH_2\!=\!CHOR + CH_3NO_2 + H_2O \longrightarrow CH_3\overset{\underset{\displaystyle |}{OH}}{C}HCH_2NO_2 + ROH$$

$$CH_2\!=\!CHOR + HCN + H_2O \longrightarrow CH_3\overset{\underset{\displaystyle |}{OH}}{C}HCN + ROH$$

The reaction of a vinyl ether with carbon dioxide and a secondary amine gives a carbamic ester (246).

$$CH_2\!=\!CHOR + CO_2 + (CH_3)_2NH \longrightarrow (CH_3)_2NCOO\overset{\underset{\displaystyle |}{OR}}{C}HCH_3$$

Manufacture. The principal manufacturers of vinyl ethers are BASF, GAF, and Union Carbide. The first two utilize vinylation of alcohols, whereas the last reportedly uses cracking of acetals.

German vinyl ether plants were described in detail at the end of World War II and variations of these processes are still in use. Vinylation of alcohols from methyl to butyl was carried out under pressure: typically 2–2.3 MPa (20–22 atm) and 160–165°C for methyl, and 0.4–0.5 MPa (4–5 atm) and 150–155°C for isobutyl. An unpacked tower, operating continuously, produced about 300 t/month, with yields of 90–95% (247).

High boiling alcohols were vinylated at atmospheric pressure. The Germans used a tower packed with Raschig rings and filled with an alcohol containing 1–5% of KOH at 160–180°C. Acetylene was recycled continuously up through the tower. The heat of reaction, about 125 kJ/mol (30 kcal/mol), was removed by cooling coils. Fresh alcohol and catalyst were added continuously at the top and withdrawn at the bottom. Yields of purified, distilled product were described as quantitative (248).

Shipment, Storage, and Prices. Methyl vinyl ether is available in tank cars or cylinders, while the other vinyl ethers are available in tank cars, tank wagons, or drums. Mild steel, stainless steel, and phenolic-coated steel are suitable for shipment and storage. If protected from air, moisture, and acidic contamination, vinyl ethers are stable for years. United States bulk prices in 1991 for methyl vinyl ether, ethyl vinyl ether, and butyl vinyl ether were listed as about $5.78/kg, $6.28/kg, and $6.08/kg, respectively.

Specifications and Analytical Methods. Vinyl ethers are usually specified as 98% minimum purity, as determined by gas chromatography. The principal impurities are the parent alcohols, limited to 1.0% maximum for methyl vinyl ether and 0.5% maximum for ethyl vinyl ether. Water (by Karl-Fischer titration) ranges from 0.1% maximum for methyl vinyl ether to 0.5% maximum for ethyl vinyl ether. Acetaldehyde ranges from 0.1% maximum in ethyl vinyl ether to 0.5% maximum in butyl vinyl ether.

Health and Safety Factors.   Because of their high vapor pressures (methyl vinyl ether is a gas at ambient conditions), the lower vinyl ethers represent a severe fire hazard and must be handled accordingly. Contact with acids can initiate violent polymerization and must be avoided. Although vinyl ethers form peroxides more slowly than saturated ethers, distillation residues must be handled with caution.

Inhalation should be avoided. A group of six rats that were exposed to 64,000 ppm of methyl vinyl ether in air for 4 h were anesthetized. All recovered and appeared normal after 72 h. One died after 96 h. The others survived the two-week observation period without noticeable effect.

The lower vinyl ethers do not appear to be skin irritants or sensitizers.

Oral toxicity is very low: isobutyl vinyl ether has LD_{50} of 17 mL/kg for white rats.

Uses.   Union Carbide consumes its vinyl ether production in the manufacture of glutaraldehyde [*111-30-8*]. BASF and GAF consume most of their production as monomers (see VINYL POLYMERS). In addition to the homopolymers, the copolymer of methyl vinyl ether with maleic anhydride is of particular interest.

BIBLIOGRAPHY

"Acetylene-Derived Chemicals" in *ECT* 3rd ed., Vol. 1, pp. 244–276, by Eugene V. Hort, GAF Corporation.

1. S. A. Miller, *Acetylene, Its Properties, Manufacture and Uses,* Vol. 1, Academic Press, Inc., New York, 1965, pp. 24–28, 42–44.
2. M. J. Haley with T. Ball and S. Yoshikawa, *Acetylene,* CEH Product Review, Chemical Economics Handbook, SRI International, Menlo Park, Calif., Oct. 1988, p. 300.5000x.
3. *Ibid.*, p. 300.5000y.
4. R. E. Colborn and K. P. C. Vollhardt, *J. Am. Chem. Soc.* **108,** 5470 (1986).
5. U.S. Pat. 4,241,230 (Dec. 23, 1980), B. M. Drinkard (to Mobil Oil Corp.).
6. Jpn. Kokai 78 59,605 (May 29, 1978), Y. Nambu and C. Fujii (to Denki Kaguku Kogyo K.K.).
7. Ref. 2, p. 300.5000s.
8. W. Reppe and co-workers, *Ann.* **596,** 2 (1955).
9. J. V. Nef, *Ann.* **308,** 277 (1899).
10. Ref. 8, p. 29.
11. L. Henry, *Ber.* **5,** 449 (1872).
12. U.S. Pat. 2,563,325 (Aug. 7, 1951), F. Fahnoe (to GAF Corp.).
13. U.S. Pat. 2,641,615 (June 9, 1953), R. F. Kleinschmidt (to GAF Corp.).
14. J. C. Sauer, *Org. Synth. Coll. Vol.* **4,** 813 (1963).
15. V. Wolf, *Chem. Ber.* **86,** 735 (1953).
16. V. Wolf, *Chem. Ber.* **87,** 668 (1954).
17. L. Henry, *Ber.* **6,** 728 (1873); **8,** 398 (1875).
18. A. Kirrmann, *Bull. Soc. Chim. Fr.* **39**(4), 698 (1926).
19. L. Henry, *Ber.* **17,** 1132 (1884).
20. Ref. 8, pp. 57–59.
21. F. J. McQuillin and W. O. Ord, *J. Chem. Soc.,* 2906 (1959).
22. Ref. 8, p. 61.
23. G. F. Hennion and W. S. Murray, *J. Am. Chem. Soc.* **64,** 1220 (1942).
24. U.S. Pat. 3,637,813 (Jan. 25, 1972), G. F. D'Allelio.
25. E. Cherbuliez, M. Gowhari, and J. Rabinowitz, *Helv. Chim. Acta* **47,** 2098 (1964).

26. U.S. Pat. 3,783,016 (Jan. 1, 1974), D. I. Randall and C. Vogel (to GAF Corp.).
27. Ref. 8, p. 69.
28. P. Chini, A. Santambrogio, and N. Palladino, *J. Chem. Soc. C*, 830 (1967).
29. W. Schulz, U. Rosenthal, D. Braun, and D. Walther, *Z. Chem.* **27,** 264 (1987).
30. W. E. Daniels, *J. Org. Chem.* **29,** 2936 (1964).
31. L. A. Akopyan and co-workers, *Polym. Sci. USSR* (in English) **17**(5), 1231 (1975).
32. M. G. Akhmerov, D. Usupov, and A. Kuchkarov, *Uzb. Khim. Zh.* **1979**(4), 59 (1979); *Chem. Abstr.* **92,** 58871 (1980).
33. U.S. Pat. 2,238,471 (Apr. 15, 1941), E. Keyssner and E. Eichler (to GAF Corp.).
34. R. L. Salvador and D. Simon, *Can. J. Chem.* **44**(21), 2570 (1966).
35. R. Lespieau and P. L. Viguier, *C. R. Acad. Sci.* **146,** 294 (1908).
36. I. G. Ali-Zade and co-workers, *Dokl. Akad. Nauk. SSSR* **173,** 89 (1967); *Chem. Abstr.* **67,** 32723 (1967).
37. L. Brandsma, *Preparative Acetylenic Chemistry,* Elsevier, Amsterdam, the Netherlands, 1971, pp. 166–168.
38. U.S. Pat. 3,257,465 (June 21, 1966), M. W. Leeds and H. L. Komarowski (to Cumberland Chemical Corp.).
39. Brit. Pat. 968,928 (Sept. 9, 1964), M. E. Chiddix and O. F. Hecht (to GAF Corp.).
40. U.S. Pat. 3,272,854 (Sept. 13, 1966), R. A. Covey, A. E. Smith, and W. L. Hubbard (to Uniroyal Corp.).
41. U.S. Pat. 2,778,830 (Jan. 22, 1957), H. Pasedach and M. Seefelder (to BASF A.G.).
42. U.S. Pat. 3,923,870 (Dec. 2, 1975), W. Singer (to Troy Chemical Corp.).
43. G. Y. Yositsch, *Zh. Russ. Fiz. Khim. Ova.* **38,** 252 (1906).
44. A. W. Johnson, *J. Chem. Soc.*, 1009 (1946).
45. G. Dupont, R. Dulou, and G. Lefebvre, *Bull. Soc. Chim. Fr.,* 816 (1954).
46. Ref. 8, p. 56.
47. U.S. Pat. 3,083,235 (Mar. 26, 1963), D. J. Mann, D. D. Perry, and R. M. Dudak (to Thiokol Chemical Corp.).
48. U.S. Pat. 2,941,010 (June 14, 1960), D. J. Mann, D. D. Perry, and R. M. Dudak (to Thiokol Chemical Corp.).
49. V. Wolf, *Chem. Ber.* **88,** 717 (1955).
50. I. Heilbron, E. R. Jones and F. Sondheimer, *J. Chem. Soc.,* 604 (1949).
51. V. Franzen, *Ann.* **587,** 131 (1954).
52. A. W. Johnson, *J. Chem. Soc.,* 1011 (1946).
53. A. W. Johnson, *J. Chem. Soc.,* 1014 (1946).
54. U.S. Pat. 3,746,726 (July 17, 1973), F. Reicheneder and K. Dury (to BASF A.G.).
55. U.S. Pat. 3,054,739 (Sept. 18, 1962), F. Reicheneder and K. Dury (to BASF A.G.).
56. K. Dury, *Angew. Chem.* **72,** 864 (1960).
57. H. Kleinert and H. Fuerst, *J. Pract. Chem.* **36,** 252 (1967).
58. L. H. Smith, *Synthetic Fiber Developments in Germany*, Textile Research Institute, New York, 1946, pp. 534–541.
59. Ger. Pat. 1,074,569 (Feb. 4, 1960), H. Pasedach and D. Ludsteck (to BASF A.G.).
60. Ger. Pat. 32,828 (May 15, 1965), H. Kleinert; *Chem. Abstr.* **63,** 17901 (1965).
61. Ref. 8, p. 51.
62. Ref. 8, pp. 45–47.
63. Y. K. Yur'ev, I. K. Korobitsyna, and E. G. Brige, *Dokl. Akad. Nauk SSSR* **62,** 625 (1948); *Chem. Abstr.* **43,** 5003 (1949).
64. U.S. Pat. 2,542,417 (Feb. 20, 1951), R. F. Kleinschmidt (to GAF Corp.).
65. Ger. Pat. 229,689 (Nov. 13, 1985), H. Drevs and R. S. Koernig; *Chem. Abstr.* **104,** 225336 (1986).
66. U.S. Pat. 2,421,650 (June 3, 1947), W. Reppe, C. Schuster, and E. Weiss (to GAF Corp.).
67. C. J. S. Appleyard and J. F. C. Gartshore, "Manufacture of 1,4-Butynediol at I. G.

Ludwigshafen," *BIOS Report 367, Item 22; OTS Report PB 28556*, U.S. Department of Commerce.

68. D. L. Fuller, A. O. Zoss, and H. M. Weir, "The Manufacture of Butynediol from Acetylene and Formaldehyde," *FIAT Report No. 926; OTS Report PB 80334*, U.S. Department of Commerce, 1946.
69. U.S. Pat. 3,560,576 (Feb. 2, 1971), J. R. Kirschner to E.I. du Pont de Nemours & Co., Inc).
70. Ger. Offen. 2,357,751 (Aug. 7, 1975), K. Baer and co-workers (to BASF A.G.)
71. Ger. Offen. 2,314,693 (Oct. 10, 1974), W. Reiss and co-workers (to BASF A.G.).
72. Ger. Offen. 2,240,401 (Mar. 7, 1974), H. Pasedach and H. Kroesche (to BASF A.G.).
73. U.S. Pat. 3,920,759 (Nov. 18, 1975), E. V. Hort (to GAF Corp.).
74. U.S. Pat. 4,117,248 (Sept. 26, 1978), J. L. Prater and R. L. Hedworth (to GAF Corp.).
75. U.S. Pat. 4,119,790 (Oct. 10, 1978), E. V. Hort (to GAF Corp.).
76. Ref. 1, pp. 476–542.
77. U.S. Pat. 2,906,614 (Sept. 29, 1959), T. R. Hopkins and J. W. Pullen (to Spencer Chemical Co.).
78. G. Griner, *C. R. Acad. Sci.* **116**, 723 (1893).
79. C. Prevost, *C. R. Acad. Sci.* **183**, 1292 (1926).
80. U.S. Pat. 2,877,205 (Mar. 10, 1959), J. Lal (to Justi and Son, Inc.).
81. U.S. Pat. 2,840,598 (June 24, 1958), H. Schwartz (to Vineland Chemical Co.).
82. U.S. Pat. 2,980,649 (Apr. 18, 1961), J. R. Caldwell and R. Gilkey (to Eastman Kodak Co.).
83. O. Bayer and co-workers, *Angew. Chem.* **62**, 61 (1950).
84. C. S. Marvel and C. H. Young, *J. Am. Chem. Soc.* **73**, 1066 (1951).
85. Jpn. Pat. 3164 (Apr. 25, 1958), T. Haya, M. Sato, and M. Yoshida (to Mitsubishi Co.); *Chem. Abstr.* **53**, 6083 (1959).
86. G. Lefebvre, G. Dupont, and R. Dulou, *C. R. Acad. Sci.* **229**, 222 (1949).
87. K. C. Brannock and G. R. Lappin, *J. Org. Chem.* **21**, 1366 (1956).
88. Ger. Pat. 855,864 (Nov. 17, 1952), A. Seib (to BASF A.G.).
89. U.S. Pat. 3,240,702 (Mar. 15, 1966), R. F. Monroe (to The Dow Chemical Company).
90. J. M. Bobbit, L. H. Amundsen, and R. I. Steiner, *J. Org. Chem.* **25**, 2230 (1960).
91. A. Valette, *Ann. Chim.* **3**(12), 644 (1948).
92. Ger. Pat. 857,369 (Nov. 27, 1952), H. Krzikalla and E. Woldan (to BASF A.G.).
93. P. Kurtz, *Ann.* **572**, 49, 69 (1951).
94. U.S. Pat. 4,633,015 (Dec. 30, 1986), A. S. C. Chan and D. E. Morris (to Monsanto Co.).
95. L. E. Craig, R. M. Elofson, and I. J. Ressa, *J. Am. Chem. Soc.* **72**, 3277 (1960).
96. U.S. Pat. 3,661,980 (May 9, 1972), W. Himmele, W. Qulla, and R. Prinz (to BASF A.G.).
97. U.S. Pat. 3,859,369 (Jan. 7, 1975), H. B. Copelin (to E.I. du Pont de Nemours & Co., Inc.).
98. Ger. Offen. 3,334,589 (Apr. 4, 1985), R. Schalenbach and H. Waldmann (to Bayer A.G.).
99. M. F. Abidova, A. S. Sultanov, and N. A. Savel'eva, *Katal. Pererab. Uglevodorodnogo. Syr'ya* **1971**(5), 175 (1971); *Chem. Abstr.* **79**, 125755 (1973).
100. N. O. Brace, *J. Am. Chem. Soc.* **77**, 4157 (1955).
101. Ger. Pat. 961,353 (Apr. 4, 1957), H. Friederich (to BASF A.G.).
102. U.S. Pat. 2,912,439 (Nov. 10, 1959), R. H. Hasek and J. E. Hardwicke (to Eastman Kodak Co.).
103. Ger. Pat. 1,094,732 (Dec. 15, 1960), H. Frensch (to Hoechst A.G.).
104. N. Clauson-Kaas, *Acta Chem. Scand.* **15**, 177 (1961); *Chem. Abstr.* **57**, 4619 (1962).
105. U.S. Pat. 2,833,787 (May 6, 1958), G. J. Carlson and co-workers (to Shell Oil Co.).
106. U.S. Pat. 3,284,419 (Nov. 8, 1966), F. G. Helfferich (to Shell Oil Co.).
107. Ger. Pat. 833,963 (Mar. 13, 1952), E. Bauer (to BASF A.G.).
108. Ref. 8, p. 142.

109. U.S. Pat. 2,779,700 (Jan. 29, 1957), P. Robitschek and C. T. Bean (to Hooker Chemical Co.).
110. U.S. Pat. 2,740,771 (Apr. 3, 1958), R. I. Longley, Jr., E. C. Chapin, and R. F. Smith (to Monsanto Chemical Co.).
111. F. J. McQuillin and W. O. Ord, *J. Chem. Soc.*, 2906 (1959).
112. I. A. Makolkin and co-workers, *Tr. Mosk. Inst. Nar. Khoz.* **1968**(46), 3 (1968); *Chem. Abstr.* **71,** 54383 (1969).
113. J. W. Copenhaver and M. H. Bigelow, *Acetylene and Carbon Monoxide Chemistry,* Reinhold, New York, 1949, pp. 131–133.
114. P. W. Feit, *Chem. Ber.* **93,** 116 (1960).
115. U.S. Pat. 2,961,471 (Nov. 22, 1960), E. V. Hort (to GAF Corp.).
116. H. Tieckelmann in R. A. Abromovitch, ed., *Pyridine and Its Derivatives,* Suppl. Part III, John Wiley & Sons, Inc., New York, 1974, pp. 670–673.
117. P. J. Dekkers, *Recl. Trav. Chim. Pays-Bas* **9,** 92 (1890).
118. C. Harries, *Ber.* **35,** 1187 (1902).
119. J. Baeseken, *Recl. Trav. Chim. Pays-Bas* **34,** 100 (1915).
120. J. Hamonet, *C. R. Acad. Sci.* **132,** 632 (1905).
121. R. Lespieau, *C. R. Acad. Sci.* **150,** 1761 (1910).
122. W. Griehl and G. Schnock, *Faserforsch. Textiltech.* **8,** 408 (1957); *Chem. Abstr.* **52,** 11781 (1958).
123. S. Sarel, L. A. Pohoryles, and R. Ben-Shoshan, *J. Org. Chem.* **24,** 1873 (1959).
124. Ger. Pat. 800,662 (Nov. 27, 1950), H. Krzikalla and K. Merkel (to BASF A.G.).
125. R. Riemschneider and W. M. Schneider, *Monatsh. Chem.* **90,** 510 (1959).
126. J. W. Hill and W. H. Carothers, *J. Am. Chem. Soc.* **57,** 925 (1935).
127. U.S. Pat. 2,870,097 (Jan. 20, 1959), D. B. Pattison (to E.I. du Pont de Nemours & Co., Inc.).
128. Ger. Pat. 800,398 (Nov. 2, 1950), B. Christ (to BASF A.G.).
129. V. V. Korshak and I. A. Gribova, *Izv. Akad. Nauk SSSR Otd. Tekh. Nauk,* 670 (1954); *Chem. Abstr.* **49,** 10893 (1955).
130. U.S. Pat. 2,222,302 (Nov. 19, 1940), W. Schmidt and F. Manchen (to GAF Corp.).
131. A. Muller and W. Vane, *Ber.* **77B,** 669 (1944).
132. S. K. Kang, W. S. Kim, and B. H. Moon, *Synthesis* **1985**(12), 1161 (1985).
133. Ref. 8, p. 178.
134. I. Geiman and co-workers, *Khim Geterotsickl Soedin.* **1981**(4), 448 (1981); *Chem. Abstr.* **95,** 80604 (1981).
135. G. A. Kliger and co-workers, *Otkrytiya Izobret.* **1988**(23), 106 (1988); *Chem. Abstr.* **109,** 212804 (1988).
136. R. E. Walkup and S. Searles, Jr., *Tetrahedron* **41**(1), 101 (1985).
137. Ger. Offen. 2,824,908 (Dec. 20, 1979), W. Mesch (to BASF A.G.).
138. Y. K. Yur'ev and N. G. Medovshchikov, *J. Gen. Chem. USSR* **9,** 628 (1939); *Chem. Abstr.* **33,** 7779 (1939).
139. *Chem. Eng.* **82**(7), 55 (1975).
140. *Chem. Mark. Rep.* **207**(11), 5 (1975).
141. M. J. Haley with W. E. Cox and S. Yoshikawa, *1,4-Butanediol,* CEH Product Review, Chemical Economics Handbook, SRI International, Menlo Park, Calif., Oct. 1988, p. 621.5030c.
142. *Chem. Eng. News* **53**(7), 8 (1975).
143. *Chem. Eng. News* **54**(9), 11 (1976).
144. Ref. 141, pp. 621.5030f,g.
145. Ref. 141, p. 621.5030l.
146. R. Fittig and M. B. Chanlaroff, *Ber.* **226,** 331 (1884).
147. S. Minoda and M. Miyajima, *Hydrocarbon Process.* **49**(11), 176 (1970).

148. K. Saotome and K. Sato, *Bull. Chem. Soc. Jpn.* **39**(3), 485 (1966).
149. H. C. Brown and K. A. Keblys, *J. Org. Chem.* **31**(2), 485 (1966).
150. Ref. 8, pp. 191–194.
151. Y. K. Yur'ev, E. G. Vendel'shtein, and L. A. Zinov'eva, *Zhur. Obshch. Khim.* **22**, 509 (1952); *Chem. Abstr.* **47**, 2747 (1953).
152. Swiss Pat. 264,598 (Jan. 16, 1950), G. Bischoff; *Chem. Abstr.* **45**, 1622 (1951).
153. J. E. Livak and co-workers, *J. Am. Chem. Soc.* **67**, 2218 (1945).
154. Ger. Pat. 810,025 (Aug. 6, 1951), C. Schuster and A. Simon (to BASF A.G.).
155. H. Hart and O. E. Curtis, Jr., *J. Am. Chem. Soc.* **78**, 112 (1956).
156. U.S. Pats. 2,993,891 (July 25, 1961), 3,030,361 (Apr. 17, 1962), 3,031,446 (Apr. 24, 1962), H. W. Zimmer and J. M. Holbert (to Chattanooga Medicine Co.).
157. Jpn. Pat. 8,271 (Sept. 24, 1956), M. Ohta (to Mitsubishi Co.,); *Chem. Abstr.* **52**, 11904 (1958).
158. V. M. Markovich and co-workers, *Khim. Farm. Zh.* **16**, 1491 (1982); *Chem. Abstr.* **98**, 125367 (1983).
159. Indian Pat. 160,027 (June 20, 1987), A. Prakosh and R. S. Prasad (to Reckitt and Colman of India Ltd.).
160. G. I. Nikishin and co-workers, *Izv. Akad. Nauk SSSR Otd. Tekh. Nauk,* 146 (1962); *Chem. Abstr.* **57**, 16390 (1962).
161. D. Elad and R. D. Youssefyeh, *Chem. Commun.* **1965**(1), 7 (1965).
162. W. E. Truce and C. E. Olson, *J. Am. Chem. Soc.* **74**, 4721 (1952).
163. Ger. Pat. 1,026,297 (Mar. 20, 1958), N. von Kutepow (to BASF A.G.); *Chem. Abstr.* **54**, 9768 (1960).
164. D. J. Cram and H. Steinberg, *J. Am. Chem. Soc.* **76**, 3630 (1954).
165. U.S. Pat. 2,778,852 (Jan. 22, 1957), K. Adam and H. G. Trieschmann (to BASF A.G.).
166. Ger. Pat. 804,567 (Apr. 26, 1951), H. Kaltschmitt and A. Tartter (to BASF A.G.).
167. V. Y. Kortun, Z. M. Kol'tsova, and V. G. Yashunskii, *Zh. Prikl. Khim. (Leningrad)* **51**, 1919 (1978); *Chem. Abstr.* **89**, 196931 (1978).
168. U.S. Pat. 2,819,304 (Jan. 7, 1958), W. Reppe, H. Friederich, and H. Laib (to BASF A.G.).
169. Ger. Pat. 917,665 (Oct. 14, 1954), H. Haussmann and G. Grafinger (to BASF A.G.).
170. H. Plieninger, *Chem. Ber.* **83**, 265 (1950).
171. U.S. Pat. 2,603,658 (July 15, 1952), F. Hanusch (to BASF A.G.).
172. W. Reppe and co-workers, *Ann.* **596**, 198 (1955).
173. T. C. Bruice and J. J. Bruno, *J. Am. Chem. Soc.* **83**, 3494 (1961).
174. A. L. Dounce, R. H. Wardlow, and R. Connor, *J. Am. Chem. Soc.* **57**, 2556 (1935).
175. S. M. McElvain and J. F. Vozza, *J. Am. Chem. Soc.* **71**, 896 (1949).
176. E. Spath and J. Lintner, *Ber.* **69**, 2727 (1936).
177. K. Hatada and Y. Ono, *Bull. Chem. Soc. Jpn.* **50**, 2517–2521 (1977) (in English).
178. U.S. Pat. 4,824,967 (Apr. 25, 1989), K. C. Liu and P. D. Taylor (to GAF Corp.).
179. A. O. Zoss and D. L. Fuller, "The Manufacture of γ-Butyrolactone," PB 60902, U.S. Department of Commerce Office of Technical Services, Oct. 1946.
180. A. E. Favorskii and M. Skossarewsky, *Zh. Russ. Fiz. Khim. Ova.* **32**, 652 (1900); *Bull. Soc. Chim. Fr.* **26**, 284 (1901).
181. G. F. Hennion and co-workers, *J. Org. Chem.* **21**, 1142 (1956).
182. U.S. Pat. 2,882,287 (Apr. 14, 1959), D. C. Rowlands and W. H. Gillen (to Air Reduction Co.).
183. C. D. Hurd and W. D. McPhee, *J. Am. Chem. Soc.* **71**, 398 (1949).
184. U.S. Pat. 2,798,885 (July 9, 1957), H. Ensslin and K. Meier (to Ciba Pharmaceutical Products Inc.).
185. I. N. Nazarov and G. A. Schvekhgeimer, *Zh. Obshch. Khim.* **29**, 463 (1959); *Chem. Abstr.* **53**, 21661 (1959).
186. G. F. Hennion and K. W. Nelson, *J. Am. Chem. Soc.* **79**, 2142 (1957).

187. Y. Pasternak, *Bull. Soc. Chim. Fr.* **1963**(8–9), 1719 (1963).
188. T. A. Favorskaya, *J. Gen. Chem. USSR* **9**, 386 (1939); *Chem. Abstr.* **33**, 9281 (1939).
189. T. A. Favorskaya, *J. Gen. Chem. USSR* **10**, 461 (1940); *Chem. Abstr.* **34**, 7845 (1940).
190. U.S. Pat. 2,250,558 (July 29, 1941), T. H. Vaughn (to Union Carbide Corp.).
191. U.S. Pat. 3,388,181 (June 11, 1968), H. D. Anspon (to GAF Corp.).
192. R. J. Tedeschi and G. Clark, Jr., *J. Org. Chem.* **27**, 4323 (1962).
193. N. C. Rose, *J. Chem. Educ.* **43**(6), 324 (1966).
194. I. N. Nazarov and L. D. Bergel'son, *Zh. Obshch. Khim.* **27**, 1540 (1957); *Chem. Abstr.* **52**, 3660 (1958).
195. G. F. Hennion and G. M. Wolf, *J. Am. Chem. Soc.* **62**, 1368 (1940).
196. Ger. Offen. 3,633,033 (Apr. 7, 1988), G. Thelen and H. W. Voges (to Huels A.G.).
197. G. A. Chukhadzhyan and co-workers, *Zh. Org. Khim.* **8**(3), 476 (1972); *Chem. Abstr.* **77**, 4739 (1972).
198. E. R. H. Jones, G. H. Whitham, and M. C. Whiting, *J. Chem. Soc.,* 4628 (1957).
199. T. Nogi and J. Tsuji, *Tetrahedron* **25**(17), 4099 (1969).
200. H. A. Stansbury and W. R. Proops, *J. Org. Chem.* **27**, 320 (1962).
201. S. S. Dehmlow and E. V. Dehmlow, *Justus Liebigs Ann. Chem.* **10**, 1753 (1973).
202. Ref. 8, p. 36.
203. R.-R. Lii and S. I. Miller, *J. Am. Chem. Soc.* **95**(5), 1602 (1973).
204. G. Saucy and co-workers, *Helv. Chim. Acta* **42**, 1945 (1959).
205. A. V. Shchelkunov, A. Ashirbekova, and V. M. Rofman, *Deposited Doc.* **1980**, 5311 (1980).
206. U.S. Pat. 3,082,260 (Mar. 19, 1963), R. J. Tedeschi, A. W. Casey, and J. P. Russell (to Air Reduction Co.).
207. M. A. Dzhragatspanyan, A. G. Mirzkhanyan, and L. A. Ustynyuk, *Arm. Khim. Zh.* **36**, 476, 547 (1983); *Chem. Abstr.* **100**, 22301, 22302 (1984).
208. U.S. Pat. 3,283,014 (Nov. 1, 1966), A. Balducci and M. de Malde (to SNAM S.p.A.).
209. A. Heath, *Chem. Eng.* **80**(22), 48 (1973).
210. U.S. Pat. 1,941,108 (Dec. 26, 1923), W. Reppe (to I. G. Farbenind. A.G.).
211. U.S. Pat. 1,959,927 (May 22, 1934), W. Reppe (to I. G. Farbenind. A.G.).
212. O. Solomon, C. Ionescu, and I. Ciuta, *Chem. Tech. Leipzig* **9**, 202 (1957); *Chem. Abstr.* **51**, 15493 (1957).
213. Shansi University, *Hua Hsueh Tung Pao,* 21 (1977); *Chem. Abstr.* **87**, 134899 (1977).
214. C. Gardner and co-workers, *J. Chem. Soc.,* 789 (1949).
215. Fr. Pat. 865,354 (May 3, 1940), H. Weese, G. Hecht, and W. Reppe (to I. G. Farbenind. A.G.).
216. J. Wislicenus, *Ann.* **192**, 106 (1878).
217. I. W. J. Still, J. N. Reed, and K. Turnbull, *Tetrahedron Lett.* **17**, 1481 (1979).
218. A. Franke, F. F. Frickel, R. Schlecker, and P. C. Theime, *Synthesis,* 712 (1979).
219. A. Skrabal and R. Skrabal, *Z. Phys. Chem. A* **181**, 449 (1938).
220. E. Levas, *C. R. Acad. Sci.* **228**, 1443 (1949).
221. F. Kipnis, H. Solonay, and J. Ornfelt, *J. Am. Chem. Soc.* **73**, 1783 (1951).
222. U.S. Pat. 2,061,946 (Nov. 24, 1936), E. Kuehn and H. Hopff (to I. G. Farbenind. A.G.).
223. W. Reppe and co-workers, *Ann.* **601**, 109 (1956).
224. L. Summers, *Chem. Rev.* **55**, 317 (1955).
225. U.S. Pat. 2,433,890 (Jan. 6, 1948), O. W. Cass (to E.I. du Pont de Nemours & Co., Inc.).
226. U.S. Pat. 4,489,011 (Dec. 18, 1984), S. S. M. Wang (to Merrell Dow Pharm., Inc.).
227. S. Siggia and R. L. Edsberg, *Anal. Chem.* **20**, 762 (1948).
228. U.S. Pat. 2,165,962 (July 1, 1939), M. Mueller-Conradi and K. Pieroh (to I. G. Farbenind. A.G.).
229. U.S. Pats. 2,487,525 (Apr. 8, 1949) and 2,502,433 (Apr. 4, 1950), J. W. Copenhaver (to GAF Corp.).

230. R. I. Hoaglin and D. H. Hirsch, *J. Am. Chem. Soc.* **71,** 3468 (1949).
231. U.S. Pat. 2,527,533 (Oct. 31, 1950), J. W. Copenhaver (to GAF Corp.).
232. R. I. Hoaglin, D. G. Kubler, and A. E. Montagna, *J. Am. Chem. Soc.* **80,** 5460 (1958).
233. M. S. Nieuwenhuizen, A. P. G. Kieboom, and H. Van Bekkum, *Synthesis,* 712 (1981).
234. R. I. Longley, Jr., and W. S. Emerson, *J. Am. Chem. Soc.* **72,** 3079 (1950).
235. Jpn. Kokai Tokkyo Koho 59,108,734 (June 23, 1984) (to Daicel Ind., Ltd.).
236. U.S. Pat. 2,546,018 (Mar. 20, 1951), C. W. Smith and co-workers (to Shell Development Co.).
237. U.S. Pat. 2,546,019 (Mar. 20, 1951), C. W. Smith and co-workers (to Shell Development Co.).
238. M. Yamauchi, S. Katayama, O. Baba, and T. Watanabe, *J. Chem. Soc. Chem. Commun.,* 281 (1983).
239. R. W. Aben and H. W. Scheeren, *J. Chem. Soc. Perkin Trans.* **1,** 3132 (1979).
240. U.S. Pat. 3,019,231 (Jan. 30, 1962), W. J. Peppel and J. D. Watkins (to Jefferson Chemical Co.).
241. W. H. Watanabe and L. E. Conlon, *J. Am. Chem. Soc.* **79,** 2828 (1957).
242. U.S. Pat. 2,560,219 (July 10, 1951), S. A. Glickman (to GAF Corp.).
243. U.S. Pat. 3,050,565 (Aug. 21, 1962), P. O. Tawney (to Uniroyal Corp.).
244. U.S. Pat. 2,768,975 (Oct. 30, 1956), R. S. Schiefelbein (to Jefferson Chemical Co.).
245. U.S. Pat. 2,736,743 (Feb. 28, 1956), C. J. Schmidle and R. C. Mansfield (to Rohm and Haas Corp.).
246. Y. Yoshida and S. Inoue, *Chem. Lett.* **11,** 1375 (1977).
247. S. A. Miller, *Acetylene, Its Properties, Manufacture and Uses,* Vol. 2, Academic Press, Inc., New York, 1965, pp. 199–202.
248. J. W. Copenhaver and M. H. Bigelow, *Acetylene and Carbon Monoxide Chemistry,* Reinhold Publishing Corp., New York, 1949, pp. 37–38.

EUGENE V. HORT
PAUL TAYLOR
GAF Corporation

ACETYLENE DICHLORIDE (1,2-DICHLOROETHYLENE). See CHLOROCARBONS AND CHLOROHYDROCARBONS.

ACETYLENE TETRABROMIDE (1,1,2,2-TETRABROMO-ETHANE). See BROMINE COMPOUNDS.

ACETYLENE TETRACHLORIDE (1,1,2,2-TETRACHLORO-ETHANE). See CHLOROCARBONS AND CHLOROHYDROCARBONS.

ACETYLENIC ALCOHOLS. See ACETYLENE-DERIVED CHEMICALS.

ACETYLENIC GLYCOLS. See ACETYLENE-DERIVED CHEMICALS.

ACETYLENIC POLYMERS.   See ELECTRICALLY CONDUCTIVE POLYMERS.

ACETYL FLUORIDE.   See FLUORINE COMPOUNDS, ORGANIC.

ACETYLIDES.   See HYDROCARBONS, ACETYLENE; CARBIDES.

ACETYL PEROXIDE.   See PEROXIDES AND PEROXY COMPOUNDS, ORGANIC.

ACETYLSALICYLIC ACID.   See ANALGESICS, ANTIPYRETICS, AND ANTI-INFLAMMATORY AGENTS; SALICYLIC ACID AND RELATED COMPOUNDS.

ACID RAIN.   See AIR POLLUTION; ATMOSPHERIC MODELS; ENVIRONMENTAL IMPACT.

ACIDS, CARBOXYLIC.   See CARBOXYLIC ACIDS.

ACONITIC ACID.   See CITRIC ACID.

ACQUIRED IMMUNODEFICIENCY SYNDROME, AIDS.   See ANTIVIRAL AGENTS; IMMUNOTHERAPEUTIC AGENTS.

ACRIDINE.   See PYRIDINE AND PYRIDINE DERIVATIVES.

ACRIDINE DYES.   See DYES AND DYE INTERMEDIATES.

ACROLEIN AND DERIVATIVES

Acrolein (2-propenal [107-02-08]), C_3H_4O, is the simplest unsaturated aldehyde (CH_2=CHCHO). The primary characteristic of acrolein is its high reactivity due to conjugation of the carbonyl group with a vinyl group. Controlling this reactivity to give the desired derivative is the key to its usefulness. Acrolein now finds commercial utility in several major products as well as a number of smaller volume products. More than 80% of the refined acrolein that is produced today goes into the synthesis of methionine. Much larger quantities of crude acrolein are produced as an intermediate in the production of acrylic acid. More than 85% of the acrylic acid produced worldwide is by the captive oxidation of acrolein.

Several review articles (1–5) and a book (6) have been published on the preparation, reactions, and uses of acrolein.

Acrolein is a highly toxic material with extreme lacrimatory properties. At room temperature acrolein is a liquid with volatility and flammability somewhat similar to acetone; but unlike acetone, its solubility in water is limited. Commercially, acrolein is always stored with hydroquinone and acetic acid as inhibitors. Special care in handling is required because of the flammability, reactivity, and toxicity of acrolein.

The physical and chemical properties of acrolein are given in Table 1.

Table 1. Properties of Acrolein

Property	Value
Physical properties	
molecular formula	C_3H_4O
molecular weight	56.06
specific gravity at 20/20°C	0.8427
coefficient of expansion at 20°C, vol/°C	0.00140
boiling point, °C	
at 101.3 kPa[a]	52.69
at 1.33 kPa[a]	−36
vapor pressure at 20°C, kPa[a]	29.3
heat of vaporization at 101.3 kPa[a], kJ/kg[b]	93
critical temperature, °C	233
critical pressure, MPa[c]	5.07
critical volume, mL/mol	189
freezing point, °C	−87.0
solubility at 20°C, % by wt	
in water	20.6
water in	6.8
refractive index, n_D^{20}	1.4013
viscosity at 20°C, mPa·s (= cP)	0.35
heat capacity (specific), kJ/(kg·K)[b]	
liquid (17–44°C)	0.396
gas (27°C)	0.221
liquid density at 20°C, kg/L[d]	0.8412
Chemical properties	
flash point, open cup	−18°C
closed cup	−26°C
flammability limits in air, vol %	
upper	31
lower	2.8
autoignition temperature in air	234°C
heat of combustion at 25°C, kJ/kg[b]	5383
heat of polymerization (vinyl), kJ/mol[b]	71–80
heat of condensation (aldol), kJ/mol[b]	42

[a]To convert kPa to mm Hg, multiply by 7.5.
[b]To convert kJ to kcal, divide by 4.184.
[c]To convert MPa to psi, multiply by 145.
[d]To convert kg/L to lb/gal, multiply by 8.345.

Manufacture

The Reaction. Acrolein has been produced commercially since 1938. The first commercial processes were based on the vapor-phase condensation of acetaldehyde and formaldehyde (1). In the 1940s a series of catalyst developments based on cuprous oxide and cupric selenites led to a vapor-phase propylene oxidation route to acrolein (7,8). In 1959 Shell was the first to commercialize this propylene oxidation to acrolein process. These early propylene oxidation catalysts were capable of only low per pass propylene conversions (ca 15%) and therefore required significant recycle of unreacted propylene (9–11).

In 1957 Standard Oil of Ohio (Sohio) discovered bismuth molybdate catalysts capable of producing high yields of acrolein at high propylene conversions (>90%) and at low pressures (12). Over the next thirty years much industrial and academic research and development was devoted to improving these catalysts, which are used in the production processes for acrolein, acrylic acid, and acrylonitrile. All commercial acrolein manufacturing processes known today are based on propylene oxidation and use bismuth molybdate based catalysts.

Many key improvements and enhancements to the bismuth molybdate based propylene oxidation catalysts have occurred over the past thirty years. These are outlined in the following tabulation.

Year	Catalyst	Company	Refs.
1957	BiMo	Sohio	12
1959	BiMoFe	Knapsack	13
1964	BiMoFeNiCo	Nippon Kayaku	14,15
1965	BiMoFeNiCoP	Nippon Kayaku	16
1970	BiMoFeNiCoP K	Nippon Kayaku	17
1970	BiMoFeCo W Si K	Nippon Shokubai	18,19

Today the most efficient catalysts are complex mixed metal oxides that consist of Bi, Mo, Fe, Ni, and/or Co, K, and either P, B, W, or Sb. Many additional combinations of metals have been patented, along with specific catalyst preparation methods. Most catalysts used commercially today are extruded neat metal oxides as opposed to supported impregnated metal oxides. Propylene conversions are generally better than 93%. Acrolein selectivities of 80 to 90% are typical.

With the maturing of the propylene oxidation catalyst area, attention in the 1980s was more focused on reaction process related improvements.

The catalytic vapor-phase oxidation of propylene is generally carried out in a fixed-bed multitube reactor at near atmospheric pressures and elevated temperatures (ca 350°C); molten salt is used for temperature control. Air is commonly used as the oxygen source and steam is added to suppress the formation of flammable gas mixtures. Operation can be single pass or a recycle stream may be employed. Recent interest has focused on improving process efficiency and minimizing process wastes by defining process improvements that use recycle of process gas streams and/or use of new reaction diluents (20–24).

The reaction is very exothermic. The heat of reaction of propylene oxidation to acrolein is 340.8 kJ/mol (81.5 kcal/mol); the overall reactions generate approximately 837 kJ/mol (200 kcal/mol). The principal side reactions produce acrylic acid, acetaldehyde, acetic acid, carbon monoxide, and carbon dioxide. A variety of other aldehydes and acids are also formed in small amounts. Proprietary processes for acrolein manufacture have been described (25,26).

Product Recovery. The reactor effluent gases are cooled to condense and separate the acrolein from unreacted propylene, oxygen, and other low-boiling components (predominantly nitrogen). This is commonly accomplished in two absorption steps where (1) aqueous acrylic acid is condensed from the reaction effluent and absorbed in a water-based stream, and (2) acrolein is condensed and absorbed in water to separate it from the propylene, nitrogen, oxygen, and carbon oxides. Acrylic acid may be recovered from the aqueous product stream if desired. Subsequent distillation refining steps separate water and acetaldehyde from the crude acrolein. In another distillation column, refined acrolein is recovered as an azeotrope with water. A typical process flow diagram is given in Figure 1.

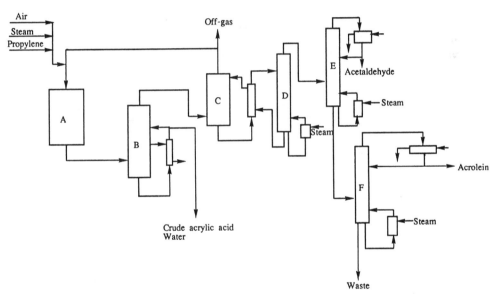

Fig. 1. A typical process flowsheet for acrolein manufacture. A, Fixed-bed or fluid-bed reactor; B, quench cooler; C, absorber; D, stripper; E and F, fractionation stills.

Economic Aspects

Presently, worldwide refined acrolein nameplate capacity is about 113,000 t/yr. Degussa has announced a capacity expansion in the United States by building a 36,000 t/yr acrolein plant in Theodore, Ala. to support their methionine business (27). The key producers of refined acrolein are as noted in Table 2.

Of these producers, Atochem, Degussa, and Daicel are reported to be in the merchant acrolein business. Union Carbide supplies only the acrolein derivative markets. Rhône-Poulenc also produces acrolein, primarily as a nonisolated intermediate to make methionine. A number of other small scale plants are located worldwide which also produce acrolein as an intermediate to make methionine.

Table 2. Refined Acrolein Producers

Producer	Annual nameplate capacity, 10^3 t/yr[a]
United States	
Union Carbide	36
Western Europe	
Degussa (Germany)	30
Atochem (France)	30
Japan	
Daicel	10
Ohita Chem.	4.5
Sumitomo	2.3

[a]Estimated (5).

The significance of industrial acrolein production may be clearer if one considers the two major uses of acrolein—direct oxidation to acrylic acid and reaction to produce methionine via 3-methylmercaptopropionaldehyde. In acrylic acid production, acrolein is not isolated from the intermediate production stream. The 1990 acrylic acid production demand in the United States alone accounted for more than 450,000 t/yr (28), with worldwide capacity approaching 1,470,000 t/yr (29). Approximately 0.75 kg of acrolein is required to produce one kilogram of acrylic acid. The methionine production process involves the reaction of acrolein with methyl mercaptan. Worldwide methionine production was estimated at about 170,000 t/yr in 1990 (30). (See ACRYLIC ACID AND DERIVATIVES; AMINO ACIDS, SURVEY.)

Specifications and Analysis

Acrolein is produced according to the specifications in Table 3. Acetaldehyde and acetone are the principal carbonyl impurities in freshly distilled acrolein. Acrolein dimer accumulates at 0.50% in 30 days at 25°C. Analysis by two gas chromatographic methods with thermal conductivity detectors can determine all significant impurities in acrolein. The analysis with Porapak Q, 175–300 μm (50–80 mesh), programmed from 60 to 250°C at 10°C/min, does not separate acetone, propionaldehyde, and propylene oxide from acrolein. These separations are made with 20% Tergitol E-35 on 250–350 μm (45–60 mesh) Chromosorb W, kept at 40°C until acrolein elutes and then programmed rapidly to 190°C to elute the remaining components.

Alternatively a bonded poly(ethylene glycol) capillary column held at 35°C for 5 min and programmed to 190°C at 8°C/min may be employed to determine all components but water. The Karl-Fischer method for water gives inaccurate results.

Hydroquinone can be determined spectrophotometrically at 292 nm in methanol after a sample is evaporated to dryness to remove the interference of

Table 3. Specifications for Acrolein

Requirement	Limit
acrolein, wt %, min	95.5
total carbonyl compounds other than acrolein, wt %, max	1.5
hydroquinone, wt %	0.10–0.25
water, wt %, max	3.0
specific gravity, 20°/20°C	0.842–0.846
pH of 10% solution in water at 25°C, max	6.0

acrolein. An alternative method is high performance liquid chromatography on 10-μm LiChrosorb RP-2 at ambient temperature with 2.0 mL/min of 20%(v/v) 2,2,4-trimethylpentane, 79.20% chloroform, and 0.80% methanol with uv detection at 292 nm.

Reactions and Derivatives

Acrolein is a highly reactive compound because both the double bond and aldehydic moieties participate in a variety of reactions.

Oxidation. Acrolein is readily oxidized to acrylic acid, $C_3H_4O_2$ [79-10-7], by passing a gaseous mixture of acrolein, air, and steam over a catalyst composed primarily of molybdenum and vanadium oxides (see ACRYLIC ACID AND DERIVATIVES). This process has been reviewed in a number of articles (31–33).

$$CH_2{=}CH{-}CHO \xrightarrow{\text{cat.}} CH_2{=}CHCOOH$$

Virtually all of the acrylic acid produced in the United States is made by the oxidation of propylene via the intermediacy of acrolein.

Reduction. Because of a lack of discrimination between the double bond and carbonyl moieties, direct hydrogenation of acrolein leads to the production of mixtures containing propyl alcohol, C_3H_8O [71-23-8], propionaldehyde, C_3H_6O [123-38-6], and allyl alcohol, C_3H_6O [107-18-16]. Both the carbonyl (34–40) and olefin (41–45) moieties may be selectively reduced (35,43,46–65).

$$CH_2{=}CH{-}CHO \longrightarrow CH_2{=}CH{-}CH_2OH$$
$$CH_2{=}CH{-}CHO \longrightarrow CH_3CH_2CHO$$

The vapor-phase reduction of acrolein with isopropyl alcohol in the presence of a mixed metal oxide catalyst yields allyl alcohol in a one-pass yield of 90.4%, with a selectivity (60) to the alcohol of 96.4%. Acrolein may also be selectively reduced to yield propionaldehyde by treatment with a variety of reducing reagents.

Reactions with Alcohols. The addition of alcohols to acrolein may be catalyzed by acids or bases. By the judicious choice of reaction conditions the regioselectivity of the addition may be controlled and alkoxy propionaldehydes, acrolein acetals, or alkoxypropionaldehyde acetals produced in high yields (66).

$$
R'OH + CH_2{=}CH{-}CHO \longrightarrow
\begin{cases}
CH_2{=}CH{-}CH\big({\scriptstyle OR' \atop OR'}\big) \\[4pt]
R'OCH_2CH_2CHO \\[4pt]
R'OCH_2CH_2CH\big({\scriptstyle OR' \atop OR'}\big)
\end{cases}
$$

Table 4 lists a variety of alkoxypropionaldehydes and certain of their properties (67). Alcohols up to n-butyl have been added to acrolein in this fashion. Methyl, ethyl, and allyl alcohols react with ease, while the addition of hexyl or octyl alcohol proceeds in low yields. Although the alkoxypropionaldehydes have found only limited industrial utility, it is anticipated that they will find use as replacements for more toxic solvents. Furthermore, the alkoxypropionaldehydes may readily be reduced to the corresponding alkoxypropanols, which may also have desirable properties as solvents.

Acrolein acetals have also been prepared in high yields (66). The formation of the acetal requires the careful control of reaction conditions to avoid additions to the double bond. Table 5 lists a variety of acrolein acetals that have been prepared and their boiling points (68).

The addition of certain glycols and polyols to acrolein leads to the production of cyclic acetal derivatives.

$$
CH_2{=}CH{-}CHO +
\begin{matrix} R{-}\overset{\displaystyle OH}{\underset{\displaystyle }{CH}} \\ R{-}\underset{\displaystyle OH}{\overset{\displaystyle }{CH}} \end{matrix} R
\longrightarrow
CH_2{=}CH{-}CH\begin{matrix} O \\ O \end{matrix}\begin{matrix} R \\ R \end{matrix}
$$

Cyclic acrolein acetals are, in general, easily formed, stable compounds and have been considered as components in a variety of polymer systems. Table 6 lists a variety of previously prepared cyclic acrolein acetals and their boiling points (69).

Reactions of acrolein with alcohols producing high yields of alkoxypropionaldehyde acetals are also known. Examples of these are displayed in Table 7 (70). The alkoxypropionaldehyde acetals may be useful as solvents or as intermediates in the synthesis of other useful compounds.

Addition of Mercaptans.   One of the largest uses of acrolein is the production of 3-methylmercaptopropionaldehyde [3268-49-3], which is an intermediate in the synthesis of D,L-methionine [59-51-8], an important chicken feed supplement.

$$CH_2{=}CH{-}CHO + CH_3SH \longrightarrow CH_3SCH_2CH_2CHO$$

3-Methylmercaptopropionaldehyde is also used to make the methionine hydroxy analog $CH_3SCH_2CH_2CH(OH)COOH$ [583-91-5], which is used commercially as an

Table 4. Alkoxypropionaldehydes from Acrolein

Compound added	Structure	Product CAS Registry Number	Molecular formula	bp, °C	Pressure at bp, kPa[a]
CH_3OH	$CH_3OCH_2CH_2CHO$	[2806-84-0]	$C_4H_8O_2$	49	6.7
C_2H_5OH	$C_2H_5OCH_2CH_2CHO$	[2806-85-1]	$C_5H_{10}O_2$	57	5.3
$n\text{-}C_3H_7OH$	$n\text{-}C_3H_7OCH_2CH_2CHO$	[19790-53-5]	$C_6H_{12}O_2$	88	12
$iso\text{-}C_3H_7OH$	$iso\text{-}C_3H_7OCH_2CH_2CHO$	[39963-51-4]	$C_6H_{12}O_2$	45	2
$n\text{-}C_4H_9OH$	$n\text{-}C_4H_9OCH_2CH_2CHO$	[13159-34-2]	$C_7H_{14}O_2$	60	2
$CH_2{=}CHCH_2OH$	$CH_2{=}CHCH_2OCH_2CH_2CHO$	[44768-60-7]	$C_6H_{10}O_2$	55	1.9
$ClCH_2CH_2OH$	$ClCH_2CH_2OCH_2CH_2CHO$	[5422-33-3]	$C_5H_9ClO_2$	75.5	0.67
$CH_2{=}C(CH_3)CH_2OH$	$CH_2{=}C(CH_3)CH_2OCH_2CH_2CHO$	[76618-56-9]	$C_7H_{12}O_2$	62	1.2

[a]To convert kPa to mm Hg, multiply by 7.5.

Table 5. Acrolein Acetals

Compound added	Structure	Product CAS Registry Number	Molecular formula	bp, °C	Pressure at bp, kPa[a]
C_2H_5OH	$CH_2{=}CHCH(OC_2H_5)_2$	[3054-95-3]	$C_7H_{14}O_2$	63	12.1
CH_3OH	$CH_2{=}CHCH(OCH_3)_2$	[6044-68-4]	$C_5H_{10}O_2$	40	16
$n\text{-}C_3H_7OH$	$CH_2{=}CHCH(OC_3H_7)_2$	[20615-55-8]	$C_9H_{18}O_2$	87.5–88	
$iso\text{-}C_3H_7OH$	$CH_2{=}CHCH(Oiso\text{-}C_3H_7)_2$	[14091-80-6]	$C_9H_{18}O_2$	54	1.6
$n\text{-}C_4H_9OH$	$CH_2{=}CHCH(OC_4H_9)_2$	[45094-50-6]	$C_{11}H_{22}O_2$	39	1.6
$C_6H_5CH_2OH$	$CH_2{=}CHCH(OCH_2C_6H_5)_2$	[40575-57-3]	$C_{17}H_{18}O_2$	120	6.7×10^{-4}
$CH_2{=}CHCH_2OH$	$CH_2{=}CHCH(OCH_2CH{=}CH_2)_2$	[3783-83-3]	$C_9H_{14}O_2$	75	3.7

[a]To convert kPa to mm Hg multiply by 7.5.

239

Table 6. Acrolein Cyclic Acetals

Compound added	Structure	Product CAS Registry Number	Molecular formula	bp, °C	Pressure at bp, kPa[a]
$HOCH_2CH_2OH$		[3984-22-3]	$C_5H_8O_2$	115.5–116.5	
$HOCH_2CH(OH)CH_2OH$		[4313-32-0]	$C_6H_{10}O_3$	80	0.4
$C(CH_2OH)_4$		[78-19-3]	$C_{11}H_{16}O_4$	142–143 m.p. 41–42	1.6
		b	$C_{16}H_{24}O_5$	198–199	0.4
$HOCH_2CH_2CH_2OH$		[5935-25-1]	$C_6H_{10}O_2$	65–66	5.9

[a]To convert kPa to mm Hg, multiply by 7.5.
[b]No CAS Registry Number has been assigned.

Table 7. Alkoxypropionaldehyde Acetals

Compound added	Structure	Product CAS Registry Number	Molecular formula	bp, °C	Pressure at bp, kPa[a]
C_2H_5OH	$C_2H_5OCH_2CH_2CH(OC_2H_5)_2$	[7789-92-6]	$C_9H_{20}O_3$	69–70	1.3
CH_3OH	$CH_3OCH_2CH_2CH(OCH_3)_2$	[14315-97-0]	$C_6H_{14}O_3$	94–95	18.9
n-C_3H_7OH	$C_3H_7OCH_2CH_2CH(OC_3H_7)_2$	[53963-14-7]	$C_{12}H_{26}O_3$	109	1.6
iso-C_3H_7OH	$C_3H_7OCH_2CH_2CH(Oiso$-$C_3H_7)_2$	[89769-16-4]	$C_{12}H_{26}O_3$	89	1.5
$C_6H_5CH_2OH$	$C_6H_5OCH_2CH_2CH(OCH_2C_6H_5)_2$	b	$C_{24}H_{26}O_3$	243–246	0.07
$CH_2{=}CHCH_2OH$	$CH_2{=}CHCH_2OCH_2CH_2CH(OCH_2CH{=}CH_2)_2$	[8431-07-1]	$C_{12}H_{20}O_3$	113–114	1.3
n-C_4H_9OH	$C_4H_9OCH_2CH_2CH(OC_4H_9)_2$	[53963-15-8]	$C_{15}H_{32}O_3$	143–144	1.3
n-$C_5H_{11}OH$	$C_5H_{11}OCH_2CH_2CH(OC_5H_{11})_2$	[53963-17-6]	$C_{18}H_{38}O_3$	153–155	0.13
$ClCH_2CH_2OH$	$ClCH_2CH_2OCH_2CH_2CH(OCH_2CH_2Cl)_2$	[688-78-8]	$C_9H_{17}Cl_3O_3$	160–162	0.7
C_2H_5SH	$C_2H_5SCH_2CH_2CH(SC_2H_5)_2$	[19157-17-6]	$C_9H_{20}S_3$	143	1.3
C_2H_5SH + HCl	$ClCH_2CH_2CH(SC_2H_5)_2$	[19157-16-5]	$C_7H_{15}ClS_2$	115–117	1.5
$CH_2{=}CHCH_2OCH_2CH_2CHO$ + C_2H_5OH	$CH_2{=}CHCH_2OCH_2CH_2CH(OC_2H_5)_2$	[107023-55-2]	$C_{10}H_{20}O_3$	86	1.7
C_2H_5OH + HCl	$ClCH_2CH_2CH(OC_2H_5)_2$	[35573-93-4]	$C_7H_{15}ClO_2$	58–62	1.1
CH_3OH + HCl	$ClCH_2CH_2CH(OCH_3)_2$	[35502-06-8]	$C_5H_{11}ClO_2$	45	1.6
C_2H_5OH + HBr	$BrCH_2CH_2CH(OC_2H_5)_2$	[59067-07-1]	$C_7H_{15}BrO_2$	80–90	2.7
n-C_3H_7OH + HCl	$ClCH_2CH_2CH(OC_3H_7)_2$	[35502-07-9]	$C_9H_{19}ClO_2$	87	2.7
iso-C_4H_9OH + HCl	$ClCH_2CH_2CH(Oiso$-$C_4H_9)_2$	[35502-09-1]	$C_{11}H_{23}ClO_2$	105	0.6

[a]To convert kPa to mm Hg, multiply by 7.5.
[b]No CAS Registry Number has been assigned.

241

effective source of methionine activity (71). All commercial syntheses of methionine and methionine hydroxy analog are based on the use of acrolein as a raw material. More than 170,000 tons of this amino acid are produced yearly (30) (see AMINO ACIDS). One method for the preparation of methionine from acrolein via 3-methylmercaptopropionaldehyde is as follows.

Methyl mercaptan adds to acrolein in nearly quantitative yields in the presence of a variety of basic catalysts (72,73). Other alkylmercaptopropionaldehydes produced by the reaction of acrolein with a mercaptan are known. Table 8 lists a variety of these and their boiling points (74).

Table 8. Alkylmercaptopropionaldehydes from Addition of Mercaptans to the Acrolein Double Bond

Compound added		Product			
	Structure	CAS Registry Number	Molecular formula	bp, °C	Pressure at bp, kPa^a
CH_3SH	$CH_3SCH_2CH_2CHO$	[3268-49-3]	C_4H_8OS	54–56	1.5
C_2H_5SH	$C_2H_5SCH_2CH_2CHO$	[5454-45-5]	$C_5H_{10}OS$	63–65	1.5
n-C_3H_7SH	n-$C_3H_7SCH_2CH_2CHO$	[44768-66-3]	$C_6H_{12}OS$	75	0.86
$C_6H_5CH_2SH$	$C_6H_5CH_2SCH_2CH_2CHO$	[16979-50-3]	$C_{10}H_{12}OS$	142.3	0.8
$CH_3C(O)SH$	$CH_3C(O)SCH_2CH_2CHO$	[53943-93-4]	$C_5H_8O_2S$	89	1.5
CF_3SH	$CF_3SCH_2CH_2CHO$	[58019-54-6]	$C_4F_3H_5OS$	46.5	2.7

^{a}To convert kPa to mm Hg, multiply by 7.5.

Reaction with Ammonia.   Although the liquid-phase reaction of acrolein with ammonia produces polymers of little interest, the vapor-phase reaction, in the presence of a dehydration catalyst, produces high yields of β-picoline [108-99-6] and pyridine [110-86-4] in a ratio of approximately 2/1.

β-Picoline may serve as an important source of nicotinic acid [59-67-6] for dietary supplements. A variety of substituted pyridines may be prepared from acrolein (75–83).

Diels-Alder Reactions.   Acrolein may participate in Diels-Alder reactions as the dieneophile or as the diene (84–89).

Acrolein as Dienophile. The participation of acrolein as the dienophile in Diels-Alder reactions is, in general, an exothermic process. Dienes such as cyclopentadiene and 1-diethylamino-1,3-butadiene react rapidly with acrolein at room temperature.

Several Diels-Alder reactions in which acrolein participates as the dienophile are of industrial significance. These reactions involve butadiene or substituted butadienes and yield the corresponding 1,2,5,6-tetrahydrobenzaldehyde derivative (THBA); examples are given in Table 9 (90). These products have found use in the epoxy and perfume/fragrance industries.

Many other acrolein derivatives produced via Diels-Alder reactions are classified as flavors and fragrances. Among those of commercial interest are lyral, $C_{13}H_{22}O_2$, (1) [31906-04-4] (91,92) and myrac aldehyde, $C_{13}H_{20}O$, (2) [80450-04-0] (92,93).

(1)   (2)

Table 9. Products of Dienes Added to Acrolein[a]

THBA Product[b]	CAS Registry Number	Time, h	bp, °C	Pressure at bp, kPa[c]
THBA	[100-50-5]	1	51–52	1.7
2,5-*endo*-methylene-THBA	[5453-80-5]	several[d]	70–72	2.7
4-methyl-THBA	[7560-64-7]	3	70–71	1.9
2,5-*endo*-ethylene-THBA	[40570-95-4]	8	84–85	1.6

[a]Reaction at 100°C unless otherwise noted; yields are 90–95%.
[b]THBA from butadiene; 2,5-*endo*-methylene from 1,3-cyclopentadiene; 4-methyl-THBA from 2-methylbutadiene; 2,5-*endo*-ethylene from 1,3-cyclohexadiene.
[c]To convert kPa to mm Hg, multiply by 7.5.
[d]At 25°C.

Acrolein as Diene. An industrially useful reaction in which acrolein participates as the diene is that with methyl vinyl ether. The product, methoxydihydropyran, is an intermediate in the synthesis of glutaraldehyde [111-30-8].

In addition to its principal use in biocide formulations (94), glutaraldehyde has been used in the film development and leather tanning industries (95). It may be converted to 1,5-pentanediol [111-29-5] or glutaric acid [110-94-1].

   Production of Acrolein Dimer.   Acting as both the diene and dienophile, acrolein undergoes a Diels-Alder reaction with itself to produce acrolein dimer, 3,4-dihydro-2-formyl-2*H*-pyran, $C_6H_8O_2$ [100-73-2]. At room temperature the rate of dimerization is very slow. However, at elevated temperatures and pressures the dimer may be produced in single-pass yields of 33% with selectivities greater than 95%.

Acrolein dimer may be easily hydrated to α-hydroxyadipaldehyde, $C_6H_{10}O_3$, [141-31-1] which may then be reduced to 1,2,6-hexanetriol [106-69-4]. 1,2,6-Hexanetriol, $C_6H_{14}O_3$, is used in a variety of pharmaceutical and cosmetics industry applications and is currently viewed as an alternative to glycerol as a humectant.

   Polymerization.   In the absence of inhibitors, acrolein polymerizes readily in the presence of anionic, cationic, or free-radical agents. The resulting polymer is an insoluble, highly cross-linked solid with no known commercial use.

   Copolymers, including one obtained by the oxidative copolymerization of acrolein with acrylic acid, a product of commercial interest, are known. There is a great variety of potential acrolein copolymers; however, significant commercial uses have not been developed. The possible application of polyacroleins or copolymers as polymeric reagents, polymeric complexing agents, and polymeric carriers has been recognized. Several articles give preparative methods as well as properties of the homopolymers and copolymers of acrolein (4).

Direct Uses of Acrolein

Because of its antimicrobial activity, acrolein has found use as an agent to control the growth of microbes in process feed lines, thereby controlling the rates of plugging and corrosion (see WASTES, INDUSTRIAL).

   Acrolein at a concentration of <500 ppm is also used to protect liquid fuels against microorganisms. The dialkyl acetals of acrolein are also useful in this application. In addition, the growth of algae, aquatic weeds, and mollusks in recirculating process water systems is also controlled by acrolein. Currently, acrolein is used to control the growth of algae in oil fields and has also been used as an H_2S scavenger (96). The ability to use acrolein safely in these direct applications is a prime concern and is a deterrent to more widespread use.

Health and Safety

Toxicity. The most frequently encountered hazards of acrolein are acute toxicity from inhalation and ocular irritation (97). Because of its high volatility, even a small spill can lead to a dangerous situation. Acrolein is highly irritating and a potent lacrimator. The odor threshold (50%), 0.23 mg/m^3 [0.09 ppm (v/v)], is close to the OSHA permissible exposure limit (PEL) and ACGIH TLV-TWA8, 0.25 mg/m^3 (0.1 ppm). The odor threshold for 100% recognition, 0.5 mg/m^3 (0.2 ppm), is above the TLV (98) but the OSHA and ACGIH short-term exposure limit (STEL) is 0.8 mg/m^3 (0.3 ppm), so perception of acrolein will generally provide adequate warning.

Concentrations of acrolein vapor as low as 0.6 mg/m^3 (0.25 ppm) may irritate the respiratory tract, causing coughing, nasal discharge, chest discomfort or pain, and difficulty with breathing (99). A concentration of 5–10 mg/m^3 (2–4 ppm) is intolerable to most individuals in a minute or two (97) and is close to the concentration considered immediately dangerous to life and health (100). At higher concentrations there may be lung injury from inhaled acrolein, and prolonged exposure may be fatal. In a short time, exposure to 25 mg/m^3 (10 ppm) or more is lethal to humans (101).

Acrolein vapor is highly irritating to the eyes, causing pain or discomfort in the eye, profuse lacrimation, involuntary blinking, and marked reddening of the conjunctiva. Splashes of liquid acrolein will produce a severe injury to the eyelids and conjunctiva and chemical burns of the cornea.

A small amount of acrolein may be fatal if swallowed. It produces burns of the mouth, throat, esophagus, and stomach. Signs and symptoms of poisoning may include severe pain in the mouth, throat, chest, and abdomen; nausea; vomiting, which may contain blood; diarrhea; weakness and dizziness; and collapse and coma (99).

Acrolein is highly toxic by skin absorption. Brief contact may result in the absorption of harmful and possibly fatal amounts of material. Skin contact causes severe local irritation and chemical burns. Poly(vinyl chloride) coated protective gloves should be used (99).

There is no specific antidote for acrolein exposure. Treatment of exposure should be directed at the control of symptoms and the clinical condition. Most of the harmful effects of acrolein result from its highly irritating and corrosive properties.

Chronic Exposure. Chronic human exposure is unlikely due to the lack of tolerance to acrolein. There is no evidence that acrolein is a human carcinogen (102), and inadequate animal data preclude any evaluation of its carcinogenicity (103). Acrolein has shown low, borderline, or moderate mutagenicity in bioassays, depending on the test system and frame of reference (104,105). Animal studies that gave little indication of teratogenicity of acrolein are insufficient for determining whether acrolein is a teratogen (106). Some embryotoxicity over a narrow dose range with acrolein administered by injection indicates that acrolein is quite embryotoxic (106).

Flammability. Acrolein is very flammable; its flash point is <0°C, but a toxic vapor cloud will develop before a flammable one. The flammable limits in air are 2.8% and 31.0% lower and upper explosive limits, respectively by volume.

Acrolein is only partly soluble in water and will cause a floating fire, so alcohol type foam should be used in firefighting. The vapors are heavier than air and can travel along the ground and flash back from an ignition source.

Reactivity. Acrolein is a highly reactive chemical, and contamination of all types must be avoided. Violent polymerization may occur by contamination with either alkaline materials or strong mineral acids. Contamination by low molecular weight amines and pyridines such as α-picoline is especially hazardous because there is an induction period that may conceal the onset of an incident and allow a contaminant to accumulate unnoticed. After the onset of polymerization the temperature can rise precipitously within minutes.

Acrolein reacts slowly in water to form 3-hydroxypropionaldehyde and then other condensation products from aldol and Michael reactions. Water dissolved in acrolein does not present a hazard. The reaction of acrolein with water is exothermic and the reaction proceeds slowly in dilute aqueous solution. This will be hazardous in a two-phase adiabatic system in which acrolein is supplied from the upper layer to replenish that consumed in the lower, aqueous, layer. The rate at which these reactions occur will depend on the nature of the impurities in the water, the volume of the water layer, and the rate of heat removal. Thus a water layer must be avoided in stored acrolein.

Dimerization of acrolein is very slow at ambient temperatures but it can become a runaway reaction at elevated temperature (ca 90°C), a consideration in developing protection against fire exposure of stored acrolein.

Storage and Handling

The following cautions should be observed: Do not destroy or remove inhibitor. Do not contaminate with alkaline or strongly acidic materials. Do not store in the presence of a water layer. In the event of spillage or misuse that cause a release of product vapor to the atmosphere, thoroughly ventilate the area, especially near floor levels where vapors will collect.

Acrolein produced in the United States is stabilized against free-radical polymerization by 1000–2500 ppm of hydroquinone and is protected somewhat against base-catalyzed polymerization by about 100 ppm of acetic acid. To ensure stability, the pH of a 10% v/v solution of acrolein in water should be below 6.

Since the principal hazard of contamination of acrolein is base-catalyzed polymerization, a "buffer" solution to shortstop such a polymerization is often employed for emergency addition to a reacting tank. A typical composition of this solution is 78% acetic acid, 15% water, and 7% hydroquinone. The acetic acid is the primary active ingredient. Water is added to depress the freezing point and to increase the solubility of hydroquinone. Hydroquinone (HQ) prevents free-radical polymerization. Such polymerization is not expected to be a safety hazard, but there is no reason to exclude HQ from the formulation. Sodium acetate may be included as well to stop polymerization by very strong acids. There is, however, a temperature rise when it is added to acrolein due to catalysis of the acetic acid–acrolein addition reaction.

Materials of Construction. Suitable materials of construction are steel, stainless steel, and aluminum 3003. Galvanized steel should not be used. Plastic tanks and lines are not recommended.

Storage tanks should have temperature monitoring with alarms to detect the onset of reactions. The design should comply with all applicable industry, federal, and local codes for a class 1B flammable liquid. The storage temperature should be below 37.8°C. Storage should be under an atmosphere of dry nitrogen and should vent vapors from the tank to a scrubber or flare.

Spill Disposal. In treatment of spills or wastes the suppression of vapors is the first concern and the aquatic toxicity to plants, fish, and microorganisms is the second. Normal procedures for flammable liquids should also be carried out.

Even small spills and leaks (<0.45 kg) require extreme caution. Unless the spill is contained in a fume hood, do not remain in or enter the area unless equipped with full protective equipment and clothing. Self-contained breathing apparatus should be used if the odor of acrolein or eye irritation is sensed. Small spills may be covered with absorbant, treated with aqueous alkalies, and flushed with water.

Acrolein is very highly toxic to fish and to the microorganisms in a biological wastewater treatment plant. Avoid drainage to sewers or to natural waters. Safe, practical methods have been devised to handle contained spills of liquid acrolein. These entail covering the acrolein with 15 cm of 3M ATC foam to suppress evaporation followed by either (a) removing some of the foam and igniting the vapors to destroy most of the acrolein under controlled burning conditions, or (b) polymerizing the acrolein by the addition of a dilute (5 to 10%) aqueous sodium carbonate solution. In situations where the foam covering and controlled burning are not feasible, the acrolein spill may be covered uniformly with dry sodium carbonate amounting to 60–120 kg/m^3 (0.5–1 lb/gal) of acrolein, followed by dilution with 5 to 10 volumes of water per volume of acrolein. This procedure effectively destroys the acrolein by polymerization but leaves water-insoluble residue. More water (ca 20 volumes) is needed to get a solution or fine suspension of polymer. Other alkalies, such as dilute aqueous sodium hydroxide, will serve the same purpose but the polymerization is more violent than when the sodium carbonate is used.

Government Regulations

The Comprehensive Environmental Response, Compensation, and Liability Act of 1980 (CERCLA) requires notification to the National Response Center of releases of quantities of hazardous substances equal to or greater than the reportable quantity (RQ) in 40 CFR 302.4, which is one pound (0.454 kg).

The Superfund Amendments and Reauthorization Act of 1986 (SARA) Title III requires emergency planning based on threshold planning quantities (TPQ) and release reporting based on RQs in 40 CFR part 355 (used for SARA 302, 303, and 304). The TPQ for acrolein is 500 lb (227 kg), and its RQ is 1 lb (0.454 kg). SARA also requires submission of annual reports of release of toxic chemicals that appear on the list in 40 CFR 372.65 (for SARA 313). Acrolein appears on that list. This information must be included in all MSDSs that are copied and distributed for acrolein.

Acrolein is a DOT Flammable Liquid having subsidiary DOT hazard classifications of Poison B and Corrosive Material. It is also an inhalation hazard that falls under the special packaging requirements of 49 CFR 173.3a.

BIBLIOGRAPHY

"Acrolein" in *ECT* 1st ed., Vol. 1, pp. 173–175, by R. L. Hasche, Tennessee Eastman Corporation; in *ECT* 1st ed; Suppl. 1, pp. 1–18, by H. R. Guest and H. A. Stansbury, Jr., Union Carbide Chemicals Company; "Acrolein and Derivatives" in *ECT* 2nd ed., Vol. 1, pp. 255–274, by H. R. Guest, B. W. Kiff, and H. A. Stansbury, Jr., Union Carbide Chemicals Company; in *ECT* 3rd ed., Vol. 1, pp. 277–297, by L. G. Hess, A. N. Kurtz, and D. B. Stanton, Union Carbide Corporation.

1. H. Schulz and H. Wagner, *Angew. Chem.* **62,** 105–118 (1950).
2. S. A. Ballard, H. de ViFinch, B. P. Geyer, G. W. Hearne, C. W. Smith, and R. R. Whetstone, *World Petroleum Congress Proceedings of the 4th Congress, Rome, 1955,* Sect. 4, Part C, pp. 141–154.
3. W. M. Weigert and H. Haschke, *Chem. Ztg.* **98,** 61–69 (1974).
4. R. C. Schulz in J. I. Kroschwitz, ed., *Encyclopedia of Polymer Science and Engineering,* 2nd ed., Vol. 1, Wiley-Interscience, New York, 1985, pp. 160–169.
5. T. Ohara, T. Sato, N. Shimizu, G. Prescher, H. Schwind, and O. Weiberg, *Ullman's Encyclopedia of Industrial Chemistry,* 5th ed., Vol. A1, 1985, pp. 149–160.
6. C. W. Smith, ed., *Acrolein,* John Wiley & Sons, Inc., New York, 1962.
7. U.S. Pat. 2,383,711 (Aug. 28, 1945), A. Clark and R. S. Shutt (to Battelle Memorial Institute).
8. U.S. Pat. 2,593,437 (Apr. 22, 1952), E. P. Goodings and D. J. Hadley (to Distillers Co., Ltd.).
9. U.S. Pat. 2,451,485 (Oct. 19, 1948), G. W. Hearne and M. L. Adams (to Shell Development Co.).
10. U.S. Pat. 2,486,842 (Nov. 1, 1949), G. W. Hearne and M. L. Adams (to Shell Development Co.).
11. U.S. Pat. 2,606,932 (Aug. 12, 1952), R. M. Cole, C. L. Cunn, and G. J. Pierotti (to Shell Development Co.).
12. U.S. Pat. 2,941,007 (June 14, 1960), J. L. Callahan, R. W. Foreman, and F. Veatch (to Standard Oil Co., Ohio).
13. U.S. Pat. 3,171,859 (Mar. 2, 1965), K. Sennewald, K. Gehramann, W. Vogt, and S. Schaefer (to Knapsack-Griesheim, A.G.).
14. U.S. Pat. 3,454,630 (July 8, 1969), G. Yamaguchi and S. Takenaka (to Nippon Kayaku Co., Ltd.).
15. U.S. Pat. 3,576,764 (Apr. 27, 1971), G. Yamaguchi and S. Takenaka (to Nippon Kayaku Co., Ltd.).
16. U.S. Pat. 3,522,299 (July 28, 1970), S. Takenaka and G. Yamaguchi (to Nippon Kayaku Co., Ltd.).
17. U.S. Pat. 3,959,384 (May 25, 1976), S. Takenaka, Y. Kido, T. Shimabara, and M. Ogawa (to Nippon Kayaku Co., Ltd.).
18. U.S. Pat. 3,825,600 (July 23, 1974), T. Ohara, M. Ueshima, and I. Yanagisawa (to Nippon Shokubai K.K.).
19. U.S. Pat. 3,907,712 (Sept. 23, 1975), T. Ohara, M. Ueshima, and I. Yanagisawa (to Nippon Shokubai K.K.).
20. U.S. Pat. 4,147,885 (Apr. 3, 1979), N. Shimizu, I. Yanagisawa, M. Takata, and T. Sato (to Nippon Shokubai K.K.).
21. U.S. Pat. 4,031,135 (June 21, 1977), H. Engelbach and co-workers (to BASF Aktiegesellschaft).
22. Eur. Pat. Appl. 253,409 (July 17, 1987), W. Etzkorn and G. Harkreader (to Union Carbide Corp.).
23. Eur. Pat. Appl. 257,565 (Aug. 20, 1987), W. Etzkorn and G. Harkreader (to Union Carbide Corp.).

24. Eur. Pat. Appl. 293,224 (May 27, 1988), M. Takata, M. Takamura, S. Uchida, and M. Sasaki (to Nippon Shokubai K.K.).
25. G. E. Schaal, *Hydrocarbon Process.* **52**, 218 (1973).
26. W. Weigert, *Chem. Eng.* **80**, 68 (1973).
27. *Chem. Eng. News* 19 (Feb. 15, 1988).
28. *Chem. Mark. Rep.* 54 (Nov. 12, 1990).
29. *Chem. Mark. Rep.* 22 (Mar. 12, 1990).
30. *Chem. Mark. Rep.* 20 (June 25, 1990).
31. Jpn. Kokai Tokkyo Koho JP 63/146841 AZ [88/146841] (June 18, 1988), K. Sarumaru and T. Shibano (to Mitsubishi Petrochemical Co. Ltd.).
32. J. B. Black, J. D. Scott, E. M. Serwicka, and J. B. Goodenough, *J. Catal.* **106**, 16–22 (1987).
33. E. M. Serwicka, J. B. Black, and J. B. Goodenough, *J. Catal.* **106**, 23–37 (1987).
34. V. Kijenski, M. Glinski, and J. Reinhercs, *Stud. Surf. Sci. Catal.* **41**, 231–240 (1988).
35. Y. Nagase and K. Wada, *Ibaraki Daigaku Kogakubu Kenkyu Shuho* **33**, 223–228 (1985).
36. Y. Nagase, H. Hattori, and K. Tanabe, *Chem. Lett.* **10**, 1615–1618 (1983).
37. T. H. Vanderspurt, *Ann. N.Y. Acad. Sci.* **333**, 155–164 (1980).
38. U.S. Pat. 4,127,508 (Nov. 28, 1978), T. H. Vanderspurt (to Celanese Corp.).
39. Ger. Offen. DE2734811 (Feb. 9, 1978), T. H. Vanderspurt (to Celanese Corp.).
40. T. Nakano, S. Umano, Y. Kino, and Y. M. Ishii, *J. Org. Chem.* **53**, 3752–3757 (1988).
41. M. A. Aramendia, V. Borau, and co-workers, *React. Kinet. Catl. Lett.* **36**, 251–256 (1988).
42. L. M. Ryzhenko and A. D. Shebaldova, *Khim. Tekhnol. Elementoorg. Soedin. Polim.* 14–19 (1984).
43. D. L. Reger, M. M. Habib, and D. V. Fauth, *Tetrahedron Lett.* 115–116 (1979).
44. M. Terassawa, K. Kaneda, T. Imanaka, and S. Tera, *J. Catal.* **51**, 406–421 (1978).
45. J. A. Cabello, J. M. Campello, A. Garcia, D. Luna, and co-workers, *Bull. Soc. Chim. Belg.* **93**, 857–862 (1984).
46. K.-J. Yang and C. S. Chein, *Inorg. Chem.* **26**, 2732–2733 (1987).
47. Z. Poltarzewski, S. Galvagno, R. Pietropaolo, and P. Staiti, *J. Catal.* **102**, 190–198 (1986).
48. M. Funakoshi, H. Komiyama, and H. Inoue, *Chem. Lett.* 245–248 (1985).
49. A. D. Shabaldova, V. N. Kravtsova, N. M. Sorokina, and co-workers, in *Nukleofil'nye Reacts. Karbonil'nykh Soedin.* (conference proceedings, Saratov, USSR) 87–89 (1982).
50. G. P. Pez and R. A. Grey, *Fund. Res. Homogenous Catal.* **4**, 97–116 (1984).
51. J. M. Campello, A. Garcia, D. Luna, and J. M. Marinas, *React. Kinet. Catal. Lett.* **21**, 209–212 (1982).
52. Jpn. Pat. Jp 57/91743 AZ [82/917343] (June 8, 1982) (to Agency of Ind. Sci. & Tech.).
53. G. V. Kudryavtsev, A. Yu Stakheev, and G. V. Lisichkin, *Zh. Vses. Khim. Ova.* **27**, 232–233 (1982).
54. J. M. Campello, A. Garcia, D. Luna, and J. M. Marinas, *Bull. Soc. Chim. Belg.* **91**, 131–142 (1982).
55. R. A. Grey, G. P. Pez, and A. Wallo, *J. Am. Chem. Soc.* **103**, 7536–7542 (1981).
56. K. Murata, A. Matsuda, *Bull. Chem. Soc. Jap.* **54**, 1989–1900 (1981).
57. K. Kaneda, M. Terasawa, T. Imanaka, and co-workers, *Fund. Res. Homogenous Catal.* **3**, 671–690 (1979).
58. Y. Nagase, *Ibaraki Daigaku Kogakubu Kenkyu Shuho* **33**, 217–221 (1985).
59. Eur. Pat. Appl. EP 183225 Al (June 4, 1986), Y. Shimasaki, Y. Hino, and M. Ueshima (to Nippon Shokubai Kagaku Kogyo Co., Ltd.).
60. A. Alba, M. A. Aramendia, V. Borau, C. Jimenez, and co-workers, *React. Kinet. Catal. Lett.* **25**, 45–50 (1984).
61. V. P. Kukolev, N. A. Balyushima, and G. H. Chukhadzhyan, *Arm. Khim. Zh.* **35**, 688–690 (1982).

62. G. Horanyi and K. Torkos, *J. Electoanal. Chem. Interfacial Electrochem.* **136,** 301 (1982).

63. U.S. Pat. 4,292,452A (Sept. 29, 1981), R. J. Lee, D. H. Meyer, and D. M. Senneke (to Standard Oil Co.).

64. R. W. Hoffman and T. Herold, *Chem. Ber.* **114,** 375–383 (1981).

65. Y. Nagse and T. Washiyama, *Ibaraki Daigaku Kogakubu Kenkyu Shuho* **27,** 171–178 (1979).

66. G. V. Kryshtal, D. Dvorak, Z. Arnold, and L. A. Yanovskaya, *Isv. Akad. Nauk. SSSR, Ser. Khim.* **4,** 921–923 (1986).

67. Ref. 6, p. 140 and references therein.

68. Ref. 6, p. 122 and references therein.

69. Ref. 6, pp. 124–125 and references therein.

70. Ref. 6, p. 130 and references therein.

71. U.S. Pat. 4,353,924 (Oct. 12, 1982), J. W. Beher, D. L. Mansfield, and D. J. Weinkauff (to Monsanto Company).

72. Rom. Pat. RO 85095B (Oct. 30, 1984), A. M. Pavlouschi, L. Levinta, and G. H. Gross (to Combinatul Petrochimic, Pitesti).

73. Jpn. Kokai Tokkyo Koho JP 55/16135 [80/16135] (Apr. 30, 1980), (to Anahi Chemical Industry Co., Ltd.).

74. Ref. 6, p. 118 and references therein.

75. J. Viala and M. Santelli, *Synthesis* 395–397 (1988).

76. Eur. Pat. Appl. EP 299362 A1 (Jan. 18, 1989), K. Nagao (to Osaka Soda Co., Ltd.).

77. Ger. Offen. DE 3634259 A1 (Apr. 21, 1988), W. Hoelderich, N. Goetz, and G. Fouquet (to BASF A.G.).

78. Ger. Offen. DE 3634975 Al (Apr. 30, 1987), R. J. Doehner, Jr. (to American Cyanamid Co.).

79. Ger. Offen. DE 3337569 Al (Apr. 25, 1985), T. Dockner, H. Hagen, and H. Krug (to BASF A.G.).

80. J. I. Grayson and R. Dinkel, *Helv. Chim. Acta* **67,** 2100–2110 (1984).

81. C. Wang and Y. Li, *Yiyao Gongye* **6,** 1–6 (1984).

82. Eur. Pat. Appl. EP75727 A2 (Apr. 6, 1983), J. I. Grayson and R. Dinkel (to Lonza A.G.).

83. A. T. Soldatenkov and co-workers, *Zh. Org. Khim.* **16,** 188–194 (1980).

84. R. Baker and M. J. Crimmin, *J. Chem. Soc. Perkin Trans.* **1,** 1264–1267 (1979).

85. Eur. Pat. Appl. EP43507 A2 (Jan. 13, 1982), K. Bruns and T. N. Dang (to Henkel K. - G.A.A.).

86. G. A. Trofimov, V. I. Lavrov, and L. N. Parshina, *Zh. Org. Khim.* **17,** 1716–1720 (1981).

87. Jpn. Kokai Tokkyo Koho JP 61/161241 AZ [86/161241] (July 21, 1986), K. Inoue, H. Takeda, and M. Kobayashi (to Mitsubishi Rayon Co., Ltd.).

88. Jpn. Kokai Tokkyo Koho JP 62/141097 AZ [87/141097] (June 24, 1987), N. Tanaka, H. Takada, M. Oku, and A. Kimura (to Koa Corp.).

89. K. G. Akopyan, A. P. Sayadyan, A. G. Dzhomardyan, and co-workers, *Prom-st. Stroit. Arkhit. Arm.* 34–35 (1988).

90. Ref. 6, pp. 216–219 and references therein.

91. U.S. Pat. 4,007,137 (Feb. 8, 1977), J. M. Sanders, W. L. Schreiber, and J. B. Hall (to International Flavors & Fragrances, Inc.).

92. U.S. Pat. 4,107,217 (Aug. 15, 1978), W. L. Schreiber and A. O. Pittet (to International Flavors and Fragrances, Inc.).

93. Ger. Offen., DE2,643,062 (Apr. 14, 1977), J. M. Sanders and co-workers (to International Flavors and Fragrances, Inc.).

94. U.S. Pat. 4,244,876 (Jan. 13, 1981), G. H. Warner, L. F. Theiling, and M. G. Freid (to Union Carbide Corp.).

95. U.S. Pat. 2,941,859 (June 21, 1960), M. L. Fein and E. M. Filachione (to Union Carbide Corp.).

96. Brit. Pat. Appl. GB 2023123 (Dec. 28, 1979), C. L. Kissel and F. F. Caserio (to Magna Corp.).
97. R. O. Beauchamp, D. A. Andjelkovich, A. D. Kligerman, K. T. Morgan, and H. d'A. Heck, *CRC Crit. Rev. Toxicol.* **14,** 309–380 (1985).
98. B. L. Carson, C. M. Beall, H. V. Ellis, L. H. Baker, and B. L. Herndon, *Acrolein Health Effects, NTIS PB82-161282; EPA-460/3-81-034, Gov. Rep. Announce. Index* **12,** 9–12 (1981). [A 121-page review of health effects literature primarily related to inhalation exposure.]
99. *Acrolein, Material Safety Data Sheet,* Union Carbide Chemicals and Plastics Company Inc., Specialty Chemicals Division, August 15, 1989.
100. Ref. 97, p. 339.
101. Syracuse Research Corporation, *Information Profiles on Potential Occupational Hazards,* Vol. 1, Single Chemicals Acrolein, NTIS PB81-147951, U.S. Department of Commerce, Springfield, Va., 1979, p. 11.
102. Ref. 97, p. 342.
103. Acrolein, in *IARC Monographs on the Evaluation of Carcinogenic Risk of Chemicals to Humans. Some Monomers, Plastics and Synthetic Elastomers and Acrolein,* Vol. 19, International Agency for Research on Cancer, Lyon, France, 1979, pp. 479–494.
104. Ref. 101, p. 9.
105. Ref. 98, p. 2.
106. Ref. 97, pp. 334–345, 338–339.

W. G. Etzkorn
J. J. Kurland
W. D. Neilsen
Union Carbide Chemicals & Plastics Company Inc.

ACRYLAMIDE

Acrylamide [*79-06-1*] (NIOSH No: A533250) has been commercially available since the mid-1950s and has shown steady growth since that time, but is still considered a small volume commodity. Its formula, $H_2C{=}CHCONH_2$ (2-propeneamide), indicates a simple chemical, but it is by far the most important member of the series of acrylic and methacrylic amides. Water soluble polyacrylamides (1) represent the most important applications, including potentially large uses in enhanced oil recovery as mobility control agents in water flooding, additives for oil well drilling fluids, and aids in fracturing, acidifying, and other operations. Other uses include flocculants for waste water treatment, the mining industry and various other process industries, soil stabilization, papermaking aids, and thickeners. Smaller but none the less important uses include dye acceptors; polymers for promoting adhesion; additives for textiles, paints, and cement; increasing the softening point and solvent resistance of resins; components of photopolymerizable systems; and cross-linking agents in vinyl polymers.

Physical Properties

Acrylamide is a white crystalline solid that is quite stable at ambient conditions, and, even at temperatures as high as its melting point (for 1 day in the absence of light), no significant polymer formation is observed. Above its melting point, however, liquid acrylamide may polymerize rapidly with significant heat evolution. Precautions should be taken when handling even small quantities of molten material. In addition to the solid form, a 50% aqueous solution of acrylamide is a popular commercial product today. This solution is stabilized by small amounts of cupric ion (25–30 ppm based on monomer) and soluble oxygen. Several other stabilizers are also available for the aqueous monomer solution, such as ethylenediaminetetraacetic acid (EDTA) [60-00-4] (2), ferric ion (3), and nitrite (4,5). The only effect of oxygen is to increase the induction period for polymerization (6). Iron complexes of cyanogen or thiocyanogen have proven to be useful stabilizers for salt-containing acrylamide solutions (7). The physical properties of solid acrylamide monomer are summarized in Table 1. Solubilities of acrylamide in various solvents are given in Table 2, and typical physical properties of a 50% solution in water appear in Table 3.

Table 1. Physical Properties of Solid Acrylamide Monomer[a]

Property	Value
molecular weight	71.08
melting point, °C	84.5 ± 0.3
vapor pressure, Pa[b]	
25°C	0.9
40°C	4.4
50°C	9.3
boiling point, °C	
0.27 kPa[b]	87
0.67 kPa[b]	103
1.4 kPa[b]	116.5
3.3 kPa[b]	136
heat of polymerization, kJ/mol[c]	−82.8
density, g/mL at 30°C	1.122
equilibrium moisture content, particle size 355 μm[d], at 22.8°C, 50% rh	1.7 g of water/kg of dry acrylamide
crystal system	monoclinic or triclinic
crystal habit	thin tabular to laminar
refractive indexes	
n_x	1.460 (calcd)
n_y	1.550 ± 0.003
n_z	1.581 ± 0.003
optic axial angles	2E 98°, 2V 58°
optic sign	(−)

[a]Ref. 5.
[b]To convert kPa to mm Hg, multiply by 7.5.
[c]To convert kJ/mol to kcal/mol, divide by 4.184.
[d]45 mesh.

Table 2. Solubilities of Acrylamide in Various Solvents at 30°C

Solvent	g/100 mL
acetonitrile	39.6
acetone	63.1
benzene	0.346
ethylene glycol monobutyl ether	31
chloroform	2.66
1,2-dichloroethane	1.50
dimethylformamide	119
dimethyl sulfoxide	124
dioxane	30
ethanol	86.2
ethyl acetate	12.6
n-heptane	0.0068
methanol	155
pyridine	61.9
water	215.5
carbon tetrachloride	0.038

Table 3. Physical Properties of 50% Aqueous Acrylamide Solution[a]

Property	Value
pH	5.0–6.5
refractive index range, 25°C (48–52%)	1.4085–1.4148
viscosity, mPa (= cP) at 25°C	2.71
specific gravity, at 25°C	1.0412
density, 25/4°C	1.038
crystallization point, °C	8–13
partial phase diagram	
eutectic temperature, °C	−8.9
eutectic composition, wt %	31.2
boiling point at 101.3 kPa[b], °C	99–104
vapor pressure	
at 23°C, kPa[b]	2.407
at 70°C, kPa[b]	27.93
specific heat (20–50°C), J/(g·K)[c]	3.47
heat of dilution to 20 wt %, J/g soln[c]	−4.6
heat of polymerization, kJ/mol[c]	−85.4
heat of melting (solution), melting range −17.3 to +19.7°C, J/g[c]	247.7
flammability	nonflammable

[a]Ref. 5.
[b]To convert kPa to mm Hg, multiply by 7.5.
[c]To convert J to cal, divide by 4.184.

Chemical Properties

Acrylamide, C_3H_5NO, is an interesting difunctional monomer containing a reactive electron-deficient double bond and an amide group, and it undergoes reactions typical of those two functionalities. It exhibits both weak acidic and basic properties. The electron withdrawing carboxamide group activates the double bond, which consequently reacts readily with nucleophilic reagents, eg, by addition.

$$\text{Nuc:H} + \text{CH}_2=\text{CHCONH}_2 \longrightarrow \text{NucCH}_2\text{CH}_2\text{CONH}_2$$

Many of these reactions are reversible, and for the stronger nucleophiles they usually proceed the fastest. Typical examples are the addition of ammonia, amines, phosphines, and bisulfite. Alkaline conditions permit the addition of mercaptans, sulfides, ketones, nitroalkanes, and alcohols to acrylamide. Good examples of alcohol reactions are those involving polymeric alcohols such as poly(vinyl alcohol), cellulose, and starch. The alkaline conditions employed with these reactions result in partial hydrolysis of the amide, yielding mixed carbamoylethyl and carboxyethyl products.

Some specific examples include the noncatalytic reaction of acrylamide with primary amines to produce a mono or bis product (5).

$$\text{RNH}_2 + \text{CH}_2=\text{CHCONH}_2 \longrightarrow \text{RNHCH}_2\text{CH}_2\text{CONH}_2 \longrightarrow \text{RN(CH}_2\text{CH}_2\text{CONH}_2)_2$$

Secondary amines give only a monosubstituted product. Both of these reactions are thermally reversible. The product with ammonia (3,3′,3″-nitrilotrispropionamide [2664-61-1], $C_9H_{18}N_4O_3$) (5) is frequently found in crystalline acrylamide as a minor impurity and affects the free-radical polymerization. An extensive study (8) has determined the structural requirements of the amines to form thermally reversible products. Unsymmetrical dialkyl hydrazines add through the unsubstituted nitrogen in basic medium and through the substituted nitrogen in acidic medium (9). Monoalkylhydroxylamine hydrochlorides react with preservation of the hydroxylamine structure (10). Primary nitramines combine in such a way as to keep the nitramine structure intact.

The reaction with sodium sulfite or bisulfite (5,11) to yield sodium-β-sulfopropionamide [19298-89-6] ($C_3H_7NO_4S·Na$) is very useful since it can be used as a scavenger for acrylamide monomer. The reaction proceeds very rapidly even at room temperature, and the product has low toxicity. Reactions with phosphines and phosphine oxides have been studied (12), and the products are potentially useful because of their fire retardant properties. Reactions with sulfide and dithiocarbamates proceed readily but have no applications (5). However, the reaction with mercaptide ions has been used for analytical purposes (13). Water reacts with the amide group (5) to form hydrolysis products, and other hydroxy compounds, such as alcohols and phenols, react readily to form ether compounds. Primary aliphatic alcohols are the most reactive and the reactions are complicated by partial hydrolysis of the amide groups by any water present.

Activated ketones react with acrylamide to yield adducts that frequently cyclize to lactams (14). The lactams can be hydrolyzed to yield substituted propionic acids. Chlorine and bromine react with acrylamide in aqueous solution to yield α,β-dihalopropionamide (5). Under acidic conditions, sizable quantities of acrylamide can be removed from water by chlorination (15). Hydrochloric and hydrobromic acids add to give β-halopropionamides. These adducts are also thermally reversible. A patent describes a procedure to prepare N-substituted acrylamide by direct transamidation of acrylamide (16). Dienes react with acrylamide to form Diels-Alder type adducts (17,18). Improved yields in the aza-annelation of cyclic ketones by the use of enamines and imines have been reported (19–21). Palladium reduced with borohydride (22), nickel boride (23), or rhodium carbonyl (24) reduces the double bond of acrylamide to yield propionamide, and acrylamide can be oxidized to a glycol with sodium hypochlorite using osmium tetroxide as a catalyst (25). In contrast, if osmium is not present, the attack occurs at the nitrogen to yield N-vinyl-N'-acryloylurea [19396-55-5] $(C_6H_8N_2O_2)$ (26). When treated with a strong base in an aprotic solvent, acrylamide forms a head-to-tail dimer, 3-acrylamidopropionamide [21963-06-4] $(C_6H_{10}N_2O_2)$ (27). Electrolytic reductive dimerization of acrylamide proceeds through tail-to-tail addition to yield adipamide [628-94-4], $C_6H_{12}N_2O_2$ (28).

The most important reactions of acrylamide are those that produce vinyl addition polymers (see ACRYLAMIDE POLYMERS). The initiation and termination mechanisms depend on the catalyst system, but the reaction can be started by any free-radical source. In practice, redox couples such as sodium persulfate and sodium bisulfite are commonly used, and the highest molecular weight polymers are obtained in aqueous solution, with molecular weights of several million prepared routinely. Acrylamide is remarkable for the very large value of $k_p/k_t^{1/2}$, $1.8 \times 10^4/(1.45 \times 10^7)^{1/2}$, which is a measure of chain length in the polymerization. However, it may be necessary to remove the inhibitor (cupric ions) from aqueous acrylamide solutions to obtain the desired polymerization results. Copolymers with acrylamide are also prepared with ease, although the molecular weights are consistently lower than that of polyacrylamide prepared under similar conditions. Acrylamide copolymerizes readily by a free-radical mechanism with other acrylates, methacrylates, and styrene. Acrylamide may be polymerized by a hydrogen transfer mechanism catalyzed by strong base in basic or aprotic solvents. The product is poly(β-alanine) or nylon-3 (29), which has properties similar to natural silk. This polymer, on hydrolysis, yields β-aminopropionic acid. A hydrogen transfer copolymer with acrolein has also been reported (30). A biocatalytic method of removing residual monomer from polymers that could become very important to acrylamide polymer users has been described (31).

The amide group is readily hydrolyzed to acrylic acid, and this reaction is kinetically faster in base than in acid solutions (5,32,33). However, hydrolysis of N-alkyl derivatives proceeds at slower rates. The presence of an electron-withdrawing group on nitrogen not only facilitates hydrolysis but also affects the polymerization behavior of these derivatives (34,35). With concentrated sulfuric acid, acrylamide forms acrylamide sulfate salt, the intermediate of the former sulfuric acid process for producing acrylamide commercially. Further reaction of the salt with alcohols produces acrylate esters (5). In strongly alkaline anhydrous solutions a potassium salt can be formed by reaction with potassium $tert$-butoxide in $tert$-butyl alcohol at room temperature (36).

Several other interesting reactions include acrylamide transition metal complexes (37–40), complexes with nucleosides (41) in dimethyl sulfoxide solution, and also complexes with several inorganic salts (42–44). Dehydration of acrylamide by treatment with fused manganese dioxide (45) at 500°C or with phosphorus pentoxide (46) yields acrylonitrile. Aldehydes such as formaldehyde, glyoxal, and chloral hydrate react with acrylamide under neutral and alkaline conditions, producing the corresponding N-methylolacrylamide [924-42-5] (47,48). Under acidic conditions, N,N-methylenebisacrylamide [110-26-9] ((H$_2$C=CHCONH)$_2$CH$_2$) is produced from formaldehyde and acrylamide (49). Under acidic conditions, methoylol ethers are formed from hydroxyl compounds and N-methylolacrylamide (50). Condensation products derived from N-methylolacrylamide and polyphenols have also been reported (51). Using p-toluenesulfonic acid as the catalyst in dioxane or ethyl acetate solvent, N,N'-oxydimethylenebisacrylamide [16958-71-7] (C$_8$H$_{12}$O$_3$N$_2$) has been obtained (52). These difunctional products have similar copolymerization parameters to acrylamide and are useful as cross-linking agents. Alcohols can be used to cap the methylol compound to provide the less reactive methylol ethers. Methanol is commonly employed, but, where increased compatibility with oleophilic systems is desired, one of the butanols is the preferred alcohol; oxalic acid is an example of a suitable catalyst (50). This reaction also occurs with cellulosic hydroxyls. Provided the system is not basic, the methylol derivative may also be condensed with carbamate esters (53), secondary amines (54), or phosphines (12) without involving the double bond. Acrylamido-N-glycolic acid and diacrylamidoacetic acid can be obtained from acrylamide and glyoxylic acid (55–58). N-acylacrylamides are of minor interest industrially. One member of the series, diacrylamide [20602-80-6] (C$_6$H$_7$NO$_2$), is a suspected side-reaction product in the sulfuric acid process of manufacture. It may be prepared by the reaction of acrylamide with acrylic anhydride or acryloyl chloride. A specific preparation for N-acetylacrylamide [1432-45-7] (C$_5$H$_7$NO$_2$) is the addition of ketene to acrylamide (59).

Analysis

The analysis of acrylamide monomer in water solutions containing at least 0.5% monomer is carried out by bromination (5). If the concentration is fairly high, in the 2–55% monomer range, then a refractive index method is easier (11,60,61). Polarography (62) and gas chromatography (63) can also be used to determine trace amounts of acrylamide monomer in other organic materials. For detecting small concentrations of polymer in aqueous acrylamide solutions, n-butanol addition will produce turbidity, which can then be compared to standards (5,61). Cupric ion inhibitor and other impurities in acrylamide samples can be determined by standard techniques (11,61,64). Other methods can also be employed to analyze the polyacrylamide content of monomer solutions, including turbidimetric (Hach) and colorimetric (Klett) methods (11,61). A summary of various analytical techniques for assaying acrylamide monomer are listed in Table 4.

Manufacture

The current routes to acrylamide are based on the hydration of inexpensive and readily available acrylonitrile [107-13-1] (C$_3$H$_3$N, 2-propenenitrile, vinyl cyanide, VCN, or cyanoethene) (see ACRYLONITRILE). For many years the principal process for making acrylamide was a reaction of acrylonitrile with H$_2$SO$_4$·H$_2$O followed by

Table 4. Acrylamide Assay Techniques[a]

Method	Approximate sensitivity, ppm[b]	Application	Interference	Ref.
refractive index	50,000	quality control	anything affecting refractive index	11, 61
bromate–bromide	1,000	assay product	unsaturated compounds	5
flame ionization	40	monomer in polymer		66
d-c polarization	10	assay product	alkali cations, acrylic esters	64
differential pulse polarography	>1	environment concerns	alkali cations, acrylic esters, vinyl cyanide	62
spectrophotometry	0.1	urinalysis	aldehydes, ketones, pyrroles, indoles, hydrazine, aromatic amines	64, 65
hplc	0.1	wipe and air		67
electron capture, gc	0.1 ppb	river water		63

[a]Ref. 65.
[b]Unless otherwise noted.

separation of the product from its sulfate salt using a base neutralization or an ion exclusion column (68).

$$CH_2{=}CHCN + H_2SO_4{\cdot}H_2O \longrightarrow CH_2{=}CHCONH_2{\cdot}H_2SO_4$$

This process yields satisfactory monomer, either as crystals or in solution, but it also produces unwanted sulfates and waste streams. The reaction was usually run in glass-lined equipment at 90–100°C with a residence time of 1 h. Long residence time and high reaction temperatures increase the selectivity to impurities, especially polymers and acrylic acid, which controls the properties of subsequent polymer products.

The ratio of reactants had to be controlled very closely to suppress these impurities. Recovery of the acrylamide product from the acid process was the most expensive and difficult part of the process. Large scale production depended on two different methods. If solid crystalline monomer was desired, the acrylamide sulfate was neutralized with ammonia to yield ammonium sulfate. The acrylamide crystallized on cooling, leaving ammonium sulfate, which had to be disposed of in some way. The second method of purification involved ion exclusion (68), which utilized a sulfonic acid ion-exchange resin and produced a dilute solution of acrylamide in water. A dilute sulfuric acid waste stream was again produced, and, in either case, the waste stream represented a problem as well as an increased production cost. As far as can be determined, no commercial acrylamide is produced today via this process.

Even in 1960 a catalytic route was considered the answer to the pollution problem and the by-product sulfate, but nearly ten years elapsed before a process was developed that could be used commercially. Some of the earlier attempts included hydrolysis of acrylonitrile on a sulfonic acid ion-exchange resin (69). Manganese dioxide showed some catalytic activity (70), and copper ions present in two different valence states were described as catalytically active (71), but copper metal by itself was not active. A variety of catalysts, such as Urushibara or Ullmann copper and nickel, were used for the hydrolysis of aromatic nitriles, but aliphatic nitriles did not react using these catalysts (72). Beginning in 1971 a series of patents were issued to The Dow Chemical Company (73) describing the use of copper metal catalysis. Full-scale production was achieved the same year. A solution of acrylonitrile in water was passed over a fixed bed of copper catalyst at 85°C, which produced a solution of acrylamide in water with very high conversions and selectivities to acrylamide. The heat of hydration is approximately -70 kJ/mol (-17 kcal/mol). This process usually produces no waste streams, but if the acrylonitrile feed contains other nitrile impurities, they will be converted to the corresponding amides. Another reaction that is prone to take place is the hydrolysis of acrylamide to acrylic acid and ammonia. However, this impurity can usually be kept at very low concentrations. American Cyanamid uses a similar process in both the United States and Europe, which provides for their own needs and for sales to the merchant market.

Mitsui Toatsu Chemical, Inc. disclosed a similar process using Raney copper (74) shortly after the discovery at Dow, and BASF came out with a variation of the copper catalyst in 1974 (75). Since 1971 several hundred patents have shown modifications and improvements to this technology, both homogeneous and heterogeneous, and reviews of these processes have been published (76). Nalco Chemical Company has patented a process based essentially on Raney copper catalyst (77) in both slurry and fixed-bed reactors and produces acrylamide monomer mainly for internal uses. Other producers in Europe, besides Dow and American Cyanamid, include Allied Colloids and Stockhausen, who are believed to use processes similar to the Raney copper technology of Mitsui Toatsu, and all have captive uses. Acrylamide is also produced in large quantities in Japan. Mitsui Toatsu and Mitsubishi are the largest producers, and both are believed to use Raney copper catalysts in a fixed bed reactor and to sell into the merchant market.

In 1985 Nitto Chemical Industry started using microorganisms for making acrylamide from acrylonitrile using an enzymatic hydration process (77,78). The reaction is catalyzed by nitrile hydralase, a nitrilasically active enzyme produced by organisms such as Corynebacterium N-774 strain, *Bacillus, Bacteridium, Micrococcus, Nocardia,* and *Pseudomonas.* This is one of the initial uses of biocatalysis in the manufacture of commodity chemicals in the petrochemical industry. There are certainly other bioprocesses in use for fine chemicals in the amino acid area, as well as fermentation processes. Improved bacterial strains and cells immobilized in acrylamide gels as well as methods of concentrating the dilute product solutions are subjects of more recent patents (79–81). The most recent release indicates a switch to *Rhodococcus rhodochrous* bacteria, which will increase their capacity from 6,000 to 20,000t/yr (82). The reaction is run at 0–15°C and a pH of 7–9 and gives almost complete conversions with very small amounts of by-products such as acrylic acid.

Acrylamide and its derivatives have been prepared by many other routes (4). The reactions of acryloyl chloride and acrylic anhydride with ammonia are classical methods. Primary and secondary amines may be used in place of ammonia to obtain N-substituted derivatives. Acryloyl isocyanate has been hydrolyzed to acrylamide but yields are poor. Exhaustive amination of methyl acrylate with ammonia yields 3,3′,3″-nitrilotrispropionamide [2664-61-1] ($C_9H_{18}N_4O_3$). This compound can be thermally decomposed to acrylamide by heating to 208–230°C at 2 kPa (15 mm Hg). Similarly, Michael-type addition products of alkylamines or aliphatic alcohols and methyl acrylate react with ammonia to give the corresponding β-substituted propionamides. These compounds may also be thermally decomposed to yield acrylamide. The Michael-type addition products of methyl acrylate and aliphatic amines may react further to give N-alkyl or N,N-dialkyl propionamide derivatives that can be thermally decomposed to mono- or dialkylsubstituted acrylamides, respectively. N-Substituted acrylamides may also be prepared from acetylene, carbon monoxide, and an amine using an iron or nickel carbonyl catalyst. However, the best route to mono-N-alkylsubstituted acrylamides is the Ritter reaction. This reaction is used to prepare diacetoneacrylamide [2873-97-4] ($C_9H_{15}NO_2$), 2-acrylamido-2-methylpropanesulfonic acid [15214-89-8] ($C_7H_{13}NO_4S$), N-isopropylacrylamide [2210-25-5] ($C_6H_{11}NO$), N-tert-butylacrylamide [107-58-4] ($C_7H_{13}NO$), and other N-alkyl acrylamides in which the carbon attached to the nitrogen is usually tertiary (83,84).

Specifications

The 50% aqueous acrylamide is the preferred form because it eliminates the handling of solids and because its cost is lower. This is a result of the new manufacturing method put into effect in 1971. The aqueous form is applicable to nearly all the end uses of acrylamide when volume is taken into account. Aqueous acrylamide is shipped in tank trucks, rail cars, or drums, but small samples can also be obtained. The solution should be kept in stainless steel or in tanks coated with plastic resin (phenolic, epoxy, or polypropylene). All containers, including tank trucks and rail cars, must be rinsed prior to disposal or return. When shipping costs are an important consideration, solid acrylamide may be the desired form. Acrylamide should be stored in a well ventilated area away from sunlight. The temperature should be under 30°C, and under these conditions no change of quality should be noticed for at least 3 months. Typical specifications for the 50% aqueous solution are shown in Table 5 and for the solid monomer in Table 6.

Health and Safety Considerations

Contact with acrylamide can be hazardous and should be avoided. The most serious toxicological effect of exposure to acrylamide monomer is as a neurotoxin. In contrast, polymers of acrylamide exhibit very low toxicity. Since the solid form sublimes, the solid or powder form of acrylamide is more likely to be a problem than the aqueous form because of possible exposure to dusts and vapors. An

Table 5. Typical Specifications for 50% Aqueous Solutions[a]

Property	Limit
assay, wt %	48–52
pH	5.0–6.5
polymer, ppm, max (BOM)[b]	100
Cu^{2+} inhibitor, ppm, max (BOM)[b]	25
color	water clear

[a]Refs. 11, 61.
[b]Based on monomer.

Table 6. Typical Specifications for Crystalline Acrylamide Monomer

appearance	white, free flowing crystal
assay, %, min	98
water, %, max	0.8
iron, as Fe^{0}, ppm, max	15
color, 20% soln, max, APHA	50
water insoluble, %, max	0.2
butanol insoluble, %, max	1.5

important characteristic of the toxicity of acrylamide monomer is that the signs and symptoms of exposure to toxic levels may be slow in developing and can occur after ingestion of small amounts over a period of several days or weeks. It is therefore important that people who have been exposed to acrylamide be monitored by a qualified physician. Signs and symptoms include increased sweating of hands and feet, numbness or tingling of the extremities, or even paralysis of the arms and legs. Acrylamide is readily absorbed through unbroken skin, and the signs are the same as with ingestion. Acute dermal LD_{50} is 2250 mg/kg for rabbit (85). Eye contact can produce conjunctival irritation and slight corneal injury and can lead to systemic exposure if contact is prolonged and/or repeated. Inhalation of vapors, dusts, and/or mists can result in serious injury to the nervous system, but again, symptoms may be slow in developing. Since the symptoms for minor repeated exposures over a long period of time are similar to those for gross human exposure, the development of such symptoms can be a signal that severe damage has already occurred. There are no reliable "early warning" signals of damaging exposure to toxic levels of acrylamide monomer, so it is imperative that all handling procedures be designed to prevent human contact. In a long-term study, rats that received relatively low concentrations of acrylamide monomer in the drinking water showed an increase in several types of malignant tumors (86). Suitable respirators and clothing that consists of a head covering, long-sleeved coverall, impervious gloves, and rubber footwear are recommended to avoid contact (61).

A large number of research studies have been published, many of which were released by government agencies (87–94). A threshold limit value (TLV) of 0.03 mg/m^3 (skin) has been set by the American Conference of Governmental Industrial Hygienists (ACGIH). ACGIH also categorizes acrylamide as A2 (suspect human carcinogen). Several studies demonstrate that acrylamide is biodegradable (95), and the hydrolysis of acrylamide proceeds readily both in rivers and in soils (96,97). Bioconcentration of acrylamide probably will not occur because of the ease of biodegradation and the high water solubility of this material. Acrylamide shows low acute toxicity to fish (61,98). Other derivatives, such as *N*-methylolacrylamide, are neurotoxins in their own right, but the LD_{50} is much higher than for acrylamide. Toxicity data for acrylamide and several derivatives are listed in Table 7.

Table 7. Toxicity of Acrylamide and Derivatives

Compound	CAS Registry Number	Molecular formula	Animal	Oral LD_{50}, g/kg	Ref.
acrylamide	[79-06-1]	C_3H_5NO	mouse	0.17	85
N-methylolacrylamide	[924-42-5]	$C_4H_7NO_2$	mouse	0.42	99
N,N'-methylenebisacrylamide	[110-26-9]	$C_7H_{10}N_2O_2$	rat	0.39	100
N-isobutoxymethylacrylamide	[16669-59-3]	$C_8H_{15}NO_2$	rat	1.0	101
N,N-dimethylacrylamide	[2680-03-7]	C_5H_9NO	rat	0.316	102
2-acrylamido-2-methylpropane-sulfonic acid	[15214-89-8]	$C_7H_{13}NO_4S$	rat	1.41	103

Handling of dry acrylamide is hazardous primarily from its dust and vapor, and this is a significant problem, especially in the course of emptying bags and drums. This operation should be carried out in an exhaust hood with the operator wearing respiratory and dermal protection. Waste air from the above mentioned ventilation should be treated by a wet scrubber before purging to the open air, and the waste water should be fed to an activated sludge plant or chemical treatment facility. Solid acrylamide may polymerize violently when melted or brought into contact with oxidizing agents. Storage areas for solid acrylamide monomer should be clean and dry and the temperature maintained at 10–25°C, with a maximum of 30°C.

The 50% aqueous product is the most desirable where water can be tolerated in the process. Employees should not be permitted to work with acrylamide until thoroughly instructed and until they can practice the required precautions and safety procedures. Anyone handling acrylamide should practice strict personal cleanliness and strict housekeeping at all times. This should include wearing a complete set of clean work clothes each day and the removal of contaminated clothing immediately. If contact is made, the affected skin area should be washed thoroughly with soap and water and contaminated clothing should be replaced. When contact can occur, such as in maintenance and repair operations or connection and disconnection during transport, protective equipment should be used. This should include impervious gloves and footwear to protect the skin, and suitable eye protection such as chemical worker's goggles. If exposure to the face is possible, a face shield should be used in addition to the goggles. Food, candy, tobacco, and beverages should be banned from areas where acrylamide is being handled, and workers should wash hands and face thoroughly with soap and water before eating or drinking. The need for good personal hygiene and housekeeping to prevent exposure cannot be overemphasized.

Aqueous solutions of 50% acrylamide should be kept between 15.5 and 38°C with a maximum of 49°C. Below 14.5°C acrylamide crystallizes from solution and separates from the inhibitor. Above 50°C the rate of polymer buildup becomes significant. Suitable materials of construction for containers include stainless steel (304 and 316) and steel lined with plastic resin (polypropylene, phenolic, or epoxy). Avoid contact with copper, aluminum, their alloys, or ordinary iron and steel.

Disposal of small amounts of acrylamide may be done by biodegradation in a conventional secondary sewage treatment plant, but any significant amounts should be avoided. Such waste material should not be allowed to get into a municipal waste treatment or landfill operation unless all appropriate precautions have been taken. When the disposal of large quantities is necessary, the supplier should be contacted. Containers that have been used for acrylamide should be thoroughly rinsed and then disposed of in an appropriate manner. In any disposal of waste materials, all of the applicable federal, state, and local statutes, rules, and regulations should be followed. Persons contemplating large-scale use of acrylamide monomer should consult the manufacturers at an early stage in the planning to ensure that their facilities and operations are adequate. Many companies refuse to supply to operations that, in their opinion, are unsafe.

Economic Aspects

The largest production of acrylamide is in Japan; the United States and Europe also have large production facilities. Some production is carried out in the Eastern Bloc countries, but details concerning quantities or processes are difficult to obtain. The principal producers in North America are The Dow Chemical Company, American Cyanamid Company, and Nalco Chemical Company (internal use); Dow sells only aqueous product and American Cyanamid sells both liquid and solid monomer. In Europe, Chemische Fabrik Stockhausen & Cie, Allied Colloids, The Dow Chemical Company, and Cyanamid BV are producers; Dow and American Cyanamid are the only suppliers to the merchant market, and crystalline monomer is available from American Cyanamid. For Japan, producers are Mitsubishi Chemical Industries, Mitsui Toatsu, and Nitto Chemical Industries Company (captive market). Crystals and solutions are available from Mitsui Toatsu and Mitsubishi, whereas only solution monomer is available from Nitto.

Estimated production capacity for the Japanese producers is 77,000 t/yr; for the American producers, about 70,000 t/yr; and for the European producers about 50,000 t/yr (104). The list prices for the monomer have increased dramatically over the past 15 years, according to the *Chemical Marketing Reporter* (105). In 1975 the price for 50% solution was $0.903/kg, compared to $1.68/kg in December 1990 (100% basis, FOB plant). The solid crystalline monomer always demands a premium price because of the added cost of production, and sold in December 1990 for

$2.27/kg compared to $1.09/kg in 1975. There are at least 35 suppliers of acrylamide monomer; most of them obviously are repackagers.

BIBLIOGRAPHY

"Acrylamide" in *ECT* 2nd ed., Vol. 1, pp. 274–284, by Norbert M. Bikales and Edwin R. Kolodny, American Cyanamid Company; in *ECT* 3rd ed., Vol. 1, pp. 298–311, by D. C. MacWilliams, Dow Chemical U.S.A.

1. D. C. MacWilliams, in R. H. Yocum and E. B. Nyquist, eds., *Functional Monomers,* Vol. 1, Marcel Dekker, Inc., New York, 1973, pp. 1–197.
2. U.S. Pat. 2,917,477 (Dec. 15, 1959), T. J. Suen and R. L. Webb (to American Cyanamid Co.).
3. E. Collinson and F. S. Dainton, *Nature* **177,** 1224 (1956).
4. U.S. Pat. 2,758,135 (Aug. 7, 1956), M. L. Miller (to American Cyanamid Co.).
5. *Chemistry of Acrylamide, Bulletin PRC 109,* Process Chemicals Department, American Cyanamid Co., Wayne, N.J., 1969.
6. J. P. Friend and A. E. Alexander, *J. Polym. Sci. Part A-1* **6,** 1833 (1968).
7. Ger. Pat. 1,030,826 (May 29, 1958), H. Wilhelm (to BASF A.G.).
8. A. LeBerre and A. Delaroix, *Bull. Soc. Chim. Fr.* **11**(2), 2639 (1974); *Chem. Abstr.* **82,** 97302 (1975); earlier paper in *Bull. Soc. Chim. Fr.* **2**(2), 640 (1973) is very significant.
9. A. LeBerre and C. Porte, *Bull. Soc. Chim. Fr.* **7–8**(2), 1627 (1975); *Chem. Abstr.* **84,** 58149 (1976).
10. U.S. Pat. 3,778,464 (Dec. 11, 1973), P. Klemchuck.
11. *Aqueous Acrylamide, Forms 260-951-88, Analytical Method PAA 46,* Chemicals and Metals Department, The Dow Chemical Company, Midland, Mich., 1976.
12. U.S. Pat. 3,699,192 (Oct. 17, 1972), P. Moretti (to U.S. Oil Company, Inc.).
13. The Dow Chemical Company, unpublished results.
14. D. Elad and D. Ginsberg, *J. Chem. Soc.* 4137 (1953).
15. B. T. Croll, G. M. Srkell, and R. P. J. Hodge, *Water Res.* **8,** 989 (1974).
16. Ger. Offen. 3,128,574 (Jan. 27, 1983), K. Laping, O. Petersen, K. H. Heinemann, H. Humbert, and F. Henn (to Deutsche Texaco A.G.).
17. J. S. Meek, R. T. Mernow, D. E. Ramey, and S. J. Cristol, *J. Am. Chem. Soc.* **73,** 5563 (1951).
18. A. I. Naimushin and V. V. Simonov, *Zh. Obshch. Khim.* **47,** 862 (1977); *Chem. Abstr.* **87,** 38678m (1977).
19. G. Stork, *Pure Appl. Chem.* **17,** 383 (1968).
20. I. Ninomiya, T. Naito, S. Higuchi, and T. Mori, *J. Chem. Soc. D* **9,** 457 (1971).
21. U.S. Pat. 4,198,415 (Apr. 15, 1980), N. J. Bach and E. C. Kornfeld (to Eli Lilly and Co.).
22. T. W. Russell and D. M. Duncan, *J. Org. Chem.* **39,** 3050 (1974).
23. T. W. Russell, R. C. Hoy, and J. E. Cornelius, *J. Org. Chem.* **37,** 3552 (1972).
24. T. Kitamura, N. Sakamoto, and T. Joh, *Chem. Lett.* **2**(4), 379 (1973).
25. U.S. Pat. 3,846,478 (Nov. 5, 1974), R. W. Cummins (to FMC Corp.).
26. U.S. Pat. 3,332,923 (July 25, 1967), L. D. Moore and R. P. Brown (to Nalco Chemical Co.).
27. A. Leoni and S. Franco, *Macromol. Synth.* **4,** 125 (1972).
28. U.S. Pats. 3,193,476 and 3,193,483 (July 6, 1965), M. M. Baizer (to Monsanto Co.).
29. D. S. Breslow, G. E. Hulse, and A. S. Matlack, *J. Am. Chem. Soc.* **79,** 3760 (1957).
30. N. Yamashita, M. Yoshihara, and T. Maeshima, *J. Polym. Sci. Part B* **10,** 643 (1972).
31. Eur. Pat. Appl. 272025 A2 (June 22, 1988), D. Byrom and M. A. Carver; Eur. Pat. Appl. EP 272026 A2 (June 22, 1988), M. A. Carver and J. Hinton (to Imperial Chemical Ind.).

32. Jpn. Kokai 76 86412 (July 29, 1976), F. Matsuda and T. Takazo (to Mitsui Toatsu Chem. Co.).
33. G. A. Chubarov, S. M. Danov, and V. I. Logutov, *Zh. Prikl. Khim. Leningrad* **52**, 2564 (1979); *Chem. Abstr.* **92**, 163293m (1980).
34. A. Conix, G. Smets, and J. Moens, *Ric. Sci. Suppl.* **25**, 200 (1954); *Chem. Abstr.* **54**, 11545e (1960).
35. T. Azuma and N. Ogata, *J. Polym. Sci. Polym. Chem. Ed.* **13**, 1959 (1975).
36. U.S. Pat. 3,084,191 (Apr. 2, 1963), J. R. Stephens (to American Cyanamid Co.).
37. M. F. Farona, W. T. Ayers, B. G. Ramsey, and J. G. Grasselli, *Inorg. Chim. Acta* **3**, 503 (1969).
38. J. Reedijk, *Inorg. Chim. Acta* **5**, 687 (1971).
39. A. Samantaray, P. K. Panda, and B. K. Mohapatra, *J. Indian Chem. Soc.* **57**, 430 (1980).
40. M. S. Barvinok and L. V. Mashkov, *Zh. Neorg. Khim.* **25**, 2846 (1980).
41. V. I. Bruskov and V. N. Bushuev, *Biofizika* **22**(1), 26 (1977); *Chem. Abstr.* **87**, 39783d (1977).
42. T. O. Osmanov, V. F. Gromov, P. M. Khomikovskii, and A. D. Abkin, *Polym. Sci. USSR* **22**, 739 (1980); **21**, 1948 (1979).
43. T. Asakara and N. Yoda, *J. Polym. Sci. Part A-1* **6**, 2477 (1968).
44. T. Asakara, K. Ikeda, and N. Yoda, *J. Polym. Sci. Part A-1* **6**, 2489 (1968).
45. U.S. Pat. 2,373,190 (Apr. 10, 1945), F. E. King (to B. F. Goodrich Co.).
46. C. Moureau, *Bull. Soc. Chim. Fr.* **9**, 417 (1973).
47. U.S. Pat. 3,064,050 (Nov. 13, 1962), K. W. Saunders and L. L. Lento, Jr. (to American Cyanamid Co.).
48. H. Fener and V. E. Lynch, *J. Am. Chem. Soc.* **75**, 5027 (1953).
49. U.S. Pat. 2,475,846 (July 12, 1949), L. A. Lindberg (to American Cyanamid Co.).
50. Ger. Offen. 2,310,516 (Sept. 19, 1974), K. Fischer and H. Petersen (to BASF A.G.).
51. T. Araki, C. Terunuma, K. Tanigawa, and N. Ando, *Kobunshi Ronbunshu* **31**, 309 (1974).
52. Jpn. Kokai 75 82008 (July 3, 1975), K. Yamamoto and co-workers (to Mitsui Toatsu Chemicals, Inc.).
53. Jpn. Kokai 74 26235 (Mar. 8, 1974), S. Kumi and co-workers (to Dainippon Ink and Chemicals, Inc.).
54. E. Mueller, K. Dinges, and W. Ganlich, *Makromol. Chem.* **57**, 27 (1962).
55. U.S. Pat. 3,185,539 (May 25, 1965), R. K. Madison and W. J. Van Loo, Jr. (to American Cyanamid Co.).
56. Jpn. Pat. 15,816 (Aug. 5, 1964), T. Oshima and M. Suzuki (to Sumitomo Chemical Co., Ltd.).
57. U.S. Pat. 3,422,139 (Jan. 14, 1969), P. Talet and R. Behar (to Nobel-Bozel).
58. Fr. Pat. 1,406,594 (July 23, 1965), P. Talet and R. Behar (to Nobel-Bozel).
59. R. E. Dunbar and G. C. White, *J. Org. Chem.* **23**, 915 (1958).
60. *Acrylamide-50 Handling and Storage Procedures, PRC 22B,* American Cyanamid Co., Wayne, N.J., 1980.
61. *Aqueous Acrylamide, Forms 260-951-88, Analytical Method PAA 44,* Chemical and Metals Department, The Dow Chemical Company, Midland, Mich., 1976.
62. S. R. Betso and J. D. McLean, *Anal. Chem.* **48**, 766 (1976).
63. B. T. Croll and G. M. Simkins, *Analyst* **97**, 281 (1972).
64. M. V. Norris, in F. D. Snell and C. L. Hilton, eds., *Encyclopedia of Industrial Chemical Analysis,* Vol. 4, Wiley-Interscience, New York, 1967, pp. 160–168.
65. D. C. MacWilliams, D. C. Kaufman, and B. F. Waling, *Anal. Chem.* **37**, 1546 (1965); A. R. Mattocks, *Anal. Chem.* **40**, 1347 (1968).
66. B. T. Croll, *Analyst (London)* **96**, 67 (1971).
67. *HPLC Determinations of Acrylamide in Water and Air Samples, Analytical Method*

PAA 58,61 in Forms 260-951-88, Chemicals and Metals Department, The Dow Chemical Company, Midland, Mich., 1981.
68. U.S. Pat. 2,734,915 (Feb. 14, 1956), G. D. Jones (to The Dow Chemical Company).
69. U.S. Pat. 3,041,375 (June 26, 1962), S. N. Heiny (to The Dow Chemical Company).
70. M. J. Sook, E. J. Forbes, and G. M. Khan, *Chem. Commun.* (5), 121 (1966).
71. U.S. Pat. 3,381,034 (Apr. 30, 1968), J. L. Greene and M. Godfrey (to Standard Oil Co., Ohio).
72. K. Watanabe, *Bull. Chem. Soc. Jap.* **37,** 1325 (1964); *Chem. Abstr.* **62,** 2735b (1965).
73. U.S. Pat. 3,597,481 (Aug. 3, 1971), B. A. Tefertiller and C. E. Habermann (to The Dow Chemical Company); U.S. Pat. 3,631,104 (Dec. 28, 1971), C. E. Habermann and B. A. Tefertiller (to The Dow Chemical Company); U.S. Pat. 3,642,894 (Feb. 15, 1972), C. E. Habermann, R. E. Friedrich, and B. A. Tefertiller (to The Dow Chemical Company); U.S. Pat. 3,642,643 (Feb. 15, 1972), C. E. Habermann (to The Dow Chemical Company); U.S. Pat. 3,642,913 (Mar. 7, 1972), C. E. Habermann (to The Dow Chemical Company); U.S. Pat. 3,696,152 (Oct. 3, 1972), C. E. Habermann and M. R. Thomas (to The Dow Chemical Company); U.S. Pat. 3,758,578 (Sept. 11, 1973), C. E. Habermann and B. A. Tefertiller (to The Dow Chemical Company); U.S. Pat. 3,767,706 (Oct. 23, 1972), C. E. Habermann and B. A. Tefertiller (to The Dow Chemical Company).
74. Brit. Pat. 1,324,509 (July 25, 1973) (to Mitsui Toatsu Chemicals, Inc.).
75. Ger. Offen. 2,320,060 (Nov. 7, 1974), T. Dockner and R. Platz (to BASF A.G.).
76. E. Otsuka and co-workers, *Chem. Econ. Eng. Rev.* **7**(4), 29 (1975).
77. Brit. Pat. 2,018,240 (Oct. 17, 1979), I. Watanabe (to Nitto Chemical).
78. U.S. Pat. 4,343,900 (Aug. 10, 1982), I. Watanabe (to Nitto Chemical).
79. Fr. Demande 2,488,908 (Feb. 26, 1982), I. Watanabe and co-workers (to Nitto Chemical).
80. U.S. Pat. 4,390,631 (June 28, 1983), I. Watanabe and co-workers (to Nitto Chemical).
81. U.S. Pat. 4,414,331 (Nov. 8, 1983), I. Watanabe and co-workers (to Nitto Chemical).
82. *Chem. Eng.* **97**(7), 19–21 (July 1990).
83. J. J. Ritter and P. P. Minieri, *J. Am. Chem. Soc.* **70,** 4045 (1948).
84. H. Plant and J. J. Ritter, *J. Am. Chem. Soc.* **73,** 4076 (1951).
85. *Aqueous Acrylamide, Form No. 192-466-76,* Chemicals and Metals Department, The Dow Chemical Company, Midland, Mich., 1976.
86. K. A. Johnson, S. J. Gorzinski, K. M. Bodner, R. A. Campbell, C. H. Wolf, M. A. Friedman, and R. W. Mast, *Toxicol. Appl. Pharmacol.* **85,** 154–168 (1986).
87. L. N. Davis, P. R. Durkin, P. H. Howard, and J. Saxena, *Investigation of Selected Potential Environmental Contaminants; Acrylamides;* EPA Report No. 560/2-76-008, 1976.
88. *Criteria for Recommended Standard Occupational Exposure to Acrylamide,* U.S. Department of Health, Education, and Welfare, Washington, D.C., 1976.
89. *Environmental and Health Aspects of Acrylamide, A Comprehensive Bibliography of Published Literature 1930 to April 1980,* EPA Report No. 560/7-81-006, 1981.
90. *Assessment of Testing Needs; Acrylamide,* EPA Report No. 560/11-80-016, 1980.
91. J. Going and K. Thomas, *Sampling and Analysis of Selected Toxic Substances; Task I Acrylamide,* EPA Report No. 560/13-79-013, 1979.
92. E. J. Conway, R. J. Petersen, R. F. Colingsworth, J. G. Craca, and J. W. Carter, *Assessment of the Need for a Character of Limitations on Acrylamide and Its Components,* EPA MRI Project No. 4308-N, 1979.
93. H. A. Tilson, *Neurobehav. Toxicol. Tetratol.* **3,** 445 (1981).
94. P. M. Edwards, *Br. J. Ind. Med.* **32,** 31 (1975).
95. B. T. Croll, G. M. Arkell, and R. P. J. Hodge, *Water Res.* **8,** 989 (1974).
96. M. J. Hynes and J. A. Pateman, *J. Gen. Microbiol.* **63,** 317 (1970).
97. H. M. Abdelmagid and M. A. Tabatabai, *J. Environ. Qual.* **11,** 701 (1982).

98. Krautter and co-workers, *Environ. Toxicol. Chem.* **5,** 373–377 (1986).
99. *N-Methylolacrylamide, PRC 14,* Process Chemicals Department, American Cyanamid Co., Wayne, N.J., 1972.
100. *N,N'-Methylenebisacrylamide, PRT 47A,* American Cyanamid Co., Wayne, N.J., 1978.
101. *N-(iso-Butoxymethyl)acrylamide, PRT 126,* Process Chemicals Department, American Cyanamid Co., Wayne, N.J., 1977.
102. *N,N-Dimethylacrylamide, Technical Bulletin,* Alcolac, Inc., Baltimore, Md., 1977.
103. *AMPS Monomer,* Lubrizol Corp., Wickliffe, Ohio, 1981.
104. *Directory of Chemical Producers,* SRI International, Menlo Park, Calif., 1988.
105. *Chem. Mark. Rep.* **238**(23), 36 (1990).

C. E. HABERMANN
Dow Chemical, USA

ACRYLAMIDE POLYMERS

Acrylamide [*79-06-1*] (CH_2=CHC—NH_2, with a carbonyl O double-bonded to the C) polymerizes in the presence of free-radical initiators to form polyacrylamide [*9003-05-8*] chains with the structure

$$(-CH_2-CH-)_n$$
$$\mid$$
$$CONH_2$$

In this article the term acrylamide polymer refers to all polymers which contain acrylamide as a major constituent. Consequently, acrylamide polymers include functionalized polymers prepared from polyacrylamide by postreaction and copolymers prepared by polymerizing acrylamide (2-propenamide, C_3H_5NO) with one or more comonomers.

Since the 1960s the industrial use of nonionic, anionic, cationic, and amphoteric acrylamide polymers has increased steadily because of their unique chemical and physical properties, low relative toxicity, and low cost. The primary market areas for these polymers are water treatment, papermaking, mineral processing, enhanced oil recovery, and superabsorbents. Enhanced oil recovery represents a large potential market, but low oil prices limited the growth of this market during the 1980s. In 1990 oil prices increased and are expected to continue increasing during the decade. Environmental concerns have contributed to a rapid growth in the use of polyacrylamide flocculants in effluent treatment and sludge dewatering. High molecular weight cationic polyacrylamides have begun to play a prominent role in these applications. Improvements in characterization techniques such as infrared spectroscopy, nmr spectroscopy, gel-permeation chromatography, and sedimentation field flow fractionation have led to a better understanding of the influence of chain microstructure and molecular weight distribution on performance.

Manufacturing processes have been improved by use of on-line computer control and statistical process control leading to more uniform final products. Production methods now include inverse (water-in-oil) suspension polymerization, inverse emulsion polymerization, and continuous aqueous solution polymerization on moving belts. Conventional azo, peroxy, redox, and gamma-ray initiators are used in batch and continuous processes. Recent patents describe processes for preparing transparent and stable microlatexes by inverse microemulsion polymerization. New methods have also been described for reducing residual acrylamide monomer in finished products.

Physical Properties

Solid Polymer. Completely dry polyacrylamide is a brittle white solid. Commercially available dry polyacrylamide powders are typically dried under mild conditions and usually contain 5–15% water. The powders are hygroscopic, and generally become increasingly hygroscopic as the ionic character of the polymer increases. Cationic polymers are particularly hygroscopic.

The physical properties of nonionic polyacrylamide are listed in Table 1. Many values of the glass-transition temperature (T_g) have been published. The measured value is highly sensitive to water, and is also dependent on the presence of functionality along the backbone. The tacticity and linearity of the polymer is claimed to be dependent on the polymerization temperature. Syndiotacticity is favored at low temperatures (13). Linear polymer chains are obtained below 50°C, but branching begins to occur as temperature is increased above this level (14). Weight loss on heating in helium or nitrogen occurs with loss of water up to 250°C (15). Above this temperature, weight loss occurs primarily with ammonia gas evolution (16).

Table 1. Physical Properties of Solid Polyacrylamide

Property	Value	Reference
density, g/cm^3	1.302	1
glass-transition temp, °C	188	2
critical surface tension, mN/m (= dyn/cm)	.35–40	3
chain structure	mainly heterotactic linear	4
	or branched, some	5
	head-to-head addition	6,7
crystallinity	amorphous (high mol wt)	8
solvents	water, ethylene glycol, formamide	9
nonsolvents	ketones, hydrocarbons, ethers, alcohol	10
fractionation solvents	water–methanol	11
gases evolved on combustion in air	H_2, CO, CO_2, NH_3, nitrogen oxides	12

Dry polyacrylamides are commercially available as nondusting powders and as spherical beads. Products can contain small amounts of additives which aid in both the stability and dissolution of the polymers in water. Most polyacrylamide powders will develop full viscosity in water with mild agitation. The length of time is dependent on the polymer molecular weight and the hardness of the water. The high shear sensitivity of high molecular weight polymer solutions can result in polymer degradation.

Polymers in Solution.   Polyacrylamide is soluble in water at all concentrations, temperatures, and pH values. An extrapolated theta temperature in water is approximately $-40°C$ (17). Insoluble gel fractions are sometimes obtained owing to cross-link formation between chains or to the formation of imide groups along the polymer chains (18). In very dilute solution, polyacrylamide exists as unassociated coils which can have an ellipsoidal or beanlike structure (19). Large aggregates of polymer chains have been observed in hydrolyzed polyacrylamides (20) and in copolymers containing a small amount of hydrophobic groups (21).

In general nonionic polyacrylamides do not interact strongly with neutral inorganic salts. Certain salts can, however, alter the hydrogen bonding between primary amide groups and water in individual chains. For example, potassium iodide [7681-11-0] causes a slight increase in solution viscosity, indicating coil expansion and an increase in hydrodynamic volume (22). Anionic polyacrylamides possessing carboxyl groups have complex interactions with many divalent cations, such as calcium and magnesium found in hard water and in the injection water used in enhanced oil recovery (23,24). Interactions with trivalent ions such as Al^{3+} or Cr^{3+} are very strong and can cause anionic polyacrylamides to gel. Usually increased water hardness reduces the observed viscosity of an anionic polyacrylamide at a given polymer concentration.

Flow Properties.   In water, high molecular weight polyacrylamide forms viscous homogeneous solutions. The measured viscosities increase slowly with increasing polymer concentration and then increase very rapidly after a critical concentration is reached, at which point the polymer chains start to overlap and interpenetrate. The critical concentration decreases with increasing shear rate because the linear and flexible entangled molecules stretch in the direction of shear stress. Normal stresses, in the direction perpendicular to flow, are observed for high molecular weight polyacrylamide solutions. These normal stresses are more pronounced for polymers with a very broad molecular weight distribution. Viscosities and viscoelastic behavior decrease with increasing temperature. In some cases a marked viscosity decrease with time is observed in solutions stored at constant temperature and zero shear. The decrease may be due to changes in polymer conformation. The rheological behavior of pure polyacrylamides over wide concentration ranges has been reviewed (5).

At concentrations well below the critical concentration for chain overlap, linear, high molecular weight polyacrylamides can reduce resistance to flow. Introduction of ionic groups into the polymer chain increases the effective polymer chain length and increases the efficiency of drag reduction. The high molecular weight portion of the molecular weight distribution is the most efficient in reducing friction. Drag reduction effects are not affected by salts if nonionic polyacrylamide is employed. If the polymer possesses carboxyl groups, however, the effectiveness decreases with increasing salt concentration. A recent review describes the desirable flow enhancing properties of polyacrylamides and other polymers (19).

In packed beds of particles possessing small pores, dilute aqueous solutions of hydrolyzed polyacrylamide will sometimes exhibit dilatant behavior instead of the usual shear thinning behavior seen in simple shear or Couette flow. In elongational flow, such as flow through porous sandstone, flow resistance can increase with flow rate due to increases in elongational viscosity and normal stress differences. The increase in normal stress differences with shear rate is typical of isotropic polymer solutions. Normal stress differences of anisotropic polymers, such as xanthan in water, are shear rate independent (25,26).

Interest in polyacrylamides exhibiting associative thickening behavior has increased recently due to the ability of these hydrophobically modified polymers to develop desirable flow properties and performance (27–29). Hydrophobically modified polyacrylamides produce a very small increase in pressure drop with increasing shear rate in flow through narrow tubes. These polymers are often of low molecular weight and possess 1.0 mol % or less hydrophobic groups along the polymer backbone. Shear degradation is usually minimized because of the low molecular weight. The hydrophobic groups dissociate at high shear rates, and then associate again at low shear rates.

Chemical Properties

The preparation of polyacrylamides and postpolymerization reactions on polyacrylamides are usually conducted in water. Reactions on the amide groups of polyacrylamides are often more complicated than reactions of simple amides because of neighboring group effects. Reaction rates, for example, can differ considerably.

Postreactions of polyacrylamide to introduce anionic, cationic, or other functional groups are often attractive from a cost standpoint. This approach can suffer, however, from side reactions resulting in cross-linking or the introduction of unwanted functionality, such as carboxyl groups from hydrolysis. In addition, postreaction on high molecular weight polyacrylamides can be difficult because of the high viscosity of even fairly dilute aqueous polyacrylamide solutions.

Hydrolysis. Polyacrylamide hydrolyzes rapidly under alkaline conditions at relatively low temperatures. Generally, higher temperatures are required to hydrolyze polyacrylamide under neutral or acidic conditions. The structure of the hydrolyzed polyacrylamide can vary considerably and depends on the conditions employed. ^{13}C-nmr studies (30–32) have shown that the sequence distributions of carboxyl and amide groups differ depending on both the pH of the hydrolysis reaction and the concentration of electrolyte. These more recent nmr studies corroborate previous studies (33–35) which indicated that hydrolysis at low pH results in blocks of carboxyl groups, while hydrolysis at high pH results in a more random distribution of carboxyl groups. Under neutral conditions, a mixture of the two distributions is obtained (32).

Alkaline hydrolysis of polyacrylamide can be expressed in terms of three rate constants, k_0, k_1, and k_2, where the subscript indicates the number of neighboring carboxyl groups next to the amide group undergoing hydrolysis. The rate constant k_0 is larger than k_1, which in turn is much larger than k_2. A negatively charged carboxyl group produces a retardation effect on the hydrolysis of a neighboring amide group. Two neighboring negatively charged carboxyl groups severely retard hydrolysis of an amide. Consequently, hydrolysis of about 70% can be obtained in the presence of excess hydroxide ion, leaving the remainder of the amides surrounded by two carboxyl groups. Lower levels of carboxyl groups are often desirable in many commercial applications. A convenient way to prepare a polyacrylamide with about 30 mol % hydrolysis is to heat an aqueous polyacrylamide solution containing excess sodium carbonate (36).

Hydrolysis of cationic polyacrylamides prepared from copolymerization of acrylamide and cationic ester monomer can occur under very mild conditions. A substantial loss in cationicity can cause a significant loss in performance in many applications. Copolymers [69418-26-4] of acrylamide and acryloxyethyltrimethylammonium chloride [44992-01-0], $C_8H_{16}NO_2(Cl)$, for instance, lose cationicity rapidly at alkaline pH (37).

$$-(CH_2-CH-CH_2-CH)_{\overline{m}} \xrightarrow{OH^-} $$

The reaction rate increases with pH. ^{13}C-nmr spectra indicate the presence of imide structures in the degraded product (38). The degree of cationicity loss is dependent on the number of cationic units having neighboring amide groups along the polymer chain (39). The percentage of esters cleaved increases as the number of esters with neighboring amides increases. Esters with no neighboring amides, such as those found in homopolymers of acryloxyethyltrimethylammonium chloride, are difficult to hydrolyze.

Mannich Reaction.   Aminomethylation of polyacrylamide with formaldehyde [50-00-0] and a secondary amine to produce a Mannich polyacrylamide has been extensively studied (40).

$$-(CH_2-CH)_{\overline{m}} + R_2NH + CH_2{=}O \longrightarrow -(CH_2-CH)_{\overline{m}} + H_2O$$

The yields of this reaction are typically 40–80%. ^{13}C-nmr studies (41) indicate that the reaction is a second-order process between polyacrylamide and dimethylaminomethanol, which is one of the equilibrium products formed in the reaction between formaldehyde and dimethylamine [124-40-3], C_2H_7N. The Mannich reaction is reversible. Extensive dialysis of Mannich polyacrylamides removes all of the dimethylaminomethyl substituents (42).

High molecular weight Mannich polyacrylamides are sold as aqueous solutions containing approximately 7% polymer or less, because of the high viscosity of their aqueous solutions and their tendency to cross-link on standing. The cross-linking reaction can be controlled by addition of formaldehyde scavengers such as guanidine nitrate [506-93-4] and urea [57-13-6].

Mannich polyacrylamides can react with alkylating agents such as methyl chloride [74-87-3], CH_3Cl, methyl bromide [74-83-9], CH_3Br, and dimethyl sulfate [77-78-1], $C_2H_6SO_4$, to produce quaternary ammonium bases which retain cationic charge at alkaline pH (40,42,43).

Sulfomethylation. The reaction of formaldehyde and sodium bisulfite [7631-90-5] with polyacrylamide under alkaline conditions to produce sulfo-methylated polyacrylamides has been known for many years (44–46). A more recent publication (47) suggests, however, that the expected sulfomethyl substitution is not obtained under the previously described strongly alkaline conditions of pH 10–12. This ^{13}C-nmr study indicates that hydrolysis of polyacrylamide occurs and the resulting ammonia reacts with the $NaHSO_3$ and formaldehyde. A recent patent claims a new high pressure, high temperature process at slightly acid pH for preparation of sulfomethylated polyacrylamide (48).

Methylol Formation. Polyacrylamide reacts with formaldehyde to form an N-methylol derivative. The reaction is conducted at about pH 7–8.8 in order to avoid cross-linking, which will occur at lower pH. The copolymer can also be prepared by copolymerizing acrylamide with commercially available N-methylolacrylamide [924-42-5], $C_4H_7NO_2$. These derivatives are useful in several mining applications (49,50). They are also useful as chemical grouts.

Reaction with Other Aldehydes. Polyacrylamide reacts with glyoxal [107-22-2], $C_2H_2O_2$, under mild alkaline conditions to yield a polymer with pendant aldehyde functionality.

$$-(CH_2-CH)_{\overline{m}} + O{=}C-C{=}O \xrightarrow{\text{base}} -(CH_2-CH)_{\overline{m}}$$
$$\quad\quad\quad | \quad\quad\quad\quad | \quad | \quad\quad\quad\quad\quad\quad\quad |$$
$$\quad\quad\quad C{=}O \quad\quad\quad H \quad H \quad\quad\quad\quad\quad C{=}O$$
$$\quad\quad\quad | \quad\quad\quad\quad\quad\quad\quad\quad\quad\quad\quad\quad\quad |$$
$$\quad\quad\quad NH_2 \quad\quad\quad\quad\quad\quad\quad\quad\quad\quad NHCH{-}CHO$$
$$\quad\quad\quad\quad\quad\quad\quad\quad\quad\quad\quad\quad\quad\quad\quad\quad\quad\quad\quad |$$
$$\quad\quad\quad\quad\quad\quad\quad\quad\quad\quad\quad\quad\quad\quad\quad\quad\quad\quad\quad OH$$

This reaction can be controlled by varying the pH and reaction temperature. Cross-linking is a competing reaction. In a typical preparation a 10% aqueous solution of a low molecular weight polyacrylamide reacts with glyoxal at pH 8–9 at room temperature. As the reaction proceeds, solution viscosity increases slowly and then more rapidly as the level of functionalization and cross-linking increases. When the desired extent of reaction is achieved, the reaction is acidified to a pH below 6 to slow the reaction down to a negligible rate. These glyoxalated polyacrylamides are used as paper additives for improvement of wet strength (51,52).

A similar reaction occurs when polyacrylamide is mixed with glyoxilic acid [298-12-4], $C_2H_2O_3$, at pH about 8. This reaction produces a polymer with the $CONHCH(OH)COOH$ functionality, which has found application in phosphate ore processing (53).

Transamidation. Polyacrylamide reacts with primary amines such as hydrazine [302-01-2], N_2H_4, (54) and hydroxylamine [7803-49-8], NH_3O, (55–57) to form substituted amides with loss of ammonia.

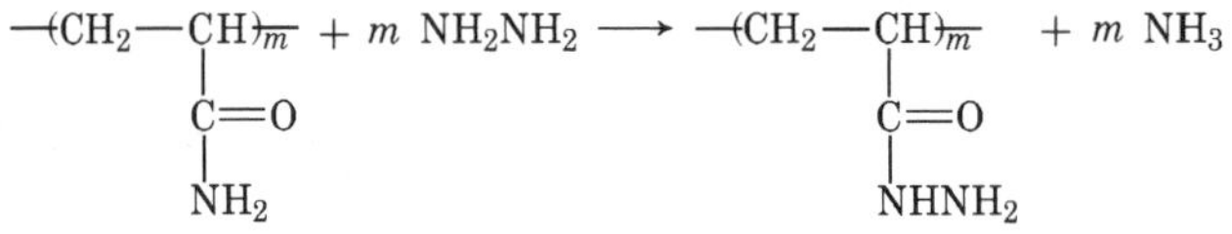

$$\text{--(CH}_2\text{---CH)}_{\overline{m}} + m\ NH_2NH_2 \longrightarrow \text{--(CH}_2\text{---CH)}_{\overline{m}} + m\ NH_3$$
$$\underset{NH_2}{\overset{C=O}{|}} \qquad\qquad\qquad \underset{NHNH_2}{\overset{C=O}{|}}$$

This reaction, conducted in alkaline solution, also produces carboxyl groups by hydrolysis of the amide (54). Recent work on the reaction of polyacrylamide with hydroxylamine indicates that maximum conversion to the hydroxamate functionality ($-CONHOH$) takes place at a pH >12 (57). Apparently, this reaction of hydroxylamine at high pH, where it is a free base, is faster than the hydrolysis of the amide by hydroxide ion. Previous studies on the reaction of hydroxylamine with low molecular weight amides indicated that a pH about 6.5 was optimum (55).

Hoffman Degradation. Polyacrylamide reacts with alkaline sodium hypochlorite [7681-52-9], NaOCl, or calcium hypochlorite [7778-54-3], $Ca(OCl)_2$, to form a polymer with primary amine groups (58). Optimum conditions for the reaction include a slight molar excess of sodium hypochlorite, a large excess of sodium hydroxide, and low temperature (59). Cross-linking sometimes occurs if the polymer concentration is high. High temperatures can result in chain scission.

Reaction with Chlorine. Polyacrylamide reacts with chlorine under acid conditions to form reasonably stable N-chloroamides. The polymers are water soluble and can provide good wet strength and wet web strength in paper (60).

Commercial Polymerization and Processing

Polyacrylamides are manufactured by free-radical polymerization of acrylamide to form chains of the structure shown, where n can range from several up to 400,000.

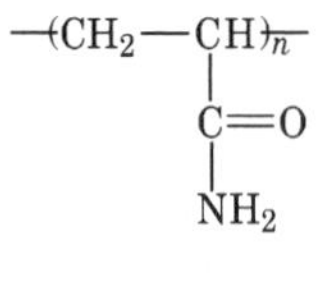

$$\text{--(CH}_2\text{---CH)}_{\overline{n}}\text{--}$$
$$\underset{NH_2}{\overset{C=O}{|}}$$

Suitable initiators include azo compounds, organic peroxides, redox systems, light with sensitizers, x rays or gamma rays, persulfates, and electroinitiation (5,61). The heat evolved during polymerization is 82.8 kJ/mol (19.8 kcal/mol) (62). The rate of polymerization is typically proportional to the square root of the initiator concentration (63). The rate is proportional to the monomer concentration to the power 1.2–1.5 (63). The rate constant at 25°C and pH 1 for propagation k_p is $(1.7 \pm 0.3) \times 10^4$ L/(mol·s). The ratio $k_p/k_t^{1/2}$ is almost always independent of pH and is equal to 4.2 ± 0.2 (64). This number is very high and explains why very high molecular weights are possible.

Termination is primarily by disproportionation (65). Additives such as inorganic salts, detergents, complexing agents, and organic solvents can either increase or decrease the rate of polymerization and alter the polymerization kinetics (66,67). Molecular weights of 4×10^7 can be obtained if polymerization occurs at low temperature and very low initiator concentrations. For most industrial applications, somewhat lower molecular weights are sufficient. Chain-transfer agents can be added to control molecular weight and prevent cross-linking. Table 2 lists chain-transfer constants for polymerization of acrylamide in water. 2-Propanol is commonly used for molecular weight control.

Table 2. Chain-Transfer Constants for Polymerization of Acrylamide in Water

Transfer species	Temperature, °C	Chain-transfer constant $\times 10^4$	Reference
monomer	25	0.0786 ± 0.0107	68
monomer	40	0.120 ± 0.0388	68
monomer	60	0.6	69
polymer	<50	near zero	68
$(CH_3)_2CHOH$	50	19	70
HSO_3^-	75	1700	71
H_2O_2	25	5	72
$K_2S_2O_8$	40	26.3 ± 7.08	68
iron(III) salts	25	very high	73
copper(II) salts	25	very high	
H_2O	25	near zero	68

Acrylamide copolymerizes with many vinyl comonomers readily. The copolymerization parameters in the Alfrey-Price scheme are $Q = 0.23$ and $e = 0.54$ (74). The effect of temperature on reactivity ratios is small (75). Solvents can produce apparent reactivity ratio differences in copolymerizations of acrylamide with polar monomers (76). Copolymers obtained from acrylamide and weak acids such as acrylic acid have compositions that are sensitive to polymerization pH. Reactivity ratios for acrylamide and many comonomers can be found in reference 77. Reactivity ratios of acrylamide with commercially important cationic monomers are given in Table 3.

Copolymers of acrylamide and acryloyloxyethyltrimethylammonium chloride have become increasingly preferred due to the favorable reactivity ratios between these two monomers, which result in copolymers with a uniform composition.

Solution Polymerization. Plant scale polymerizations in water are conducted either adiabatically or isothermally. Molecular weight control, exotherm control, and reduction of residual monomer are factors which limit the types of initiators employed. Commercially available high molecular weight solution polyacrylamides are usually manufactured and sold at about 5% solids so that the viscosities permit the final product to be pumped easily.

In a typical adiabatic polymerization, approximately 20 wt % aqueous acrylamide is charged into a stainless steel reactor equipped with agitation, condenser,

Table 3. Acrylamide Monomer (M_1) Reactivity Ratios[a]

Comonomer M_2	CAS Registry Number	Molecular formula	r_1	r_2	Reference
acryloyloxyethyltrimethylammonium chloride	[44992-01-0]	$C_8H_{16}NO_2 \cdot Cl$	0.64	0.48	78
			0.33 ± 0.09	0.40 ± 0.11	79
N,N-diallyldimethylammonium chloride	[7398-69-8]	$C_8H_{16}N \cdot Cl$	6.7	0.58	78
			6.4 ± 0.4[b]	0.06 ± 0.03[b]	79
N,N-dimethylaminoethyl methacrylate	[2867-47-2]	$C_8H_{15}NO_2$	0.26	1.66	78
N,N-dimethylaminopropylacrylamide	[3845-76-9]	$C_8H_{16}N_2O$	0.47	0.95	78
N,N-dimethylaminopropyl methacrylate	[20602-77-1]	$C_9H_{17}NO_2$	0.47	1.10	78
methacrylolyloxyethyltrimethylammonium chloride	[5039-78-1]	$C_9H_{18}NO_2 \cdot Cl$	0.27	1.71	78
methacrylamidopropyltrimethylammonium chloride	[51410-72-1]	$C_{10}H_{21}N_2O \cdot Cl$	0.57	1.13	78

[a] At 40°C unless otherwise noted.
[b] At 60°C.

and cooling jacket or coils. To initiate the polymerization, an aqueous solution of sodium bisulfite [7631-90-5] is added, followed by the addition of a solution of ammonium persulfate [7727-54-0], $N_2H_8S_2O_8$. As the polymerization proceeds, the temperature rises to about 90°C, and then begins to fall at the end of the polymerization. The molecular weight obtained depends primarily on the initiator concentration employed.

Isothermal polymerizations are carried out in thin films so that heat removal is efficient. In a typical isothermal polymerization, aqueous acrylamide is sparged with nitrogen for one hour at 25°C and EDTA ($C_{10}H_{16}N_2O_8$) is then added to complex the copper inhibitor. Polymerization can then be initiated as above with the ammonium persulfate–sodium bisulfite redox couple. The batch temperature is allowed to rise slowly to 40°C and is then cooled to maintain the temperature at 40°C. The polymerization is complete after several hours, at which time additional sodium bisulfite is added to reduce residual acrylamide.

Solution polyacrylamides can also be prepared at high polymer solids by radiation processes (80,81). Polyacrylamides with molecular weights up to 20 million can be prepared by irradiation of acrylamide and comonomers in a polyethylene bag with cobalt-60 gamma radiation at dose rates of 120–200 J/kg·h. The total dose of radiation is controlled to avoid cross-linking.

Polymerization on Moving Belts. Dry polyacrylamides are sometimes preferred, particularly when transportation distances are long. Care must be taken during drying to avoid forming insoluble polymer and to prevent molecular weight decreases (82). A continuous process has been developed for preparing dry polyacrylamides which consists of polymerizing aqueous acrylamide on a moving belt and drying the resulting polymer (82–86). In one such process (82) an aqueous solution of acrylamide and azobisisobutyronitrile [78-67-1], $C_8H_{12}N_4$, is pumped onto a concave, moving belt along with an aqueous solution containing a redox initiator system. The polymerization proceeds as the solution moves along the belt, reaching temperatures up to about 95°C. The belt speed is controlled so that the polymerization is complete when the polymer reaches the end of the belt. At the end of the belt the polymer gel is sliced into small granules and dried in an oven. The dried polymer is then passed through a grinder to produce the optimum particle size for handling and use.

Several recent patents describe improvements in the basic belt process. In one case a higher solids polymerization is achieved by cooling the starting monomer until some monomer crystallizes and then introducing the resulting monomer slurry onto the belt as above. The latent heat of fusion of the monomer crystals absorbs some of the heat of polymerization, which otherwise limits the solids content of the polymerization (87). In another patent a concave belt is described which becomes flat near the end. This change leads to improved release of polymer (88).

Another method used commercially is to grind the gel and extract the water (and residual monomer) with methanol. The methanol is dried and recycled. The dry polymer is ground and screened.

Dry Bead Process. Dry polyacrylamides can also be prepared in the form of dry beads with particle sizes ranging from about 100 to 2000 micrometers (89). These free-flowing beads are formed by azeotropically distilling water from inverse suspension polyacrylamides, collecting the particles by filtration, and further drying the particles in a fluid bed drier for short times. The resulting particles can be dissolved readily in water in the same manner as other dry polyacrylamides. The size and shape of the particles prepared in the suspension polymerization process are a function of the types and amounts of surfactants and additives employed. Typically, 0.03 to 0.2 wt % (based on water plus polymer) of an oil-soluble polymeric surfactant is used to obtain the desired particle size. Greater amounts of surfactants lead to smaller particles, which are more useful in the inverse suspension form than in the dry form because of dusting problems inherent in small-particle-size dry powders. Certain water soluble ionic organic compounds have been found to be effective in improving the stability of the particles and providing a narrower particle size distribution when used in conjunction with the polymeric stabilizers. In the absence of the stabilizer, irregularly shaped, unstable particles can result. The choice of the stabilizer is dependent on the charge of the polyacrylamide being produced.

Inverse Emulsion Process. A method of avoiding the high solution viscosities of water-soluble polymers comprises emulsifying the aqueous monomer solution in oil containing surfactants, homogenizing the mixture to form a water-in-oil emulsion and then polymerizing the monomers in the emulsion. For use, the emulsion is inverted in water, releasing the polymer. The original patent (90) illustrated a process for production of dry solid polymers. Considerable attention has been given to processes where the inverse emulsion polymerization produces finely divided particles which are small enough to resist settling and can be sold as is (91). Molecular weights can be over two million and are frequently as high as twenty million. A typical process for production of an anionic polyacrylamide follows (92). To a stirred tank are added, with cooling, 50% aqueous acrylamide, acrylic acid, and deionized water to give monomer concentration of 40% by weight. Ammonia is added to the monomer mixture while the temperature is maintained at 25°C. When the pH reaches 6, this mixture is pumped into another tank containing an oil and a sorbitan oleate surfactant. The resulting mixture is homogenized, oxygen removed by nitrogen sparging, polymerization initiated with a redox initiator, and the batch temperature controlled by cooling and the rate of initiator addition. At the end of the polymerization, an additional surfactant is added which enables the emulsion to invert rapidly into water.

Recent patents and publications describe process improvements. Conversions can be followed by on-line hplc (93). The enzyme amidase can be used to reduce residual monomers (94–96). A hydrogenation process for reduction of acrylamide in emulsions containing more that 5% residual monomer has been patented (95). Biodegradable oils have been developed (97).

Microemulsion Polymerization. Polyacrylamide microemulsions are low viscosity, nonsettling, clear, thermodynamically stable water-in-oil emulsions with particle sizes less than about 100 nm (98–100). They were developed to try to overcome the inherent settling problems of the larger particle size, conventional inverse emulsion polyacrylamides. To achieve the smaller microemulsion particle size, increased surfactant levels are required, making this system more expensive than inverse emulsions. Acrylamide microemulsions form spontaneously when the correct combinations and types of oils, surfactants, and aqueous monomer solutions are combined. Consequently, no homogenization is required. Polymerization of acrylamide microemulsions is conducted similarly to conventional acrylamide inverse emulsions. To date, polyacrylamide microemulsions have not been commercialized, although work has continued in an effort to exploit the unique features of this technology (100).

Uses

Polyacrylamides are classified according to weight-average molecular weight ($\overline{M}_w$) as follows:

high	15×10^6,
low	2×10^5,
very low	2×10^3.

Most uses for high molecular weight polyacrylamides in water treating, mineral processing, and paper manufacture are based on the ability of these polymers to flocculate small suspended particles by charge neutralization and bridging (101). Low molecular weight polymers are employed as dispersants (qv), crystal growth modifiers, or selective mineral depressants. In oil recovery, polyacrylamides adjust the rheology of injected water so that the polymer solution moves uniformly through the rock pores sweeping the oil ahead of it. Other applications such as superabsorbents and soil modification rely on the very hydrophilic character of polyacrylamides.

Water Treating. Municipal sludge plants and industrial waste treatment plants produce sludge which can be dewatered very efficiently by acrylamide polymers. Polyacrylamide flocculants can provide clean filtrates, higher filter cake solids and reduced sludge volume, and improve incineration and composting efficiency. Many municipal sludge treatment plants use pressure filtration for better water removal. These sludges are very hydrophilic, are negatively charged, and are difficult to dewater. Cationic polyacrylamides are being used to an increasingly greater extent in these applications, particularly, acrylamide–acryloxyethyltrimethylammonium chloride copolymers [69418-26-4]. Anionic polyacrylamides are also used extensively, alone or in combination with cationic polymers.

Paper mill effluents contain large amounts of degraded lignins, other colored material, and a high BOD. Cationic polyacrylamides are now being used to clarify these effluents. Sources of drinking water may contain trihalomethane precursors. These are refractory organic matter such as lignins, which in the presence of chlorine will form trihalomethanes which are suspected carcinogens. Highly cationic polymers in the presence of alum [17927-65-0], $Al_2(SO_4)_3 \cdot xH_2O$, are found to be effective in removing these precursors.

Anionic and nonionic polyacrylamides effectively remove suspended solids such as silt and clay from potable water. Suppliers provide special grades which meet EPA/FDA regulations for residual acrylamides. A recent publication (102) states that hydrolyzed polyacrylamides with narrow interchain charge distributions provide better performance in flocculation of clay. These polymers were prepared by alkaline hydrolysis. (See FLOCCULATING AGENTS.)

Industrial plants are lowering discharges by recycling process streams. Low molecular weight polyacrylates and anionic polyacrylamides are used as dispersants to inhibit scale formation on metal surfaces by calcium carbonate [471-34-1], $CaCO_3$; gypsum [7778-18-9], $CaSO_4 \cdot 2H_2O$; silica magnesium compounds; and other substances (103–105). The polymers alter the structures of the crystals during crystal growth and/or affect the rate of crystallization. The amount of scale formation on metal surfaces is strongly affected by the polymer molecular weight and the functional groups present on the polymer chains (106).

Mineral Processing.   Both synthetic and natural hydrophilic polymers are used in the mineral processing industry as flocculants and flotation modifiers. Most synthetic polymers in use are based on acrylamide alone or with anionic or cationic comonomers. Nonionic polymers are effective as flocculants for the insoluble gangue minerals in the acid leaching of copper and uranium (107,108), for thickening of iron ore slimes (109), and for thickening of gold flotation tailings (110). In some uranium leach operations, a low molecular weight cationic polymer with a relatively low charge density is used along with the nonionic polymer to improve supernatant clarity. Anionic polymers are extensively used in the industry. They are used as flocculants for insoluble residues formed in cyanide leaching of gold (111). Acrylamide–acrylic acid copolymers are used for thickening of copper, lead, and zinc concentrates in flotation of sulfide ores. These copolymers, containing from 50–100% carboxylate groups, are used to flocculate fine iron oxide particles in the manufacture of alumina from bauxite at high pH (112). Hydroxamated polyacrylamides, prepared from nonionic polyacrylamide and hydroxylamine salts, are also effective in this Bayer process (113). Copolymers [40623-73-2] of acrylamide and acrylamido-2-methylpropanesulfonic acid [15214-89-8] have been patented as phosphate slime dewatering aids (114).

Low molecular weight polyacrylamide derivatives with mineral specific functionalities have been developed as highly selective depressants. The depressants have certain ecological advantages over natural depressants such as starches and guar gums. The depressants provide efficient mineral recovery without flocculation. Partially hydrolyzed polyacrylamides with molecular weights of 7,000 to 85,000 can be used in sylvanite (KCl) recovery (115). Polymers having the functionality $-CONHCH_2OH$ are efficient modifiers in hematite–silica separations (116). Polymers containing the $-CONHCH(OH)COOH$ functionality provide excellent selectivity in separation of apatite from siliceous gangue in phosphate benefication. Valuable sulfide minerals containing copper and nickel can be separated effectively from gangue sulfide minerals such as pyrite in froth flotation processes when acrylamide–allylthiourea copolymers are added to depress the pyrite (117).

Acrylamide copolymers are effective iron ore pellet binders (118). When the ore slurry in water has a pH above 8, anionic polymers are effective. If the ore is acid washed to remove manganese, then a cationic polymer is effective.

Paper Manufacture.   Polyacrylamides are used as wet-end additives to promote drainage of water from the cellulose web, to retain white pigments and clay fillers in the sheet, to promote sheet uniformity, and to provide dry tensile strength improvements (119). Cationic polyacrylamides which have reacted with glyoxal are used to promote wet strength (120). These wet-strength resins have been used in paper towels. Recently, these glyoxalated polymers have been modified so that they can be used in toilet tissue. These polymers provide an initial high wet tensile strength with rapid tensile strength decay in water so that sewers may not become clogged. Anionic polyacrylamides have been used with alum to increase dry strength (121). Primary amide functionality promotes strong interfiber bonds between cellulose fibers. Sometimes paper mills use dry-strength additives so that recycled fiber, groundwood, thermomechanical pulp, and other low cost fiber can be used to produce liner board and other paper grades which must meet ICC requirements for Mullen burst strength and Concorra crush strength. Details on paper manufacture can be found in reference 122. All additives used for manufacture of food grade papers are subject to FDA regulations, and are listed in the *Code of Federal Regulations,* paragraphs 176.170, 176.180, 178.3400, and 178.3650 (1988).

Enhanced Oil Recovery.   Polymer flooding is a potentially important use for anionic polyacrylamides having molecular weights greater than 5 million and carboxyl contents of about 30%. The ionic groups provide the proper viscosity and mobility ratio for efficient displacement. The anionic charge prevents excessive adsorption onto negatively charged pores in reservoir rock. Viscosity loss is observed in brines particularly when calcium ion is present. A primary advantage of anionic polyacrylamides is low cost (123). Profile modification is a process where flooding water is diverted from zones with high permeability to other zones of lower permeability containing oil. For this, polymeric hydrogels are used (21). Metals such as chromium and aluminum can be injected with anionic polyacrylamides to cross-link the polymers in more permeable reservoir zones prior to the water flood (124–126).

Polyacrylamides are used in many other oilfield applications. These include cement additives for fluid loss control in well cementing operations (127), viscosity control additives for drilling muds (128), and fracturing fluids (129). Copolymers [40623-73-2] of acrylamide and acrylamidomethylpropanesulfonic acid do not degrade with the high concentrations of acids used in acid fracturing.

Hydrophobically Associating Polymers.   Extensive research in the 1980s focused on acrylamide copolymers containing small amounts of hydrophobic side chains. At zero or low shear rates, the apparent viscosity can be very large because of association of the hydrophobic groups between chains. In oil reservoir conditions, the polymers tolerate high salt concentrations while providing proper viscosifying properties (130). These associative thickeners are also used in coatings (131) and in oil spill clean-up (132). Reference 133 gives more information about associative polymers.

Superabsorbents.   Water-swellable polymers are used extensively in consumer articles and for industrial applications. Most of these polymers are cross-linked acrylic copolymers of metal salts of acrylic acid and acrylamide or other monomers such as 2-acrylamido-2-methylpropanesulfonic acid. These hydrogel forming systems can have high gel strength as measured by the shear modulus (134). Sometimes inorganic water-insoluble powder is blended with the polymer to increase gel strength (135). Patents describe processes for making cross-linked polyurethane foams which contain superabsorbent polymers (136,137).

Other Applications.   Polyacrylamides are used in many additional applications including soil modification (138), dust control (139,140), humidity control

(141), protein purification (142), removal of barium from wastewater (143), and removal of arsenic from hydrocarbons (144). Polyacrylamides have been used for many years in sugar manufacture and textile treatment.

Analytical Methods

A survey of characterization methods for linear and branched nonionic polyacrylamides is given in reference 5. Gel-permeation chromatography, diffusion and sedimentation, intrinsic viscosity, and light scattering are discussed emphasizing some difficulties encountered in obtaining consistent data. Characterization of anionic polyacrylamides used in oil recovery by ir spectroscopy, ^{13}C-nmr, tga, and x-ray diffraction is described in reference 8.

Characterization of molecular weight is difficult. New methods for characterizing anionic polyacrylamides include band sedimentation and low angle light scattering (145), hydrodynamic chromatography (146), field flow fractionation (147), and packed bed chromatography (148,149). Cationic polyacrylamides are difficult to analyze by size-exclusion chromatography because they adsorb irreversibly to column packings. Cationic functionalized silica beads have been used to overcome this problem (150).

Two-dimensional nmr spectroscopy has led to a much better understanding of biopolymers and their function (151). This technique has been applied to polyacrylamide for absolute assignments of proton and carbon spectra at the tetrad level (152).

Specifications, Shipping and Storage

Before polyacrylamides are sold, the amount of residual acrylamide is determined. In one method, the monomer is extracted from the polymer and the acrylamide content is determined by hplc (153). A second method is based on analysis by cationic exchange chromatography (154). For dry products the particle size distribution can be quickly determined by use of a shaker and a series of test sieves. Batches with small particles can present a dust hazard. The percentage of insoluble material is determined in both dry and emulsion products.

Polyacrylamide powders are typically shipped in moisture-resistant bags or fiber packs. Emulsion and solution polymers are sold in drums, tote bins, tank trucks, and tank cars. The transportation of dry and solution products is not regulated in the United States by the Department of Transportation, but emulsions require a DOT NA 1693 label.

Under normal conditions, dry polymers are stable for one year or more. The emulsion and solution products have somewhat shorter shelf lives.

Safety and Health

Dry anionic and nonionic polyacrylamides have acute oral (rat) and dermal (rabbit) LD_{50} values of greater than 2.5 and greater than 10.0 g/kg, respectively.

Dry cationic polyacrylamides have acute oral (rat) and dermal (rabbit) LD_{50} values of greater than 5.0 and greater than 2.0 g/kg, respectively. Emulsion nonionic, anionic, and cationic polyacrylamides have both acute oral (rat) and dermal (rabbit) LD_{50} values of greater than 10 g/kg. Dry nonionic and cationic material caused no skin and minimal eye irritation during primary irritation studies with rabbits. Dry anionic polyacrylamide did not produce any eye or skin irritation in laboratory animals. Emulsion nonionic polyacrylamide produced severe eye irritation in rabbits, while anionic and cationic material produced minimal eye irritation in rabbits. Emulsion nonionic, anionic, and cationic polyacrylamide produced severe, irreversible skin irritation when tested in rabbits which had the test material held in skin contact by a bandage for 24 h. This represents an exaggeration of spilling the product in a boot for several hours. When emulsion nonionic, cationic, and anionic polyacrylamides were tested un-

Table 4. Suppliers of Polyacrylamide

Region	Companies
United States	Allied Colloids, Inc.
	American Cyanamid Company
	Aqua Ben Corporation
	Betz Laboratories, Inc.
	Calgon Corporation (Merck & Co.)
	Chemtall, Inc. (SNF Floerger)
	Dearborn Chemical Company (W. R. Grace & Company)
	The Dow Chemical Company
	Drew Chemical Corporation (Ashland Chemical, Inc.)
	Exxon Chemical Company
	Hercules, Inc.
	Nalco Chemical Company
	Polypure, Inc.
	Secodyne, Inc.
	Stockhausen, Inc.
Europe	Allied Colloids, Ltd.
	American Cyanamid Company
	BASF AG
	Chemische Fabrik Stockhausen & Cie
	The Dow Chemical Company
	SNF Floerger (France)
	Kemira Oy (Finland)
	Rohm GmbH
	Rhône-Poulenc Specialties Chimiques (France)
Japan	Dai-Ichi Kogyo Seiyaku Company, Ltd.
	Kurita Water Industries, Ltd.
	Kyoritsu Yuki Company, Ltd.
	Mitsubishi Chemical Industries Company, Ltd.
	Mitsui-Cyanamid, Ltd.
	Sankyo Kasei Company, Ltd.
	Sanyo Chemical Industries, Ltd.
	Takenaka Komuten Company, Ltd.
	Toa Gosei Chemical Industry Company, Ltd.

der conditions representing spilling of product on clothing, only mild skin irritation was noted. Polyacrylamides are used safely for numerous indirect food packaging applications, potable water, and direct food applications.

Economic Aspects

Worldwide, there are many sellers of polyacrylamides. Some of these are repackagers. Suppliers are listed in Table 4. Selling prices for polyacrylamides vary considerably depending on the product form (solution, emulsion, dry), type (anionic, nonionic, cationic), and other factors. Prices can range from as low as about $2/kg for simple dry nonionic polyacrylamides to $10/kg and more for highly charged cationic polymers. In many applications, such as sludge dewatering in waste treatment, the need for increased performance has lead to increased functionalization (eg, higher cationicity) and increased cost. In Western Europe it was estimated for 1988 that 47×10^6 kg of polyacrylamides were used, with the primary outlets being water treating and paper manufacture (155). Probably a large amount was exported to Africa for mining. In Japan the major consumption has been in paper manufacture and water treatment. Table 5 gives an estimated breakdown of acrylamide consumption for polymerization to anionic, nonionic, and cationic polyacrylamides. The majority of polymer for EOR was exported (155).

Table 5. Consumption of Acrylamide for Polyacrylamides, 10^3 t

Year	Paper	Water treatment	Mining	EOR	Other	Total
		Western Europe				
1979	4	1	8.5	3.5	4	*21*
1985	3.5	9	6	1	3.5	*23*
1988	6	17	12	1	11	*47*
		Japan				
1987	28	20		2	5	*55*
1988	28	22		2	5	*57*
1989	29	24		3	5	*61*

BIBLIOGRAPHY

"Acrylamide Polymers" in *ECT* 3rd ed., Vol. 1, pp. 312–330, by J. D. Morris and R. J. Penzenstadler, The Dow Chemical Company.

1. O. G. Lewis, *Physical Constants of Linear Homopolymers,* Springer-Verlag, New York, 1968, p. 23.
2. J. Klein and R. Hietzmann, *Makromol. Chem.* **179,** 1895 (1978).
3. J. Brandrup and E. H. Immergut, eds., *Polymer Handbook,* 2nd ed., John Wiley & Sons, Inc., New York, 1975, p. III-222.

4. J. E. Lancaster and M. N. O'Connor, *J. Polym. Sci. Polym. Lett. Ed.* **20**, 547 (1982).

5. W. M. Kulicke, R. Kniewske, and J. Klein, *Prog. Polym. Sci.* **8**, 376 (1982).

6. S. Sawant and H. Morawetz, *J. Polym. Sci. Polym. Lett. Ed.* **20**, 385–388 (1982).

7. S. Sawant and H. Morawetz, *Macromolecules* **17**, 2427–2431 (1984).

8. S. J. Guerrero, P. Boldarino, and J. A. Zurimendi, *J. Appl. Polym. Sci.* **30**, 962 (1985).

9. Ref. 5, p. 383.

10. Ref. 3, p. IV-245.

11. Ref. 3, p. IV-183.

12. *MAGNIFLOC 905N Flocculant, Material Safety Data,* American Cyanamid Company, Wayne, N.J., 1989.

13. F. A. Bovey, *High Resolution NMR of Macromolecules,* Academic Press, New York, 1972, p. 147.

14. Ref. 5, p. 379.

15. E. H. Gleason, M. L. Miller, and G. F. Sheats, *J. Polym. Sci.* **38**, 133 (1959).

16. Ref. 8, p. 960.

17. A. Silberberg, J. Eliassaf, and A. Katchalsky, *J. Polym. Sci.* **23**, 259 (1957).

18. J. J. Kozakiewicz, American Cyanamid Company, Stamford, Conn., unpublished results, 1987.

19. W. M. Kulicke, M. Koetter, and H. Graeger, *Advances in Polymer Science 89,* Springer-Verlag Berlin, Heidelberg, 1989, p. 32.

20. W. P. Shyluk and F. S. Stow, Jr., *J. Appl. Polym. Sci.* **13**, 1023 (1969).

21. U.S. Pat. 4,861,499 (Aug. 29, 1989), R. G. Ryles and R. E. Neff (to American Cyanamid Company).

22. M. Leca, *Polym. Bull.* **16**, 537–543 (1986).

23. G. Muller, J. P. Laine, and J. C. Fenyo, *J. Polym. Sci. Chem.* **17**, 659–672 (1979).

24. F. Schwartz and J. Francois, *Makromol. Chem.* **182**, 2775 (1981).

25. W. M. Kulicke, *Polymer Preprints* **30**, 379 (1989).

26. W. M. Kulicke, N. Boese, and M. Bouldin in D. N. Schulz and G. A. Stahl, eds., *Water-Soluble Polymers for Petroleum Recovery,* Plenum Publishing Corp., New York, 1988, pp. 1–18.

27. U.S. Pat. 4,730,028 (Mar. 8, 1988), J. Bock and P. L. Valint (to Exxon Research and Engineering Co.).

28. J. K. Borchardt in J. K. Borchardt and T. F. Yen, eds., *Enhanced Oil Recovery and Production Simulation* (ACS Symposium Ser. Vol. 396), American Chemical Society, Washington, D.C., 1989.

29. S. Evani and G. D. Rose, *Polym. Mater. Sci. Eng.* **57**, 477–481 (1987).

30. F. Halverson, J. E. Lancaster, and M. N. O'Connor, *Macromolecules* **18**, 1139 (1985).

31. N. D. Truong, J. C. Galin, J. Francois, and Q. T. Pham, *Polymer* **27**, 459 (1986).

32. K. Ysuda, K. Okajima, and K. Kamide, *Polymer Journal* **20**, 1101 (1988).

33. H. Kheradmand, J. Francois, and V. Plazanet, *Polymer* **29**, 860 (1988).

34. J. Moens and G. Smets, *J. Polym. Sci.* **23**, 931 (1957).

35. G. Smets and A. M. Hesbain, *J. Polym. Sci.* **40**, 217 (1959).

36. U.S. Pat. 3,022,279 (Feb. 20, 1962), A. C. Profitt (to The Dow Chemical Company).

37. R. Aksberg and L. Wagberg, *J. Appl. Polym. Sci.* **38**, 297 (1989).

38. F. Lafuma and G. Durand, *Polym. Bull.* **21**, 315 (1989).

39. D. R. Draney, S. Y. Huang, J. J. Kozakiewicz, and D. W. Lipp, *American Chemical Society Polymer Preprints* **31**, 500 (1990).

40. M. Tramonti, L. Angolini, and N. Ghedini, *Polymer* **29**, 775 (1980).

41. C. J. McDonald and R. M. Beaver, *Macromolecules* **12**, 203 (1979).

42. R. H. Pelton, *J. Polym. Sci. Polym. Chem. Ed.* **22**, 3955 (1984).

43. U.S. Pat. 4,137,165 (Jan. 20, 1979), A. T. Coscia and N. D. O'Connor (to American Cyanamid Company).

44. U.S. Pat. 3,332,927 (July 25, 1967), M. F. Hoover (to Calgon Corporation).
45. A. M. Schiller and T. J. Suen, *Ind. Eng. Chem.* **48,** 2132 (1956).
46. U.S. Pat. 3,979,348 (Sept. 7, 1976), E. G. Ballweber and K. G. Phillips (to Nalco Chemical Company).
47. D. P. Bakalik and D. J. Kowalski, *Polym. Mater. Sci. Eng.* **57,** 845 (1987).
48. U.S. Pat. 4,762,894 (Aug. 9, 1988), D. W. Fong and D. J. Kowalski (to Nalco Chemical Company).
49. U.S. Pat. 4,282,087 (Aug. 4, 1989), R. M. Goodman and H. P. Panzer (to American Cyanamid Company).
50. U.S. Pat. 4,360,425 (Nov. 23, 1982), S. K. Lim and R. M. Goodman (to American Cyanamid Company).
51. U.S. Pat. 4,605,702 (Aug. 12, 1986), G. J. Guerro, R. J. Proverb, and R. F. Tarvin (to American Cyanamid Company).
52. U.S. Pat. 4,603,176 (July 29, 1986), D. W. Bjorkquist and W. M. Schmitt (to The Proctor and Gamble Company).
53. D. R. Nagaraj, A. S. Rothenberg, D. W. Lipp, and H. P. Panzer, *Int. J. of Miner. Process.* **20,** 291 (1987).
54. S. Machida, *Tappi* **52,** 1734 (1969).
55. U.S. Pat. 4,587,306 (May 6, 1986), L. Vio and G. Meunier (to Societe Nationale Elf Aquitaine).
56. A. J. Domb, E. G. Cravalho, and R. Langer, *J. Poly. Sci. Polym. Chem. Ed.* **26,** 2623 (1988).
57. World Pat. 88/06602 (Sept. 7, 1988), A. J. Domb and co-workers (to Massachusetts Institute of Technology).
58. H. Tanaka and R. Senju, *Bull. Chem. Soc. Jpn.* **49**(10), 2821 (1976).
59. H. Tanaka, *J. Polym. Sci. Polym. Chem. Ed.* **17,** 1239 (1979).
60. Eur. Pat. Appl. 289,823 (May 5, 1987), Y. L. Fu, S. Y. Huang, and R. W. Dexter (to American Cyanamid Company) .
61. G. S. Misra and U. D. N. Bajpai, *Prog. Polym. Sci.* **8**(1–2), 61–131 (1982).
62. F. S. Dainton, K. Irvin, and D. A. G. Wamsley, *Trans. Faraday Soc.* **56,** 1784 (1960).
63. T. J. Suen, Y. Jen, and J. V. Lockwood, *J. Polym. Sci.* **31,** 481 (1958).
64. D. J. Currie, F. S. Dainton, and W. S. Watt, *Polymer* **6,** 451 (1965).
65. T. J. Suen, A. M. Schiller, and W. N. Russel, *J. Appl. Polym. Sci.* **3,** 126 (1960).
66. V. F. Kurenkov and V. A. Myagchenkov, *Eur. Polym. J.* **16,** 1229 (1980).
67. J. Barton, V. Juranicova, and V. Vaskova, *Makromol. Chem.* **186,** 1935 (1985).
68. S. M. Shawki and A. E. Hamielec, *J. Appl. Polym. Sci.* **23,** 3341 (1979).
69. Ref. 3, p. II-59.
70. T. J. Suen, A. M. Schiller, and W. N. Russell, in *Polymerization and Polycondensation Processes* (Adv. Chem. Ser. 34), American Chemical Society, Washington, D.C., 1963, p. 217.
71. Ref. 63, p. 493.
72. F. S. Dainton and M. Tordoff, *Trans. Faraday Soc.* **53,** 499 (1957).
73. E. A. S. Cavell and I. T. Gilson, *J. Polym. Sci. Part A-1* **4,** 541 (1966).
74. R. Z. Greenley, *J. Macromol. Sci. Chem.* **14,** 427, 445 (1980).
75. S. M. Shawki and A. E. Hamielec, *J. Appl. Polym. Sci.* **23,** 3155 (1979).
76. S. Plochocka, *J. Macromol. Sci. Rev. Macromol. Chem.* **20**(1), 67 (1979).
77. W. M. Thomas and D. W. Wang in J. I. Kroschwitz, ed., *Encyclopedia of Polymer Science and Engineering,* 2nd. ed., Vol. 1, Wiley-Interscience, New York, 1985, p. 182.
78. H. Tanaka, *J. Polym. Sci. Polym. Chem. Ed.* **24,** 29–36 (1986).
79. W. Baade, D. Hunkler, and A. E. Hamielec, *J. Appl. Polym. Sci.* **38,** 185 (1989).
80. U.S. Pat. 4,024,040 (May 17, 1977), C. J. Phalangas, A. J. Restaino, and H. B. Yun (to Hercules, Inc.).

81. U.S. Pat. 4,115,339 (Sept. 19, 1978), A. J. Restaino (to Hercules, Inc.).
82. U.S. Pat. 4,138,839 (Feb. 6, 1979), P. H. Landolt and L. N. Allen (to American Cyanamid Company).
83. U.S. Pat. 4,132,844 (Jan. 2, 1979), M. W. Coville (to American Cyanamid Company).
84. Fr. Pats. 2,428,053 and 2,428,054 (June 9, 1978), J. Boutin and J. Neel (to Rhône-Poulenc Industries).
85. U.S. Pat. 4,762,862 (Aug. 9, 1988), A. Yada and co-workers (to Dai-ichi Kogyo Seiyaku Co.).
86. Ger. Pat. DE 3,621,429 (Jan. 8, 1987), J. M. Lucas and A. C. Pericone (to Milchem, Inc.); *Chem. Abstr.* **106,** 138949e (1987).
87. Aust. Pat. AU8818380 (Jan. 5, 1989), W. B. Davies (to American Cyanamid Company).
88. U.S. Pat. 4,857,610 (Aug. 15, 1989), M. Chemlir and J. Pauen (to Chemische Fabrik Stockhausen GmbH).
89. Eur. Pat. EP 102,760 (Aug. 9, 1982), P. Flesher and A. S. Allen (to Allied Colloids, Ltd.).
90. U.S. Pat. 3,284,393 (Nov. 8, 1966), J. W. Vanderhoff and R. M. Wiley (to The Dow Chemical Company).
91. U.S. Pat. 3,826,771 (July 30, 1974), D. R. Anderson and A. J. Frisque (to Nalco Chemical Company).
92. U.S. Pat. 4,439,332 (Mar. 27, 1984), S. Frank, A. T. Coscia, and J. M. Schmitt (to American Cyanamid Company).
93. Ref. 79, p. 187.
94. U.S. Pat. 4,742,114 (May 3, 1988), R. L. Wetgrove and R. W. Kaesler (to Nalco Chemical Company).
95. U.S. Pat. 4,427,821 (Nov. 22, 1982), D. W. Fong and R. J. Allain (to Nalco Chemical Company).
96. Eur. Pat. EP 272,026 (Apr. 12, 1987), M. A. Carver and J. Hinton (to Imperial Chemical Industries).
97. U.S. Pat. 4,824,894 (Apr. 25, 1987), R. Schnee, A. S. Cordialo, and J. Masanek (to Rohm GmbH Chemische Fabrik).
98. U.S. Pat. 4,021,364 (May 3, 1977), P. Speiser and G. Birrenbach (to Prof. Dr. Peter Speiser, Forsch. Switzerland).
99. F. Candau, in M. A. El-Nokaly, ed., *Polymer Association Structures: Microemulsions and Liquid Crystals* (ACS Symposium Ser. 384), American Chemical Society, Washington, D.C., 1989, pp. 47–61.
100. World Pat. WO 10274 (June 17, 1988), F. Candau and P. Buchert (to Norsolor).
101. F. Mabire, R. Audebert, and C. Quivoron, *J. Colloid Interface Sci.* **97,** 120 (1984).
102. K. Takada and J. Ryugo, *Kobunshi Ronbunshu* **46**(3), 145 (1989); *Chem. Abstr.* **111**(2), 8178v (1989).
103. U.S. Pat. 4,499,022 (Feb. 12, 1985), W. F. Massler III and Z. Amjad (to the B.F. Goodrich Company).
104. Fr. Pat. 2,604,712 (Apr. 8, 1988), C. Roque and A. Ribba (to Institut Francois du Petrole).
105. Jpn. Pat. J61/107997 (May 26, 1986) (Kurita Water Industries, Ltd.); *Chem. Abstr.* **105,** 232,183b (1986).
106. Z. Amjad, *J. Colloid Interface Sci.* **123,** 523–536 (1988).
107. J. M. W. Mackenzie, *Eng. Min. J.* 80–87 (Oct. 1980).
108. J. P. MacDonald, P. L. Mattison, and J. M. W. MacKenzie, *J.S. Afr. Inst. Min. Metall.* **81,** 303 (1981).
109. *Superfloc 16 Plus Flocculant,* American Cyanamid Company, Wayne, N.J., 1986.
110. *Mining Chemicals Handbook,* Rev. ed., American Cyanamid Company, Wayne, N.J., 1986.
111. D. L. Dauplaise and M. F. Werneke, *Proceedings of the Consolidation and Dewatering*

      of Fine Particles Conference, University of Alabama, Tuscaloosa, Ala., 1982, pp. 90–113.

112. H. I. Heitner, T. Foster, and H. P. Panzer in J. I. Kroschwitz, ed., *Encyclopedia of Polymer Science and Engineering,* 2nd ed., Vol. 9, Wiley-Interscience, New York, 1987, p. 830.

113. U.S. Pat. 4,767,540 (Aug. 30, 1988), D. P. Spitzer and W. S. Yen (to American Cyanamid Company).

114. U.S. Pat. 4,342,653 (Aug. 3, 1982), F. Halverson (to American Cyanamid Company).

115. U.S. Pat. 4,533,465 (Aug. 6, 1985), R. M. Goodman and S. K. Lim (to American Cyanamid Company).

116. Ref. 53, p. 293.

117. U.S. Pat. 4,866,150 (Sept. 12, 1989), D. W. Lipp and D. R. Nagaraj (to American Cyanamid Company).

118. Eur. Pat. EP 288,150 (Mar. 27, 1987), A. Allen (to Allied Colloids, Ltd.).

119. U.S. Pat. 4,824,523 (June 1, 1987), L. Wagberg and T. Linndstrum (to Svenska Traforskningsinstitutet).

120. U.S. Pat. 3,556,932 (Jan. 19, 1971), A. T. Coscia and L. Williams (to American Cyanamid Company).

121. W. F. Reynolds and R. B. Wasser in J. P. Casey, ed., *Pulp and Paper Chemistry and Chemical Technology,* John Wiley & Sons, Inc., New York, 1981, pp. 1447–1470.

122. J. P. Casey, ed., *Pulp and Paper Chemistry and Chemical Technology,* 4 vols., John Wiley & Sons, Inc., New York, 1987, pp. 1–2609.

123. J. K. Borchardt, "Oil Field Applications," in J. I. Kroschwitz, ed., *Encyclopedia of Polymer Science and Engineering,* 2nd ed., Vol. 10, Wiley-Interscience, New York, 1987, p. 350.

124. U.S. Pat. 4,744,419 (May 17, 1988), R. Sydansk and P. Argabright (to Marathon Oil Company).

125. U.S. Pat. 4,783,492 (Nov. 8, 1988), H. Dovan and R. Hutchins (to Union Oil Company of Calif.).

126. U.S. Pat. 4,606,407 (Aug. 8, 1986), P. Shu (to Mobil Oil Corp.).

127. U.S. Pat. 4,806,164 (Feb. 21, 1989), L. Brothers (to Halliburton Co.).

128. U.S. Pat. 4,423,199 (Dec. 27, 1983), C. J. Chang and T. E. Stevens (to Rohm and Haas Co.).

129. U.S. Pat. 4,417,989 (Nov. 29, 1983), W. D. Hunter (to Texaco Development Corp.).

130. J. Bock and co-workers, "Hydrophobically Associating Polymers," in G. A. Stahl and D. N. Schulz, eds., *Water-Soluble Polymers for Petroleum Recovery,* Plenum Press, New York, 1988, pp. 147–160.

131. U.S. Pat. 4,722,962 (Feb. 2, 1988), G. D. Shay (to DeSoto, Inc.).

132. U.S. Pat. 4,734,205 (Mar. 29, 1988), D. F. Jacques and J. Bock (to Exxon Research and Engineering Co.).

133. A. Karunasena, R. G. Brown, and J. E. Glass, in J. E. Glass, ed., *Polymers in Aqueous Media: Performance Through Association* (Adv. Chem. Ser. 223), American Chemical Society, Washington, D.C., 1989.

134. U.S. Pat. 4,654,039 USRE 32649 (Apr. 19, 1988), K. Brandt, T. Inglin, and S. Goldman (to The Proctor and Gamble Co.).

135. U.S. Pat. 4,500,670 (Feb. 19, 1985), M. J. McKinley and D. P. Sheridan (to The Dow Chemical Company).

136. U.S. Pat. 4,725,628 (Feb. 16, 1988), C. Garvey, J. Pazos, and G. Ring (to Kimberly Clark Corp.).

137. U.S. Pat. 4,725,629 (Feb. 16, 1988), C. Garvey and J. Pazos (to Kimberly Clark Corp.).

138. U.S. Pat. 4,797,145 (Jan. 10, 1989), G. A. Wallace and A. Wallace.

139. U.S. Pat. 4,417,992 (Nov. 29, 1983), B. R. Bhattacharyya and W. J. Roe (to Nalco Chemical Co.).
140. U.S. Pat. 4,801,635 (Jan. 31, 1989), J. K. Zinkan and L. U. Koenig, Jr. (to Zinkan Enterprises).
141. U.S. Pat. 4,683,258 (July 28, 1987), H. Itch and co-workers (to Mitsui Toatsu Chemical Company, Inc.).
142. U.S. Pat. 4,716,219 (Dec. 29, 1987), B. Eggiman, E. Hochuli, and A. Schacher (to Hoffman-LaRoche, Inc.).
143. U.S. Pat. 4,491,523 (Jan. 1, 1985), R. W. Foreman (to Park Chemical Co.).
144. U.S. Pat. 4,541,918 (Sept. 17, 1985), P. R. Stapp (to Phillips Petroleum Co.).
145. G. Holzwarth, L. Soni, and D. N. Schulz, *Macromolecules* **19**, 422 (1986).
146. D. A. Hoagland, K. A. Larson, and R. K. Prud'homme, in H. G. Barth, eds., *Modern Methods of Particle Size Analyses,* John Wiley & Sons, Inc., New York, 1984.
147. J. J. Kirkland and W. W. Yau, *J. Chromatogr.* **353**, 95–107 (1986).
148. D. A. Hoagland and R. K. Prud'homme, *J. Appl. Polym. Sci.* **36**, 935 (1988).
149. U.S. Pat. 4,532,043 (July 30, 1985), R. K. Prud'homme, F. W. Stanley, Jr., and Martin A. Langhorst (to The Dow Chemical Company).
150. M. Stickler and F. Eisenbeiss, *Eur. Polym. J.* **20**, 849 (1984).
151. H. Kessler, M. Gehrke, and C. Griesinger, *Angew. Chem., Int. Ed. Engl.* **27**, 490–536 (1988).
152. K. Hikichi, M. Ikura, and M. Yosuda, *Polym. J.* **29**, 851 (1988).
153. *Methods Exam. Waters Assoc. Matl. 1988,* United Kingdom Dept. of the Environment; *Chem. Abstr.* **108**, 197583v (1988).
154. G. Schmoetzer, *Chromatographia* **4**, 391–395 (1971); *Chem. Abstr.* **76**, 25744c (1972).
155. *Chemical Economics Handbook, CEH 80 Database,* SRI International, Menlo Park, Calif., Oct. 10, 1989.

DAVID LIPP
JOSEPH KOZAKIEWICZ
American Cyanamid Company

ACRYLIC ACID AND DERIVATIVES

The term acrylates includes derivatives of both acrylic (CH_2=CHCOOH) and methacrylic acids (CH_2=C(CH_3)COOH). This article discusses the preparation, properties, and reactions of acrylic acid monomers only (see METHACRYLIC ACID AND DERIVATIVES). Acrylic acid (propenoic acid) was first prepared in 1847 by air oxidation of acrolein (1). Interestingly, after use of several other routes over the past half century, it is this route, using acrolein from the catalytic oxidation of propylene, that is currently the most favored industrial process. Polymerization of acrylic esters has been known for just over a century (2). However, it was not until 1930 that the technical difficulties of their manufacture and polymerization were overcome (3). The rate of consumption of acrylates grew between 10 and 20% annually during the late 1970s. Growth fluctuated with the economy in the early 1980s with some announced capacity increases being delayed until later in the decade. Although growth in the ester markets has dropped appreciably, new applications for polymers of acrylic acid in the superabsorbent and detergent fields surged in the late 1980s. Current annual growth is ~3–5%; nonester markets contribute more than the more mature ester area. Manufacture continues to be concentrated in the United States, Europe, and Japan. Worldwide capacity for acrylic acid in the early 1990s will reach about 1.8×10^9 kg per year if all currently announced expansions are completed.

Area	Capacity, 10^3t
United States	682
Europe	636
Far East	500

Far East capacity is primarily in Japan but includes Taiwan, Korea, and China.

Acrylates are primarily used to prepare emulsion and solution polymers. The emulsion polymerization process provides high yields of polymers in a form suitable for a variety of applications. Acrylate polymer emulsions were first used as coatings for leather in the early 1930s and have found wide utility as coatings, finishes, and binders for leather, textiles, and paper. Acrylate emulsions are used in the preparation of both interior and exterior paints, floor polishes, and adhesives. Solution polymers of acrylates, frequently with minor concentrations of other monomers, are employed in the preparation of industrial coatings. Polymers of acrylic acid can be used as superabsorbents in disposable diapers, as well as in formulation of superior, reduced-phosphate-level detergents.

The polymeric products can be made to vary widely in physical properties through controlled variation in the ratios of monomers employed in their preparation, cross-linking, and control of molecular weight. They share common qualities of high resistance to chemical and environmental attack, excellent clarity, and attractive strength properties (see ACRYLIC ESTER POLYMERS). In addition to acrylic acid itself, methyl, ethyl, butyl, isobutyl, and 2-ethylhexyl acrylates are manufactured on a large scale and are available in better than 98–99% purity (4). They usually contain 10–200 ppm of hydroquinone monomethyl ether as polymerization inhibitor.

Physical Properties

Physical properties of acrylic acid and representative derivatives appear in Table 1. Table 2 gives selected properties of commercially important acrylate esters, and Table 3 lists the physical properties of many acrylic esters.

Acrylic acid is a moderately strong carboxylic acid. Its dissociation constant is 5.5×10^{-5}. Vapor pressure as a function of temperature is given in Table 4 for

Table 1. Physical Properties of Acrylic Acid Derivatives

Property	Acrylic acid	Acrolein	Acrylic anhydride	Acryloyl chloride	Acrylamide
molecular formula	$C_3H_4O_2$	C_3H_4O	$C_6H_6O_3$	C_3H_3OCl	C_3H_5ON
CAS Registry Number	[79-10-7]	[107-02-8]	[2051-76-5]	[814-68-6]	[79-06-1]
melting point, °C	13.5	−88			84.5
boiling point[a], °C	141	52.5	38[b]	75	125[c]
refractive index[d], n_D	1.4185[e]	1.4017	1.4487	1.4337	
flash point, Cleveland open cup, °C	68				
density[d], g/mL	1.045[e]	0.838		1.113	1.122[f]

[a] At 101.3 kPa = 1 atm unless otherwise noted.
[b] At 0.27 kPa.
[c] At 16.6 kPa.
[d] At 20°C, unless otherwise noted.
[e] At 25°C.
[f] At 30°C.

Table 2. Properties of Commercially Important Acrylate Esters[a]

Property	Methyl	Ethyl	n-Butyl	Isobutyl	2-Ethylhexyl
solubility at 23°C, parts per 100 of solvent					
in water	5	1.5	0.2	0.2	0.01
of water in ester	2.5	1.5	0.7	0.6	0.15
heat of vaporization, kJ/g[b]	0.39	0.35	0.19	0.30	0.25
specific heat, J/g·C[b]	2.00	1.97	1.93	1.93	1.93
boiling points[c] of azeotropes					
with water, °C	71	81.1	94.5		
water content, %	7.2	15	40		
with methanol, °C	62.5	64.5			
methanol content, %	54	84.4			
with ethanol, °C	73.5	77.5			
ethanol content, %	42.4	72.7			
with n-butanol, °C			119		
butanol content, %			89		
heat of polymerization, kJ/mol[b]	78.7	77.8	77.4		

[a]Refs. 4–5.
[b]To convert J to cal, divide by 4.184.
[c]At 101.3 kPa = 1 atm.

acrylic acid and four important esters (4,16–18). The lower esters form azeotropes both with water and with their corresponding alcohols.

Reactions

Acrylic acid and its esters may be viewed as derivatives of ethylene, in which one of the hydrogen atoms has been replaced by a carboxyl or carboalkoxyl group. This functional group may display electron-withdrawing ability through inductive effects of the electron-deficient carbonyl carbon atom, and electron-releasing effects by resonance involving the electrons of the carbon–oxygen double bond. Therefore, these compounds react readily with electrophilic, free-radical, and nucleophilic agents.

Carboxylic Acid Functional Group Reactions.   Polymerization is avoided by conducting the desired reaction under mild conditions and in the presence of polymerization inhibitors. Acrylic acid undergoes the reactions of carboxylic acids and can be easily converted to salts, acrylic anhydride, acryloyl chloride, and esters (16–17).

Salts are made by reaction of acrylic acid with an appropriate base in aqueous medium. They can serve as monomers and comonomers in water-soluble or water-dispersible polymers for floor polishes and flocculants.

Acrylic anhydride is formed by treatment of the acid with acetic anhydride or by reaction of acrylate salts with acryloyl chloride. *Acryloyl chloride* is made by reaction of acrylic acid with phosphorous oxychloride, or benzoyl or thionyl chloride. Neither the anhydride nor the acid chloride is of commercial interest.

Esters. Most acrylic acid is used in the form of its methyl, ethyl, and butyl esters. Specialty monomeric esters with a hydroxyl, amino, or other functional group are used to provide adhesion, latent cross-linking capability, or different solubility characteristics. The principal routes to esters are direct esterification with alcohols in the presence of a strong acid catalyst such as sulfuric acid, a soluble sulfonic acid, or sulfonic acid resins; addition to alkylene oxides to give hydroxyalkyl acrylic esters; and addition to the double bond of olefins in the presence of strong acid catalyst (19,20) to give ethyl or secondary alkyl acrylates.

$$CH_2\!=\!CHCOOH + ROH \xrightarrow{H^+} CH_2\!=\!CHCOOR + H_2O$$

$$CH_2\!=\!CHCOOH + \overset{O}{\overbrace{CH_2\!-\!CH_2}} \longrightarrow CH_2\!=\!CHCOOCH_2CH_2OH$$

$$CH_2\!=\!CHCOOH + CH_2\!=\!CH_2 \xrightarrow{H^+} CH_2\!=\!CHCOOCH_2CH_3$$

Acrylic esters may be saponified, converted to other esters (particularly of higher alcohols by acid catalyzed alcohol interchange), or converted to amides by aminolysis. Transesterification is complicated by the azeotropic behavior of lower acrylates and alcohols but is useful in preparation of higher alkyl acrylates.

Amides. Reaction of acrylic acid with ammonia or primary or secondary amines forms amides. However, acrylamide (qv) is better prepared by controlled hydrolysis of acrylonitrile (qv). Esters can be obtained by carrying out the nitrile hydrolysis in the presence of alcohol.

$$CH_2\!=\!CHC\!\equiv\!N \xrightarrow[H_2O]{H_2SO_4} CH_2\!=\!CHCO\overset{+}{N}H_3HSO_4^- \xrightarrow[ROH,\ H^+]{H_2O} CH_2\!=\!CH\overset{\overset{\displaystyle O}{\|}}{C}OH(R)$$

Unsaturated Group Reactions. In addition to a comprehensive review of these reactions (16), there are excellent texts (17,18). Free-radical-initiated polymerization of the double bond is the most common reaction and presents one of the more troublesome aspects of monomer manufacture and purification.

Substituted ring compounds are formed readily by Diels-Alder reactions.

$$CH_2\!=\!CHCOOR + CH_2\!=\!CH\!-\!CH\!=\!CH_2 \longrightarrow \hexagon\!-\!COOR$$

Additions. Halogens, hydrogen halides, and hydrogen cyanide readily add to acrylic acid to give the 2,3-dihalopropionate, 3-halopropionate, and 3-cyanopropionate, respectively (21).

On storage or at elevated temperatures, acrylic acid dimerizes to give 3-acryloxypropionic acid [*24615-84-7*], $C_6H_8O_4$.

$$2\ CH_2\!=\!CHCOOH \longrightarrow CH_2\!=\!CHCOOCH_2CH_2COOH$$

Although the reaction is second order in acrylic acid concentration, the rate of dimer formation for neat acrylic acid available commercially is quite ade-

Table 3. Physical Properties of Acrylic Esters[a], CH_2=CHCOOR

Compound	Molecular formula	CAS Registry Number	Boiling point °C	kPa[b]	Refractive index n_D^{20}	Density d_4^{20}
n-Alkyl esters[c]						
methyl	$C_4H_6O_2$	[96-33-3]	80	101	1.4040	0.9535
ethyl	$C_5H_8O_2$	[140-88-5]	43	13.7	1.4068	0.9234
propyl	$C_6H_{10}O_2$	[925-60-0]	44	5.3	1.4130	0.9078
butyl	$C_7H_{12}O_2$	[141-32-2]	35	1.1	1.4190	0.8998
pentyl	$C_8H_{14}O_2$	[2998-23-4]	48	0.9	1.4240	0.8920
hexyl	$C_9H_{16}O_2$	[2499-95-8]	40	0.15	1.4280	0.8882
heptyl	$C_{10}H_{18}O_2$	[2499-58-3]	57	0.13	1.4311	0.8846
octyl	$C_{11}H_{20}O_2$	[2499-59-4]	57	0.007	1.4350	0.8810
nonyl	$C_{12}H_{22}O_2$	[2664-55-3]	76	0.03	1.4375	0.8785
decyl	$C_{13}H_{24}O_2$	[2156-96-9]	120	0.67	1.4400	0.8781
dodecyl	$C_{15}H_{28}O_2$	[2156-97-0]	120	0.11	1.4440	0.8727
tetradecyl	$C_{17}H_{32}O_2$	[21643-42-5]	138	0.05	1.4468	0.8700
hexadecyl	$C_{19}H_{36}O_2$	[13402-02-3]	170	0.20	1.4470 (30°C)	0.8620 (30°C)
Secondary and branched-chain alkyl esters[d]						
isopropyl	$C_6H_{10}O_2$	[689-12-3]	52	13.7	1.4060	0.8932
isobutyl	$C_7H_{12}O_2$	[106-62-8]	62	6.7	1.4150	0.8896
sec-butyl	$C_7H_{12}O_2$	[2998-08-5]	60	6.7	1.4140	0.8914
2-ethylhexyl	$C_{11}H_{20}O_2$	[103-11-7]	85	1.07	1.4365	0.8852
Esters of olefinic alcohols[e]						
allyl	$C_6H_8O_2$	[999-55-3]	47	5.33	1.4320	0.9441
2-methylallyl	$C_7H_{10}O_2$	[818-67-7]	68	6.67	1.4372	0.9285
Aminoalkyl esters[f]						
2-(dimethylamino)ethyl	$C_7H_{13}O_2N$	[2439-35-2]	61	1.47	0.9434	1.4375
2-(diethylamino)ethyl	$C_9H_{17}O_2N$	[2426-54-2]	70	0.67	0.9251	1.4425

	CAS	Formula			n	d
Esters of ether alcohols[c]						
2-methoxyethyl	[3121-67-7]	$C_6H_{10}O_3$	59	1.60	1.4272	1.0131
2-ethoxylethyl	[106-74-1]	$C_8H_{12}O_3$	78	3.07	1.4282	0.9819
Cycloalkyl esters[g]						
cyclohexyl	[3066-71-5]	$C_9H_{14}O_2$	75	1.47	1.4600	0.9796
4-methylcyclohexyl	[16491-65-9]	$C_{10}H_{16}O_2$	55	0.27	1.4550	0.9537
Esters of halogenated alcohols[h]						
2-bromoethyl	[4823-47-6]	C_5H_7Br	53	5	1.4770	1.4774
2-chloroethyl	[2206-89-5]	C_5H_7Cl	74	29	1.4477	
Glycol diacrylates[i]						
ethylene glycol (monoester)	[818-61-1]	$C_5H_8O_3$	40	0.001	1.4482 (25°C)	
ethylene glycol	[2274-11-5]	$C_8H_{10}O_4$	70	0.13	1.4529	
propylene glycol	[999-61-1]	$C_9H_{12}O_4$	63	0.04	1.4470	
1,3-propanediol	[25151-33-1]	$C_9H_{12}O_4$	65	<0.13	1.4529	
1,4-butanediol	[31442-13-4]	$C_{10}H_{14}O$	83	0.04	1.4538	
diethylene glycol	[4074-88-8]	$C_{10}H_{14}O_5$	94	0.03	1.4572	
1,5-pentanediol	[36840-85-4]	$C_{11}H_{16}O_4$	94	0.04	1.4551	
1,10-decanediol	[13048-45-5]	$C_{16}H_{26}O_4$	145	0.01		

[a] In most cases, the references include additional examples of the class of alcohols. Nitroalkyl esters are also known (6–7).
[b] To convert kPa to mm Hg, multiply by 7.5.
[c] Ref. 8.
[d] Ref. 9.
[e] Ref. 10.
[f] Ref. 11.
[g] Ref. 12.
[h] Refs. 13–15.
[i] Ref. 5.

Table 4. Vapor Pressures of Acrylic Acid and Important Esters[a], kPa[b]

Temperature, °C	Acrylic acid	Methyl acrylate	Ethyl acrylate	Butyl acrylate	2-Ethylhexyl acrylate
−20		0.85	0.31		
−10		1.72	0.61		
0		3.12	1.16	0.15	
10		5.40	2.20	0.28	
20		9.09	3.93	0.53	
30	0.8	14.5	6.73	0.97	
40	1.4	22.7	10.9	1.71	
50	2.4	34.0	16.8	2.84	0.13
60	4.0	49.6	25.3	4.53	0.25
70	6.6	70.7	37.3	7.13	0.44
80	10		53.3	10.9	0.75
90	16		74.7	16.0	1.23
100	24			23.06	1.97
110	35			32.8	3.06
120	49			45.6	4.62
130	69			61.9	6.79
140	99			82.3	10.1
150					14.4
160					20.0
170					28.0
180					37.3
190					50.0
200					65.6
210					85.3

[a]Ref. 4.
[b]To convert kPa to mm Hg, multiply by 7.5.

quately expressed by

$$\text{rate} = 3.58 \times 10^{17} \exp(-10{,}500/T)$$

over the first several percent conversion (5), where rate is in ppm/day and T is the Kelvin temperature. Since this rate is approximately 100 ppm/day at 20°C, significant dimer can build up on prolonged storage. The reaction is accelerated by addition of strong acids, bases, or large amounts of water (several wt %). However, water at concentrations as low as 0.1–0.2 wt %, present in commercially available acrylic acid, has negligible effect on dimer formation (5,22). Continuation of this reaction leads to a distribution of polyester oligomers formed by successive additions across the double bond (5). Acrylic acid can be regenerated by thermal or acid catalyzed cracking of these oligomers (23). Cracking of the corresponding esters gives acrylic acid and acrylic ester.

$$CH_2{=}CHCOOCH_2CH_2COOR \longrightarrow CH_2{=}CHCOOH + CH_2{=}CHCOOR$$

Michael condensations are catalyzed by alkali alkoxides, tertiary amines, and quaternary bases and salts. Active methylene compounds and aliphatic nitro compounds add to form β-substituted propionates. These addition reactions are frequently reversible at high temperatures. Exceptions are the tertiary nitro adducts which are converted to olefins at elevated temperatures (24).

$$CH_3CHNO_2CH_3 + CH_2{=}CHCOOC_2H_5 \longrightarrow CH_3{-}\underset{\underset{NO_2}{|}}{\overset{\overset{CH_3}{|}}{CH}}CH_2CH_2COOC_2H_5$$

$$CH_3{-}\underset{\underset{NO_2}{|}}{\overset{\overset{CH_3}{|}}{CH}}CH_2CH_2COOC_2H_5 \longrightarrow CH_2{=}\overset{\overset{CH_3}{|}}{C}CH_2CH_2COOC_2H_5 + HNO_2$$

The addition of alcohols to form the 3-alkoxypropionates is readily carried out with strongly basic catalyst (25). If the alcohol groups are different, ester interchange gives a mixture of products. Anionic polymerization to oligomeric acrylate esters can be obtained with appropriate control of reaction conditions. The 3-alkoxypropionates can be cleaved in the presence of acid catalysts to generate acrylates (26). Development of transition-metal catalysts for carbonylation of olefins provides routes to both 3-alkoxypropionates and 3-acryloxypropionates (27,28). Hence these are potential intermediates to acrylates from ethylene and carbon monoxide.

Additions of mercaptans with alkaline catalysts give 3-alkylthiopropionates (29). In the case of hydrogen sulfide, the initially formed 3-mercaptopropionate reacts with a second molecule of acrylate to give a 3,3′-thiodipropionate (30,31).

$$H_2S + 2\, CH_2{=}CHCOOR \longrightarrow S(CH_2CH_2COOR)_2$$

Polythiodipropionic acids and their esters are prepared from acrylic acid or an acrylate with sulfur, hydrogen sulfide, and ammonium polysulfide (32). These polythio compounds are converted to the dithio analogs by reaction with an inorganic sulfite or cyanide.

$$2\, CH_2{=}CHCOOCH_3 + S_x + H_2S \xrightarrow{(NH_4)_2S_x}$$
$$S_x(CH_2CH_2COOCH_3)_2 \xrightarrow{NaCN} CH_3OOCCH_2CH_2SSCH_2CH_2COOCH_3$$

Ammonia and amines add to acrylates to form β-aminopropionates, which add easily to excess acrylate to give tertiary amines. The reactions are reversible (33).

$$NH_3 + CH_2{=}CHCOOR \longrightarrow H_2NCH_2CH_2COOR \xrightarrow{CH_2{=}CHCOOR}$$
$$HN(CH_2CH_2COOR)_2 \xrightarrow{CH_2{=}CHCOOR} N(CH_2CH_2COOR)_3$$

Aqueous ammonia and acrylic esters give tertiary amino esters, which form the corresponding amide upon ammonolysis (34). Modern methods of molecular quantum modelling have been applied to the reaction pathway and energetics for several nucleophiles in these Michael additions (35,36).

Acrylic esters dimerize to give the 2-methylene glutaric acid esters catalyzed by tertiary organic phosphines (37) or organic phosphorous triamides, phosphonous diamides, or phosphinous amides (38). Yields of 75–80% dimer, together with 15–20% trimer, are obtained. Reaction conditions can be varied to obtain high yields of trimer, tetramer, and other polymers.

Manufacture

Various methods for the manufacture of acrylates are summarized in Figure 1, showing their dependence on specific raw materials. For a route to be commercially attractive, the raw material costs and utilization must be low, plant investment and operating costs not excessive, and waste disposal charges minimal.

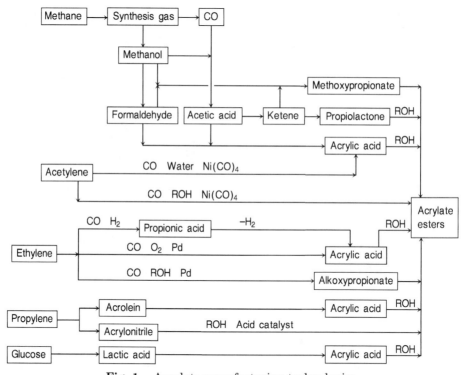

Fig. 1. Acrylate manufacturing technologies.

After development of a new process scheme at laboratory scale, construction and operation of pilot-plant facilities to confirm scale-up information often require two or three years. An additional two to three years is commonly required for final design, fabrication of special equipment, and construction of the plant. Thus, projections of raw material costs and availability five to ten years into the future become important in adopting any new process significantly different from the current technology.

In the 1980s cost and availability of acetylene have made it an unattractive raw material for acrylate manufacture as compared to propylene, which has been readily available at attractive cost (see ACETYLENE-DERIVED CHEMICALS). As a consequence, essentially all commercial units based on acetylene, with the exception of BASF's plant at Ludwigshafen, have been shut down. All new capacity recently brought on stream or announced for construction uses the propylene route. Rohm and Haas Co. has developed an alternative method based on alkoxycarbonylation of ethylene, but has not commercialized it because of the more favorable economics of the propylene route.

Propylene requirements for acrylates remain small compared to other chemical uses (polypropylene, acrylonitrile, propylene oxide, 2-propanol, and cumene for acetone and phenol). Hence, cost and availability are expected to remain attractive and new acrylate capacity should continue to be propylene-based until after the turn of the century.

Propylene Oxidation. The propylene oxidation process is attractive because of the availability of highly active and selective catalysts and the relatively low cost of propylene. The process proceeds in two stages giving first acrolein and then acrylic acid (39) (see ACROLEIN AND DERIVATIVES).

$$CH_2=CHCH_3 + O_2 \longrightarrow CH_2=CHCHO + H_2O$$

$$CH_2=CHCHO + \tfrac{1}{2} O_2 \longrightarrow CH_2=CHCOOH$$

Single-reaction-step processes have been studied. However, higher selectivity is possible by optimizing catalyst composition and reaction conditions for each of these two steps (40,41). This more efficient utilization of raw material has led to two separate oxidation stages in all commercial facilities. A two-step continuous process without isolation of the intermediate acrolein was first described by the Toyo Soda Company (42). A mixture of propylene, air, and steam is converted to acrolein in the first reactor. The effluent from the first reactor is then passed directly to the second reactor where the acrolein is oxidized to acrylic acid. The products are absorbed in water to give about 30–60% aqueous acrylic acid in about 80–85% yield based on propylene.

Japan Catalytic Chemical Co. (43) and Mitsubishi Petrochemical Co. (44) offer licenses to their acrylate manufacturing technology (including high quality catalysts). Thus, although most manufacturers have also developed their own catalyst and process technologies, many have also taken licenses from these companies and either constructed entire plants based on those disclosures or combined that technology with their own developments for their operating plants.

Catalysts. Catalyst performance is the most important factor in the economics of an oxidation process. It is measured by activity (conversion of reactant), selectivity (conversion of reactant to desired product), rate of production (production of desired product per unit of reactor volume per unit of time), and catalyst life (effective time on-stream before significant loss of activity or selectivity).

Early catalysts for acrolein synthesis were based on cuprous oxide and other heavy metal oxides deposited on inert silica or alumina supports (39). Later, catalysts more selective for the oxidation of propylene to acrolein and acrolein to acrylic acid were prepared from bismuth, cobalt, iron, nickel, tin salts, and molybdic, molybdic phosphoric, and molybdic silicic acids. Preferred second-stage catalysts generally are complex oxides containing molybdenum and vanadium. Other components, such as tungsten, copper, tellurium, and arsenic oxides, have been incorporated to increase low temperature activity and productivity (39,45,46).

Catalyst performance depends on composition, the method of preparation, support, and calcination conditions. Other key properties include, in addition to chemical performance requirements, surface area, porosity, density, pore size distribution, hardness, strength, and resistance to mechanical attrition.

Patents claiming specific catalysts and processes for their use in each of the two reactions have been assigned to Japan Catalytic (45,47–49), Sohio (50), Toyo Soda (51), Rohm and Haas (52), Sumitomo (53), BASF (54), Mitsubishi Petrochemical (56,57), Celanese (55), and others. The catalysts used for these reactions remain based on bismuth molybdate for the first stage and molybdenum vanadium oxides for the second stage, but improvements in minor component composition and catalyst preparation have resulted in yields that can reach the 85–90% range and lifetimes of several years under optimum conditions. Since plants operate under more productive conditions than those optimum for yield and life, the economically most attractive yields and productive lifetimes may be somewhat lower.

Oxidation Step. A review of mechanistic studies of partial oxidation of propylene has appeared (58). The oxidation process flow sheet (Fig. 2) shows equipment and typical operating conditions. The reactors are of the fixed-bed shell-and-tube type (about 3–5 m long and 2.5 cm in diameter) with a molten salt coolant on the shell side. The tubes are packed with catalyst, a small amount of inert material at the top serving as a preheater section for the feed gases. Vaporized propylene is mixed with steam and air and fed to the first-stage reactor. The feed composition is typically 5–7% propylene, 10–30% steam, and the remainder air (or a mixture of air and absorber off-gas) (56,57,59,60).

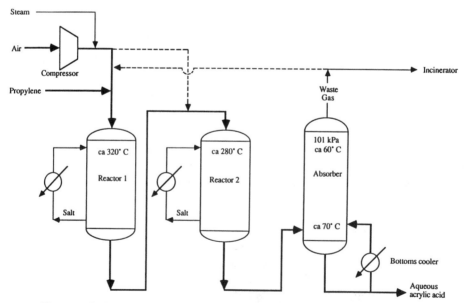

Fig. 2. Oxidation process. To convert kPa to mm Hg, multiply by 7.5.

The preheated gases react exothermically over the first-stage catalyst with the peak temperature in the range of 330–430°C, depending on conditions and catalyst selectivity. The conversion of propylene to waste gas (carbon dioxide and carbon monoxide) is more exothermic than its conversion to acrolein. At the end of the catalyst bed the temperature of the mixture drops toward that of the molten salt coolant.

If necessary, first-stage reactor effluent may be further cooled to 200–250°C by an interstage cooler to prevent homogeneous and unselective oxidation of acrolein taking place in the pipes leading to the second-stage reactor (56,59).

The acrolein-rich gaseous mixture containing some acrylic acid is then passed to the second-stage reactor, which is similar to the first-stage reactor, but packed with a catalyst designed for selective conversion of acrolein to acrylic acid. Here, the temperature peaks in the range of 280–360°C, again depending on conditions. The temperature of the effluent from the second-stage reactor again approximates that of the salt coolant. The heat of reaction is recovered as steam in external waste-heat boilers.

The process is operated at the lowest temperature consistent with high conversion. Conversion increases with temperature; the selectivity generally decreases only with large increases in temperature. Catalyst life also decreases with increasing temperatures. Catalysts are designed to give high performance over a range of operating conditions permitting gradual increase of salt temperature over the operating life of the catalysts to maintain productivity and selectivity near the initial levels, thus compensating for gradual loss of catalyst activity.

The gaseous reactor effluent from the second-stage oxidation is fed to the bottom of the aqueous absorber and cooled from about 250°C to less than 80°C by contact with aqueous acrylic acid. The gas passes through the absorber to complete the recovery of product. The water is fed to the top of the absorber at about 30–60°C to minimize acrylic acid losses and the absorber off-gas is sent to a flare or to a furnace to convert all residual organic material to waste gas. Some of the absorber off-gas may be recycled to the first-stage reactor feed to allow achievement of optimum oxygen-to-propylene ratio at reduced steam levels (59). If the resulting oxygen level is too low for best performance in the second-stage oxidation, an interstage feed of supplemental air (or air plus steam) may be introduced (56). The aqueous effluent from the bottom of the absorber is 30–60% acrylic acid depending on whether off-gas recycle with low steam feed or air feed with higher steam level is chosen. This is sent to the separations section for recovery. The overall yield of acrylic acid in the oxidation reaction steps is in the range of 75–86%, depending on the catalysts, conditions, and age of catalyst employed.

Acrylic Acid Recovery.   The process flow sheet (Fig. 3) shows equipment and conditions for the separations step. The acrylic acid is extracted from the absorber effluent with a solvent, such as butyl acetate, xylene, diisobutyl ketone, or mixtures, chosen for high selectivity for acrylic acid and low solubility for water and by-products. The extraction is performed using 5–10 theoretical stages in a tower or centrifugal extractor (46,61–65).

The extract is vacuum-distilled in the solvent recovery column, which is operated at low bottom temperatures to minimize the formation of polymer and dimer and is designed to provide acrylic acid-free overheads for recycle as the extraction solvent. A small aqueous phase in the overheads is mixed with the

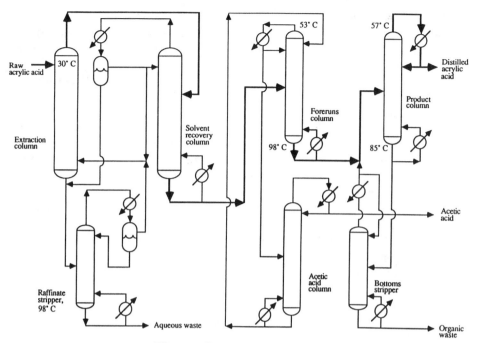

Fig. 3.   Separations process.

raffinate from the extraction step. This aqueous material is stripped before disposal both to recover extraction solvent values and minimize waste organic disposal loads.

It is possible to dispense with the extraction step if the oxidation section is operated at high propylene concentrations and low steam levels to give a concentrated absorber effluent. In this case, the solvent recovery column operates at total organic reflux to effect azeotropic dehydration of the concentrated aqueous acrylic acid. This results in a reduction of aqueous waste at the cost of somewhat higher energy usage.

The bottoms from the solvent recovery (or azeotropic dehydration column) are fed to the foreruns column where acetic acid, some acrylic acid, and final traces of water are removed overhead. The overhead mixture is sent to an acetic acid purification column where a technical grade of acetic acid suitable for ester manufacture is recovered as a by-product. The bottoms from the acetic acid recovery column are recycled to the reflux to the foreruns column. The bottoms from the foreruns column are fed to the product column where the glacial acrylic acid of commerce is taken overhead. Bottoms from the product column are stripped to recover acrylic acid values and the high boilers are burned. The principal losses of acrylic acid in this process are to the aqueous raffinate and to the aqueous layer from the dehydration column and to dimerization of acrylic acid to 3-acryloxypropionic acid. If necessary, the product column bottoms stripper may include provision for a short-contact-time cracker to crack this dimer back to acrylic acid (60).

In any case, mild conditions and short residence times to minimize dimer formation are maintained throughout the separations section. In addition, free-

radical polymerization inhibitors are fed to each unit to prevent polymer formation and resulting equipment failure. The glacial acrylic acid produced at this
stage of the process is typically better than 99.5% pure, with the principal
contaminants being water and acetic acid at about 0.1–0.2 wt %, acetic acid at
about 0.1 wt %, and acrylic acid dimer at 0.1–0.5 wt % depending on storage time
after distillation. Propionic acid is present at 0.02–0.04 wt %. The monomethyl
ether of hydroquinone is added as storage and shipping stabilizer at 0.02 wt %
(200 ppm). Low concentrations of aldehydes, primarily furfuraldehyde, but including acetaldehyde, acrolein, and benzaldehyde, may be present in commercial
glacial acrylic acid. These impurities may have an adverse effect in achievement
of very high molecular weight polymers. Further distillation or chemical treatment prior to final distillation is employed to remove carbonyl impurities. Effective agents include hydrazine, amino acids, and alkyl or aryl amines (66–68).

Esterification. The process flow sheet (Fig. 4) outlines the process and
equipment of the esterification step in the manufacture of the lower acrylic esters
(methyl, ethyl, or butyl). For typical art, see references 69–74. The part of the flow
sheet containing the dotted lines is appropriate only for butyl acrylate, since the
lower alcohols, methanol and ethanol, are removed in the wash column. Since the
butanol is not removed by a water or dilute caustic wash, it is removed in the
azeotrope column as the butyl acrylate azeotrope; this material is recycled to the
reactor.

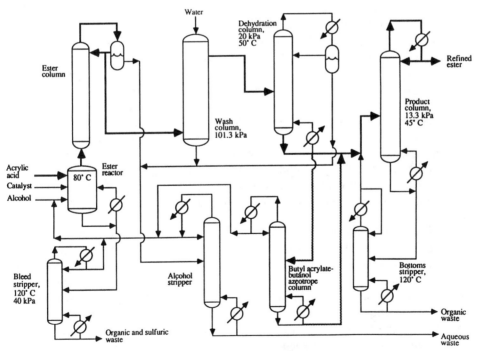

Fig. 4. Esterification process. To convert kPa to mm Hg, multiply by 7.5.

Acrylic acid, alcohol, and the catalyst, eg, sulfuric acid, together with the
recycle streams are fed to the glass-lined ester reactor fitted with an external
reboiler and a distillation column. Acrylate ester, excess alcohol, and water of

esterification are taken overhead from the distillation column. The process is operated to give only traces of acrylic acid in the distillate. The bulk of the organic distillate is sent to the wash column for removal of alcohol and acrylic acid; a portion is returned to the top of the distillation column. If required, some base may be added during the washing operation to remove traces of acrylic acid.

A continuous bleed is taken from the reactor to remove high boilers. Values contained in this bleed are recovered in the bleed stripper and the distillate from this operation is recycled to the esterification reactor. The bleed stripper residue is a mixture of high boiling organic material and sulfuric acid, which is recovered for recycle in a waste sulfuric acid plant.

If a waste sulfuric acid regeneration plant is not available, eg, as part of a joint acrylate–methacrylate manufacturing complex, the preferred catalyst for esterification is a sulfonic acid type ion-exchange resin. In this case the residue from the ester reactor bleed stripper can be disposed of by combustion to recover energy value as steam.

The wet ester is distilled in the dehydration column using high reflux to remove a water phase overhead. The dried bottoms are distilled in the product column to provide high purity acrylate. The bottoms from the product column are stripped to recover values and the final residue incinerated. Alternatively, the bottoms may be recycled to the ester reactor or to the bleed stripper.

Conventional polymerization inhibitors are fed to each of the distillation columns. The columns are operated under reduced pressure to give low bottom temperatures and minimize polymerization.

The aqueous layer from the ester column distillate, the raffinate from washing the ester, and the aqueous phase from the dehydration step are combined and distilled in the alcohol stripper. The wet alcohol distillate containing a low level of acrylate is recycled to the esterification reactor. The aqueous column bottoms are incinerated or sent to biological treatment. Biological treatment is common.

Process conditions for methyl acrylate are similar to those employed for ethyl acrylate. However, in the preparation of butyl acrylate the excess butanol is removed as the butanol–butyl acrylate azeotrope in the azeotrope column.

The esters are produced in minimum purity of 99.5%. The yield, based on acrylic acid, is in the range of about 95–98% depending on the ester and reaction conditions. Monomethyl ether of hydroquinone (10–100 ppm) is added as polymerization inhibitor and the esters are used in this form in most industrial polymerizations.

Acetylene-Based Routes. Walter Reppe, the father of modern acetylene chemistry, discovered the reaction of nickel carbonyl with acetylene and water or alcohols to give acrylic acid or esters (75,76). This discovery led to several processes which have been in commercial use. The original Reppe reaction requires a stoichiometric ratio of nickel carbonyl to acetylene. The Rohm and Haas modified or semicatalytic process provides 60–80% of the carbon monoxide from a separate carbon monoxide feed and the remainder from nickel carbonyl (77–78). The reactions for the synthesis of ethyl acrylate are

$$4\ C_2H_2 + 4\ C_2H_5OH + Ni(CO)_4 + 2\ HCl \longrightarrow 4\ CH_2{=}CHCOOC_2H_5 + H_2 + NiCl_2$$

$$C_2H_2 + C_2H_5OH + 0.05\ Ni(CO)_4 + 0.8\ CO + 0.1\ HCl \longrightarrow$$
$$CH_2{=}CHCOOC_2H_5 + 0.05\ NiCl_2 + 0.05\ H_2$$

The stoichiometric and the catalytic reactions occur simultaneously, but the catalytic reaction predominates. The process is started with stoichiometric amounts, but afterward, carbon monoxide, acetylene, and excess alcohol give most of the acrylate ester by the catalytic reaction. The nickel chloride is recovered and recycled to the nickel carbonyl synthesis step. The main by-product is ethyl propionate, which is difficult to separate from ethyl acrylate. However, by proper control of the feeds and reaction conditions, it is possible to keep the ethyl propionate content below 1%. Even so, this is significantly higher than the propionate content of the esters from the propylene oxidation route.

Other by-products formed are relatively easy to separate, including esters of higher unsaturated monobasic acids (alkyl 3-pentenoate and 3,5-heptadienoate) (5) and esters of multiply-unsaturated dibasic acids, eg, suberates.

The reaction is initiated with nickel carbonyl. The feeds are adjusted to give the bulk of the carbonyl from carbon monoxide. The reaction takes place continuously in an agitated reactor with a liquid recirculation loop. The reaction is run at about atmospheric pressure and at about 40°C with an acetylene:carbon monoxide mole ratio of 1.1:1 in the presence of 20% excess alcohol. The reactor effluent is washed with nickel chloride brine to remove excess alcohol and nickel salts and the brine–alcohol mixture is stripped to recover alcohol for recycle. The stripped brine is again used as extractant, but with a bleed stream returned to the nickel carbonyl conversion unit. The neutralized crude monomer is purified by a series of continuous, low pressure distillations.

The modified Reppe process was installed by Rohm and Haas at their Houston plant in 1948 and later expanded to a capacity of about 182×10^6 kg/yr. Rohm and Haas started up a propylene oxidation plant at the Houston site in late 1976. The combination of attractive economics and improved product purity from the propylene route led to a shutdown of the acetylene-based route within a year.

Reppe's work also resulted in the high pressure route which was established by BASF at Ludwigshafen in 1956. In this process, acetylene, carbon monoxide, water, and a nickel catalyst react at about 200°C and 13.9 MPa (2016 psi) to give acrylic acid. Safety problems caused by handling of acetylene are alleviated by the use of tetrahydrofuran as an inert solvent. In this process, the catalyst is a mixture of nickel bromide with a cupric bromide promotor. The liquid reactor effluent is degassed and extracted. The acrylic acid is obtained by distillation of the extract and subsequently esterified to the desired acrylic ester. The BASF process gives acrylic acid, whereas the Rohm and Haas process provides the esters directly.

The BASF plant has a capacity of about 91 million kg/yr and is still in operation. However, they have constructed several plants in Germany and in the United States during the late 1970s and late 1980s, all of which are based on propylene oxidation. They have shut down a smaller Reppe unit in the United States, and it is expected that, as they bring on new propylene-based capacity in Germany during the early 1990s, the acetylene process will be phased out.

Nickel carbonyl is volatile, has little odor, and is extremely toxic. Symptoms of dangerous exposure may not appear for several days. Effective medical treatment should be started immediately. The plant should be designed to ensure containment of nickel carbonyl and to prevent operator contact.

All other organic waste-process and vent streams are burned in a flare, in an incinerator, or in a furnace where fuel value is recovered. Wastewater streams are handled in the plant biological treatment area.

Although some very minor manufacturers of acrylic acid may still use hydrolysis of acrylonitrile (see below), essentially all other plants worldwide use the propylene oxidation process.

Acrylonitrile Route.   This process, based on the hydrolysis of acrylonitrile (79), is also a propylene route since acrylonitrile (qv) is produced by the catalytic vapor-phase ammoxidation of propylene.

$$CH_2=CHCH_3 + NH_3 + \tfrac{3}{2} O_2 \longrightarrow CH_2=CHCN + 3 H_2O$$

The yield of acrylonitrile based on propylene is generally lower than the yield of acrylic acid based on the direct oxidation of propylene. Hence, for the large volume manufacture of acrylates, the acrylonitrile route is not attractive since additional processing steps are involved and the ultimate yield of acrylate based on propylene is much lower. Hydrolysis of acrylonitrile can be controlled to provide acrylamide rather than acrylic acid, but acrylic acid is a by-product in such a process (80).

The sulfuric acid hydrolysis may be performed as a batch or continuous operation. Acrylonitrile is converted to acrylamide sulfate by treatment with a small excess of 85% sulfuric acid at 80–100°C. A hold-time of about one hour provides complete conversion of the acrylonitrile. The reaction mixture may be hydrolyzed and the aqueous acrylic acid recovered by extraction and purified as described under the propylene oxidation process prior to esterification. Alternatively, after reaction with excess alcohol, a mixture of acrylic ester and alcohol is distilled and excess alcohol is recovered by aqueous extractive distillation. The ester in both cases is purified by distillation.

Important side reactions are the formation of ether and addition of alcohol to the acrylate to give 3-alkoxypropionates. In addition to high raw material costs, this route is unattractive because of large amounts of sulfuric acid–ammonium sulfate wastes.

Ketene Process.   The ketene process based on acetic acid or acetone as the raw material was developed by B. F. Goodrich (81) and Celanese (82). It is no longer used commercially because the intermediate β-propiolactone is suspected to be a carcinogen (83). In addition, it cannot compete with the improved propylene oxidation process (see KETENES AND RELATED SUBSTANCES).

Ethylene Cyanohydrin Process.   This process, the first for the manufacture of acrylic acid and esters, has been replaced by more economical ones. During World War I, the need for ethylene as an important raw material for the synthesis of aliphatic chemicals led to development of this process (16) in both Germany, in 1927, and the United States, in 1931.

In the early versions, ethylene cyanohydrin was obtained from ethylene chlorohydrin and sodium cyanide. In later versions, ethylene oxide (from the direct catalytic oxidation of ethylene) reacted with hydrogen cyanide in the presence of a base catalyst to give ethylene cyanohydrin. This was hydrolyzed and converted to acrylic acid and by-product ammonium acid sulfate by treatment with about 85% sulfuric acid.

Losses by polymer formation kept the yield of acrylic acid to 60–70%.

Preferably, esters were prepared directly by a simultaneous dehydration–esterification process.

The process has historic interest. It was replaced at the Rohm and Haas Company by the acetylene-based process in 1954, and in 1970 at Union Carbide by the propylene oxidation process.

Other Syntheses. Acrylic acid and other unsaturated compounds can also be made by a number of classical elimination reactions. Acrylates have been obtained from the thermal dehydration of hydracrylic acid (3-hydroxypropanoic acid [503-66-2]) (84), from the dehydrohalogenation of 3-halopropionic acid derivatives (85), and from the reduction of dihalopropionates (2). These studies, together with the related characterization and chemical investigations, contributed significantly to the development of commercial organic chemistry.

Vapor-Phase Condensations of Acetic Acid or Esters with Formaldehyde. Addition of a methylol group to the α-carbon of acetic acid or esters, followed by dehydration, gives the acrylates.

$$CH_3COOH(R) + CH_2O \longrightarrow CH_2{=}CHCOOH(R) + H_2O$$

The reaction is generally carried out at atmospheric pressure and at 350–400°C. A variety of catalysts, eg, bases and metal salts and oxides on silica or alumina–silicates, have been patented (86–91). Conversions are in the 30–70% range and selectivities in the 60–90% range, depending on the catalyst and the ratio of formaldehyde to acetate.

The procedure is technically feasible, but high recovery of unconverted raw materials is required for the route to be practical. Its development depends on the improvement of catalysts and separation methods and on the availability of low cost acetic acid and formaldehyde. Both raw materials are dependent on ample supply of low cost methanol.

Although the rapid cost increases and shortages of petroleum-based feedstocks forecast a decade ago have yet to materialize, shift to natural gas or coal may become necessary in the new century. Under such conditions, it is possible that acrylate manufacture via acetylene, as described above, could again become attractive. It appears that condensation of formaldehyde with acetic acid might be preferred. A coal gasification complex readily provides all of the necessary intermediates for manufacture of acrylates (92).

Oxidative Carbonylation of Ethylene—Elimination of Alcohol from β-Alkoxypropionates. Spectacular progress in the 1970s led to the rapid development of organotransition-metal chemistry, particularly to catalyze olefin reactions (93,94). A number of patents have been issued (28,95–97) for the oxidative carbonylation of ethylene to provide acrylic acid and esters. The procedure is based on the palladium catalyzed carbonylation of ethylene in the liquid phase at temperatures of 50–200°C. Esters are formed when alcohols are included. Anhydrous conditions are desirable to minimize the formation of by-products including acetaldehyde and carbon dioxide (see ACETALDEHYDE).

During the reaction, the palladium catalyst is reduced. It is reoxidized by a co-catalyst system such as cupric chloride and oxygen. The products are acrylic acid in a carboxylic acid–anhydride mixture or acrylic esters in an alcoholic solvent. Reaction products also include significant amounts of 3-acryloxypropionic acid [*24615-84-7*] and alkyl 3-alkoxypropionates, which can be converted thermally to the corresponding acrylates (23,98). The overall reaction may be represented by:

$$CH_2{=}CH_2 + CO + HOR + \tfrac{1}{2}\,O_2 \longrightarrow CH_2{=}CHCOOR + H_2O$$

It is preferrable to carry out the reactions to give an intermediate β-alkoxypropionate. When the reaction is carried out in ethanolic ethyl β-ethoxypropionate as solvent, and a trace of mercury(II) is included in the catalyst system, the yield of intermediate ethyl β-ethoxypropionate rises to 95–97%; the principal by-product is the corresponding β-chloropropionate ester. Acid-catalyzed thermal cracking of the mixture at 120–150°C gives very high yields of ethyl acrylate (26). Although yields are excellent, the reaction medium is extremely corrosive, so high cost materials of construction are necessary. In addition, the high cost of catalyst and potential toxicity of mercury require that the inorganic materials be recovered quantitatively from any waste stream. Hence, high capital investment, together with continued favorable costs for propylene, have prevented commercialization of this route.

The elimination of alcohol from β-alkoxypropionates can also be carried out by passing the alkyl β-alkoxypropionate at 200–400°C over metal phosphates, silicates, metal oxide catalysts (99), or base-treated zeolites (98). In addition to the route via oxidative carbonylation of ethylene, alkyl β-alkoxypropionates can be prepared by reaction of dialkoxy methane and ketene (100).

Dehydrogenation of Propionates. Oxidative dehydrogenation of propionates to acrylates employing vapor-phase reactions at high temperatures (400–700°C) and short contact times is possible. Although selective catalysts for the oxidative dehydrogenation of isobutyric acid to methacrylic acid have been developed in recent years (see METHACRYLIC ACID AND DERIVATIVES) and a route to methacrylic acid from propylene to isobutyric acid is under pilot-plant development in Europe, this route to acrylates is not presently of commercial interest because of the combination of low selectivity, high raw material costs, and purification difficulties.

Liquid-Phase Oxidation of Acrolein. As discussed before, the most attractive process for the manufacture of acrylates is based on the two-stage, vapor-phase oxidation of propylene. The second stage involves the oxidation of acrolein. Considerable art on the liquid-phase oxidation of acrolein (17) is available, but this route cannot compete with the vapor-phase technology.

Specialty Acrylic Esters

Higher alkyl acrylates and alkyl-functional esters are important in copolymer products, in conventional emulsion applications for coatings and adhesives, and as reactants in radiation-cured coatings and inks. In general, they are produced in direct or transesterification batch processes (17,101,102) because of their relatively low volume.

Direct, acid catalyzed esterification of acrylic acid is the main route for the manufacture of higher alkyl esters. The most important higher alkyl acrylate is 2-ethylhexyl acrylate prepared from the available oxo alcohol 2-ethyl-1-hexanol (see ALCOHOLS, HIGHER ALIPHATIC). The most common catalysts are sulfuric or toluenesulfonic acid and sulfonic acid functional cation-exchange resins. Solvents are used as entraining agents for the removal of water of reaction. The product is washed with base to remove unreacted acrylic acid and catalyst and then purified by distillation. The esters are obtained in 80–90% yield and in excellent purity.

Transesterification of a lower acrylate ester and a higher alcohol (102,103) can be performed using a variety of catalysts and conditions chosen to provide acceptable reaction rates and to minimize by-product formation and polymerization.

$$CH_2=CHCOOR + R'OH \xrightarrow{H^+} CH_2=CHCOOR' + ROH$$

Pure dry reactants are needed to prevent catalyst deactivation; effective inhibitor systems are also desirable as well as high reaction rates, since many of the specialty monomers are less stable than the lower alkyl acrylates. The alcohol–ester azeotrope (8) should be removed rapidly from the reaction mixture and an efficient column used to minimize reactant loss to the distillate. After the reaction is completed, the catalyst may be removed and the mixture distilled to obtain the ester. The method is particularly useful for the preparation of functional monomers which cannot be prepared by direct esterification. Dialkylaminoethyl acrylic esters are readily prepared by transesterification of the corresponding dialkylaminoethanol (102,103). Catalysts include strong acids and tetraalkyl titanates for higher alkyl esters; and titanates, sodium phenoxides, magnesium alkoxides, and dialkyltin oxides, as well as titanium and zirconium chelates, for the preparation of functional esters. Because of loss of catalyst activity during the reaction, incremental or continuous additions may be required to maintain an adequate reaction rate.

Hydroxyethyl and 2-hydroxypropyl acrylates are prepared by the addition of ethylene oxide or propylene oxide to acrylic acid (104,105).

The reactions are catalyzed by tertiary amines, quaternary ammonium salts, metal salts, and basic ion-exchange resins. The products are difficult to purify and generally contain low concentrations of acrylic acid and some diester which should be kept to a minimum since its presence leads to product instability and to polymer cross-linking.

Analytical Methods

Chemical assay is preferably performed by gas–liquid chromatography (glc) or by the conventional methods for determination of unsaturation such as bromination or addition of mercaptan, sodium bisulfite, or mercuric acetate.

Acidity is determined by glc or titration, and the dimer content of acrylic acid by glc or a saponification procedure. The total acidity is corrected for the dimer acid content to give the value for acrylic acid.

Storage and Handling

Acrylic acid and esters are stabilized with minimum amounts of inhibitors consistent with stability and safety. The acrylic monomers must be stable and there should be no polymer formation for prolonged periods with normal storage and shipping (4,106). The monomethyl ether of hydroquinone (MEHQ) is frequently used as inhibitor and low inhibitor grades of the acrylate monomers are available for bulk handling. MEHQ at 10–15 ppm is generally adequate for the esters, but a higher concentration (200 ppm) is needed for acrylic acid.

The effectiveness of phenolic inhibitors is dependent on the presence of oxygen and the monomers must be stored under air rather than an inert atmosphere. Temperatures must be kept low to minimize formation of peroxides and other products. Moisture may cause rust-initiated polymerization.

Acrylic acid has a relatively high freezing point (13°C) and the inhibitor may not be distributed uniformly between phases when frozen acid is partially thawed. If the liquid phase is inadequately inhibited, it could polymerize and initiate violent polymerization of the entire mass. Provisions should be made to maintain the acid as a liquid. High temperatures should be avoided because of dimer formation. If freezing should occur, melting should take place at room temperature (25°C); material should not be withdrawn until the total is thawed and well-mixed to provide good distribution of the inhibitor and dissolved oxygen. No part of the mass should be subjected to elevated temperatures during the melting process.

Dimer formation, which is favored by increasing temperature, generally does not reduce the quality of acrylic acid for most applications. The term dimer includes higher oligomers formed by further addition reactions and present in low concentrations relative to the amount of dimer (3-acryloxypropionic acid). Glacial acrylic acid should be stored at 16–29°C to maintain high quality.

The acrylic esters may be stored in mild or stainless steel, or aluminum. However, acrylic acid is corrosive to many metals and can be stored only in glass, stainless steel, aluminum, or polyethylene-lined equipment. Stainless steel types 316 and 304 are preferred materials for acrylic acid.

For most applications, the phenolic inhibitors do not have to be removed. The low-inhibitor grades of acrylic monomers are particularly suitable for the manufacture of polymer without pretreatment. Removal of inhibitor from ester is best done by adsorption with ion-exchange resins or other adsorbents. Phenolic inhibitors may be removed from esters with an alkaline brine wash, generally a solution containing 5% caustic and 20% salt. Vigorous agitation during washing should be avoided to prevent the formation of emulsions. The washed monomers may be used without drying in emulsion processes. Washed uninhibited monomers are less stable and should be used promptly. They should be stored under refrigeration, but should not be permitted to freeze because of the danger of explosive polymerization during thawing. The heat of polymerization is approximately 75 kJ/mol (18 kcal/mol).

The relatively low flash points of some acrylates create a fire hazard. Also, the ease of polymerization must be borne in mind in all operations. The lower and upper explosive limits for methyl acrylate are 2.8 and 25 vol %, respectively.

Corresponding limits for ethyl acrylate are 1.8 vol % and saturation, respectively. All possible sources of ignition of monomers must be eliminated.

Health and Safety Factors

The toxicity of common acrylic monomers has been characterized in animal studies using a variety of exposure routes. Toxicity varies with level, frequency, duration, and route of exposure. The simple higher esters of acrylic acid are usually less absorbed and less toxic than lower esters. In general, acrylates are more toxic than methacrylates. Data appear in Table 5.

Table 5. Acute Toxicity of Acrylic Acid and Esters

Monomer	Methyl acrylate	Ethyl acrylate	Butyl acrylate	Acrylic acid
rat oral LD$_{50}$, g/kg	0.3	0.8–1.8	3.7–8.1	0.3–2.5
rabbit dermal LD$_{50}$, g/kg	1.3	1.2–3.0	1.7–5.7	0.3–1.6
rat inhalation LC$_{50}$, ppm	750–1350	2180	2370	1200–4000
rabbit eye irritation	severe to corrosive	severe to corrosive	slight to moderate	corrosive
rabbit skin irritator	severe irritation	severe irritation	moderate	corrosive
odor threshold, ppb	2.3–4.8	0.5	35[a]	94[b]
TLV/TWA[c,d], mg/m^3	35	20	52	5.9
ppm	10, skin	5	10	2, skin
STEL[d], mg/m^3		100		
ppm		25		
		A–2 suspect human carcinogen		

[a]May have delayed eye irritation.
[b]Vapor exposure can cause irreversible eye damage.
[c]Ref. 107.
[d]Ref. 83.

With respect to acute toxicity, based on lethality in rats or rabbits, acrylic monomers are slightly to moderately toxic. Mucous membranes of the eyes, nose, throat, and gastrointestinal tract are particularly sensitive to irritation. Acrylates can produce a range of eye and skin irritations from slight to corrosive depending on the monomer.

Full eye protection should be worn whenever handling acrylic monomers; contact lenses must never be worn. Prolonged exposure to liquid or vapor can result in permanent eye damage or blindness. Excessive exposure to vapors causes nose and throat irritation, headaches, nausea, vomiting, and dizziness or drowsiness (solvent narcosis). Overexposure may cause central nervous system depression. Both proper respiratory protection and good ventilation are necessary wherever the possibility of high vapor concentration arises.

Swallowing acrylic monomers may produce severe irritation of the mouth, throat, esophagus, and stomach, and cause discomfort, vomiting, diarrhea, dizziness, and possible collapse.

Skin redness and from slight to corrosive irritation is caused by direct contact. Acrylic acid is more corrosive than esters. The monomers not only irritate the skin, but may also be absorbed through the skin. Therefore, gloves and protective clothing and shoes or boots should be used in addition to eye (or full face) protective equipment. Upon contact, the skin should be flushed with copious amounts of water; follow-up medical attention should be sought. Medical attention should also be obtained if any of the earlier mentioned symptoms appear.

Repeated exposures to acrylic monomers can produce allergic dermatitis (or skin sensitization) resulting in rash, itching, or swelling. After exposure to one monomer, this dermatitis may arise upon subsequent exposure to the same or even a different acrylic monomer.

Repeated exposures of animals to high (near-lethal) concentrations of vapors result in inflammation of the respiratory tract, as well as degenerative changes in the liver, kidneys, and heart muscle. These effects arise at concentrations far above those causing irritation. Such effects have not been reported in humans. The low odor threshold and irritating properties of acrylates cause humans to leave a contaminated area rather than tolerate the irritation.

Current TLV/TWA values are provided in *Material Safety Data Sheets* provided by manufacturers upon request. Values for 1989 (83,107) appear in Table 5.

Acrolein, acrylamide, hydroxyalkyl acrylates, and other functional derivatives can be more hazardous from a health standpoint than acrylic acid and its simple alkyl esters. Furthermore, some derivatives, such as the alkyl 2-chloroacrylates, are powerful vesicants and can cause serious eye injuries. Thus, although the hazards of acrylic acid and the normal alkyl acrylates are moderate and they can be handled safely with ordinary care to industrial hygiene, this should not be assumed to be the case for compounds with chemically different functional groups (see INDUSTRIAL HYGIENE AND PLANT SAFETY; TOXICOLOGY).

In 1983 the National Toxicology Program (NTP) reported that ethyl acrylate produced tumors in the rodent forestomach after gavage (forced feeding via stomach tube) for 2 yr. The response occurred only at the site of contact after lifetime exposure to levels that were both irritating and ulcerating to that tissue. Based on this study, both the NTP and the International Agency for Research on Cancer (IARC) concluded that there was sufficient evidence for carcinogenicity of ethyl acrylate in experimental animals and by extension classified ethyl acrylate as possibly carcinogenic in humans. Several other studies of simple acrylate esters using alternate exposure methods failed to show evidence of oncogenic response.

BIBLIOGRAPHY

"Acrylic and Methacrylic Acids" in *ECT* 1st ed., Vol. 1, pp. 176–179, by F. J. Glavis, Rohm & Haas Company; "Acrylic Resins and Plastics" in *ECT* 1st ed., Vol. 1, pp. 180–184, by E. H. Kroeker, Rohm & Haas Company; "Acrylic Acid and Derivatives" in *ECT* 2nd ed., Vol. 1, pp. 285–313, by F. J. Glavis and E. H. Specht, Rohm & Haas Company; "Acrylic Acid and Derivatives" in *ECT* 3rd ed., Vol. 1, pp. 330–354, by J. W. Nemec and W. Bauer, Jr., Rohm and Haas Company.

1. J. Redtenbacher, *Ann.* **47,** 125 (1843).
2. W. Caspary and B. Tollens, *Ann.* **167,** 240 (1873).
3. S. Hochheiser, *Rohm and Haas,* University of Pennsylvania Press, Philadelphia, 1986, pp. 31ff.
4. *Storage and Handling of Acrylic and Methacrylic Esters and Acids, Bulletin 84C7,* Rohm and Haas Co., Philadelphia, Pa., 1987; *Acrylic and Methacrylic Monomers— Specifications and Typical Properties, Bulletin 84C2,* Rohm and Haas Co., Philadelphia, Pa., 1986; *Rocryl Specialty Monomers—Specifications and Typical Properties, Bulletin 77S2,* Rohm and Haas Co., Philadelphia, Pa., 1989.
5. Rohm and Haas Company, Philadelphia, Pa.
6. N. S. Marans and R. P. Zelinski, *J. Am. Chem. Soc.* **72,** 2125 (1950).
7. U.S. Pat. 2,967,195 (Jan. 3, 1961), M. H. Gold (to Aerojet-General Corp.).
8. C. E. Rehberg and C. H. Fisher, *J. Am. Chem. Soc.* **66,** 1203 (1944); *Ind. Eng. Chem.* **40,** 1429 (1948).
9. C. E. Rehberg, W. A. Faucette, and C. H. Fisher, *J. Am. Chem. Soc.* **66,** 1723 (1944).
10. C. E. Rehberg and C. H. Fisher, *J. Org. Chem.* **12,** 226 (1947).
11. C. E. Rehberg and W. A. Faucette, *J. Am. Chem. Soc.* **71,** 3164 (1949).
12. C. E. Rehberg and W. A. Faucette, *J. Org. Chem.* **14,** 1094 (1949).
13. C. E. Rehberg, M. B. Dixon, and W. A. Faucette, *J. Am. Chem. Soc.* **72,** 5199 (1950).
14. D. W. Coddington, T. S. Reid, A. H. Ahlbrecht, C. H. Smith, Jr., and D. R. Usted, *J. Polym. Sci.* **15,** 515 (1955).
15. W. Postelnek, L. E. Coleman, and A. M. Lovelace, *Fortschr. Hochpolym. Forsch.* **1,** 75 (1958).
16. E. H. Riddle, *Monomeric Acrylic Esters,* Reinhold Publishing Co., New York, 1954.
17. H. Rauch-Puntigam and T. Volker, *Acryl- und Methacrylverbindungen,* Springer-Verlag, Berlin, 1967.
18. M. Sittig, *Vinyl Monomers and Polymers,* Noyes Development Corp., Park Ridge, N.J., 1966.
19. U.S. Pat. 3,703,539 (Nov. 21, 1962), B. A. Di Liddo (to B. F. Goodrich Co.).
20. U.S. Pat. 4,490,553 (Dec. 25, 1984), J. D. Chase and W. W. Wilkison (to Celanese Corporation).
21. R. Mozingo and L. A. Patterson, *Org. Synth. Coll. Vol. 3,* 576 (1955).
22. F. M. Wampler III, *Plant/Operations Progress* **1**(3), 183–189 (1988).
23. U.S. Pat. 3,888,912 (June 10, 1975), M. D. Burguette (to Minnesota Mining and Manufacturing Co.).
24. U.S. Pat. 3,642,843 (Feb. 15, 1972), J. W. Nemec (to Rohm and Haas Co.).
25. C. E. Rehberg, M. B. Dixon, and C. H. Fisher, *J. Am. Chem. Soc.* **68,** 544 (1946); **69,** 2966 (1947); C. E. Rehberg and M. B. Dixon, *J. Am. Chem. Soc.* **72,** 2205 (1950).
26. U.S. Pat. 3,227,746 (Jan. 4, 1966), F. Knorr and A. Spes (to Wacker-Chemie G.m.b.H.).
27. D. M. Fenton and K. L. Olivier, *Chem. Technol.* 220 (Apr. 1972).
28. U.S. Pat. 3,987,089 (Oct. 19, 1976), F. L. Slejko and J. S. Clovis (to Rohm and Haas Co.).
29. C. D. Hurd and L. L. Gershbein, *J. Am. Chem. Soc.* **69,** 2328 (1947).
30. L. L. Gershbein and C. D. Hurd, *J. Am. Chem. Soc.* **69,** 241 (1947).
31. E. A. Fehnel and M. Carmack, *Org. Syn.* **30,** 65 (1950).
32. U.S. Pat. 3,769,315 (Oct. 30, 1973), R. L. Keener and H. Raterink (to Rohm and Haas Co.).
33. S. M. McElvain and G. Stork, *J. Am. Chem. Soc.* **68,** 1049 (1946).
34. U.S. Pat. 2,580,832 (Jan. 1, 1952), E. W. Pietrusza (to Allied Chemical Co.).
35. C. B. Frederick and C. H. Reynolds, *Toxicol. Lett.* **47,** 241–247 (1989).

36. R. Osman, K. Namboodiri, H. Weinstein, and J. R. Rabinowitz, *J. Am. Chem. Soc.* **110,** 1701–1707 (1988).
37. U.S. Pat. 3,074,999 (Jan. 22, 1963), M. B. Rauhut and H. Currier (to American Cyanamid Co.).
38. U.S. Pats. 3,342,853 and 3,342,854 (Sept. 19, 1967), J. W. Nemec and co-workers (to Rohm and Haas Co.).
39. C. R. Adams, *Chem. Ind.* **26,** 1644 (1970).
40. S. Sakuyama, T. O'Hara, N. Shimizu, and K. Kubota, *Chem. Technol.* 350 (June 1973).
41. T. O'Hara and co-workers, "Acrylic Acid and Derivatives" in *Ullmanns Encyclopedia of Industrial Chemistry,* 5th ed., Vol. A1, VCH, Verlagsgesellschaft mbH, Weinheim 1985, pp. 161–176.
42. *Hydrocarbon Process.* **48**(11), 145 (1969).
43. *Hydrocarbon Process.* **60**(11), 124 (1981).
44. *Hydrocarbon Process.* **68**(11), 91 (1989).
45. U.S. Pat. 3,475,488 (Oct. 28, 1969), N. Kurata, T. Ohara, and K. Oda (to Nippon Shokubai Kagaku Kogyo Co., Ltd.).
46. Brit. Pats. 915,799 and 915,800 (Jan. 16, 1963), D. J. Hadley and R. H. Jenkins (to Distillers Co., Ltd.).
47. U.S. Pats. 4,203,906 (May 20, 1980) 4,256,753 (Mar. 17, 1981), M. Takada, H. Uhara, and T. Sato (to Nippon Shokobai Kogaku Kogyo Co., Ltd.).
48. U.S. Pat. 4,537,874 (Aug. 27, 1985), T. Sato, M. Takata, M. Ueshima, and I. Nagai (to Nippon Shokubai Kogaku Co., Ltd.).
49. U.S. Pat. 4,438,217 (Mar. 20, 1984), M. Takata, R. Aoki, and T. Sato (to Nippon Shokubai Kogaku Co., Ltd.).
50. U.S. Pats. 2,881,212 (Apr. 7, 1959) and 3,087,964 (Apr. 30, 1963), J. D. Idol, J. L. Callahan, and R. W. Foreman (to Standard Oil Co., Ohio).
51. Jpn. Pat. 43-13609 (June 8, 1968), M. Izawa and co-workers (to Toyo Soda Manufacturing Co.).
52. U.S. Pats. 3,441,613 (Apr. 29, 1969) and 3,527,716 (Sept. 8, 1970), J. W. Nemec and F. W. Schlaefer (to Rohm and Haas Co.).
53. U.S. Pat. 4,092,354 (May 30, 1978), T. Shiraishi, T. Kechiwada, and Y. Nagaoka (to Sumitomo Chemical Co., Ltd.).
54. U.S. Pat. 3,527,797 (Sept. 8, 1970), R. Krabetz, H. Engelbach, and H. Zinke-Allmang (to Badische Anilin- und Soda-Fabrik A.G.).
55. U.S. Pats. 3,939,096 (Feb. 17, 1976) and 3,962,322 (June 8, 1976), P. C. Richardson (to Celanese Corp.).
56. U.S. Pat. 4,365,087 (Dec. 21, 1982), K. Kadowacki, K. Sarumaru, and T. Shibano (to Mitsubishi Petrochemical Co., Ltd.).
57. U.S. Pat. 4,356,114 (Oct. 26, 1982), K. Kadowacki, K. Sarumaru, and Y. Tanaka (to Mitsubishi Petrochemical Co., Ltd.).
58. T. P. Snyder and C. G. Hill, Jr., *Catal. Review—Sci. Eng.* **31,** 43–95 (1989). A current review with leading references to much of the literature of 1975–1990 in partial oxidation of propylene.
59. U.S. Pat. 4,147,885 (Apr. 3, 1979), N. Shimezur, I. Yonagisawa, M. Takata, and T. Sato (to Nippon Shokubai Kogaku Kogyo Co., Ltd.).
60. U.S. Pat. 4,317,926 (Mar. 2, 1982), T. Sato, M. Baba, and M. Okane (to Nippon Shokubai Kogaku Co., Ltd.).
61. Brit. Pat. 997,325 (July 7, 1965), F. C. Newman (to Distillers Co., Ltd.).
62. Jpn. Pat. 49-18728 (June 4, 1971) (to Toa Gosei Chemical Industry).
63. W. Krolikowski, *Soc. Plast. Eng. J.* 1031 (Sept. 1964).
64. U.S. Pat. 3,968,153 (July 6, 1976), T. Ohrui, T. Sakahibara, Y. Aono, M. Kato, H. Takao, and M. Ayano (to Sumitomo Chemical Co.).

65. U.S. Pat. 3,962,074 (June 8, 1976), W. K. Schropp (to Badische Anilin- und Soda-Fabrik A.G.).

66. U.S. Pat. 3,725,208 (Apr. 3, 1973), S. Maezawa, H. Yoshikawa, K. Sakamoto, J. Fugii, and M. Hashimoto (to Nippon Kayaku Co.).

67. U.S. Pat. 3,893,895 (July 8, 1975), J. Dehnert, A. Kleeman, T. Lussling, E. Noll, H. Schaefer, and G. Schreyer (to Deutsche Gold und Silber Scheideanstalt).

68. U.S. Pat. 4,358,347 (Nov. 9, 1982), B. Mettetal and R. Kolonko (to The Dow Chemical Company).

69. U.S. Pat. 3,914,290 (Oct. 21, 1975), S. Otsuki and I. Miyanohara (to Rohm and Haas Co.).

70. U.S. Pat. 2,916,512 (Dec. 8, 1959), G. J. Fischer and A. F. McLean (to Celanese Corp.).

71. U.S. Pat. 3,087,962 (Apr. 30, 1963), N. M. Bortnick (to Rohm and Haas Co.).

72. U.S. Pat. 3,882,167 (May 6, 1975), E. Lohmar, A. Ohorodnik, K. Gehrman, and P. Stutzke (to Hoechst A.G.).

73. U.S. Pat. 2,947,779 (Aug. 2, 1960), J. D. Idol, R. W. Foreman, and F. Veach (to Standard Oil Co. Ohio).

74. Brit. Pat. 923,595 (Apr. 18, 1963), F. J. Bellringer, C. J. Brown, and P. B. Brindley (to Distillers Co. Ltd.).

75. W. Reppe, *Justus Liebigs Ann. Chem.* **582,** 1 (1953).

76. U.S. Pat. 3,023,327 (Feb. 27, 1962), W. Reppe and R. Stadler (to Badische Anilin- und Soda Fabrik A.G.).

77. U.S. Pats. 2,582,911 (Jan. 15, 1952) and 2,613,222 (Oct. 7, 1952) and 2,773,063 (Dec. 4, 1956), H. T. Neher, E. H. Specht, and A. Neuman (to Rohm and Haas Co.).

78. M. Salkind, E. H. Riddle, and R. W. Keefer, *Ind. Eng. Chem.* **51,** 1232, 1328 (1959).

79. *Hydrocarbon Process.* **44**(11), 169 (1965).

80. U.S. Pat. 2,734,915 (Feb. 14, 1956), G. D. Jones (to The Dow Chemical Company).

81. U.S. Pats. 2,356,459 (Aug. 22, 1944) and 2,361,036 (Oct. 24, 1944), E. F. King, and U.S. Pat. 3,002,017 (Sept. 26, 1961), N. Wearsch and A. J. De Paola (to B. F. Goodrich Co.).

82. U.S. Pat. 3,069,433 (Dec. 18, 1962), K. A. Dunn (to Celanese Corp.).

83. *Threshold Limit Values and Biological Exposure Indices for 1989–1990,* American Conference of Governmental Industrial Hygienists, Cincinnati, Ohio, 1989.

84. F. K. Beilstein, *Ann. Chem.* **122,** 372 (1862).

85. W. von Schneider and E. Erlenmeyer, *Ber.* **3,** 340 (1870).

86. U.S. Pat. 2,821,543 (Jan. 28, 1958), R. W. Etherington (to Celanese Corp. of America).

87. U.S. Pat. 3,014,958 (Dec. 26, 1961), T. A. Koch and I. M. Robinson (to E. I. du Pont de Nemours & Co., Inc.).

88. U.S. Pats. 3,578,702 (May 11, 1971) and 3,574,703 (Apr. 13, 1971), T. C. Snapp, Jr., A. E. Blood, and H. J. Hagemeyer, Jr. (to Eastman Kodak Co.).

89. U.S. Pats. 3,840,587 and 3,840,588 (Oct. 8, 1974), A. J. C. Pearson (to Monsanto Co.).

90. U.S. Pat. 3,933,888 (Jan. 20, 1976), F. W. Schlaefer (to Rohm and Haas Co.).

91. U.S. Pat. 4,490,476 (Dec. 25, 1984), R. J. Piccolini and M. J. Smith (to Rohm and Haas Co.).

92. J. Haggin, *Chem. Eng. News,* 7–13 (May 19, 1986).

93. R. F. Heck, *Organotransition Metal Chemistry,* Academic Press, Inc., New York, 1974, Chapt. IX.

94. E. I. Becker and M. Tsutsui, eds., *Organometallic Reactions,* Vol. 3, Wiley-Interscience, New York, 1972.

95. U.S. Pats. 3,346,625 (Oct. 10, 1967), D. M. Fenton and K. L. Olivier; 3,397,225 (Aug. 13, 1968), D. M. Fenton; 3,349,119 (Oct. 24, 1967); and 3,381,030 (Apr. 30, 1968), D. M. Fenton and K. L. Olivier (to Union Oil Co. of California).

96. U.S. Pats. 3,920,736 (Nov. 18, 1975) and 3,876,694 (Apr. 8, 1975), W. Gaenzler (to Rohm G.m.b.H. Chemische Fabrik).

97. U.S. 3,579,568 (May 18, 1971), R. F. Heck and P. M. Henry (to Hercules Inc.).
98. U.S. Pat. 4,814,492 (Mar. 21, 1989), E. C. Nelson (to Texaco, Inc.).
99. U.S. Pat. 3,022,339 (Feb. 20, 1962), E. Enk and F. Knoerr (to Wacker Chemie G.m.b.H.).
100. U.S. Pat. 4,827,021 (May 2, 1989), G. C. Jones, W. D. Nottingham, and P. W. Raynolds (to Eastman Kodak Co.).
101. U.S. Pat. 2,917,538 (Dec. 15, 1959), R. L. Carlyle (to The Dow Chemical Company).
102. P. L. De Beneville, L. S. Luskin, and H. J. Sims, *J. Org. Chem.* **23**, 1355 (1958).
103. U.S. Pat. 4,777,265 (Oct. 11, 1988), F. Merger and co-workers (to BASF AG).
104. U.S. Pat. 2,484,487 (Oct. 11, 1949), J. R. Caldwell (to Eastman Kodak Co.).
105. U.S. Pat. 3,059,024 (Oct. 16, 1962), A. Goldberg, J. Fertig, and H. Stanley (to National Starch and Chemical Corp.).
106. L. S. Kirch, J. A. Kargol, J. W. Magee and W. S. Stuper, *Plant/Operations Progress* **1**(4) 270–274 (1988).
107. "Air Contaminants—Permissible Exposure Limit," *Title 29 Code of Federal Regulations Part CFR 1910.1000*, OSHA, 1989, p. 3112.

WILLIAM BAUER, JR.
Rohm and Haas Company

ACRYLIC AND MODACRYLIC FIBERS.   See FIBERS, ACRYLIC.

ACRYLIC ESTER POLYMERS

SURVEY

Acrylic esters are represented by the generic formula

The nature of the R group determines the properties of each ester and the polymers it forms. Polymers of this class are amorphous and distinguished by their water-clear color and their stability on aging. Acrylic monomers are extremely versatile building blocks. They are relatively moderate to high boiling liquids that readily polymerize or copolymerize with a variety of other monomers. Copolymers with methacrylates, vinyl acetate, styrene, and acrylonitrile are commercially significant. Polymers designed to fit specific application requirements ranging from soft, tacky adhesives to hard plastics can be tailored from these versatile monomers. Although the acrylics have been higher in cost than many other common monomers, they find use in high quality products where their unique characteristics and efficiency offset the higher cost.

Historically, the development of the acrylates proceeded slowly; they first received serious attention from Otto Rohm. Acrylic acid (propenoic acid) was first prepared by the air oxidation of acrolein in 1843 (1,2). Methyl and ethyl acrylate were prepared in 1873, but were not observed to polymerize at that time (3). In 1880 poly(methyl acrylate) was reported by G. W. A. Kahlbaum, who noted that on dry distillation up to 320° C the polymer did not depolymerize (4). Rohm observed the remarkable properties of acrylic polymers while preparing for his doctoral dissertation in 1901; however, a quarter of a century elapsed before he was able to translate his observations into commercial reality. He obtained a U.S. patent on the sulfur vulcanization of acrylates in 1912 (5). Based on the continuing work in Rohm's laboratory, the first limited production of acrylates began in 1927 by the Rohm and Haas Company in Darmstadt, Germany (6). Use of this class of compounds has grown from that time to a total U.S. consumption in 1989 of approximately 400,000 metric tons. Total worldwide consumption is probably twice that.

Physical Properties

To a large extent, the properties of acrylic ester polymers depend on the nature of the alcohol radical and the molecular weight of the polymer. As is typical of polymeric systems, the mechanical properties of acrylic polymers improve as molecular weight is increased; however, beyond a critical molecular weight, which often is about 100,000 to 200,000 for amorphous polymers, the improvement is slight and levels off asymptotically.

Glass Transition. The glass-transition temperature T_g reflects the mechanical properties of polymers over a specified temperature range. Below T_g polymers are stiff, hard, brittle, and glasslike; above T_g, if the molecular weight is high enough, they are relatively soft, limp, stretchable, and can be somewhat elastic. At even higher temperatures they flow and are tacky. Methods used to determine glass-transition temperatures and the reported values for a large number of polymers may be found in references 7–9. Values for the T_g of common acrylate homopolymers are found in Table 1.

The glass transition, unlike a true thermodynamic transition, takes place over a temperature range of several degrees and is dependent on both the experimental method and the time scale used for its determination. Below the transition temperature the majority of the polymer chains have a relatively fixed configuration, and little translation or rotation of polymer chains takes place. Above the glass-transition temperature the polymer chain has sufficient thermal energy for rotational motion or considerable torsional oscillation; thus the glass-transition temperature marks the onset of segmental mobility. During the transition there is no significant absorption of latent heat, but for most polymers there is an increase in the specific volume, coefficient of expansion, compressibility, specific heat, and refractive index. The transition is not sharp and T_g is taken as the midpoint of the

Table 1. Physical Properties of Acrylic Polymers

Polymer	Monomer molecular formula	CAS Registry Number	T_g, °C[a]	Density, g/cm^3 [b]	Refractive index, n_D
methyl acrylate	$C_4H_6O_2$	[9003-21-8]	6	1.22	1.479
ethyl acrylate	$C_5H_8O_2$	[9003-32-1]	-24	1.12	1.464
propyl acrylate	$C_6H_{10}O_2$	[24979-82-6]	-45		
isopropyl acrylate	$C_6H_{10}O_2$	[26124-32-3]	-3	1.08	
n-butyl acrylate	$C_7H_{12}O_2$	[9003-49-0]	-50	1.08	1.474
sec-butyl acrylate	$C_7H_{12}O_2$	[30347-35-4]	-20		
isobutyl acrylate	$C_7H_{12}O_2$	[26335-74-0]	-43		
tert-butyl acrylate	$C_7H_{12}O_2$	[25232-27-3]	43		
hexyl acrylate	$C_9H_{16}O_2$	[27103-47-5]	-57		
heptyl acrylate	$C_{10}H_{18}O_2$	[29500-72-9]	-60		
2-heptyl acrylate	$C_{10}H_{18}O_2$	[61634-83-1]	-38		
2-ethylhexyl acrylate	$C_{11}H_{20}O_2$	[9003-77-4]	-65		
2-ethylbutyl acrylate	$C_9H_{16}O_2$	[39979-32-3]	-50		
dodecyl acrylate	$C_{15}H_{28}O_2$	[26246-92-4]	-30		
hexadecyl acrylate	$C_{19}H_{36}O_2$	[25986-78-1]	35		
2-ethoxyethyl acrylate	$C_7H_{12}O_3$	[26677-77-0]	-50		
isobornyl acrylate	$C_{13}H_{20}O_2$	[30323-87-6]	94		
cyclohexyl acrylate	$C_9H_{14}O_2$	[27458-65-7]	16		

[a]Refs. 7 and 10.
[b]Ref. 11.

temperature interval over which the discontinuity occurs. The physical properties of amorphous acrylic polymers, which are dependent on the segmental relaxation rate, evidence a principal change in the glass-transition region. Chemical reactivity, mechanical and dielectric relaxation, viscous flow, load-bearing capacity, hardness, tack, heat capacity, refractive index, thermal expansivity, creep, crystallization, and diffusion differ markedly below and above the transition region.

The T_g of a polymer can be altered by the copolymerization of two or more monomers. The approximate T_g value for copolymers can be calculated from a knowledge of the weight fraction W of each monomer type and the T_g (in degees kelvin) of each homopolymer (12).

$$\frac{1}{T_g} = \frac{W_1}{T_{g1}} + \frac{W_2}{T_{g2}} + \frac{W_3}{T_{g3}} + \cdots + \frac{W_n}{T_{gn}}$$

Table 1 shows that most acrylics have low glass-transition temperatures. Therefore, in copolymers they tend to soften and flexibilize the overall composition. Plasticizers also lower the transition temperature. However, unlike incorporated acrylic comonomers, they can be lost through volatilization or extraction.

Much more information can be obtained by examining the mechanical properties of a viscoelastic material over an extensive temperature range. A convenient nondestructive method is the measurement of torsional modulus. A number of instruments are available (13–18). More details on use and interpretation of these measurements may be found in references 8 and 19–25. An increase in modulus value means an increase in polymer hardness or stiffness. The various regions of elastic behavior are shown in Figure 1. Curve A of Figure 1 is that of a soft polymer, curve B of a hard polymer. To a close approximation both are transpositions of each other on the temperature scale. A copolymer curve would fall between those of the homopolymers, with the displacement depending on the amount of hard monomer in the copolymer (26–28).

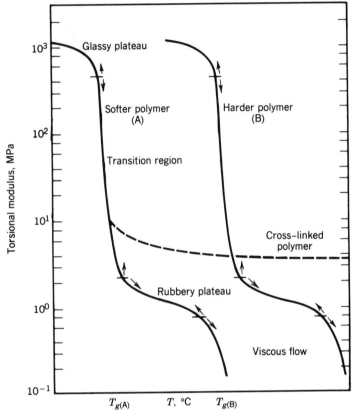

Fig. 1. Modulus–temperature curve of amorphous and cross-linked acrylic polymers. To convert MPa to kg/cm^2, multiply by 10.

Cross-linking of a polymer elevates and extends the rubbery plateau; little effect on T_g is noted until extensive cross-linking has been introduced (23,25,28). A cross-link joins more than two primary polymer chains together. In practice, cross-linking of acrylic polymers is used to decrease thermoplasticity and solubility and increase resilience. In some instances cross-linking moieties are used in reactions of a polymer with a substrate (20). The chemistry of cross-linking is described in references 11 and 29–38.

Molecular Weight.   The values of the mechanical properties of polymers increase as the molecular weight increases. However, beyond some critical molecular weight, often about 100,000 to 200,000 for amorphous polymers, the increase in property values is slight and levels off asymptotically. As an example, the glass-transition temperature of a polymer usually follows the relationship

$$T_g = T_{gi} - k/M_n$$

where T_{gi} is the glass-transition temperature at infinite molecular weight and M_n the number-average molecular weight. The value of k for acrylate polymers is about 2×10^5 (39). A detailed discussion of the effect of molecular weight on the properties of a polymer may be found in reference 40.

Mechanical and Thermal Properties. The first member of the acrylate series, poly(methyl acrylate), has little or no tack at room temperature; it is a tough, rubbery, and moderately hard polymer. Poly(ethyl acrylate) is more rubberlike, considerably softer, and more extensible. Poly(butyl acrylate) is softer still, and much tackier. This information is quantitatively summarized in Table 2 (41). In the n-alkyl acrylate series, the softness increases through n-octyl acrylate. As the chain length is increased beyond n-octyl, side-chain crystallization occurs and the materials become brittle (42); poly(n-hexadecyl acrylate) is hard and waxlike at room temperature but is soft and tacky above its softening point.

Table 2. Mechanical Properties of Acrylic Polymers

Polyacrylate	Elongation, %	Tensile strength, kPa[a]
methyl	750	6895
ethyl	1800	228
butyl	2000	21

[a]To convert kPa to psi, multiply by 0.145.

Unless subjected to extreme conditions, acrylic polymers are durable and degrade slowly. In contrast to methacrylate polymers, which depolymerize on strong heating, poly(methyl acrylate) when pyrolyzed *in vacuo* from 292 to 399°C yields only a small quantity of monomer (43–44). Oxidative degradation of acrylic polymers can occur by the combination of oxygen with free radicals generated in the polymer to form hydroperoxides. The rate of oxidation is quite slow unless extreme conditions with oxygen under high pressure and high temperatures are employed (45–47).

Solution Properties. Typically, if a polymer is soluble in a solvent, it is soluble in all proportions. As solvent evaporates from the solution, no phase separation or precipitation occurs. The solution viscosity increases continually until a coherent film is formed. The film is held together by molecular entanglements and secondary bonding forces. The solubility of the acrylate polymers is affected by the nature of the side group. Polymers that contain short side chains are relatively polar and are soluble in polar solvents such as ketones, esters, or ether alcohols. As the side chain increases in length the polymers are less polar and dissolve in relatively nonpolar solvents, such as aromatic or aliphatic hydrocarbons.

The quantitative treatment of solubility is based on the familiar free energy equation governing mutual miscibility:

$$\Delta F = \Delta H_m - T\Delta S_m$$

Solubility occurs when the free energy of mixing is negative. On solution there is more movement of polymer chains; therefore, entropy increases as a polymer dissolves. Because the entropy term is always positive and large, the sign and magnitude of the heat term is the deciding factor in determining the sign of the free energy change. When the difference between the solubility parameters of two substances is small, the heat of mixing must be small; the free energy tends to be negative and solubility should occur. A polymer dissolves in a solvent or is compatible with another polymer if the solubility parameters and polarities are similar (38,48–53). Representative solubility parameter values are given in Table 3.

Table 3. Solubility Parameters of Acrylic Homopolymers Calculated by Small's Method[a]

Polymer	$(J/cm^3)^{1/2}$ [b]
methyl acrylate	4.7
ethyl acrylate	4.5
n-butyl acrylate	4.3

[a]Refs. 23 and 53.
[b]To convert $(J/cm^3)^{1/2}$ to $(cal/cm^3)^{1/2}$, divide by 2.05.

Polymer solution viscosity is dependent on the concentration of the solvent, the molecular weight of the polymer, the polymer composition, the solvent composition, and the temperature. More extensive information on the properties of polymer solutions may be found in references 9 and 54–56.

Chemical Properties

Under conditions of extreme acidity or alkalinity, acrylic ester polymers can be made to hydrolyze to poly(acrylic acid) or an acid salt and the corresponding alcohol. However, acrylic polymers and copolymers have a greater resistance to both acidic and alkaline hydrolysis than competitive poly(vinyl acetate) and vinyl acetate copolymers. Even poly(methyl acrylate), the most readily hydrolyzed polymer of the series, is more resistant to alkali than poly(vinyl acetate) (57). Butyl acrylate copolymers are more hydrolytically stable than ethyl acrylate copolymers (58).

Acrylic polymers are fairly insensitive to normal uv degradation since the primary uv absorption of acrylics occurs below the solar spectrum (59). The incorporation of absorbers, such as o-hydroxybenzophenone [117-99-7] , further improves the uv stability (59). Under normal use conditions acrylic polymers have superior resistance to degradation and show remarkable retention of their original properties.

Both side-chain and main-chain scission products are observed when polyacrylates are irradiated with gamma radiation (60). The nature of the alkyl side group affects the observed ratio of these two processes (61,62).

Acrylic Ester Monomers

The physical properties of the principal commercial acrylic esters are given in Table 4. A more comprehensive listing of physical properties, including other less common acrylates, is provided in the article ACRYLIC ACID AND DERIVATIVES.

There are currently two principal processes used for the manufacture of monomeric acrylic esters: the semicatalytic Reppe process and the propylene oxidation process. The newer propylene oxidation process is preferred because of economy and safety. In this process acrolein [107-02-8] is first formed by the catalytic oxidation of propylene vapor at high temperature in the presence of steam. The acrolein is then oxidized to acrylic acid [79-10-7].

$$CH_2{=}CHCH_3 + O_2 \xrightarrow{\text{catalyst}} CH_2{=}CHCHO + H_2O$$

$$2\ CH_2{=}CHCHO + O_2 \xrightarrow{\text{catalyst}} 2\ CH_2{=}CHCOOH$$

Both one-step and two-step oxidation processes are known. A number of catalyst systems are known; most use a molybdenum compound as the main component. The acrylic acid is esterified with alcohol to the desired acrylic ester in a separate process (63–66).

In normal practice, inhibitors such as hydroquinone (HQ) [123-31-9] or the monomethyl ether of hydroquinone (MEHQ) [150-76-5] are added to acrylic monomers to stabilize them during shipment and storage. Uninhibited acrylic monomers should be used promptly or stored at 10°C or below for no longer than a few weeks. Improperly inhibited monomers have the potential for violent polymerizations. HQ and MEHQ require the presence of oxygen to be effective inhibitors; therefore, these monomers should be stored in contact with air and not under inert atmosphere. Because of the low concentration of inhibitors present in most commercial grades of acrylic monomers (generally less than 100 ppm), removal before use is not normally required. However, procedures for removal of inhibitors are available (67).

The common acrylic ester monomers are combustible liquids. Commercially, acrylic monomers are shipped with DOT red labels in bulk quantities, tank cars, or tank trucks. Mild steel is the usual material of choice for the construction of bulk storage facilities for acrylic monomers. Moisture must be excluded to avoid rusting of the tanks and contamination of the monomers. Copper or copper alloys must not be allowed to contact acrylic monomers intended for use in polymerization because copper is an inhibitor (67).

Numerous methods for the determination of monomer purity, including procedures for the determination of saponification equivalent and bromine num-

Table 4. Physical Properties of Acrylic Monomers

Acrylate	CAS Registry Number	Molecular weight	bp, °C[a]	d^{25}, g/cm^3	Flash point, °C[b]	Water solubility, g/100 g H$_2$O	Heat of evaporation, J/g[c]	Specific heat, J/g·K[c]
methyl	[96-33-3]	86	79–81	0.950	10	5	385	2.01
ethyl	[140-88-5]	100	99–100	0.917	10	1.5	347	1.97
n-butyl	[141-32-2]	128	144–149	0.894	39	0.2	192	1.92
isobutyl	[106-63-8]	128	61–63[d]	0.884	42	0.2	297	1.92
t-butyl	[1663-39-4]	128	120	0.879	19	0.2		
2-ethylhexyl	[103-11-7]	184	214–220	0.880	90[e]	0.01	255	1.92

[a]At 101.3 kPa unless otherwise noted.
[b]Tag open cup unless otherwise noted.
[c]To convert J to cal, divide by 4.184.
[d]At 6.7 kPa = 50 mm Hg.
[e]Cleveland open cup.

ber, specific gravity, refractive index, and color, are available from manufacturers (68–70). Concentrations of minor components are determined by iodimetry or colorimetry for HQ or MEHQ, by the Karl-Fisher method for water, and by turbidity measurements for trace amounts of polymer. Gas–liquid chromatography is widely used both for a direct determination of monomer quality and for identification and determination of minor components.

The toxicities of acrylic monomers range from moderate to slight. In general, they can be handled safely and without difficulty by trained personnel following established safety practices. Table 5 summarizes investigations of the toxicity of the common acrylic monomers in animals under acute toxicity conditions (67).

Table 5. Toxicities of Acrylic Monomers

| Monomer | Acute oral LD$_{50}$, rats, mg/kg | Acute precutaneous LD$_{50}$, rabbits, mg/kg | Inhalation | |
			LC$_{50}$, rats, mg/L	TLV, ppm
methyl acrylate	300	1235	3.8	10
ethyl acrylate	760	1800	7.4	25
butyl acrylate	3730	3000	5.3	

Liquid methyl and ethyl acrylate are moderately toxic when taken internally, brought into contact with the eyes, or absorbed through the skin. Eye contact is the most serious hazard. Vapors of methyl and ethyl acrylate are moderately lacrimatory and irritating to the respiratory membranes. Threshold limit values (TLV) have been adopted by OSHA, restricting prolonged low level exposures to vapors of methyl and ethyl acrylate in industrial situations (Table 5). Repeated exposure of animals to the vapors of methyl or ethyl acrylate at lethal or near lethal concentrations produces degenerative changes in the liver, kidneys, and heart muscle. However, their obvious odors and irritating effects reduce the likelihood of significant exposure. The higher acrylates have similar but more moderate toxicities. Certain individuals have specific allergic reactions to acrylic monomers, including eye irritation, headache, or skin eruptions. Ethyl acrylate appears on the California Safe Drinking Water and Toxic Enforcement Act of 1986 (Proposition 65) as a known carcinogen (71).

In normal practice, good ventilation to reduce exposure to vapors, splash-proof goggles to avoid eye contact, and protective clothing to avoid skin contact are required for the safe handling of acrylic monomers. A more extensive discussion of these factors should be consulted before handling these monomers (67).

Radical Polymerization

Usually, free-radical initiators such as azo compounds or peroxides are used to initiate the polymerization of acrylic monomers. Photochemical (72–74) and radiation-initiated (75) polymerizations are also well known. At a constant temperature, the initial rate of the bulk or solution radical polymerization of acrylic monomers is first order with respect to monomer concentration and one-half order with respect to the initiator concentration. Rate data for polymerization of several common acrylic monomers initiated with 2,2'-azobisisobutyronitrile (AIBN) [78-67-1] have been determined and are shown in Table 6. The table also includes heats of polymerization and volume percent shrinkage data.

Table 6. Polymerization Data for Acrylic Ester Monomers in Solution[a]

Acrylate	Concentration, solvent	k_{sp}, L/mol·h[b]	Heat, kJ/mol[c]	Shrinkage, vol %
methyl	3 M, methyl propionate	250	78.7	24.8
ethyl	3 M, benzene	313	77.8	20.6
butyl	1.5 M, toluene	324	77.4	15.7

[a]Ref. 76.
[b]At 44.1°C.
[c]To convert kJ to kcal divide by 4.184.

Acrylate and methacrylate polymerizations are accompanied by the liberation of a considerable amount of heat and a substantial decrease in volume. Both of these factors strongly influence most manufacturing processes. Excess heat must be dissipated to avoid uncontrolled exothermic polymerizations. In general, the percentage of shrinkage decreases as the size of the alcohol substituent increases; on a molar basis, the shrinkage is relatively constant (77).

The free-radical polymerization of acrylic monomers follows a classical chain mechanism in which the chain-propagation step entails the head-to-tail growth of the polymeric free radical by attack on the double bond of the monomer.

$$R'-CH_2CH \cdot \quad + \quad CH_2{=}CH \quad \longrightarrow \quad R'-CH_2CH-CH_2-CH \cdot$$
$$\underset{COOR}{|} \qquad\qquad \underset{COOR}{|} \qquad\qquad\qquad \underset{COOR}{|} \quad \underset{COOR}{|}$$

Chain termination can occur by either combination or disproportionation, depending on the conditions of the process (78,79).

Some details of the chain-initiation step have been elucidated. With an oxygen radical-initiator such as the t-butoxyl radical, both double bond addition and hydrogen abstraction are observed. Hydrogen abstraction is observed at the ester alkyl group of methyl acrylate. Double bond addition occurs in both a head-to-head and a head-to-tail manner (80).

Acrylate polymerizations are markedly inhibited by oxygen; therefore, considerable care is taken to exclude air during the polymerization stages of manufacturing. This inhibitory effect has been shown to be caused by copolymerization of oxygen with monomer, forming an alternating copolymer (81,82).

$$R'-CH_2CH \cdot \quad + \quad O_2 \overset{fast}{\longrightarrow} R'-CH_2CHOO-CH_2CHOO \cdot$$
$$\underset{COOR}{|} \qquad\qquad\qquad\qquad \underset{COOR}{|} \qquad \underset{COOR}{|}$$

In the presence of any substantial amount of oxygen this reaction is extremely rapid, but the terminal peroxy radical formed reacts slowly with monomer and has a relatively rapid termination rate.

$$R'-CH_2CHOO-CH_2CHOO\cdot + CH_2{=}CH \xrightarrow{\text{slow}} R'-CH_2CHOO-CH_2CH\cdot$$
$$\underset{COOR}{|} \quad \underset{COOR}{|} \quad\quad \underset{COOR}{|} \quad\quad\quad \underset{COOR}{|} \quad \underset{COOR}{|}$$

The overall effect, aside from the change in the polymer composition, is a decrease in the rate of monomer reaction, the kinetic chain length, and the polymer molecular weight (83).

The vast majority of all commercially prepared acrylic polymers are copolymers of an acrylic ester monomer with one or more different monomers. Copolymerization greatly increases the range of available polymer properties and has led to the development of many different resins suitable for a broad variety of applications. Several review articles are available (84,85).

In general, acrylic ester monomers copolymerize readily with each other or with most other types of vinyl monomers by free-radical processes. The relative ease of copolymerization for 1:1 mixtures of acrylate monomers with other common monomers is presented in Table 7. Values above 25 indicate that good copolymerization is expected. Low values can often be offset by a suitable adjustment in the proportion of comonomers or in the method of their introduction into the polymerization reaction (86).

Table 7. Relative Ease of Copolymer Formation for 1:1 Ratios of Acrylic and Other Monomers, $\dfrac{r(\text{smaller})}{r(\text{larger})} \times 100$

	Monomer 1		
Monomer 2	Methyl acrylate	Ethyl acrylate	Butyl acrylate
acrylonitrile, [107-13-1]	53	46	74
butadiene, [106-99-0]	66	4.7	8.1
methyl methacrylate, [80-62-6]	50.3	30.6	14.6
styrene, [100-42-5]	21	16	26
vinyl chloride, [75-01-4]	2.7	2.1	1.6
vinylidene chloride, [75-35-4]	100	52	55
vinyl acetate, [108-05-4]	1.1	0.7	0.6

The $r(\text{smaller})/r(\text{larger})$ values listed in Table 7 correspond to the reactivity ratios for monomer 1, r_1, and monomer 2, r_2. These are defined as

$$r_1 = k_{11}/k_{12}$$
$$r_2 = k_{22}/k_{21}$$

For a growing radical chain that has monomer 1 at its radical end, its rate constant for combination with monomer 1 is designated k_{11} and with monomer 2, k_{12}. Similarly, for a chain with monomer 2 at its growing end, the rate constant for combination with monomer 2 is k_{22} and with monomer 1, k_{21}. The reactivity ratios may be calculated from Price-Alfrey Q and e values, which are given in Table 8 for the more important acrylic esters (87). The sequence distributions of numerous acrylic copolymers have been determined experimentally utilizing nmr techniques (88,89). Several review articles discuss copolymerization (84,85).

Table 8. Q and e Values for Acrylic Monomers[a]

Monomer	Q	e
methyl acrylate	0.44	+0.60
ethyl acrylate	0.41	+0.46
butyl acrylate	0.30	+0.74
isobutyl acrylate	0.41	+0.34
2-ethylhexyl acrylate	0.14	+0.90

[a]Ref. 88.

Small amounts of specially functionalized monomers are often copolymerized with acrylic monomers in order to modify or improve the properties of the polymer. These functional monomers can bring about improvements either directly or by providing sites for further reaction with metal ions, cross-linkers, or other compounds and resins. Table 9 lists some of the more common functional monomers used in the preparation of acrylic copolymers.

Bulk Polymerization. The bulk polymerization of acrylic monomers is characterized by a rapid acceleration in the rate and the formation of a cross-linked insoluble network polymer at low conversion (90,91). Such network polymers are thought to form by a chain-transfer mechanism involving abstraction of the hydrogen alpha to the ester carbonyl in a polymer chain followed by growth of a branch radical. Ultimately, two of these branch radicals combine (91). Commercially, the bulk polymerization of acrylic monomers is of limited importance.

Solution Polymerization. The solution polymerization of acrylic monomers to form soluble acrylic polymers is an important commercial process. In general, the polyacrylate esters of the lower alcohols are soluble in aromatic hydrocarbons, esters, ketones, and chlorohydrocarbons. They are insoluble or only slightly soluble in aliphatic hydrocarbons, ethers, and alcohols. The higher poly(alkyl acrylates) are generally insoluble in oxygenated organic solvents and soluble in both aliphatic and aromatic hydrocarbons and in chlorohydrocarbons. Cost, toxicity, flammability, volatility, and chain-transfer activity are the primary considerations in the selection of a suitable solvent. Chain transfer to solvent is an important factor in controlling the molecular weight of polymers prepared by this method. The chain-transfer constants C_s for poly(ethyl acrylate) in various solvents are listed in Table 10.

The type of initiator utilized for a solution polymerization depends on several factors, including the solubility of the initiator, the rate of decomposition of the initiator, and the intended use of the polymeric product. The amount of initiator used may vary from a few hundredths to several percent of the monomer weight. As the amount of initiator is decreased, the molecular weight of the polymer is increased as a result of initiating fewer polymer chains per unit weight of monomer, and thus the initiator concentration is often used to control molecu-

Table 9. Functional Monomers for Copolymerization with Acrylic Monomers

Monomer	Structure	CAS Registry Number	Molecular formula
Carboxyl			

methacrylic acid	$CH_2{=}CCOOH$
	       |
	      CH_3
		[79-41-4]	$C_4H_6O_2$

acrylic acid	$CH_2{=}CHCOOH$	[79-10-7]	$C_3H_4O_2$
itaconic acid	CH_2COOH | $CH_2{=}CCOOH$	[97-65-4]	$C_5H_6O_4$
Amino			
t-butylaminoethyl methacrylate	CH_3 | $CH_2{=}CCOO(CH_2)_2NHC(CH_3)_3$	[24171-27-5]	$C_{10}H_{19}NO_2$
dimethylaminoethyl methacrylate	CH_3 | $CH_2{=}CCOO(CH_2)_2N(CH_3)_2$	[2867-47-2]	$C_8H_{15}NO_2$
Hydroxyl			
2-hydroxyethyl methacrylate	CH_3 | $CH_2{=}CCOOCH_2CH_2OH$	[868-77-9]	$C_6H_{10}O_3$
2-hydroxyethyl acrylate	$CH_2{=}CHCOOCH_2CH_2OH$	[818-61-1]	$C_5H_8O_3$
N-Hydroxymethyl			
N-hydroxymethyl acrylamide	$CH_2{=}CHCONHCH_2OH$	[924-42-5]	$C_4H_7NO_2$
N-hydroxymethyl methacrylamide	CH_3 | $CH_2{=}CCONHCH_2OH$	[923-02-4]	$C_5H_9NO_2$
Oxirane			
glycidyl methacrylate	CH_3 O | / \\ $CH_2{=}CCOOCH_2CH{-}CH_2$	[106-91-2]	$C_7H_{10}O_3$
Multifunctional			
1,4-butylene dimethacrylate	CH_3 | $(CH_2{=}CCOOCH_2)_2$	[2082-81-7]	$C_{12}H_{18}O_4$

Table 10. Chain-Transfer Constants to Common
Solvents for Poly(ethyl acrylate)[a]

Solvent	$C_s \times 10^5$
benzene	5.2
toluene	26.0
isopropyl alcohol	260
isobutyl alcohol	46.5
chloroform	14.9
carbon tetrachloride	15.5

[a]Refs. 79, 92, and 93.

lar weight. Organic peroxides, hydroperoxides, and azo compounds are the initiators of choice for the preparations of most acrylic solution polymers and copolymers.

The molecular weight of a polymer can be controlled through the use of a chain-transfer agent, as well as by initiator concentration and type, monomer concentration, and solvent type and temperature. Chlorinated aliphatic compounds and thiols are particularly effective chain-transfer agents used for regulating the molecular weight of acrylic polymers (94). Chain-transfer constants (C_s at 60°C) for some typical agents for poly(methyl acrylate) are as follows (87):

carbon tetrabromide	0.41
ethanethiol	1.57
butanethiol	1.69

Solution polymerizations of acrylic esters are usually conducted in large stainless steel, nickel, or glass-lined cylindrical kettles designed to withstand at least 446 kPa (65 psi). An anchor-type agitator is suitable for solutions with viscosity up to 1.0 Pa·s (10 P); a slow-speed ribbon agitator with close clearance to the walls is required for solutions of higher viscosity. In large kettles, turbine agitators with swept-back blades are often used. Typically, production kettles are fitted with a jacket for steam or hot water heating and for cooling, a reflux condenser, inlets for addition of the reaction ingredients, sightglasses, a thermometer, and a rupture disk. A bottom valve is used for discharging the finished product to receiving tanks or drums.

Since acrylic polymerizations liberate considerable heat, violent or runaway reactions are avoided by gradual addition of the reactants to the kettle. Usually the monomers are added by a gravity feed from weighing or measuring tanks situated close to the kettle. The rate of monomer addition is adjusted to permit removal of heat with full flow of water in the condenser and a partial flow in the cooling jacket. Flow in the jacket can be increased to control the polymerization in cases of erroneous feed rates or other unexpected circumstances. A supply of inhibitor is kept on hand to stop the polymerization if the cooling becomes inadequate.

Initiators, usually from 0.02 to 2.0 wt % of the monomer of organic peroxides or azo compounds, are dissolved in the reaction solvents and fed separately to the kettle. Since oxygen is often an inhibitor of acrylic polymerizations, its presence is undesirable. When the polymerization is carried out below reflux temperatures, low oxygen levels are obtained by an initial purge with an inert gas such as carbon dioxide or nitrogen. A blanket of the inert gas is then maintained over the polymerization mixture. The duration of the polymerization is usually 24 h (95).

A typical process for the preparation of a 94.8% ethyl acrylate–5.2% acrylic acid copolymer as an approximately 39% solution in ethyl acetate is given below:

Reactor charge	Parts
ethyl acetate	61.4
benzoyl peroxide	0.1
Monomer charge	
ethyl acrylate	36.5
acrylic acid	2.0

The solvent and initiator are charged to the reactor and heated to reflux (ca 80°C). Forty percent of the monomer charge is then added. The remainder of the monomer is added in four equal increments at 24, 50, 79, and 110 min after addition of the initial monomer charge. The reaction mixture is kept at reflux overnight, then cooled and packaged (96).

Acrylic resins are shipped in drum, tank truck, and tank car quantities. Storage tanks, usually of 20 to 40 m³ capacity, are constructed of stainless steel, although tanks of mild steel can be used if provisions are made to keep them free of water. Moisture in mild steel tanks causes rusting and discoloration of the polymer solution. Typically, storage tanks are equipped with a bottom outlet and are dished or sloped to permit complete drainage. Most are also equipped with a manhole so that they can be entered for cleaning. Because the viscosity of most acrylic solution polymers varies with temperature, temperature must be controlled by means of tank location, insulation, and heating or cooling devices. Steel pipe with steel or stainless steel valves is used for the resin transfer lines (97).

Emulsion Polymerization.   Emulsion polymerization is the most important industrial method for the preparation of acrylic polymers. The principal markets for aqueous dispersion polymers made by emulsion polymerization of acrylic esters are the paint, paper, adhesives, textile, floor polish, and leather industries, where they are used principally as coatings or binders. Copolymers of either ethyl acrylate or butyl acrylate with methyl methacrylate are most common.

Most of the lower alkyl acrylates readily polymerize in water in the presence of a surfactant and a water-soluble initiator. The final product is an opaque, gray, or milky-white dispersion or latex of high molecular weight polymer at a concentration of 30–60% in water. The particle size of acrylic copolymer dispersions ranges from about 0.1 μm to about 1.0 μm. These emulsion polymerizations are usually rapid and give high molecular weight polymers at high concentration and low viscosity. Difficulties in agitation, heat transfer, and material transfer, which are often encountered in the handling of viscous polymer solutions, are greatly decreased with aqueous dispersions. In addition, the safety hazards and the expense of flammable solvents are eliminated.

The surfactants used in the emulsion polymerization of acrylic monomers are classified as anionic, cationic, or nonionic. Anionic surfactants, such as salts of alkyl sulfates and alkylarene sulfates and phosphates, or nonionic surfactants, such as alkyl or aryl polyoxyethylenes, are most common (87,98–101). Mixed anionic–nonionic surfactant systems are also widely utilized (102–105).

Water-soluble peroxide salts, such as ammonium or sodium persulfate, are the usual initiators. The initiating species is the sulfate radical anion generated from either the thermal or redox cleavage of the persulfate anion. The thermal dissociation of the persulfate anion, which is a first-order process at constant temperature (106), can be greatly accelerated by the addition of certain reducing agents or small amounts of polyvalent metal salts, or both (87). By using redox initiator systems, rapid polymerizations are possible at much lower temperatures (25–60°C) than are practical with a thermally initiated system (75–90°C).

Industrial acrylic emulsion polymerizations are usually conducted by batch processes in jacketed stainless steel or glass-lined kettles designed to withstand an internal pressure of at least 446 kPa (65 psi). Glass-lined equipment is preferred for easier cleaning. Agitators are constructed from the same materials as the reactor. Versatility in controlling agitation is provided by use of a variable speed drive. A baffle is sometimes used in the kettle to improve mixing; however, excessive shear must be avoided to minimize the formation of coagulum. The temperature of the reactants is controlled by circulating steam and cold water through the jacket. A schematic diagram of a typical plant installation is given in Figure 2. A filtered feed line for emulsified monomer is installed to enter through the top of the kettle. Separate feed lines are also provided in the top of the reactor for adding aqueous solutions of initiators and/or activators. Additional equipment includes a temperature recorder, manometer, sightglass, and emergency stack equipped with a rupture disk.

Monomer emulsions are prepared in separate stainless steel emulsification tanks that are usually equipped with a turbine agitator, manometer level gage, cooling coils, a sprayer inert gas, temperature recorder, rupture disk, flame arrester, and various nozzles for charging the ingredients. Monomer emulsions are commonly fed continuously to the reactor throughout the polymerization.

A simple stainless steel drumming tank is used to receive the polymerized emulsion and hold it until it is packaged into drums or tank cars. This tank is also employed to adjust the solids content and pH; for the addition of preservatives, stabilizers, and thickeners; and for blending operations. A paddle-type low speed agitator is used for mixing. A cooling jacket on the drumming tank permits discharge of hot dispersion from the reactor, thus increasing the productivity of the kettle. The cooled dispersion is passed through a coarse filter prior to packaging (87).

Numerous recipes have been published, primarily in the patent literature, that describe the preparation of acrylate and methacrylate homopolymer and copolymer dispersions (107,108). A typical process for the preparation of a 50%

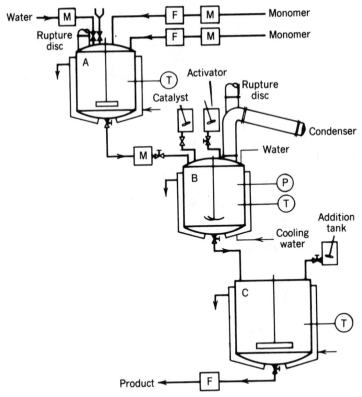

Fig. 2. Emulsion polymerization plant. A, Emulsion feed tank; B, polymerization reactor; C, drumming tank; F, filter; M, meter; P, pressure gauge; T, temperature indication.

methyl methacrylate, 49% butyl acrylate, and 1% methacrylic acid terpolymer as an approximately 45% dispersion in water begins with the following charges:

Monomer emulsion charge	Parts
deionized water	13.65
sodium lauryl sulfate	0.11
methyl methacrylate	22.50
butyl acrylate	22.05
methacrylic acid	0.45

Initiator charge	
ammonium persulfate	0.23

Reactor charge	
deionized water	30.90
sodium lauryl sulfate	0.11

The monomer emulsion charge is prepared by adding the listed ingredients in the order given while maintaining good agitation. The reactor charge is heated with good agitation under a nitrogen atmosphere to 85°C; then the initiator charge is added to the reactor and the monomer emulsion feed is begun. The monomer emulsion is fed uniformly over 2.5 h while maintaining 85°C. After the addition is complete, the temperature is raised to 95°C to complete the conversion of monomer. The product is then cooled to room temperature, filtered, and packaged.

Acrylic dispersion polymers are shipped in bulk or in drums. Tank trucks and tank cars used for bulk shipment are constructed of stainless or resin-coated steel and are insulated to prevent freezing. Filament-wound glass-fiber-reinforced polyester tanks are recommended for storage because of their relatively low cost, ease of installation, and chemical resistance. Usually storage tanks are located in an enclosed and heated environment to prevent freezing during cold weather. Dispersion polymers are subject to the various instability problems common to all colloidal systems, such as sedimentation, skinning (surface film), gritting (solid within the dispersion), gumming (deposits on walls), and sponging (formation of an aerogel). Undesirable changes may be caused by time, drift in pH, evaporation, high or low temperature, shear and turbulence, and foaming. Oxidative degradation is not usually encountered with acrylic dispersion polymers, but bacterial attack is common and is avoided by pH adjustment, addition of bactericidal agents, and careful housekeeping (95). Recent advances in the industrial aspects of emulsion polymerization can be found in reference 109.

Suspension Polymerization. Suspension polymerization yields polymer in the form of tiny beads, which are primarily used as molding powders and ion-exchange resins. Most suspension polymers prepared as molding powders are poly(methyl methacrylate); copolymers containing up to 20% acrylate for reduced brittleness and improved processibility are also common.

In a suspension polymerization monomer is suspended in water as 0.1–5-mm droplets, stabilized by protective colloids or suspending agents. Polymerization is initiated by a monomer-soluble initiator and takes place within the monomer droplets. The water serves as both the dispersion medium and a heat-transfer agent. Particle size is controlled primarily by the rate of agitation and the concentration and type of suspending aids. The polymer is obtained as small beads about 0.1–5 mm in diameter, which are isolated by filtration or centrifugation.

Suitable protective colloids for the preparation of acrylic suspension polymers include cellulose derivatives, polyacrylate salts, starch, poly(vinyl alcohol), gelatin, talc, clay, and clay derivatives (95). These materials are added to prevent the monomer droplets from coalescing during polymerization (110). Thickeners such as glycerol, glycols, polyglycols, and inorganic salts are also often added to improve the quality of acrylic suspension polymers (95). Other constituents may be added to assist in the formation of uniform beads or to influence the use properties of the polymers through plasticization or cross-linking. These include lubricants, such as lauryl or cetyl alcohol and stearic acid, and cross-linking monomers such as di- or trivinylbenzene, diallyl esters of dibasic acids, and glycol dimethacrylates.

Initiators of suspension polymerization are organic peroxides or azo compounds that are soluble in the monomer phase but insoluble in the water phase.

The amount of initiator influences both the polymerization rate and the molecular weight of the product (95).

Because the polymerization occurs totally within the monomer droplets without any substantial transfer of materials between individual droplets or between the droplets and the aqueous phase, the course of the polymerization is expected to be similar to bulk polymerization. Accounts of the quantitative aspects of the suspension polymerization of methyl methacrylate generally support this model (95,111,112). Developments in suspension polymerization, including acrylic suspension polymers, have been reviewed (113,114).

Graft Copolymerization.   Graft copolymers are formed by attaching one polymer as a branch to the chain of another polymer. This method requires the generation of radical sites on the first polymer onto which monomer of a second type is polymerized. The presence of distinct but chemically bonded segments of two polymers often confers interesting and useful properties. Graft polymerizations of acrylic monomers onto both synthetic (115,116) and natural polymers (117–120) are reported. The radical sites for grafting are generated by a variety of methods including chemical (121–123), photochemical (124,125), radiation (126), and mechanical mastication (127). The synthesis and characterization of graft copolymers, including acrylic graft copolymers, has been reviewed (128,129).

Radiation-Induced Polymerization.   The uv cure of coatings, printing ink, and photoresists has become increasingly important in recent years because it offers the economic advantage of requiring less energy than thermal cure and the ecological advantage of reduced solvent emissions. A chief drawback of this technology is its inability to cure highly pigmented films because the pigments reflect and absorb uv radiation and prevent adequate light penetration.

Ultraviolet curable coatings (130) and printing inks (131) are typically composed of the following ingredients: pigment, monomer, polymer, photoinitiator, and inhibitor. The formulation is applied to a substrate as a thin film, then cured (polymerized) rapidly by exposure to uv radiation. The polymers utilized often contain unsaturation in order to co-cure with the monomer. The function of the photoinitiator is to absorb the radiation and initiate the free-radical polymerization. Because of their rapid cure rates, the high boiling acrylate monomers, both monofunctional and multifunctional, are widely used as the solvent media in uv curable systems. Methacrylate functionality is often used as the co-curing unsaturation site on the polymer (132). Methacrylate functional polymers and acrylate monomers also find use in uv curing photoresist applications (133). References 132, 134, 135, and 136 are reviews of the state of the art in uv cure technology, including the use of acrylate monomers.

Electron-beam curing of coatings is becoming an area of active interest and increasing industrial importance. This technology uses formulations very similar to those employed by the uv coatings industry, including the widespread use of higher boiling acrylate monomers. A good description and overview of this technique can be found in reference 137 (see also RADIATION CURING).

Ionic Polymerization

Acrylate monomers do not generally polymerize by a cationic mechanism. However, the anionic polymerization of acrylic monomers to stereoregular or block copolymers is well known. These polymerizations are conducted in organic solvents, primarily using organometallic compounds as initiators. Currently this technology is of minor commercial significance, but stereoregular forms of numerous polyacrylates have been prepared and characterized. These include poly(*t*-butyl acrylate) (138–141), poly(isopropyl acrylate) (142), and poly(isobutyl acrylate) (143,144). Carefully controlled reaction conditions are usually required to obtain polymers with some measurable degree of crystallinity. In nonpolar solvents the anionic polymerization of acrylates generally yields isotactic polymer, whereas in polar solvents syndiotactic polymerization is favored. The physical and chemical properties of the various forms are often quite different. A general review covers these and other aspects of the anionic polymerization of acrylates (145).

Initiation of these anionic polymerizations is considered to take place via a Michael reaction:

$$R_3C^-M^+ + CH_2{=}CHCOOR \longrightarrow R_3C{-}CH_2{-}\overset{-}{C}H{-}COOR$$

Propagation occurs by head-to-tail addition of monomer:

$$R_3C{-}CH_2{-}\overset{-}{C}H{-}COOR + CH_2{=}CHCOOR \longrightarrow R_3C{-}CH_2{-}\underset{\underset{COOR}{|}}{C}H{-}CH_2{-}\underset{\underset{COOR}{|}}{\overset{-}{C}}H$$

Despite numerous efforts, there is no generally accepted theory explaining the causes of stereoregulation in acrylic and methacrylic anionic polymerizations. Complex formation with the cation of the initiator (146) and enolization of the active chain end are among the more popular hypotheses (147). Unlike free-radical polymerizations, copolymerizations between acrylates and methacrylates are not observed in anionic polymerizations; however, good copolymerizations within each class are reported (148).

A brief review has appeared covering the use of metal-free initiators in living anionic polymerizations of acrylates and a comparison with Du Pont's group-transfer polymerization method (149). Tetrabutylammonium thiolates run room temperature polymerizations to quantitative conversions yielding polymers of narrow molecular weight distributions in dipolar aprotic solvents. Block copolymers are accessible through sequential monomer additions (149–151) and interfacial polymerizations (152,153).

Economic Aspects

In 1989 the U.S. production of acrylic ester monomers was ca 450,000 t. This represents about 45% of the worldwide production; Western Europe (ca 35%) and Japan (ca 15%) account for most of the remainder. Essentially all of this was converted to acrylic polymers and copolymers. The U.S. production is principally from four companies:

Producer	Capacity, t
Rohm and Hass	280,000
Hoechst Celanese	195,000
BASF	105,000
Union Carbide	100,000

The historical trend in U.S. production of the various acrylic monomers is shown in Table 11.

Table 11. U.S. Production of Acrylic Monomers, 10^3 t[a]

Monomer	1969	1975	1980	1984
methyl	14	20	22	28
ethyl	91	109	136	138
n-butyl	29	81	145	192
2-ethylhexyl	15	15	31	39
other	4	9	23	27
acrylate esters, total	*153*	*234*	*357*	*424*

[a]Ref. 154.

In 1989 the U.S. prices for ethyl and butyl acrylate were in the range of $1.20 to 1.50/kg. Lower volume acrylates generally ranged in price from $3.00 to 5.00/kg. Prices for acrylic polymers are generally one and one-half to three times the monomer costs.

Synthetic emulsion polymers account for approximately 70% of the U.S. consumption of acrylate monomers. Major end uses for these latex polymers are coatings (32%), textiles (17%), adhesives (7%), paper (5%), and floor polishes (3%). The U.S. producers of acrylic copolymer emulsions include Rohm and Haas, Reichhold, National Starch, Union Carbide, Air Products, Unocal, B. F. Goodrich, and H. B. Fuller.

Solution polymers are the second most important use for acrylic monomers, accounting for about 12% of the monomer consumption. The major end use for these polymers is in coatings, primarily industrial finishes. Other uses of acrylic monomers include graft copolymers, suspension polymers, and radiation curable inks and coatings.

Analytical Test Methods and Specifications

Emulsion Polymers. Acrylic emulsion polymers are usually characterized by their composition, solids content, viscosity, pH, particle size distribution, glass-transition temperature, minimum film-forming temperature, and surfactant type. Where applicable, details of these characteristics are similar to those described for solution polymers. Particle size is most easily evaluated by photon correlation spectroscopy and sedimentation field flow fractionation (155) or more quantitatively by observation with an electron microscope (156). Minimum film-forming temperatures are determined by casting a film of the dispersion on a variable temperature bar and observing the minimum temperature at which a continuous film is obtained (157). A dispersion is classified as anionic, cationic, or nonionic depending primarily on the type of surfactant used in its preparation. In addition, acrylic dispersions are often evaluated by their mechanical (primarily shear), freeze–thaw, and thermal stability, their tendency to form sediment on long term standing, and their compatibility with other dispersions, salts, surfactants, and pigments. Details on methods for determining all of the properties are reported (87).

Solution Polymers. Acrylic solution polymers are usually characterized by their composition, solids content, viscosity, molecular weight, glass-transition temperature, and solvent. The compositions of acrylic polymers are most readily determined by physicochemical methods such as spectroscopy, pyrolytic gas–liquid chromatography, and refractive index measurements (97,158). The solids content of acrylic polymers is determined by dilution followed by solvent evaporation to constant weight. Viscosities are most conveniently determined with a Brookfield viscometer, molecular weight by intrinsic viscosity (158), and glass-transition temperature by calorimetry.

Health and Safety Factors

Acrylic polymers are considered to be nontoxic. In fact, the FDA allows certain acrylate polymers to be used in the packaging and handling of food. However, care must be exercised because additives or residual monomers present in various types of polymers can display toxicity. For example, some acrylic latex dispersions can be mild skin or eye irritants. This toxicity is usually ascribed to the surfactants in the latex and not to the polymer itself.

Potential health and safety problems of acrylic polymers occur in their manufacture (159). During manufacture, considerable care is exercised to reduce the potential for violent polymerizations and to reduce exposure to flammable and potentially toxic monomers and solvents. Recent environmental legislation governing air quality has resulted in completely closed kettle processes for most acrylic polymerizations. Acrylic solution polymers are treated as flammable mixtures. Dispersion polymers are nonflammable.

Uses

The combination of durability and clarity and the ability to tailor molecules relatively easily to specific applications have made acrylic esters prime candidates for numerous and diverse applications. At normal temperatures the polyacrylates are soft polymers and therefore tend to find use in applications that require flexibility or extensibility. However, the ease of copolymerizing the softer acrylates with the harder methacrylates, styrene, acrylonitrile, and vinyl acetate, allows the manufacture of products that range from soft rubbers to hard nonfilm-forming polymers.

Coatings. Their excellent clarity, toughness, color retention, uv stability, and chemical inertness make acrylic ester emulsion polymers prime paint binders. Acrylics are widely used in all types of paint formulations: interior and exterior; flats, semigloss, and gloss; and primers and topcoats. Widely used copolymer systems include all-acrylic (acrylate–methacrylate), vinyl–acrylic (vinyl acetate–acrylic), and styrene–acrylic. The all-acrylic copolymers are generally favored for exterior applications because of their excellent durability (160). Typical paint formulations are given in reference 76, and general techniques for manufacturing acrylic latex paints are given in reference 161. Water-borne acrylic latex coatings compete for the protection of structural steel (162) (see COATING PROCESSES).

Both acrylic emulsion and solution polymers are widely used in industrial finishing, including factory finished wood (163,164), metal furniture and containers (165), and can and coil coating (166). Thermosetting coatings based on acrylic polymers cross-linked with melamines, epoxies, and isocyanates are common. Acrylic coatings are applied to surfaces in various ways: spray, roll dip, and curtain coating are most common. New coating methods applicable to acrylics include radiation cure of acrylic monomers (167–172); acrylic powder coatings, a coating where the pigment and the binder are applied as powders (173–176); and electrode deposition of acrylic emulsion polymers (177–179). High solids and low emission industrial finishing systems are described in reference 180, and a review of the chemistry of water-based acrylic polymers used for industrial finishing appears in reference 181. A thorough technical discussion and review of the place of acrylic emulsions in metal coil coatings may be found in reference 182. A summary of the properties and applications of pigmented and clear lacquers based on solution acrylics is presented in reference 183. Clear acrylic solution coatings are used to protect the luster of polished zinc, copper, and brass (184). The technology and formulation of high-solids solution acrylic resins as baking enamels have been reviewed (185) (see COATINGS).

Hydrophobe-modified copolymers of acrylate esters with acrylic or methacrylic acid are finding increasing use as high quality thickeners for both trade sales and industrial paints (186). Formulations thickened with these unique water-soluble polymers show excellent flow and leveling characteristics.

Textiles.   The uses of acrylic emulsion polymers in textiles are numerous and diverse. Large volume uses include binders for fiberfill and nonwoven fabrics, textile bonding or lamininating, flocking, backcoating, and pigment printing binders. All-acrylic binders are most common; however, some vinyl–acrylics are also used. Self-cross-linking polymers containing copolymerized N-methylolacrylamide are common. The cross-linking improves washing and dry cleaning durability as well as the overall strength of the binder (187). Acrylics are favored because of their resistance to discoloration and for the soft feel they can provide.

As binders for fiberfill and nonwovens, the emulsions are applied to a loose web or mat, then heated to form a film that sticks the loose fibers together. Polyester (188–191), glass (192), and rayon (193) mats are bonded in this manner for a variety of end uses including quilting, clothing, disposable diapers and towels, filters, and roofing (see NONWOVEN FABRICS).

The exceptional resistance of acrylic to uv light, heat, ozone-yellowing, water, stiffening on aging, and dry cleaning make acrylics ideal as a textile coating material where the use demands are severe: automotive upholstery, window drapes, pile fabrics used for outerwear, and boots or boot linings (194). Automotive and furniture upholstery fabrics are often back-coated with thin films of acrylic polymers to improve dimensional handling properties, prevent pattern distortion, prevent unraveling, and minimize seam slippage. Frothed and foamed back-coatings eliminate the strike-through problems, yield a softer fabric, and save on energy costs (195). Crushed acrylic latex foams are used as backings for drapery fabric. The foam protects the drapery from sunlight; it stabilizes the fabric, improves drape, and provides a softer hand than conventional backcoatings (196) (see FOAMED PLASTICS). Acrylics are also recommended for fabric-to-fabric, fabric-to-foam, and fabric-to-nonwoven bonding and find some use as carpetbacking (197).

The preparation of flocked fabric using acrylic adhesives is detailed in reference 198. In flocking, cut fibers are bonded to an adhesive-coated fabric to achieve both a decorative and a functional effect. Acrylics can be tailored to provide the unique balance between softness and durability required for this application. Textiles can be durably printed using acrylic emulsion polymers as binders for pigments. The printed fabrics are highly resistant to washing, dry cleaning, and wear, and are soft and pleasant to the touch (35,199). Printing paste formulations and technology can be found in reference 200.

Another textile use of acrylic polymers is fabric finishing, to impart a desired hand or feel, or to aid soil release, or for permanent-press features. Copolymers of acrylate esters with acrylic or methacrylic acid serve as thickeners for a variety of textile coating formulations (see TEXTILES, FINISHING).

Adhesives. Acrylic emulsion and solution polymers form the basis of a variety of adhesive types. The principal use is in pressure-sensitive adhesives, where a film of a very low T_g ($< -20°C$) acrylic polymer or copolymer is used on the adherent side of tapes, decals, and labels. Acrylics provide a good balance of tack and bond strength with exceptional color stability and resistance to aging (201,202). Acrylics also find use in numerous types of construction adhesive formulations and as film-to-film laminating adhesives (qv).

Paper. Various paper and board mills have adopted acrylic–vinyl acetate copolymer emulsion polymers as pigment binders for their coated grade papers and boards based on their cost–performance balance. The advantages over the more widely used styrene–butadiene copolymers are higher brightness, opacity, and coating solids, combined with improved adhesive ability (203,204). Acrylic ester polymers have been compared to natural rubber, butadiene–acrylonitrile, and butadiene–styrene as paper saturants (205). Acrylic emulsions are offered that interact with clay in starch–latex-pigmented coatings (206). Acrylic emulsions find some use in size-press applications (207) and in beater addition (208) (see PAPER MAKING MATERIALS AND ADDITIVES).

Other Applications. Acrylic ester emulsions are used for leather finishing (209) and as acrylic polymer–leather composites. Acrylics can be used in every stage of pigskin leather production from tanning to finishing. The use of specially designed acrylics imparts uniformity, break improvement, better durability, and surface resistance without impairing the beautiful aesthetic qualities of pigskin (210) (see LEATHER).

Acrylic modifiers for cement impact strength and adhesion to substrates are discussed in reference 211. Both water-soluble acrylic and acrylic emulsion polymers are used in the ceramic industry as temporary binders, deflocculants, and additive components in ceramic bodies and glazes (212) (see CERAMICS).

Acrylics are used as binders for both aqueous and solvent-based caulks and sealants (qv) (213,214). Acrylic elastomeric roof mastics are used to protect polyurethane foam from uv radiation damage. The water-based formulation is easy to apply and is extremely resistant to chemical and uv radiation attack, impacts such as hailstones, and temperature change (215). Powdered acrylic polymers serve as processing aids and plate-out scavengers for both plasticized and unplasticized poly(vinyl chloride) in the manufacture of blown film and thin-gauge calendered film. This polymer type produces changes in the melt viscosity that make it possible to calender smooth, thick vinyl sheet (216). An acrylic emulsion that forms a thin insoluble film over leaves and fruit was used to control "Greasy Spot," a disease that can cause serious leaf-spotting and subsequent leaf fall in citrus trees (217). A comprehensive formulation guide for acrylic floor polishes (qv) is available (218).

Polyacrylate elastomers find limited use in hydraulic systems and gasket applications because of their superior heat resistance compared to the nitrile rubbers (219,220). Ethylene–acrylate copolymers were introduced in 1975. The applications include transmission seals, vibration dampers, dust boots, and steering and suspension seals. Further details and performance comparisons with other elastomers are given in reference 221 (see also ELASTOMERS, SYNTHETIC, ACRYLIC ELASTOMERS).

BIBLIOGRAPHY

"Acrylic Ester Polymers, Survey" in *ECT* 3rd ed., Vol. 1, pp. 386–408, by Benjamin B. Kine and Ronald W. Novak, Rohm and Haas Company.

1. J. Redtenbacher, *Ann.* **47,** 125 (1843).
2. Fr. Englehorn, *Berichte* **13,** 433 (1880); L. Balbiano and A. Testa, *Berichte* **13,** 1984 (1880).
3. W. Caspary and B. Tollens, *Ann.* **167,** 241 (1873).
4. G. W. A. Kahlbaum, *Berichte* **13,** 2348 (1880).
5. U.S. Pat. 1,121,134 (Dec. 15, 1914), O. Rohm.
6. E. H. Riddle, *Monomeric Acrylic Esters,* Reinhold Publishing Corp., New York, 1954.
7. J. Brandrup and E. H. Immergut, *Polymer Handbook,* 2nd ed., Interscience Publishers, Wiley-Interscience, New York, 1975.
8. H. B. Burrell, *Off. Dig. Fed. Soc. Paint Technol.* 131 (Feb. 1962).
9. D. W. van Krevelen, *Properties of Polymers,* Elsevier, Amsterdam, 1972.
10. J. L. Gardon in N. M. Bikales, ed., *Encyclopedia of Polymer Science and Technology,* Vol. 3, Interscience Publishers, a division of John Wiley & Sons, Inc., New York, 1965, pp. 833–862.
11. T. G. Fox, Jr., *Bull. Am. Phys. Soc.* 1(3), 123 (1956).
12. J. A. Shetter, *J. Polym. Sci. Part B* **1,** 209 (1963).
13. L. E. Nielsen, *Mechanical Properties of Polymers*, Van Nostrand Reinhold Co., Inc., New York, 1962, p. 122.
14. I. Williamson, *Br. Plast.* **23,** 87 (1950).
15. R. F. Clark, Jr., and R. M. Berg, *Ind. Eng. Chem.* **34,** 1218 (1942).
16. *ASTM Standards, ASTM D1043-61T,* Vol. 27, American Society for Testing and Materials, Philadelphia, Pa., 1964.
17. S. D. Gehman, D. E. Woodford, and C. S. Wikinson, Jr., *Ind. Eng. Chem.* **39,** 1108 (1947).
18. *ASTM Standards, ASTM D1053-61,* Vol. 19, American Society for Testing and Materials, Philadelphia, Pa., 1974.
19. A. C. Nuessle and B. B. Kine, *Ind. Eng. Chem.* **45,** 1287 (1953).
20. A. C. Nuessle and B. B. Kine, *Am. Dyest. Rep.* **50**(26), 13 (1961).

21. V. J. Moser, *Am. Dyest. Rep.* **53**(38), 11 (1964).
22. M. K. Lindemann, *Appl. Polym. Symp.* **10,** 73 (1969).
23. W. H. Brendley, Jr., *Paint Varn. Prod.* **63**(3), 23 (1973).
24. R. Zdanowski and G. L. Brown, Jr., *Resin Review,* Vol. 9, No. 1, Rohm and Haas Co., Philadelphia, Pa., 1959, p. 19.
25. R. Bakule and J. M. Blickensderfer, *Tappi* **56**(4), 70 (1973).
26. A. V. Tobolsky, *J. Polym. Sci. Polym. Symp.* **9,** 157 (1975).
27. L. J. Hughes and G. E. Britt, *J. Appl. Polym. Sci.* **5**(15), 337 (1961).
28. S. Krause and N. Roman, *J. Polym. Sci. Part A* **3,** 1631 (1965).
29. H. L. Gerhart, *Off. Dig. Fed. Soc. Paint Technol.* 680 (June 1961).
30. R. M. Christenson and D. P. Hart, *Off. Dig. Fed. Soc. Paint Technol.* 684 (June 1961).
31. H. A. Vogel and H. G. Brittle, *Off. Dig. Fed. Soc. Paint Technol.* 699 (June 1961).
32. J. D. Murdock and G. H. Segall, *Off. Dig. Fed. Soc. Paint Technol.* 709 (June 1961).
33. J. D. Petropoulous, C. Frazier, and L. E. Cadwell, *Off. Dig. Fed. Soc. Paint Technol.* 719 (1961).
34. D. G. Applegath, *Off. Dig. Fed. Soc. Paint Technol.* 737 (June 1961).
35. U.S. Pat. 2,886,474 (May 12, 1959), B. B. Kine and A. C. Nuessle (to Rohm and Haas Co.).
36. U.S. Pat. 2,923,653 (Feb. 2, 1960), N. A. Matlin and B. B. Kine (to Rohm and Haas Co.).
37. D. H. Klein, *J. Paint Technol.* **42**(545), 335 (1970).
38. S. H. Rider and E. E. Hardy, *Polymerization and Polycondensation Processes* (Adv. Chem. Ser. No. 34), American Chemical Society, Washington, D.C., 1962.
39. R. H. Wiley and G. M. Braver, *J. Polym. Sci.* **3,** 647 (1948).
40. J. R. Martin, J. F. Johnson, and A. R. Cooper, *J. Macrol. Sci. Rev. Macromol. Chem.* **8,** 57 (1972).
41. A. S. Craemer, *Kunststoffe* **30,** 337 (1940).
42. B. E. Rehberg and C. H. Fisher, *Ind. Eng. Chem.* **40,** 1429 (1948).
43. G. G. Cameron and D. R. Kane, *Makromol. Chem.* **109,** 194 (1967).
44. L. Gunawan and J. K. Haken, *J. Poly. Sci. Poly. Chem. Ed.* **23,** 2539 (1985).
45. A. R. Burgess, *Chem. Ind.* 78 (1952).
46. R. Stelle and H. Jacobs, *J. Appl. Polym. Sci.* **2**(4), 86 (1959).
47. B. G. Achhammer, *Mod. Plast.* **35,** 131 (1959).
48. K. L. Hoy, *J. Paint Technol.* **43,** 76 (1970).
49. A. F. M. Martin, *Handbook of Solubility Parameters and Other Cohesion Parameters,* CRC Press, Inc., Boca Raton, Fla., 1983.
50. J. H. Hildebrand and R. L. Scott, *The Solubility of Non-Electrolytes,* 3rd ed., Rheinhold Publishing Corp., New York, 1949.
51. P. A. Small, *J. Appl. Chem.* **3,** 71 (1953).
52. H. Burrell, *Off. Dig. Fed. Soc. Paint Technol.* 726 (Oct. 1955).
53. J. L. Gardon, *J. Paint Technol.* **38,** 43 (1966).
54. A. Rudin and H. K. Johnston, *J. Paint Technol.* **43**(559), 39 (1971).
55. T. P. Forbarth, *Chem. Eng.* **69**(5), 96 (1962).
56. M. Salkind, E. H. Riddle, and R. W. Keefer, *Ind. Eng. Chem.* **51,** 1328 (1959).
57. H. Warson, *The Applications of Synthetic Resin Emulsions,* Ernest Benn Ltd., London, 1972.
58. R. F. B. Davies and G. E. J. Reynolds, *J. Appl. Polym. Sci.* **12,** 47 (1968).
59. A. R. Burgess, *Chem. Ind.* 78 (1952).
60. M. Tabata, G. Nilsson, and A. Lund, *J. Polym. Sci. Polym. Chem. Ed.* **21,** 3257 (1983).
61. R. K. Graham, *J. Polym. Sci.* **38,** 209 (1959).
62. W. Burlant, J. Minsch, and C. Taylor, *J. Polym. Sci. Part A* **2,** 57 (1964).
63. S. Sakuyama, T. Ohara, N. Shimixu, and K. Kubota, *Chem. Technol.* 350 (June 1973).
64. F. T. Maler and W. Bayer, *Encycl. Chem. Process Des.* **1,** 401 (1976).

65. D. J. Hadley and E. M. Evans, *Propylene and Its Industrial Derivatives,* John Wiley & Sons, Inc., New York, 1973, pp. 416–497.
66. U.S. Pat. 3,875,212 (1975), T. Ohrui (to Sumitomo Chemical).
67. *Storage and Handling of Acrylic and Methacrylic Esters and Acids, CM-17,* Rohm and Haas Co., Philadelphia, Pa.
68. *Analytical Methods for the Acrylic Monomers, CM-18,* Rohm and Haas Co., Philadelphia, Pa.
69. *Acrylate Monomers, Bulletin F-40252,* Union Carbide Corp., New York.
70. *Celanese Acrylates, Product Manual N-70-1,* Celanese Chemical Co., New York.
71. State of California Health and Welfare Agency Safe Drinking Water and Toxic Enforcement Act of 1986.
72. M. H. Mackoy and H. W. Melville, *Trans. Faraday Soc.* **45,** 323 (1949).
73. G. M. Burnett and L. D. Loan, *International Symposium on Macromolecular Chemistry, Prague,* 1957, p. 113.
74. U.S. Pats. 2,367,660 and 2,367,661 (Jan. 23, 1945), C. L. Agre (to E. I. du Pont de Nemours & Co., Inc.).
75. A. Chapiro, *Radiative Chemistry of Polymeric Systems,* Wiley-Interscience, New York, 1972.
76. *Preparation, Properties and Uses of Acrylic Polymers, CM-19,* Rohm and Haas Co., Philadelphia, Pa.
77. T. G. Fox, Jr., and R. Loshock, *J. Am. Chem. Soc.* **75,** 3544 (1953).
78. J. L. O'Brien, *J. Am. Chem. Soc.* **77,** 4757 (1955).
79. E. P. Bonsall, *Trans. Faraday Soc.* **49,** 686 (1953).
80. P. G. Griffiths and co-workers, *Tetrahedron Lett.* **23,** 1309 (1982).
81. G. V. Schulz and G. Henrici, *Makromol. Chem.* **18/19,** 473 (1956).
82. F. R. Mayo and A. A. Miller, *J. Am. Chem. Soc.* **80,** 2493 (1956).
83. M. M. Mogilevich, *Russ. Chem. Rev.* **48,** 199 (1979).
84. D. A. Tirrell in J. I. Kroschwitz, ed., *Encyclopedia of Polymer Science and Engineering,* Wiley-Interscience, New York, 1986, pp. 192–233.
85. O. Yoshiahi and T. Imato, *Rev. Phys. Chem. Jpn.* **42**(1), 34 (1972).
86. F. W. Billmeyer, Jr., *Textbook of Polymer Chemistry,* Interscience Publishers, a division of John Wiley & Sons, Inc., New York, 1957.
87. *Emulsion Polymerization of Acrylic Monomers, CM-104,* Rohm and Haas Co., Philadelphia, Pa.
88. J. J. Uibel and F. J. Dinon, *J. Polym. Sci. Polym. Chem. Ed.* **21,** 1773 (1983).
89. F. A. Bovey, *High Resolution NMR of Macromolecules,* Academic Press, New York, 1972.
90. M. S. Matheson, E. E. Aner, E. B. Bevilacqua, and E. J. Hart, *J. Am. Chem. Soc.* **73,** 5395 (1951).
91. T. G. Fox and R. Gratch, *Ann. N.Y. Acad. Sci.* **57,** 367 (1953).
92. L. Maduga, *An. Quim.* **65,** 993 (1969).
93. P. G. Griffiths, E. Rizzardo, and D. H. Solomon, *J. Macromol. Sci. Chem.* **17,** 45 (1982).
94. J. G. Kloosterboer and H. L. Bressers, *Polym. Bull.* **2,** 205 (1982).
95. *The Manufacture of Acrylic Polymers, CM-107,* Rohm and Haas Co., Philadelphia, Pa.
96. *The Manufacture of Ethyl Acrylate–Acrylic Acid Copolymers, TMM-48,* Rohm and Haas Co., Philadelphia, Pa.
97. *Bulk Storage and Handling of Acryloid Coating Resins, C-186,* Rohm and Haas Co., Philadelphia, Pa.
98. B. B. Kine and G. H. Redlich, in D. T. Wasan, ed., *Surfactants in Chemical/Process Engineering,* Marcel Dekker, Inc., New York, 1988, p. 163.
99. A. E. Alexander, *J. Oil Colour Chem. Assoc.* **42,** 12 (1962).
100. C. E. McCoy, Jr., *Off. Dig. Fed. Soc. Paint Technol.* **35,** 327 (1963).
101. A. F. Sirianni and R. D. Coleman, *Can. J. Chem.* **42,** 682 (1964).

102. G. L. Brown, *Off. Dig. Fed. Soc. Paint Technol.* **28,** 456 (1956).
103. Brit. Pat. 940,366 (Oct. 30, 1963), P. R. van Ess (to Shell Oil).
104. U.S. Pat. 3,080,333 (Mar. 5, 1963), R. J. Kray and C. A. DeFazio (to Celenese).
105. A. R. M. Azad, R. M. Fitch, and J. Ugelstad, in *Colloidal Dispersions Micellar Behavior* (Sympos. Ser. No. 9), American Chemical Society, Washington, D.C., 1980.
106. H. Fikentscher, H. Gerrens, and H. Schuller, *Angew. Chem.* **72,** 856 (1960).
107. U.S. Pat. 3,458,466 (July 29, 1969), W. J. Lee (to The Dow Chemical Company).
108. U.S. Pat. 3,344,100 (Sept. 26, 1967), F. J. Donat and co-workers (to B. F. Goodrich Co.).
109. H. Warson, *Makromol. Chem. Suppl.* **10/11,** 265 (1985).
110. G. F. D'Alelio, *Fundamental Principles of Polymerization,* John Wiley & Sons, Inc., New York, 1952.
111. G. S. Whitby and co-workers, *J. Polym. Sci.* **16,** 549 (1955).
112. B. N. Rutovshii and co-workers, *J. Appl. Chem. USSR* **26,** 397 (1953).
113. H. Warson, *Polym. Paint Colour J.* **178,** 625 (1988).
114. H. Warson, *Polym. Paint Colour J.* **178,** 865 (1988).
115. L. J. Hughes and G. L. Brown, *J. Appl. Polym. Sci.* **7,** 59 (1963).
116. B. N. Kishore and co-workers, *J. Polym. Sci. Polym. Chem. Ed.* **24,** 2209 (1986).
117. G. Graczyk and V. Hornof, *J. Macromol. Sci. Chem.* **12,** 1633 (1988).
118. O. Y. Mansour and A. B. Moustafa, *J. Polym. Sci. Polym. Chem. Ed.* **13,** 2795 (1975).
119. T. Nagabhushanam and M. Santoppa, *J. Polym. Sci. Polym. Chem. Ed.* **14,** 507 (1976).
120. L. Zhi-Chong and co-workers, *J. Macromol. Sci. Chem.* **12,** 1487 (1988).
121. C. E. Brockway and K. B. Moser, *J. Polym. Sci. Part A-1* **1,** 1025 (1963).
122. G. Smets, A. Poot, and G. L. Dunean, *J. Polym. Sci.* **54,** 65 (1961).
123. C. H. Bamford and E. F. T. White, *Trans. Faraday Soc.* **52,** 719 (1956).
124. J. A. Hicks and H. W. Melville, *Nature (London)* **171,** 300 (1953).
125. I. Sahata and D. A. I. Goring, *J. Appl. Poly. Sci.* **20,** 573 (1976).
126. R. K. Graham, M. J. Gluchman, and M. J. Kampf, *J. Polym. Sci.* **38,** 417 (1959).
127. D. J. Angier, E. D. Farlie, and W. F. Watson, *Trans. Inst. Rubber Ind.* **34,** 8 (1958).
128. W. J. Burlant and A. S. Hoffman, *Block and Graft Polymers,* Reinhold Publishing Corp., New York, 1960.
129. R. J. Ceresa, *Block and Graft Copolymers,* Butterworth, Inc., Washington, D.C., 1962.
130. D. McGinniss, in *Ultraviolet Light Induced Reactions in Polymers* (Sympos. Ser. No. 25), American Chemical Society, Washington, D.C., 1976.
131. J. W. Vanderhoff, in ref. 130.
132. S. P. Pappas, ed., *U.V. Curing: Science and Technology,* Technology Marketing Corp., Stamford, Conn., 1978.
133. U.S. Pat. 3,418,295 (Dec. 24, 1968), A. C. Schwenthaler (to E. I. du Pont de Nemours & Co., Inc.).
134. V. D. McGinnis, *SME Technical Paper* (Ser.) *FC 76-486,* 1976.
135. J. V. Koleche, *Photochemistry of Cycloaliphatic Epoxides and Epoxy Acrylates,* Vol. I., ASTM International, 1989.
136. *The Curing of Coatings with Ultra-Violet Radiation, D8667 G.D.,* Tioxide of Canada, Sorel, P.Q., Canada.
137. T. A. Du Plessis and G. De Hollain, *J. Oil Colour Chem. Assoc.* **62,** 239 (1979).
138. M. L. Miller and C. E. Rauhut, *J. Polym. Sci.* **38,** 63 (1959).
139. B. Garret, *J. Am. Chem. Soc.* **81,** 1007 (1959).
140. A. Kawasahi and co-workers, *Makromol. Chem.* **49,** 76 (1961).
141. G. Smets and W. Van Hurnbeeck, *J. Polym. Sci. Part A-1* **1,** 1227 (1963).
142. W. E. Goode, R. P. Fellman, and F. H. Owens, in C. G. Overberger, ed., *Macromolecular Synthesis,* Vol. 1, John Wiley & Sons, Inc., New York, 1963, p. 25.
143. J. Furukawa, T. Tsuruta, and T. Makimoto, *Makromol. Chem.* **42,** 162 (1960).
144. T. Makmoto, T. Tsuruta, and J. Furukawa, *Makromol. Chem.* **50,** 116 (1961).
145. Y. Heimei and co-workers, in N. A. J. Platzer, ed., *Appl. Polym. Symp.* (26), 39 (1975).

146. D. E. Glusker and co-workers, *J. Polym. Sci.* **49**, 315 (1961).

147. D. J. Cram and K. R. Kopecky, *J. Am. Chem. Soc.* **81**, 2748 (1959).

148. H. Yuki and co-workers, in O. Vogl and J. Furukawan, eds., *Ionic Polymerization,* Marcel Dekker, Inc., New York, 1976.

149. M. T. Reetz, *Angew. Chem. Int. Ed. Engl.* **27**, 994 (1988).

150. V. V. Korshak and co-workers, *Makromol. Chem. Suppl.* **6**, 55 (1984).

151. C. P. Bosmyak, I. W. Parsons, and J. N. Hay, *Polymer* **21**, 1488 (1980).

152. F. L. Keohan and co-workers, *J. Polym. Sci. Polym. Chem. Ed.* **22**, 679 (1984).

153. Y. D. Lee and H. B. Tsai, *Makromol. Chem.* **190**, 1413 (1989).

154. *Synthetic Organic Chemicals, U.S. Production and Sales, 1970–1986,* U.S. International Trade Commission, Washington, D.C.

155. E. A. Collines and co-workers, *J. Coating Technol.* **47**, 35 (1975).

156. E. B. Bradford and J. W. Vanderhoff, *J. Polym. Sci. Part C* **1**, 41 (1963).

157. T. F. Protxman and G. L. Brown, *J. Appl. Polym. Sci.* **81** (1960).

158. P. W. Allen, *Technique of Polymers Characterization,* Butterworths, London, 1959; *Dilute Solution Properties of Acrylic and Methacrylic Polymers, SP-160,* Rohm and Haas Co., Philadelphia, Pa.

159. M. Harmon and J. King, U.S. NTIS AD Rep. AD-A0I7443, 142 (1974).

160. R. E. Harren, A. Mercurio, and J. D. Scott, *Aust. Oil Colour Chem. Assoc. Proc. News* 17 (Oct. 1977).

161. *Resin Review,* Vol. 18, Rohm and Haas Co., Philadelphia, Pa., 1968.

162. R. N. Washburne, *Am. Paint Coatings J.* **67**, 40 (1983).

163. R. P. Hopkins, E. W. Lewandowski, and T. E. Purcell, *J. Paint Technol.* **44**, 85 (1972).

164. T. E. Purcell, *Am. Paint J.* (June 1972).

165. Technical Practices Committee, *Materials Performance,* Vol. 15, National Assoc. of Corrosion Engineers, Houston, Tex., 1976, pp. 4, 13.

166. K. E. Buffington, *Maint. Eng.* (Mar. 1976).

167. I. K. Schahidi, J. C. Trebellas, and J. A. Vona, *Paint Varn. Prod.* **64**, 39 (1974).

168. C. B. Rybyn and co-workers, *J. Paint Technol.* **46**, 60 (1974).

169. *Multifunctional Acrylates in Radiation Curing,* Celanese Co., New York.

170. C. H. Carder, *Paint Varn. Prod.* **64**, 19 (1974).

171. *Materials for Photocurable Coatings and Inks,* Union Carbide Corp., New York.

172. R. A. Hickner and L. M. Ward, *Paint Varn. Prod.* **64**, 27 (1974).

173. F. Wingler and co-workers, *Farbe + Lack* **78**, 1063 (1972).

174. R. Muller, *Farbe + Lack* **78**, 1070 (1972).

175. J. K. Rankin, *J. Oil Colour Chem. Assoc.* **56**, 112 (1973).

176. F. A. Kyrlova, *Lakokras. Mater. Ikh Primen.* **3**, 20 (1969).

177. J. B. Zicherman, *Am. Paint J.* (May 1, 1972).

178. U.S. Pat. 1,535,228 (Aug. 16, 1968), K. Shibayama (to Eiki Jidaito).

179. F. Beck and co-workers, *Farbe + Lack* **73**, 298 (1967).

180. A. Mercurio and S. N. Lewis, *J. Paint Technol.* **47**, 37 (1975).

181. W. H. Brendley, Jr., and E. C. Carl, *Paint Varn. Prod.* **63**, 23 (1973).

182. M. R. Yunaska and J. E. Gallagher, *Resin Review,* Vol. 19, Rohm and Haas Co., Philadelphia, Pa., 1969, pp. 1–3.

183. G. Allyn, *Materials and Methods,* Reinhold Publishing Co., New York, 1956.

184. *Protecting the Surfaces of Copper and Copper-Base Alloys,* International Copper Research Association, New York.

185. R. R. Kuhn, N. Roman, and J. D. Whiteman, *Mod. Paint Coatings* **71**, 50 (1981).

186. J. C. Thiabault, P. R. Sperry, and E. J. Schaller, in J. E. Glass, ed., *Water-Soluble Polymers,* American Chemical Society, Washington, D.C., 1986, p. 375.

187. N. Sutterlin, *Makromol. Chem. Suppl.* **10/11**, 403 (1985).

188. V. J. Moser, *Resin Review,* Vol. 14, Rohm and Haas Co., Philadelphia, Pa., 1964, p. 25.

189. D. I. Lunde, *Nonwoven Fabrics Forum, Clemson Univ., June 15–17,* 1982.

190. U.S. Pat. 3,157,562 (Nov. 17, 1964), B. B. Kine, V. J. Moser, and H. A. Alps (to Rohm and Haas Co.).
191. U.S. Pat. 3,101,292 (Aug. 20, 1963), B. B. Kine and N. A. Matlin (to Rohm and Haas Co.).
192. J. R. Lawrence, *Resin Review,* Vol. 6, Rohm and Haas Co., Philadelphia, Pa., 1956, p. 12.
193. U.S. Pat. 2,931,749 (Apr. 5, 1960), B. B. Kine and N. A. Matlin (to Rohm and Haas Co.).
194. G. C. Kantner, *Text. World* **132,** 89 (1982).
195. G. C. Kantner, *Am. Text.* **12,** 30 (Feb. 1983).
196. L. Thompson and H. Mayfield, *AATCC National Technical Conference,* 1974, p. 258.
197. F. X. Chancler and J. G. Brodnyan, *Resin Review,* Vol. 23, Rohm and Haas Co., Philadelphia, Pa., 1973, p. 3.
198. V. J. Moser and D. G. Strong, *Am. Dyest. Rep.* **55,** 52 (1966).
199. U.S. Pat. 2,883,304 (Apr. 21, 1959), B. B. Kine and A. C. Nuessle (to Rohm and Haas Co.).
200. Brit. Pat. 1,011,041 (1962), K. Craemer (to Badische Anilin).
201. K. F. Foley and S. G. Chu, *Adhes. Age* 24 (Sept. 1986).
202. D. Satas, *Adhes. Age* 28 (Aug. 1988).
203. J. J. Latimer, *Pulp Pap. Can.* **82,** 83 (1981).
204. J. E. Young and J. J. Latimer, *South. Pulp Pap. J.* **45,** 5 (1982).
205. P. J. McLaughlin, *Tappi* **42,** 994 (1959).
206. L. Mlynar and R. W. McNamec, Jr., *Tappi* **55,** 359 (1972).
207. F. L. Schucker, *Resin Review,* Vol. 7, Rohm and Haas Co., Philadelphia, Pa., 1957, p. 20.
208. H. C. Adams, T. J. Drennen, and L. E. Kelley, *Tappi* **48,** 486 (1965).
209. J. A. Handscomb, *J. Soc. Leather Trades' Chem.* **43,** 237 (1959).
210. W. C. Prentiss, *J. Am. Leather Chem. Assoc.* **71,** 54 (1976).
211. J. A. Lavelle and P. E. Wright, *Resin Review,* Vol. 24, Rohm and Haas Co., Philadelphia, Pa., 1974, p. 3.
212. J. R. Johnson, *Resin Review,* Vol. 11, Rohm and Haas Co., Philadelphia, Pa., 1961, p. 3.
213. P. E. Wright and H. C. Young, *Resin Review,* Vol. 24, Rohm and Haas Co., Philadelphia, Pa., 1974, p. 17.
214. P. H. Dougherty and H. T. Freund, *Resin Review,* Vol. 17, Rohm and Haas Co., Philadelphia, Pa., 1967, p. 3.
215. L. S. Frankel, D. A. Perry, and J. J. Lavelle, *Resin Review,* Vol. 32, Rohm and Haas Co., Philadelphia, Pa., 1982, p. 1.
216. J. T. Lutz, Jr., *Resin Review,* Vol. 19, Rohm and Haas Co., Philadelphia, Pa., 1969, p. 18.
217. M. Cohen, *Proc. Fl. State Hortic. Soc.* **72,** 56 (1959).
218. *Resin Review,* Vol. 16, Rohm and Haas Co., Philadelphia, Pa., 1966, p. 12.
219. T. M. Vial, *Rubber Chem. Technol.* **44,** 344 (1971).
220. P. Fram, in N. M. Bikales, ed., *Encyclopedia of Polymer Science and Technology,* Vol. 1, Interscience Publishers, a division of John Wiley & Sons, Inc., New York, 1964, pp. 226–246.
221. D. L. Schultz, *Rubber World* **182,** 51 (May 1980).

RONALD W. NOVAK
Rohm and Haas Company

2-CYANOACRYLIC ESTER POLYMERS

The polymers of the 2-cyanoacrylic esters, more commonly known as the alkyl 2-cyanoacrylates, are hard glassy resins that exhibit excellent adhesion to a wide variety of materials. The polymers are spontaneously formed when their liquid precursors or monomers are placed between two closely fitting surfaces. The spontaneous polymerization of these very reactive liquids and the excellent adhesion properties of the cured resins combine to make these compounds a unique class of single-component, ambient-temperature-curing adhesives of great versatility. The materials that can be bonded run the gamut from metals, plastics, most elastomers, fabrics, and woods to many ceramics.

The utility of these adhesives arises from the electron-withdrawing character of the groups adjacent to the polymerizable double bond, which accounts for both the extremely high reactivity or cure rate and their polar nature, which enables the polymers to adhere tenaciously to many diverse substrates.

The polymers first synthesized (1,2) were reported to be clear glassy resins. This original work involved a thermal polymerization, and it was not until the early 1950s that scientists at Eastman Kodak discovered the rapid room-temperature cure and excellent adhesion of these materials quite by accident. While determining the refractive index of a freshly prepared monomer, they discovered that the glass prisms of the refractometer had become tightly bonded. Further work led to the discovery that many other substrates became bonded in the same manner. This resulted in the commercialization in 1958 of Eastman 910, the first in what would become a large family of 2-cyanoacrylic ester adhesives. At present, a number of manufacturers in the United States, Europe, Japan, and elsewhere market extended lines of these adhesives all over the world. Some of the major producers and their trademarks include Loctite (Prism and Superbonder), Toagosei (Aron Alpha, Krazy Glue), Henkel (Sicomet), National Starch (Permabond), Sumitomo (Cyanobond), Three Bond (Super Three), and Alpha Giken (Alpha Ace, Alpha Techno).

Physical Properties

The physical properties of the monomers must be discussed along with those of the cured polymers because consideration of one without the other presents an incomplete picture. The 2-cyanoacrylic ester monomers are all thin, water-clear liquids with viscosities of 1–3 mPa·s (= cP). Although a number of the esters have been prepared and characterized, only a relative few are of any significant commerical interest, and, of those, the methyl and ethyl esters by far predominate. The physical properties of the principal monomers are included in Table 1.

The base monomers are too thin for convenient use and therefore are generally formulated with stabilizers, thickeners, and property-modifying additives. The viscosities of such formulated adhesives can range from that of the base monomer (for wicking grades) to thixotropic gels with viscosities of 20,000 to 50,000 mPa·s (= cP) for larger gaps. The liquid products are characterized by sharp, lacrimatory, faintly sweet odors, except for several recent variants. The

Table 1. Properties of Common Cyanoacrylate[a] Monomers, ROOCC=CH$_2$

$$\underset{\underset{\text{ROOCC}=\text{CH}_2}{|}}{\text{CN}}$$

Property	Methyl	Ethyl	Isopropyl	Allyl	n-Butyl	Isobutyl	Methoxyethyl	Ethoxyethyl
							R	
molecular formula	C$_5$H$_5$O$_2$N	C$_6$H$_7$O$_2$N	C$_7$H$_9$O$_2$N	C$_7$H$_7$O$_2$N	C$_8$H$_{11}$O$_2$N	C$_8$H$_{11}$O$_2$N	C$_7$H$_9$O$_3$N	C$_8$H$_{11}$O$_3$N
CAS Registry Number	[137-05-3]	[7085-85-0]	[10586-17-1]	[7324-02-9]	[6066-65-1]	[1069-55-2]	[27816-23-5]	[21982-91-8]
odor	very sharp, lacrimatory →		← sharp acrylic odor, lacrimatory →			→	← virtually odorless →	
viscosity, mPa·s (= cP)	2.2	1.9	2.1	2.0	2.1	2.0	2.6	5.0
density, g/cm^3	1.10	1.05	1.01	1.05	0.98	0.99	1.06	1.07
boiling point, °C at kPa[b]	48–49	54–56	53–56	78–82	83–84	71–73	96–100	104–106
	0.33–0.36	0.21–0.40	0.27–0.33	0.80	0.40	0.25–0.29	0.35–0.44	0.67
refractive index, n_D, 20°C	1.4406	1.4349	1.4291	1.4565	1.4330	1.4352		1.4470
heat of polymerization, kJ/mol[c]	57.7	58.1	67.8	63 ± 4	63 ± 1	66.9		
flash point, °C	83	83		82	85	93		130

[a] Alkyl cyanoacrylates or alkyl 2-cyano-2-propenoates.
[b] To convert kPa to mm Hg multiply by 7.5.
[c] To convert kJ to kcal divide by 4.184.

substitution of an alkoxyalkyl ester side chain for the normal alkyl group renders these products nearly odor free as well as slightly less effective as adhesives.

The outstanding characteristic of the 2-cyanoacrylic esters is their high reactivity. The liquid monomers and adhesives will polymerize nearly instantaneously via an anionic mechanism when brought into contact with any weakly basic surface. Even the presence of a weakly basic substance such as adsorbed surface moisture is adequate to initiate the curing reaction. The adhesives cure very rapidly and are potentially hazardous skin bonders because of the presence of moisture and protein in the skin. The adhesive bonding process is accomplished by placing a small drop or bead on one surface and quickly bringing the mating part in contact with light pressure and holding it in place for a period of several seconds to several minutes. The rapid polymerization will cause rapid fixturing and adhering of the mating parts. As one might expect, conditions of low humidity and/or acidic groups on the surface can slow or inhibit the cure reaction.

Polymerization

The basic polymerization reaction is described by the following equations.

Initiation

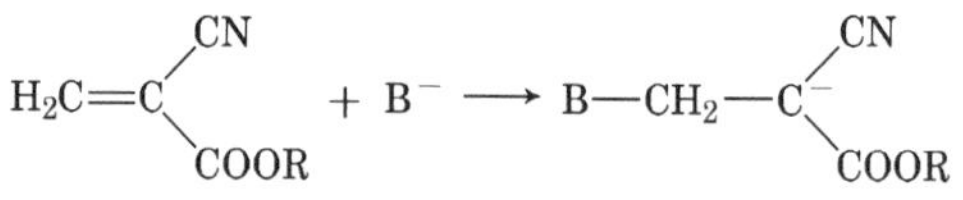

$$H_2C{=}C\begin{smallmatrix}CN\\COOR\end{smallmatrix} + B^- \longrightarrow B{-}CH_2{-}C^-\begin{smallmatrix}CN\\COOR\end{smallmatrix}$$

Propagation

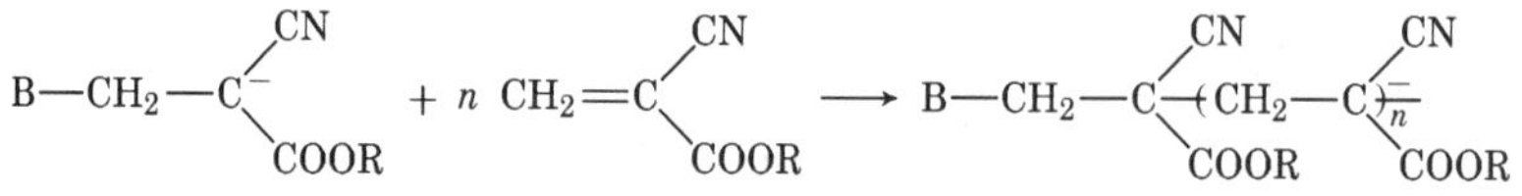

$$B{-}CH_2{-}C^-\begin{smallmatrix}CN\\COOR\end{smallmatrix} + n\ CH_2{=}C\begin{smallmatrix}CN\\COOR\end{smallmatrix} \longrightarrow B{-}CH_2{-}C{-}(CH_2{-}C)_n^-\begin{smallmatrix}CN\ \ CN\\COOR\ \ COOR\end{smallmatrix}$$

Termination

$$B{-}CH_2{-}C{-}(CH_2{-}C)_n^-\begin{smallmatrix}CN\ \ CN\\COOR\ \ COOR\end{smallmatrix} + H^+ \longrightarrow B{-}CH_2{-}(CH_2{-}C)_n{-}H\begin{smallmatrix}CN\ \ CN\\COOR\ \ COOR\end{smallmatrix}$$

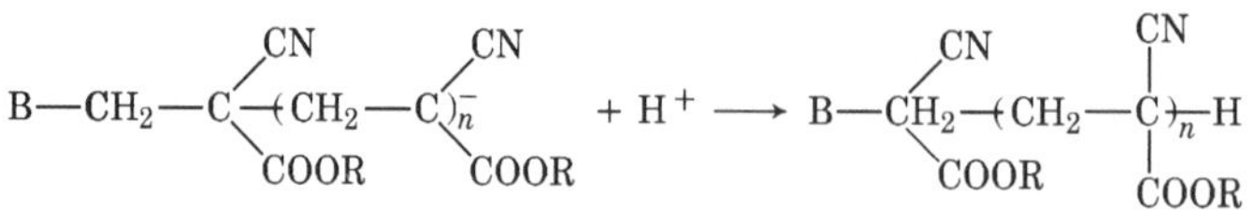

The reaction proceeds until all available monomer has reacted or until it is terminated by an acidic species. The gel point in these *in situ* polymerizations, as represented by the time of fixture, occurs within several seconds on strongly catalytic surfaces such as thermoset rubbers to several minutes on noncatalytic surfaces.

When the surface conditions are acidic or the ambient humidity is low enough to affect the cure significantly, a surface accelerator may be used to promote the reaction. Available from most manufacturers, these basic solutions may be dip, wipe, or spray applied. Recently, new additive chemistry has been developed that accelerates the cure under adverse conditions without the need for a separate accelerator.

The bulk physical properties of the polymers of the 2-cyanoacrylic esters appear in Table 2. All of these polymers are soluble in N-methylpyrrolidinone, N,N-dimethylformamide, and nitromethane. The adhesive bonding properties of typical formulated adhesives are listed in Table 3.

The cured polymers are hard, clear, and glassy thermoplastic resins with high tensile strengths. The polymers, because of their highly polar structure, exhibit excellent adhesion to a wide variety of substrate combinations. They tend to be somewhat brittle and have only low to moderate impact and peel strengths. The addition of fillers such as poly(methyl methacrylate) (PMMA) reduces the brittleness somewhat. Newer formulations are now available that contain dissolved elastomeric materials of various types. These rubber-modified products have been found to offer adhesive bonds of considerably improved toughness (3,4).

The structure–property relationships for the lower cyanoacrylic ester polymers generally indicate that cure rates, tensile strength, tensile shear strengths, and hardness vary inversely with increasing ester chain length and that glass-transition temperatures (T_g) and adhesive bond service temperatures decrease with increasing ester chain lengths.

Several special-property cyanoacrylic esters that offer variations on the general theme are also available. Allyl esters give bond properties similar to the ethyl or isopropyl esters, but are reported to cross-link by a free-radical mechanism through the allyl group, providing increased thermal resistance. This reaction is very sluggish and requires long exposures at high temperatures to proceed. The alkoxyalkyl esters, several of which are now commercially available, are practically odorless because of the inclusion of an alkoxy group on the ester side chain. These modifications also impart slightly reduced adhesive performance.

Manufacture and Processing

The cyanoacrylic esters are prepared via the Knoevenagel condensation reaction (5), in which the corresponding alkyl cyanoacetate reacts with formaldehyde in the presence of a basic catalyst to form a low molecular weight polymer. The polymer slurry is acidified and the water is removed. Subsequently, the polymer is cracked and redistilled at a high temperature onto a suitable stabilizer combination to prevent premature repolymerization. Strong protonic or Lewis acids are normally used in combination with small amounts of a free-radical stabilizer.

The above batch process has undergone numerous refinements to improve yields, processing characteristics, purity, and storage stability, but it remains the standard method of manufacture for these products. Recently a continuous process has been reported by Bayer AG (6) wherein the condensation is carried out in an extruder. The by-products are removed in a degassing zone, and the molten polymer, mixed with stabilizers, is subsequently cracked to yield raw monomer.

Adhesives formulated from the 2-cyanoacrylic esters typically contain stabilizers and thickeners, and may also contain tougheners, colorants, and other special property-enhancing additives. Both anionic and free-radical stabilizers

Table 2. Cured Bulk Properties of Common 2-Cyanoacrylic Esters

Property	Monomer type							
	Methyl	Ethyl	Isopropyl	Allyl	Butyl	Isobutyl	Methoxyethyl	Ethoxyethyl
softening point, °C (Vicat)	165	126	154	78	165	197		52
melting point, °C	205		179			192	165	103
refractive index, n_D, 20°C	1.45	1.45	1.45			1.26	1.4	1.48
dielectric constant[a] at 1 MHz	3.34	3.908	3.8	3.3	5.4			
dissipation factor[a] at 1 MHz			2.04	0.02				
volume resistivity, MΩ·mm		3×10^{15}	9×10^{12}	7×10^{14}	5.37×10^{9}			
tensile strength[b], steel–steel, MPa[c]	31	27.6	20.7			20.4	24.5	30.3
elongation, %	<2	<2	<2	10		<2		
flexural modulus[d], MPa[c]	3400	2069		1752				
hardness (Rockwell)	M 65	M 58	R 18					

[a] ASTM D150.
[b] ASTM D638.
[c] To convert MPa to psi multiply by 145.
[d] ASTM D790.

Table 3. Adhesive Bond Properties of 2-Cyanoacrylic Esters with Metals and Various Polymeric Materials

	Ester type					
	Methyl	Ethyl[a]	Butyl	Isobutyl	Methoxyethyl	Ethoxyethyl
set time, s						
steel	20	10	30	20	15	5
aluminum	3	10	5	20	15	5
nitrile rubber	5	3	5	5	5	3
neoprene rubber	5	3	5	5	5	3
ABS	20	10	20	20	5	3
polystyrene	20	10		20		20
polycarbonate	20	10	20	20	60	20
PMMA	10	5		10		15
PVC	5	3	2	5		10
nylon	30	15		30		10
phenolic resin	5	3	30	5	25	5
bond strength[b], kPa[c]						
steel	206	172	151	96	206	165
aluminum	186	158	151	96	138	117
ABS	48	48	96	48	48	48
polystyrene	34	34	83	34		34
polycarbonate	69	69	90	69		41
PMMA	48	48	62	41		48
PVC	96	96	62	83	55	69
nylon	76	76	62	34		41
phenolic resin	69	76	90	62	62	55

[a]Set times for allyl esters are similar to those for ethyl esters, as are bond strengths to steel, aluminum, ABS, PS, and PC.
[b]According to ASTM D1002.
[c]To convert kPa to psi multiply by 0.145.

are required, since the monomer will polymerize via both mechanisms. The anionic stabilizers that have been reported include acidic gases such as NO, SO_2, SO_3, BF_3, and HF (7–9). Strong protonic acids such as the aliphatic and aromatic sulfonic acids (10), and even strong mineral acids, have been used at low concentrations. Combinations of nonvolatile strong acids with gaseous stabilizers (11) have demonstrated synergistic improvements.

Although the anionic polymerization mechanism is the predominant one for the cyanoacrylic esters, the monomer will polymerize free-radically under prolonged exposure to heat or light. To extend the usable shelf life, free-radical stabilizers such as quinones or hindered phenols are a necessary part of the adhesive formulation.

Economic Aspects

Production of the 2-cyanoacrylic ester adhesives on a worldwide basis is estimated to be approximately 2400 metric tons. This amounts to only 0.02% of the total volume of adhesives produced but about 3% of the dollar volume.

Because of the high costs of raw materials and the relatively complex synthesis, the 2-cyanoacrylic esters are moderately expensive materials when considered in bulk quantities. Depending on the quantity and the specific ester or formulation involved, the prices for cyanoacrylic ester adhesives can range from approximately $30/kg to over $1000/kg. For these reasons, as well as several technical factors related to handling and performance, cyanoacrylic ester adhesives are best suited to small bonding applications, very often where single drops or small beads are adequate for bonding. In such cases the cost of the adhesive becomes inconsequential compared to the value of the service it performs, and these adhesives become very economical to use.

Specifications and Standards

A wide variety of adhesives based on the cyanoacrylic esters are now available worldwide. Product distinctions are based on ester type and specific performance attributable to that ester, formulation, viscosity, and cure speed. A number of special performance grades have appeared that maximize performance in specific areas, sometimes quite significantly. Adhesive grades are now available with improved thermal resistance; improved peel strengths, impact resistance, and toughness; and low odor, low blooming characteristics. New formulations contain cure additives that reduce the sensitivity of cure to acidic surfaces and low atmospheric humidity. More recently, surface primers have been introduced that have dramatically improved bond strengths to traditionally hard-to-bond surfaces such as polyolefins and other low energy polymers.

The adhesives are routinely tested for specific performance and compliance to many user specifications with different requirements. Government purchases are covered in the Mil Standard A-46050C (12).

Analytical and Test Methods

The routine compositional and functional testing done on the adhesives includes gas chromatographic testing for purity, potentiometric titrations for acid stabilizer concentrations, accelerated thermal stability tests for shelf life, fixture time cure speed tests, and assorted ASTM tests for tensile shear strengths, peel and impact strengths, and hot strengths.

Health and Safety Factors

The 2-cyanoacrylic esters have sharp, pungent odors and are lacrimators, even at very low concentrations. These esters can be irritating to the nose and throat at concentrations as low as 3 ppm; eye irritation is observed at levels of 5 ppm (13). The TLV for methyl 2-cyanoacrylate is 2 ppm and the short-term exposure limit is 4 ppm (14). Good ventilation when using the adhesives is essential.

Eye and skin contact should be avoided because of the adhesive's rapid tissue-bonding capabilities. In case of eye or skin contact, the affected area should be flushed with copious amounts of water. Especially with eye contact, the bonded area should not be forced apart, since this will produce more damage than the initial bonding. Medical attention is recommended. Soaking in warm water will gradually weaken and release the bond. Contact through clothing may produce a rapid exotherm and the clothing involved should be flooded with water. Efforts to pull the involved clothing away may result in skin damage.

The cured 2-cyanoacrylic ester polymers are relatively nontoxic. Oral doses of 6400 mg/kg failed to kill laboratory rats. Mild skin irritation was observed with guinea pigs, but there was no evidence of sensitization or absorption through the skin (15).

Both the liquid and cured 2-cyanoacrylic esters support combustion. These adhesives should not be used near sparks, heat, or open flame, or in areas of acute fire hazard. Highly exothermic polymerization can occur from direct addition of catalytic substances such as water, alcohols, and bases such as amines, ammonia, or caustics, or from contamination with any of the available surface activator solutions.

Uses

The combination of fast room-temperature cure and excellent adhesion to numerous substrates makes the adhesives of the 2-cyanoacrylic esters a natural choice in the assembly of small close-fitting parts in a number of diverse market areas. Some of the market segments served by these versatile materials include automotive, electronic, sporting goods, toys, hardware, morticians, law enforcement, cosmetics, jewelry, and medical devices. Although they are not approved for use in the United States, their strong tissue bonding characteristics have led to their use as chemical sutures and hemostatic agents in other countries around the world.

The diverse nature of the cyanoacrylate adhesives applications illustrates vividly that there is no truly typical application. The number of applications in which these adhesives are used is being expanded daily as technological improvements continue to broaden their capabilities.

Examples of some of the current uses for these adhesives include retention of automotive gaskets, weather stripping, side trim, and wiring harnesses; athletic shoes; swim masks; shotgun recoil pads; arrow feathers; dolls and doll furniture; circuit board component mounting and wire tacking; and lipstick tube and compact mirrors. Medical devices include balloon catheters and tubing sets; the adhesives are used by morticians to seal eyes and lips.

Law enforcement agencies make use of the volatility and reactivity of the vapors to develop fingerprints not retrievable with conventional methods.

The new technical developments have made possible quick bonding to woods, papers, and porous surfaces. Polyolefins, which comprise approximately 50% of the U.S. thermoplastic production, are now bondable with the cyanoacrylic ester adhesives. These new capabilities are sure to provide for continued market growth in the years ahead.

BIBLIOGRAPHY

"Acrylic Ester Polymers, 2-Cyanoacrylic Ester Polymers" in *ECT* 3rd ed., Vol. 1, pp. 408–413, by H. W. Coover, Jr., and J. M. McIntire, Tennessee Eastman Company.

1. U.S. Pat. 2,467,926 (Apr. 19, 1949), A. E. Ardis (to BF Goodrich Co.).
2. U.S. Pat. 2,467,927 (Apr. 19, 1949), A. E. Ardis (to BF Goodrich Co.).
3. U.S. Pat. 4,012,945 (July 23, 1978), E. R. Gleave (to Loctite Corp.).
4. U.S. Pat. 4,440,910 (Apr. 3, 1984), J. T. O'Connor (to Loctite Corp.).
5. U.S. Pat. 2,721,858 (Oct. 25, 1955), F. Joyner and G. Hawkins (to Eastman Kodak).
6. Ger. Pat. 3,320,796 (Dec. 13, 1984), H. Waniczek and H. Bartl (to Bayer AG).
7. U.S. Pat. 2,794,788 (June 4, 1957), H. W. Coover, Jr., and N. H. Shearer (to Eastman Kodak Co.).
8. U.S. Pat. 2,796,251 (July 24, 1956), F. B. Joyner and N. H. Shearer (to Eastman Kodak).
9. U.S. Pat. 3,557,185 (Jan. 19, 1971), K. Ito and K. Kondo (to Toa Gosei).
10. U.S. Pat. 3,652,635 (Mar. 28, 1972), S. Kawamura and co-workers (to Toa Gosei).
11. Brit. Pat. GB 2,107,328B (Apr. 27, 1983), L. Lizardi, B. Malofsky, J. C. Liu, and C. Mariotti (to Loctite Corp.).
12. *Adhesives, Cyanoacrylate, Rapid Room Temperature Curing, Solventless,* Army Materials and Mechanics Research Center, Watertown, Mass.
13. W. A. McGee, F. A. Oglesby, R. L. Raleigh, and W. D. Fassett, *Am. Ind. Hyg. Assoc. J.* **29,** 558–561 (1968).
14. *Threshold Limit Values for Chemical Substances in Work Room Air,* American Conference of Government Industrial Hygienists, Cincinnati, Ohio, 1975.
15. W. Thomsen, in G. Schneberger, *Adhesives in Manufacturing,* Marcel Dekker, Inc., New York, 1983, p. 305.

J. T. O'Connor
Loctite Corporation

ACRYLIC FIBERS.   See Fibers, acrylic.

ACRYLONITRILE

Prior to 1960, acrylonitrile [*107-13-1*] (also called acrylic acid nitrile, propylene nitrile, vinyl cyanide, propenoic acid nitrile) was produced commercially by processes based on either ethylene oxide and hydrogen cyanide or acetylene and hydrogen cyanide. The growth in demand for acrylic fibers, starting with the introduction of Orlon by Du Pont around 1950, spurred efforts to develop improved process technology for acrylonitrile manufacture to meet the growing market (see Fibers, acrylic). This resulted in the discovery in the late 1950s by Sohio (1) and also by Distillers (2) of a heterogeneous vapor-phase catalytic process for acrylonitrile by selective oxidation of propylene and ammonia, commonly referred to as the propylene ammoxidation process. Commercial introduction of this lower cost process by Sohio in 1960 resulted in the eventual displacement of all other acrylonitrile manufacturing processes. Today over 90% of the approximately 4,000,000 metric tons produced worldwide each year use the Sohio-developed ammoxidation process. Acrylonitrile is among the top 50 chemicals produced in the United States as a result of the tremendous growth in its use as a starting material for a wide range of chemical and polymer products. Acrylic fibers remain the largest use of acrylonitrile; other significant uses are in resins and nitrile elastomers and as an intermediate in the production of adiponitrile and acrylamide.

Physical Properties

Acrylonitrile (C_3H_3N, mol wt = 53.064) is an unsaturated molecule having a carbon–carbon double bond conjugated with a nitrile group. It is a polar molecule because of the presence of the nitrogen heteroatom. There is a partial shift in the bonding electrons toward the more electronegative nitrogen atom, as represented by the following heterovalent resonance structures.

$$CH_2{=}CH{-}C{\equiv}N: \longleftrightarrow CH_2{=}CH{-}\overset{+}{C}{=}\overset{-}{\underset{..}{N}}: \longleftrightarrow \overset{+}{C}H_2{-}CH{=}C{=}\overset{-}{\underset{..}{N}}:$$

Tables 1 and 2 list some physical properties and thermodynamic information, respectively, for acrylonitrile (3–5).

Table 1. Physical Properties of Acrylonitrile

Property	Value
appearance/odor	clear, colorless liquid with faintly pungent odor
boiling point, °C	77.3
freezing point, °C	−83.5
density, 20°C, g/cm^3	0.806
volatility, 78°C, %	>99
vapor pressure, 20°C, kPa^a	11.5
vapor density (air = 1)	1.8
solubility in water, 20°C, wt %	7.3
pH (5% aqueous solution)	6.0–7.5
critical values	
temperature, °C	246
pressure, MPa^b	3.54
volume, cm^3/g	3.798
refractive index, n_D^{25}	1.3888
dielectric constant, 33.5 MHz	38
ionization potential, eV	10.75
molar refractivity (D line)	15.67
surface tension, 25°C, mN/m (= dyn/cm)	26.6
dipole moment, C·m^c	
liquid	1.171×10^{-29}
vapor	1.294×10^{-29}
viscosity, 25°C, mPa·s (= cP)	0.34

^{a}To convert kPa to mm Hg multiply by 7.5.
^{b}To convert MPa to psi multiply by 145.
^{c}To convert C·m to debye, divide by 3.336×10^{-30}.

Table 2. Thermodynamic Data[a]

Property	Value
flash point, °C	0
autoignition temperature, °C	481
flammability limits in air, 25°C, vol %	
lower	3.0
upper	17.0
free energy of formation, ΔG_g°, 25°C, kJ/mol	195
enthalpy of formation, 25°C, kJ/mol	
ΔH_g°	185
ΔH_l°	150
heat of combustion, liquid, 25°C, kJ/mol	1761.5
heat of vaporization, 25°C, kJ/mol	32.65
molar heat capacity, kJ/(kg·K)	
liquid	2.09
gas at 50°C, 101.3 kPa[b]	1.204
molar heat of fusion, kJ/mol	6.61
entropy, S, gas at 25°C, 101.3 kPa[b], kJ/(mol·K)	274

[a]To convert kJ to kcal divide by 4.184.
[b]101.3 kPa = 1 atm.

Acrylonitrile is miscible in a wide range of organic solvents, including acetone, benzene, carbon tetrachloride, diethyl ether, ethyl acetate, ethylene cyanohydrin, petroleum ether, toluene, some kerosenes, and methanol. Compositions of some common azeotropes of acrylonitrile are given in Table 3. Table 4 presents the solubility of acrylonitrile in water as a function of temperature (6). Vapor–liquid equilibria for acrylonitrile in combination with acetonitrile, acrolein, HCN, and water have been published (6–9). Table 5 gives the vapor pressure of acrylonitrile over aqueous solutions.

Acrylonitrile has been characterized using infrared, Raman, and ultraviolet spectroscopies, electron diffraction, and mass spectroscopy (10–18).

Table 3. Azeotropes of Acrylonitrile

Azeotrope	Boiling point, °C	Acrylonitrile concentration, wt %
tetrachlorosilane	51.2	89
water	71.0	88
isopropyl alcohol	71.6	56
benzene	73.3	47
methanol	61.4	39
carbon tetrachloride	66.2	21
chlorotrimethylsilane	57.0	7

Table 4. Solubilities of Acrylonitrile in Water

Temperature, °C	Acrylonitrile in water, wt %	Water in acrylonitrile, wt %
−50		0.4
−30		1.0
0	7.1	2.1
10	7.2	2.6
20	7.3	3.1
30	7.5	3.9
40	7.9	4.8
50	8.4	6.3
60	9.1	7.7
70	9.9	9.2
80	11.1	10.9

Table 5. Acrylonitrile Vapor Pressure over Aqueous Solutions at 25°C

Acrylonitrile, wt %	Vapor pressure, kPa[a]
1	1.3
2	2.9
3	5.3
4	6.9
5	8.4
6	10.0
7	10.9

[a]To convert kPa to mm Hg multiply by 7.5.

Chemical Properties

Acrylonitrile undergoes a wide range of reactions at its two chemically active sites, the nitrile group and the carbon–carbon double bond. Detailed descriptions of specific reactions have been given (19,20). Acrylonitrile polymerizes readily in the absence of a hydroquinone inhibitor, especially when exposed to light. Polymerization is initiated by free radicals, redox catalysts, or bases and can be carried out in the liquid, solid, or gas phase. Homopolymers and copolymers are most easily produced using liquid-phase polymerization (see ACRYLONITRILE POLYMERS). Acrylonitrile undergoes the reactions typical of nitriles, including hydration with sulfuric acid to form acrylamide sulfate ($C_3H_5NO \cdot H_2SO_4$ [*15497-99-1*]), which can be converted to acrylamide (C_3H_5NO [*79-06-1*]) by neutralization with a base; and complete hydrolysis to give acrylic acid ($C_3H_4O_2$ [*79-10-7*]). Acrylamide (qv) is also formed directly from acrylonitrile by partial hydrolysis using copper-based catalysts (21–24); this has become the preferred commercial route for acrylamide production. Industrially important acrylic esters can be formed by reaction of acrylamide sulfate with organic alcohols. Methyl acrylate

($C_4H_6O_2$ [*96-33-3*]) has been produced commercially by the alcoholysis of acryl-amide sulfate with methanol. Reactions at the activated double bond of acryloni-trile include Diels-Alder addition to dienes, forming cyclic products; hydrogena-tion over metal catalysts to give propionitrile (C_3H_5N [*107-12-0*]) and propylamine (C_3H_9N [*107-10-8*]); and the industrially important hydrodimerization to produce adiponitrile ($C_6H_8N_2$ [*111-69-3*]) (25–27). Other reactions include addition of halo-gens across the double bond to produce dihalopropionitriles, and cyanoethylation by acrylonitrile of alcohols, aldehydes, esters, amides, nitriles, amines, sulfides, sulfones, and halides.

Manufacturing and Processing

Acrylonitrile is produced in commercial quantities almost exclusively by the vapor-phase catalytic propylene ammoxidation process developed by Sohio (28).

$$C_3H_6 + NH_3 + \tfrac{3}{2}\, O_2 \xrightarrow{\text{catalyst}} C_3H_3N + 3\, H_2O$$

A schematic diagram of the commercial process is shown in Figure 1. The commer-cial process uses a fluid-bed reactor in which propylene, ammonia, and air contact a solid catalyst at 400–510°C and 49–196 kPa (0.5–2.0 kg/cm^2) gauge. It is a single-pass process with about 98% conversion of propylene, and uses about 1.1 kg propylene per kg of acrylonitrile produced. Useful by-products from the process are HCN (about 0.1 kg per kg of acrylonitrile), which is used primarily in the manufacture of methyl methacrylate, and acetonitrile (about 0.03 kg per kg of acrylonitrile), a common industrial solvent. In the commercial operation the hot reactor effluent is quenched with water in a countercurrent absorber and any unreacted ammonia is neutralized with sulfuric acid. The resulting ammonium sulfate can be recovered and used as a fertilizer. The absorber off-gas containing primarily N_2, CO, CO_2, and unreacted hydrocarbon is either vented directly or first passed through an incinerator to combust the hydrocarbons and CO. The acrylonitrile-containing solution from the absorber is passed to a recovery col-umn that produces a crude acrylonitrile stream overhead that also contains HCN. The column bottoms are passed to a second recovery column to remove water and produce a crude acetonitrile mixture. The crude acetonitrile is either incinerated or further treated to produce solvent quality acetonitrile. Acrylic fiber quality (99.2% minimum) acrylonitrile is obtained by fractionation of the crude acryloni-trile mixture to remove HCN, water, light ends, and high boiling impurities. Disposal of the process impurities has become an increasingly important aspect of the overall process, with significant attention being given to developing cost-effective and environmentally acceptable methods for treatment of the process waste streams. Current methods include deep-well disposal, wet air oxidation, ammonium sulfate separation, biological treatment, and incineration (29).

   Although acrylonitrile manufacture from propylene and ammonia was first patented in 1949 (30), it was not until 1959, when Sohio developed a catalyst capable of producing acrylonitrile with high selectivity, that commercial manu-facture from propylene became economically viable (1). Production improvements

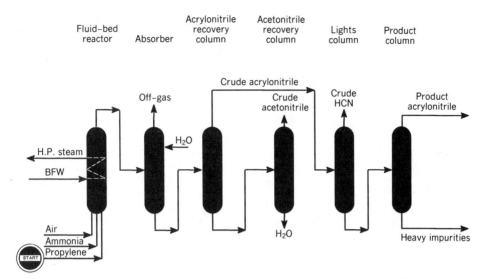

Fluid–bed   Acrylonitrile   Acetonitrile
reactor   recovery   recovery   Lights   Product
                 Absorber   column   column   column   column

Fig. 1.   Process flow diagram of the commercial propylene ammoxidation process for acrylonitrile. BFW is boiler feed water.

over the past 30 years have stemmed largely from development of several generations of increasingly more efficient catalysts. These catalysts are multicomponent mixed metal oxides mostly based on bismuth–molybdenum oxide. Other types of catalysts that have been used commercially are based on iron–antimony oxide, uranium–antimony oxide, and tellurium–molybdenum oxide.

Fundamental understanding of these complex catalysts and the surface-reaction mechanism of propylene ammoxidation has advanced substantially since the first commercial plant began operation. Mechanisms for selective ammoxidation of propylene over bismuth molybdate and antimonate catalysts are shown in Figures 2 and 3. The rate-determining step is abstraction of an α-hydrogen of propylene by an oxygen in the catalyst to form a π-allyl complex on the surface (31–33). Lattice oxygens from the catalyst participate in further hydrogen abstraction, followed by oxygen insertion to produce acrolein in the absence of ammonia, or nitrogen insertion to form acrylonitrile when ammonia is present (34–36). The oxygens removed from the catalyst in these steps are replenished by gas-phase oxygen, which is incorporated into the catalyst structure at a surface site separate from the site of propylene reaction. In the ammoxidation reaction, ammonia is activated by an exchange with O^{2-} ions to form isoelectronic NH^{2-} moieties according to the following:

$$NH_3 + O^{2-} \longrightarrow NH^{2-} + H_2O$$

These are the species inserted into the allyl intermediate to produce acrylonitrile.

The active site on the surface of selective propylene ammoxidation catalyst contains three critical functionalities associated with the specific metal components of the catalyst (37–39): an α-H abstraction component such as Bi^{3+}, Sb^{3+}, or Te^{4+}; an olefin chemisorption and oxygen or nitrogen insertion component such

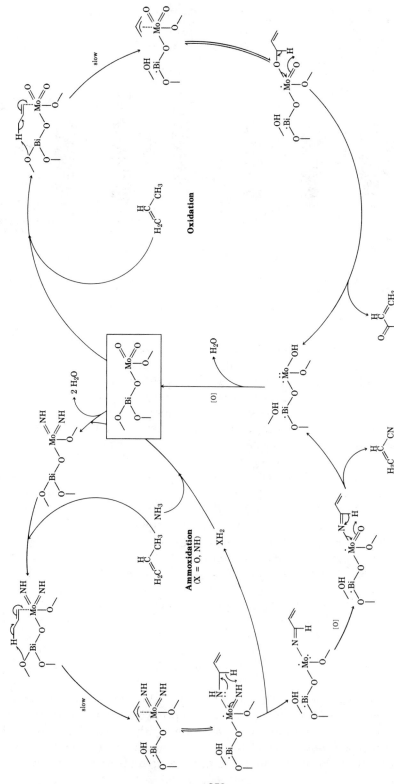

Fig. 2. Mechanism of selective ammoxidation and oxidation of propylene over bismuth molybdate catalysts. (31).

358

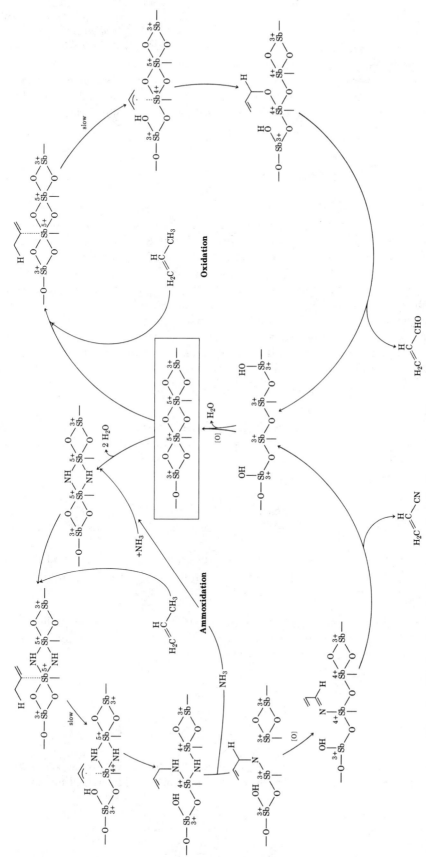

Fig. 3. Mechanism of selective ammoxidation and oxidation of propylene over antimonate catalysts (31).

as Mo^{6+} or Sb^{5+}; and a redox couple such as Fe^{2+}/Fe^{3+} or Ce^{3+}/Ce^{4+} to enhance transfer of lattice oxygen between the bulk and surface of the catalyst. The surface and solid-state mechanisms of propylene ammoxidation catalysis have been determined using Raman spectroscopy (40,41), neutron diffraction (42–44), x-ray absorption spectroscopy (45,46), x-ray diffraction (47–49), pulse kinetic studies (36), and probe molecule investigations (50).

Obsolete Acrylonitrile Processes

Processes rendered obsolete by the propylene ammoxidation process (51) include the ethylene cyanohydrin process (52–54) practiced commercially by American Cyanamid and Union Carbide in the United States and by I. G. Farben in Germany. The process involved the production of ethylene cyanohydrin by the base-catalyzed addition of HCN to ethylene oxide in the liquid phase at about 60°C. A typical base catalyst used in this step was diethylamine. This was followed by liquid-phase or vapor-phase dehydration of the cyanohydrin. The liquid-phase dehydration was performed at about 200°C using alkali metal or alkaline earth metal salts of organic acids, primarily formates and magnesium carbonate. Vapor-phase dehydration was accomplished over alumina at about 250°C.

$$C_2H_4O + HCN \xrightarrow[\text{catalyst}]{\text{base}} HOC_2H_4CN \xrightarrow[-H_2O]{\text{catalyst}} C_3H_3N$$

A second commercial route to acrylonitrile used by Du Pont, American Cyanamid, and Monsanto was the catalytic addition of HCN to acetylene (55).

$$C_2H_2 + HCN \xrightarrow{\text{catalyst}} C_3H_3N$$

The reaction occurs by passing HCN and a 10:1 excess of acetylene into dilute hydrochloric acid at 80°C in the presence of cuprous chloride as the catalyst.

These processes use expensive C_2 hydrocarbons as feedstocks and thus have higher overall acrylonitrile production costs compared to the propylene-based process technology. The last commercial plants using these process technologies were shut down by 1970.

Other routes to acrylonitrile, none of which achieved large-scale commercial application, are acetaldehyde and HCN (56), propionitrile dehydrogenation (57,58), and propylene and nitric oxide (59,60):

$$CH_3CHO + HCN \longrightarrow CH_3CH(OH)CN \xrightarrow{-H_2O} C_3H_3N$$

$$CH_3CH_2CN \longrightarrow C_3H_3N + H_2$$

$$4\,C_3H_6 + 6\,NO \longrightarrow 4\,C_3H_3N + 6\,H_2O + N_2$$

Numerous patents have been issued disclosing catalysts and process schemes for manufacture of acrylonitrile from propane. These include the direct heterogeneously catalyzed ammoxidation of propane to acrylonitrile using mixed metal oxide catalysts (61–64).

$$C_3H_8 + NH_3 + 2\ O_2 \xrightarrow{\text{catalyst}} C_3H_3N + 4\ H_2O$$

A two-step process involving conventional nonoxidative dehydrogenation of propane to propylene in the presence of steam, followed by the catalytic ammoxidation to acrylonitrile of the propylene in the effluent stream without separation, is also disclosed (65).

$$C_3H_8 \xrightarrow{\text{catalyst}} C_3H_6 + H_2 \xrightarrow[+NH_3]{+\ ^{3\!/_2}\ O_2} C_3H_3N + 3\ H_2O + H_2$$

Because of the large price differential between propane and propylene, which has ranged from \$155/t to \$355/t between 1987 and 1989, a propane-based process may have the economic potential to displace propylene ammoxidation technology eventually. Methane, ethane, and butane, which are also less expensive than propylene, and acetonitrile have been disclosed as starting materials for acrylonitrile synthesis in several catalytic process schemes (66,67).

Economic Aspects

The propylene-based process developed by Sohio was able to displace all other commercial production technologies because of its substantial advantage in overall production costs, primarily due to lower raw material costs. Raw material costs less by-product credits account for about 60% of the total acrylonitrile production cost for a world-scale plant. The process has remained economically advantaged over other process technologies since the first commercial plant in 1960 because of the higher acrylonitrile yields resulting from the introduction of improved commercial catalysts. Reported per-pass conversions of propylene to acrylonitrile have increased from about 65% to over 80% (28,68–70).

More than half of the worldwide acrylonitrile production is situated in Western Europe and the United States (Table 6). In the United States, production is dominated by BP Chemicals, with more than a third of the domestic capacity (Table 7). Nearly one-half of the U.S. production was exported in 1988 (Table 8), with most going to Japan and the Far East. The export market has been an increasingly important outlet for U.S. acrylonitrile producers, with the percentage of U.S. production exported growing from around 10% in the mid-1970s to 53% in 1987 and 43% in 1988. Japanese and Far East producers have not been able to satisfy the increasing domestic demand in recent years because of higher propylene costs relative to the United States. This makes it more economical to import acrylonitrile from the United States than to install new domestic production. Table 9 provides a breakdown of worldwide demand between 1976 and 1988. Growth in demand has averaged about 3.6% per year between 1984 and 1988. Projections are for a 3% per year growth rate through 1993 (72).

Table 6. Worldwide Acrylonitrile Production in 1988[a]

Region	Production, 10^3 t
Western Europe	1,200
United States	1,170
Japan	600
Far East	200
Mexico	60

[a]Ref. 71.

Table 7. U.S. Acrylonitrile Producers[a]

Company	Approximate capacity[b], 10^3 t/y
BP Chemicals	430
Monsanto Chemical Company	220
Sterling Chemical Company	220
E. I. du Pont de Nemours & Co., Inc.	175
American Cyanamid Company	160

[a]Ref. 72.
[b]As of March 6, 1989.

Table 8. U.S. Acrylonitrile Exports[a], 10^3 t

Destination	1988	1987
Far East/Asia	176	189
Japan	92	186
Western Europe	109	86
Mexico/Canada	71	60
South America	34	48
Middle East/Africa	28	41
total export	*510*	*610*

[a]Ref. 73.

Table 9. Worldwide Acrylonitrile Demand, 10^3 t/yr

Region	1988	1985	1980	1976
Western Europe	1,200	1,140	880	880
Japan	680	635	510	570
United States	660	640	660	590
Far East	560	385	270	200
Mexico/South America	250	200	130	81
total	*3,350*	*3,000*	*2,450*	*2,321*

Analysis, Testing, Standards, and Quality Control

Standard test methods for chemical analysis have been developed and published (74). Included is the determination of commonly found chemicals associated with acrylonitrile and physical properties of acrylonitrile that are critical to the quality of the product (75–77). These include determination of color and chemical analyses for HCN, quinone inhibitor, and water. Specifications appear in Table 10.

Table 10. Commercial Specifications for Acrylonitrile

Parameter	Specification
acetone, ppm, max	300
acetonitrile, ppm, max	500
aldehydes, ppm, max	100
color, APHA, max	15
distillation range, °C, min, ibp	74.5
°C, max, 97%	78.5
HCN, ppm, max	10
inhibitor, hydroquinone monomethyl ether, ppm	35–50
iron, ppm, max	0.10
nonvolatile matter, ppm, max	100
peroxides, ppm, max	0.5
pH, 5% aqueous	6.0–7.5
refractive index, n_D^{25}	1.3882–1.3892
water, wt %, max	0.50
purity, wt %, min	99.0

Storage and Transport

Acrylonitrile must be stored in tightly closed containers in cool, dry, well-ventilated areas away from heat, sources of ignition, and incompatible chemicals. Storage vessels, such as steel drums, must be protected against physical damage, with outside detached storage preferred. Storage tanks and equipment used for transferring acrylonitrile should be electrically grounded to reduce the possibility of static spark-initiated fire or explosion.

Acrylonitrile is transported by rail car, barge, and pipeline. Department of Transportation (DOT) regulations require labeling acrylonitrile as a flammable liquid and poison. Transport is regulated under DOT 49 CFR 172.101. Bill of lading description is: Acrylonitrile, Flammable Liquid, Poison B, UN 1093 "RQ."

Health and Safety Factors

Acrylonitrile is highly toxic if ingested, with an acute LDL_0 value for laboratory rats of 113 mg/kg (78,79). It is moderately toxic if inhaled (rat $LCL_0 = 500$ ppm/4 h) and it is extremely irritating and corrosive to skin and eyes. Ingestion may cause gastrointestinal disturbances and inhalation may result in respiratory tract irritation. Symptoms of overexposure include irritation, nausea, vomiting, and diarrhea and may include headache, weakness, shortness of breath, dizziness, collapse, loss of consciousness, respiratory arrest, and death. Acrylonitrile is readily absorbed through the skin, and contact may result in inflammation, itching, reddening, and blistering. These symptoms may be delayed. First-aid treatment for inhalation or ingestion involves inhalation of amyl nitrite and induced vomiting if the victim is conscious.

Acrylonitrile is categorized as a cancer hazard by OSHA. It has been determined to be carcinogenic to laboratory animals and mutagenic in both mammalian and nonmammalian tests. Genetic transformations and damage have been reported in tissue cultures exposed to acrylonitrile. Animal tests show that it is a reproductive toxicant only at maternally toxic doses. Permissible exposure limits for acrylonitrile in the United States are 2 ppm for an 8-h time-weighted average concentration and 10 ppm as the ceiling concentration for a 15-min period.

Acrylonitrile will polymerize violently in the absence of oxygen if initiated by heat, light, pressure, peroxide, or strong acids and bases. It is unstable in the presence of bromine, ammonia, amines, and copper or copper alloys. Neat acrylonitrile is generally stabilized against polymerization with trace levels of hydroquinone monomethyl ether and water.

Acrylonitrile is combustible and ignites readily, producing toxic combustion products such as hydrogen cyanide, nitrogen oxides, and carbon monoxide. It forms explosive mixtures with air and must be handled in well-ventilated areas and kept away from any source of ignition, since the vapor can spread to distant ignition sources and flash back.

Federal regulations (40 CFR 261) classify acrylonitrile as a hazardous waste and it is listed as Hazardous Waste Number U009. Disposal must be in accordance with federal (40 CFR 262, 263, 264), state, and local regulations only at properly permitted facilities. It is listed as a toxic pollutant (40 CFR 122.21) and introduction into process streams, storm water, or waste water systems is in violation of federal law. Strict guidelines exist for clean-up and notification of leaks and spills. Federal notification regulations require that spills or leaks in excess of 100 lb (45.5 kg) be reported to the National Response Center. Substantial criminal and civil penalties can result from failure to report such discharges into the environment.

Uses

Worldwide consumption of acrylonitrile increased 52% between 1976 and 1988, from 2.5×10^6 to 3.8×10^6 t/yr. The trend in consumption over this time period is shown in Table 11 for the principal uses of acrylonitrile: acrylic fiber, acrylonitrile–butadiene–styrene (ABS) resins, adiponitrile, nitrile rubbers, elastomers, and styrene–acrylonitrile resins (SAN). Since the 1960s acrylic fibers have remained the major outlet for acrylonitrile production in the United States and especially in Japan and the Far East. Acrylic fibers always contain a comonomer. Fibers containing 85 wt % or more acrylonitrile are usually referred to as acrylics whereas fibers containing 35 to 85 wt % acrylonitrile are termed modacrylics (see

FIBERS, ACRYLIC). Acrylic fibers are used primarily for the manufacture of apparel, including sweaters, fleece wear, and sportswear, as well as for home furnishings, including carpets, upholstery, and draperies. Demand is largely subject to trends in the fashion industry, with the popularity of bulky sweaters, fleece wear, and sportswear helping to increase U.S. consumption to record levels in the 1980s. Acrylic fibers consume about 65% of the acrylonitrile produced worldwide, while in the United States acrylic fiber consumes only about 45% of the acrylonitrile used domestically. Growth in demand for acrylic fibers in the 1990s is expected to be modest, between 2 and 3% per year, primarily from overseas markets. Domestic demand is expected to be flat.

Table 11. Worldwide Acrylonitrile Uses and Consumption, 10^3 t

Use	1988	1985	1980	1976
acrylic fibers	2,520	2,410	2,040	1,760
ABS resins	550	435	300	270
adiponitrile	310	235	160	90
other (including nitrile rubber, SAN resin, acrylamide, and barrier resins)	460	390	240	420

Significant growth in acrylonitrile end use has come from ABS and SAN resins and adiponitrile (see ACRYLONITRILE POLYMERS). ABS resins are second to acrylic fibers as an outlet for acrylonitrile. These resins normally contain about 25% acrylonitrile and are characterized by their chemical resistance, mechanical strength, and ease of manufacture. Consumption of ABS resins increased significantly in the 1980s with its growing application as a specialty performance polymer in construction, automotive, machine, and appliance applications. Opportunities still exist for ABS resins to continue to replace more traditional materials for packaging, building, and automotive components. SAN resins typically contain between 25 and 30% acrylonitrile. Because of their high clarity, they are used primarily as a substitute for glass in drinking cups and tumblers, automobile instrument panels, and instrument lenses. Together, ABS and SAN resins account for about 20% of domestic acrylonitrile consumption. The largest increase among the end uses for acrylonitrile over the past 10 years has come from adiponitrile, which has grown to become the third largest outlet for acrylonitrile. It is used by Monsanto as a precursor for hexamethylenediamine (HMDA, $C_6H_{16}N_2$ [124-09-4]) and is made by a proprietary acrylonitrile electrohydrodimerization process (25). HMDA is used exclusively for the manufacture of nylon-6,6. The growth of this acrylonitrile outlet in recent years stems largely from replacement of adipic acid ($C_6H_{10}O_4$ [124-04-9]) with acrylonitrile in HDMA production rather than from a significant increase in nylon-6,6 demand. A nonelectrochemical catalytic route has also been developed for acrylonitrile dimerization to adiponitrile (26,27,80,81). This technology, if it becomes commercial, can provide additional replacement opportunity for acrylonitrile in nylon manufacture. The use of acrylonitrile for HMDA production should continue to grow at a faster rate than the other outlets for acrylonitrile, but it will not approach the size of the acrylic fiber market for acrylonitrile consumption.

Acrylamide (qv) is produced commercially by heterogeneous copper-catalyzed hydration of acrylonitrile (21–24). Acrylamide is used primarily in the form of a polymer, polyacrylamide, in the paper and pulp industry and in waste water treatment as a flocculant to separate solid material from waste water streams (see ACRYLAMIDE POLYMERS). Other applications include mineral processing, coal processing, and enhanced oil recovery in which polyacrylamide solutions were found effective for displacing oil from rock. The latter use peaked in the early 1980s, coinciding with the peak in world oil prices. However, the decline in world oil prices since 1980 has limited application of acrylamide for enhanced oil recovery to a few test oil fields in the United States. Any future applications of polymer flooding for enhanced oil recovery will require an increase in oil prices to the levels experienced in the early 1980s. Other growth markets for acrylamide, including use in binders, adhesives, and absorbents, will not replace lost demand for enhanced oil recovery application.

Nitrile rubber finds broad application in industry because of its excellent resistance to oil and chemicals, its good flexibility at low temperatures, high abrasion and heat resistance (up to 120°C), and good mechanical properties. Nitrile rubber consists of butadiene–acrylonitrile copolymers with an acrylonitrile content ranging from 15 to 45% (see ELASTOMERS, SYNTHETIC, NITRILE RUBBER). In addition to the traditional applications of nitrile rubber for hoses, gaskets, seals, and oil well equipment, new applications have emerged with the development of nitrile rubber blends with poly(vinyl chloride) (PVC). These blends combine the chemical resistance and low temperature flexibility characteristics of nitrile rubber with the stability and ozone resistance of PVC. This has greatly expanded the use of nitrile rubber in outdoor applications for hoses, belts, and cable jackets, where ozone resistance is necessary.

Other acrylonitrile copolymers have found specialty applications where good gas-barrier properties are required along with strength and high impact resistance. Examples of these are BP Chemicals' Barex 210 acrylonitrile–methyl acrylate–butadiene copolymer and Monsanto's Lopac styrene-containing nitrile copolymer. These barrier resins compete directly in the alcoholic and other beverage bottle market with traditional glass and metal containers as well as with poly(ethylene terephthalate) (PET) and PVC in the beverage bottle market (see BARRIER POLYMERS). Other applications include food, agricultural chemicals, and medical packaging. Total acrylonitrile consumption for barrier resin applications is small, consuming less than about 1% of the total U.S. acrylonitrile production. Projections of a significant growth in demand for nitrile barrier resins remain unfulfilled because of an FDA ban in 1977 on the use of acrylonitrile-based copolymers in beverage bottles. Although the ban was lifted in 1982 and limits were set on acrylonitrile exposure in beverage and food packaging applications, it is unlikely that acrylonitrile copolymers will penetrate the current plastic bottle market dominated by PET.

A growing specialty application for acrylonitrile is in the manufacture of carbon fibers (qv). They are produced by pyrolysis of oriented polyacrylonitrile fibers and are used to reinforce composites (qv) for high performance applications in the aircraft, defense, and aerospace industries. These applications include rocket engine nozzles, rocket nose cones, and structural components for aircraft and orbital vehicles where light weight and high strength are needed. Other small specialty applications of acrylonitrile are in the production of fatty amines, ion-exchange resins, and fatty amine amides used in cosmetics, adhesives, corrosion inhibitors, and water treatment resins. Examples of these specialty amines include 2-acrylamido-2-methylpropanesulfonic acid ($C_7H_{13}NSO_4$ [15214-89-8]), 3-methoxypropionitrile (C_4H_7NO [110-67-8]), and 3-methoxypropylamine ($C_4H_{11}NO$ [5332-73-0]).

BIBLIOGRAPHY

"Acrylonitrile" in *ECT* 1st ed., Vol. 1, pp. 184–189, by H. S. Davis, American Cyanamid Company; in *ECT* 2nd ed., Vol. 1, pp. 338–351, by W. O. Fugate, American Cyanamid Company; in *ECT* 3rd ed., Vol. 1, pp. 414–426, by Louis T. Groet, Badger, B. V.

1. U.S. Pat. 2,904,580 (Sept. 15, 1959), J. D. Idol (to The Standard Oil Co.).
2. Brit. Pat. 876,446 (Oct. 3, 1959) and U.S. Pat. 3,152,170 (Oct. 6, 1964), J. L. Barclay, J. B. Bream, D. J. Hadley, and D. G. Stewart (to Distillers Company Ltd.).
3. M. A. Dalin, I. K. Kolchin, and B. R. Serebryakov, *Acrylonitrile,* Technomic, Westport, Conn., 1971, pp. 161–162.
4. R. M. Paterson, M. I. Bornstein, and E. Garshick, *Assessment of Acrylonitrile as a Potential Air Pollution Problem,* GCA-TR-75-32-G(6), GCA Corporation, 1976.
5. *IARC Monographs on the Evaluation of the Carcinogenic Risk of Chemicals to Humans,* International Agency for Research on Cancer, Vol. 19, Feb. 1979.
6. Ref. 3, p. 166.
7. N. M. Sokolov, *Rev. Chim.* **20,** 169 (1969).
8. N. M. Sokolov, *Proc. Int. Symp. Distill.* **3,** 110 (1969).
9. N. M. Sokolov, N. N. Sevryugova, and N. M. Zhavoronkor, *Theor. Osn. Khim. Tekhnol.* **3,** 449 (1969).
10. T. Fukuyama and K. Kuchitsu, *J. Mol. Struct.* **5,** 131 (1970).
11. *The Chemistry of Acrylonitrile,* 1st ed., American Cyanamid Company, New York, 1951, pp. 14–15.
12. *EPA/NIH Mass Spectral Data Base,* Vol. 1, U.S. National Bureau of Standards, Washington, D.C., 1978, p. 5.
13. M. C. L. Gerry, K. Yamada, and G. Winnewisser, *J. Phys. Chem. Ref. Data* **8,** 107 (1979).
14. S. Suzer and L. Andrews, *J. Phys. Chem.* **93,** 2123 (1989).
15. A. R. H. Cole and A. A. Green, *J. Mol. Spectrosc.* **48,** 246 (1973).
16. V. I. Khvostenko, I. I. Furlei, V. A. Mazunov, and R. S. Rafikov, *Dokl. Akad. Nauk SSSR* **213,** 1364 (1973).
17. J. A. Nuth and S. Glicker, *J. Quant. Spectosc. Radiat. Trans.* **28,** 223 (1982).
18. G. Cazzoli and Z. Kisiel, *J. Mol. Spectrosc.* **130,** 303 (1988).
19. Ref. 3, pp. 120–159.
20. Ref. 11, pp. 21–51.
21. U.S. Pats. 3,597,481 (Aug. 3, 1971), 3,631,104 (Dec. 28, 1971), Re. 31,430 (Oct. 25, 1983), B. A. Tefertiller and C. E. Habermann (to The Dow Chemical Company).
22. U.S. Pat. 4,048,226 (Sept. 13, 1977), W. A. Barber and J. A. Fetchin (to American Cyanamid Co.).
23. U.S. Pat. 4,086,275 (Apr. 25, 1978), K. Matsuda and W. A. Barber (to American Cyanamid Co.).
24. U.S. Pat. 4,178,310 (Dec. 11, 1979), J. A. Fetchin and K. H. Tsu (to American Cyanamid Co.).

25. U.S. Pat. 3,193,480 (July 6, 1965), M. M. Baizer, C. R. Campbell, R. H. Fariss, and R. Johnson (to Monsanto Chemical Co.).
26. U.S. Pat. 3,529,011 (Sept. 15, 1970), J. W. Badham (to Imperial Chemical Industries Austalia Ltd.).
27. Eur. Pat. Appl. E.P. 314,383 (May 3, 1989), G. Shaw and J. Lopez-Merono (to Imperial Chemical Industries PLC).
28. J. L. Callahan, R. K. Grasselli, E. C. Milberger, and H. A. Strecker, *Ind. Eng. Chem. Prod. Res. Dev.* **9**, 134 (1970).
29. *Chem. Eng. News* **67**(2), 23 (1989).
30. U.S. Pat. 2,481,826 (Sept. 13, 1949), J. N. Cosby (to Allied Chemical & Dye Corp.).
31. J. D. Burrington, C. T. Kartisek, and R. K. Grasselli, *J. Catal.* **87**, 363 (1984).
32. C. R. Adams and T. J. Jennings, *J. Catal.* **2**, 63 (1963).
33. C. R. Adams and T. J. Jennings, *J. Catal.* **3**, 549 (1964).
34. G. W. Keulks, *J. Catal.* **19**, 232 (1970).
35. G. W. Keulks and L. D. Krenzke, *Proceedings of the International Congress on Catalysis, 6th, 1976,* The Chemical Society, London, 1977, p. 806; *J. Catal.* **61**, 316 (1980).
36. J. F. Brazdil, D. D. Suresh, and R. K. Grasselli, *J. Catal.* **66**, 347 (1980).
37. R. K. Grasselli, J. F. Brazdil, and J. D. Burrington, *Proceedings of the International Congress on Catalysis, 8th, 1984,* Verlag Chemie, Weinheim, 1984, Vol. V, p. 369.
38. R. K. Grasselli, *Applied Catal.* **15**, 127 (1985).
39. R. K. Grasselli, *React. Kinet. Catal. Lett.* **35**, 327 (1987).
40. J. F. Brazdil, L. C. Glaeser, and R. K. Grasselli, *J. Catal.* **81**, 142 (1983).
41. L. C. Glaeser, J. F. Brazdil, M. A. Hazle, M. Mehicic, and R. K. Grasselli, *J. Chem. Soc. Faraday Trans. 1,* **81**, 2903 (1985).
42. R. G. Teller, J. F. Brazdil, and R. K. Grasselli, *Acta Cryst.* C **40**, 2001 (1984).
43. R. G. Teller, J. F. Brazdil, R. K. Grasselli, R. T. L. Corliss, and J. Hastings, *J. Solid State Chem.* **52**, 313 (1984).
44. R. G. Teller, J. F. Brazdil, R. K. Grasselli, and W. Yelon, *J. Chem. Soc. Faraday Trans. 1,* **81**, 1693 (1985).
45. M. R. Antonio, R. G. Teller, D. R. Sandstrom, M. Mehicic, and J. F. Brazdil, *J. Phys. Chem.* **92**, 2939 (1988).
46. M. R. Antonio, J. F. Brazdil, L. C. Glaeser, M. Mehicic, and R. G. Teller, *J. Phys. Chem.* **92**, 2338 (1988).
47. J. F. Brazdil and R. K. Grasselli, *J. Catal.* **79**, 104 (1983).
48. J. F. Brazdil, L. C. Glaeser, and R. K. Grasselli, *J. Phys. Chem.* **87**, 5485 (1983).
49. A. W. Sleight, in J. J. Burton and R. L. Garten, eds., *Advanced Materials in Catalysis,* Academic Press, New York, 1977, pp. 181–208.
50. R. K. Grasselli and J. D. Burrington, *Adv. Catal.* **30**, 133 (1981).
51. K. Weissermel and H. J. Arpe, *Industrial Organic Chemistry,* A. Mullen, trans., Verlag Chemie, New York, 1978, pp. 266–267.
52. U.S. Pat. 2,690,452 (Sept. 28, 1954), E. L. Carpenter (to American Cyanamid Co.).
53. U.S. Pat. 2,729,670 (Jan. 3, 1956), P. H. DeBruin (to Stamicarbon N.V.).
54. *Chem. Eng. News,* **23**(20), 1841 (Oct. 25, 1945).
55. D. J. Hadley and E. G. Hancock, eds., *Propylene and Its Industrial Derivatives,* Halsted Press, a division of John Wiley & Sons, Inc., New York, 1973, p. 418.
56. K. Sennewald, *World Petroleum Congress Proceedings, 5th, 1959,* Section IV, Paper 19, pp. 217–227.
57. U.S. Pat. 2,554,482 (May 29, 1951), N. Brown (to E. I. du Pont de Nemours & Co., Inc.).
58. U.S. Pat. 2,385,552 (Sept. 25, 1945), L. R. U. Spence and F. O. Haas (to Rohm and Haas Co.).
59. U.S. Pat. 2,736,739 (Feb. 28, 1956), D. C. England and G. V. Mock (to E. I. du Pont de Nemours & Co., Inc.).

60. U.S. Pat. 3,184,415 (May 18, 1965), E. B. Huntley, J. M. Kruse, and J. W. Way (to E. I. du Pont de Nemours & Co., Inc.).
61. U.S. Pats. 4,783,545 (Nov. 8, 1988), 4,837,233 (June 6, 1989), and 4,871,706 (Oct. 3, 1989), L. C. Glaeser, J. F. Brazdil, and M. A. Toft (to The Standard Oil Co.).
62. Brit. Pats. 1,336,135 (Nov. 7, 1973), N. Harris and W. L. Wood; 1,336,136 (Nov. 7, 1973), N. Harris (to Power-Gas Ltd.).
63. U.S. Pat. 3,833,638 (Sept. 3, 1974), W. R. Knox, K. M. Taylor, and G. M. Tullman (to Monsanto Co.).
64. U.S. Pats. 4,849,537 (July 18, 1989) and 4,849,538 (July 18, 1989), R. Ramachandran, D. L. MacLean, and D. P. Satchell, Jr. (to The BOC Group, Inc.).
65. U.S. Pat. 4,609,502 (Sept. 2, 1986), S. Khoobiar (to The Halcon SD Group, Inc.).
66. U.S. Pat. 3,751,443 (Aug. 7, 1973), K. E. Khchelan, O. M. Revenko, A. N. Shatalova, and E. G. Gelperina.
67. J. Perkowski, *Przem. Chem.* **51,** 17 (1972).
68. U.S. Pat. 4,746,753 (May 24, 1988), J. F. Brazdil, D. D. Suresh, and R. K. Grasselli (to The Standard Oil Co. (Ohio)).
69. U.S. Pat. 4,503,001 (Mar. 5, 1985), R. K. Grasselli, A. F. Miller, and H. F. Hardman (to The Standard Oil Co. (Ohio)).
70. U.S. Pat. 4,228,098 (Oct. 14, 1980), K. Aoki, M. Honda, T. Dozono, and T. Katsumata (to Asahi Kasei Kogyo Kabushiki Kaisha).
71. *Chem. Eng. News,* 36 (June 19, 1989).
72. *Chem. Mark. Rep.* **235,** 50 (1989).
73. *Propylene Newsletter,* DeWitt & Company, Houston, Tex., March 15, 1989.
74. Ref. 3, pp. 163–165.
75. *Annual Book of ASTM Standards, E 1178-87,* American Society for Testing and Materials, Philadelphia, Pa., 1988.
76. *Annual Book of ASTM Standards, E 203-75,* American Society for Testing and Materials, Philadelphia, Pa., 1986.
77. *Annual Book of ASTM Standards, E 299-84,* American Society for Testing and Materials, Philadelphia, Pa., 1984.
78. *Material Safety Data Sheet Number 1386,* BP Chemicals Inc., Cleveland, Ohio, 1989.
79. *Assessment of Human Exposures to Atmospheric Acrylonitrile,* SRI International, Menlo Park, Calif., 1979.
80. U.S. Pat. 3,489,789 (Jan. 13, 1970), R. A. Dewar and M. A. Riddolls (to Imperial Chemical Industries Australia Ltd.).
81. U.S. Pat. 3,549,685 (Dec. 22, 1970), J. W. Badham, P. J. Gregory, and J. B. Glen (to Imperial Chemical Industries Australia Ltd.).

JAMES F. BRAZDIL
BP Research

ACRYLONITRILE POLYMERS

SURVEY AND SAN

Acrylonitrile (AN), C_3H_3N, first became an important polymeric building block in the 1940s. Although it had been discovered in 1893 (1), its unique properties were not realized until the development of nitrile rubbers during World War II (see ELASTOMERS, SYNTHETIC, NITRILE RUBBER) and the discovery of solvents for the homopolymer with resultant fiber applications (see FIBERS, ACRYLIC) for textiles and carbon fibers (qv). As a comonomer, acrylonitrile (qv) contributes hardness, rigidity, solvent and light resistance, gas impermeability, and the ability to orient. These properties have led to many copolymer application developments since 1950.

The utility of acrylonitrile [107-13-1] in thermoplastics was first realized in its copolymer with styrene, C_8H_8 [100-42-5], in the late 1950s. Styrene–acrylonitrile (SAN) copolymers [9003-54-7] have superior properties to polystyrene in the areas of toughness, rigidity, and chemical and thermal resistance (2), and, consequently, many commercial applications for them have developed. These optically clear materials containing between 15 and 35% AN can be readily processed by extrusion and injection molding, but they lack real impact resistance.

The subsequent development of ABS (acrylonitrile–butadiene–styrene) resins [9003-56-9], which contain an elastomeric component within a SAN matrix, further boosted commercial application of the basic SAN copolymer as a portion of these rubber-toughened thermoplastics (see ACRYLONITRILE POLYMERS, ABS RESINS). When SAN is grafted onto a butadiene-based rubber, and optionally blended with additional SAN, the two-phase thermoplastic ABS is produced. ABS has the useful SAN properties of rigidity and resistance to chemicals and solvents, while the elastomeric component contributes real impact resistance. Because ABS is a two-phase system and each phase has a different refractive index, the final ABS is normally opaque. A clear ABS can be made by adjusting the refractive indexes through the inclusion of another monomer such as methyl methacrylate. ABS is a very versatile material and modifications have brought out many specialty grades such as clear ABS and high temperature and flame retardant grades. Saturated hydrocarbon elastomers or acrylic elastomers (3,4) can be used instead of those based on butadiene, C_4H_6 [106-99-0].

In the late 1960s a new class of AN copolymers and multipolymers was introduced that contain $\geq 60\%$ acrylonitrile. These are commonly known as barrier resins and have found their greatest acceptance where excellent barrier properties toward gases (5), chemicals, and solvents are needed. They may be processed into bottles, sheets, films, and various laminates, and have found wide usage in the packaging industry (see BARRIER POLYMERS).

Acrylonitrile has found its way into a great variety of other polymeric compositions based on its polar nature and reactivity, imparting to other systems some or all of the properties noted above. Some of these areas include adhesives and binders, antioxidants, medicines, dyes, electrical insulations, emulsifying agents, graphic arts, insecticides, leather, paper, plasticizers, soil-modifying agents, solvents, surface coatings, textile treatments, viscosity modifiers, azeotropic distillations, artificial organs, lubricants, asphalt additives, water-soluble polymers, hollow spheres, cross-linking agents, and catalyst treatments (6).

SAN Physical Properties and Test Methods

SAN resins possess many physical properties desired for thermoplastic applications. They are characteristically hard, rigid, and dimensionally stable with load bearing capabilities. They are also transparent, have high heat distortion temperatures, possess excellent gloss and chemical resistance, and adapt easily to conventional thermoplastic fabrication techniques (7).

SAN polymers are random linear amorphous copolymers. Physical properties are dependent on molecular weight and the percentage of acrylonitrile. An increase of either generally improves physical properties, but may cause a loss of processability or an increase in yellowness. Various processing aids and modifiers can be used to achieve a specific set of properties. Modifiers may include mold release agents, uv stabilizers, antistatic aids, elastomers, flow and processing aids, and reinforcing agents such as fillers and fibers (7).

Methods for testing and some typical physical properties are listed in Table 1.

Table 1. Properties of Injection-Molded Commercial SAN Resins[a]

Property	Monsanto Lustran-35	Dow Tyril-880	ASTM Method
tensile strength, MPa[b]	79.4	82.1	D638
ultimate elongation, %	3.0	3.0	D638
modulus of elasticity, GPa[c]	3.45	3.86	D638
Izod impact strength, J/m[d]	24.0	26.7	D256
hardness—Rockwell M	83	80	D785
deflection temperature, °C	104.4	103.3	D648, annealed
Vicat softening temperature, °C	111.1	111.1	D1525
melt-flow rate, g/10 min	7.0	3.0	D1238, cond 1
coefficient of linear thermal expansion, cm/(cm·°C)	6.8×10^{-5}	6.6×10^{-5}	D696
flammability, cm/min		2.0	D635
specific heat, J/(g·K)[e]		1.3	Dow test
dielectric constant, kHz (MHz)		3.18 (3.02)	D150
dissipation factor, kHz (MHz)		0.007 (0.012)	D150
refractive index, n_D	1.57	1.565	D542
water absorption, % in 24 h	0.25	0.35	D570
specific gravity	1.07	1.08	D792
mold shrinkage, cm/cm	0.003–0.004	0.003–0.007	

[a]Data taken from Monsanto and Dow product data sheets.
[b]To convert MPa to psi multiply by 145.
[c]To convert GPa to psi multiply by 145,000.
[d]To convert J/m to ftlb/in. divide by 53.39.
[e]To convert J to cal divide by 4.184.

The properties of SAN resins are dependent on their acrylonitrile content. Both melt viscosity and hardness increase with increasing acrylonitrile level. Unnotched impact and flexural strengths depict dramatic maxima at ca 87.5 mol % (78 wt %) acrylonitrile (8). With increasing acrylonitrile content, copolymers show continuous improvements in barrier properties and chemical and uv resistance, but thermal stability deteriorates (9). The glass-transition temperature (T_g) of SAN varies nonlinearly with acrylonitrile content, showing a maximum at 50 mol % AN. The alternating SAN copolymer has the highest T_g (10,11). The fatigue resistance of SAN increases with AN content to a maximum at 30 wt %, then decreases with higher AN levels (12). The effect of acrylonitrile incorporation on SAN resin properties is shown in Table 2.

Table 2. Compositional Effects on SAN Physical Properties[a]

AN, wt %	Tensile strength, MPa[b]	Elongation, %	Impact strength, J/m notch[c]	Heat distortion, temp., °C	Solution viscosity, mPa·s (= cP)
5.5	42.27	1.6	26.6	72	11.1
9.8	54.61	2.1	26.0	82	10.7
14.0	57.37	2.2	27.1	84	13.0
21.0	63.85	2.5	27.1	88	16.5
27.0	72.47	3.2	27.1	88	25.7

[a]Ref. 13.
[b]To convert MPa to psi multiply by 145.
[c]To convert J/m to ftlb/in. divide by 53.39.

SAN Chemical Properties and Analytical Methods

SAN resins show considerable resistance to solvents and are insoluble in carbon tetrachloride, ethyl alcohol, gasoline, and hydrocarbon solvents. They are swelled by solvents such as benzene, ether, and toluene. Polar solvents such as acetone, chloroform, dioxane, methyl ethyl ketone, and pyridine will dissolve SAN (14). The interactions of various solvents and SAN copolymers containing up to 52% acrylonitrile have been studied along with their thermodynamic parameters, ie, the second virial coefficient, free-energy parameter, expansion factor, and intrinsic viscosity (15).

The properties of SAN are significantly altered by water absorption (16). The equilibrium water content increases with temperature while the time required decreases. A large decrease in T_g can result. Strong aqueous bases can degrade SAN by hydrolysis of the nitrile groups (17).

The molecular weight of SAN can be easily determined by either intrinsic viscosity or size-exclusion chromatography (sec). Relationships for both multipoint and single point viscosity methods are available (18,19). Two intrinsic viscosity and molecular weight relationships for azeotropic copolymers have been given (20,21):

$$[\eta] = 3.6 \times 10^{-4} \ M_w^{0.62} \ \text{dL/g in MEK at } 30°C$$

$$[\eta] = 2.15 \times 10^{-4} \ M_w^{0.68} \ \text{dL/g in THF at } 25°C$$

Chromatographic techniques are readily applied to SAN for molecular weight determination. Size-exclusion chromatography or gel permeation chromatography (gpc) (22) columns and conditions have been described for SAN (23). Chromatographic detector differences have been shown to be of the order of only 2–3% (24). High pressure precipitation chromatography can achieve similar molecular weight separation (25). Liquid chromatography (lc) can be used with sec-fractioned samples to determine copolymer composition (26). Thin layer chromatography will also separate SAN by compositional (monomer) variations (25).

Residual monomers in SAN have been a growing environmental concern and can be determined by a variety of methods. Monomer analysis can be achieved by polymer solution or directly from SAN emulsions (27) followed by "head space" gas chromatography (gc) (28,29). Liquid chromatography (lc) is also effective (30).

SAN Manufacture

The reactivities of acrylonitrile and styrene radicals toward their monomers are quite different, resulting in SAN copolymer compositions that vary from their monomer compositions (31). Further complicating the reaction is the fact that AN is soluble in water (see ACRYLONITRILE) and slightly different behavior is observed between water-based emulsion and suspension systems and bulk or mass polymerizations (32). SAN copolymer compositions can be calculated from copolymerization equations (33) and published reactivity ratios (34). The difference in radical reactivity causes the copolymer composition to drift as polymerization proceeds, except at the azeotrope composition where copolymer composition matches monomer composition. Figure 1 shows these compositional variations (35). When SAN copolymer compositions vary significantly, incompatibility results, causing loss of optical clarity, mechanical strength, and moldability, as well as heat, solvent, and chemical resistance (36). The termination step has been found to be controlled by diffusion even at low conversions, and the termination rate constant varies with acrylonitrile content. The average half-life of the radicals increases with styrene concentration from 0.3 s at 20 mol % to 6.31 s with pure styrene (37). Further complicating SAN manufacture is the fact that both the heat (38,39) and rate (40) of copolymerization vary with monomer composition.

The early kinetic models for copolymerization, Mayo's terminal mechanism (41) and Alfrey's penultimate model (42), did not adequately predict the behavior of SAN systems. Copolymerizations in DMF and toluene indicated that both penultimate and antepenultimate effects had to be considered (43,44). The resulting reactivity model is somewhat complicated, since there are eight reactivity ratios to consider.

The first quantitative model, which appeared in 1971, also accounted for possible charge-transfer complex formation (45). Deviation from the terminal model for bulk polymerization was shown to be due to antepenultimate effects (46). More recent work with numerical computation and ^{13}C-nmr spectroscopy data on SAN sequence distributions indicates that the penultimate model is the most appropriate for bulk SAN copolymerization (47,48). A kinetic model for azeotropic SAN copolymerization in toluene has been developed that successfully predicts conversion, rate, and average molecular weight for conversions up to 50% (49).

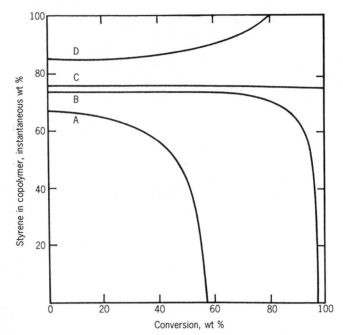

Fig. 1. Approximate compositions of styrene–acrylonitrile copolymers formed at different conversions starting with various monomer mixtures (35): S/AN = A, 65/35; B, 70/30; C, 76/24; D, 90/10.

An emulsion model that assumes the locus of reaction to be inside the particles and considers the partition of AN between the aqueous and oil phases has been developed (50). The model predicts copolymerization results very well when bulk reactivity ratios of 0.32 and 0.12 for styrene and acrylonitrile, respectively, are used.

Commercially, SAN is manufactured by three processes: emulsion, suspension, and continuous bulk.

Emulsion Process. The emulsion polymerization process utilizes water as a continuous phase with the reactants suspended as microscopic particles. This low viscosity system allows facile mixing and heat transfer for control purposes. An emulsifier is generally employed to stabilize the water insoluble monomers and other reactants, and to prevent reactor fouling. With SAN the system is composed of water, monomers, chain-transfer agents for molecular weight control, emulsifiers, and initiators. Both batch and semibatch processes are employed. Copolymerization is normally carried out at 60 to 100°C to conversions of ~97%. Lower temperature polymerization can be achieved with redox-initiator systems (51).

Figure 2 shows a typical batch or semibatch emulsion process (52). A typical semibatch emulsion recipe is shown in Table 3 (53).

The initial charge is placed in the reactor, purged with an inert gas such as N$_2$, and brought to 80°C. The initiator is added, followed by addition of the remaining charge over 100 min. The reaction is completed by maintaining agitation at 80°C for 1 h after monomer addition is complete. The product is a free-flowing white latex with a total solids content of 35.6%.

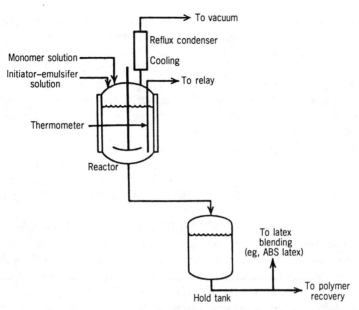

Fig. 2. Styrene–acrylonitrile batch emulsion process (52).

Table 3. Semibatch Mode Emulsion Recipe for SAN Copolymers

Ingredient	Parts
Initial reactor charge	
acrylonitrile	90
styrene	111
Na alkanesulfonate (emulsifier)	63
$K_2S_2O_8$ (initiator)	0.44
4-(benzyloxymethylene)cyclohexene (mol wt modifier)	1
water	1400
Addition charge	
acrylonitrile	350
styrene	1000
Na alkanesulfonate (emulsifier)	15
$K_2S_2O_8$ (initiator)	4
4-(benzyloxymethylene)cyclohexene (mol wt modifier)	10
water	1600

Compositional control for other than azeotropic compositions can be achieved with both batch and semibatch emulsion processes. Continuous addition of the faster reacting monomer, styrene, can be practiced for batch systems, with the feed rate adjusted by computer through gas chromatographic monitoring during the course of the reaction (54). A calorimetric method to control the monomer feed rate has also been described (8). For semibatch processes, adding the monomers at a rate that is slower than copolymerization can achieve equilibrium. It has been found that constant composition in the emulsion can be achieved after ca 20% of the monomers have been charged (55).

Residual monomers in the latex are avoided either by effectively reacting the monomers to polymer or by physical or chemical removal. The use of *tert*-butyl peroxypivalate as a second initiator toward the end of the polymerization or the use of mixed initiator systems of $K_2S_2O_8$ and *tert*-butyl peroxybenzoate (56) effectively increases final conversion and decreases residual monomer levels. Spray devolatilization of hot latex under reduced pressure has been claimed to be effective (56). Residual acrylonitrile also can be reduced by postreaction with a number of agents such as monoamines (57) and dialkylamines (58), ammonium–alkali metal sulfites (59), unsaturated fatty acids or their glycerides (60,61), their aldehydes, esters of olefinic alcohols, cyanuric acid (62,63), and myrcene (64).

The copolymer latex can be used "as is" for blending with other latexes, such as in the preparation of ABS, or the copolymer can be recovered by coagulation. The addition of electrolyte or freezing will break the latex and allow the polymer to be recovered, washed, and dried. Process refinements have been made to avoid the difficulties of fine particles during recovery (65–67).

The emulsion process can be modified for the continuous production of latex. One such process (68) uses two stirred-tank reactors in series, followed by insulated hold-tanks. During continuous operation, 60% of the monomers are continuously charged to the first reactor with the remainder going into the second reactor. Surfactant is added only to the first reactor. The residence time is 2.5 h for the first reactor where the temperature is maintained at 65°C for 92% conversion. The second reactor is held at 68°C for a residence time of 2 h and conversion of 95%.

Suspension Process.   Like the emulsion process, water is the continuous phase for suspension polymerization, but the resultant particle size is larger, well above the microscopic range. The suspension medium contains water, monomers, molecular weight control agents, initiators, and suspending aids. Stirred reactors are used in either batch or semibatch mode. Figure 3 illustrates a typical suspension manufacturing process while a typical batch recipe is shown in Table 4 (69). The components are charged into a pressure vessel and purged with nitrogen. Copolymerization is carried out at 128°C for 3 h and then at 150°C for 2 h. Steam stripping removes residual monomers (70), and the polymer beads are separated by centrifugation for washing and final dewatering.

Compositional control in suspension systems can be achieved with a corrected batch process. A suspension process has been described where styrene monomer is continuously added until 75–85% conversion, and then the excess acrylonitrile monomer is removed by stripping with an inert gas (71,72).

Elimination of unreacted monomers can be accomplished by two methods: dual initiators to enhance conversion of monomers to product (73–75) and steam stripping (70,76). Several process improvements have been claimed for dewatering beads (77), to reduce haze (78–81), improve color (82–86), remove monomer (87,88), and maintain homogeneous copolymer compositions (71,72,89).

Continuous Bulk Process.   The continuous bulk process has several advantages including high space-time yield, and good quality products uncontaminated with residual ingredients such as emulsifiers or suspending agents.

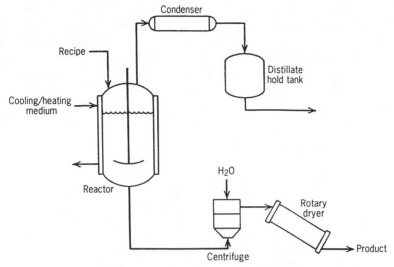

Fig. 3.   Styrene–acrylonitrile suspension process (69).

Table 4. Batch Mode Recipe for SAN Copolymers[a]

Ingredient	Parts
acrylonitrile	30
styrene	70
dipentene (4-isopropenyl-1-methylcyclohexene)	1.2
di-*tert*-butyl peroxide	0.03
acrylic acid–2-ethylhexyl acrylate (90:10) copolymer	0.03
water	100

[a]Ref. 69.

SAN manufactured by this method generally has superior color and transparency and is preferred for applications requiring good optical properties. It is a self-contained operation without waste treatment or environmental problems since the products are either polymer or recycled back to the process.

In practice, the continuous bulk polymerization is rather complicated. Because of the high viscosity of the copolymerizing mixture, complex machinery is required to handle mixing, heat transfer, melt transport, and devolatilization. In addition, considerable time is required to establish steady-state conditions in both a stirred tank reactor and a linear flow reactor. Thus system start-up and product grade changes produce some off-grade or intermediate grade products. Copolymerization is normally carried out between 100 and 200°C. Solvents are used to reduce viscosity or the conversion is kept to 40–70%, followed by devolatilization to remove solvents and monomers. Devolatilization is carried out from 120 to 260°C under vacuum at less than 20 kPa (2.9 psi). The devolatilized melt is then fed through a strand die, cooled, and pelletized.

A schematic of a continuous bulk SAN polymerization process is shown in Figure 4 (90). The monomers are continuously fed into a screw reactor where copolymerization is carried out at 150°C to 73% conversion in 55 min. Heat of polymerization is removed through cooling of both the screw and the barrel walls. The polymeric melt is removed and fed to the devolatilizer to remove unreacted monomers under reduced pressure (4 kPa or 30 mm Hg) and high temperature (220°C). The final product is claimed to contain less than 0.7% volatiles. Two devolatilizers in series are found to yield a better quality product as well as better operational control (91,92).

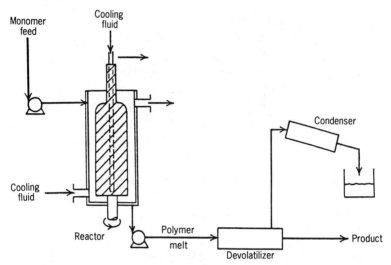

Fig. 4.   Styrene–acrylonitrile continuous mass process (90).

Two basic reactor types are used in the continuous bulk process, the stirred tank reactor (93) and the linear flow reactor. The stirred tank reactor consists of a horizontal cylinder chamber equipped with various agitators (94,95) for mixing the viscous melt and an external cooling jacket for heat removal. With adequate mixing, the composition of the melt inside the reactor is homogeneous. Operation at a fixed conversion, with monomer make-up added at an amount and ratio equal to the amount and composition of copolymer withdrawn, produces a fixed composition copolymer. The two types of linear flow reactors employed are the screw reactor (90) and the tower reactor (95). A screw reactor is composed of two concentric cylinders. The reaction mixture is conveyed toward the outlet by rotating the inner screw, which has helical threads, while heat is removed from both cylinders. A tower reactor with separate heating zones has a scraper agitator in the upper zone, while the lower portion generates plug flow. In the linear flow reactors the conversion varies along the axial direction as does the copolymer composition, except where operating at the azeotrope composition. A stream of monomer must be added along the reactor to maintain SAN compositional homogeneity at high conversions. A combined stirred tank followed by a linear flow reactor process has been disclosed (95). Through continuous recycle copolymerization, a copolymer of identical composition to monomer feed can be achieved, regardless of the reactivity ratios of the monomers involved (96).

The devolatilization process has been developed in many configurations. Basically, the polymer melt is subjected to high temperatures and low pressures to remove unreacted monomer and solvent. A two-stage process using a tube and shell heat exchanger with enlarged bottom receiver to vaporize monomers has been described (92). A copolymer solution at 40–70% conversion is fed into the first-stage exchanger and heated to 120–190°C at a pressure of 20–133 kPa and then discharged into the enlarged bottom section to remove at least half of the unreacted acrylonitrile. The product from this section is then charged to a second stage and heated to 210–260°C at <20 kPa. The devolatilized product contains ~1% volatiles. Preheating the polymer solution and then flashing it into a multipassage heating zone at lower pressure than the preheater produces essentially volatile-free product (91,97). SAN can be steam stripped to quite low monomer levels in a vented extruder which has water injected at a pressure greater than the vapor pressure of water at that temperature (98).

A twin-screw extruder is used to reduce residual monomers from ca 50 to 0.6%, at 170°C and 3 kPa with a residence time of 2 min (94). In another design, a heated casing encloses the vented devolatilization chamber, which encloses a rotating shaft with specially designed blades (99,100). These continuously regenerate a large surface area to facilitate the efficient vaporization of monomers. The devolatilization equipment used for the production of polystyrene and ABS is generally suitable for SAN production.

Processing. SAN copolymers may be processed using the conventional fabrication methods of extrusion, blow molding, injection molding, thermoforming, and casting. Small amounts of additives, such as antioxidants, lubricants, and colorants, may also be used. Typical temperature profiles for injection molding and extrusion of predried SAN resins are as follows (101).

Molding Temperatures

cylinder	193–288°C
mold	49–88°C
melt	218–260°C

Extrusion Temperatures

hopper zone	water-cooled
rear zone	177–204°C
middle zone	210–232°C
torpedo zone and die	204–227°C

Other Copolymers

Acrylonitrile copolymerizes readily with many electron-donor monomers other than styrene. Hundreds of acrylonitrile copolymers have been reported, and a comprehensive listing of reactivity ratios for acrylonitrile copolymerizations is readily available (34,102). Copolymerization mitigates the undesirable properties of acrylonitrile homopolymer, such as poor thermal stability and poor processability. At the same time, desirable attributes such as rigidity, chemical resistance, and excellent barrier properties are incorporated into melt-processable resins.

Barex 210, a commercial high barrier resin produced by BP America, is a copolymer of acrylonitrile and methyl acrylate [96-33-3]. This resin is an excellent example of the use of acrylonitrile to provide gas barrier, chemical resistance, high tensile strength, and stiffness, and utilization of a comonomer to provide thermal stability and processability. In addition, modification with an elastomer provides toughness and impact strength. This material has a unique combination of useful packaging qualities, including transparency, and is an excellent barrier to permeation by gases, organic solvents, and most essential oils. As a result, it prevents the migration of volatile flavors and odors from packaged foods and toiletry products (103) while also providing protection from atmospheric oxygen.

Barex 210 extruded sheet can be easily thermoformed into lightweight, rigid containers (103,104). Packages can be printed, laminated, or metallized. Recent developments in extrusion and injection blow molding (103,105), laminated film structures (103,106), and coextrusion (103,107) have led to packaging uses for a variety of products. Barex 210 is especially well-suited for bottle production. This acrylonitrile copolymer also provides a good example of the dependence of properties on the degree and temperature of orientation (108,109). Figure 5 illustrates the improvement in tensile strength, elongation, and the ability to absorb impact energy due to orientation (109). Tensile strength and impact strength increase with the extent of stretching, and decrease with the orientation temperature, and oxygen permeability decreases with orientation.

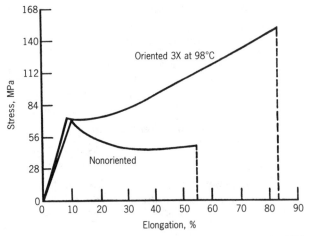

Fig. 5.   Stress elongation of Barex 210 sheet (109). To convert MPa to psi multiply by 145.

Acrylonitrile–methyl acrylate–indene terpolymers, by themselves, or in blends with acrylonitrile–methyl acrylate copolymers, exhibit even lower oxygen and water permeation rates than the indene-free copolymers (110,111). Terpolymers of acrylonitrile with indene and isobutylene also exhibit excellent barrier properties (112), and permeation of gas and water vapor through acrylonitrile–styrene–isobutylene terpolymers is also low (113,114).

Copolymers of acrylonitrile and methyl methacrylate (115) and terpolymers of acrylonitrile, styrene, and methyl methacrylate (116,117) are used as barrier polymers. Acrylonitrile copolymers and multipolymers containing butyl acrylate (118–121), ethyl acrylate (122), 2-ethylhexyl acrylate (118,121,123,124), hydroxyethyl acrylate (120), vinyl acetate (119,125), vinyl ethers (125,126), and vinylidene chloride (121,122,127–129) are also used in barrier films, laminates, and coatings. Environmentally degradable polymers useful in packaging are prepared from polymerization of acrylonitrile with styrene and methyl vinyl ketone (130). Table 5 gives the structures, formulas, and CAS Registry Numbers for several comonomers of acrylonitrile.

Table 5. Monomers Commonly Copolymerized with Acrylonitrile

Monomer	Molecular formula	Structural formula	CAS Registry Number
methyl methacrylate	$C_5H_8O_2$	$CH_2{=}C(CH_3)COOCH_3$	[80-62-6]
methyl acrylate	$C_4H_6O_2$	$CH_2{=}CHCOOCH_3$	[96-33-3]
indene	C_9H_8		[95-13-6]
isobutylene	C_4H_8	$CH_2{=}C(CH_3)_2$	[115-11-7]
butyl acrylate	$C_7H_{12}O_2$	$CH_2{=}CHCOOC_4H_9$	[141-32-2]
ethyl acrylate	$C_5H_8O_2$	$CH_2{=}CHCOOC_2H_5$	[140-88-5]
2-ethylhexyl acrylate	$C_{11}H_{20}O_2$	$CH_2{=}CHCOOC_8H_{17}$	[103-11-7]
hydroxyethyl acrylate	$C_5H_8O_3$	$CH_2{=}CHCOOC_2H_4OH$	[818-61-1]
vinyl acetate	$C_4H_6O_2$	$CH_2{=}CHOOCCH_3$	[108-05-4]
vinylidene chloride	$C_2H_2Cl_2$	$CH_2{=}C(Cl)_2$	[75-35-4]
methyl vinyl ketone	C_4H_6O	$CH_2{=}CHCOCH_3$	[78-94-4]
α-methylstyrene	C_9H_{10}	$CH_2{=}C(CH_3)C_6H_5$	[98-83-9]
vinyl chloride	C_2H_3Cl	$CH_2{=}CHCl$	[75-01-4]
4-vinylpyridine	C_7H_7N	$CH_2{=}CHC_5H_4N$	[100-43-6]
acrylic acid	$C_3H_4O_2$	$CH_2{=}CHCOOH$	[79-10-7]

Although the arrangement of monomer units in acrylonitrile copolymers is usually random, alternating or block copolymers may be prepared by using special techniques. For example, the copolymerization of acrylonitrile, like that of other vinyl monomers containing conjugated carbonyl or cyano groups, is changed in the presence of certain Lewis acids. Effective Lewis acids are metal compounds with nontransition metals as central atoms, including alkylaluminum halides, zinc halides, and triethylaluminum. The presence of the Lewis acid increases the tendency of acrylonitrile to alternate with electron-donor molecules, such as styrene, α-methylstyrene, and olefins (131–135). This alternation is often attributed to a ternary molecular complex or charge-transfer mechanism, where complex formation with the Lewis acid increases the electron-accepting ability of acrylonitrile, which results in the formation of a molecular complex between the acrylonitrile–Lewis acid complex and the donor molecule. This ternary molecular complex polymerizes as a unit to yield an alternating polymer. Cross-propagation and complex radical mechanisms have also been proposed (136).

A number of methods such as ultrasonics (137), radiation (138), and chemical techniques (139–141), including the use of polymer radicals, polymer ions, and organometallic initiators, have been used to prepare acrylonitrile block copolymers (142). Block comonomers include styrene, methyl acrylate, methyl methacrylate, vinyl chloride, vinyl acetate, 4-vinylpyridine, acrylic acid, and *n*-butyl isocyanate.

Acrylonitrile has been grafted onto many polymeric systems. In particular, acrylonitrile grafting has been used to impart hydrophilic behavior to starch (143–145) and polymer fibers (146). Exceptional water absorption capability results from the grafting of acrylonitrile to starch, and the use of 2-acrylamido-2-methylpropanesulfonic acid [*15214-89-8*] along with acrylonitrile for grafting results in copolymers that can absorb over 5000 times their weight of deionized water (147).

Economic Aspects

Since its introduction in the 1950s, SAN has shown steady growth. The combined properties of SAN copolymers such as optical clarity, rigidity, chemical and heat resistance, high tensile strength, and flexible molding characteristics, along with reasonable price have secured their market position. The largest portion of SAN ($\geq 80\%$) is incorporated into ABS resins, and their markets are inexorably joined. Both of these polymeric materials have shown similar responses during the 1980s to market conditions such as recessions and oil prices (148).

There are two major producers of SAN resin in the United States, Monsanto Chemical Company and The Dow Chemical Company, which market these materials under the names of Lustran and Tyril, respectively. Some typical physical properties of these have been shown in Table 1. Production figures for SAN and ABS for the 1980s are shown in Table 6 (148).

Table 6. U.S. Production of SAN and ABS Resins,[a]
10^3 Metric Tons[b]

Year	SAN	ABS
1979	56.2	956
1980	50.3	773
1981	48.5	791
1982	41.3	626
1983	42.2	857
1984	44.9	965
1985	39.5	923
1986	41.7	996
1987	57.1	1022
1988	67.1	1098

[a]Ref. 148.
[b]Dry-weight basis.

Health and Toxicology

SAN resins themselves appear to pose few health problems in that they have been approved by the FDA for beverage bottle use (149). The main concern is that of toxic residuals, eg, acrylonitrile, styrene, or other polymerization components such as emulsifiers, stabilizers, or solvents. Each component must be treated individually for toxic effects and safe exposure level.

Acrylonitrile is believed to behave similarly to hydrogen cyanide (enzyme inhibition of cellular metabolism) (150) and is believed to be a potential carcinogen (151). It can also affect the cardiovascular system and kidney and liver functions (150). Further information on the toxicology and human exposure to acrylonitrile is available (152–154) (see ACRYLONITRILE).

Styrene toxicity is regarded to be relatively low. It is an irritant to the eyes and respiratory tract, and while prolonged exposure to the skin may cause irritation, styrene is unlikely to be absorbed through the skin in harmful amounts. The American Conference of Government Industrial Hygienists (ACGIH) threshold limit value (TLV) for styrene is 50 ppm time-weighted average (TWA) (155). More information on human exposure to styrene in the workplace is available (156,157) (see STYRENE).

Uses

Useful properties of acrylonitrile copolymers, such as rigidity, gas barrier, chemical and solvent resistance, and toughness, are dependent upon the acrylonitrile content in the copolymers. The choice of the composition of SAN copolymers is dictated by their particular applications and performance requirements. The well-balanced and unique properties possessed by these copolymers have led to broad usage in a wide variety of applications.

Because of their excellent barrier properties, acrylonitrile polymers and copolymers have been widely used in films and laminates for packaging (158–162). In addition to laminates (163–167), SAN copolymers are used in membranes (168–171), controlled-release formulations (172,173), polymeric foams (174,175), fire-resistant compositions (176,177), ion-exchange resins (178), reinforced paper (179), concrete and mortar compositions (180,181), safety glass (182), solid ionic conductors (183), negative resist materials (184), electrophotographic toners (185), and optical recording materials (186). SAN copolymers are also used as coatings (187), dispersing agents for colorants (188), carbon-fiber coatings for improved adhesion (189), and synthetic wood pulp (190). SAN copolymers have been blended with aromatic polyesters to improve hydrolytic stability (191), with methyl methacrylate polymers to form highly transparent resins (192), and with polycarbonate to form toughened compositions with good impact strength (193–195). Table 7 lists the most common uses of SAN copolymers in major industrial markets (7,101). Some important modifications of SAN copolymers are listed in Table 8.

Acrylonitrile has contributed the desirable properties of rigidity, high temperature resistance, clarity, solvent resistance, and gas impermeability to many polymeric systems. Its availability, reactivity, and low cost ensure a continuing market presence and provide potential for many new applications.

Table 7. SAN Copolymer Uses[a]

Application	Articles
appliances	air conditioner parts, decorated escutcheons, washer and dryer instrument panels, washing machine filter bowls, refrigerator shelves, meat and vegetable drawers and covers, blender bowls, mixers, lenses, knobs, vacuum cleaner parts, humidifiers, and detergent dispensers
automotive	batteries, bezels, instrument lenses, signals, glass-filled dashboard components, and interior trim
construction	safety glazing, water filter housings, and water faucet knobs
electronic	battery cases, instrument lenses, cassette parts, computer reels, and phonograph covers
furniture	chair backs and shells, drawer pulls, and caster rollers
housewares	brush blocks and handles, broom and brush bristles, cocktail glasses, disposable dining utensils, dishwasher-safe tumblers, mugs, salad bowls, carafes, serving trays, and assorted drinkware, hangers, ice buckets, jars, and soap containers
industrial	batteries, business machines, transmitter caps, instrument covers, and tape and data reels
medical	syringes, blood aspirators, intravenous connectors and valves, petri dishes, and artificial kidney devices
packaging	bottles, bottle overcaps, closures, containers, display boxes, films, jars, sprayers, cosmetics packaging, liners, and vials
custom molding	aerosol nozzles, camera parts, dentures, disposable lighter housings, fishing lures, pen and pencil barrels, sporting goods, toys, telephone parts, filter bowls, tape dispensers, terminal boxes, toothbrush handles, and typewriter keys

[a]Refs. 7 and 101.

Table 8. Modified SAN Copolymers

Modifier	Remarks	Reference
polybutadiene	ABS, impact resistant	a
EPDM rubber[b]	impact and weather resistant	196,197
polyacrylate	impact and weather resistant	198,199
poly(ethylene-co-vinyl acetate) (EVA)	impact and weather resistant	200
EPDM + EVA	impact and weather resistant	201
silicones	impact and weather resistant	202
chlorinated polyethylene	impact and weather resistant and flame retardant	203
polyester, cross-linked	impact resistant	204
poly(α-methylstyrene)	heat resistant	205
poly(butylene terephthalate)	wear and abrasion resistant	206
ethylene oxide–propylene oxide copolymers	used as lubricants to improve processability	207
sulfonation	hydrogels of high water absorption	208
glass fibers	high tensile strength and hardness	209

[a]See ACRYLONITRILE POLYMERS, ABS RESINS.
[b]Ethylene–propylene–diene monomer rubber.

BIBLIOGRAPHY

"Acrylonitrile Polymers (Survey and SAN)" in *ECT* 3rd ed., Vol. 1, pp. 427–442, F. M. Peng, Monsanto Company.

1. *The Chemistry of Acrylonitrile*, 2nd ed., American Cyanamid Co., Petrochemical Div., New York, 1959, p. ix.
2. Ref. 1, p. 29.
3. J. L. Ziska, "Olefin-Modified Styrene–Acrylonitrile" in R. Juran, ed., *Modern Plastics Encyclopedia 1989*, **65**(11), McGraw-Hill, Inc., New York, p. 105.
4. D. M. Bennett, "Acrylic–Styrene–Acrylonitrile" in R. Juran, ed., *Modern Plastics Encyclopedia 1989*, **65**(11), McGraw-Hill, Inc., New York, p. 96.
5. S. P. Nemphos and Y. C. Lee, *Am. Chem. Soc., Div. Org. Coat. Plast. Chem. Pap.* **33**(2), 618 (1973).
6. Ref. 1, pp. 39–61.
7. F. L. Reithel, "Styrene–Acrylonitrile (SAN)," in R. Juran, ed., *Modern Plastics Encyclopedia 1989*, **65**(11), McGraw-Hill, Inc., New York, p. 105.
8. B. N. Hendy in N. A. J. Platzer, ed., *Copolymers, Polyblends, and Composites* (Adv. Chem. Ser. 142), American Chemical Society, Washington, D.C., 1975, p. 115.
9. W. J. Hall and H. K. Chi, *SPE National Technical Conference: High Performance Plastics,* Cleveland, Ohio, Oct. 5–7, 1976, pp. 1–5.
10. N. W. Johnston, *Polym. Prepr. Am. Chem. Soc.* **14**(1), 46 (1973).
11. N. W. Johnston, *J. Macromol. Sci. Rev. Macromol. Chem.* **C14,** 215 (1976).
12. J. A. Sauer and C. C. Chen, *Polym. Eng. Sci.* **24,** 786 (1984).
13. A. W. Hanson and R. L. Zimmerman, *Ind. Eng. Chem.* **49,** 1803 (1957).
14. Brit. Pat. 590,247 (July 11, 1947) (to Bakelite Ltd.).
15. R. F. Blanks and B. N. Shah, *Polym. Prepr. Am. Chem. Soc.* **17**(2), 407 (1976).
16. S. A. Jabarin and E. A. Lofgren, *Polym. Eng. Sci.* **26,** 405 (1986).
17. M. Kopic, F. Flajsman, and Z. Janovic, *J. Macromol. Sci. Chem.* **A24**(1), 17 (1987).
18. H. U. Khan and G. S. Bhargava, *J. Polym. Sci., Polym. Lett. Ed.* **22**(2), 95 (1984).
19. H. U. Khan, V. K. Gupta, and G. S. Bhargava, *Polym. Commun.* **24**(6), 191 (1983).
20. V. H. Gerrens, H. Ohlinger, and R. Fricker, *Makromol. Chem.* **87,** 209 (1965).
21. Y. Shimura, I. Mita, and H. Kambe, *J. Polym. Sci., Polym. Lett. Ed.* **2,** 403 (1964).
22. A. R. Cooper, "Molecular Weight Determination," in J. I. Kroschwitz, ed., *Encyclopedia of Polymer Science and Engineering*, 2nd ed., Vol. 10, Wiley-Interscience, New York, 1987, pp. 1–19.
23. L. H. Garcia-Rubio, J. F. MacGregor and A. E. Hamielec, *Polym. Prepr. Am. Chem. Soc.* **22**(1), 292 (1981).
24. K. Tsuchida, *Nempo—Fukui-ken Kogyo Shikenjo,* **1980,** 74 (1981).
25. G. Gloeckner and R. Konigsveld, *Makromol. Chem. Rapid Commun.* **4,** 529 (1983).
26. G. Gloeckner, *Pure Appl. Chem.* **55,** 1553 (1983).
27. M. Alonso, F. Recasens, and L. Puiganer, *Chem. Eng. Sci.* **41,** 1039 (1986).
28. G. Hempel and U. Ruedt, *Dtsch. Lebensm.-Rundsch.* **84,** 239 (1988).
29. I. Bruening, *Bol. Tec. PETROBRAS,* **26,** 299 (1983).
30. K. Hidaka, *Shokuhin Eiseigaku Zasshi* **22,** 536 (1981).
31. R. G. Fordyce and E. C. Chapin, *J. Am. Chem. Soc.* **69,** 581 (1947).
32. W. V. Smith, *J. Am. Chem. Soc.* **70,** 2177 (1948).
33. P. J. Flory, *Principles of Polymer Chemistry,* Cornell Univ. Press, Ithaca, N.Y., 1957, Chapter V.
34. L. J. Young, in J. Brandrup and E. H. Immergut, eds., *Polymer Handbook,* 2nd ed., John Wiley and Sons, Inc., New York, 1975, pp. II-128–II-139; see also R. Z. Greenley in

J. Brandup and E. H. Immergut, eds., *Polymer Handbook,* 3rd ed., Wiley-Interscience, New York, 1989, pp. II-165–II-171.

35. C. H. Basdekis, *ABS Plastics,* Reinhold Publishing Corp., New York, 1964, p. 47.
36. Brit. Pat. 1,328,625 (Aug. 30, 1973) (to Daicel Ltd.).
37. R. V. Kucher, L. N. Anisimova, Yu. S. Zaitsev, N. B. Lachinov, and V. P. Zubov, *Polym. Sci. USSR* **20,** 2793 (1978).
38. M. Suzuki, H. Miyama, and S. Fujimoto, *Bull. Chem. Soc. Jpn.* **35,** 60 (1962).
39. H. Miyama and S. Fujimoto, *J. Polym. Sci.* **54,** S32 (1961).
40. G. Mino, *J. Polym. Sci.* **22,** 369 (1956).
41. F. R. Mayo and F. M. Lewis, *J. Am. Chem. Soc.* **66,** 1594 (1944).
42. E. T. Mertz, T. Alfrey, and G. Goldfinger, *J. Polym. Sci.* **1,** 75 (1946).
43. A. Guyot and J. Guillot, *J. Macromol. Sci. Chem.* **A1,** 793 (1967).
44. A. Guyot and J. Guillot, *J. Macromol. Sci. Chem.* **A2,** 889 (1968).
45. J. A. Seiner and M. Litt, *Macromolecules* **4,** 308 (1971).
46. B. Sandner and E. Loth, *Faserforsch Textiltech.* **27,** 571 (1976).
47. D. J. T. Hill, J. H. O'Donnell, and P. W. O'Sullivan, *Macromolecules* **15,** 960 (1982).
48. G. S. Prementine and D. A. Tirrell, *Macromolecules* **20,** 3034 (1987).
49. C. C. Lin, W. Y. Chiu, and C. T. Wang, *J. Appl. Polym. Sci.* **23,** 1203 (1978).
50. T. Kikuta, S. Omi, and H. Kubota, *J. Chem. Eng. Jpn.* **9,** 64 (1976).
51. Brit. Pat. 1,093,349 (Nov. 29, 1967), T. M. Fisler (to Ministerul Industriei Chimice).
52. U.S. Pat. 3,772,257 (Nov. 13, 1973), G. K. Bochum, P. K. Hurth Efferen, J. M. Liblar, and H. J. K. Hurth (to Knapsack A. G.).
53. U.S. Pat. 4,439,589 (Mar. 27, 1984), H. Alberts, R. Schubart, and A. Pischtschan (to Bayer A.G.).
54. A. Guyot, J. Guillot, C. Pichot, and L. R. Gurerrero in D. R. Bassett and A. E. Hamielec, eds., *Emulsion Polymers and Emulsion Polymerization* (Symp. Ser. 165), American Chemical Society, Washington, D.C., 1981, p. 415.
55. J. Snuparek, Jr. and F. Krska, *J. Appl. Polym. Sci.* **21,** 2253 (1977).
56. Ger. Pat. 141,314 (Apr. 23, 1980), H. P. Thiele and co-workers (to VEB Chemische Werke Buna).
57. Pol. Pat. PL 129,792 (Nov. 15, 1985), P. Penczek, E. Wardzinska, and G. Cynkowska (to Instytut Chemii Przemyslowej).
58. U.S. Pat. 4,251,412 (Feb. 17, 1981), G. P. Ferrini (to B. F. Goodrich Co.).
59. U.S. Pat. 4,255,307 (Mar. 10, 1981), J. R. Miller (to B. F. Goodrich Co.).
60. U.S. Pat. 4,228,119 (Oct. 14, 1980), I. L. Gomez and E. F. Tokas (to Monsanto Co.).
61. U.S. Pat. 4,215,024 (July 29, 1980), I. L. Gomez and E. F. Tokas (to Monsanto Co.).
62. U.S. Pat. 4,215,085 (July 29, 1980), I. L. Gomez (to Monsanto Co.).
63. U.S. Pat. 4,275,175 (June 23, 1981), I. L. Gomez (to Monsanto Co.).
64. U.S. Pat. 4,252,764 (Feb. 24, 1981), E. F. Tokas (to Monsanto Co.).
65. U.S. Pat. 3,248,455 (Apr. 26, 1966), J. E. Harsch and co-workers (to U.S. Rubber Co.).
66. Brit. Pat. 1,034,228 (June 29, 1966) (to U.S. Rubber Co.).
67. U.S. Pat. 3,249,569 (May 3, 1966), J. Fantl (to Monsanto Co.).
68. U.S. Pat. 3,547,857 (Dec. 15, 1970), A. G. Murray (to Uniroyal, Inc.).
69. Brit. Pat. 971,214 (Sept. 30, 1964) (to Monsanto Co.).
70. U.S. Pat. 4,193,903 (Mar. 18, 1980), B. E. Giddings, E. Wardlow, Jr., and B. L. Mehosky (to The Standard Oil Co. (Ohio)).
71. U.S. Pat. 3,738,972 (June 12, 1973), K. Moriyama and T. Osaka (to Daicel Ltd.).
72. Brit. Pat. 1,328,625 (Aug. 30, 1973), K. Moriyama and S. Takahashi (to Daicel Ltd.).
73. Jpn. Pat. 80 000,725 (Jan. 7, 1980), S. Kato and M. Astumi (to Denki Kagaku Kogyo K.K.).
74. Jpn. Pat. 78 082,892 (July 21, 1978), S. Kato and M. Momoka (to Denki Kagaku Kogyo K.K.).

75. Jpn. Pat. 79 020,232 (July 20, 1979), S. Kato and M. Momoka (to Denki Kagaku Kogyo K.K.).
76. Jpn. Pat. 79 119,588 (Sept. 17, 1979), K. Kido, H. Wakamori, G. Asai, and K. Kushida (to Kureha Chemical Industry Co., Ltd.).
77. Jpn. Pat. 82 167,303 (Oct. 15, 1982) (to Toshiba Machine Co.).
78. U.S. Pat. 3,198,775 (Aug. 3, 1965), R. E. Delacretaz, S. P. Nemphos, and R. L. Walter (to Monsanto Co.).
79. U.S. Pat. 3,258,453 (June 28, 1966), H. K. Chi (to Monsanto Co.).
80. U.S. Pat. 3,287,331 (Nov. 22, 1966), Y. C. Lee and L. P. Paradis (to Monsanto Co.).
81. U.S. Pat. 3,681,310 (Aug. 1, 1972), K. Moriyama and T. Moriwaki (to Daicel Ltd.).
82. U.S. Pat. 3,243,407 (Mar. 29, 1966), Y. C. Lee (to Monsanto Co.).
83. U.S. Pat. 3,331,810 (July 18, 1967), Y. C. Lee (to Monsanto Co.).
84. U.S. Pat. 3,331,812 (July 18, 1967), Y. C. Lee and S. P. Nemphos (to Monsanto Co.).
85. U.S. Pat. 3,356,644 (Dec. 5, 1967), Y. C. Lee (to Monsanto Co.).
86. U.S. Pat. 3,491,071 (Jan. 20, 1970), R. Lanzo (to Montecatini Edison SpA).
87. Jpn. Pat. 83 103,506 (June 20, 1983) (to Mitsubishi Rayon, Ltd.).
88. Jpn. Pat. 88 039,908 (Feb. 20, 1988) (to Mitsui Toatsu Chemicals, Inc.).
89. Belg. Pat. 904,985 (Oct. 16, 1986), S. Ikuma (to Mitsubishi Monsanto Chemical Co.).
90. U.S. Pat. 3,141,868 (July 21, 1964), E. P. Fivel (to Resines et Verms Artificiels).
91. U.S. Pat. 3,201,365 (Aug. 17, 1965), R. K. Charlesworth, W. Creck, S. A. Murdock, and K. G. Shaw (to The Dow Chemical Company).
92. U.S. Pat. 2,941,985 (June 21, 1960), J. L. Amos and C. T. Miller (to The Dow Chemical Company).
93. U.S. Pat. 3,031,273 (Apr. 24, 1962), G. A. Latinen (to Monsanto Co.).
94. U.S. Pat. 2,745,824 (May 15, 1956), J. A. Melchore (to American Cyanamid Co.).
95. Jpn. Pat. 73 021,783 (Mar. 19, 1973), H. Sato, I. Nagai, T. Okamoto, and M. Inoue (to Toray, K. K. Ltd.).
96. R. L. Zimmerman, J. S. Best, P. N. Hall, and A. W. Hanson in R. F. Gould, ed., *Polymerization and Polycondensation Processes* (Adv. Chem. Ser. 34), American Chemical Society, Washington, D.C., 1962, p. 225.
97. Jpn. Pat. 87 179,508 (Aug. 6, 1987), N. Ito and co-workers (to Mitsui Toatsu Chemicals, Inc.).
98. Jpn. Pat. 83 037,005 (Mar. 4, 1983) (to Mitsui Toatsu Chemicals, Inc.).
99. U.S. Pat. 3,067,812 (Dec. 11, 1962), G. A. Latinen and R. H. M. Simon (to Monsanto Co.).
100. U.S. Pat. 3,211,209 (Oct. 12, 1965), G. A. Latinen and R. H. M. Simon (to Monsanto Co.).
101. F. M. Peng, "Acrylonitrile Polymers" in J. I. Kroschwitz, ed., *Encyclopedia of Polymer Science and Engineering,* 2nd ed., Vol. 1, Wiley-Interscience Inc., New York, 1985, p. 463.
102. F. M. Peng, *J. Macromol. Sci. Chem.* **A22,** 1241 (1985).
103. S. Woods, *Canadian Packaging* **39**(9), 48 (1986).
104. P. R. Lund and T. J. Bond, *Soc. Plast. Eng. Tech. Pap.* **24,** 61 (1978).
105. R. C. Adams and S. J. Waisala, *TAPPI Pap. Synth. Conf. Prepr.,* Atlanta, Ga., **79** (1975).
106. Jpn. Pat. 85 097,823 (May 31, 1985) (to Mitsui Toatsu Chemicals, Inc.).
107. U.S. Pat. 4,452,835 (June 5, 1984), G. Vasudevan (to Union Carbide Corp.).
108. Ger. Pat. 2,656,993 (June 22, 1978), R. E. Isley (to The Standard Oil Co. (Ohio)).
109. J. A. Carlson, Jr. and L. Borla, *Mod. Plast.* **57**(6), 117 (1980).
110. U.S. Pat. 3,926,871 (Dec. 16, 1975), L. W. Hensley and G. S. Li (to The Standard Oil Co. (Ohio)).

111. U.S. Pat. 4,195,135 (Mar. 25, 1980), G. S. Li and J. F. Jones (to The Standard Oil Co. (Ohio)).
112. U.S. Pat. 3,997,709 (Dec. 14, 1976), W. Y. Aziz, L. E. Ball, and G. S. Li (to The Standard Oil Co. (Ohio)).
113. Fr. Pat. 2,207,938 (June 21, 1974) (to Polysar Ltd.).
114. Can. Pat. 991,787 (June 22, 1976), M. H. Richmond and H. G. Wright (to Polysar Ltd.).
115. U.S. Pat. 4,301,112 (Nov. 17, 1981), M. M. Zwick (to American Cyanamid Co.).
116. U.S. Pat. 4,025,581 (May 24, 1977), J. A. Powell and A. Williams (to Rohm and Haas Co.).
117. Jpn. Pat. 79 058,794 (May 11, 1979), H. Furukawa and S. Matsumura (to Kanegafuchi Chemical Industry Co., Ltd.).
118. Fr. Pat. 2,389,644 (Dec. 1, 1978), P. Hubin-Eschger (to ATO-Chimie S.A.).
119. Jpn. Pat. 75 124,970 (Oct. 1, 1975), A. Kobayashi and M. Ohya (to Kureha Chemical Industry Co., Ltd.).
120. Jpn. Pat. 87 013,425 (Jan. 22, 1987), M. Fujimoto, T. Yamashita, and T. Matsumoto (to Kanebo NSC K.K.).
121. U.S. Pat. 4,000,359 (Dec. 28, 1976), W. A. Watts and J. L. Wang (to Goodyear Tire and Rubber Co.).
122. U.S. Pat. 3,832,335 (Sept. 27, 1974), J. W. Bayer (to Owens-Illinois, Inc.).
123. Jpn. Pat. 74 021,105 (May 29, 1974), M. Takahashi, K. Yanagisawa, and T. Mori (to Sekisui Chemical Co., Ltd.).
124. U.S. Pat. 3,959,550 (May 25, 1976), M. S. Guillod and R. G. Bauer (to Goodyear Tire and Rubber Co.).
125. Fr. Pat. 2,041,137 (Mar. 5, 1971), Q. A. Trementozzi (to Monsanto Co.).
126. Ger. Pat. 2,134,814 (May 31, 1972), T. Yamawaki, M. Hayashi, and K. Endo (to Mitsubishi Chemical Industries Co., Ltd.).
127. U.S. Pat. 3,725,120 (Apr. 3, 1973), C. A. Suter (to Goodyear Tire and Rubber Co.).
128. Ger. Pat. 1,546,809 (Jan. 31, 1974), D. S. Dixler (to Air Products and Chemicals, Inc.).
129. Jpn. Pat. 87 256,871 (Nov. 9, 1987), H. Sakai and T. Kotani (to Asahi Chemical Industry Co., Ltd.).
130. Ger. Pat. 2,436,137 (Feb. 13, 1975), B. N. Hendy (to Imperial Chemical Industries Ltd.).
131. S. Yabumoto, K. Ishii, and K. Arita, *J. Polym. Sci.* **A1,** 1577 (1969).
132. N. G. Gaylord, S. S. Dixit, and B. K. Patnaik, *J. Polym. Sci. B,* **9,** 927 (1971).
133. N. G. Gaylord, S. S. Dixit, S. Maiti, and B. K. Patnaik, *J. Macromol. Sci. Chem.* **6,** 1495 (1972).
134. K. Arita, T. Ohtomo, and Y. Tsurumi, *J. Polym. Sci., Polym. Lett. Ed.* **19,** 211 (1981).
135. C. D. Eisenbach and co-workers. *Angew. Makromol. Chem.* **145/146,** 125 (1986).
136. H. Hirai, *J. Polym. Sci. Macromol. Rev.* **11,** 47 (1976).
137. A. Henglein, *Makromol. Chem.* **14,** 128 (1954).
138. A. Chapiro, *J. Polym. Sci.* **23,** 377 (1957).
139. A. D. Jenkins, *Pure Appl. Chem.* **46,** 45 (1976).
140. H. Craubner, *J. Polym. Sci., Polym. Chem. Ed.* **18,** 2011 (1980).
141. I. G. Krasnoselskaya and B. L. Erusalimskii, *Vysokomol. Soedin., Ser. B* **29,** 442 (1987).
142. A. Noshay and J. E. McGrath, *Block Copolymers: Overview and Critical Survey,* Academic Press, Inc., New York, 1977.
143. J. E. Turner, M. Shen, and C. C. Lin, *J. Appl. Polym. Sci.* **25,** 1287 (1980).
144. E. I. Stout, D. Trimnell, W. M. Doane, and C. R. Russell, *J. Appl. Polym. Sci.* **21,** 2565 (1977).
145. W. P. Lindsay, *TAPPI Annual Meeting Preprints,* Atlanta, Ga., 1977, p. 203.
146. F. Sundardi, *J. Appl. Polym. Sci.* **22,** 3163 (1978).
147. U.S. Pat. 4,134,863 (Jan. 16, 1979), G. F. Fanta, E. I. Stout, and W. M. Doane (to U.S. Dept. of Agriculture).

148. *Chem. Eng. News* **67**(25), 45 (June 19, 1989).
149. United States Food and Drug Administration, *Fed. Regist.* **52**(173), 33802–33803 (Sept. 8, 1987).
150. United States Food and Drug Administration, *Fed. Regist.,* 4510–26, **43**(11), 2586 (Jan. 17, 1978).
151. National Institute for Occupational Safety and Health, "A Recommended Standard for Occupational Exposure to Acrylonitrile," DHEW Publ. No. 78–116, U.S. Government Printing Office Washington, D.C., 1978.
152. N. I. Sax and R. J. Lewis, Sr., *Dangerous Properties of Industrial Materials,* 7th ed., Van Nostrand Reinhold, 1989.
153. R. E. Lenga, *The Sigma-Aldrich Library of Chemical Safety Data,* Sigma-Aldrich Corp., Milwaukee, Wis., 1985.
154. M. Sittig, ed., *Priority Toxic Pollutants,* Noyes Data Corporation, Park Ridge, N.J., 1980.
155. "Threshold Limit Values for Chemical Substances in the Work Environment Adopted by ACGIH for 1985–86," American Conference of Government Industrial Hygienists, Cincinnati, Ohio, 1985.
156. "Criteria for Recommended Standard Occupational Exposure to Styrene", U.S. Department of Health and Human Services (NIOSH), Washington, D.C. Rep. 83–119, pp. 18, 227, 1983; available from NTIS, Springfield, Va.
157. J. Santodonato and co-workers, *Monograph on Human Exposure to Chemicals in the Work Place; Styrene, PB86–155132,* Syracuse, N.Y., July 1985.
158. Jpn. Pat. 76 000,581 (Jan. 6, 1976), M. Nishizawa and co-workers (to Toray Industries, Inc.).
159. Jpn. Pat. 82 185,144 (Nov. 15, 1982) (to Gunze Ltd.).
160. Jpn. Pat. 81 117,652 (Sept. 16, 1981) (to Asahi Chemical Industry Co., Ltd.).
161. U.S. Pat. 4,389,437 (June 21, 1983), G. P. Hungerford (to Mobil Oil Corp.).
162. Jpn. Pat. 88 041,139 (Feb. 22, 1988) (to Mitsui Toatsu Chemicals, Inc.).
163. Jpn. Pat. 83 119,858 (July 16, 1983) (to Toyobo Co., Ltd.).
164. Jpn. Pat. 83 183,465 (Oct. 26, 1983) (to Dainippon Printing Co., Ltd.).
165. Jpn. Pat. 84 012,850 (Jan. 23, 1984) (to Asahi Chemical Industry Co., Ltd.).
166. Jpn. Pat. 85 009,739 (Jan. 18, 1985) (to Toyo Seikan Kaisha, Ltd.).
167. Eur. Pat. Appl. EP 138,194 (Apr. 24, 1985), J. H. Im and W. E. Shrum (to The Dow Chemical Company).
168. U.S. Pat. 4,364,759 (Dec. 21, 1982), A. A. Brooks, J. M. S. Henis, and M. K. Tripodi (to Monsanto Co.).
169. Jpn. Pat. 84 202,237 (Nov. 16, 1984) (to Toyota Central Research and Development Laboratories, Inc.).
170. Jpn. Pat. 85 202,701 (Oct. 14, 1985), T. Kawai and T. Nogi (to Toray Industries, Inc.).
171. H. Kawato, M. Kakimoto, A. Tanioka, and T. Inoue, *Kenkyu Hokoku—Asahi Garasu Kogyo Gijutsu Shoreikai* **49,** 77 (1986).
172. Eur. Pat. Appl. EP 141,584 (May 15, 1985), R. W. Baker (to Bend Research, Inc.).
173. Y. S. Ku and S. O. Kim, *Yakhak Hoechi* **31**(3), 182 (1987).
174. U.S. Pat. 4,330,635 (May 18, 1982), E. F. Tokas (to Monsanto Co.).
175. Ger. Pat. 3,523,612 (Jan. 23, 1986), N. Sakata and I. Hamada (to Asahi Chemical Industry Co., Ltd.).
176. Jpn. Pat. 84 024,752 (Feb. 8, 1984) (to Kanebo Synthetic Fibers, Ltd.).
177. Ger. Pat. 3,512,638 (Feb. 27, 1986), H. J. Kress and co-workers (to Bayer A.G.).
178. Jpn. Pat. 88 089,403 (Apr. 20, 1988) (to Sumitomo Chemical Co., Ltd.).
179. Jpn. Pat. 82 059,075 (Dec. 13, 1982) (to Nichimen Co., Ltd.).
180. S. Milkov and T. Abadzhieva, *Fiz. Khim. Mekh.* **11,** 81 (1983).
181. A. M. Gadalla and M. E. El-Derini, *Polym. Eng. Sci.* **24,** 1240 (1984).

182. Ger. Pat. 2,652,427 (May 26, 1977), G. E. Cartier and J. A. Snelgrove (to Monsanto Co.).
183. Jpn. Pat. 82 137,359 (Aug. 24, 1982) (to Nippon Electric Co., Ltd.).
184. Jpn. Pat. 83 001,143 (Jan. 6, 1983) (to Fujitsu Ltd.).
185. Jpn. Pat. 88 040,169 (Feb. 20, 1988), Y. Takahashi, M. Nakamura, Y. Kitahata, and K. Maeda (to Sharp Corp.).
186. Jpn. Pat. 86 092,453 (May 10, 1986), Y. Ichihara and Y. Uratani (to Mitsubishi Petrochemical Co., Ltd.).
187. Braz. Pat. 87 818 (Dec. 22, 1987), T. P. Christini (to E. I. du Pont de Nemours & Co., Inc.).
188. Jpn. Pat. 84 018,750 (Jan. 31, 1984) (to Sanyo Chemical Industries, Ltd.).
189. Jpn. Pat. 84 080,447 (May 9, 1984) (to Mitsubishi Rayon Co., Ltd.).
190. P. Albihn and J. Kubat, *Plast. Rubber Process. Appl.* **3**(3), 249 (1983).
191. U.S. Pat. 4,327,012 (Apr. 27, 1982), G. Salee (to Hooker Chemicals and Plastics Corp.).
192. Jpn. Pat. 83 217,536 (Dec. 17, 1983) (to Asahi Chemical Industry Co., Ltd.).
193. Eur. Pat. Appl. EP 96,301 (Dec. 21, 1983), H. Peters and co-workers (to Bayer A.G.).
194. Y. Fujita and co-workers, *Kobunshi Ronbunshu,* **43**(3), 119 (1986).
195. H. Takahashi and co-workers, *J. Appl. Polym. Sci.* **36,** 1821 (1988).
196. Ger. Pat. 2,830,232 (Mar. 29, 1979), W. J. Peascoe (to Uniroyal, Inc.).
197. Jpn. Pat. 79 083,088 (July 2, 1979), S. Ueda, K. Tazaki, H. Kitayama, and I. Kuribayashi (to Asahi-Dow Ltd.).
198. Ger. Pat. 2,826,925 (Jan. 17, 1980), J. Swoboda, G. Lindenschmidt, and C. Bernhard (to BASF A.G.).
199. U.S. Pat. 3,944,631 (Mar. 16, 1976), A. J. Yu and R. E. Gallagher (to Stauffer Chemical Co.).
200. H. Bartl and co-workers, Paper 15, *ACS National Meeting, Division of Industrial Engineering Chemistry,* Atlanta, Ga., Mar. 29–Apr. 3, 1981.
201. Jpn. Pat. 79 083,049 (July 2, 1979), S. Ueda, K. Tazaki, H. Kitayama, and N. Asamizu (to Asahi-Dow Ltd.).
202. U.S. Pat. 4,071,577 (Jan. 31, 1978), J. R. Falender, C. M. Mettler, and J. C. Saam (to Dow Corning Corp.).
203. *Mod. Plast.* **60**(1), 92 (1983).
204. U.S. Pat. 4,224,207 (Sept. 23, 1980), J. C. Falk (to Borg-Warner Corp.).
205. U.S. Pat. 4,169,195 (Sept. 25, 1979), M. K. Rinehart (to Borg-Warner Corp.).
206. Jpn. Pat. 81 016,541 (Feb. 17, 1981) (to Asahi-Dow Ltd.).
207. Ger. Pat. 2,916,668 (Nov. 13, 1980), J. Hambrecht, G. Lindenschmidt, and W. Regel (to BASF A.G.).
208. Jpn. Pat. 80 157,604 (Dec. 8, 1980) (to Daicel Chemical Industries, Ltd.).
209. T. C. Wallace, *SPE Regional Technical Conference,* Las Vegas, Nev., Sept. 16–18, 1975.

General References

Reference 101 is a general reference.
C. H. Bamford and G. E. Eastmond, "Acrylonitrile Polymers" in N. M. Bikales, ed., *Encyclopedia of Polymer Science and Technology,* Vol. 1, Interscience Publishers, a Division of John Wiley & Sons, Inc., New York, 1964, pp. 374–425.

LAWRENCE E. BALL
BENEDICT S. CURATOLO
BP Research

ABS RESINS

Acrylonitrile–butadiene–styrene (ABS) polymers [9003-56-9] are composed of elastomer dispersed as a grafted particulate phase in a thermoplastic matrix of styrene and acrylonitrile copolymer (SAN) [9003-54-7]. The presence of SAN grafted onto the elastomeric component, usually polybutadiene or a butadiene copolymer, compatabilizes the rubber with the SAN component. Property advantages provided by this graft terpolymer include excellent toughness, good dimensional stability, good processability, and chemical resistance. Property balances are controlled and optimized by adjusting elastomer particle size, morphology, microstructure, graft structure, and SAN composition and molecular weight. Therefore, although the polymer is a relatively low cost engineering thermoplastic the system is structurally complex. This complexity is advantageous in that altering these structural and compositional parameters allows considerable versatility in the tailoring of properties to meet specific product requirements. This versatility may be even further enhanced by adding various monomers to raise the heat deflection temperature, impart transparency, confer flame retardancy, and, through alloying with other polymers, obtain special product features. Consequently, research and development in ABS systems is active and continues to offer promise for achieving new product opportunities.

Physical Properties

The range of properties typically available for general purpose ABS is illustrated in Table 1 (1). Numerous grades of ABS are available including new alloys and specialty grades for high heat, plating, flaming-retardant, or static dissipative product requirements (1,2). Reference 1 discusses stress–strain behavior, creep, stress relaxation, and fatigue in ABS materials.

Impact Resistance.　Toughness is a primary consideration in the selection of ABS for many applications. ABS is structured to dissipate the energy of an impact blow through shear and dilational modes of deformation. Upon impact, the particulate rubber phase promotes both the initiation and termination of crazes. Crazes are regions of considerable strength that contain both voids and polymer fibrils oriented in the stress direction. Crazes are terminated by mutual interference or are stopped by other rubber particles, thereby dissipating energy without the formation of a crack, which would lead to catastrophic failure. Shear deformation also contributes to stress relaxation. The behavior of the rubber phase is understood from analyses of the stress distribution surrounding the particulate rubber phase. The rubber component may exist in a state of triaxial tension due to the higher rate of thermal contraction of the rubber compared to styrene–acrylonitrile copolymer upon cooling after molding. Crazes, in general, are initiated at local points of stress concentration. Mechanisms and the factors affecting fracture toughness have been discussed in detail in the literature (3–12).

　　The inherent ductility of the matrix phase depends on the composition of the SAN copolymer and is reported to increase with increasing acrylonitrile content (3). Controlling rubber particle size, distribution, and microstructure are impor-

Table 1. Material Properties of General Purpose and Heat Distortion Resistant ABS[a]

Properties	ASTM Method	High impact	Medium impact	Heat resistant
notched Izod impact at RT, J/m[b]	D256	347–534	134–320	107–347
tensile strength, MPa[c]	D638	33–43	30–52	41–52
tensile modulus, GPa[d]	D638	1.7–2.3	2.1–2.8	2.1–2.6
flexural modulus, GPa[d]	D790	1.7–2.4	2.2–3.0	2.1–2.8
elongation to yield, %	D638	2.8–3.5	2.3–3.5	2.8–3.5
Rockwell hardness	D785	80–105	105–112	100–111
heat deflection[e], °C at 1820 kPa[f]	D648	96–102	93–104	104–116
heat deflection[e], °C at 455 kPa[f]	D648	99–107	102–107	110–118
Vicat softening pt, °C	D1525	91–106	94–107	104–118
coefficient of linear thermal expansion, $\times 10^5$ cm/cm·°C	D696	9.5–11.0	7.0–8.8	6.5–9.2
dielectric strength, kV/mm	D149	16–31	16–31	14–35
dielectric constant, $\times 10^6$ Hz	D150	2.4–3.8	2.4–3.8	2.4–3.8

[a]Ref. 1.
[b]To convert J/m to ft·lb/in. divide by 53.4.
[c]To convert MPa to psi multiply by 145.
[d]To convert GPa to psi multiply by 145,000.
[e]Annealed.
[f]To convert kPa to psi multiply by 0.145.

tant in optimizing impact strength. Good adhesion between the rubber and the matrix phase is also essential and is achieved by an optimized graft structure (3,8,10,13). Typically, toughness is increased by increasing the rubber content and the molecular weight of the ungrafted SAN.

Rheology. Effects of structure of ABS on viscosity functions can be distinguished by considering effects at lower shear rates (<10/s) vs higher shear rates. At higher shear rates melt viscosity is primarily determined by ungrafted SAN structure and the percentage of graft phase. The modulus curves correspond in their shape to that of the ungrafted SAN component, and the rubber particle type and concentration have little effect on the temperature dependence of the viscosity function (14). The extrudate swell, however, becomes smaller with increasing rubber concentration (15).

By contrast, the graft phase structure has a marked effect on viscosity at small deformation rates. The long time relaxation spectra are affected by rubber particle–particle interactions (16,17), which are strongly dependent on particle size, grafting, morphology, and rubber content. Depending on particle surface area, a minimum amount of graft is needed to prevent the formation of three-dimensional networks of associated rubber particles (17). Thus at low shear rates ABS can behave similarly to a cross-linked rubber; the network structure, however, is dissolved by shearing forces. Extensive studies on the viscoelastic properties of ABS in the molten state have been reported (14–21). Effects of lubricants and other nonpolymeric components have also been described (22).

Gloss. Surface gloss values can be achieved ranging from a very low matte finish at <10% (60° Gardner) to very high gloss in excess of 95%. Gloss is dependent on the specific grade and the mold or polishing roll surface.

Electrical Properties. (See Table 1.) A new family of ABS products exhibiting electrostatic dissipative properties without the need for nonpolymeric additives or fillers (carbon black, metal) is now also commercially available (2).

Thermal Properties. ABS is also used as a base polymer in high performance alloys. Most common are ABS–polycarbonate alloys which extend the property balance achievable with ABS to offer even higher impact strength and heat resistance (2).

Color. ABS is sold as an unpigmented powder, unpigmented pellets, precolored pellets matched to exacting requirements, and "salt-and-pepper" blends of ABS and color concentrate. Color concentrates can also be used for on-line coloring during molding.

Chemical Properties

The behavior of ABS may be inferred from consideration of the functional groups present within the polymer.

Chemical Resistance. The term chemical resistance is generally used in an applications context and refers to resistance to the action of solvents in causing swelling or stress cracking as well as to chemical reactivity. In ABS the polar character of the nitrile group reduces interaction of the polymer with hydrocarbon solvents, mineral and vegetable oils, waxes, and related household and commercial materials. Good chemical resistance provided by the presence of acrylonitrile as a comonomer combined with relatively low water absorptivity (<1%) results in high resistance to staining agents (eg, coffee, grape juice, beef blood) typically encountered in household applications (23).

Like most polymers, ABS undergoes stress cracking when brought into contact with certain chemical agents under stress (23,24). Injection molding conditions can significantly affect chemical resistance, and this sensitivity varies with the ABS grade. Certain combinations of melt temperature, fill rate, and packing pressure can significantly reduce stress cracking resistance, and this effect is interactive in complex ways with the imposed stress level that the part is subjected to in service. Both polymer orientation and stress appear to be considerations; thus critical strains can be higher in the flow direction (25). Consequently, all media to be in contact with the ABS part during service should be evaluated under anticipated end-use conditions.

Processing Stability. Processing can influence resultant properties by chemical and physical means (26,27). Degradation of the rubber and matrix phases has been reported under very severe conditions (28). Morphological changes may become evident as agglomeration of dispersed rubber particles during injection molding at higher temperatures (28). Physical effects such as orientation and molded-in stress can have marked effects on mechanical properties. Thus the proper selection and control of process variables are important to maintain optimum performance in molded parts. Antioxidants (qv) added at the compounding step have been shown to help retention of physical properties upon processing (26).

Appearance changes evident under certain processing conditions include color development (26), changes in gloss (28), and splaying. Discoloration may be minimized by reducing stock temperatures during molding or extrusion. Splaying is the formation of surface imperfections elongated in the direction of flow and is typically caused by moisture, occluded air, or gaseous degradation products; proper drying conditions are essential to prevent moisture-induced splay.

Techniques for evaluating processing stability and mechanochemical effects include using a Brabender torque rheometer (29,30), injection molding (26,28), capillary rheometry (26,28), and measuring melt index as a function of residence time (26).

Thermal Oxidative Stability. ABS undergoes autoxidation and the kinetic features of the oxygen consumption reaction are consistent with an autocatalytic free-radical chain mechanism. Comparisons of the rate of oxidation of ABS with that of polybutadiene and styrene–acrylonitrile copolymer indicate that the polybutadiene component is significantly more sensitive to oxidation than the thermoplastic component (31–33). Oxidation of polybutadiene under these conditions results in embrittlement of the rubber because of cross-linking; such embrittlement of the elastomer in ABS results in the loss of impact resistance. Studies have also indicated that oxidation causes detachment of the grafted styrene–acrylonitrile copolymer from the elastomer which contributes to impact deterioration (34).

Examination of oven-aged samples has demonstrated that substantial degradation is limited to the outer surface (34), ie, the oxidation process is diffusion limited. Consistent with this conclusion is the observation that oxidation rates are dependent on sample thickness (32). Impact property measurements by high speed puncture tests have shown that the critical thickness of the degraded layer at which surface fracture changes from ductile to brittle is about 0.2 mm. Removal of the degraded layer restores ductility (34). Effects of embrittled surface thickness on impact have been studied using ABS coated with styrene–acrylonitrile copolymer (35).

Antioxidants have been shown to improve oxidative stability substantially (36,37). The use of rubber-bound stabilizers to permit concentration of the additive in the rubber phase has been reported (38–40). The partitioning behavior of various conventional stabilizers between the rubber and thermoplastic phases in model ABS systems has been described and shown to correlate with solubility parameter values (41). Pigments can adversely affect oxidative stability (32). Test methods for assessing thermal oxidative stability include oxygen absorption (31,32,42), thermal analysis (43,44), oven aging (34,45,46), and chemiluminescence (47,48).

Photooxidative Stability. Unsaturation present as a structural feature in the polybutadiene component of ABS (also in high impact polystyrene, rubber-modified PVC, and butadiene-containing elastomers) also increases liability with regard to photooxidative degradation (49–51). Such degradation only occurs in the outermost layer (52,53), and impact loss upon irradiation can be attributed to embrittlement of the rubber and possibly to scission of the grafted styrene–acrylonitrile copolymer (49,54). Oxidative degradation induced by prior processing may affect photosensitivity (49,55). Appearance changes such as yellowing are also induced by irradiation and caused by chromophore formation in both the polybutadiene and styrene–acrylonitrile copolymer components (49,56). Comparative data on ABS with other acrylic-based plastics have been reported (57).

Applications involving extended outdoor exposure, especially in direct sunlight, require protective measures such as the use of stabilizing additives, pigments, and protective coatings and film. Light stabilizers provide some measure of protection (58,59) as illustrated by the very successful use of ABS in interior automotive trim. Effects of polymer-bound stabilizers have been described (60). Pigments can significantly enhance stability. Paints are also highly effective in minimizing weather degradation (61). A particularly effective technique for sheet products is the lamination during extrusion of an acrylic film to ABS. Test methods for assessing light stability include outdoor exposure (62) and accelerated testing (62,63). Reactivity toward singlet oxygen has been reported (64). The current trend in accelerated light-aging for ABS in automotive applications is the use of xenon arc testing.

Flammability. The general purpose grades are usually recognized as 94 HB according to the requirements of Underwriters' Laboratories UL94 and also meet the requirements, dependent on thickness, of the Motor Vehicle Safety Standard 302. Flame-retardant (FR) grades (V0, V1, and V2) are also available which meet Underwriters' UL 94/94 5V and Canadian Standards Association (CSA) requirements beginning at a minimum thickness of 1.57 mm. Flame retardancy is achieved by utilizing halogen in combination with antimony oxide or by alloys with PVC or PC. A new FR grade utilizing polymer-bound bromine has been developed to avoid additive bloom and toxicity (65).

Polymerization

In all manufacturing processes, grafting is achieved by the free-radical copolymerization of styrene and acrylonitrile monomers in the presence of an elastomer. Ungrafted styrene–acrylonitrile copolymer is formed during graft polymerization and/or added afterwards.

Mechanism. A grafting mechanism involving the direct reaction of initiator radicals with the elastomer is supported by reports that percentage grafting is dependent on the nature of the initiator (66–69). Initiators which generate radicals effective in abstraction from the rubber backbone (eg, oxy vs carbon radicals) are generally more efficient in promoting graft formation. However, kinetic data and the fact that grafting does occur in reactions initiated by carbon radicals (eg, from 2,2′-azobisisobutyronitrile) suggest that grafting by copolymerization may also occur (66) and proceed concurrently. The degree of grafting is a function of factors including the 1,2-vinyl content of the polybutadiene, monomer concentration, extent of conversion, initiator, temperature, and mercaptan concentration (66–71).

Monomer compositional drifts may also occur due to preferential solution of the styrene in the rubber phase or solution of the acrylonitrile in the aqueous phase (72). In emulsion systems, rubber particle size may also influence graft structure so that the number of graft chains per unit of rubber particle surface area tends to remain constant (73). Factors affecting the distribution (eg, core-shell vs "wart-like" morphologies) of the grafted copolymer on the rubber particle surface have been studied in emulsion systems (74). Effects due to preferential solvation of the initiator by the polybutadiene have been described (75,76).

In addition to graft copolymer attached to the rubber particle surface, the formation of styrene–acrylonitrile copolymer occluded within the rubber particle may occur. The mechanism and extent of occluded polymer formation depends on the manufacturing process. The factors affecting occlusion formation in bulk (77) and emulsion processes (78) have been described. The use of block copolymers of styrene and butadiene in bulk systems can control particle size and give rise to unusual particle morphologies (eg, coil, rod, capsule, cellular) (77).

Manufacturing

There are three commercial processes for manufacturing ABS: emulsion, mass, and mass-suspension.

Emulsion Process.   The emulsion (79,80) ABS process involves two steps, production of a rubber latex and subsequent polymerization of styrene and acrylonitrile in the presence of the rubber latex to produce an ABS latex. This latex is then processed to isolate the ABS resin (81,82).

The rubber latex is usually produced in batch reactors. The rubber can be polybutadiene [9003-17-2] or a copolymer of 1,3-butadiene [106-99-0] and either acrylonitrile [107-13-1] or styrene [100-42-5]. The latex normally has a polymer content of approximately 30 to 50%; most of the remainder is water. In addition to the monomers, the polymerization ingredients include an emulsifier, a polymerization initiator, and usually a chain-transfer agent for molecular weight control.

After the rubber latex is produced, it is subjected to further polymerization in the presence of styrene (C_8H_8) and acrylonitrile (C_3H_3N) monomers to produce the ABS latex. This can be done in batch, semibatch, or continuous reactors. The other ingredients required for this polymerization are similar to those required for the rubber latex reaction.

The ABS polymer is recovered through coagulation of the ABS latex. Coagulation is usually achieved by the addition of an agent to the latex which destabilizes the emulsion. The resulting slurry can then be filtered or centrifuged to recover the ABS resin. The wet resin is dried to a low moisture content. A variety of dryers can be used for ABS, including tray, fluid bed, and rotary kiln type dryers.

The emulsion process for making ABS has been commercially practiced since the early 1950s. Its advantage is the capability of producing ABS with a wide range of compositions, particularly higher rubber contents than are possible with the other processes. Typically, the rubber content percentage can be varied from approximately 10 to over 90. Mixing and transfer of the heat of reaction in an emulsion polymerization is achieved more easily than in the mass polymerization process because of the low viscosity and good thermal properties of the water phase. The energy requirements for the emulsion process are generally higher than for the other processes because of the energy usage in the polymer recovery area. The emulsion process has a greater wastewater treatment demand than the other processes because of the quantity of water used.

Mass Process.   In the mass (or bulk) (83) ABS process the polymerization is conducted in a monomer medium rather than in water. This process usually consists of a series of two or more continuous reactors. The rubber used in this process is most commonly a solution-polymerized linear polybutadiene (or copolymer containing sytrene), although some mass processes utilize emulsion-polymerized ABS with a high rubber content for the rubbery component (84). If a linear rubber is used, a solution of the rubber in the monomers is prepared for feeding to the reactor system. If emulsion ABS is used as the source of rubber, a dispersion of the ABS in the monomers is usually prepared after the water has been removed from the ABS latex.

If a linear rubber is used as a feedstock for the mass process (85), the rubber becomes insoluble in the mixture of monomers and SAN polymer which is formed in the reactors, and discrete rubber particles are formed. This is referred to as phase inversion since the continuous phase shifts from rubber to SAN. Grafting of some of the SAN onto the rubber particles occurs as in the emulsion process. Typically, the mass-produced rubber particles are larger (0.5 to 5 μm) than those of emulsion-based ABS (0.1 to 1 μm) and contain much larger internal occlusions of SAN polymer. The reaction recipe can include polymerization initiators, chain-transfer agents, and other additives. Diluents are sometimes used to reduce the viscosity of the monomer and polymer mixture to facilitate processing at high conversion. The product from the reactor system is devolatilized to remove the unreacted monomers and is then pelletized. Equipment used for devolatilization includes single- and twin-screw extruders, and flash and thin film evaporators. Unreacted monomers are recovered for recycle to the reactors to improve the process yield.

The mass ABS process was originally adapted from the mass polystyrene process (86). Mass-produced ABS typically has very good unpigmented color and is usually somewhat translucent, which may reduce the concentration of colorants required. The mass-produced grafted rubber typically is more efficient at impact modification than emulsion-grafted rubber; however, the extent of rubber incorporation is limited to approximately 15% because of viscosity limitations in the process. The surface gloss of the mass-produced ABS is generally lower than that of emulsion ABS due to the presence of the larger rubber particles.

Mass-Suspension Process. The mass-suspension process (87) utilizes a mass reaction to produce a partially converted mixture of polymer and monomer and then employs a suspension reaction technique (88) to complete the polymerization. This is a batch process. The mass reaction is the same as that described above for mass polymerization using a linear rubber, and the rubber particles formed during phase inversion are also similar to those formed in the mass process. When the conversion of the monomers is approximately 15 to 30% complete, the mixture of polymer and unreacted monomers is suspended in water with the introduction of a suspending agent. The reaction is continued until a high degree of monomer conversion is attained. Unreacted monomers are stripped from the product before the slurry is centrifuged and dried. The mass-suspension product is in the form of small beads (typically 100 to 500 μm in diameter). The morphology and properties of the mass-suspension product are similar to those of the mass-polymerized product. The mass-suspension process retains some of the process advantages of the water-based emulsion process, such as lower viscosity in the reactor and good heat removal capability.

Compounding. ABS either is sold as an unpigmented product, in which case the customer may add pigments during the forming process, or it is colored by the manufacturer prior to sale. Much of the ABS produced by the mass process is sold unpigmented. If colorants, lubricants, stabilizers, or alloying resins are added to the product, a compounding operation is required. ABS is compounded on a range of equipment, including batch and continuous melt mixers, and both single- and twin-screw extruders. In the compounding step, more than one type of ABS may be employed (ie, emulsion and mass-produced) to obtain an optimum balance of properties for a specific application. Products can also be made in the compounding process by combining emulsion ABS having a high rubber content with mass or suspension polymerized SAN.

Analysis

Analytical investigations may be undertaken to identify the presence of an ABS polymer, characterize the polymer, or identify nonpolymeric ingredients. Fourier transform infrared (ftir) spectroscopy is the method of choice to identify the presence of an ABS polymer and determine the acrylonitrile–butadiene–styrene ratio of the composite polymer (89,90). Confirmation of the presence of rubber domains is achieved by electron microscopy. Comparison with available physical property data serves to increase confidence in the identification or indicate the presence of unexpected structural features. Identification of ABS via pyrolysis gas chromatography (91) and dsc (92) has also been reported.

Detailed compositional and molecular weight analyses involve: determining the percentage of grafted rubber; determining the molecular weight and distribution of the grafted SAN and the ungrafted SAN; and determining compositional data on the grafted rubber, the grafted SAN, and the ungrafted SAN. This information is provided by a combination of phase-separation and instrumental techniques. Separation of the ungrafted SAN from the graft rubber is accomplished by ultracentrifugation of ABS dispersions (93,94) which causes sedimentation of the grafted rubber. Cleavage of the grafted SAN from the elastomer is achieved using oxidizing agents such as ozone [10028-15-6] (94,95), potassium permanganate [7722-64-7] (96), or osmium tetroxide [20816-12-0] with tert-butylhydroperoxide [75-91-2] (97). Chromatographic and spectroscopic analyses of the isolated fractions provide structural data on the grafted and ungrafted SAN components (98). Information on the microstructure of the rubber is provided by analysis of the cleavage products derived from the substrate (94,96). The extraction of ungrafted rubber has also been reported (99).

Additional information on elastomer and SAN microstructure is provided by ^{13}C-nmr analysis (100). Rubber particle composition may be inferred from glass-transition data provided by thermal or mechanochemical analysis. Rubber particle morphology as obtained by transmission or scanning electron microscopy (101) is indicative of the ABS manufacturing process (77). (See Figs. 1 and 2.)

The isolation and/or identification of nonpolymerics has been described, including analyses for residual monomers (90,102,103) and additives (90,104–106). The determination of localized concentrations of additives within the phases of ABS has been reported; the partitioning of various additives between the elastomeric and thermoplastic phases of ABS has been shown to correlate with solubility parameter values (41).

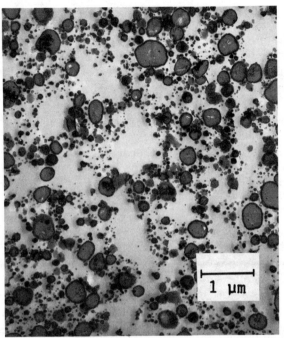

Fig. 1. Transmission electron micrograph of ABS produced by an emulsion process. Staining of the rubber bonds with osmium tetroxide provides contrast with the surrounding SAN matrix phase.

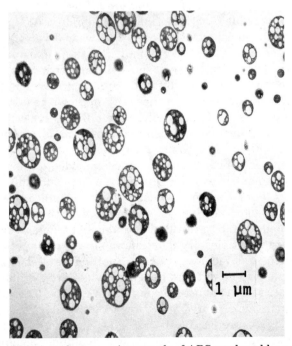

Fig. 2. Transmission electron micrograph of ABS produced by a mass process. The rubber domains are typically larger in size and contain higher concentrations of occluded SAN than those produced by emulsion technology.

Processing

Good thermal stability plus shear thinning allow wide flexibility in viscosity control for a variety of processing methods. ABS exhibits non-Newtonian viscosity behavior. For example, raising the shear rate one decade from 100/s to 1000/s (typical in-mold shear rates) reduces the viscosity by 75% on a general purpose injection molding grade. Viscosity can also be reduced by raising melt temperature; typically increasing the melt temperature 20 to 30°C within the allowable processing range reduces the melt viscosity by about 30%. ABS can be processed by all the techniques used for other thermoplastics: compression and injection molding, extrusion, calendering, and blow-molding (see PLASTICS PROCESSING). Clean, undegraded regrind can be reprocessed in most applications (plating excepted), usually at 20% with virgin ABS. Postprocessing operations include cold forming; thermoforming; metal plating; painting; hot stamping; ultrasonic, spin, and vibrational welding; and adhesive bonding.

Material Handling and Drying.   Although uncompounded powders are available from some suppliers, most ABS is sold in compounded pellet form. The pellets are either precolored or natural to be used for in-house coloring using dry or liquid colorants or color concentrates. These pellets have a variety of shapes including diced cubes, square and cylindrical strands, and spheroids. The shape and size affect several aspects of material handling such as bulk density, feeding of screws, and drying (qv). Very small particles called fines can be present as a carryover from the pelletizing step or transferring operations; these tend to congregate at points of static charge build up. Certain additives can be used to control static charges on pellets (107).

ABS is mildly hygroscopic. The moisture diffuses into the pellet and moisture content is a reversible function of relative humidity. At 50% relative humidity typical equilibrium moisture levels can be between 0.3 and 0.6% depending on the particular grade of ABS. In very humid situations moisture content can be double this value. Although there is no evidence that this moisture causes degradation during processing, drying is required to prevent voids and splay (108) and achieve optimum surface appearance. Drying down to 0.1% is usually sufficient for general purpose injection molding and 0.05% for critical applications such as plating. For nonvented extrusion and blow-molding operations a maximum of 0.02% is required for optimum surface appearance.

Desiccant hot air hopper dryers are recommended, preferably mounted on the processing equipment. Tray driers are not recommended, but if used the pellet bed should be no more than 5 cm deep. Many variables affect drying rates (109,110); the pellet temperature has a stronger effect than the dew point. Most pellet drying problems can be a result of actual pellet temperatures being too low in the hopper. Large particles dry much more slowly than pellets, thus regrind should be protected from moisture regain. Supplier data sheets should be consulted for specific drying conditions. Several devices are available commercially for analytically determining moisture contents in ABS pellets (111–113). Alternatives to pellet drying are vented injection molding (114) and cavity-air pressurization (counterpressure) (115).

Injection Molding.   *Equipment.*   Although plunger machines can be used, the better choice is the reciprocating screw injection machine because of better melt homogeneity. Screws with length-to-diameter ratios of 20:1 and a compression ratio of 2–3:1 are recommended. General purpose screws vary significantly in number and depth of the metering flights; long and shallow metering zones can create melt temperature override which is particularly undesirable with flame-retardent (FR) grades of ABS. Screws with a generous transition length perform best because of better melting rate control (116). Good results have been realized with a long transition "zero-meter" screw design (117). Some comments on the performance of general purpose and two-stage vented screws used for coloring with concentrates is given in reference 118. Guidelines for nozzle and nonreturn valve selection as well as metallurgy are given in references 119 and 120. Gas-nitrided components should be avoided; ion-nitrided parts are acceptable.

A variety of mold types can be used: two plate, three plate, stack, or runnerless. Insulated runner molds are not recommended. If heated torpedoes are used with hot manifold molds, they should be made from a good grade of stainless steel and not from beryllium copper. Molds are typically made from P-20, H-13, S-7, or 420 stainless; chrome or electroless plating is recommended for use with FR grades of ABS. Mold cavities should be well vented (0.05 mm deep) to prevent gas burns. Polished, full round, or trapezoidal runners are recommended; half or quarter round runners are not. Most conventional gating techniques are acceptable (119,120). On polished molds a draft angle of 0.5° is suggested to ease part ejection; side wall texturing requires an additional 1° per 0.025 mm of texture depth. Mold shrinkage is typically in the range of 0.5 to 0.9% (0.005 to 0.009 cm/cm) depending on grade, and the shrinkage value for a given grade can vary much more widely than this because of the design of parts and molding conditions.

Processing Conditions. Certain variables should be monitored, measured, and recorded to aid in reproducibility of the desired balance of properties and appearance. The individual ABS suppliers provide data sheets and brochures specifying the range of conditions that can be used for each product. Relying on machine settings is not adequate. Identical cylinder heater settings on two machines can result in much different melt temperatures. Therefore, melt temperatures should be measured with a fast response hand pyrometer on an air shot recovered under normal screw rpm and back-pressure. Melt temperatures range from 218 to 268°C depending on the grade. Generally, the allowable melt temperature range within a grade is at least 28°C. Excessive melt temperatures cause color shift, poor gloss control, and loss of properties. Similarly, a fill rate setting of 1 cm/s ram travel will not yield the same mold filling time on two machines of different barrel size. Fill time should be measured and adjusted to meet the requirements of getting a full part, and to take advantage of shear thinning without undue shear heating and gas burns. Injection pressure should be adjusted to get a full part free of sinks and good definition of gloss or texture. Hydraulic pressures of less than 13 MPa (1900 psi) usually suffice for most molding. Excessive pressure causes flash and can result in loss of some properties. Mold temperatures for ABS range from 27 to 66°C (60 to 82°C for high heat grades). The final properties of a molded part can be influenced as much by the molding as by the grade of ABS selected for the application (121). The factors in approximate descending order of importance are polymer orientation, heat history, free volume, and molded-in stress. Izod impact strength can vary severalfold as a function of melt temperature and fill rate because of orientation effects, and the response curve is ABS grade dependent (122). The effect on tensile strength is qualitatively the same, but the magnitude is in the range of 5 to 10%. Modulus effects are minimal. Orientation distribution in the part is very sensitive to the flow rate in the mold; therefore, fill rate and velocity-to-pressure transfer point are important variables to control (123). Dart impact is also sensitive to molding variables, and orientation and thermal history can also be key factors (124). Heat deflection temperature can be influenced by packing pressure (125) because of free volume considerations (126). The orientation on the very surface of the part results from an extensionally stretching melt front and can have deleterious effects on electroplate adhesion and paintability. A phenomenon called the mold-surface-effect, which involves grooving the nonappearance half of the mold, can be employed to reduce unwanted surface orientation on the noncorresponding part surface (127–129). Other information regarding the influence of processing conditions on part quality are given in references 130–134.

Part Design.   For optimum economics and production cycle time, wall thicknesses for ABS parts should be the minimum necessary to satisfy service strength requirements. The typical design range is 0.08 to 0.32 cm, although parts outside this range have been successfully molded. A key principle that guides design is avoiding stress concentrators such as notches and sharp edges. Changes in wall thickness should be gradual, sharp corners should be avoided, and generous radii (25% of the wall thickness) used at wall intersections with ribs and bosses. To avoid sinks, rib thickness should be between 50 and 75% of the nominal wall. Part-strength at weld lines can be diminished; thus welds should be avoided if possible or at least placed in noncritical areas of the part (135). Because of polymer orientation, properties such as impact strength vary from point to point on the same part and with respect to the flow direction (121). Locations of highest Izod impact strength can be points of lowest dart impact strength because of the degree and direction of orientation. ABS suppliers can provide assistance with design of parts upon inquiry and through design manuals (136). There are a number of special considerations when designing parts for metal plating to optimize the plating process, plate deposition uniformity, and final part quality (137). ABS parts can be also designed for solid–solid or solid–foam co-injection molding (138) and for gas-assisted-injection molding (139).

Extrusion.   *Equipment.*   Since moisture removal is even more critical with extrusion than injection molding, desiccant hot-air hopper drying of the pellets to 0.02% moisture is essential for optimum properties and appearance. The extruder requirements are essentially the same for pipe, profile, or sheet. Two-stage vented extruders are preferred since the improved melting control and volatile removal can provide higher rates and better surface appearance. Barrels are typically 24:1 minimum L/D for single-stage units and 24 or 36:1 for two-stage vented units. The screws are typically 2:1 to 2.5:1 compression ratio and single lead, full flighted with a 17.7° helix angle. Screen packs (20–40 mesh = 840–420 μm) are recommended.

For sheet, streamlined coat-hanger type dies are preferred over the straight manifold type. Typically, three highly polished and temperature controlled rolls are used to provide a smooth sheet surface and control thickness (140). Special embossing rolls can be substituted as the middle roll to impart a pattern to the upper surface of the sheet. ABS and non-ABS films can be fed into the polishing rolls to provide laminates for special applications, eg, for improved weatherability, chemical resistance, or as decoration. Two rubber pull rolls, speed synchronzied with the polishing rolls, are located far enough downstream to allow sufficient cooling of the sheet; finally, the sheet goes into a shear for cutting into lengths for shipping.

Pipe can be sized using internal mandrels with air pressure contained by a downstream plug or externally using a vacuum bushing and tank. Cooling can be done by immersion, cascade, or mist. Water temperatures of 41 to 49°C at the sizing zone reduce stresses. Foamcore pipe has increased in market acceptance significantly over the last few years, and cooling unit lengths must be longer than for solid pipe. Drawdown should not exceed 10 to 15%.

Profile dies can be flat plates or the streamline type. Flat plate dies are easy to build and inexpensive but can have dead spots that cause hang-up, polymer degradation, and shutdowns for cleaning. Streamlined, chrome-plated dies are more expensive and complicated to build but provide for higher rates and long runs. The land length choice represents a tradeoff; long lands give better quality profile and shape retention but have high pressure drops that affect throughput. Land length to wall thickness ratios are typically 10:1. Drawdown can be used to compensate for die swell but should not exceed 25% to minimize orientation. Sizing jigs vary in complexity depending on profile design; water mist, fog, or air cooling can be used. The latter gives more precise sizing. Also, water immersion vacuum sizing can be used. Accurate, infinitely adjustable speed control is important to the takeoff end equipment to guarantee dimensional control of the profile.

With sheet or pipe, multilayer coextrusion can be used. Solid outer–solid core coextrusion can place an ABS grade on the outside that has special attributes such as color, dullness, chemical resistance, static dissipation, or fire-retardancy over a core ABS that is less expensive or even regrind. Composites can be created in which the core optimizes desired physical properties such as modulus, whereas the outer layer optimizes surface considerations not inherent in the core material. Solid outer–foam core can provide composites with significant reductions in specific gravity (0.7). Dry blowing agents can be "dusted" onto the pellets or liquid agents injected into the first transition section of the extruder.

Extrusion processing conditions vary depending on the ABS grade and application; vendor bulletins should be consulted for details. Information for assistance in troubleshooting extrusion problems can be found in reference 141.

Calendering. The rheological characteristics of the sheet extrusion grades of ABS easily adapt them to calendering to produce film from 0.12 to 0.8 mm thick for vacuum forming or as laminates for sheet. The advantages of this process over extrusion are the capability for thinner gauge product and quick turnaround for short runs.

Blow Molding. Although ABS has been blow molded for over 20 years, this processing method has been gaining popularity recently for a variety of applications (142). Better blow-molding grades of ABS are being provided by tailoring the composition and rheological characteristics specifically to the process. Whereas existing polyolefin equipment can often be easily modified and adjusted to mold ABS, there are some key requirements that require attention.

Pellet predrying is required down to 0.02 to 0.03% moisture. High shear polyolefin screws must be replaced with low shear 2.0:1 to 2.5:1 screws with L/D ratios of 20:1 to 24:1 to keep the melt temperature in the 193 to 221°C optimum range. The land length of the tooling can be reduced to 3:1–5:1 because ABS shows less die swell; this also helps to reduce the melt pressure resulting from the higher viscosity. The accumulator tooling should be streamlined to reduce hang-up and improve re-knit, and be capable of handling the higher pressures required with large programmed parisons. Mold temperatures of 77 to 88°C provide good surface finish. It is recommended that the material vendor be consulted to confirm equipment capability and provide safety and processing information (143).

Secondary Operations. *Thermoforming.* ABS is a versatile thermo-forming material. Forming techniques in use are positive and negative mold vacuum forming, bubble and plug assist, snapback and single- or twin-sheet pressure forming (144). It is easy to thermoform ABS over the wide temperature range of 120 to 190°C. As-extruded sheet should be wrapped to prevent scuffing and moisture pickup. Predrying sheet that has been exposed to humid air prevents surface defects; usually 1 to 3 h at 70–80°C suffices. Thick sheet should be heated slowly to prevent surface degradation and provide time for the core temperature to reach the value needed for good formability. Relatively inexpensive tooling can be made from wood, plaster, epoxies, thermoset materials, or metals. Tools should have a draft angle of 2° to 3° on male molds and ½° to 1° on female molds. More draft may be needed on textured molds. Vacuum hole diameters should not exceed 50% of the sheet thickness. Mold design should allow for 0.003 to 0.008 cm/cm mold shrinkage; exact values depend on mold configuration, the material grade, and forming conditions. Maximum depth of draw is usually limited to part width in simple forming, but more sophisticated forming techniques or relaxed wall uniformity requirements can allow greater draw ratios. Some definitions for draw ratios are given in reference 145. Pressure forming, with well-designed tools, can make parts approaching the appearance and detailing obtained by injection molding. Additional information on pressure forming is given in reference 146.

Cold Forming. Some ABS grades have ductility and toughness such that sheet can be cold formed from blanks 0.13–6.4 mm thick using standard metal-working techniques. Up to 45% diameter reduction is possible on the first draw; subsequent redraws can yield 35%. Either aqueous or nonaqueous lubrication is required. More details are available in reference 147.

Other Operations. *Metallizing.* ABS can be metallized by electroplating, vacuum deposition, and sputtering. Electroplating (qv) produces the most robust coating; progress is being made on some of the environmental concerns associated with the chemicals involved by the development of a modified chemistry. An advantage to sputtering is that any metal can be used, but wear resistance is not as good as with electroplating. Attention must be paid to the molding and handling of the ABS parts since contamination can affect plate adhesion, and surface defects are magnified after plating. Also, certain aspects of part design become more important with plating; these are covered in references 128 and 137. (See also ELECTROLESS PLATING; METALLIC COATINGS.)

Fastening, Bonding, and Joining. Often parts can be molded with various snap-fit designs (148) and bosses to receive rivets or self-tapping screws. Thermal-welding techniques that are easily adaptable to ABS are spin welding (149), hot plate welding, hot gas welding, induction welding, ultrasonic welding, and vibrational welding (150,151). ABS can also be nailed, stapled, and riveted. There are a variety of adhesives and solvent cements for bonding ABS to itself or other materials such as wood, glass, and metals; for more information, contact the material or adhesives suppliers. Joining ABS with materials of different coefficients of thermal expansion requires special considerations when wide temperature extremes are encountered. An excellent review of joining methods for plastics is given in reference 152.

Applications

Its broad property balance and wide processing window has allowed ABS to become the largest selling engineering thermoplastic. ABS enjoys a unique position as a "bridge" polymer between commodity plastics and other higher performance engineering thermoplastics. Table 2 summarizes estimates for 1988 regional consumption of ABS resins by major use.

Table 2. Markets for ABS Plastics by Region in 1988[a], 10^3 t

	United States and Canada	Western Europe	Japan	Total	%
transportation	139	120	96	355	25
appliances	95	97	117	309	22
business machines	124	60	99	283	20
pipe and fittings	91	23		114	8
other	98	144	102	344	25
total	*547*	*444*	*414*	*1405*	*100*

[a]Ref. 153.

In 1988 the largest market for ABS resins worldwide was transportation. Uses are numerous and include both interior and exterior applications. Interior injection-molded applications account for the greatest volume. General purpose and high heat grades have been developed for automotive instrument panels, consoles, door post covers, and other interior trim parts. ABS resins are considered by many the preferred material for components situated above the "waistline" of the car. Exterior applications include radiator grilles, headlight housings, and extruded/thermoformed fascias for large trucks. ABS plating grades also account for significant ABS sales and include applications such as knobs, light bezels, mirror housings, grilles, and decorative trim. Appliances were the second largest market segment for ABS. The majority of this consumption was for major appliances; extruded/thermoformed door and tank liners lead the way. Transparent ABS grades are also used in refrigerator crisper trays. Other applications in the appliance market include injection-molded housings for kitchen appliances, power tools, vacuum sweepers, sewing machines, and hair dryers.

A large "value-added" market for ABS is business machines and other electrical and electronic equipment. Although general purpose injection-molding grades meet the needs of applications such as telephones and micro floppy disk covers, significant growth exists in more demanding flame-retardant applications such as computer housings and consoles.

Pipe and fittings remain a significant market for ABS, particularly in North America. ABS foam core technology allows ABS resin to compete effectively with PVC in the primary drain-waste and vent (DWV) pipe market.

Other uses of ABS include consumer and industrial applications such as luggage, toys, medical devices, furniture, shower stalls, and bathroom fixtures.

Economic Aspects

Capacity. Estimated ABS capacity worldwide in 1989 is given in Table 3. Accurate ABS capacity figures are difficult to obtain because significant production capability is considered "swing" and can be used to manufacture polystyrene or SAN as well as ABS. The United States has the largest ABS nameplate production capacity of any country at 867×10^3 tons accounting for approximately 25% of the world's capacity. Three producers account for over 50% of the world's capacity; GE Plastics is the largest (Table 4).

Table 3. Worldwide Capacity for ABS Plastic in 1989[a], 10^3 t

Americas	
United States	867
Canada	70
other	157
total	*1,094*
Europe	
Benelux	282
Germany	210
France	60
Eastern Europe	78
other	270
total	*837*
Pacific	
Japan	753
Taiwan	502
Korea	160
other	92
total	*1,507*
Worldwide total	*3,438*

[a]Ref. 154.

Table 4. World Capacity of Leading ABS Producers[a]

Company	1990 Capacity, 10^3 t
GE Plastics	740
Chi Mei Industrial	550
Monsanto	547

[a]Ref. 154.

Price. The price history of ABS in the United States is presented in Table 5 for the period from 1979 to 1987. The almost continual rise in prices during this period not only reflects the overall increase in cost of labor and feedstocks, but the changing mix of ABS resins and blends towards higher value, higher performance applications.

Table 5. U.S. Unit Sales Values for ABS Resins[a]

Year	Value, $/kg	Year	Value, $/kg
1979	1.17	1984	1.85
1980	1.41	1985	1.78
1981	1.63	1986	1.63
1982	1.74	1987	1.74
1983	1.78		

[a]Ref. 155.

Although ABS resins have a long history by industry standards, the products are anything but mature. ABS resins and blends are, and are expected to remain, the engineering thermoplastics of choice for a wide array of markets.

BIBLIOGRAPHY

"Acrylonitrile Polymers, ABS Resins" in *ECT* 3rd ed., Vol. 1, pp. 442–456, by G. A. Morneau, W. A. Pavelich, and L. G. Roettger, Borg-Warner Chemicals.

1. C. T. Pillichody and P. D. Kelley in I. I. Rubin, ed., *Handbook of Plastic Materials and Technology,* John Wiley & Sons, Inc., New York, 1990, Chapt. 3.
2. R. D. Leaversuch, *Mod. Plast.* **66**(1), 77 (1989).
3. H. Kim, H. Keskkula, and D. R. Paul, *Polymer* **31**, 869 (1990).
4. G. H. Michler, *J. Mater. Sci.* **25**, 2321 (1990).
5. E. M. Donald and E. J. Kramer, *J. Mater. Sci.* **17**, 1765 (1982).
6. L. V. Newmann and J. G. Williams, *J. Mater. Sci.* **15**, 773 (1980).
7. C. B. Bucknall, *Toughened Plastics,* Applied Science Publishers, London, 1977.
8. M. Rink, T. Ricco, W. Lubert, and A. Pavan, *J. Appl. Polym. Sci.* **22**, 429 (1978).
9. J. Mann and G. R. Williamson in R. N. Haward, ed., *The Physics of Glassy Polymers,* John Wiley & Sons, Inc., New York, 1973, Chapt. 8.
10. H. Keskkula, *Appl. Polym. Symp.* **15**, 51 (1970).
11. J. A. Schmitt, *J. Polym. Sci.* **C30**, 437 (1970).
12. S. L. Rosen, *Polym. Eng. Sci.* **7**, 115 (1967).
13. L. Bohn, Angew. *Makromol. Chem.* **20**, 129 (1971).
14. A. Zosel, *Rheol. Acta* **11**, 229 (1972).
15. H. Munstedt, *Polym. Eng. Sci.* **21**, 259 (1981).
16. Y. Aoki and K. Nakayama, *Polym. J.* **14**, 951 (1982).
17. Y. Aoki, *Macromolecules* **20**, 2208 (1987).
18. Y. Aoki, *J. Non-Newtonian Fluid Mech.* **22**, 91 (1986).
19. M. G. Huguet and T. R. Paxton, *Colloidal and Morphological Behavior of Block and Graft Copolymers,* Plenum, New York, 1971, pp. 183–192.
20. T. Masuda and co-workers, *Pure Appl. Chem.* **56**, 1457 (1984).
21. A. Casale, A. Moroni, and C. Spreafico in *Copolymers, Polyblends, and Composites* (Adv. in Chem. Ser. No. 142), American Chemical Society, Washington, D.C., 1975, p. 172.

22. L. L. Blyler Jr., *Polym. Eng. Sci.* **14**(11), 806 (1974).
23. D. M. Kulich, P. D. Kelley, and J. E. Pace in J. I. Kroschwitz, ed., *Encyclopedia of Polymer Science and Engineering,* 2nd ed., Vol. 1, Wiley-Interscience, New York, 1985, p. 396.
24. F. M. Smith, *Manufacture of Plastics,* Vol. 1, Reinhold Publishing Corp., New York, 1964, p. 443.
25. D. L. Fulkner, *Polym. Eng. Sci.* **24**, 1174 (1984).
26. J. M. Heaps, *Rubber Plast. Age,* 967 (1968).
27. T. H. Rogers and R. B. Roennau, *Chem. Eng. Progress* **62**(11), 94 (1966).
28. A. Casale and O. Salvatore, *Polym. Eng. Sci.* **15**, 286 (1975).
29. W. I. Congdon, H. E. Bair, and S. K. Khanna, *Org. Coatings Plast. Chem.* **40**, 739 (1979).
30. M. L. Heckaman, *Soc. Plast. Eng. Tech. Pap.* **18**, 512 (1972).
31. B. D. Gesner, *J. Appl. Polym. Sci.* **9**, 3701 (1965).
32. P. G. Kelleher, *J. Appl. Polym. Sci.* **10**, 843 (1966).
33. J. Shimada, *J. Appl. Polym. Sci.* **12**, 655 (1968).
34. M. D. Wolkowicz and S. Gaggar, *Polym. Eng. Sci.* **21**, 571 (1981).
35. P. So and L. J. Broutman, *Polym. Eng. Sci.* **22**, 888 (1982).
36. B. Gilg, H. Muller, and K. Schwarzenbach. Paper presented at *Advances in Stabilization and Controlled Degradation of Polymers,* New Paltz, N.Y., June 1982.
37. J. Shimada, K. Kabuki, and M. Ando, *Rev. Electr. Commun. Lab.* **20**, 564 (1972).
38. F. Gugumus in G. Scott, ed., *Developments in Polymer Stabilization,* Vol. 1, Elsevier Applied Science Publishers, Ltd., London 1979, p. 319.
39. G. Scott, *Developments in Polymer Stabilization,* Vol. 4, Elsevier Applied Science Publishers, Ltd., London, 1981, p. 181.
40. Eur. Pats. 0109008 and 0108396, J. C. Wozny (to Borg-Warner Corp.).
41. D. M. Kulich and M. D. Wolkowicz in *Rubber-Toughened Plastics* (Adv. in Chem. Ser. No. 222), American Chemical Society, Washington, D.C., 1989, p. 329.
42. B. D. Gesner, *SPE J.* **25**, 73 (1969).
43. J. Kovarova, L. Rosik, and J. Pospisil, *Polym. Mater. Sci. Eng.* **58**, 215 (1988).
44. F. Gugumus in G. Scott, ed., *Developments in Polymer Stabilization,* Vol. 8, Elsevier, New York, 1987, p. 243.
45. D. M. Chang, *Org. Coatings Plast. Chem.* **44**, 347 (1981).
46. M. G. Wygoski, *Polym. Eng. Sci.* **16**, 265 (1976).
47. L. Zlatkevich in P. Klemchuk, ed., *Polymer Stabilization and Degradation* (ACS Symp. Ser. No. 280), American Chemical Society, Washington, D.C., 1985, Chapt. 27.
48. L. Zlatkevich, *J. Polym. Sci. Polym. Phys. Ed.* **25**, 2207 (1987).
49. G. Scott and M. Tahan, *Eur. Polym. J.* **13**, 981 (1977).
50. M. Tahan, *Weathering of Plastics and Rubber, International Symposium of the Institute of Electrical Engineers, London, June 1976,* Chamelon Press, Ltd., London, 1976, p. A2.1.
51. J. B. Adeniyi, *Eur. Polym. J.* **20**, 291 (1984).
52. E. Priebe, P. Simak, and G. Stange, *Kunststoffe* **62**, 105 (1972).
53. T. Hirai, *Jpn. Plast.* 23, Oct. 1970.
54. M. Ghaemy and G. Scott, *Polym. Degrad. Stab.* **3**, 233 (1981).
55. J. B. Adeniyi and E. G. Kolawole, *Eur. Polym. J.* **20**, 43 (1984).
56. R. D. Deanin, I. S. Rabinovic, and A. Llompart in *Multicomponent Polymer Systems* (Adv. in Chem. Ser. No. 99), American Chemical Society, Washington, D.C., 1971, p. 229.
57. A. Blaga and R. S. Yamasaki, *Durab. of Build. Mater.* **4**, 21 (1986).
58. J. Shimada and K. Kabuki, *J. Appl. Polym. Sci.* **12**, 671 (1968).
59. T. Kurumada, H. Ohsawa, and T. Yamazaki, *Polym. Degrad. Stab.* **19**, 263 (1987).

60. E. G. Kolawole and J. B. Adeniyi, *Eur. Polym. J.* **18,** 469 (1982).
61. T. R. Bullet and P. R. Mathews, *Plast. Polym.* **39,** 200 (1971).
62. A. Davis and D. Gordon, *J. Appl. Polym. Sci.* **18,** 1159 (1974).
63. P. G. Kelleher, D. J. Boyle, and R. J. Miner, *Mod. Plast.* 189 (Sept. 1969).
64. M. L. Kaplan and P. G. Kelleher, *J. Polym. Sci. Part A-1* **8,** 3163 (1970).
65. *Product Information,* GE Plastics, Pittsfield, Mass.
66. R. A. Hayes and S. Futamura, *J. Polym. Sci. Polym. Chem. Ed.* **19,** 985 (1981).
67. A. Brydon, G. M. Burnett, and C. G. Cameron, *J. Polym. Sci. Polym. Chem. Ed.* **12,** 1011 (1974).
68. P. W. Allen, G. Ayrey, and C. G. Moore, *J. Polym. Sci.* **36,** 55 (1959).
69. J. L. Locatelli and G. Riess, *Angew. Makromol. Chem.* **32,** 117 (1973).
70. G. Reiss and J. L. Locatelli in *Copolymers, Polyblends, and Composites* (Adv. in Chem. Ser. No. 142), American Chemical Society, Washington, D.C., 1975, p. 186.
71. B. Chauvel and J. C. Daniel in *Copolymers, Polyblends, and Composites* (Adv. in Chem. Ser. No. 142), American Chemical Society, Washington, D.C., 1975, p. 159.
72. J. L. Locatelli and G. Riess, *Angew. Makromol. Chem.* **27,** 201 (1972).
73. C. F. Parsons and E. L. Suck, Jr., *Multicomponent Polymer Systems* (Adv. in Chem. Ser. No. 99), American Chemical Society, Washington, D.C., 1971, p. 340.
74. J. Stabenow and F. Haaf, *Angew. Makromol. Chem.* **29–30,** 1 (1973).
75. J. L. Locatelli and G. Riess, *J. Polym. Sci. Polym. Chem. Ed.* **2,** 3309 (1973).
76. J. L. Locatelli and G. Riess, *Angew. Makromol. Chem.* **32,** 101 (1973).
77. A. Echte in *Rubber-Toughened Plastics* (Adv. in Chem. Ser. No. 222), American Chemical Society, Washington, D.C., 1989, p. 15.
78. E. Beati, M. Pegoraro, E. Pedemonte, *Angew. Makrom. Chem.* 149, 55 (1987).
79. G. Odian, *Principles of Polymerization,* McGraw-Hill, Inc. New York, 1970, Chapt. 4.
80. P. J. Flory, *Principles of Polymer Chemistry,* Cornell University Press, Ithaca, N.Y., 1953, pp. 203–217.
81. U.S. Pat. 2,820,773 (Jan. 21, 1958), C. W. Childers and C. F. Fisk (to United States Rubber Co.).
82. U.S. Pat. 3,238,275 (Mar. 1, 1966), W. C. Calvert (to Borg-Warner Corp.).
83. Ref. 79, Chapt. 3.
84. U.S. Pat. 3,950,455 (Apr. 13, 1976), T. Okamoto and co-workers (to Toray Industries, Inc.).
85. U.S. Pat. 3,660,535 (May 2, 1972), C. R. Finch and J. E. Knutzsch (to The Dow Chemical Company).
86. U.S. Pat. 2,694,692 (Nov. 16, 1954), J. L. Amos, J. L. McCurdy, and O. R. McIntire (to The Dow Chemical Company).
87. U.S. Pat. 3,515,692 (June 2, 1970), F. E. Carrock and K. W. Doak (to Dart Industries, Inc.).
88. F. Rodriguez, *Principles of Polymer Systems,* McGraw-Hill, Inc., New York, 1970, Chapt. 5.
89. J. Haslam, H. A. Willis, and D. C. M. Squirrell, *Identification and Analysis of Plastics,* 2nd ed., Heyden Book Co. Inc., Philadelphia, 1980, Chapt. 8.
90. J. C. Cobbler and G. E. Stobbe in F. D. Snell and L. S. Ettre, eds., *Encyclopedia of Industrial Chemical Analysis,* Vol. 18, Wiley-Interscience, New York, 1973, p. 332.
91. T. Okumoto and T. Tadaoki, *Nippon Kagaku Kaishi* **1,** 71 (1972).
92. K. Sircar and T. Lamond, *Thermochim. Acta* **7,** 287 (1973).
93. B. D. Gesner, *J. Polym. Sci. Part A* **3,** 3825 (1965).
94. L. D. Moore, W. W. Moyer, and W. J. Frazer, *Appl. Polym. Symp.* **7,** 67 (1968).
95. J. Tsurugi, T. Fukumoto, and K. Ogawa, *Chem. High Polym. (Tokyo)* **25,** 116 (1968).
96. H. Shuster, M. Hoffmann, and K. Dinges, *Angew. Makrom. Chem.* **9,** 35 (1969).
97. D. Kranz, K. Dinges, and P. Wendling, *Angew. Makrom. Chem.* **51,** 25 (1976).

98. D. Kranz, H. V. Pohl, and H. Baumann, *Angew. Makrom. Chem.* **26,** 67 (1972).

99. R. R. Turner, D. W. Carlson, and A. G. Altenau, *J. Elastomer Plast.* **6,** 94 (1974).

100. L. W. Jelinski and co-workers, *J. Polym. Sci. Polym. Chem. Ed.* **20,** 3285 (1982).

101. V. G. Kampf and H. Shuster, *Angew. Makrom. Chem.* **14,** 111 (1970).

102. D. Simpson, *Br. Plast.* 78 (1968).

103. L. I. Petrova, M. P. Noskova, V. A. Balandine, and Z. G. Guricheva, *Gig. Sanit.* **37,** 62 (1972).

104. T. R. Crompton, *Chemical Analysis of Additives in Plastics,* Pergamon Press, New York, 1971.

105. N. E. Skelly, J. D. Graham, and Z. Iskandarani, *Polym. Mater. Sci. Eng.* **59,** 23 (1988).

106. R. Yoda, *Bunseki* **1,** 29 (1984).

107. R. J. Pierce and J. W. Bozzelli, Paper presented at the 45th Annual Technical Conference of the Society of Plastics Engineers, May 1987, p. 19.

108. L. W. Fritch, Paper presented at the 33rd Annual Technical Conference of the Society of Plastics Engineers, May 1975, p. 70.

109. L. W. Fritch, *Plast. Technol.* 69 (1980).

110. *Cycolac Brand ABS Pellet Drying,* Technical Publication SR-601A, GE Plastics, 1989.

111. J. W. Bozzelli, B. J. Furches, and S. L. Janiki, *Mod. Plast.* 7 (1988).

112. *Product Bulletin: Micro Moisture II,* ZARAD Technology Inc., Cary, Ill.

113. *Product Bulletin E38855,* DuPont 903 Moisture Evolution Analyzer, DuPont Company, Instrument Systems, Wilmington, Del.

114. B. Miller, *Plast. World* 51 (1987).

115. H. Lord, Paper presented at the 36th Annual Technical Conference of the Society of Plastics Engineers, May 1978, p. 83.

116. R. E. Nunn, *Injection Molding Handbook,* Van Nostrand Reinhold Co., New York, 1986, Chapt. 3.

117. B. Miller, *Plast. World* **40**(3), 34 (1982).

118. B. Furches and J. Bozzelli, Paper presented at the 45th Annual Technical Conference of the Society of Plastics Engineers, May 1987, p. 6.

119. *Cycolac Brand ABS Resin Injection Molding,* Technical Publication CYC-400, GE Plastics, Pittsfield, Mass., 1990.

120. *Molding Flame Retardent ABS,* Technical Publication P-408, GE Plastics, Pittsfield, Mass., 1989.

121. L. W. Fritch, *Injection Molding Handbook,* Van Nostrand Reinhold Co., New York, 1986, Chapt. 19.

122. L. W. Fritch, *Plast. Eng.* 43 (1989).

123. L. W. Fritch, Paper presented at the 45th Annual Technical Conference of the Society of Plastics Engineers, May 1987, p. 218.

124. L. W. Fritch, Paper presented at the 40th Annual Technical Conference of the Society of Plastics Engineers, May 1982, p. 332.

125. L. W. Fritch, Paper presented at the 5th Pacific Area Technical Conference of the Society of Plastics Engineers, Feb. 1980.

126. S. Gaggar and J. Wilson, Paper presented at the 40th Annual Technical Conference of the Society of Plastics Engineers, May 1982, p. 157.

127. L. W. Fritch, Paper presented at the 37th Annual Technical Conference of the Society of Plastics Engineers, May 1979, p. 15.

128. L. W. Fritch, *Prod. Finish. Cincinnati* **48**(5), 42 (1984).

129. L. W. Fritch, *Plast. Machin. Equip.* 43 (1988).

130. R. M. Criens and H. Mosle, Paper presented at the 42nd Annual Technical Conference of the Society of Plastics Engineers, May 1984, p. 587.

131. J. W. Bozzelli and P. A. Tiffany, Paper presented at the 44th Annual Technical Conference of the Society of Plastics Engineers, May 1986, p. 120.

132. U. Wolfel and G. Menges, Paper presented at the 45th Annual Technical Conference of the Society of Plastics Engineers, May 1987, p. 292.
133. S. M. Janosz, Paper presented at the 45th Annual Technical Conference of the Society of Plastics Engineers, May 1987, p. 323.
134. H. Cox and C. Mentzer, *Polym. Eng. Sci.* **26,** 488 (1986).
135. G. Brewer, Paper presented at the 45th Annual Technical Conference of the Society of Plastics Engineers, May 1987, p. 252.
136. *Cycolac Brand ABS Resin Design Guide,* Technical Publication CYC-350, GE Plastics, Pittsfield, Mass., 1990.
137. *Cycolac Brand ABS Electroplating,* Technical Publication 402, GE Plastics, Pittsfield, Mass., 1990.
138. M. Snyder, *Plast. Machin. Equip.* 50 (1988).
139. K. C. Rusch, Paper presented at the 45th Annual Technical Conference of the Society of Plastics Engineers, May 1987, p. 1014.
140. W. Virginski, Paper presented at the 46th Annual Technical Conference of the Society of Plastics Engineers, May 1988, p. 205.
141. *Plast. World,* 28 (1987).
142. L. E. Ferguson and R. J. Brinkmann, Paper presented at the 45th Annual Technical Conference of the Society of Plastics Engineers, May 1987, p. 866.
143. *Cycolac Brand ABS—General Purpose Blow Molding Grades,* Technical Publication SR-616, GE Plastics, Pittsfield, Mass., 1989.
144. *Thermoforming Cycolac Brand ABS,* Technical Publication P-406, GE Plastics, Pittsfield, Mass., 1989.
145. J. L. Throne, Paper presented at the 45th Annual Technical Conference of the Society of Plastics Engineers, May 1987, p. 412.
146. N. Nichols and G. Kraynak, *Plast. Technol.* 73 (1987).
147. R. Royer, Paper presented at the Regional Conference of the Society of Plastics Engineers, Quebec Section, 1968, p. 43.
148. G. Trantina and M. Minnicheli, Paper presented at the 45th Annual Technical Conference of the Society of Plastics Engineers, May 1987, p. 438.
149. T. L. La Bounty, Paper presented at the 43rd Annual Technical Conference of the Society of Plastics Engineers, May 1985, p. 855.
150. *Cycolac Brand ABS Resin—Assembly Techniques,* Technical Publication CYC-352, GE Plastics, Pittsfield, Mass., 1990.
151. H. Potente and H. Kaiser, Paper presented at the 47th Annual Technical Conference of the Society of Plastics Engineers, May 1989, p. 464.
152. V. K. Stokes, Paper presented at the 47th Annual Technical Conference of the Society of Plastics Engineers, May 1989, p. 442.
153. *Chemical Economics Handbook,* SRI International, Menlo Park, Calif., 1989, 580.0180D.
154. *World Petrochemicals,* SRI International, Menlo Park, Calif., 1990, WORL 2-16.
155. *Synthetic Organic Chemicals,* United States Production and Sales, USITC Publication, 1989.

Donald M. Kulich
John E. Pace
Leroy W. Fritch, Jr.
Angelo Brisimitzakis
GE Plastics

ACTINIDES AND TRANSACTINIDES

ACTINIDES

The actinide elements are a group of chemically similar elements with atomic numbers 89 through 103 and their names, symbols, atomic numbers, and discoverers are given in Table 1 (1–3) (see RADIOACTIVITY, NATURAL; THORIUM AND THORIUM COMPOUNDS; URANIUM AND URANIUM COMPOUNDS; PLUTONIUM AND PLUTONIUM COMPOUNDS; NUCLEAR REACTORS; and RADIOISOTOPES).

Each of the elements has a number of isotopes (2,4), all radioactive and some of which can be obtained in isotopically pure form. More than 200 in number and mostly synthetic in origin, they are produced by neutron or charged-particle induced transmutations (2,4). The known radioactive isotopes are distributed among the 15 elements approximately as follows: actinium and thorium, 25 each; protactinium, 20; uranium, neptunium, plutonium, americium, curium, californium, einsteinium, and fermium, 15 each; berkelium, mendelevium, nobelium, and lawrencium, 10 each. There is frequently a need for values to be assigned for the atomic weights of the actinide elements. Any precise experimental work would require a value for the isotope or isotopic mixture being used, but where there is a purely formal demand for atomic weights, mass numbers that are chosen on the basis of half-life and availability have customarily been used. A list of these is provided in Table 1.

Thorium and uranium have long been known, and uses dependent on their physical or chemical, not on their nuclear, properties were developed prior to the discovery of nuclear fission. The discoveries of actinium and protactinium were among the results of the early studies of naturally radioactive substances. The first transuranium element, synthetic neptunium, was discovered during an investigation of nuclear fission, and this event rapidly led to the discovery of the next succeeding element, plutonium. The realization that plutonium as ^{239}Pu undergoes fission with slow neutrons and thus could be utilized in a nuclear weapon supplied the impetus for its thorough investigation. This research has provided the background of knowledge and techniques for the production and identification of nine more actinide elements (and six transactinide elements).

Thorium, uranium, and plutonium are well known for their role as the basic fuels (or sources of fuel) for the release of nuclear energy (5). The importance of the remainder of the actinide group lies at present, for the most part, in the realm of pure research, but a number of practical applications are also known (6). The actinides present a storage-life problem in nuclear waste disposal and consideration is being given to separation methods for their recovery prior to disposal (see HAZARDOUS WASTE TREATMENT; NUCLEAR REACTORS, WASTE MANAGEMENT).

Source

Only the members of the actinide group through Pu have been found to occur in nature (2,3,7,8). Actinium and protactinium are decay products of the naturally

Table 1. The Actinide Elements

Atomic number	Element	CAS Registry Number	Symbol	Atomic weight[a]	Discoverers and date of discovery
89	actinium	[7440-34-8]	Ac	227	A. Debierne, 1899
90	thorium	[7440-29-1]	Th	232	J. J. Berzelius, 1828
91	protactinium	[7440-13-3]	Pa	231	O. Hahn and L. Meitner, 1917, and F. Soddy and J. A. Cranston, 1917
92	uranium	[7440-61-1]	U	238	M. H. Klaproth, 1789
93	neptunium	[7439-99-8]	Np	237	E. M. McMillan and P. H. Abelson, 1940
94	plutonium	[7440-07-5]	Pu	242	G. T. Seaborg, E. M. McMillan, J. W. Kennedy, and A. C. Wahl, 1940–1941
95	americium	[7440-35-9]	Am	243	G. T. Seaborg, R. A. James, L. O. Morgan, and A. Ghiorso, 1944–1945
96	curium	[7440-51-9]	Cm	248	G. T. Seaborg, R. A. James, and A. Ghiorso, 1944
97	berkelium	[744-40-6]	Bk	249	S. G. Thompson, A. Ghiorso, and G. T. Seaborg, 1949
98	californium	[7440-71-3]	Cf	249	S. G. Thompson, K. Street, Jr., A. Ghiorso, and G. T. Seaborg, 1950
99	einsteinium	[7429-92-7]	Es	254	A. Ghiorso, S. G. Thompson, G. H. Higgins, G. T. Seaborg, M. H. Studier, P. R. Fields, S. M. Fried, H. Diamond, J. F. Mech, G. L. Pyle, J. R. Huizenga, A. Hirsch, W. M. Manning, C. I. Browne, H. L. Smith, and R. W. Spence, 1952
100	fermium	[7440-72-4]	Fm	257	A. Ghiorso, S. G. Thompson, G. H. Higgins, G. T. Seaborg, M. H. Studier, P. R. Fields, S. M. Fried, H. Diamond, J. F. Mech, G. L. Pyle, J. R. Huizenga, A. Hirsch, W. M. Manning, C. I. Browne, H. L. Smith, and R. W. Spence, 1953
101	mendelevium	[7440-11-1]	Md	258	A. Ghiorso, B. G. Harvey, G. R. Choppin. S. G. Thompson, and G. T. Seaborg, 1955
102	nobelium	[10028-14-5]	No	259	A. Ghiorso, T. Sikkeland, J. R. Walton, and G. T. Seaborg, 1958
103	lawrencium	[22537-19-5]	Lr	260	A. Ghiorso, T. Sikkeland, A. E. Larsh, and R. M. Latimer, 1961

[a]Mass number of longest lived or most available isotope.

occurring uranium isotope ^{235}U, but the concentrations present in uranium minerals are small and the methods involved in obtaining them from the natural source are very difficult and tedious in contrast to the relative ease with which the elements can be synthesized. Thorium and uranium occur widely in the earth's crust in combination with other elements, and, in the case of uranium, in significant concentrations in the oceans. The extraction of these two elements from their ores has been studied intensively and forms the basis of an extensive technology. Neptunium (^{239}Np and ^{237}Np) and plutonium (^{239}Pu) are present in trace amounts in nature, being formed by neutron reactions in uranium ores. Longer lived ^{244}Pu, possibly from a primordial source, has been found in very small concentration (1 part in 10^{18}) in the rare earth mineral bastnasite [12172-82-6] (8). Mining these elements from these sources is not feasible because the concentrations involved are exceedingly small. Thus, with the exceptions of uranium and thorium, the actinide elements are synthetic in origin for practical purposes, ie, they are products of nuclear reactions. High neutron fluxes are available in modern nuclear reactors, and the most feasible method for preparing actinium, protactinium, and most of the actinide elements is through the neutron irradiation of elements of high atomic number (3,9).

Actinium can be prepared by the transmutation of radium,

$$^{226}Ra + n \longrightarrow {}^{227}Ra + \gamma; \quad {}^{227}Ra \xrightarrow[41.2 \text{ min}]{\beta^-} {}^{227}Ac$$

and gram amounts have been obtained in this way. The actinium is isolated by means of solvent extraction or ion exchange.

Protactinium can be produced in the nuclear reactions

$$^{230}Th + n \longrightarrow {}^{231}Th + \gamma; \quad {}^{231}Th \xrightarrow[25.6 \text{ h}]{\beta^-} {}^{231}Pa$$

However, the quantity of ^{231}Pa produced in this manner is much less than the amount (more than 100 g) that has been isolated from the natural source. The methods for the recovery of protactinium include coprecipitation, solvent extraction, ion exchange, and volatility procedures. All of these, however, are rendered difficult by the extreme tendency of protactinium(V) to form polymeric colloidal particles composed of ionic species. These cannot be removed from aqueous media by solvent extraction; losses may occur by adsorption to containers; and protactinium may be adsorbed by any precipitate present.

Kilogram amounts of neptunium (^{237}Np) have been isolated as a by-product of the large-scale synthesis of plutonium in nuclear reactors that utilize ^{235}U and ^{238}U as fuel. The following transmutations occur:

$$^{238}U + n \longrightarrow {}^{237}U + 2n; \quad {}^{237}U \xrightarrow[6.75 \text{ d}]{\beta^-} {}^{237}Np$$

and

$$^{235}U + n \longrightarrow {}^{236}U + \gamma; \quad {}^{236}U + n \longrightarrow {}^{237}U + \gamma; \quad {}^{237}U \xrightarrow{\beta^-} {}^{237}Np$$

The wastes from uranium and plutonium processing of the reactor fuel usually contain the neptunium. Precipitation, solvent extraction, ion exchange, and volatility procedures (see DIFFUSION SEPARATION METHODS) can be used to isolate and purify the neptunium.

Plutonium as the important isotope ^{239}Pu is prepared in ton quantities in nuclear reactors. It is produced by the following reactions, wherein the excess neutrons produced by the fission of ^{235}U are captured by ^{238}U to yield ^{239}Pu.

$$^{235}U + n \longrightarrow \text{fission products} + 2.5\,n + 200\text{ MeV}$$

$$^{238}U + n \longrightarrow {}^{239}U \xrightarrow[23.5\text{ min}]{\beta^-} {}^{239}Np \xrightarrow[2.3\text{ d}]{\beta^-} {}^{239}Pu$$

The plutonium usually contains isotopes of higher mass number (Fig. 1). A variety of industrial-scale processes have been devised for the recovery and purification of plutonium. These can be divided, in general, into the categories of precipitation, solvent extraction, and ion exchange.

The isotope ^{238}Pu, produced in kilogram quantities by the reactions

$$^{237}Np + n \longrightarrow {}^{238}Np \quad \text{and} \quad {}^{238}Np \xrightarrow[2.1\text{d}]{\beta^-} {}^{238}Pu$$

is an important fuel for isotopically powered energy sources used for terrestrial and extraterrestrial applications.

Kilogram quantities of americium as ^{241}Am can be obtained by the processing of reactor-produced plutonium. Much of this material contains an appreciable proportion of ^{241}Pu, which is the parent of ^{241}Am. Separation of the americium is effected by precipitation, ion exchange, or solvent extraction.

The nuclear reaction sequences of neutron captures and beta decays involved in the preparation of the actinide elements by means of the slow neutron irradiation of ^{239}Pu are indicated in Figure 1. The irradiations can be performed by placing the parent material in the core of a high-neutron-flux reactor where fluxes of neutrons in excess of 10^{14} neutrons/(cm^2·s) may be available. Figure 2 gives an indication of the time required for typical preparation of various heavy isotopes from ^{239}Pu as the starting material. For example, beginning with 1 kg of ^{239}Pu, about 1 mg of ^{252}Cf would be present after 5–10 yr of continuous irradiation at a neutron flux of 3×10^{14} neutrons/(cm^2·s). Much larger quantities can be produced by irradiating larger quantities of plutonium in production reactors, followed by irradiation of the curium thus produced in higher-neutron-flux reactors, ca 10^{15} neutrons/(cm^2·s), such as those at the Savannah River Plant in South Carolina and the High Flux Isotopes Reactor (HFIR) at the Oak Ridge National Laboratory (ORNL) in Tennessee. Such programs have led to the production of kilogram quantities of curium (^{244}Cm and heavier isotopes), gram quantities of californium, 100-mg quantities of berkelium, and milligram quantities of einsteinium (6,10). The elements 95 to 100 are also produced in increasing quantities by nuclear power reactors.

Ion exchange (qv; see also CHROMATOGRAPHY) is an important procedure for the separation and chemical identification of curium and higher elements. This technique is selective and rapid and has been the key to the discovery of the

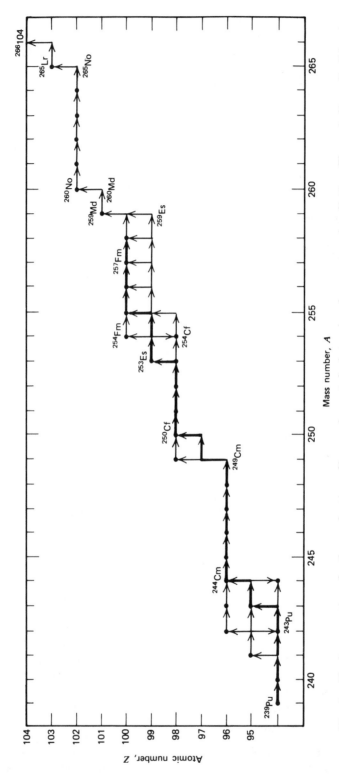

Fig. 1. Nuclear reactions for the production of heavy elements by intensive slow neutron irradiation. The main line of buildup is designated by heavy arrows. The sequence above ^{258}Fm represents predictions.

416

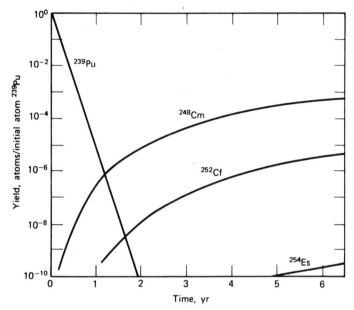

Fig. 2. Production of heavy nuclides by the irradiation of ^{239}Pu at a flux of 3×10^{14} neutrons/(cm^2·s).

transcurium elements, in that the elution order and approximate peak position for the undiscovered elements were predicted with considerable confidence (9). Thus the first experimental observation of the chemical behavior of a new actinide element has often been its ion-exchange behavior—an observation coincident with its identification. Further exploration of the chemistry of the element often depended on the production of larger amounts by this method. Solvent extraction is another useful method for separating and purifying actinide elements.

There are many similarities in the chemical properties of the lanthanide elements (see LANTHANIDES) and those of the actinides, especially with elements in the same oxidation state. A striking example of this resemblance is furnished by their ion-exchange behavior. Figure 3 shows the comparative elution data for tripositive actinide and lanthanide ions obtained by the use of the ion-exchange resin Dowex-50 (a copolymer of styrene and divinylbenzene with sulfonic acid groups) and the eluting agent ammonium α-hydroxyisobutyrate [2539-76-6]. In this system, which is used for illustration because of its historical importance, the elutions occur in the inverse order of atomic number. The elution sequence depends on a balance between the adherence to the resin and the stability of the complex ion formed with the eluting agent and may be correlated with the variation of ionic radius with atomic number.

Actinide ions of the III, IV, and VI oxidation states can be adsorbed by cation-exchange resins and, in general, can be desorbed by elution with chloride, nitrate, citrate, lactate, α-hydroxyisobutyrate, ethylenediaminetetraacetate, and other anions (11,12).

Ion-exchange separations can also be made by the use of a polymer with exchangeable anions; in this case, the lanthanide or actinide elements must be

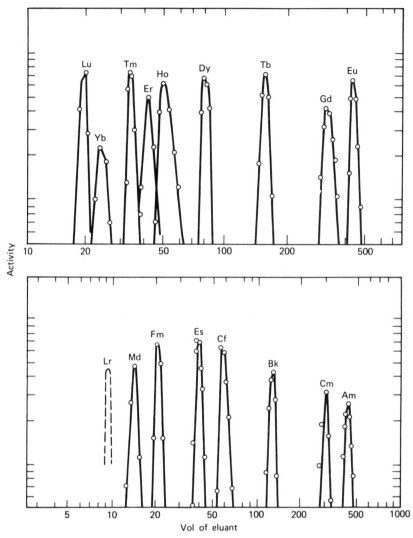

Fig. 3.   The elution of tripositive actinide and lanthanide ions. Dowex-50 ion-exchange resin was used with ammonium α-hydroxyisobutyrate as the eluant. The position predicted for short-lived lawrencium is indicated by a broken line.

initially present as complex ions (11,12). The anion-exchange resins Dowex-1 (a copolymer of styrene and divinylbenzene with quaternary ammonium groups) and Amberlite IRA-400 (a quaternary ammonium polystyrene) have been used success-fully. The order of elution is often the reverse of that from cationic-exchange resins.

Extraction chromatography (see ADSORPTION), in which the organic ex-tractant is adsorbed on the surfaces of a fine, porous powder placed in a column, offers another excellent method for separating the actinide elements from each other. Useful cation extracting agents include bis(2-ethylhexyl)phosphoric acid [298-07-7], mono(2-ethylhexyl)phenylphosphonic acid ester [1518-07-6], and n-tri-butyl phosphate [126-73-8] (12). Excellent anion extracting agents include tertiary amines such as tricapryl amine [1116-76-3] or trilauryl amine [102-87-4], or quaternary amines such as tricaprylmethyl ammonium chloride [5137-55-3] (nitrate, thiocyanate) (12). Satisfactory supporting agents can be found in commercially available diatomaceous earths or silica microspheres.

It is possible to prepare very heavy elements in thermonuclear explosions, owing to the very intense, although brief (order of a microsecond), neutron flux furnished by the explosion (3,13). Einsteinium and fermium were first produced in this way; they were discovered in the fallout materials from the first thermonuclear explosion (the "Mike" shot) staged in the Pacific in November 1952. It is possible that elements having atomic numbers greater than 100 would have been found had the debris been examined very soon after the explosion. The preparative process involved is multiple neutron capture in the uranium in the device, which is followed by a sequence of beta decays. For example, the synthesis of ^{255}Fm in the Mike explosion was via the production of ^{255}U from ^{238}U, followed by a long chain of short-lived beta decays,

$$^{255}\text{U} \xrightarrow{-\beta^-} {}^{255}\text{Np} \xrightarrow{-\beta^-} {}^{255}\text{Pu} \xrightarrow{-\beta^-} \cdots \longrightarrow {}^{255}\text{Fm}$$

all of which occur after the neutron reactions are completed.

The process of neutron irradiation in high-flux reactors cannot be used to prepare the elements beyond fermium (^{257}Fm), except at extremely high neutron fluxes, because some of the intermediate isotopes that must capture neutrons have very short half-lives that preclude the necessary concentrations. Transfermium elements are prepared in charged-particle bombardments (2,3). Such syntheses are characterized by the limited availability of target materials of high atomic number, the small reaction yields, and the difficulties intrinsic in the isolation of very short-lived substances. Nonchemical separations of short-lived isotopes from the target materials are carried out during bombardments. Numerous isotopes of mendelevium, nobelium, and lawrencium (the heaviest actinide elements) are produced by bombardment with heavy ions. Despite the fact that these are usually produced on a "one-atom-at-a-time" basis, the chemical properties of these elements have been studied using the tracer technique. Cyclotrons can be used to accelerate heavy ions; in addition, linear accelerators designed for this express purpose are in operation in several laboratories throughout the world.

Isotopes sufficiently long-lived for work in weighable amounts are obtainable, at least in principle, for all of the actinide elements through fermium (100); these isotopes with their half-lives are listed in Table 2 (4). Not all of these are available as individual isotopes. It appears that it will always be necessary to study the elements above fermium by means of the tracer technique (except for some very special experiments) because only isotopes with short half-lives are known.

Experimental Methods of Investigation

All of the actinide elements are radioactive and, except for thorium and uranium, special equipment and shielded facilities are usually necessary for their manipulation (9,14,15). On a laboratory scale, enclosed containers (gloved boxes) are

Table 2. Long-Lived Actinide Nuclides Suitable for Investigation

Element	Isotope	CAS Registry Number	Half-life
actinium-227	^{227}Ac	[14952-40-0]	21.8 yr
thorium-232	^{232}Th	[7440-29-1]	1.41×10^{10} yr
protactinium-231	^{231}Pa	[14331-85-2]	3.25×10^4 yr
uranium-238	^{238}U^a	[24678-82-8]	4.47×10^9 yr
neptunium-236	^{236}Np^b	[15770-36-4]	1.55×10^5 yr
neptunium-237	^{237}Np	[13994-20-2]	2.14×10^6 yr
plutonium-238	^{238}Pu	[13981-16-3]	87.8 yr
plutonium-239	^{239}Pu	[15117-48-3]	24,150 yr
plutonium-240	^{240}Pu	[14119-33-6]	6,540 yr
plutonium-241	^{241}Pu	[14119-32-5]	14.9 yr
plutonium-242	^{242}Pu	[13982-10-0]	3.87×10^5 yr
plutonium-244	^{244}Pu	[14119-34-7]	8.3×10^7 yr
americium-241	^{241}Am	[14596-10-2]	433 yr
americium-242	^{242}Am	[13981-54-9]	152 yr
americium-243	^{243}Am	[14993-75-0]	7,400 yr
curium-242	^{242}Cm	[15510-73-3]	163.0 d
curium-243	^{243}Cm	[15757-87-6]	30 yr
curium-244	^{244}Cm	[13981-15-2]	18.1 yr
curium-245	^{245}Cm	[15621-76-8]	8,540 yr
curium-246	^{246}Cm	[15757-90-1]	4,800 yr
curium-247	^{247}Cm	[15758-32-4]	1.6×10^7 yr
curium-248	^{248}Cm	[15758-33-5]	3.6×10^5 yr
curium-250	^{250}Cm^c	[15743-88-6]	1.1×10^4 yr
berkelium-247	^{247}Bk^b	[15752-38-2]	1,380 yr
berkelium-249	^{249}Bk	[14900-25-5]	320 d
californium-249	^{249}Cf	[15237-97-5]	350 yr
californium-250	^{250}Cf	[13982-11-1]	13.1 yr
californium-251	^{251}Cf	[15765-19-2]	898 yr
californium-252	^{252}Cf	[13981-17-4]	2.6 yr
einsteinium-253	^{253}Es	[15840-02-5]	20.5 d
einsteinium-254	^{254}Es	[15840-03-6]	276 d
einsteinium-255	^{255}Es	[15840-04-7]	40 d
fermium-257	^{257}Fm	[15750-26-2]	100 d

^{a}Natural mixture (^{238}U, 99.3%; ^{235}U, 0.72%; and ^{234}U, 0.006%). Half-life given is for the major constituent ^{238}U.
^{b}Available so far only in trace quantities from charged particle irradiations.
^{c}Available only in very small amounts from neutron irradiations in thermonuclear explosions.

generally used for safe handling of these substances. In some work, all operations are performed by remote control. Neptunium in the form of the long-lived isotope ^{237}Np is relatively convenient to work with in chemical investigations. Because of the existence of large quantities of the fissionable isotope ^{239}Pu, the physiological toxicity of plutonium deserves emphasis. In this form plutonium is a dangerous poison by reason of its intense alpha radioactivity [1.4×10^8 alpha particles/(mg·min)] and its physiological behavior. Ingested plutonium may be transferred to the bone and, over a period of time, give rise to neoplasms.

The study of the chemical behavior of concentrated preparations of short-lived isotopes is complicated by the rapid production of hydrogen peroxide in aqueous solutions and the destruction of crystal lattices in solid compounds. These effects are brought about by heavy recoils of high energy alpha particles released in the decay process.

Most chemical investigations with plutonium to date have been performed with ^{239}Pu, but the isotopes ^{242}Pu and ^{244}Pu (produced by intensive neutron irradiation of plutonium) are more suitable for such work because of their longer half-lives and consequently lower specific activities. Much work on the chemical properties of americium has been carried out with ^{241}Am, which is also difficult to handle because of its relatively high specific alpha radioactivity, about 7×10^9 alpha particles/(mg·min). The isotope ^{243}Am has a specific alpha activity about twenty times less than ^{241}Am and is thus a more attractive isotope for chemical investigations. Much of the earlier work with curium used the isotopes ^{242}Cm and ^{244}Cm, but the heavier isotopes offer greater advantages because of their longer half-lives. The isotope ^{248}Cm, which can be obtained in relatively high isotopic purity as the alpha-particle decay daughter of ^{252}Cf, is the most practical for chemical studies. Berkelium (as ^{249}Bk) and californium (as a mixture of the isotopes ^{249}Cf, ^{250}Cf, ^{251}Cf, and ^{252}Cf) are available as the result of intensive neutron irradiation of lighter elements. The best isotope for the study of californium is ^{249}Cf, which can be isolated in pure form through its beta-particle-emitting parent, ^{249}Bk. The isotope ^{253}Es (half-life, 20 d), also a product from such intensive neutron irradiation, is used to study the chemical properties of einsteinium. The isotope ^{254}Es (half-life, 276 d) is more useful for work with macroscopic quantities but is produced in much smaller amounts than ^{253}Es. Weighable amounts of berkelium, californium, and einsteinium are difficult to handle because of their intense radioactivity. Spontaneous fission is a mode of decay for ^{252}Cf (half-life, 2.6 yr), 1 μg of which emits approximately 2×10^8 neutrons/min, and the chief mode of decay of ^{254}Cf (half-life, 56 d), 1 μg of which emits approximately 8×10^{10} neutrons/min. Californium produced in the highest flux reactors unfortunately contains ^{252}Cf, which makes it very difficult to handle. In work with more than a few micrograms of ^{252}Cf it is necessary to do all manipulations by remote control, which is exceedingly cumbersome on such a small scale; therefore, ^{249}Cf is generally used.

Special techniques for experimentation with the actinide elements other than Th and U have been devised because of the potential health hazard to the experimenter and the small amounts available (15). In addition, investigations are frequently carried out with the substance present in very low concentration as a radioactive tracer. Such procedures continue to be used to some extent with the heaviest actinide elements, where only a few score atoms may be available; they were used in the earliest work for all the transuranium elements. Tracer studies offer a method for obtaining knowledge of oxidation states, formation of complex ions, and the solubility of various compounds. These techniques are not applicable to crystallography, metallurgy, and spectroscopic studies.

Microchemical or ultramicrochemical techniques are used extensively in chemical studies of actinide elements (16). If extremely small volumes are used, microgram or lesser quantities of material can give relatively high concentrations in solution. Balances of sufficient sensitivity have been developed for quantitative measurements with these minute quantities of material. Since the amounts of material involved are too small to be seen with the unaided eye, the actual chemical work is usually done on the mechanical stage of a microscope, where all of the essential apparatus is in view. Compounds prepared on such a small scale are often identified by x-ray crystallographic methods.

Position in the Periodic Table and Electronic Structure

Prior to 1944 the location of the heaviest elements in the periodic table had been a matter of question, and the elements thorium, protactinium, and uranium were commonly placed immediately below the elements hafnium, tantalum, and tungsten. In 1944, on the basis of earlier chemical studies of neptunium and plutonium, the similarity between the actinide and the lanthanide elements was recognized (1,7,14). The intensive study of the heaviest elements shows a series of elements similar to the lanthanide series, beginning with actinium (Fig. 4). Corresponding pairs of elements show resemblances in spectroscopic and magnetic behavior that arise because of the similarity of electronic configurations for the ions of the homologous elements in the same state of oxidation, and in crystallographic properties, owing to the near matching of ionic radii for ions of the same charge. The two series are not, however, entirely comparable. A difference lies in the oxidation states. The tripositive state characteristic of lanthanide elements does not appear in aqueous solutions of thorium and protactinium and does not

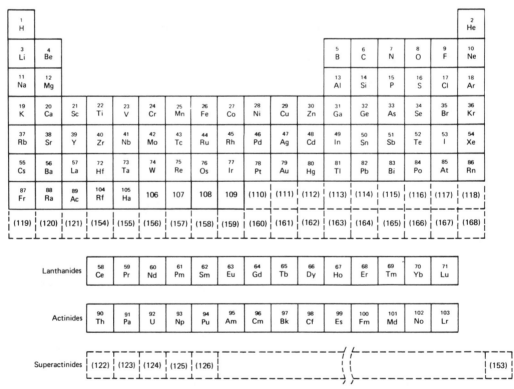

Fig. 4. Futuristic periodic table showing predicted locations of a large number of transuranium elements (atomic numbers in parentheses).

become the most stable oxidation state in aqueous solution until americium is reached. The elements uranium through americium have several oxidation states, unlike the lanthanides. These differences can be interpreted as resulting from the proximity in the energies of the $7s$, $6d$, and $5f$ electronic levels.

Table 3 presents the actual or predicted electronic configurations of the actinide elements (2,14). Similar information for the lanthanide elements is given for purposes of comparison (14). As indicated, fourteen $4f$ electrons are added in the lanthanide series, beginning with cerium and ending with lutetium; in the actinide elements, fourteen $5f$ electrons are added, beginning, formally, with thorium and ending with lawrencium. In the cases of actinium, thorium, uranium, americium, berkelium, californium, and einsteinium the configurations were determined from an analysis of spectroscopic data obtained in connection with the

Table 3. Electronic Configurations for Gaseous Atoms of Lanthanide and Actinide Elements

Atomic number	Element	CAS Registry Number	Electronic configuration[a]
57	lanthanum	[7439-91-0]	$5d6s^2$
58	cerium	[7440-45-1]	$4f5d6s^2$
59	praseodymium	[7440-10-0]	$4f^36s^2$
60	neodymium	[7440-00-8]	$4f^46s^2$
61	promethium	[7440-12-2]	$4f^56s^2$
62	samarium	[7440-19-9]	$4f^66s^2$
63	europium	[7440-53-1]	$4f^76s^2$
64	gadolinium	[7440-54-2]	$4f^75d6s^2$
65	terbium	[7440-27-9]	$4f^96s^2$
66	dysprosium	[7429-91-6]	$4f^{10}6s^2$
67	holmium	[7440-60-0]	$4f^{11}6s^2$
68	erbium	[7440-52-0]	$4f^{12}6s^2$
69	thulium	[7440-30-4]	$4f^{13}6s^2$
70	ytterbium	[7440-64-4]	$4f^{14}6s^2$
71	lutetium	[7439-94-3]	$4f^{14}5d6s^2$
89	actinium		$6d7s^2$
90	thorium		$6d^27s^2$
91	protactinium		$5f^26d7s^2$
92	uranium		$5f^36d7s^2$
93	neptunium		$5f^46d7s^2$
94	plutonium		$5f^67s^2$
95	americium		$5f^77s^2$
96	curium		$5f^76d7s^2$
97	berkelium		$5f^97s^2$
98	californium		$5f^{10}7s^2$
99	einsteinium		$5f^{11}7s^2$
100	fermium		$5f^{12}7s^2$
101	mendelevium		$(5f^{13}7s^2)$
102	nobelium		$(5f^{14}7s^2)$
103	lawrencium		$(5f^{14}6d7s^2$ or $5f^{14}7s^27p)$

[a]Beyond xenon.

[b]Beyond radon. The configurations enclosed in parentheses are predicted.

measurement of the emission lines from neutral and charged gaseous atoms. The knowledge of the electronic structures for protactinium, neptunium, plutonium, curium, and fermium results from atomic beam experiments (15).

Measurements of paramagnetic susceptibility, paramagnetic resonance, light absorption, fluorescence, and crystal structure, in addition to a consideration of chemical and other properties, have provided a great deal of information about the electronic configurations of the aqueous actinide ions and of actinide compounds. In general, all of the electrons beyond the radon core in the actinide compounds and in aqueous actinide ions are in the $5f$ shell. There are exceptions, such as U_2S_3, and subnormal compounds, such as Th_2S_3, where $6d$ electrons are present.

Properties

The close chemical resemblance among many of the actinide elements permits their chemistry to be described for the most part in a correlative way (13,14,17,18).

Oxidation States.   The oxidation states of the actinide elements are summarized in Table 4 (12–14). The most stable states are designated by bold face type and those which are very unstable are indicated by parentheses. These latter states do not exist in aqueous solutions and have been produced only in solid compounds. The IV state of curium is limited to CmO_2 and CmF_4 (solids) and a complex ion stable in highly concentrated cesium fluoride solution, whereas the IV state of californium is limited to CfO_2 and CfF_4 (solids), and double salts such as $7NaF{\cdot}6CfF_4$ (solid). In the second half of the series the II state first appears in the form of solid compounds at californium and becomes successively more stable in proceeding to nobelium. The II state is observed in aqueous solution for mendelevium (and presumably for fermium) and is the most stable state for nobelium. Americium(II), observed only in solid compounds, and berkelium(IV) show the stability of the half-filled $5f$ configuration ($5f^7$) and nobelium(II) shows the stability of the full $5f$ configuration ($5f^{14}$). The greater tendency toward the II state in the actinides, as compared to the lanthanides, is a result of the increasing binding of the $5f$ (and $6d$) electrons upon approaching the end of the actinide series.

The actinide elements exhibit uniformity in ionic types. In acidic aqueous solution, there are four types of cations, and these and their colors are listed in Table 5 (12–14,17). The open spaces indicate that the corresponding oxidation

Table 4. The Oxidation States of the Actinide Elements

Atomic number and element														
89	90	91	92	93	94	95	96	97	98	99	100	101	102	103
Ac	Th	Pa	U	Np	Pu	Am	Cm	Bk	Cf	Es	Fm	Md	No	Lr
						(2)			(2)	(2)	2	2	**2**	
3	(3)	(3)	3	3	3	**3**	**3**	**3**	**3**	3	3	3	3	**3**
	4	4	4	4	4	4	4	4	(4)					
		5	5	**5**	5	5								
			6	6	6	6								
				7	(7)									

Table 5. Ion Types and Colors for Actinide Ions

Element	M^{3+}	M^{4+}	MO_2^+	MO_2^{2+}	MO_5^{3-}
actinium	colorless				
thorium		colorless			
protactinium		colorless	colorless		
uranium	red	green	color unknown	yellow	
neptunium	blue to purple	yellow-green	green	pink to red	dark green
plutonium	blue to violet	tan to orange-brown	rose	yellow to pink-orange	dark green
americium	pink or yellow	color unknown	yellow	rum-colored	
curium	pale green	color unknown			
berkelium	green	yellow			
californium	green				

states do not exist in aqueous solution. The wide variety of colors exhibited by actinide ions is characteristic of transition series of elements. In general, protactinium(V) polymerizes and precipitates readily in aqueous solution and it seems unlikely that ionic forms are present in such solutions.

Corresponding ionic types are similar in chemical behavior, although the oxidation–reduction relationships and therefore the relative stabilities differ from element to element. The ions MO_2^+ and MO_2^{2+} are stable with respect to their binding of oxygen atoms and remain unchanged through a great variety of chemical treatment. They behave as single entities with properties intermediate to singly or doubly charged ions and ions of similar size but of higher charge. The VII oxidation states found for neptunium and plutonium are probably in the form of ions of the type MO_5^{3-} in alkaline aqueous solution; in acid solution these elements in the VII oxidation state readily oxidize water.

The reduction potentials for the actinide elements are shown in Figure 5

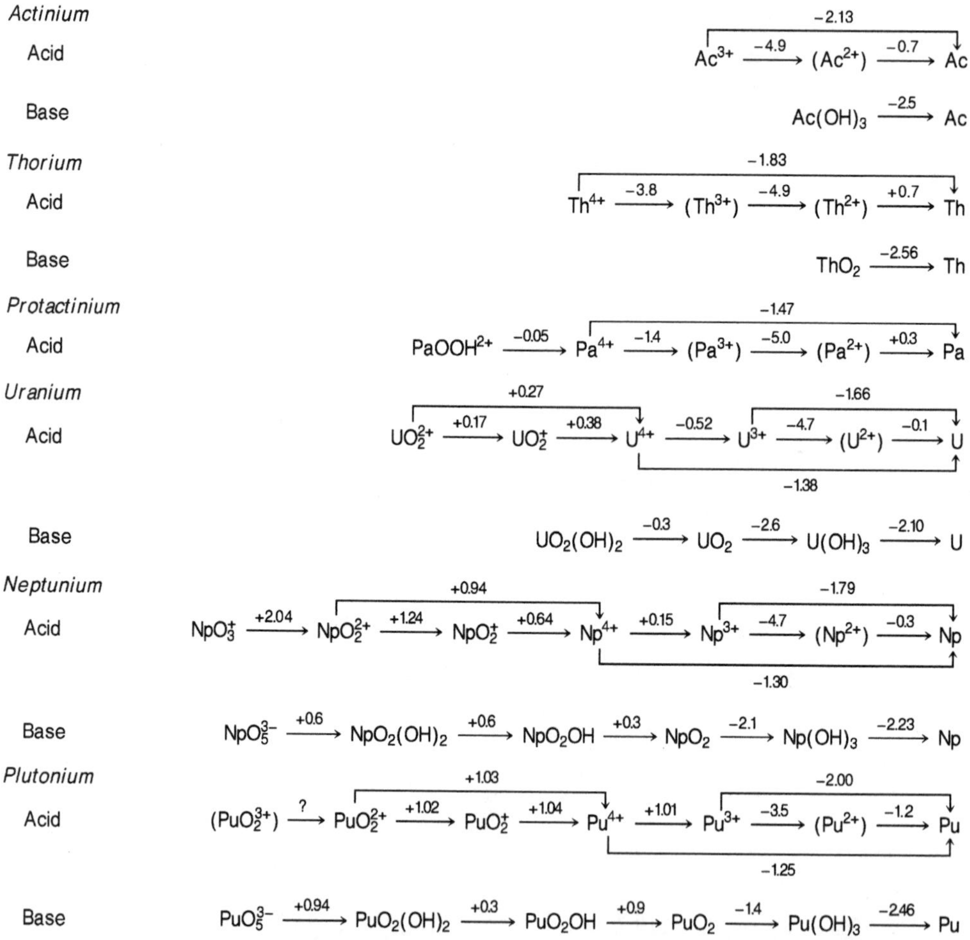

Fig. 5. Standard (or formal) reduction potentials of actinium and the actinide ions in acidic (pH 0) and basic (pH 14) aqueous solutions (values are in volts vs standard hydrogen electrode) (19).

(12–14,17,20). These are formal potentials, defined as the measured potentials corrected to unit concentration of the substances entering into the reactions; they are based on the hydrogen-ion–hydrogen couple taken as zero volts; no corrections are made for activity coefficients. The measured potentials were established by cell, equilibrium, and heat of reaction determinations. The potentials for acid solution were generally measured in 1 M perchloric acid and for alkaline solution in 1 M sodium hydroxide. Estimated values are given in parentheses.

The $M^{4+} \leftrightharpoons M^{3+}$ and $MO_2^{2+} \leftrightharpoons MO_2^+$ couples are readily reversible, and reactions are rapid with other one-electron reducing or oxidizing agents that involve no bond changes. The rate varies with reagents that normally react by two-electron or bond-breaking changes. The $MO_2^+ \leftrightharpoons M^{3+}$, $MO_2^{2+} \leftrightharpoons M^{3+}$, $MO_2^+ \leftrightharpoons M^{4+}$, and $MO_2^{2+} \leftrightharpoons M^{4+}$ couples are not reversible, presumably because of slowness introduced in the making and breaking of oxygen bonds.

Table 6 presents a summary of the oxidation–reduction characteristics of actinide ions (12–14,17,20). The disproportionation reactions of UO_2^+, Pu^{4+}, PuO_2^+,

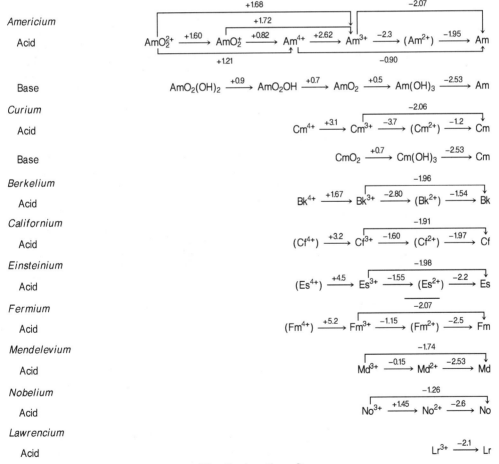

Fig. 5 (*continued*).

Table 6. Stability of Actinide Ions in Aqueous Solution

Ion	Stability
Md^{2+}	stable to water, but readily oxidized
No^{2+}	stable
Ac^{3+}	stable
U^{3+}	aqueous solutions evolve hydrogen on standing
Np^{3+}	stable to water, but readily oxidized by air to Np^{4+}
Pu^{3+}	stable to water and air, but easily oxidized to Pu^{4+}; oxidizes slightly under the action of its own alpha radiation (in form of ^{239}Pu)
Am^{3+}	stable; difficult to oxidize
Cm^{3+}	stable
Bk^{3+}	stable; can be oxidized to Bk^{4+}
Cf^{3+}	stable
Es^{3+}	stable
Fm^{3+}	stable
Md^{3+}	stable, but rather easily reduced to Md^{2+}
No^{3+}	easily reduced to No^{2+}
Lr^{3+}	stable
Th^{4+}	stable
Pa^{4+}	stable to water, but readily oxidized
Pa^{5+}	stable; hydrolyzes readily
U^{4+}	stable to water, but slowly oxidized by air to UO_2^{2+}
Np^{4+}	stable to water, but slowly oxidized by air to NpO_2^+
Pu^{4+}	stable in concentrated acid, eg, 6 M HNO_3, but disproportionates to Pu^{3+} and PuO_2^{2+} at lower acidities
Am^{4+}	known in solution only as complex fluoride and carbonate ions
Cm^{4+}	known in solution only as complex fluoride ion
Bk^{4+}	marginally stable; easily reduced to Bk^{3+}
UO_2^+	disproportionates to U^{4+} and UO_2^{2+}; most stable at pH 2–4
NpO_2^+	stable; disproportionates only at high acidities
PuO_2^+	always tends to disproportionate to Pu^{4+} and PuO_2^{2+} (ultimate products); most stable at very low acidities
AmO_2^+	disproportionates in strong acid to Am^{3+} and AmO_2^{2+}; reduces fairly rapidly under the action of its own alpha radiation at low acidities (in form of ^{241}Am)
UO_2^{2+}	stable; difficult to reduce
NpO_2^{2+}	stable; easy to reduce
PuO_2^{2+}	stable; easy to reduce; reduces slowly under the action of its own alpha radiation (in form of ^{239}Pu)
AmO_2^{2+}	easy to reduce; reduces fairly rapidly under the action of its own alpha radiation (in form of ^{241}Am)
NpO_5^{3-}	observed only in alkaline solution
PuO_5^{3-}	observed only in alkaline solution; oxidizes water

and AmO_2^+ are very complicated and have been studied extensively. In the case of plutonium, the situation is especially complex: four oxidation states of plutonium [(III), (IV), (V), and (VI)] can exist together in aqueous solution in equilibrium with each other at appreciable concentrations.

Hydrolysis and Complex Ion Formation. Hydrolysis and complex ion formation are closely related phenomena (13,14).

Of the actinide ions, the small, highly charged M^{4+} ions exhibit the greatest degree of hydrolysis and complex ion formation. For example, the ion Pu^{4+} hydrolyzes extensively and also forms very strong anion complexes. The hydrolysis of Pu^{4+} is of special interest in that polymers that exist as positive colloids can be produced; their molecular weight and particle size depend on the method of preparation. Polymeric plutonium with a molecular weight as high as 10^{10} has been reported.

The degree of hydrolysis or complex ion formation decreases in the order $M^{4+} > MO_2^{2+} > M^{3+} > MO_2^+$. Presumably the relatively high tendency toward hydrolysis and complex ion formation of MO_2^{2+} ions is related to the high concentration of charge on the metal atom. On the basis of increasing charge and decreasing ionic size, it could be expected that the degree of hydrolysis for each ionic type would increase with increasing atomic number. For the ions M^{4+} and M^{3+}, beginning at about uranium, such a regularity of hydrolytic behavior is observed, but for the remaining two ions, MO_2^+ and MO_2^{2+}, the degree of hydrolysis decreases with increasing atomic number, thus indicating more complicated factors than simple size and charge.

The extensive hydrolysis of protactinium in its V oxidation state makes the chemical investigation of protactinium extremely difficult. Ions of protactinium(V) must be held in solution as complexes, eg, with fluoride ion, to prevent hydrolysis.

The tendency toward complex ion formation of the actinide ions is determined largely by the factors of ionic size and charge. Although there is variation within each of the ionic types, the order of complexing power of different anions is, in general, fluoride > nitrate > chloride > perchlorate for mononegative anions and carbonate > oxalate > sulfate for dinegative anions. The actinide ions form somewhat stronger complex ions than homologous lanthanide ions.

Actinide ions form complex ions with a large number of organic substances (12). Their extractability by these substances varies from element to element and depends markedly on oxidation state. A number of important separation procedures are based on this property. Solvents that behave in this way are tributyl phosphate, diethyl ether [60-29-7], ketones such as diisopropyl ketone [565-80-5] or methyl isobutyl ketone [108-10-1], and several glycol ether type solvents such as diethyl Cellosolve [629-14-1] (ethylene glycol diethyl ether) or dibutyl Carbitol [112-73-2] (diethylene glycol dibutyl ether).

A number of organic compounds, eg, acetylacetone [123-54-6] and cupferron [135-20-6], form compounds with aqueous actinide ions (IV state for reagents mentioned) that can be extracted from aqueous solution by organic solvents (12). The chelate complexes are especially noteworthy and, among these, the ones formed with diketones, such as 3-(2-thiophenoyl)-1,1,1-trifluoroacetone [326-91-0] ($C_4H_3SCOCH_2COCF_3$), are of importance in separation procedures for plutonium.

Metallic State. The actinide metals, like the lanthanide metals, are highly electropositive (13,14,17). They can be prepared by the electrolysis of molten salts or by the reduction of a halide with an electropositive metal, such as calcium or barium. Their physical properties are summarized in Table 7 (13,14,17,21). Metallic protactinium, uranium, neptunium, and plutonium have complex structures

Table 7. Properties of Actinide Metals

Element	Melting point, °C	Heat of vaporization, ΔH_v, kJ/mol (kcal/mol)	Boiling point, °C	Phase	Range of stability, °C	Crystal structure Symmetry	a_0	b_0	c_0	Density, g/mL, at T, °C
actinium	1100 ± 50	293 (70)		α		FC cubic	0.5311			10.07, 25
thorium	1750	564 (130)	3850	α	RT–1360	FC cubic	0.5086			11.724, 25
				β	1360–1750	BC cubic	0.411			
protactinium	1575			α	RT–1170	tetragonal	0.3929		0.3241	15.37, 25
				β	1170–1575	BC cubic	0.381			
uranium	1132	446.4 (106.7)	3818	α	RT–668	orthorhombic	0.2854	0.5869	0.4956	18.97, 25
				β	668–774	tetragonal	1.0759		0.5656	18.11, 720
				γ	774–1132	BC cubic	0.3525			18.06, 805
neptunium	637 ± 2	418 (100)	3900	α	RT–280 ± 5	orthorhombic	0.4721	0.4888	0.4887	20.45, 25
				β	280 ± 5–577 ± 5	tetragonal	0.4895		0.3386	19.36, 313
				γ	577 ± 5–637 ± 2	BC cubic	0.3518			18.04, 600
plutonium	646	333.5 (79.7)	3235	α	RT–122	monoclinic	0.6183	0.4822 $\beta = 101.8°$	1.0963	19.86, 21
				β	122–207	BC monoclinic	0.9284	1.0463 $\beta = 92.13°$	0.7859	17.70, 190
				γ	207–315	orthorhombic	0.3159	0.5768	1.0162	17.13, 235
				δ	315–457	FC cubic	0.4637			15.92, 320
				δ′	457–479	tetragonal	0.3326		0.4463	16.01, 460
				ε	479–640	BC cubic	0.3636			16.48, 490
americium	1173	230 (55)	2011	α	RT–658	hexagonal	0.3468		1.1241	13.67, 20
				β	793–1004	FC cubic	0.4894			13.65, 20
				γ	1050–1173					
curium	1345	386 (92.2)	3110	α	below 1277	hexagonal	0.3496		1.1331	13.51, 25
				β	1277–1345	FC cubic	0.5039			12.9, 25
berkelium	1050			α	below 930	hexagonal	0.3416		1.1069	14.78, 25
				β	930–986	FC cubic	0.4997			13.25, 25
californium	900 ± 30			α	below 900	hexagonal	0.339		1.101	15.1, 25
				β		FC cubic	0.575			8.70, 25
einsteinium	860 ± 30			α	below 860	hexagonal				
				β		FC cubic	0.575			8.84

that have no counterparts among the lanthanide metals. Plutonium metal has very unusual metallurgical properties. It is known to exist in six allotropic modifications between room temperature and its melting point. One of the most interesting features of plutonium metal is the contraction undergone by the δ and δ' phases with increasing temperature. Also noteworthy is the fact that for no phase do both the coefficient of thermal expansion and the temperature coefficient of resistivity have the conventional sign. The resistance decreases if the phase expands on heating. Americium is the first actinide to show resemblance in crystal structure to the lanthanide metals.

With respect to chemical reactivity, the actinide metals resemble the lanthanide metals more than metals of the $5d$ elements such as tantalum, tungsten, rhenium, osmium, and iridium. A wide range of intermetallic compounds has been observed and characterized, including compounds or alloys with members of groups IB, IIA, IIIA, IVA, VIII, VA, and the VIB chalcogenides (13,17). The $5f$ electrons in the lighter actinides are not as localized as the $4f$ electrons in the lanthanides and, with energies close to those of the $6d$ and $7s$ electrons, participate actively in bonding. The participation in bonding by $5f$ electrons is apparently more prominent in the early actinides than in the heavier actinides, where a localized behavior becomes evident. The behavior of the $5f$ electrons makes the actinide metals and their metallic compounds different in their behavior than the transition and lanthanide metals and their compounds.

Solid Compounds. The tripositive actinide ions resemble tripositive lanthanide ions in their precipitation reactions (13,14,17,20,22). Tetrapositive actinide ions are similar in this respect to Ce^{4+}. Thus the fluorides and oxalates are insoluble in acid solution, and the nitrates, sulfates, perchlorates, and sulfides are all soluble. The tetrapositive actinide ions form insoluble iodates and various substituted arsenates even in rather strongly acid solution. The MO_2^+ actinide ions can be precipitated as the potassium salt from strong carbonate solutions. In solutions containing a high concentration of sodium and acetate ions, the actinide MO_2^{2+} ions form the insoluble crystalline salt $NaMO_2(O_2CCH_3)_3$. The hydroxides of all four ionic types are insoluble; in the case of the MO_2^{2+} ions, compounds of the type exemplified by sodium diuranate ($Na_2U_2O_7$) can be precipitated from alkaline solution. The NpO_5^{3-} and PuO_5^{3-} anions, which seem to exist in alkaline solution, form insoluble compounds with several di- and tripositive cations. Peroxide solutions react with actinide ions, particularly M^{4+} ions, to form complex peroxy compounds in solution, and such compounds can be precipitated even from moderately acid solutions. Inorganic anions, eg, sulfate, nitrate, and chloride, are often incorporated in the solid peroxy compounds.

Thousands of compounds of the actinide elements have been prepared, and the properties of some of the important binary compounds are summarized in Table 8 (13,17,18,22). The binary compounds with carbon, boron, nitrogen, silicon, and sulfur are not included; these are of interest, however, because of their stability at high temperatures. A large number of ternary compounds, including numerous oxyhalides, and more complicated compounds have been synthesized and characterized. These include many intermediate (nonstoichiometric) oxides, and besides the nitrates, sulfates, peroxides, and carbonates, compounds such as phosphates, arsenates, cyanides, cyanates, thiocyanates, selenocyanates, sulfites, selenates, selenites, tellurates, tellurites, selenides, and tellurides.

Table 8. Properties and Crystal Structure Data for Important Actinide Binary Compounds

Compound	Color	Melting point, °C	Symmetry	Space group or structure type	Lattice parameters				Density, g/mL
					a_0, nm	b_0, nm	c_0, nm	Angle, deg	
AcH_2	black		cubic	fluorite ($Fm3m$)	0.5670				8.35
ThH_2	black		tetragonal	$F4_1/mmm$	0.5735		0.4971		9.50
Th_4H_{15}	black		cubic	$I\bar{4}3d$	0.911				8.25
$\alpha\text{-}PaH_3$	gray		cubic	$Pm3n$	0.4150				10.87
$\beta\text{-}PaH_3$	black		cubic	$\beta\text{-W}$	0.6648				10.58
$\alpha\text{-}UH_3$	?		cubic	$Pm3n$	0.4160				11.12
$\beta\text{-}UH_3$	black		cubic	$\beta\text{-W}$ ($Pm3n$)	0.6645				10.92
NpH_2	black		cubic	fluorite	0.5343				10.41
NpH_3	black		trigonal	$P\bar{3}c1$	0.651		0.672		9.64
PuH_2	black		cubic	fluorite	0.5359				10.40
PuH_3	black		trigonal	$P\bar{3}c1$	0.655		0.676		9.61
AmH_2	black		cubic	fluorite	0.5348				10.64
AmH_3	black		trigonal	$P\bar{3}c1$	0.653		0.675		9.76
CmH_2	black		cubic	fluorite	0.5322				10.84
CmH_3	black		trigonal	$P\bar{3}c1$	0.6528		0.6732		10.06
BkH_2	black		cubic	fluorite	0.5248				11.57
BkH_3	black		trigonal	$P\bar{3}c1$	0.6454		0.6663		10.44
Ac_2O_3	white		hexagonal	La_2O_3 ($P\bar{3}m1$)	0.407		0.629		9.19
Pu_2O_3	?		cubic	$Ia3$ (Mn_2O_3)	1.103				10.20
Pu_2O_3	black	2085	hexagonal	La_2O_3	0.3841		0.5958		11.47
Am_2O_3	tan		hexagonal	La_2O_3	0.3817		0.5971		11.77
Am_2O_3	reddish brown		cubic	$Ia3$	1.103				10.57
Cm_2O_3	white to faint tan	2260	hexagonal	La_2O_3	0.3792		0.5985		12.17
Cm_2O_3			monoclinic	$C2/m$ (Sm_2O_3)	1.4282	0.3641	0.8883	$\beta = 100.29$	11.90
Cm_2O_3	white		cubic	$Ia3$	1.1002				10.80
Bk_2O_3	light green		hexagonal	La_2O_3	0.3754		0.5958		12.47
Bk_2O_3	yellow-green		monoclinic	$C2/m$	1.4197	0.3606	0.8846	$\beta = 100.23$	12.20
Bk_2O_3	yellowish brown		cubic	$Ia3$	1.0887				11.66
Cf_2O_3	pale green		hexagonal	La_2O_3	0.372		0.596		12.69
Cf_2O_3	lime green		monoclinic	$C2/m$	1.4121	0.3592	0.8809	$\beta = 100.34$	12.37
Cf_2O_3	pale green		cubic	$Ia3$	1.083				11.39
Es_2O_3	white		hexagonal	La_2O_3	0.37		0.60		12.7
Es_2O_3	white		monoclinic	$C2/m$	1.41	0.359	0.880	$\beta = 100$	12.4
Es_2O_3	white		cubic	$Ia3$	1.0766				11.79

Compound	Color	m.p. (°C)	Crystal system	Structure	a	b	c	angle	density
ThO_2	white	ca 3050	cubic	fluorite	0.5597				10.00
PaO_2	black		cubic	fluorite	0.5509				10.45
UO_2	brown to black	2875	cubic	fluorite	0.5471				10.95
NpO_2	apple green		cubic	fluorite	0.5425				11.14
PuO_2	yellow-green to brown	2400	cubic	fluorite	0.53960				11.46
AmO_2	black		cubic	fluorite	0.5374				11.68
CmO_2	black		cubic	fluorite	0.5358				11.92
BkO_2	yellowish-brown		cubic	fluorite	0.5332				12.31
CfO_2	black		cubic	fluorite	0.5310				12.46
Pa_2O_5	white		cubic	fluorite-related	0.5446 or 0.5492				11.14
Np_2O_5	dark brown		monoclinic	$P2_{1}/c$	0.4183	0.6584	0.4086	$\beta = 90.32$	8.18
$\alpha\text{-}U_3O_8$	black-green	1150 (dec)	orthorhombic	$C2mm$	0.6716	1.1960	0.4147		8.39
$\beta\text{-}U_3O_8$	black-green		orthorhombic	$Cmcm$	0.7069	1.1445	0.8303		8.32
$\gamma\text{-}UO_3$	orange	650 (dec)	orthorhombic	$Fddd$	0.981	1.993	0.971		7.80
$AmCl_2$	black		orthorhombic	$Pbnm$ ($PbCl_2$)	0.8963	0.7573	0.4532		6.78
$CfCl_2$	red-amber		?						
$AmBr_2$	black		tetragonal	$SrBr_2$ ($P4/n$)	1.1592		0.7121		7.00
$CfBr_2$	amber		tetragonal	$SrBr_2$	1.1500		0.7109		7.22
ThI_2	gold		hexagonal	$P6_{3}/mmc$	0.397		3.175		7.45
AmI_2	black	ca 700	monoclinic	EuI_2 ($P2_{1}/c$)	0.7677		0.7925	$\beta = 98.46$	6.60
CfI_2	violet		hexagonal	CdI_2 ($P\bar{3}m1$)	0.456		0.699		6.63
CfI_2	violet		rhombohedral	$CdCl_2$ ($R\bar{3}m$)	0.743	0.8311		$\alpha = 36$	6.58
AcF_3	white		trigonal	LaF_3 ($P\bar{3}c1$)	0.741		0.753		7.88
UF_3	black	>1140 (dec)	trigonal	LaF_3	0.718		0.7348		8.95
NpF_3	purple		trigonal	LaF_3	0.7129		0.7288		9.12
PuF_3	purple	1425	trigonal	LaF_3	0.7092		0.7254		9.33
AmF_3	pink	1393	trigonal	LaF_3	0.7044		0.7225		9.53
CmF_3	white	1406	trigonal	LaF_3	0.7014		0.7194		9.85
BkF_3	yellow-green		orthorhombic	YF_3 ($Pnma$)	0.670		0.441		9.70
BkF_3	yellow-green		trigonal	LaF_3	0.697	0.709	0.714		10.15
CfF_3	light green		orthorhombic	YF_3	0.6653		0.4393		9.88
CfF_3	light green		trigonal	LaF_3	0.6945	0.7039	0.7101		10.28
$AcCl_3$	white		hexagonal	UCl_3 ($P6_{3}/m$)	0.762		0.455		4.81
UCl_3	green	835	hexagonal	$P6_{3}/m$	0.7443		0.4321		5.50
$NpCl_3$	green	ca 800	hexagonal	UCl_3	0.7413		0.4282		5.60

433

Table 8 (*continued*)

Compound	Color	Melting point, °C	Symmetry	Space group or structure type	a_0, nm	b_0, nm	c_0, nm	Angle, deg	Density, g/mL
$PuCl_3$	emerald green	760	hexagonal	UCl_3	0.7394		0.4245		5.71
$AmCl_3$	pink or yellow	715	hexagonal	UCl_3	0.7382		0.4214		5.87
$CmCl_3$	white	695	hexagonal	UCl_3	0.7374		0.4185		5.95
$BkCl_3$	green	603	hexagonal	UCl_3	0.7382		0.4127		6.02
$\alpha\text{-}CfCl_3$	green	545	orthorhombic	$TbCl_3$ ($Cmcm$)	0.3859	1.1748	0.8561		6.07
$\beta\text{-}CfCl_3$	green		hexagonal	UCl_3	0.7379		0.4090		6.12
$EsCl_3$	white to orange		hexagonal	UCl_3	0.740		0.407		6.20
$AcBr_3$	white		hexagonal	UBr_3 ($P6_3/m$)	0.806		0.468		5.85
UBr_3	red	730	hexagonal	$P6_3/m$	0.7936		0.4438		6.55
$NpBr_3$	green		hexagonal	UBr_3	0.7916		0.4390		6.65
$NpBr_3$	green		orthorhombic	$TbCl_3$ ($Cmcm$)	0.4109	1.2618	0.9153		6.67
$PuBr_3$	green	681	orthorhombic	$TbCl_3$	0.4097	1.2617	0.9147		6.72
$AmBr_3$	white to pale yellow		orthorhombic	$TbCl_3$	0.4064	1.2661	0.9144		6.85
$CmBr_3$	pale yellow-green	625 ± 5	orthorhombic	$TbCl_3$	0.4041	1.2700	0.9135		6.85
$BkBr_3$	light green		monoclinic	$AlCl_3$ ($C2/m$)	0.723	1.253	0.683	$\beta = 110.6$	5.604
$BkBr_3$	light green		orthorhombic	$TbCl_3$	0.403	1.271	0.912		6.95
$BkBr_3$	yellow green		rhombohedral	$FeCl_3$ ($R\bar{3}$)	0.766			$\alpha = 56.6$	5.54
$CfBr_3$	green		monoclinic	$AlCl_3$	0.7214	1.2423	0.6825	$\beta = 110.7$	5.673
$CfBr_3$	green		rhombohedral	$FeCl_3$	0.758			$\alpha = 56.2$	5.77
$EsBr_3$	straw		monoclinic	$AlCl_3$	0.727	1.259	0.681	$\beta = 110.8$	5.62
PaI_3	black		orthorhombic	$TbCl_3$ ($Cmcm$)	0.433	1.40	1.002		6.69
UI_3	black		orthorhombic	$TbCl_3$	0.4328	1.3996	0.9984		6.76
NpI_3	brown		orthorhombic	$TbCl_3$	0.430	1.403	0.995		6.82
PuI_3	green		orthorhombic	$TbCl_3$	0.4326	1.3962	0.9974		6.92
AmI_3	pale yellow	ca 950	hexagonal	BiI_3 ($R\bar{3}$)	0.742		2.055		6.35
AmI_3	yellow		orthorhombic	$PuBr_3$	0.428	1.394	0.9974		6.95
CmI_3	white		hexagonal	BiI_3	0.744		2.040		6.40
BkI_3	yellow		hexagonal	BiI_3	0.7584		2.087		6.02
CfI_3	red-orange		hexagonal	BiI_3	0.7587		2.081		6.05
EsI_3	amber to light yellow		hexagonal	BiI_3	0.753		2.084		6.18
ThF_4	white	1068	monoclinic	UF_4 ($C2/c$)	1.300	1.099	0.860	$\beta = 126.4$	6.20
PaF_4	reddish-brown		monoclinic	UF_4	1.288	1.088	0.849	$\beta = 126.4$	6.38
UF_4	green		monoclinic	$C2/c$	1.2803	1.0792	0.8372	$\beta = 126.3$	6.73
NpF_4	green	960	monoclinic	UF_4	1.268	1.066	0.834	$\beta = 126.3$	6.86

Compound	Color	mp	Crystal system	Space group	a	b	c	Angle (°)	Density
PuF_4	brown	1037	monoclinic	UF_4	1.260	1.057	0.828	$\beta = 126.3$	7.05
AmF_4	tan		monoclinic	UF_4	1.256	1.058	0.825	$\beta = 125.9$	7.23
CmF_4	light gray-green		monoclinic	UF_4	1.250	1.049	0.818	$\beta = 126.1$	7.36
BkF_4	pale yellow-green		monoclinic	UF_4	1.2396	1.0466	0.8118	$\beta = 126.3$	7.55
CfF_4	light green		monoclinic	UF_4	1.2327	1.0402	0.8113	$\beta = 126.4$	7.57
$\alpha\text{-}ThCl_4$	white		orthorhombic		1.118	0.593	0.909		4.12
$\beta\text{-}ThCl_4$	white	770	tetragonal	UCl_4 ($I4_1/amd$)	0.8473		0.7468		4.60
$PaCl_4$	greenish-yellow		tetragonal	UCl_4	0.8377		0.7481		4.72
UCl_4	green	590	tetragonal	$I4_1/amd$	0.8296		0.7481		4.89
$NpCl_4$	red-brown	518	tetragonal	UCl_4	0.8266		0.7475		4.96
$\alpha\text{-}ThBr_4$	white		tetragonal	$I4_1/a$	0.6737		1.3601		5.94
$\beta\text{-}ThBr_4$	white		tetragonal	UCl_4	0.8931		0.7963		5.77
$PaBr_4$	orange-red		tetragonal	UCl_4	0.8824		0.7957		5.90
UBr_4	brown	519	monoclinic	$2/c/\text{-}$	1.092	0.869	0.705	$\beta = 93.15$	
$NpBr_4$	dark red	464	monoclinic	$2/c/\text{-}$	1.089	0.874	0.705	$\beta = 94.19$	
ThI_4	yellow	556	monoclinic	$P2_1/n$	1.3216	0.8068	0.7766	$\beta = 98.68$	6.00
PaI_4	black								
UI_4	black								
PaF_5	white		tetragonal	$\bar{I}42d$	1.153		0.519		5.81
$\alpha\text{-}UF_5$	grayish white		tetragonal	$I4/m$	0.6512		0.4463		6.47
$\beta\text{-}UF_5$	pale yellow		tetragonal	$\bar{I}42d$	1.1456		0.5196		
NpF_5			tetragonal	$I4/m$	0.653		0.445		
$PaCl_5$	yellow	306	monoclinic	$C2/c$	0.800	1.142	0.843	$\beta = 106.4$	3.81
$\alpha\text{-}UCl_5$	brown		monoclinic	$P2_1/n$	0.799	1.069	0.848	$\beta = 91.5$	
$\beta\text{-}UCl_5$	red-brown		triclinic	$P\bar{1}$	0.707	0.965	0.635	$\alpha = 89.10$, $\beta = 117.36$, $\gamma = 108.54$	
$\alpha\text{-}PaBr_5$	orange-brown		monoclinic	$P2_1/c$	1.264	1.282	0.992	$\beta = 108$	
$\beta\text{-}PaBr_5$	brown		monoclinic	$P2_1/n$	0.9385	1.2205	0.895	$\beta = 91.1$	
UBr_5			monoclinic	$P2_1/n$					
PaI_5	black		orthorhombic		0.698	0.2160	2.130		
UF_6	white	64.02^a	orthorhombic	$Pnma$	0.9900	0.8962	0.5207		5.060
NpF_6	orange	55	orthorhombic	$Pnma$	0.9909	0.8997	0.5202		5.026
PuF_6	reddish-brown	52	orthorhombic	$Pnma$	0.995	0.902	0.526		4.86
UCl_6	dark green	178	hexagonal	$P\bar{3}m1$	1.09		0.603		3.62

[a]At 151.6 kPa, to convert kPa to atm, divide by 101.3.

Hundreds of actinide organic derivatives, including organometallic compounds, are known (12,19,23). A number of interesting actinide organometallic compounds of the π-bonded type have been synthesized and characterized (see ORGANOMETALLICS). The triscyclopentadienyl compounds, although more covalent than the analogous lanthanide compounds, are highly ionic and include UCp_3, $NpCp_3$, $PuCp_3$, $AmCp_3$, $CmCp_3$, $BkCp_3$, and $CfCp_3$ ($Cp = C_5H_5$); each, except the uranium compound, is relatively stable and appreciably volatile but is sensitive to air (23). The tetrakiscyclopentadienyl complexes ($ThCp_4$, $PaCp_4$, UCp_4, and $NpCp_4$) are, like the Cp_3 complexes, soluble in organic solvents and moderately air-sensitive but not appreciably volatile. A number of triscyclopentadienyl actinide halides, of the general formula MCp_3X (M = Th, U, Np), are known, and these are soluble in a range of organic solvents, are more stable to heat than the tetrakis complexes, and are moderately air-sensitive. The pentamethylcyclopentadienides of thorium and uranium also exist. Of special interest are the cyclooctatetraene [629-20-1] (COT) complexes, including the bis compounds $Th(COT)_2$, $Pa(COT)_2$, $U(COT)_4$, $Np(COT)_4$, and $Pu(COT)_4$, and substituted derivatives of these (23). Characterized by monoclinic symmetry with a π-binding sandwich structure involving $5f$ electron orbitals, the prototype compound involving uranium is known as uranocene, in view of the analogy to ferrocene. The compounds are air-sensitive, are only sparingly soluble in organic solvents, and can be sublimed in vacuum. A few are very stable to air. The actinides also form tetraallyl complexes, $M(C_3H_5)_4$, but these are stable only at low temperatures. Many organoactinide complexes with σ and π ligands of the type Cp_3M-R (R = alkyl, aryl, or alkynyl) are known, as well as a number of borohydride compounds. Additional solid organoactinide complexes include alkoxides, dialkylamides, chelates such as β-diketones and β-ketoesters, β-hydroxyquinolines, N-nitroso-N-phenylhydroxylamines (cupferron type), tropolones, N,N-dialkyldithiocarbamates, and phthalocyanines; many of these are useful for separation schemes.

Crystal Structure and Ionic Radii. Crystal structure data have provided the basis for the ionic radii (coordination number = CN = 6), which are summarized in Table 9 (13,14,17). For both M^{3+} and M^{4+} ions there is an actinide contraction, analogous to the lanthanide contraction, with increasing positive charge on the nucleus.

As a consequence of the ionic character of most actinide compounds and of the similarity of the ionic radii for a given oxidation state, analogous compounds are generally isostructural. In some cases, eg, UBr_3, $NpBr_3$, $PuBr_3$, and $AmBr_3$, there is a change in structural type with increasing atomic number, which is consistent with the contraction in ionic radius. The stability of the MO_2 structure (fluorite type) is especially noteworthy, as is shown by the existence of such compounds as PaO_2, AmO_2, CmO_2, and CfO_2 despite the instability of the IV oxidation state of these elements in solution. The actinide contraction and the isostructural nature of the compounds constitute some of the best evidence for the transition character of this group of elements.

Absorption and Fluorescence Spectra. The absorption spectra of actinide and lanthanide ions in aqueous solution and in crystalline form contain narrow bands in the visible, near-ultraviolet, and near-infrared regions of the spectrum (13,14,17,24). Much evidence indicates that these bands arise from electronic transitions within the $4f$ and $5f$ shells in which the $4f^n$ and $5f^n$ configura-

Table 9. Ionic Radii of Actinide and Lanthanide Elements

No. of 4f or 5f electrons	Lanthanide series				Actinide series			
	Element	Radius, nm	Element	Radius, nm	Element	Radius, nm	Element	Radius, nm
0	La^{3+}	0.1032			Ac^{3+}	0.112	Th^{4+}	0.094
1	Ce^{3+}	0.101	Ce^{4+}	0.087	(Th^{3+})	(0.108)	Pa^{4+}	0.090
2	Pr^{3+}	0.099	Pr^{4+}	0.085	(Pa^{3+})	0.104	U^{4+}	0.089
3	Nd^{3+}	0.0983			U^{3+}	0.1025	Np^{4+}	0.087
4	Pm^{3+}	0.097			Np^{3+}	0.101	Pu^{4+}	0.086
5	Sm^{3+}	0.0958			Pu^{3+}	0.100	Am^{4+}	0.085
6	Eu^{3+}	0.0947			Am^{3+}	0.0980	Cm^{4+}	0.084
7	Gd^{3+}	0.0938			Cm^{3+}	0.0970	Bk^{4+}	0.083
8	Tb^{3+}	0.0923	Tb^{4+}	0.076	Bk^{3+}	0.0960	Cf^{4+}	0.0821
9	Dy^{3+}	0.0912			Cf^{3+}	0.0950	Es^{4+}	0.081
10	Ho^{3+}	0.0901			Es^{3+}	0.0940		
11	Er^{3+}	0.0890						
12	Tm^{3+}	0.0880						
13	Yb^{3+}	0.0868						
14	Lu^{3+}	0.0861						

tions are preserved in the upper and lower states for a particular ion. In general, the absorption bands of the actinide ions are some ten times more intense than those of the lanthanide ions. Fluorescence, for example, is observed in the trichlorides of uranium, neptunium, americium, and curium, diluted with lanthanum chloride (15).

Practical Applications

The practical use of three actinide nuclides, ^{239}Pu, ^{235}U, and ^{233}U, as nuclear fuel is well known (5,9). When a neutron of any energy strikes the nucleus of one of these nuclides, each of which is capable of undergoing fission with thermal (essentially zero energy) neutrons, the fission reaction can occur in a self-sustaining manner. A controlled self-perpetuating chain reaction using such a nuclear fuel can be maintained in such a manner that the heat energy can be extracted or converted by conventional means to electrical energy. The complete utilization of nonfissionable ^{238}U (through conversion to fissionable ^{239}Pu) and nonfissionable ^{232}Th (through conversion to fissionable ^{233}U) can be accomplished by breeder reactors.

In addition, three other actinide nuclides (^{238}Pu, ^{241}Am, and ^{252}Cf) have other practical applications (6). One gram of ^{238}Pu produces approximately 0.56 W of thermal power, primarily from alpha-particle decay, and this property has been used in space exploration to provide energy for small thermoelectric power units (see THERMOELECTRIC ENERGY CONVERSION). The most noteworthy example of this latter type of application is a radioisotopic thermoelectric generator left on the moon. It produced 73 W of electrical power to operate the scientific experiments of the Apollo lunar exploration, and was fueled with 2.6 kg of the plutonium isotope in the form of plutonium dioxide, PuO_2. Similar generators powered the instrumentation for other Apollo missions, the Viking Mars lander, and the Pioneer and Voyager probes to Jupiter, Saturn, Uranus, Neptune, and beyond. Americium-241 has a predominant gamma-ray energy of 60 keV and a long half-life of 433 yr for decay by the emission of alpha particles, which makes it particularly useful for a wide range of industrial gaging applications, the diagnosis of thyroid disorders, and for smoke detectors. When mixed with beryllium it generates neutrons at the rate of 1.0×10^7 neutrons/(s·g) ^{241}Am. The mixture is designated ^{241}Am–Be and a large number of such sources are in worldwide daily use in oil-well logging operations to find how much oil a well is producing in a given time span. Californium-252 is an intense neutron source: 1 g emits 2.4×10^{12} neutrons/s. This isotope is being tested for applications in neutron activation analysis, startup sources for nuclear reactors, neutron radiography, portable sources for field use in mineral prospecting and oil-well logging, and in airport neutron-activation detectors for nitrogenous materials (ie, explosives). Both ^{238}Pu and ^{252}Cf are being studied for possible medical applications: the former as a heat source for use in heart pacemakers and heart pumps and the latter as a neutron source for irradiation of certain tumors for which gamma-ray treatment is relatively ineffective.

TRANSACTINIDES

The elements beyond the actinides in the periodic table can be termed the transactinides. These begin with the element having atomic number 104 and extend, in principle, indefinitely. Although only six such elements, numbers 104–109, were definitely known in 1991, there are good prospects for the discovery of a number of additional elements just beyond number 109 or in the region of larger atomic numbers. They are synthesized by the bombardment of heavy nuclides with heavy ions.

As indicated in Figure 4, the early transactinide elements find their place back in the main body of the periodic table. The discoverers of the currently known transactinide elements, suggested names and symbols, and dates of discovery are listed in Table 10 (19). Because there are competing claims for the discovery of these elements, the two groups of discoverers in each case have suggested names for elements 104 and 105. In the case of elements 106–109, names for the elements have not been suggested in order to avoid another duplication.

Study of the chemical properties of element 104 has confirmed that it is indeed homologous to hafnium as demanded by its position in the periodic table (20). Chemical studies have been made for element 105, showing some similarity to tantalum (25); no chemical studies have been made for elements 106–109. Such studies are very difficult because the longest-lived isotope of 104 (^{261}104) has a half-life of only about 1 min, of 105 (^{262}105) a half-life of about 40 s, of 106 (^{263}106) a half-life of about 1 s, and of elements 107–109 half-lives in the range of milliseconds.

On the basis of the simplest projections it is expected that the half-lives of the elements beyond element 109 will become shorter as the atomic number is increased, and this is true even for the isotopes with the longest half-life for each element. This is illustrated by Figure 6, in which the half-lives of the longest-lived

Table 10. The Transactinide Elements

Atomic number	Element	Symbol	Atomic weight[a]	Discoverers and date of discovery
104	rutherfordium [53850-36-5] (U.S.A.)	Rf	261	A. Ghiorso, M. Nurmia, J. Harris, K. Eskola, and P. Eskola, 1969
104	kurchatovium (USSR)	Ku		G. N. Flerov, Yu. Ts. Oganesyan, Yu. V. Lobanov, V. I. Kuznetsov, V. A. Druin, V. P. Perelygin, K. A. Gavrihov, S. P. Tretiakova, and V. M. Plotko, 1964
105	hahnium [53850-35-4] (U.S.A.)	Ha	262	A. Ghiorso, M. Nurmia, K. Eskola, J. Harris, and P. Eskola, 1970
105	nielsbohrium (USSR)	Ns		G. N. Flerov, Yu. Ts. Oganesyan, Yu. V. Lobanov, Yu. A. Lazarev, and S. P. Tretiakova, 1970
106	[54038-81-2]		263	A. Ghiorso, J. M. Nitschke, J. R. Alonso, C. T. Alonso, M. Nurmia, G. T. Seaborg, E. K. Hulet, and R. W. Lougheed, 1974
106				Yu. Ts. Oganesyan, Yu. P. Tretiakov, A. S. Iljinov, A. G. Demin, A. A. Pleve, S. P. Tretiakova, V. M. Plotko, M. P. Ivanov, N. A. Danilov, Yu. S. Korotkin, and G. N. Flerov, 1974
107			262	G. Münzenberg, S. Hofmann, F. P. Hessberger, W. Reisdorf, K. H. Schmidt, J. H. R. Schneider, W. F. W. Schneider, P. Armbruster, C. C. Sahm, and B. Thuma, 1981
107				Y. T. Oganessian, A. G. Demin, N. A. Danilov, M. P. Ivanov, A. S. Iljinov, N. N. Kolesnikov, B. N. Markov, V. M. Plotko, S. P. Tretiakova, and G. N. Flerov, 1976
108			265	G. Münzenberg, P. Armbruster, H. Folger, F. P. Hessberger, S. Hofmann, J. Keller, K. Poppensieker, W. Reisdorf, K. H. Schmidt, H. J. Schött, M. E. Leino, and R. Hingmann, 1984
109			264	G. Münzenberg, P. Armbruster, F. P. Hessberger, S. Hofmann, K. Poppensieker, F. P. Hessberger, S. Hofmann, K. Poppensieker, W. Reisdorf, J. H. R. Schneider, K. H. Schmidt, C. C. Sahm, and D. Vermeulen, 1982

[a]Mass number of longest lived isotopes.

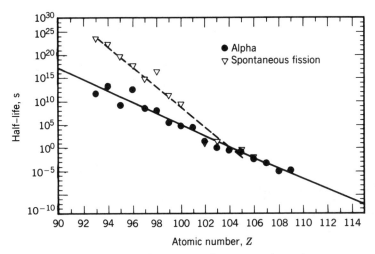

Fig. 6. Longest-lived isotopes of transuranium elements.

isotopes (for alpha and spontaneous fission decay) of transuranium elements are plotted against the increasing atomic number. Thus, if this rate of decrease could be extrapolated to ever-increasing atomic numbers, it would present somewhat dismal future prospects for heavier transuranium elements, but other factors have entered the picture in recent years. These have led to an optimism concerning the prospects for the synthesis and identification of elements beyond the observed upper limit of the periodic table, elements that have come to be referred to as superheavy elements.

Complicated theoretical calculations, based on filled shell (magic number) and other nuclear stability considerations, have led to extrapolations to the far transuranium region (2,26,27). These suggest the existence of closed nucleon shells at $Z = 114$ (proton number) and $N = 184$ (neutron number) that exhibit great resistance to decay by spontaneous fission, the main cause of instability for the heaviest elements. Earlier considerations had suggested a closed shell at $Z = 126$, by analogy to the known shell at $N = 126$, but this is not now considered to be important.

Table 11 illustrates the known closed proton and neutron shells and the predicted closed nuclear shells (shown in parentheses) that might be important in stabilizing the superheavy elements. Included by way of analogy are the long-known closed electron shells observed in the buildup of the electronic structure of atoms. These correspond to the noble gases, and the extra stability of these closed shells is reflected in the relatively small chemical reactivity of these elements. The predicted (in parentheses) closed electronic structures occur at $Z = 118$ and $Z = 168$.

Enhancing the prospects for the actual synthesis and identification of super-heavy nuclei is the fact that the calculations show the doubly magic nucleus $^{298}114$ not to be a single long-lived specimen but to be a part of a rather large "island of stability" in a "sea of spontaneous fission" (2,26,27). The grid lines of Figure 7 show "magic" numbers of protons or neutrons giving rise to exceptional stability. The doubly magic region at 82 protons and 126 neutrons is shown by a

Table 11. Closed Proton (Z) and Neutron (N) Shells with Closed Electron (Noble Gas) Shells for Comparison[a]

Z	N	e^-
2(He)	2(^{4}He)	2(He)
8(O)	8(^{16}O)	10(Ne)
20(Ca)	20(^{40}Ca)	18(Ar)
28(Ni)	28(^{56}Ni)	36(Kr)
50(Sn)	50(^{88}Sr)	54(Xe)
82(Pb)	82(^{140}Ce)	86(Rn)
	126(^{208}Pb)	
(114)	(184)	(118)
(126)		
		(168)

[a]Predicted shells shown in parentheses.

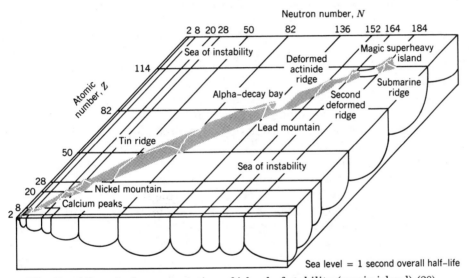

Fig. 7. Allegorical representation of island of stability (magic island) (28).

mountain; a predicted doubly magic but less stable region at 114 protons and 184 neutrons is shown by a hill at the island of stability. The ridges depict areas of enhanced stability due to a single magic number. Other calculations suggest that there should be stabilizing, deformed nuclear shells (or subshells) at lower neutron numbers, such as $N = 162$.

The effects of a rather distinct deformed shell at $N = 152$ were clearly seen as early as 1954 in the alpha-decay energies of isotopes of californium, einsteinium, and fermium. In fact, a number of authors have suggested that the entire transuranium region is stabilizied by shell effects with an influence that increases markedly with atomic number. Thus the effects of shell substructure lead to an increase in spontaneous fission half-lives of up to about 15 orders of magnitude for the heavy transuranium elements, the heaviest of which would otherwise have half-lives of the order of those for a compound nucleus (10^{-14} s or less) and not of milliseconds or longer, as found experimentally. This gives hope for the synthesis and identification of several elements beyond the present heaviest (element 109) and suggest that the peninsula of nuclei with measurable half-lives may extend up to the island of stability at $Z = 114$ and $N = 184$.

Turning to consideration of electronic structure, upon which chemical properties must be based, modern high speed computers have made possible the calculation of such structures (13,26,27). The calculations show that elements 104 through 112 are formed by filling the $6d$ electron subshell, which makes them, as expected, homologous in chemical properties with the elements hafnium ($Z = 72$) through mercury ($Z = 80$). Elements 113 through 118 result from the filling of the $7p$ subshell and are expected to be similar to the elements thallium ($Z = 81$) through radon ($Z = 86$). Thus these calculations are consistent with the periodic table shown in Figure 4, which shows the filling of the electron subshells extending beyond element 118.

The calculations indicate that the $8s$ subshell should fill at elements 119 and 120, thus making these an alkali and alkaline earth metal, respectively. Next, the calculations point to the filling, after the addition of a $7d$ electron at element 121 of the inner $5g$ and $6f$ subshells, 32 places in all, which the author has termed the superactinide elements and which terminates at element 153. This is followed by the filling of the $7d$ subshell (elements 154 through 162) and $8p$ subshell (elements 163 through 168).

Actually, more careful calculations have indicated that the picture is not this simple. The calculations indicate that other electrons ($8p$ and $7d$) in addition to those identified in the above discussion enter prominently as early as element 121, and other anomalies may enter as early as element 103, thus further complicating the picture. These perturbations, caused by spin-orbit splitting, become especially significant beyond the superactinide series, and lead to predictions of chemical properties that are not consistent, element by element, with those suggested by Figure 4. The largest deviations occur far beyond the region of expected nuclear stability, the only region where it might be possible to synthesize or find superheavy elements.

Thus it can be seen that elements in and near the island of stability based on element 114 can be predicted to have chemical properties as follows. Element 114 should be a homologue of lead, that is, should be eka-lead, and element 112 should be eka-mercury, element 110 should be eka-platinum, etc (26,27). If there is an island of stability at element 126, this element and its neighbors should have chemical properties like those of the actinide and lanthanide elements (26).

Attempts to synthesize and identify superheavy elements, through bombardments of a wide range of heavy nuclides with a wide range of heavy ions, have so far been unsuccessful (27). Unfortunately, the yield of the desired product nuclei is predicted to be very small because the overwhelming proportion of the nuclear reactions lead to fission and other reactions, rather than to the desired synthesis of superheavy nuclei through amalgamation of the heavy ion projectile and the target nucleus. An inherent difficulty in the synthesis of the superheavy nuclei situated near the center of the island of stability is the simultaneous requirement that there be a sufficient number of neutrons as well as protons in the product nucleus. The desired product nuclei have larger ratios of neutrons to protons than the constituent projectiles and target nuclei, and hence, somewhat unusual nuclear reactions are required. There is no way to know, prior to the actual experimental attempts, which nuclear reactions will be most effective. There are a number of aspects of the dynamics of the reaction of the heavy incident nuclear projectile with the heavy target nucleus that are not understood and that might present serious or fatal impediments to the production of super-heavy nuclei; eg, the coulomb repulsive force exerted on a heavy projectile may be too large relative to the nuclear attractive force to allow the desired amalgama-tion reaction to occur, and nuclear matter may be too viscous to allow the required ready amalgamation of the incident heavy ion with the heavy target nucleus.

If the half-life of a superheavy nucleus should happen to be as long as a few times 10^8 yr (now considered to be very unlikely), this would be long enough to allow the isotope to survive and still be present on the earth (as in the case of ^{235}U, which has a half-life of 7×10^8 yr), provided that it was initially present as a result of the cosmic nuclear reactions that led to the creation of the solar system. Every attempt to find evidence in nature, direct or indirect, of the superheavy elements associated with the island of stability centered around element 114 has been inconclusive. Because of the physical limitations inherent in any experimental technique, it is not possible to say that such superheavy elements do not exist in nature. The results of such searches establish that the concentration, if they are present, is extremely small, eg, much less than one part in 10^{12} parts of ore. Searches have also been made in cosmic rays, meteorites, and moon rocks, with generally negative results except for some indirect evidence of former presence in meteorites during the early history of the meteorite's life. There was some evi-dence found in the Allende meteorite, which fell in Mexico in 1969; the evidence for such an extinct superheavy element (now considered to be unlikely) was the observation of a unique composition of xenon isotopes that might have been formed from decay by spontaneous fission with a half-life of 10^7–10^9 yr (29). The postulated current synthesis of a broad range of chemical elements, possibly including superheavy elements, in stars might enhance the prospects for finding even shorter-lived superheavy elements in cosmic rays (30); elements as heavy as uranium have apparently been found in cosmic rays emanating from such stars.

BIBLIOGRAPHY

"Actinides" in *ECT* 2nd ed., Vol. 1, pp. 351–371, by G. T. Seaborg, U.S. Atomic Energy Commission; "Actinides and Transactinides," in *ECT* 3rd ed., Vol. 1, pp. 456–488, by Glenn T. Seaborg, University of California, Berkeley.

1. G. T. Seaborg and J. J. Katz, eds., *The Actinide Elements, National Nuclear Energy Series,* Div. IV, 14A, McGraw-Hill Book Co., Inc., New York, 1954; G. T. Seaborg, ed., *Transuranium Elements: Products of Modern Alchemy,* Benchmark Papers, Dowden, Hutchinson & Ross, Inc., Stroudsburg, Pa., 1978.
2. *Transurane, Gmelins Handbuch der anorganischen Chemie,* Part A, *The Elements,* Verlag Chemie GmbH, Weinheim/Bergstrasse, 1972–1973, A1,I, 1973, A1,II, 1974.
3. E. K. Hyde, I. Perlman, and G. T. Seaborg, *Nuclear Properties of the Heavy Elements,* Vol. II, *Detailed Radioactivity Properties,* Prentice-Hall, Inc., Englewood Cliffs, N.J., 1964.

4. C. M. Lederer, J. M. Hollander, and I. Perlman, *Table of Isotopes, Sixth Edition,* John Wiley & Sons, Inc., New York, 1967.

5. *Proceedings of the First United Nations International Conference on the Peaceful Uses of Atomic Energy, Geneva, 1955,* United Nations, New York, 1955; *Proceedings of the Second United Nations International Conference on the Peaceful Uses of Atomic Energy, Geneva, 1958,* United Nations, New York, 1958; *Proceedings of the Third United Nations International Conference on the Peaceful Uses of Atomic Energy, Geneva, 1964,* United Nations, New York, 1964; *Proceedings of the Fourth United Nations International Conference on the Peaceful Uses of Atomic Energy, Geneva, 1971,* United Nations, New York, and IAEA, Vienna, 1972.

6. G. T. Seaborg, *Nucl. Appl. Technol.* **9,** 830 (1970).

7. G. T. Seaborg, J. J. Katz, and W. M. Manning, eds., *The Transuranium Elements: Research Papers, National Nuclear Energy Series,* Div. IV, 14B, McGraw-Hill Book Co., Inc., New York, 1949.

8. D. C. Hoffman, F. O. Lawrence, J. L. Mewherter, and F. M. Rourke, *Nature (London)* **234,** 132 (1971).

9. G. T. Seaborg, *Man-Made Transuranium Elements,* Prentice-Hall, Inc., Englewood Cliffs, N.J., 1963.

10. J. L. Crandall, *Production of Berkelium and Californium, Proceedings of the Symposium Commemorating the 25th Anniversary of Elements 97 and 98 held on January 20, 1975, Lawrence Berkeley Laboratory, Report LBL-4366.* Available as *TID 4500-R64* from National Technical Information Center, Springfield, Va., 1975.

11. *Series on Radiochemistry,* National Academy of Sciences. Reports available from Office of Technical Services, Department of Commerce, Washington, D.C.: P. C. Stevenson and W. E. Nervik, *Actinium* (with scandium, yttrium, rare earths), *NAS-NS-3020;* E. Hyde, *Thorium, NAS-NS-3004;* H. W. Kirby, *Protactinium, NAS-NS-3016;* J. E. Gindler, *Uranium, NAS-NS-3050;* G. A. Burney and R. M. Harbour, *Neptunium, NAS-NS-3060;* G. H. Coleman and R. W. Hoff, *Plutonium, NAS-NS-3058;* R. A. Penneman and R. K. Keenan, *Americium and Curium, NAS-NS-3006;* G. H. Higgins, *The Transcurium Elements, NAS-NS-3031.*

12. *Transurane, Gmelins Handbuch der anorganischen Chemie,* Part D, *Chemistry in Solution,* Springer-Verlag, Berlin, Heidelberg, New York, 1975. D1, D2.

13. C. Keller, *The Chemistry of the Transuranium Elements,* Verlag Chemie GmbH, 1971.

14. J. J. Katz and G. T. Seaborg, *The Chemistry of the Actinide Elements,* Methuen & Co., Ltd., London and John Wiley & Sons, Inc., New York, 1957.

15. Ref. 2, A2.

16. G. T. Seaborg, *The Transuranium Elements,* Yale University Press, New Haven, Conn., 1958.

17. J. C. Bailor, Jr., J. Emeleus, R. Nyholm, and A. F. Trotman-Dickenson, eds., *Comprehensive Inorganic Chemistry,* Vol. 5, *Actinides,* Pergamon Press, Oxford, New York, 1973.

18. Ref. 2, Index, 1979.

19. J. J. Katz, G. T. Seaborg, and L. R. Morss, eds., *The Chemistry of the Actinide Elements,* 2nd ed., Chapman and Hall, London, 1986.

20. W. Müller and R. Lindner, eds., *Transplutonium 1975, 4th International Transplutonium Element Symposium, Proceedings of the Symposium at Baden Baden September 13–17, 1975;* W. Müller and H. Blank, eds., *Heavy Element Properties, 4th International Transplutonium Element Symposium, 5th International Conference on Plutonium and Other Actinides 1975, Proceedings of the Joint Session of the Baden Baden Meetings September 13, 1975,* North-Holland Publishing Co., Amsterdam, American Elsevier Publishing Co., Inc., New York.

21. Ref. 2, Parts B1, B2, B3, *The Metals and Alloys,* 1976, 1977.

22. Ref. 2, Part C, *The Compounds,* 1972.
23. E. C. Baker, G. W. Halstead, and K. N. Raymond, *Struct. Bonding (Berlin)* **25,** 23–68 (1976); T. J. Marks and A. Streitwieser, Jr., Chapt. 22 of ref. 19; T. J. Marks, Chapt. 23 of ref. 19.
24. W. T. Carnall and H. M. Crosswhite, Chapt. 16 of ref. 19.
25. K. E. Gregorich, R. A. Henderson, D. M. Lee, M. J. Nurmia, R. M. Chasteler, H. L. Hall, D. A. Bennett, C. M. Gannett, R. B. Chadwick, J. D. Leyba, D. C. Hoffman, and G. Herrmann, *Radiochim. Acta* **43,** 223 (1988).
26. G. T. Seaborg, *Ann. Rev. Nucl. Sci.* **18,** 53 (1968); O. L. Keller, Jr., and G. T. Seaborg, *Ann. Rev. Nucl. Sci.* **27,** 139 (1977).
27. G. Hermann, *Superheavy Elements, International Review of Science, Inorganic Chemistry,* Series 2, Vol. 8, Butterworths, London, and University Park Press, Baltimore, Md., 1975; G. T. Seaborg and W. Loveland, *Contemp. Physics* **28,** 233 (1987).
28. K. E. Gregorich and G. T. Seaborg, LBL Report 27947; *Radioanal. Nucl. Chem.* **142,** 27 (1990).
29. E. Anders, H. Huguchi, J. Gros, H. Takahashi, and J. W. Morgan, *Science* **190,** 1262 (1975).
30. V. Trimble, *Rev. Mod. Phys.* **47,** 877 (1975).

GLENN T. SEABORG
University of California, Berkeley

ACTIVATED SLUDGE.   See WATER, SEWAGE.

ACTIVATION ANALYSIS.   See FINE ART EXAMINATION AND CONSERVATION; NONDESTRUCTIVE TESTING.

ADAMANTANE.   See HYDROCARBONS, SURVEY.

ADHESION.   See ADHESIVES.

ADHESIVES

An *adhesive* is a material capable of holding together solid materials by means of surface attachment. *Adhesion* is the physical attraction of the surface of one material for the surface of another. An *adherend* is the solid material to which the adhesive adheres and the *adhesive bond* or *adhesive joint* is the assembly made by joining adherends together by means of an adhesive. *Practical adhesion* is the physical strength of an adhesive bond. It primarily depends on the forces of adhesion, but its magnitude is determined by the physical properties of the adhesive and the adherend, as well as the engineering of the adhesive bond.

The *interphase* is the volume of material in which the properties of one substance gradually change into the properties of another. The interphase is useful for describing the properties of an adhesive bond. The *interface,* contained within the interphase, is the plane of contact between the surface of one material and the surface of another. Except in certain special cases, the interface is imaginary. It is useful in describing surface energetics.

Theories of Adhesion

There is no unifying theory of adhesion describing the relationship between practical adhesion and the basic intermolecular and interatomic interactions which take place between the adhesive and the adherend either at the interface or within the interphase. The existing adhesion theories discussed below are, for the most part, rationalizations of observed phenomena, although in some cases, predictions regarding the relative ranking of practical adhesion can actually be made.

Diffusion Theory.   The diffusion theory of adhesion is mostly applied to polymers. It assumes mutual solubility of the adherend and adhesive to form a true interphase. The solubility parameter, the square root of the cohesive energy density of a material, provides a measure of the intermolecular interactions occurring within the material. Thermodynamically, solutions of two materials are most likely to occur when the solubility parameter of one material is equal to that of the other. Thus, the observation that "like dissolves like." In other words, the adhesion between two polymeric materials, one an adherend, the other an adhesive, is maximized when the solubility parameters of the two are matched; ie, the best practical adhesion is obtained when there is mutual solubility between adhesive and adherend. The diffusion theory is not applicable to substantially dissimilar materials, such as polymers on metals, and is normally not applicable to adhesion between substantially dissimilar polymers.

Electrostatic Theory.   The basis of the electrostatic theory of adhesion is the differences in the electronegativities of adhering materials (1,2). If two materials having measurably different electronegativities are brought into contact, electron transfer from the material of lower electronegativity to that of higher electronegativity can occur. This transfer then forms a double layer of charge across an interface and a net attraction results. (The electrostatic theory of adhesion is generally regarded as being incomplete.)

Surface Energetics and Wettability Theory.   The surface energetics and wettability theory of adhesion is concerned with the effect of intermolecular and interatomic forces on the surface energies of the adhesive and the adherend and the interfacial energy between the two. The distances over which interatomic and intermolecular forces operate are on the order of 10^{-7} cm. For these forces (and hence adhesion) to have any measurable value, the adhesive must come into intimate contact with the adherend; the surface of the adherend must be completely "wetted" by the adhesive. Surface energy, the excess energy that a material possesses because of its surface, can be understood by examining a molecule within the bulk of a material, where the molecule experiences intermolecular bonding in essentially all directions. In contrast, a molecule at a surface has intermolecular bonds only in the direction of the bulk material and in the same plane as the surface. There are no intermolecular forces above the surface and the lack of such forces causes a change in the physical properties of the layers of molecules immediately adjacent to the surface. In general, the density of the surface layers is less than that of the bulk. Because the distances between molecules are increased, the intermolecular energy is increased, giving rise to excess or surface energy. Interfacial energy, which is usually smaller in magnitude than surface energy, is found when a material is in contact with a dissimilar substance. This energy arises because the intermolecular forces in one medium are not necessarily matched by the intermolecular forces in another.

For liquids, the quantities surface energy and surface tension, usually given the symbol γ, are numerically the same and are given in units of millijoules per square meter. Liquid surface energy, or surface tension, is easily observed in that liquids appear to have a skin. Surface energies of common organic liquids vary from about 10 mJ/m^2 up to about 65 mJ/m^2 at room temperature. The room-temperature surface energy of water is approximately 72 mJ/m^2.

For a solid surface, surface energy and surface tension are not the same. Thermodynamically, the surface energy can be defined as a change in a free energy function with a change in the surface area of a material. The surface energy of solids cannot be measured directly. However, the value for most inorganic solids is thought to be in the hundreds of mJ/m^2, and for polymers, to be approximately the same as that of organic liquids.

The wettability theory of adhesion is inextricably related to the study of contact angles of liquids on solid surfaces. A force balance at the point of contact between the liquid and the solid can be written (3)

$$\gamma_{LV} \cos \theta = \gamma_{SV} - \gamma_{SL} \tag{1}$$

where γ_{LV} is the liquid–vapor interfacial tension, θ is the contact angle, γ_{SV} is the solid–vapor interfacial tension, and γ_{SL} is the solid–liquid interfacial tension. These parameters are all defined in Figure 1.

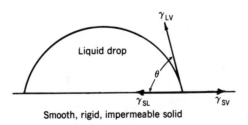

Fig. 1. Schematic of a liquid drop on a solid surface showing the contact angle, θ, as well as the liquid–solid interfacial energy, γ_{SL}, the liquid–vapor interfacial tension, γ_{LV}, and the solid–vapor interfacial energy, γ_{SV}.

W_A, the work of adhesion, may be defined in terms of the surface energies of the adhesive and the adherend (4)

$$W_A = \gamma_1 + \gamma_2 - \gamma_{12} \tag{2}$$

and the Young-Dupre equation allows one to determine the work of adhesion

$$W_A = \gamma_{\mathrm{LV}} (1 + \cos \theta) \qquad (3)$$

from a simple measurement of the contact angle and the liquid surface tension. Much of the experimental work in adhesion science has centered around the relationship between practical adhesion and the work of adhesion. Recent investigations indicate that practical adhesion can be related to the work of adhesion times a function describing the energy dissipation mechanisms within an adhesive bond (5).

The surface energy of polymers is an important consideration in the wettability theory of adhesion (6–8). The critical wetting tension, γ_{C}, of a solid surface provides a criterion for the surface's complete wetting. For complete wetting, the liquid surface tension of the liquid that is wetting should be less than the critical wetting tension of the solid surface wetted. Critical wetting tension is determined by the measurement of the equilibrium contact angles of various liquids on a single surface to generate a $\cos \theta$ versus γ_{LV} plot. This plot, extrapolated to $\cos \theta = 1$, defines the surface's critical wetting tension. A situation where $\gamma_{\mathrm{LV}} \leq \gamma_{\mathrm{C}}$ is necessary but not sufficient for good adhesion.

Also important in the wettability theory of adhesion is bonding across the interface, which can come about through hydrogen bonding, Lewis acid–base or other donor–acceptor interactions, or covalent bonding. Bonding at an interface is exemplified by the action of coupling agents such as certain silanes, zirconates, titanates, and chrome complexes. These agents form a covalent bond between an inorganic substrate and an organic overcoat. The titanates have also been used to modify the rheology of inorganic particulates dispersed in organic media. The silanes have found wide use in adhesion building situations, most notably that of fiberglass composites and especially in improving environmental resistance of adhesive bonds (see SILICON COMPOUNDS, SILYLATING AGENTS).

Mechanical Interlocking Theory. If an adhesive and an adherend, differing substantially in physical properties, meet at a sharp interface, there is an abrupt plane of stress transfer under load. In addition, there is minimal interfacial area between the two, and hence minimal bonding opportunities. If instead of a sharp interface, there is a microscopically rough surface to which the adhesive can be applied, then there is the possibility of mechanical interlocking. A microscopically rough surface provides more points of stress transfer under load as well as more surface area for possible interfacial bonding, than does a sharp one. A rough surface also provides the possibility of a lock and key effect. If the adhesive has wet the adherend so thoroughly that the adhesive has flowed completely around asperities on the adherend surface, then the adhesive would have to pass physically through the adherend asperities in order to be removed. This is much the same situation as when a key is turned in a lock. The asperities within the lock cannot be passed over by the key and hence the key cannot be removed. The adhesive's viscosity, as well as the time of contact of the adhesive on the adherend, play a substantial role in determining how well a mechanically roughened surface is adhered to by an adhesive.

Guidelines for Good Adhesion. The various adhesion theories can be used to formulate guidelines for good adhesion:

1. An adhesive should possess a liquid surface tension that is less than the critical wetting tension of the adherend's surface.
2. The adherend should be mechanically rough enough so that the asperities on the surface are on the order of, or less than, one micrometer in size.
3. The adhesive's viscosity and application conditions should be such that the asperities on the adherend's surface are completely wetted.
4. If an adverse environment is expected, covalent bonding capabilities at the interface should be provided.

For good adhesion, the adhesive and the adherend should, if possible, display mutual solubility to the extent that both diffuse into one another, providing an interphasal zone.

History of Adhesive Bonding

One of the earliest records of the use of adhesive bonding is a mural discovered in a sepulcher in ancient Thebes which depicts veneering of furniture by means of a hot-melt adhesive. Veneering was discussed in *The Natural History* of Pliny the Elder, and adhesive bonding is mentioned in the Bible. The oldest forms of adhesives were those of natural origin, eg, glues that are made from the collagen-containing parts of animals, or from animal blood. Casein, or milk, glues have been known since the 9th century and exist to this date in a form not substantially different from that discussed by Theophilus. Some of the more ancient resins used to form adhesives are materials such as bitumen, shellac, and pitch.

Some important advances in adhesive technology include (9):

The synthesis in 1869 of nitrocellulose [*9004-70-0*], the first synthetic adhesive.

The synthesis in 1912 of phenol–formaldehyde resins by Baekeland, which forms the basis of many modern day adhesives.

The synthesis in 1928 of polychloroprene or Neoprene, which has found much use in high strength elastomeric adhesives.

The 1930s marked the development of pressure-sensitive adhesive tapes. Even though adhesives were used in bonding wooden aircraft in the 1900s, the first metal bonding adhesive was developed for aircraft in 1941 by Nicholas de Bruyne of the Aerotech Company. A significant development in the late 1940s was the advent of the first epoxy resin adhesives followed in the 1950s and 1960s by the cyanoacrylate adhesives which cure under the influence of moisture. Anaerobic, as well as silicone, adhesives were developed. High temperature resistant adhesives, such as the aromatic polyimides, were formulated in the United States in the late 1960s, in response to the supersonic transport program. From the late 1940s through the 1980s, the most significant advances in adhesive bonding technology involved the marriage of different types of polymers to form hybrid adhesives. Most notable is the use of elastomeric materials in combination with high strength, cross-linked polymers such as epoxies or phenolics. Recently, work in the area of adhesives has centered on modification of high temperature materials and elastomers or other substances. Post-It Notes, which use a specific type of microstructured adhesive, are an example of the continually expanding applications of adhesives. Advances have also been made in the methods of adhesives application. Application methods range from easy-to-mix containers to robots for high-volume adhesive application.

Advantages and Disadvantages of Using Adhesives

Adhesive Advantages.   Adhesive bonding technology is but one means of joining adherends. Other joining technologies include the use of mechanical fasteners eg, screws, nuts and bolts, rivets, and nails. In comparison to other methods of joining, adhesives provide several advantages. First, a properly applied adhesive provides a joint having a more uniform stress distribution under load than a mechanical fastener which requires a hole in the adherend. A hole acts as a stress concentrator; the presence of a hole causes the overall joint strength to be less than expected from the materials' properties in the assembly under load. Second, adhesives provide the ability to bond dissimilar materials such as metals without facilitation of corrosion. If two metals which are susceptible to galvanic corrosion are joined by means of a metallic mechanical fastener, the more active metal corrodes more rapidly than if left unjoined. If this assembly is, instead, made by an adhesive, the adhesive acts as an electrical insulator protecting the more active material from the less active one. Third, using an adhesive to make an assembly increases fatigue resistance. This factor is especially important in the aircraft industry, where adhesives are routinely used to increase resistance to fatigue crack propagation in comparison to the use of mechanical fasteners. Fourth, adhesive joints can be made of heat or shock sensitive materials. Use of these materials is precluded by mechanical fasteners because of the shock of putting the fastener through the assembly. Fifth, adhesive joining can bond and seal simultaneously. In attempting to create a pressure vessel, a fuselage for an aircraft, or similar application, an adhesive provides a bond which is strong enough to hold major loads and also seals the assembly against the ingress of water or the egress of air or pressurized gases. And sixth, use of an adhesive to form an assembly usually results in a weight reduction in comparison to mechanical fasteners since adhesives, for the most part, have densities which are substantially less than that of metals.

Adhesive Disadvantages.   There are some limitations in using adhesives to form assemblies. The major limitation is that the adhesive joint is formed by means of surface attachment and is, therefore, sensitive to the substrate surface condition. If the surface is substantially contaminated by oils or dirt then the adhesive does not adhere well and the durability of the adhesive bond is decreased. Fasteners present no such problem. Thus, concomittant to the increased useage of adhesives, a technology of substrate surface preparation has developed. If the substrate has a layer of extraneous material on the surface, then the adhesive bond is more likely to fail in service because of the ingress of moisture into the adhesive bond. Coupling agents can play a role in improving durability under these conditions.

Another limitation of adhesive bonding is the lack of a nondestructive quality control procedure. When a rivet or a mechanical fastener has not been properly installed, it is normally obvious. However, an adhesive which is totally enclosed by the adherends is very difficult to inspect. A number of nondestructive testing (qv) techniques, eg, ultrasonic scanning and x-ray penetration, have been developed, but these methods cannot predict the strength of an adhesive bond. Rather, they provide information as to the presence of voids within the bond. Adhesive joining can also lead to slower processing than mechanical fasteners. This depends substantially on the type of adhesive that is being used. Finally, adhesive joining is still somewhat limited because most designers of assemblies are simply not familiar with the engineering characteristics of adhesives. Because an adhesive bond depends upon surface attachment for its strength, the adhesive must be in contact with a substantially larger area than that occupied by a mechanical fastener. Adhesive joints need to be designed to use adhesive bonding advantages, rather than as a direct replacement to mechanical fasteners.

Mechanical Tests of Adhesive Bonds

The three principal forces to which adhesive bonds are subjected are illustrated in Figure 2. Figure 2a shows a shear force in which one adherend is forced past the other whereas Figure 2b illustrates peeling. In the peeling situation shown, at least one of the adherends is flexible enough to be bent away from the adhesive bond. Cleavage is shown in Figure 2c. The cleavage force and the peeling force are very similar to one another, but the former applies when the adherends are nondeformable and the latter when the adherends are deformable.

The principal type of shear test specimen used in the industry, the lap shear specimen, is 2.54 cm wide and has a 3.23-cm^2 overlap bonded by the adhesive. Adherends are chosen according to the industry: aluminum for aerospace, steel for automotive, and wood for construction applications. Adhesive joints made in this fashion are tested to failure in a tensile testing machine. The temperature of test, as well as the rate of extension, are specified. Results are presented in units of pressure, where the area of the adhesive bond is considered to be the area over which the force is applied. Although the 3.23-cm^2 overlap is taken as a standard in the industry, many applications use far larger overlap areas and thus must be tested in such fashion.

Peel tests are accomplished using many different geometries. In the simplest peel test, the T-peel test, the adherends are identical in size, shape, and thickness. Adherends are attached at their ends to a tensile testing machine and then separated in a "T" fashion. The temperature of the test, as well as the rate of adherend separation, is specified. The force required to open the adhesive bond is measured and the results are reported in terms of newtons per meter (pounds per inch, ppi). There are many other peel test configurations, each dependent upon the adhesive application. Such tests are well described in the ASTM literature.

Fracture mechanics (qv) tests are typically used for structural adhesives. Thus, tests such as the double cantilever beam test (Fig. 2c), in which two thick adherends joined by an adhesive are broken by cleavage, provide information

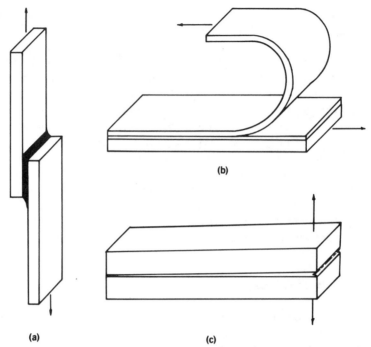

Fig. 2.   Illustrations of forces to which adhesive bonds are subjected. (**a**) A standard lap shear specimen where the black area shows the adhesive. The adherends are usually 25 mm wide and the lap area is 312.5 mm^2. The arrows show the direction of the normal application of load. (**b**) A peel test where the loading configuration, shown by the arrows, is for a 180° peel test. (**c**) A double cantilever beam test specimen used in the evaluation of the resistance to crack propagation of an adhesive. The normal application of load is shown by the arrows. This load is applied by a tensile testing machine or other mechanical means of holding open the end of the specimen.

relating to structural flaws. Results can be reported in a number of ways. The most typical uses a quantity known as the strain energy release rate, G_c, given in energy per unit area.

Table 1 provides the approximate load bearing capabilities of various adhesive types. Because the load-bearing capabilities of an adhesive are dependent upon the adherend material, the loading rate, temperature, and design of the adhesive joint, wide ranges of performance are listed.

Chemistry and Uses of Adhesives

Although materials such as Portland cement (see CEMENT), solder (see SOLDER AND BRAZING ALLOYS), and silicates can be considered to be adhesives, this discussion only includes organic materials such as those that form the materials presented in Table 1.

Structural Adhesives.   A structural adhesive is a resin system, usually a thermoset, that is used to bond high strength materials in such a way that the bonded joint is able to bear a load in excess of 6.9 MPa (1,000 psi) at room

Table 1. Load-Bearing Capabilities of Adhesives[a]

Adhesive type	Shear load, MPa[b]	Peel load, N/m[c]
pressure sensitive	0.005–0.02[d]	300–600
rubber based	0.3–7	1000–7000
emulsion	10–14	
hot melt	1–15	1000–5000
natural product (structural)	10–14	
polyurethane	6–17	2000–10,000
acrylic	6–20	900–6000
epoxy	14–50	700–18,000
phenolic	14–35	700–9,000
polyimide	13–17	350–1760

[a]Load bearing capabilities are dependent upon the adherend, joint design, rate of loading, and temperature. Values given represent the type of adherends normally used at room temperature. Lap shear values approximate those obtainable from an overlap of 3.2 cm^2.
[b]To convert from MPa to psi, multiply by 145.
[c]To convert from N/m to ppi, divide by 175.
[d]Pressure sensitive adhesives normally are rated in terms of shear holding power, ie, time to fail in minutes under a constant load.

temperature. Structural adhesives are the strongest form of adhesive and are meant to hold loads permanently. They exist in a number of forms. The most common form is the two-part adhesive, widely available as a consumer product. The next most familiar is that which is obtained as a room temperature curing liquid. Less common are primer–liquid adhesive combinations which cure at room temperature. Structural adhesive pastes which cure at 120°C are widely available in the industrial market. Paste adhesives are applied by methods such as troweling or by means of a spatula. Adhesives requiring the highest technology are obtained in film form: the material is usually available as a roll in which the adhesive has been cast onto a release liner. The film adhesive is first removed from its release liner and then smoothed onto the parts to be bonded. These parts are mated and the adhesive is cured in an autoclave where both pressure and temperature can be controlled.

Epoxy Resins. The chemistry of epoxy resin adhesives is quite varied. However, the most widely used is that formed from the reaction of 4,4'-isopropylidene diphenol (bisphenol A) [80-05-7], $C_{15}H_{16}O_2$, and epichlorohydrin [106-89-8], C_3H_5ClO. This epoxy resin is more commonly known as the diglycidyl ether of bisphenol A. Its chemistry can be extended by further reaction with the bisphenol A starting material, creating resins of varying molecular weights that have either oxirane or hydroxyl end functionality. The formulation of most epoxy adhesives is based upon proper selection of various molecular weights of the diglycidyl ether of bisphenol A and its chain-extended product (see EPOXY RESINS).

Common room temperature curing chemistries for epoxy resins include reaction with aliphatic amines or mercaptans to generate amino alcohols and mercaptoether alcohols, respectively. A common high temperature curing method utilizes anhydride. Both the mercaptan and the anhydride addition reactions are usually catalyzed. One of the more common catalysts for this reaction is tris(dimethylamino)phenol [31194-38-4], $C_{12}H_{21}N_3O$. Etherification, involving the reaction of alcohols that are often present in epoxy resins with an oxirane to create ether alcohols, is another common curing chemistry. Although this reaction does not typically take place at room temperature, it does proceed readily in the presence of amine catalysts at temperatures in excess of 120°C. Imidazoles, which react by an anionic polymerization mechanism with the oxirane at elevated temperatures, are also used in epoxy adhesive chemistry.

Adhesives which are meant to cure at temperatures of 120 or 171°C require curatives which are latent at room temperature, but react quickly at the cure temperatures. Dicyanodiamide [461-58-5], $C_2H_4N_4$, is one such latent curative for epoxy resins. It is insoluble in the epoxy at room temperature but rapidly solubilizes at elevated temperatures. Other latent curatives for 171°C are complexes of imidazoles with transition metals, complexes of Lewis acids (eg, boron trifluoride and amines), and diaminodiphenylsulfone, which is also used as a curing agent in high performance composites. For materials which cure at lower temperatures (120°C), these curing agents can be made more soluble by alkylation of dicyanodiamide. Other materials providing latency at room temperature but rapid cure at 120°C are the blocked isocyanates, such as the reaction products of toluene diisocyanate and amines. At 120°C the blocked isocyanate decomposes to regenerate the isocyanate and liberate an amine which can initiate polymerization of the epoxy resin. Materials such as Monuron can also be used to accelerate the cure of dicyanodiamide so that it takes place at 120°C.

The two-part epoxy adhesive, readily available in hardware stores or other consumer outlets, comes in two tubes. One tube contains the epoxy resin, the other contains an amine hardener. Common diamine room temperature epoxy curing agents are materials such as the polyamides, available under the trade name Versamid. These polyamides are the reaction products of dimer acids and aliphatic diamines such as diethylenetriamine [111-40-0], $C_4H_{13}N_3$. Other room temperature curing agents are triethylenetetraamine [112-24-3], $C_6H_{18}N_4$, and the polypropylene glycol diamines, known as the Jeffamines. The physical properties of room temperature curing epoxies depend heavily upon the chemical structure of the curing agent. If there is a high cross-link density because of polyamino functionality, the resultant cured epoxy resin is likely to be brittle. If instead, there is a long distance between the amine functionalities in the curing agent, the cured resin is likely to be more flexible.

Epoxy structural adhesives are used in an extraordinarily wide range of applications. They are available in essentially all of the forms discussed above, except for primer–liquid combinations or as room temperature curing liquids. The highest technology application for epoxies is in aerospace structural adhesive bonding. Epoxy resin structural adhesives are used for secondary structure in the fuselage, wings, and control surfaces of aircraft. Of specific interest is the formation of honeycomb structures in which a structural film adhesive is used to bond metallic or nonmetallic honeycomb face sheets, generating a very lightweight, yet stiff, structure. Epoxy structural adhesives are also used in the automotive industry for such applications as heming flange bonding or stiffener bonding.

Phenolic Resins.   Phenolic resins (qv) are formed by the reaction of phenol [108-95-2], C_6H_6O, and formaldehyde [50-00-0], CH_2O. If basic conditions and an excess of formaldehyde are used, the result is a resole phenolic resin, which will cure by itself liberating water. If an acid catalyst and an excess of phenol are used, the result is a novolac phenolic resin, which is not self-curing. Novolac phenolic resins are typically formulated to contain a curing agent which is most often a material known as hexamethylenetetraamine [100-97-0], $C_6H_{12}N_4$. Phenolic resin adhesives are found in film or solution form and normally require cures of at least 170°C. Volatiles are eliminated during the curing process for most phenolic-based adhesives. The cure is performed under very high applied pressures typically exceeding 300–700 kPa (44–100 psi) and bonding is usually accomplished in a hot press or in an autoclave.

Phenolic resins are the oldest form of synthetic structural adhesives. Usage ranges from bonding automobile and other types of brake linings to aerospace applications. These adhesives have a reputation for providing the most durable structural bonds to aluminum. Because of volatiles, however, and the need for high pressures, the phenolic resins are used less as adhesives than the epoxy resins.

Acrylic Adhesives. Acrylic structural adhesives can be classified into three major types: the surface-activated acrylics (anaerobics), the surface-activated second-generation acrylics, and the cyanoacrylates.

Anaerobic structural adhesives are typically formulated from acrylic monomers such as methyl methacrylate [80-62-6], $C_5H_8O_2$, and methacrylic acid [79-41-4], $C_4H_6O_2$ (see ACRYLIC ESTER POLYMERS). Very often, cross-linking agents such as dimethacrylates are also added. A peroxide, such as cumene hydroperoxide [80-15-9], is used as the polymerization initiator. Anaerobics polymerize by a redox mechanism which takes place between the hydroperoxide and iron oxide containing surfaces. Materials such as saccharin [81-07-2] and *N,N'*-dimethyl-*p*-toluidine [99-97-8] are used to catalyze this reaction. The surface-activated acrylics have found use as thread locking adhesives and are widely used as such in machinery construction, automotive engines, and similar applications. Surface-activated second-generation acrylic adhesives will be discussed later under elastomer modification of structural adhesives.

Cyanoacrylate adhesives (Super-Glues) are materials which rapidly polymerize at room temperature. The standard monomer for a cyanoacrylate adhesive is ethyl 2-cyanoacrylate [7085-85-0], which readily undergoes anionic polymerization. Very rapid cure of these materials has made them widely used in the electronics industry for speaker magnet mounting, as well as for wire tacking and other applications requiring rapid assembly. Anionic polymerization of a cyanoacrylate adhesive is normally initiated by water. Therefore, atmospheric humidity or the surface moisture content must be at a certain level for polymerization to take place. These adhesives are not cross-linked as are the surface-activated acrylics. Rather, the cyanoacrylate material is a thermoplastic, and thus, the adhesives typically have poor temperature resistance.

High Temperature Resistant Adhesives. Structural adhesives having high temperature resistance are used in the aerospace and other industries wherever light weight, thermally resistant joints are needed. Such joints are often found in automotive, jet, and rocket engines. Resins having high glass-transition temperatures and high cross-link densities, that are highly aromatic in character, are used to generate these adhesives. In general, the polymer backbone is formed by multiple rather than single bonds. One example is a polyimide, formed by the reaction of an aromatic anhydride and an aromatic amine. Telechelic polyimides having terminal norbornyl or acetylenic groups can also be generated in order to provide a cross-linking resin. Polyphenylquinoxalines have been investigated as high temperature resistant structural adhesives and bismaleimides, using either Michael addition or Diels-Alder curing chemistry, have recently found their way into this technology.

Elastomeric Modified Adhesives.   The major characteristic of the resins discussed above is that after cure, or after polymerization, they are extremely brittle. Thus, the utility of unmodified common resins as structural adhesives would be very limited. For highly cross-linked resin systems to be useful structural adhesives, they have to be modified to ensure fracture resistance. Modification can be effected by the addition of an elastomer which is soluble within the cross-linked resin. Modification of a cross-linked resin in this fashion generally decreases the glass-transition temperature but increases the resin flexibility, and thus increases the fracture resistance of the cured adhesive. Recently, structural adhesives have been modified by elastomers which are soluble within the uncured structural adhesive, but then phase separate during the cure to form a two-phase system. The matrix properties are mostly retained: the glass-transition temperature is only moderately affected by the presence of the elastomer, yet the fracture resistance is substantially improved.

The earliest known elastomeric modification of a structural adhesive occurred during World War II, when it was demonstrated that phenolic resin adhesives could be modified with poly(vinyl acetal) resins (10). Aluminum sheet was coated using a phenolic resin and then particles of poly(vinyl acetal) [*26591-54-8*] were sprinkled onto the uncured phenolic resin. The bond was closed, cured at high temperature and high pressure, and a fracture-resistant bond having improved performance was obtained. Phenolic resins have also been modified by acrylonitrile–butadiene elastomers, which provide higher peel strength of the cured adhesive than poly(vinyl acetal) modification. In both cases, flexibilization is presumably effected because the elastomer is soluble within the phenolic resin. This is fairly clearly demonstrated in the case of the acrylonitrile–butadiene elastomer modification because the glass-transition temperature of these phenolic materials is typically below room temperature.

Elastomer modifiers have also been used to improve the performance of epoxy-based structural adhesives. In much the same fashion as with the phenolic resins, butadiene–acrylonitrile elastomers modify epoxies for improvement in peel strength, eg, high molecular weight butadiene–nitrile rubbers containing carboxyl groups have been used as flexibilizing agents. Epoxy resins have also been modified by elastomers which phase separate in order to provide a toughened epoxy system. Such elastomers are based upon butadiene–nitrile rubbers that are low in molecular weight and are carboxy or amino terminated. Prereaction of these elastomers with an epoxy resin forms a copolymer which is soluble in the resin before cure, but insoluble after cure. Formulators of modified epoxy resin adhesives must also be careful to control the molecular weight between cross-links as the mechanism of toughening by phase-separated elastomers requires a certain ductility of the resin. Materials such as bisphenol A can be included as a chain extension agent or judicious choices of the curing agent used in the formulation must be made. Other elastomers have also been used to modify epoxy structural adhesives. Notable in this technology is a phase-separated acrylic elastomer.

Acrylic structural adhesives have been modified by elastomers in order to obtain a phase-separated, toughened system. A significant contribution in this technology has been made in which acrylic adhesives were modified by the addition of chlorosulfonated polyethylene to obtain a phase-separated structural adhesive (11). Such adhesives also contain methyl methacrylate, glacial methacrylic acid, and cross-linkers such as ethylene glycol dimethacrylate [97-90-5]. The polymerization initiation system, which includes cumene hydroperoxide, N,N'-dimethyl-p-toluidine, and saccharin, can be applied to the adherend surface as a primer, or it can be formulated as the second part of a two-part adhesive. Modification of cyanoacrylates using elastomers has also been attempted: copolymers of acrylonitrile, butadiene, and styrene; ethylene copolymers with methylacrylate; or copolymers of methacrylates with butadiene and styrene have been used. However, because of the extreme reactivity of the monomer, modification of cyanoacrylate adhesives is very difficult and material purity is essential in order to be able to modify the cyanoacrylate without causing premature reaction.

Urethane Adhesives.   Urethane structural adhesives are those based upon the reaction of a diisocyanate and a diol (see URETHANE POLYMERS). The cross-linking reaction is between an isocyanate and carbamate to form an allophanate or between an isocyanate and a urea to form a biuret. The diols are typically polyethers such as polypropylene or polytetramethylene oxide glycol. Polyester polyols, such as polycaprolactone diol, are also used. Polyethylene and similiar glycols are not used because of their affinity for water. Toluene diisocyanate [1321-38-6] and diphenylmethane 4,4'-diisocyanate [101-68-8] are used to make polyurethane structural adhesives but because of the toxicity of these aromatic isocyanates, they are not used in low molecular weight form. Rather, a higher molecular weight material having lower volatility, and hence less possibility of user sensitization, is formed by reaction of the diisocyanate and a polyol. Lower molecular weight polyols that are liquid at room temperature and hence easy to mix with the diisocyanate are used. Other diisocyanates include the aliphatic hydrogenated diphenylmethane diisocyanate and isophorone diisocyanate [4098-71-9], used in applications where clear products are needed.

The physical properties of polyurethane adhesives result from a special form of phase separation which occurs in the cross-linked polyurethane structure. The urethane portions of polyurethanes tend to separate from the polyol portion of the resin, providing good shear strength, good low temperature flexibility, and high peel strength. Catalysts such as dibutyltin dilaurate [77-58-7], stannous octoate [1912-83-0], 1,4-diazabicyclo[2.2.2]octane [280-57-9], and mercury compounds are used to promote the reaction of isocyanates and polyols. Polyurethanes are mostly used as two-part structural adhesives. In one part the material is mostly the diisocyanate or the diisocyanate prepolymer while the second part is mostly the polyol. Various viscosity modifiers such as inorganic fillers or other polymers are also used. Additionally, polyurethanes can be made into one-part reactive systems obtained by using catalysts which unblock at cure temperatures or by providing a moisture curing mechanism. Reaction of isocyanates and moisture liberates amines which can then react with isocyanates to provide a polyurea chain extension. Structural adhesives can also be formulated using blocked isocyanates. That is, isocyanates can react with low molecular weight compounds materials which will to form thermally revert back to the isocyanate at a certain temperature.

Physical Properties of Structural Adhesives. Table 1 lists the ranges of performance which can be expected from structural adhesives. Those materials providing highest temperature performance are also those which are most likely to be the most brittle. Room temperature curing adhesives, in general, provide the lowest performance. The best overall performance is from those obtained using the 120°C curing epoxy-based, rubber-modified, film adhesive; these are the materials of choice for demanding applications such as aerospace structural adhesive bonding.

Natural-Product-Based Adhesives. *Protein-Based Adhesives.* Protein-based adhesives are normally used as structural adhesives; they are all polyamino acids that are derived from blood, fish skin, casein [9000-71-9], soybeans, or animal hides, bones, and connective tissue (collagen). Setting or cross-linking methods typically used are insolubilization by means of hydrated lime and denaturation. Denaturation methods require energy which can come from heat, pressure, or radiation, as well as chemical denaturants such as carbon disulfide [75-15-0] or thiourea [62-56-6]. Complexing salts such as those based upon cobalt, copper, or chromium have also been used. Formaldehyde and formaldehyde donors such as hexamethylenetetraamine can be used to form cross-links. Removal of water from a protein will also often denature the material.

A typical formula for a protein-based adhesive includes a natural protein that has been solubilized by means of sodium hydroxide and then dispersed in water. This ionized protein then is mixed with a defoamer, hydrated lime which acts as the cross-linking agent, sodium silicate, various chemical denaturants, and biocides. These last materials are added because proteins are also nutrients for microbes. Fillers are also added in order to modify viscosity. Depending upon the formulation, the pot life of such an adhesive can be from several hours to several days. The adhesive is typically applied by a roll coater and is normally heat cured in a press. The formula just described is typical of those used for the manufacture of plywood where the critical features include the moisture content of the wood, lack of foam in the coating, a material which is roll coatable, appropriate shelf life, and finally, resistance to moisture. The moisture content is critical to adhesive bond performance in that the physical properties of wood (qv) are substantially affected by the amount of water it contains. If plywood is made in such a fashion that the wood is either very dry or very wet during bonding, the plywood, when it changes moisture content during use, will change its size enough that it may delaminate. Protein-based adhesives are normally ranked by their resistance to moisture. Blood or blood–soybean-blend adhesives are considered to be the most water resistant, followed by casein or casein–soybean blends. The adhesive of least water resistance is based upon soybeans or animal hide glue.

Starch-Based Adhesives. Starches in the form of amylose [9005-82-7] and amylopectin [9037-22-3], both branched carbohydrates, are obtained from plants by hot water leeching of roots and seeds. The resultant starch (qv) is a granular, semicrystalline material, which must be cooked in order to be dispersed in water and hence used as an adhesive. No true aqueous solutions form upon heating.

Rather the mixture is a dispersion of various portions of the starch molecules. Starch can be chemically modified in a number of fashions. One way is to modify starch in water in the presence of a mineral acid leading to a composition known as thin boiling starch. Treatment using sodium hypochlorite [7681-52-9] or other oxidizing media leads to oxidized starches in which the association of the chains is reduced. Dextran [9004-54-0] is obtained by roasting acidified, dried starch. Starch can also be esterified to give various chemical materials. Typical starch-based adhesives include various additives and modifiers. One modifier is a plasticizer, which can be used to flexibilize the starch adhesive. Preservatives are also added to prevent these materials from providing nutrients for microbes. Fillers, such as kaolin [1332-58-7] clay, and calcium carbonate [471-34-1], are added both to modify the viscosity of the starch adhesive and to reduce the cost of the material. Borax [1303-96-4], which can act as a viscosity modifier, is very often used because it increases the tack, and acts as an antimicrobial. Starch-based adhesives find most use in the area of binding paper (qv). They have been used as gum label and envelope adhesives, and, in certain formulations, provide the adhesive for the bottom of paper grocery bags.

Cellulosics.   Cellulosic adhesives are obtained by modification of cellulose [9004-34-6] (qv) which comes from cotton linters and wood pulp. Cellulose can be nitrated to provide cellulose nitrate [9004-70-0], which is soluble in organic solvents. When cellulose nitrate is dissolved in amyl acetate [628-63-7], for example, a general purpose solvent-based adhesive which is both waterproof and flexible is formed. Cellulose esterification leads to materials such as cellulose acetate [9004-35-7], which has been used as a pressure-sensitive adhesive tape backing. Cellulose can also be ethoxylated, providing hydroxyethylcellulose which is useful as a thickening agent for poly(vinyl acetate) emulsion adhesives. Etherification leads to materials such as methylcellulose [9004-67-5] which are soluble in water and can be modified with glyceral [56-81-5] to produce adhesives used as wallpaper paste (see CELLULOSE ESTERS; CELLULOSE ETHERS).

Tackifying Resins.   Tackifying resins have found great use in modifying a number of different types of adhesives. Abietic acid [514-10-3] and pimaric acid [127-27-5], known as rosin acids, are obtained from pine tree sap. These acids are not used in their natural form; rather they are modified by a number of techniques, such as heating to high temperatures to induce disproportionation, reaction with alcohols to provide an esterified product, or reaction with various catalysts either to hydrogenate or to polymerize the material. Aromatic resins, for example the coumarone–indene resins, are obtained from natural product streams, such as coal, petroleum, or wood tar. Chemicals such as indene [95-13-6] or methylindene are polymerized with styrene [100-42-5] or methylstyrene in the presence of a Lewis acid to provide aromatic tackifying resins. Aliphatic hydrocarbon tackifying resins are obtained from polymerization products of *cis-* and *trans*-piperylene (1,3-pentadiene) and isoprene and dicylopentadiene; terpene resins are obtained from turpentine and citrus peels. Additionally, natural products such as α-pinene, β-pinene, and dipentene are polymerized in the presence of aluminum chloride to provide terpene resins. The uses of tackifying resins are discussed below.

Pressure-Sensitive Adhesives.   A pressure-sensitive adhesive, a material which adheres with no more than applied finger pressure, is aggressively and permanently tacky. It requires no activation other than the finger pressure, exerts a strong holding force, and should be removeable from a smooth surface without leaving a residue.

Applications and Formulation. Pressure-sensitive adhesives are most widely used in the form of adhesive tapes. These tapes are used for an extraordinary number of applications: masking, medical applications, electrical insulation, assembly, packaging, and other applications. The application governs the choice of tape backing and the adhesive formulation. A transparent backing having a relatively weak adhesive is used for paper mending; a filament filled backing having an aggressive adhesive is used for packaging applications. Pressure-sensitive adhesives are also obtainable in aerosol form for use in various graphic arts applications.

The general formula for a pressure-sensitive adhesive includes an elastomeric polymer, a tackifying resin, any necessary fillers, various antioxidants and stabilizers, if needed, and cross-linking agents. In formulating a pressure-sensitive adhesive, a balance of three physical properties needs to be taken into account: shear strength, peel strength, and tack. The shear strength or shear holding power of the adhesive is typically measured by hanging a weight on the end of a piece of tape and measuring the time to failure. Tack is the technical term applied to quantify the sticky feel of the material. In general, the shear strength and the tack of a pressure-sensitive adhesive increase and then go through a maximum as a function of the amount of tackifying resin added. The peel strength usually increases with the amount of tackifying resin. The shear holding power often depends upon the mode of cross-linking. Thus, a balance of properties appropriate to the application is obtained by controlling the rubber-to-resin ratio as well as the level and type of cross-linking agent.

The most widely, and perhaps the earliest, elastomer used to formulate pressure-sensitive adhesives is natural rubber. In 1845 a patent was granted for a formulation including India rubber, gum of southern pine, balsam of Peru, and ground litharge (12). This formulation provided a material akin to a pressure-sensitive adhesive. Other elastomers which have been used are the butyl rubbers, poly(vinyl ether)s, acrylics (especially those having long chain alkyl groups), and silicones. The modes of tackifying depend upon the elastomers. Natural rubber, butyl rubber, and acrylic materials are typically tackified by the addition of a tackifying agent of the type described above. Block copolymers such as the styrene–butadiene–styrene block or the styrene–isoprene–styrene block must be treated so that the continuous phase is the tackified material. Thus, the choice of tackifying resin depends upon the solubility of the tackifying agent in the continuous phase. Vinyl ethers are tackified by adding lower molecular weight poly(vinyl ether). Silicone pressure-sensitive adhesives are typically tackified by the addition of silicone gum and silicone resin to the silicone elastomer.

Natural-rubber-based pressure-sensitive adhesives can be cured by standard rubber curatives, eg, sulfur plus an accelerator (see RUBBER, NATURAL); butyl rubber adhesives can be cured by *p*-quinone dioxime plus oxidizers; and acrylics can be cured by the addition of difunctional acrylics which form cross-links upon curing. Silicones are typically cross-linked by means of peroxide cures. In order to form tapes, pressure-sensitive adhesives can be applied to substrates by several methods. The most common is to apply adhesive out of solvent. However, considering the increased regulation of solvents, the more environmentally safe method is to coat pressure-sensitive adhesives either as hot-melts or as water-based emulsions.

Hot-Melt Adhesives. Hot-melt adhesives are 100% nonvolatile thermoplastic materials that can be heated to a melt and then applied as a liquid to an adherend. The bond is formed when the adhesive resolidifies. The oldest example of a hot-melt adhesive is sealing wax.

Formulations. Hot-melt adhesive formulation involves providing the best physical properties over as large a temperature range as possible. This type of adhesive is generally useful in the temperature range where the material is either leathery or rubbery, ie, between the glass-transition temperature and the melt temperature. Hot-melt adhesives are based on thermoplastic polymers that may be compounded or uncompounded: ethylene–vinyl acetate copolymers, paraffin waxes, polypropylene, phenoxy resins, styrene–butadiene copolymers, ethylene–ethyl acrylate copolymers, and low , and low density polypropylene are used in the compounded state; polyesters, polyamides, and polyurethanes are used in the mostly uncompounded state.

The most widely used thermoplastic polymer is the ethylene–vinyl acetate copolymer, which is obtainable in a wide range of molecular weights as well as in a variety of compositions. Often flexibilizers or plasticizers are added in order to improve both the mechanical shock resistance and the thermal properties of the adhesive. Polybutenes, phthalates, and tricresyl phosphate have been used as plasticizers. Tackifying agents can also be added. Because hot-melt adhesives are frequently ethylene-based, they are subject to oxidation if, as in a typical situation, the adhesive sits in an applicator for long periods before use. Thus, antioxidants such as hindered phenols are often used, as are fillers. Fillers are added to opacify or to modify the adhesive's flow characteristics, as well as to reduce cost. Wax is also a very important component. Wax alters surface characteristics by decreasing both the liquid adhesive's surface tension and its viscosity in the melt. Upon solidification, however, the wax acts to increase the strength of the adhesive. Both paraffin and microcrystalline wax are used (see WAXES).

In the area of molecularly designed hot-melt adhesives, the most widely used resins are the polyamides (qv), formed upon reaction of a diamine and a dimer acid. Dimer acids (qv) are obtained from the Diels-Alder reaction of unsaturated fatty acids. Linoleic acid is an example. Judicious selection of diamine and diacid leads to a wide range of adhesive properties. Typical shear characteristics are in the range of thousands of kilopascals and are dependent upon temperature. Although hot-melt adhesives normally become quite brittle below the glass-transition temperature, these materials can often attain physical properties that approach those of a structural adhesive. These properties severely degrade as the material becomes liquid above the melt temperature.

Hot-melt adhesives, normally applied from an automatic applicator or a hand-held gun, are extraordinarily useful materials because of rapid bonding characteristics. Problems in using hot-melt adhesives are usually associated with the lack of high temperature performance because, being thermoplastic, they tend to creep under load. Many of these problems can be solved by adding some degree of cross-linking, which must take place after the adhesive has been applied, not at the temperature at which the adhesive is kept liquid inside the applicator.

Moisture-curing urethanes have been attempted as cross-linking agents. One of the largest applications is in the paperback book industry in the area of book binding. Hot-melt adhesives have also been used in shoe and furniture manufacture where, in the latter case, the problems of creep do enter into the application consideration. Box and carton sealing is a very large use area, as is that of corrugated paper manufacture. Consumer applications are growing also.

Solvent- and Emulsion-Based Adhesives. *Solvent-Based Adhesives.* Solvent-based adhesives, as the name implies, are materials that are formed by solution of a high molecular weight polymer in an appropriate solvent. Although the polymer is usually an elastomer, there are a number of cases in which the adhesive contains a nonelastomeric polymer such as poly(vinyl chloride). These adhesives can be divided into three categories: the contact bond, the reactivatable, and the solvent weld adhesive. The contact bond adhesive is applied by spray or roll coating on both sides of an adherend combination. After some portion of the solvent evaporates, the adherends are joined and the adhesive rapidly bonds or knits to itself. There is a window of time in which there is just enough tack exhibited by the adhesive to bond. The reactivatable adhesive is applied to both sides of an adherend combination and the solvent is allowed to evaporate completely. The adhesive, nontacky and dry to the touch, can be reactivated by wiping the surface with solvent. In the solvent weld adhesive, the polymer is more than likely a nonelastomeric material and the solvent mix is chosen to dissolve the adherend to which the bond is to be made. After a portion of the adherend dissolves, a solution forms between the high molecular weight, polymeric adhesive and the adherend at the surface. This type of adhesion is best explained by diffusion theory.

A rubber-based adhesive can be a very complex mixture of ingredients. The main component is an elastomer from which the adhesive derives most of its strength. Tackifiers are often added both to provide tack and to increase the autohesion or knitting characteristics; plasticizers are often added to make the adhesive more permanently soft; and pigments and fillers are used to change color, control viscosity, and reduce cost. Solvents, a key portion of solvent-based adhesives, reduce viscosity to allow application and modify the green strength or knitting characteristics. Curing systems, which can build heat resistance and increase the shear strength of the material, are often added also. Metal oxides are frequently used because they participate in the cure in the normal sense of vulcanization of elastomers. These oxides can act as acid acceptors when the base resin is a Neoprene polymer. Antioxidants can also be added in order to provide stability.

Almost all common elastomers have been used in rubber-based adhesives; the required performance governs the type. A natural-rubber-based adhesive, in combination with naphtha, was originally introduced in 1791. More recently, the use of reclaimed tire rubber has been attempted. The most widely utilized elastomer is Neoprene (polychloroprene), which is especially good in contact bond adhesives when used in the presence of appropriate tackifying resins (see ELASTOMERS, SYNTHETIC, POLYCHLOROPRENE). The tackifying resins are much the same as those found in the formulation of pressure-sensitive adhesives. The application characteristics of the adhesives derive mainly from the type of solvent system employed. New regulations regarding solvent emissions prescribe the use of a latex system, which can provide many of the same properties as a solvent-based system but are often not as good overall. The first elastomer-based latex system was derived from natural rubber latex.

The largest market for elastomer-based adhesives is that of lamination (see LAMINATES). Modern office furniture having Formica as a surfacing material is made with an elastomer-based adhesive. Tile adhesives for ceramic or carpet tile are also elastomer-based and usually solvent applied. Paper adhesives and shoe manufacture also use substantial quantities of rubber-based materials. Solvent-weld adhesives are used to join plastic plumbing and to repair vinyl sheeting.

Emulsion Adhesives. The most widely used emulsion-based adhesive is that based upon poly(vinyl acetate)–poly(vinyl alcohol) copolymers formed by free-radical polymerization in an emulsion system. Poly(vinyl alcohol) is typically formed by hydrolysis of the poly(vinyl acetate). The properties of the emulsion are derived from the polymer employed in the polymerization as well as from the system used to emulsify the polymer in water. The emulsion is stabilized by a combination of a surfactant plus a colloid protection system. The protective colloids are similar to those used paint (qv) to stabilize latex. For poly(vinyl acetate), the protective colloids are isolated from natural gums and cellulosic resins (carboxymethylcellulose or hydroxyethylcellulose). The hydrolized polymer may also be used. The physical properties of the poly(vinyl acetate) polymer can be modified by changing the co-monomer used in the polymerization. Any material which is free-radically active and participates in an emulsion polymerization can be employed. Plasticizers (qv), tackifiers, viscosity modifiers, solvents (added to coalesce the emulsion particles), fillers, humectants, and other materials are often added to the adhesive to meet specifications for the intended application. Because the presence of foam in the bond line could decrease performance of the adhesion joint, agents that control the amount of air entrapped in an adhesive bond must be added. Biocides are also necessary: many of the materials that are used to stabilize poly(vinyl acetate) emulsions are natural products. Poly(vinyl acetate) adhesives known as "white glue" or "carpenter's glue" are available under a number of different trade names. Applications are found mostly in the area of adhesion to paper and wood (see VINYL POLYMERS).

Economic Aspects

Although the manufacture and sale of adhesives is a worldwide enterprise, the adhesives business can be characterized as a fragmented industry. The 1987 Census of Manufacturers obtained reports from 712 companies in the United States, each of which considers itself to be in the adhesives or sealants business (13); only 275 of these companies had more than 20 employees. The total value of material shipped by these companies approached 4.7 billion dollars, 3 billion of which were attributed to materials classified as adhesives. Adhesive types and the corresponding dollar value of material in 1987 are given in Table 2. Phenolics, poly(vinyl acetate) adhesives, rubber cements, and hot-melt adhesives are the leading products in terms of monetary value. These products are used primarily in the wood, paper, and packaging industries. The annual growth rate of the adhesives market is 2.3% (14) and individual segments of the market are expected to

Table 2. Value of Shipments Produced by Adhesives Suppliers in 1987[a]

Adhesive type	Value, $ $\times$ 10^6
natural product adhesives	154
epoxy adhesives	162
phenolics (including resorcinol-based)	219
urea and modified urea	147
poly(vinyl acetate) emulsion	407
poly(vinyl chloride)	32
acrylic	123
cyanoacrylate	38
urethane	93
hot melts	376
rubber and synthetic resin combinations	346
rubber cement emulsion	55
rubber cement (solvent-based)	128

[a]Ref. 13.

grow faster than this rate. For example, environmentally forgiving materials, such as emulsion-based or hot-melt adhesives, are in the faster growth category.

Market analysis predicts that the demand for adhesives will increase to ten billion dollars by 1993, given a manufacturing volume of 4.76 million metric tons (14). The cost of specific adhesive materials depends upon the specifications of the final product, the manufacturing process, and the marketing strategies used by individual companies. For example, given two epoxy adhesives having similar formulations, one could be sold at a relatively low price to a general consumer market where exacting manufacturing conditions are not required, and the other could be sold at a substantially higher price into an aerospace market where the manufacturing conditions must be exact and performance must meet demanding specifications.

Most adhesive manufacturers have a broad line of products often spanning the entire range of available materials. For example, the National Starch and Chemical Corporation, now a subsidiary of Unilever, markets adhesives which range from hot-melts to dextrin [9004-53-9] based materials under brand names such as Duro-Tak, Panel Master, Instant-Lok and Bondmaster. The 3M Company, which is also a multinational organization, markets all forms of adhesives from the well-known Scotch Pressure-Sensitive Tapes to Scotch-Grip and Scotch-Weld adhesives. Ciba-Geigy, another multinational company, sells structural adhesives under the Araldite brand name. A large number of adhesive manufacturers sell into specialty markets. An example is the Loctite Corporation, which sells cyanoacrylates, acrylics, and anaerobic adhesives under a variety of brand names such as Loctite and Permatex. Consumer needs are filled by such companies as Borden, which manufactures and sells the well known Elmer's Glue, a synthetic resin emulsion. Krazy Glue is a cyanoacrylate adhesive marketed by the Krazy Glue Company; Duro Super Glue is sold by Loctite Corporation. Companies such as H. B. Fuller manufacture many adhesives commonly found in hardware stores.

Listings of brand names of adhesive products (15) and adhesive manufacturers (16) are available.

Health and Safety

Health and safety information is available from the manufacturer of every adhesive sold in the United States. The toxicology of a particular adhesive is dependent upon its components, which run the gamut of polymeric materials from natural products which often exhibit low toxicity to isocyanates which can cause severe allergic reactions. Toxicological information may be found in articles discussing the manufacture of the specific chemical compounds that comprise the adhesives.

BIBLIOGRAPHY

1. B. V. Derjaguin and V. P. Smilga, *J. Appl. Phys.* **38,** 4609 (1967).
2. B. V. Derjaguin and Yu. P. Toporov, in K. L. Mittal, ed., *Physicochemical Aspects of Polymer Surfaces,* Plenum Press, New York, 1981, pp. 605–612.
3. T. Young, *Trans. R. Soc. (London)* **95,** 65 (1805).
4. A. Dupre, *Theorie Mechanique de la Chaleur,* Paris (1869).
5. A. N. Gent and G. R. Hamed, "Adhesion," in J. I. Kroschwitz, ed., *Encyclopedia of Polymer Science and Engineering,* Vol 1, John Wiley & Sons, New York, 1985, pp. 476–578.
6. H. W. Fox and W. A. Zisman, *J. Colloid Sci.* **5,** 514 (1950).
7. H. W. Fox and W. A. Zisman, *J. Colloid Sci.* **7,** 109, 428 (1952).
8. W. A. Zisman, in F. M. Fowkes, ed., *Contact Angle, Wettability and Adhesion,* (Adv. Chem. Ser. No. 43), American Chemical Society, Washington, D.C., 1964, p. 1.
9. J. Delmonte, *The Technology of Adhesives,* Reinhold, New York, 1947.
10. Brit. Pat. 577,823 (June 3, 1946), N. A. DeBruyne.
11. U.S. Pat. 3,890,407 (June 17, 1975), P. C. Briggs, Jr., and L. C. Muschiatti (to E. I. Du Pont de Nemours and Company).
12. U.S. Pat. 3,965 (1845), W. H. Shecut and H. H. Day.
13. *Adhesives Age* **33**(2), 32 (1990).
14. *Adhesives Age* **32**(9), 46 (1989).
15. A. Sweum, ed., *Adhesives Red Book,* Communication Channels, Inc., Atlanta, Ga, 1982.
16. *Adhesives Age* **32**(6), 33 (1989).

General References

"Adhesion and Bonding;" "Adhesive Compositions," Vol. 1 (1985). "Pressure Sensitive Adhesives and Products," Vol. 3 (1988) in J. I. Kroschwitz, ed., *Encyclopedia of Polymer Science and Engineering*, John Wiley & Sons, Inc., New York.
I. M. Skeist, ed., *Handbook of Adhesives,* 3rd ed. Van Nostrand-Reinhold, New York, 1990. A basic resource for practitioners of this technology.
D. Satas, ed., *Handbook of Pressure Sensitive Adhesive Technology,* Van Nostrand-Reinhold, New York, 1989.
R. D. Adams and W. C. Wake, *Structural Adhesive Joints in Engineering,* Elsevier, New York, 1984.
S. R. Hartshorn, ed., *Structural Adhesives: Chemistry and Technology,* Plenum, New York, 1986.

N. J. De Lollis, *Adhesives, Adherends, Adhesion,* Krieger Publishing Co., Melbourne, Fla., 1980.

A. Pizzi, ed., *Wood Adhesives: Chemistry and Technology,* Marcel Dekker, New York, 1983.

W. C. Wake, ed., *Adhesion and the Formulation of Adhesives,* Elsevier Publishing Co., New York, 1982.

S. Wu, *Polymer Interface and Adhesion,* Marcel Dekker, New York, 1982. A basic textbook covering surface effects on polymer adhesion.

K. W. Allen, ed., *Adhesion,* Vols. 1–14, Elsevier Science Publishers Ltd., Barking, UK, Latest volume is 1990. An annual volume containing recently presented research results.

ALPHONSUS V. POCIUS
The 3M Company

ADIPIC ACID

Adipic acid, hexanedioic acid, 1,4-butanedicarboxylic acid, mol wt 146.14, $HOOCCH_2CH_2CH_2CH_2COOH$ [124-04-9], is a white crystalline solid with a melting point of about 152°C. Little of this dicarboxylic acid occurs naturally, but it is produced on a very large scale at several locations around the world. The majority of this material is used in the manufacture of nylon-6,6 polyamide [32131-17-2], which is prepared by reaction with 1,6-hexanediamine [124-09-4]. W. H. Carothers' research team at the Du Pont Company discovered nylon in the early 1930s (1), and the fiftieth anniversary of its commercial introduction was celebrated in 1989. Growth has been strong and steady during this period, resulting in an adipic acid demand of nearly two billion metric tons per year worldwide in 1989. The large scale availability, coupled with the high purity demanded by the polyamide process, has led to the discovery of a wide variety of applications for the acid.

Chemical and Physical Properties

Adipic acid is a colorless, odorless, sour tasting crystalline solid. Its fundamental chemical and physical properties are listed in Table 1. Further information may be obtained by referring to studies of infrared and Raman spectroscopy of adipic acid crystals (11,12), ultraviolet spectra of solutions (13), and specialized thermodynamic properties (14,4). Solubility and solution properties are described in Table 2. The crystal morphology is monoclinic prisms strongly influenced by impurities (21). Both process parameters (22) and additives (21) profoundly affect

Table 1. Physical and Chemical Properties of Adipic Acid

Property	Value	Reference
molecular formula	$C_6H_{10}O_4$	
molecular weight	146.14	
melting point, °C	152.1 ± 0.3	2
specific gravity	1.344 at 18°C (sol)	3
	1.07 at 170°C (liq)	4
coefficient of cubical expansion, K^{-1}	4.0×10^{-4} at 35–150°C (sol)	4
	10.3×10^{-4} at 155–168°C (liq)	5
vapor density, air $= 1$	5.04	
vapor pressure, Pa[a]		
solid at °C		6
18.5	9.7	
32.7	19.3	
47.0	38.0	
liquid at °C		7
205.5	1,300	
216.5	2,000	
244.5	6,700	
265.0	13,300	
specific heat, kJ/kg·K[b]	1.590 (solid state)	8
	2.253 (liquid state)	9,8
	1.680 (vapor, 300°C)	
heat of fusion, kJ/kg[b]	115	
entropy of fusion, J/mol·K[b]	79.8	4,10
heat of vaporization, kJ/kg[b]	549	
melt viscosity, mPa·s ($=$ cP)	4.54 at 160°C	
	2.64 at 193°C	
heat of combustion, kJ/mol[b]	2800	10

[a]To convert Pa to mm Hg divide by 133.3.
[b]To convert J to cal divide by 4.184.

crystal morphology in the crystallization of adipic acid, an industrially significant process. Aqueous solutions of the acid are corrosive and their effect on
various steel alloys have been tested (23). Generally, austenitic stainless steels
containing nickel and molybdenum and over 18% chromium are resistant. Data
on twenty metals were summarized in one survey (24). Bulk and handling properties of adipic acid are summarized in Table 3.

Chemical Reactions

Adipic acid undergoes the usual reactions of carboxylic acids, including esterification, amidation, reduction, halogenation, salt formation, and dehydration.
Because of its bifunctional nature, it also undergoes several industrially significant polymerization reactions.

　　Esterification.　Esters and polyesters comprise the second most important
class of adipic acid derivatives, next to polyamides. The acid readily reacts with

Table 2. Solution Properties of Adipic Acid

Property	Value		Reference
heat of solution in H_2O, kJ/kg[a]	214 at 10–20°C		15
	241 at 90–100°C		
dissociation constant in H_2O	K_1	K_2	16,17
at 25°C	3.7×10^{-5}	3.86×10^{-6}	
at 50°C	3.29×10^{-5}	3.22×10^{-6}	
at 74°C	2.90×10^{-5}	2.55×10^{-6}	
solubility in H_2O, g/100 g H_2O			18
at 15°C	1.42		
at 40°C	4.5		
at 60°C	18.2		
at 80°C	73		
at 100°C	290		
pH of aqueous solutions			19
0.1 wt %	3.2		
0.4 wt %	3.0		
1.2 wt %	2.8		
2.5 wt %	2.7		
solubility in organic solvents at 25°C			
very soluble in	methanol, ethanol		
soluble in	acetone, ethyl acetate		
very slightly soluble in	cyclohexane, benzene		
distribution coefficient			
organic solvents vs H_2O	$D, \dfrac{\text{wt \% in } H_2O}{\text{wt \% in solvent}}$		
CCl_4, $CHCl_3$, C_6H_6	>10		20
disopropyl ketone	4.8		
butyl acetate	2.9		
ethyl ether	2.2		
methyl isobutyl ketone	1.2		
ethyl acetate	0.91		
methyl propyl ketone	0.55		
methyl ethyl ketone	0.50		
cyclohexanone	0.32		
n-butanol	0.31		

[a]To convert kJ to kcal divide by 4.184.

Table 3. Bulk Phase Handling Properties of Adipic Acid

Property	Value	Reference
bulk density[a], kg/m^3	640–800	19
flash point, Cleveland open cup, °C	210	5
flash point, closed cup, °C	196	
autoignition temperature, °C	420	5
dust cloud ignition temperature, °C	550	
minimum explosive concentration (dust in air), kg/m^3	0.035	25,26
minimum dust cloud ignition energy, J[b]	6.0×10^{-2}	27
maximum rate of pressure rise, MPa[c]/s	18.6	

[a]A function of particle size.
[b]To convert J to cal divide by 4.184.
[c]To convert MPa to psi multiply by 145.

alcohols to form either the mono- or diester. Although the reaction usually is acid-catalyzed, conversion may be enhanced by removal of water as it is produced. The methyl ester is an industrially important material, because it is a distillable derivative which provides a means of separating or purifying acid mixtures. Recent modifications of adipic acid manufacturing processes have included methanol esterification of the dicarboxylic acid by-product mixture. Thus glutaric acid [110-94-1] and succinic acid [110-15-6] can be recovered upon hydrolysis, or disposed of as the esters (28). Monomethyl adipate can be electrolyzed as the salt to give dimethyl sebacate [106-79-6] (Kolbe synthesis) (29), an important ten-carbon diacid. Diesters from moderately long-chain (eight or ten carbon) alcohols are also an important group, finding use as plasticizers, eg, for PVC resins. Table 4 lists the boiling points of several representative adipate esters. Reactions with diols (especially ethylene glycol) give polyesters, also important as plasticizers in special applications. In another important use of adipate esters, low molecular weight polyesters terminated in hydroxyl groups react with polyisocyanates to give polyurethane resins. Polyurethanes consumed about 4% of adipic acid production in the United States in 1986 (30).

Table 4. Esters of Adipic Acid

Ester	CAS Registry Number	Pressure, kPa[a]	Boiling point, °C
monomethyl	[627-91-8]	1.3	158
dimethyl	[627-93-0]	1.7	115
monoethyl	[626-86-8]	0.9	160
diethyl	[141-28-6]	1.7	127
di-*n*-propyl	[106-19-4]	1.5	151
di-*n*-butyl	[105-99-7]	1.3	165
di-2-ethylhexyl	[103-23-1]	0.67	214
di-*n*-nonyl	[151-32-6]	0.67	230
di-*n*-decyl	[105-97-5]	0.67	244
di-tridecyl	[16958-92-2]	101.3	349
octyl decyl	[110-29-2]	0.67	235
di-(2-butoxyethyl)	[141-18-4]	0.53	215

[a]To convert kPa to mm Hg multiply by 7.5.

Salt Formation. Salt-forming reactions of adipic acid are those typical of carboxylic acids. Alkali metal salts and ammonium salts are water soluble; alkaline earth metal salts have limited solubility (see Table 5). Salt formation with amines and diamines is discussed in the next section.

Amidation. Heating of the diammonium salt or reaction of the dimethyl ester with concentrated ammonium hydroxide gives adipamide [628-94-4], mp 228°C, which is relatively insoluble in cold water. Substituted amides are readily formed when amines are used. The most industrially significant reaction of adipic acid is its reaction with diamines, specifically 1,6-hexanediamine. A water-soluble polymeric salt is formed initially upon mixing solutions of the two materials; then heating with removal of water produces the polyamide, nylon-6,6. This reaction

Table 5. Solubility of Adipic Acid Salts

Salt	CAS Registry Number	Temperature, °C	Solubility, g/100 g H_2O
disodium (hemihydrate)	[7486-38-6]	14	59
dipotassium	[19147-16-1]	15	65
diammonium	[3385-41-9]	14	40
calcium			
(monohydrate)	[18850-78-7]	13	4
(anhydrous)	[22322-28-7]	100	1
barium			
(monohydrate)		12	12
(anhydrous)	[60178-88-0]	100	7

has been studied extensively, and the literature contains hundreds of references to it and to polyamide product properties (31).

$$n \text{ HOOC(CH}_2)_4\text{COOH} + n \text{ H}_2\text{N(CH}_2)_6\text{NH}_2 \longrightarrow [^-\text{OOC(CH}_2)_4\text{COO}\overset{-}{\overset{+}{\text{N}}}\text{H}_3\text{(CH}_2)_6\overset{+}{\text{N}}\text{H}_3]_n \longrightarrow$$

$$(2n - 1)\text{H}_2\text{O} + \text{HO}[\overset{\overset{\text{O}}{\|}}{\text{C}}\text{(CH}_2)_4\overset{\overset{\text{O}}{\|}}{\text{C}}\text{NH(CH}_2)_6\text{NH]}_n\text{H}$$

Reduction. Hydrogenation of dimethyl adipate over Raney-promoted copper chromite at 200°C and 10 MPa produces 1,6-hexanediol [629-11-8], an important chemical intermediate (32). Promoted cobalt catalysts (33) and nickel catalysts (34) are examples of other patented processes for this reaction. An earlier process, which is no longer in use, for the manufacture of the 1,6-hexanediamine from adipic acid involved hydrogenation of the acid (as its ester) to the diol, followed by ammonolysis to the diamine (35).

Cyclization/Dehydration. Heating above the melting point results in elimination of water and formation of a linear or polymeric anhydride [2035-75-8], not the cyclic anhydride as produced in the case of glutaric anhydride [108-55-4] and succinic anhydride [108-30-5]. Decarboxylation occurs at temperatures above 230–250°C, leaving cyclopentanone [120-92-3] as the chief product, bp 131°C. This reaction is catalyzed by metals such as calcium (36) or barium (37). Behavior of adipic acid upon Curie-point pyrolysis has been reviewed; mass spectroscopy was used to analyze the anhydrides, cyclic ketones, and rearranged fragments (38). Cyclization of the esters is accomplished by standard condensation chemistry with basic reagents. For example, cyclization via the acyloin condensation occurs in the presence of sodium metal, producing 2-hydroxycyclohexanone [533-60-8] (39).

Miscellaneous Reactions. Conversion of the acid to the acid chloride is accomplished using standard laboratory techniques. The resulting acid chloride frequently is used in subsequent synthesis reactions. An example is the laboratory synthesis of nylon-6,6 via the nylon rope trick, in which the diamine reacts with adipoyl chloride [111-50-2] in a two-phase system. Polyamide produced at the interface may be pulled continuously from the open vessel in a startling demonstration of polymerization chemistry (40). The acid–nitrile interchange is another unique reaction, in which a mixture of adipic acid and adiponitrile [111-69-3] are heated together, producing an equilibrium mixture containing significant amounts of 5-cyanopentanoic acid [5264-33-5]. This material is a precursor to caprolactam [105-60-2] and may be isolated from the reaction mixture by a number of methods, including esterification and hydrogenation (41).

Manufacture and Processing

Several general reviews of adipic acid manufacturing processes have been published since it became of commercial importance in the 1940s (42–46), including a very thorough report based on patent studies (47). Adipic acid historically has been manufactured predominantly from cyclohexane [110-82-7] and, to a lesser extent, phenol [108-95-2]. During the 1970s and 1980s, however, much research has been directed to alternative feedstocks, especially butadiene [106-99-0] and cyclohexene [110-83-8], as dictated by shifts in hydrocarbon markets. All current industrial processes use nitric acid [7697-37-2] in the final oxidation stage. Growing concern with air quality may exert further pressure for alternative routes as manufacturers seek to avoid NO_x abatement costs, a necessary part of processes that use nitric acid.

Since adipic acid has been produced in commercial quantities for almost 50 years, it is not surprising that many variations and improvements have been made to the basic cyclohexane process. In general, however, the commercially important processes still employ two major reaction stages. The first reaction stage is the production of the intermediates cyclohexanone [108-94-1] and cyclohexanol [108-93-0], usually abbreviated as KA, KA oil, ol-one, or anone-anol. The KA (ketone, alcohol), after separation from unreacted cyclohexane (which is recycled) and reaction by-products, is then converted to adipic acid by oxidation with nitric acid. An important alternative to this use of KA is its use as an intermediate in the manufacture of caprolactam, the monomer for production of nylon-6 [25038-54-4]. The latter use of KA predominates by a substantial margin on a worldwide basis, but not in the United States.

PREPARATION OF KA BY OXIDATION OF CYCLOHEXANE

There are three main variations to the basic cyclohexane oxidation process pioneered by Du Pont in the 1940s. The first, which can be termed metal-catalyzed oxidation, is the oldest process still in use and forms the base for the other two. It employs a cyclohexane-soluble catalyst, usually cobalt naphthenate [61789-51-3] or cobalt octoate [136-52-7], and moderate temperatures (150–175°C) and pressures (800–1200 kPa = 115–175 psi). Air is fed to each of a series of stirred tank reactors or to a column reactor which contains numerous reaction stages, along with cyclohexane. The catalyst, at 0.3–3 ppm based on cyclohexane feed, is usually premixed by injection into the feed stream, though it is not uncommon to divide the catalyst stream into many separate additions to each of the series reactors. The conversion of cyclohexane to oxidized products is 3–8 mol %, which is quite low compared to most important industrial processes. There are claims of commercial processes operating as low as 1 mol % conversion (48), which translates to 99% of the feed material being recovered and recycled to the oxidation reactors. Low conversion is the major factor in achieving high selectivities to ketone (K) and alcohol (A) (and to cyclohexylhydroperoxide [766-07-4] discussed below). This is so because the intermediates of interest (K, A, and cyclohexylhydroperoxide) are all much more easily oxidized than is cyclohexane (49,50). Selectivities vary inversely and linearly with conversion, ranging from around 90 mol % at 1–2 mol % conversion to 65–70 mol % at 8 mol % conversion. Table 6 illustrates the range of reaction conditions to be found in the patent literature.

Because the process operates at such low conversion of cyclohexane per pass through the oxidation reactors, large quantities of unreacted cyclohexane must be recovered by distillation of the oxidizer effluent. This, and the increase in energy prices in the 1970s, has resulted in considerable attention being given to the energy conservation schemes employed in recovering the cyclohexane. Examples of techniques used in energy conservation are process–process heat interchange, high efficiency packed distillation columns, and use of the "pinch-point" technique in designing recovery steps. Contacting the final crude KA oil with water or solutions of caustic soda, or both, for removal of mono- and dibasic acid impurities also can be considered an energy conservation technique since this treatment can eliminate the final steam stripper often used to purify the crude KA oil.

Regardless of the techniques used to purify the KA oil, several waste streams are generated during the overall oxidation–separation processes and must be disposed of. The spent oxidation gas stream must be scrubbed to remove residual cyclohexane, but afterwards will still contain CO, CO_2, and volatile hydrocarbons (especially propane, butane, and pentane). This gas stream is either burned and the energy recovered, or it is catalytically abated. There are usually several aqueous waste streams arising from both water generated by the oxidation reactions and wash water. The principal hydrocarbon constituents of these aqueous wastes are the C_1–C_6 mono- and dibasic acids, but also present are butanol [71-36-3], pentanol [71-41-0], ε-hydroxycaproic acid [1191-25-9] , and various lactones and diols (71,72). The spent caustic streams contain similar components in addition to the caustic values. These streams can be burned for recovery of sodium carbonate or sold directly as a by-product for use in the paper industry. The most concentrated waste stream is one often called still bottoms, heavy ends, or nonvolatile residue. It comes from the final distillation column in which the KA oil is steam-stripped overhead. The tails stream from this column contains most of the nonvolatile by-products, as well as metals and residues from the catalysts and from corrosion. Both the metals and acid content may be high enough to dictate that this stream be classified as a hazardous waste. It usually is burned and the energy used to generate steam (73). Much effort has gone into recovering valuable materials from it over the years, including adipic acid, which may be present in as much as 3–4% of the cyclohexane oxidized (74). It has potential as a feedstock in the production of monobasic acids, polyester polyols, butanediol, and maleic acid (75,76). The frequency of fugitive emissions from cyclohexane oxidation plants has been reviewed (77).

High Peroxide Process.   An alternative to maximizing selectivity to KA in the cyclohexane oxidation step is a process which seeks to maximize cyclo-

Table 6. Reaction Conditions for Air Oxidation of Cyclohexane

Process and company	Temperature, °C	Pressure, MPa[a]	Catalyst or additive	Reactor type	Cyclohexane conversion, mol %	KA yield, mol %	Reference
metal-catalyzed							
Du Pont	170	1.1	Co	column	6	76	51
Stamicarbon	155	0.9	Co	tank	4	77	52
high peroxide							
BASF							53–55
oxidation	145	1.1	none	tank	3	83	
deperoxidation	125		Co/NaOH				
Du Pont							56–58
oxidation	160	1.0	Co	column	4	82	
deperoxidation	120		Co, Cr				
Rhône Poulenc							59–62
oxidation	175	1.8	none	tank	4	84	
deperoxidation	115		Cr, V, Mo				
Stamicarbon							48,63,64
oxidation	160	1.3	none	tank	3	86	
deperoxidation	100		Co/NaOH				
boric acid							
Halcon	165	1.0	H_3BO_3	tank	3	87	65,66
ICI	165	1.0	H_3BO_3	tank	5	85	67
IFP	165	1.2	H_3BO_3		12	85	68,69
Monsanto	165	1.0	H_3BO_3	tank	4	87	70

[a]To convert MPa to psi multiply by 145.

473

hexylhydroperoxide, also called P or CHHP. This peroxide is one of the first intermediates produced in the oxidation of cyclohexane. It is produced when a cyclohexyl radical reacts with an oxygen molecule (78) to form the cyclohexylhydroperoxy radical. This radical can extract a hydrogen atom from a cyclohexane molecule, to produce CHHP and another cyclohexyl radical, which extends the free-radical reaction chain.

The peroxide can be converted to KA easily, and in high yield, in a number of ways; thus maximization of CHHP, at high yield, gives a process with high yield to KA. Techniques employed to produce high CHHP yield include drastically cutting or eliminating metal catalysts in the oxidation step, minimizing cyclohexane conversion, passivating reactor walls, lowering reaction temperature (to as low as 140°C), adding water to the reaction mix to extract acid catalysts from the cyclohexane phase, and adding metal-chelating agents to the reaction mix. Optimization of this process can produce CHHP in a proportion as high as 75% of the reaction products (59). The CHHP then can be converted to KA by any of the following methods: decomposing it with homogeneous or heterogeneous catalysts from the group Co, Cr, Mo, V, Cu, or Ru; dehydrating it by treatment with caustic soda (which preferentially gives K); or hydrogenating it (which preferentially gives A). KA is separated from the reaction mixture in a manner similar to the conventional process. It may be possible, however, to avoid a final steam distillation of the KA overhead if the tails stream from the distillation train is sufficiently clean. This could result from a high yield process that employs thorough water and caustic washing. Figure 1 illustrates schematically the high peroxide process practiced by Stamicarbon (60).

Borate-Promoted Oxidation.   Another alternative to the basic cyclohexane oxidation process is one which maximizes only the yield of A. This process uses boric acid as an additive to the cyclohexane stream as both a promoter and an esterifying agent for the A that is produced. Metaboric acid [10043-35-3] is fed to the first series oxidizer as a slurry in cyclohexane to give a molar ratio of boron:cyclohexane of around 1.5:100. No other metal catalyst is used. Esterifying the A effectively shields it from overoxidation and thus allows the attainment of very high yields (ca 90%) (65). The ratio A:K in the final product can exceed 10:1. The process was developed in the mid-1960s by a number of companies, including Halcon/Scientific Design (79,80), Institute Francais Petrole (68,81), and Stamicarbon (82). The process was licensed and commercialized by several companies in the decade following its development, including Monsanto, ICI, and Bayer. The major drawback to the process is the need to hydrolyze the

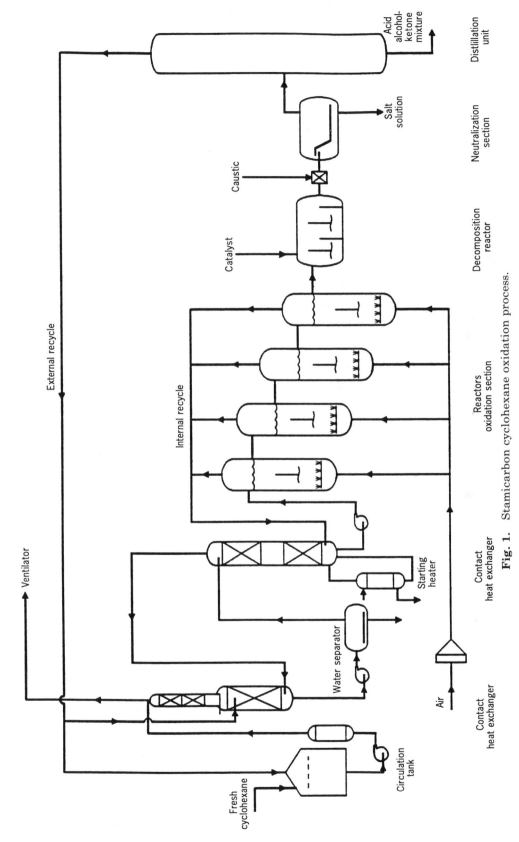

Fig. 1. Stamicarbon cyclohexane oxidation process.

borate ester in order to recover A. This is an energy-intensive step and can be quite a mechanical nuisance because of the requirement for handling boric acid solids. Without careful attention to energy conservation and engineering, the savings that accrue from the high yield can be more than offset. The process does, because of its high yield, offer advantages in waste minimization and product purity. It does, however, introduce boron into the waste streams.

Other Routes to KA.   Phenol [108-95-2] may be hydrogenated to KA in very high yield, typically 97–99% (83). Depending on catalyst selection and operating conditions, the ratio of K to A in the product can be varied over a large range. If the product is to be further oxidized with nitric acid to make adipic acid, then A would be the preferred choice since its reaction with nitric acid results in lower consumption of nitric than does K. If caprolactam is to be produced, then the process would be designed to produce relatively more K, since this is the starting material for the production of caprolactam. Because of the high yield to KA the purification is relatively simple, consisting mainly of removing the small amount of unreacted phenol via ion exchange (84). Economics in recent years has dictated against this process because of the relatively high cost of phenol compared to cyclohexane. Typical reaction conditions are 140°C and 400 kPa (58 psi) using a heterogeneous nickel on silica catalyst (83).

Cyclohexene, produced from the partial hydrogenation of benzene [71-43-2], also can be used as the feedstock for A manufacture. Such a process involves selective hydrogenation of benzene to cyclohexene, separation of the cyclohexene from unreacted benzene and cyclohexane (produced from over-hydrogenation of the benzene), and hydration of the cyclohexene to A. Asahi has obtained numerous patents on such a process and is in the process of commercialization (85,86). Indicated reaction conditions for the partial hydrogenation are 100–200°C and 1–10 kPa (0.1–1.5 psi) with a Ru or zinc-promoted Ru catalyst (87–90). The hydration reaction uses zeolites as catalyst in a two-phase system. Cyclohexene diffuses into an aqueous phase containing the zeolites and there is hydrated to A. The A then is extracted back into the organic phase. Reaction temperature is 90–150°C and reactor residence time is 30 min (91–94).

ARCO has developed a coproduct process which produces KA along with propylene oxide [75-56-9] (95–97). Cyclohexane is oxidized as in the high peroxide process to maximize the quantity of CHHP. The reactor effluent then is concentrated to about 20% CHHP by distilling off unreacted cyclohexane and cosolvent *tert*-butyl alcohol [75-65-0]. This concentrate then is contacted with propylene [115-07-1] in another reactor in which the propylene is epoxidized with CHHP to form propylene oxide and KA. A molybdenum catalyst is employed. The product ratio is about 2.5 kg of KA per kilogram of propylene oxide.

NITRIC ACID OXIDATION OF CYCLOHEXANOL(ONE)

Although many variations of the cyclohexane oxidation step have been developed or evaluated, technology for conversion of the intermediate ketone–alcohol mixture to adipic acid is fundamentally the same as originally developed by Du Pont in the early 1940s (98,99). This step is accomplished by oxidation with 40–60% nitric acid in the presence of copper and vanadium catalysts. The reaction proceeds at high rate, and is quite exothermic. Yield of adipic acid is 92–96%, the major by-products being the shorter chain dicarboxylic acids, glutaric and succinic acids, and CO_2. Nitric acid is reduced to a combination of NO_2, NO, N_2O, and N_2. Since essentially all commercial adipic acid production arises from nitric acid oxidation, the trace impurities patterns are similar in the products of most manufacturers.

Chemistry. Papers addressing the mechanism of nitric acid oxidation began appearing in the mid 1950s (100). Then, a series of reports beginning in 1962 described the mechanism of the oxidation in considerable detail (101–105). The reaction pathway diagram shown in Figure 2 is based on these and other studies of nitric acid oxidation chemistry. A key intermediate in the reaction sequence is 2-oximinocyclohexanone [24858-28-4], , produced via nitrosation of cyclohexanone. Nitrous acid [7782-77-6] is produced during the conversion of cyclohexanol to the ketone, and also upon oxidation of aldehyde and alcohol impurities usually accompanying the KA and arising in the cyclohexane oxidation step. The nitric acid oxidation chemistry is controlled by nitrous acid, which is in equilibrium with NO, NO_2, HNO_3, and H_2O in the reacting mixture. Total inhibition of reaction can be achieved by incorporating a small amount of urea [57-13-6], which effectively scavenges (106) nitrous acid from the mixture. Further nitration leads to 2-nitro-2-nitrosocyclohexanone [23195-89-3], which is converted via hydrolytic cleavage of the ring to 6-nitro-6-hydroximinohexanoic acid (nitrolic acid) [1069-46-1]. Of all the intermediates shown in Figure 2, the nitrolic acid is the only one of sufficient stability to be isolable under very mild conditions. It is hydrolyzed to adipic acid in one of the slowest steps in the sequence. Further hydrolysis produces adipic acid. Nitrous oxide (N_2O) is formed by further reaction of the nitrogen-containing products of nitrolic acid hydrolysis. The NO and NO_2 are reabsorbed and converted back to nitric acid, but N_2O cannot be recovered in this manner, and thus is the major nitric acid derived by-product of the process.

About 60–70% of the reaction occurs as in path 1 in Figure 2, the remainder by other pathways. About 20% of the reaction occurs by the vanadium oxidation of 1,2-dioxygenated intermediates (path 2 in Fig. 2). This chemistry has been discussed in detail (104,105). This path is noteworthy since it does not produce the nonrecoverable nitrous oxide. The other reactions shown in Figure 2 occur to varying degrees, depending on either an excess or deficiency of nitrous acid, arising from variations in reaction conditions. These lead to varying yields of the lower dicarboxylic acids. Yield of monobasic and dibasic acid by-products also is a function of the purity of the KA feed. A distinguishing characteristic for several of the commercial processes is the degree to which the intermediate KA is refined, prior to feeding it to nitric acid oxidation.

Process Description. In a typical industrial adipic acid plant, as schematically illustrated in Figure 3, the KA mixture reacts in reactor A with 45–55% nitric acid containing copper (0.1–0.5%) and vanadium (0.02–0.1%) catalyst (107,108). Design of the oxidation reactor for optimum yield and heat removal has been the subject of considerable research and development over the years of use of this process (109). The reaction occurs at 60–90°C and 0.1–0.4 MPa (14–58 psi). It is very exothermic (6280 kJ/kg = 1500 kcal/kg), and can reach an autocatalytic runaway state at temperatures above about 150°C. Control is achieved by limiting the KA feed to a large excess of nitric acid in a stirred tank or circulating loop reactor. Two stages of oxidation are sometimes employed to achieve improved product quality (110). Oxides of nitrogen are removed by bleaching with air in column C, then water is removed by vacuum distillation in column E.

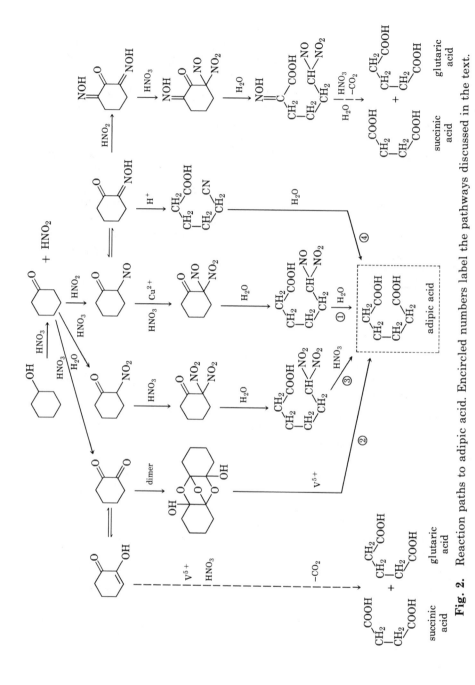

Fig. 2. Reaction paths to adipic acid. Encircled numbers label the pathways discussed in the text.

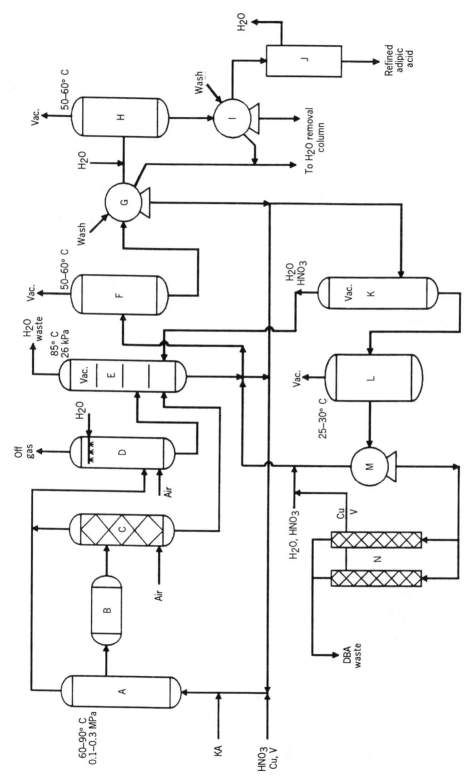

Fig. 3. Typical nitric acid oxidation process. A, reactor; B, optional cleanup reactor; C, bleacher; D, NO_x absorber; E, concentrating still; F, crude crystallizer; G, centrifuge or filter; H, refined crystallizer; I, centrifuge or filter; J, dryer; K, purge evaporator; L, purge crystallizer; M, centrifuge or filter; N, ion-exchange beds; DBA = dibasic acids.

The concentrated stream, nominally adipic acid and lower dibasic acid co-products in 35–50% HNO_3 (organic-free basis), is then cooled and crystallized (F). Crude adipic acid product is removed via filtration or centrifugation (G), and the mother liquor is returned to the oxidizer. Further refining is required to achieve polymer grade material, usually by recrystallization from water. Residual lower dibasic acids, nitrogen-containing impurities, and metals are removed in this step. Additional purification steps occasionally are employed, including slurry washing, further recrystallization, and charcoal treatment. The bleacher off-gas, containing NO and NO_2, reacts with air and is reabsorbed as nitric acid for reuse (D).

In order to control the concentration of lower dibasic acid by-products in the system, a portion of the mother liquor stream is diverted to a purge treatment process. Following removal of nitric acid by distillation (Fig. 3, K), copper and vanadium catalyst are recovered by ion-exchange treatment (Fig. 3, N). This area of the process has received considerable attention in recent years as companies strive to improve efficiency and reduce waste. Patents have appeared describing addition of SO_2 to improve ion-exchange recovery of vanadium (111), improved separation of glutaric and succinic acids by dehydration and distillation of anhydrides (112), formation of imides (113), improved nitric acid removal prior to dibasic acid recovery (114), and other claims (115).

Because of the highly corrosive nature of the nitric acid streams, adipic acid plants are constructed of stainless steel, or titanium in the more corrosive areas, and thus have high investment costs.

Wastes and Emissions.   Nitric acid oxidation may be used to recover value from waste streams generated in the cyclohexane oxidation portion of the process, such as the water wash (116) and nonvolatile residue (76) streams. The nitric acid oxidation step produces three major waste streams: an off-gas containing oxides of nitrogen and CO_2; water containing traces of nitric acid and organics from the water removal column; and a dibasic acid purge stream containing adipic, glutaric, and succinic acids. The off-gas usually is passed through a reducing-flame burner to the atmosphere, or it may be oxidized back to NO_x at 1000–1300°C and recovered as nitric acid, as claimed in a patent (117). The overhead water stream usually is treated (eg, neutralization, biotreatment) and reused or disposed of. The dibasic acid stream usually is either burned or disposed of by deepwell injection or biotreatment. However, as more uses for these acids are discovered, the necessity for their disposal diminishes. The principal emissions of concern from these processes are related to nitric acid, either as the various oxides of nitrogen or as a very dilute solution of the acid itself. The fate of these waste streams varies widely, subject to the usually very complex environmental and regulatory situations at each individual manufacturing site. These issues are now a prime consideration, equal to economics, in the design of chemical processing systems in the petrochemical industry (118).

OTHER ROUTES TO ADIPIC ACID

A number of processes for producing adipic acid without producing the intermediates K and A, and from feedstocks other than cyclohexane and phenol, have been investigated. None has been employed at a commercial scale. A one-step air oxidation process, first researched by Halcon (119,120) and Gulf (121) in the 1960s, and developed by Asahi and others (122–126) in the 1970s, uses an acetic acid [64-19-7] solvent for the cyclohexane. High concentrations of soluble cobalt catalyst (60–300 ppm) are used, along with cyclohexanone or acetaldehyde [75-07-0] promoter. Yields to adipic acid of 70–75% are reported at cyclohexane conversions of 50–75%. Reaction temperature is a moderate 70–100°C. References to air oxidation processes have continued to appear through the 1980s (127–130).

It has been known since the early 1950s that butadiene reacts with CO to form aldehydes and ketones that could be treated further to give adipic acid (131). Processes for producing adipic acid from butadiene and carbon monoxide [630-08-0] have been explored since around 1970 by a number of companies, especially ARCO, Asahi, BASF, British Petroleum, Du Pont, Monsanto, and Shell. BASF has developed a process sufficiently advanced to consider commercialization (132). There are two main variations, one a carboalkoxylation and the other a hydrocarboxylation. These differ in whether an alcohol, such as methanol [67-56-1], is used to produce intermediate pentenoates (133), or water is used for the production of intermediate pentenoic acids (134). The former is a two-step process which uses high pressure, >31 MPa (306 atm), and moderate temperatures (100–150°C) (132–135). Butadiene, CO, and methanol react in the first step in the presence of cobalt carbonyl catalyst and pyridine [110-86-1] to produce methyl pentenoates. A similar second step, but at lower pressure and higher temperature with rhodium catalyst, produces dimethyl adipate [627-93-0]. This is then hydrolyzed to give adipic acid and methanol (135), which is recovered for recycle. Many variations to this basic process exist. Examples are ARCO's palladium/copper-catalyzed oxycarbonylation process (136–138), and Monsanto's palladium and quinone [106-51-4] process, which uses oxygen to reoxidize the by-product hydroquinone [123-31-9] to quinone (139).

Other processes explored, but not commercialized, include the direct nitric acid oxidation of cyclohexane to adipic acid (140–143), carbonylation of 1,4-butanediol [110-63-4] (144), and oxidation of cyclohexane with ozone [10028-15-5] (145–148) or hydrogen peroxide [7722-84-1] (149–150). Production of adipic acid as a by-product of biological reactions has been explored in recent years (151–156).

Storage, Handling, and Shipping

When dispersed as a dust, adipic acid is subject to normal dust explosion hazards. See Table 3 for ignition properties of such dust–air mixtures. The material is an irritant, especially upon contact with the mucous membranes. Thus protective goggles or face shields should be worn when handling the material. Prolonged contact with the skin should also be avoided. Eye wash fountains, showers, and washing facilities should be provided in work areas. However, *MSDS Sheet 400* (5) reports that no acute or chronic effects have been observed.

The material should be stored in corrosion-resistant containers, away from alkaline or strong oxidizing materials. In the event of a spill or leak, nonsparking equipment should be used, and dusty conditions should be avoided. Spills should be covered with soda ash, then flushed to drain with large amounts of water (5).

Adipic acid is shipped in quantities ranging from 22.7 kg (50-lb bags) to 90.9 t

(200,000-lb hopper cars). Upon long standing, the solid material tends to cake, dependent on such factors as initial particle size and moisture content. Shipping data in the United States are "Adipic Acid," *DOT-ID NA 9077, DOT Hazard Class ORM-E*. It is regulated only in packages of 2.3 t (5,000 lb) or more (hopper cars and pressure-differential cars and trucks) (157).

Economic Aspects

The continuing pursuit of a wide variety of alternate manufacturing processes indicates an effort by competitors to position themselves to take advantage of potential shifts in petrochemical feedstock prices. A large number of the reports concern the use of C_4 feedstocks, notably butadiene, although several major modifications to cyclohexane-based or benzene-based processes are included. The continued buildup of capacity in nylon-6,6 intermediates, especially in the Far East, attests to the confidence in continued growth by the major participants. Although the nylon-6 [25038-54-4] market currently is larger in Europe, both markets will share in the growth, especially in the developing areas of the world. The emergence of new polymers for specialized applications may tend to limit growth in certain areas. For example, polypropylene may take a significant share of the lower cost carpet market. Specialized polyamides such as nylon-4,6 [24936-71-8] have now appeared, although this one consumes adipic acid.

Adipic acid is a very large volume organic chemical. Worldwide production in 1986 reached 1.6×10^6 t (3.5×10^9 lb) (158) and in 1989 was estimated at more than 1.9×10^6 t (Table 7). It is one of the top fifty (159) chemicals produced in the United States in terms of volume, with 1989 production estimated at 745,000 t (160). Growth rate in demand in the United States for the period 1988–1993 is estimated at 2.5% per year based on 1987–1989 (160). Table 7 provides individual capacities for U.S. manufacturers. Western European capacity is essentially equivalent to that in the United States at 800,000 t/yr. Demand is highly cyclic (161), reflecting the automotive and housing markets especially. Prices usually follow the variability in crude oil prices. Adipic acid for nylon takes about 60% of U.S. cyclohexane production; the remainder goes to caprolactam for nylon-6, export, and miscellaneous uses (162). In 1989 about 88% of U.S. adipic acid production was used in nylon-6,6 (77% fiber and 11% resin), 3% in polyurethanes, 2.5% in plasticizers, 2.7% miscellaneous, and 4.5% exported (160).

The outlook as of late 1990 was positive, especially on the worldwide scene. Significant expansions or introduction of new capacity were announced in Western Europe, Russia, and the Far East. Several large petrochemical complexes involving nylon-6,6 facilities have been announced in the USSR during the late 1980s (163). The expansion is continuing; a 120,000-t/yr adipic acid plant was included in a large complex announced for the 1990s at Yarkova, in Western Siberia (164). A 40,000-t/yr expansion also was reported by BASF, due to begin production in 1990 (165). New technology for adipic acid manufacture was developed by BASF, but they have not yet announced plans to commercialize it (166).

Because of projected nylon-6,6 growth of 4–10% (167) per year in the Far East, several companies have announced plans for that area. A Rhône-Poulenc/Oriental Chemical Industry joint venture (Kofran) announced a 1991 startup for a

Table 7. Worldwide Adipic Acid Capacities[a]

Company	Location	Capacity, 10^3 t/yr
North America		
Allied-Signal	Hopewell, Va.	14
E. I. du Pont de Nemours & Co.	Orange, Tex.	180
E. I. du Pont de Nemours & Co.	Victoria, Tex.	320
Monsanto	Pensacola, Fla.	290
Du Pont of Canada	Maitland, Ont.	109
subtotal		*913*
Western Europe		
UCB-Ptal SA (Belgium)	Oostende, W. Vlaanderen	27
Rhône-Poulenc (France)	Chalampe, Haut-Rhin	230
BASF (Germany)	Ludwigshafen	200
Bayer (Germany)	Leverkusen	40
ICI (United Kingdom)	Wilton, Cleveland	300
subtotal		*797*
Far East		
Asahi Chemical (Japan)	Nobeaka	70
Kanto Denka Kogyo (Japan)	Shibukara	10
others less than 10,000 t each	Japan	9
subtotal		*89*
remaining areas of the world		
Rhodia SA (Brazil)	Paulinia, Sao Paulo	55
Liaoyang Pet. Fiber (China)	Liaoyang	55
Poland State Complexes	three sites	17
subtotal		*127*
worldwide total		*1938*

[a]Ref. 158 and news release updates since 1987.

50,000-t/yr plant in Onsan, South Korea (168,169). Asahi announced plans for a 15,000-t/yr expansion of adipic acid capacity at their Nobeoka complex in late 1989, accompanied by a 60,000-t/yr cyclohexanol plant at Mizushima based on their new cyclohexene hydration technology (170). In early 1990 the Du Pont Company announced plans for a major nylon-6,6 complex for Singapore, including a 90,000-t/yr adipic acid plant due to start up in 1993 (167). Plans or negotiations for other adipic acid capacity in the area include Formosa Plastics (Taiwan) (171) and BASF-Hyundai Petrochemical (South Korea) (167). Adipic acid is a truly worldwide commodity. An average of 20,000 t/yr was exported from the United States in the years 1980–1986 (172). Western European exports have been 40,000–45,000 t/yr in the same period (173). Japan has been a net importer of the acid.

Specifications and Analysis

Quality Specifications. Because of the extreme sensitivity of polyamide synthesis to impurities in the ingredients (eg, for molecular-weight control, dye receptivity), adipic acid is one of the purest materials produced on a large scale.

In addition to food-additive and polyamide specifications, other special requirements arise from the variety of other applications. Table 8 summarizes the more
important specifications. Typical impurities include monobasic acids arising from
the air oxidation step in synthesis, and lower dibasic acids and nitrogenous
materials from the nitric acid oxidation step. Trace metals, water, color, and oils
round out the usual specification lists.

Table 8. Quality Specifications

| Parameter | Application | | Page reference[a] |
	Food grade	Other	
melting range	151.5–154.0°C		519
assay	99.6% min.		11
water	0.2% max.		552
residue on ignition	20.0 ppm max.		11
arsenic (as As)	3.0 ppm max.		464
heavy metals (as Pb)	10.0 ppm max		11,513
iron (as Fe)		2.0 ppm	
ICV color		5.0 max	
caproic acid		10.0 ppm	
succinic acid		50.0 ppm	
nitrogen		3 ppm	
hydrocarbon oil		10 ppm	

[a]Refers to pages in reference 174.

Analytical Procedures. Standard methods for analysis of food-grade
adipic acid are described in the Food Chemicals Codex (see references in Table 8).
Classical methods are used for assay (titration), trace metals (As, heavy metals as
Pb), and total ash. Water is determined by Karl-Fisher titration of a methanol
solution of the acid. Determination of color in methanol solution (APHA, Hazen
equivalent, max. 10), as well as iron and other metals, are also described elsewhere (175). Other analyses frequently are required for resin-grade acid. For
example, hydrolyzable nitrogen (NH_3, amides, nitriles, etc) is determined by
distillation of ammonia from an alkaline solution. Reducible nitrogen (nitrates
and nitroorganics) may then be determined by adding DeVarda's alloy and continuing the distillation. Hydrocarbon oil contaminants may be determined by ir
analysis of halocarbon extracts of alkaline solutions of the acid.

    Monobasic acids are determined by gas chromatographic analysis of the free
acids; dibasic acids usually are derivatized by one of several methods prior to
chromatographing (176,177). Methyl esters are prepared by treatment of the sample with BF_3–methanol, H_2SO_4–methanol, or tetramethylammonium hydroxide.
Gas chromatographic analysis of silylation products also has been used extensively. Liquid chromatographic analysis of free acids or of derivatives also has
been used (178). More sophisticated hplc methods have been developed recently to
meet the needs for trace analyses in the environment, in biological fluids, and
other sources (179,180). Mass spectral identification of both dibasic and monobasic acids usually is done on gas chromatographically resolved derivatives.

Toxicity, Safety, and Industrial Hygiene

Adipic acid is relatively nontoxic; no OSHA PEL or NIOSH REL have been established for the material. Airborne exposure should be limited to 10 mg/m^3 (total dust), the ACGIH TLV-TWA for an organic nuisance dust (5). Toxicity in laboratory animals based on exposure to adipic acid has been reported (181).

eye, rabbit (eye irritant)	20 mg/24 h (SEV)
oral, rat	LDL$_0$: 3600 mg/kg
intraperitoneal, rat	LD$_{50}$: 275 mg/kg
oral, mouse	LD$_{50}$: 1900 mg/kg
intraperitoneal, mouse	LD$_{50}$: 275 mg/kg
intravenous, mouse	LD$_{50}$: 680 mg/kg

Adipic acid is excreted essentially unmetabolized in human urine, based on tests with a series of dicarboxylic acids (182). However, adipic acid may be produced via liver metabolism of longer chain diacids, as observed in a recent study with rats (183). The acid has achieved "generally recognized as safe" (GRAS) status from the U.S. Food and Drug Administration for use as a direct ingredient in food for such uses as acidulant, leavening agent, or pH control agent (184). The sodium salt [23311-84-4] has not achieved GRAS status. Maximum permissible usage of the acid in foods was studied with respect to toxicity and teratological and mutagenicity effects (185). No mutagenic or teratological activity was observed (186). Recommended maximum concentration in water reservoirs is 2 mg/L (5).

Adipic acid is an irritant to the mucous membranes. In case of contact with the eyes, they should be flushed with water. It emits acrid smoke and fumes on heating to decomposition. It can react with oxidizing materials, and the dust can explode in admixture with air (see Table 3). Fires may be extinguished with water, CO_2, foam, or dry chemicals.

Environmental Aspects. Airborne particulate matter (187) and aerosol (188) samples from around the world have been found to contain a variety of organic monocarboxylic and dicarboxylic acids, including adipic acid. Traces of the acid found in Southern California air were related both to automobile exhaust emission (189) and, indirectly, to cyclohexene as a secondary aerosol precursor (via ozonolysis) (190). Dibasic acids (eg, succinic acid) have been found even in such unlikely sources as the Murchison meteorite (191). Public health standards for adipic acid contamination of reservoir waters were evaluated with respect to toxicity, odor, taste, transparency, foam, and other criteria (192). Biodegradability of adipic acid solutions was also evaluated with respect to BOD/theoretical oxygen demand ratio, rate, lag time, and other factors (193).

Uses

About 85% of U.S. adipic acid production is used captively by the producer, almost totally in the manufacture of nylon-6,6 (194). The remaining 15% is sold in the merchant market for a large number of applications. These have been developed as a result of the large scale availability of this synthetic petrochemical commodity. Prices for 1960–1989 for standard resin-grade material have paralled raw material and energy costs (petroleum and natural gas) as follows: 57–71 ¢/kg (1960s); 40–42 ¢/kg (1970–1973); increasing to 88–90 ¢/kg (1974–1979); then to $1.15–1.26/kg (1980s) (194).

Polyamides. In 1988, 77% of U.S. demand for adipic acid was for nylon-6,6 fiber, while 11% was used in nyon-6,6 resins (195). In Western Europe only about 66% was for polyamide, because of the stronger competition from nylon-6. The fiber applications include carpets (67%), apparel (13%), tire cord (7%), and miscellaneous (13%). Nylon-6,6 resins were distributed between injection molding (85%) for such applications as automotive and electrical parts and for extrusion resins (15%) for strapping, film, and wire and cable.

Polyurethanes. About 3% of the U.S. polyurethanes market in 1988 was derived from the condensation product of polyisocyanates with low molecular weight polyadipates having hydroxyl end groups (195). In 1986 this amounted to 29,000 t, or 4% of total adipic acid consumption. The percentage was similar in Western Europe. About 90% of these adipic acid containing polyurethanes are used in flexible or semirigid foams and elastomers, with the remainder used in adhesives, coatings, and spandex fibers.

Plasticizers. About 2.5% of U.S. adipic acid consumed in 1988 was used in two basic types of adipic ester based plasticizers (195). Simple adipate esters prepared from C_8–C_{13} alcohols are used especially as PVC plasticizers (qv). For special applications requiring low volatility or extraction resistance, polyester derivatives of diols or polyols are preferred.

Adiponitrile Synthesis. 1,6-Hexanediamine, the second ingredient in the production of nylon-6,6 polyamide, is prepared by hydrogenation of adiponitrile [*111-69-3*]. In several large plants around the world, the nitrile is produced from adipic acid by dehydration of the ammonium salt. A significant percentage of adipic acid capacity in Western Europe currently is used for this purpose. The technology has been reviewed in the literature (196). Although this was the primary adiponitrile process for several years, new processes based on propylene and butadiene have supplanted this technology in the United States. For several years, Du Pont operated a process based on the chlorination and cyanation of butadiene, but this was shut down in 1983 (197,198). Du Pont produces adiponitrile at two large U.S. plants and one French joint venture by direct nickel(0)-catalyzed homogeneous hydrocyanation of butadiene (199). Monsanto and Asahi developed and practiced the electrolytic coupling of acrylonitrile process, used in the United States, Western Europe, and Japan (200,201).

Miscellaneous Applications. About 2% of U.S. consumption in 1988 was distributed among several other applications, amounting to several thousand tons each (195). Wet-strength resins based on polyamide–epichlorohydrin products consumed about 6800 t in 1986. Unsaturated polyester resins (3600 t in 1986) are used in surface coatings, flexible alkyd resins (qv), coil coatings, and other coatings because of their curing properties. Adipic acid also is used as a food acidulant in jams, jellies, and gelatins. Although it has only 2% of the acidulant market, 3200 t were used for this purpose in 1989 (202). The synthetic lubricant market consumed about 1800 t as the C_{8-13} adipate esters in 1986, for gas turbines, compressors, and military jet engines. An environmentally significant use of the acid, and especially its dibasic acid by-products, is as a buffer in the scrubbing operation of power plant flue gas desulfurization (203–205). Adipoyl chloride is occasionally used as a softening agent for leather.

BIBLIOGRAPHY

"Adipic Acid" under "Acids, Dicarboxylic" in *ECT* 1st ed., Vol. 1, pp. 153–154, by P. F. Bruins, Polytechnic Institute of Brooklyn; "Adipic Acid" in *ECT* 2nd ed., Vol. 1, pp. 405–421, by W. L. Standish and S. V. Abramo, E. I. du Pont de Nemours & Co., Inc.; in *ECT* 3rd ed., Vol. 1, pp. 510–531, by D. C. Danly and C. R. Campbell, Monsanto Chemical Intermediates Company.

1. E. Bolton, *Ind. Eng. Chem.* **34,** 53 (1942).
2. H. Serwy, *Bull. Soc. Chim. Belg.* **42,** 483 (1933).
3. Armour Research Foundation, *Anal. Chem.* **20,** 385 (1948).
4. S. Khetarpal, L. Krishan, and H. Bhatnager, *Ind. J. Chem., Sect. A.* **19A,** 516–519 (1986).
5. P. Igoe, D. Wilson, and W. Silverman, *Material Safety Data Sheet No. 400* in *Genum's Reference Collection,* Genum Publishing Corp., Schenectady, N.Y., 1989.
6. A. Granovskaya, *Zh. Fiz. Khim.* **21,** 967 (1947).
7. F. Kraft and H. Noerdlinger, *Ber. Dtsch. Chem. Ges.* **22,** 818 (1889).
8. A. Van Dooren and B. Mueller, *Thermochim. Acta* **54**(1–2), 115–129 (1982).
9. P. E. Verkade, H. Hartman, and J. Coops, *Recl. Trav. Chim. Pays-Bas* **45,** 380 (1926).
10. I. Contineanu, E. Corlateanu, J. Herscovici, and I. Dumitri, *Rev. Chim. (Bucharest)* **31,** 763–764 (1980).
11. S. Kahnyakina and G. Puchkovskaya, *Zh. Prikl. Spektrosk.* **34,** 885–891 (1981).
12. Y. Morechal, *Can. J. Chem.* **63,** 1684–1688 (1985).
13. K. Urano, K. Kawamoto, and K. Hayoshi, *Yosui To Haisui* **23**(2), 196–202 (1981).
14. A. Babinkov and co-workers, *Termodin. Organ. Sordin. Gor'kii* **1979**(8), 28–33 (1979).
15. A. Apelblat, *J. Chem. Thermodyn.* **18,** 351–357 (1986).
16. I. Jones and R. Soper, *J. Chem. Soc.* **1936,** 135 (1936).
17. J. Burgot, *Talanta* **25,** 233–235 (1978).
18. A. Apelblat and E. Manzurola, *J. Chem. Thermodyn.* **19,** 317–320 (1987).
19. *Adipic Acid, Product Bulletin E-99079-1,* E. I. du Pont de Nemours & Co., Inc., 1989.
20. C. S. Marvel and J. C. Richards, *Anal. Chem.* **21,** 1480 (1949).
21. K. Chow, J. Go, M. Mehdizadeh, and D. Grant, *Int. J. Pharm.* **20**(1–2), 3–24 (1984).
22. L. Hus, C. Chang, J. Beddow, and A. Vetter, *Proc. Tech. Prog. Int. Powder and Bulk Solids Handling and Processing, Atlanta, Georgia, May 24–26, 1983,* Books Demand UMI, Ann Arbor, Mich., 1983, pp. 52–66.
23. O. Georgescu, S. Ivascon, and M. Apostolescu, *Rev. Chim. (Bucharest)* **36,** 839–844 (1985).
24. *Corrosion Data Survey, Metals Section,* 5th ed. and 6th ed., National Association of Corrosion Engineers, Houston, Texas, 1974 and 1985, p. 6 and p. 4.
25. G. Lunn, *J. Hazard. Mater.* **17**(2), 207–213 (1988).
26. E. Scholl and co-workers, *Inst. Explos. Sprengtech. Bergbau-Versuchstrecke FRG, STF-Rep.* **1979**(2), 99 (1979).
27. D. Felstead, R. Rogers, and D. Young, *Conf. Ser.—Inst. Phys. (Electrostatics), ICI, United Kingdom* **66,** 105–110 (1983).
28. U.S. Pat. 4,375,552 (Mar. 1, 1983), V. Kuceski (to C. P. Hall Co.).
29. M. Seko, A. Yomiyama, and T. Isoya, *Chem. Econ. Eng. Rev.* **11**(9), 48–50 (1979).
30. R. T. Gerry, "Adipic Acid" in *Chemical Economics Handbook, Marketing Research Report,* SRI International, Menlo Park, Calif., 1987, p. 608.5032M.

31. M. I. Kohan, *Nylon Plastics*, John Wiley & Sons, Inc., New York, 1973, pp. 14–73.
32. Ger. Offen. 3,510,876 (Oct. 2, 1986), W. Hoelderich and co-workers (to Badische Anilin- und Soda-Fabrik A. G.).
33. Jpn. Kokai Tokkyo Koho 6105,036 (Jan. 10, 1986), K. Tsukada, N. Fukuoka, and I. Kinoshita (to Kao Corp.).
34. Jpn. Kokai Tokkyo Koho 80 04,090 (Jan. 29, 1980), J. Kanetaka and S. Mori (to Mitsubishi Petrochemical Co.).
35. Fr. Pat. 1,509,288 (Jan. 12, 1968), P. Volpe and W. Humphrey (to Celanese Corp.).
36. W. Hentzchel and J. Wislicenus, *Justus Leibigs Ann. Chem.* **275,** 312 (1983).
37. G. Vavon and A. Apchie, *Bull. Soc. Chim. Fr.* **43,** 667 (1928).
38. J. Dullinga, N. Nibbering, and A. Boerboom, *J. Chem. Soc. Perkin Trans.* **2,** 1065–1076 (1984).
39. J. Sheehan, R. O'Neill, and M. White, *J. Am. Chem. Soc.* **72,** 3376 (1950).
40. P. Morgan, *J. Chem. Educ.* **36**(4), 182 (1959).
41. Ger. Offen. 3,235,938 (Apr. 21, 1983), K. Kimura and T. Isoya (to Asahi Chemical Industries, Ltd.).
42. I. V. Berezin, E. T. Denisov, and N. M. Emanuel, *The Oxidation of Cyclohexane,* Pergamon Press, Oxford, England, 1965.
43. S. A. Miller, *Chemical and Process Engineering,* **1969,** 63 (June 1969).
44. Tamarapu Sridhar, *Mass Transfer in Cyclohexane Oxidation,* Ph.D. Thesis, Dept. of Chemical Engineering, Monash Univ., Australia, 1978.
45. V. D. Luedeke, "Adipic Acid" in *Encyclopedia of Chemical Process and Design,* J. McKetta and W. Cunningham, eds., Vol. 2, Marcel Dekker, Inc., New York, 1977, pp. 128–146.
46. A. K. Suresh, T. Shidhar, and O. E. Potter, *AIChE J.* **34**(1), 55–93 (1988).
47. Y. C. Yen and S. Y. Wu, *Nylon 6,6, PEP Report 54B,* SRI International, Menlo Park, Calif., 1987.
48. U.S. Pat. 4,238,415 (Dec. 9, 1980), W. O. Bryan (to Stamicarbon N. V.).
49. L. Bateman, H. Hughes, and A. L. Morris, *Faraday Discuss. Chem. Soc.* **1953**(14), 190–199 (1953).
50. D. G. Hendry and co-workers, *J. Org. Chem.* **41,** 2 (1976).
51. U.S. Pat. 3,530,185 (Sept. 22, 1970), K. Pugi (to E. I. du Pont de Nemours & Co., Inc.).
52. *Hydrocarbon Process.* **48,** 163 (1969).
53. U.S. Pat. 4,163,027 (July 31, 1979), P. Magnussen, G. Herrman, and E. Frommer (to Badische Anilin- und Soda-Fabrik A. G.).
54. Ger. Offen. 3,328,771 (Feb. 28, 1985), Stoessel and co-workers (to Badische Anilin- und Soda-Fabrik A. G.).
55. U.S. Pat. 4,704,476 (Nov. 3, 1987), J. Hartig, G. Herrman, and E. Lucas (to Badische Anilin- und Soda-Fabrik A. G.).
56. U.S. Pat. 3,957,876 (May 18, 1976), M. Rapoport and J. O. White (to E. I. du Pont de Nemours & Co., Inc.).
57. U.S. Pat. 3,987,100 (Oct. 19, 1976), W. J. Barnette, D. L. Schmitt, and J. O. White (to E. I. du Pont de Nemours Co., Inc.).
58. U.S. Pat. 4,465,861 (Aug. 14, 1984), J. Hermolin (to E. I. du Pont de Nemours & Co., Inc.).
59. U.S. Pat. 3,925,316 (Dec. 9, 1975), J. C. Brunie, N. Creene, and F. Maurel (to Rhône-Poulenc S. A.).
60. Adapted from L. L. van Dierendonck and J. A. de Leeuw den Bouter, *PT/Procestechniek* **39**(3), 44–48 (1984).
61. U.S. Pat. 3,923,895 (Dec. 2, 1975), M. Costantini, N. Creene, M. Jouffret, and J. Nouvel (to Rhône-Poulenc S. A.).
62. U.S. Pat. 3,927,105 (Dec. 16, 1975), J. C. Brunie and N. Creene (to Rhône-Poulenc S. A.).
63. U.S. Pat. 4,326,085 (Apr. 20, 1982), M. De Cooker (to Stamicarbon N. V.).

64. Eur. Pat. 092,867 (Nov. 2, 1983), J. G. Housmans and co-workers (to Stamicarbon N.V.).
65. U.S. Pat. 3,932,513 (Jan. 13, 1976), J. L. Russell (to Halcon International, Inc.).
66. U.S. Pat. 3,796,761 (Aug. 18, 1971), Marcell and co-workers (to Halcon International, Inc.).
67. Brit. Pat. 1,590,958 (June 10, 1981), J. F. Risebury (to Imperial Chemical Industries, Ltd.).
68. J. Alagy and co-workers, *Hydrocarbon Process.* **47**(12), 131 (1968).
69. H. Van Landeghem, *Ind. Eng. Chem. Process. Des. Dev.* **13,** 317 (1974).
70. U.S. Pat. 3,895,067 (Jan. 12, 1973), G. H. Mock and co-workers (to Monsanto Co.).
71. E. F. J. Duynstee and co-workers, *Recl. Trav. Chim. Pays-Bas* **89,** 769–780 (1970).
72. Ref. 50, pp. 1–5.
73. C. T. Chi and J. H. Lester, *Presentation to MCA Waste Minimization Workshop,* New Orleans, La., Nov. 11–13, 1987.
74. U.S. Pat. 3,260,743 (July 12, 1966), W. B. Hogeman (to Monsanto Co.); U.S. Pat. 3,365,490 (Jan. 23, 1968), W. J. Arthur and L. S. Scott (to E. I. du Pont de Nemours & Co., Inc.); U.S. Pat. 3,969,465 (July 13, 1976), J. K. Brunner (to Badische Anilin- und Soda-Fabrik A. G.); U.S. Pat. 4,105,856 (Aug. 8, 1978), C. A. Newton (to El Paso Products, Co.); K. J. Mehta and co-workers, *Chem. Eng. World* **24**(30), 63–65 (1989).
75. U.S. Pat. 3,993,691 (Nov. 23, 1976), J. K. Brunner (to Badische Anilin- und Soda-Fabrik A. G.); U.S. Pat. 4,166,056 (Aug. 28, 1979), K. P. Satterly and F. E. Livingston (to Witco Chemical Corp.); U.S. Pat. 4,233,408 (Nov. 11, 1980), K. P. Satterly and F. E. Livingston (to Witco Chemical Corp.).
76. Jpn. Pat. 82-041456-B (Sept. 3, 1982), (to Sumitomo Chemical Co.).
77. B. Harris and B. Tichenor, *Proceedings of 74th Annual Meeting,* Air Pollution Control Association, Pittsburgh, Pa., 1981, Vol. 3, paper 81–41.5.
78. Ref. 43, p. 69.
79. U.S. Pat. 3,243,449 (Mar. 29, 1966), C. N. Winnick (to Halcon International, Inc.).
80. *Eur. Chem. News* **15,** 22 (May 2, 1969).
81. *Eur. Chem. News* **11,** 32 (June 9, 1967).
82. U.S. Pat. 3,287,423 (Nov. 22, 1966), J. Steeman and J. von den Hoff (to Stamicarbon, N.V.).
83. U.S. Pat. 2,794,056 (May 28, 1957), L. O. Winstrom (to Allied Chemical Co.).
84. Brit. Pat. 979,268 (Jan. 1, 1965), J. G. Mather and F. G. Webster (to Imperial Chemical Industries, Ltd.).
85. *Japan Chem. Week,* 5 (Oct. 29, 1987).
86. *Comline Chemicals and Materials,* Comline News Service, Tokyo 107, Japan, Feb. 22, 1988.
87. Jpn. Pat. 59184138 (Oct. 19, 1984), O. Mitsui and Y. Fukuoka (to Asahi Chemical Industry, Ltd.).
88. Jpn. Pat. 59186929 (Oct. 23, 1984), O. Mitsui and Y. Fukuoka (to Asahi Chemical Industry, Ltd.).
89. Jpn. Pat. 61050930 (Mar. 13, 1986), H. Nagahara and Y. Fukuoka (to Asahi Chemical Industry, Ltd.).
90. Jpn. Pat. 62045541 (Feb. 27, 1987), H. Nagahara and M. Konishi (to Asahi Chemical Industry, Ltd.).
91. Jpn. Pat. 60104029 (June 8, 1985), Y. Fukuoka and O. Mitsui (to Asahi Chemical Industry, Ltd.).
92. Fr. Pat. 2,554,440 (May 10, 1985), O. Mitsui and Y. Fukuoka (to Asahi Chemical Industry, Ltd.).
93. Eur. Pat. 162,475 (Nov. 11, 1985), M. Tojo and Y. Fukuoka (to Asahi Chemical Industry, Ltd.).
94. Ger. Offen. 3,441,072 (May 23, 1985), O. Mitsui and Y. Fukuoka (to Asahi Chemical Industry, Ltd.).

95. U.S. Pat. 3,987,115 (Oct. 19, 1976), J. G. Zajacek and F. J. Hilbert (to Atlantic Richfield Co.).
96. U.S. Pat. 4,080,387 (Mar. 21, 1978), J. C. Jubin, I. E. Katz, and R. G. Tave (to Atlantic Richfield Co.).
97. T. T. Shih and W. J. Klingebiel, paper presented to *The First Shanghai International Symposium on Technology of Petroleum and Petrochemical Industry,* May 16–20, 1989, Shanghai, China.
98. U.S. Pat. 2,557,282 (1951), C. Hamblett and A. Mac Alevy (to E. I. du Pont de Nemours & Co., Inc.).
99. U.S. Pat. 2,703,331 (1953), M. Goldbeck and F. Johnson (to E. I. du Pont de Nemours & Co., Inc.).
100. H. Godt and J. Quinn, *J. Am. Chem. Soc.* **78,** 1461–1464 (1956).
101. I. Lubyanitski, R. Minati, and M. Furman, *Russ. J. Phys. Chem. (Engl. trans.),* **32,** 294–297 (1962).
102. I. Lubyanitski, *Zh. Obshch. Khim.* **36,** 343 (1962).
103. I. Lubyanitski, *Zh. Prikl. Khim. (Leningrad)* **36,** 819–823 (1963).
104. D. van Asselt and W. van Krevelen, *Chem. Eng. Sci.* **18,** 471–483 (1963).
105. D. van Asselt and W. van Krevelen, *Recl. Trav. Chim. Pays-Bas* **82,** 51–56, 429–437, 438–449 (1963).
106. U.S. Pat. 3,758,564 (Sept. 11, 1973), D. Davis (to E. I. du Pont de Nemours & Co., Inc.).
107. U.S. Pat. 3,564,051 (1971), E. Haarer and G. Wenner (to Badische Anilin- und Soda-Fabrik A. G.).
108. Brit. Pat. 1,092,603 (1969), G. Riegelbauer, A. Wegerich, A. Kuerzinger, and E. Haarer (to Badische Anilin- und Soda-Fabrik A. G.).
109. T. Hearfield, *Chem. Eng. (London)* **1980** (361), 625–627 (1980).
110. U.S. Pat. 3,359,308 (Dec. 19, 1967), O. Sampson (to E. I. du Pont de Nemours & Co., Inc.).
111. Eur. Pat. Appl. 122-249A1 (Oct. 17, 1984), C. Hsu and D. Laird (to Monsanto Co.).
112. U.S. Pat. 4,254,283 (Mar. 3, 1981), G. Mock (to Monsanto Co.).
113. Ger. Offen. 3,002,256 (July 30, 1981), W. Rebafka, G. Heilen, and W. Klink (to Badische Anilin- und Soda-Fabrik A. G.).
114. U.S. Pat. 4,014,903 (Mar. 29, 1977), W. Moore (to Allied Chemical Corp.).
115. Brit. Pat. 1,480,480 (July 20, 1977), A. Bowman (to Imperial Chemical Industries, Ltd.).
116. U.S. Pat. 4,227,021 (1980), O. Grosskinsky and co-workers (to Badische Anilin- und Soda-Fabrik A. G.).
117. Jpn. Kokai Tokkyo Koho, JP61–257940 (Nov. 15, 1986), T. Sakamoto, H. Suga, and T. Sakasegawa (to Asahi Chemical Industries Co., Ltd.).
118. S. Fathi-Afshar and J. Yang, *Chem. Eng. Sci.* **40,** 781–797 (1985).
119. Brit. Pat. 956,779 (Apr. 29, 1964) (to Halcon International, Inc.).
120. Brit. Pat. 956,780 (Apr. 29, 1964) (to Halcon International, Inc.).
121. U.S. Pat. 3,231,608 (Jan. 22, 1966), J. Kollar (to Gulf Research and Development Corp.).
122. Jpn. Pat. 45-16444 (June 8, 1970), G. Inoue and co-workers (to Asahi Chemical Industries, Ltd.).
123. Jpn. Pat. 50-116415 (Sept. 11, 1975), S. Furuhashi (to Asahi Chemical Industries, Ltd.).
124. Jpn. Pat. 51-29427 (Mar. 12, 1976), M. Nishino and co-workers (to Toray Industries).
125. U.S. Pat. 4,032,569 (June 28, 1977), A. Onopchenko and co-workers (to Gulf Research and Development Corp.).
126. K. Tanaka, *Chem. Technol.* **4**(9), 555 (1974).
127. Jpn. Pat. 58-021642 (Feb. 8, 1983), M. Suematsu and K. Nakaoka (to Toray Industries).
128. H. Shen and H. Weng, *Ind. Eng. Chem. Res.* **27,** 2254–2260 (1988).
129. E. Sorribes, J. Navarro, A. Romero, and L. Jodra, *Rev. R. Acad. Cienc. Exactas, Fis. Nat. Madrid* **81**(1), 233–235 (1987).

130. D. Rao and R. Tirukkoyilur, *Ind. Eng. Chem. Process Des. Dev.* **25**(1), 299–304 (1986).
131. D. Forster and J. F. Roth, eds., *Homogeneous Catalysis II* (Advances in Chemistry Series 132), American Chemical Society, Washington, D.C. 1974; H. Adkins and co-workers, *J. Org. Chem.* **17**, 980–987 (1952); U.S. Pat. 2,729,651 (Jan. 3, 1956), W. Reppe (to Badische Anilin- und Soda-Fabrik A. G.); USSR Pat. 198,324 (June 28, 1967), N. S. Imyanitov and co-workers; U.S. Pat. 3,509,209 (Apr. 28, 1970), D. M. Fenton (to Union Oil of California); U.S. Pat. 3,876,695 (Apr. 8, 1975), N. Von Kutepow (to Badische Anilin- und Soda-Fabrik A. G.).
132. *Chem. Eng. News,* 14 (May 25, 1987).
133. S. Hosaka and co-workers, *Tetrahedron* **27**, 3821–3829 (1971); W. E. Billeys and co-workers, *Chem. Commun.* 1067–1068 (1971).
134. Ger. Offen. 2,630,086 (Jan. 12, 1978), H. Schneider and co-workers (to Badische Anilin- und Soda-Fabrik A. G.); U.S. Pat. 4,316,047 (Feb. 16, 1982), R. Kummer and co-workers (to Badische Anilin- und Soda-Fabrik, A. G.).
135. U.S. Pat. 4,169,956 (Oct. 2, 1979), R. Kummer and co-workers (to Badische Anilin- und Soda-Fabrik A. G.); U.S. Pat. 4,171,451 (Oct. 16, 1979), R. Kummer and co-workers (to Badische Anilin- und Soda-Fabrik A. G.); U.S. Pat. 4,360,695 (Nov. 23, 1982), P. Magnussen and co-workers (to Badische Anilin- und Soda-Fabrik A. G.).
136. U.S. Pat. 4,171,450 (Oct. 16, 1979), H. S. Kesling, Jr., and co-workers (to Atlantic Richfield Co.).
137. U.S. Pat. 4,166,913 (Sept. 4, 1979), H. S. Kesling, Jr., and co-workers (to Atlantic Richfield Co.).
138. U.S. Pat. 4,195,184 (Mar. 25, 1980), H. S. Kesling, Jr., and co-workers (to Atlantic Richfield Co.).
139. U.S. Pat. 4,575,562 (Mar. 11, 1986), C. K. Hsu and co-workers (to Monsanto Co.).
140. U.S. Pat. 3,306,932 (Feb. 28, 1967), D. D. Davis (to E. I. du Pont de Nemours & Co., Inc.).
141. U.S. Pat. 3,654,355 (Nov. 19, 1969), W. H. Mueller and co-workers (to Monsanto Co.).
142. U.S. Pat. 3,636,100 (Jan. 18, 1972), W. H. Mueller and co-workers (to Monsanto Co.).
143. U.S. Pat. 3,636,101 (Jan. 18, 1972), T. F. Doumani (to Union Oil Co. of California).
144. Brit. Pat. 1,278,353 (June 21, 1972), H. Arnold and co-workers (to Monsanto Co.).
145. Brit. Pat. 1,239,224 (July 14, 1971), C. Gardner (to Imperial Chemical Industries, Ltd.).
146. U.S. Pat. 3,607,926 (Sept. 21, 1971), R. D. Smetana (to Texaco, Inc.).
147. Fr. Add. 96,191 (May 19, 1972), C. Gardner (to Imperial Chemical Industries, Ltd.); Jpn. Pat. 56-5374 (Feb. 4, 1981), S. Miyazaki (to Agency of Industrial Sciences and Technology).
148. Brit. Pat. 1,361,749 (July 31, 1974), S. D. Razumovskii and co-workers (to USSR).
149. Fr. Pat. 2,140,088 (Dec. 1, 1973), O. Grosskinsky, G. Herrmann, and R. Kaiser (to Badische Anilin- und Soda-Fabrik A. G.).
150. Jpn. Pat. 54-135720 (Oct. 22, 1979), Y. Ishii (to Yasutaka).
151. U.S. Pat. 3,912,586 (Oct. 14, 1975), H. Kaneyuki and co-workers (to Mitsui Petrochemical Industries).
152. Jpn. Pat. 57-129694 (Aug. 11, 1982), H. Nakano and co-workers (to Dainippon Ink and Chemicals).
153. U.S. Pat. 4,400,468 (Aug. 23, 1983), M. Faber (to Hydrocarbon Research).
154. Jpn. Pat. 58-149687 (Sept. 6, 1983), T. Minoda, T. Oomori, and H. Narishima (to Nissan Chemical Industries).
155. Eur. Pat. 74,169 (Mar. 16, 1983), P. C. Maxwell (to Celanese Corp.).
156. *Chem. Econ. Eng. Rev.* 35–36 (Jan./Feb. 1986).
157. Hazardous Materials Table, *Code of Federal Regulations 49CFR 172.101* (revised Nov. 1989).
158. Ref. 30, p. 608.5031C.
159. *Chem. Eng. News* **67**(15), 12 (1989).

160. *Chem. Mark. Rep.* **236**(15), 54 (1989).
161. *Chem. Mark. Rep.* **230**(14), 58 (1989).
162. *Chem. Mark. Rep.* **236**(17), 50 (1989).
163. *Chem. Econ. Eng. Rev.* **17**(6), 45 (1985).
164. *Eur. Chem. News* **50**(1329), 25 (1988).
165. *Eur. Chem. News* **52**(1370), 10 (1989).
166. *Chem. Eng. News* **65**(21), 14 (1987).
167. *Chem. Week* **146**(2), 11 (1990).
168. *Agence Economique and Financiere,* 7 (Nov. 1, 1987).
169. *Chem. Eng. News* **67**(49), 15 (1989).
170. *Jpn. Chem. Week* 2–3 (April 27, 1989).
171. *Eur. Chem. News* **50**(1329), 4 (1988).
172. *U.S. Exports, EM546,* U.S. Dept. of Commerce, Bureau of Census, data for 1986.
173. Ref. 30, p. 608.5031A.
174. *Food Chemicals Codex,* 3rd ed., National Academy of Sciences, National Academy Press, Washington, D.C., 1981.
175. R. Keller in F. Snell and C. Hilton, eds., *Encyclopedia of Industrial Chemicals Analysis,* Vol. 4, Wiley-Interscience, New York, 1967, p. 408–423.
176. "Chromatography" in R. Freis and J. Lawrence, eds., *Derivatization in Analytical Chemistry,* Vol. 1, Plenum Press, New York, 1981.
177. J. Drozd, *Chemical Derivatization in Gas Chromatography,* Elsevier, Amsterdam, 1980.
178. R. Schwarzenbach, *J. Chromatogr.* **251,** 339–358 (1982).
179. M. Gennaro and co-workers, *Ann. Chim. (Rome)* **78**(3–4), 137–152 (1988).
180. G. Lippe and co-workers, *Clin. Biochem.* **20**(4), 275–279 (1987).
181. N. Sax, *Dangerous Properties of Industrial Materials,* Van-Nostrand Reinhold Co., New York, 1984, p. 141.
182. J. Svendsen, L. Sydnes, and J. Whist, *Spectrosc. Inst. J.* **3**(4–5), 380–386 (1984).
183. J. Vamecq, J. Draye, and J. Brison, *Am. J. Physiol.* **256**(4, pt. 1), G680–688 (1989).
184. *Federal Register* **47**(123) (June 25, 1982).
185. Y. Hirayama, *Shokuhin Eisei Kenkyu* **33,** 852–855 (1983).
186. H. Shimuzu and co-workers, *Sangyo Igaku* **27,** 400–419 (1985); see also D. Guest and co-workers, eds., *Patty's Industrial Hygiene and Toxicology,* 3rd ed., Vol. 2C, Wiley-Interscience, New York, 1982, p. 4945.
187. Y. Yokouchi and Y. Ambi, *Atmos. Environ.* **20,** 1727–1734 (1986).
188. R. Ferek, A. Lazrus, P. Haagenson, and J. Winchester, *Environ. Sci. Technol.* **17,** 315–324 (1983).
189. K. Kawamura and I. Kaplan, *Environ. Sci. Technol.* **21,** 105–110 (1987).
190. B. Appel and co-workers, *Environ. Sci. Technol.* **13,** 98–104 (1979).
191. E. Pelzer, J. Bada, G. Schlesinger, and S. Miller, *Adv. Space. Res.* **4**(12), 69–74 (1984).
192. Y. Novikov and co-workers, *Gig. Sanit.* **1983**(9), 72–75 (1983).
193. K. Urano and Z. Kato, *J. Hazard Mater.* **13**(2), 147–159 (1986).
194. Ref. 30, p. 608.5032C,Q.
195. *Chem. Mark. Rep.* 54 (Oct. 9, 1989).
196. J. Szymanowski and A. Sobczynska, *Przem. Chem.* **66,** 373–377 (1987).
197. U.S. Pat. 2,680,761 (1952), R. Halliwell (to E. I. du Pont de Nemours & Co., Inc.).
198. U.S. Pat. 2,518,608 (1947), M. Farlow (to E. I. du Pont de Nemours & Co., Inc.).
199. *Eur. Chem. News* **23**(2), 17 (1973); U.S. Pats. 3,496,217; 3,496,218; 3,766,237; 3,526,654; 3,542,847; 3,536,748; W. Drinkard and co-workers (to E. I. du Pont de Nemours & Co., Inc.).
200. M. M. Baizer and D. E. Danly, *Chem. Technol.* **10**(10), 161–164, 302–311 (1980).
201. M. Kato, *Nikkakyo Geppo* **26,** 561–567 (1973).
202. *CPI Purchasing,* 31 (Aug. 1989).

203. U.S. Pat. 4,423,018 (Dec. 27, 1983), D. Danly and J. Lester (to Monsanto Co., now dedicated to the public).
204. S. Litherland and co-workers, *Energy Research Abstr.* **10**(8), Abstr. No. 12145 (1985).
205. C. Chi and J. Lester, Jr., *CHEMTECH,* 308 (May 1990).

General References

Y. C. Yen and S. Y. Wu, *Nylon-6,6, Report No. 54B, Process Economics Program,* SRI International, Menlo Park, Calif., Jan. 1987, pp. 1–148.
V. D. Luedeke, "Adipic Acid" in *Encyclopedia of Chemical Processing and Design,* J. McKetta and W. Cunningham, eds., Vol. 2, Marcel Dekker, Inc., New York, 1977, pp. 128–146.

DARWIN D. DAVIS
DONALD R. KEMP
E. I. du Pont de Nemours & Co., Inc.

ADRENALINE.   See EPINEPHRINE; HORMONES; PSYCHOPHARMACOLOGICAL AGENTS.

ADSORBENTS.   See ADSORPTION, LIQUID SEPARATION; ALUMINUM COMPOUNDS, ALUMINA; CARBON, ACTIVATED CARBON; ION EXCHANGE; MOLECULAR SIEVES; SILICON COMPOUNDS; SYNTHETIC, INORGANIC SILICATES.

ADSORPTION

Adsorption is the term used to describe the tendency of molecules from an ambient fluid phase to adhere to the surface of a solid. This is a fundamental property of matter, having its origin in the attractive forces between molecules. The force field creates a region of low potential energy near the solid surface and, as a result, the molecular density close to the surface is generally greater than in the bulk gas. Furthermore, and perhaps more importantly, in a multicomponent system the composition of this surface layer generally differs from that of the bulk gas since the surface adsorbs the various components with different affinities. Adsorption may also occur from the liquid phase and is accompanied by a similar change in composition, although, in this case, there is generally little difference in molecular density between the adsorbed and fluid phases.

The enhanced concentration at the surface accounts, in part, for the catalytic activity shown by many solid surfaces, and it is also the basis of the application of adsorbents for low pressure storage of permanent gases such as methane. However, most of the important applications of adsorption depend on the selectivity, ie, the difference in the affinity of the surface for different components. As a result of this selectivity, adsorption offers, at least in principle, a relatively straightforward means of purification (removal of an undesirable trace component from a fluid mixture) and a potentially useful means of bulk separation.

Fundamental Principles

Forces of Adsorption. Adsorption may be classified as chemisorption or physical adsorption, depending on the nature of the surface forces. In physical adsorption the forces are relatively weak, involving mainly van der Waals (induced dipole–induced dipole) interactions, supplemented in many cases by electrostatic contributions from field gradient–dipole or –quadrupole interactions. By contrast, in chemisorption there is significant electron transfer, equivalent to the formation of a chemical bond between the sorbate and the solid surface. Such interactions are both stronger and more specific than the forces of physical adsorption and are obviously limited to monolayer coverage. The differences in the general features of physical and chemisorption systems (Table 1) can be understood on the basis of this difference in the nature of the surface forces.

Table 1. Parameters of Physical Adsorption and Chemisorption

Parameter	Physical adsorption	Chemisorption
heat of adsorption (ΔH)	low, < 2 or 3 times latent heat of evaporation	high, > 2 or 3 times latent heat of evaporation
specificity	nonspecific	highly specific
nature of adsorbed phase	monolayer or multilayer, no dissociation of adsorbed species	monolayer only may involve dissociation
temperature range	only significant at relatively low temperatures	possible over a wide range of temperature
forces of adsorption	no electron transfer, although polarization of sorbate may occur	electron transfer leading to bond formation between sorbate and surface
reversibility	rapid, nonactivated, reversible	activated, may be slow and irreversible

Heterogeneous catalysis generally involves chemisorption of the reactants, but most applications of adsorption in separation and purification processes depend on physical adsorption. Chemisorption is sometimes used in trace impurity removal since very high selectivities can be achieved. However, in most situations the low capacity imposed by the monolayer limit and the difficulty of regenerating the spent adsorbent more than outweigh this advantage. The higher capacities achievable in physical adsorption result from multilayer formation and this is obviously critical in such applications as gas storage, but it is also an important consideration in most adsorption separation processes since the process cost is directly related to the adsorbent capacity.

In very small pores the molecules never escape from the force field of the pore wall even at the center of the pore. In this situation the concepts of monolayer and multilayer sorption become blurred and it is more useful to consider adsorption simply as pore filling. The molecular volume in the adsorbed phase is similar to that of the saturated liquid sorbate, so a rough estimate of the saturation capacity can be obtained simply from the quotient of the specific micropore volume and the molar volume of the saturated liquid.

Selectivity. Selectivity in a physical adsorption system may depend on differences in either equilibrium or kinetics, but the great majority of adsorption separation processes depend on equilibrium-based selectivity. Significant kinetic selectivity is in general restricted to molecular sieve adsorbents—carbon molecular sieves, zeolites, or zeolite analogues. In these materials the pore size is of molecular dimensions, so that diffusion is sterically restricted. In this regime small differences in the size or shape of the diffusing molecule can lead to very large differences in diffusivity. In the extreme limit one species (or one class of compounds) may be completely excluded from the micropores, thus giving a highly selective molecular sieve separation. The most important example of such a process is the separation of linear hydrocarbons from their branched and cyclic isomers using a 5A zeolite adsorbent. A second example, where the difference in diffusivities is less extreme but still large enough to produce an efficient separation, is air separation over carbon molecular sieve or 4A zeolite, in which oxygen, the faster diffusing component, is preferentially adsorbed.

A degree of control over the kinetic selectivity of molecular sieve adsorbents can be achieved by controlled adjustment of the pore size. In a carbon sieve this may be accomplished by adjusting the burn-out conditions or by controlled deposition of an easily crackable hydrocarbon. In a zeolite, ion exchange offers the simplest possibility but controlled silanation or boration has also been shown to be effective in certain cases (1).

Control of equilibrium selectivity is generally achieved by adjusting the balance between electrostatic and van der Waals forces. This may be accomplished by changing the chemical nature of the surface and also, to a lesser extent, by adjusting the pore size. In carbon adsorbents surface oxidation offers a simple and effective way of introducing surface polarity and thus modifying the selectivity. One example is shown in Figure 1. On an untreated carbon adsorbent n-hexane is adsorbed more strongly than sulfur dioxide, whereas on an oxidized surface the relative affinities are reversed. With zeolite adsorbents, changing the nature of the exchangeable cation by ion exchange or adjusting the silicon–aluminum ratio of the framework, which determines the cation density, are the most common approaches. In some instances the aluminum-free zeolite analogue (a porous crystalline silicate) may be prepared with the same channel geometry but with a nonpolar surface.

Adsorption on a nonpolar surface such as pure silica or an unoxidized carbon is dominated by van der Waals forces. The affinity sequence on such a surface generally follows the sequence of molecular weights since the polarizabil-

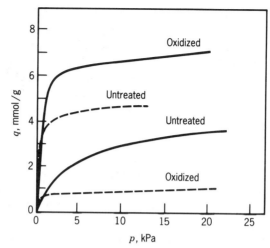

Fig. 1. Equilibrium isotherms for adsorption on activated carbon at 298 K showing the effect of surface modification (2). ——, SO_2; - - -, n-hexane. To convert kPa to torr multiply by 7.5.

ity, which is the main factor governing the magnitude of the van der Waals interaction energy, is itself roughly proportional to the molecular weight.

Hydrophilic and Hydrophobic Surfaces.   Water is a small, highly polar molecular and it is therefore strongly adsorbed on a polar surface as a result of the large contribution from the electrostatic forces. Polar adsorbents such as most zeolites, silica gel, or activated alumina therefore adsorb water more strongly than they adsorb organic species, and, as a result, such adsorbents are commonly called hydrophilic. In contrast, on a nonpolar surface where there is no electrostatic interaction water is held only very weakly and is easily displaced by organics. Such adsorbents, which are the only practical choice for adsorption of organics from aqueous solutions, are termed hydrophobic.

The most common hydrophobic adsorbents are activated carbon and silicalite. The latter is of particular interest since the affinity for water is very low indeed; the heat of adsorption is even smaller than the latent heat of vaporization (3). It seems clear that the channel structure of silicalite must inhibit the hydrogen bonding between occluded water molecules, thus enhancing the hydrophobic nature of the adsorbent. As a result, silicalite has some potential as a selective adsorbent for the separation of alcohols and other organics from dilute aqueous solutions (4).

Capillary Condensation.   The equilibrium vapor pressure in a pore or capillary is reduced by the effect of surface tension. As a result, liquid sorbate condenses in a small pore at a vapor pressure that is somewhat lower than the saturation vapor pressure. In a porous adsorbent the region of multilayer physical adsorption merges gradually with the capillary condensation regime, leading to upward curvature of the equilibrium isotherm at higher relative pressure. In the capillary condensation region the intrinsic selectivity of the adsorbent is lost, so in separation processes it is generally advisable to avoid these conditions. However, this effect is largely responsible for the enhanced capacity of macroporous desiccants such as silica gel or alumina at higher humidities.

Practical Adsorbents

To achieve a significant adsorptive capacity an adsorbent must have a high specific area, which implies a highly porous structure with very small micropores. Such microporous solids can be produced in several different ways. Adsorbents such as silica gel and activated alumina are made by precipitation of colloidal particles, followed by dehydration (see ALUMINUM COMPOUNDS, ALUMINUM OXIDE; SILICA, AMORPHOUS SILICA). Carbon adsorbents are prepared by controlled burn-out of carbonaceous materials such as coal, lignite, and coconut shells (see CARBON, ACTIVATED CARBON). These procedures generally yield a fairly wide distribution of pore size (Fig. 2). The crystalline adsorbents (zeolite and zeolite analogues) are different in that the dimensions of the micropores are determined by the crystal structure and there is therefore virtually no distribution of micropore size (see MOLECULAR SIEVES). Although structurally very different from the crystalline adsorbents, carbon molecular sieves also have a very narrow distribution of pore size. The adsorptive properties depend on the pore size and the pore size distribution as well as on the nature of the solid surface. A simple classification of some of the common adsorbents according to these features is as follows:

Surface polarity	Pore size distribution	
	Narrow	Broad
polar	zeolites (Al rich)	activated alumina silica gel
nonpolar	carbon molecular sieves silicalite	activated carbon

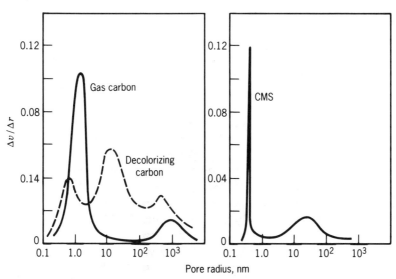

Fig. 2. Pore size distribution of typical samples of activated carbon (small pore gas carbon and large pore decolorizing carbon) and carbon molecular sieve (CMS). $\Delta v/\Delta r$ represents the increment of specific micropore volume for an increment of pore radius.

Despite the difference in the nature of the surface, the adsorptive behavior of the molecular sieve carbons resembles that of the small pore zeolites. As their name implies, molecular sieve separations are possible on these adsorbents based on the differences in adsorption rate, which, in the extreme limit, may involve complete exclusion of the larger molecules from the micropores.

Important properties and a number of applications of several commercial adsorbents are summarized in Tables 2–4.

Amorphous Adsorbents.   The amorphous adsorbents (silica gel, activated alumina, and activated carbon) typically have specific areas in the 200–1000-m^2/g range, but for some activated carbons much higher values have been achieved ($\sim$1500 m^2/g). The difficulty is that these very high area carbons tend to lack physical strength and this limits their usefulness in many practical applications. The high area materials also contain a large proportion of very small pores, which renders them unsuitable for applications involving adsorption of large molecules. The distinction between gas carbons, used for adsorption of low molecular weight permanent gases, and liquid carbons, which are used for adsorption of larger molecules such as color bodies from the liquid phase, is thus primarily a matter of pore size.

In a typical amorphous adsorbent the distribution of pore size may be very wide, spanning the range from a few nanometers to perhaps one micrometer. Since different phenomena dominate the adsorptive behavior in different pore size ranges, IUPAC has suggested the following classification:

micropores, <2-nm diameter

mesopores, 2–50-nm diameter

macropores, >50-nm diameter

This division is somewhat arbitrary since it is really the pore size relative to the size of the sorbate molecule rather than the absolute pore size that governs the behavior. Nevertheless, the general concept is useful. In micropores (pores which are only slightly larger than the sorbate molecule) the molecule never escapes from the force field of the pore wall, even when in the center of the pore. Such pores generally make a dominant contribution to the adsorptive capacity for molecules small enough to penetrate. Transport within these pores can be severely limited by steric effects, leading to molecular sieve behavior.

The mesopores make some contribution to the adsorptive capacity, but their main role is as conduits to provide access to the smaller micropores. Diffusion in the mesopores may occur by several different mechanisms, as discussed below. The macropores make very little contribution to the adsorptive capacity, but they commonly provide a major contribution to the kinetics. Their role is thus analogous to that of a super highway, allowing the adsorbate molecules to diffuse far into a particle with a minimum of diffusional resistance.

Crystalline Adsorbents.   In the crystalline adsorbents, zeolites and zeolite analogues such as silicalite and the microporous aluminum phosphates, the dimensions of the micropores are determined by the crystal framework and there is therefore virtually no distribution of pore size. However, a degree of control can sometimes be exerted by ion exchange, since, in some zeolites, the exchangeable cations occupy sites within the structure which partially obstruct the pores. The

Table 2. Properties and Applications of Amorphous Adsorbents

Adsorbent	Pore diameter, nm	Particle density, g/cm^3	Specific area, m^2/g	Applications
activated carbon (large pore)	1–10^3 (broad range)	0.6–0.8	200–600	water purification, sugar decolorizing
activated carbon (small pore)	1–10	0.5–0.9	400–1200	removal of light organics
carbon molecular sieve	0.4–0.5, 10–10^2 (bimodal)	0.9–1.0	100–300	air separation (N$_2$ production)
silica gel (high area)	2–10	1.09	800	
silica gel (low area)	10–50	0.62	300	general purpose desiccants
activated alumina	2–10	1.2–1.3	300–400	

Table 3. Properties and Application of Polymeric Adsorbents

Type of polymer	Representative commercial product	Properties	Applications
sulfonated styrene–divinylbenzene copolymers[a] with various degrees of cross-linking	Dowex-50 Amberlite IR120B	pore diameter, porosity, density, etc vary with degree of hydration or dehydration	sugar separations[b], eg, various fructose, glucose
macroreticular sulfonated styrene–divinylbenzene	Diaion HPK-25	porosity ~ 0.33, microparticles ~ 80-µm diameter	removal of NH$_3$ or light amines[c]

[a]Ion exchanged to Ca^{2+} form.
[b]Liquid-phase operation.
[c]Gas or liquid phase.

Table 4. Properties and Applications of Crystalline Adsorbents[a]

Structure	Cation	Typical formula of unit cell or pseudocell	Window	Effective channel diameter, nm	Applications
4A	Na^+	$Na_{12}[(AlO_2)_{12}(SiO_2)_{12}]$	obstructed 8-ring	0.38	desiccant; CO_2 removal; air separation (N_2)
5A	Ca^{2+}	$Ca_5Na_2[(AlO_2)_{12}(SiO_2)_{12}]$	free 8-ring	0.44	linear paraffin separation; air separation (O_2)
3A	K^+	$K_{12}[(AlO_2)_{12}(SiO_2)_{12}]$	obstructed 8-ring	0.29	drying of reactive gases
13X	Na^+	$Na_{86}[(AlO_2)_{86}(SiO_2)_{106}]$	12-ring (free)	0.84	air separation (O_2), removal of mercaptans
10X	Ca^{2+}	$Ca_{43}[(AlO_2)_{86}(SiO_2)_{106}]$	12-ring (obstructed)	0.80	
SrBaX	Sr^{2+}, Ba^{2+}	$Sr_{21}Ba_{22}[(AlO_2)_{86}(SiO_2)_{106}]$	12-ring (obstructed)	0.80	separation of C_8 aromatics
KY	K^+	$K_{56}[(AlO_2)_{56}(SiO_2)_{136}]$	12-ring (free)	0.80	
Mordenite	H^+	$H_8[(AlO_2)_8(SiO_2)_{40}]$	12-ring (free)	0.70	trapping of Kr from nuclear off-gas
	Ag^+	$Ag_8[(AlO_2)_8(SiO_2)_{40}]$	12-ring (free)	0.70	trapping of CH_3I from nuclear off-gas
AgX	Ag^+	$Ag_{86}[(AlO_2)_{86}(SiO_2)_{106}]$	12-ring (free)	0.84	removal of organics in aqueous systems
silicalite/HZSM5		$(SiO_2)_{96}$	10-ring	0.60	

[a]Structural details can be found in refs. 5 and 6. A simplified description is given in ref. 7.

500

crystals of these materials are generally quite small (1–5 μm) and they are aggregated with a suitable binder (generally a clay) and formed into macroporous particles having dimensions large enough to pack directly into an adsorber vessel. Such materials therefore have a well-defined bimodal pore size distribution with the intracrystalline micropores (a few tenths of a nanometer) linked together through a network of macropores having a diameter of the same order as the crystal size ($\sim$1 μm).

Desiccants. A solid desiccant is simply an adsorbent which has a high affinity and capacity for adsorption of moisture so that it can be used for selective adsorption of moisture from a gas (or liquid) stream. The main requirements for an efficient desiccant are therefore a highly polar surface and a high specific area (small pores). The most widely used desiccants (qv) are silica gel, activated alumina, and the aluminum rich zeolites (4A or 13X). The equilibrium adsorption isotherms for moisture on these materials have characteristically different shapes (Fig. 3), making them suitable for different applications.

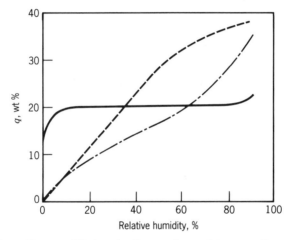

Fig. 3. Adsorption equilibrium isotherms for moisture on three commercial adsorbents: pelletized 4A zeolite (——), silica gel (- - -), and a typical activated alumina (–·—·–).

The zeolites have high affinity and high capacity at low partial pressures, shown by the nearly rectangular form of the isotherm. This makes them useful desiccants where a very low humidity or dew point is required. The 3A zeolite is a molecular sieve desiccant, since its micropores are small enough to exclude most molecules other than water. It is therefore useful for drying reactive gases. The major disadvantage of zeolite desiccants is that a high temperature is required for regeneration (>300°C), which makes their use uneconomic when only a moderately low dew point is required.

Considerable variation in the moisture isotherm for alumina can be obtained by different preparation and pretreatment. However, in general, the initial slope of the isotherm is not as steep as that of a zeolite, indicating a lower moisture affinity at low partial pressure, but the capacity at high humidities is often higher than that of a zeolitic adsorbent. Regeneration temperatures are typically in the 250–350°C range. Alumina adsorbents are also more robust than zeolites and less sensitive to deactivation by organics, but they are generally less suitable than the zeolites where very low humidity is the primary requirement.

The isotherm for silica gel is more nearly linear over a wide range of partial pressure, although the affinity for moisture is lower than that of either alumina or the zeolites. However, a correspondingly lower regeneration temperature is also required. This can be as low as 120°C, making silica gel the most suitable candidate for pressure swing driers, desiccant cooling systems (8), and other applications where low grade heat is used for regeneration of the adsorbent.

Loaded Adsorbents.   Where highly efficient removal of a trace impurity is required it is sometimes effective to use an adsorbent preloaded with a reactant rather than rely on the forces of adsorption. Examples include the use of zeolites preloaded with bromine to trap traces of olefins as their more easily condensible bromides; zeolites preloaded with iodine to trap mercury vapor, and activated carbon loaded with cupric chloride for removal of mercaptans.

Adsorption Equilibrium

Henry's Law.   Like any other phase equilibrium, the distribution of a sorbate between fluid and adsorbed phases is governed by the principles of thermodynamics. Equilibrium data are commonly reported in the form of an isotherm, which is a diagram showing the variation of the equilibrium adsorbed-phase concentration or loading with the fluid-phase concentration or partial pressure at a fixed temperature. In general, for physical adsorption on a homogeneous surface at sufficiently low concentrations, the isotherm should approach a linear form, and the limiting slope in the low concentration region is commonly known as the Henry's law constant. The Henry constant is simply a thermodynamic equilibrium constant and the temperature dependence therefore follows the usual van't Hoff equation:

$$\lim_{p \to 0} \left(\frac{\partial q}{\partial p} \right)_T \equiv K' = K'_0 e^{-\Delta H_0 / RT} \tag{1}$$

in which $-\Delta H_0$ is the limiting heat of adsorption at zero coverage. Since adsorption, particularly from the vapor phase, is usually exothermic, $-\Delta H_0$ is a positive quantity and K' therefore decreases with increasing temperature. A corresponding dimensionless Henry constant (K) may also be defined, based on the ratio of adsorbed and fluid-phase concentrations:

$$\lim_{c \to 0} \left(\frac{\partial q}{\partial c} \right)_T \equiv K = K_0 e^{-\Delta U_0 / RT} \tag{2}$$

Since, for an ideal vapor phase, $p = cRT$, these quantities are related by

$$K = RTK'; \qquad -\Delta H_0 = -\Delta U_0 + RT \tag{3}$$

Henry's law corresponds physically to the situation in which the adsorbed phase is so dilute that there is neither competition for surface sites nor any significant interaction between adsorbed molecules. At higher concentrations both of these effects become important and the form of the isotherm becomes more complex. The isotherms have been classified into five different types (9) (Fig. 4). Isotherms for a microporous adsorbent are generally of type I; the more complex forms are associated with multilayer adsorption and capillary condensation.

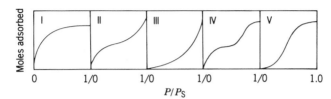

Fig. 4. The Brunaner classification of isotherms (I–V).

Langmuir Isotherm. Type I isotherms are commonly represented by the ideal Langmuir model:

$$\frac{q}{q_s} = \frac{bp}{1 + bp} \tag{4}$$

where q_s is the saturation limit and b is an equilibrium constant which is directly related to the Henry constant ($K' = bq_s$). The Langmuir model was originally developed to represent monolayer adsorption on an ideal surface, for which q_s corresponds to the monolayer coverage. However, in applying this model to physical adsorption on a microporous solid, the saturation limit becomes the quantity of sorbate required to fill the micropore volume. This expression is of the correct form to represent a type I isotherm, since at low pressure it approaches Henry's law while at high pressure it tends asymptotically to the saturation limit. The equilibrium constant b ($= K'/q_s$) decreases with increasing temperature (eq. 1); therefore, for a given pressure range, the isotherm approaches rectangular or irreversible form at low temperatures (large b) and linear form at high temperatures (small b).

Although very few systems conform accurately to the Langmuir model, this model provides a simple qualitative representation of the behavior of many systems and it is therefore widely used, particularly for adsorption from the vapor phase. According to the Langmuir model the heat of adsorption should be independent of loading, but this requirement is seldom fulfilled in practice. Both increasing and decreasing trends are commonly observed (Fig. 5). For a polar sorbate on a polar adsorbent (ie, a system in which electrostatic forces are dominant) a decreasing trend is normally observed, since the relative importance of the electrostatic contribution declines at high loadings as a result of preferential occupation of the most favorable sites and consequent screening of cations. In contrast, where van der Waals forces are dominant (nonpolar sorbates), a rising trend of heat of adsorption with loading is generally observed. This is commonly attributed to the effect of intermolecular attractive forces, but other explanations are also possible (11). In homologous series such as the linear paraffins the heat of adsorption increases linearly with carbon number (12).

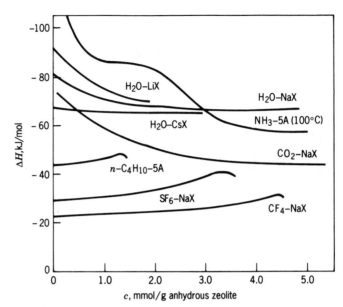

Fig. 5. Variation of isosteric heat of adsorption with adsorbed phase concentration. Reprinted from ref. 10, courtesy of Marcel Dekker, Inc. To convert kJ to kcal divide by 4.184.

Freundlich Isotherm.   The isotherms for some systems, notably hydrocarbons on activated carbon, conform more closely to the Freundlich equation:

$$q = bp^{1/n} \qquad (n > 1.0) \tag{5}$$

Although the Freundlich expression does not reduce to Henry's law at low concentrations, it often provides a good approximation over a wide range of conditions. This form of equation can be explained as resulting from energetic heterogeneity of the surface. Superposition of a set of Langmuir isotherms with different b values (corresponding to sites of different energy) yields an expression of this form.

Adsorption of Mixtures.   The Langmuir model can be easily extended to binary or multicomponent systems:

$$\frac{q_1}{q_{s1}} = \frac{b_1 p_1}{1 + b_1 p_1 + b_2 p_2 + \cdots} ; \quad \frac{q_2}{q_{s2}} = \frac{b_2 p_2}{1 + b_1 p_1 + b_2 p_2; + \cdots} \tag{6}$$

Thermodynamic consistency requires $q_{s1} = q_{s2}$, but this requirement can cause difficulties when attempts are made to correlate data for sorbates of very different molecular size. For such systems it is common practice to ignore this requirement, thereby introducing an additional model parameter. This facilitates data fitting but it must be recognized that the equations are then being used purely as a convenient empirical form with no theoretical foundation.

Equation 6 shows that the adsorption of component 1 at a partial pressure p_1 is reduced in the presence of component 2 as a result of competition for the available surface sites. There are only a few systems for which this expression (with $q_{s1} = q_{s2} = q_s$) provides an accurate quantitative representation, but it provides useful qualitative or semiquantitative guidance for many systems. In particular, it has the correct asymptotic behavior and provides explicit recognition of the effect of competitive adsorption. For example, if component 2 is either strongly adsorbed or present at much higher concentration than component 1, the isotherm for component 1 is reduced to a simple linear form in which the apparent Henry's law constant depends on p_2:

$$q_1 \simeq \left(\frac{b_1 q_{s1}}{1 + b_2 p_2} \right) p_1 \qquad (7)$$

For an equilibrium-based separation, a convenient measure of the intrinsic selectivity of the adsorbent is provided by the separation factor (α_{12}), which is defined by analogy with the relative volatility as

$$\alpha_{12} = \frac{(X_1/Y_1)}{(X_2/Y_2)} \qquad (8)$$

where X and Y refer to the mole fractions in the adsorbed and fluid phases, respectively, at equilibrium. For a system that obeys the Langmuir model (eq. 6) it is evident that $\alpha_{12} = b_1/b_2$ and is thus independent of concentration. The Langmuir isotherm is therefore sometimes referred to as the constant separation factor model.

The assumption of a constant separation factor is often a reasonable approximation for preliminary process design but this assumption is often violated in real systems, where some variation of the separation factor with composition is common and more extreme variations involving azeotrope formation ($\alpha = 1.0$ at a particular composition) and selectivity reversal (α varying from greater than 1.0 to less than 1.0 with changing composition) are not uncommon. There have been many attempts to improve the correlation of equilibrium data by using more complex expressions, one of the more widely used being the Langmuir-Freundlich or loading ratio correlation (13):

$$\frac{q_1}{q_s} = \frac{b_1 p_1^{n_1}}{1 + b_1 p_1^{n_1} + b_2 p_2^{n_2} + \cdots} ; \quad \frac{q_2}{q_s} = \frac{b_2 p_2^{n_2}}{1 + b_1 p_1^{n_1} + b_2 p_2^{n_2} + \cdots} \qquad (9)$$

This has the advantage that the expressions for the adsorbed-phase concentration are simple and explicit, and, as in the Langmuir expression, the effect of competition between sorbates is accounted for. However, the expression does not reduce to Henry's law in the low concentration limit and therefore violates the requirements of thermodynamic consistency. Whereas it may be useful as a basis for the correlation of experimental data, it should be treated with caution and should not be used as a basis for extrapolation beyond the experimental range.

Ideal Adsorbed Solution Theory.   Perhaps the most successful approach to the prediction of multicomponent equilibria from single-component isotherm data is ideal adsorbed solution theory (14). In essence, the theory is based on the assumption that the adsorbed phase is thermodynamically ideal in the sense that the equilibrium pressure for each component is simply the product of its mole fraction in the adsorbed phase and the equilibrium pressure for the pure component *at the same spreading pressure*. The theoretical basis for this assumption and the details of the calculations required to predict the mixture isotherm are given in standard texts on adsorption (7) as well as in the original paper (14). Whereas the theory has been shown to work well for several systems, notably for mixtures of hydrocarbons on carbon adsorbents, there are a number of systems which do not obey this model. Azeotrope formation and selectivity reversal, which are observed quite commonly in real systems, are not consistent with an ideal adsorbed phase and there is no way of knowing a priori whether or not a given system will show ideal behavior.

Adsorption Kinetics

Intrinsic Kinetics.   Chemisorption may be regarded as a chemical reaction between the sorbate and the solid surface, and, as such, it is an activated process for which the rate constant (k) follows the familiar Arrhenius rate law:

$$k = k_0 e^{-E/RT} \tag{10}$$

Depending on the temperature and the activation energy (E), the rate constant may vary over many orders of magnitude.

In practice the kinetics are usually more complex than might be expected on this basis, since the activation energy generally varies with surface coverage as a result of energetic heterogeneity and/or sorbate–sorbate interaction. As a result, the adsorption rate is commonly given by the Elovich equation (15):

$$q = \frac{1}{k'} \ln(1 + k''t) \tag{11}$$

where k' and k'' are temperature-dependent constants.

In contrast, physical adsorption is a very rapid process, so the rate is always controlled by mass transfer resistance rather than by the intrinsic adsorption kinetics. However, under certain conditions the combination of a diffusion-controlled process with an adsorption equilibrium constant that varies according to equation 1 can give the appearance of activated adsorption.

As illustrated in Figure 6, a porous adsorbent in contact with a fluid phase offers at least two and often three distinct resistances to mass transfer: external film resistance and intraparticle diffusional resistance. When the pore size distribution has a well-defined bimodal form, the latter may be divided into macropore and micropore diffusional resistances. Depending on the particular system and the conditions, any one of these resistances may be dominant or the overall rate of mass transfer may be determined by the combined effects of more than one resistance.

External Fluid Film Resistance.   A particle immersed in a fluid is always surrounded by a laminar fluid film or boundary layer through which an adsorbing

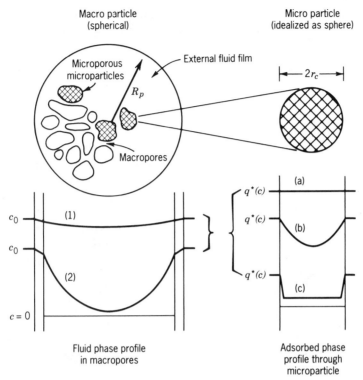

Fig. 6. Concentration profiles through an idealized biporous adsorbent particle showing some of the possible regimes. (1) + (a) rapid mass transfer, equilibrium throughout particle; (1) + (b) micropore diffusion control with no significant macropore or external resistance; (1) + (c) controlling resistance at the surface of the microparticles; (2) + (a) macropore diffusion control with some external resistance and no resistance within the microparticle; (2) + (b) all three resistances (micropore, macropore, and film) significant; (2) + (c) diffusional resistance within the macroparticle and resistance at the surface of the microparticle with some external film resistance.

or desorbing molecule must diffuse. The thickness of this layer, and therefore the mass transfer resistance, depends on the hydrodynamic conditions. Mass transfer in packed beds and other common contacting devices has been widely studied. The rate data are normally expressed in terms of a simple linear rate expression of the form

$$\frac{\partial q}{\partial t} = k_f a(c - c^*) \tag{12}$$

and the variation of the mass transfer coefficient (k_f) with the hydrodynamic conditions is generally accounted for in terms of empirical correlations of the general form

$$\mathrm{Sh} \equiv \frac{2k_f R}{D_m} = f(\mathrm{Re}, \mathrm{Sc}) \tag{13}$$

where Re and Sc are the (particle-based) Reynolds and Schmidt numbers. One of the most widely used correlations, applicable to both gas and liquid systems over a wide range of conditions, is (16)

$$\text{Sh} = \frac{2k_f R}{D_m} = 2.0 + 1.1\, \text{Sc}^{1/3}\text{Re}^{0.6} \tag{14}$$

Macropore Diffusion.   Transport in a macropore can occur by several different mechanisms, the most important of which are bulk molecular diffusion, Knudsen diffusion, surface diffusion, and Poiseuille flow. In liquid systems bulk molecular diffusion is generally dominant, but in the vapor phase the contributions from Knudsen and surface diffusion may be large or even dominant. The contribution from Poiseuille flow, ie, forced flow through the pore under the influence of the pressure gradient, is generally relatively minor since pressure gradients are usually kept small. However, this is not true in the pressurization and blowdown steps of a pressure swing process, where the contribution from Poiseuille flow can be dominant.

A molecule colliding with the pore wall is reflected in a specular manner so that the direction of the molecule leaving the surface has no correlation with that of the incident molecule. This leads to a Fickian mechanism, known as Knudsen diffusion, in which the flux is proportional to the gradient of concentration or partial pressure. The Knudsen diffusivity (D_K) is independent of pressure and varies only weakly with temperature:

$$D_K = 9700\rho\sqrt{T/M} \qquad (\text{cm}^2/\text{s}) \tag{15}$$

where ρ is the pore radius (cm) and M the molecular weight. Knudsen diffusion becomes dominant when collisions with the pore wall occur more frequently than collisions between diffusing molecules, ie, when the pore diameter is smaller than the mean free path. Since the mean free path varies inversely with pressure there is a gradual transition from the molecular to the Knudsen regime as the pressure is reduced, but the pressure at which this occurs depends on the pore size. At atmospheric temperature and pressure Knudsen diffusion is dominant in pores of less than about 10 nm diameter. In the intermediate region, which spans the range of the macropores in many commercial adsorbents, both mechanisms are of comparable significance.

The combined effects of Knudsen and molecular diffusion may be estimated approximately from the reciprocal addition rule:

$$\frac{1}{\epsilon_p D_p} = \frac{\tau}{\epsilon_p}\left(\frac{1}{D_K} + \frac{1}{D_m}\right) \tag{16}$$

The factor ϵ_p takes account of the fact that diffusion occurs only through the pore and not through the matrix; τ is a tortuosity factor which accounts for the increased path length and reduced concentration gradient arising from the random orientation of the pores as well as any other geometric effects. In a typical adsorbent $\tau \sim 3.0$ and $\epsilon_p \sim 0.3$, so the effect of these two factors is to reduce the diffusivity by about one order of magnitude relative to the value for a straight cylindrical capillary.

Micropore Diffusion.   In very small pores in which the pore diameter is not much greater than the molecular diameter the diffusing molecule never escapes from the force field of the pore wall. Under these conditions steric effects and the effects of nonuniformity in the potential field become dominant and the Knudsen mechanism no longer applies. Diffusion occurs by an activated process involving jumps from site to site, just as in surface diffusion, and the diffusivity becomes strongly dependent on both temperature and concentration.

The true driving force for any diffusive transport process is the gradient of chemical potential rather than the gradient of concentration. This distinction is not important in dilute systems where thermodynamically ideal behavior is approached. However, it becomes important at higher concentration levels and in micropore and surface diffusion. To a first approximation the expression for the diffusive flux may be written

$$J = -Bq\ \partial\mu/\partial z \tag{17}$$

where q is the concentration in the adsorbed phase. Assuming an ideal vapor phase, the expression for the chemical potential is

$$\mu = \mu^\circ + RT \ln p \tag{18}$$

where

$$\frac{\partial\mu}{\partial z} = RT \frac{d \ln p}{dq} \frac{\partial q}{\partial z} \tag{19}$$

Combining equations 17 and 18 yields, for the Fickian diffusivity (defined by $J = -D\ \partial q/\partial z$),

$$D = D_0 \frac{d \ln p}{d \ln q} \tag{20}$$

where $D_0 = BRT$. If the equilibrium relation is linear, $d \ln p/d \ln q = 1.0$ and $D \rightarrow D_0$. At higher concentrations the equilibrium relationship is nonlinear, and as a result the diffusivity is generally concentration dependent. For the special case of a Langmuir isotherm (eq. 4), $d \ln p/d \ln q = (1 - q/q_s)^{-1}$ so

$$D = \frac{D_0}{1 - q/q_s} \tag{21}$$

A rapid increase in diffusivity in the saturation region is therefore to be expected, as illustrated in Figure 7 (17). Although the corrected diffusivity (D_0) is, in principle, concentration dependent, the concentration dependence of this quantity is generally much weaker than that of the thermodynamic correction factor ($d \ln p/d \ln q$). The assumption of a constant corrected diffusivity is therefore an

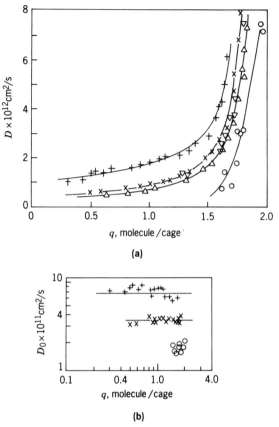

Fig. 7. Variation of (**a**) intracrystalline diffusivity and (**b**) corrected diffusivity (D_0) with sorbate concentration for n-heptane in a commercial sample of 5A zeolite crystals: $\bigcirc$, 409 K; $\triangle, \triangledown$, 439 K (ads, des); $\times$, 462 K; $+$, 491 K. Reproduced by permission of National Research Council of Canada from ref. 17.

acceptable approximation for many systems. More detailed analysis shows that the corrected diffusivity is closely related to the self-diffusivity or tracer diffusivity, and at low sorbate concentrations these quantities become identical.

The temperature dependence of the corrected diffusivity follows the usual Eyring expression

$$D_0 = D_\infty e^{-E/RT} \tag{22}$$

in which E is the activation energy or the energy barrier between adjacent sites. In small pore zeolites and carbon molecular sieves the dominant contribution to this energy is the repulsive interaction encountered by the molecule in penetrating the concentration in the pore. As a result, the activation energy shows a well-defined correlation with the molecule diameter and the size of the micropore, as illustrated in Figure 8.

Sorption Rates in Batch Systems. Direct measurement of the uptake rate by gravimetric, volumetric, or piezometric methods is widely used as a means of

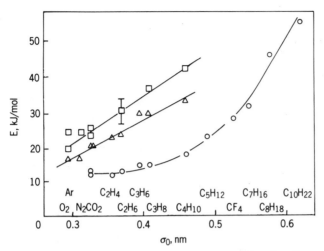

Fig. 8. Variation of activation energy with kinetic molecular diameter for diffusion in 4A zeolite (□), 5A zeolite (○), and carbon molecular sieve (MSC-5A) (△). Kinetic diameters are estimated from the van der Waals co-volumes. From ref. 7. To convert kJ to kcal divide by 4.184.

measuring intraparticle diffusivities. Diffusive transport within a particle may be represented by the Fickian diffusion equation, which, in spherical coordinates, takes the form

$$\frac{\partial q}{\partial t} = D\left(\frac{\partial^2 q}{\partial r^2} + \frac{2}{r}\frac{\partial q}{\partial r}\right) \tag{23}$$

For a step change in sorbate concentration at the particle surface ($r = R$) at time zero, assuming isothermal conditions and diffusion control, the expression for the uptake curve may be derived from the appropriate solution of this differential equation:

$$\frac{m_t}{m_\infty} = 1 - \frac{6}{\pi^2}\sum_{n=1}^{\infty}\frac{1}{n^2}\,e^{-n^2\pi^2 Dt/R^2} \tag{24}$$

or, in the initial region,

$$\frac{m_r}{m_\infty} = \frac{6}{\sqrt{\pi}}\left(\frac{Dt}{R^2}\right)^{1/2} \tag{25}$$

The time constant R^2/D, and hence the diffusivity, may thus be found directly from the uptake curve. However, it is important to confirm by experiment that the basic assumptions of the model are fulfilled, since intrusions of thermal effects or extraparticle resistance to mass transfer may easily occur, leading to erroneously low apparent diffusivity values.

    In certain adsorbents, notably partially coked zeolites and some carbon molecular sieves, the resistance to mass transfer may be concentrated at the surface of the particle, leading to an uptake expression of the form

$$\frac{m_t}{m_\infty} = 1 - e^{-k_s t} \tag{26}$$

in place of equation 24. The difference between surface resistance and intraparticle diffusion control is easily apparent from the form of the uptake curves (see Fig. 9). Since both D and k_s are generally concentration dependent, it is preferable to make differential measurements over small concentration steps in order to simplify the interpretation of the experimental data.

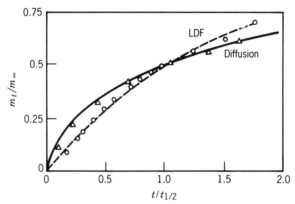

Fig. 9. Uptake curves for N_2 in two samples of carbon molecular sieve showing conformity with diffusion model (eq. 24) for sample 1 ($\triangle$), and with surface resistance model (eq. 26) for sample 2 ($\bigcirc$); LDF = linear driving force. Data from ref. 18.

For a macroporous sorbent the situation is slightly more complex. A differential balance on a shell element, assuming diffusivity transport through the macropores with rapid adsorption at the surface (or in the micropores), yields

$$\epsilon_p \frac{\partial c}{\partial t} + (1 - \epsilon_p) \frac{\partial q}{\partial t} = \epsilon_p D_p \left(\frac{\partial^2 c}{\partial r^2} + \frac{2}{r} \frac{\partial c}{\partial r} \right) \tag{27}$$

Assuming a linear equilibrium relationship (over the range of the small concentration step, $q^* = Kc$), this becomes

$$\frac{\partial c}{\partial t} = \frac{\epsilon_p D_p}{1 + (1 - \epsilon_p)K} \left(\frac{\partial^2 c}{\partial r^2} + \frac{2}{r} \frac{\partial c}{\partial r} \right) \tag{28}$$

which is of the same form as equation 23 but with an effective diffusivity given by

$$D = \frac{\epsilon_p D_p}{1 + (1 - \epsilon_p)K} \tag{29}$$

For adsorption from the vapor phase, K may be very large (sometimes as high as 10^7) and then clearly the effective diffusivity is very much smaller than the pore diffusivity. Furthermore, the temperature dependence of K follows equation 2, giving the appearance of an activated diffusion process with $E \approx (-\Delta U)$.

As a result of these difficulties the reported diffusivity data show many apparent anomalies and inconsistencies, particularly for zeolites and other microporous adsorbents. Discrepancies of several orders of magnitude in the diffusivity values reported for a given system under apparently similar conditions are not uncommon (18). Since most of the intrusive effects lead to erroneously low values, the higher values are probably the more reliable.

Adsorption Column Dynamics

In most adsorption processes the adsorbent is contacted with fluid in a packed bed. An understanding of the dynamic behavior of such systems is therefore needed for rational process design and optimization. What is required is a mathematical model which allows the effluent concentration to be predicted for any defined change in the feed concentration or flow rate to the bed. The flow pattern can generally be represented adequately by the axial dispersed plug-flow model, according to which a mass balance for an element of the column yields, for the basic differential equation governing the dynamic behavior,

$$-D_L \frac{\partial^2 c_i}{\partial z^2} + \frac{\partial}{\partial z}(vc_i) + \frac{\partial c_i}{\partial t} + \left(\frac{1-\epsilon}{\epsilon}\right)\frac{\partial \overline{q}_i}{\partial t} = 0 \tag{30}$$

The term $\partial \overline{q}_i / \partial t$ represents the overall rate of mass transfer for component i (at time t and distance z) averaged over a particle. This is governed by a mass transfer rate expression which may be thought of as a general functional relationship of the form

$$\frac{\partial \overline{q}}{\partial t} = f(c_i, c_j, \ldots, q_i, q_j, \ldots) \tag{31}$$

This rate equation must satisfy the boundary conditions imposed by the equilibrium isotherm and it must be thermodynamically consistent so that the mass transfer rate falls to zero at equilibrium. It may be a linear driving force expression of the form

$$\frac{\partial \overline{q}_i}{\partial t} = k_s(q_i^* - \overline{q}_i) \tag{32}$$

where $q_i^*(c_i, c_j, \ldots)$ represents the equilibrium adsorbed phase concentration of component i, or it may be a set of diffusion equations with their associated boundary conditions.

For an isothermal system the simultaneous solution of equations 30 and 31, subject to the boundary conditions imposed on the column, provides the expressions for the concentration profiles $c_i(z,t)$, $\bar{q}_i(z,t)$ in both phases. If the system is nonisothermal, an energy balance is also required and since, in general, both the equilibrium concentration and the rate coefficients are temperature dependent, all equations are coupled. Analytical solutions are possible only for the simpler cases: single-component isothermal systems with linear or rectangular equilibrium isotherms. In the general case of a multicomponent nonisothermal system, numerical solutions offer the only practical approach.

The form of the response for an adiabatic three-component system (two adsorbable components in an inert carrier) is illustrated in Figure 10. In general, if there are n components (counting both heat and nonadsorbing species as components), the response contains $(n - 1)$ transitions or mass transfer zones, separated by $(n - 2)$ plateaus between the initial and final states. When the change imposed at the column inlet involves an increase in the concentration of the more strongly adsorbed species, the concentration at the intermediate plateau will exceed both its initial concentration and its final steady-state concentration. This phenomenon, known as roll-up, results from displacement by the more strongly adsorbed species, which travels more slowly through the column.

Equilibrium Theory. The general features of the dynamic behavior may be understood without recourse to detailed calculations since the overall pattern of the response is governed by the form of the equilibrium relationship rather than

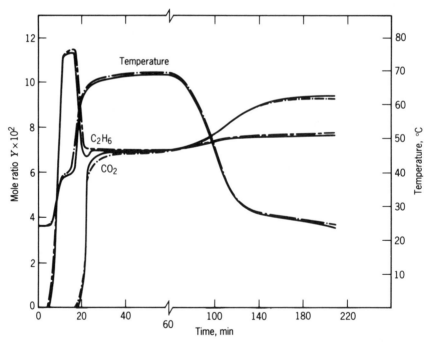

Fig. 10. Comparison of theoretical (——) and experimental (–··–·) concentration and temperature breakthrough curves for sorption of C_2H_6–CO_2 mixtures from a N_2 carrier on 5A molecular sieve. Feed: 10.5% CO_2, 7.03% C_2H_6 (molar basis) at 24°C, 116.5 kPa (1.15 atm). Column length, 48 cm. Theoretical curves were calculated numerically using the linear driving force model with a Langmuir equilibrium isotherm. Experimental data are from ref. 19. From ref. 20, courtesy of Pergamon Press.

by kinetics. Kinetic limitations may modify the form of the concentration profile but they do not change the general pattern. To illustrate the different types of transition, consider the simplest case: an isothermal system with plug flow involving a single adsorbable species present at low concentration in an inert carrier, for which equation 30 reduces to

$$v \frac{\partial c}{\partial z} + \frac{\partial c}{\partial t} + \left(\frac{1 - \epsilon}{\epsilon} \right) \frac{\partial \overline{q}}{\partial t} = 0 \tag{33}$$

Assuming local equilibrium, $\overline{q} = f(c)$ where this function represents the isotherm equation, this becomes

$$\frac{v}{1 + ((1 - \epsilon)/\epsilon) f'(c)} \frac{\partial c}{\partial z} + \frac{\partial c}{\partial t} = 0 \tag{34}$$

where $f'(c) = dq^*/dc$ is simply the slope of the equilibrium isotherm at concentration c.

Equation 34 has the form of the kinematic wave equation and represents a transition traveling with the wave velocity w, given by

$$w = \left(\frac{\partial z}{\partial t} \right)_c = \frac{v}{1 + ((1 - \epsilon)/\epsilon) f'(c)} \tag{35}$$

For a linear system $f'(c) = K$, so the wave velocity becomes independent of concentration and, in the absence of dispersive effects such as mass transfer resistance or axial mixing, a concentration perturbation propagates without changing its shape. The propagation velocity is inversely dependent on the adsorption equilibrium constant.

For a nonlinear system the behavior depends on the shape of the isotherm. If the isotherm is unfavorable (Fig. 11), $f'(c)$ increases with concentration so that w

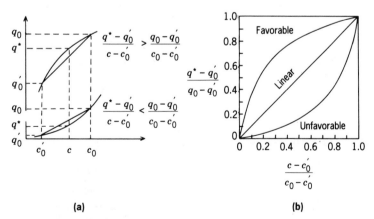

(a)   (b)

Fig. 11. (a) Equilibrium isotherm and (b) dimensionless equilibrium diagram showing favorable, linear, and unfavorable isotherms.

decreases with concentration. This leads to a spreading profile, as illustrated in Fig. 12b. However, if the isotherm is favorable (in the direction of the concentration change), an entirely different situation arises. Then $f'(c)$ decreases with concentration so that w increases with concentration. This would lead to the physically unreasonable overhanging profiles shown in Figure 12a. In fact, what happens is that the continuous solution is replaced by the equivalent shock transition so that response becomes a shock wave which propagates at a steady velocity w' given by

$$w' = \frac{v}{1 + ((1 - \epsilon)/\epsilon)(\Delta q/\Delta c)} \tag{36}$$

where $\Delta q/\Delta c$ represents the ratio of the concentration changes in the adsorbed and fluid phases.

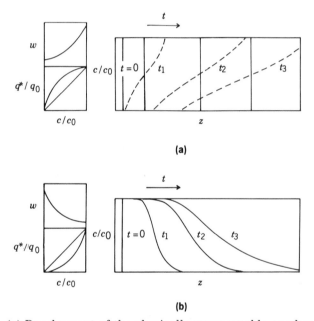

Fig. 12. (a) Development of the physically unreasonable overhanging concentration profile and the corresponding shock profile for adsorption with a favorable isotherm and (b) development of the dispersive (proportionate pattern) concentration profile for adsorption with an unfavorable isotherm (or for desorption with a favorable isotherm). From ref. 7.

Constant Pattern Behavior.   In a real system the finite resistance to mass transfer and axial mixing in the column lead to departures from the idealized response predicted by equilibrium theory. In the case of a favorable isotherm the shock wave solution is replaced by a constant pattern solution. The concentration profile spreads in the initial region until a stable situation is reached in which the mass transfer rate is the same at all points along the wave front and exactly matches the shock velocity. In this situation the fluid-phase and adsorbed-phase profiles become coincident, as illustrated in Figure 13. This represents a stable situation and the profile propagates without further change in shape—hence the term constant pattern. The form of the concentration profile under constant pattern conditions may be easily deduced by integrating the mass transfer rate expression subject to the condition $c/c_0 = q/q_0$, where q_0 is the adsorbed phase concentration in equilibrium with c_0.

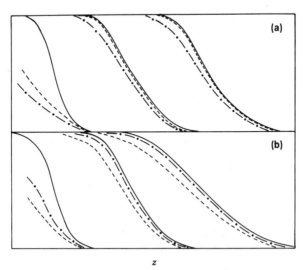

Fig. 13. Schematic diagram showing (**a**) approach to constant pattern behavior for a system with a favorable isotherm and (**b**) approach to proportionate pattern behavior for a system with an unfavorable isotherm. y axis: c/c_0, ——; $\bar{q}/q_0$, - - -; c^*/c_0, —·—. From ref. 7.

The distance required to approach the constant pattern limit decreases as the mass transfer resistance decreases and the nonlinearity of the equilibrium isotherm increases. However, when the isotherm is highly favorable, as in many adsorption processes, this distance may be very small, a few centimeters to perhaps a meter.

Length of Unused Bed. The constant pattern approximation provides the basis for a very useful and widely used design method based on the concept of the length of unused bed (LUB). In the design of a typical adsorption process the basic problem is to estimate the size of the adsorber bed needed to remove a certain quantity of the adsorbable species from the feed stream, subject to a specified limit (c') on the effluent concentration. The length of unused bed, which measures the capacity of the adsorber which is lost as a result of the spread of the concentration profile, is defined by

$$\text{LUB} = (1 - q'/q_0)L = (1 - t'/\bar{t})L \tag{37}$$

where q' is the capacity at the break time t' and $\bar{t}$ is the stoichiometric time (see Fig. 14). The values of t', $\bar{t}$, and hence the LUB are easily determined from an

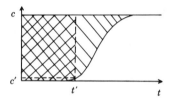

Fig. 14. Sketch of breakthrough curve showing break time t' and the method of calculation of the stoichiometric time $\bar{t}$ and LUB. From ref. 7. ▨ = the integral of equation 38; ▩ = the integral of equation 39.

experimental breakthrough curve since, by overall mass balance,

$$\bar{t} = \frac{L}{v}\left[1 + \left(\frac{1-\epsilon}{\epsilon}\right)\left(\frac{q_0}{c_0}\right)\right] = \int_0^\infty \left(1 - \frac{c}{c_0}\right) dt \tag{38}$$

$$t' = \frac{L}{v}\left[1 + \left(\frac{1-\epsilon}{\epsilon}\right)\left(\frac{q'}{c_0}\right)\right] = \int_0^{t'} \left(1 - \frac{c}{c_0}\right) dt \tag{39}$$

Under constant pattern conditions the LUB is independent of column length although, of course, it depends on other process variables. The procedure is therefore to determine the LUB in a small laboratory or pilot-scale column packed with the same adsorbent and operated under the same flow conditions. The length of column needed can then be found simply by adding the LUB to the length calculated from equilibrium considerations, assuming a shock concentration front.

One potential problem with this approach is that heat loss from a small scale column is much greater than from a larger diameter column. As a result, small columns tend to operate almost isothermally whereas in a large column the system is almost adiabatic. Since the temperature profile in general affects the concentration profile, the LUB may be underestimated unless great care is taken to ensure adiabatic operation of the experimental column.

Proportionate Pattern Behavior. If the isotherm is unfavorable, the stable dynamic situation leading to constant pattern behavior can never be achieved. The situation is shown in Figure 13b. The equilibrium adsorbed-phase concentration lies above rather than below the actual adsorbed-phase profile. As the mass transfer zone progresses through the column it broadens, but the limiting situation, which is approached in a long column, is simply local equilibrium at all points ($c = c^*$) and the profile therefore continues to spread in proportion to the length of the column. This difference in behavior is important since the LUB approach to design is clearly inapplicable under these conditions.

Favorable and unfavorable equilibrium isotherms are normally defined, as in Figure 11, with respect to an increase in sorbate concentration. This is, of course, appropriate for an adsorption process, but if one is considering regeneration of a saturated column (desorption), the situation is reversed. An isotherm which is favorable for adsorption is unfavorable for desorption and vice versa. In most adsorption processes the adsorbent is selected to provide a favorable adsorption isotherm, so the adsorption step shows constant pattern behavior and proportionate pattern behavior is encountered in the desorption step.

Detailed Modelling Results. The results of a series of detailed calculations for an ideal isothermal plug-flow Langmuir system are summarized in Figure 15. The solid lines show the form of the theoretical breakthrough curves for adsorption and desorption, calculated from the following set of model equations and expressed in terms of the dimensionless variables ζ, τ, and β:

Differential Balance for Column

$$v\frac{\partial c}{\partial z} + \frac{\partial c}{\partial t} + \left(\frac{1-\epsilon}{\epsilon}\right)\frac{\partial \overline{q}}{\partial t} = 0 \tag{40}$$

Rate Equation

$$\frac{\partial \overline{q}}{\partial t} = k(q^* - \overline{q}) \tag{41}$$

Equilibrium Isotherm

$$\frac{q^*}{q_s} = \frac{bc}{1+bc} \tag{42}$$

Initial Conditions

$$\overline{q}(z, 0) = c(z, 0) = 0 \quad \text{(adsorption)}$$
$$\overline{q}(z, 0) = q_0, \, c(z, 0) = c_0 \quad \text{(desorption)} \tag{43}$$

Boundary Conditions

$$c(0, t) = c_0 \quad \text{(adsorption)}$$
$$c(0, t) = 0 \quad \text{(desorption)} \tag{44}$$

Bed Length Parameter

$$\zeta = \frac{z}{v}\left(\frac{q_0}{c_0}\right)\left(\frac{1-\epsilon}{\epsilon}\right)$$

Dimensionless Time

$$\tau = k\left(t - \frac{z}{v}\right) \tag{45}$$

Nonlinearity Parameter

$$\beta = 1 - \frac{q_0}{q_s} = \frac{1}{1+bc_0}$$

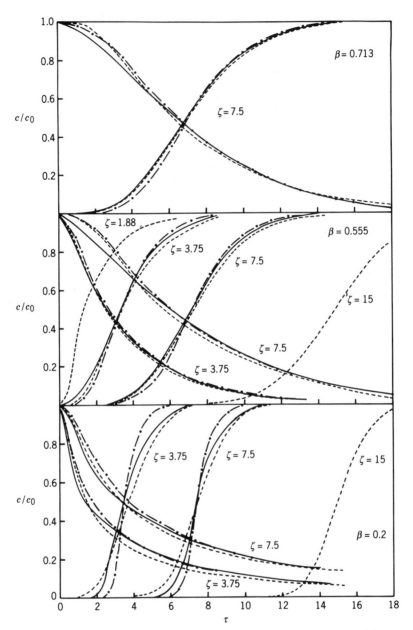

Fig. 15. Theoretical breakthrough curves for a nonlinear (Langmuir) system showing the comparison between the linear driving force (———), pore diffusion (- - -), and
intracrystalline diffusion (—·—) models based on the Glueckauf approximation (eqs. 40–45).
From ref. 7.

Also shown are the corresponding curves calculated for the same system
assuming a diffusion model in place of the linear rate expression. For
intracrystalline diffusion $k = 15D_0/r_c^2$, whereas for macropore diffusion $k = (15\epsilon_p D_p/R_p^2)(c_0/q_0)$, in accordance with the Glueckauf approximation (21).

For linear or moderately nonlinear systems ($\beta \rightarrow 1.0$) there is little difference in the response curves for all three models, thus verifying the validity of the Glueckauf approximation. Differences between the models, however, become more significant for a highly nonlinear isotherm ($\beta \rightarrow 0$). For linear or near linear systems the adsorption and desorption curves are mirror images, but as the isotherm becomes more nonlinear the adsorption and desorption curves become increasingly asymmetric. The adsorption curve approaches its limiting constant pattern form whereas the desorption curve approaches the limiting proportionate pattern form. In the long-time region the desorption curve is governed entirely by equilibrium, so that the curves for all three rate models again become coincident.

The main conclusion to be drawn from these studies is that for most practical purposes the linear rate model provides an adequate approximation and the use of the more cumbersome and computationally time consuming diffusing models is generally not necessary. The Glueckauf approximation provides the required estimate of the effective mass transfer coefficient for a diffusion controlled system. More detailed analysis shows that when more than one mass transfer resistance is significant the overall rate coefficient may be estimated simply from the sum of the resistances (7):

$$\frac{1}{kK} = \frac{R}{3k_f} + \frac{R^2}{15KD_c} + \frac{R^2}{15\epsilon_p D_p} \tag{46}$$

Adsorption Chromatography. The principle of gas–solid or liquid–solid chromatography may be easily understood from equation 35. In a linear multicomponent system (several sorbates at low concentration in an inert carrier) the wave velocity for each component depends on its adsorption equilibrium constant. Thus, if a pulse of the mixed sorbate is injected at the column inlet, the different species separate into bands which travel through the column at their characteristic velocities, and at the outlet of the column a sequence of peaks corresponding to the different species is detected. Measurement of the retention time ($\bar{t}$) under known flow conditions thus provides a simple means of determining the equilibrium constant (Henry constant):

$$\int_0^\infty \frac{ct\,dt}{\int_0^\infty c\,dt} = \bar{t} = \frac{L}{v}\left[1 + \left(\frac{1-\epsilon}{\epsilon}\right)K\right] \tag{47}$$

In an ideal system with no axial mixing or mass transfer resistance the peaks for the various components propagate without spreading. However, in any real system the peak broadens as it propagates and the extent of this broadening is directly related to the mass transfer and axial dispersion characteristics of the column. Measurement of the peak broadening therefore provides a convenient way of measuring mass transfer coefficients and intraparticle diffusivities. The simplest approach is to measure the second moments of the response peak over a range of flow rates:

$$\sigma^2 \equiv \int_0^\infty \frac{c(t - \bar{t})^2\,dt}{\int_0^\infty c\,dt} \tag{48}$$

Solution of the model equations shows that, for a linear isothermal system and a pulse injection, the height equivalent to a theoretical plate (HETP) is given by

$$H = \frac{\sigma^2}{\bar{t}^2} L = \frac{2D_L}{v} + \frac{2v}{kK}\left(\frac{\epsilon}{1-\epsilon}\right)\left[1 + \frac{\epsilon}{(1-\epsilon)K}\right]^{-2} \qquad (49)$$

where $1/kK$ is the overall mass transfer resistance defined by equation 46.

For liquid systems D_L/v is approximately independent of velocity, so that a plot of H versus v provides a convenient method of determining both the axial dispersion and mass transfer resistance. For vapor-phase systems at low Reynolds numbers D_L is approximately constant since dispersion is determined mainly by molecular diffusion. It is therefore more convenient to plot H/v versus $1/v^2$, which yields D_L as the slope and the mass transfer resistance as the intercept. Examples of such plots are shown in Figure 16.

Applications

The applications of adsorbents are many and varied and may be classified as nonregenerative uses, in which the adsorbent is used once and discarded, and regenerative applications, in which the adsorbent is used repeatedly in a cyclic manner involving sequential adsorption and regeneration steps.

Nonregenerative Uses

Desiccant in dual pane windows
Odor removal in health care products
Desiccant in refrigeration and air conditioning systems
Cigarette filters

Regenerative Uses

Water purification (some systems)
Removal of trace impurities from gases or liquid streams
Bulk separations (gas or liquid)
Low pressure storage of methane
Desiccant cooling (open-cycle air conditioning)

In terms of tonnage usage, some of the nonregenerative applications, notably as desiccants in dual pane windows, in cigarette filters, and in water purification, are surprisingly important, but most of the important chemical engineering applications are regenerative since, with a few notable exceptions, the cost of the adsorbent is too great to allow nonregenerative use.

The application of adsorbents (generally high area activated carbon) as a means of storing methane fuel (natural gas) at relatively high density under moderate pressure is relatively new. With current technology, capacities of about 180 m^3 STP/m^3 at 3 MPa (30 atm) pressure are achievable (24) but somewhat

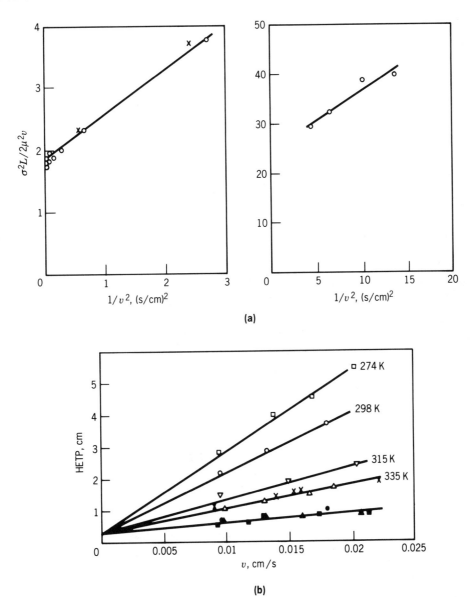

Fig. 16. Plots showing **(a)** variation of $(\sigma^2 L/2\mu^2 v)$ with $1/v^2$ for O_2 (left plot, $\times$, 0.84–0.72 mm = 20–25 mesh; $\bigcirc$, 0.42–0.29 mm = 40–50 mesh) and N_2 (right plot, on 3.2-mm pellets) in Bergbau-Forschung carbon molecular sieve and **(b)** variation of HETP with liquid velocity (interstitial) for fructose (solid symbols), and glucose (open symbols) in a column packed with KX zeolite crystals. From refs. 22 and 23.

higher capacities are needed to compete with liquid fuels for motor vehicles. However, depending on the cost of crude oil and the potential for improvement of the adsorbent, this technology could become important in the future.

Open-cycle desiccant cooling is another area of emerging technology (8). Rather than cooling and dehumidifying by mechanical work, as in a conventional air conditioning system, in an open-cycle desiccant system dehumidification is achieved directly, while cooling is achieved by controlled evaporation. The energy input is in the form of the heat required to regenerate the desiccant. A significant advantage of this system is that it can be designed to operate with a low regeneration temperature, thus making it possible to utilize low grade heat or even solar heat to drive the system.

Adsorption Separation and Purification Processes. The main area of current application of adsorption is in separation and purification processes. Many different ways of operating such processes have been devised and it is helpful to consider the various systems according to the mode of fluid–solid contact (see Fig. 17). In a cyclic batch process at least two beds are employed and each bed is successively saturated with the preferentially adsorbed species (or class of species) during the adsorption step and then regenerated during a desorption step in which the direction of mass transfer is reversed to remove the adsorbed species from the bed. In the continuous countercurrent process the adsorbent can be regarded as circulating continuously between the adsorption and desorption beds, in both of which fluid and solid contact in countercurrent flow. More commonly, as in the Sorbex type of process, the adsorbent is not physically circulated but the same effect is achieved in a fixed adsorbent bed equipped with multiple inlet and outlet ports to which the fluid streams are directed in sequence. Such systems can achieve a close approximation to countercurrent flow without the problems inherent in circulating the solid adsorbent.

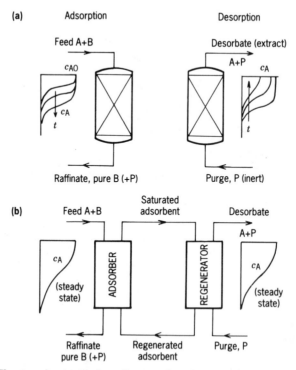

Fig. 17. The two basic modes of operation for an adsorption process: (**a**) cyclic batch system; (**b**) continuous countercurrent system with adsorbent recirculation. From ref. 7.

However, the system is relatively expensive, so it is generally used only for difficult separations (low separation factor) which cannot be carried out efficiently in a simple batch process.

The other major difference between adsorption processes lies in the method by which the adsorbent bed is regenerated. The advantages and disadvantages of three different methods—temperature swing, pressure swing, and displacement—are summarized in Table 5. For efficient removal of trace impurities it is normally essential to use a highly selective adsorbent on which the sorbate is strongly held. Temperature swing regeneration is therefore generally used in such applications. However, in bulk separations all three regeneration methods are widely used.

Table 5. Factors Governing Choice of Regeneration Method[a]

Method	Advantages	Disadvantages
thermal swing	good for strongly adsorbed species; small change in T gives large change in q^*	thermal aging of adsorbent
	desorbate may be recovered at high concentration	heat loss means inefficiency in energy usage
		unsuitable for rapid cycling, so adsorbent cannot be used with maximum efficiency
	gases and liquids	in liquid systems the latent heat of the interstitial liquid must be added
pressure swing	good where weakly adsorbed species is required at high purity	very low P may be required
		mechanical energy more expensive than heat
	rapid cycling—efficient use of adsorbent	desorbate recovered at low purity
displacement desorption	good for strongly held species	product separation and recovery needed (choice of desorbent is crucial)
	avoids risk of cracking reactions during regeneration	
	avoids thermal aging of adsorbent	

[a]Ref. 7.

Process Design. As with any chemical engineering process, the choice of process type and the details of the design are dictated primarily by economic considerations, subject to the overriding requirements of safety and reliability.

Although the principles of adsorption processes are well understood, most practical designs still rely on a good deal of empiricism since factors such as the aging and deterioration of an adsorbent under practical operating conditions are difficult to predict except from experience.

In general the sophistication of the design procedures is closely related to the level of sophistication of the process. In the simple thermal swing, batch type processes for removal of trace impurities the beds are generally sized on the basis of equilibrium capacity and LUB with a suitable allowance for aging of the adsorbent. The desorption or regeneration temperature is generally selected on the basis of equilibrium data as the minimum temperature which will allow the required specification of the purity of the raffinate product to be easily met. Use of a higher regeneration temperature generally gives a purer product but only at the cost of increased energy consumption and reduction of the service life of the adsorbent. The quantity of purge gas is estimated in two ways: from the overall heat balance with the required temperature rise and from the mass balance with the assumption of equilibrium at the bed outlet. Depending on the particular system and the process conditions, either of these considerations may be limiting. In general, when regeneration is carried out by purging at atmospheric pressure, the purge requirement is determined by the heat balance, but when the regeneration is carried out at elevated pressure, the mass balance is often the major constraint.

The design of pressure swing systems is also largely dependent on the scale-up of pilot-plant units, although, with such systems, the use of a detailed numerical simulation to guide the optimization of the operating conditions is more common. This is true also of countercurrent and simulated countercurrent processes where the initial design is commonly based on a simple McCabe-Thiele diagram (25) (see ADSORPTION, LIQUID SEPARATION).

NOTATION

a	ratio of external surface area to particle volume
b	Langmuir equilibrium constant
B	mobility
c	sorbate concentration in fluid phase
c_0	initial value of c
D	diffusivity
D_e	effective diffusivity
D_m	molecular diffusivity
D_K	Knudsen diffusivity
D_L	axial dispersion coefficient
D_0	corrected diffusivity
D_∞	preexponential factor
D_p	pore diffusivity
E	activation energy
$-\Delta H_0$	limiting heat of adsorption
K	dimensionless equilibrium constant
K_0	preexponential factor
K'	Henry's law constant

K'_0	preexponential factor
k	rate constant
k_0	preexponential factor
k', k''	constants in Elovich equation
k_f	fluid film mass transfer coefficient
k_s	surface mass transfer coefficient
L	bed length
M	molecular weight
m_t	mass adsorbed (or desorbed) at time t
m_∞	mass adsorbed (or desorbed) at $t \to \infty$
p	partial pressure of sorbate
q	adsorbed phase concentration
q_0	initial value of q
$\bar{q}$	adsorbed phase concentration averaged over a particle
q^*	equilibrium value of q
q_s	saturation limit in Langmuir expression
r	radial coordinate
R	radius of adsorbent particle
R	gas constant
t	time
$\bar{t}$	mean residence time or stoichiometric time
t'	break time
T	temperature
$-\Delta U$	change of interval energy on adsorption
v	interstitial fluid velocity
w	wave velocity
w'	shock velocity
z	distance
α	separation factor
ϵ	voidage of adsorbent bed
ϵ_p	porosity of particle
μ	chemical potential
ρ	mean pore radius
σ^2	variance of pulse response
τ	tortuosity factor

BIBLIOGRAPHY

"Adsorptive separation, introduction" in *ECT* 3rd ed., Vol. 1, pp. 531–544, by Theodore Vermeulen, University of California, Berkeley; "Adsorption, theoretical" in *ECT* 2nd ed., Vol. 1, pp. 421–459, by Sydney Ross, Rensselaer Polytechnic Institute; "Adsorption, theoretical" in *ECT* 1st ed., Vol. 1, pp. 206–222, by P. H. Emmett, Mellon Institute of Industrial Research.

1. A. Thijs, G. Peters, E. F. Vansant, I. Verhaert, and P. deBievre, *J. Chem. Soc. Faraday Trans. 1* **79**, 2821 (1983).
2. Y. Matsumura, *Proceedings of the First Indian Carbon Conference, New Delhi,* December 1982, pp. 99–106.
3. E. M. Flanigen and co-workers, *Nature* **271**, 512 (1978).
4. S. M. Klein and W. H. Abraham, *AIChE Symp. Ser.* **79**(230), 53 (1984).
5. D. W. Breck, *Zeolite Molecular Sieves,* John Wiley & Sons, Inc., New York, 1974.

6. W. M. Meier and D. H. Olson, *Atlas of Zeolite Structure Types,* Juris Druck and Verlag AG, Zurich, 1978.
7. D. M. Ruthven, *Principles of Adsorption and Adsorption Processes,* Wiley-Interscience, New York, 1984.
8. T. R. Penney and I. Maclaine-Cross, *Proceedings, Desiccant Cooling and Dehumidification Workshop, June 10–11, 1986, Chattanooga, Tenn.,* Sponsored by Electric Power Research Institute, Gas Research Institute, and Tennessee Valley Authority.
9. S. Brunauer, L. S. Deming, W. E. Deming, and J. E. Teller, *J. Am. Chem. Soc.* **62,** 1723 (1940).
10. D. M. Ruthven, *Sep. Purif. Methods* **5**(2), 189 (1976).
11. D. M. Ruthven and K. F. Loughlin, *J. Chem. Soc. Faraday Trans. 1* **68,** 696 (1972).
12. A. V. Kiselev and K. D. Shcherbakova in "Molecular Sieves," *Proceedings 1st International Zeolite Conference, London, 1967,* Society of Chemical Industry, London, 1968.
13. R. Sips, *J. Chem. Phys.* **16,** 490 (1948).
14. A. L. Myers and J. M. Prausnitz, *AIChE J.* **11,** 121 (1965).
15. P. G. Ashmore, *Catalysis and Inhibition of Chemical Reactions,* Butterworths, London, 1963, p. 164.
16. N. Wakao and T. Funazkri, *Chem. Eng. Sci.* **33,** 1375 (1978).
17. I. H. Doetsch, D. M. Ruthven, and K. F. Loughlin, *Can. J. Chem.* **52,** 2717 (1974).
18. J. A. Dominguez, D. Psaris, and A. I. La Cava, *AIChE Symp. Ser.* **84**(264), 73 (1988).
19. D. Basmadjian and D. W. Wright, *Chem. Eng. Sci.* **36,** 937 (1981).
20. A. I. Liapis and O. K. Crosser, *Chem. Eng. Sci.* **37,** 958 (1982).
21. E. Glueckauf, *Trans. Faraday Soc.* **51,** 1540 (1955).
22. D. M. Ruthven, N. S. Raghavan, and M. M. Hassan, *Chem. Eng. Sci.* **41,** 1325 (1986).
23. C. B. Ching and D. M. Ruthven, *Zeolites* **8,** 68 (1988).
24. S. S. Barton, J. A. Holland, and D. F. Quinn, *Proceedings of the 2nd International Conference on Adsorption, Santa Barbara, May 1986,* Engineering Foundation, New York, 1987, p. 99.
25. D. M. Ruthven and C. B. Ching, *Chem. Eng. Sci.* **44,** 1011 (1989).

General References

D. M. Ruthven, *Principles of Adsorption and Adsorption Processes,* Wiley-Interscience, New York, 1984.
M. Suzuki, *Adsorption Engineering,* Kodansba-Elsevier, Tokyo, 1990.
R. T. Yang, *Gas Separation by Adsorption Processes,* Butterworths, Stoneham, Mass., 1987.
P. Wankat, *Large Scale Adsorption and Chromatography,* CRC Press, Boca Raton, Fla., 1986.
A. E. Rodrigues and D. Tondeur, eds., *Percolation Processes,* NATO ASI No. 33, Sijthoff & Noordhoff, Alpen aan den Rijn, 1980.
A. E. Rodrigues, M. D. Le van, and D. Tondeur, *Adsorption: Science and Technology,* NATO ASI E158, Kluwer, Amsterdam, 1989.
N. Wakao, *Heat and Mass Transfer in Packed Beds,* Gordon & Breach, New York, 1982.
M. Smisek and S. Cerny, *Active Carbon,* Elsevier, Amsterdam, 1970.
T. Vermeulen, M. D. LeVan, N. K. Hiester, and G. Klein, "Adsorption and Ion Exchange," Section 16 of *Perry's Chemical Engineers' Handbook,* 6th ed., McGraw-Hill Book Co., New York, 1984.

Douglas M. Ruthven
University of New Brunswick, Canada

ADSORPTION, GAS SEPARATION

Gas-phase adsorption is widely employed for the large-scale purification or bulk separation of air, natural gas, chemicals, and petrochemicals (Table 1). In these uses it is often a preferred alternative to the older unit operations of distillation and absorption.

Table 1. Commercial Adsorption Separations

Separation[a]	Adsorbent
Gas bulk separations	
normal paraffins, isoparaffins, aromatics	zeolite
N_2/O_2	zeolite
O_2/N_2	carbon molecular sieve
CO, CH_4, CO_2, N_2, Ar, NH_3/H_2	zeolite, activated carbon
acetone/vent streams	activated carbon
C_2H_4/vent streams	activated carbon
H_2O/ethanol	zeolite
Gas purifications	
H_2O/olefin-containing cracked gas, natural gas, air, synthesis gas, etc	silica, alumina, zeolite
CO_2/C_2H_4, natural gas, etc	zeolite
organics/vent streams	activated carbon, others
sulfur compounds/natural gas, hydrogen, liquified petroleum gas (LPG), etc	zeolite
solvents/air	activated carbon
odors/air	activated carbon
NO_x/N_2	zeolite
SO_2/vent streams	zeolite
Hg/chlor–alkali cell gas effluent	zeolite

[a]Ref. 1.

An adsorbent attracts molecules from the gas, the molecules become concentrated on the surface of the adsorbent, and are removed from the gas phase. Many process concepts have been developed to allow the efficient contact of feed gas mixtures with adsorbents to carry out desired separations and to allow efficient regeneration of the adsorbent for subsequent reuse. In nonregenerative applications, the adsorbent is used only once and is not regenerated.

Most commercial adsorbents for gas-phase applications are employed in the form of pellets, beads, or other granular shapes, typically about 1.5 to 3.2 mm in diameter. Most commonly, these adsorbents are packed into fixed beds through which the gaseous feed mixtures are passed. Normally, the process is conducted in a cyclic manner. When the capacity of the bed is exhausted, the feed flow is stopped to terminate the loading step of the process, the bed is treated to remove the adsorbed molecules in a separate regeneration step, and the cycle is then repeated.

The growth in both variety and scale of gas-phase adsorption separation processes, particularly since 1970, is due in part to continuing discoveries of new, porous, high-surface-area adsorbent materials (particularly molecular sieve zeolites) and, especially, to improvements in the design and modification of adsorbents. These advances have encouraged parallel inventions of new process concepts. Increasingly, the development of new applications requires close cooperation in adsorbent design and process cycle development and optimization.

Adsorption Principles

The design and manufacture of adsorbents for specific applications involves manipulation of the structure and chemistry of the adsorbent to provide greater attractive forces for one molecule compared to another, or, by adjusting the size of the pores, to control access to the adsorbent surface on the basis of molecular size. Adsorbent manufacturers have developed many technologies for these manipulations, but they are considered proprietary and are not openly communicated. Nevertheless, the broad principles are well known.

The attention of this article is focused on physical adsorption, which involves relatively weak intermolecular forces, because most commercial applications of adsorption rely on this phenomenon alone. Chemisorption is discussed only briefly in some sections on specific applications.

Adsorption Forces. Coulomb's Law allows calculations of the electrostatic potential resulting from a charge distribution, and of the potential energy of interaction between different charge distributions. Various elaborate computations are possible to calculate the potential energy of interaction between point charges, distributed charges, etc. See reference 2 for a detailed introduction.

An electric dipole consists of two equal and opposite charges separated by a distance. All molecules contain atoms composed of positively charged nuclei and negatively charged electrons. When a molecule is placed in an electric field between two charged plates, the field attracts the positive nuclei toward the negative plate and the electrons toward the positive plate. This electrical distortion, or polarization of the molecule, creates an electric dipole. When the field is removed, the distortion disappears, and the molecule reverts to its original condition. This electrical distortion of the molecule is called induced polarization; the dipole formed is an induced dipole.

The magnitude of the induced dipole moment depends on the electric field strength in accord with the relationship $\mu_i = \alpha F$, where μ_i is the induced dipole moment, F is the electric field strength, and the constant α is called the polarizability of the molecule. The polarizability is related to the dielectric constant of the substance. Group-contribution methods (2) can be used to estimate the polarizability from knowledge of the number of each type of bond within the molecule, eg, the polarizability of an unsaturated bond is greater than that of a saturated bond.

The total potential energy of adsorption interaction may be subdivided into parts representing contributions of the different types of interactions between adsorbed molecules and adsorbents. Adopting the terminology of Barrer (3), the total energy Φ_{Total} of interaction is the sum of contributions resulting from dispersion energy Φ_D, close-range repulsion Φ_R, polarization energy Φ_P, field–dipole interaction $\Phi_{F-\mu}$, field gradient–quadrupole interaction $\Phi_{\delta F-Q}$, and adsorbate–adsorbate interactions, denoted self-potential Φ_{SP}:

$$\Phi_{Total} = \underbrace{\Phi_D + \Phi_R + \Phi_P}_{\text{nonspecific}} + \underbrace{\Phi_{F-\mu} + \Phi_{\delta F-Q}}_{\text{specific}} + \underbrace{\Phi_{SP}}_{\text{adsorbate–adsorbate}}$$

The Φ_D and Φ_R terms always contribute, regardless of the specific electric charge distributions in the adsorbate molecules, which is why they are called nonspecific. The third nonspecific Φ_P term also always contributes, whether or not the adsorbate molecules have permanent dipoles or quadrupoles; however, for adsorbent surfaces which are relatively nonpolar, the polarization energy Φ_P is small.

The $\Phi_{F-\mu}$ and $\Phi_{\delta F-Q}$ terms are specific contributions, which are significant when adsorbate molecules possess permanent dipole and quadrupole moments. In the absence of these moments, these terms are zero, as is true also if the adsorbent surface has no electric fields, a completely nonpolar adsorbent.

Finally, the Φ_{SP} term is the contribution resulting from interactions between adsorbate molecules. At low coverages of the adsorbent by adsorbate molecules, this contribution approaches zero, and at high coverage it often causes a noticeable increase in the heat of adsorption.

The $\Phi_D + \Phi_R$ (dispersion plus repulsion) terms are known as the London or van der Waals forces. Spherical, nonpolar molecules are well described by the familiar Lennard-Jones 6–12 potential equation:

$$\Phi_D + \Phi_R = 4\epsilon[-(\sigma/r)^6 + (\sigma/r)^{12}]$$

where r is the intermolecular separation distance, and σ (length units) and ϵ (energy units) are constants characteristic of the colliding molecules. Values of force constants σ and ϵ have been compiled (2).

These forces arise from the fact that each molecule contains atoms having a nucleus and surrounded by a cloud of electrons. The electron cloud fluctuates and is nonsymmetrical at various instants in time. Although a nonpolar neutral molecule has no net permanent charge or dipole, these fluctuating electron distributions provide fluctuating dipoles in each molecule. These fluctuating dipoles interact to generate forces between molecules or between adsorbed molecules and adsorbent surfaces. These contributions to the potential energy of adsorption are present even if the adsorbed molecules are nonpolar and even if the adsorbent structure contains no strong electrostatic fields.

The contribution Φ_P is due to the polarization of the molecules by electric fields on the adsorbent surface, eg, electric fields between positively charged cations and the negatively charged framework of a zeolite adsorbent. The attractive interaction between the induced dipole and the electric field is called the polarization contribution. Its magnitude is dependent upon the polarizability α of the molecule and the strength of the electric field F of the adsorbent (4): $\Phi_P = -\frac{1}{2}\alpha F^2$.

The first of the two specific interaction terms $\Phi_{F-\mu}$ is due to the attractive interaction between the permanent dipole moment μ of a molecule and the electric field on the adsorbent surface (4):

$$\Phi_{F-\mu} = -F\mu \cos \Theta$$

where Θ is the dipole–axis/field angle.

The other specific interaction term $\Phi_{\delta F-Q}$ is due to the attractive interaction between the permanent quadrupole moment Q of the molecule and the electric field gradient on the adsorbent surface (4):

$$\Phi_{\delta F-Q} = \tfrac{1}{2}Q \, dF/dr$$

The final contribution, the self-potential term Φ_{SP}, is the sum of all the above interactions of adsorbed molecules with each other.

Finally, an analysis of the energies of adsorption on many practical polar and nonpolar adsorbents has shown not only that the magnitude of the Φ_P term depends directly upon the polarizability α, but also that the sum of all of the nonspecific terms taken together, ie, $\Phi_D + \Phi_R + \Phi_P$, increases monotonically, with increasing α (4).

Adsorption Selectivities. For a given adsorbent, the relative strength of adsorption of different adsorbate molecules depends on the relative magnitudes of the polarizability α, dipole moment μ, and quadrupole moment Q of each. These properties of some common molecules are given in Table 2. Often, just the consideration of the values of α, μ, and Q allows accurate qualitative predictions to be made of the relative strengths of adsorption of given molecules on an adsorbent or of the best adsorbent type (polar or nonpolar) for a particular separation.

Table 2. Electrostatic Properties of Some Common Molecules

Molecule	Polarizability $\alpha \times 10^{40}$, $C^2 \cdot m^2/J^a$	Dipole moment $\mu \times 10^{30}$, $C \cdot m^b$	Quadrupole moment $Q \times 10^{40}$, $C \cdot m^2 {}^c$
Ar	1.83	0.00	0.00
H_2	0.90	0.00	2.09
N_2	0.78	0.00	-4.91
O_2	1.77	0.00	-1.33
CO	2.19	0.37	-6.92
CO_2	3.02	0.00	-13.71
CS_2	9.41	0.00	12.73
N_2O	3.32	0.54	-12.02
NH_3	2.67	5.10	-7.39
C_2H_6	4.97	0.00	-3.32
C_6H_6	11.49	0.00	-30.7
HCl	2.94	3.57	13.28

^{a}To convert $C^2 \cdot m^2/J$ to cm^3, divide by 1.113×10^{-16}.
^{b}To convert $C \cdot m$ to debyes, divide by 3.336×10^{-30}.
^{c}To convert $C \cdot m^2$ to Buckinghams, divide by 3.336×10^{-40}.

For example, the strength of the electric field F and field gradient ($\delta F = dF/dr$) of the highly polar cationic zeolites is strong. For this reason, nitrogen is more strongly adsorbed than is oxygen on such adsorbents, primarily because of the stronger quadrupole of N_2 compared to O_2.

In contrast, nonpolar activated carbon adsorbents lack strong electric fields and field gradients. Such adsorbents adsorb O_2 slightly more strongly than N_2, because of the slightly higher polarizability of O_2. Relative selectivities on nonpolar adsorbents often parallel the relative volatilities of the same compounds. Compounds with higher boiling points are more strongly adsorbed. In this case, the higher boiling O_2 (bp ~ 90 K) is more strongly adsorbed than is N_2 (bp ~ 77 K).

The polarizabilities of molecules in a homologous series increase steadily with increasing numbers of atoms. Therefore, the relative strengths of adsorption also increase (along with the boiling points).

For a given adsorbate molecule, the relative strength of adsorption on different adsorbents depends largely on the relative polarizability and electric field strengths of adsorbent surfaces. On the one hand, water molecules, with relatively low polarizability but a strong dipole and moderately strong quadrupole moment, are strongly adsorbed by polar adsorbents (eg, cationic zeolites), but only weakly adsorbed by nonpolar adsorbents (eg, silicalite or nonoxidized forms of activated carbon). On the other hand, saturated hydrocarbons with low molecular weight have greater polarizabilities than does water, but no dipoles and only weak quadrupoles. These molecules are adsorbed less strongly than water on polar adsorbents, but more strongly than water on nonpolar adsorbents. Therefore, polar adsorbents are often called hydrophilic adsorbents and nonpolar adsorbents are called hydrophobic adsorbents.

Isotherms and Isobars. The graphical presentation of the equilibrium adsorbate loading vs adsorbate pressure (or concentration) at constant temperature (Fig. 1) is an adsorption isotherm (1). A graph of the adsorbate loading vs

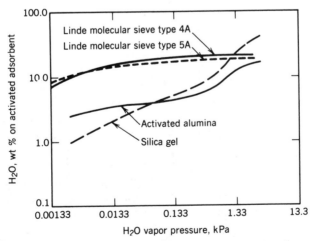

Fig. 1. Water isotherms for various adsorbents (1). Activation conditions: Linde molecular sieves, 350°C and <1.33 Pa; activated alumina, 350°C and <1.33 Pa; silica gel, 175°C and <1.33 Pa. To convert kPa to mm Hg, divide by 0.133.

temperature at constant adsorbate pressure (Fig. 2) is an adsorption isobar (1). The greater the strength of adsorption, the greater is the adsorbate loading at a given temperature and partial pressure of the adsorbate up to the point where the maximum adsorption capacity of the adsorbent has been attained.

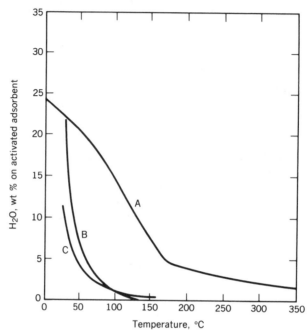

Fig. 2. Water isobars for various adsorbents: Equilibrium H_2O capacity vs temperature for three adsorbents (1); $p_{H_2O} = 1.33$ kPa (10 mm Hg) at 12°C and 101.3 kPa (1 atm). Activation conditions: A, Linde molecular sieve, 350°C and <1.33 Pa; B, activated alumina, 350°C and <1.33 Pa; C, silica gel, 175°C and <1.33 Pa. To convert Pa to mm Hg, multiply by 0.0075.

The strength of adsorption of unsaturated hydrocarbons by a polar adsorbent (zeolite) is much greater than for saturated hydrocarbons, and increases with increasing carbon number (Fig. 3) (5). This observation may be understood as a consequence of the increasing polarizability of molecules with increasing numbers of bonds and the presence of dipole and stronger quadrupole moments in the unsaturated hydrocarbons compared to the saturated hydrocarbons.

Heats of Adsorption. Physical adsorption processes are exothermic, ie, they release heat. Because the entropy change ΔS on adsorption is negative (adsorbed molecules are more ordered than in the gas phase) and the free energy change ΔG must be negative for adsorption to be favored, thermodynamics ($\Delta G = \Delta H - T \Delta S$) requires the enthalpy change ΔH on adsorption (heat of adsorption) to be negative (exothermic). Adsorption strengths thus decrease with increasing temperature.

The integral heat of adsorption is the total heat released when the adsorbate loading is increased from zero to some final value at isothermal conditions. The

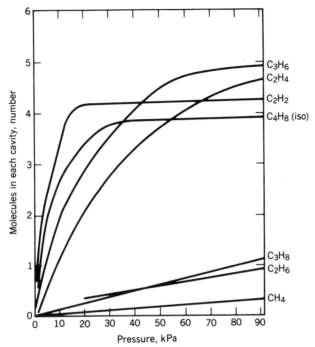

Fig. 3. Adsorption of hydrocarbons by zeolites is much greater for unsaturated hydrocarbons whose molecules contain double or triple bonds. From top to bottom, the curves show adsorption (at 150°C) of propylene, ethylene, acetylene, and isobutylene (unsaturated) and propane, ethane, and methane (saturated) (5). To convert kPa to mm Hg, multiply by 7.5. Courtesy of *Scientific American*.

differential heat of adsorption δH_{iso} is the incremental change in heat of adsorption with a differential change in adsorbate loading. This heat of adsorption δH_{iso} may be determined from the slopes of adsorption isosteres (lines of constant adsorbate loading) on graphs of $\ln P$ vs $1/T$ (Fig. 4) (6) through the Clausius-Clapeyron relationship:

$$\frac{d \ln P}{d(1/T)} = - \frac{\delta H_{iso}}{R}$$

where R is the gas constant, P the adsorbate absolute pressure, and T the absolute temperature.

Differential heats of adsorption for several gases on a sample of a polar adsorbent (natural zeolite chabazite) are shown as a function of the quantities adsorbed in Figure 5 (4). Consideration of the electrical properties of the adsorbates, included in Table 2, allows the correct prediction of the relative order of adsorption selectivity:

$$Ar < O_2 < N_2 < CO << CO_2$$

At low adsorbate loadings, the differential heat of adsorption decreases with increasing adsorbate loadings. This is direct evidence that the adsorbent surface

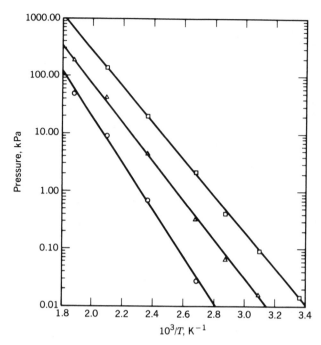

Fig. 4. Adsorption isosteres, water vapor on 4A (NaA) zeolite pellets (6). H_2O loading: □, 15 kg/100 kg zeolite; △, 10 kg/100 kg, ○, 5 kg/100 kg. To convert kPa to mm Hg, multiply by 7.5. Courtesy of Union Carbide.

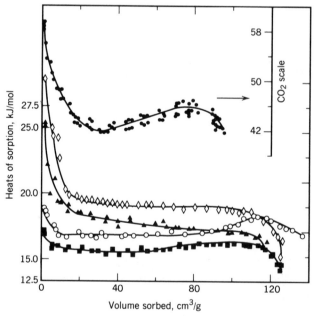

Fig. 5. Differential heats of sorption in natural chabazite (4). ▲ = N_2; ■ = Ar; ○ = O_2; ◇ = CO; ● = CO_2. See Table 2 for polarizability, dipole moment, and quadrupole moment values for the gases. Volume adsorbed is expressed as cm^3 of adsorbate as liquid. To convert kJ to kcal, divide by 4.184. Courtesy of Academic Press.

is energetically heterogeneous, ie, some adsorption sites interact more strongly with the adsorbate molecules. These sites are filled first so that adsorption of additional molecules involves progressively lower heats of adsorption.

All practical adsorbents have surfaces that are heterogeneous, both energetically and geometrically (not all pores are of uniform and constant dimensions). The degree of heterogeneity differs substantially from one adsorbent type to another. These heterogeneities are responsible for many nonlinearities, both in single component isotherms and in multicomponent adsorption selectivities.

In Figure 5, the heat of adsorption of CO_2 increases slightly at the higher adsorbate loadings. This increase is due to the increasing self-potential contribution at the higher loadings.

Isotherm Models.   Many efforts have been made over the years to develop isotherm models for data correlation and design predictions for both single component and multicomponent adsorption. Unfortunately, no single model is accurate over broad ranges of adsorbent and adsorbate types, pressures, temperatures, and loadings, especially for multicomponent systems. This is probably due to deficiencies in the models in adequately describing both the heterogeneities of the surface and the effects of the adsorbate on the properties of the adsorbent itself. Most models assume the adsorbent is inert, ie, not altered by the presence of the adsorbate molecules; however, partial changes in some adsorbent properties are commonly observed.

Nevertheless, each of the more popular isotherm models have been found useful for modeling adsorption behavior in particular circumstances. The following outlines many of the isotherm models presently available. Detailed discussions of derivations, assumptions, strengths, and weaknesses of these and other isotherm models are given in references 4 and 7–16.

Not all of the isotherm models discussed in the following are rigorous in the sense of being thermodynamically consistent. For example, specific deficiencies in the Freundlich, Sips, Dubinin-Radushkevich, Toth, and vacancy solution models have been identified (14).

The Sips and related LRC (loading ratio correlation) models fail to properly predict Henry's law behavior (as required for thermodynamic consistency) at the zero pressure limit (8). Thermodynamic inconsistency of the LRC model had also been noted by the original authors (17); nevertheless, the model has been found useful in predicting multicomponent performance from single component data and correlating multicomponent data (18). However, users of models lacking thermodynamic consistency must take due care, particularly in extrapolation beyond the range of actual experimental data.

Thermodynamically Consistent Isotherm Models.   These models include both the statistical thermodynamic models and the models that can be derived from an assumed equation of state for the adsorbed phase plus the thermodynamics of the adsorbed phase, ie, the Gibbs adsorption isotherm,

$$\left(\frac{d\Phi}{dP}\right)_T = \frac{qRT}{P}$$

where Φ is the spreading pressure, P the partial pressure of adsorbate, q the adsorbate loading x per quantity w of adsorbent $= x/w$, T the temperature, and R

the gas constant. In the following models, $\Theta = q/q_{max}$ is the fractional surface coverage, where q_{max} is the maximum loading. Constants are q_{max}, all K's, all k's, all A's, b, c, n, s, t, β, and τ. The vapor pressure of pure adsorbate is P_0.

Henry's law:	$q = KP_0$
Langmuir:	$K'P_0 = \Theta/(1 - \Theta)$
Volmer:	$bP_0 = [\Theta/(1 - \Theta)] \exp[\Theta/(1 - \Theta)]$
van der Waals:	$K''P_0 = [\Theta/(1 - \Theta)] \exp[\Theta/(1 - \Theta)] \exp[\beta/RT]$
Virial:	$K'''P_0/x = \exp[2A_1x + (\frac{3}{2})A_2x^2 + \cdots]$

The Langmuir model is discussed in reference 19; the Volmer in reference 20; and the van der Waals and virial equations in reference 8.

Statistical Thermodynamic Isotherm Models. These approaches were pioneered by Fowler and Guggenheim (21) and Hill (22). Examples of the application of this approach to modeling of adsorption in microporous adsorbents are given in references 3, 23–27. Excellent reviews have been written (4,28).

Semiempirical Isotherm Models. Some of these models have been shown to have some thermodynamic inconsistencies and should be used with due care. Nevertheless, they have each been found to be useful for data correlation and interpolation, as well as for the calculation of some thermodynamic properties.

Models Based on the Polanyi Adsorption Potential

$$A = RT \ln(P_0/P)$$

Dubinin-Radushkevich. This model (29) is the same as the more general Dubinin-Astakhov equation (30) (see below), with $n = 2$.

Dubinin-Astakhov:

$$\Theta = \exp\left[-(A/E)^n\right]$$

where n is generally between 1 and 3.

Radke-Prausnitz. This model (31) is also known as the Langmuir-Freundlich model:

$$\Theta = \frac{k'P}{[1 + (k'P)]^\tau} \quad \text{for} \quad 0 < \tau \le 1$$

Toth. This model (32) is represented as:

$$\Theta = \frac{kP}{[1 + (kP^t)^{1/t}]}$$

UNILAN. The uniform distribution, Langmuir local isotherm model (9,12):

$$\Theta = \frac{1}{2s} \ln\left[\frac{(c + Pe^s)}{(c + Pe^{-s})}\right]$$

BET. This model (33) estimates the coverage corresponding to one monolayer of adsorbate and is used to measure the surface areas of solids:

$$\Theta = \frac{b(P/P_0)}{[(1 - P/P_0)(1 - P/P_0 + bP/P_0)]}$$

Isotherm Models for Adsorption of Mixtures. Of the following models, all but the ideal adsorbed solution theory (IAST) and the related heterogeneous ideal adsorbed solution theory (HIAST) have been shown to contain some thermodynamic inconsistencies. References to the limited available literature data on the adsorption of gas mixtures on activated carbons and zeolites have been compiled, along with a brief summary of approximate percentage differences between data and theory for the various theoretical models (16). In the following the subscripts i and j refer to different adsorbates.

Markham and Benton. This model (34) is known as the extended Langmuir isotherm equation for two components, i and j:

$$\Theta_i = K_i P_i/(1 + K_i P_i + K_j P_j)$$
$$\Theta_j = K_j P_j/(1 + K_i P_i + K_j P_j)$$

Leavitt Loading Ratio Correlation (LRC) Method. The LRC model (17) for a single component i parallels Sips model (35):

$$\Theta_i = (K_i P_i)^{1/ni}/[1 + (K_i P_i)^{1/ni}]$$

but with

$$-\ln K_i = A_{1i} + A_{2i}/T$$

For the binary system of components i and j, the LRC model (17) is

$$\Theta_i = (K_i P_i)^{1/ni}/[1 + (K_i P_i)^{1/ni} + (K_j P_j)^{1/nj}]$$

Ideal Adsorbed Solution (IAS) Model. For components i and j, assuming ideal gas behavior, this model (36) is

$$\frac{\Phi A}{RT} = \int_0^{P_i^\circ} [q_i^\circ(P)] \, d(\ln P) = \int_0^{P_j^\circ} [q_j^\circ(P)] \, d(\ln P)$$

$$PY_i = P_i^\circ X_i$$
$$PY_j = P_j^\circ X_j = P_j^\circ(1 - X_i)$$

where P_i° is the vapor pressure of component i, $q_i^\circ(P)$ the equilibrium loading of pure i at pressure P, Y_i the vapor phase mole fraction of component i, and X_i the adsorbed phase mole fraction of component i. These equations are solved simultaneously to determine P_i°, P_j°, and X_i, and the following equations are used to calculate q_i, q_j, and q_total:

$$\{1/q_\text{total}\} = X_i/[(q_i)(P_i^\circ)] + X_j/[(q_j)(P_j^\circ)]$$

$$q_i = q_\text{total}X_i$$

$$q_j = q_\text{total}X_j$$

Heterogeneous Ideal Adsorbed Solution Theory (HIAST).   This IAS theory has been extended to the case of adsorbent surface energetic heterogeneity and is shown to provide improved predictions over IAST (12).

Vacancy Solution Model.   The initial model (37) considered the adsorbed phase to be a mixture of adsorbed molecules and vacancies (a vacancy solution) and assumed that nonidealities of the solution can be described by the two-parameter Wilson activity coefficient equation. Subsequently, it was found that the use of the three-parameter Flory-Huggins activity coefficient equation provided improved prediction of binary isotherms (38).

Adsorption Dynamics.   An outline of approaches that have been taken to model mass-transfer rates in adsorbents has been given (see ADSORPTION). Detailed reviews of the extensive literature on the interrelated topics of modeling of mass-transfer rate processes in fixed-bed adsorbers, bed concentration profiles, and breakthrough curves include references 16 and 26. The related simple design concepts of WES, WUB, and LUB for constant-pattern adsorption are discussed later.

Reactions on Adsorbents.   To permit the recovery of pure products and to extend the adsorbent's useful life, adsorbents should generally be inert and not react with or catalyze reactions of adsorbate molecules. These considerations often affect adsorbent selection and/or require limits be placed upon the severity of operating conditions to minimize reactions of the adsorbate molecules or damage to the adsorbents.

However, even then, gradual reactions of trace impurities in a feed stream or slowly occurring reactions that modify the adsorbent may still cause a gradual decline in the adsorbent performance, as illustrated in Figure 6 (39). To compensate, adsorbent beds are sized to account for the gradual loss in capacity and to allow their use for a given period of time. Most commonly, at the end of its useful life, the adsorbent is dumped from the beds and replaced with fresh adsorbent. However, in some cases, the process equipment is designed to allow periodic *in situ* rejuvenation of the adsorbent, eg, a periodic burning-off of coke accumulated on the adsorbent.

Adsorbent Principles

Principal Adsorbent Types.   Commercially useful adsorbents can be classified by the nature of their structure (amorphous or crystalline), by the sizes of their pores (micropores, mesopores, and macropores), by the nature of their surfaces (polar, nonpolar, or intermediate), or by their chemical composition. All

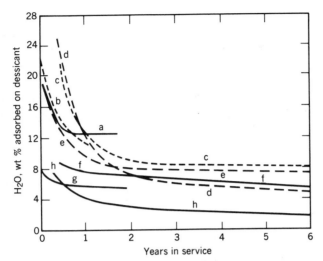

Fig. 6. Adsorption capacity of various dessicants vs years of service in dehydrating high pressure natural gas (39). a, Alumina H-151, gas ~27°C and 123 kPa, from oil and water separators; b, silica gel, gas ~38°C and 145 kPa, from oil absorption plant; c, sorbead, 136-kPa gas from absorption plant; regeneration gas inlet temperature 243°C (maximum allowable dew point −6.7°C); d, sorbead, 40-kPa gas containing propane; regeneration gas temperature 177°C (maximum allowable dew point −34°C); e, sorbead, 1950–1956 data; f, activated alumina, same gas as for Curve d; g, activated bauxite(florite), residue gas from gasoline absorption plant; h, activated alumina, same gas for for Curve c. Courtesy of Gulf Publishing Company.

of these characteristics are important in the selection of the best adsorbent for any particular application.

However, the size of the pores is the most important initial consideration because, if a molecule is to be adsorbed, it must not be larger than the pores of the adsorbent. Conversely, by selecting an adsorbent with a particular pore diameter, molecules larger than the pores may be selectively excluded, and smaller molecules can be allowed to adsorb.

Pore size is also related to surface area and thus to adsorbent capacity, particularly for gas-phase adsorption. Because the total surface area of a given mass of adsorbent increases with decreasing pore size, only materials containing micropores and small mesopores (nanometer diameters) have sufficient capacity to be useful as practical adsorbents for gas-phase applications. Micropore diameters are less than 2 nm; mesopore diameters are between 2 and 50 nm; and macropores diameters are greater than 50 nm, by IUPAC classification (40).

The practical adsorbents used in most gas phase applications are limited to the following types, classified by their amorphous or crystalline nature.

Amorphous: silica gel, activated alumina, activated carbon, and molecular sieve carbons.

Crystalline: molecular sieve zeolites, and related molecular sieve materials that are not technically zeolites, eg, silicalite, $AlPO_4$s, SAPOs, etc.

Typical pore size distributions for these adsorbents have been given (see ADSORPTION). Only molecular sieve carbons and crystalline molecular sieves have large pore volumes in pores smaller than 1 nm. Only the crystalline molecular sieves have monodisperse pore diameters because of the regularity of their crystalline structures (41).

Activated carbons are made by first preparing a carbonaceous char with low surface area followed by controlled oxidation in air, carbon dioxide, or steam. The pore-size distributions of the resulting products are highly dependent on both the raw materials and the conditions used in their manufacture, as may be seen in Figure 7 (42).

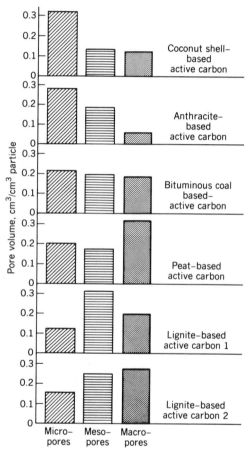

Fig. 7.   Pore size distribution in some active carbons obtained using different precursors (42). Courtesy of Marcel Dekker Publishing Company.

Assuming the pores are large enough to admit the molecules of interest, the most important consideration is the nature of the adsorbent surface, because this characteristic controls adsorption selectivity.

Practical adsorbents may also be classified according to the nature of their surfaces.

Highly polar: molecular sieve zeolites with high aluminum and cation contents.

Moderately polar: crystalline molecular sieves with low aluminum and low cation contents, silica gel, activated alumina, activated carbons with highly oxidized surfaces, crystalline molecular sieve $AlPO_4$s.

Nonpolar: silicalite, F-silicalite, other high silica content crystalline molecular sieves, activated carbons with reduced surfaces.

Adsorption Properties. Typical adsorption isotherms for water on various adsorbents are given in Figure 1, and the corresponding isobars in Figure 2. Not only do the more highly polar molecular sieve zeolites adsorb more water at lower pressures than do the moderately polar silica gel and alumina gel, but they also hold onto the water more strongly at higher temperatures. For the same reason, temperatures required for thermal regeneration of water-loaded zeolites is higher than for less highly polar adsorbents.

Isotherms for H_2O and n-hexane adsorption at room temperature and for O_2 adsorption at liquid oxygen temperature on 13X (NaX) zeolite and on the crystalline SiO_2 molecular sieve silicalite are are shown in Figure 8 (43). Silicalite

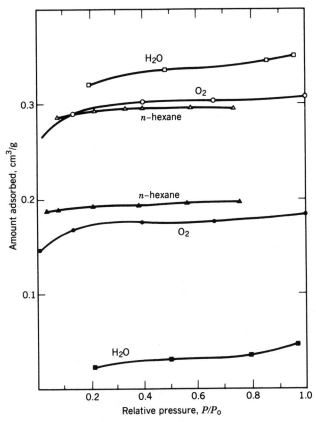

Fig. 8. Adsorption isotherms of H_2O, O_2, and n-hexane on zeolite NaX (open symbols) and silicalite (filled symbols). Oxygen is at $-183°C$ and water and n-hexane (C_6H_{14}) at RT. Volume adsorbed is expressed as cm^3 of adsorbate as liquid. Courtesy of *Nature, London* (43).

adsorbs water very weakly. Further modification of silicalite by fluoride incorporation provides an extremely hydrophobic adsorbent, shown in Figure 9 (44). These examples illustrate the broad range of properties of crystalline molecular sieves.

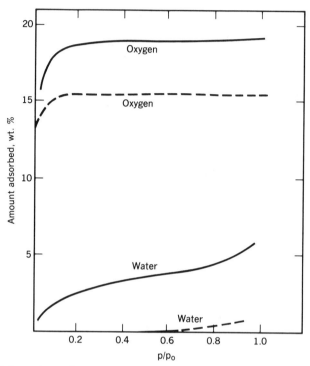

Fig. 9. Adsorption of oxygen (90 K) and water (RT) on silicalite (———) and F-silicalite (–––) (44).

Activated carbons contain chemisorbed oxygen in varying amounts unless special care is taken to eliminate it. Desired adsorption properties often depend upon the amount and type of chemisorbed oxygen species on the surface. Therefore, the adsorption properties of an activated carbon adsorbent depend on its prior temperature and oxygen-exposure history. In contrast, molecular sieve zeolites and other oxide adsorbents are not affected by oxidizing or reducing conditions.

This principle is illustrated in Figure 10 (45). Water adsorption at low pressures is markedly reduced on a poly(vinylidene chloride)-based activated carbon after removal of surface oxygenated groups by degassing at 1000°C. Following this treatment, water adsorption is dominated by capillary condensation in mesopores, and the size of the adsorption–desorption hysteresis loop increases, because the pore volume previously occupied by water at the lower pressures now remains empty until the water pressure reaches pressures ($\sim$0.3 to 0.4 times the vapor pressure) at which capillary condensation can occur.

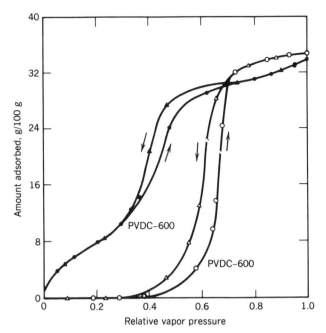

Fig. 10.   Adsorption (●, ○)–desorption (▲, △) isotherms of water vapor on poly(vinylidene chloride) (PVDC) carbon before (filled symbols) and after (open symbols) outgassing at 1000°C (46). Courtesy of *Carbon*.

Typical adsorption isotherms for light hydrocarbons on activated carbon prepared from coconut shells are shown in Figure 11 (46). The polarizabilities and boiling points of these compounds increase in the order

$$CH_4 < C_2H_4 < C_2H_6 < C_3H_6 < C_3H_8$$

The relative strengths of adsorption of these compounds follow the same order, as expected for a nonpolar adsorbent, except that C_3H_6 was adsorbed more strongly than C_3H_8. This result indicates that the surface is weakly polar and that specific (dipole–field and quadrupole–field gradient) contributions to the adsorption potential alter the expected order slightly. This situation may also be due to chemisorbed oxygen species on the surface.

Physical Properties.   Physical properties of importance include particle size, density, volume fraction of intraparticle and extraparticle voids when packed into adsorbent beds, strength, attrition resistance, and dustiness. These properties can be varied intentionally to tailor adsorbents to specific applications (See ADSORPTION LIQUID SEPARATION; ALUMINUM COMPOUNDS, ALUMINUM OXIDE; CARBON, ACTIVATED CARBON; ION EXCHANGE; MOLECULAR SIEVES; and SILICON COMPOUNDS, SYNTHETIC INORGANIC SILICATES).

Deactivation.   Adsorbent degradation by chemical attack or physical damage is not reversible. Acids or acid gases can react with adsorbents with alkaline surface chemistry, eg, some zeolites, and cause loss of adsorption capacity. Other adsorbents, such as silica gel, are sensitive to alkalies. The constant thermal

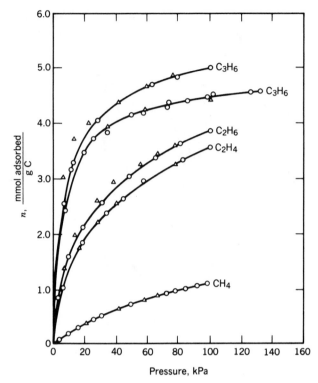

Fig. 11.   Adsorption isotherms for hydrocarbons on activated coconut-shell carbon at 25°C (46). ○, Adsorption; △, desorption. To convert kPa to mm Hg, multiply by 7.5. Courtesy of *Industrial and Engineering Chemistry*.

expansion and contraction in temperature swing adsorption (TSA) processes can cause damage to the internal pore and/or crystal structure. Activated alumina and silica gel can be dehydrated by excessive temperatures. When water is present, hydrothermal cycling can cause explosive steam release that physically damages some adsorbents. Some types of silica gel are susceptible to breakup caused by water droplets; special decrepitation-resistant grades are available.

Adsorption Processes

Adsorption processes are often identified by their method of regeneration. Temperature-swing adsorption (TSA) and pressure-swing adsorption (PSA) are the most frequently applied process cycles for gas separation. Purge-swing cycles and nonregenerative approaches are also applied to the separation of gases. Special applications exist in the nuclear industry. Others take advantage of reactive sorption. Most adsorption processes use fixed beds, but some use moving or fluidized beds.

TEMPERATURE SWING

A temperature-swing or thermal-swing adsorption (TSA) cycle is one in which desorption takes place at a temperature much higher than adsorption. The principal application is for separations in which contaminants are present at low concentration, ie, for purification. The TSA cycles are characterized by low residual loadings and high operating loadings. Figure 12 depicts the isotherms for the two temperatures of a TSA cycle. The available operating capacity is the difference between the loadings X_1 and X_2. These high adsorption capacities for low concentrations mean that cycle times are long, hours to days, for reasonably sized beds. This long cycle time is fortunate, because packed beds of adsorbent respond slowly to changes in gas temperature. A purge and/or vacuum removes the thermally desorbed components from the bed, and cooling returns the bed to adsorption condition. Systems in which species are strongly adsorbed are especially suited to TSA. Such applications include drying, sweetening, CO_2 removal, and pollution control.

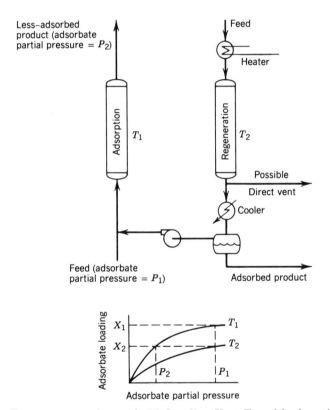

Fig. 12. Temperature-swing cycle (4). Loading X_1 at T_1 and feed partial pressure P_1; X_2 at the higher T_2 and the lower P_2 needed in the product (1).

Principles. In a TSA cycle, two processes occur during regeneration, heating and purging. Heating must provide adequate thermal energy to raise the adsorbate, adsorbent and adsorber temperature, desorb the adsorbate, and make up for heat losses. Regeneration is heating-limited (or stoichiometric-limited) when transfer of energy to the system is limiting. Equilibrium determines the maximum capacity of the purge gas to transfer the desorbed material away.

Regeneration is stripping-limited (or equilibrium-limited) when transferring adsorbate away is limiting.

Heating occurs by either direct (external heat exchange to the purge gas) or, less commonly, indirect (heating elements, coils or panels, inside the adsorber) contact of the adsorbent by the heating medium. Direct heating is simpler and is invariably used for stripping-limited regeneration. Microwave fields (47) and dielectric fields (48) are alternate methods for supplying indirect heating. The complexity of indirect heating limits its use to heating-limited regeneration where purge gas is in short supply. Coils or panels can supply indirect cooling as well. The use of steam for the regeneration of activated carbon is a combination of thermal desorption and purge displacement; direct heating is supplied by water adsorption.

Another process for thermal cycling is parametric pumping. In the direct mode of parametric pumping, a single adsorbent bed is indirectly heated and cooled while the fluid feed is pumped forward and backward through the column from reservoirs at each end. For a binary fluid, one component concentrates in one reservoir and one in the other. In the recuperative mode of parametric pumping, the heating and cooling takes place outside the adsorbent column.

Steps. Thermal-swing cycles have at least two steps, adsorption and heating. A cooling step is also normally used after the heating step. A portion of the feed or product stream can be utilized for heating, or an independent fluid can be used. Easily condensable contaminants may be regenerated with noncondensable gases and recovered by condensation. Water-immiscible solvents are stripped with steam, which may be condensed and separated from the solvent by decantation. Fuel and/or air may be used when the impurities are to be burned or incinerated.

The highest regeneration temperatures are the most efficient for desorption. However, heater cost, metallurgy, and the thermal stability of the adsorbent and the fluids must be considered. Silica gel requires the lowest temperatures and the lowest amount of heat of any commercial adsorbent. Activated carbons and aluminas can tolerate the highest temperatures. Although thermal-swing regeneration can be done at the same pressure as adsorption, lowering the pressure can achieve better desorption and is often used; such cycles are actually a hybrid of PSA and TSA. The heating gas is normally used for the cooling step. Rather than cooling the bed, adsorption can sometimes be started on a hot bed. If certain criteria are met (49), the dynamic adsorption performance does not depend on cooling.

Flow Sheet. The most common processing scheme is a pair of fixed-bed adsorbers alternating between the adsorption step and the regeneration steps. An example is given in Figure 12. However, the variations possible to achieve special needs are endless. The flow directions can be varied. Single beds provide interrupted flow, but multiple beds can ensure constant flow. Beds can be configured in lead–trim, parallel trains, series cool–heat, or closed-loop (1). Regeneration may even be *ex situ* rather than *in situ*.

The normal flow direction through a fixed bed is usually in a vertical direction. The mechanical complexities required for horizontal- or annular-flow beds often outweigh the decrease in pressure drop achieved. Because allowable velocities for crushing exceed those for lifting, the cycle step with the highest pressure drop should be downward. All other flows can then be in the same direction as the limiting flow (cocurrent) or in the opposite direction (countercurrent). Each combination of flow directions for heating and cooling produces a different residual of adsorbate (Fig. 13).

Although most applications of fixed bed have multiple adsorber beds to treat continuous streams, batch operation using a single adsorber bed is an alternative. For purification applications, where one vessel can contain enough adsorbent to provide treatment for days, weeks, or even months, the cost savings and simplicity often justify the inconvenience of stopping adsorption treatment periodically for a short regeneration.

When the mass transfer zone is a major portion of an adsorbent bed, the equilibrium capacity is poorly utilized. A lead–trim configuration uses the adsorbent more fully. The feed flows successively through a lead bed and then a trim bed. The lead bed is nearly exhausted before it is taken out of service to be regenerated. When a lead bed is removed from adsorption, the trim bed becomes the lead, and a fully regenerated bed becomes the new trim bed.

When large flows are to be treated, designing and building a single adsorber vessel large enough to treat the entire stream is not practical. Instead, the feed flow is split equally between parallel beds and/or trains of adsorbers. This provides the additional advantage of a convenient method of turning down the process to save on utilities.

At the start of the cooling step, the adsorber vessel is a large heat sink containing valuable energy: the sum of all of the sensible heats of the adsorbent, the vessel, and any internals. Using three adsorber beds—one on adsorption, one on heating, and one on cooling—the purge gas flows in series first to cool a hot bed and then to heat a spent bed. Thus all of the heat from the bed being cooled is recovered.

Thermal energy can be conserved by using a thermal-pulse cycle. When desorption is heat-limited, only a short soak time at temperature completes regeneration. The entire adsorbent bed need not be at desorption temperature before beginning the cooling step. Only a pulse of heating gas that contains the heat of desorption is required to move through the bed, desorbing the adsorbate until it exhausts its thermal energy as it reaches the outlet. Because temperature fronts spread as they move through packed beds, a small excess of heat is added to the stoichiometric quantity to ensure that the outlet reaches the desired level before being cooled.

When the gas available for regeneration is in short supply, the regeneration steps are often carried out in a closed loop. This recycle of the bed effluent back to the inlet has the advantage of concentrating the impurity and making it easier to separate by condensation or other recovery means. Heating is usually accomplished with a semiclosed loop which has a constant fresh gas makeup and a bleed to draw off the desorbed material. However, contaminant is at a higher level than in an open loop and product purity is harder to achieve.

Drying. The single most common gas phase application for TSA is drying. The natural gas, chemical, and cryogenics industries all use zeolites, silica gel, and activated alumina to dry streams. Adsorbents are even found in mufflers.

Zeolites, activated alumina, and silica gel have all been used for drying of pipeline natural gas. Alumina and silica gel have the advantage of having higher

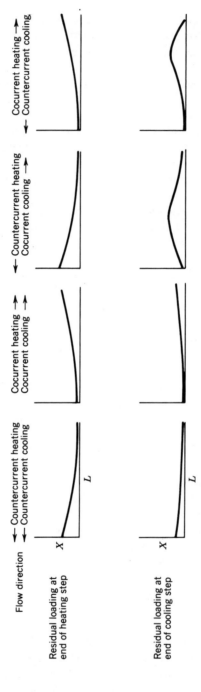

Fig. 13. Shape of residual loading gradient (124). L is bed length. Courtesy of *Chemical Engineering*.

equilibrium capacity and of being more easily regenerated with waste level heat (50–52). However, the much lower dewpoint and longer life attainable with 4A makes zeolites the predominant adsorbent. Special acid-resistant zeolites are used for natural gas containing large amounts of acid gases, such as CO_2 and H_2S.

The low dewpoint that can be achieved with zeolites is especially important when drying feed streams to cryogenic processes to prevent freeze-up at process temperatures. Natural gas is dried before liquefaction to liquefied natural gas (LNG), both in peak demand and in base load facilities (53,54). Zeolites have largely replaced silica gel and activated alumina in drying natural gas for ethane recovery utilizing the cryogenic turboexpander process, and for helium recovery (54). The air to be cryogenically distilled into N_2, O_2 and argon must be purified of both water and CO_2. This purification is accomplished with 13X zeolites (53).

The 4A zeolite, silica gel, and activated alumina all find applications drying synthesis gas, inert gas, hydrocracker gas, rare gases, and reformer recycle H_2. Cracked gas before low temperature distillation for olefin production is a reactive stream. The 3A or pore-closed 4A zeolite size selectively adsorbs water but excludes the hydrocarbons, thus preventing coking (52). This molecular sieving also prevents coadsorption of hydrocarbons which would otherwise be lost during desorption with the water. Small pore zeolites are also applied to the drying of ethylene, propylene, and acetylene as they are drawn from salt cavern, or conventional, storage (54). When industrial gases containing Cl_2, SO_2 and HCl are dried, acid-resistant zeolites are used.

A recently developed drying application for zeolites is the prevention of corrosion in mufflers (52,55). Internal corrosion in mufflers is caused primarily by the condensation of water and acid as the system cools. A unique UOP zeolite adsorption system takes advantage of the natural thermal cycling of an automotive exhaust system to desorb the water and acid precursors.

Sweetening.   Another significant purification application area for adsorption is sweetening. Hydrogen sulfide, mercaptans, organic sulfides and disulfides, and COS need to be removed to prevent corrosion and catalyst poisoning. They are to be found in H_2, natural gas, deethanizer overhead, and biogas. Often adsorption is attractive because it dries the stream as it sweetens.

In the sweetening of wellhead natural gas to prevent pipeline corrosion, 4A zeolites allow sulfur compound removal without CO_2 removal (to reduce shrinkage), or the removal of both to upgrade low thermal content gas. When minimizing the formation of COS during desulfurization is desirable, calcium-exchanged zeolites are commonly used because they are less catalytically active for the reaction of CO_2 with H_2S to form COS and water. Natural gas for steam–methane reforming in ammonia production must be sweetened to protect the sulfursensitive, low temperature shift catalyst. Zeolites are better than activated carbon because mercaptans, COS, and organic sulfides are also removed (54). Many refinery H_2 streams require H_2S and water removal by 4A and 5A zeolites to prevent poisoning of catalysts such as those in catalytic reformers.

Other Separations.   Other TSA applications range from CO_2 removal to hydrocarbon separations, and include removal of air pollutants and odors, and purification of streams containing HCl and boron compounds. Because of their high selectivity for CO_2 and their ability to dry concurrently, 4A, 5A, and 13X zeolites are the predominant adsorbents for CO_2 removal by temperature-swing processes. The air fed to an air separation plant must be H_2O- and CO_2-free to prevent fouling of heat exchangers at cryogenic temperatures; 13X is typically used here. Another application for 4A-type zeolite is for CO_2 removal from base-load and peak-shaving natural gas liquefaction facilities.

The removal of volatile organic compounds (VOC) from air is most often accomplished by TSA. Air streams needing treatment can be found in most chemical and manufacturing plants, especially those using solvents. At concentrations from 500 to 15,000 ppm, recovery of the VOC from steam used to regenerate activated carbon adsorbent thermally is economically justified. Concentrations above 15,000 ppm are typically in the explosive range and require the use of inert gas rather than air for regeneration. Below about 500 ppm, recovery is not economically justifiable, but environmental concerns often dictate adsorptive recovery followed by destruction. Activated carbon is the traditional adsorbent for these applications, which represent the second largest use for gas-phase carbons. New forms of activated carbon, such as carbon fabrics (56) and adsorbent wheels (57), have been introduced to reduce the airflow pressure drop, which can result in large utility consumptions (see CARBON, ACTIVATED CARBON).

A number of inorganic pollutants are removable by TSA processes. One of the major pollutants requiring removal is SO_2 from flue gases and from sulfuric acid plant tail gases. The Sulfacid and Hitachi fixed-bed processes, the Sumitomo and BF moving-bed processes, and the Westvaco fluidized-bed process all use activated carbon adsorbents for proven SO_2 removal (58). Zeolites with high acid resistance, such as mordenite and clinoptilolite, have proven to be effective adsorbents for dry SO_2 removal from sulfuric acid tail gas (59), and special zeolite adsorbents have been incorporated into the UOP PURASIV S process for this application (54).

Zeolites have also proven applicable for removal of nitrogen oxides (NO_x) from wet nitric acid plant tail gas (59) by the UOP PURASIV N process (54). The removal of NO_x from flue gases can also be accomplished by adsorption. The Unitaka process utilizes activated carbon with a catalyst for reaction of NO_x with ammonia, and activated carbon has been used to convert NO to NO_2, which is removed by scrubbing (58). Mercury is another pollutant that can be removed and recovered by TSA. Activated carbon impregnated with elemental sulfur is effective for removing Hg vapor from air and other gas streams; the Hg can be recovered by *ex situ* thermal oxidation in a retort (60). The UOP PURASIV Hg process recovers Hg from chlor–alkali plant vent streams using more conventional TSA regeneration (54). Mordenite and clinoptilolite zeolites are used to remove HCl from Cl_2, chlorinated hydrocarbons, and reformer catalyst gas streams (61). Activated aluminas are also used for such applications, and for the adsorption of fluorine and boron–fluorine compounds from alkylation (qv) processes (50).

PRESSURE SWING

A pressure-swing adsorption (PSA) cycle is one in which desorption takes place at a pressure much lower than adsorption. Its principal application is for bulk separations where contaminants are present at high concentration. The PSA cycles are characterized by high residual loadings and low operating loadings.

Figure 14 shows the operating loading $(X_1 - X_2)$ that derives from the partial pressure at feed conditions and the lower pressure P_2 at the end of desorption. These low adsorption capacities for high concentrations mean that cycle times must be short, seconds to minutes, for reasonably sized beds. Fortunately, packed beds of adsorbent respond rapidly to changes in pressure. A purge usually removes the desorbed components from the bed, and the bed is returned to adsorption condition by repressurization. Applications may require additional steps. Systems with weakly adsorbed species are especially suited to PSA adsorption. The applications of PSA include drying, upgrading of H_2 and fuel gases, and air separation. Several broad reviews of PSA have been written (62–64).

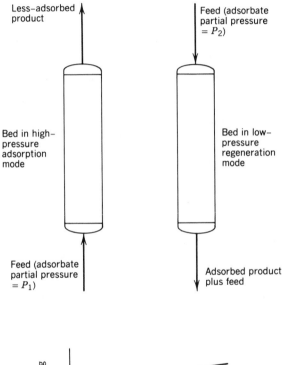

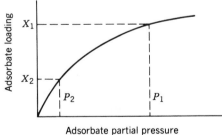

Fig. 14. Pressure-swing cycle (1).

Principles. In a PSA cycle two processes occur during regeneration, depressurizing, and purging. Depressurization must provide adequate reduction in the partial pressure of the adsorbates to allow desorption. Enough purge gas must flow through the adsorbent to transfer the desorbed material away. Equilibrium determines the maximum capacity of the gas to accomplish this. These cycles operate at nearly constant temperature and require no heating or cooling steps. They utilize the exothermic heat of adsorption remaining in the adsorbent to supply the endothermic heat of desorption. Pressure-swing cycles are classified as PSA, VSA (vacuum-swing adsorption), PSPP (pressure-swing parametric pumping) or RPSA (rapid pressure-swing adsorption). PSA swings between a high superatmospheric and a low superatmospheric pressure, and VSA swings from a superatmospheric pressure to a subatmospheric pressure. Otherwise, the principles involved are the same.

The other means of accomplishing pressure cycling of an adsorbent is parametric pumping, in which a single adsorbent bed is alternately pressurized with forward flow and depressurized with backward flow through the column from reservoirs at each end. Like TSA parametric pumping, one component concentrates in one reservoir and one in the other. As the name implies, pressure-swing parametric pumping embodies pressure changes that are more than pressuring and depressurizing a bed of adsorbent. Significant pressure gradients occur in the bed, much as temperature gradients are imposed in TSA parametric pumping. These gradients are especially critical to the way that RPSA cycles operate and result in much smaller adsorbent beds and simpler processes (57).

In most applications of adsorption, the separation is carried out by adsorbing the more strongly adsorbed species from the less strongly adsorbed. These separations are thus equilibrium-limited. However, an adsorptive separation can also be based on a rate- or kinetically-limited system (65). Slightly larger molecules diffuse more slowly through a microporous adsorbent with properly selected pore diameter. Therefore, in a rapidly cycling process such as PSA, smaller molecules can be preferentially adsorbed even in the absence of any equilibrium selectivity. Indeed rate-limited PSA has preferentially adsorbed oxygen from air on 4A zeolite when the equilibrium selectivity favors N_2 adsorption (52).

Steps. A pressure-swing cycle has at least three steps: adsorption, blowdown, and repressurization. Although not always necessary, a purge step is normally used. In finely tuned processes, cocurrent depressurization and pressure-equalization steps are frequently added.

At the completion of adsorption, the less selectively adsorbed components have been recovered as product. However, a significant quantity of the weakly adsorbed species are held up in the bed, especially in the void spaces. A cocurrent depressurization step reduces the bed pressure by allowing flow out of the bed cocurrently to feed flow and thus reduces the amount of product retained in the voids (holdup), improving product recovery, and increases the concentration of the more strongly adsorbed components in the bed. The purity of the more selectively adsorbed species has been shown to depend strongly on the cocurrent depressurization step for some applications (66). A cocurrent depressurization step is optional because a countercurrent one always exists. Criteria have been developed to indicate when the use of both is justified (67).

None of the selectively adsorbed components is removed from the adsorption vessel until the countercurrent depressurization (blowdown) step. During this step, the strongly adsorbed species are desorbed and recovered at the adsorption inlet of the bed. The reduction in pressure also reduces the amount of gas in the bed. By extending the blowdown with a vacuum (ie, VSA), the productivity of the cycle can be greatly increased.

Additional stripping of the adsorbates from the adsorbent and purging of them from the voids is accomplished by the purge step. This step can occur concurrently with the end of the blowdown or be carried out afterward. This step is accomplished with a flow of product into the product end to provide a low residual of the selectively adsorbed components at the effluent end of the bed.

The repressurization step returns the adsorber to feed pressure and completes the steps of a PSA cycle. Pressurization is carried out with product and/or feed. Pressurizing with product is done countercurrent to adsorption so that purging of the product end continues; indeed it may be merely a continuation of the purge step but with the bed exit valve closed. Pressurizing with feed cocurrent to adsorption in effect begins adsorption without producing any product.

Pressure equalization steps are used to conserve gas and compression energy. They are applied to reduce the quantity of feed or product gas needed to pressurize the beds. Portions of the effluent gas during depressurization, blowdown, and purge can be used for repressurization.

Flow Sheet. The most common processing scheme has two or three fixed-bed adsorbers alternating between the adsorption step and the desorption steps. The simplest two-bed configuration is illustrated in Figure 14. However, the variations possible to achieve special separations are endless. Single beds with external surge vessels provide continuous flow; multiple beds are used to accommodate additional steps. An example of the bed sequencing needed for multiple steps in a four-bed PSA is shown in Figure 15. Beds can be configured in series or parallel to accomplish coproduction.

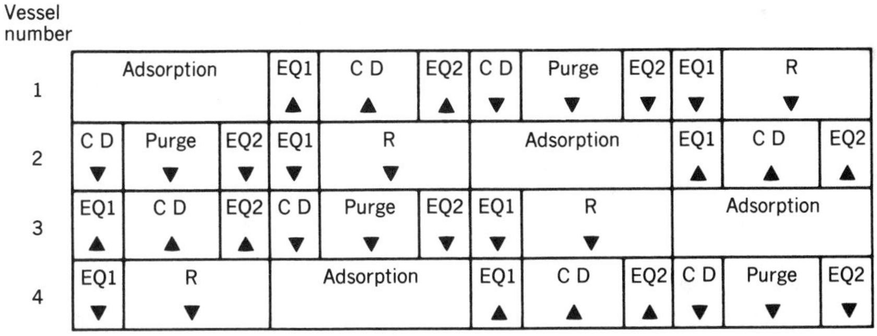

Vessel number

1	Adsorption		EQ1 ▲	C D ▲	EQ2 ▲	C D ▼	Purge ▼	EQ2 ▼	EQ1 ▼	R ▼	
2	C D ▼	Purge ▼	EQ2 ▼	EQ1 ▼	R ▼		Adsorption		EQ1 ▲	C D ▲	EQ2 ▲
3	EQ1 ▲	C D ▲	EQ2 ▲	C D ▼	Purge ▼	EQ2 ▼	EQ1 ▼	R ▼	Adsorption		
4	EQ1 ▼	R ▼		Adsorption		EQ1 ▲	C D ▲	EQ2 ▲	C D ▼	Purge ▼	EQ2 ▼

Fig. 15. Four-bed PSA system cycle sequence chart (64). EQ, equalization; C D▲, cocurrent depressurization; C D▼, countercurrent depressurization; R, repressurization; ▲, cocurrent flow; ▼, countercurrent flow. Courtesy of American Institute of Chemical Engineers.

The flow directions in a PSA process are fixed by the composition of the stream. The most common configuration is for adsorption to take place up-flow. All gases with compositions rich in adsorbate are introduced into the adsorption inlet end, and so effluent streams from the inlet end are rich in adsorbate. Similarly, adsorbate-lean streams to be used for purging or repressurizing must flow into the product end.

Because RPSA is applied to gain maximum product rate from minimum adsorbent, single beds are the norm. In such cycles where the steps take only a few seconds, flows to and from the bed are discontinuous. Therefore, surge vessels are usually used on feed and product streams to provide uninterrupted flow. Some RPSA cycles incorporate delay steps unique to these processes. During these steps, the adsorbent bed is completely isolated; and any pressure gradient is allowed to dissipate (68). The UOP Polybed PSA system uses five to ten beds to maximize the recovery of the less selectively adsorbed component and to extend the process to larger capacities (69).

Purifications.   The major purification applications for PSA are for hydrogen, methane, and drying. One of the first commercial uses was for gas drying in which the original two-bed Skarstrom cycle was used. This cycle uses adsorption, countercurrent blowdown, countercurrent purge, and cocurrent repressurization to produce a dry air stream with less than 1 ppm H_2O (70). About half of all dryers of instrument air use a PSA cycle similar to this one, most commonly using activated alumina or silica gel (71). Zeolites are used to obtain the lowest possible dewpoints. Some applications for drying air do not require a low level of H_2O, but only a significant lowering of the dew point. The pneumatic compressor systems used in vehicle air-brakes are an example; when a 10–30-K dew point depression is needed for higher discharge air temperatures in the presence of compressor oil, zeolites have been demonstrated to have an advantage over activated alumina and silica gel (72).

High purity H_2 is needed for applications such as hydrogenation, hydrocracking, and ammonia and methanol production. As a significant source of such gas, PSA is able to produce purities as high as 99.9999% using technologies such as the UOP Polybed approach (69). Most H_2 purification by PSA is associated with steam reforming of natural gas and with ethylene-plant and refinery off-gas streams (62). Hydrogen is also available in coke-oven gas, cracked ammonia, and coal-gasification gas. The contaminants that have to be removed by PSA include carbon oxides, N_2, O_2, Ar, NH_3, CH_4, and heavier hydrocarbons. To remove these components, adsorbent beds are compounded of activated carbon, zeolites, and carbon molecular sieves.

Bulk Separations.   Air separation, methane enrichment, and iso-/normal separations are the principal bulk separations for PSA. Others are the recovery of CO and CO_2.

The PSA process is used to separate air into N_2 and O_2. Many companies market systems for PSA O_2; zeolites 5A, 13X, clinoptilolite and mordenite, and carbon molecular sieves are commonly used in PSA, VSA, and RPSA cycles. The product purity ranges from 85 to 95% (limited by the argon, which remains with the O_2). About two-thirds of the O_2 produced is employed for electric furnace steel, with lesser amounts for waste water treating and solid waste and kilns (62). Smaller production units are used for patients requiring respiratory inhalation therapy in the hospital and at home (64) and for pilots on board aircraft (73). Enriched air, 25 to 55% O_2, used to enhance combustion, chemical reactions, and ozone production can be produced by tuning PSA processes (63). High purity, up to 99.99%, N_2 is produced by PSA and VSA cycles with zeolites and carbon molecular sieves (74). The major use for the N_2 is inert blanketing, such as in metal heat-treating furnaces; small units are used to purge aircraft fuel tanks (52) and in the food and beverage industry.

The upgrading of methane to natural gas pipeline quality is another significant PSA separation area. Methane is recovered from fermentation gases of landfills and wastewater purification plants and from poor-quality natural gas wells and tertiary oil recovery when CO_2 is the major bulk contaminant. Fermentation gases are saturated with water and contain "garbage" components such as sulfur and halogen compounds, alkanes, and aromatics (75). These impurities must first be removed by TSA using activated carbon or carbon molecular sieves. The CO_2 is then selectively adsorbed in a PSA cycle using either zeolites or silica gel in an equilibrium separation, or carbon molecular sieve in a kinetic-assisted equilibrium separation (76,77).

One version of the UOP IsoSiv process uses PSA to separate normal paraffins from branched and cyclic hydrocarbons in the C_5 to C_9 range. Zeolite 5A is used because its pores can size-selectively adsorb straight-chain molecules while excluding branched and cyclic species. The normal hydrocarbon fraction has better than 95% purity, and the higher octane isomer fraction contains less than 2% normal hydrocarbons (64).

PURGE SWING

A purge-swing adsorption cycle is one in which desorption takes place at the same temperature and total pressure as adsorption. Regeneration is accomplished either by partial-pressure reduction by an inert gas purge or by adsorbate displacement by an adsorbable gas. Its major application is for bulk separations when contaminants are at high concentration. Like PSA, purge cycles are characterized by high residual loadings, low operating loadings, and short cycle times (minutes). Mixtures of weakly adsorbed components are especially suited to purge-swing adsorption. Applications include the separation of normal from branched and cyclic hydrocarbons, gasoline vapor recovery, and bulk drying of organics.

Principles. Purging must provide adequate reduction in the partial pressure of the adsorbates to allow desorption. With enough purge volume, loadings as high as the loading X_1 in equilibrium with the feed partial pressure P_1 can be achieved, as shown in Figure 16. Reduction in partial pressure operates analogously to the reduction in system pressure in PSA cycles. Equilibrium determines the maximum capacity of the gas to purge the adsorbate. These cycles operate adiabatically at nearly constant inlet temperature and require no heating or cooling steps. As with PSA, purge processes utilize the exothermic heat of adsorption remaining in the adsorbent to supply the endothermic heat of desorption. Purge cycles are divided into two categories, inert purge and displacement purge. In inert-purge stripping, inert refers to the fact that the purge gas is not appreciably adsorbable at the cycle conditions. Inert purging desorbs the adsorbate solely by partial pressure reduction.

In displacement-purge stripping, displacement refers to the displacing action of the purge gas caused by its ability to adsorb at the cycle conditions. This competitive adsorption tends to desorb the adsorbate in addition to the partial

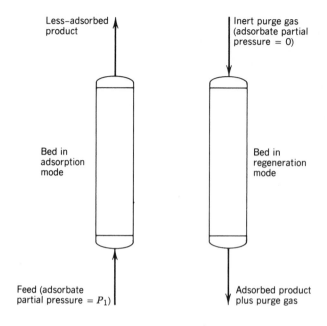

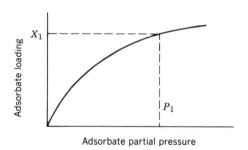

Fig. 16.   Inert-purge cycle (1).

pressure reduction of dilution. Displacement purging is not as dependent on the heat of adsorption remaining on the adsorbent, because the adsorption of purge gas can release much or all of the energy needed to desorb the adsorbate. The adsorbate must be more selectively adsorbed than the displacement purge so that it can desorb purge fluid during the adsorption step. The displacement purge gas composition must be carefully selected, because it contaminates both the product stream and the recovered adsorbate and requires distillation as illustrated in Figure 17. The displacement purge is more efficient for less selective adsorbate–adsorbent systems; systems with high equilibrium loading of adsorbate require more purging (78).

   Steps.   A purge-swing cycle usually has two steps, adsorption and purge. Sometimes, a cocurrent purge is added. After the adsorption step has been completed and the less selectively adsorbed components have been recovered, an appreciable amount of product is still stored in the bed. A purge cocurrent to feed can increase recovery by displacing the fluid held in the voids.

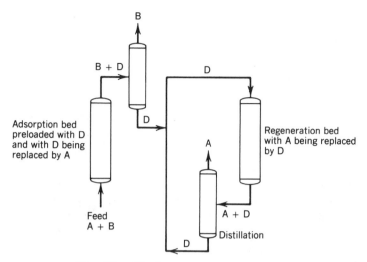

Fig. 17. Displacement-purge cycle (1).

The more selectively adsorbed components are stripped from the adsorbent bed during the countercurrent purge step. By purging into the product end of the bed, a lower residual loading of the selectively adsorbed species can be achieved in the portion of the adsorber that determines product quality.

Flow Sheet. Most purge-swing applications use two fixed-bed adsorbers to provide a continuous flow of feed and product (Fig. 16). Single beds are used when the flow to be treated is intermittent or cyclic. Because the purge flow is invariably greater than that of adsorption, purge is carried out in the down-flow direction to prevent bed lifting, and adsorption is up-flow.

Applications. Several purge-swing processes for the separation of $C_{10}–C_{18}$ iso- from normal paraffins have been commercialized: Exxon's Ensorb, UOP's IsoSiv, Texaco Selective Finishing (TSF), Leuna Werke's Parex, and the Shell Process (52). All of these processes take advantage of the molecular size selectivity of 5A zeolite, but vary in the purge fluid. Ammonia is used in a displacement-purge cycle in Ensorb. Normal paraffins or light naphtha with a carbon number of two to four less than the feed stream are used for the displacement purge for TSF, Parex, and the Shell Process (79). One version of UOP purge IsoSiv for C_5 to C_9 naphtha employs H_2 in an inert-purge cycle (64). UOP also developed a similar process, OlefinSiv, for the separation of isobutylene from normal butenes using a size-selective zeolite with displacement purge (80).

Since 1971 all U.S. automobile models must have canisters of activated carbon to control gasoline vapors. Any gasoline vapors from the carburetor or the gas tank during running, from the tank during diurnal cycling, and from carburetor hot-soak losses are adsorbed by the carbon and held until they can be regenerated. The vapors are desorbed by an inert purge of air and are drawn into the carburetor as fuel when the engine is running (81). This gas-phase use for activated carbon is the third largest after solvent recovery and air purification.

Another use of inert-purge regeneration is UOP's Adsorptive Heat Recovery (AHR) drying system (82). The AHR system has been commercialized for drying azeotropic ethanol to be blended with gasoline into "gasohol." The process uses a closed loop of N_2 as the inert purge to desorb the water. The AHR process can be economically superior to azeotropic distillation for drying feeds with as much as 20% H_2O.

NONREGENERATIVE PROCESSES

Gas-phase adsorption can also be used when regenerating the adsorbent is not practical. Most of these applications are used where the facilities to effect a regeneration are not justified by the small amount of adsorbent in a single unit. Nonregenerative adsorbents are used in packaging, dual-pane windows, odor removal, and toxic chemical protection.

Applications. Silica gel is the adsorbent most commonly used as a desiccant in packaging. Activated carbon is used in packaging and storage to adsorb other chemicals for preventing the tarnishing of silver, retarding the ripening or spoiling of fruits, "gettering" (scavenging) outgassed solvents from electronic components, and removing odors.

Adsorbents are used in dual-pane windows to prevent fogging between the sealed panes that could result from the condensation of water or the solvents used in the sealants (83). Synthetic zeolites (3A, 4A, 13X) or, less frequently, blends of zeolites with silica gel are installed in the spacing strips in double-glazed windows to adsorb water during initial dry-down and any in-leakage and to adsorb organic solvents emitted from the sealants during their cure. The adsorbent or mix of adsorbents applied depends on the sealant system and filling gas used.

The largest use of activated carbon is for the purification of air streams. Much of this carbon is used to treat recirculated air in large occupied enclosures, such as office buildings, apartments, and plants. The carbon is incorporated into thin filterlike frames to treat the large volumes of air with low pressure drop. Odors are also removed from smaller areas by activated carbon filters in kitchen hoods, air conditioners, and electronic air purifiers. On a smaller scale, gas masks containing carbon or carbon impregnated with promoters are used to protect wearers from odors and toxic chemicals. The smallest scale carbon filters are those used in cigarettes. In addition to protection from hazardous chemicals in industry, activated carbon gas masks can protect against gas-warfare chemicals. Activated carbon fibers (qv) have been formed into fabrics for clothing to protect against vesicant and percutaneous chemical vapors (84).

REACTIVE ADSORPTION

Although chemisorbents are not used as extensively as physical adsorbents, a number of commercially significant processes employ chemisorption for gas purification.

Iron Sponge. An old method for removal of sulfur compounds involves contacting gases containing H_2S and H_2O with α- or γ-ferric oxide monohydrates at approximately 38°C to adsorb the sulfur in the form of ferric sulfide, followed by periodic reoxidation of the surface to form elemental sulfur and to "revivify" the ferric oxides (85). The iron sponge is reused in this cycle until buildup of sulfur in its pores reduces its effectiveness and it is replaced with fresh adsorbent. The process is most efficient when the treated gases contain oxygen to allow continuous revivification. Spent adsorbents may be regenerated, eg, by oxidation of the sulfur to SO_2 to be fed to a sulfuric acid plant or by solvent extraction with carbon disulfide, and reused.

Mercury Removal. Trace amounts of mercury found in natural gas in some parts of the world are known to cause significant pinhole corrosion damage to aluminum heat exchanger surfaces in cryogenic coldboxes upstream of liquefied natural gas LNG plants (86). Mercury can be removed from such streams, and other industrial gases, down to low concentrations by treatment in an *ex situ* TSA regenerative process using an activated carbon adsorbent containing sulfur, and allowing reactions involving the formation of mercuric sulfide to remove mercury from the gas. Alternatively, the mercury can be removed by a newly developed adsorbent that may be employed in either nonregenerative or TSA regenerative process cycles (87).

Nuclear Waste Management. Separation of radioactive wastes provides a number of relatively small scale but vitally important uses of gas-phase purification applications of adsorption. Such applications often require extremely high degrees of purification because of the high toxicity of many radioactive elements.

Delay for Decay. Nuclear power plants generate radioactive xenon and krypton as products of the fission reactions. Although these products are trapped inside the fuel elements, portions can leak out into the coolant (through fuel cladding defects) and can be released to the atmosphere with other gases through an air ejector at the main condenser.

To prevent such release, off gases are treated in Charcoal Delay Systems, which delay the release of xenon and krypton, and other radioactive gases, such as iodine and methyl iodide, until sufficient time has elapsed for the short-lived radioactivity to decay. The delay time is increased by increasing the mass of adsorbent and by lowering the temperature and humidity; for a boiling water reactor (BWR), a typical system containing 21 t of activated carbon operated at 255 K, at 500 K dewpoint, and 101 kPa (15 psia) would provide about 42 days holdup for xenon and 1.8 days holdup for krypton (88). Humidity reduction is typically provided by a combination of a cooler–condenser and a molecular sieve adsorbent bed.

If the spent fuel is processed in a nuclear fuel reprocessing plant, the radioactive iodine species (elemental iodine and methyl iodide) trapped in the spent fuel elements are ultimately released into dissolver off gases. The radioactive iodine may then be captured by chemisorption on molecular sieve zeolites containing silver (89).

Other Applications. Many applications of adsorption involving radioactive compounds simply parallel similar applications involving the same compounds in nonradioactive forms, eg, radioactive carbon-14, or deuterium- or tritium-containing versions of CO_2, H_2O, hydrocarbons. For example, molecular sieve zeolites are commonly employed for these separations, just as for the corresponding nonradioactive uses.

MOVING AND FLUIDIZED BEDS

Most adsorption systems use stationary-bed adsorbers. However, efforts have been made over the years to develop moving-bed adsorption processes in which the adsorbent is moved from an adsorption chamber to another chamber for regeneration, with countercurrent contacting of gases with the adsorbents in each chamber. Union Oil's Hypersorption Process (90) is an example. However, this process proved uneconomical, primarily because of excessive losses resulting from adsorbent attrition.

The commercialization by Kureha Chemical Co. of Japan of a new, highly attrition-resistant, activated-carbon adsorbent as Beaded Activated Carbon (BAC) allowed development of a process employing fluidized-bed adsorption and moving-bed desorption for removal of volatile organic carbon compounds from air. The process has been marketed as GASTAK in Japan and as PURASIV HR (91) in the United States, and is now marketed as SOLDACS by Daikin Industries, Ltd.

The discovery (92) that the graphite coating of molecular sieves can dramatically improve their attrition resistance without significantly impairing adsorption performance should allow the extension of moving-bed technology to bulk gas separations (93).

Design Methods

Design techniques for gas-phase adsorption range from empirical to theoretical. Methods have been developed for equilibrium, for mass transfer, and for combined dynamic performance. Approaches are available for the regeneration methods of heating, purging, steaming, and pressure swing. Several broad reviews have been published on analytical equations describing adsorption (94,95), of experimental adsorption equilibrium and kinetic data (96), on theoretical models for adsorption processes (97,98), and on adsorption design considerations (1).

Adsorption.   In the design of the adsorption step of gas-phase processes, two phenomena must be considered, equilibrium and mass transfer. Sometimes adsorption equilibrium can be regarded as that of a single component, but more often several components and their interactions must be accounted for. Design techniques for each phenomenon exist as well as some combined models for dynamic performance.

Equilibrium.   Among the aspects of adsorption, equilibrium is the most studied and published. Many different adsorption equilibrium equations are used for the gas phase; the more important have been presented (see section on Isotherm Models). Equally important is the adsorbed phase mixing rule that is used with these other models to predict multicomponent behavior.

Many simple systems that could be expected to form ideal liquid mixtures are reasonably predicted by extending pure-species adsorption equilibrium data to a multicomponent equation. The potential theory has been extended to binary mixtures of several hydrocarbons on activated carbon by assuming an ideal mixture (99) and to hydrocarbons on activated carbon and carbon molecular sieves, and to O_2 and N_2 on 5A and 10X zeolites (100). Mixture isotherms predicted by IAST agree with experimental data for methane + ethane and for ethylene + CO_2 on activated carbon, and for CO + O_2 and for propane + propylene on silica gel (36). A statistical thermodynamic model has been successfully applied to equilibrium isotherms of several nonpolar species on 5A zeolite, to predict multicomponent sorption equilibria from the Henry constants for the pure components (26). A set of equations that incorporate surface heterogeneity into the IAST model provides a means for predicting multicomponent equilibria, but the agreement is only good up to 50% surface saturation (9).

For most models of adsorptive equilibrium, however, the coefficients derived from pure species are not adequate to predict multicomponent equilibrium for nonideal mixtures. Fitting the systems ethane + ethylene + propane on 5A zeolite and $H_2S + CO_2$ on H-mordenite required using binary parameters with the IAST or the real adsorbed solution theory models (101). A coalescing factor applied to the potential theory did collapse all isotherms to a single curve for activated carbon, zeolites, and silica gel. A binary interaction parameter that is a function of the coalescing factor was needed to gain agreement with binary data (102). For the multicomponent system of H_2, CO, CH_4, CO_2, and H_2S on activated carbon, an interaction parameter was required in the extended Langmuir equation to predict multicomponent equilibrium (103). Cross-correlation coefficients were necessary to apply a statistical model to three nonideal ternary zeolite systems (104). An activity coefficient whose composition dependence is described by the Wilson equation has been added to the vacancy solution model (VSM) to fit data for hydrocarbons on activated carbon and $O_2 + N_2$ on 10X zeolite (37). Activity coefficients of the adsorbate–adsorbate interactions or treatment of the surface as heterogeneous are correlative methods that allow extension of the IAST to binary adsorption (10).

Mass Transfer. The degree of approach to equilibrium that can be achieved in adsorption is determined by the mass-transfer rates. One useful design concept is the mass-transfer zone (MTZ), an extension of the ion-exchange zone method (105). Figure 18**b** is a depiction of the adsorbate loading in a fixed bed during adsorption. The ordinate is loading (X) and the abscissa is distance (L) from the inlet of the bed. Between the inlet and the exhaustion point (L_e), the loading is in equilibrium with the feed gas, and this section is called the equilibrium section. From the breakthrough point (L_b) to the outlet of the bed, the adsorbate loading is still at the residual loading level and is unused bed. Mass transfer between the gas and the adsorbent is occurring between the breakthrough and exhaustion points, and so this zone is called the mass-transfer zone (MTZ). The length of the bed, L_b to L_e, is called the mass-transfer zone length (MTZL). The MTZL is usually correlated to flow rate or flow velocity (106).

Most dynamic adsorption data are obtained in the form of outlet concentrations as a function of time as shown in Figure 18**a**. The area iebai measures the removal of the adsorbate, as would the stoichiometric area idcai, and is used to calculate equilibrium loading. For constant pattern adsorption, the breakthrough time (Θ_b), and the stoichiometric time (Θ_s), are used to calculate LUB as $(1 - \Theta_b/\Theta_s)L_{bed}$ (107). This LUB concept is commonly used for drying and desulfurization design in the natural gas industry and for air prepurification before cryogenic distillation.

Another way of subdividing the bed is illustrated in Figure 18**b**. If the mass-transfer resistance were negligible, the MTZ would become a square or stoichiometric front along the line dsc. The area febgf represents used adsorbent loading, while the area ehbe between the potential loading and the actual loading curve eb is unused. By material balance, the area fdcgf up to the stoichiometric front would also represent the used capacity. Therefore, areas febgf and fdcgf are equal. This

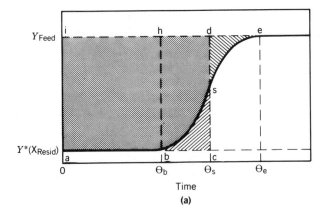

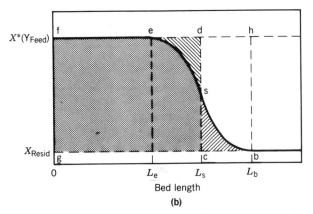

Fig. 18.   (a) Time trace of adsorbate composition in an adsorber effluent during adsorption. (b) Adsorbate loading along the flow axis of an adsorber during adsorption (1).

portion of the bed up to the stoichiometric point, s, is called the weight of equivalent equilibrium section (WES). The rest of the adsorbent from the stoichiometric point to the breakthrough point is termed the weight of unused bed (WUB), because it is equivalent to a bed with no usable capacity in the stoichiometric interpretation. Adsorption beds can thus be sized by combining a WES calculated from equilibrium data and a WUB derived from kinetic data.

Dynamic Performance.   Most models do not attempt to separate the equilibrium behavior from the mass-transfer behavior. Rather they treat adsorption as one dynamic process with an overall dynamic response of the adsorbent bed to the feed stream. Although numerical solutions can be attempted for the rigorous partial differential equations, simplifying assumptions are often made to yield more manageable calculating techniques.

The *J*-function is a definite integral of an expression including I_0, the modified Bessel function of the first kind. *J*-function curves use stoichiometric time and the number of theoretical stages as the two parameters to fit breakthrough curves and extend to other conditions. These curves have been approximated for use on PC microcomputers (108). A phenomenological model requires the determination of two parameters, a transfer coefficient, and a linear isotherm constant, from a complete breakthrough curve. The solution to the model is in an infinite series form, which is calculable by a hand-held calculator or personal computer (109). Another method separates the equilibrium from the kinetic effects by constructing effective equilibrium curves. Because the solution to the model involves nonlinear algebraic or differential equations, graphs called solution charts are used to predict breakthrough fronts (110). Theoretical stages form the essence of the discrete cell model graphical procedures, which are applied to flat isotherms and incorporate pore diffusivity and axial dispersion (98). Another solution technique is the use of fast Fourier transforms. Linear isotherms are required, but their applicability for predicting breakthrough curves has been demonstrated for isothermal and nonisothermal adsorbers (111). Another model, with a solution in infinite series form, incorporates separate mass-transfer coefficients for external film, macropore, and micropore resistances (112). Techniques have also been developed to predict breakthrough from fluidized beds. The behavior of organic solvents adsorbed from air on activated carbon was shown to exhibit breakthrough times that can be correlated to the adsorption capacity and the amount of bed expansion (113).

Some specific design methods have been developed for particular applications. Several procedures have been published for the design of gas dryers. The J-function has been applied to silica gel dryers after a correlated correction factor accounted for the nonisothermality (114). In other work on drying with activated alumina and silica gel, constant-pattern LUBs were shown applicable for designs at H_2O contents of less than 0.003 kg/kg air (115). Equilibrium and kinetic parameters for H_2O on activated alumina were determined for a more rigorous nonisothermal model that predicted adiabatic behavior (116). Breakthrough times for several organic vapors on activated carbon respirator cartridges have been found to be predictable by using a theory of statistical moments. In one study, equilibrium data was correlated by the Potential Theory and breakthrough was calculated using the normal probability distribution curve (117). In another, the equilibrium data was represented by a Freundlich equation (118). For heavy hydrocarbon recovery from natural gas on silica gel, the equilibrium data were fit to a Freundlich isotherm and the breakthrough composition was found to have a power dependence on the extent of adsorptive saturation of the adsorbent (111). A WUB design approach was found to predict breakthrough for several organics on activated carbon using a Potential Theory equilibrium curve (119). Correlations of equilibrium capacity and WUB were also found applicable in the removal of H_2S from natural gas using 5A zeolite (120).

Regeneration. In recent years, considerable effort has been expended to better understand and quantify the process of regeneration. Methods are available to predict thermal, purge, and steaming requirements. Models are available to simulate all of the regeneration types, temperature, pressure, and purge swings.

Thermal Requirements. When a temperature-swing cycle is heating limited, the regeneration design is only concerned with transferring energy to the system. Charts for isothermal, linear-isotherm adsorption that were derived by Hougen and Marshall (121) from earlier work on heat transfer from a gas to fixed beds (122) can be reapplied to heat transfer for heating-limited regeneration when the heat of adsorption is negligible (123). This approach has been expanded to include both a correction of one dimensionless time per dimensionless bed length for heat losses and another correction particular to the H_2O–4A zeolite system studied (124). Because cooling is a heating-limited step, it can be calculated by the modified Hougen and Marshall method. As mentioned in the discussion of thermal swing, the cooling step can be performed under some process conditions by starting adsorption with a hot adsorbent; performance is not affected.

Purge Requirements. The amount of purge gas needed in stripping-limited regeneration is similar to that for purge regeneration, but it differs primarily in the temperature at which the isothermal desorption occurs. For a pressure-swing process, the theoretical minimum volumetric purge-to-feed ratio is the ratio of the purge pressure to the feed pressure (123), and one model shows the optimum ratio to be the minimum purge volume that can be used with the given cyclic steady-state conditions (125). For a thermal-swing process, the minimum purge-to-adsorbent ratio is the ratio of the heat capacity of the solid to that of the gas (126). Specific design purge data have been published for purge-swing activated carbon automotive evaporative emissions control (81) and for pressure-swing drying of pneumatic system air (72). The pneumatic system process exhibits an optimum purge ratio for maximizing the attainable dewpoint depression. An isothermal purge-swing model that uses Langmuir equilibrium to simulate adsorbent performance has been presented (127).

Steaming Requirements. The steaming of fixed beds of activated carbon is a combination of thermal swing and displacement purge swing. The exothermic heat released when the water adsorbs from the vapor phase is much higher than is possible with heated gas purging. This cycle has been successfully modeled by equilibrium theory (128).

Temperature Swing. Several fairly comprehensive reviews of thermal-swing adsorption models are found in the literature (96,97,129,130). Many of these models have been used to carry out parametric analyses with a goal of energy minimization. A nonisothermal model for single components using equilibrium theory demonstrated that efficiency improves with increased purge contact time and high heat capacity purge gas but is minimally affected by initial bed loading (131), and the model defined conditions under which the desorption can be continued with a cold stream without additional overall purge gas. That work also introduced the concept that minimum thermal energy is required at the characteristic temperature, that temperature at which the slope of the equilibrium isotherm is equal to the ratio of the heat capacity at the adsorbent to that of the purge gas. A nonequilibrium, mechanistic model with a multicomponent VSM model of equilibrium (132) and a nonequilibrium nonadiabatic computer model with a Langmuir-like isotherm (130) reached similar conclusions on optimization. Another nonequilibrium nonadiabatic computer model with an Antoine equation isotherm was used with supporting data to demonstrate that significant energy can be saved by proper timing of the cooling step (133). Most modeling has assumed that the purge gas is clean even though solvent recovery processes use a recirculated stream when an inert purge gas is employed. Using the method of characteristics and a Freundlich isotherm, an equilibrium theory modeled the incorporation of closed-loop heating and cooling steps (134).

Pressure Swing. Design equations have been developed to predict temperature rise, minimum bed length to retain the heat front, minimum purge rate, and effluent composition (135). A nonequilibrium, nonisothermal simulation program with a Freundlich isotherm equation was found to agree with data for drying with silica gel (136). A somewhat simpler isothermal model using an isotherm approximated by two straight lines successfully calculated the volumetric purge-to-feed ratio needed to achieve varying product dryness using silica gel (137). An adiabatic equilibrium model with a Langmuir isotherm was used to study the blowdown step of a cycle removing CO_2 on activated carbon and 5A zeolite (138). Changing to an isothermal assumption introduced significant errors into the results. The countercurrent pressurization step was investigated with an isothermal equilibrium model using a Langmuir isotherm for O_2 production from air with 5A zeolite (139). The model predicted the dependence of O_2 concentration on countercurrent pressure and was used to study other parameters. An isothermal model with linear isotherms and component-specific pore diffusivity was used and compared to data for the kinetic-limited separation of air by RS-10 zeolite (140). The simulations agreed well with the experimental parametric studies of time and pressure of feed, blowdown, purge, and pressurization. An equilibrium model was formulated to simulate RPSA using a Freundlich isotherm for separation of N_2 and CH_4 (141). Pressure responses, flow rates, and compositions compared favorably as a function of feed pressure, cycle frequency, and product rate. A nonequilibrium, nonisothermal model for RPSA was developed using a linear isotherm and Darcy's law for pressure drop (142). The model predicted performance in agreement with previous data (57) for air separation on 5A zeolite.

Pressure Drop. The prediction of pressure drop in fixed beds of adsorbent particles is important. When the pressure loss is too high, costly compression may be increased, adsorbent may be fluidized and subject to attrition, or the excessive force may crush the particles. As discussed previously, RPSA relies on pressure drop for separation. Because of the cyclic nature of adsorption processes, pressure drop must be calculated for each of the steps of the cycle. The most commonly used pressure drop equations for fixed beds of adsorbent are those of Ergun (143), Leva (144), and Brownell and co-workers (145). Each of these correlations uses a particle Reynolds number ($Re = D_pG/\mu$) and friction factor (f) to calculate the pressure drop (ΔP) per unit length (L) by the equation

$$\frac{\Delta P}{L} = \frac{fG^2}{2g_c D_p \rho}$$

where D_p is the particle diameter, G the mass flux, μ the gas viscosity, and ρ the gas density. The methods differ in their definition of D_p and f. For up-flow in fixed-bed adsorbers, fluidization occurs when the pressure drop just balances the weight, corrected by any buoyancy:

$$\frac{\Delta P}{L} = \frac{(1 - \epsilon)(\rho_s - \rho)g}{g_c}$$

where ρ_s is the density of the solid. For down-flow in packed beds, the potential for crushing the adsorbent must be checked. Two forces act to crush the particles, pressure drop and the weight of the bed. The sum of these two ($\Delta P + (1 - \epsilon)\rho_s L$) should be kept less than that which is known to cause adsorbent damage.

Future Directions

Advances in fundamental knowledge of adsorption equilibrium and mass transfer will enable further optimization of the performance of existing adsorbent types. Continuing discoveries of new molecular sieve materials will also provide adsorbents with new combinations of useful properties. New adsorbents and adsorption processes will be developed to provide needed improvements in pollution control, energy conservation, and the separation of high value chemicals. New process cycles and new hybrid processes linking adsorption with other unit operations will continue to be developed.

Fundamentals. Marked improvements in the prediction of multicomponent equilibria from single-component data will be achieved by developing more realistic theoretical models that provide for nonideal adsorbate phases and heterogeneities of surface energetics and geometries, and that allow for the effect of adsorbates on adsorbent properties. Molecular modeling and molecular-dynamic simulations of adsorption phenomena on high speed computers will improve dramatically over the next 5 to 10 years. These improvements will be driven by the need to better predict multicomponent adsorption behavior and to develop better tools for the design of adsorbents with desired properties. Improvements in the understanding of mass-transfer rate processes in adsorption will allow the development of new separation processes based on differences in the relative rates of adsorption.

New Adsorbent Materials. Silicalite and other hydrophobic molecular sieves, the new family of $AlPO_4$ molecular sieves, and steadily increasing families of other new molecular sieves (including structures with much larger pores than those now commercially available), as well as new carbon molecular sieves and pillared interlayer clays (PILCS), will become more available for commercial applications, including adsorption. Adsorbents with enhanced performance, both highly selective physical adsorbents and easily regenerated, weak chemisorbents will be developed, as will new rate-selective adsorbents.

Process Concepts. Hybrid systems involving gas-phase adsorption coupled with catalytic processes and with other separations processes (especially distillation and membrane systems) will be developed to take advantage of the unique features of each. The roles of adsorption systems will be to efficiently achieve very high degrees of purification; to lower fouling contaminant concentrations to very low levels in front of membrane and other separations processes; or to provide unique separations of azeotropes, close-boiling isomers, and temperature-sensitive or reactive compounds.

Applications. Both industrial emissions reduction and indoor air-pollution abatement uses will grow. For example, the development of adsorbents with higher capacity for removal of radon from humid air could allow the development of a one-bed, delay-for-decay system in which radon adsorbs, decays to lead, and is precipitated onto the adsorbent.

Many existing applications involve small adsorption systems for home and automobile applications, eg, refrigerant drying in automobile air conditioners, dual-pane window desiccants, medical oxygen systems, and muffler corrosion protection. Such small adsorption systems will continue to be developed for new uses in indoor air pollution and odor abatement and for the enhancement of the performance of other equipment and appliances. For example, adsorption-based control of the composition of air in refrigerators can provide improvements in the storage of fruits and vegetables.

Design Methods. Improvements in the ability to predict multicomponent equilibrium and mass-transfer rate performance will allow significant improvements in the design of new adsorption systems and in the energy efficiency of existing systems.

Computer Systems. Improved "smart" control systems based on new computer capabilities (microprocessors, artificial intelligence, expert systems) will be used increasingly in adsorption systems to provide more efficient operation. For example, the adjustment of the adsorption–regeneration cycle to account for changing feed compositions and flow rates can significantly reduce energy consumption by carrying out regenerations only when needed. Enhanced computer capabilities will also allow coupling of more sophisticated equilibrium models with more exact models for adsorption dynamics to provide improved design tools.

BIBLIOGRAPHY

"Adsorptive Separation, Gases" in *ECT* 3rd ed., Vol. 1, pp. 544–581, by D. B. Broughton, UOP Process Division, UOP Inc.

1. G. E. Keller, II, R. A. Anderson, and C. M. Yon, in R. W. Rousseau, ed., *Handbook of Separation Process Technology,* John Wiley & Sons, Inc., New York, 1987, 644–696.
2. J. O. Hirschfelder, C. F. Curtiss, and R. B. Bird, *Molecular Theory of Gases and Liquids,* John Wiley & Sons, Inc., New York, 1954, p. 215, 949, 1110.
3. R. M. Barrer and D. E. W. Vaughan, *J. Phys. Chem. Solids* **32,** 731 (1971).
4. R. M. Barrer, *Zeolites and Clay Minerals as Adsorbents and Catalysts,* Academic Press, London, 1978, p. 164, 174, 185.
5. D. W. Breck and J. V. Smith, *Sci. Am.,* 8 (Jan. 1959).
6. Data from Union Carbide Molecular Sieves, UOP, Tarrytown, N.Y.
7. W. A. Steele, "The Physical Adsorption of Gases on Solids," *Adv. Colloid Interface Sci.* **1,** 3–78 (1967). (Review with 360 refs.).
8. D. M. Ruthven, *Principles of Adsorption and Adsorption Processes,* John Wiley & Sons, Inc., New York, 1984, Chapt. 3, 4, p. 108.
9. A. L. Myers in A. L. Myers and G. Belfort, eds., *Fundamentals of Adsorption,* Engineering Foundation, New York, 1984, pp. 365–381.
10. A. L. Myers in A. I. Liapis, ed., *Fundamentals of Adsorption,* Engineering Foundation, New York, 1987, pp. 3–25.
11. A. L. Myers, *NATO ASI Ser., Ser. E,* **158** (*Adsorpt. Sci. Technol.*), 15–36 (1989).
12. D. P. Valenzuela and A. L. Myers, *Adsorption Equilibrium Data Handbook,* Prentice Hall, Engelwood Cliffs, N.J., 1989.
13. D. P. Valenzuela and A. L. Myers, *Sep. Purif. Methods* **13**(2), 153–183 (1984).
14. O. Talu and A. L. Myers, *AIChE J.* **34,** 1887–1893 (1988).
15. *Ibid.,* 1931–1932 (1988).
16. R. T. Yang, *Gas Separation by Adsorption Processes,* Butterworths, Boston, 1987, p. 86.
17. C. M. Yon and P. H. Turnock, *AIChE Symp. Ser.* **67**(117), 75 (1971).
18. R. T. Maurer in J. R. Katzer, ed., *Molecular Sieves—II* (ACS Symp. Ser. 40) American Chemical Society, Washington, D.C., 1977, p. 379.
19. I. Langmuir, *J. Am. Chem. Soc.* **40,** 1361 (1918).
20. M. Volmer, *Z. Phys. Chem.* **115,** 253 (1925).

21. R. H. Fowler and E. A. Guggenheim, *Statistical Thermodynamics,* Cambridge University Press, Cambridge, 1939.

22. T. L. Hill, *Introduction to Statistical Thermodynamics,* Addison-Wesley, Reading, Mass., 1960.

23. V. A. Bakaev, *Dokl. Akad. Nauk SSSR* **167,** 369 (1967).

24. L. Riekert, *Adv. Catal.* **21,** 287 (1970).

25. P. Brauer, A. Lopatkin, and G. Ph. Stepanez in E. M. Flanigen and L. B. Sand, eds., *Molecular Sieve Zeolites, Adv. in Chem 102,* American Chemical Society, Washington, D.C., 1971, p. 97.

26. D. M. Ruthven, K. F. Loughlin, and K. A. Holborrow, *Chem. Eng. Sci.* **28,** 701 (1973).

27. D. M. Ruthven, *Nat. Phys. Sci.* **232**(29), 10 (1971).

28. Ref. 8, p. 75ff.

29. M. M. Dubinin and L. V. Radushkevich, *Dokl. Akad. Nauk SSSR, Ser. Khim.* **55,** 331 (1947).

30. M. M. Dubinin and V. A. Astakhov, *Izv. Akad. Nauk. SSSR, Ser. Khim.* **71,** 5 (1971).

31. C. J. Radke and J. M. Prausnitz, *Ind. Eng. Chem. Fundam.* **11,** 445 (1972); *AIChE J.* **18,** 761 (1972).

32. J. Toth, *Acta. Chim. Acad. Sci. Hung.* **69,** 311 (1971).

33. S. Brunauer, P. H. Emmett, and E. Teller, *J. Am. Chem. Soc.* **60,** 309 (1938).

34. E. C. Markham and A. F. Benton, *J. Am. Chem. Soc.* **53,** 497 (1931).

35. R. Sips, *J. Chem. Phys.* **16,** 490 (1948).

36. A. L. Myers and J. M. Prausnitz, *AIChE J.* **11,** 121 (1965).

37. S. Suwanayuen and R. P. Danner, *AIChE J.* **26,** 68, 76 (1980).

38. T. W. Cochran, R. L. Kabel, and R. P. Danner, *AIChE J.* **31,** 268 (1985).

39. A. Kohl and F. Riesenfeld, *Gas Purification,* 4th ed., Gulf Publishing Co., Houston, Tex., 1985, p. 651.

40. K. S. W. Sing and co-workers, *Pure Appl. Chem.* **57,** 603 (1985).

41. D. W. Breck, *Zeolite Molecular Sieves—Structure, Chemistry, and Use,* John Wiley & Sons, Inc., New York, 1974.

42. R. C. Bansal, J.-B. Donnet, and F. Stoeckli, *Active Carbon,* Marcel Dekker, New York, 1988, p. ix.

43. E. M. Flanigen and co-workers, *Nature (London)* **271,** 512 (1978).

44. E. M. Flanigen and R. L. Patton, UOP, Tarrytown, N.Y., private communication.

45. R. C. Bansal, T. L. Dhami, and S. Parkash, *Carbon* **16,** 389 (1978).

46. W. K. Lewis, E. R. Gilliland, B. Chertow, and W. P. Cadogan, *Ind. Eng. Chem.* **42,** 1326 (1950).

47. M. Benchanaa, M. Lallemant, M. H. Simonet-Grange, and G. Bertrand, *Thermochim. Acta.* **152,** 43–51 (1989).

48. H. R. Burkholder, G. E. Fanslow, and D. D. Bluhm, *Ind. Eng. Chem. Fundam.* **25,** 414–416 (1986).

49. D. Basmadjian, *Can. J. Chem. Eng.* **53,** 234–238 (1975).

50. B. Crittenden, *Chem. Eng.* **452,** 21–24 (1988).

51. K. P. Goodboy and H. L. Fleming, *Chem. Eng. Progr.* **80,** 63–68 (1984).

52. D. M. Ruthven, *Chem. Eng. Progr.* **84,** 42–50 (1988).

53. H. L. Brooking and D. C. Walton, *Chem. Eng.* **257,** 13–17 (1972).

54. R. A. Anderson in ref. 18, pp. 637–649.

55. S. R. Dunne, *Automotive Corrosion and Prevention Conf. Proc.* Society of Automotive Engineers, Warrendale, Pa., 1989, 165–173.

56. R. E. Kenson and J. F. Jackson, Prepared Paper, Air Pollution Control Association, Annual Mtg., 1988.

57. G. E. Keller, II, and R. J. Jones in W. H. Flank, ed., *Adsorption and Ion Exchange with Synthetic Zeolites, Am. Chem. Soc. Symp. Ser. 135,* American Chemical Society, Washington, D.C., 1980, pp. 275–286.

58. H. Juentgen, *Carbon* **15**, 273–283 (1977).
59. J. R. Kiovsky, P. B. Koradia, and D. S. Hook, *Chem. Eng. Progr.* **72**, 98–103 (1976).
60. W. D. Lovett and F. T. Cunniff, *Chem. Eng. Progr.* **70**, 43–47 (1974).
61. A. Dyer, *An Introduction to Zeolite Molecular Sieves,* John Wiley & Sons, Inc., New York, 1988, 102–105.
62. J. R. Martin, C. F. Gotzmann, F. Notaro, and H. A. Stewart, *Adv. Cryog. Eng.* **31**, 1071–1086 (1986).
63. S. Sircar in A. E. Rodrigues, M. D. LeVan, and D. Tondeur, eds., *Adsorption: Science and Technology,* Kluwer Academic Publishers, Dordrecht, Netherlands, 1988, 285–321.
64. R. T. Cassidy and E. S. Holmes, *AIChE Symp. Ser.* **80**, 68–75 (1984).
65. Z. J. Pan, R. T. Yang, and J. A. Ritter in G. E. Keller, II, and R. T. Yang, eds., *New Directions in Sorption Technology,* Butterworths, Boston, 1988.
66. P. Cen and R. T. Yang, *Ind. Eng. Chem. Fundam.* **25**, 758–767 (1986).
67. S. S. Suh and P. C. Wankat, *AIChE J.* **35**, 523–526 (1989).
68. G. E. Keller, II, in T. E. White, Jr., C. M. Yon, and E. H. Wagener, eds., *Industrial Gas Separations, Am. Chem. Soc. Symp. Ser. 223,* American Chemical Society, Washington, D.C., 1983, pp. 145–169.
69. R. T. Cassidy in ref. 57, pp. 248–259.
70. C. W. Skarstrom in N. N. Li, ed., *Recent Developments in Separation Science,* Vol. 2, CRC Press, Boca Raton, Fla., 1975, pp. 95–106.
71. J. W. Armond in R. P. Townsend, ed., *The Properties and Applications of Zeolites,* The Chemical Society, London, 1980, pp. 92–102.
72. J. P. Ausikaitis in ref. 18, pp. 681–695.
73. J. B. Tedor, T. C. Horch, and T. J. Dangieri, *SAFE J.* **12**, 4–9 (1982).
74. M. Kawai and T. Kaneko, *Gas Sep. Purif.* **3**, 2–6 (1989).
75. R. Kumar and J. K. VanSloun, *Chem. Eng. Progr.* **85**, 34–40 (1989).
76. E. Richter, *Erdoel Kohle, Erdgas, Petrochem.* **40**, 432–438 (1987).
77. A. Kapoor and R. T. Yang, *Chem. Eng. Sci.* **44**, 1723–1733 (1989).
78. S. Sircar and R. Kumar, *Ind. Eng. Chem. Proc. Des. Dev.* **24**, 358–364 (1985).
79. R. T. Yang, *Gas Separation by Adsorption Processes,* Butterworths, Stoneham, Mass., 1987.
80. M. S. Adler and D. R. Johnson, *Chem. Eng. Progr.* **75**, 77–79 (1979).
81. P. J. Clarke, J. E. Gerrard, C. W. Skarstrom, J. Vardi, and D. T. Wade, *SAE Trans.* **76**, 824–842 (1968).
82. D. R. Garg and C. M. Yon, *Chem. Eng. Progr.* **82**, 54–60 (1986).
83. J. P. Ausikaitis, *Glass Dig.* **61**, 69–78 (1982).
84. R. N. Macnair and G. N. Arons in P. N. Cheremisinoff and F. Eleerbusch, eds., *Carbon Adsorption Handbook,* Ann Arbor Science, Ann Arbor, Mich., 1978, 819–859.
85. Ref. 39, p. 421.
86. M. D. Bingham, *Field Detection and Implications of Mercury in Natural Gas, Soc. Petrol. Engrs. Production Engrg.,* 120–124 (May, 1990).
87. J. Markovs, UOP, Tarrytown, N.Y., private communication; U.S. Pat. 4,874,525 (Oct. 17, 1989).
88. J. T. Collins, M. J. Bell, and W. M. Hewitt in A. A. Moghissi and co-workers, eds., *Nuclear Power Waste Technology,* American Society of Mechanical Engineers, New York, 1978, Chapt. 4.
89. D. W. Holladay, *A Literature Survey: Methods for the Removal of Iodine Species from Off-Gases and Liquid Waste Streams of Nuclear Power and Nuclear Fuel Reprocessing Plants, with Emphasis on Solid Sorbents,* Report ORNL/TM-6350 (January, 1979), p. 46, available from National Technical Information Service, Springfield, Va.
90. C. Berg, *Pet. Refiner* **30**(9), 241 (Sept. 1951).
91. "Beaded Carbon Ups Solvent Recovery," *Chem. Eng.* **84**(18) (Aug. 29, 1977).

92. U.S. Pat. 4,526,877 (July 2, 1985), A. Acharya and W. E. BeVier.
93. G. E. Keller, III, *Separations: New Directions for an Old Field* (AIChE Monogr. Ser. 17) American Institute of Chemical Engineers, 1987, p. 83.
94. C. Huang and J. R. Fair, *AIChE J.* **34**, 1861–1877 (1988).
95. S. Sircar and A. L. Myers, *Ads. Sci. Technol.* **2**, 69–87 (1985).
96. M. S. Ray, *Sep. Sci. Tech.* **18**, 95–120 (1983).
97. J. W. Carter in R. P. Townsend, ed., *The Properties and Applications of Zeolites,* The Chemical Society, London, 1980, pp. 76–91.
98. D. D. Do, *AIChE J.* **31**, 1328–1337 (1985).
99. R. J. Grant and M. Manes, *Ind. Eng. Chem. Fundam.* **5**, 490–498 (1966).
100. S. J. Doong and R. T. Yang, *Ind. Eng. Chem. Res.* **27**, 630–635 (1988).
101. G. Gamba, R. Rota, G. Storti, S. Carra, and M. Morbidelli, *AIChE J.* **35**, 959–966 (1989).
102. S. D. Mehta and R. P. Danner, *Ind. End. Chem. Fundam.* **24**, 325–330 (1985).
103. J. A. Ritter and R. T. Yang, *Ind. Eng. Chem. Res.* **26**, 1679–1686 (1987).
104. R. Rota, G. Gamba, R. Paludetto, S. Carra, and M. Morbidelli, *Ind. Eng. Chem. Res.* **27**, 848–851 (1988).
105. A. S. Michaels, *Ind. Eng. Chem.* **44**, 1922–1930 (1952).
106. H. M. Barry, *Chem. Eng.* **67**, 105–120 (1960).
107. J. J. Collins, *Chem. Eng. Progr. Symp. Ser.* **63**, 31–35 (1967).
108. S. L. Forbes and D. W. Underhill, *JAPCA* **36**, 61–64 (1986).
109. R. Mohilla, J. Argelan, and R. Szolcsanyi, *Int. Chem. Eng.* **27**, 723–729 (1987).
110. D. Basmadjian and C. Karayannopoulos, *Ind. Eng. Chem. Proc. Des. Dev.* **24**, 140–149 (1985).
111. C. L. Humphries, *Hydrocarbon Process.* **45**, 88–95 (1966).
112. P. I. Cen and R. T. Yang, *AIChE J.* **32**, 1635–1641 (1986).
113. H. Hori, I. Tanaka, and T. Akiyama, *JAPCA* **38**, 269–271 (1988).
114. H. Lee and W. P. Cummings, *Chem. Eng. Progr. Symp. Series* **63**(74), 42–49 (1967).
115. L. C. Eagleton and H. Bliss, *Chem. Eng. Progr.* **49**, 543–548 (1953).
116. J. W. Carter, *Br. Chem. Eng.* **14**, 303–306 (1969).
117. O. Grubner and W. A. Burgess, *Environ Sci. Technol.* **15**, 1346–1351 (1981).
118. Y. E. Yoon and J. H. Nelson, *Am. Ind. Hyg. Assoc. J.* **45**, 517–524 (1984).
119. L. A. Jonas and J. A. Rehrmann, *Carbon* **11**, 59–64 (1973).
120. C. W. Chi and H. Lee, *AIChE Symp. Ser.* **69**, 95–101 (1973).
121. O. A. Hougen and W. K. Marshall, *Chem. Eng. Progr.* **43**, 197–208 (1947).
122. C. C. Furnas, *Trans. Am. Inst. Chem. Eng.* **24**, 142–193 (1930).
123. G. M. Lukchis, *Chem. Eng.* **80**, (13), 111–116; (16), 83–87; (18), 83–90 (1973).
124. C. W. Chi, *AIChE Symp. Ser.* **74**, 42–46 (1977).
125. R. P. Underwood, *Chem. Eng. Sci.* **41**, 409–411 (1986).
126. R. Kumar and G. L. Dissinger, *Ind. Eng. Chem. Proc. Des. Dev.* **25**, 456–464 (1986).
127. I. Zwiebel, R. L. Gariepy, and J. J. Schnitzer, *AIChE J.* **18**, 1139–1147 (1972); **20**, 915–923 (1974).
128. A. Jedrzejak and M. Paderewski, *Int. Chem. Eng.* **28**, 707–712 (1988).
129. J. L. Bravo, Report of DOE Contract No. DE-AS07-831D12473, 1984, 150–181.
130. J. M. Schork and J. R. Fair, *Ind. Eng. Chem. Res.* **27**, 457–469 (1988).
131. D. Basmadjian, K. D. Ha, and C. Y. Pan, *Ind. Eng. Chem. Proc. Des. Dev.* **14**, 328–340 (1975).
132. C. Huang and J. R. Fair, *AIChE J.* **35**, 1667–1677 (1989).
133. M. M. Davis and M. D. LeVan, *Ind. Eng. Chem. Res.* **28**, 778–785 (1989).
134. A. Jedrzejak, *Chem. Eng. Technol.* **11**, 352–358 (1988).
135. D. H. White, Jr., and P. G. Barkley, *Chem. Eng. Progr.* **85**, 25–33 (1989).
136. K. Chihara and M. Suzuki, *J. Chem. Eng. Japan* **16**, 293–299 (1983).

137. J. W. Carter and M. L. Wyszynski, *Chem. Eng. Sci.* **38,** 1093–1099 (1983).

138. R. Kumar, *Ind. Eng. Chem. Res.* **28,** 1677–1683 (1989).

139. J. L. Liow and C. N. Kenney, *AIChE J.* **36,** 53–65 (1990).

140. H. Shin and K. S. Knaebel, *AIChE J.* **34,** 1409–1416 (1988).

141. P. H. Turnock and R. H. Kadlec, *AIChE J.* **17,** 335–342 (1971).

142. S. J. Doong and R. T. Yang, *AIChE Symp. Ser.* **84,** 145–154 (1988).

143. S. Ergun, *Chem. Eng. Progr.* **48,** 89–94 (1952).

144. M. Leva, *Chem. Eng.* **56,** 115–117 (1949).

145. L. E. Brownell, H. S. Dombrowski, and C. A. Dickey, *Chem. Eng. Progr.* **46,** 415–422 (1950).

JOHN D. SHERMAN
CARMEN M. YON
UOP

ADSORPTION, LIQUID SEPARATION

Nearly every chemical manufacturing operation requires the use of separation processes to recover and purify the desired product. In most circumstances, the efficiency of the separation process has a significant impact on both the quality and the cost of the product (1). Liquid-phase adsorption has long been used for the removal of contaminants present at low concentrations in process streams. In most cases, the objective is to remove a specific feed component; alternatively, the contaminants are not well defined, and the objective is the improvement of feed quality defined by color, taste, odor, and storage stability (2–5) (see WASTES, INDUSTRIAL; WATER, INDUSTRIAL WATER TREATMENT).

In contrast to trace impurity removal, the use of adsorption for bulk separation in the liquid phase on a commercial scale is a relatively recent development. The first commercial operation occurred in 1964 with the advent of the UOP Molex process for recovery of high purity n-paraffins (6–8). Since that time, bulk adsorptive separation of liquids has been used to solve a broad range of problems, including individual isomer separations and class separations. The commercial availability of synthetic molecular sieves and ion-exchange resins and the development of novel process concepts have been the two significant factors in the success of these processes. This article is devoted mainly to the theory and operation of these liquid-phase bulk adsorptive separation processes.

Adsorbate–Adsorbent Interactions

An adsorbent can be visualized as a porous solid having certain characteristics. When the solid is immersed in a liquid mixture, the pores fill with liquid, which at equilibrium differs in composition from that of the liquid surrounding the particles. These compositions can then be related to each other by enrichment factors that are analogous to relative volatility in distillation. The adsorbent is selective for the component that is more concentrated in the pores than in the surrounding liquid.

The choice of separation method to be applied to a particular system depends largely on the phase relations that can be developed by using various separative agents. Adsorption is usually considered to be a more complex operation than is the use of selective solvents in liquid–liquid extraction (see EXTRACTION, LIQUID–LIQUID), extractive distillation, or azeotropic distillation (see DISTILLATION, AZEOTROPIC AND EXTRACTIVE). Consequently, adsorption is employed when it achieves higher selectivities than those obtained with solvents.

A significant advantage of adsorbents over other separative agents lies in the fact that favorable equilibrium-phase relations can be developed for particular separations; adsorbents can be produced that are much more selective in their affinity for various substances than are any known solvents. This selectivity is particularly true of the synthetic crystalline zeolites containing exchangeable cations. These zeolites became available in the early 1960s under the name of molecular sieves (qv) (9).

An example of unique selectivity is provided by the use of 5A molecular sieves for the separation of linear hydrocarbons from branched and cyclic types. In this system only the linear molecules can enter the pores; others are completely excluded because of their larger cross section. Thus the selectivity for linear molecules with respect to other types is infinite. In the more usual case, all the feed components access the selective pores, but some components of the mixture are adsorbed more strongly than others. A selectivity between the different components that can be used to accomplish separation is thus established.

Another example of unique selectivities is the separation of olefins from paraffins in feed mixtures containing about five successive molecular sizes, eg, C_{10} to C_{14}. Liquid–liquid extraction might be considered for this separation. However, polar solvents give solubility patterns of the type shown in Figure 1. Each olefin is more soluble than the paraffin of the same chain length, but the solubility of both species declines as chain length increases. Thus, in a broad-

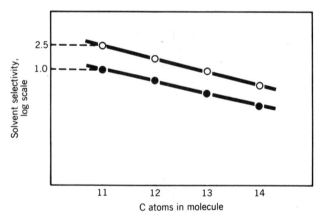

Fig. 1. Liquid–liquid extraction selectivity: ○, olefins; ●, paraffins.

boiling mixture, solubilities of paraffins and olefins overlap and separation becomes impossible. In contrast, the relative adsorption of olefins and paraffins from the liquid phase on the adsorbent used commercially for this operation is shown in Figure 2. Not only is there selectivity between an olefin and paraffin of the same chain length, but also chain length has little effect on selectivity. Consequently, the complete separation of olefins from paraffins becomes possible.

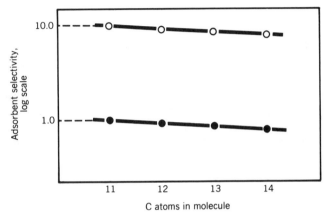

Fig. 2.   Liquid-phase selectivity of UOP Olex adsorbent: ○, olefins; ●, paraffins.

Unique adsorption selectivities are employed in the separation of C_8 aromatic isomers, a classical problem that cannot be easily solved by distillation, crystallization, or solvent extraction (10). Although p-xylene [106-42-3] can be separated by crystallization, its recovery is limited because of the formation of eutectic with m-xylene [108-58-3]. However, either p-xylene, m-xylene, o-xylene [95-47-6], or ethylbenzene [100-41-4] can be extracted selectively by suitable modification of zeolitic adsorbents.

Literature dealing with adsorbent–adsorbate interactions in liquid phase is largely confined to patents (11–43). Although theoretical consistency tests exist for such data (44), the search for an adsorbent of suitable selectivity remains an art.

Practical Adsorbents

The search for a suitable adsorbent is generally the first step in the development of an adsorption process. A practical adsorbent has four primary requirements: selectivity, capacity, mass transfer rate, and long-term stability. The requirement for adequate adsorptive capacity restricts the choice of adsorbents to microporous solids with pore diameters ranging from a few tenths to a few tens of nanometers.

Traditional adsorbents such as silica [7631-86-9], SiO_2; activated alumina [1318-23-6], Al_2O_3; and activated carbon [7440-44-0], C, exhibit large surface areas and micropore volumes. The surface chemical properties of these adsorbents make them potentially useful for separations by molecular class. However, the micropore size distribution is fairly broad for these materials (45). This characteristic makes them unsuitable for use in separations in which steric hindrance can potentially be exploited (see ALUMINUM COMPOUNDS, ALUMINA; SILICON COMPOUNDS, SYNTHETIC INORGANIC SILICATES).

Typical nonsieve, polar adsorbents are silica gel and activated alumina. Equilibrium data have been published on many systems (11–16,46,47). The order of affinity for various chemical species is: saturated hydrocarbons < aromatic hydrocarbons = halogenated hydrocarbons < ethers = esters = ketones < amines = alcohols < carboxylic acids. In general, the selectivities are parallel to those obtained by the use of selective polar solvents; in hydrocarbon systems, even the magnitudes are similar. Consequently, the commercial use of these adsorbents must compete with solvent-extraction techniques.

The principal nonpolar-type adsorbent is activated carbon. Equilibrium data have been reported on hydrocarbon systems, various organic compounds in water, and mixtures of organic compounds (11,15,16,46,47). With some exceptions, the least polar component of a mixture is selectively adsorbed; eg, paraffins are adsorbed selectively relative to olefins of the same carbon number, but dicyclic aromatics are adsorbed selectively relative to monocyclic aromatics of the same carbon number (see CARBON, ACTIVATED CARBON).

Polymeric resins [*81133-25-7*], $-(C_{10}H_{10})_n-$, are widely used in the food and pharmaceutical industries as cation–anion exchangers for the removal of trace components and for some bulk separations, such as fructose from glucose (48). These resins are primarily attractive for aqueous-phase separations and offer a fairly wide potential range of surface chemistries to fit a number of separation needs. For example, polymeric resins are effective in partitioning by size and molecular weight and may also be effective in ion exclusion (see ION EXCHANGE).

In contrast to these adsorbents, zeolites offer increased possibilities for exploiting molecular-level differences among adsorbates. Zeolites are crystalline aluminosilicates containing an assemblage of SiO_4 and AlO_4 tetrahedra joined together by oxygen atoms to form a microporous solid, which has a precise pore structure (49). Nearly 40 distinct framework structures have been identified to date. Table 1 and Figure 3 summarize some of those structures that have been widely used in the chemical industry. The versatility of zeolites lies in the fact that widely different adsorptive properties may be realized by the appropriate control of the framework structure, the silica-to-alumina ratio (Si/Al), and the cation form. For example, zeolite A, shown in Figure 4, has a three-dimensional isotropic channel structure constricted by an eight-membered oxygen ring. Its effective pore size can be controlled at about 0.3, 0.4, and 0.45 nm by exchanging

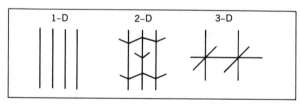

Fig. 3.   Schematic diagram of molecular sieve pore structure. See Table 1.

Table 1. Molecular Sieve Pore Structures

Common name	Ring size, number of atoms	Free aperture, nm	Pore structure[a]	CAS Registry Number	Formula
faujasite	12	0.74	3-D	[12173-28-3]	$(Ca, Mg, Na_2, K_2)_{29.5}[(AlO_2)_{59}(SiO_2)_{133}] \cdot 235H_2O$
mordenite	8	0.29 × 0.57	1-D	[12173-98-7]	$Na_8[(AlO_2)_8(SiO_2)_{40}] \cdot 24H_2O$
	12	0.67 × 0.7	1-D		
L	12	0.71	1-D		$K_9[(AlO_2)_9(SiO_2)_{27}] \cdot 22H_2O$
ZSM-5	10	0.54 × 0.56	1-D	[58339-99-4]	$(Na, TPA^b)_3 [(AlO_2)_3(SiO_2)_{93} \cdot 16H_2O]$
	10	0.51 × 0.56	1-D		
Erionite	8	0.36 × 0.52	2-D	[12150-42-8]	$(Ca, Mg, Na_2, K_2)_{4.5}[(AlO_2)_9(SiO_2)_{27}] \cdot 27H_2O$
A	8	0.42	3-D		$Na_{12} [(AlO_2)_{12}(SiO_2)_{12}] \cdot 27H_2O$

[a]See Figure 3.
[b]TPA = tetrapropylammonium.

577

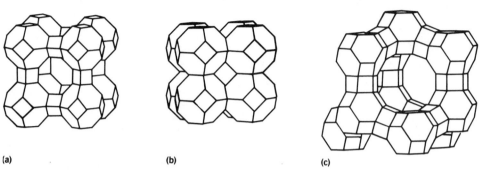

(a)   (b)   (c)

Fig. 4. Three zeolites with the same structural polyhedron, cubo-octahedrons. (a) Type A, $Na_{12}[(AlO_2)_{12}(SiO_2)_{12}]\cdot 27H_2O$; (b) sodalite [*1302-90-5*]; (c) faujasite (Type X, Y), where X = $Na_{86}[(AlO_2)_{86}(SiO_2)_{106}]\cdot 264H_2O$; Y = $Na_{56}[(AlO_2)_{56}(SiO_2)_{136}]\cdot 250H_2O$

with potassium, sodium, and calcium, respectively. The potassium form, with 0.3-nm pores, is used for removing water from olefinic hydrocarbons. The sodium form can be used to efficiently remove water from nonreactive hydrocarbons, such as alkanes. The substitution of calcium can provide a pore size that will admit *n*-paraffins and exclude other hydrocarbons.

Large-pore zeolites, X, Y, and mordenites, have pores defined by 12-membered oxygen rings with a free diameter of 0.74 nm. The framework structure of X and Y faujasites sketched in Figure 4 consists of a total of 192 SiO_2 and AlO_2 units. The Si/Al (atomic) ratio for X is generally between 1.0 and 1.5, whereas for Y it is between 1.5 and 3.0. With suitable procedures, Y can be dealuminated to make ultrastable Y with Si/Al ratios exceeding 100. Adsorption properties of faujasites are strongly dependent on not only the cation form, but also the Si/Al ratio. The flexibility provided by faujasites in the adsorption of C_8 aromatics is shown in Table 2. The selectivity order, from the most selectively adsorbed to the least selectively adsorbed, can be changed significantly by the choice of zeolite properties.

Table 2. Selectivity of Zeolites in C_8H_{10} Aromatic Systems

	Adsorbent[a]			
	No. 1	No. 2	No. 3	No. 4
p-xylene	1	2	3	4
ethylbenzene	2	1	4	3
m-xylene	3	3	1	2
o-xylene	4	4	2	1

[a]Key: 1 = most selectively adsorbed, 4 = least selectively adsorbed.

The separation of fructose from glucose illustrates the interaction between the framework structure and the cation (Fig. 5) (50). Ca^{2+} is known to form complexes with sugar molecules such as fructose. Thus, Ca–Y shows a high selectivity for fructose over glucose. However, Ca–X does not exhibit high selectivity. On the other hand, K–X shows selectivity for glucose over fructose. This

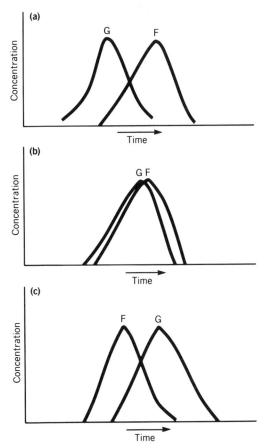

Fig. 5.   Fructose–glucose separation on faujasite adsorbents. (**a**) Ca–Y adsorbent; (**b**) Ca–X adsorbent; (**c**) K–X adsorbent.

polar nature of faujasites and their unique shape-selective properties, more than the molecular-sieving properties, make them most useful as practical adsorbents.

Polymeric cation-exchange resins are also used in the separation of fructose from glucose. The UOP Sarex process has employed both zeolitic and polymeric resin adsorbents for the production of high fructose corn syrup (HFCS). The operating characteristics of these two adsorbents are substantially different and have been compared in terms of fundamental characteristics such as capacity, selectivity, and adsorption kinetics (51).

The zeolite and the resin adsorbents show different adsorption isotherm characteristics, particularly at higher concentration (51). The resin adsorbent isotherm is slightly concave upward, whereas the zeolite isotherm is linear, or even slightly concave downward. Resins, therefore, have an advantage in a UOP Sarex operation that involves high feed-solids concentration.

In addition to the fundamental parameters of selectivity, capacity, and mass-transfer rate, other more practical factors, namely, pressure drop characteristics and adsorbent life, play an important part in the commercial viability of a practical adsorbent.

Pressure Drop Characteristics.   Ion-exchange resins are compressible and exhibit a characteristic stress–strain relationship, as shown in Figure 6. In addition, they undergo shrinking and swelling as a result of osmotic pressure variation resulting from concentration changes. Zeolite adsorbents are rigid and do not exhibit much strain with pressure. When resins are used as adsorbents, the implications of shrinking and swelling and compressibility must be considered to ensure safe operation below the design pressure drop. The pressure drop can often become a major factor in the determination of maximum throughput.

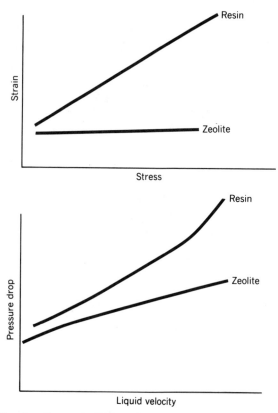

Fig. 6.   Compressibility characteristics of adsorbents.

Adsorbent Life.   Long term stability under rugged operating conditions is an important characteristic of an adsorbent. By their nature zeolites are not stable in an aqueous environment and must be specially formulated to enhance their stability in order to obtain several years of service. Polymeric resins do not suffer from dissolution problems. However, they are prone to chemical attack (52).

Commercial Processes

Industrial-scale adsorption processes can be classified as batch or continuous (53,54). In a batch process, the adsorbent bed is saturated and regenerated in a cyclic operation. In a continuous process, a countercurrent staged contact between the adsorbent and the feed and desorbent is established by either a true or a simulated recirculation of the adsorbent.

The efficiency of an adsorption process is significantly higher in a continuous mode of operation than in a cyclic batch mode (55). In a batch chromatographic operation, the liquid composition at a given level in the bed undergoes a cyclic change with time, and large portions of the bed do not perform any useful function at a given time. In continuous operation, the composition at a given level is invariant with time, and every part of the bed performs a useful function at all times. The height equivalent of a theoretical plate (HETP) in a batch operation is roughly three times that in a continuous mode. For difficult separations, batch operation may require 25 times more adsorbent inventory and twice the desorbent circulation rate than does a continuous operation. In addition, in a batch mode, the four functions of adsorption, purification, desorption, and displacement of the desorbent from the adsorbent are inflexibly linked, whereas a continuous mode allows more degrees of freedom with respect to these functions, and thus a better overall operation.

Continuous Countercurrent Processes

The need for a continuous countercurrent process arises because the selectivity of available adsorbents in a number of commercially important separations is not high. In the *p*-xylene system, for instance, if the liquid around the adsorbent particles contains 1% *p*-xylene, the liquid in the pores contains about 2% *p*-xylene at equilibrium. Therefore, one stage of contacting cannot provide a good separation, and multistage contacting must be provided in the same way that multiple trays are required in fractionating materials with relatively low volatilities.

The hypersorption process (56) developed by Union Oil Company in the early 1950s for the recovery of propane and heavier components from natural gas is the earliest example of large-scale countercurrent adsorption processes. This process used an activated carbon adsorbent flowing as a dense bed continuously downward through a rising gas stream. A unit built for the Dow Chemical Company had a gas capacity of more than 500,000 m^3/day. However, this process proved to be less economical than cryogenic distillation and is no longer in operation. A number of commercial moving-bed designs exist mainly for ion exchange. A good review of these designs can be found in reference 57.

Since the 1960s the commercial development of continuous countercurrent processes has been almost entirely accomplished by using a flow scheme that simulates the continuous countercurrent flow of adsorbent and process liquid without the actual movement of the adsorbent. The idea of a simulated moving bed (SMB) can be traced back to the Shanks system for leaching soda ash (58).

Such a concept was originally used in a process developed and licensed by UOP under the name UOP Sorbex (59,60). Other versions of the SMB system are also used commercially (61). Toray Industries built the Aromax process for the production of *p*-xylene (20,62,63). Illinois Water Treatment and Mitsubishi have commercialized SMB processes for the separation of fructose from dextrose (64–66). The following discussion is based on the UOP Sorbex process.

MOVING-BED OPERATION

A hypothetical moving-bed system and a liquid-phase composition profile are shown in Figure 7. The adsorbent circulates continuously as a dense bed in a closed cycle and moves up the adsorbent chamber from bottom to top. Liquid streams flow down through the bed countercurrently to the solid. The feed is assumed to be a binary mixture of A and B, with component A being adsorbed selectively. Feed is introduced to the bed as shown.

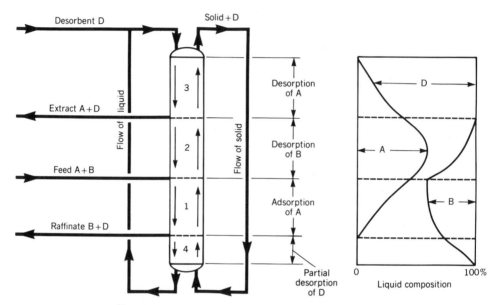

Fig. 7. Adsorptive separation with moving bed.

Desorbent D is introduced to the bed at a higher level. This desorbent is a liquid of different boiling point from the feed components and can displace feed components from the pores. Conversely, feed components can displace desorbent from the pores with proper adjustment of relative flow rates of solid and liquid.

Raffinate product, consisting of the less strongly adsorbed component B mixed with desorbent, is withdrawn from a position below the feed entry. Only a portion of the liquid flowing in the bed is withdrawn at this point; the remainder continues to flow into the next section of the bed. Extract product, consisting of the more strongly adsorbed component A mixed with desorbent, is withdrawn from the bed; again, only a portion of the flowing liquid in the bed is withdrawn, and the remainder continues to flow into the next bed section.

The positions of introduction and withdrawal of net streams divide the bed into four zones, each of which performs a different function as described below.

Zone 1. The primary function of this zone is to adsorb A from the liquid. The solid entering at the bottom carries only B and D in its pores. As the liquid stream flows downward, countercurrent to this solid, component A is transferred from the liquid stream into the pores of the solid. At the same time, component D is desorbed (transferred from the pores to the liquid stream) to make room for A.

Zone 2. The primary function of this zone is to remove B from the pores of the solid. When the solid arrives at the fresh feed point, the pores contain the quantity of A that was adsorbed in Zone 1. However, the pores also contain a large quantity of B, because the solid has just been in contact with fresh feed. The liquid entering the top of Zone 2 contains no B, only A and D. As the solid moves upward, countercurrent to this stream, B is gradually displaced from the pores and is replaced by A and D. Thus, when the solid arrives at the top of Zone 2, the pores contain only A and D. By proper regulation of the liquid rate in Zone 2, B can be desorbed completely from the pores. This B desorption can be accomplished without simultaneously desorbing all of A, because A is more strongly adsorbed than B.

Zone 3. The function of this zone is to desorb A from the pores. The solid entering the zone carries A and D in the pores; the liquid entering the top of the zone consists of pure D. As the solid rises, A in the pores is displaced by D.

Zone 4. The purpose of this zone is to act as a buffer to prevent component B, which is at the bottom of Zone 1, from passing into Zone 3, where it would contaminate extracted component A. When the adsorbent leaves Zone 3, the pores are completely filled with desorbent. The liquid entering the top of Zone 4 is of raffinate composition and contains B and D. If the flow rate in Zone 4 is properly regulated, component B will be readsorbed completely from the liquid, preventing its entry into Zone 3, where it would contaminate the product A.

Difficulties of Moving-Bed Operation. The use of a moving bed introduces the problem of mechanical erosion of the adsorbent. Obtaining uniform flow of both solid and liquid in beds of large diameter is also difficult. The performance of this type of operation can be greatly impaired by nonuniform flow of either phase.

The use of a series of fluidized beds may be considered when solid overflows from each bed to the next. However, this arrangement involves a sacrifice in mass-transfer efficiency because the number of theoretical equilibrium trays cannot exceed the number of physical beds. In contrast, the flow through dense and fixed beds of adsorbent, as practiced in chromatography, can provide hundreds of theoretical trays in beds of modest length. Another disadvantage of a fluidized-bed operation is the large-sized equipment required to contain a given inventory of adsorbent (see FLUIDIZATION; MASS TRANSFER; REACTOR TECHNOLOGY). In view of these difficulties, only a few fluidized-bed operations are practiced commercially. The Purasiv HR system using beaded activated carbon for the recovery of solvent is one example. This process uses a staged fluidized bed for adsorption and a moving bed for regeneration (67).

SIMULATED MOVING-BED OPERATION

In the moving-bed system of Figure 7, solid is moving continuously in a closed circuit past fixed points of introduction and withdrawal of liquid. The same results can be obtained by holding the bed stationary and periodically moving the positions at which the various streams enter and leave. A shift in the positions of the introduction of the liquid feed and the withdrawal in the direction of fluid flow through the bed simulates the movement of solid in the opposite direction.

Of course, moving the liquid feed and withdrawal positions continuously is impractical. However, approximately the same effect can be produced by providing multiple liquid-access lines to the bed and periodically switching each stream to the adjacent line. Functionally, the adsorbent bed has no top or bottom and is equivalent to an annular bed. Therefore, the four liquid-access positions can be moved around the bed continually, always maintaining the same distance between the various streams.

The commercial application of this concept (68) is portrayed in Figure 8, which shows the adsorbent as a stationary bed. A liquid circulating pump is provided to pump liquid from the bottom outlet to the top inlet of the adsorbent chamber. A fluid-directing device known as a rotary valve (69,70) is provided. The rotary valve functions on the same principle as a multiport stopcock in directing each of several streams to different lines. At the right-hand face of the valve, the four streams to and from the process are continuously fed and withdrawn. At the left-hand face of the valve, a number of lines are connected that terminate in distributors within the adsorbent bed.

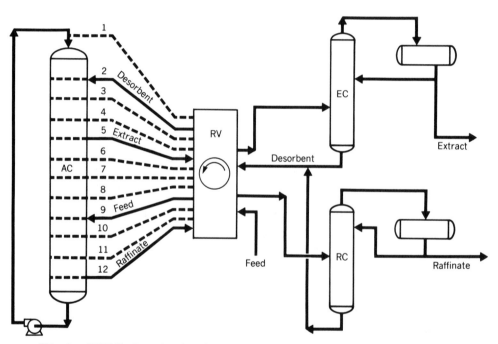

Fig. 8.   UOP Sorbex simulated moving bed for adsorptive separation. AC = adsorbent chamber, RV = rotary valve, EC = extract column, RC = raffinate column.

At any particular moment, only four lines from the rotary valve to the adsorbent chamber are active. Figure 8 shows the flows at a time when lines 2, 5, 9, and 12 are active. When the rotating element of the rotary valve is moved to its next position, each net flow is transferred to the adjacent line; thus, desorbent enters line 3 instead of line 2, extract is drawn from 6 instead of 5, feed enters 10 instead of 9, and raffinate is drawn from 1 instead of 12.

Figure 7 shows that in the moving-bed operation, the liquid flow rate in each of the four zones is different because of the addition or withdrawal of the various steams. In the simulated moving-bed of Figure 8, the liquid flow rate is controlled by the circulating pump. At the position shown in Figure 8, the pump is between the raffinate and desorbent ports, and therefore should be pumping at a rate appropriate for Zone 4. However, after the next switch in position of the rotary valve, the pump is between the feed and raffinate ports, and should therefore be pumping at a rate appropriate for Zone 1. Stated briefly, the circulating pump must be programmed to pump at four different rates. The control point is altered each time an external stream is transferred from line 12 to line 1.

To complete the simulation, the liquid-flow rate relative to the solid must be the same in both the moving-bed and simulated moving-bed operations. Because the solid is physically stationary in the simulated moving-bed operation, the liquid velocity relative to the vessel wall must be higher than in an actual moving-bed operation.

The primary control variables at a fixed feed rate, as in the operation pictured in Figure 8, are the cycle time, which is measured by the time required for one complete rotation of the rotary valve (this rotation is the analog of adsorbent circulation rate in an actual moving-bed system), and the liquid flow rate in Zones 2, 3, and 4. When these control variables are specified, all other net rates to and from the bed and the sequence of rates required at the liquid circulating pump are fixed. An analysis of sequential samples taken at the liquid circulating pump can trace the composition profile in the entire bed. This profile provides a guide to any changes in flow rates required to maintain proper performance before any significant effect on composition of the products has appeared. Various aspects of process control are described in the patent literature (71–73).

Temperature and pressure are not considered as primary operating variables: temperature is set sufficiently high to achieve rapid mass-transfer rates, and pressure is sufficiently high to avoid vaporization. In liquid-phase operation, as contrasted to vapor-phase operation, the required bed temperature bears no relation to the boiling range of the feed, an advantage when heat-sensitive stocks are being treated.

MODELING OF UOP SORBEX SYSTEMS

The theoretical performance of the commercial simulated moving-bed operation is practically identical to that of a system in which solids flow continuously as a dense bed countercurrent to liquid. A model in which the flows of solid and liquid are continuous, as shown in Figure 7, is therefore adequate.

The operation is modeled in terms of theoretical equilibrium trays having the same significance as in fractionating columns (see DISTILLATION). Solid and liquid are assumed to flow continuously through hypothetical well-mixed theoretical trays in which equilibrium is attained. The number of theoretical trays has no relation to the number of bed segments in the actual operation. Each segment can be equivalent to many theoretical trays. The number is determined by bed height, mass-transfer coefficient, and flow rates. Axial mixing is generally of much greater significance in liquid than in vapor systems because of the greater mass of process fluid in the voids relative to that in the pores. To allow for the effect of axial mixing in the liquid phase, a second parameter is introduced into the model by assuming that the solid entrains a certain quantity of liquid from deck to deck.

The relationship of this type of model to a true differential analysis has been discussed for the case of linear equilibrium and first-order kinetics (74,75). A minor extension of this work leads to the following relations for a bed section in which flow rates of solid and liquid are constant. For the number of theoretical trays, n,

$$n = KkH/Lz \tag{1}$$

where K is the adsorption equilibrium constant; k the mass-transfer coefficient; H the bed height; L the net liquid-flow rate; and $z = j\ln[j/(j-1)]$, where j is an integer. The entrainment factor per unit liquid flow rate is given by

$$e = KkD°/L^2 \tag{2}$$

where $D°$ is the axial diffusion coefficient. Equations 1 and 2 are approximations that apply for efficient beds at reasonable operating conditions. These equations appear to be ambiguous because the values of n and e are different, depending on the component being considered. However, this ambiguity is not of great practical significance because each zone of the system is performing a function that is critical with respect to only one component. Values of n and e corresponding to the equilibrium properties of the critical component in each zone should be used.

The model of theoretical equilibrium trays with entrainment is readily treated by computer with methods analogous to those used for the design of fractionating columns.

McCABE-THIELE DIAGRAM

The McCabe-Thiele approach has been developed to describe the Sorbex process (76). Two feed components, A and B, with a suitable adsorbent and a desorbent, C, are separated in an isothermal continuous countercurrent operation. If A is the more strongly adsorbed component and the system is linear and noninteracting, the flows in each section of the process must satisfy the following constraints for complete separation of A from B:

Section	Condition
IV	$S/(D + F - E - R) > K_{CB}$
I	$S/(D + F - E) > K_{BA}$
II	$S/(D - E) < K_{AB}$
III	$S/D < K_{CA}$

The required direction of the net flow of each component is illustrated in Figure 9. The UOP Sorbex process has four flow-rate variables (S/F, D/F, E/F, and R/F) and four inequality constraints, one for each section of the bed. Once the equilibrium is fixed, the only remaining degree of freedom is the margin by which the inequality constraints are fulfilled. Once that is decided, the inequality constraints become four equations that define all flow-rate ratios for the system. Once the flow rates are fixed, a preliminary estimate of the number of theoretical stages in each section may be obtained by a McCabe-Thiele diagram, shown in Figure 10.

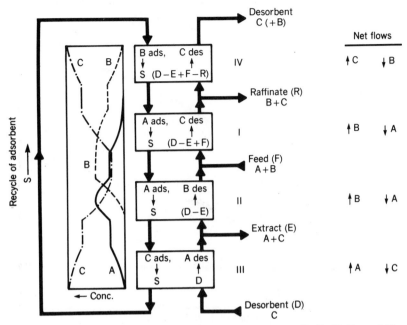

Fig. 9. Schematic diagram of a UOP Sorbex process. D, E, F, R, and S represent flow rates for desorbent, extract, feed, raffinate, and net solids, respectively.

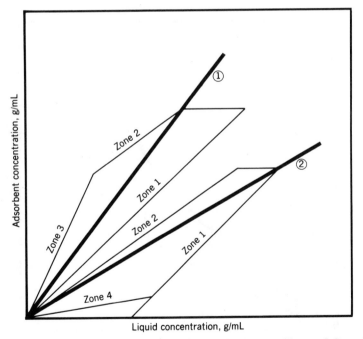

Fig. 10. UOP Sorbex operation with linear isotherms. Slope of ① $= K_1$; slope of ② $= K_2$. Conditions for separation: $K_1 > K_2$, $L_3/S \geq K_1$, $K_2 \leq L_2/S \leq K_1$, $K_2 \leq L_1/S \leq K_1$, $L_4/S \leq K_2$, $L_1 - L_2 = F$; where $K =$ adsorption coefficient, $L =$ net liquid flow rate, $S =$ net solids flow rate, and $F =$ feed flow rate.

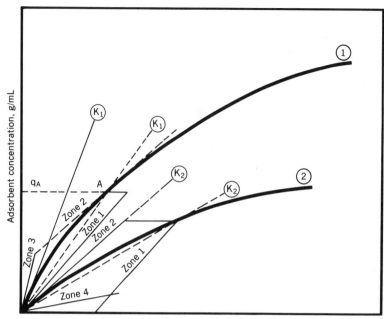

Fig. 11. UOP Sorbex operation with convex isotherms. Conditions for separation: at point A, slope of ① $= K_2$, $K_1 > K_2$, $L_3/S \geq K_1$, $K_2 \leq L_2/S \leq K_T \leq K_1$, $K_2' \leq L_1/S \leq K_1'$, $L_4/S \leq K_2'$, $L_1 - L_2 = F$; where K, L, S, and F are as defined in Figure 10.

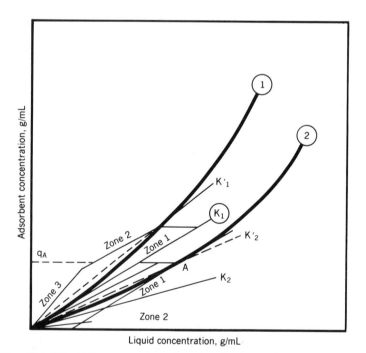

Fig. 12. Conditions for separation: At point A, slope of ② $= K_1$, $K_1 > K_2$, $L_3/S > K_1'$, $K_2 \leq K_T \leq L_1/S \leq K_1$, $K_2' \leq L_2/S \leq K_1'$, $L_4/S \leq K_2'$; where K, L, and S are as defined in Figure 10. Adsorbent concentration of component 2 less than q_A.

McCabe-Thiele diagrams for nonlinear and more practical systems with pertinent inequality constraints are illustrated in Figures 11 and 12. The convex isotherms are generally observed for zeolitic adsorbents, particularly in hydrocarbon separation systems, whereas the concave isotherms are observed for ion-exchange resins used in sugar separations.

UOP Sorbex Applications

The first UOP Sorbex process was licensed in 1962 as a UOP Molex process to separate n-paraffins from branched paraffins, cyclic paraffins, and aromatics. This plant started up in 1964, and its products were used for the manufacture of biodegradable detergents. Since then, about 80 units have been put on-stream in a variety of applications that produce in excess of 8 million t/yr of products. The extent of commercialization, as of July 1989, is shown in Table 3. The problems associated with large-scale commercial operation have been solved, as is evidenced by the satisfactory operation of a single p-xylene unit with a capacity of nearly 400,000 t/yr. The performance of the rotary valves has been excellent, and they have required only routine maintenance. The largest adsorbent bed in commercial use has a diameter of 6.7 m and is performing as well as a pilot plant with a bed diameter of 0.1 m. This performance is achieved by paying particular attention to the internal design of large adsorbent chambers to ensure good distribution and uniform flow of liquids (77–79). Evolutionary improvements in the UOP Sorbex process design have reduced product impurities from tenths of a percent to a few hundred parts per million. The following examples show various applications of continuous bulk adsorptive separation processes operating in liquid phase.

Table 3. UOP Sorbex Processes for Commodity Chemicals

UOP Processes	Separation	Licensed units
Parex	p-xylene from C_8 aromatics	44
Molex	n-paraffins from branched and cyclic hydrocarbons	21
Olex	olefins from paraffins	6
Cymex	p- or m-cymene from cymene isomers	1
Cresex	p- or m-cresol from cresol isomers	1
Sarex	fructose from dextrose plus polysaccharides	5
total		*78*

n-Paraffin Separation. The UOP Molex process is used to separate n-paraffins from branched paraffins and aromatics in a variety of applications. In the C_5–C_6 range, the Molex process is used to enhance the octane number of gasoline. As illustrated in Figure 13, the UOP Molex process is integrated with a catalytic isomerization process, UOP Penex, to achieve maximum isomerization octane (80). This octane upgrade is a result of the separation of the low octane n-paraffins from the high octane branched paraffins and the recycle of the

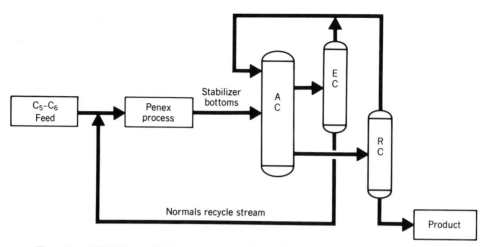

Fig. 13. UOP Penex-Molex process. AC = adsorbent chamber, EC = extract column, RC = raffinate column.

n-paraffins to UOP Penex for further conversion. The magnitude of the octane increase for the UOP Penex-Molex process over the once-through UOP Penex process depends on the C_5–C_6 content of the feed. Typically, the product octanes are 4 to 5 numbers higher, for instance, 83.5 RON (research octane number) for the UOP Penex process and 88.5 RON for the UOP Penex-Molex process.

Linear paraffins in the C_{10} to C_{15} range are used for the production of alcohols and plasticizers and biodegradable detergents of the linear alkylbenzene sulfonate and nonionic types (see ALCOHOLS; PLASTICIZERS; SURFACTANTS). Here the UOP Molex process is used to extract n-paraffins from a hydrotreated kerosine (6–8).

The fermentation of n-paraffins in the C_{10} to C_{23} range for protein production has provided a new outlet for these hydrocarbons (see FOODS, NONCONVENTIONAL). Because it operates in liquid phase, the UOP Molex process can readily accomplish the separation of n-paraffins from such a wide boiling feedstock.

Olefin–Paraffin Separation.   The catalytic dehydrogenation of n-paraffins offers a route to the commercial production of linear olefins. Because of limitations imposed by equilibrium and side reactions, conversion is incomplete. Therefore, to obtain a concentrated olefin product, the olefins must be separated from the reactor effluent (81–85), and the unreacted n-paraffins must be recycled to the catalytic reactor for further conversion.

The performance of the adsorptive section of such a combination in a commercial installation is shown in Table 4. The feedstock includes C_{11}–C_{14} components, and olefins are recovered at about 94% efficiency.

The olefin product contains 1.1% of residual n-paraffins. Essentially similar results have been obtained in commercial operations on C_8–C_{10} and C_{15}–C_{18} feedstocks. The desorbents used are generally hydrocarbon mixtures of lower boiling range than the feed components. The concentrated olefin stream may then be used for production of detergent alcohols.

p-Xylene Separation. p-Xylene finds wide use as a precursor for the production of polyester fibers and plastics. Before the advent of adsorptive tech-

Table 4. Commercial Operation of Linear C_{11}–C_{14} Olefin Extraction

Component	Feed, wt %	Extract, wt %	Raffinate, wt %
n-olefins	9.0	96.2	0.6
n-paraffins	90.1	1.1	98.5
other components[a]	0.9	2.7	0.9
total	*100.0*	*100.0*	*100.0*

[a]Including aromatics and branched-chain aliphatics.

niques, the p-xylene was commonly separated from hydrocarbon mixtures by crystallization. However, recovery was limited to 55 to 60% because of the formation of eutectic mixtures.

Since 1971 mainly adsorptive separation processes are used to obtain high purity p-xylene (55,84–86). A typical commercial process for the separation of p-xylene from other C_8 aromatics produces about 99.8% purity p-xylene at greater than 95% recovery.

Ethylbenzene Separation. Ethylbenzene [*100-41-4*], which is primarily used in the production of styrene, is difficult to separate from mixed C_8 aromatics by fractionation. A column of about 350 trays operated at a reflux:feed ratio of 20 is required. No commercial adsorptive unit to accomplish this separation has yet been installed, but the operation has been performed successfully in pilot plants (see Table 5). About 99% of the ethylbenzene in the feed was recovered at a purity of 99.7%. This operation, the UOP Ebex process, requires about 40% of the energy that is required by fractional distillation.

Table 5. Ethylbenzene Separation, Pilot-Plant Scale

Component	Feed, wt %	Ethylbenzene, wt %	Residue, wt %
ethylbenzene	30.5	99.7	0.4
p-xylene	12.9	0.1	18.4
m-xylene	35.8	0.1	51.3
o-xylene	20.8	0.1	29.9
total	*100.0*	*100.0*	*100.0*

Fructose–Dextrose Separation. Fructose–dextrose separation is an example of the application of adsorption to nonhydrocarbon systems. An aqueous solution of the isomeric monosaccharide sugars, $C_6H_{12}O_6$, fructose and dextrose (glucose), accompanied by minor quantities of polysaccharides, is produced commercially under the designation of "high" fructose corn syrup by the enzymatic conversion of cornstarch. Because fructose has about double the sweetness index of dextrose, the separation of fructose from this mixture and the recycling of dextrose for further enzymatic conversion to fructose is of commercial interest (see SUGAR; SWEETENERS).

The UOP Sarex process has been used since 1978 for the separation of high purity fructose from a mixture of fructose, glucose, and polysaccharides (87,88). The pilot-plant performance of fructose–glucose separation is given in Table 6.

Table 6. Separation of Fructose from High Fructose Corn Syrup, Pilot-Plant Scale

Component	Feed, wt %[a]	Extract, wt %[a]	Raffinate, wt %[a]
fructose	41.9	95.0	7.0
dextrose	53.1	5.0	84.7
other saccharides	5.0	0.0	8.3
total	*100.0*	*100.0*	*100.0*

[a]Dry basis.

Aromatic and Nonaromatic Hydrocarbon Separation. Aromatics are partially removed from kerosines and jet fuels to improve smoke point and burning characteristics. This removal is commonly accomplished by hydroprocessing, but can also be achieved by liquid–liquid extraction with solvents, such as furfural, or by adsorptive separation. Table 7 shows the results of a simulated moving-bed pilot-plant test using silica gel adsorbent and feedstock components mainly in the C_{10}–C_{15} range. The extent of extraction does not vary greatly for each of the various species of aromatics present. Silica gel tends to extract all aromatics from nonaromatics (89).

Table 7. Separation of Aromatics from Nonaromatics in C_{10}–C_{15} Light Cycle Oil

Components	Feed, vol %	Extract, vol %	Raffinate, vol %	Extraction, %
aromatics				
alkylbenzenes	17.9	34.3	7.5	76
tetralins, indanes	8.3	14.8	4.2	71
indenes	1.7	3.4	0.6	80
naphthalenes	13.6	30.0	1.6	86
acenaphthenes, biphenyls	4.4	9.1	0.8	89
tricyclic aromatics	1.7	3.3	0.2	92
total aromatics	*47.6*	*95.0*	*14.9*	*82*
nonaromatics	52.4	5.0	85.1	4
total	*100.0*	*100.0*	*100.0*	*41*

Citric Acid Separation. Citric acid [77-92-9] and other organic acids can be recovered from fermentation broths using the UOP Sorbex technology (90–92). The conventional means of recovering citric acid is by a lime and sulfuric acid process in which the citric acid is first precipitated as a calcium salt and then reacidulated with sulfuric acid. However, this process generates significant by-products and thus can become inefficient.

UOP has developed a UOP Sorbex process for the recovery and purification of citric acid from fermentation broths. The process provides technical-grade citric acid, $C_6H_8O_7$, which can be further recrystallized to obtain food-grade citric acid (qv).

Separation of Fatty Acids. Tall oil is a by-product of the pulp and paper manufacturing process and contains a spectrum of fatty acids, such as palmitic, stearic, oleic, and linoleic acids, and rosin acids, such as abietic acid. The conventional refining process to recover these fatty acids involves intensive distillation under vacuum. This process does not yield high purity fatty acids, and moreover, a significant degradation of fatty acids occurs because of the high process temperatures. These fatty and rosin acids can be separated using a UOP Sorbex process (93–99) (Tables 8 and 9).

Table 8. UOP Sorbex Separation of Fatty Acids from Rosin Acids in Distilled Tall Oil

Composition	Feed, wt %	Extract, wt %	Raffinate, wt %
rosin acid	33.0	1.50	94.50
linoleic acid	32.7	48.50	2.00
oleic acid	32.75	48.508	2.00
neutrals	1.50	1.50	1.50

Table 9. UOP Sorbex Separation of Saturated and Unsaturated Tall Oil Fatty Acids

Composition	Feed, wt %	Extract, wt %	Raffinate, wt %
oleic acid	48.55	92.00	5.00
linoleic acid	48.55	5.00	92.00
rosin acid	1.50	1.50	1.50
neutrals	1.50	1.50	1.50

Cyclic-Batch Processes

Continuous processes have wide application in different areas of the chemical industry. The separation efficiency of a continuous process is generally higher than that of a batch or cyclic-batch process. However, in some applications the cyclic-batch process may be preferred because of the complexity of design and the difficulty of controlling the continuous processes. Examples of commercial cyclic-batch adsorption processes operating in liquid phase include the UOP methanol recovery (UOP MRU) and oxygenate removal (UOP ORU) processes, which separate oxygenates from C_4 hydrocarbons; the UOP Cyclesorb process, which separates fructose from glucose; and ion-exclusion processes for recovering sucrose from molasses.

UOP MRU–ORU Processes. Methyl *tert*-butyl ether [*1634-04-4*] (MTBE) is an additive that is used to increase the octane value of gasoline. In environmentally sensitive locations where a minimum oxygen content for gasoline is specified, MTBE can provide the desired oxygen content while increasing octane and maintaining compatibility with the hydrocarbon components of gasoline. The MTBE synthesis typically requires an excess of methanol [*67-56-1*] relative to isobutylene [*115-11-7*] to achieve as high a conversion as possible. Downstream of the reactor, this excess methanol must be separated from the MTBE and unreacted C_4 hydrocarbons so that it can be recycled back to the reactor. A two-stage water wash is conventionally used for methanol recovery. The UOP MRU process offers an alternative means of recovering the 3 to 4% methanol contained in the C_4 stream (100).

In most cases, the linear olefins in the C_4 stream are subsequently sent to alkylation or polymerization processes, both of which are sensitive to trace oxygenates. The UOP ORU process can be used to remove trace dimethyl ether, *tert*-butyl alcohol, and water down to concentrations of a few parts per million. Figure 14 shows how the UOP MRU and UOP ORU processes are integrated with the MTBE and downstream plants. The UOP MRU process is a multibed thermal swing adsorption process that operates in a cyclic-batch adsorption mode. Methanol is preferentially adsorbed over the less polar hydrocarbons. The desorption step is accomplished by a temperature swing using the reactor feed. The UOP ORU process involves liquid-phase adsorption of the remaining trace oxygenates from the effluent C_4 stream followed by vapor-phase desorption using an external regenerant. Both processes are in commercial use.

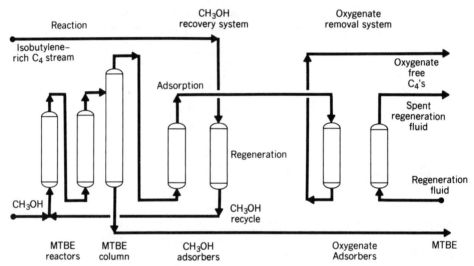

Fig. 14.   UOP raffinate process treatment process. MTBE = methyl *tert*-butyl ether.

UOP Cyclesorb Process.   The UOP Cyclesorb process was developed by UOP in the mid-1980s as an alternative to the UOP Sarex process. The UOP Cyclesorb process is simpler in design and yet attains reasonable efficiency of separation. The UOP Cyclesorb is a continuous recycle chromatographic process, wherein a series of chromatographic columns are used to develop the separation of fructose from glucose, and a series of internal recycle streams of impure and dilute portions of the chromatograph are used to improve the efficiency. The following process description is based on UOP patent literature (101,102).

A schematic diagram of a six-vessel UOP Cyclesorb process is shown in Figure 15. The UOP Cyclesorb process has four external streams: feed and desorbent enter the process, and extract and raffinate leave the process. In addition, the process has four internal recycles: dilute raffinate, impure raffinate, impure extract, and dilute extract. Feed and desorbent are fed to the top of each column, and the extract and raffinate are withdrawn from the bottom of each column in a predetermined sequence established by a switching device, the UOP

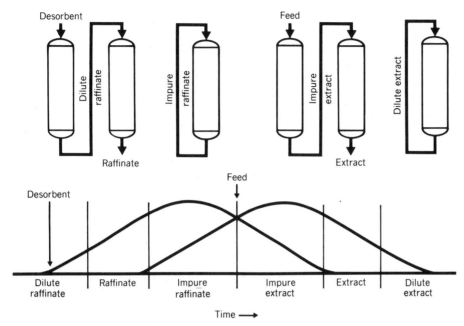

Fig. 15.   UOP Cyclesorb process flow.

rotary valve. The flow of the internal recycle streams is from the bottom of a column to the top of the same column in the case of dilute extract and impure raffinate and to the top of the next column in the case of dilute raffinate and impure extract.

Such a flow establishes a chromatographic profile in each column that is moving from top to bottom. Also, the profile is staggered from one column to the other as each column is in a different stage of development of the chromatogram. The concentration at any given point in a column varies with time, but the variation is more gradual than can be expected in a batch chromatographic process. Therefore, all portions of the column perform a useful function at any given time.

Ion-Exclusion Processes for Sucrose.   Molasses, which is a by-product of raw cane or beet sugar manufacturing processes, is a heavy, viscous liquid that is separated from the final low grade massecuite from which no more sugar can be crystallized by the usual methods. Molasses has a reasonably high sugar content. The recovery of sucrose from molasses has been the object of intense investigation for more than 50 years. In 1953, the ion-exclusion process was introduced by the Dow Chemical Company (103). This process, which was developed to separate the ionic from the nonionic constituents of molasses, was based on the fact that, under equilibrium conditions, certain ion-exchange resins have a different affinity for nonionic species than for ionic species.

The ion-exclusion process for sucrose purification has been practiced commercially by Finn Sugar (104). This process operates in a cyclic-batch mode and provides a sucrose product that does not contain the highly molassogenic salt impurities and thus can be recycled to the crystallizers for additional sucrose recovery.

Liquid Chromatography

Conventional liquid chromatography has not attained great commercial significance in the area of large-scale bulk separations from the liquid phase. In 1952 the Sun Oil Company announced such a process, designated as the Arosorb process, for separating aromatic hydrocarbons from naphthas with a silica gel (105). However, the Arosorb process did not attain wide commercial usage, largely because of the simultaneous development of efficient liquid–liquid extraction processes for the same application.

However, chromatographic processes still have a considerable applicability (106) (see ANALYTICAL METHODS). For instance, in small-scale operations, the greater simplicity of the chromatograph may more than compensate economically for the larger adsorbent inventory and desorbent usage. Chromatography may also be advantageous when it is required to separate several pure products from a single feed stream. A simulated moving-bed system can yield only two well-separated fractions from a single feed stream.

The separating power of a chromatographic process arises from the development of many theoretical plates to achieve adsorption equilibrium within a column of moderate length. Even though the separation factor between two components may be small, any desired resolution may be achieved with sufficient theoretical plates.

The height equivalent to a theoretical plate (HETP) for a chromatographic column is approximated by the van Deemter equation (107):

$$\text{HETP} = A + B/v + Cv$$

where v is the interstitial velocity, and A, B, and C are related to such factors as particle size, bed porosity, and adsorption equilibrium constant. To obtain a small HETP, and therefore an efficient separation process, uniform packing is essential. Such packing requires careful loading procedures to eliminate variation in voidage and stratification of particle sizes and shapes across the cross section of the bed.

In analytical chromatography, the primary objective is to maximize the resolution between two components subject to some restrictions on the maximum time of elution. As a result, the feed pulse loading is minimized, and the number of theoretical plates is maximized. In preparative chromatography, the objective is to maximize production rate as well as reduce capital and operating costs at a given separation efficiency. The adsorption column is therefore commonly run under overload conditions with a finite feed pulse width. The choice of operating conditions for preparative chromatography has been discussed (53). In production chromatography, the optimal pulse sequence occurs when the successive pulses of feed are introduced at intervals such that the feed components are just resolved both within a given sample and between adjacent samples.

Gas versus Liquid Adsorption

The question of whether adsorption should be done in the gas or liquid phase is an interesting one. Often the choice is clear. For example, in the separation of nitrogen from oxygen, liquid-phase separation is not practical because of low temperature requirements. In $C_{10}-C_{14}$ olefin separation, a gas-phase operation is not feasible because of reactivity of feed components at high temperatures. Also, in the case of substituted aromatics separation, such as p-xylene from other C_8 aromatics, the inherent selectivities of individual components are so close to one another that a simulated moving-bed operation in liquid phase is the only practical choice.

However, in some cases, the answer is not clear. A variety of factors need to be taken into consideration before a clear choice emerges. For example, UOP's Molex and IsoSiv processes are used to separate normal paraffins from non-normals and aromatics in feedstocks containing C_5-C_{20} hydrocarbons, and both processes use molecular sieve adsorbents. However, Molex operates in simulated moving-bed mode in liquid phase, and IsoSiv operates in gas phase, with temperature swing desorption by a displacement fluid. The following comparison of UOP's Molex and IsoSiv processes indicates some of the primary factors that are often used in decision making:

Factor	Molex	IsoSiv
adsorbent/extract	1	2
desorbent/extract	1	3
tolerance to feed contaminants	low	high

Outlook

Liquid adsorption processes hold a prominent position in several applications for the production of high purity chemicals on a commodity scale. Many of these processes were attractive when they were first introduced to the industry and continue to increase in value as improvements in adsorbents, desorbents, and process designs are made. The UOP Parex process alone has seen three generations of adsorbent and four generations of desorbent. Similarly, liquid adsorption processes can be applied to a much more diverse range of problems than those presented in Table 3.

A surprisingly large number of important industrial-scale separations can be accomplished with the relatively small number of zeolites that are commercially available. The discovery, characterization, and commercial availability of new zeolites and molecular sieves are likely to multiply the number of potential solutions to separation problems. A wider variety of pore diameters, pore geometries, and hydrophobicity in new zeolites and molecular sieves as well as more precise control of composition and crystallinity in existing zeolites will help to broaden the applications for adsorptive separations and likely lead to improvements in separations that are currently in commercial practice.

The value of many chemical products, from pesticides to pharmaceuticals to high performance polymers, is based on unique properties of a particular isomer from which the product is ultimately derived. For example, trisubstituted aromatics may have as many as 10 possible geometric isomers whose ratio in the mixture is determined by equilibrium. Often the purity requirement for the desired product includes an upper limit on the content of one or more of the other isomers. This separation problem is a complicated one, but one in which adsorptive separation processes offer the greatest chances for success.

BIBLIOGRAPHY

"Adsorptive separation, liquids," in *ECT* 3rd ed., Vol. 1, pp. 563–581, D. B. Broughton, UOP Process Division, UOP, Inc.

1. *Separation and Purification: Critical Needs and Opportunities,* National Research Council Report, National Academy Press, 1987.
2. C. L. Mantell, *Adsorption,* 2nd ed., McGraw-Hill, Inc., New York, 1951.
3. V. R. Deitz, *Bibliography of Solid Adsorbents,* N.B.S. Circular 566, National Bureau of Standards, Washington, D.C., 1956.
4. T. Vermeulen, N. K. Heister, and G. Klein, *Perry's Chemical Engineers Handbook,* 4th ed., McGraw-Hill, Inc., New York, 1963, Sect. 16.
5. *Adsorption Handbook,* Pittsburgh Chemical Company, Activated Carbon Division, 1961.
6. D. B. Broughton, *Chem. Eng. Prog.* **64,** 60 (1968).
7. D. B. Broughton and A. G. Lickus, *Pet. Refiner* **40**(5), 173 (1961).
8. D. B. Carson and D. B. Broughton, *Pet. Refiner* **38**(4), 130 (1959).
9. D. W. Breck, *Zeolite Molecular Sieves,* John Wiley & Sons, Inc., New York, 1974.
10. L. Berg, *Chem. Eng. Prog.* **65**(9), 52 (1969).
11. A. E. Herschler and T. S. Mertes, *Ind. Eng. Chem.* **47,** 193 (1955).
12. D. Haresnape, F. A. Fidler, and R. A. Lowry, *Ind. Eng. Chem.* **41,** 2691 (1949).
13. B. J. Mair and M. Shamaiengar, *Anal. Chem.* **30,** 276 (Feb. 1958).
14. B. J. Mair, A. L. Gaboriault, and F. D. Rossini, *Ind. Eng. Chem.* **39,** 1072 (1947).
15. S. Eagle and J. W. Scott, *Ind. Eng. Chem.* **42,** 1287 (1950).
16. A. E. Hirschler and S. Amon, *Ind. Eng. Chem.* **30,** 276 (Feb. 1958).
17. U.S. Pat. 3,133,126 (May 12, 1964), R. N. Fleck and C. G. Wright (to Union Oil Company).
18. Brit. Pat. 1,108,305 (Apr. 3, 1968), D. W. Peck, R. R. Gentry, and H. E. Frite (to Union Carbide Corp.).
19. U.S. Pat. 3,843,518 (Oct. 22, 1974), E. M. Magee and F. J. Healy (to Esso Research & Engineering Company).
20. U.S. Pat. 3,761,533 (Sept. 25, 1973), S. Otani and co-workers (to Toray Industries Inc.).
21. U.S. Pat. 3,686,343 (Aug. 22, 1972), R. Bearden and R. J. De Feo, Jr. (to Esso Research & Engineering Company).
22. U.S. Pat. 3,724,170 (Apr. 3, 1973), P. T. Allen, B. M. Drinkard, and E. H. Vager (to Mobil Oil Corp.).
23. U.S. Pat. 3,626,020 (Dec. 7, 1971), R. W. Neuzil (to Universal Oil Products Co.).
24. U.S. Pat. 3,558,730 (Jan. 26, 1971), R. W. Neuzil (to Universal Oil Products Co.).
25. U.S. Pat. 3,558,732 (Jan. 26, 1971), R. W. Neuzil (to Universal Oil Products Co.).
26. U.S. Pat. 3,663,638 (May 16, 1972), R. W. Neuzil (to Universal Oil Products Co.).
27. U.S. Pat. 3,686,342 (Aug. 22, 1972), R. W. Neuzil (to Universal Oil Products Co.).
28. U.S. Pat. 3,734,974 (May 22, 1973), R. W. Neuzil (to Universal Oil Products Co.).
29. U.S. Pat. 3,706,813 (Dec. 19, 1972), R. W. Neuzil (to Universal Oil Products Co.).
30. U.S. Pat. 3,851,006 (Nov. 26, 1974), A. J. de Rosset and R. W. Neuzil (to Universal Oil Products Co.).
31. U.S. Pat. 3,698,157 (Oct. 17, 1972), P. T. Allen and B. M. Drinkard (to Mobil Oil Corp.).
32. U.S. Pat. 3,917,734 (Nov. 4, 1975), A. J. de Rosset (to Universal Oil Products Co.).
33. U.S. Pat. 3,665,046 (May 23, 1972), A. J. de Rosset (to Universal Oil Products Co.).
34. U.S. Pat. 3,510,423 (May 5, 1973), R. W. Neuzil (to Universal Oil Products Co.).
35. U.S. Pat. 3,723,561 (Mar. 27, 1973), J. W. Priegnitz (to Universal Oil Products Co.).
36. U.S. Pat. 3,851,006 (Nov. 26, 1974), A. J. de Rosset (to Universal Oil Products Co.).
37. F. Wolf and K. Pilchowski, *Chem. Technol.* **23**(11), (1971) (in Ger.).

38. R. M. Moore and J. R. Katzer, *AIChE J.* **18,** 816 (1972).
39. C. N. Satterfield and C. S. Cheng, *AIChE J.* **18,** 710 (1972).
40. F. Wolf, K. Pilchowski, K. H. Mohrmann, and E. Hause, *Chem. Technol.* **27**(12), (1975) (in Ger.).
41. S. K. Suri and V. Ramkrishna, *Trans. Faraday Soc.* **65**(6), 1960 (1969).
42. J. F. Walter and E. B. Stuart, *AIChE J.* **10,** 889 (1964).
43. U.S. Pat. 3,929,669 (Dec. 30, 1975), D. H. Rosback and R. W. Neuzil (to Universal Oil Products Co.).
44. S. Sircar and A. L. Meyers, *AIChE J.* **17,** 186 (1971).
45. R. T. Yang, *Gas Separation by Adsorption Processes,* Butterworth, London, 1986.
46. E. Heftmann, ed., *Chromatography,* Van Nostrand-Reinhold, New York, 1975.
47. J. J. Kipling, *Adsorption from Solutions of Non-Electrolytes,* Academic Press, Inc., New York, 1965.
48. F. C. Nachod and J. Schubert, *Ion-Exchange Technology,* Academic Press, Inc., New York, 1956.
49. R. M. Barrer, *Zeolites and Clay Minerals as Sorbents and Molecular Sieves,* Academic Press, Inc., London, 1978.
50. J. A. Johnson and A. R. Oroskar, "Sorbex Technology for Industrial Scale Separation," in H. G. Karge and J. Weitkamp, eds., *Zeolites as Catalysts, Sorbents, and Detergent Builders,* Elsevier Science Publishers BV, Amsterdam, 1989.
51. C. Ho, C. B. Ching, and D. M. Ruthven, *Ind. Eng. Chem. Res.* **26,** 1407 (1987).
52. S. A. Fisher and G. Otten, *Proceedings of the 42nd International Water Conference,* Engineering Society of Western Pennsylvania, Pittsburgh, 1981.
53. D. M. Ruthven, *Principles of Adsorption and Adsorption Processes,* John Wiley & Sons, Inc., New York, 1984.
54. G. E. Keller II, in T. E. Whyte and co-workers, eds., *Industrial Gas Separation* (ACS Symposium Series No. 223), American Chemical Society, Washington, D.C., 1983.
55. D. B. Broughton, R. W. Neuzil, J. M. Pharis, and C. S. Brearly, *Chem. Eng. Prog.* **66**(9), 70 (1970).
56. C. Berg, *Trans. Am. Inst. Chem. Eng.* **42,** 665 (1946).
57. P. C. Wankat, *Large Scale Adsorption and Chromatography,* CRC Press, Boca Raton, Fla., 1986.
58. R. E. Treybal, *Mass Transfer Operations,* 3rd ed. McGraw Hill, New York, 1980.
59. D. B. Broughton, H. J. Bieser, and M. C. Anderson, *Pet. Int. (Milan),* **23**(3), 91 (1976) (in English).
60. D. B. Broughton, H. J. Bieser, and R. A. Persak, *Pet. Int. (Milan),* **23**(5), 36 (1976) (in English).
61. P. E. Barker and G. Gavelson, *Separation and Purification Methods* **17,** 1 (1988).
62. S. Otani and co-workers, *Chem. Econ. Eng. Rev.* **3**(6), 56 (1971).
63. S. Otani, *Chem. Eng.* **80**(9), 106 (1973).
64. *Making Waves in Liquid Processing,* Illinois Water Treatment Company, IWT Adsep System, Rockford, Ill., 1984, **VI** (1).
65. Tetsua Hirota, *Sugar Azucar* (Jan. 1980).
66. Advertisement, *Sugar Azucar* (March 1980).
67. *Chem. Eng.* 39, (Aug. 29, 1977).
68. U.S. Pat. 2,985,589 (May 23, 1961), D. B. Broughton and C. G. Gerhold (to Universal Oil Products Co.).
69. U.S. Pat. 3,040,777 (June 26, 1962), D. B. Carson (to Universal Oil Products Co.).
70. U.S. Pat. 3,192,954 (July 6, 1965), C. G. Gerhold and D. B. Broughton (to Universal Oil Products Co.).
71. U.S. Pat. 3,268,604 (Aug. 23, 1966), D. M. Boyd (to Universal Oil Products Co.).
72. U.S. Pat. 3,268,603 (Aug. 23, 1966), D. M. Boyd (to Universal Oil Products Co.).
73. U.S. Pat. 3,131,232 (Apr. 28, 1964), D. B. Broughton (to Universal Oil Products Co.).

74. T. Miyauchi and T. Vermeulen, *Ind. Eng. Chem. Fundam.* **2,** 304 (1963).
75. S. Hartland and J. C. Mecklenburgh, *Chem. Eng. Sci.* **21,** 1209 (1966).
76. D. M. Ruthven and C. B. Ching, *Chem. Eng. Sci.* **44,** 1011 (1989).
77. U.S. Pat. 3,208,833 (Sept. 28, 1964), D. B. Carson (to Universal Oil Products Co.).
78. U.S. Pat. 3,214,247 (Oct. 26, 1964), D. B. Broughton (to Universal Oil Products Co.).
79. U.S. Pat. 3,523,762 (Aug. 11, 1970), D. B. Broughton (to Universal Oil Products Co.).
80. R. J. Schmidt, B. H. Johnson, and J. A. Weiszmann, 1987 National Petroleum Refiners Association Annual Meeting, San Antonio, Tex., Mar. 1987.
81. D. B. Broughton and R. C. Berg, *Hydrocarbon Process.* **48**(6), 115 (1969).
82. D. B. Broughton and R. C. Berg, National Petroleum Refiners Association 1969 Annual Meeting, Mar. 23, 1969, technical paper AM-69-38.
83. J. A. Johnson, S. R. Raghuram, and P. R. Pujado, "Olex: A Process for Producing High Purity Olefins," presented at the AIChE Summer National Meeting, Minneapolis, Minn., Aug. 1987.
84. G. Koenig, *Erdoel Kohle* **26,** 323 (1973) (in German).
85. D. P. Thornton, *Hydrocarbon Process.* **49**(11), 151 (1970).
86. F. H. Adams, *Eur. Chem. News* **62** (Oct. 13, 1972).
87. R. W. Neuzil and R. A. Jensen, 85th National Meeting of the AIChE, Philadelphia, Pa., June, 1978.
88. A. J. de Rosset, R. W. Neuzil, and D. J. Korous, *Ind. Eng. Chem. Proc. Des. Dev.* **15,** 261 (1978).
89. D. B. Broughton and L. C. Hardison, "Hydrocarbon-Type Separation by Unisorb," Paper presented at the 27th Midyear Meeting of the American Petroleum Institute, Division of Refining, San Francisco, Calif., May 15, 1962.
90. U.S. Pat. 4,720,579 (Jan. 19, 1988), S. Kulprathipanja (to UOP, Inc.).
91. U.S. Pat. 4,851,573 (June 25, 1989), S. Kulprathipanja, J. Priegnitz, and A. R. Oroskar (to UOP, Inc.).
92. U.S. Pat. 4,851,574 (July 25, 1989), S. Kulprathipanja (to UOP, Inc.).
93. U.S. Pat. 4,529,551 (1985), M. T. Cleary, S. Kulprathipanja, and R. W. Neuzil (to UOP, Inc.).
94. U.S. Pat. 4,534,900 (1985), M. T. Cleary (to UOP, Inc.).
95. U.S. Pat. 4,495,094 (1985), M. T. Cleary (to UOP, Inc.).
96. U.S. Pat. 4,495,106 (1985), M. T. Cleary and W. C. Laughlin (to UOP, Inc.).
97. U.S. Pat. 4,521,343 (1985), T. H. Chao and M. T. Cleary (to UOP, Inc.).
98. S. A. Gembicki, S. M. Shah, and M. T. Cleary, *Pulp Chemicals Association,* Pine Mountain, Ga., Mar. 10, 1983.
99. R. W. Johnson and E. Fritz, *Fatty Acids in Industry,* Marcel Dekker, New York, 1989.
100. A. Benchikha and D. R. Garg, "The C4 Raffinate Treatment Process; Methanol Recovery/Oxygenate Removal," Presentation at HUELS MTBE Symposium in Marl, West Germany, Sept. 6, 1988.
101. U.S. Pat. 4,402,832 (Sept. 6, 1983), C. G. Gerhold (to UOP, Inc.).
102. U.S. Pat. 4,478,721 (Oct. 23, 1984), C. G. Gerhold (to UOP, Inc.).
103. M. Wheaton and W. C. Bauman, *I&EC Eng. and Proc. Dev.* **45,** 228 (1953).
104. H. Hongisto and H. Heikkila, *Sugar Azucar* **56,** 60 (Mar. 1978).
105. W. H. Davis, J. I. Harper, and E. R. Weatherley, *Pet. Refiner* **31,** 109 (1952).
106. C. J. King, *Separation Processes,* McGraw Hill, Inc., New York, 1971.
107. J. J. van Deemter, F. J. Zuiderweg, and A. Klinberg, *Chem. Eng. Sci.* **5,** 271 (1956).

STANLEY A. GEMBICKI
ANIL R. OROSKAR
JAMES A. JOHNSON
UOP

ADSORPTIVE SEPARATION.   See ADSORPTION; ADSORPTION, GAS SEPARATION; ADSORPTION, LIQUID SEPARATION.

ADVANCED CERAMICS

ELECTRONIC CERAMICS

Electronic ceramics is a generic term describing a class of inorganic, nonmetallic materials utilized in the electronics industry. Although the term electronic ceramics, or electroceramics, includes amorphous glasses and single crystals, it generally pertains to polycrystalline inorganic solids comprised of randomly oriented crystallites (grains) intimately bonded together. This random orientation of small, micrometer-size crystals results in an isotropic ceramic possessing equivalent properties in all directions. The isotropic character can be modified during the sintering operation at high temperatures or upon cooling to room temperature by processing techniques such as hot pressing or poling in an electric or magnetic field (see CERAMICS AS ELECTRICAL MATERIALS).

The properties of electroceramics are related to their ceramic microstructure, ie, the grain size and shape, grain–grain orientation, and grain boundaries, as well as to the crystal structure, domain configuration, and electronic and defect structures. Electronic ceramics are often combined with metals and polymers to meet the requirements of a broad spectrum of high technology applications, computers, telecommunications, sensors (qv), and actuators. Roughly speaking, the multibillion dollar electronic ceramics market can be divided into six equal parts as shown in Figure 1. In addition to SiO_2-based optical fibers and

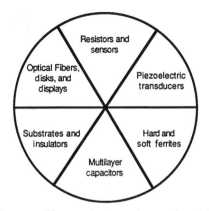

Fig. 1.   Electronic ceramics market (1).

Table 1. Electronic Ceramic Functions and Products

Function	Material	Products[a]
insulators	porcelain, glass, steatite	high voltage insulation
packaging	Al_2O_3, BeO, AlN	IC substrates, packages (MMCs)
capacitors (energy storage)	$BaTiO_3$, $SrTiO_3$, TiO_2	multilayer and barrier layer capacitors
piezoelectrics	$Pb(Zr_{1-x}Ti_x)O_3$, SiO_2 (quartz)	vibrators, oscillators, filters, motors, and actuators
magnetics	$Mn_{1-x}Zn_xFe_2O_4$, $Ni_{1-x}Zn_xFe_2O_4$	inductors, transformers, memory devices
semiconductors	$(Ba,La)TiO_3$, V_2O_3, $Fe_{2-x}Ti_xO_3$, $ZnO-Bi_2O_3$, $MgCr_2O_4-TiO_2$, CdS, SiC	PTC, NTC-thermistors, varistors, pH sensor, humidity sensor, solar cells, electric heater
conductors	RuO_2, $NaAl_{11}O_{17}$, $Zr_{1-2x}Y_{2x}O_{2-x}$, $YBa_2Cu_3O_{7-\delta}$	resistors (thick film), solid electrolytes, oxygen sensors, superconductors

[a]MMC = multicomponent components; PTC = positive temperature coefficient; NTC = negative temperature coefficient.

displays, electronic ceramics encompass a wide range of materials and crystal structure families (see Table 1) used as insulators, capacitors, piezoelectrics (qv), magnetics, semiconductor sensors, conductors, and the recently discovered high temperature superconductors. The broad scope and importance of the electronic ceramics industry is exemplified in Figure 2, which schematically displays electroceramic components utilized in the automotive industry. Currently, the growth of the electronic ceramic industry is driven by the need for large-scale integrated circuitry giving rise to new developments in materials and processes. The development of multilayer packages for the microelectronics industry, composed of multifunctional three-dimensional ceramic arrays called monolithic ceramics (MMC), continues the miniaturization process begun several decades ago to provide a new generation of robust, inexpensive products.

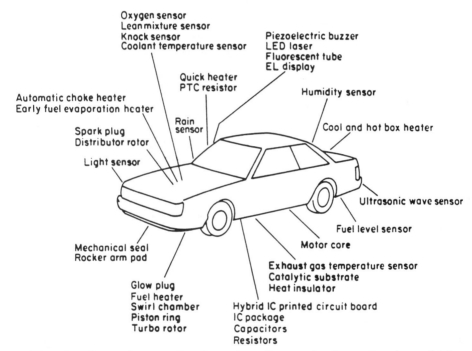

Fig. 2. Electronic ceramics for automotive applications. Courtesy of Nippon Denso, Inc.

Structure–Property Relations

An overview of the atomistic and electronic phenomena utilized in electroceramic technology is given in Figure 3. More detailed discussions of compositional families and structure–property relationships can be found in other articles. (See, for example, FERROELECTRICS, MAGNETIC MATERIALS, and SUPERCONDUCTING MATERIALS.)

Multilayer capacitors, piezoelectric transducers, and positive temperature coefficient (PTC) thermistors make use of the ferroelectric properties of barium

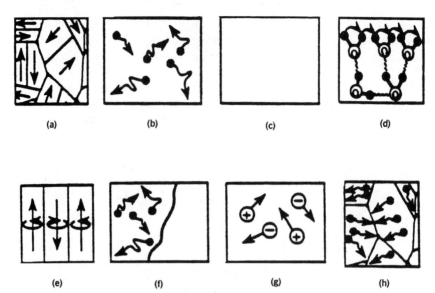

Fig. 3. An overview of atomistic mechanisms involved in electroceramic components and the corresponding uses: (**a**) ferroelectric domains: capacitors and piezoelectrics, PTC thermistors; (**b**) electronic conduction: NTC thermistor; (**c**) insulators and substrates; (**d**) surface conduction: humidity sensors; (**e**) ferrimagnetic domains: ferrite hard and soft magnets, magnetic tape; (**f**) metal–semiconductor transition: critical temperature NTC thermistor; (**g**) ionic conduction: gas sensors and batteries; and (**h**) grain boundary phenomena: varistors, boundary layer capacitors, PTC thermistors.

titanate (IV) [*12047-27-7*], BaTiO$_3$, and lead zirconate titanate [*12626-81-2*]. On cooling from high temperature, these ceramics undergo phase transformations to polar structures having complex domain patterns. Large peaks in the dielectric constant accompany the phase transitions where the electric dipole moments are especially responsive to electric fields. As a result, modified compositions of barium titanate (qv), BaTiO$_3$, are widely used in the multilayer capacitor industry and most piezoelectric transducers are made from lead zirconate titanate, PbZr$_{1-x}$Ti$_x$O$_3$, (PZT) ceramics. Applying a large dc field (poling) aligns the domains and makes the ceramic piezoelectric. The designation PZT is a registered trademark of Vernitron, Inc.

Similar domain phenomena are observed in ferrimagnetic oxide ceramics such as manganese ferrite [*12063-10-4*], MnFe$_2$O$_4$, and BaFe$_{11}$O$_{17}$, but the underlying mechanism is different. The unpaired spins of Fe^{3+} and Mn^{2+} ions give rise to magnetic dipole moments which interact via neighboring oxygen ions through a super-exchange mechanism. The magnetic dipoles are randomly oriented in the high temperature paramagnetic state, but on cooling through the Curie temperature, T_C, align to form magnetic domains within the ceramic grains. The peak in the magnetic permeability at T_C is analogous to the peak in the dielectric constant of ferroelectric ceramics. Domain walls move easily in soft ferrites (qv) like MnFe$_2$O$_4$ and γ-Fe$_2$O$_3$, which are used in transformers and magnetic tape. In barium ferrite [*11138-11-7*], the spins are firmly locked to the hexagonal axis, making it useful as a permanent magnet.

Several kinds of conduction mechanisms are operative in ceramic thermistors, resistors, varistors, and chemical sensors. Negative temperature coefficient (NTC) thermistors make use of the semiconducting properties of heavily doped transition metal oxides such as n-type $Fe_{2-x}Ti_xO_3$ and p-type $Ni_{1-x}Li_xO$. Thick film resistors are also made from transition-metal oxide solid solutions. Glass-bonded $Bi_{2-2x}Pb_{2x}Ru_2O_{7-x}$ having the pyrochlore [12174-36-6] structure is typical.

Phase transitions are involved in critical temperature thermistors. Vanadium , VO_2, and vanadium trioxide [1314-34-7], V_2O_3, have semiconductor–metal transitions in which the conductivity decreases by several orders of magnitude on cooling. Electronic phase transitions are also observed in superconducting ceramics like $YBa_2Cu_3O_{7-x}$, but here the conductivity increases sharply on cooling through the phase transition.

Ionic conductivity is used in oxygen sensors and in batteries (qv). Stabilized zirconia, $Zr_{1-x}Ca_xO_{2-x}$, has a very large number of oxygen vacancies and very high O^{2-} conductivity. β-Alumina [12005-48-0], $NaAl_{11}O_{17}$, is an excellent cation conductor because of the high mobility of Na^+ ions. Ceramics of β-alumina are used as the electrolyte in sodium–sulfur batteries.

Surface conduction is monitored in most humidity sensors through the use of porous ceramics of $MgCr_2O_4$–TiO_2 that adsorb water molecules which then dissociate and lower the electrical resistivity.

Grain boundary phenomena are involved in varistors, boundary layer capacitors, and PTC thermistors. The formation of thin insulating layers between conducting grains is crucial to the operation of all three components. The reversible electric breakdown in varistors has been traced to quantum mechanical tunneling through the thin insulating barriers. In a $BaTiO_3$–PTC thermistor, the electric polarization associated with the ferroelectric phase transition neutralizes the insulating barriers, causing the ceramic to lose much of its resistance below T_C. Boundary layer capacitors have somewhat thicker barriers which cannot be surmounted, and hence the ceramic remains an insulator. However, the movement of charges within the conducting ceramic grains raises the dielectric constant and increases the capacitance.

Lastly, the importance of electroceramic substrates and insulators should not be overlooked. Here one strives to raise the breakdown strength by eliminating the interesting conduction mechanisms just described. Spark plugs, high voltage insulators, and electronic substrates and packages are made from ceramics like alumina, mullite [55964-99-3], and porcelain [1332-58-7].

Electroceramic Processing

Fabrication technologies for all electronic ceramic materials have the same basic process steps, regardless of the application: powder preparation, powder processing, green forming, and densification.

Powder Preparation. The goal in powder preparation is to achieve a ceramic powder which yields a product satisfying specified performance standards. Examples of the most important powder preparation methods for electronic ceramics include mixing/calcination, coprecipitation from solvents, hydrothermal processing, and metal organic decomposition. The trend in powder synthesis is toward powders having particle sizes less than 1 μm and little or no hard agglomerates for enhanced reactivity and uniformity. Examples of the four basic methods are presented in Table 2 for the preparation of $BaTiO_3$ powder. Reviews of these synthesis techniques can be found in the literature (2,5).

The mixing of components followed by calcination to the desired phase(s) and then milling is the most widely used powder preparation method (2). Mixing/calcination is straightforward, and in general, the most cost effective use of capital equipment. However, the high temperature calcination produces an agglomerated powder which requires milling. Contamination from grinding media and mill lining in the milling step can create defects in the manufactured product in the form of poorly sintered inclusions or undesirable compositional modification. Furthermore, it is difficult to achieve the desired homogeneity, stoichiometry, and phases for ceramics of complex composition.

Coprecipitation is a chemical technique in which compounds are precipitated from a precursor solution by the addition of a precipitating agent, for example, a hydroxide (5). The metal salt is then calcined to the desired phase. The advantage of this technique over mixing/calcination techniques is that more intimate mixing of the desired elements is easily achieved, thus allowing lower calcination temperatures. Limitations are that the calcination step may once again result in agglomeration of fine powder and the need for milling. An additional problem is that the ions used to provide the soluble salts (eg, chloride from metal chlorides) may linger in the powder after calcination, affecting the properties in the sintered material.

Hydrothermal processing uses hot (above 100°C) water under pressure to produce crystalline oxides (6). This technique has been widely used in the formation process of Al_2O_3 (Bayer Process), but not yet for other electronic powders. The situation is expected to change, however. The major advantage of the hydrothermal technique is that crystalline powders of the desired stoichiometry and phases can be prepared at temperatures significantly below those required for calcination. Another advantage is that the solution phase can be used to keep the particles separated and thus minimize agglomeration. The major limitation of hydrothermal processing is the need for the feedstocks to react in a closed system to maintain pressure and prevent boiling of the solution.

Metal organic decomposition (MOD) is a synthesis technique in which metal-containing organic chemicals react with water in a nonaqueous solvent to produce a metal hydroxide or hydrous oxide, or in special cases, an anhydrous metal oxide (7). MOD techniques can also be used to prepare nonoxide powders (8,9). Powders may require calcination to obtain the desired phase. A major advantage of the MOD method is the control over purity and stoichiometry that can be achieved. Two limitations are atmosphere control (if required) and expense of the chemicals. However, the cost of metal organic chemicals is decreasing with greater use of MOD techniques.

Powder Processing.   A basic guideline of powder manufacturing is to do as little processing as possible to achieve the targeted performance standards (see POWDERS, HANDLING). Ceramic powder fabrication is an iterative process during which undesirable contaminants and defects can enter into the material at any stage. Therefore, it is best to keep the powder processing scheme as simple as

Table 2. Methods Used to Prepare BaTiO₃ Electronic Ceramic Powders

Method	Reaction	Particle size
mixing/calcination[a]	$BaCO_3 + TiO_2 \xrightarrow{\Delta T} BaTiO_3 + CO_2 \uparrow$	1 μm to 100s of μm
coprecipitation[b]	$Ba^{2+} + TiO^{2+} + 2\, C_2O_4^{2-} \xrightarrow{H_2O} BaTi(C_2O_4)_2 \cdot 4H_2O \xrightarrow{\Delta T}$ $BaTiO_3 + 4\, H_2O \uparrow + 4\, CO_2 \uparrow$	if calcined, mean size of ≈ 0.5 μm after milling
hydrothermal[c]	$Ba^{2+} + TiO_2 + H_2O \xrightarrow[\Delta T, P]{OH^-} BaTiO_3 + 2\, H_2O$	nanosize to 50 μm
metal organic decomposition[c]	$Ba(i\text{-}OC_3H_7)_2 + Ti((OC_4H_9)_4 + 3\, H_2O \rightarrow BaTiO_3 + 2\, C_3H_7OH + 4\, C_4H_9OH$	5.0–35.0 nm, depending upon calcination conditions

[a]Ref. 2.
[b]Ref. 3.
[c]Ref. 4.

607

possible to maintain flexibility. Uncontrollable factors such as changes in the characteristics of as-received powders must be accommodated in the processing from batch to batch of material. Keeping the processing simple is not always possible: the more complex the material system, the more complex the processing requirements.

A fundamental requirement in powder processing is characterization of the as-received powders (10–12). Many powder suppliers provide information on tap and pour densities, particle size distributions, specific surface areas, and chemical analyses. Characterization data provided by suppliers should be checked and further augmented where possible with in-house characterization. Uniaxial characterization compaction behavior, in particular, is easily measured and provides data on the nature of the agglomerates in a powder (13,14).

Milling is required for most powders, either to reduce particle size or to aid in the mixing of component powders (15). Commonly employed types of comminution include ball milling, and vibratory, attrition, and jet milling, each possessing advantages and limitations for a particular application. For example, ball milling is well-suited to powder mixing but is rather inefficient for comminution.

Green Forming.   Green forming is one of the most critical steps in the fabrication of electronic ceramics. The choice of green forming technique depends on the ultimate geometry required for a specific application. There are many different ways to form green ceramics, several of which are summarized in Table 3. Multilayer capacitors require preparation and stacking of two-dimensional ceramic sheets to obtain a large capacitance in a small volume. Techniques used to prepare two-dimensional sheets of green ceramic, including tape casting, (16–22) are discussed later under processing of multilayer ceramics. Manufacturing methods for ceramic capacitors have been reviewed (23).

Table 3. Green Forming Procedures for Electronic Ceramics

Green forming method	Geometries	Applications
uniaxial pressing	disks, toroids, plates	disk capacitors, piezo transducers, magnets
cold isostatic pressing	complex and simple	spark plugs, ZrO_2–O_2 sensors
colloidal casting	complex shapes	crucibles, porcelain insulators
extrusion	thin sheets (>80 μm), rods, tubes, honeycomb substrates	substrates, thermocouple insulator, catalytic converters, PTC thermistor heaters
injection molding	small complex shapes (<1.0 cm)	ZrO_2–O_2 sensors

Uniaxial pressing is the method most widely used to impart shape to ceramic powders (24). Binders, lubricants, and other additives are often incorporated into ceramic powders prior to pressing to provide strength and assist in particle compaction (25). Simple geometries such as rectangular substrates for integrated circuit (IC) packages can be made by uniaxial pressing (see INTEGRATED CIRCUITS).

More complex shapes can be made by cold isostatic pressing (CIP). CIP uses deformable rubber molds of the required shape to contain the powder. The application of isostatic pressure to the mold suspended in a pressure transfer media, such as oil, compacts the powder. CIP is not as easily automated as uniaxial pressing, but has found wide application in the preparation of more complex shapes such as spark plug insulators (26).

Slip or colloidal casting has been used to make complex shapes in the whiteware industry for many years (24). Other work has shown that colloidal casting can be used to produce electronic ceramic materials having outstanding strength because hard agglomerates can be eliminated in the suspension processing (27–29). Colloidal casting uses a porous mold in which the fine particles in a colloidal suspension accumulate because of capillary forces at the wall surface of the mold. Relatively dense packing of the particles, to approximately 60% of theoretical density, can be achieved. More importantly, hard aggregates can be eliminated from the colloid by suitable powder selection and processing. Drying of the resulting material may not be trivial and sections greater than about ~1.25 cm thick are sometimes difficult to obtain.

In addition to being the preferred forming technique for ceramic rods and tubes, extrusion processes are used to fabricate the thick green sheets used in many electronic components (24,30,31). The smallest thickness for green sheets prepared by extrusion techniques is about 80 μm. Organic additives similar to those used in tape casting are employed to form a high viscosity plastic mass that retains its shape when extruded. The extrusion apparatus, schematically shown in Figure 4, consists of a hopper for introduction of the plasticized mass, a de-airing chamber, and either a screw-type or plunger-type transport barrel in which the pressure is generated for passage of the plastic mass through a die of the desired geometry. The plastic mass is extruded onto a carrier belt and passed through dryers to relax the plastic strain remaining after extrusion. The green sheet can be stamped or machine diced to form disks, wafers, or other platelike shapes.

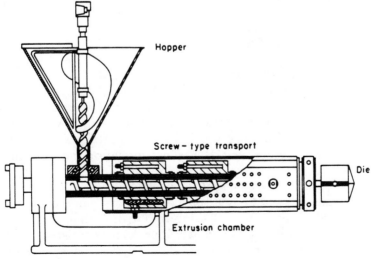

Fig. 4. Schematic of extrusion type apparatus for green sheet fabrication.

Injection molding is particularly suited to mass production of small complex shapes with relatively small (<1.0 cm) cross sections (32–34). Powders are mixed using thermoplastic polymers and other organic additives. A molten mass composed of the ceramic and a thermoplastic binder system are injected via a heated extruder into a cooled mold of desired shape. The organic is burned out and the ceramic consolidated. Machining fragments from the green ceramic can be recycled because the thermoplastic polymers can be reversibly heated. Molds can be relatively expensive so injection molding is best suited to the preparation of a large number of single parts. Because of the high organic content required, organic removal is not trivial. Green sections greater than 1.0 cm thick require slow heating rates during burnout to avoid bloating and delamination of the green ceramic.

Densification.   Densification generally requires high temperatures to eliminate the porosity in green ceramics. Techniques include pressureless sintering, hot-pressing, and hot isostatic pressing (HIP). Pressureless sintering is the most widely used because of ease of operation and economics. Hot-pressing is limited to relatively simple shapes whereas more complex shapes can be consolidated using HIP (35). Sintering is used for most oxide electronic ceramics. Hot-pressing and HIP, which employ pressure and high temperatures, are used to consolidate ceramics in which dislocation motion (leading to pore elimination) is sluggish. Both techniques are particularly useful for nonoxide materials such as silicon nitride [*12033-89-5*] and silicon carbide [*409-21-2*] (35,36) (see CARBIDES; NITRIDES).

Special precautions are often used in the sintering of electronic ceramics. Heating rates and hold times at maximum temperature are critical to microstructural development and grain size control. Sintering cycles may include intermediate temperature annealing or controlled cooling to relieve residual strains or avoid deleterious phase transformations. Atmosphere control may be important to prevent loss of volatile components or avoid reduction reactions. In continuous production, sequential burnout (organics) and sintering may take place in the same furnace, requiring complex temperature cycles even for relatively simple devices. Complex devices such as thick film circuits and monolithic multicomponent ceramics may require many sequential fabrication and sintering steps.

Processing of Multilayer Ceramics

Rapid advances in integrated circuit technology have led to improved processing and manufacturing of multilayer ceramics (MLC) especially for capacitors and microelectronic packages. The increased reliability has been the result of an enormous amount of research aimed at understanding the various microstructural–property relationships involved in the overall MLC manufacturing process. This includes powder processing, thin sheet formation, metallurgical interactions, and testing.

Presently, multilayer capacitors and packaging make up more than half the electronic ceramics market. For multilayer capacitors, more than 20 billion units are manufactured a year, outnumbering by far any other electronic ceramic component. Multilayer ceramics and hybrid packages consist of alternating layers of dielectric and metal electrodes, as shown in Figures 5 and 6, respectively.

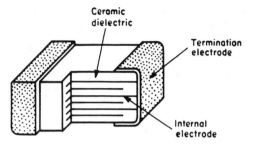

Fig. 5. Schematic cross section of a conventional MLC capacitor.

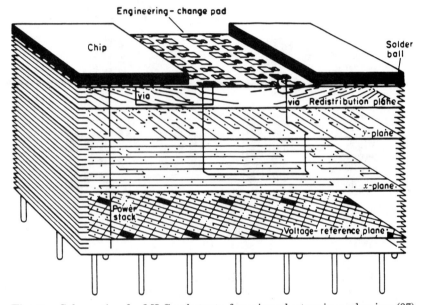

Fig. 6. Schematic of a MLC substrate for microelectronic packaging (37).

The driving force for these compact configurations is miniaturization. For capacitors, the capacitance (C) measured in units of farads, F, is

$$C = \frac{\epsilon_0 A K}{t}$$

where K is the dielectric constant (unitless); ϵ_0 the permittivity of free space = 8.85×10^{-12} F/m; A the electrode area, m^2; and t the thickness of dielectric layer, m. Thus C increases with increasing area and number of layers and decreasing thickness. Typical thicknesses range between 15 and 35 μm. Similarly, for substrate packages, the multilayer configuration incorporates transversely integrated conductor lines and vertical conducting paths (vias) allowing for numerous interconnects to components throughout the device system and power distribution in a relatively small space. MLC substrates capable of providing 12,000 electrical connections containing 350,000 vias are currently manufactured (38,39).

A number of processing steps, shown in Figure 7, are used to obtain the multilayer configuration(s) for the ceramic–metal composites. The basic process steps are slip preparation, green tape fabrication, via-hole punching (packages), printing of internal electrodes or metallization, stacking and laminating, dicing or dimensional control, binder burnout, sintering, end termination, and encapsulation. After each processing step, quality control in the form of nondestructive physical and electrical tests ensures a uniform end-product.

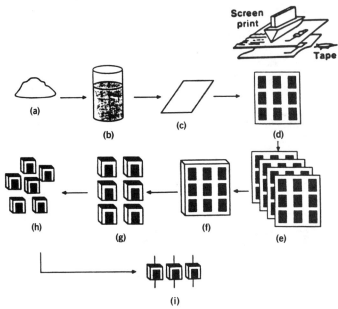

Fig. 7. Fabrication process for MLC capacitors. Steps are (**a**) powder; (**b**) slurry preparation; (**c**) tape preparation; (**d**) electroding; (**e**) stacking; (**f**) lamination; (**g**) dicing; (**h**) burnout and firing; and (**i**) termination and lead attachment.

The basic building block, the ceramic green sheet, starts using a mixture of dielectric powder suspended in an aqueous or nonaqueous liquid system or vehicle comprised of solvents, binders, plasticizers, and other additives to form a slip that can be cast in thin, relatively large area sheets. The purpose of the binder (20,000–30,000 molecular weight polymers) is to bind the ceramic particles together to form flexible green sheets. Electrodes are screened on the tape using an appropriate paste of metal powders. Solvents play a number of key roles, ranging from deagglomeration of ceramic particles to control the viscosity of the cast slip, to formation of microporosity in the sheet as the solvent evaporates. Plasticizers, ie, small to medium sized organic molecules, decrease cross-linking between binder molecules, imparting greater flexibility to the green sheet. Dispersants, typically 1,000 to 10,000 molecular weight polymer molecules, are added to slips to aid in the de-agglomeration of powder particles, allowing for higher green densities in the cast tape. Several review articles on the functional additives in tape cast systems are available (16,17,25,40–44). The resulting slip should have pseudoplastic rheological behavior so that the slip flows during high shear rate casting operations, but displays little or no flow afterward, thus maintaining tape dimension (45).

There are several methods to make large ceramic sheets for MLC manufacturing (17–23). The methods include glass, belt and carrier film casting, and wet lay down techniques. The relative advantages and limitations of each technique have been reviewed (46). The two most commonly employed techniques, belt casting and doctor blading, are depicted schematically in Figure 8.

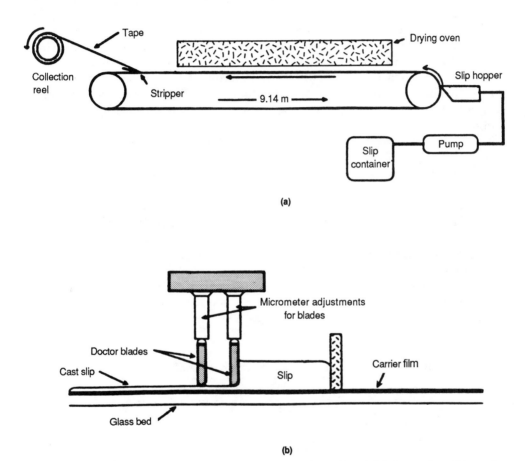

Fig. 8. Schematic of methods for MLC manufacturing; (**a**) belt casting; (**b**) carrier film casting using a doctor blade.

Metallization of the green sheets is usually carried out by screen printing, whereby a suitable metal ink consisting of metal powders dispersed in resin and solvent vehicles is forced through a patterned screen. Palladium [7440-05-3] and silver–palladium (Ag:Pd) alloys are the most common form of metallization; tungsten [7440-33-7] and molybdenum [7439-98-7] are used for high (>1500°C) temperature MLCs (47–52). Following screening, the metallized layers are stacked and laminated to register (align) and fuse the green sheets into a monolithic component. Proper registration is crucial to achieve and maintain capacitance design (MLC capacitors) and for proper via-hole placement in MLC packages.

Sintering is the most complex process in MLC fabrication. Ideally, the binder burnout and sintering steps are performed during the same temperature cycle and in the same atmosphere. Most binders burn out by 500°C, well before pore closure in the densification of most ceramics. Sintering behavior of the many different MLC components must be reconciled to achieve a dense material. Internal metallization and the dielectric must co-fire in a single process. Firing temperatures are related to material composition and can be adjusted using additives. Densification rates are related both to the process temperature and to particle characteristics (size, size distribution, and state of agglomeration). Thus, the burnout and sintering conditions depend heavily on the system.

After densification, external electrode termination and leads are attached for MLC capacitor components, and pin module assembly and IC chip joining is carried out for MLC packages. The devices are then tested to ensure performance and overall reliability.

Thick Film Technology

Equally important as tape casting in the fabrication of multilayer ceramics is thick film processing. Thick film technology is widely used in microelectronics for resistor networks, hybrid integrated circuitry, and discrete components, such as capacitors and inductors along with metallization of MLC capacitors and packages as mentioned above.

In principle, the process is equivalent to the silk-screening technique whereby the printable components, paste or inks, are forced through a screen with a rubber or plastic squeegee (see Fig. 7). Generally, stainless steel or nylon

Table 4. Components of Thick Film Compositions[a]

Component	Composition
functional phase	
conductor	Au, Pt/Au
	Ag, Pd/Ag
	Cu, Ni
resistor	RuO_2
	$Bi_2Ru_2O_7$
	LaB_6
dielectric	$BaTiO_3$
	glass
	glass–ceramic
	Al_2O_3
binder	glass: borosilicates, aluminosilicates
	oxides: CuO, CdO
vehicle	volatile phase: terpineol, mineral spirits
	nonvolatiles: ethyl cellulose, acrylates

[a]Ref. 53.

filament screens are masked using a polymeric material forming the desired printed pattern in which the composition is forced through to the underlying substrate.

Thick film compositions possess three parts: (1) functional phase, (2) binder, and (3) vehicle. The functional phase includes various metal powders for conductors, electronic ceramics for resistors, and dielectrics for both capacitors and insulation. Examples of typical components for thick film compositions are given in Table 4. The binder phase, usually a low ($<1000°C$) melting glass adheres the fired film to the substrate whereas the fluid vehicle serves to temporarily hold the unfired film together and provide proper rheological behavior during screen printing. Thick film processing for hybrid integrated circuits typically takes place below 1000°C providing flexible circuit designs.

Current and Future Developments in Multilayer Electronic Ceramics

Advances in the field of electronic ceramics are being made in new materials, novel powder synthesis methods, and in ceramic integration. Monolithic multicomponent components (MMC) take advantage of three existing technologies: (1) thick film methods and materials, (2) MLC capacitor processes, and (3) the concept of cofired packages as presented in Figure 9. Figure 10 shows an exploded view of a monolithic multicomponent ceramic substrate.

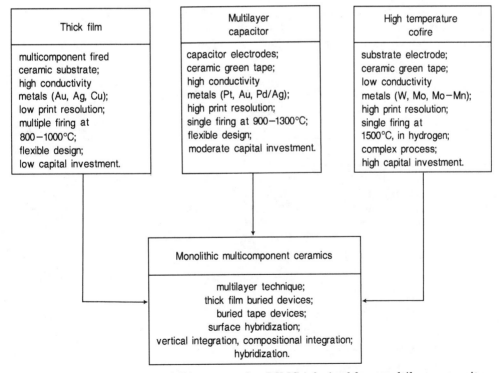

Fig. 9. Monolithic multilayer ceramics (MMCs) derived from multilayer capacitor, high temperature cofire, and thick film technologies.

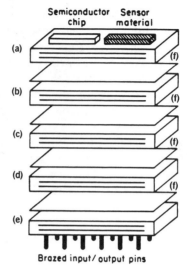

Fig. 10. Exploded view of a monolithic multicomponent ceramic substrate. Layers (**a**) signal distribution; (**b**) resistor; (**c**) capacitor; (**d**) circuit protection; and (**e**) power distribution are separated by (**f**) barrier layers.

New materials for packaging include aluminum nitride [*24304-00-5*], AlN, silicon carbide [*409-21-2*], SiC, and low thermal expansion glass-ceramics, replacing present day alumina packaging technology. As shown in Table 5, these new materials offer significant advantages to meeting the future requirements of the microelectronics industry. Properties include higher thermal conductivity,

Table 5. Properties of High Performance Ceramic Substrates[a]

Properties	AlN	SiC	Glass-ceramics	90% Al_2O_3
thermal conductivity, W/(m·K)	230	270	5	20
thermal expansion coefficient, RT − 400°C × 10^{-7}/°C[b]	43	37	30–42	67
dielectric constant at 1 MHz	8.9	42	3.9–7.8	9.4
flexural strength, kg/cm^2	3500–4000	4500	1500	3000
thin film metals	Ti/Pd/Au Ni/Cr/Pd/Au	Ti/Cu	Cr/Cu, Au	Cr/Cu
thick film metals	Ag–Pd Cu	Au Ag–Pd	Au, Cu, Ag–Pd	Ag–Pd Cu, Au
cofired metals	W	Mo	Au–Cu, Ag–Pd	W, Mo
cooling capability, °C/W				
air	6	5	60	30
water[c]	<1	<1	<1	<1

[a]Ref. 39.
[b]RT = room temperature.
[c]External cooling.

lower dielectric constant, cofire compatibility, and related packaging characteristics such as thermal expansion matching of silicon and high mechanical strength as compared to Al_2O_3.

Greater dimensional control and thinner tapes in multilayer ceramics are the driving forces for techniques to prepare finer particles. Metal organic decomposition and hydrothermal processing are two synthesis methods that have the potential to produce submicrometer powders having low levels of agglomeration to meet the demand for more precise tape fabrication.

As stated above, the development of multifunctional MLCs based on existing technologies offers excellent growth potential since MMCs combine the possibilities of both the high cofire (packaged) substrates and burial of surface devices (54–57). Burial of surface devices promises gains in both circuit density and device hermiticity, leading to increased reliability. Processing trade-offs are expected since current electronic materials for multilayer applications (capacitors, transducers, sensors) are densified at very different firing temperatures. Consequently, integrated components will likely be of lower tolerance and limited range, at least in the early developmental stages. Current efforts have been directed toward incorporation of multilayer capacitor-type power planes and burial of thick film components, including resistors and capacitors. The latter processing technology offers more immediate possibilities as it is developed to cofire at conventional thick film processing temperatures for which a wide range of materials exist.

The continuing miniaturization of electronic packaging should see the replacement of components and processes using such thin film technologies developed for semiconductors as sputtering, chemical vapor deposition, and sol–gel (see SOL–GEL TECHNOLOGY; THIN FILMS) (58,59). Sputtering is the process whereby a target material is bombarded by high energy ions which liberate atomic species from the target for deposition on a substrate. Chemical vapor deposition (CVD) involves a gaseous stream of precursors containing the reactive constituents for the desired thin film material, generally reacted on a heated substrate. The more recent process for thin films, sol–gel, uses a nonaqueous solution of metal–organic precursor. Through controlled hydrolyses, a thin, adherent film is pre-

Table 6. Current and Future Developments in Thin Film Electronic Ceramics[a]

Material	Application	Methods
PT, PZT, PLZT	nonvolatile memory, ir, pyroelectric detectors, electro–optic waveguide, and spatial light modulators	sol–gel, sputtering
diamond (C)	cutting tools, high temperature semiconductors, protective optical coatings	chemical vapor deposition (CVD)
SiO_2, $BaTiO_3$	capacitors	sol–gel, sputtering, chemical vapor deposition (CVD)
1:2:3 superconductors	squids, nmr, interconnects	

[a]Refs. 58 and 59.

pared by dip-coating or spin-coating. The dried "gel" film is then crystallized and densified through heat treatments. Both existing and future developments of thin film electronic ceramics and methods are presented in Table 6.

BIBLIOGRAPHY

1. *Japan Electronics Almanac,* Dempa Publications, Inc., Tokyo, 1986, p. 412.
2. D. W. Johnson in G. Y. Chin, ed., *Advances in Powder Technology,* American Society for Metals, Metals Park, Ohio, 1982, pp. 23–37.
3. K. Osseo-Asare, F. J. Arriagada, and J. H. Adair, "Solubility Relationships in the Coprecipitation Synthesis of Barium Titanate: Heterogeneous Equilibria in the Ba–Ti–C_2O_4–H_2O System," in G. L. Messing, E. R. Fuller, Jr., and Hans Hausin, eds., *Ceramic Powder Science,* Vol. 2, 1987, pp. 47–53.
4. D. Miller, J. H. Adair, W. Huebner, and R. E. Newnham, "A Comparative Assessment of Chemical Synthesis Techniques for Barium Titanate," Paper, 88th Annual Meeting of the American Ceramic Society, Pittsburgh, Pa., April 27–30, 1987.
5. B. J. Mulder, *Am. Ceram. Soc. Bull.* **49**(11), 990–993 (1970).
6. E. P. Stambaugh and J. F. Miller, "Hydrothermal Precipitation of High Quality Inorganic Oxides," in S. Somiya, ed., *Proceedings of First International Symposium on Hydrothermal Reactions,* Gakujutsu Bunken Fukyu-kai (c/o Tokyo Institute of Technology), Tokyo, Japan, 1983, pp. 859–872.
7. K. S. Mazdiyasni, C. T. Lynch, and J. S. Smith, *J. Ceram. Soc.* **48**(7), 372–375 (1965).
8. R. R. Wills, R. A. Markle, and S. P. Mukherjee, *Am. Ceram. Soc. Bull.* **62**(8), 904–911 (1983).
9. R. West, X.-H. Zhang, I. P. Djurovich, and H. Stuger, "Crosslinking of Polysilanes as Silicon Carbide Precursors," in L. L. Hench and D. R. Ulrich, eds., *Science of Ceramic Chemical Processing,* John Wiley & Sons, New York, 1986, pp. 337–344.
10. K. K. Verna and A. Roberts in G. Y. Onoda, Jr., and L. L. Hench, eds., *Ceramic Processing Before Firing,"* John Wiley & Sons, Inc., New York, 1978, pp. 391–407.
11. J. H. Adair, A. J. Roese, and L. G. McCoy, "Particle Size Analysis of Ceramic Powders," in K. M. Nair, ed., *Advances in Ceramics,* Vol. 2, The American Ceramic Society, Columbus, Ohio, 1984.
12. J. W. McCauley, *Am. Chem. Soc. Bull.* **63**(2), 263–265 (1984).
13. G. L. Messing, C. J. Markhoff, and L. G. McCoy, *Am. Ceram. Soc. Bull.* **61**(8), 857–860 (1982).
14. D. E. Niesz and R. B. Bennett, in ref. 10, pp. 61–73.
15. C. Greskovich, "Milling" in F. F. Y. Wang, ed., *Treatise on Materials Science and Technology,* Vol. 9, Academic Press, New York, 1976.
16. R. E. Mistler, D. J. Shanefield, and R. B. Runk, in ref. 10, pp. 411–448.
17. J. C. Williams, "Doctor-Blade Process," in F. F. Y. Wang, ed., *Treatise on Materials Science and Technology,* Vol. 9, Academic Press, New York, 1976.
18. U.S. Pat. 3,717,487 (1973) (to Sprague Electric Company).
19. B. Schwartz and D. L. Wilcox, *Ceramic Age,* 40–44 (June 1967).
20. R. B. Runk and M. J. Andrejco, *Am. Ceram. Soc. Bull.* **54**(2), 199–200 (1975).
21. C. Wentworth and G. W. Taylor, *Am. Ceram. Soc. Bull.* **46**(12), 1186–1193 (1967).
22. R. E. Mistler, *Am. Ceram. Soc. Bull.* **52**(11), 850–854 (1973).
23. J. M. Herbert, *Methods of Preparation, Ceramic Dielectrics and Capacitors,* Gordon and Breach Science Publishers, New York, 1985, Chapt. 3.
24. F. H. Norton, *Forming Plastic Masses, Fine Ceramics: Technology and Applications,* Robert E. Krieger Publishing, Huntington, NY, 1978, Chapt. 10.

25. T. Morse, *Handbook of Organic Additives for Use in Ceramic Body Formulation,* Montana Energy and MHD Research and Development Institute, Inc., Butte, Mont., 1979.
26. D. B. Quinn, R. E. Bedford, and F. L. Kennard, "Dry-Bag Isostatic Pressing and Contour Grinding of Technical Ceramics," in J. A. Mangels and G. L. Messing, eds., *Advances in Ceramics,* Vol. 9 (Forming of Ceramics), 1984, pp. 4–31.
27. I. A. Aksay, F. F. Lange, and B. I. Davis, *J. Am. Ceram. Soc.* **66**(10), C190–C192 (1983).
28. F. F. Lange, B. I. Davis, and E. Wright, *J. Am. Ceram. Soc.* **69**(1), 66–69 (1986).
29. I. A. Aksay and C. H. Schilling, in ref. 26, pp. 85–93.
30. G. N. Howatt, R. G. Breckenridge, and J. M. Brownlow, *J. Am. Ceram. Soc.* **30**(8), 237–242 (1947).
31. J. J. Thompson, *Am. Ceram. Soc. Bull.* **42**(9), 480–481 (1963).
32. J. A. Mangels and W. Trela, in ref. 26, pp. 85–93.
33. T. J. Whalen and C. F. Johnson, *Am. Ceram. Soc. Bull.* **60**(2), 216–220 (1981).
34. M. J. Edirisinghe and J. R. G. Evans, *Int. J. High Technol. Ceram.* **2**(1), 1–31 (1986).
35. R. R. Wills, M. C. Brockway, and L. G. McCoy, "Hot Isostatic Pressing of Ceramic Materials," in R. F. Davis, H. Palmour III, and R. L. Porter, eds., *Materials Science Research,* Vol. 17 (Emergent Process Methods for High-Technology Ceramics), Plenum Press, New York, 1984.
36. M. H. Leipold, "Hot Pressing," in F. F. Y. Wang, ed., *Treatise on Materials Science and Technology,* Vol. 9 (Ceramic Fabrication Processes), Academic Press, New York, 1976.
37. A. J. Blodgett, Jr., *Sci. Am.* **249**(1), 86–96 (1983).
38. R. R. Tummala and E. J. Rymaszewski, *Microelectronics Packaging Handbook,* Van Nostrand Reinhold, New York, 1989.
39. R. R. Tummala, *Am. Ceram. Soc. Bull.* **67**(4), 752–758 (1988).
40. D. J. Shanefield and R. S. Mistler, *Am. Ceram. Soc. Bull.* **53**(5), 416–420 (1974).
41. D. J. Shanefield and R. S. Mistler, *Am. Ceram. Soc. Bull.* **53**(8), 564–568 (1974).
42. R. A. Gardner and R. W. Nufer, *Solid State Technol.* (May 8–13, 1974).
43. A. G. Pincus and L. E. Shipley, *Ceram. Ind.* **92**(4), 106–110 (1969).
44. N. Sarkar and G. K. Greminger, Jr. *Am. Ceram. Soc. Bull.* **62**(11), 1280–1284 (1983).
45. G. Y. Onoda, Jr., in ref. 10, pp. 235–251.
46. J. H. Adair, D. A. Anderson, G. O. Dayton, and T. R. Shrout, *J. Mater. Ed.* **9**(1,2), 71–118 (1987).
47. D. A. Chance, *Met. Trans.* **1**, 685–694 (March 1970).
48. I. Burn and G. H. Maher, *J. Mater. Sci.* **10**, 633–640 (1975).
49. U.S. Pat. 4,075,681 (Feb. 1978), M. J. Popowich.
50. T. L. Rutt and J. A. Syne, "Fabrication of Multilayer Ceramic Capacitor by Metal Impregnation," *IEEE Trans. Parts Hybrids Packag.,* **PHP-9,** 144–147 (1973).
51. D. A. Chance and D. L. Wilcox, *Met. Trans.* **2**, 733–741 (March 1971).
52. D. A. Chance and D. L. Wilcox, *Proc. IEEE* **59**(10), 1455–1462 (1971).
53. L. M. Levinson, *Electronic Ceramics,* Marcel Dekker, Inc., New York, 1988, Chapt. 6.
54. K. Utsumi, Y. Shimada, and H. Takamizawa, "Monolithic Multicomponent Ceramic (MMC) Substrate," in K. A. Jackson, R. C. Pohanka, D. R. Ulhmann, and D. R. Ulrich, eds., *Electronic Packaging Materials Science,* Materials Research Society, Pittsburgh, Pa., 1986, pp. 15–26.
55. W. A. Vitriol and J. I. Steinberg, "Development of a Low Fire Cofired Multilayer Ceramic Technology," 1983, pp. 593–598.
56. H. T. Sawhill and co-workers, "Low Temperature Co-Firable Ceramics with Co-Fired Resistors," International Society of Hybrid Microelectronics Proceedings, 1986, pp. 473–480.
57. C. C. Shiflett, D. B. Buchholz, and C. C. Faudskar, "High-Density Multilayer Hybrid Circuits Made with Polymer Insulating Layers (Polyhic's)," Society of Hybrid Microelectronics Proceedings, 1980, pp. 481–486.

58. S. L. Swartz, "Topics in Electronic Ceramics," *IEEE Trans. Elect. Insul.* Digest on Dielectrics **25**, 935–987 (Oct. 1990).
59. C. P. Poole, Jr., T. Datta, and H. A. Farach, *Copper Oxide Superconductors,* John Wiley & Sons, New York, 1988.

General references

R. C. Buchanan, ed., *Ceramic Materials for Electronics,* Marcel Dekker, Inc., New York, 1986.
L. M. Levinson, ed., *Electronic Ceramics,* Marcel Dekker, Inc., New York, 1988.
B. Jaffe, W. R. Cook, Jr., and H. Jaffe, *Piezoelectric Ceramics,* Academic Press, New York, 1971.

ROBERT E. NEWNHAM
THOMAS R. SHROUT
Pennsylvania State University

STRUCTURAL CERAMICS

Advanced structural ceramics are those ceramics intended for use as load-bearing members. They are materials that combine the properties and advantages of traditional ceramics (qv), such as chemical inertness, high temperature capability, and hardness, with the ability to carry a significant mechanical stress. Like all ceramics, they are inorganic and nonmetallic; in addition, they are often multicomponent and/or multiphased materials having complex crystal structures. These materials are usually intended to be fully dense and to have tight dimensional tolerances. In addition to being designed to withstand substantially higher levels of mechanical and thermal stress, there are other important features which make advanced structural ceramics different from traditional ones. Starting powders, compositions, processing, and resulting microstructure must be carefully controlled to provide required levels of performance. Consequently, advanced structural ceramics are more expensive than traditional ceramics.

Most of the advanced structural ceramics under development today are based on silicon nitride [*12033-89-5*], Si_3N_4; silicon carbide [*409-21-2*], SiC; zirconia [*1314-23-4*], ZrO_2; or alumina [*1344-28-1*], Al_2O_3. In addition, materials such as titanium diboride [*12045-63-5*], TiB_2; aluminum nitride [*24304-00-5*], AlN; silicon aluminum oxynitride [*52935-33-8*], SiAlON; and some other ceramic carbides and nitrides are often classified as advanced or high tech ceramics because of processing methods or applications. Ceramic matrix composites are also receiving increasing attention as advanced structural ceramics (see COMPOSITE MATERIALS, CERAMIC MATRIX). Monolithic silicon nitride, silicon carbide, and zirconia each represent a family of materials rather than a single species. A wide range of microstructures and properties can be tailored within each family, through compositional or processing modifications, in order to optimize materials performance for specific applications.

Physical Properties

Advanced structural ceramics typically possess some combination of high temperature capabilities, high strength, toughness or flaw tolerance, high hardness, mechanical strength retention at high temperatures, wear resistance, corrosion resistance, thermal shock resistance, creep resistance, and long term durability. Figure 1 shows the typical stress and temperature ranges of application for SiC, Si_3N_4, and ZrO_2. Zirconia ceramics, which find application under conditions of high stress and moderately high (up to 600°C) temperatures, have the highest low temperature strength. Although the low temperature strength of silicon nitride is less than that for zirconia, silicon nitride maintains these strength properties up to approximately 1200°C. Silicon carbide is somewhat weaker than silicon nitride over the entire temperature range, but maintains good strength and creep resistance at even higher (1500°C) temperatures.

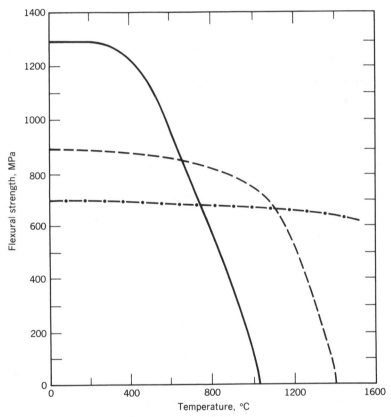

Fig. 1.   Stress and temperature ranges of application for ZrO_2 (——), Si_3N_4 (‑‑‑), and SiC (‑·‑·‑) advanced structural ceramics. To convert MPa to psi, multiply by 145.

Processing and Fabrication Technology

The relationship between processing and properties is especially critical for advanced structural ceramics because subsequent successful operation in severe environments often requires carefully controlled compositions and microstructures. Fabrication generally takes place in four steps: powder processing, consolidation/forming, densification, and finishing. Starting powders must be chemically pure and fine grained. Then, depending on the forming and densification processes to be used and the final properties and microstructure desired, the powders may be mixed with various additives. For example, additives can be used to improve the flowability of dry powders to make mold filling easier. Plasticizers may be added to improve the formability of powder blends for some shape forming operations. Binders are almost always added to powder blends, especially those intended for dry forming, in order to improve adherence of the fine powder particles and impart strength to the green part. (Green refers to powder compacts, formed by any process, which have not yet been subjected to a densification operation.) Sintering additives are also necessary for covalent materials, particularly nonoxides, to enhance densification rates.

Once the powder has been processed and the composition set, several techniques, including dry pressing (uniaxial or isostatic), slip casting, plastic forming (extrusion or injection molding), and tape casting, can be used for forming. In a dry pressing operation, powders are fed into a die or cavity and compacted under pressure. In general, pressure may be divided into two categories: uniaxial (unipressing) and isostatic (isopressing). For unipressing the powder is fed into a die and pressure is externally applied along a single axis, which limits this technique to relatively simple shapes. For isopressing, powder is fed into a compressible mold or bag and pressure is applied uniformly from all directions by a liquid or gas medium. Slip casting involves adding the ceramic powder to a liquid medium that is typically aqueous to produce a slurry which is then poured into a porous mold. In time, water from the slurry is absorbed by the mold and a solid ceramic results. This technique is most suitable for low volume production of relatively simple shapes and traditionally has been used for hollow or tubular components. Advances in slurry compositions, mold materials, and mold design have resulted in faster casting times however, permitting thicker cross-section parts and more complex shapes.

For plastic forming techniques the ceramic powder is combined with plasticizers such as thermoplastic resins and other additives to make a mixture which is deformable under pressure. This mixture is then heated slightly, facilitating plasticity, and either forced through dies (extrusion) or into molds (injection molding). Extrusion is a continuous, high volume process, but is limited to shapes having a constant cross section, eg, rods and tubes. Injection molding is a high volume process capable of producing complex shapes, but tooling costs for molds can be very high.

Regardless of the consolidation method used, the formed or green part must generally undergo a burn-out step prior to densification in order to remove the binders, plasticizers, and other decomposable additives which were added in the forming step. In the case of dry pressing usually only small amounts (several volume percent) of binders are added, whereas for injection-molded components, the additives can comprise 30–40 vol % of the green part. Additives are typically organic compounds that decompose at temperatures less than 700°C. However, in order to ensure complete removal of additives without disruption to the part, eg, swelling or cracking, burn-out must be done slowly and under carefully controlled conditions. Burn-out may be a separate step or it may be incorporated in the early stages of the densification procedure.

High temperature consolidation techniques such as conventional sintering, reaction sintering, hot-pressing, and hot isostatic pressing, are generally necessary for advanced structural ceramics. Sintering involves subjecting a powder compact to high temperature without application of pressure and, especially in the case of nonoxides, usually requires additives to promote densification and/or to inhibit grain growth. Achieving full density may be more difficult than in pressure-assisted methods and the composition and quantity of densification aids required may degrade the material's high temperature properties. Reaction sintering or reaction bonding involves the infiltration of a powder bed by the appropriate gaseous constituent or molten material at elevated temperatures to produce the desired composition. Complex shapes can be formed using the starting powder and little or no shrinkage occurs during the reaction. However, residual unreacted starting material and/or residual porosity in the final product can be a problem.

Hot-pressing is the simultaneous application of uniaxial high pressure and temperature to the powder. Fully dense materials can be formed, often having high strength, but the application of uniaxial pressure limits this technique to simple shapes. In the case of hot isostatic pressing (HIP), pressure is applied to the powder compact equally in all directions through the use of a compressed gas. Fabrication of more complex shapes, not possible using hot-pressing, can be accomplished, and densities approaching theoretical are often achieved. However, this technique requires that the material be either sintered to closed porosity prior to application of isostatic pressure or encased in a compressible can of metal or molten glass.

Microwave sintering of powders, a relatively new technique, has the advantage of providing more uniform heating of the component, because it does not rely on conduction and convection. Other routes to forming and densification which avoid traditional powder processing are also under development. Sol–gel processing, chemical vapor deposition (CVD) and organometallic polymer pyrolysis are examples.

Although some postforming grinding and machining are often necessary, the intrinsic hardness of advanced structural ceramics makes them difficult and costly to machine. In addition, grinding can introduce surface flaws which may serve as failure sites. Thus, forming processes producing near net-shape components, such as injection molding and hot isostatic pressing, are desirable because their usage reduces the amount of postforming machining required.

Applications

Throughout the development of structural ceramics the focus has been on applications for gas turbine, diesel, and spark-ignited engines. The ability of ceramics to function at higher temperatures than superalloys, and to do so without cooling, has been a particularly important driving force. Ceramics utilization in heat engines can lead to reduced fuel consumption and increased performance through higher engine operating temperatures, the elimination of mechanical losses resulting from cooling, lower inertia, and reduced friction. In addition to the properties critical to mechanical performance, advanced structural ceramics have the significant advantage of not requiring imported, strategic metals (eg, Ni, Co, Cr) for their fabrication.

Advanced structural ceramics are also under investigation for use in numerous other high performance applications including antifriction roller and ball bearings, metal-cutting and shaping tools, hot extrusion and hot forging dies, industrial wear parts (eg, sand-blast nozzles, pump seals, thread guides, chute liners), and various military applications (eg, armor, radomes, ir domes, gun barrel liners). More comprehensive works on the processing, properties, and applications of advanced ceramics are available (1–4).

Silicon Carbide Structural Ceramics

Silicon carbide (see CARBIDES, SILICON CARBIDE) has been a candidate material for structural ceramic applications far longer than other materials. Properties such as the relatively low thermal expansion, high strength-to-weight ratio, high stiffness, high thermal conductivity, hardness, erosion and corrosion resistance, and most importantly, the maintenance of strength as high as 1650°C, have led to a wide range of applications. Additionally, it is possible to produce both large quantities of pure silicon carbide powders and required component shapes. An in-depth review of SiC structural ceramics is available (5).

Material System.   Silicon carbide occurs in a variety of polymorphic crystalline forms, generally designated β-SiC for the cubic form and α-SiC for the hexagonal and rhombohedral varieties. The alpha form appears most stable at temperatures above 2000°C, whereas the cubic β-SiC is the most common product when silicon carbide is produced at lower temperatures (4). Most silicon carbide powder is produced by the Acheson process involving the reduction of high purity silica sand surrounding an electrically heated core of petroleum coke or anthracite coal (6). The reaction is carried out at about 2400°C in the core for as long as 36 h where the higher temperature α-SiC is formed. The product is separated based on purity (determined by crystal color) and ground. For structural ceramic applications this material must be milled, often to submicrometer sizes, and chemically cleaned of impurities. The large scale of this process leads to a relatively low cost for such a high purity raw material.

α-SiC can also be produced directly in the desired purity by the plasma gas-phase reaction of species such as silane [7803-62-5], SiH_4, and methane [74-82-8], CH_4 (7). β-SiC powders can be produced by the same gas-phase reaction at lower temperature (1500–1600°C) or by polymer decomposition reactions (8).

Fabrication Technology.   Silicon carbides for structural application can be classed as reaction-bonded, liquid-phase sintered, and solid-state sintered. Reaction-bonded SiCs are actually a composite of a continuous SiC matrix having 5 to 20% silicon [7440-21-3], metal filling the remaining volume (9). To form this material, a preform of powder containing carbon [7440-44-0] added either as a powder or as the decomposition product of a carbon source resin, is infiltrated with silicon at about 1500°C either through direct contact or using silicon vapor. The silicon reacts with the carbon preform to form a bridging structure of more SiC. Excess silicon remains, filling the residual pore space and giving a fully dense product having structural integrity to 1370°C. Silicon melts at 1410°C. The preform can be fabricated by any of the traditional ceramic shaping processes (5,10). The silicon carbide powder utilized for the preform does not require the submicrometer particle sizes and the purity of other forms of dense SiC, although finer sized SiC preforms tend to give a stronger product (11). Reaction bonding also leads to little dimensional change (<1%) from the preform, allowing large shapes and complex shapes having tight dimensional tolerances to be fabricated. Moreover, the low (1500°C) temperatures employed during reaction bonding, combined with the flexibility in powder size and purity, provide a good quality product at a reasonable cost.

A desire for a higher strength form of dense SiC suitable for applications at temperatures above 1300–1400°C led to the development of liquid-phase sintered material. This process involves reaction of an oxide additive (typically 1 to 2% Al_2O_3) and the silica present on the surface of SiC grains to create a melt during densification above 2000°C. A simultaneous application of pressure through hot-pressing is often required to achieve a dense product and the resultant hot-pressed SiC exhibits excellent room temperature strength and wear resistance. However, the hot-pressing process is limited to the production of simple shapes, and the presence of the aluminosilicate phase tends to decrease elevated temperature strength and oxidation resistance (12). Liquid-phase sintered SiC usage has therefore been limited for applications requiring complex shapes and long term exposure to high temperatures.

A new type of liquid-phase sintered SiC using yttria [1314-36-9], Y_2O_3, as the oxide additive and submicrometer SiC powder for enhanced densification, produces a material which can be densified without the application of pressure (13). This material, sintered from cold isostatically pressed billets, appears to be comparable to silicon nitride in strength and fracture toughness.

Two forms of dense high purity SiC resulted from a discovery (14) that simultaneous additions of carbon and boron [7440-42-4] allow silicon carbide densification without the application of pressure. One is a sinterable material based on β-SiC raw materials and about 1% carbon and 0.5% boron (15). Control of powder reactivity and densification conditions is critical because densification temperatures approach the region for transformation to the alpha form of SiC resulting in excessive grain growth rates and resultant lowering of final strength. The other product results from the same additive system, but is based on the more readily available α-SiC (16). Ultrafine SiC powder is utilized both to increase the driving force for densification and to minimize diffusion distances (17). There has been extensive analysis of the role of these two additives in densification (18–21). The carbon appears to react with surface oxygen present on the SiC grains, forming volatile CO and thereby purifying the surfaces. The role of boron is more difficult to define, but it is believed to promote both volume diffusion and grain boundary diffusion.

Both α- and β-sintered silicon carbides have been fabricated by all of the available ceramic forming methods. Sintering is performed in vacuum or in an argon atmosphere (22). Shrinkages during densification are typically 15 to 20 linear percent. The resultant material is about 98% dense having small isolated porosity and residual carbon particles. Analysis indicates no presence of second phases along the grain boundaries. This leads to a material with excellent retention of strength at temperatures in excess of 1500°C and superior resistance to elevated temperature oxidation (23,24).

A postsintering HIP treatment has been used to further improve the strength of sintered silicon carbide (25). The component is sintered until all surface porosity is closed, about 95% dense or higher, then subjected to a combination of temperature and argon gas pressure to achieve final densification. Finer grain sizes are achieved in the final body because sintering temperatures tend to be lower. The resultant material shows good density uniformity, even in thick cross-section components, and density levels over 99% of theoretical density. If densification is performed properly, a significant increase and improved uniformity in strength results (26).

Properties of Dense Silicon Carbide. Properties of the SiC structural ceramics are shown in Table 1. These properties are for representative materials. Variations can exist within a given form depending on the manufacturer. Figure 2 shows the flexure strength of the SiC as a function of temperature. Sintered or sinter/HIP SiC is the preferred material for applications at temperatures over 1400°C and the liquid-phase densified materials show best performance at low temperatures. The reaction-bonded form is utilized primarily for its ease of manufacture and not for superior mechanical properties.

Table 1. Properties of Silicon Carbide Ceramics

Property	Material densification mode				
	Reaction-bonded	Sintered alpha[a]	Sintered beta	Hot-pressed (Al_2O_3)[b]	Sintered (Y_2O_3)[c]
density, kg/m^3	3.1	3.1	3.0	3.3	3.2
hardness, kg/mm^2	1620	2800		2400	
flexure strength, MPa[d] at 25°C	245	460	490	702	917
Young's modulus, GPa[e]	383	410	372	446	
Poisson's ratio, GPa[e]	0.24	0.14	0.16	0.17	
thermal expansion coeff, $\times 10^{-6}/°C$	4.8	4.02	4.4	4.6	
thermal conductivity, W/(m·K) at 25°C	135	126	71	80	

[a]Material subjected to a postsintering HIP treatment; sinter/HIP alpha SiC has a density of 3.2 g/mL and a flexural strength of 530 MPa at 25°C.
[b]Contains about 2% alumina.
[c]Contains some yttria, see text.
[d]To convert MPa to psi, multiply by 145.
[e]To convert GPa to psi, multiply by 145,000.

Flexural stress SiC rupture curves are shown in Figure 3 (27). All the forms tend to be fairly resistant to time-dependent failure by elevated temperature creep. In addition, SiC shows outstanding resistance to oxidation even at 1200°C as a result of formation of a protective high purity silica surface layer (28).

The elastic modulus and thermal expansion properties are dominated by the characteristics of the SiC crystal itself and the thermal conductivity or thermal diffusivity of silicon carbides tends to be substantially higher than those of other structural ceramics. Thermal diffusivity as a function of temperature is shown in Figure 4 (29). These values tend to be sensitive to the form of silicon carbide, but all values drop significantly as temperature increases. The combination of a high

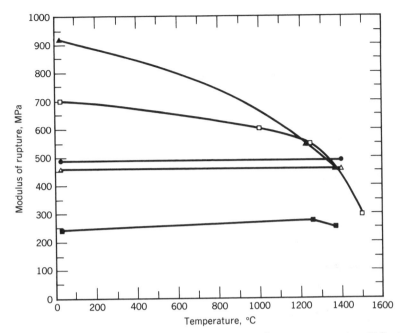

Fig. 2. Strength as a function of temperature for representative SiC structural ceramics: ▲, sintered (Y_2O_3 added); □, hot-pressed (2% Al_2O_3); ●, sintered beta; △, sintered alpha; and ■, reaction-bonded. To convert MPa to psi, multiply by 145.

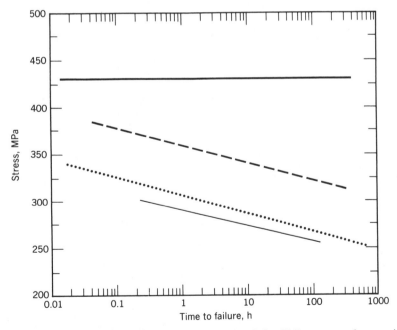

Fig. 3. Stress rupture behavior in air at 1200°C for SiC structural ceramics: ——, hot-pressed; ‒‒‒, reaction-bonded; ⋯, sintered alpha; ——, sintered beta. To convert MPa to psi, multiply by 145.

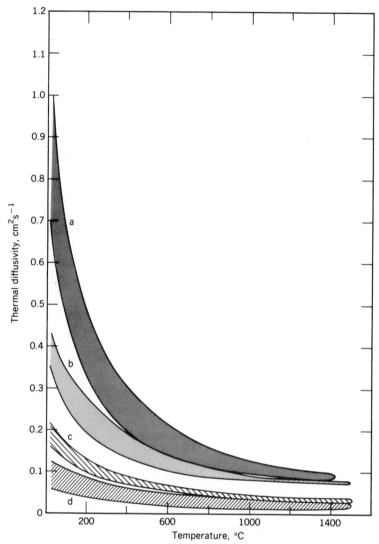

Fig. 4. Thermal diffusivity of silicon-based structural ceramics: (a) reaction-bonded SiC; (b) hot-pressed and sintered SiC; (c) hot-pressed Si_3N_4 (1% MgO, 8% Y_2O_3); (d) RS–Si_3N_4 (density is 2.1–2.9 g/mL).

elastic modulus and moderate thermal expansion coefficient result in SiC being susceptible to damage by thermal shock. Resistance to thermal shock is significantly lower than that of silicon nitride, but higher than that of the high expansion zirconia structural ceramics. Thermal shock behavior is also very application-dependent. For example, very rapid temperature changes can lead to a preference of Si_3N_4 over SiC, whereas during moderate rates of temperature change the high thermal conductivity of SiC can lead to a superior performance.

Fracture toughness of SiC tends to be lower than that of other structural ceramics leading to some concern about the application of SiC in certain heat engines, such as turbine rotors which may be susceptible to impact from foreign objects (30). The yttria liquid-phase sintered SiC is, however, reported to be comparable to other structural ceramics in fracture toughness. Erosion and corrosion characteristics have not been measured as extensively as other mechanical properties. Wear and coefficient of friction measurements have mostly been application specific, but point out the importance of surface preparation and characterization. Published erosion results show good resistance to angular particle or slurry erosion. Reaction-bonded SiC tends to be the most susceptible to erosive wear because of preferential wear of surface connected free silicon grains (31). Reaction-bonded SiC also appears much less resistant to acids, alkali, and high temperature combustion products than the single-phase sintered material (32). In contact with sodium sulfate, or acidic or basic coal slags from coal gasification, SiC tends to corrode slightly in a pitting reaction. In basic coal slag reactions at temperatures from 1000 to 1300°C, the reaction involves dissolution of the protective silica oxidation layer followed by reaction with Fe or Ni to form low-melting point silicides (33). Sintered silicon carbide has also been shown to corrode at elevated temperature in hydrogen-containing atmospheres. The reaction appears to be a decarburization of the SiC, particularly at grain boundaries, resulting in silicon rich regions and some grain fallout (34). Corrosion from sodium silicate glass vapors and particulates has demonstrated that both sintered and reaction-bonded SiC corrode through passive oxidation followed by dissolution of the oxide coating. The silicon component in reaction-bonded SiC was oxidized more rapidly than the SiC phase (35).

Applications of Silicon Carbide. Silicon carbides are used more for the low temperature wear properties than for the high temperature behavior. Applications such as sand blasting nozzles, automotive water pump seals, bearings, pump components, and extrusion dies utilize the high hardness, abrasion resistance, and corrosion resistance of silicon carbide (4,5,32). Elevated temperature structural applications range from rocket nozzle throats to furnace rollers and the combination of high thermal conductivity and high temperature strength and stability make silicon carbide heat exchanger tubes and diffusion furnace components feasible.

Most engine applications involve auxiliary components such as turbocharger rotors, valve train parts to reduce friction losses, piston wrist pins, and precombustion chambers. Application of SiC for pistons and cylinder liners has been demonstrated, but the high thermal conductivity makes SiC use more difficult than that of other structural ceramics. However, high thermal conductivity and strength at high temperatures make SiC the material of choice for combustors. Well-developed fabrication technology and lower raw material cost have also resulted in the use of SiC for many hot path stationary gas turbine components. SiC turbine rotors and vanes have also been demonstrated, but material strength considerations have often resulted in the selection of Si_3N_4.

Future applications may involve use of SiC as substrates for silicon chips, making use of the high thermal conductivity of SiC and its close thermal expansion match to silicon. The low density and high stiffness of silicon carbides may also result in applications in space. One such application is for space-based mirrors, making use of the high degree of surface polish possible on dense SiC.

Silicon Nitride Structural Ceramics

Silicon nitride (see NITRIDES) is a key material for structural ceramic applications in environments of high mechanical and thermal stress such as in vehicular propulsion engines. Properties which make this material uniquely suitable are high mechanical strength at room and elevated temperatures, good oxidation and creep resistance at high temperatures, high thermal shock resistance, excellent abrasion and corrosion resistance, low density, and, consequently, a low moment of inertia. Additionally, silicon nitride is made from abundant raw materials.

Material System. There are two basic techniques for the industrial synthesis of Si_3N_4 powder, although other methods are available (36). The older and most widely used method is the nitridation of silicon. Silicon is heated in a nitrogen [7727-37-9] atmosphere at temperatures of 1100–1450°C in the presence of an iron catalyst (37). The purity of the product depends on the purity of the starting materials, the amount of catalyst used, and the extent to which the catalyst is removed (38). The other commercial process is a type of ammonolysis where silicon tetrachloride [10026-04-7] or a silane reacts with liquid ammonia [7664-41-7] at low temperatures. The silicon compound is dissolved in an aromatic solvent such as toluene and silicon imide is formed at the interface between the liquid ammonia and the organic phase. The silicon imide is separated and thermally converted to crystalline silicon nitride (34) resulting in a powder of high purity.

Silicon nitride exists in two hexagonal crystallographic modifications designated as the α- and β-phases (40). The latter is prevalent at high temperatures. Metallic impurities are deleterious to Si_3N_4 thermomechanical properties and in the purest powders their total concentration does not exceed 100 ppm. Tolerance levels for the various metals vary, but any contaminant must be homogeneously dispersed in the powder rather than present in discrete particles. Alkalies and metals forming low melting glasses are wholly unacceptable because they may cause failure at high temperatures; Al and Mg are not as problematic because they are frequently used as sintering aids; transition metals such as Fe, Ni, or Cr can adversely affect the material strength (36).

Fabrication Technology. A variety of simple and complex-shaped dense parts are made from Si_3N_4 powders by ceramic processing techniques. Inasmuch as silicon nitride is a covalently bonded compound having a low diffusion coefficient, sintering aids are used to achieve complete densification. Aids are typically oxides such as Al_2O_3, Y_2O_3, ZrO_2, MgO, lanthanide oxides, and, at times, AlN as well. They are added singly or in combinations in amounts of several weight percent and may reach as high as 15 wt % of the matrix. The α-Si_3N_4 crystallites, which typically comprise over 90% of the starting powder material, dissolve in the liquid phase formed by the reaction of the sintering additives with the silica layer present on the surface of the silicon nitride particles, then reprecipitate as β-Si_3N_4 (41). This α- to β-phase transformation is facilitated by the presence of liquid phase as well as β-crystal nuclei (42). It does not appear to be reversible. The strongest ceramics are realized when all the pores are eliminated and full conversion to β-Si_3N_4 takes place. A dense crystalline matrix consisting of rodlike grains characteristic of the β-phase is formed (43). The crystallites are surrounded by a thin intergranular amorphous or crystalline phase which forms upon cooling (44).

Both composition and the quantity of sintering additives profoundly affect the properties of silicon nitride ceramics (43). Additives facilitate densification, serve as the strength limiting factor at high temperatures, and may also adversely affect oxidation resistance. Thus sintering additives are kept to a minimum. The specific additive used depends on the ceramic's ultimate application. For uses at lower (up to 1000°C) temperatures, magnesia–magnesia/alumina combinations are frequently employed; in the intermediate (up to 1200°C) range yttria–alumina formulations are usually preferred; for the applications in which structural integrity and performance are required at temperatures up to 1400°C, yttria alone is used.

Si_3N_4 powder is typically mixed with the appropriate amount of sintering additives and an organic binder. This mixture then undergoes extensive comminution, generally by milling, often using silicon nitride grinding media. Powder processing for the most critical applications is frequently done in clean room environments (see CLEAN ROOM TECHNOLOGY).

The shaping of these fine, submicrometer powders into complex components and their subsequent consolidation into dense ceramic parts of ideally zero porosity is a major technological challenge. The parts formed need to be consolidated to near-net shape because Si_3N_4 machining requires expensive diamond grinding. Additionally, Si_3N_4 dissociates at or near the typical densification temperatures used in the fabrication of structural ceramics and, therefore, special measures have to be taken to preserve the compositional integrity of the material.

Parts of simple geometries can be readily made by uniaxial die pressing or cold isostatic pressing and densified by sintering or hot isostatic pressing (HIP). Typical sintering temperatures are in the range of 1700–2000°C, depending on the composition and concentration of the sintering aid, and in order to prevent silicon nitride decomposition, the parts are usually embedded in a protective powder such as boron or silicon nitride and/or placed in a closed vessel (45). An overpressure of 0.1–10 MPa (15–1500 psi) of nitrogen is usually maintained during sintering. Satisfactory products can be obtained at atmospheric pressures, particularly using silicon nitride formulations designed for applications at lower temperatures. Parts having densities over 99% of theoretical can be made by this technique and shrinkage can be controlled to achieve near net shape fabrication.

Using hot-pressing, shaping and densification occur in a single process step. The temperatures are in the range of 1650–1800°C and applied pressures are from 30–40 MPa (4000–6000 psi) (45), resulting in parts of high quality. This method is limited to simple shapes and low production volumes, however, and the process may also impart anisotropic characteristics to the material (46).

There are two operational variations in the HIP process as applied to silicon nitride parts. In one, shaped specimens are encapsulated in a glass and then hot pressed isostatically (46,47); in the other, called sinter-HIP, the parts are first presintered to closed porosity ($<7\%$) and then hot isostatically pressed directly, without encapsulation (45,48–50). The latter technique, although a two-step process, offers the advantage of eliminating both the risk of the diffusion of deleterious glass components into the work pieces and the need for a post-HIP decapsulation step. Process conditions for both variations are similar: temperatures range from 1700–2000°C and pressures from 100–200 MPa (45). The quality of the resulting products is high. Densities approach theoretical values and mechanical properties are comparable to those of hot-pressed components.

Injection molding and slip casting are used for making complex-shaped silicon nitride structural ceramic components. Injection molding is a high volume production technique (51). To impart the appropriate rheological characteristics the powder is mixed with significant (10–15 wt %) amounts of organic binder components which then need to be removed from the part prior to densification. This process requires careful time-temperature control and may take from a few days to even weeks (52). Complex parts such as turbine rotors and turbochargers have been fabricated having adequate green strength and the requisite microstructure for densification by HIP processing into high quality ceramics (53–57).

Slip casting, used increasingly for silicon nitride parts, has the advantage that only small amounts of chemical additives are employed and therefore, binder burn-out problems are basically eliminated (46). The biggest disadvantage is the time it takes for the part to form. Tens of hours may be required to cast a gas turbine engine rotor and during that time the stability of the slip suspension may change (58). Therefore, pressure-assisted casting is being used to accelerate part formation (58,59) reducing the casting time to about an hour. After removal from the mold, the part needs to be dried carefully and slowly to prevent crack formation (58). Automated process methodology is used to shape tens of thousands of silicon nitride turbochargers per month by pressure-assisted slip casting. These are then densified either by HIP or gas overpressure sintering.

An alternative technique for the fabrication of complex Si_3N_4 parts is reaction bonding. A shape is formed from silicon powder by any of the aforementioned methods and the resulting green compact, having a typical density that is 60–70% of the theoretical value of Si, is then nitrided by reaction with nitrogen at 1300–1400°C to form reaction-bonded silicon nitride (RBSN). The reaction, carried out over a 3 to 10 day period, is exothermic and needs to be carefully controlled so as not to exceed, even locally, the melting point of silicon (1410°C). The final product is not fully dense; it has a residual connected porosity of about 12–30% (45,49). This process allows for real net shape fabrication of complex components because the part does not shrink during the reaction. Precise dimensional tolerances can be obtained without postfabrication machining. Another advantage is that no sintering aids are used so that RBSN parts retain their strength even at high temperatures and they are resistant to creep. However, the low density of RBSN materials results in typical strength values that are only a fraction of fully dense silicon nitride ceramics. The high porosity RBSN materials also have lower oxidation resistance and are more brittle than dense silicon nitride (45,46,49,60). Modifications to the reaction-bonding process material density include postsintering (SRBSN) (45) or post-HIP (HIPRBSN) (48). Property improvements have been realized, but this approach does not appear to offer quality advantages over the more direct processes.

Properties.   Properties of structural silicon nitride ceramics are given in Table 2. These values represent available, well-tested materials. However, test methodology and the quality of the specimens, particularly their surface finish, can affect the measured values. Another important material property is tensile

Table 2. Properties of Silicon Nitride Ceramics

Property	Material densification mode			
	Reaction-bonded[a]	Sintered[b]	Hot-pressed isostatically[c]	Hot-pressed[d]
density, kg/m^3	2.5	3.26	3.23	3.2
elastic modulus, GPa[e]	180	300	310	310
hardness, kg/mm^2	1350	1370	1620	1800
flexure strength, MPa[f] at				
ambient temperature	340	700	900	700
1000°C		600		610
1200°C		480		570
1370°C		210	580	310
fracture toughness, MPa√m	3–4	4.6	4.7–5.5	4.9
thermal expansion coeff, 25–1000°C, 10^{-6}/°C	3	3.9	3.9	3.5
thermal conductivity, W/(m·K) at 25°C	12	32	38	32

[a]Coors Ceramics Company, Bulletin #980.
[b]GTE Laboratories AY-6, 6 wt % yttria + 2 wt % alumina (43).
[c]Norton NT154, 4 wt % yttria (61).
[d]Norton NC 132, 1 wt % magnesia (46).
[e]To convert GPa to psi, multiply by 145 × 10^3.
[f]To convert MPa to psi, multiply by 145.

strength. Values obtained on Norton's NT154 material are: 750 MPa at RT, 500 MPa at 1200°C, and 350 MPa (50,000 psi) at 1400°C (62).

As noted, the oxidation resistance of silicon nitride ceramics depends on the type and concentration of the sintering aids. In materials designed for high temperature applications the specific weight gain resulting from oxidation upon a 500-h air exposure at 1200°C and 1350°C is about 1–2 g/m^2 and 2–4 g/m^2, respectively. The kinetics of the oxidation process have been investigated (63,64) as has the corrosion resistance (65). Corrosion resistance is also dependent on material formulation and density.

Applications. Silicon nitride is the leading material for components in advanced automotive, diesel, and gas turbine engines. The range of potentially useful ceramic parts includes both static structural components and dynamic ones such as turbocharger rotors, gasifier turbine rotors, valves, valve guides, valve seats, piston components, cam followers, fuel injector links, and bearings. Some of these parts have been commercialized, others are being evaluated.

The first commercial use of silicon nitride ceramics for automotive applications was in glow plugs to reduce engine startup wait times for light duty diesel engines. More recently, Si_3N_4 hot plugs have been installed in similar engines. These plugs also reduce engine emissions and noise (66). Silicon nitride turbocharger rotors are being used (67,68) in Japanese production cars sold in Japan. 1990 production rates were about 20,000/month and are projected to reach 30,000/month in 1991. A limited production run of a U.S. model was also equipped with a ceramic turbocharger (60). The primary advantage is the low density of the material and a consequent lower moment of inertia leading to faster engine response and to a decrease in turbocharger lag. Introduction of such turbochargers into Japanese light duty diesel engines is probable (66). Silicon nitride cam followers are being introduced into heavy-duty diesel engines in the United States because of their superior wear performance (69); an additional advantage is a reduction or even elimination of the need for forced lubrication and for costly lubrication channels. Ceramic fuel injector links are being incorporated into heavy-duty diesel engines, also because of superior wear resistance (70). A silicon nitride exhaust port liner is being used in sports cars because it results in faster heat up of the catalytic converter and consequent reduction in hydrocarbon and oxide emissions (70). A Si_3N_4 rocker arm wear pad has been introduced (67).

The lower inertia of silicon nitride valves allows them to follow the valve-lifting cam more closely resulting in a more stable operation which could increase the engine speed by up to 1000 rpm (68). This advantage was utilized effectively in stock car racing in which engines equipped with silicon nitride valves outperformed standard engines in many competitions (71). In conventional applications the use of ceramic valves offers the potential for increased fuel efficiency by reducing the spring load resulting in a lower camshaft torque. Large silicon nitride components, including valves about 50 cm in diameter and 50 cm high, have been fabricated and are being evaluated in coal gasification plants in Japan (72).

The emerging field of gas turbine engines is a technology in which silicon nitride ceramics serve as enabling materials. The automotive version of this engine is designed to provide the powertrain for the next generation of passenger cars (73). Car manufacturers in the United States, Germany, and Japan are actively involved in the development of such engines and there are several variations in its design; one of the common features is that the engine is intended to operate at a gas inlet temperature of 1375°C. The only suitable materials for rotating components under these conditions are silicon nitride ceramics. Gas turbine rotors which successfully passed spin tests at maximum speed both at room temperature and at 1395°C have been developed (73).

A car powered with a gas turbine engine having ceramic rotors has been undergoing successful road tests in Germany for several years (74) and a similar engine of larger capacity is being developed in Japan for stationary use for cogeneration of electric power. Another application for silicon nitride rotors is in auxiliary power unit (APU) engines (75).

In other useful applications, silicon nitride bearings have been found to offer excellent performance; silicon nitride cutting tool inserts are a commercial product; wear parts such as sand blast nozzles, seals, and die liners are also commercially produced; the superior performance of heat exchangers has been demonstrated and there are also military applications.

Zirconia Structural Ceramics

Zirconia ceramics represent a fairly new class of advanced structural materials (see ZIRCONIUM AND ZIRCONIUM COMPOUNDS). Their potential use in structural applications was first realized in the mid-1970s. Since then numerous publications have appeared devoted entirely to these materials (76–81).

Material System. Pure zirconia at atmospheric pressures exhibits three well-defined crystalline polymorphs: the monoclinic, tetragonal, and cubic phases. The monoclinic phase is stable up to about 1170°C where it transforms to the tetragonal phase. At 2370°C the tetragonal phase transforms to the cubic phase which exists up to 2680°C, the zirconia melting point (82). On cooling through the tetragonal-to-monoclinic transformation temperature, a large volume increase (3–5%) occurs. This change is sufficient to cause cracking. Thus, fabrication of large components of pure zirconia is not possible. The transformation volume expansion can be used to advantage, however, by the addition of cubic stabilizing oxides, most commonly magnesia, calcium oxide [1305-78-8], CaO, and yttria. These oxides can stabilize the relatively weak cubic form down to room temperature. Moreover, if insufficient stabilizing oxide is added, and the material is properly processed, zirconia particles can be retained in the metastable tetragonal form at room temperature. These materials are referred to as partially stabilized zirconia (PSZ) ceramics.

During application of stress, eg, in the region of a propagating crack, metastable tetragonal particles transform to the stable monoclinic phase. The resulting volume expansion places the region around these particles, ie, adjacent to the crack, in compression and crack propagation is retarded until the applied stress is increased. The extra work required to move the crack through the matrix can lead to increases in strength, toughness, and resistance to thermal shock. The phases present in these ceramics, their amount, size, and distribution, can be controlled to produce materials having a range of properties tailored for specific applications.

The transformation is believed to occur by a diffusionless shear process (83). It is often referred to as martensitic transformation, having a thermal hysteresis between the cooling and heating cycles. The transformation is dependent on particle size; finer particles transforming at a lower temperature than coarser particles. Transformation toughening can also result upon incorporation of fine zirconia particles into another matrix such as Al_2O_3 (84). These materials are called zirconia toughened ceramics (ZTC). A third type of transformation-toughened ceramic material is formed using a low concentration of yttria in zirconia and a very fine grain size (85–87). An approximately 100% tetragonal zirconia polycrystalline (TZP) ceramic results.

Fabrication Technology. Stabilizing additives must be uniformly distributed within the starting powders for zirconia ceramics. Homogeneous distribution can be attained by controlled coprecipitation of hydroxides which are then decomposed by calcination yielding powders of fine particle sizes. Active sinterable powders are produced commercially, usually by hydrolysis of a mixture of $ZrOCl_2$ and YCl_3 to precipitate the mixed hydroxide. The method produces a powder having a very fine (about 0.3 μm) particle size. Alternative methods of fabricating fine active powders include CVD and hydrothermal oxidation. Preparation of zirconia powders is covered in the literature (78,88–90).

Zirconia powders may be shaped using techniques such as slip casting, dry pressing, and injection molding. The ceramics may be densified by sintering, hot pressing, or hot isostatic pressing, provided the thermal treatment of the material is appropriate to develop the desired microstructure. Other fabrication methods, such as microwave sintering, are also under investigation. Reaction sintering has been used to produce microstructures of zirconia particles in various ceramic matrices as well (91,92).

Partially Stabilized Zirconia.   PSZ is comprised of a cubic zirconia matrix having a fine dispersion of tetragonal particles. Stabilizing additives are on the order of several weight percent of MgO, CaO, or Y_2O_3 to produce the appropriate microstructures. Powders are first sintered at an appropriate temperature, solution annealed in the single-phase cubic region of the zirconia phase diagram, and then heat-treated (aged) in the two-phase tetragonal + cubic region to nucleate and grow tetragonal precipitates within the cubic matrix. A critical size range, submicrometer to several micrometers, exists for stress-induced transformation of tetragonal zirconia particles. If the material is aged too long and the precipitates grow larger than the critical size, particles spontaneously transform to the monoclinic phase upon cooling to room temperature; if the particles are smaller than the critical size, transformation does not occur. The critical size limit depends on the matrix constraint and the composition of the zirconia. As the stabilizing oxide content is increased, the chemical free energy associated with the phase transformation decreases and hence larger particles can be induced to remain in the metastable tetragonal form.

Tetragonal Zirconia Polycrystal.   TZP ceramics may be produced from compositions stabilized using Y_2O_3 (2–4 mol %) or cerium(IV) oxide [1306-38-3], CeO_2 (9–14 mol %) by sintering in the single-phase tetragonal region of the phase diagram (87,93). In order to retain the tetragonal phase to room temperature, the grain size must be kept very small (usually <1 μm). Each grain is restrained by surrounding grains from transforming to the stable monoclinic form. TZP materials exhibit exceptionally high fracture strength values and high toughness. Yttria TZP, however, exhibits a serious decrease in strength when aged in air between 150° and 300°C (94–97). Because of fine grain size and the presence of a grain boundary phase, TZP ceramics exhibit pseudosuperplasticity at 1200°C with extension of >100% measured in tension (98,99). This feature provides opportunities for shape forming of this material.

Zirconia Toughened Ceramics.   Zirconia particles can be embedded in host matrices to form a variety of transformation-toughened ceramics (100). Hosts include Al_2O_3 (84,101), $β''$-alumina (102), mullite [55964-99-3] (103,104,91), Si_3N_4 (105–109), SiAlON (110), cordierite [12182-53-5] (111), glass ceramics (112), TiB_2 (113,114), MgO (115), and molybdenum silicide [12136-78-6] $MoSi_2$ (116). Requirements are that the host matrix not react with the ZrO_2 and that the matrix have a sufficiently high elastic modulus to retain the ZrO_2 in the tetragonal state. The zirconia particles can then transform as they do in PSZ or TZP materials. Optimum toughness and strength result using very fine (usually <1 μm) zirconia. ZTC are generally formed by either mixing powders of zirconia and the matrix and sintering or chemically preparing powders of mixed composition by coprecipitation. Mixed powder synthesis yields intergranular particles whereas the chemical route yields intragranular ones. Mechanical properties are optimized by maintaining a very well dispersed zirconia phase and avoiding particle growth during sintering. However, sintering temperatures must be high enough to achieve full density.

Toughening Mechanisms. The mechanics of tetragonal-to-monoclinic transformation can effect the strength and toughness of the ceramic.

Stress-Induced Transformation Toughening. The stress field of a crack can initiate martensitic transformation (117–121); the transformed particles expand against the matrix, resulting in compressive stress on the crack surface. This stress acts to reduce and eventually stop propagation of the crack. The particles which have transformed in the vicinity of the crack comprise what is called a process zone. This zone tends to shield the crack tip from applied stress.

Compressive Surface Layers. Spontaneous tetragonal-to-monoclinic transformation of zirconia particles may take place at or near the surface of a macroscopic part as a result of the absence of the hydrostatic constraint near the free surface. The particles expand and induce a compressive strain. This compressive surface layer leads to high fracture strength and in some cases the strength may be doubled. Surface grinding has been found to be the most effective method of inducing this transformation.

Microcrack Toughening. Toughening can result from both residual and stress induced microcracks. Residual microcracks are formed when the retained tetragonal particles are larger than some critical size causing them to spontaneously transform on cooling. The stresses generated around the transformed particles induce microcracks between the particles and the matrix. The microcracks may extend in the stress field of the propagating crack, or deflect the propagating crack, thereby absorbing or dissipating fracture energy and, hence, increasing the toughness of the ceramic. However, although *fracture toughness* is increased, the presence of either stress-induced or residual microcracks can result in a significant reduction in the *fracture strength*. Because of increased fracture toughness, microstructures having residual microcracks are useful in situations requiring resistance to thermal shock. These materials must be carefully processed to produce particles large enough to transform but small enough to cause only limited microcrack development.

The microstructure of zirconia materials can be designed to yield optimum combinations of fracture strength and toughness by controlling the relative amount of stress-induced transformation toughening and microcracking (122). The temperature dependence of mechanical properties is also correlated with different toughening mechanisms. Transformation toughening provides high strength and toughness at low and intermediate temperatures but its effectiveness decreases as temperature increases. Microcrack toughening is a less effective toughening mechanism but it is essentially temperature-independent.

Crack Deflection. Crack deflection can result when particles transform ahead of a propagating crack. The crack can be deflected by the localized residual stress field which develops as a result of phase transformation. The force is effectively reduced on the deflected portion of the propagating crack resulting in toughening of the part.

Properties. Transformation toughened ceramics have excellent strength and toughness at low and intermediate temperatures. Compared to SiC and Si_3N_4, ZrO_2-toughened ceramics can withstand significantly higher applied stress at room temperature (see Fig. 1), but SiC and Si_3N_4 have much greater high temperature potential. Zirconia ceramics have limited high ($>800-1000°C$) temperature capability for two reasons: creep rates are high compared to nonoxide ceramics; and the contribution from the transformation toughening mechanism decreases as the temperature increases. That is, as the tetragonal phase becomes more stable, the driving force for the transformation decreases. Table 3 lists the properties of zirconia ceramics each of which is a family of materials. Specific properties are a function of amount and type of stabilizing agent, processing conditions utilized, and resulting microstructure.

Table 3. Properties of Zirconia Ceramics

Property	PSZ[a]	TZP[b]	ZTA[c]
density, kg/m^3	5.7	6.0	4.2
hardness, kg/mm^2	1000	1300	1600
flexural strength, MPa[d]	300–700	1000–2500	400–900
fracture toughness, $MPa\sqrt{m}$	4–8	5–15	5–10
elastic modulus, GPa[e]	200	200	340
thermal expansion coeff, $\times 10^{-6}/K$	9–10	10–11	8–9
thermal conductivity, $W/(m\cdot K)$	2.0–2.5	2.7	7–10
maximum service temperature, °C	950	500	1700

[a]Properties of PSZ depend on whether CaO, MgO, or Y_2O_3 is used as the stabilizing agent.
[b]Properties of TZP depend on whether CeO_2 or Y_2O_3 is used as the stabilizing agent.
[c]Properties of zirconia toughened alumina, ZTA, depend on the specific microstructure and the proportions of zirconia and alumina.
[d]To convert MPa to psi, multiply by 145.
[e]To convert GPa to psi, multiply by 145×10^3.

TZP materials have exceptionally high fracture strength values: strengths greater than 1000 MPA are consistently achieved and values over 2000 MPa have been reported. Toughness is generally greater than 5 $MPa\sqrt{m}$ (123). Yttria TZP, however, exhibits a serious decrease in strength when aged in air between 150° and 300°C (94–97). The effect appears to be related to water vapor in the air reacting with the Y_2O_3–ZrO_2 at the ceramic surface, promoting the tetragonal-to-monoclinic transformation, and forming microcracks. There are indications that this problem can be avoided or at least minimized by achieving a suitably fine grain size (0.2–0.6 μm), by adding finely dispersed alumina to the yttria–zirconia (124), or by substituting CeO_2 for Y_2O_3 as the stabilizing oxide (125). Addition of alumina (~ 20 wt % Al_2O_3) to Y–TZP inhibits grain growth and transformation of the tetragonal particles. The toughness of Ce–TZP (>30 $MPa\sqrt{m}$) can be higher than that of Y–TZP (15–20 $MPa\sqrt{m}$) but the strength is relatively low, 500–1000 MPa for Ce–TZP vs 1500–2000 MPa for Y–TZP (125). A limiting factor in the application of zirconia ceramics is the decrease in properties that result as temperatures increase. Several possible strategies exist for improving the high temperature capabilities of zirconia ceramics (126). SiC whisker reinforced zirconia ceramics have also been investigated (127–129).

Applications. One of the most demanding applications for zirconia ceramics is in automotive engine parts, particularly for the diesel engine (130).

Applications attempt to exploit its low thermal conductivity and/or the wear-resistance characteristics. One approach utilizes ceramic liners or inserts (eg, piston crowns, head face plates, and piston liners) attached to metal engine components. PSZ is a favored material for this approach, not only because it has low thermal conductivity and is a good insulator, but more importantly, because its high thermal expansion coefficient is close to that of cast iron. This compatibility facilitates attachment and reduces the possibility of failure during engine cycling. Other engine applications for zirconia include components which are limited by wear, particularly in the valve train, such as cams, cam followers, tappets, and exhaust valves.

Alumina–zirconia ceramics have superior strength, toughness, and wear resistance when compared to conventional alumina and these composite ceramics have found use as cutting tool tips and abrasion wheels. Applications include scissors and shears for cutting of difficult materials such as Kelvar, and cutting and slitting of industrial materials, such as magnetic tape, plastic film, and paper items. The fracture toughness and thermal shock resistance of transformation-toughened PSZ has made it a leading candidate for both wire drawing and hot extrusion dies. Seals in valves, chemical pumps, and abrasive slurry pumps and impellers are being made of zirconia ceramics. In some applications involving abrasive slurries, PSZ materials can be more wear resistant than silicon carbide. Components requiring long life under low load conditions, such as thread guides and bearings and guides for dot matrix printers can also be made successfully from zirconia. MgO–PSZ has found the widest commercial use because of the range of tailored microstructures which can be produced.

Zirconia also has suitable properties for thermal barrier coatings, for turbine rotors for example, because of its high thermal expansion coefficient, low thermal conductivity, good chemical stability, and thermal shock resistance (131–133). Plasma-sprayed zirconia compositions have been investigated and the most durable coatings were found to be formed from a partially stabilized zirconia composition. A major problem encountered with such coatings is corrosive attack by the mineral constituents in fuel oil leading to destabilization of the tetragonal zirconia to give the monoclinic form.

In all applications involving zirconia, the thermal instability of the tetragonal phase presents limitations especially for prolonged use at temperatures greater than $\sim 1000°C$ or uses involving thermal cycling. Additionally, the sensitivity of Y–TZP ceramics to aqueous environments at low temperatures has to be taken into account. High raw material costs have precluded some applications particularly in the automotive industry.

Environmental Aspects

Exposure limits for silicon carbide and powders of zirconium compounds (including zirconium dioxide) have been established by ACGIH. TLV–TWA's are 10 mg/m^3 and 5 mg/m^3, respectively. OSHA guidelines for zirconium compounds call for a PEL of 5 mg/m^3. There are no exposure limits for silicon nitride powder, but prudent practice suggests a TLV–TWA of 0.1 mg/m^3. The solid ceramics present no apparent health hazard. In machining such ceramics, however, care should be taken to prevent inhalation of respirable particles in amounts in excess of established limits. Disposal should be in approved landfills; the materials are inert and should pose no danger to the environment.

Economic Aspects

The production of advanced ceramic mechanical and wear components in Japan in 1989 is estimated at one billion U.S. dollars. Silicon nitride powder production amounted to about ten million U.S. dollars, about one tenth of the total nonoxide ceramic powder production. It is projected that in the year 2000 the market for these ceramic components will be in the $2.7–4.2 billion range (134). In Western Europe the market for mechanoceramics in 1989 was estimated at $200 million U.S. dollars and projected to grow modestly in the near future. The amount of silicon nitride and silicon carbide sold was $32 and $21 million, respectively (135). It is projected that in the United States the market for structural ceramics in automotive applications will grow in the year 2000 to $820 million (in 1990 dollars) from the current estimate of $81 million for 1990. In the same time interval the market for wear parts and other industrial advanced ceramics (excluding cutting tools) is expected to grow to $720 million from its 1990 value of $150 million. The corresponding figures for the aerospace and defense-related advanced ceramics market are $445 and $80 million, respectively (136). Structural ceramics producers are given in Table 4.

Table 4. Producers of Structural Ceramics

Location, company	Ceramic type
Australia	
Nilcra	ZrO_2
Europe	
ASEA	Si_3N_4
ESK	SiC, Si_3N_4
Feldmühle[a]	ZrO_2
Japan	
Kyocera Corp.[a]	SiC, Si_3N_4, ZrO_2
NGK Insulators	SiC, Si_3N_4
NTK Technical Ceramics	Si_3N_4
United States	
Carborundum Co.	SiC
Ceramatec	ZrO_2
Coors[b]	ZrO_2
Corning	ZrO_2
Dow Chemical	ZrO_2
WR Grace	ZrO_2
GTE	Si_3N_4
Norton Co.	SiC, Si_3N_4, ZrO_2
Zircoa[b]	ZrO_2

[a]Kycocera Corp. licenses Feldmühle and Max Planck Institute zirconia technology.
[b]Largest U.S. suppliers of ZrO_2 ceramics.

Premium zirconia powders cost from U.S. $55 to $100/kg. Suppliers include Ferro and Z-Tech (a subsidiary of ICI); Japanese sources of powders include Toyo Soda Manufacturing Company (now Tosoh) and Daichi Kigenso.

BIBLIOGRAPHY

1. W. Bunk and H. Hausner, eds., *Proc. of 2nd Int. Symp. on Ceramic Materials and Components for Engines,* Verlag Deutsche Keramizche Gesellschaft, Bad Honeff, 1986.
2. P. F. Becher, M. V. Swain, and S. Somiya, eds., *Advanced Structural Ceramics, Materials Research Society Symp. Proc.* Vol. 78, Materials Research Society, Pittsburgh, Pa., 1987.
3. J. B. Wachtman, Jr., ed., *Structural Ceramics: Treatise on Materials Science and Technology,* Vol. 29, Academic Press, New York, 1989.
4. G. L. Leatherman and R. N. Katz, *Superalloys, Supercomposites and Superceramics,* Academic Press, New York, 1989, p. 671.
5. M. Srinivasan in Ref. 3, p. 99.
6. N. Ault and J. Crowe, *Am. Ceram. Soc. Bull.* **68,** 1062 (1989).
7. K. Kajima, H. Noguchi, and M. Konishi, *J. Mater. Sci.* **24,** 2929 (1989).
8. P. Shaffer, *Ceram. Eng. Sci. Proc.* **6,** 1289 (1985).
9. C. Forrest, P. Kennedy, and J. Shennan, *Spec. Ceram.* **5,** 99 (1972).
10. F. Kennard, *Ceram. Eng. Sci. Proc.* **7,** 1095 (1986).
11. P. Kennedy in S. Hampshire, ed., *Non-Oxide Technical and Engineering Ceramics,* Elsevier Applied Science, London, 1986, p. 301.
12. G. Weaver and B. Olson in R. Marshal, J. Faust, Jr., and C. Ryan, eds., *Silicon Carbide—1973,* University of South Carolina Press, Columbia, S.C., 1974, p. 367.
13. R. Storm, *Sagamore Army Materials Research, 37th Conf. Proc.* 1990.
14. S. Prochazka in J. Burke, A. Gorum, and R. Katz, eds., *Ceramics for High Performance Applications,* Brook Hill, Chestnut Hill, Mass., 1974, p. 239.
15. S. Prochazka, *Spec. Ceram.* **6,** 171 (1975).
16. Y. Murata and R. Smoak in S. Somiya and S. Saito, eds., *Proc. Int. Symp. of Factors in Densification and Sintering of Oxide and Nonoxide Ceramics,* Gakujutsu Bunken Fukyu-Kai, Tokyo, 1979, p. 382.
17. W. Bocke, H. Landfermann and H. Hausner, *Powder Metal. Int.* **13**(1), 37 (1981).
18. W. Bocker and H. Hausner, *Powder Metal. Int.* **10**(2), 87 (1978).
19. E. Maddrell, *J. Mater. Sci. Lett.* **6,** 486 (1987).
20. T. Mizutani, M. Hayashi, and A. Tsuge, *J. Ceram Soc. Jpn. Int. Ed.* **96,** 211 (1988).
21. R. Hamminger, *J. Am. Ceram. Soc.* **72**(9), 1741 (1989).
22. S. Prochazka, C. Johnson, and R. Giddings in Ref. 16, p. 366.
23. J. Coppola and co-workers in Ref. 16, p. 400.
24. L. Ogbuji, *Ceram. Int.* **12,** 173 (1986).
25. G. Watson, T. Moore, and M. Millard, *Am. Ceram. Soc. Bull.* **64**(9), 1253 (1985).
26. K. Hunold, *Powder Metal. Int.* **21**(3), 22 (1989).
27. R. Katz and G. Quinn in F. Riley, ed., *Progress in Nitrogen Ceramics,* Martinus Nijhoff Publishers, The Hague, Netherlands, 1983, p. 491.
28. D. Larsen and J. Adams, *Technical Report AFWAL-TR-83-414,* April 1984.
29. D. Larsen and J. Adams in Ref. 27, p. 695.
30. J. Cuccio and co-workers, *Proc. 27th Automotive Technol. Dev. Contractors' Coord. Mtg.,* Publication P-230, Society of Automotive Engineers, Warrendale, Pa., April 1990, p. 335.
31. D. Shetty, I. Wright, and A. Clauer, *Wear* **79,** 275 (1982).

32. S. Lasday, *Ind. Heat.* 35 (Aug. 1990).
33. M. Ferber and co-workers, *J. Am. Ceram. Soc.* **68**(4), 191 (1985).
34. G. Fischman and S. Brown, *Mater. Sci. Eng.* **71**, 295 (1985).
35. D. Butt, J. Mecholsky, and V. Goldfarb, *J. Am. Ceram. Soc.* **72**(9), 1628 (1989).
36. S. Natansohn in A. Vary and J. Snyder, eds., *Nondestructive Testing of High Performance Ceramics,* The American Ceramic Society, Westerville, Ohio, 1987, p. 73.
37. A. Atkinson, A. Moulson, and E. W. Roberts, *J. Am. Ceram. Soc.* **59**, 285 (1976).
38. W. Rhodes and S. Natansohn, *Am. Ceram. Soc. Bull.* **68**, 804 (1989).
39. U.S. Pat. 4,405,589 (Sept. 20, 1983), T. Iwai, T. Kawahito, and T. Yamada (to Ube Industries, Ltd.).
40. E. Turkdogan, P. Bills, and V. Tippett, *J. Appl. Chem.* **8**, 296 (1958).
41. F. Lange, *Int. Metals Rev.* **25**, 1 (1980).
42. S. Natansohn and V. Sarin in H. Hausner, G. Messing, and S. Hirano, eds., *Ceramic Powder Processing Science,* Deutsche Keramische Gesellschaft e.V. Köln, Germany, 1989, p. 433.
43. C. Quackenbush and J. Smith, *Paper No. 84-GT-228,* American Society of Mechanical Engineers, New York, 1984.
44. D. Clarke and G. Thomas, *J. Am. Ceram. Soc.* **61**, 114 (1978).
45. G. Ziegler, *Z. Werkstofftech.* **14**, 189 (1983).
46. M. Torti in ref. 3, p. 161.
47. H. Larker in F. Riley, ed., *Progress in Nitrogen Ceramics,* Martinus Nijhoff, Boston, Mass., 1983, p. 717.
48. G. Ziegler and G. Woetting, *Int. J. High Technol. Ceram.* **1**, 31 (1985).
49. G. Woetting and G. Ziegler, *Sprechsaal* **119**, 555 (1986).
50. G. Ziegler, J. Heinrich, and G. Woetting, *J. Mater. Sci.* **22**, 3041 (1987).
51. G. Bandyopadhyay and K. French, *J. Eng. for Gas Turbines and Power* **108**, 536 (1986).
52. C. Quackenbush, K. French, and J. Neil, *Automotive Technol. Dev. Contractors' Coord. Mtg., 19th Summary Report,* U.S. Dept. of Energy, Washington, D.C., 1981, p. 424.
53. J. Neil and co-workers, *Paper No. 82-GT-252,* American Society of Mechanical Engineers, New York, 1982.
54. G. Bandyopadhyay and co-workers in J. Tennery, ed., *Proc. 3rd. Int. Symp. on Ceramics Materials and Components for Engines,* American Ceramic Society, Westerville, Ohio, 1989, p. 1397.
55. J. Neil and co-workers, *Proc. 27th Automotive Technol. Dev. Contractors' Coord. Mtg.,* Society of Automotive Engineers, Warrendale, Pa., 1990, p. 303.
56. G. Bandyopadhyay and co-workers, *Paper No. 90-GT-47,* American Society of Mechanical Engineers, New York, 1990.
57. J. Neil and co-workers, *Preprints of the 28th Annual Automotive Technol. Dev. Contractors' Coord. Mtg.* (Oct. 1990, Dearborn, Mich.), Society of Automotive Engineers, Warrendale, Pa., 1990.
58. J. Pollinger and B. Busovne, *Proc. 27th Automotive Technol. Dev. Contractors' Coord. Mtg.,* Society of Automotive Engineers, Warrendale, Pa., 1990, p. 357.
59. B. Mc Entire and co-workers, *Proc. 27th Automotive Technol. Dev. Contractors' Coord. Mtg.,* Society of Automotive Engineers, Warrendale, Pa., 1990, p. 341.
60. G. L. Leatherman and R. N. Katz, *Superalloys, Supercomposites, and Superceramics,* Academic Press, New York, 1989, p. 671.
61. B. Mc Entire and co-workers, *Preprints of the 28th Annual Automotive Technol. Dev. Contractors' Coord. Mtg.* (Oct. 1990, Dearborn, Mich.), Society of Automotive Engineers, Warrendale, Pa., 1990.
62. N. Hecht and co-workers, *Preprints of the 28th Annual Automotive Technol. Dev. Contractors' Coord. Mtg.,* (Oct. 1990, Dearborn, Mich.), Society of Automotive Engineers, Warrendale, Pa. 1990.

63. J. Smith and C. Quackenbush, *Am. Ceram. Soc. Bull.* **59,** 533 (1980).
64. S. Natansohn in V. Tennery, ed., *Proc. 3rd Int. Symp. on Ceramics Materials and Components for Engines,* The American Ceramic Society, Westerville, Ohio, 1989, p. 27.
65. *Technical Bull. on Characteristics of Kyocera Technical Ceramics,* Kyocera Corporation, 1988.
66. R. Katz in Ref. 3, p. 1.
67. J. Wachtman, Jr., and co-workers, *Japanese Structural Ceramics Research and Development, Technical Assessment Report,* Science Applications International Corporation, McLean, Va., 1989, Chapt. 5.
68. C. Lewis, *Mater. Eng.* 30 (May 1989).
69. K. Weber and N. Hakim, *1st Int. Ceramic Science and Technology Congress, Paper 4-SI-89C,* (Anaheim, Calif., October, 1989).
70. R. Katz, *7th Cimtec World Ceramic Congress, Paper B4.1-L01,* (Montecatini Terme, Italy, July 1990).
71. R. Larsen, *1st Int. Ceramic Science and Technology Congress, Paper 9-SI-89C* (Anaheim, Calif., Oct. 1989).
72. H. Yamaguchi and co-workers, *1st Int. Ceramic Science and Technology Congress, Paper 15-SI-89C,* (Anaheim, Calif., October 1989).
73. H. Helms and co-workers, *Preprints of the 28th Annual Automotive Technol. Development Contractors' Coord. Mtg.* (Dearborn, Mich., Oct. 1990).
74. K. D. Moergenthaler, *Preprints of the 28th Annual Automotive Technol. Development Contractors' Coord. Mtg.* (Dearborn, Mich., Oct. 1990).
75. T. Bornemisza, *7th Cimtec World Ceramic Congress, Paper B4.1-L05,* (Montecatini Terme, Italy, July 1990).
76. A. H. Heuer and L. W. Hobbs, eds., *Science and Technology of Zirconia I, Adv. in Ceram. 3,* American Ceramic Society, Columbus, Ohio, 1981.
77. N. Claussen, M. Rühle, and A. H. Heuer, eds., *Science and Technology of Zirconia II, Adv. in Ceram. 12,* American Ceramic Society, Columbus, Ohio, 1984.
78. R. Stevens, *Zirconia and Zirconia Ceramics,* 2nd ed., Magnesium Elektron Publication No. 113, 1986.
79. Reference 2, pp. 3–172.
80. S. Somiya, N. Yamamoto, and H. Yanagida, eds., *Science and Technology of Zirconia III, Adv. in Ceram. 24,* American Ceramic Society, Columbus, Ohio, 1988.
81. W. Roger Cannon in Ref. 3, p. 195.
82. M. Yoshimura, *Am. Ceram. Soc. Bull.* **67**(12), 1950 (1988).
83. R. Garvie, R. Hannink, and R. Pascoe, *Nature (London)* **258,** 703 (1975).
84. N. Claussen, *J. Am. Ceram. Soc.* **59**(1–2), 49 (1976).
85. T. Gupta and co-workers, *J. Mater. Sci.* **12**(12), 2421 (1977).
86. T. Gupta, F. Lange, and J. Bechtold, *J. Mater. Sci.* **13,** 1464 (1978).
87. T. Gupta, *Sci. Sintering,* **10,** 2421 (1978).
88. M. van de Graaf and A. Burggraaf in Ref. 77, p. 744.
89. D. Clough, *Ceram. Eng. Sci. Proc.* **6**(9–10), 1244 (1985).
90. W. Zevert and co-workers, *J. Mater. Sci.* **25**(8), 3449 (1990).
91. M. Hölmstrom, T. Chartier, and P. Boch, *Mater. Sci. Eng. A* **109,** 105 (1989).
92. Y. Shen and R. Brook, *Sci. Sintering* **17**(1–2), 35 (1985).
93. I. Nettleship and R. Stevens, *Int. J. High Technol. Ceram.* **3,** 1 (1987).
94. K. Tsukuma, Y. Kubota, and Tsukidate in Ref. 77, p. 382.
95. T. Sato and co-workers, *J. Am. Ceram. Soc.* **68**(12), C320 (1985).
96. N. Hecht, S. Jang, and D. McCullum in Ref. 80, p. 133.
97. M. Matsui and co-workers in Ref. 80, p. 607.
98. F. Wakai, *Br. Ceram. Trans. J.* **88**(6), 205 (1989).

99. Y. Yoshizawa and T. Sakuma, *J. Am. Ceram. Soc.* **73**(10), 3069 (1990).
100. N. Claussen in Ref. 77, p. 325.
101. J. Wang and R. Stevens, *J. Mater. Sci.* **24**, 3421 (1989).
102. L. Viswanathan, Y. Ikuma, and A. V. Virkar, *J. Mater. Sci.* **18**, 109 (1983).
103. J. Moya and M. Osendi, *J. Mater. Sci.* **19**, 2909 (1984).
104. Q.-M. Yuan, J.-Q. Tan, and Z.-G. Jin, *J. Am. Ceram. Soc.* **69**(3), 265 (1986).
105. N. Claussen and J. Jahn, *J. Am. Ceram. Soc.* **61**(1–2), 94 (1978).
106. F. Lange, L. Falk, and B. Davis, *J. Mater. Res.* **2**(1), 66 (1987).
107. S. Kobayashi and S. Wada in Ref. 80, p. 127.
108. A. Tjernlund and co-workers in Ref. 80, p. 1015.
109. T. Ekström, L. Falk, and E. Knutsonwedel, *J. Mater. Sci. Lett.* **9**(7), 823 (1990).
110. S. Hirano, T. Hayashi, and T. Nakashima, *J. Mater. Sci.* **24**(10), 3712 (1989).
111. I. Wadsworth, J. Wang, and R. Stevens, *J. Mater. Sci.* **25**(9), 3982 (1990).
112. G. Leatherman and M. Tomozawa, *J. Mater. Sci.* **25**(10), 4488 (1990).
113. T. Watanabe and K. Shoubu, *J. Am. Ceram. Soc.* **68**(2), C34 (1985).
114. K. Shobu and co-workers in Ref. 80, 1091.
115. Y. Ikuma, W. Komatsu, and S. Yaegashi, *J. Mater. Sci. Lett.* **4**, 63 (1985).
116. J. Petrovic and R. Honnell, *J. Mater. Sci.* **25**(10), 4453 (1990).
117. D. Porter and A. Heuer, *J. Am. Ceram. Soc.* **60**(3–4), 183 (1977).
118. A. Evans and A. Heuer, *J. Am. Ceram. Soc.* **63**(5–6), 241 (1980).
119. F. Lange, *J. Mater. Sci.* **17**(1), 225 (1982).
120. A. Evans in Ref. 77, p. 193.
121. *J. Am. Ceram. Soc.* **69**(3,7) (1986). These two issues of the Journal contain a collection of articles devoted to the subject of transformation toughening in ZrO_2 containing ceramics.
122. M. Rühle, N. Claussen, and A. Heuer, *J. Am. Ceram. Soc.* **69**(3), 195 (1986).
123. T. Masaki, *J. Am. Ceram. Soc.* **69**(8), 638 (1986).
124. K. Tsukuma, K. Ueda, and M. Skimada, *J. Am. Ceram. Soc.* **68**(1), C4 (1985).
125. K. Tsukuma, T. Takahata, and M. Shiomi in Ref. 80, p. 721.
126. N. Claussen, *Mater. Sci. Eng.* **71**, 23 (1985).
127. N. Claussen, K. Weisskopf, and M. Rühle, *J. Am. Ceram. Soc.* **69**(3), 288 (1986).
128. N. Claussen and G. Petzow in R. E. Tressler, G. L. Messing, C. G. Pantano, and R. E. Newnham, eds., *Tailoring Multiphase and Composite Ceramics, Mater. Sci. Res.* **20**, 1986, p. 649.
129. E. Lucchini and S. Maschio, *J. Mater. Sci. Lett.* **9**(4), 417 (1990).
130. U. Dworak and co-workers in Ref. 2, p. 480.
131. R. Bratton and S. Lau in Ref. 1, p. 226.
132. R. Miller, *Surf. Coat Technol.* **30**(1), 1 (1987).
133. R. Vincenzini, *Ind. Ceram.* **10**(3), 113 (1990).
134. A. Kato and H. Yoshida, *7th Cimtec World Ceramic Congress, Paper* B6-L02, (Montecatini Terme, Italy, July 1990).
135. H. Reh, *7th Cimtec World Ceramic Congress, Paper B6-L04* (Montecatini Terme, Italy, July 1990).
136. R. Spriggs, R. Katz, and S. Hellem, *7th Cimtec World Ceramic Congress, Paper B6-L03*, (Montecatini Terme, Italy, July 1990).

Marina R. Pascucci
Jeffrey T. Neil
Samuel Natansohn
GTE Laboratories Incorporated

AERATION

BIOTECHNOLOGY

The supply of oxygen to a growing biological species, aeration, in aerobic bio-reactors is one of the most critical requirements in biotechnology. It was one of the biggest hurdles that had to be overcome in designing bioreactors (fermenters) capable of turning penicillin from a scientific curiosity to the first major antibiotic (1). Aeration is usually accomplished by transferring oxygen from the air into the fluid surrounding the biological species, from where it is in turn transferred to the biological species itself. The rate at which oxygen is demanded by the biological species in a bioreactor depends very significantly on the species, on its concentration, and on the concentration of the other nutrients in the surrounding fluid (1,2) (see CELL CULTURE TECHNOLOGY). There is no unique set of units used to define this rate requirement, but some typical figures are given in Table 1. The very wide range is noteworthy; during the course of a batch bioreaction, oxygen demand often passes through a marked maximum when the species is most biologically active (1).

Table 1. Oxygen Demands of Biological Species

Biological species	kg $O_2/(m^3 \cdot h)$	Reference
bacteria/yeasts	1 to 7	1,3
plant cells	0.03 to 0.3	4
seed priming[a]	1 to 8 $\times$ 10^{-2}	5
mammalian cells[b]	2 to 10 $\times$ 10^{-3}	6

[a]Based on a seed density of 100 kg/m^3.
[b]Based on a cell density of 10^{12} cells/m^3.

The main reason for the importance of aeration lies in the limited solubility of oxygen in water, a value which decreases in the presence of electrolytes and other solutes and as temperature increases. A typical value for the solubility of oxygen (the equilibrium saturation concentration) in water in the presence of air at atmospheric pressure at 25°C is about 0.008 kg O_2/m^3 (= 8 parts per million = 0.25 mmol/L). Thus, for a yeast or bacterial bioreaction demanding oxygen at the rates given in Table 1, all oxygen is utilized in about 10 to 40 s (3,7).

In addition to each bioreaction demanding oxygen at a different rate, there is a unique relationship for each between the rate of reaction and the level of dissolved oxygen (1,8). A typical generalized relationship is shown in Figure 1 for a particular species, eg, *Penicillium chrysogenum* or yeast. The shape of the curve is such that a critical oxygen concentration, C_{crit}, can be defined above which the rate of the bioreaction is independent of oxygen concentration, ie, zero order with respect to oxygen. The typical values given in Table 2 indicate that the critical

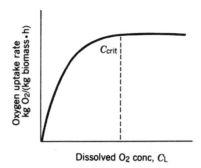

Fig. 1. The relationship between rate of oxygen uptake and dissolved oxygen, concentration where C_{crit} is the critical oxygen concentration.

Table 2. Critical Oxygen Concentrations at 30°C[a]

Organism	C_{crit}, (kg O_2/m^3) × 10^4
Azotobacter vinelandi	6 to 16
Pseudomonas denitrificans	3
Penicillium chrysogenum	3
Aspergillus oryzae	6.5

[a]Ref. 1.

concentration is usually on the order of 1 to 20% of the oxygen saturation value. Thus for each species, oxygen should be transferred rapidly enough to allow the oxygen demand to be met throughout the volume in which the bioreaction is occurring using a level of dissolved oxygen above the critical value. Failure to do so leads to a reduction in the overall rate and possibly a change of bioreaction to a different and unwanted metabolic pathway. If oxygen concentration falls low enough, an anaerobic reaction may develop. Yeast fermentations are a particularly good example of such possibilities (1,8).

In the cases of pellets (9), flocs (10), and immobilized cells and enzymes (11), aeration becomes more complex (1). Here it is the level of oxygen at the active biological site within the solid particle or aggregate rather than the level of dissolved oxygen in the surrounding fluid that determines the overall rate of reaction. Indeed, in certain cases such as penicillin fermentations, pellets form of such a size that the center of the pellet becomes inactive (12). A similar effect is observed when biofilms form on cooling surfaces. At a certain thickness the oxygen concentration at the base of the film becomes zero, ie, that region becomes anaerobic, and the biological species, eg *Pseudomonas fluorescens,* at the surface which is being fouled dies and sloughs off (13).

Principles of Oxygen Transfer

The Basic Mass Transfer Steps. Figure 2 shows the steps through which oxygen must pass in moving from air (or oxygen-enriched air) to the reaction site in a biological species (1,14). The steps consist of transport through the gas film

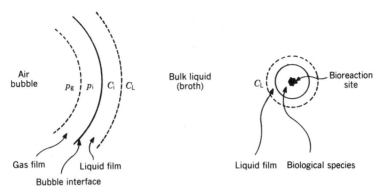

Fig. 2. Steps by which oxygen is transferred from the gas phase to the biological reaction site. Terms are defined in the text.

inside the bubble, across the bubble–liquid interface, through the liquid film around the bubble, across the well-mixed bulk liquid (broth), through the liquid film around the biological species, and finally transport within the species (eg, cell, seed, microbial floc) to the bioreaction site. Each step offers a resistance to oxygen transfer. In the last step, the resistance to transport is negligible for freely suspended bacteria or cells which are extremely small, but for immobilized cells and biological flocs and pellets the rate of oxygen diffusion through their structure may be rate-limiting. In the more complex situation, the rate of oxygen diffusion to the active sites depends on mass transfer through the external boundary layer followed by diffusion through the solid as governed by Fick's law. The theory is essentially the same as that applied to chemical catalysis (qv) and is described very well in reference 1, and other standard texts.

The rate-limiting step typically occurs at the air–liquid interface and, for biological species without diffusion limitations, the overall relationship can be simply written at steady state as

oxygen transfer rate from air (OTR)

= oxygen uptake rate by biological species (OUR)

Provided this equality is satisfied and the dissolved oxygen concentration in the well-mixed liquid is greater than the critical concentration throughout the bioreactor, then the maximum oxygen demand of the species should be met satisfactorily. Design of the bioreactor must ensure that the above requirements are achieved economically and without damaging the biological species.

When the oxygen demand is low, eg, at the start of a batch fermentation when the amount of biomass is small, the oxygen uptake rate at the bioreaction site is the limiting step and the level of dissolved oxygen is high (near saturation). When the oxygen demand becomes high, however, the rate-limiting step is that associated with transfer across the bubble–liquid interface and this rate of transfer is critically dependent on the fluid flow at the gas–liquid interface as well as the concentration of oxygen in both the gas and the liquid phases. The liquid-phase oxygen concentration is now low, but it should not be allowed to fall below the critical value. It is when this gas–liquid mass transfer step is rate-limiting that the biggest demands are made on the bioreactor. Then there is the greatest difficulty in getting enough oxygen into the suspending fluid (broth) to satisfy the oxygen demand of the biological species.

The Basic Mass Transfer Relationship. The basic principles which underlie oxygenation (aeration) are exactly the same as those which determine the rate of transfer of any sparingly soluble gas (oxygen) from the gas stream (air) to the unsaturated liquid (broth). The rate at which this transfer takes place is dependent on four principal parameters (1,14,15). The first is the area of contact between the gas and the liquid. In most bioreactors, this is provided by dispersing air in the suspending fluid (medium or broth) to give a large specific area of contact as implied by Figure 2. However, there are also other methods, especially in wastewater treatment, by which large specific areas of contact are produced. Solid supports are used on which films of the irrigated, biologically active species grow, eg, trickling bed filters and rotating disk contactors (11,14). The other three mass transfer rate parameters are the driving force available (ie, the difference in concentration of oxygen in the two phases); the two-phase fluid dynamics (including the effect of viscosity); and the chemical composition of the liquid.

The first assumption in all such physical mass transfer processes is that equilibrium exists at the interface between the two phases. This assumption implies that, at the interface, the concentration of the gas in the liquid, C_i, is equal to its solubility at its partial pressure in the gas phase, p_i. Since, for sparingly soluble gases such as oxygen, there is a direct proportionality between the two,

$$p_i = HC_i \tag{1}$$

where H is the Henry's law constant. The second assumption is that as the gas is absorbed, its concentration falls progressively, from a high in the bulk concentration in the gas phase, p_g, to that at the interface, p_i, then again from that on the liquid side of the interface, C_i, to a low in the bulk liquid, C_L (see Fig. 2).

The rate of mass transfer, J, is then assumed to be proportional to the concentration differences existing within each phase, the surface area between the phases, A, and a coefficient (the gas or liquid film mass transfer coefficient, k_g or k_L, respectively) which relates the three. Thus

$$J = k_g A(p_g - p_i) \tag{2}$$

$$= k_L A(C_i - C_L) \tag{3}$$

Equations 2 and 3 also stand as the definitions of k_g and k_L. For sparingly soluble gases, however, $p_i \simeq p_g$ so that from equation 1,

$$C_i = p_i/H = p_g/H = C_g^* \tag{4}$$

and

$$J = k_L A(C_g^* - C_L) \tag{5}$$

where C_g^* is the solubility of oxygen in the broth that is in equilibrium with the gas phase partial pressure of oxygen. Thus the aeration rate per unit volume of bioreactor, N, is given by

$$N = J/V = k_L (A/V)(C_g^* - C_L) = k_L a(C_g^* - C_L) \tag{6}$$

where a is the interfacial area per unit volume. At steady state, N must equal the OUR, which should be that demanded by the biological species in the range $C_L > C_{crit}$.

The Driving Force for Mass Transfer. The rate of mass transfer increases as the driving force, $C_g^* - C_L$, is increased. C_g^* can be enhanced as follows. From Dalton's law of partial pressures

$$p_g = P_g y \tag{7}$$

where y is the mole (volume) fraction of oxygen in the gas phase and P_g is the total pressure, ie, back pressure plus static head. C_g^* (eqs. 4 and 5) is also a function of composition and temperature, decreasing with both. C_L can be low provided it is above C_{crit}. However, in a very large scale bioreactor where circulation times are of the same order as the time for total oxygen depletion (3,7), the average value may need to be kept well above C_{crit} so that local values below it may be avoided. There are very few studies on the variations of C_L in any bioreactor (16,17), partly because of the difficulty of measuring it. C_L is usually measured by an oxygen electrode which gives C_L/C_g^* as a percentage and this value is sensitive to the local velocity over the probe, particularly if the liquid is viscous (18).

In a stirred bioreactor the liquid is generally considered well-mixed, ie, C_L is spatially constant. The gas phase too may be well-mixed (19) so that

$$p_g = \text{constant} = (p_g)_{out} \tag{8}$$

where $(p_g)_{out}$ is the partial pressure of oxygen in the exit gas and

$$C_g^* = (p_g)_{out}/H \tag{9}$$

This situation is most probable with fairly intense agitation and on the small scale. On the other hand, for ease of application, it is often assumed that no oxygen is utilized. This is the so-called no-depletion model. In this case

$$C_g^* = (p_g)_{in}/H \tag{10}$$

Again, it is most reasonable on the small scale, but only with low $k_L a$ values. Most of the data leading to $k_L a$ values in the literature (especially those in the review in reference 20) were obtained by making this assumption.

For large scale bioreactors (21), especially those of the air lift type (22), the gas phase is best considered as being in plug flow, so that a log mean value of driving force is obtained:

$$\Delta C_{lm} = \frac{\Delta C_{in} - \Delta C_{out}}{\ln(\Delta C_{in}/\Delta C_{out})} \tag{11}$$

where

$$\Delta C_{in} = [(p_g)_{in}/H] - C_L \tag{12}$$

and

$$\Delta C_{out} = [(p_g)_{out}/H] - C_L \tag{13}$$

The Mass Transfer Coefficient, $k_L a$. Because of the interaction between k_L and a when air is dispersed in the media, the two have rarely been measured separately. When they have been measured, k_L has been found to be dependent on the relative velocity between the phases, the level of turbulence in the liquid, and the size of bubbles. However, it is a relatively weak function of all of those. In addition, it appears that impurities, whether in solution or suspension, reduce the ease with which oxygen can pass across the interface (23) and, consequently, reduce the value of k_L.

The specific surface, a, is also relatively insensitive to the fluid dynamics, especially in low viscosity broths. On the other hand, it is quite sensitive to the composition of the fluid, especially to the presence of substances which inhibit coalescence. In the presence of coalescence inhibitors, the Sauter mean bubble size, d_B, is significantly smaller (24), and, especially in stirred bioreactors, bubbles very easily circulate with the broth. This leads to a large hold-up, ie, increased volume fraction of gas phase, ϵ_H. d_B, ϵ_H, and a are all related

$$a = 6\epsilon_H/d_B \tag{14}$$

Therefore, a values in the presence of coalescence inhibitors are much higher than without them (25), the changes in a outweighing changes in k_L. Thus it is found that coalescence inhibitors such as electrolytes and low molecular weight alcohols increase $k_L a$ by up to an order of magnitude compared to water (26). Coalescence promoters such as antifoams generally lower $k_L a$ by somewhat similar amounts (27,28). In addition, neither d_B nor k_L is a strong function of the fluid dynamics, but the amount of air recirculation is, so $k_L a$ and ϵ_H are both related to the fluid dynamics in similar ways (29).

Increases in broth viscosity significantly reduce $k_L a$ and cause bubble size distributions to become bimodal (30). Overall, $k_L a$ decreases approximately as the square root of the apparent broth viscosity (31). $k_L a$ can also be related to temperature by the relationship (32)

$$k_L a = (k_L a)_{20} \theta^{(T-20)} \tag{15}$$

where $(k_L a)_{20}$ is the value at 20°C, $\theta \simeq 1.022$, and T is the temperature, °C; ie, $k_L a$ increases by about 2.5% per °C.

The Measurement of $k_L a$. There are two main methods for measuring $k_L a$, the unsteady-state method, and the steady-state method. In the most common unsteady-state method, the level of dissolved oxygen is first reduced to zero, either by bubbling through nitrogen or by adding sodium sulfite (20). Then, the increase in dissolved oxygen concentration as a function of time is followed using an oxygen electrode. One of the assumptions set out earlier about the extent of gas-phase mixing has to be made and the choice can be very significant, giving k_La values differing by a factor of 3 for the same raw data (33). If two oxygen electrodes, one in the liquid phase and one in the gas, are employed, this difficulty can be eliminated, but the technique is quite difficult to use (34,35). In addition, the electrode response time requires consideration (18,20) as does the quality of liquid-phase mixing (22). Most significantly, it should again be emphasized that k_La is very dependent on composition. For example, the k_La of distilled water measured by nitrogen degassing is significantly less than that measured by sulfite deoxygenation. This difference occurs because the sulfite acts as a coalescence inhibitor, greatly enhancing hold-up, and therefore a and k_La. For stirred bioreactors, the unsteady-state technique can be adapted to give k_La values for real bioreactions (36). Though the method has the same inherent weaknesses as all the dynamic techniques, its advantage is that the k_La is determined for the system of interest. These experimental values may well be very different from those predicted by literature correlations based on water or idealized liquids. This point is likely to be particularly valid for mycelial fermentations for example. Additionally, the presence of antifoam can make a dramatic difference (28).

Steady-state techniques are better carried out on real bioreactions. In most cases however, and especially for batch systems, it is necessary to carry out an oxygen balance on the air in order to determine the mass of oxygen utilized and thus the mass flux represented in equation 6. Assumptions must again be made about the gas- and liquid-phase mixing. The significance of the assumptions is less than in the unsteady-state, except on the large scale where the well-mixed liquid assumption breaks down (16,17). No really satisfactory way of overcoming this problem has yet been proposed. Given suitable instrumentation, the steady-state method of measurement for the bioreaction of interest is the recommended technique. However, it is not suitable in systems having very low oxygen demands, where the extent of oxygen utilization is insufficient to give an accurate measure of the drop in oxygen concentration. In that case, the unsteady-state method (36) is best, as exemplified by its successful use with seed priming bioreactors (5). Recently, some new techniques have been proposed, but they are not yet well established (37–39).

Aeration in Bioreactors

A huge variety of bioreactors has been developed and a thorough review is available (40). It is not feasible to consider them all and large numbers are only curiosities. A useful subdivision has been made into three generic types involving the way in which air is dispersed to give the desired specific surface area. These are bioreactors driven by rotating agitators (stirred tanks), bioreactors driven by gas compression (bubble columns/loop fermenters), and bioreactors driven by circulating liquid (jet loop reactors) (41). The first two are the most important.

Stirred Tank Bioreactors. Traditionally, stirred tanks have been the most common types of bioreactors for aerobic processes and they remain so even in the face of newer designs. One of the main reasons is their extreme flexibility. Operational designs using controlled air flow rates up to about 1 vvm (volume of air/min per unit volume of fermentation fluid) and variable speed motors capable of transmitting powers up to about 3 W/kg with control down to close to zero are suitable for almost any bioreaction. These tanks are also relatively insensitive to fill, ie, to the proportion of liquid added to the bioreactor, and are therefore quite satisfactory for fed batch operations. Control of dissolved oxygen can be carried out by either altering aeration rate and/or agitator power input (via speed control). They are also capable of handling relatively satisfactorily broths which become significantly viscous during the course of a fermentation. In that case design for good bulk mixing (homogenization) may be the most demanding task that the agitator is required to carry out (42).

Until recently most industrial scale, and even bench scale, bioreactors of this type were agitated by a set of Rushton turbines having about one-third the diameter of the bioreactor (43) (Fig. 3). In this system, the air enters into the lower agitator and is dispersed from the back of the impeller blades by gas-filled or ventilated cavities (44). The presence of these cavities causes the power drawn by the agitator, ie, the power required to drive it through the broth, to fall and this has important consequences for the performance of the bioreactor with respect to aeration (35). k_La has been related to the power per unit volume, P/V, in W/m^3 and to the superficial air velocity, v_s, in m/s (20), where v_s is the air flow rate per cross-sectional area of bioreactor. This relationship in water is

$$k_La = 2.6 \times 10^{-2} \left(\frac{P}{V}\right)^{0.4} v_s^{0.5} \tag{16}$$

and for electrolyte solutions:

$$k_La = 2.0 \times 10^{-3} \left(\frac{P}{V}\right)^{0.7} v_s^{0.2} \tag{17}$$

Each equation is independent of impeller type. As pointed out earlier, the absolute k_La values vary considerably from liquid to liquid. However, similar relationships have been found for other fluids, including fermentation broths, and also for hold-up, ϵ_H. Therefore, loss of power reduces the ability of the Rushton turbines to transfer oxygen from the air to the broth.

There are two other special features of Rushton turbines (45). First, there is their radial flow. As these turbines disperse the gas–liquid mixture, they drive it radially outward at each agitator leading to rather poor top-to-bottom mixing. Second, if the air flow rate is too high, they can no longer disperse the air and the impellers are said to become flooded (46). New impellers have been introduced which are better able to handle the large quantities of air needed in large scale fermentations and which do not lose power so significantly. These are still radial flow impellers and the best example is the Scaba impeller (43) (Fig. 4a). Other types having good power characteristics and improved top-to-bottom bulk blending of nutrients (including air) are the Lightnin' A315 (Fig. 4b) and the Prochem Maxflo T (Fig. 4c). The Ekato Intermig (not shown) is another type proving popular in Europe (43). Though these impellers are not yet very well characterized (43,48), it is quite possible that they will take over from Rushton turbines in

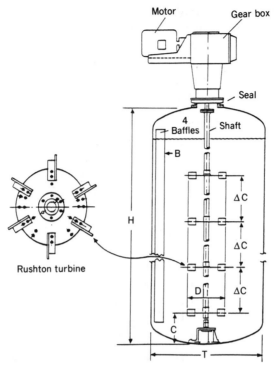

Fig. 3. A large scale fermenter agitated by Rushton turbines (43) where $B/T = 0.1$, $H/T \simeq 3.3$, $D/T \simeq 0.35$, $C/T = 0.25$, and $\Delta C/T = 0.51$.

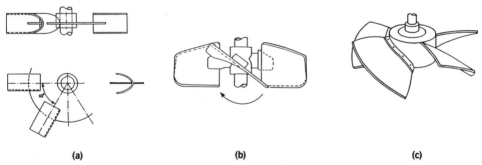

Fig. 4. Examples of alternative agitators: (**a**) Scaba 6 SRGT, (**b**) Lightnin' A 315, and (**c**) Prochem Maxflo T (43,47).

the future. Indeed, as the interaction between fluid dynamics, biological performance, and aeration becomes better understood, it is quite likely that agitator types for specific bioreactions will be chosen on a well-reasoned basis. It is also interesting to note that the reported improvements achieved by these newer agitators have been attributed to better bulk blending compared to Rushton turbines rather than to improved aeration (42).

Bubble Column and Loop Bioreactors. Air driven bioreactors are said to offer these advantages (40): no opening for a shaft is required and therefore they are less likely to become contaminated; they are less likely to damage bio-reactions involving fragile material such as plant (4) or mammalian cells (49); and they are very simple to operate on the very small scale and more economic on the very large scale where huge agitators and motors would otherwise be required (50). Examples are given in Figure 5.

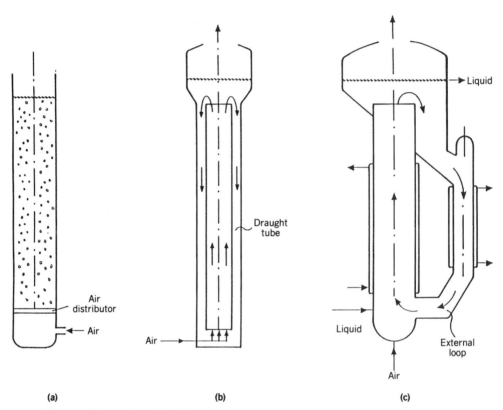

Fig. 5.   Examples of air driven bioreactors: (**a**) bubble column, (**b**) draught tube, and (**c**) external loop.

The bubble column is clearly the simplest of these bioreactors to construct. However, because of its rather ill-defined liquid circulation, air-lift reactors having either internal (draught tube) or external (loop) circulation of broth have been introduced. The major disadvantages of all three types are the poor capability of handling very viscous fermentations, especially those having a yield stress; the inflexibility, especially of the air-lift types, which only work well using a fill closely matched to the size of the bioreactor (this match affects both circulation rates, mixing and mass transfer, and bubble disengagement); and lack of indepen-dent control of dO_2 and mixing, since both are closely linked to the aeration rate. Reference 51 presents a very interesting literature review and analysis of bubble columns. In contrast to stirred tank bioreactors, the bubble size may in certain cases be very dependent on the way the air is introduced, ie, on the type of sparger employed. For systems having hindered coalescence, very fine (0.5 to 1 mm diameter) bubbles can be formed by using porous disks provided the superficial gas velocity is less than about 10^{-2} m/s. These conditions lead to very high k_La values. However, unless coalescence is very repressed, the bubble size grows within about 0.5 to 1 m to give k_La values similar to those found in coarse bubble systems. There may also be disengagement problems if coalescence does not occur.

Coarse bubble systems are typically found with orifice, perforated disks, or pipe spargers. Under most realistic conditions, bubbles of about 4–6 mm are formed. For a wide range of sizes

$$k_La = 0.32v_s^{0.7} \tag{18}$$

Figure 6 enables a comparison to be made of k_La values in stirred bioreactors and bubble columns (51). It can be seen that bubble columns are at least as energy-efficient as stirred bioreactors in coalescing systems and considerably more so when coalescence is repressed at low specific power inputs (gas velocities).

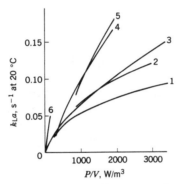

Fig. 6. A comparison of k_La values (51). Represented are: 1, stirred bioreactor using water, $v_s = 0.02$ m/s, k_La (eq. 16); 2, stirred bioreactor using water, $v_s = 0.04$ m/s, k_La (eq. 16); 3, bubble column using water, k_La (eq. 18); 4, stirred bioreactor using salt water, $v_s = 0.02$ m/s, k_La (eq. 17); 5, stirred bioreactor using salt water, $v_s = 0.04$ m/s, k_La (eq. 17); and 6, bubble column using salt water (noncoalescing).

It is also interesting to use Figure 6 to make a comparison of different aeration devices on the basis of energy-efficiency. From equation 6 and assuming a constant driving force

$$N \propto \text{OTR} \propto k_La \tag{19}$$

and from Figure 6

$$k_La \propto (P/V)^{0.6} \tag{20}$$

Therefore

$$\frac{\text{OTR}}{P/V} \propto \left(\frac{P}{V}\right)^{-0.4} \tag{21}$$

or

$$\frac{(OTR)V}{P} \propto \left(\frac{P}{V}\right)^{-0.4} \tag{22}$$

so that

$$\left(\frac{\text{kg } O_2}{\text{kW·h}}\right) \propto \left(\frac{P}{V}\right)^{-0.4} \tag{23}$$

From equation 23, it can be seen that the higher the power input per unit volume, the lower the oxygen transfer efficiency. Therefore, devices should be compared at equal transfer rates. All devices become less energy efficient as rates of transfer increase (3).

External and internal loop air-lifts and bubble column reactors containing a range of coalescing and non-Newtonian fluids, have been studied (52,53). It was shown that there are distinct differences in the characteristics of external and internal loop reactors (54). Overall, in this type of equipment

$$k_L a \propto \mu_a^{-0.9} \tag{24}$$

showing a greater fall as viscosity increases than in stirred tank bioreactors ($k_L a \propto \mu_a^{-0.5}$) and supporting the contention that such devices are unsuitable for viscous broths. The complete lack of oxygen transfer found in the downcomer (52) is also a significant factor and could easily lead to values of $C_L < C_{crit}$ being found in that region.

Applications to Different Biological Species

Mycelial Fermentations. Mycelial fermentations typically become viscous and shear thinning and difficult to mix (47). For such systems $k_L a$ and mixing are inextricably interlinked. Agitators that give better bulk blending per unit of energy can also give higher $k_L a$ values by involving more of the fermenter volume in the mass transfer process (42). Agitation levels link with mycelial structure and if pelleted growth can be encouraged, for example, using *Penicillium chrysogenum* (12), viscosity does not significantly increase and high levels of $k_L a$ can be maintained at satisfactory specific power inputs. Difficulties may arise, however, if the pellets are too large, as a result of diffusion resistance within them (1), leading to oxygen starvation of the potentially active sites at the center of the pellet (9). Though concern has been expressed about high power levels affecting mycelial length and structure, the effect has been found to be rather small where careful experiments have been conducted. Neither has a strong link been established between morphology and productivity (55).

Xanthan Gum Fermentations. Though xanthan gum fermentations become very viscous, the oxygen demand at this stage in a batch fermentation appears to be rather low and not rate limiting. Bulk blending and pH control is of more significance (56). Stirred bioreactors are therefore preferable using large impeller to tank diameter ratios.

High Oyxgen Demanding Fermentations. High oxygen demanding fermentations often require a higher level of oxygen over relatively short periods of time. The use of enriched air, or pure oxygen, and back pressure for a short period to enhance C_g^* as well as maximizing power and aeration rate to give the highest $k_L a$ may be worth considering. In such circumstances the flexibility of the stirred bioreactor is a distinct advantage.

Mammalian Cell Culture. Air-lift (57) and bubble column bioreactors (49) have been considered necessary for handling fragile mammalian cells because of the low oxygen demands (6). However, standard stirred bioreactors have been found not to damage hybridoma cells even at rather high (0.25 kW/m^3) power inputs using head space (nonbubbling) aeration on a small scale (58). The use of enriched air and back pressure might also extend the range of sizes of bioreactors operated in this mode. The successful use of stirred bioreactors at very low sparge rates has also been reported for such systems (59).

Plant Cell Culture. Air-lift bioreactors have been favored for plant cell systems since these cultures were first studied (4). However, they can give rise to problems resulting from flotation of the cells to form a meringue on the top. It is interesting to note that some reports indicate that stirred bioreactors do not damage such cells (4).

Seed Priming Bioreactors. Seed priming is a new technique enabling seeds suspended in an osmotica to imbibe moisture and thus be brought to the point of germination (5). However, germination does not occur. Subsequently, on sowing, germination is very rapid and synchronous. While the process, which takes up to about 14 days, is taking place, the seeds require oxygen. Both bubble column (60) and stirred bioreactors (5) have been used successfully, although the former requires high air rates to keep seeds in suspension (60). In addition, some seeds such as onions, appear to have a critical oxygen concentration greater than saturation with respect to air which may be due to diffusion limitations within the seed. In that case enriched air must be used (60) and, in order for the process to be economic, stirred bioreactors are appropriate.

Single-Cell Protein. Systems involving single-cell proteins are often very large throughput, continuous processing operations such as the Pruteen process developed by ICI. These are ideal for air-lift bioreactors of which the pressure cycle fermenter is a special case (50).

Biological Aerobic Wastewater Treatment. Biological aerobic wastewater treatment is a rather specialized biotechnical application (61). The activated sludge process consists of an aerated bioreactor to which the basic principles of oxygen transfer discussed here apply. Either aerated agitators or air spargers (diffusers) are used. Where the effluent has especially high oxygen demands, however, pure oxygen or oxygen-enriched air is employed. Two examples are the UNOX process (62) which involves agitation and the ICI deep-shaft process (equivalent to a loop fermenter) which is very effective where space is a premium (63). For maintaining the aeration of large quantities of relatively pure water having a low oxygen demand, where space is not a limitation, such as in reservoirs, simple plunging jets having low $k_L a$ values but very high energy efficiency are suitable (27). There is also a range of specialized devices in which the area for oxygen transfer is achieved by the use of extended solid surfaces on which a biofilm, irrigated by the water being treated, grows, eg, trickle beds and rotating disk contactors (61). Also, three-phase systems (fluidized beds) have been developed (11). These are bubble columns in which the air flow keeps inert solids such as sand in suspension giving a large surface area on which the biofilm can grow (11).

NOMENCLATURE

Symbol	Definition	SI units
a	interfacial area of air per unit volume of liquid	m^2/m^3
A	interfacial area available for mass transfer	m^2
C_{crit}	critical dissolved oxygen concentration	kg/m^3
C_i	concentration of oxygen in the liquid phase at the interface	kg/m^3
C_L	concentration of oxygen in the bulk liquid	kg/m^3
C_g^*	saturation concentration of oxygen in the liquid for an oxygen partial pressure p_g	kg/m^3
d_B	Sauter mean bubble size	m
H	Henry's law constant	$kPa \cdot m^3/kg$
J	oxygenation rate	kg/s
k_g	gas film mass transfer coefficient	$kg/(s \cdot m^2 \cdot kPa)$
k_L	liquid film mass transfer coefficient	m/s
N	oxygen mass flux	$kg/(m^3 \cdot s)$
P	power imparted to the liquid	W
P_g	total pressure of the gas phase	kPa
p_g	partial pressure of oxygen in the bulk of the gas phase	kPa
p_i	partial pressure of oxygen in the gas phase at the interface	kPa
T	temperature	$°C$
v_s	superficial gas velocity	m/s
y	volume (mole) fraction of oxygen in gas phase	dimensionless
ϵ_H	hold-up	dimensionless
θ	temperature coefficient = 1.07	dimensionless
μ_a	apparent dynamic viscosity	$kg/(m \cdot s)$

Subscripts

in	air entering the bioreactor
out	gas leaving the bioreactor
lm	ln mean

BIBLIOGRAPHY

1. J. E. Bailey and D. F. Ollis, *Biochemical Engineering Fundamentals*, 2nd ed., McGraw-Hill Book Co., New York, 1986.
2. K. Kargi and M. Moo-Young in M. Moo-Young, ed., *Comprehensive Biotechnology*, Vol. 2, Pergamon Press, Oxford, 1985, Chapt. 2.

3. K. van't Riet, *Trends Biotechnol.* **1**, 113 (1983).
4. A. H. Scragg and co-workers, *Proceedings of the International Conference on Bioreactors and Biotransformations,* NEL/Elsevier, London, 1987, p. 12.
5. A. W. Nienow and P. A. Brocklehurst, *Proceedings of the International Conference on Bioreactors and Biotransformations,* NEL/Elsevier, London, 1987, p. 52.
6. M. Lavery and A. W. Nienow, *Biotechnol. Bioeng.* **30**, 368 (1987).
7. A. P. J. Sweere, K. Ch. M. M. Luyben, and N. W. F. Kossen, *Enzyme Microb. Technol.* **9**, 386 (1987).
8. B. Atkinson and F. Mavituna, *Biochemical Engineering and Biotechnology Handbook,* Macmillan, London, 1983.
9. B. Metz and N. W. F. Kossen, *Biotechnol. Bioeng.* **19**, 781 (1977).
10. B. Atkinson and I. S. Daoud, *Adv. Biochem. Eng.* **4**, 42 (1976).
11. C. Webb in C. F. Forster and D. A. J. Wase, eds., *Environmental Biotechnology,* Ellis Horwood, Chichester, 1987, Chapt. 9.
12. K. Schugerl and co-workers, *Proceedings of the 2nd International Conference on Bioreactor Fluid Dynamics,* British Hydromechanics Research Association, Cranfield, UK, 1988, p. 229.
13. J. A. Howell and B. Atkinson, *Water Research* **10**, 304 (1976).
14. A. W. Nienow, in ref. 11, Chapt. 10.
15. P. V. Danckwerts, *Gas–Liquid Reactions,* McGraw-Hill Book Co., New York, 1970.
16. N. M. G. Oosterhuis and N. W. F. Kossen, *Biotechnol. Bioeng.* **26**, 546 (1984).
17. R. Manfredini, V. Cavallera, L. Marini, and G. Donati, *Biotechnol. Bioeng.* **25**, 3115 (1983).
18. Y. H. Lee and G. T. Tsao, *Adv. Biochem. Eng.* **13**, 35 (1979).
19. N. P. D. Dang, D. A. Karrer, and I. J. Dunn, *Biotechnol. Bioeng.* **19**, 953 (1977).
20. K. van't Riet, *Ind. Eng. Chem. (Proc. Des. Dev.)* **18**, 357 (1979).
21. D. I. C. Wang and co-workers, *Fermentation and Enzyme Technology,* Wiley-Interscience, New York, 1979.
22. G. Andre, C. W. Robinson, and M. Moo-Young, *Chem. Eng. Sci.* **38**, 1845 (1983).
23. J. T. Davies, *Turbulence Phenomena,* Academic Press, New York, 1972.
24. J. C. Lee and D. Meyrick, *Trans. Inst. Chem. Eng.* **48**, T37 (1970).
25. A. Prins and K. van't Riet, *Trends in Biotechnol.* **5**, 296 (1987).
26. M. Zlokarnik, *Adv. Biochem. Eng.* **8**, 134 (1978).
27. J. A. C. van de Donk, R. G. J. M. Lans, and J. M. Smith, *Proceedings of the Third European Conference on Mixing,* British Hydromechanics Research Association, Cranfield, 1979, p. 289.
28. B. C. Buckland and co-workers, in ref. 12, p. 1.
29. S. P. S. Andrews, *Trans Inst. Chem. Eng.* **60**, 3 (1982).
30. V. Machon, J. Vlcek, A. W. Nienow, and J. Solomon, *Chem. Eng. J.* **14**, 67 (1980).
31. A. D. Hickman, and A. W. Nienow, *Proceedings of First International Conference on Bioreactor Fluid Dynamics,* British Hydromechanics Research Association, Cranfield, UK, 1986, p. 301.
32. M. L. Jackson, and C. C. Shen, *AIChEJ.* **24**, 63 (1978).
33. C. M. Chapman, L. G. Gibilaro, and A. W. Nienow, *Chem. Eng. Sci.* **37**, 891 (1983).
34. S. N. Davies and co-workers, *Proceedings of the Fifth European Conf. on Mixing,* British Hydromechanics Research Association, Cranfield, UK, 1985, p. 27.
35. A. W. Nienow and co-workers, in ref. 12, p. 159.
36. B. Bandyapadhyay, A. E. Humphrey, and H. Taguchi, *Biotechnol. Bioeng.* **9**, 533 (1967).
37. A. D. Hickman, *Proceedings of the Sixth European Conference on Mixing,* AIDIC/British Hydromechanics Research Association, Cranfield, UK, 1988, p. 369.
38. M. Zlokarnik, *Adv. Biochem. Eng.* **11**, 157 (1979).
39. H. N. Chang, B. Halard, and M. Moo-Young. *Biotechnol. Bioeng.* **34**, 1147 (1989).

40. K. Schugerl, *Int. Chem. Eng.* **22**, 591 (1982).
41. H. Blenke, *Adv. Biochem. Eng.* **13**, 121 (1979).
42. B. C. Buckland and co-workers, *Biotechnol. Bioeng.* **31**, 737 (1988).
43. A. W. Nienow, *Chem. Eng. Progr.* **86**, 61 (1990).
44. M. M. C. G. Warmoeskerken and J. M. Smith, in ref. 12, p. 179.
45. J. C. Middleton in N. Harnby, M. F. Edwards, and A. W. Nienow, eds, *Mixing in the Process Industries,* Butterworths, London, 1985, Chapt. 17.
46. A. W. Nienow, M. M. C. G. Warmoeskerken, J. M. Smith, and M. Konno, in ref. 34, p. 143.
47. A. W. Nienow, *Trends Biotechnol.* **8**, 224 (1990).
48. G. J. Balmer, I. P. T. Moore, and A. W. Nienow in C. S. Ho and J. Y. Oldshue, *Biotechnology Processes,* AIChE, New York, 1987, p. 116.
49. A. Handa-Corrigan, A. N. Emery, and R. E. Spier, *Enzyme Microb. Tech.* **11**, 230 (1989).
50. S. R. L. Smith, *Philos. Trans. Roy. Soc. (London), Ser. B.* **290**, 341 (1980).
51. J. J. Heinen and K. van't Riet, *Proceedings of the Fourth European Conf. on Mixing,* British Hydromechanics Research Association, Cranfield, UK, 1982, p. 195.
52. M. K. Popovic and C. W. Robinson, *AIChEJ.* **35**, 393 (1989).
53. M. Y. Chisti and M. Moo-Young, *J. Chem. Tech. Biotechnol.* **42**, 211 (1988).
54. M. Y. Chisti, *Airlift Bioreactors,* Elsevier Applied Science, New York, 1989.
55. M. T. Belmar Campero and C. R. Thomas, in ref. 12, p. 215.
56. E. Galindo, A. W. Nienow, and R. S. Badham, in ref. 12, 1988, p. 65.
57. L. A. Wood and P. W. Thompson, in ref. 31, p. 157.
58. S. Oh, A. W. Nienow, M. Al-Rubeai, and A. N. Emery, *J. Biotechnol.* **12**, 45 (1989).
59. M. P. Backer and co-workers, *Biotechnol. Bioeng.* **32**, 993 (1988).
60. W. Bujalski, A. W. Nienow, and D. Gray, *Ann. Appl. Biology,* **115**, 171, (1989).
61. C. F. Forster and D. W. M. Johnstone, in ref. 11, Chapt. 1.
62. A. J. Blatchford, E. M. Tramontini, and A. J. Griffiths, *Water Pollut. Control* **81**, 601 (1982).
63. J. Walker and G. W. Wilkinson, *Ann. N.Y. Acad. Sci.* **326**, 181 (1979).

ALVIN W. NIENOW
University of Birmingham (UK)

WATER TREATMENT

Aeration for water treatment, the transfer of oxygen [7782-44-7], O_2, from air to water [7732-18-5], is well-studied (1–10). The basic purpose of aeration, which is used primarily for the treatment of wastewater, is to improve water quality for subsequent usage (see WATER). Aeration can bring about the physical removal of taste and odor producing substances such as hydrogen sulfide [7783-06-4], H_2S, and other volatiles as well as the chemical removal of metals (iron, manganese), gases (hydrogen sulfide), and other compounds (organics and inorganics) through oxidation and settling (11). Additionally, aeration is used extensively for the biological oxidation of both domestic and industrial organic wastes.

The function of aeration in a wastewater treatment system is to maintain an aerobic condition. Water, upon exposure to air, tends to establish an equilibrium concentration of dissolved oxygen (DO). Oxygen absorption is controlled by gas solubility and diffusion at the gas–liquid interface. Mechanical or artificial aeration may be utilized to speed up this process. Agitating the water, creating drops or a thin layer, or bubbling air through water speeds up absorption because each increases the surface area at the interface.

Oxygen Solubility

The solubility of a gas in water is affected by temperature, total pressure, the presence of other dissolved materials, and the molecular nature of the gas. Oxygen solubility is inversely proportional to the water temperature and, at a given temperature, directly proportional to the partial pressure of the oxygen in contact with the water. Under equilibrium conditions, Henry's law applies

$$C = HP$$

where C is the concentration (solubility) of the oxygen in the water, P is the partial pressure of the oxygen, and H is the Henry's law constant. The partial pressure of oxygen in air is assumed to be more or less constant at 21% by volume at sea level. The value decreases at higher altitudes, however.

Dissolved matter lowers oxygen solubility. At 20°C and 101.3 kPa (1 atm), the equilibrium concentration of dissolved oxygen in seawater is 7.42 mg/L. It is 9.09 mg/L in chloride-free water and 9.17 mg/L in clean water. This lessening of oxygen solubility is of importance to wastewater treatment. The solubility of atmospheric oxygen in a domestic sewage is much less than in distilled water (12).

Diffusion

The driving force in diffusion involves differences in the concentration of the diffusing substance. The molecular diffusion of a gas into a liquid is dependent on the characteristics of the gas and the liquid, the temperature of the liquid, the concentration deficit, the gas to liquid contact area, and the period of contact. Diffusion may be expressed by Fick's law (13,14):

$$\frac{dm}{dt} = -DA\,\frac{dc}{dx}$$

where dm/dt is the mass transfer rate by diffusion in time, dc/dx is the concentration gradient, A is the area of contact per unit time, and D is the diffusion coefficient. Thus, achieving saturation and the rate of oxygenation are not independent of one another because the further the system is from saturation, the greater the driving force.

One important factor affecting oxygen transfer from air to water is the resistance generated by the film at the air–water interface (Fig. 1) (15). The interface exhibits properties that are strikingly different from those of the main body of water or air. The degree of resistance depends on the presence of surface active materials at the interface (16). Minute amounts of materials such as soap, detergent, or organic acids are capable of causing considerable additional resistance to gas–liquid transfer (17).

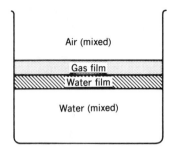

Fig. 1.   Schematic representation of the two films at the air–water interface.

Two-Film Theory.   The two-film theory is one of the best analyses of the gas absorption process in water (5,6). This theory involves consideration of the resistance to mass transfer that is offered by the two films at the air–water interface. Components are defined in Figure 2. The films are postulated to be present regardless of the type of aerator used. All the gas diffusing into the water has to diffuse through both films, the characteristics of which depend on the vapor pressure of the gas and solubility of the gas in water. Resistance to diffusion of a gas molecule results from collisions with molecules in the gas or liquid film through which diffusion is taking place. Molecules are much closer together in liquids than in a gas so that the resistance to diffusion is higher in the water film, although resistance may vary depending on the intensity of turbulence and bubble shearing. Resistance is also increased by increased film thickness. The film thickness of quiescent water has been calculated to be about 0.2 cm (6,18,19) whereas a falling drop of water traveling at a very high velocity is reported to have a film thickness of about 0.003 cm.

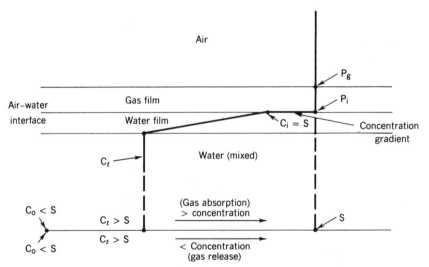

Fig. 2.   Schematic representation for the two-film theory of gas transfer; P_g = partial pressure of gas; P_i = partial pressure of the gas at the interface; C_t = concentration of gas at time t; C_i = initial concentration of gas at the interface; C_0 = initial concentration of gas at $t = 0$; and S = gas saturation. Adapted from refs. 19 and 20.

In the case of a less soluble gas such as oxygen, diffusion occurs so slowly through the liquid film that only a small concentration difference is required to overcome the resistance of the gas film. Thus the liquid film at the interface is considered to be very close to oxygen saturation and it is not necessary to consider gas film resistance in the calculation (14).

The most basic oxygen transfer equation (6) states that the amount of diffusion is proportional to the surface of the interface (6,20):

$$\frac{dw}{Ad\theta} = k_g(P_g - P_i) = k_L(C_i - C_L)$$

where w is the weight of solute, θ the time in hours, $dw/d\theta$ the rate of absorption, A the area of liquid–gas interface, k_g the diffusion coefficient through the gas film, k_L the diffusion coefficient through the liquid film, P the concentration of oxygen in the gas, and C the concentration of oxygen in the liquid; the subscripts g, i, and L apply to conditions in the main body of the gas, at the interface, and in the liquid, respectively. For oxygen, because of the negligible resistance exhibited by the gas film, k_L represents the overall gas transfer coefficient, $k_L = K_L$, so the equation may be rewritten

$$\frac{dw}{dt} = k_L A(C_i - C_L) = K_L A(C_i - C_L)$$

and C_i represents the saturation concentration in the liquid at the partial pressure of oxygen, P_g, for any given temperature.

The diffusion coefficient k_L depends upon the characteristics of the absorption process. Reducing the thickness of the surface films increases the coefficient and correspondingly speeds up the absorption rate. Therefore, agitation of the liquid increases diffusion through the liquid film and a higher gas velocity past the liquid surface could cause more rapid diffusion through the gas film.

Methods for Determining Oxygen Demand

Several methods have been developed to estimate the oxygen demand in waste water treatment systems. Commonly used laboratory methods are biochemical oxygen demand (BOD), chemical oxygen demand (COD), total oxygen demand (TOD), total organic carbon (TOC), and theoretical oxygen demand (ThOD).

Biochemical Oxygen Demand. 5-Days BOD or BOD_5, is the measurement of dissolved oxygen used in the biochemical degradation of organic wastes. BOD_5 measurement is widely used in wastewater and water treatment fields. The BOD_5, however, has a number of limitations; the 5-day test period is inconvenient in terms of the day-to-day operations of treatment plants; the test requires a high concentration of live, acclimated seed bacteria, a skilled technician, and a laboratory in which to monitor the samples; only biodegradable wastes are accounted for; nonbiodegradable toxics require pretreatment; and the effect of nitrifying bacteria must be reduced.

Oxygen is used in these microbiolreactions to degrade substrates, in this case organic wastes, to produce energy required for cell synthesis and for respiration. A minimum residual of 0.5 to 2.0 mg/L DO is usually maintained in the reactors to prevent oxygen depletion in the treatment systems.

Chemical Oxygen Demand. The COD test is used to measure the oxygen equivalent of the organic matter that can be oxidized by using strong oxidizing agents in an acid medium at high temperatures. The COD of a wastewater is, in general, higher than the BOD because more materials can be oxidized chemically than biologically. It is possible to correlate COD with BOD_5 for many wastewaters, however, and a COD can be very useful. It can be determined in 3 h. TOC, TOD, and ThOD are less frequently employed but can be substituted for COD.

Aerating Systems

Aerators designed to facilitate the transfer of oxygen from air to water increase interfacial area by producing liquid turbulence and circulation. There are four basic types of aerators summarized in Table 1.

Table 1. Aerator Specifications

Aerator type	Materials of construction	Oxygen transfer efficiency, OTE, %	Oxygen transfer rate, OTR, g/(W·h)
diffused aerator			
coarse bubble	noncorrosive metal or plastic	4–20	0.73–1.09
fine bubble	ceramic, cast iron, or plastic	10–30	1.09–1.52
static tube	noncorrosive metal or plastic	7–20	1.09–1.58
jet aerator	noncorrosive metal	10–25	1.52–2.13
mechanical aerator			
low speed surface	large diameter turbine having gear reducers (fixed bridge or platform mounted)		1.22–2.74
high speed surface	noncorrosive metal small diameter propeller having direct drive		1.22–1.52
rotary brush and disk	noncorrosive metal gear reducers or direct drive		1.52–2.11
horizontally mixing aspirating aerator	noncorrosive metal having hollow or solid shaft, small marine propeller at high speed, direct drive		0.61–1.82

Diffused Aerators. Diffused aerators are of two types, a coarse bubble and a fine bubble. The air is compressed to the bottom of a container or holding system such as a pond or tank, forced through pipes, and released through a diffuser. The function of the diffuser is to uniformly distribute and break up the air into bubbles. Two such systems are shown in Figure 3a and 3b. Another version of a compressed air aerator utilizes spargers (Fig. 3**c**) where air is piped under rotating turbines. The turbines circulate the water and break up the bubbles. Static tube aerators are also used (Fig. 3**d**). The air is supplied by tubes at the bottom of the holding system and intimately mixed with water before being released at the top.

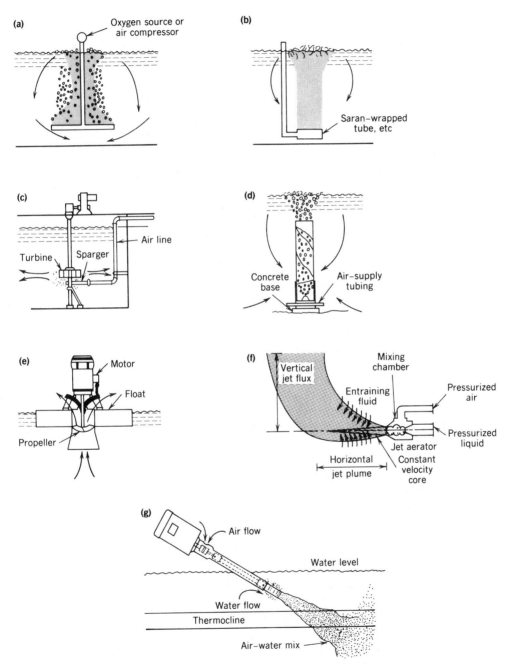

Fig. 3. Commercial aerating systems. Diffused aerators utilizing (**a**), (**b**) pipes, (**c**) a sparger, and (**d**) underwater air-supply tubes; (**e**) a mechanical aerator; (**f**) a jet aerator; and (**g**) a horizontally mixing aspirator.

Coarse bubble aerators have the following advantages: they are non-clogging, maintain the liquid temperature, and have low maintenance costs. The disadvantages are their high initial cost and low oxygen transfer rate; and they may foul. The advantages of fine bubble aerators include their good operational flexibility and fairly good mixing ability. They also maintain liquid temperature. Disadvantages are the high initial and maintenance costs because of the required air filter and other auxiliary equipment.

Static tube aerators are economically attractive and have a high transfer efficiency. They are well-suited for lagoon applications. On the other hand, they are poor mixers and are not recommended for use when the sludge concentration is over 3000 mg/L.

Mechanical Aerators. Mechanical aerators are modular in design and built using electric motors. This type of aerator is also known as a surface splasher because it pumps water vertically into the air. During this process the water is broken up into small droplets allowing exchange of oxygen from the air to the water. One type of mechanical aerator is shown in Figure 3e.

Mechanical aerators may be of the low speed surface, high speed surface, or rotary brush and disk variety. Low speed surface aerators have both high pumping and good oxygen transfer rates. The disadvantages include poor thermal retention as well as both high initial and high maintenance costs. The gear reducers can also cause problems. High speed surface aerators have the advantage of having a fairly low initial cost and of being portable and flexible in operation. However, these aerators have poor thermal retention, high maintenance cost, and poor oxygen transfer rate. They are also poor mixers. Advantages of rotary brush and disk aerators include good oxygen transfer rate and good mixing ability. Disadvantages are that some require gear reducers and all have a high power requirement, high cost, and require a special tank design.

Jet Aerators. Jet aerators are a cross between the diffused and mechanical aerators. Air and water are pumped separately under the water surface into a mixing chamber and ejected as a jet at the bottom of the tank or pond (Fig. 3f). Jet aerators are suited for deep tanks and have only moderate cost. Disadvantages include high operational costs, limitations caused by tank geometries, and nozzles that can clog. Additionally, they require blowers, pumps, and pretreated waste.

Horizontally Mixing Aspirator Aerators. An aerator using a horizontally mixing aspirator has a marine propeller, submerged under water, attached to a solid or a hollow shaft. The other end of the shaft is out of the water and attached to an electric motor. When the propeller is rotated at high velocity, at either 1800 or 3600 rpm, a pressure drop develops around the propeller. Air is then aspirated under the water and mixed with the water, and moved out. This type of aerator, shown in Figure 3g, is very efficient in mixing wastewater.

Horizontally mixing aspirating aerators produce no splashing and have no tank limitations. They are low in initial cost, are very good mixers, and have good thermal retention. Additionally, they are portable, have operational flexibility, and generate less noise than many aerators. Disadvantages are primarily their poor oxygen transfer rate coupled with some mechanical problems.

Several combinations of the aerators described are also available commercially. More information is available in references 21–23.

Oxygen Requirements

The oxygen requirements of wastewater depend upon the treatment system utilized. For lagoons having a flow-through system without sludge recycle, there is not only carbonaceous and ammonical BOD, but also a major portion of the sludge settles and is subsequently digested. Recommended for lagoon systems is a minimum of 2 kg of oxygen per kg of BOD_5 loaded at actual lagoon working conditions. To treat the ammonical BOD effectively, a minimum of 2.0 mg/L dissolved oxygen (DO) should be maintained in the lagoon at all times.

Activated sludge systems (ASS) vary considerably depending on cell residence time. The oxygen demand can range from 1.0 to 2.0 kg of oxygen per kg of BOD_5 removed. If nitrification occurs, however, additional oxygen should be provided at the rate of 4.6 kg O_2 per kg of ammonia removed. When diffused air systems are used, 0.01 to 0.06 m^3 (0.5 to 2.0 ft^3) per 3.785 L of wastewater or 34 to 51 m^3 (1200 to 1800 ft^3) of air per 0.45 kg of BOD loaded should be provided (21).

Oxygen Transfer Rate

Oxygen transfer rate (OTR) is estimated by the following standard procedure (12), where the rate of mass transfer per unit volume of liquid is taken to be directly proportional to the driving force of the system

$$\text{OTR} = K_L a V (C_\infty^* - C)$$

where OTR is the oxygen transfer occuring in the liquid volume V; $K_L a$ is the overall volumetric transfer coefficient in clean water based on the liquid film resistance; C_∞^* is the dissolved oxygen concentration at saturation, approached at infinite aeration time; and C is the dissolved oxygen concentration (DO).

The value of the saturation concentration, C_∞^*, is the spatial average of the value determined from a clean water performance test and is not corrected for gas-side oxygen depletion; therefore $K_L a$ is an apparent value because it is determined on the basis of an uncorrected C_∞^*. A true volumetric mass transfer coefficient can be evaluated by correcting for the gas-side oxygen depletion. However, for design purposes, C_∞^* can be estimated from the surface saturation concentration and effective saturation depth by

$$C_\infty^* = C_{sT}^* \left(\frac{P_1 - P_{vT} + \gamma_{wT} d_e}{P_s - P_{vT}} \right)$$

where C_{sT}^* is the tabular value of DO surface saturation concentration at temperature T, standard total pressure $P_s = 101.3$ kPa (1 atm), and 100% relative humidity; C_∞^* is the spatial average dissolved oxygen saturation concentration approached at infinite aeration time at temperature T; d_e is the effective saturation depth; P_b is the atmospheric pressure; P_{vT} is the saturated vapor pressure of water at temperature T; and γ_{wT} is the mass density of water at temperature T.

The effective saturation depth, d_e, represents the depth of water under which the total pressure (hydrostatic plus atmospheric) would produce a saturation concentration equal to C_∞^* for water in contact with air at 100% relative humidity. This can be calculated using the above equation, based on a spatial average value of $C_\infty^* T$, measured by a clean water test. For design purposes, d_e, can be estimated from clean water test results on similar systems, and it can range from 5 to 50% of tank liquid depth. Effective depth values for coarse bubble diffused air, fine bubble diffused air, and low speed surface aerators are 26 to 34%, 21 to 44%, and 5 to 7%, of the liquid depth, respectively.

The oxygen transfer rate for aerators is normally reported at standard conditions. Thus, in order to make meaningful comparisons, the ORT under working or field conditions should be adjusted to standard conditions oxygen requirement for treatment (SORT) by means of

$$SORT = ORT \div \alpha \left[\frac{\beta C_{walt} - C_L}{9.09} \right] \theta^{(T-20)}$$

where β is the salinity–surface tension correction factor = (wastewater C_∞^*)/(clean water C^*); C_{walt} is the oxygen saturation concentration for wastewater at a given temperature and altitude (mg/L) (see Table 2); C_L is the residual oxygen concentration (mg/L); T is the operating temperature of the wastewater (C); α is the oxygen transfer correction factor = (wastewater $K_L a$)/(clean water $K_L a$). For domestic waste α is usually about 0.8 to 0.85.

Table 2. Altitude Correction Factors for Oxygen Transfer

Elevation, m	Correction factor	Elevation, m	Correction factor
sea level	1.000	1524	0.830
305	0.950	1829	0.800
610	0.925	2134	0.765
914	0.900	2438	0.730
1219	0.860		

The SORT value when divided by the standard conditions oxygen transfer rate (OTR) given for a mechanical aerator yields the oxygen concentration required for treatment.

$$\frac{SORT}{24(OTR)}$$

When using a compressed air or diffused aeration system, the SORT has to be converted to a standard volume of air required per minute. This conversion can be accomplished through the equation

$$SVM = \frac{SORT}{OTE \times m \times 1440(min/day)}$$

where SVM is the standard volume of air required per minute, OTE is the oxygen transfer efficiency (see Table 1), and m is a correction factor. When SORT is in units of kg/day and $m = 1.157$ kg/m^3, SVM is in cubic meters; when SORT is in lb/day and $m = 0.047$ lb/ft^3, SVM is in cubic feet.

BIBLIOGRAPHY

1. W. E. Adeney and H. G. Becker, *Phil. Mag. S* **225,** 317–337 (1919).
2. W. E. Adeney and H. G. Becker, *Phil. Mag. S* **232,** 335–404 (1920).
3. W. E. Adeney, A. G. G. Leonard, and A. Richardson, *Phil. Mag.* **45,** 836–845 (1923).
4. W. E. Adeney, *Philos. Mag.* 1140–1148 (1926).
5. W. G. Whitman, *Chem. Metall. Eng.* **29**(4), 146–148 (1923).
6. W. K. Lewis and W. G. Whitman, *Ind. Eng. Chem.* **16,** 1215–1220 (1924).
7. H. W. Streeter, C. T. Wright, and R. W. Kehr, *Sewage Works J.* **8,** 282–316 (1936).
8. W. W. Eckenfelder, Jr., *Sewage Works J.* **24,** 1221–1228 (1952).
9. H. R. King, *Sewage Ind. Wastes* **27,** 894–908 (1955).
10. H. R. King, *Sewage Ind. Wastes* **27,** 1007–1026 (1955).
11. W. F. Langelier, *J. AWWA* **24,** 62–72 (1932).
12. L. C. Brown and C. R. Baillod, M. ASCE Modeling and Interpreting Oxygen Transfer Data, **108,** EE4, 17240, 607–627 (1982).
13. C. J. Geankoplis, *Transport Process and Unit Operations,* 2nd ed., Allyn & Bacon, Newton, Mass., pp. 373.
14. P. D. Haney, *J. AWWA* **46,** 353–378 (1954).
15. W. H. Bartholomew, E. O. Karow, and M. R. Sfat, *Ind. Eng. Chem.* **42,** 1801–1809 (1950).
16. W. W. Eckenfelder, Jr., and E. L. Barnhart, *AIChE J.* **7,** 631–634 (1961).
17. R. W. Kehr, *Sewage Ind. Wastes* **10,** 228 (1938).
18. J. R. Baylis, *Elimination of Taste and Odor in Water,* McGraw-Hill Book Co., New York, 1935, pp. 304–307.
19. M. H. Hutchinson and T. K. Sherwood, *Ind. Eng. Chem.* **29,** 836 (1937).
20. A. Pasveer, *Sewage Ind. Wastes* **27,** 1130–1146 (1955).
21. Metcalf and Eddy, *Wastewater Engineering Treatment Disposal Reuse,* 2nd ed., 1979.
22. Aeration, Oxygen Transfer Modeling, MOP, WPCF, No. FD-13,15 (1985).
23. F. N. Kemmer, *Nalco Water Handbook,* McGraw-Hill Book Co., New York, 1988.

RICHARD RAJENDEN
Aeromix Systems, Inc.

AEROSOLS

Classically, aerosols are particles or droplets that range from about 0.15 to 5 μm in size and are suspended or dispersed in a gaseous medium such as air. However, the term aerosol, as used in this discussion, identifies a large number of products which are pressure-dispensed as a liquid or semisolid stream, a mist, a fairly dry to wet spray, a powder, or even a foam. This definition of aerosol focuses on the container and the method of dispensing, rather than on the form of the product.

Aerosol technology may be defined as involving the development, preparation, manufacture, and testing of products that depend on the power of a liquefied or compressed gas to expel the contents from a container. This definition can be extended to include the physical, chemical, and toxicological properties of both the finished aerosol system and the propellants.

The aerosol container has enjoyed commercial success in a wide variety of product categories. Insecticide aerosols were introduced in the late 1940s. Additional commodities, including shave foams, hair sprays, antiperspirants, deodorants, paints, spray starch, colognes, perfumes, whipped cream, and automotive products, followed in the 1950s. Medicinal metered-dose aerosol products have also been developed for use in the treatment of asthma, migraine headaches, and angina.

The production of aerosols has increased both in the United States and worldwide, in spite of a substantial decline during the middle to late 1970s when the use of chlorofluorocarbons (see FLUORINE COMPOUNDS, ORGANIC) as propellants was seriously restricted. Hydrocarbons (qv) have since replaced the chlorofluorocarbons as propellants and aerosols continue to be used by the general public. Not only did the total number of aerosol units produced increase during the 1980s, but each year new applications for the aerosol system were introduced. As seen in Table 1, in 1989 U.S. aerosol production reached a new high (2912 million units) accompanied by growth in all areas except animal products and the miscellaneous category. Although the personal products category continued to grow in 1990, total aerosol production declined.

Personal products are the fastest growing segment of the aerosol industry and represent the largest of the categories. An increase in the use of hair spray as well as deodorants and antiperspirants accounts for the major growth. Increase in industrial aerosol automotive products was also large in the 1980s.

Advantages of Aerosol Packaging

Aerosol products are hermetically sealed, ensuring that the contents cannot leak, spill, or be contaminated. The packages can be considered to be tamper-proof. They deliver the product in an efficient manner generating little waste, often to sites of difficult access. By control of particle size, spray pattern, and volume delivered per second, the product can be applied directly without contact by the user. For example, use of aerosol pesticides can minimize user exposure and aerosol first-aid products can soothe without applying painful pressure to a wound. Spray contact lens solutions can be applied directly and aerosol lubri-

Table 1. Production of Aerosols in the United States[a], 1982–1990, Millions of Units

Aerosol product category	1973[b]	1982	1983	1984	1985	1986	1987	1988	1989	1990
personal products[c]	1495	628	747	832	879	952	964	1100	1015	1050
household products	690	556	586	603	630	635	640	650	680	680
automotive, industrial	176	286	303	337	342	375	379	440	475	415
paints, finishes	270	295	300	307	290	297	307	331	350	350
insect sprays	135	166	180	184	190	193	190	190	197	190
food products	94	132	132	138	140	136	140	157	175	175
animal products	18	20	19	16	15	22	22	8	8	8
miscellaneous	24	13	15	15	23	18	80	30	12	15
Total	*2902*	*2096*	*2282*	*2432*	*2509*	*2628*	*2722*	*2906*	*2912*	*2883*

[a]Ref. 2.
[b]Highest annual total production prior to the 1980s is given for comparison.
[c]Medical aerosols are included in the personal products category and in 1989 amounted to approximately 60 million units.

cants (qv) can be used on machinery in operation. Some preparations, such as stable foams, can only be packaged as aerosols. Spray shaving creams and furniture polish are examples of stable foams.

The use of metered-dose valves in aerosol medical applications permits an exact dosage of an active drug to be delivered to the respiratory system where it can act locally or be systemically absorbed. For example, inhalers prescribed for asthmatics produce a fine mist that can penetrate into the bronchial tubes (see ANTIASTHMATIC AGENTS).

Formulation of Aerosols

Aerosols are unique. The various components are all part of the product, and in the aerosol industry, the formulating chemist must be familiar with the entire package assembly and each of its components. All aerosols consist of product concentrate, propellant, container, and valve (including an actuator and dip tube). There are many variations of these components, and only when each component is properly selected and assembled does a suitable aerosol product result. A typical aerosol system is shown in Figure 1.

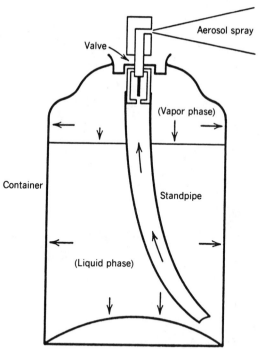

Fig. 1. Solution-type aerosol system in which internal pressure is typically 240 kPa at 21°C. To convert kPa to psi, multiply by 0.145.

The aerosol formulater must be knowledgeable about the availability and usage of propellents, various valves and containers, including pressure limitations and construction features, as well as any other components necessary for the product concentrate system. In contrast, the formulation of a nonaerosol emulsion is not, to any great extent, affected by either the container or package closure. Nonaerosol products packaged using a pump may be an exception, however, as this means of closure can also play a role in product formulation.

Product Concentrate. An aerosol's product concentrate contains the active ingredient and any solvent or filler necessary. Various propellant and valve systems, which must consider the solvency and viscosity of the concentrate–propellant blend, may be used to deliver the product from the aerosol container. Systems can be formulated as solutions, emulsions, dispersions, or pastes.

Solutions. To deliver a spray, the formulated aerosol product should be as homogeneous as possible. That is, the active ingredients, the solvent, and the propellant should form a solution. Because the widely used halocarbon and hydrocarbon propellants do not always have the desired solubility characteristics for all the components in the product concentrate, special formulating techniques using solvents such as alcohols (qv), acetone (qv), and glycols (qv), are employed.

The rate of spray is determined by propellant concentration, the solvent used, and valve and vapor pressure. The pressure must be high enough to deliver the product at the desired rate under the required operating conditions. For example, a windshield ice remover that is likely to be used around 0°C must be formulated to provide an adequate pressure at that temperature. Spray dryness or wetness and droplet size depend upon propellant concentration.

Generally, aerosol packaging consists of many delicately balanced variables. Even hardware design plays an important part. For example, valves that produce considerable breakup are used for the warm sensation desired in some personal products.

Emulsions. Aerosol emulsions (qv) may be oil in water (o/w), such as shaving creams, or water in oil (w/o), such as air fresheners and polishes. These aerosols consist of active ingredients, an aqueous or nonaqueous vehicle, a surfactant, and a propellant, and can be emitted as a foam or as a spray.

Foams. Systems that dispense foams (qv) are generally o/w emulsions, although nonaqueous solvents can also be used as the external phase. When the propellant is a hydrocarbon such as an isobutane–propane blend, as little as 3–4% in a 90–97% emulsion concentrate is sufficient to produce a suitable foam. Although the majority of the propellant is emulsified, some vaporizes and is present in the head space. The resultant pressure is generally on the order of 276 kPa (40 psig). When the valve is depressed, the pressure forces the emulsion up the dip tube and out of the container. Depending on the formulation, either a stable foam, such as would be expected in a shaving cream, or a quick-breaking foam, which collapses in a relatively short period of time, appears. The propellant is an important part of the formulation and is generally considered to be part of the immiscible phase. When the propellant is included in the internal phase, a foam is emitted; when the propellent is in the external phase, the product is dispensed as a spray. Figure 2 illustrates these situations.

When the propellant is in the internal phase (Fig. 2**a**), the propellant vapor, upon discharge, must pass through the emulsion formulation in order to escape into the atmosphere. In traveling through this emulsion, the trapped propellant forms a foam matrix. These systems, are typically oil-in-water emulsions.

An emulsion system in which the propellant is in the external or continuous

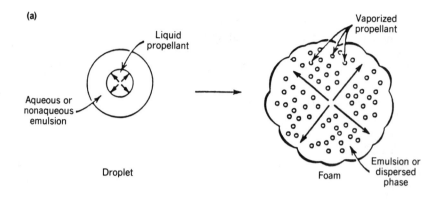

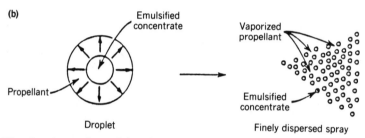

Fig. 2. Aerosol emulsion droplets containing propellant (**a**) in the internal phase with subsequent formation of aerosol foam and (**b**) in the external phase with subsequent formation of a wet spray.

phase is shown in Figure 2**b**. As the liquefied propellant vaporizes, it escapes directly into the atmosphere, leaving behind droplets of the formulation which are emitted as a wet spray. This system is typical of many water-based aerosols or w/o emulsions.

Extended stability testing is a necessity for emulsion systems in metal containers because of the corrosion potential of water. In most cases where a stable emulsion exists, there is less corrosion potential in a w/o system because the water is the internal phase.

Quick-breaking foams consist of a miscible solvent system such as ethanol (qv) [64-17-5] and water, and a surfactant that is soluble in one of the solvents but not in both. These foams are advantageous for topical application of pharmaceuticals because, once the foam hits the affected area, the foam collapses, delivering the product to the wound without further injury from mechanical dispersion. This method is especially useful for treatment of burns. Some personal products such as nail polish remover and after-shave lotion have also been formulated as quick-breaking foams.

Two advantages of foam systems over sprays (qv) are the increased control of the area to which the product is delivered and the decreased incidence of airborne particle release.

Sprays.   Aerosol spray emulsions are of the water-in-oil type. The preferred propellant is a hydrocarbon or mixed hydrocarbon–hydrofluorocarbon. About 25 to 30% propellant, miscible with the oil, remains in the external phase of the emulsion. When this system is dispensed, the propellant vaporizes, leaving behind droplets of the w/o emulsion (Fig. 2**b**). A vapor tap valve, which tends to produce finely dispersed particles, is employed. Because the propellant and the product concentrate tend to separate on standing, products formulated using this system, such as pesticides and room deodorants, must be shaken before use.

Dispersions.   In a powder aerosol the powder is dispersed or suspended in propellant using dispersants (qv) (oily vehicles) and suspending agents. Moisture content should be below 300 ppm and compacting, agglomeration, and sedimentation need to be minimized so that a fine powder can be uniformly dispensed without clogging of the valve. Powders must have a particle size of less than 40 μm (325 mesh screen) to pass through the valve orifices. Sedimentation rate can be substantially reduced by adjusting the density of either the propellant or the powder. Techniques include using a mixture of propellants of varying densities and adding an inert powder to the active ingredients. The use of surfactants (qv) as dispersing agents can also serve to lubricate the valve to prevent its sticking.

Pharmaceutical powder aerosols have more stringent requirements placed upon the formulation regarding moisture, particle size, and the valve. For metered-dose inhalers, the dispensed product must be delivered as a spray having a relatively small (3–6 μm) particle size so that the particles can be deposited at the proper site in the respiratory system. On the other hand, topical powders must be formulated to minimize the number of particles in the 3–6-μm range because of the adverse effects on the body if these materials are accidently inhaled.

Pastes.   Aerosols utilizing a paste as the product concentrate base differ from other formulations in that the product and the propellant do not come in contact with one another. The paste is placed in a bag that is attached to the valve system and fitted into the container. The propellant is then placed between the bag and the outer wall so that the propellant presses against the outside of the bag, dispensing the contents through the valve.

Propellants.   The propellant, said to be the heart of an aerosol system, maintains a suitable pressure within the container and expels the product once the valve is opened. Propellants may be either a liquefied halocarbon, hydrocarbon, or halocarbon–hydrocarbon blend, or a compressed gas such as carbon dioxide (qv), nitrogen (qv), or nitrous oxide.

Liquefied Gas Propellants.   One of the advantages in using a liquefied gas propellant is that the pressure in the aerosol container remains constant until the contents are completely expelled. The disadvantages are primarily ones of safety and environmental impact.

Chlorofluorocarbons (CFCs).   Prior to 1978 most aerosol products contained chlorofluorocarbon propellants. Since that time, the use of chlorinated fluorocarbons for aerosols has been seriously curtailed. These compounds have been implicated in the depletion of the ozone (qv) layer and are considered to be greenhouse gases (see AIR POLLUTION; ATMOSPHERIC MODELS).

The 1990 Clean Air Act regulates the production and use of CFCs, hydrochlorocarbons, hydrochlorofluorocarbons (HCFCs), and hydrofluorocarbon (HFC) substitutes. CFC and halon (Class I substances) usage is to be phased out in steps until total phaseout occurs on January 1, 2000. HCFC (Class II substance) production is currently frozen and a total HCFC production ban becomes effective on January 1, 2030. The EPA may, however, grant small and limited exceptions for Class I and II production schedules. Bans for HCFC usage in aerosols and foams begin January 1, 1994, although some safety and medical aerosol products, and foams used for insulation are exempted. In addition, warning labels must be used for CFCs and halons on all transport containers and all products containing these substances by May 15, 1993. Warning labels must also be used by that date for HCFCs on all transport containers and for all products containing or made with these substances if there exist suitable alternatives.

In the United States, use of CFC propellants, designated as Propellants 11, 12, and 114, is strictly limited to specialized medicinal aerosol products such as metered-dose inhalers. The physical properties and chemical names of these propellents are given in Table 2.

Table 2. Physical Properties of Chlorofluorocarbon and Hydrocarbon Propellants

Property	Propellant					
	11	12	114	A-108	A-31	A-17
chemical name	trichloro-monofluoro-methane	dichloro-difluoro-methane	1,2-dichloro-1,1,2,2-tetra-fluoroethane	propane	isobutane	n-butane
CAS Registry Number	[75-69-4]	[75-71-8]	[76-14-2]	[74-98-6]	[75-28-5]	[106-97-8]
formula	CCl_3F	CCl_2F_2	$CClF_2CClF_2$	C_3H_8	$HC(CH_3)_3$	C_4H_{10}
molecular weight	137.4	120.9	170.9	44.1	58.1	58.1
boiling point, °C	23.8	−29.8	3.6	−42.2	−10.2	−0.6
vapor pressure, kPa[a]						
21°C	194	585	190	846	315	214
54°C	269	1349	507	1893	763	556
liquid density, g/mL, 21°C	1.485	1.325	1.468	0.5005[b]	0.5788[b]	0.5571[b]
flammability limit, vol % in air	nonflam-mable	nonflam-mable	nonflam-mable	2.3–7.3	1.8–8.4	1.6–6.5

[a]To convert kPa to psi, multiply by 0.145.
[b]At 68°C.

Fluorocarbons are assigned numbers based on their chemical composition and, in general, these numbers are preceded by a manufacturer's trademark. In the numbering system, the digit on the right denotes the number of fluorine atoms in the compound; the second digit from the right indicates the number of hydrogen atoms plus 1; and the third digit from the right indicates the number of carbon atoms less 1. In the case of isomers, each has the same number. The most symmetrical is indicated by the number without any letter following it. As the isomers become more unsymmetrical, the letters a, b, and c are appended. If a molecule is cyclic, the number is preceded by C.

Hydrocarbons. Hydrocarbons such as propane, butane, and isobutane, which find use as propellants, are assigned numbers based upon their vapor pressure in psia at 21°C. For example, as shown in Table 2, aerosol-grade propane is known as A-108, n-butane as A-17. Blends of hydrocarbons, eg, A-46, and blends of hydrocarbons and hydrochlorocarbons or HCFCs are also used. The chief problem associated with hydrocarbon propellants is their flammability.

Hydrocarbons have, for the most part, replaced CFCs as propellants. Most personal products such as hair sprays, deodorants, and antiperspirants, as well as household aerosols, are formulated using hydrocarbons or some form of hydro-carbon–halocarbon blend. Blends provide customized vapor pressures and, if halocarbons are utilized, a decrease in flammability. Some blends form azeotropes which have a constant vapor pressure and do not fractionate as the contents of the container are used.

Hydrofluorocarbons and Hydrochlorofluorocarbons. The properties of HFC and HCFC propellants are given in Table 3. Propellant 22 is nonflammable and can be mixed to form nonflammable blends. Some of these propellants are scheduled for phase-out by 2015–2030.

Table 3. Properties of Hydrofluorocarbon and Hydrochlorofluorocarbon Propellants

	Propellant		
Property	22	142b	152a
chemical name	chlorodifluoro-methane	1-chloro-1,1-difluoro-ethane	1,1-difluoro-ethane
CAS Registry Number	[75-45-6]	[75-68-3]	[75-37-6]
formula	$CHClF_2$	$C_2H_3ClF_2$	$C_2H_4F_2$
molecular weight	86.5	100.5	66.1
boiling point, °C	−40.8	−9.44	−23.0
vapor pressure, kPa[a]			
21°C	834	200	434
54°C	2048	669	1220
density, g/mL at 21°C	1.21	1.12	0.91
solubility in water, wt % at 21°C	3.0	0.5	1.7
Kauri-butanol value	25	20	11
flammability limit, vol % in air	nonflammable	6.3–14.8	3.9–16.9
flash point, °C	none	none	< −50°

[a]To convert kPa to psi, multiply by 0.145.

Compressed Gas Propellants. The compressed gas propellants, so named because they are gaseous in conventional aerosol containers, are nontoxic, nonflammable, low in cost, and very inert. When used in aerosols, however, the pressure in the container drops as the contents are depleted. Although the problem is lessened when the contents are materials in which the propellant is somewhat soluble, this pressure drop may cause changes in the rate and characteristics of the aerosol spray. A compressed gas aerosol system is illustrated in Figure 3.

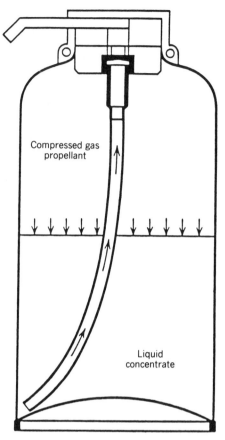

Fig. 3.   Compressed gas aerosol using an insoluble gas as propellant.

Considerable developmental effort is being devoted to aerosol formulations using the compressed gases given in Table 4. These propellants are used in some food and industrial aerosols. Carbon dioxide and nitrous oxide, which tend to be more soluble, are often preferred. When some of the compressed gas dissolves in the product concentrate, there is partial replenishment of the headspace as the gas is expelled. Hence, the greater the gas solubility, the more gas is available to maintain the initial conditions.

Compressed gas systems were originally developed simply to provide a means of expelling a product from its container when the valve was depressed. Semisolid products such as a cream, ointment, or caulking compound are dispensed as such. A liquid concentrate and a compressed gas propellant (Fig. 3) produce a spray when a mechanical breakup actuator is used. Nitrogen, insoluble in most materials, is generally used as the propellant.

Aerosols using an insoluble gas are not intended to be shaken before use. Shaking causes some of the propellant to be dispersed in the liquid concentrate. Although the product may then be dispersed to a greater extent, greater loss of propellant also results. If enough propellant is lost, the product will become inoperative.

Table 4. Properties of the Compressed Gas Propellants

Property	Carbon dioxide	Nitrous oxide	Nitrogen
CAS Registry Number	[124-38-9]	[10024-97-2]	[7727-37-9]
formula	CO_2	N_2O	N_2
molecular weight	44.0	44.0	28.0
boiling point, °C	−78	−88	−196
critical temperature, °C	31	37	−111
vapor pressure, kPa^a at 21°C	5772	4966	
solubility in water, vol gas/vol liq at 21°C and 101.3 kPa^a	0.82	0.6	0.016
flammability limits, vol % in air	nonflammable	nonflammable	nonflammable
flash point, °C	none	none	none

^{a}To convert kPa to psi, multiply by 0.145.

When gases that are somewhat soluble in a liquid concentrate are used, both concentrate and dissolved gas are expelled. The dissolved gas then tends to escape into the atmosphere, dispersing the liquid into fine particles. The pressure within the container decreases as the product is dispersed because the volume occupied by the gas increases. Some of the gas then comes out of solution, partially restoring the original pressure. This type of soluble compressed gas system has been used for whipped creams and toppings and is ideal for use with antistick cooking oil sprays. It is also used for household and cosmetic products either where hydrocarbon propellants cannot be used or where hydrocarbons are undesirable.

Other Propellants. Dimethyl ether (DME) [115-10-6] is finding use as an aerosol propellant. DME is soluble in water, as shown in Table 5. Although this solubility reduces DME's vapor pressure in aqueous systems, the total aerosol solvent content may be lowered by using DME as a propellant. The chief disadvantage is that DME is flammable and must be handled with caution.

Containers. Aerosol containers, made to withstand a certain amount of pressure, vary in both size and materials of construction. They are manufactured from tin-plated steel, aluminum, and glass. The most popular aerosol container is the three-piece tin-plated steel container. Glass containers, which are usually plastic coated, generally have thicker walls than conventional glass jars. They are limited to a maximum size of 120 mL and are used for pharmaceutical and cosmetic aerosols.

Steel. The steel container's most usual form is cylindrical with a concave (or flat) bottom and a convex top dome with a circular opening finished to receive a valve with a standard 2.54-cm opening. The three pieces (body, bottom, and top) are produced separately and joined by high speed manufacturing. The size of the container is described by its diameter and height to top seam, in that order. Hence a 202 × 509 container is 54.0 mm (2²⁄₁₆ in.) in diameter by 141.3 mm (5⁹⁄₁₆ in.) high. Tables of available sizes and overflow volumes and suggested fill levels can be readily obtained from manufacturers.

Tin-plated steel was long the mainstay of the U.S. aerosol industry and still

Table 5. Properties of Dimethyl Ether Propellant

Property	Dimethyl ether
CAS Registry Number	[115-10-6]
formula	CH_3OCH_3
molecular weight	46.07
boiling point, °C	−24.8
vapor pressure, kPa[a]	
21°C	434
54°C	1200
density, g/mL at 21°C	0.66
solubility in water, wt % at 21°C and autogenous pressure	34
Kauri-butanol value	60
flammability limits in air, vol %	3.4–1.8
flash point, °C	−41

[a]To convert kPa to psi, multiply by 0.145.

represents a very large volume. The tin coating provides both protective internal and external surfaces and the means for soldering the flat body plate into a leakproof cylinder. Both tin and lead solders are used, and both frequently contain small (5%) amounts of other metals, such as antimony and silver, for additional strength. In some products, the tin coating is removed sacrificially by the product; attack on steel is avoided only as long as some of the tin remains.

Mass production welding processes have been developed for the steel container side walls and have largely replaced the soldered side seam. The welding process has a slight financial advantage because it eliminates the need for tin. In addition, the welded joint is esthetically more desirable, it is less than one-half the width of the soldered joint, and it does not weaken during prolonged storage at elevated temperatures.

In order to increase resistance of the container to the effect of the product or to protect the product from the tin plating, an inert, internal organic coating can be applied.

Aluminum. The majority of aluminum containers are of monobloc (one-piece) construction, impact extruded from a slug of lubricated aluminum alloy. These containers are widely used for many products and are available in a vast array of heights and diameters. Because these containers lend themselves to additional shaping, many unusual shapes can be found in the marketplace. They may also be coated after the extrusion process.

Aluminum containers are recommended for many applications because of the very hard, corrosion-resistant oxide coating. They are deficient in only one respect: once the protective skin has been penetrated, aluminum corrosion accelerates.

Two-piece aluminum containers are also available. These consist of impact-extruded upper shells having a seamed-on aluminum bottom. The valve opening is machined from the solid aluminum rather than rolled. All coatings must be applied to the can body after the forming operation.

Glass. Glass containers present completely different design considerations from metal ones. They are totally nonreactive with the product, free of potentially leaking seams (the valve joint may be an exception), transparent or opaque as desired, and can be beautiful in design. The larger sizes attract attention on store shelves. On the other hand, they can break in manufacture, in shipping, and in usage. Although the extent and hazards of breakage can be reduced by means of a thick vinyl coating, glass containers are heavy, making them most costly to ship. Glass containers are processed more slowly and have higher scrap rates, and they are generally limited to low pressure product systems.

Valves. The dispensing valve and actuator serve to close the opening through which the product and frequently the propellant entered the container, to retain the pressure within the container and to dispense the product in the precise form and dosage intended by the manufacturer and expected by the consumer. An aerosol valve shown in Figure 4, consists of seven components. Many variations exist both for special purposes and to avoid existing patents.

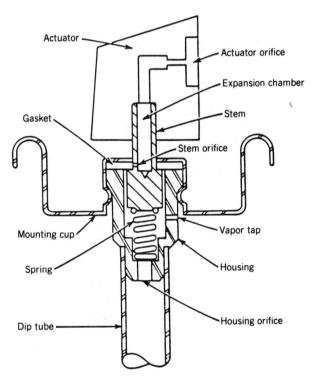

Fig. 4. Aerosol valve components.

The *mounting cup* (ferrule for bottle valves) mechanically joins the valve to the container. The mounting cup may be made from a variety of materials, but is typically tin-plated steel coated on the underside. It contains the gasket which provides the seal. Soft gasketing material is applied wet and bonded in place or, more frequently in larger cans, cut rubber, polyethylene, or polypropylene gaskets are used.

The *housing* physically holds the valve pieces together by means of a mechanical lock (crimp) and fits into the pedestal of the mounting cup. It is made from any of a number of common thermoplastics and contains the metering orifices for both the liquid and vapor phases of the effluent. Many valves do not meter vapor; the flow of the liquid is controlled by other means. A vapor tap may also be present to reduce the flammability of the product when it is emitted or to produce a finer, drier spray.

The plastic *stem* is the movable segment of the valve. It provides the opening mechanism and usually contains another metering orifice as shown.

The *spring* ensures a solid closing action and is usually wound from stainless steel wire. The *dip tube* conducts the product from the container to the valve. It is usually extruded from polyethylene or polypropylene and has an inside diameter of over 2.54 mm, although it can be provided in capillary sizes having diameters down to 0.25 mm. These small tubes are used to reduce flow rate and may function in place of the liquid metering orifice in the valve housing.

The *actuator* contains the final orifice and a finger pad or mechanical linkage for on–off control. The spray pattern is largely affected by the construction of the actuator, particularly by the chamber preceding the orifice. Actuators are often termed mechanical breakup and nonmechanical breakup depending upon the complexity of this chamber. Mechanical breakup actuators are of more expensive two-piece construction. Actuators are usually molded from polyethylene or polypropylene; the breakup insert may be almost any material, including metal.

Valves may differ depending upon the form of the product. Manufacturers stock a wide array of standard components. In addition, most are willing to produce unique combinations at no additional charge if no major tooling changes are required. The largest numbers of aerosol valves are used to produce sprays. These are actuated by either tilting forward or depressing vertically. Foam valves differ from spray valves primarily in the actuator, which is relatively wide open. Small actuator orifices throw the foam and are used for products such as rug shampoos and decorative snow. Internally, foam valves have no vapor tap and contain relatively large orifices.

Metering valves are used extensively in the pharmaceutical field for inhalers and other products requiring controlled dosage delivery. They typically deliver 50 to 150 mg of product per stroke with good repeatability. The metering is achieved by plugging the body orifice with the downward stroke of the stem, allowing only the product in the housing to escape.

Filtered valves contain a fine internal filter, typically below the body orifice. This filter prevents clogging by the debris sometimes found in product and package. The use of filtration is recommended with any valve systems containing body, stem, or actuator orifices of 0.25-mm (or smaller) diameter unless exceptional care is taken in the cleaning of product and package components. Valves containing these small orifices are used for products propelled by compressed gas.

Codispensing valve (and container) systems are used for products that consist of two reactive ingredients that must be kept separate until dispensed. They can deliver the product components in stoichiometrically correct amounts, thoroughly mixing them in the process. Successful use of these systems requires proper metering throughout the life of the container, safety considerations (should the internal seals between reactants fail), acceptably low permeation across internal membranes, end-use excess of the preferred component, and the usual product–package compatibility. The codispensing valves are the most technically demanding systems in the marketplace.

Filling of Aerosols

All aerosols are produced by either a cold filling or pressure filling process. The cold fill process is used for some aerosols which contain a metered-dose valve, although pressure filling is also adaptable to metered valves. Generally a concentrate is prepared which is filled into the aerosol container and then the valve is added. For the most part, pressure filling is carried out either by an under-the-cap filler or through the valve. If an under-the-cap filler is used, a vacuum is drawn, the propellant is added (under the valve cup), and then the valve is sealed in place. Where filling is done through the valve stem, the product is first filled into the container and a valve is crimped into place and at the same time a vacuum is drawn in the can. The propellant filler then forms a seal around the head of the can and under high pressure the propellant is forced through the actuator and valve stem into the container. The contents are then checked for leaks and an overcap is added to complete the process.

Barrier Packs. Barrier packs utilize a plastic bag or piston to separate the product from the propellant. Many designs have been submitted for these systems, but only two basic ones are commercially available. In barrier packs using a plastic bag (sepro containers) there is never any contact between product and propellant, unless one or the other permeates through the bag or the bag breaks. In those using a piston, there may be some seepage of propellant between the wall of the piston and the wall of the container; however, some of the newer designs greatly limit the extent of this seepage.

Piston Packs. A piston barrier pack, shown in Figure 5, consists of a free piston fitted into a container. It is used to package viscous, semisolid products such as cheese spreads, cake-decorating icings, pharmaceutical ointments, caulking compounds, grease and lubricants, shave gels, and other household and industrial items. The use of liquefied gas propellants is advantageous in piston packs because a constant pressure is maintained. These gases cannot be used with all products however, because there may be some seepage of vapor through the piston or between the walls of the piston and the container.

The product is filled through the 2.54-cm opening of the metal container and occupies the space above the piston. The valve is then sealed into place. After the propellant is added through a small hole in the bottom of the can, the opening is sealed using a rubber plug and pressurized either with nitrogen to about 620 kPa (90 psig) or with about 5 to 10 g of hydrocarbon propellant. Valves with relatively large openings, such as foam valves, are frequently used. Products packaged using this system are semisolid and viscous; they are dispensed as a lazy stream rather than as a foam or spray.

Sepro Container. The Sepro container consists of a collapsible plastic bag fitted into a standard three-piece, tin-plated container such as a 202 × 214, 202 × 406, or 202 × 509 can. The product is placed within the bag, and the propellant is

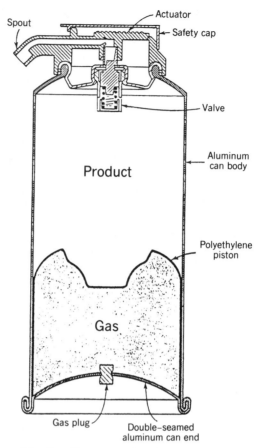

Fig. 5. Piston-type barrier pack system.

added through the bottom of the container, which is fitted with a one-way valve. There is no limitation on the viscosity of the product but compatibility with the plastic bag must be considered. A free-flowing liquid can be dispensed either as a stream or a fine spray, depending on the type of valve employed. A viscous material is often dispensed as a stream. This system has been used for caulking compounds, postfoaming gels, and depilatories.

Economic Aspects

In 1989 the aerosol industry in the United States consisted of about 250 companies that ranged from propellant, valve, and overcap, manufacturers to contract product fillers and product marketers. The industry employed over 50,000 people in 1989 and retail sales amounted to over $10 billion.

There was a small drop in aerosol compounds in 1990 (2.88 billion units) as compared to 1989 (2.91 billion units). Aerosol valve and container production remained high. 3.13 billion valves were produced in 1990. 320 million of the containers shipped in 1990 were aluminum.

BIBLIOGRAPHY

"Aerosols" in *ECT* 2nd ed., Vol. 1, pp. 470–480, by M. S. Sage, Sage Laboratories, Inc.; in *ECT* 3rd ed., Vol. 1, pp. 582–597, by Antoine Kawam and John B. Flynn, The Gillette Company.

1. *Aerosol Guide,* 6th ed., Aerosol Division, Chemical Specialties Manufacturers Association, Washington, D.C., 1971.
2. Annual Aerosol Survey, Chemical Specialties Manufacturers Association, Washington, D.C., 1990; *Chem. Week* 18 (May 8, 1991).

General References

D. P. Dunn, *Aerosol Age* **33,** 38 (Jan. 1988).
Handbook of Pharmaceutical Excipients, American Pharmaceutical Association/The Pharmaceutical Society of Great Britain, Washington, D.C. and London, 1986, pp. 19, 99, 101, 145, 240, 333.
Link Labs, *Aerosol Age* **32,** 30–32 (Sept. 1987).
M. A. Johnsen, *The Aerosol Handbook,* 2nd ed., Wayne Dorland Company, Mendham, N.J., 1982.
P. A. Sanders, *Principles of Aerosol Technology,* Van Nostrand Reinhold Co., Inc., New York, 1970.
P. A. Sanders, *Handbook of Aerosol Technology,* 2nd ed., Van Nostrand Reinhold Co., Inc., New York, 1979.
J. J. Sciarra and L. Stoller, *The Science and Technology of Aerosol Packaging,* John Wiley & Sons, Inc., New York, 1974.
J. J. Sciarra and A. J. Cutie, in H. A. Lieberman, M. M. Rieger, and G. S. Banker, eds., *Pharmaceutical Dosage Forms: Disperse Systems,* Vol. 2, Marcel Dekker, Inc., New York, 1989, pp. 417–460.
J. J. Sciarra and A. J. Cutie, in A. R. Gennaro, ed., *Remington's Pharmaceutical Sciences,* 18th ed., Mack Publishing Company, Easton, Pa., 1990.

JOHN J. SCIARRA
Sciarra Aeromed Development Corporation

AGAR. See GUMS.

AGAVE. See FIBERS, VEGETABLE.

AGRICULTURAL CHEMICALS. See FERTILIZERS; FUNGICIDES; HERBICIDES; INSECT CONTROL TECHNOLOGY; SOIL CHEMISTRY OF PESTICIDES.

AIR CONDITIONING

In the past 100 years, air conditioning has progressed from its beginnings in the refrigeration industry (see REFRIGERATION AND REFRIGERANTS) to being an indispensable factor in society. Applications range from providing human comfort to controlling the conditions essential for the production of many chemical products.

The design of air conditioning systems is a highly specialized branch of engineering. A number of excellent information sources regarding air conditioning are available (1).

Basic Principles

Thermodynamic principles govern all air conditioning processes (see HEAT EX-CHANGE TECHNOLOGY, HEAT TRANSFER). Of particular importance are specific thermodynamic applications both to equipment performance which influences the energy consumption of a system and to the properties of moist air which determine air conditioning capacity. The concentration of moist air defines a system's load.

Thermodynamics.   Many definitions and formulations exist for the laws of thermodynamics, a detailed treatment of which may be found in standard engineering texts (2). Definitions that apply best to air conditioning are as follows:

First Law. This is the law of conservation of energy which states that the flow of energy into a system must equal the flow of energy out of the same system minus the energy that remains inside the system boundary. For an open system in which the energy flows are not time dependent and in which there is no accumulation of energy in the system, the first law may be written as

$$\sum_{\text{in}} \dot{Q} + \sum_{\text{in}} \dot{m}_i \left(h_i + Z_i + \frac{V_i^2}{2g} \right) = \sum_{\text{out}} \dot{W} + \sum_{\text{out}} \dot{m}_j \left(h_j + Z_j + \frac{V_j^2}{2g} \right)$$

where $\dot{Q}$ = rate of heat transfer to the system, $\dot{m}$ = mass flow rate, h = enthalpy of substance, Z = elevation of boundary above a horizontal reference, V = velocity of substance, g = gravitational acceleration, and $\dot{W}$ = work done by the system.

Open steady-flow systems, which include almost all air conditioning processes, follow this law. Examples include the energy flows in a cooling and dehumidifying coil or an evaporative cooling system.

Second Law. This law defines the maximum theoretical performance for air conditioning equipment and provides a means of identifying energy losses in a system. It states that no heat engine operating in a closed cycle may produce work when communicating with a single temperature source. Air conditioning is the result of a heat engine operating in reverse. This means that work is added to the system and there must always be at least two temperatures, a low temperature source from which heat is received and a high temperature sink to which heat is rejected.

The Carnot cycle is formulated directly from the second law of thermodynamics. It is a perfectly reversible, adiabatic cycle consisting of two constant entropy processes and two constant temperature processes. It defines the ultimate efficiency for any process operating between two temperatures. The coefficient of performance (COP) of the reverse Carnot cycle (refrigerator) is expressed as

$$\text{COP} = \frac{\text{useful effect}}{\text{work input}} = \frac{T_1}{T_2 - T_1}$$

where T_1 = absolute temperature of the cold source and T_2 = absolute temperature of the hot sink.

In some applications, large quantities of waste or low cost heat are generated. The absorption cycle can be directly powered from such heat. It employs two intermediate heat sinks. Its theoretical coefficient of performance is described by

$$\text{COP}_a = \frac{T_1(T_2 - T_{S1})}{T_2(T_{S2} - T_1)}$$

where T_{S1} = absolute temperature of one intermediate heat sink (condenser), and T_{S2} = absolute temperature of the other intermediate heat sink (absorber).

Real processes always involve losses and irreversibilities and thus deviate from theory. Typical inefficiencies arise from temperature differences between the air stream and the heat exchange fluid, friction between moving parts, fluid pressure drops through heat exchangers and ducts, and pressure drops through pipes and valves. The fewer the inefficiencies, the more closely the process approaches the theoretical limit. Good texts on air conditioning (see general references or reference 1, provide a more detailed description of actual processes.

A simple cooling cycle serves to illustrate the concepts. Figure 1 shows a temperature–entropy plot for an actual refrigeration cycle. Gas at state 1 enters the compressor and its pressure and temperature are increased to state 2. There is a decrease in efficiency represented by the increase in entropy from state 1 to state 2 caused by friction, heat transfer, and other losses in the compressor. From state 2 to states 3 and 4 the gas is cooled and condensed by contact with a heat sink. Losses occur here because the refrigerant temperature must always be above the heat sink temperature for heat transfer to take place. The liquid refrigerant is commonly expanded adiabatically from state 4 to state 5 by a throttling device; thus there is a further loss in efficiency. Heat is added to the refrigerant from state 5 to state 1 and the cycle is repeated. The loss in this process again results from the temperature difference between the heat source and the refrigerant. Process A–B–C–D represents the Carnot cycle. Because the area inside the cycle boundary represents the work added, it is evident that the actual refrigeration cycle requires more work and thus has a considerably lower COP than the ideal cycle.

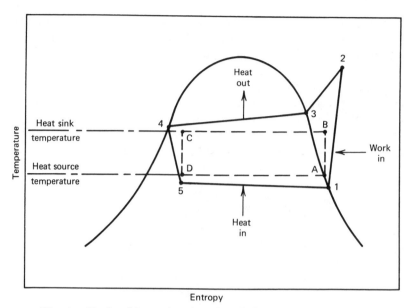

Fig. 1. Real refrigeration (——) and Carnot (– – –) cycles.

As illustrated both by the second law of thermodynamics and the example in Figure 1, the temperature difference between the heat source and sink needs to be minimized in order to increase the efficiency of the process. This means rejecting heat to the sink having the lowest possible temperature while obtaining heat from the source having the highest. Increased efficiency results from using the largest heat exchangers consistent with economy to minimize temperature differences between the working fluid and the source and sink. Lack of maintenance, especially in regard to heat exchangers, significantly increases the temperature differentials and the energy consumption. Other considerations resulting from the second law include minimizing fluid pressure losses by using the largest practical air ducts and fluid piping and utilizing the most efficient mechanical devices obtainable.

Psychrometrics.   Psychrometrics is the branch of thermodynamics that deals specifically with moist air, a binary mixture of dry air and water vapor. The properties of moist air are frequently presented on psychrometric charts such as that shown in Figure 2 for the normal air conditioning range at atmospheric pressure. Similar charts exist for temperatures below 0°C and above 50°C as well as for other barometric pressures. All mass properties are related to the mass of the dry air.

The following quantities are found on a psychrometric chart:

Dry-bulb Temperature (DB) (Abscissa).   DBT is the temperature of a gas or mixture of gases indicated by an accurate thermometer after correction for radiation effect.

Dew-point Temperature (DPT).   DPT is the temperature at which the condensation of water vapor in a space begins for a given state of humidity and pressure as the temperature is reduced. It is the temperature corresponding to saturation (100% rh) for a given absolute humidity at constant pressure.

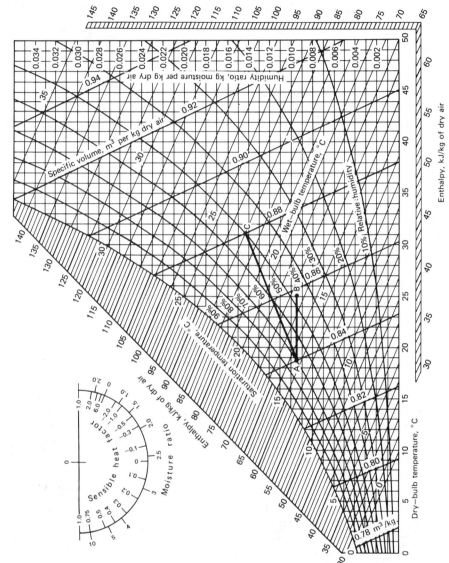

Fig. 2. Psychrometric chart at atmospheric pressure, 101.3 kPa (1 atm)(3). To convert kJ to kcal, divide by 4.184. Courtesy of Business News Publishing Co.

Enthalpy.   Enthalpy is the thermodynamic property of a substance defined as the sum of its internal energy plus the quantity Pv/J, where P = pressure of the substance, v = its specific volume, and J = the mechanical equivalent of heat. Enthalpy is also known as total heat and heat content.

Humidity Ratio (Ordinate).   The humidity ratio is the weight of water vapor in the air per unit weight of dry air.

Relative Humidity (rh).   Relative humidity is the ratio of the mole fraction of water vapor present in the air to the mole fraction of water vapor present in saturated air at the same temperature and barometric pressure; it approximately equals the ratio of the partial pressure (or density) of the water vapor in the air to the saturation pressure (or density) of water vapor at the same temperature.

Saturation Temperature.   The temperature at which the water vapor in moist air is in equilibrium with liquid water.

Sensible Heat Factor.   The ratio of the change in sensible (constant moisture content) cooling enthalpy to the change in total cooling enthalpy.

Specific Volume.   The volume of air per unit mass.

Wet-bulb Temperature.   The equilibrium temperature which air attains if adiabatically saturated by water from a condensed phase.

The psychrometric chart may be used to determine the change in properties of air required for a condition or a process. For example, point A on Figure 2 has a dry-bulb temperature of 18.8°C and a relative humidity of 70%. Further, the air has a wet-bulb temperature of 15.2°C, an enthalpy of 61 kJ/kg (14.6 kcal/kg), an absolute humidity of 0.0095 kg/kg, and a specific volume of 0.84 m³/kg. Sensible heating (no moisture addition) may be represented by the line from point A to point B. This causes the relative humidity to decrease to 48% and the enthalpy to increase to 68 kJ/kg (16.3 kcal/kg). The amount of energy required to raise the dry-bulb temperature from 18.8°C to 25°C is 7 kJ/kg (1.7 kcal/kg) of dry air. If the air mass or flow rate were known, the total energy required could be determined. Cooling and dehumidifying is represented by the line from point C to point A. This increases the relative humidity and requires the removal of 26 kJ/kg (6.2 kcal/kg) of dry air. The amount of moisture removed is 0.0047 kg/kg of dry air. Constructing a line parallel to A–C but through point 0 on the nomograph in the upper left of Figure 2 reveals that the sensible heat factor is 0.5. Other air conditioning processes, except those involving substantial pressure changes, can be plotted on a psychrometric chart although the process may not always be a straight line.

Design Conditions

Fundamental to the design of any air conditioning system is the determination of the operating conditions of temperature and humidity. Worker comfort must also be considered.

Process Requirements.   Typical inside dry-bulb temperatures and relative humidities used for preparing, processing, and manufacturing various products, and for storing both raw and finished goods, are listed in Table 1. In some instances, the conditions have been compromised for the sake of worker comfort and do not represent the optimum for the product. In others, the conditions listed have no effect on the product or process other than to increase worker efficiency.

Table 1. Typical Industrial Inside-Design Conditions[a]

Industry	Process	Dry-bulb temperature, °C	Rh, %
abrasives	manufacture	24–27	45–50
bakery	dough mixer	24–27	40–50
	fermenting	24–28	70–75
	proof box	33–36	80–85
	bread cooler	21–27	80–85
	cold room	4–7	
	make-up room	26–28	65–70
	cake mixing	35–41	
	crackers and biscuits	15–18	50
	wrapping	15–18	60–65
	storage:		
	dried ingredients	21	55–65
	fresh ingredients	0–7	80–85
	flour	21–24	50–65
	shortening	7–21	55–60
	sugar	27	35
	water	0–2	
	wax paper	21–27	40–50
brewery	storage:		
	hops	−1–0	55–60
	grain	27	60
	liquid yeast	0–1	75
	lager	0–2	75
	ale	4–7	75
	fermenting cellar:		
	lager	4–7	75
	ale	13	75
	racking cellar	0–2	75
candy (chocolate)	candy centers	27–29	40–50
	hand-dipping room	15–18	50–55
	enrobing room	24–27	55–60
	enrobing:		
	loading end	27	50
	enrober	32	13
	stringing	21	40–50
	tunnel	4–7	dp–40
	packing	18	55
	pan specialty room	21–24	45
	general storage	18–21	40–50
candy (hard)	manufacturing	24–27	30–40
	mixing and cooling	24–27	40–45
	tunnel	13	dp–55
	packing	18–24	40–45
	storage	18–24	45–50
	drying: jellies, gums	49–65	15
	cold room: marshmallow	24–27	45–50
chewing gum	manufacturing	25	33
	rolling	20	63
	stripping	22	53

Table 1. (*Continued*)

Industry	Process	Dry-bulb temperature, °C	Rh, %
	breaking	23	47
	wrapping	23	58
ceramics	refractory	43–65	50–90
	molding room	27	60–70
	clay storage	15–27	35–65
	decal and decorating	24–27	45–50
cereal	packaging	24–27	45–50
cosmetics	manufacturing	18–21	
distilling	storage:		
	grain	15	35–40
	liquid yeast	0–1	
	manufacturing	15–24	45–60
	aging	18–22	50–60
electrical products	electronic and x ray: coils and transmission winding	22	15
	tube assembly	20	40
	electrical installations: manufacturing and laboratory	21	50–55
	thermostat assembly and calibration	24	50–55
	humidistat assembly and calibration	24	50–55
	close-tolerance assembly	22	40–45
	meter assembly test	23–24	60–63
	switchgear:		
	fuse and cut-out assembly	23	50
	capacitor winding	23	50
	paper storage	23	50
	conductor wrapping	24	65–70
	lightning arrestor	20	20–40
	circuit breaker: assembly and test	24	30–60
	rectifiers: process selenium and copper oxide plates	23	30–40
furs	drying	43	
	shock treatment	−8	
	storage	4–10	55–65
	cutting	comfort	
	vinyl-laminating room	13	15
leather	drying:		
	vegetable-tanned	21	75
	chrome-tanned	49	75
	storage	10–15	40–60
lenses (optical)	fusing	comfort	
	grinding	27	80
matches	manufacturing	22–23	50
	drying	21–24	40
	storage	15–17	50
munitions	metal percussion elements:		
	drying parts	88	
	drying paints	43	
	black-powder drying	52	

Table 1. (*Continued*)

Industry	Process	Dry-bulb temperature, °C	Rh, %
	condition and load powder-type fuse	21	40
	load tracer pellets	27	40
pharmaceutical	powder storage:		
	before manufacturing	21–27	30–35
	after manufacturing	24–27	15–35
	milling room	27	35
	tablet compressing	21–27	40
	tablet coating	27	35
	effervescent: tablet and powder	32	15
	hypodermic tablet	24–27	30
	colloids	21	30–50
	cough syrup	27	40
	glandular products	26–27	5–10
	ampule manufacturing	27	35
	gelatin capsule	78	40–50
	capsule storage	24	35–40
	microanalysis	comfort	
	biological manufacturing	27	35
	liver extract	21–27	20–30
	serums; animal room	comfort	
photo material	drying	−7 to 52	40–80
	cutting and packing	18–24	40–70
	storage: film base, film paper, coated paper	21–24	40–65
	safety film	15–27	45–50
	nitrate film	4–10	40–50
plastic	manufacturing:		
	thermosetting compounds	27	25–30
	cellophane	24–27	45–65
plywood	hot press: resin	32	60
	cold press	32	15–25
precision machining	spectrographic analysis	comfort	
	gear matching and assembly	24–27	35–40
	storage:		
	gasket	38	50
	cement and glue	18	40
	machines:		
	gauging, assembly	comfort	
	adjusting precision	comfort	
	parts	comfort	
	honing	24–27	35–45
printing	multicolor lithographing:		
	pressroom	24–27	46–48
	stockroom	23–27	49–51
	sheet and web printing	comfort	
	storage, folding, etc	comfort	
refrigeration equipment	valve manufacturing	24	40
	compressor assembly	21–24	30–45
	refrigerator assembly	comfort	

Table 1. (*Continued*)

Industry	Process	Dry-bulb temperature, °C	Rh, %
	testing	18–28	47
rubber-dipped goods	manufacturing	32	
	cementing	27	25–30
	surgical articles	24–32	25–30
	storage before manufacturing	15–24	40–50
	laboratory (ASTM standard)	23	50
textiles	cotton:		
	opening, picking	21–24	55–70
	carding	28–30	50–55
	drawing and roving	27	55–60
	ring spinning:		
	conventional	27–29	60–70
	long-draft	27–29	
	frame spinning	27–29	55–60
	spooling, warping	26–27	60–65
	weaving	26–27	70–85
	cloth room	24	65–70
	combing	24	55–65
	linen:		
	carding, spinning	24–27	60
	weaving	27	80
	woolens:		
	picking	27–29	60
	carding	27–29	65–70
	spinning	27–29	50–60
	dressing	24–27	60
	weaving:		
	light goods	27–29	55–70
	heavy goods	27–29	60–65
	drawing	24	50–60
	worsteds:		
	carding, combing, and gilling	27–29	60–70
	storage	21–29	75–80
	drawing	27–29	50–70
	cap spinning	27–29	50–55
	spooling, winding	24–27	55–60
	weaving	27	50–60
	finishing	24–27	60
	silk:		
	preparation and dressing	27	60–65
	weaving, spinning	27	65–70
	throwing	27	60
	rayon:		
	spinning	27–32	50–60
	throwing	27	55–60
	weaving:		
	regenerated	27	50–60
	acetate	27	55–60
	spun rayon	27	80

Table 1. (*Continued*)

Industry	Process	Dry-bulb temperature, °C	Rh, %
	picking	24–27	50–60
	carding, roving, drawing	27–32	50–60
	knitting: viscose or cuprammonium	27–29	65
	synthetic-fiber preparation and weaving:		
	viscose	27	60
	celanese	27	70
	nylon	27	50–60
tobacco	cigar and cigarette manufacturing	21–24	55–65
	softening	32	85–88
	stemming, stripping	24–29	75
	storage and preparation	26	70
	conditioning	24	75
	packing and shipping	24	60

[a]Listed conditions are typical; final design conditions are established by customer requirements.

Specific inside design conditions are required in industrial applications for one or more of the following reasons:

A constant temperature is required for close-tolerance measuring, gauging, machining, or grinding operations, to prevent expansion and contraction of machine parts, machined products, and measuring devices. In this instance a constant temperature is normally more important than the temperature level. Relative humidity is secondary in importance but should not go above 45% to minimize formation of a surface moisture film.

Some nonhygroscopic materials such as metals, glass, and plastics, have the ability to capture water molecules within microscopic surface crevices, thus forming an invisible, noncontinuous surface film. The density of the film increases as the relative humidity increases. Thus, relative humidity must be held below the critical point at which metals may etch or at which the electrical resistance of insulating materials is significantly decreased.

Where highly polished surfaces are manufactured or stored for short intervals between different phases of processing, relative humidity and temperature are both maintained constant to minimize surface moisture films. If these surfaces are shipped or stored for extended intervals, protective coverings or coatings may be required.

The temperature and humidity should be maintained at comfort conditions consistent with the operator's expected level of activity in order to minimize perspiration. Constant temperature and humidity may also be required in machine rooms to prevent the etching or corrosion of machine parts. If perspiration causes only minor damage to the product and results in few rejects, then inside design conditions at 27°C and 40% rh are satisfactory. Where even small amounts of perspiration cause extreme damage to precision-machined parts and result in a high amount of rejects, inside design conditions of 21°C and 40% rh are recommended.

Control of relative humidity is needed to maintain the strength, pliability, and moisture regain of hygroscopic materials such as textiles and paper. Humidity control may also be required in some applications to reduce the effect of static electricity. Temperature and/or relative humidity may also have to be controlled in order to regulate the rate of chemical or biochemical reactions, such as the drying of varnishes, the application of sugar coatings, the preparation of synthetic fibers and other chemical compounds, or the fermentation of yeast.

Human Comfort. ASHRAE has extensively researched the effect of air conditioning on human comfort. The more practical results are summarized below; reference 4 contains a complete discussion.

Thermal comfort may be defined as "that condition of mind in which satisfaction is expressed with the thermal environment" (4). It is thus defined by a statistically valid sample of people under very specific and controlled conditions. No single environment is satisfactory for everybody, even if all wear identical clothing and perform the same activity. The comfort zone specified in ASHRAE Standard 55 (5) is based on 90% acceptance, or 10% dissatisfied.

Recent experiments (4) have shown that there are no significant age or gender-related differences in thermal environment preference when all other factors such as weight of clothing and activity level are the same. Whereas people often accept thermal environments outside of their comfort range, there is no evidence that they adapt to these other conditions. Their environmental preference does not change. Similarly there is no evidence that there is any seasonal or circadian rhythm influence on a person's thermal preference.

Local areas of thermal discomfort, ie, one part of the body warmer or cooler than preferred, may cause a person to be uncomfortable when the overall temperature and humidity would normally produce a sensation of thermal comfort. Some causes of this are nonuniform thermal radiation, such as hot or cold windows, walls, panels, floors, and ceilings. Experiments show that people are more sensitive to the asymmetry caused by a warm overhead surface than by a cold vertical surface. However, the percentage of people dissatisfied begins to rise rapidly once the radiant surface temperature rises more than a few degrees above the air temperature. Comfort charts are normally based on an environment where the mean radiant temperature is the same as the air temperature. This means that if there are significant surfaces such as radiant ceilings or large windows outside the range of the air temperature, the conditions for comfort may have to be adjusted. Drafts, normally felt by the local cooling effect of the air moving past the body, are another cause of local thermal discomfort. They are one of the most annoying factors in offices and often result in complaints and demands for higher ambient temperatures. Drafts can come from improperly designed air distribution systems as well as from localized surfaces such as cold windows or walls.

Figure 3 shows the winter and summer comfort zones plotted on the coordinates of the ASHRAE psychrometric chart. These zones should provide acceptable conditions for room occupants wearing typical indoor clothing who are at or near sedentary activity. Figure 3 applies generally to altitudes from sea level to 2150 m and to the common case for indoor thermal environments where the temperature of the surfaces (t_r) approximately equals air temperature (t_a) and the air velocity is less than 0.25 m/s. A wide range of environmental applications is covered by ASHRAE Comfort Standard 55 (5). Offices, homes, schools, shops, theaters, and many other applications are covered by this specification.

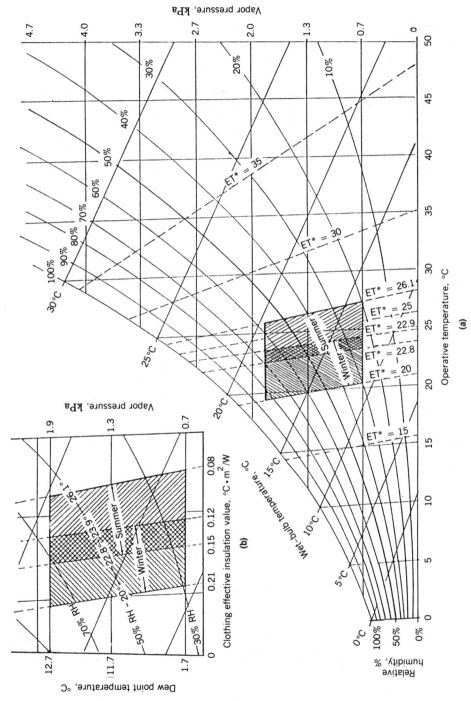

Fig. 3. Comfort zones at 6% of population predicted dissatisfied from ref. 4. RH lines are valid only when the air temperature equals the average temperature of the surfaces. (**a**) Operative temperature range where ET* is effective temperature as defined in text. (**b**) Comfort zone detail. To convert kPa to mm Hg, multiply by 7.5.

697

Effective temperature (ET*) is a single number representing those combinations of temperature and humidity which are equivalent in terms of comfort. It is defined as the dry-bulb temperature of the environment at 50% relative humidity. Standard effective temperature loci for normally clothed, sedentary persons are plotted on Figure 3. The sensation of comfort depends in part upon the wetness of one's skin. Thus, as a person becomes more active the effective temperature lines become more horizontal and the influence of relative humidity is more pronounced.

When air movement, clothing, or activity are not as specified in the definition of ET*, Figure 4, derived from an equation developed by Fanger (6), may be used. Knowledge of the energy expended during the course of routine physical activities is also necessary, since the production of body heat increases in propor-

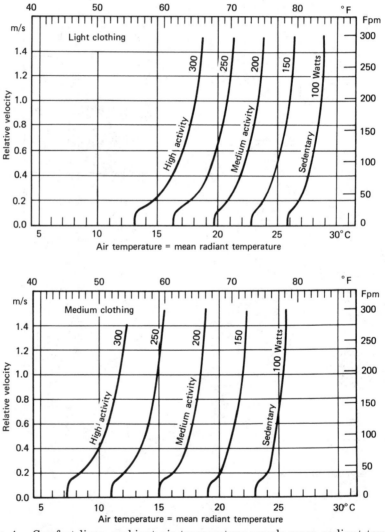

Fig. 4.   Comfort lines, ambient air temperature equals mean radiant temperature (4). To convert watts to kcal/min, multiply by 0.143.

tion to exercise intensity. Table 2 presents probable metabolic rates (or the energy cost) for various activities. However, for higher activity levels, the values given could be in error by as much as 50%. Engineering calculations should allow for this. The activity level of most people is a combination of activities or work–rest periods. A weighted average metabolic rate is generally satisfactory, provided that activities alternate several times per hour.

Table 2. Metabolic Rate at Different Activities

Activity	Metabolic rate, W^a
resting	
seated, quiet	100
standing, relaxed	120
walking	
on the level: 0.9 m/s (2 mph)	200
1.35 m/s (3 mph)	260
1.8 m/s (4 mph)	380
miscellaneous occupations	
bakery, eg, cleaning tins, packing boxes	140–200
brewery, eg, filling bottles, loading beer boxes onto belt	120–240
foundry work	
using a pneumatic hammer	300–340
tending furnaces	500–700
general laboratory work	140–180
machine work	
light, eg, electrical industry	200–240
heavy, eg, steel work	350–450
shop assistant	200
teacher	160
vehicle driving	
car	150
heavy vehicle	320
domestic work	
house cleaning	200–340
cooking	160–200
washing by hand and ironing	200–360
office work	
typing	120–140
miscellaneous office work	110–130
drafting	110–130
leisure activities	
calisthenics exercise	300–400
dancing, social	240–440
tennis, singles	360–460
squash, singles	500–720
golf, swinging and walking	140–260
golf, swinging and golf cart	140–180

^{a}Ranges are given for those activities which may vary considerably from one place of work or leisure to another, or when performed by different people. Some occupational and leisure time activities are difficult to evaluate because of differences in exercise intensity and body position. To convert W to kcal/h, divide by 1.162.

Equipment Size Requirements.   Determining the proper size of air conditioning and heating equipment requires detailed study and calculation. A comprehensive statement of requirements and allowable variations must be supplied so that the best comfort conditioning system choice can be made. With such information it is possible to estimate not only equipment size, but also yearly energy requirements using any of several comprehensive computer (7) or manual methods. Different systems can be designed and compared to ensure that the owner has a cost-effective yet energy-efficient system.

An analysis of the building structure must be conducted to determine the effects of heat gain from the sun. This analysis includes building orientation, type of construction, surrounding vegetation and structures, and reflective surfaces. People, lighting, machinery loads, and heat gains from chemical processes are evaluated as well as hooded (ventilated) processes, and lengths of operation. Outdoor air required both for ventilation to remove odors and contaminants, and for replacement of air exhausted through hoods must be conditioned. ASHRAE Standard 62 (8) provides recommendations for minimum outdoor air requirements; local codes should be investigated. A building often requires cooling at low outdoor temperatures as a result of high internal heat gains. Interior portions of many buildings require cooling during occupied hours throughout the year. The use of outdoor air, when its temperature and dew point are suitable (economizer cycle), is an efficient means of air conditioning. Care must be taken, however, in applying economizer cycles to areas where close humidity control is required because additional humidification or dehumidification may be needed.

The following information is customarily required for thorough design of new air conditioning systems or renovation of existing systems.

Temperature and/or humidity to be maintained
    Allowable seasonal variations
    Permissible control tolerance
Outdoor conditions to be assumed for design extremes under which the plant must operate
Architectural plans and details of building construction (if original plans are not available, the building must be measured carefully and details of construction must be determined by inspection)
    Orientation of the building
    Neighboring structures
    Special zoning requirements based on load concentrations, and differences in conditions required for various processes
    Glass areas
        Type of glass
        Shading devices
        Reveals and overhangs
Sensible heat gains
    Power equipment
        Usage factor
        Percent loaded
        Hours used
        Rated power requirement

Lighting
    Usage factor
    Hours used
    Auxiliaries
    Rated power requirement
Miscellaneous, eg, ovens, exposed steam pipes, use of exhaust hoods
Temperature of product entering space
    Product temperature above space temperature (resulting in a heat gain to the space)
    Product temperature below space temperature (producing a credit to sensible-heat gain or a heating requirement)
Latent heat gains
    Evaporation from wet surfaces due to process
    Migration of water vapor through building materials (especially important in low dew-point application)
    Water vapor from moist product
Sensible and latent heat gains
    People
        Degree of activity
        Time of occupancy
        Number
    Gas burning equipment
        Usage factor
        Heating value of gas
        Use of hoods
    Equipment of evaporating water
        Usage factor
        Capacity
        Energy required to operate
    Chemical and biological reactions
    Infiltration of air
        Frequency of door opening
        Window cracks
        Porosity of building structure
    Steam or water released
Ventilation air
    For human occupancy
    Toxic fume and smoke dilution
    Odor dilution
    Offsetting exhaust hood requirements
Energy sources
    Availability and cost
    Options
Heat recovery possibilities
    Process streams
    Ventilation
    Equipment
    Lighting

Air conditioning system
Refuse

Air Conditioning and Humidification Systems

Air conditioning may involve heating or cooling air, humidifying or drying it, and the control of chemical impurities to maintain the desired space conditions. Proper controls and energy conserving practices are also important to air conditioning and humidification.

Typical Air Conditioning Systems. Two broad categories of air conditioning systems exist, unitary and applied. Unitary systems are self-contained units that are "off the shelf." They use electricity for cooling, and may use electricity, natural gas, fuel oil, or propane for heating. Heat rejected during the cooling cycle is dissipated to the outdoors. Multiple unitary systems may be employed to provide greater overall reliability and to permit individual control of various sections of a plant. A typical unit for rooftop mounting is shown in Figure 5. It contains means for heating, ventilating, and cooling.

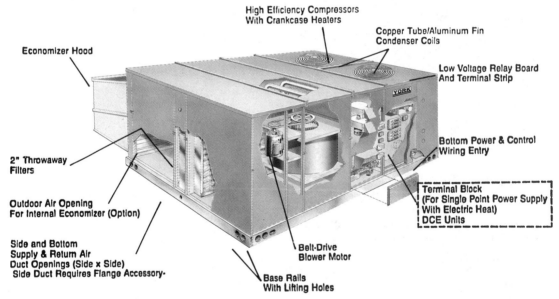

Fig. 5.   Rooftop unitary system. Courtesy of York International Corporation.

More flexibility is obtained with applied equipment (Fig. 6), which is normally used to condition a relatively large area of a plant. This is usually part of a "field erected" system. In applied systems, outdoor air for ventilation or cooling (economizer cycle) is drawn through a preconditioning or preheat coil and mixed with air returned from the conditioned space. Dampers regulate the relative amounts of outdoor and recycled air for temperature control. The air is filtered before passing into the conditioning section which contains cooling and dehumidifying coils, air washers for humidity control, and heating coils. Bypass of return air may be included for temperature control. The refrigerating effect is provided in one of several ways. Well water may be employed if available in sufficient quality and quantity and at a suitable temperature. More commonly, refrigerating machines or "chillers" are used. In small systems, reciprocating compressors are employed and the refrigerant may be directly admitted to the cooling coil. In larger applications, water is chilled and circulated through the central station unit. For applications in excess of 350 kW (1.2×10^6 Btu/h), reciprocating compressors may be replaced by centrifugal systems. Most systems are electrically powered; however, steam or gas turbines are used occasionally. Absorption chillers are frequently used when a suitable supply of "waste" heat is available. Low pressure steam, hot water, and process streams may provide the motive force. Solar heated water is also finding application (9) (see SOLAR ENERGY).

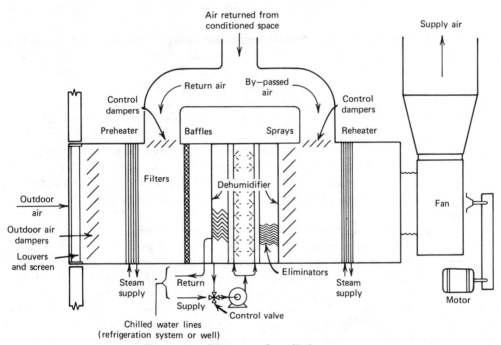

Figure 6.   Diagram of applied system.

Humidification.   For winter operation, or for special process requirements, humidification may be required (see SIMULTANEOUS HEAT AND MASS TRANSFER). Humidification can be effected by an air washer which employs direct water sprays (see EVAPORATION). Regulation is maintained by cycling the water sprays or by temperature control of the air or water. Where a large humidification capacity is required, an ejector which directly mixes air and water in a nozzle may be employed. Steam may be used to power the nozzle. Live low pressure steam can also be released directly into the air stream. Capillary-type humidifiers employ wetted porous media to provide extended air and water contact. Pan-type humidifiers are employed where the required capacity is small. A water filled pan is located on one side of the air duct. The water is heated electrically or by steam. The use of steam, however, necessitates additional boiler feed water treatment and may add odors to the air stream. Direct use of steam for humidification also requires careful attention to indoor air quality.

Dehumidification.   Dehumidification may be accomplished in several ways (see DRYING). Moderate changes in humidity can be made by exposing the air stream to a surface whose temperature is below the dew point of the air. The air is cooled and releases a portion of its moisture. Closed cycle air conditioning systems normally effect dehumidification also. The cooled air may require re-heating to attain the desired dry-bulb temperature if there is insufficient sensible load in the space.

Another method of moderate dehumidification is by direct contact between the air and cold water using open circuit equipment. An air washer or capillary humidifier maintaining the water at a cold temperature is an example.

Some industrial processes produce predominately latent air conditioning loads. Others dictate very low humidities and when the dew point falls below $0°C$, freezing becomes a major concern. Dehydration equipment, using solid sorbents such as silica gel and activated alumina, or liquid sorbents such as lithium chloride brine and triethylene glycol, may be used. The process is exothermic and may require cooling the exiting air stream to meet space requirements. Heat is also required for reactivation of the sorbent material.

Solid sorbent materials have the ability to adsorb water vapor until an equilibrium condition is attained. The total weight of water that can be adsorbed in a particular material is a function of the temperature of the material and of the relative humidity of the air (see ADSORPTION). To regenerate the sorbent, its temperature must be raised or the relative humidity lowered. The solid sorbents most commonly used are silica (qv), alumina (see ALUMINUM COMPOUNDS), and molecular sieves (qv).

Liquid sorbent materials in aqueous solutions reduce the vapor pressure relative to that of pure water. Such a solution, if of proper strength, causes moisture to condense from the air even though the solution temperature is above the dew point of the air. Brines such as calcium chloride, lithium chloride, lithium bromide, and calcium bromide may be used singly or in combination. They are generally somewhat corrosive in nature. Triethylene glycol, and to a lesser extent diethylene glycol and ethylene glycol, are also used (see GLYCOLS). Brines should not have a solidification curve too near the working range; they must be odorless, relatively noncorrosive, chemically stable, and reasonable in cost. The most serious hazards of application are corrosion and carry-over from equipment into the room. Regeneration of the brine is generally accomplished by boiling the excess water out of the brine and exhausting it to another air stream. The solution is then recooled to a temperature suitable for dehumidification. Equipment is available to make this process continuous.

Chemical Neutralization.   Spray-type air washers are used extensively for removal or neutralization of noxious components from large volumes of air, particularly exhaust air streams. Appropriate reagents are sprayed into the washer to purify the air by neutralization, eg, sodium hydroxide solution is used if the air contains acidic gases. The solution must be continuously reconcentrated and any precipitated salts removed. The contact efficiency of such washers is high, and the simple construction provides easy maintenance and constant efficiency (see AIR POLLUTION CONTROL METHODS).

Evaporative Cooling.   Evaporative air cooling equipment deposits water directly into the air stream through evaporation (see HEAT EXCHANGE TECHNOLOGY; SIMULTANEOUS HEAT AND MASS TRANSFER). These systems are employed where the application has a high sensible heat load and requires final design relative humidities of 50% or greater, or where the entering air's relative humidity is very low. Evaporative cooling and humidification are often accomplished by the same equipment, depending on the relative temperatures of the air and water.

There are several basic types of evaporative cooling devices. Among them are spray air washers, cell washers, and wetted media air coolers. Intimate contact between the spray water and the flowing air causes heat and mass transfer between the air and the water. Cell washers obtain intimate air–water contact by passing the air through cells packed with glass, metal, or fiber screens. Wetted media coolers contain evaporative pads, made usually of aspen wood fibers, and a water circulating pump to lift the sump water to a distributing system from which it runs down through the pads and back into the sump. Washers are commonly available from 1 to 118 m³/s (2000–250,000 ft³/min) capacity depending on the type; however, there is no limit to sizes that can be constructed. Air velocity, air dry-bulb, air wet-bulb, water spray density, spray pressure, and other design factors must be considered for each application.

Continued satisfactory performance of any evaporative cooling device depends largely on a regular cleaning and inspection schedule. The frequency of this maintenance varies with operating conditions; however, a weekly inspection is common practice. A spray system requires the most attention: partially clogged nozzles are indicated by a rise in spray pressure; eroded orifices by a fall in pressure. Strainers can minimize these problems. For continuous operation a bypass around the strainer or duplex strainers is required. Air washer tanks should be drained and dirt deposits removed at regular intervals. Eliminators and baffles should be inspected periodically and repainted to prevent damage by corrosion. A small amount of water, depending on the hardness of the makeup water, should be bled-off to maintain an acceptable concentration of solids according to recommendations of a water treatment specialist (see WATER, INDUSTRIAL WATER TREATMENT). In the case of cell-type washers, a differential pressure gage to measure the air flow resistance across the cells can be used to determine the need to clean the media. In all washers, proper water treatment must be maintained to prevent the growth of bacteria, fungi, and other microorganisms.

Air Conditioning Control.   When sized to meet design conditions, a heating or cooling system normally operates over a wide range of temperatures and loads; thus proper control becomes important. Controls may range from a single thermostat to complex computer systems. The general references provide several texts on these subjects. The system must be adjusted and maintained in order to provide operation for many years. The simplest control system which will produce the necessary results is best. The design professional should have the information noted in the sections on air conditioning equipment requirements for all anticipated operating conditions. For energy efficiency the specifications should contain wide tolerances.

Energy Conservation.   The design of systems that conserve energy requires knowledge of the building, its operating schedule, and the systems to be installed (see ENERGY MANAGEMENT). The following approaches lead to reduced energy consumption:

Use Equipment Only When Needed.   Start morning warm-up no earlier than necessary and do not use outside air for ventilation until the building is occupied. Use minimum amounts of outdoor air according to reference 8. Supply heat at night only to maintain a temperature above 13°C.

Supply Heating and Cooling from the Most Efficient Source.

Sequence Heating and Cooling.   Do not supply both at the same time. The zoning and system selection should eliminate or at least minimize simultaneous heating and cooling.

Provide Only the Heating or Cooling Actually Needed.   Generally, the supply temperature of hot and cold air, or water, should be reset according to actual need. This is especially important on systems or zones that allow simultaneous heating and cooling.

Uses of Air Conditioning in Industry

Many industrial processes require accurate environmental control. Examples include: chemical reactions and processes that are affected by atmospheric conditions; biochemical reactions; quality, uniformity, and standardization of certain products; factors such as rate of crystallization and size of crystals; product moisture content or regain; deliquescence, lumping, and caking of hygroscopic materials; expansion and contraction of machines and products; physical, chemical, and biological cleanliness; effects of static electricity; odors and fumes; conditions in storage and packaging; quality of painted and lacquered finishes; simulation of stratosphere or space conditions; and productivity and comfort of workers. Controlled atmospheric conditions are especially important to the textile, pharmaceutical, food processing, explosives, and photographic materials industries. Analytical laboratories, clean rooms, and computer control rooms also require air conditioned environments.

Synthetic Fibers.   In the synthetic textile industry, air conditioning is used to achieve uniform quality and viscosity for spinning; to control the rate of reaction and coagulation; to control toxic fumes and evaporation from acid baths; to prevent stretching during the winding of wet threads; to control regain; and to prevent crystal formation on threads and machines. Because of toxic fumes in some rayon processes, air conditioning with a large amount of outdoor air and extensive exhaust is a hygienic necessity. In the mechanical handling of the finished synthetic yarns, in throwing, weaving, and knitting operations, air conditioning is necessary for quality and production control (see TEXTILES).

Pharmaceuticals and Biologicals.   Air conditioning is an important processing requirement in the production of many pharmaceutical materials, as in the manufacture of pills and capsules and in the packaging of the finished product to maintain constancy in formula, quality, and dosage. It helps to achieve constant production rates, cleanliness, and purity of product; to prevent lumping, caking, and sticking; to reduce diffusion of material into the air; to remove noxious fumes and gases; and to produce the desired polish on coated pills (see PHARMACEUTICALS).

Rubber. In the rubber industry, air conditioning provides uniform performance in drying and shortens the drying period, controls oxidation, eliminates blisters in dipping operations, preserves tensile strength, minimizes explosion hazards from static electricity, and reduces the concentration of toxic fumes (see RUBBER COMPOUNDING).

Photographic Materials. Air conditioning is an essential element in the processing of photographic materials to control the moisture regain of film and minimize static discharge, thus reducing fire hazard and preventing fogging and streaking. Air conditioning assures dust-free air and provides ideal conditions for packaging and storage, thereby reducing production losses. Careful control of temperature, humidity, cleanliness, and ventilation is practical and essential (see PHOTOGRAPHY).

Explosives. The munitions industry employs air conditioning to control uniformity in the manufacture and loading of various explosive mixtures, to control drying and moisture content, to minimize static discharges, to reduce the hazards of fire and explosion, to remove and neutralize toxic fumes, to remove and recover dust or solvents from manufacturing or loading processes, and to provide proper atmospheric conditions for the storage of raw materials or finished product (see EXPLOSIVES).

Breweries. Air conditioning and the extensive use of refrigeration are necessary to provide controlled temperature in wort cooling, fermentation, storage, and final packaging of the finished beer. Sanitation and removal of carbon dioxide are important aspects of this application (see BEER).

Food Processing. Air conditioning, including drying and freezing, has many applications in food processing. Examples are food dehydration, blast freezing, smoke houses, storage facilities, canned and dried foods, frozen foods, meatpacking, chewing gum manufacturing, concentrated fruit juices, and hard and chocolate candy (see FOOD PROCESSING).

Metal Industry. Air conditioning plays an important part in cupola and blast furnaces and in Bessemer converters. It is used to supply air of proper moisture content and to increase worker comfort, eg, in crane cabs, pulpits, or other individual "hot" operating spots. This is accomplished by individual or group spot-cooling. For powder metallurgy, a low temperature is needed in metal fitting and humidity control is required in metal finishing to eliminate perspiration stains, etching, and rust (see METAL TREATMENTS; METALLURGY, POWDER).

Ceramics. The drying of ceramic products before firing is controlled by air conditioning to standardize form and dimension, establish uniform drying at a controlled rate, and prevent strains that may otherwise cause cracking and crazing during firing (see CERAMICS).

Laboratories. A wide range of temperatures from -110 to $120°C$, humidities from 5 to 95%, pressures from near vacuum to many atmospheres, as well as other special ambient conditions, are not unusual for test chambers and in some instances even for complete laboratories. Requirements may be either to maintain constant conditions or to alternate high and low temperatures in conjunction with high and low humidities. Test facilities may be designed to test specifications and standards of materials and products as well as to determine environmental standards. Simulation of conditions outside Earth's atmosphere is required for testing aerospace vehicles, eg, the radiation of the sun is approximated by banks of high energy lamps on one wall; other surfaces are maintained at a temperature of $-180°C$ or below to simulate the coldness of space.

Economic Evaluation of Air Conditioning Systems

The total economic picture, including life cycle costs and energy expenditures, must be considered in selecting an efficient air conditioning system. For many systems, the ratio of annual energy usage to first cost ranges from 0.25 to 2. Thus, over its useful life, a system consumes many times its initial costs.

Comparing two or more complex alternatives is more difficult than examining equipment capacity or first cost. Characteristics of alternatives should be weighted for relative importance and measured on a common scale to allow proper evaluation. Many characteristics such as first cost, capacity, space requirement, and annual energy use can be measured objectively and used for system comparisons. Experience has shown that items such as maintenance expense, component life, and downtime can also be reliably estimated. Other factors, eg, system maintainability, flexibility, and comfort, are more arbitrary.

Life cycle cost analysis is the proper tool for evaluation of alternative systems (11,12). The total cost of a system, including energy cost, maintenance cost, interest, cash flow, equipment replacement and/or salvage value, taxes, inflation, and energy cost escalation, can be estimated over the useful life of each alternative system. A list of life cycle cost items which may be considered for each system is presented in Tables 3 and 4. Reference 14 presents a cash flow analysis which also includes factors such as energy cost escalation.

Table 3. Owning and Operating Cost Data and Summary[a]

Owning costs

Initial cost of system (amortized)	*Amortization factors*
equipment[b]	amortization period n (number of
control systems—complete	years during which initial cost is to
wiring and piping costs attributable	be recovered)
to system	interest rate, i
any increase in building construction	capital recovery factor
cost attributable to system	equivalent uniform annual cost
any decrease in building construction	
cost attributable to system	*Annual fixed charges*
installation costs	income taxes
miscellaneous	property taxes
	insurance
	rent
	Total annual fixed charges

Table 3. (*Continued*)

Operating costs

Annual energy and fuel costs
 electric energy costs
  chiller or compressor
  pumps
   chilled water
   heating water
   condenser or tower water
   well water
   boiler auxiliaries (including
    fuel oil heaters)
  fans
   condenser or tower
  inside air handling
  exhaust
  make-up air
  boiler auxiliaries and equipment
   room ventilation
  resistance heaters (primary or
   supplementary)
  heat pump
  domestic water heating
  lighting
  cooking and food service
   equipment
  miscellaneous (elevators,
   escalators, computers, etc)
 gas, oil, coal, or purchased steam costs
  on-site generation of the
   electrical power requirements
   under electric energy costs
  heating
   direct heating

ventilation
 preheaters
 reheaters
 supplementary heating (ie, oil
  preheating)
 other
domestic water heating
cooking and food service equipment
air conditioning
 absorption
 chiller or compressor
  gas and diesel engine driven
  gas turbine driven
  steam turbine driven
 miscellaneous
water
 condenser make-up water
 sewer charges
 chemicals
 miscellaneous
Total annual fuel and energy costs

Wages of engineers and operators

Annual maintenance allowances
 replacement or servicing of oil, air, or
  water filters
 contracted maintenance service
 lubricating oil and grease
 general housekeeping costs
 replacement of worn parts (labor and
  material)
 refrigerant
Total annual maintenance allowance

Summary

 equivalent uniform annual cost
 total annual fixed charges
 total annual fuel and energy costs
 wages of engineers and operators
 total annual maintenance allowance
 Total Annual Owning and Operating Costs.

[a]Ref. 13.
[b]See Table 4.

Table 4. Initial Costs[a]

energy and fuel service costs	cooling distribution equipment
fuel service, storage, handling, piping, and distribution costs	pumps, piping, piping insulation, condensate drains, etc
electrical service entrance and distribution equipment costs	terminal units, mixing boxes, diffusers, grilles, etc
total energy plant	air treatment and distribution equipment
heat producing equipment	air heaters, humidifiers, dehumidifiers, filters, etc
boilers and furnaces	fans, ducts, duct insulation, dampers, etc
steam–water converters	exhaust and return systems
heat pumps or resistance heaters	system and controls automation
make-up heaters	terminal or zone controls
heat producing equipment auxiliaries	system program control
refrigeration equipment	alarms and indicator system
compressors, chillers, or absorption units	building construction and alteration
cooling towers, condensers, well water supplies	mechanical and electric space
refrigeration equipment auxiliaries	chimneys and flues
heat distribution equipment	building insulation
pumps, reducing valves, piping, piping insulation, etc	solar radiation controls
terminal units or devices	acoustical and vibration treatment
	distribution shafts, machinery foundations, furring

[a]Ref. 13.

BIBLIOGRAPHY

"Air Conditioning" in *ECT* 1st ed., Vol. 1, pp. 238–252, by L. Macrow and U. A. Bowman, Carrier Corporation; in *ECT* 2nd ed., Vol. 1, pp. 481–501, by R. Elsea and R. C. Terwilliger, Jr., Carrier Corporation; in *ECT* 3rd ed., Vol. 1, pp. 598–624, by K. W. Cooper and R. A. Erth, York Division, Borg-Warner Corporation.

1. *ASHRAE Handbooks,* American Society of Heating, Refrigeration, and Air Conditioning Engineers, Inc., Publications Department, Atlanta, Ga. four vols.: *Fundamentals, Equipment, Systems and Applications,* and *Refrigeration.* One volume is revised each year.
2. F. C. McQuiston and J. D. Parker, *Heating, Ventilating and Air Conditioning,* 3rd ed., John Wiley & Sons, Inc., New York, 1988.
3. W. F. Stoecker, *Using SI units in Heating, Air Conditioning, and Refrigeration,* rev. ed., Business News Publishing Co., 1976.
4. Ref. 1, *Fundamentals,* 1989, Chapter 8.
5. *ASHRAE Standards,* American Society of Heating, Refrigerating, and Air Conditioning Engineers, Inc., Publications Department, Atlanta, Ga. Standard 55, 1981. Standards are upgraded on a regular basis.
6. Ref. 1, *Fundamentals,* 1989, p. 8.17.
7. *A Bibliography of Available Computer Programs in the Area of Heating, Ventilating, Air Conditioning and Refrigeration,* American Society of Heating, Refrigerating, and Air Conditioning Engineers, Inc., 1986.
8. Ref. 5, Standard 62.

9. A. B. Newton, *ASHRAE J.* **34** (Feb. 1975).
10. Ref. 1, *Equipment,* 1988.
11. R. T. Ruegg, *Life-Cycle Cost Manual for the Federal Energy Management Program,* U.S. Department of Energy Federal Energy Management program, NTIS–PR–360, May 1982.
12. R. T. Ruegg and S. R. Petersen, *Comprehensive Guide for Least-cost Energy Decisions,* NBS Special Publication 709, January 1987.
13. Ref. 1, *Systems and Applications,* 1987.
14. Ref. 1, *Systems and Applications,* Chapt. 49, 1987.

General References

ASHRAE Journal (monthly) and *ASHRAE Transactions* (semiannually) American Society of Heating, Refrigerating and Air Conditioning Engineers, Atlanta, Ga.
ASHRAE Transactions (semiannually)
Heating/Piping/Air Conditioning, (monthly) Penton Publications, Inc., Cleveland, Ohio.
R. W. Haines, *HVAC System Design Handbook,* TAB Books, Blue Ridge Summit, Pa., 1988.
B. C. Langley, *Control Systems for Air Conditioning and Refrigeration,* Prentice Hall, Englewood Cliffs, N.J. 1985.
G. W. Gupton Jr., *HVAC Controls, Operation and Maintenance,* Fairmont Press, Lilburn, Ga., 1987.
R. W. Haines, *Control Systems for Heating, Ventilating and Air Conditioning,* Van Nostrand-Reinhold, New York, 4th ed., 1987.

KENNETH W. COOPER
Poolpak, Inc.

AIR, LIQUID.   See CRYOGENICS.

AIR POLLUTION

Air pollution, as defined by textbooks published in the 1970s and 1980s, is any atmospheric condition in which substances are present in concentrations high enough above their normal ambient levels to produce a measurable effect on humans, animals, vegetation, or materials. This definition is deficient, however, because it does not include the so-called greenhouse or ozone-depleting gases which have the potential to alter the global climate and hence the global ecosystem. (The effects of these gases on humans, animals, vegetation, or materials have not been, and may never be, observed.) Therefore, in an attempt to be more comprehensive, the following definition is offered: air pollution is the presence of any substance in the atmosphere at a concentration high enough to produce an objectionable effect on humans, animals, vegetation, or materials, or to significantly alter the natural balance of any ecosystem. Substances can be solids, liquids, or gases, and can be produced by anthropogenic activities or natural sources. In this article only nonbiological material is considered and the discussion of airborne radioactive contaminants is limited to radon [10043-92-2] (see HELIUM-GROUP GASES), which is discussed in the context of indoor air pollution.

Perceptions of air pollutant effects are limited by ease of detection. Historically, odors, soiling of surfaces, and smoke-belching stacks were detectable. Three rare meteorological events, however, made it clear that air pollutants can be hazardous to human health and can even cause death at high enough concentrations. The first, a week-long air stagnation in the Meuse Valley in Belgium in 1930, led to the death of 60 people and respiratory problems for a large number of others. In 1948 similar conditions in Donora, Pennsylvania, resulted in nearly 7000 illnesses and 20 deaths, and in 1952, 4000 deaths were attributed to a four-day "killer fog" in London, England. Although these episodes dramatized the acute health effects of high concentrations of air pollutants, it was concern over longer term, chronic effects that led to the initiation of National Ambient Air Quality Standards (NAAQS) for six criteria pollutants in the United States in the early 1970s.

The original six criteria pollutants, so named because the Environmental Protection Agency is required to summarize published information on each and the summaries are called criteria documents, were: sulfur dioxide [7446-09-5], SO_2; carbon monoxide (qv) [630-08-0], CO; nitrogen dioxide [10102-44-0], NO_2; ozone (qv) [10028-15-6], O_3; suspended particulates; and nonmethane hydrocarbons, NMHC. The NMHC are now referred to as volatile organic compounds, VOC. The criteria pollutants captured the attention of regulators for several reasons: they were ubiquitous; there was substantial evidence linking them to health effects at high concentrations; three of them, O_3, SO_2, and NO_2, were also known as phytotoxins; and they were fairly easy to measure. The NMHC were dropped from the list shortly after the criteria pollutants were so designated. In the late 1970s, lead (qv) [7439-92-1], Pb, was added to the list and in 1987, so was particulate matter having an aerodynamic diameter of less than or equal to 10 μm, PM_{10}.

There have been several developments since the establishment of the criteria pollutants. In the mid-1970s it was shown that high concentrations of O_3 and sulfate haze could be transported hundreds of miles and acid deposition studies in the 1980s clearly illustrated the international and global aspects of this transport. Then stratospheric O_3 depletion and global warming became issues and air pollution was finally viewed in a global context. At the same time that the geographic scale of air pollution was expanding, the number of pollutants of concern also increased and detection capabilities improved, leading to the establishment of a hazardous air pollutant category which includes any potentially toxic substance in the air that is not a criteria pollutant.

Air Pollution Components

Air pollution can be considered to have three components: sources, transport and transformations in the atmosphere, and receptors. The source emits airborne substances that, when released, are transported through the atmosphere. Some of the substances interact with sunlight or chemical species in the atmosphere and are transformed. Pollutants that are emitted directly to the atmosphere are called primary pollutants; pollutants that are formed in the atmosphere as a result of transformations are called secondary pollutants. The reactants that undergo transformation are referred to as precursors. An example of a secondary pollutant is O_3, and its precursors are NMHC and nitrogen oxides, NO_x, a combination of nitric oxide [*10102-43-9*], NO, and NO_2. The receptor is the person, animal, plant, material, or ecosystem affected by the emissions.

Sources. There are three types of air pollution sources: point, area, and line sources. A point source is a single facility that has one or more emissions points. An area source is a collection of smaller sources such as emissions from residential heating within a particular geographic area. A line source is a one-dimensional, horizontal configuration such as a roadway. Most emissions emanate from a specific stack or vent. Emissions emanating from sources other than stacks, eg, storage piles or unpaved lots, are classified as fugitive emissions.

EPA requires that each state develop emissions inventories for all primary pollutants and precursors to secondary pollutants that are classified as criteria or hazardous air pollutants (1). In clean, rural areas, county-wide emissions totals for individual pollutant species may be all that is needed as long as emissions from large point sources are inventoried separately. For urban areas having severe air pollution problems, gridded emissions inventories are required. An area is divided into grids, typically 5 to 10 km to a side, and area- and line-source emissions are calculated for each grid. Large point sources are listed individually. Such inventories are used as inputs to sophisticated air quality models, employed to develop air pollution control strategies (see AIR POLLUTION CONTROL METHODS).

Emissions rates for a specific source can be measured directly by inserting sampling probes into the stack or vent and this has been done for most large point sources. It would be an impossible task to do for every source in an area inventory, however. Instead, emission factors, based on measurements from similar sources or engineering mass-balance calculations, are applied to most sources. An emission factor is a statistical average or quantitative estimate of the amount of a pollutant emitted from a specific source type as a function of the amount of raw material processed, product produced, or fuel consumed. Emission factors for most sources have been compiled (2). Emission factors for motor vehicles are determined as a function of vehicle model year, speed, temperature, etc. The vehicles are operated using various driving patterns on a chassis dynamometer. Dynamometer-based emissions data are used in EPA's MOBILE 4 model (3) to calculate total fleet emissions for a given roadway system.

Each year, EPA publishes a summary of air pollution emissions and air quality trends for the criteria pollutants (4). Table 1 contains the summary for 1989. U.S. emissions estimates for these pollutants are available back to 1940 (5).

Transport and Transformation. Once emitted into the atmosphere, the fate of a particular pollutant depends upon the stability of the atmosphere, which determines the concentration of the species, the stability of the pollutant in the atmosphere, which determines the persistence of the substance. Transport depends upon the stability of the atmosphere which, in turn, depends upon the ventilation. The stability of a pollutant depends on: the presence or absence of clouds, fog, or precipitation; the pollutant's solubility in water and reactivity with other atmospheric constituents (which may be a function of temperature);

Table 1. Nationwide Air Pollutant Emissions Estimates for the United States in 1989[a]

Source category	Pollutant emissions, 10^6 t/yr					
	Particulates	SO_x	NO_x	VOC	CO	Pb
transportation	1.5	1.0	7.9	6.4	40.0	2.2[b]
stationary fuel combustion	1.8	16.8	11.1	0.9	7.8	0.5[b]
industrial processes	2.7	3.3	0.6	8.1	4.6	2.3[b]
solid waste	0.3	0.0	0.1	0.6	1.7	2.3[b]
miscellaneous	1.0	0.0	0.2	2.5	6.7	
Totals[c]	7.2	21.1	19.9	18.5	60.9	7.2[b]

[a]Ref. 4.
[b]10^3 t/yr.
[c]The sums of the subcategories may not equal totals as a result of rounding.

the concentrations of other atmospheric constituents; the pollutant's stability in the presence of sunlight; and the deposition velocity of the pollutant.

Atmospheric stability can be examined utilizing the Gaussian Plume model which ignores possible transformations:

$$X(x,y,z = 0) = \frac{Q}{\pi \sigma_y \sigma_z u} \exp - \left\{ \left(\frac{H^2}{2\sigma_z^2} \right) + \left(\frac{y^2}{2\sigma_y^2} \right) \right\}$$

where X is the concentration at the receptor located on the ground ($z = 0$); Q is the pollutant release rate; σ_y and σ_z are the crosswind and vertical plume standard deviations, which are functions of the atmospheric stability and the distance downwind (x); u is the mean wind speed; H is the effective stack height, equal to the height of release only if the plume is not buoyant; and x and y are the downwind and crosswind distances. As the ventilation ie, u, σ_y, and σ_z, increases, the concentration of the pollutants decreases for a given emission rate Q. The atmospheric stability is determined by comparing the actual lapse rate to the dry adiabatic lapse rate. An air parcel warmer than the surrounding air rises and cools at the dry adiabatic lapse rate of 9.8°C/1000 m. When the actual temperature-decline-with-altitude is greater than 9.8°C/1000 m, the atmosphere is unstable, the σ's become larger, and the concentrations of pollutants lower. As the lapse rate becomes smaller, the dispersive capacity of the atmosphere declines and reaches a minimum when the lapse rate becomes positive. At that point, a temperature inversion exists. Temperature inversions form every evening in most places. However, these inversions are usually destroyed the next morning as the sun heats the earth's surface. Most episodes of high pollutant concentrations are associated with multiday inversions.

The stability or persistence of a pollutant in the atmosphere depends on the pollutant's atmospheric residence time. Mean residence times and principal atmospheric sinks for a variety of species are given in Table 2. Species like SO_2, NO_x (NO and NO_2), and coarse particles have lifetimes less than a day; thus important environmental impacts from these as primary pollutants are usually within close

Table 2. Mean Atmospheric Residence Times (τ) and Dominant Sinks of Air Pollutants

Species	CAS Registry Number	τ	Dominant sink[a]	Sink location[b]	Reference
SO_2	[7446-09-5]	0.5 days	OH	T	6
NO_x		0.5 days	OH	T	6
coarse particles (diameter >2.5 μm)		<1 day	S, P	T	7
fine particles (diameter <2.5 μm)		5 days[c]	P	T	7
O_3 (tropospheric)	[10028-15-6]	90 days[d]	NO, uv, Sr, O	T	6
CO	[630-08-0]	100 days	OH	T	6
CO_2	[124-38-9]	120 yr[e]	O	T	8
CH_4	[74-82-8]	7–10 yr	OH	T	6
$CFCl_3$ (CFC-11)	[75-69-4]	65–75 yr	uv	St	6,8
CF_2Cl_2 (CFC-12)	[75-71-8]	110–130 yr	uv	St	6,8
N_2O	[10024-97-2]	120–150 yr	uv	St	6,8
$C_2Cl_3F_3$ (CFC-113)	[76-13-1]	90 yr	uv	St	6

[a]Sinks, chemical species, or method: OH, reaction with OH radical; S, sedimentation; P, precipitation scavenging; NO, reaction with NO radical; uv, photolysis by ultraviolet radiation; Sr, destruction at surfaces; O, adsorption or destruction at oceanic surface.
[b]T = troposphere; St = stratosphere.
[c]Applies to particles released in the lower troposphere only; the most important sink is scavenging by precipitation, so in the absence of precipitation, these particles will remain suspended longer.
[d]Tropospheric residence time only; shorter lifetime applies to urban areas where NO quickly destroys O_3.
[e]Combined lifetime for atmosphere, biosphere, and upper ocean.

proximity to the emissions sources. In the presence of high concentrations of NO, the residence time of O_3 is on the order of seconds to hours. In the relatively nonpolluted environment of the free troposphere, from approximately 1500 m to the top of the troposphere ($\sim$12 km), the 90-day lifetime applies. Consequently, given the right conditions, O_3 could have important environmental impact far downwind from its source. In fact, concentrations of O_3 near the NAAQS limit have been transported from the Gulf Coast of the United States to the Northeast over a several-day period (9).

Particles having diameters less than 2.5 μm have negligible settling velocities and therefore have residence times which are considerably longer than those of larger particles. As a result, multiday transport of haze produced by fine particles over distances of more than a thousand kilometers is possible (10). The longer lifetimes of the greenhouse gases, those listed below CO in Table 2, result in the accumulation and relatively even distribution of these gases around the globe. Chlorofluorocarbons (CFCs) and nitrous oxide [10024-97-2], N_2O, are essentially inert in the troposphere and are only destroyed in the stratosphere by ultraviolet solar radiation. The photolysis products of CFCs are the reactants which are responsible for stratospheric O_3 depletion.

To determine the fate of a pollutant after it is released, both monitoring and modeling are available. Monitoring of the criteria pollutants is done routinely by state and local air pollution agencies in most large urban areas and in some other areas as well. Recommended techniques for measuring criteria and many other pollutants are available (11). Monitoring is expensive and time consuming, however, and even the most extensive urban networks are insufficient to assess accurately the geographic distribution of pollutants. Consequently, air pollution models are employed. A microscale model extends from the emission source to less than 10 km downwind. The Gaussian Plume model is an example. From 10 km to about 100 km downwind, mesoscale or urban-scale models, which are used to describe the pollutant patterns both within and downwind of urban areas, are applied. From 100 km to about 1000 km, synoptic or regional-scale models are employed. These models are used to estimate pollution patterns in areas the size of the eastern United States. Above 1000 km, global-scale or general-circulation models, which calculate the distributions of species having long atmospheric residence times such as the greenhouse gases, are used. The complexity of the model depends on both the scale of the area covered and the number and kinds of transformation processes which are included (see ATMOSPHERIC MODELS).

Receptors. The receptor can be a person, animal, plant, material, or ecosystem. The criteria and hazardous air pollutants were so designated because, at sufficient concentrations, they can cause adverse health effects to human receptors. Some of the criteria pollutants also cause damage to plant receptors. An Air Quality Criteria Document (12) exists for each criteria pollutant and these documents summarize the most current literature concerning the effects of criteria pollutants on human health, animals, vegetation, and materials. The receptors which have generated much concern regarding acid deposition are certain aquatic and forest ecosystems, and there is also some concern that acid deposition adversely affects some materials.

For visibility-reducing air pollutants, CFCs, and greenhouse gases, the receptor is the atmosphere. Visibility-reducing species alter atmospheric optical properties and CFCs alter its natural chemical composition in such a way that the atmosphere becomes more transparent to potentially harmful uv solar radiation. The greenhouse gases alter atmospheric radiative properties and consequently have the potential to alter the global heat budget.

Air Quality Management

In the United States, the framework for air quality management is the Clean Air Act (CAA), which defines two categories of pollutants: criteria and hazardous. For the criteria pollutants, the CAA requires that EPA establish NAAQS and emissions standards for some large new sources and for motor vehicles, and gives the primary responsibility for designing and implementing air quality improvement programs to the states. For the hazardous air pollutants, only emissions standards for some sources are required. The NAAQS apply uniformly across the United States whereas emissions standards for criteria pollutants depend on the severity of the local air pollution problem and whether an affected source already exists or is proposed. In addition, individual states have the right to set their own ambient air quality and emissions standards (which must be at least as stringent as the federal standards) for all pollutants and all sources except motor vehicles.

With respect to motor vehicles, the CAA allows the states to choose between two sets of emissions standards: the Federal standards or the more stringent California ones.

The two levels of NAAQS, primary and secondary, are listed in Table 3. Primary standards were set to protect public health within an adequate margin of safety; secondary standards, where applicable, were chosen to protect public welfare, including vegetation. According to the CAA, the scientific bases for the NAAQS are to be reviewed every 5 years so that the NAAQS levels reflect current knowledge. In practice, however, the review cycle takes considerably longer.

Table 3. National Ambient Air Quality Standards for Criteria Pollutants[a]

Pollutant	Primary $\mu g/m^3$	ppm	Secondary $\mu g/m^3$	ppm	Averaging time
PM_{10}	50		50		annual arithmetic mean
	150		150		24-h[b]
SO_2	80	(0.03)			annual arithmetic mean
	365	(0.14)			24-h[b]
			1300	(0.50)	3-h[b]
CO	(10)	9			8-h[b]
	(40)	35			1-h[b]
NO_2	(100)	0.053	(100)	0.053	annual arithmetic mean
Pb	1.5		1.5		maximum quarterly average
O_3	(235)	0.12	(235)	0.12	maximum daily[c] 1-h average

[a]Parenthetical value is an approximately equivalent concentration.
[b]Not to be exceeded more than once per year.
[c]Not to be exceeded on more than three days in three years.

In order to analyze trends in criteria pollutants nationwide, the U.S. EPA has established three types of monitoring systems comprised of 274 sites. The first is a network of 98 National Air Monitoring Sites (NAMS), located in areas having high pollutant concentrations and high population exposures. The system was established by regulations promulgated in 1979. In addition, EPA also regularly evaluates data from the State and Local Monitoring System (SLAMS) and from Specific Purpose Monitors (SPM). To determine if NAAQS are met, states are required to monitor the criteria pollutants' concentrations in areas that are likely to be near or to exceed the standards. If an area exceeds a NAAQS for a given pollutant, it is designated as a nonattainment area for that pollutant, and the state is required to establish a State Implementation Plan (SIP).

The SIP is a strategy designed to achieve emissions reductions sufficient to meet the NAAQS within a deadline that is determined by the severity of the local pollution problem. Areas that receive long (six years or more) deadlines must show continuous progress by reducing emissions by a specified percentage each year. For SO_2 and NO_2, the initial SIPs were very successful in achieving the NAAQS. For other criteria pollutants, particularly O_3 and to a lesser extent CO, however, many areas are starting a third round of SIP preparations with little hope of meeting the NAAQS in the near future. If a state misses an attainment deadline, fails to revise an inadequate SIP, or fails to implement SIP requirements, EPA has the authority to enforce sanctions such as banning construction of new stationary sources and withholding federal grants for highways.

In nonattainment areas, the degree of control on small sources is left to the discretion of the state and is largely determined by the degree of required emissions reductions. Large existing sources must be retrofitted with reasonable available control technology (RACT) to minimize emissions. All large new sources and existing sources that undergo major modifications must meet EPA's new source performance standards at a minimum. Additionally, in nonattainment areas, they must be designed using lowest achievable emission rate (LAER) technology, and emissions offsets must be obtained. Offsets require that emissions from existing sources within the area be reduced below legally allowable levels so that the amount of the reduction is greater than or equal to the emissions expected from the new source. RACT usually is less stringent than LAER: it may not be feasible to retrofit certain sources using the LAER technology.

In attainment areas, new large facilities must be designed to incorporate the best available control technology (BACT). Generally, BACT is more stringent than RACT and equal to or less stringent than LAER. In addition, there are also rules that specify how much deterioration in baseline air quality a new facility can cause. In no situation can the facility cause a new violation in the NAAQS.

Large sources of SO_2 and NO_x may also require additional emission reductions because of the 1990 Clean Air Act Amendments. To reduce acid deposition, the amendments require that nationwide emissions of SO_2 and NO_x be reduced on an annual basis by approximately 10 million and 2 million tons, respectively, by the year 2000.

Once a substance is designated by EPA as a hazardous air pollutant (HAP), EPA has to promulgate a NESHAP (National Emission Standard for Hazardous Air Pollutants), designed to protect public health with an ample margin of safety.

Air Pollution Issues

Photochemical Smog. Photochemical smog is a complex mixture of constituents formed when VOCs and NO_x are irradiated by sunlight. O_3, the most abundant species formed in photochemical smog, is the primary concern. Extensive studies have shown that O_3 is both a lung irritant and a phytotoxin. It is responsible for crop damage and is suspected of being a contributor to forest decline in Europe and in parts of the United States. There are, however, a multitude of other photochemical smog species that also have significant environmental consequences. The most important of these pollutants are particles, hydrogen peroxide (qv) [7722-84-1], H_2O_2, peroxyacetyl nitrate [2278-22-0], PAN, $C_2H_3NO_5$ aldehydes, and nitric acid (qv) [7697-37-2], HNO_3.

Photochemical smog is a summertime phenomenon for most parts of the United States. In the warmer parts of the country, especially in Southern California, the smog season begins earlier and lasts into the fall. Despite almost two decades of reducing VOC emissions from stationary and mobile sources and NO_x emissions from mobile sources, little or no progress has been made in reducing the number of areas in the United States designated as nonattainment for O_3. For the 1985–1987 period, there were 64 O_3-nonattainment areas, mostly urban, in the United States. This number increased to 101 areas after the anomalously hot summer of 1988, but then decreased to 96 areas in 1989. An area is classified nonattainment if the O_3 design value exceeds 0.12 ppm. The design value is equal to the 4th highest maximum daily 1-h O_3 concentration within a 3-yr period. For most of the nonattainment areas, this value falls between 0.13 to 0.16 ppm. The three areas having the highest design values for the 1987–1989 period were the Los Angeles (0.33 ppm), Houston (0.22), and New York (0.20) metropolitan areas.

There is a significant clean air background O_3 concentration, measured at pristine areas of the globe, that varies according to the season and latitude. This concentration consists of natural sources of O_3, but it undoubtedly contains some anthropogenic contribution because it may have increased since the last century (13). In the summertime in the United States, the average O_3 background is about 0.04 ppm (14). Background O_3 has four sources: intrusions of O_3-rich stratospheric air, *in situ* O_3 production from methane oxidation, the photooxidation of naturally-emitted VOCs from vegetation, and the long-range transport of O_3 formed from the photooxidation of anthropogenic VOCs and NO_x emissions. Although there are several mechanisms which transport O_3-rich air from the stratosphere into the lower troposphere, the most important appears to be associated with large-scale eddy transport that occurs in the vicinity of upper air troughs of low pressure associated with the jet stream (15). This is an intermittent mechanism, so the contribution of stratospheric to surface O_3 has a considerable temporal variation and on very rare occasions, has produced brief ground level concentrations exceeding 0.12 ppm (16). Three other mechanisms are described below.

In the presence of sunlight, hv, NO_2 photolyzes and produces O_3

$$NO_2 + hv \longrightarrow NO + O \tag{1}$$

$$O + O_2 + M \longrightarrow O_3 + M \tag{2}$$

$$NO + O_3 \longrightarrow NO_2 + O_2 \tag{3}$$

where M is any third body molecule (most likely N_2 or O_2 in the atmosphere) that remains unchanged in the reaction. This process produces a steady-state concentration of O_3 that is a function of the initial concentrations of NO and NO_2, the solar intensity, and the temperature. Although these reactions are extremely important in the atmosphere, the steady-state O_3 produced is much lower than the observed concentrations, even in clean air. In order for ozone to accumulate, there must be a mechanism that converts NO to NO_2 without consuming a molecule of O_3, as does reaction 3. Reactions involving hydroxyl radicals and hydrocarbons or VOC constitute such a mechanism. In clean air OH may be generated by

$$O_3 + hv \longrightarrow O_2 + O(^1D) \tag{4}$$

$$O(^1D) + H_2O \longrightarrow 2\ OH \tag{5}$$

where $O(^1D)$ is an excited form of an O atom that is produced from a photon at a wavelength between 280 and 310 nm. This seed OH can then produce the following chain reactions:

$$OH + CH_4 \longrightarrow H_2O + CH_3 \tag{6}$$

$$CH_3 + O_2 + M \longrightarrow CH_3O_2 + M \tag{7}$$

$$CH_3O_2 + NO \longrightarrow CH_3O + NO_2 \tag{8}$$

The NO_2 can then photolyze producing O_3 (eqs. 1 and 2) and the CH_3O radical continues to react:

$$CH_3O + O_2 \longrightarrow HCHO + HO_2 \tag{9}$$

The HO_2 radical also forms more NO_2:

$$HO_2 + NO \longrightarrow NO_2 + OH \tag{10}$$

resulting in more O_3. In addition, OH is regenerated to begin the cycle again. Further, the formaldehyde photodissociates:

$$HCHO + h\nu \xrightarrow{\ a\ } H_2 + CO \tag{11}$$

$$\xrightarrow{\ b\ } HCO + H \tag{12}$$

$$HCO + O_2 \longrightarrow HO_2 + CO \tag{13}$$

$$H + O_2 \longrightarrow HO_2 \tag{14}$$

and the HO_2 from both equations 13 and 14 can form additional NO_2. Moreover, CO can be oxidized:

$$CO + OH \longrightarrow CO_2 + H \tag{15}$$

and the H radical can form another NO_2 (eqs. 14 and 10). Thus the oxidation of one CH_4 molecule is capable of producing three O_3 molecules and two OH radicals. Although this chain reaction is less than 100% efficient, on the average, the chain results in a net production of O_3 and OH. Two examples of competing chain-terminating reactions are

$$HO_2 + HO_2 \longrightarrow H_2O_2 + O_2 \tag{16}$$

$$OH + NO_2 \xrightarrow{\ M\ } HNO_3 \tag{17}$$

In a polluted or urban atmosphere, O_3 formation by the CH_4 oxidation mechanism is overshadowed by the oxidation of other VOCs. Seed OH can be produced from reactions 4 and 5, but the photodisassociation of carbonyls and nitrous acid [7782-77-6], HNO_2, (formed from the reaction of OH + NO and other reactions) are also important sources of OH in polluted environments. An imperfect, but useful, measure of the rate of O_3 formation by VOC oxidation is the rate

Table 4. Median Concentration of the Ten Most Abundant Ambient Air Hydrocarbons in 39 U.S. Cities and Their Reactivity with Hydroxyl Radical

Compound	CAS Registry Number	Median concentration[a] ppb C	Reactivity with OH[b,c]
isopentane	[78-78-4]	45.3	494
n-butane	[106-97-8]	40.3	351
toluene	[108-88-3]	33.8	831
propane	[74-98-6]	23.5	143
ethane	[74-84-0]	23.3	36
n-pentane	[109-66-0]	22.0	480
ethylene	[74-85-1]	21.4	1013
m-xylene	[108-38-3]	18.1[d]	3117
p-xylene	[106-42-3]	18.1[d]	1818
2-methylpentane	[107-83-5]	14.9	
isobutane	[75-28-5]	14.8	325
		Biogenic species	
α-pinene	[80-56-8]		7792
isoprene	[78-79-5]		12078

[a]Ref. 17.
[b]Ref. 18.
[c]Relative to reaction of CH_4 + OH at 298°C.
[d]Combined xylene [1330-20-7] concentration.

of the initial OH–VOC reaction, shown in Table 4 relative to the OH–CH_4 rate for some commonly occurring VOCs. Also given are the median VOC concentrations. Shown for comparison are the relative reaction rates for two VOC species that are emitted by vegetation: isoprene and α-pinene. In general, internally bonded olefins are the most reactive, followed in decreasing order by terminally bonded olefins, multialkyl aromatics, monoalkyl aromatics, C_5 and higher paraffins, C_2–C_4 paraffins, benzene, acetylene, and ethane.

The reaction mechanisms by which the VOCs are oxidized are analogous to, but much more complex than, the CH_4 oxidation mechanism. The fastest reacting species are the natural VOCs emitted from vegetation. However, natural VOCs also react rapidly with O_3, and whether they are a net source or sink is determined by the natural VOC to NO_x ratio and the sunlight intensity. At high VOC/NO_x ratios, there is insufficient NO_2 formed to offset the O_3 loss. However, when O_3 reacts with the internally bonded olefinic compounds, carbonyls are formed and, the greater the sunshine, the better the chance the carbonyls will photolyze and produce OH which initiates the O_3-forming chain reactions.

Once the sun sets, O_3 formation ceases and, in an urban area, ozone is rapidly scavenged by freshly emitted NO (eq. 3). On a typical summer night, however, a nocturnal inversion begins to form around sunset, usually below a few hundred meters and consequently, the surface-based NO emissions are trapped below the top of the inversion. Above the inversion to the top of the mixed layer (usually about 1500 m), O_3 is depleted at a much slower rate. The next morning, the inversion dissipates and the O_3-rich air aloft is mixed down into the O_3-

depleted air near the surface. This process, in combination with the onset of photochemistry as the sun rises, produces the sharp increase in surface O_3 shown in Figure 1. As shown, the overnight O_3 depletion is less in the more rural areas than in a large urban area such as New York City. This is a result of the lower overnight levels of NO in rural areas. Even in the absence of NO or other O_3 scavengers (olefins, for example), O_3 decreases at night near the ground faster than aloft because of its destruction at any surface, ie, the ground, buildings, trees. At the remote mountaintop sites, Whiteface and Utsayantha, there is no overnight decrease in O_3 concentrations.

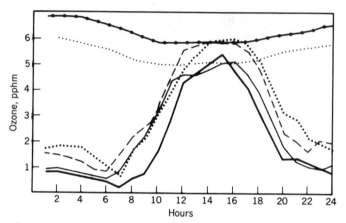

Fig. 1. Average hourly ozone concentrations during August 1, 1973 to August 17, 1973 for selected sites in New York State; site designations are: ⋯⋯ Kingston; ‑‑‑‑ Rensselaer; ⸺ Glens Falls; ⸺ New York City; ⋯⋯ Mt. Utsayantha; ‑•‑ Whiteface Mountain (19). Courtesy of Air and Waste Management Association.

Although photochemical smog is a complex mixture of many primary and secondary pollutants and involves a myriad of atmospheric reactions, there are characteristic pollutant concentration versus time profiles that are generally observed within and downwind of an urban area during a photochemical smog episode. In particular, the highest O_3 concentrations are generally found 10–100 km downwind of the urban emissions areas, unless the air is completely stagnant. This fact, in conjunction with the long lifetime of O_3 in the absence of high concentrations of NO, means that O_3 is a regional air pollution problem. In the Los Angeles basin, high concentrations of O_3 are transported throughout the basin and multiday episodes are exacerbated by the accumulation of O_3 aloft which is then mixed to the surface daily. On the east coast, a typical O_3 episode is associated with a high pressure system anchored offshore producing a southwesterly flow across the region. As a result, emissions from Washington, D.C. travel and mix with emissions from Baltimore and over a period of a few days continue traveling northeastward through Philadelphia, New York City, and Boston. Under these conditions, the highest O_3 concentrations typically occur in central Connecticut (20).

It is obvious that in order to reduce O_3 in a polluted atmosphere, reductions in the VOC and NO_x precursors are required. However, the choice of whether to control VOC or NO_x or both depends on the local VOC/NO_x ratio. At low VOC/NO_x ratios, O_3 formation is suppressed through equations 3 and 17. Consequently, in this case reducing NO_x emissions, emitted mainly as NO, reduces the amount of O_3 (eq. 3) and OH (eq. 17) scavenged, increasing the O_3 concentrations. This is illustrated in the O_3-isopleth diagrams in Figure 2. Although the four chemical mechanisms used to obtain Figure 2 give somewhat different results, the shape of the isopleths are quite similar and the key features are summarized in Figure 3. The region in the upper left is the NO_x-inhibition region where a decrease in NO_x alone results in an increase in O_3, but a decrease in VOC decreases O_3. The region at the bottom right is the hydrocarbon, HC, or VOC saturation region where reducing VOCs has no effect on the O_3 level. Here, a reduction in NO_x results in lower O_3. In the middle is the knee region, where reductions in either NO_x or VOC reduce O_3. The upper boundary of this region varies from day to day and from place to place as its location is a function of the reactivity of the VOC mix and the sunlight intensity.

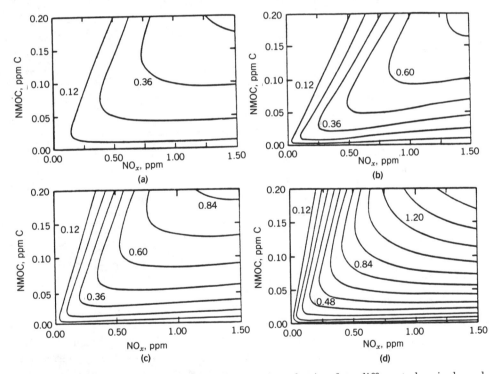

Fig. 2. Examples of ozone isopleths generated using four different chemical mechanisms (**a**) EPA; (**b**) FSM; (**c**) CBII; (**d**) ELSTAR; NMOC = nonmethane organic compounds (21). Courtesy of Pergamon Press.

As a guideline, VOC controls are generally the most efficient way to reduce O_3 in areas having a median 6 A.M. to 9 A.M. VOC/NO_x ratio of 10:1 or less; whereas areas with a higher ratio may need to consider NO_x reductions as well (23). The 1990 Clean Air Act Amendments require that O_3 nonattainment areas

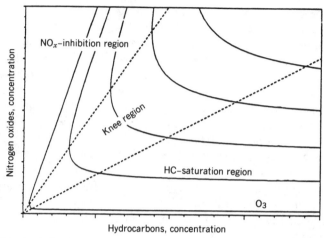

Fig. 3. Typical O_3 isopleth diagram showing the three chemical regimes; HC = hydrocarbons (22). Courtesy of Pergamon Press.

reduce both VOC and NO_x from major stationary sources unless the air quality benefits are greater in the absence of NO_x reductions. Large cities in the northeast tend to have ratios <10:1; cities in the south (Texas and eastward) tend to have ratios >10:1. Determining a workable control strategy is further complicated by the transport issue. For example, on high O_3 days in the northeast, the upwind air entering Philadelphia and New York City frequently contains O_3 already near or over the NAAQS as a result of emissions from areas to the west and south (24). Consequently, control strategies must be developed on a coordinated, multistate regional basis.

Because of the mixture of VOCs in the atmosphere, the composition of smog reaction products and intermediates is extremely complex. H_2O_2, formed via reaction 16, is important because when dissolved in cloud droplets it is an important oxidant, responsible for oxidizing SO_2 to sulfuric acid [7664-93-9], H_2SO_4, the primary cause of acid precipitation. The oxidation of many VOCs produces acetyl radicals, CH_3CO, which can react with O_2 to produce peroxyacetyl radicals, $CH_3(CO)O_2$, which react with NO_2

$$CH_3(CO)O_2 + NO_2 \rightleftharpoons CH_3C(O)O_2NO_2 \text{ (PAN)} \tag{18}$$

At high enough concentrations, PAN is a potent eye irritant and phytotoxin. On a smoggy day in the Los Angeles area, PAN concentrations are typically 5 to 10 ppb; in the rest of the United States PAN concentrations are generally a fraction of a ppb. An important formation route for formaldehyde [50-00-0], HCHO, is reaction 9. However, ozonolysis of olefinic compounds and some other reactions of VOCs can produce HCHO and other aldehydes. Aldehydes are important because they are temporary reservoirs of free radicals (see eqs. 11 and 12). HCHO is a known carcinogen. Nitric acid is formed by OH attack on NO_2 and by a dark-phase series of reactions initiated by $O_3 + NO_2$. Nitric acid is important because it is the second most abundant acid in precipitation. In addition, in Southern California it is the major cause of acid fog.

Particles are the major cause of the haze and the brown color that is often associated with smog. The three most important types of particles produced in smog are composed of organics, sulfates, and nitrates. Organic particles are formed when large VOC molecules, especially aromatics and cyclic alkenes, react with each other and form condensable products. Sulfate particles are formed by a series of reactions initiated by the attack of OH on SO_2 in the gas phase or by liquid-phase reactions. Nitrate particles are formed by

$$HNO_3(g) + NH_3(g) \rightleftharpoons NH_4NO_3(s) \tag{19}$$

or by the reactions of HNO_3 with NaCl or alkaline soil dust.

Volatile Organic Compounds (VOC).   VOCs include any organic-carbon compound that exists in the gaseous state in the ambient air. (In some of the older literature the term VOC is used interchangeably with NMHC. VOC sources may be any process or activity utilizing organic solvents, coatings, or fuel. Emissions of VOCs are important: some are toxic by themselves, and most are precursors of O_3 and other species associated with photochemical smog. As a result of control measures designed to reduce O_3, VOC emissions are declining in the United States. Between 1980–1989, nationwide VOC emissions declined 19% (4). Trends in ambient VOC concentrations cannot be determined, however, because of the lack of measurements.

Nitrogen Oxides (NO$_x$).   Most of the NO_x is emitted as NO, which is then oxidized to NO_2 in the atmosphere (see eqs. 3 and 8). All combustion processes (see COMBUSTION TECHNOLOGY) are sources of NO_x. At the high temperatures generated during combustion, some N_2 is converted to NO in the presence of O_2 and, in general, the higher the combustion temperature, the more NO_x produced. NO_2 is one of the criteria pollutants as well as a precursor to O_3, so it was the target of successful U.S. emissions reduction strategies in the 1970s and 1980s. As a result, in 1987, all areas of the United States, excepting the Los Angeles/Long Beach area, were in compliance with the NAAQS for NO_2. From 1980 to 1989 nationwide NO_x emissions and ambient concentrations declined 5% (4).

However, NO_x remains an important issue throughout the United States. In addition to being an essential ingredient of photochemical smog and a precursor to HNO_3, itself an ingredient of acid precipitation and fog, NO_2 is the only important gaseous species in the atmosphere that absorbs visible light. In high enough concentrations it can contribute to a brownish discoloration of the atmosphere.

Sulfur Oxides (SO$_x$).   The combustion of sulfur-containing fossil fuels, especially coal, is the major source of SO_x. Between 97 and 99% of the SO_x emitted from combustion sources is in the form of SO_2, a criteria pollutant. The remainder is mostly sulfur trioxide [7446-11-9], SO_3, which in the presence of atmospheric water [7732-18-5] vapor is immediately transformed into H_2SO_4, a liquid particulate. Both SO_2 and H_2SO_4 at sufficient concentrations produce deleterious effects on the respiratory system. In addition, SO_2 is a phytotoxin. As with NO_2, control strategies designed to reduce the ambient levels of SO_2 have been highly successful. In the 1960s, most industrialized urban areas in the eastern United States had an SO_2 air quality problem. By 1987 only Pittsburgh, Pa. and Steubenville, Ohio, exceeded the 24-h SO_2 NAAQS, and only Steubenville exceeded the annual NAAQS. Over the past ten years, nationwide emissions declined 10% and ambient concentrations decreased 24% (4). However, the 1990 Clean Air Act Amendments require additional SO_2 reductions because of the role that SO_2 plays in acid deposition. In addition, there is some concern over the health effects of H_2SO_4 particles, which are emitted directly from some sources as well as being formed in the atmosphere (25).

Carbon Monoxide (CO).   Carbon monoxide (qv) is emitted during any combustion process. Transportation sources account for about two-thirds of the CO emissions nationally, but, in certain areas, most of the CO comes from woodburning fireplaces and stoves. CO is absorbed through the lungs into the blood stream and reacts with hemoglobin [9034-51-9] to form carboxyhemoglobin, which reduces the oxygen carrying capacity of the blood.

Emissions of CO in the United States peaked in the late 1960s, but have decreased consistently since that time as transportation sector emissions significantly decreased. Between 1968 and 1983, CO emissions from new passenger cars were reduced by 96% (see EXHAUST CONTROL, AUTOMOTIVE). This has been partially offset by an increase in the number of vehicle-miles travelled annually. Even so, there has been a steady decline in the CO concentrations across the United States and the decline is expected to continue until the late 1990s without the implementation of any additional emissions-reduction measures. In 1989, there were still 41 U.S. urban areas that exceeded the CO NAAQS on one or more days per year, but the number of exceedances declined by about 80% from 1980 to 1989. Over the same time period, nationwide CO emissions decreased 23%, and ambient concentrations declined by 25% (4).

Particulate Matter.   In the air pollution field, the terms particulate matter, particulates, particles, and aerosols (qv) are used interchangeably and all refer to finely divided solids and liquids dispersed in the air. The original EPA primary standards were for total suspended particulates, TSP, the weight of any particulate matter collected on the filter of a high volume air sampler. On the average, these samplers collect particles that are less than about 30–40 μm in diameter, but collection efficiencies vary according to both wind direction and speed. In 1987, the term PM_{10}, particulate matter having an aerodynamic diameter of 10 μm or less, was introduced. The 10-μm diameter was chosen because 50% of the 10-μm particles deposit in the respiratory tract below the larynx during oral breathing and the fraction deposited decreases as particle diameter increases above 10 μm. Because the NAAQS standard (see Table 3) was only enacted in 1987, currently available PM_{10} data are insufficient to determine trends. However, from 1979 to 1988 TSP emissions declined 22% and ambient concentrations decreased 20% (4).

Atmospheric aerosols can be classified into three size modes: nuclei, accumulation, and large or coarse-particle modes. Characteristics are given in Figure 4. The bulk of the aerosol mass usually occurs in the 0.1- to 10-μm size range which encompasses most of the accumulation mode and part of the large-particle mode. The nuclei mode is transient as nuclei, formed by combustion, nucleation, and chemical reactions, coagulate and grow into the accumulation mode. Particles in the accumulation mode are relatively stable because they exceed the size range where coagulation is important, and they are too small to have appreciable

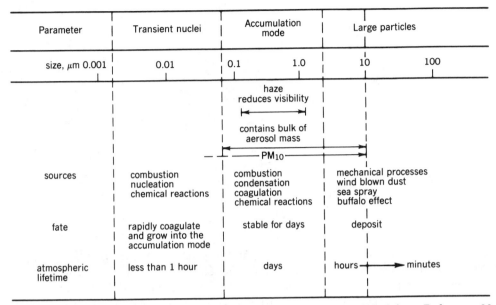

Fig. 4. Some important aerosol characteristics. Data modified from Reference 26 which was adopted in part from Reference 27.

deposition velocities. Consequently, particles "accumulate" in this mode. Particles larger than about 2.5 μm begin to have appreciable deposition velocities, so their lifetimes in the atmosphere decrease significantly as particle size increases. The sources of large particles are mostly mechanical processes.

Figure 5 shows the mass size distribution of typical ambient aerosols. Note the mass peaks in the accumulation mode and between 5 and 10 μm. The minimum in the curves at about 1–2.5 μm results from a lack of sources for these particles. Coagulation is not significant for the accumulation-mode particles, and particles produced by mechanical process are larger than 2.5 μm. Consequently, particles less than about 2.5 μm have different sources from particles greater than 2.5 μm and it is convenient to classify PM_{10} into a coarse-particulate-mass mode (CPM, dia ≥ 2.5–10 μm) and a fine-particulate-mass mode (FPM, dia < 2.5 μm). By knowing the relative amounts of CPM and FPM as well as the chemical composition of the major species, information on the PM_{10} sources can be deduced. In urban areas the CPM and FPM are usually comparable in mass; in rural areas the FPM is generally lower than in urban areas, but higher than the CPM mass. A significant fraction of the rural FPM is generally transported from upwind sources, whereas most of the CPM is generated locally.

Chemical composition data for CPM and FPM for a variety of locations are summarized in Table 5. These data illustrate several important points. First, the distributions of the PM_{10} between CPM and FPM vary from about 0.4 to 0.7. Second, the ratio of PM_{10} to TSP varies from 0.58 to 0.79. In general, both this ratio and the ratio of FPM to PM_{10} tend to be higher at rural sites, but Bermuda, because of the large influence of sea salt in the CPM, is an exception. Sulfate

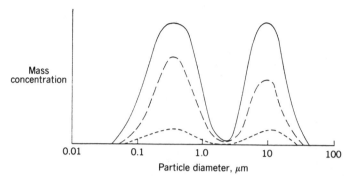

Fig. 5. Size distributions of atmospheric particles in (——) urban, (— —) rural, and (– – –) remote background areas.

(SO_4^{2-}), carbon (as organic carbon, OC, and elemental carbon, EC), and nitrate (NO_3^-) compounds generally account for 70–80% of the FPM. In the eastern United States, SO_4^{2-} compounds are the dominant species, although very little SO_4^{2-} is emitted directly into the atmosphere. Thus most of the sulfate is a secondary aerosol formed from the oxidation of SO_2, and in the eastern United States coal-burning emissions are the source of most of the SO_2.

In the atmosphere, the principal SO_2 oxidation routes include homogeneous oxidation by OH, and the heterogeneous oxidation in water droplets by H_2O_2, O_3, or, in the presence of a catalyst, O_2. Atmospheric particles which have been identified as catalysts include many metal oxides and soot. The water droplets include cloud and dew droplets as well as aerosols which contain sufficient water: under high relative-humidity conditions, hygroscopic salts deliquesce, and form liquid aerosols. Sulfuric acid is the initial SO_2 oxidation product. This rapidly reacts with any available ammonia [7664-41-7], NH_3 to form ammonium bisulfate [7803-63-6], NH_4HSO_4. If sufficient NH_3 is present, the final product is ammonium sulfate [7783-20-2], $(NH_4)_2SO_4$. In some urban areas in the western United States, NO_3^- is more abundant than SO_4^{2-}. The NO_3^- in the FPM exists primarily as ammonium nitrate [6484-52-2], NH_4NO_3, (see eq. 19). However, acidic SO_4^{2-} (H_2SO_4 or NH_4HSO_4) readily reacts with NH_4NO_3 and abstracts the NH_3, leaving behind gaseous HNO_3. Consequently, unless there is sufficient NH_3 to completely convert all of the SO_4^{2-} to $(NH_4)_2SO_4$, NH_4NO_3 does not accumulate in the atmosphere. The western U.S. cities shown in Table 5 have sufficient NH_3, largely because of the presence of animal feedlots, and NH_4NO_3 accumulates. In the eastern United States, there is generally insufficient NH_3.

Organic compounds are a major constituent of the FPM at all sites. The major sources of OC are combustion and atmospheric reactions involving gaseous VOCs. As is the case with VOCs, there are hundreds of different OC compounds in the atmosphere. A minor but ubiquitous aerosol constituent is elemental carbon. EC is the nonorganic, black constituent of soot. Combustion and pyrolysis are the only processes that produce EC, and diesel engines and wood burning are the most significant sources.

Crustal dust and water make up most of the remaining FPM mass. Crustal dust is composed of aerosolized soil and rock from the earth's crust. Although this

Table 5. Summary of Detailed Particulate Measurements at Urban and Rural Locations

Species	Urban					Rural		
	Denver, Colo.	Detroit, Mich.	Long Beach, Calif.	Claremont, Calif.	Jacksonville, Fla.	Blue Ridge Mts, Va.	Lewes, Del.	Bermuda
	Concentrations $\mu g/m^3$							
PM_{10}	39.5	52.9[a]	79.1	59.7	38.8	32.5	22.0	20.8
FPM		25.1	57.8	36.0	22.6	23.9	15.9	9.4
CPM	50.9[b]	27.8	23.1	22.9	16.2	8.6	6.1	11.4
FPM/PM_{10}		0.48[a]	0.73	0.60	0.58	0.74[a]	0.72	0.45
TSP	104.4	90.5				43.7	27.9	32.8
PM_{10}/TSP		0.58[a]				0.74	0.79	0.63
	Concentration as percentage of FPM							
sulfate compounds	14	53	7	22	41	55	50	25
nitrate compounds	20	<1	33	28	3	<1		2
organic compounds	22	27	29	29	13	24	30	15
elemental carbon	15	4	12	5	6	5	6	<1
	Concentration as percentage of CPM							
crustal species	91	73			85		48	17
sea salt							43	57
	References							
	28	29,30	31	31	32	33	34	35

[a]PM_{10} in Detroit and Blue Ridge Mts is PM_{15} which is $\cong$ to PM_{10}.
[b]CPM in Denver is dia $\geq 2.5\ \mu m$ to about 30 μm.

729

is natural material, human activities (traffic, which produces the buffalo effect by entraining street dust, construction activities, agricultural and land-use practices, etc) affect the rate at which crustal material is aerosolized. Since it is aerosolized by frictional processes, the diameter of most of the crustal dust is ≥ 2.5 μm and typically accounts for most of the CPM and particles >10 μm. In global average crustal material, the major elements contained in decreasing order are O, Si, Al, Fe, Ca, Na, K, and Mg. On the average, Si accounts for 20% of the crustal aerosol mass (36). Consequently, the crustal mass can be estimated from Si measurements alone. However, the relative amounts of the elements do vary spatially.

Lead. Lead (qv) is of concern because of its tendency to be retained by living organisms. When excessive amounts accumulate in humans, lead can inhibit the formation of hemoglobin and produce life-threatening lead poisoning. In smaller doses, lead is also suspected of causing learning disabilities in children. From 1980 to 1989, nationwide Pb emissions decreased 90%. The primary source, transportation, showed a 96% reduction (4) as a direct result of the removal of lead compounds (qv) such as tetraethyllead [78-00-2] $(C_2H_5)_4Pb$ from fuels, primarily gasoline. Trends of Pb in the ambient air have responded to the emissions reductions. In 1989 only a few isolated monitoring sites that were dominated by industrial sources experienced violations of the NAAQS (4).

Air Toxics. There are thousands of commercial chemicals used in the United States. Hundreds are emitted into the atmosphere and have some potential to adversely affect human health at certain concentrations; some are known or suspected carcinogens. Identifying all of these substances and promulgating emissions standards is beyond the present capabilities of existing air quality management programs. Consequently, toxic air pollutants (TAPs) need to be prioritized based on risk analysis, so that those posing the greatest threats to health can be regulated. Although the criteria pollutants were so designated because they can have significant public-health impacts, these materials are not considered TAPs because they are regulated elsewhere in the CAA. A distinguishing feature between TAPs and criteria pollutants is that criteria pollutants are considered to be national problems whereas TAPs are most often localized near the source of the TAP emissions.

There are three types of TAP emissions: continuous, intermittent, and accidental. Both routine emissions associated with a batch process or a continuous process that is operated only occasionally can be intermittent sources. A dramatic example of an accidental emission was the release of methyl isocyanate [624-83-9] in Bhopal, India. As a result of this accident, the U.S. Congress created Title III, a free-standing statute included in the Superfund Amendments and Reauthorization Act (SARA) of 1986. Title III provides a mechanism by which the public can be informed of the existence, quantities, and releases of toxic substances, and requires the states to develop plans to respond to accidental releases of these substances. Further, it requires anyone releasing specific toxic chemicals above a certain threshold amount to annually submit a toxic chemical release form to EPA. At present, there are 308 specific chemicals subject to Title III regulation (37).

Lists of workplace air standards for over 700 substances, many of which would be considered a TAP if present in sufficient quantity in the ambient air, are available (38). Toxicological data for these substances can be found in reference 39.

The 1970 Clean Air Act required that EPA provide an ample margin of safety to protect against hazardous air pollutants by establishing national emissions standards for certain sources. From 1970 to 1990, over 50 chemicals were considered for designation as HAPs, but EPA's review process was completed for only 28 chemicals. NESHAPs have been promulgated for only eight substances: beryllium [7440-41-7], mercury [7439-97-6], vinyl chloride [75-01-4], asbestos [1332-21-4], benzene [71-43-2], radionuclides, inorganic arsenic [7440-38-2], and coke-oven emissions (40). However, in the 1990 Clean Air Act Amendments, 189 substances are listed (Table 6) that EPA must regulate by enforcing maximum achievable control technology (MACT). The Amendments mandate that EPA issue MACT standards for all sources of the 189 substances by the year 2000. In addition, EPA must determine the risk remaining after MACT is in place and develop health-based standards that would limit the cancer risk to one case in one million exposures. EPA may add or delete substances from this list.

Because EPA was so slow in promulgating standards for HAPs prior to the 1990 Amendments, most states developed and implemented their own TAP control programs. Such programs, as well as the pollutants they regulate, differ widely from state to state. The ambient standards for a given substance are usually selected to be some small fraction of the TLV for that substance.

Odors. The 1977 Clean Air Act Amendments directed EPA to study the effects, sources, and control feasibility of odors. Although no federal legislation has been established to regulate odors, individual states have responded to odor complaints by enforcing common nuisance laws. About 50% of all citizen air pollution complaints concern odors (see ODOR MODIFICATION). A disagreeable odor is perceived as an indication of air pollution but many substances can be detected by the human olfactory system at concentrations well below those considered harmful. For example, hydrogen sulfide [7783-06-4], H_2S, can be detected by most people at 0.0047 ppm, whereas the occupation health 8-h TLV is 10 ppm. Although exposures to such odors in low concentrations may not in itself cause physical harm, the exposure can lead to nausea, loss of appetite, and other effects.

Odors are characterized by quality and intensity. Descriptive qualities such as sour, sweet, pungent, fishy, and spicy are commonly used. Intensity is determined by how much the concentration of the odoriferous substance exceeds its detection threshold (the concentration at which most people can detect an odor). Odor intensity is approximately proportional to the logarithm of the concentration. However, several factors affect the ability of an individual to detect an odor: the sensitivity of a subject's olfactory system, the presence of other masking odors, and olfactory fatigue (ie, reduced olfactory sensitivity during continued exposure to the odorous substance). In addition, the average person's sensitivity to odor decreases with age.

Visibility. Although there is no NAAQS designed to protect visual air quality, the 1977 Clean Air Act Amendments set as a national goal "the remedying of existing and prevention of future impairment of visibility in mandatory Class I Federal areas which impairment results from man-made pollution." Class I areas are certain national parks and wildernesses that were in existence in 1977. The Amendments also directed EPA to promulgate appropriate regulations to protect

Table 6. Substances Listed as Hazardous Air Pollutants as Defined by the 1990 Clean Air Act Amendments

Substance	CAS Registry Number	Substance	CAS Registry Number
acetaldehyde	[75-07-0]	hydrazine	[302-01-2]
acetamide	[60-35-5]	hydrochloric acid	[7647-01-0]
acetonitrile	[75-05-8]	hydrogen fluoride	[7664-39-3]
acetophenone	[98-86-2]	hydroquinone	[123-31-9]
2-acetylaminofluorene	[53-96-3]	isophorone	[78-59-1]
acrolein	[107-02-8]	lindane (all isomers)	[58-89-9]
acrylamide	[79-06-1]	maleic anhydride	[108-31-6]
acrylic acid	[79-10-7]	methanol	[67-56-1]
acrylonitrile	[107-13-1]	methoxychlor	[72-43-5]
allyl chloride	[107-05-1]	methyl bromide	[74-83-9]
4-aminobiphenyl	[92-67-1]	methyl chloride	[74-87-3]
aniline	[62-53-3]	methyl chloroform	[71-55-6]
o-anisidine	[90-04-0]	methyl ethyl ketone	[78-93-3]
asbestos	[1332-21-4]	methyl hydrazine	[60-34-4]
benzene	[71-43-2]	methyl iodide	[74-88-4]
benzidine	[92-87-5]	methyl isobutyl ketone	[108-10-1]
benzotrichloride	[98-07-7]	methyl isocyanate	[624-83-9]
benzyl chloride	[100-44-7]	methyl methacrylate	[80-62-6]
biphenyl	[92-52-4]	methyl tert-butyl ether	[1634-04-4]
bis(2-ethylhexyl) phthalate	[117-81-7]	4,4'-methylene bis(2-chloroaniline)	[101-14-4]
bis(chloromethyl) ether	[542-88-1]	methylene chloride	[75-09-2]
bromoform	[75-25-2]	methylene diphenyl diisocyanate	[101-68-8]
1,3-butadiene	[106-99-0]	4,4'-methylenedianiline	[101-77-9]
calcium cyanamide	[156-62-7]	naphthalene	[91-20-3]
caprolactam	[105-60-2]	nitrobenzene	[98-95-3]
captan	[133-06-2]	4-nitrobiphenyl	[92-93-3]
carbaryl	[63-25-2]	4-nitrophenol	[100-02-7]
carbon disulfide	[75-15-0]	2-nitropropane	[79-46-9]
carbon tetrachloride	[56-23-5]	N-nitroso-N-methylurea	[684-93-5]
carbonyl sulfide	[463-58-1]	N-nitrosodimethylamine	[62-75-9]
catechol	[120-80-9]	N-nitrosomorpholine	[59-89-2]
chloramben	[133-90-4]	parathion	[56-38-2]
chlordane	[57-74-9]	pentachloronitrobenzene	[82-68-8]
chlorine	[7782-50-5]	pentachlorophenol	[87-86-5]
chloroacetic acid	[79-11-8]	phenol	[108-95-2]
2-chloroacetophenone	[532-27-4]	p-phenylenediamine	[106-50-3]
chlorobenzene	[108-90-7]	phosgene	[75-44-5]
chlorobenzilate	[510-15-6]	phosphine	[7803-51-2]
chloroform	[67-66-3]	phosphorus	[7723-14-0]
chloromethyl methyl ether	[107-30-2]	phthalic anhydride	[85-44-9]
chloroprene	[126-99-8]	polychlorinated biphenyls	[1336-36-3]
cresylic acid	[1319-77-3]	1,3-propane sultone	[1120-71-4]
o-cresol	[95-48-7]	β-propiolactone	[57-57-8]
m-cresol	[108-39-4]	propionaldehyde	[123-38-6]
p-cresol	[106-44-5]	propoxur (Baygon)	[114-26-1]
cumene	[98-82-8]	propylene dichloride	[78-87-5]
2,4-D, salts and esters	[94-75-7]	propylene oxide	[75-56-9]

Table 6. (*Continued*)

Substance	CAS Registry Number	Substance	CAS Registry Number
DDE	[3547-04-4]	1,2-propylenimine	[75-55-8]
diazomethane	[334-88-3]	quinoline	[91-22-5]
dibenzofurans	[132-64-9]	quinone	[106-51-4]
1,2-dibromo-3-chloropropane	[96-12-8]	styrene	[100-42-5]
dibutylphthalate	[84-74-2]	styrene oxide	[96-09-3]
1,4-dichlorobenzene(p)	[106-46-7]	2,3,7,8-tetrachlorodibenzo-*p*-dioxin	[1746-01-6]
3,3-dichlorobenzidene	[91-94-1]	1,1,2,2-tetrachloroethane	[79-34-5]
dichloroethyl ether	[111-44-4]	tetrachloroethylene	[127-18-4]
1,3-dichloropropene	[542-75-6]	titanium tetrachloride	[7550-45-0]
dichlorvos	[62-73-7]	toluene	[108-88-3]
diethanolamine	[111-42-2]	2,4-toluenediamine	[95-80-7]
N,N-diethylaniline	[121-69-7]	2,4-toluene diisocyanate	[584-84-9]
diethyl sulfate	[64-67-5]	*o*-toluidine	[95-53-4]
3,3-dimethoxybenzidine	[119-90-4]	toxaphene	[8001-35-2]
dimethylaminoazobenzene	[60-11-7]	1,2,4-trichlorobenzene	[120-82-1]
3,3′-dimethylbenzidine	[119-93-7]	1,1,2-trichloroethane	[79-00-5]
dimethylcarbamoyl chloride	[79-44-7]	trichloroethylene	[79-01-6]
dimethylformamide	[68-12-2]	2,4,5-trichlorophenol	[95-95-4]
1,1-dimethyl hydrazine	[57-14-7]	2,4,6-trichlorophenol	[88-06-2]
dimethyl phthalate	[131-11-3]	trimethylamine	[121-44-8]
dimethyl sulfate	[77-78-1]	trifluralin	[1582-09-8]
4,6-dinitro-*o*-cresol, and salts	[534-52-1]	2,2,4-trimethylpentane	[540-84-1]
2,4-dinitrophenol	[51-28-5]	vinyl acetate	[108-05-4]
2,4-dinitrotoluene	[121-14-2]	vinyl bromide	[593-60-2]
1,4-dioxane	[123-91-1]	vinyl chloride	[75-01-4]
1,2-diphenylhydrazine	[122-66-7]	vinylidene chloride	[75-35-4]
epichlorohydrin	[106-89-8]	xylenes (isomers and mixture)	[1330-20-7]
1,2-epoxybutane	[106-88-7]	*o*-xylenes	[95-47-6]
ethyl acrylate	[140-88-5]	*m*-xylenes	[108-38-3]
ethyl benzene	[100-41-4]	*p*-xylenes	[106-42-3]
ethyl carbamate	[51-79-6]		
ethyl chloride	[75-00-3]	antimony compounds	
ethylene dibromide	[106-93-4]	arsenic compounds	
ethylene dichloride	[107-06-2]	beryllium compounds	
ethylene glycol	[107-21-1]	cadmium compounds	
ethyleneimine	[151-56-4]	chromium compounds	
ethylene oxide	[75-21-8]	cobalt compounds	
ethylene thiourea	[96-45-7]	coke oven emissions	
ethylidene dichloride	[75-34-3]	cyanide compounds	
formaldehyde	[50-00-0]	glycol ethers	
heptachlor	[76-44-8]	lead compounds	
hexachlorobenzene	[118-74-1]	manganese compounds	
hexachlorobutadiene	[87-68-3]	mercury compounds	
hexachlorocyclopentadiene	[77-47-4]	fine mineral fibers	
hexachloroethane	[67-72-1]	nickel compounds	
hexamethyl-1,6-diisocyanate	[822-06-0]	polycyclic organic matter	
hexamethylphosphoroamide	[680-31-9]	radionuclides (including radon)	
hexane	[110-54-3]	selenium compounds	

against visibility impairment in these areas. In 1981, EPA directed 36 states to amend their State Implementation Plans to develop control programs for visual impairment that could be traced to particular sources. This type of impairment is called plume blight, and it was the initial focus of EPA's effort because it involved easily identifiable sources. The 1990 Clean Air Act Amendments direct EPA to promulgate appropriate regulations to address regional haze in affected Class I areas. EPA has not dealt with a third type of visibility impairment, urban-scale haze, because the source-receptor relationships are extremely complex (41).

Visibility or visual range is the maximum distance at which a black object, a target, can be distinguished from the horizon. Under certain viewing conditions, the apparent contrast (C) between a target and the horizon decreases exponentially with the distance (x) between the target and observer (42)

$$C = C_0 e^{-b_{\text{ext}} x} \tag{20}$$

where C_0 is the contrast at $x = 0$, and b_{ext} is the extinction coefficient, a proportionality constant relating the intensity of light received by an observer from a target to the intensity of light emitted by the target. The maximum distance, $x_{\max}$, at which an observer can distinguish the target from the horizon occurs when $C = \epsilon$, where ϵ is the observer's contrast threshold. Substituting in equation 20 and rearranging:

$$x_{\max} = \frac{\ln C_0 - \ln \epsilon}{b_{\text{ext}}} = V \tag{21}$$

where V is the visibility. For a black target, $C_0 = -1$ and experiments have shown that ϵ is between 0.02 and 0.05. Using the more sensitive value for ϵ, equation 21 reduces to (43)

$$V = 3.9/b_{\text{ext}} \tag{22}$$

Equation 22 only applies under ideal conditions, ie, black target against a bright horizon, and a well-illuminated, homogeneous atmosphere. Actual conditions, such as a nonblack target, which alters the value of C_0, different viewing angles, and different illumination conditions such as cloud cover and different sun angles, are not described so simply, but they can have a profound effect on visual range. Nevertheless, b_{ext} is a useful indicator of the inverse of visual range. It is widely used as an indicator of visual air quality. The total extinction can be written as the sum of a number of components:

$$b_{\text{ext}} = b_{\text{sp}} + b_{\text{R}} + b_{\text{ap}} + b_{\text{ag}} \tag{23}$$

where b_{sp} is the light extinction due to light scattered by particles; b_{R} is Rayleigh scattering, the light scattered by air molecules, and is a function of the atmospheric pressure; b_{ap} is the light absorbed by particles; and b_{ag} is the light absorbed by gas molecules.

Rayleigh scattering accounts for only a minor part of the extinction, except on the clearest days. It is a function of atmospheric pressure alone and does not depend appreciably on the composition of the pollutant gases present. At sea level and 25°C, b_R is equal to 13.2×10^{-6} m^{-1} which, in the absence of particles and absorbing gases, corresponds to a visual range of about 300 km. Light extinction budgets (excluding b_R) for several areas of the United States are presented in Table 7.

Table 7. Light Extinction Budgets, excluding b_R, Mean Percentage Contribution

	Location		
Component	Denver[a]	Detroit[b]	Blue Ridge Mts.[c]
b_{ap}	29	8	3
b_{sp}	64	88	97
b_{ag}	7	4	$<<1$
Total	100	100	100

[a]Ref. 44.
[b]Ref. 30.
[c]Ref. 33.

In general, visibility limitations are dominated by light scattering from fine particles. The most efficient are those that are the same size (0.4–0.7 μm) as the wavelengths of visible light. As shown in Figure 5, a peak in particulate mass distribution occurs in this range, and therefore, these particles almost always dominate b_{sp}. Exceptions occur during fog, precipitation, and dust storms. On a per mass basis, the most efficient light-scattering fine particles are hygroscopic ones. Sulfate, nitrate, and ammonium particles absorb significant amounts of water at moderate to high relative humidities. As the particles increase in size, they become more efficient light scatterers. Light absorption by particles in the atmosphere results almost exclusively from elemental carbon which also scatters light. The only common light-absorbing gaseous pollutant is NO_2, which usually accounts for a few percent or less of the total extinction.

There are three scales of visual impairment: plume blight, urban-scale haze, and regional-scale haze. Plume blight occurs when a plume from a large point source travels into an otherwise clean area and interferes with viewing a particular vista. Such events can occur anywhere, but are usually most noticeable in the western United States. Plume blight is frequently associated with sulfate particles from a sulfur dioxide-emitting point source. Most large urban areas occasionally experience urban haze, but the public perception of the haze is highest in those cities having scenic mountain vistas such as Los Angeles and Denver. Most of the light extinction in urban haze can be accounted for by particles of sulfates (sulfuric acid and the ammonium salts), nitrate (as ammonium nitrate), organic carbon, and elemental carbon.

Regional haze, a haze that extends for hundreds of miles, is usually dominated by sulfates. In the Southwest, for example, occasional haze obscures scenic vistas over large areas. It is attributed to a combination of sulfates from coal-fired power plant and smelter emissions, and transport of urban, southern California haze composed mainly of carbonaceous particles, nitrates and, to a lesser extent, sulfates from southern California. In the East, a denser sulfate-dominated haze frequently extends over much of the area during the summer. In the rural West, mean sulfate concentrations are ~ 1 $\mu g/m^3$, whereas, in the rural East, they average ~ 8 $\mu g/m^3$; also, the relative humidity is generally much higher in the East. The primary source of the eastern haze is coal-burning emissions. Natural haze, caused mainly by aerosols generated from biogenic VOC emissions from vegetation, was historically cited as the regional haze in areas such as the Blue Ridge and Smoky Mountain Regions. Except on the cleanest days, however, sulfate haze now dominates natural haze in these regions (33).

Air pollutants can also cause discolorations of the atmosphere. The most common are brownish discolorations such as the "brown LA haze" and the "brown clouds" observed in Denver and elsewhere. Three factors can contribute to the brown tint. The first is the presence of nitrogen dioxide, a brownish gas, most commonly viewed in a plume. In the urban hazes, however, discoloration from NO_2 is usually overwhelmed by the effects caused by particles. Since fine particles preferentially scatter blue light in the forward direction, the light viewed through an optically thin cloud as the sun is behind the observer is deficient in the blue wavelengths and appears brown. In dense haze clouds, the preferential scattering is masked by multiple-scattering effects and the haze is seen as white. However, a cloud can appear brown along its edges where it is optically thin. If the cloud is between the observer and the sun, it appears as white, but a dense cloud in the distance against a background of a bright blue sky can appear brown through chromatic adaptation. In chromatic adaptation, the blue receptors in the human eye are desensitized by the bright blue background; as a result, the white light from the haze appears to be brown.

Acid Deposition.   Acid deposition, the deposition of acids from the atmosphere to the surface of the earth, can be dry or wet. Dry deposition involves acid gases or their precursors or acid particles coming in contact with the earth's surface and thence being retained. The principal species associated with dry acid deposition are $SO_2(g)$, acid sulfate particles, ie, H_2SO_4 and NH_4HSO_4, and $HNO_3(g)$. Measurements of dry deposition are quite sparse, however, and usually only speciated as total SO_4^{2-} and total NO_3^-. In general, dry acid deposition is estimated to be a small fraction of the total because most of the dry deposited material has been neutralized by basic gases and particles in the atmosphere. The sulfate and nitrate resulting from dry deposition, however, is estimated to be a significant fraction of the total SO_4^{2-} and NO_3^- deposition. Current spatial and temporal dry deposition data are insufficient for specific estimates. On the other hand, there are abundant data on wet acid deposition. Wet acid deposition, acid precipitation, is the process by which acids are deposited by rain or snow. The principal dissolved acids are H_2SO_4 and HNO_3. Other acids such as HCl and organic acids usually account for only a minor part of the acidity although organic acids can be significant contributors in remote areas.

Both acid particles and gases can be incorporated into cloud droplets. Particles are incorporated into droplets by: nucleation, Brownian diffusion, impaction, diffusiophoresis (transport into the droplet induced by the flux of water vapor to the same surface), thermophoresis (thermally induced transport to a cooler surface), and electrostatic transport. Advective and diffusive attachment dominate all other mechanisms for pollutant gas uptake by cloud droplets. Modeling and experimental evidence suggests that most of the H_2SO_4 is formed in cloud water droplets. SO_2 diffuses into the droplet and is oxidized to H_2SO_4 by one of several mechanisms. At a pH greater than about 5.5, oxidation of SO_2 by dissolved O_3 is the dominant reaction. At lower pH, SO_2 oxidation is dominated by the reaction with H_2O_2. Under some conditions, oxidation by O_2, catalyzed by metals or soot, may contribute to the formation of H_2SO_4. Most of the HNO_3 in precipitation results from HNO_3 that diffuses into the droplet. However, there is also observational evidence that some HNO_3 is formed in the droplets, but the mechanism has not been identified.

The pH of rainwater in equilibrium with atmospheric CO_2 is 5.6, a value frequently cited as the natural background pH. However, in the presence of other naturally occurring species such as SO_2, SO_4^{2-}, NH_3, organic acids, sea salt, and alkaline crustal dust, the natural values of unpolluted rainwater vary between 4.9 and 6.5 depending upon time and location. Across the United States, the mean annual average precipitation pH varies from 4.2 in western Pennsylvania to 5.7 in the West (see Figure 6). In general, precipitation of the lowest pH occurs in the summer. Precipitation pH is generally lowest in the eastern United States within and downwind of the largest SO_2 and NO_x emissions areas. In the East, SO_4^{2-} concentrations in precipitation are 1.5 to 2.5 times higher during the summer than in winter, but the NO_3^- values are about the same year round. Consequently, the lower pH in summer precipitation results mostly from the higher SO_4^{2-} concentrations. The equivalent ratio of sulfate to nitrate in precipitation where 1.0 represents equal concentrations is an inexact measure of the relative contribution of these two species to the acidity. In the East during the winter, this ratio ranges

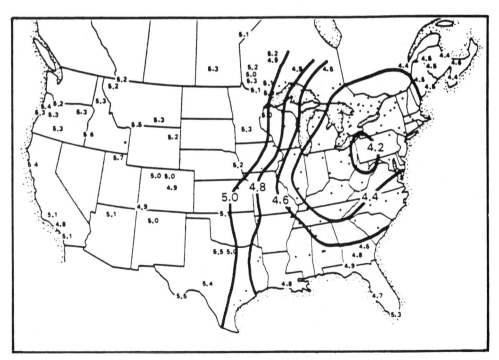

Fig. 6.   1985 annual precipitation-weighted pH (45).

from 1.0 to 2.5; during the summer, it ranges from 2.0 to 3.0. On the average in the eastern United States, about 60% of the wet-deposited acidity can be attributed to SO_4^{2-} and 40% to NO_3^- (45).

Since SO_2 and NO_2 are criteria pollutants, their emissions are regulated. In addition, for the purposes of abating acid deposition in the United States, the 1990 Clean Air Act Amendments require that nationwide SO_2 and NO_x emissions be reduced by approximately 10 million and 2 million t/yr, respectively, by the year 2000. Reasons for these reductions are based on concerns which include acidification of lakes and streams, acidification of poorly buffered soils, and acid damage to materials. An additional major concern is that acid deposition is contributing to the die-back of forests at high elevations in the eastern United States and in Europe.

Global Warming (The Greenhouse Effect).   Solar energy, mostly in the form of visible light, is absorbed by the earth's surface and reemitted as longer wavelength infrared radiation. Certain gases in the atmosphere, primarily water vapor and to a lesser degree CO_2, have the ability to absorb the outgoing ir radiation and translate it to heat. The result, a higher atmospheric equilibrium temperature than would occur in the absence of these molecules, is called the greenhouse effect. Gases that have the ability to absorb radiation and bring about temperature enhancement are called greenhouse gases. Without atmospheric water vapor and CO_2, the earth's mean surface temperature would be $-18°C$ instead of the present 17°C. However, there is concern that increasing concentrations of CO_2 and other trace greenhouse gases that result from human activities will enhance the greenhouse effect and cause global warming. Scenarios that could then result include: an alteration in existing precipitation patterns, an increase in the severity of storms, the dislocation of suitable land for agriculture, the dislocation and possible extinction of certain biological species and ecosystems, and the flooding of many coastal areas because of rising sea levels resulting from the thermal expansion of the oceans, the melting of glaciers, and, probably less so, from the melting of polar ice caps.

Measurements since 1958 clearly show that atmospheric CO_2 concentrations are increasing at the rate of about 0.3%/yr. The present concentration is ~ 350 ppm compared to the preindustrial revolution (1800) value of 285 ppm that has been estimated from ice cores. Projections based on the current rate of increase and future energy uses show that the CO_2 concentration will approach 600 ppm some time in the middle of the 21st century. In the 1970s and 1980s, the concentrations of other greenhouse gases were also discovered to be increasing. These gases include methane, nitrous oxide, tropospheric O_3, and a variety of CFCs. The greenhouse gases (not including water), concentrations, rates of increase in the atmosphere, and estimates of relative greenhouse efficiencies and atmospheric residence times are presented in Table 8. Three important features are evident. First, CO_2 is by far the most abundant anthropogenic greenhouse gas. Second, all of the other greenhouse gases are much more efficient ir-absorbers than CO_2. Third, most of the gases have very long atmospheric residence times so that even if emissions were to cease, the gases would remain in the atmosphere for decades (some for centuries). The major sources of greenhouse gases are summarized in Table 9.

From the analyses of air trapped in Antarctic and Greenland ice, the concen-

Table 8. Summary of Important Greenhouse Gases[a]

Gas	CAS Registry Number	1990 Concentrations	Concentration increases, %/yr	Greenhouse efficiency[b]	Atmospheric residence times, yrs[c]
CO_2	[124-38-9]	350 ppm	0.3	1	120[d]
CH_4	[94-82-8]	1.68 ppm	0.8	25	7–10
N_2O	[10024-97-2]	340 ppb	0.2	250	~150
O_3	[10028-15-6]	40 ppb[e]	f	3000[g]	0.4
$CFCl_3$ (CFC-11)	[75-69-4]	226 ppt	4	17500	75
CF_2Cl_2 (CFC-12)	[75-71-8]	392 ppt	4	20000	120
$CHClF_2$ (HCFC-22)	[75-45-6]	100 ppt	7	7500	20
$C_2H_2F_4$ (HFC-134a)	[811-97-2]			11600[h]	15.5[h]
$C_2Cl_3F_3$ (CFC-113)	[76-13-1]	30–70 ppt	11	22500[h]	90
CH_3CCl_3 (methylchloroform)	[71-55-6]	125 ppt	7	2500	5.5–10
CCl_4	[56-23-5]	75–100 ppt	1	12500	50

[a]Ref. 46 unless otherwise indicated.
[b]Relative to CO_2 on a volume basis. A more useful way to compare the relative warming potentials is to consider the lifetimes of the species and integrate the efficiencies over a fixed time interval, usually 100 years (see Ref. 8).
[c]Residence times may vary slightly from those in Table 2 because of the differences in the primary source.
[d]Ref. 8.
[e]Mean tropospheric background for O_3 is given. Value can be much higher in polluted areas.
[f]Small positive trend is evident (see Ref. 47).
[g]Estimated from Ref. 48.
[h]Ref. 49.

Table 9. Sources of Greenhouse Gases[a]

Gases	Sources
CO_2	fossil fuel combustion, deforestation, oceans, respiration
CH_4	wetlands, rice paddies, enteric fermentation (animals), biomass burning, termites
N_2O	natural soils, cultivated and fertilized soils, oceans, fossil fuel combustion
O_3	photochemical reactions in the troposphere, transport from stratosphere
CFC-11	use in manufacturing of foam, aerosol propellant
CFC-12	use as refrigerant, aerosol propellant, manufacturing of foams
HCFC-22	use as refrigerant, production of fluoropolymers
CFC-113	use as electronics solvent
CH_3CCl_3	use as industrial degreasing solvent
CCl_4	use as intermediate in production of CFC-11, -12, solvent

[a]Ref. 46.

trations of greenhouse gases, except for O_3, in the preindustrial atmosphere (averaged over ~1000 yr) can be estimated. The enhancement of the greenhouse effect as a result of current concentrations of gases relative to preindustrial concentrations is called the enhanced greenhouse effect or radiative forcing. Using a radiative convective model, the contributions from the various gases to the radiative forcing in the 1980s can be estimated (6). Such estimates are shown in Figure 7 where CO_2 is shown to account for about half of the radiative forcing. The relative contribution from CO_2 has been shrinking and is expected to continue to do so. Other more efficient ir-absorber species are increasing in concentration faster than CO_2.

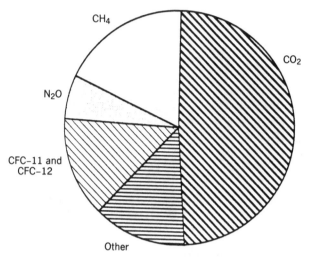

Fig. 7.   Estimates of greenhouse-gas contributions to global warming in the 1980s. Percentages of total contributions are: CO_2, 49; CH_4, 18; CFC–11 and CFC–12, 14; N_2O, 6; and others, 13 (50).

Although there is no doubt that greenhouse gas concentrations and the radiative forcing are increasing, there is no unequivocal evidence that this forcing is actually causing a net warming of the earth. Analyses of global temperature trends since the 1860s show that the global temperature has increased about 0.5–0.7°C (51,52), but this number decreases to ~0.5°C when corrections for heat-island effects are considered (53). This is in reasonable agreement with modeling results which predict a temperature increase of ~1°C (54). However, most of the temperature increase occurred prior to the increases in CO_2 (55). A detailed analysis of global temperature and CO_2 concentration time-series from 1958 to 1988, the period of atmospheric CO_2 measurements, shows an excellent positive correlation, but CO_2 changes lag temperature change by an average of 5 months (56). Thus, although there is strong evidence linking temperature and CO_2 changes, the cause and effect has not been demonstrated, and it is not clear which is the cause and which is the effect. The lack of a definitive relationship may also be obscured by changes in other factors that affect the earth's heat budget such as increased atmospheric aerosols or cloud cover and natural climatic cycles. Global circulation models (GCMs) predict that the average global temperature will increase from 2.0 to 5.5°C as CO_2 concentrations double from those of pre-industrialized levels (57,58). The temperature increases are not expected to be uniformly distributed. The uncertainties in these models are large because many of the important feedback processes involving oceans and clouds are not adequately understood to properly incorporate them into the model system (see ATMOSPHERIC MODELS).

Stratospheric O_3 Depletion. In the stratosphere, O_3 is formed naturally when O_2 is dissociated by uv solar radiation in the region 180–240 nm:

$$O_2 + uv \longrightarrow O + O \tag{24}$$

and the atomic oxygen then reacts with molecular oxygen according to equation 2. Ultraviolet radiation in the 200–300 nm region can also dissociate O_3:

$$O_3 + uv \longrightarrow O_2 + O \tag{25}$$

Equation 25 represents the reaction responsible for the removal of uv-B radiation (280–330 nm) that would otherwise reach the earth's surface. There is concern that any process that depletes stratospheric ozone will consequently increase uv-B (in the 293–320 nm region) reaching the surface. Increased uv-B is expected to lead to increased incidence of skin cancer and it could have deleterious effects on certain ecosystems. The first concern over O_3 depletion was from NO_x emissions from a fleet of supersonic transport aircraft that would fly through the stratosphere and cause reactions according to equations 3 and 26 (59):

$$NO_2 + O \longrightarrow NO + O_2 \tag{26}$$

The net effect of this sequence is the destruction of 2 molecules of O_3 as the one is lost in NO_2 formation and the O of equation 26 would have combined with O_2 to form the other. In addition, the NO acts as a catalyst. It is not consumed, and therefore can participate in the reaction sequence many times.

In the mid-1970s, it was realized that the CFCs in widespread use because of their chemical inertness, would diffuse unaltered through the troposphere and into the mid-stratosphere where they, too, would be photolyzed by uv (<240 nm) radiation. For example, CFC-12 can photolyze:

$$CF_2Cl_2 + uv \longrightarrow CF_2Cl + Cl \tag{27}$$

$$CF_2Cl + O_2 \longrightarrow CF_2O + ClO \tag{28}$$

forming Cl and ClO radicals which then react with ozone and O:

$$Cl + O_3 \longrightarrow ClO + O_2 \tag{29}$$

$$ClO + O \longrightarrow Cl + O_2 \tag{30}$$

In this sequence the Cl also acts as a catalyst and two O_3 molecules are destroyed. It is estimated that before the Cl is finally removed from the atmosphere in 1–2 yr by precipitation, each Cl atom will have destroyed approximately 100,000 O_3 molecules (60). The estimated O_3-depletion potential of some common CFCs, hydrofluorocarbons, HFCs, and hydrochlorofluorocarbons, HCFCs, are presented in Table 10. The O_3-depletion potential is defined as the ratio of the emission rate of a compound required to produce a steady-state O_3 depletion of 1% to the amount of CFC-11 required to produce the 1% depletion. The halons, bromo-chlorofluorocarbons or bromofluorocarbons that are widely used in fire extin-

Table 10. Ozone Depletion Potential Relative to CFC-11[a]

Compound	CAS Registry Number	Relative ozone depletion potential
$CFCl_3$ (CFC-11)	[75-69-11]	1.0
$CFCl_2$ (CFC-12)	[75-71-8]	0.87
$C_2Cl_3F_3$ (CFC-113)	[76-13-1]	0.76
(CFC-114)		0.56
C_2ClF_5 (CFC-115)	[76-15-3]	0.27
$CHClF_2$ (HCFC-22)	[75-45-6]	0.043
$C_2HCl_2F_3$ (HCFC-123)	[306-83-2]	0.016
C_2HClF_4 (HFCF-124)	[2837-89-0]	0.017
C_2HF_5 (HCFC-125)	[354-33-6]	0
$C_2H_2F_4$ (HFC-134a)	[811-97-2]	0
CCl_4	[56-23-5]	1.1
CH_3CCl_3	[71-55-6]	0.12

[a]Ref. 61. Results are based on a Lawrence Livermore National Laboratory 2-Dimensional Model.

guishers, are also ozone-depleting compounds. Although halon emissions, and thus the atmospheric concentrations, are much lower than the most common CFCs, halons are of concern because they are from three to ten times more destructive to O_3 than the CFCs.

The strongest evidence that stratospheric O_3 depletion is occurring comes from the discovery of the Antarctic ozone hole. In recent years during the spring, O_3 depletions of 60% integrated over all altitudes and 95% in some layers have been observed over Antarctica. During winter in the southern hemisphere, a polar vortex develops that prevents the air from outside of the vortex from mixing with air inside the vortex. The depletion begins in August, as the approaching spring sun penetrates into the polar atmosphere, and extends into October. When the hole was first observed, existing chemical models could not account for the rapid O_3 loss, but attention was soon focused on stable reservoir species for chlorine. These compounds, namely HCl and $ClNO_3$, are formed in competing reactions involving Cl and ClO that temporarily or permanently remove Cl and ClO from participating in the O_3 destruction reactions. For example,

$$Cl + CH_4 \longrightarrow HCl + CH_3 \tag{31}$$

$$ClO + NO_2 + M \longrightarrow ClNO_3 + M \tag{32}$$

where M is again any third body molecule, remaining unchanged, in the atmosphere. Within the polar vortex, temperatures as low as $-90°C$ allow the formation of polar stratospheric ice clouds. On the surfaces of the ice particles that compose these clouds, heterogeneous reactions which break down these reservoir species occur. Two important ones are

$$ClNO_3 + HCl(s) \longrightarrow Cl_2 + HNO_3(s) \tag{33}$$

$$H_2O(s) + ClNO_3 \longrightarrow HOCl + HNO_3(s) \tag{34}$$

During the polar winter night, Cl_2, HOCl, and HNO_3(s) accumulate. When sunlight returns to the polar regions, the chlorine compounds are photolyzed producing Cl and ClO. Nitrogen oxides remain sequestered, and without NO_2 to deplete ClO, massive O_3 destruction occurs until the polar vortex dissipates later in the spring and the mixing of air from lower latitudes occurs.

There is additional evidence that stratospheric O_3 concentrations have declined an average of 2.5% globally from 1969 to 1986. After data was adjusted for known cycles that cause variations in the O_3, declines were most evident during winter months (62,63). The cause of this global decrease in stratospheric O_3 is under investigation.

In 1976 the United States banned the use of CFCs as aerosol propellants. No further steps were taken until 1987 when the United States and some 50 other countries adopted the Montreal Protocol, specifing a 50% reduction of fully halogenated CFCs by 1999. In 1990, an agreement was reached among 93 nations to accelerate the discontinuation of CFCs and completely eliminate production by the year 2000. The 1990 Clean Air Act Amendments contain a phaseout schedule for CFCs, halons, carbon tetrachloride, and methylchloroform. Such steps should stop the increase of CFCs in the atmosphere but, because of the long lifetimes, CFCs will remain in the atmosphere for centuries.

Indoor Air Pollution.   Indoor air pollution, the presence of air pollutants in indoor air, is of growing concern in offices and residential buildings. Pollution in the industrial environment has been monitored and regulated for some time (see INDUSTRIAL HYGIENE AND PLANT SAFETY). Partly in response to the "energy crisis" of the early 1970s, buildings are now constructed more air-tight. Unfortunately, air-tight structures create a setting conducive to the accumulation of indoor air pollutants. Numerous sources and types of pollutants found indoor can be classified into seven categories: tobacco smoke, radon, emissions from building materials, combustion products from inside the building, pollutants which infiltrate from outside the building, emissions from products used within the home, and biological pollutants. Concentrations of the pollutants depend upon strength of the indoor sources, the ventilation rate of the building, and the outdoor pollutant concentration.

Tobacco smoke contains a variety of air pollutants. In a survey of 80 homes in an area where the outdoor TSP varied between 10–30 $\mu g/m^3$, the indoor TSP was the same, or less, in homes having no smokers. In homes having one smoker, the TSP levels were between 30–60 $\mu g/m^3$, while in homes having two or more smokers, the levels were between 60–120 $\mu g/m^3$ (64). In other studies, indoor TSP levels exceeding 1000 $\mu g/m^3$ have been found in homes with numerous smokers. In addition to TSP, burning tobacco emits CO, NO_x, formaldehyde [50-00-0], benzopyrenes, nicotine [54-11-5], phenols, and some metals such as cadmium [7440-43-9] and arsenic [7440-38-2] (65).

Radon-222 [14859-67-7], ^{222}Rn, is a naturally occurring, inert, radioactive gas formed from the decay of radium-226 [13982-63-3], ^{226}Ra. Because Ra is a ubiquitous, water-soluble component of the earth's crust, its daughter product, Rn, is found everywhere. A major health concern is radon's radioactive decay products. Radon has a half-life of 4 days, decaying to polonium-218 [15422-74-9], ^{218}Po, with the emission of an α particle. It is ^{218}Po, an α-emitter having a half-life of 3 min, and polonium-214 [15735-67-8], ^{214}Po, an α-emitter having a half-life of 1.6 $\times$

10^{-4} s, that are of most concern. Polonium-218 decays to lead-214 [15067-28-4], ^{214}Pb, a β-emitter having $t_{1/2} = 27$ min, which decays to bismuth-214 [14733-03-0], ^{214}Bi, a β-emitter having $t_{1/2} = 20$ min, which decays to ^{214}Po. Radon is an inert gas that, when inhaled, is not retained in the lungs. But the Rn daughters, when inhaled, either by themselves or attached to an airborne particle, are retained and the subsequent α-emissions irradiate the surrounding lung tissue.

Radon can enter buildings through emissions from soil, water, or construction materials. The soil route is by far the most common, and construction material the least common, although there have been isolated incidents where construction materials contained high levels of Ra. The emission rate of Rn depends on the concentration of Ra in the soil, the porosity of the soil, and the permeability of the building's foundation. For example, Rn is transported faster through cracks and sumps in the basement floor than through concrete. In the ambient air Rn concentrations are typically 9.25–37 mBq/L, whereas the mean concentration in U.S. residences is about 44 mBq/L (66). However, it is estimated that there are 1 million residences that have concentrations exceeding 0.3 Bq/L or 300 mBq/L, which is the level for remedial action recommended by the National Council on Radiation Protection and Measurements (66). The highest values ever measured in U.S. homes exceeded 37 Bq/L (67). Remedial action consists of (1) reducing the transport of Rn into the building by sealing cracks with impervious fillers and installing plastic or other barriers that have proven effective; (2) removing the daughters from the air by filtration; and (3) increasing the infiltration of outside air using an air-exchanger system.

Of the pollutants emitted from construction materials within the home, asbestos (qv) [1332-21-4] has received the most attention. Asbestos is a generic term for a number of naturally occurring fibrous hydrated silicates. By EPA's definition, a fiber is a particle that possesses a 3:1 or greater aspect ratio (length:diameter). The family of asbestos minerals is divided into two types: serpentine [12168-92-2] and amphibole. One type of serpentine, chrysotile [12001-29-5], $Mg_6Si_4O_{10}(OH)_8$, accounts for 90% of the world's asbestos production. The balance of the world's production is accounted for by two of the amphiboles: amosite [12172-73-5], $Fe_5Mg_2(Si_8O_{22})(OH)_2$, and crocidolite [12001-28-4], $Na_2(Fe^{3+})_2(Fe^{2+})_3Si_8O_{22}(OH)_2$. Three other amphiboles, anthophyllite [77536-67-5], $(Mg,Fe)_7Si_8O_{22}(OH)_2$, tremolite [77536-68-6], $Ca_2Mg_5Si_8O_{22}(OH)_2$ and actinolite [77536-66-4], $Ca_2(Mg,Fe)_5Si_8O_{22}(OH)_2$, have been only rarely mined. The asbestos minerals differ in morphology, durability, range of fiber diameters, surface properties, and other attributes that determine uses and biological effects. Known by ancients as the magic mineral because of its ability to be woven into cloth, its physical strength, and its resistance to fire, enormous heat, and chemical attack, asbestos was incorporated into many common building products (68).

All forms of asbestos were implicated in early studies linking exposure to airborne fibers and asbestosis (pulmonary interstitial fibrosis), lung cancer, and mesothelioma (a rare form of cancer of the lung or abdomen). However, most of the asbestos-related diseases are now thought to result from exposure to airborne amphiboles rather than chrysotile, the most common asbestos type, and to fibers greater than or equal to 5 μm in length (69). In the 1970s, the spray-on application of asbestos was banned and substitutes were found for many products. Nevertheless, asbestos was used liberally in buildings for several decades, and many of them are still standing. Asbestos in building materials does not spontaneously shed fibers, but when the materials become damaged by normal decay, renovation, or demolition, the fibers can become airborne and contribute to the indoor air pollution problem.

Formaldehyde (qv) [50-00-0], HCHO, another important pollutant emanating from building material, is important because of irritant effects and suspected carcinogenicity. Traces of formaldehyde can be found in the air in virtually every modern home. Mobile homes and houses insulated using urea–formaldehyde [9011-05-6] foam, an efficient insulation material that can be injected into the sidewalls of conventional homes, have the highest concentrations. In 1982, use of the foam was banned in the U.S. Higher formaldehyde emissions can occur in mobile homes using particle board held together using an urea–formaldehyde resin. This can also be a problem in a conventional house, but it is usually exacerbated in a mobile home because of the low rate of air exchange. Plywood is also a source of formaldehyde as the layers of wood are held together using a similar urea-formaldehyde resin adhesive. In general, however, particle board contains more adhesive per unit mass, so the emissions are greater. Other sources of indoor formaldehyde are paper products, carpet backing, and some fabrics.

Whenever unvented combustion occurs indoors or when venting systems attached to combustion units malfunction, a variety of combustion products will be released to the indoor environment. Indoor combustion units include: nonelectric stoves and ovens, furnaces, hot water heaters, space heaters, and wood-burning fireplaces or stoves. Products of combustion include CO, NO, NO_2, fine particles, aldehydes, polynuclear aromatics, and other organic compounds. Especially dangerous sources are unvented gas and kerosene [8008-20-6] space heaters which discharge pollutants directly into the living space. The best way to prevent the accumulation of combustion products indoors is to make sure all units are properly vented and properly maintained.

Pollutants from outdoors can also be drawn inside under certain circumstances such as incorrectly locating an air intake vent downwind of a combustion exhaust stack. High outdoor pollutant concentrations can also infiltrate buildings. Unreactive pollutants like CO diffuse through any openings in a building and pass unaltered through any air-intake system. Given sufficient time, the indoor/outdoor ratio for CO approaches 1.0 if outside air is the only CO source. For reactive species such as ozone, which is destroyed on contact with most surfaces, the indoor/outdoor ratio is usually around 0.5, but this ratio varies considerably depending on the ventilation rate and the internal surface area within the building (70).

Air contaminants are emitted to the indoor air from a wide variety of activities and consumer products, some of which are summarized in Table 11. Most indoor activities produce some types of pollutants. When using volatile products or engaging in the activities listed, care should be exercised to minimize exposure through proper use of the product and by providing adequate ventilation.

Biological indoor air pollutants include airborne bacteria, viruses, fungi, spores, molds, algae, actinomycetes, and insect and plant parts. Microorganisms many of which multiply in the presence of high humidity, can produce infections, disease, or allergic reactions; the nonviable biological pollutants can produce

Table 11. Emission Mechanisms of Indoor Air Pollutants Arising from Activities and Consumer Products[a]

Activity or product	Aerosol production		Evaporation or sublimation	Unintentional outgassing
	Intentional	Unintentional		
cleaning	×	×	×	×
painting	×		×	×
polishing	×		×	×
stripping	×	×	×	
refinishing	×		×	×
hobbies, crafts	×	×	×	×
deodorizer	×		×	
insecticide	×		×	
disinfectant	×		×	
personal grooming product	×		×	×
plastic				×

[a]Ref. 71.

allergic reactions. The most notable episode was the 1976 outbreak of Legionella (Legionnaires') disease in Philadelphia where the American Legion convention attendees inhaled Legionella virus from a contaminated central air conditioning system. A similar incident in an industrial environment occurred in 1981: more than 300 workers came down with "Pontiac fever" as a result of inhalation exposure to a similar virus aerosolized from contaminated machining fluids (72). Preventive maintenance of air management systems and increased ventilation rates reduce the concentrations of all species, and should consequently reduce the incidence of adverse affects.

BIBLIOGRAPHY

"Smokes and Fumes" in *ECT* 1st ed., Vol. 12, pp. 558–573, by G. P. Larson, Air Pollution Control District of Los Angeles County; "Smokes, Fumes, and Smog" in *ECT* 2nd ed., Vol. 18, pp. 400–415, by R. B. Engdahl, Battelle Memorial Institute; "Pollution—Air Pollution" in *ECT* 2nd ed., Supplement Vol., pp. 730–737, by G. P. Sutton, Envirotech Corp.; "Air Pollution" in *ECT* 3rd ed., Vol. 1, pp. 624–649 by P. R. Sticksel, R. B. Engdahl, Battelle Memorial Institute.

1. *Procedures for Emission Inventory Preparation, Vol. 1–4,* Pub. No. EPA 450–4–81–026A–E, U.S. Environmental Protection Agency, Research Triangle Park, N.C., 1981.
2. *Compilation of Air Pollution Emission Factors,* Pub. No. AP–42, 5th ed., U.S. Environmental Protection Agency, Research Triangle Park, N.C., 1989.
3. *Compilation of Air Pollution Emission Factors, Vol. 2. Mobile Sources,* Pub. No. AP–42, 5th ed., U.S. Environmental Protection Agency, Research Triangle Park, N.C., 1989.
4. *National Air Quality & Emissions Trends Report, 1989,* Pub. No. EPA-450/4-91-003, U.S. Environmental Protection Agency, Research Triangle Park, N.C., 1991.
5. *National Air Pollutant Emissions Estimates 1940–1986,* Pub. No. EPA-450/4-87-024, U.S. Environmental Protection Agency, Research Triangle Park, N.C., 1988.

6. V. Ramanathan and co-workers, *Rev. Geophys.* **25,** 1441 (1987).

7. P. Warneck, *Chemistry of Natural Atmospheres,* Academic Press, New York, 1988, p. 367.

8. Intergovernmental Panel on Climate Change, *Scientific Assessment of Climate Change, Section 2, Radiative Forcing of Climate,* United Nations, New York, 1990, p. 14.

9. G. T. Wolff and P. J. Lioy, *Environ. Sci. Technol.* **14,** 1257 (1980).

10. G. T. Wolff, N. A. Kelly, and M. A. Ferman, *Science* **311,** 703 (1981).

11. J. P. Lodge, ed., *Methods of Air Sampling and Analysis,* Lewis Publishers, Chelsea, Mich., 1989, 763 pp.

12. *Air Quality Criteria for Oxone and Other Photochemical Oxidants,* Publication No. EPA-600-8-84-020F, 5 vols., U.S. Environmental Protection Agency, Research Triangle Park, N.C., 1986. EPA publishes separate criteria documents for all the criteria pollutants and they are updated about every five years.

13. A. M. Hough and R. G. Derwent, *Nature* **344,** 645 (1990).

14. N. A. Kelly, G. T. Wolff, and M. A. Ferman, *Atmos. Environ.* **16,** 1077 (1978).

15. W. Johnson and W. Viezee, *Atmos. Environ.* **15,** 1309 (1981).

16. W. Attmannspacher and R. Hartmannsgruber, *Pure Appl. Geophys.* **106–108,** 1091 (1973).

17. R. L. Seila, W. A. Lonneman, and S. A. Meeks, *Determination of C_2 to C_{12} Ambient Air Hydrocarbons in 39 U.S. Cities from 1984 through 1986,* Pub. No. EPA/600/3-89/058, U.S. Environmental Protection Agency, Research Triangle Park, N.C., 1989.

18. P. Warneck, *Chemistry of Natural Atmospheres,* Academic Press, New York, 1988, pp. 721–729.

19. W. N. Stasiuk, Jr., and P. E. Coffey, *J. Air Pollut. Control Assoc.* **24,** 564 (1974).

20. G. T. Wolff and co-workers, *Environ. Sci. Technol.* **11,** 506 (1977).

21. A. M. Dunker, S. Kumar, and P. H. Berzins, *Atmos. Environ.* **18,** 311 (1984).

22. N. A. Kelly and R. G. Gunst, *Atmos. Environ.* **24A,** 2991 (1990).

23. *Catching Our Breath. Next Steps for Reducing Urban Ozone,* U.S. Office of Technology Assessment, Washington, D.C., 1989, pp. 101–102.

24. G. T. Wolff, P. J. Lioy, G. D. Wight, and R. E. Pasceri, *J. Air Pollut. Control Assoc.* **27,** 460 (1977).

25. *An Acid Aerosols Issue Paper,* Pub. No. EPA/600/8-88/005F, U.S. Environmental Protection Agency, Washington, D.C., 1989.

26. G. T. Wolff, *Ann. NY Acad. Sci.* **338,** 379 (1980).

27. K. Willeke and K. T. Whitby, *J. Air Pollut. Control Assoc.* **25,** 529 (1975).

28. R. J. Countess, G. T. Wolff, and S. H. Cadle, *J. Air Pollut. Control Assoc.* **30,** 1195 (1980).

29. G. T. Wolff and co-workers, *Atmos. Environ.* **19,** 305 (1985).

30. G. T. Wolff and co-workers, *J. Air Pollut. Control Assoc.* **32,** 1216 (1982).

31. G. T. Wolff, M. S. Ruthkosky, D. P. Stroup, and P. E. Korsog, *Atmos. Environ.* **25A** (1991).

32. G. T. Wolff and co-workers, *J. Air Waste Mgt. Assoc.* **40,** 1638 (1990).

33. M. A. Ferman, G. T. Wolff, and N. A. Kelly, *J. Air. Pollut. Control Assoc.* **31,** 1074 (1981).

34. G. T. Wolff and co-workers, *J. Air Pollut. Control Assoc.* **36,** 585 (1986).

35. G. T. Wolff and co-workers, *Atmos. Environ.* **20,** 1229 (1986).

36. M. S. Miller, S. K. Friedlander, and G. M. Hidy in G. M. Hidy, ed., *Aerosol and Atmospheric Chemistry,* Academic Press, New York, 1972, pp. 301–312.

37. P. W. Fisher, R. M. Currie, and R. J. Churchill, *J. Air Pollut. Control Assoc.* **38,** 1376 (1988).

38. *Threshold Limit Values and Biological Exposure Indices,* American Conference of Governmental and Industrial Hygienists, Cincinnati, Ohio, 1989, p. 124.

39. N. I. Sax, *Dangerous Properties of Industrial Materials,* Van Nostrand Reinhold, New York, 1979, p. 1108.
40. J. A. Cannon, *J. Air Pollut. Control Assoc.* **36,** 562 (1986).
41. J. C. Mesta in P. S. Bhardwaja, ed., *Visibility Protection Research and Policy Aspects,* Air and Waste Management Association, Pittsburgh, Pa., 1987, pp. 1–8.
42. W. E. Middleton, *Vision Through the Atmosphere,* University of Toronto Press, Toronto, Canada, 1968.
43. H. Koschmieder, *Beitr. Phys. Frein Atm.* **12,** 33 (1924).
44. P. J. Groblicki, G. T. Wolff, and R. J. Countess, *Atmos. Environ.* **15,** 2473 (1981).
45. D. Albritton and co-workers, *NAPAP Interim Assessment,* Vol. 3 (Atmosphere Process and Deposition), National Acid Precipitation Assessment Program, Washington, D.C., 1987.
46. *Policy Options for Stabilizing Global Climate,* U.S. Environmental Protection Agency, Washington, D.C., 1990.
47. A. M. Hough and G. Derwent, *Nature* **344,** 645 (1990).
48. J. Fishman, V. Ramanathan, P. J. Crutzen, and S. C. Liu, *Nature* **282,** 818 (1979).
49. D. A. Fisher and co-workers, *Nature* **344,** 513 (1990).
50. J. Hansen and co-workers, *J. Geophys. Res.* **93,** 9341 (1988).
51. P. D. Jones, T. M. L. Wigley, and P. B. Wright, *Nature* **322,** 430 (1986).
52. J. Hansen and S. Lebedeff, *J. Geophys. Res. D11,* **13,** 345 (1987).
53. T. R. Karl and P. D. Jones, *Bull. Am. Meteorol. Soc.* **70,** 265 (1989).
54. V. Ramanathan, *Science* **240,** 293 (1988).
55. R. S. Lindzen, *Bull. Am. Meteorol. Soc.* **71,** 288 (1990).
56. C. Kuo, C. Lindberg, and D. J. Thomson, *Nature* **343,** 709 (1990).
57. S. H. Schneider, *Sci. Am.* **261,** 70 (Sept. 1989).
58. J. F. B. Mitchell, C. A. Senior, and W. J. Ingram, *Nature* **341,** 132 (1989).
59. P. J. Crutzen, *Q. J. Royal Meteorol. Soc.* **96,** 320 (1970).
60. M. J. Molina and F. S. Rowland, *Nature* **249,** 810 (1974).
61. D. A. Fisher and co-workers, *Nature* **344,** 508 (1990).
62. M. B. McElroy and R. J. Salawitch, *Science* **243,** 763 (1989).
63. F. S. Rowland, *Am. Sci.* **77,** 36 (1989).
64. J. D. Spengler and co-workers, *Atmos. Environ.* **15,** 23 (1981).
65. California Department of Consumer Affairs, *Clean Your Room, Compendium on Indoor Air Pollution,* Sacramento, Calif., 1982, p. III.EI–III.E.II.
66. Mueller Associates, Inc., Syscon Corporation, and Brookhaven National Laboratory, *Handbook of Radon in Buildings,* Hemisphere Publishing Corporation, New York, 1988, p. 95.
67. H. W. Alter and R. A. Oswald, *J. Air Pollut. Control Assoc.* **37,** 227 (1987).
68. P. Brodeur, *New Yorker* **44,** 117 (Oct. 12, 1968).
69. B. T. Mossman and co-workers, *Science* **247,** 294(1990).
70. C. J. Weschler, H. C. Shields, and D. V. Naik, *J. Air Waste Manage. Assoc.* **39,** 1562 (1989).
71. *Indoor Air Pollutants,* National Academy Press, Washington, D.C., 1981, p. 101.
72. L. A. Herwaldt and co-workers, *Ann. Intern. Med.* **100,** 333 (1984).

General References

References 4, 5, 6, 7, 8, 11, 12, 45, 56, 61, 64, and 70 and the following books and reports constitute an excellent list for additional study. Reference 7 is an especially useful resource for global atmospheric chemistry.

J. H. Seinfeld, *Atmospheric Chemistry and Physics of Air Pollution,* John Wiley & Sons, Inc., New York, 1986.

B. J. Finlayson-Pitts and J. N. Pitts, Jr., *Atmospheric Chemistry Fundamentals and Experimental Techniques,* John Wiley & Sons, Inc., New York, 1986.

T. E. Graedel, D. T. Hawkins, and L. D. Claxton, *Atmospheric Chemical Compounds Sources, Occurrence and Bioassay,* Academic Press, New York, 1986.

Atmospheric Ozone 1985, World Meteorological Organization, Geneva, Switzerland (3 vols.); an excellent compendium on tropospheric and stratospheric processes.

G. T. Wolff, J. L. Hanisch, and K. Schere, eds., *The Scientific and Technical Issues Facing Post-1987 Ozone Control Strategies,* Air and Waste Management Association, Pittsburgh, Pa., 1988.

J. H. Seinfeld, "Urban Air Pollution: State of the Sciences," *Science* **243,** 745 (1989).

S. H. Schneider, "The Greenhouse Effect: Science and Policy," *Science* **243,** 771 (1989).

GEORGE T. WOLFF
General Motors Research Laboratories

AIR POLLUTION CONTROL METHODS

Air pollution (qv), recognized in the National Ambient Air Quality Standards (NAAQS) as being characterized by a time–dosage relationship, is defined as the presence in the atmosphere (or ambient air) of one or more contaminants of such quantity and duration as may be injurious to human, plants, or animal life, property, or conduct of business (1,2). Thus, air pollutants may be rendered less harmful by reducing the concentration of contaminants, the exposure time, or both.

Selection of pollution control methods is generally based on the need to control ambient air quality in order to achieve compliance with standards for criteria pollutants, or, in the case of nonregulated contaminants, to protect human health and vegetation. There are three elements to a pollution problem: a source, a receptor affected by the pollutants, and the transport of pollutants from source to receptor. Modification or elimination of any one of these elements can change the nature of a pollution problem. For instance, tall stacks which disperse effluent modify the transport of pollutants and can thus reduce nearby SO_2 deposition from sulfur-containing fossil fuel combustion. Although better dispersion aloft can solve a local problem, if done from numerous sources it can unfortunately cause a regional one, such as the acid rain now evident in the northeastern United States and Canada (see ATMOSPHERIC MODELS). References 3–15 discuss atmospheric dilution as a control measure. The better approach, however, is to control emissions at the source.

There are three main classes of pollutants: gases, particulates (which may be either liquid or solid or a combination), and odors (which may originate as gases or particulates). Although odors are controlled similarly to other pollutants, they are often discussed separately because of the different methods used for sensing and measurement (see ODOR MODIFICATION). Many effluents contain several contaminants: one or two may be present as gases; the others often exist as liquid or solid particulates of various sizes. The possibility that effluent pollutants may be present in more than one physical state must be taken into account in sampling, analysis, and control. To achieve air pollution control, reliable measurements are needed to quantify both the pollutant concentration and the contribution of individual sources. These data are necessary for designing control equipment, for monitoring emissions, and for maintaining acceptable ambient air quality.

Measurement of Air Pollution

Measurement techniques are divided into two categories: ambient and source measurement. Ambient air samples often require detection and measurement in the ppmv to ppbv (parts by volume) range, whereas source concentrations can range from tenths of a volume percent to a few hundred ppmv. Federal regulations (16–17) require periodic ambient air monitoring at strategic locations in a designated air quality control region. The number of required locations and complexity of monitoring increases with region population and with the normal concentration level of pollutants. Continuous monitoring is preferable, but for particulates one 24-h sample every sixth day may be permitted. In some extensive metropolitan sampling networks, averaged results from continuous monitors are telemetered to a single data processing center. Special problems have been investigated using portable, vehicle-carried, or airborne ambient sampling equipment. The utilization of remote-guided miniature aircraft has been reported as a practical, cost-effective ambient sampling method (18). Ambient sampling may fulfill one or more of the following objectives: (1) establishing and operating a pollution alert network, (2) monitoring the effect of an emission source, (3) predicting the effect of a proposed installation, (4) establishing seasonal or yearly trends, (5) locating the source of an undesirable pollutant, (6) obtaining permanent sampling records for legal action or for modifying regulations, and (7) correlating pollutant dispersion with meteorological, climatological, or topographic data, and with changes in societal activities.

The problems of source sampling are distinct from those of ambient sampling. Source gas may have high temperature or contain high concentrations of water vapor or entrained mist, dust, or other interfering substances so that particulates or gases may be deposited on or absorbed into the grain structure of the gas-extractive sampling probes. Depending on the objective or regulations, source sampling may be infrequent, occasional, intermittent, or continuous. Typical objectives are: (1) demonstrating compliance with regulations, (2) obtaining emission data, (3) measuring product loss or optimizing process operating variables, (4) obtaining data for engineering design, (5) determining collector efficiency or acceptance testing of purchased equipment, and (6) determining need for maintenance of process or control equipment.

Sampling of Gaseous Pollutants. Gaseous pollutant detection is dependent upon the chemistry of the material involved.

Reference methods for criteria (19) and hazardous (20) pollutants established by the US EPA include: sulfur dioxide [7446-09-5] by the West-Gaeke method; carbon monoxide [630-08-0] by nondispersive infrared analysis; ozone [10028-15-6] and nitrogen dioxide [10102-44-0] by chemiluminescence (qv); and hydrocarbons by gas chromatography coupled with flame-ionization detection. Gas chromatography coupled with a suitable detector can also be used to measure ambient concentrations of vinyl chloride monomer [75-01-4], halogenated hydrocarbons and aromatics, and polyacrylonitrile [25014-41-9] (21–22) (see CHROMATOGRAPHY; TRACE AND RESIDUE ANALYSIS).

Automated analyzers may be used for continuous monitoring of ambient pollutants and EPA has developed continuous procedures (23) as alternatives to the referenced methods. For source sampling, EPA has specified extractive sampling trains and analytical methods for pollutants such as SO_2 and SO_3 [7446-11-9] sulfuric acid [7664-93-9] mists, NO_x, mercury [7439-97-6], beryllium [7440-41-7], vinyl chloride , and VOCs (volatile organic compounds).). Some EPA New Source Performance Standards require continuous monitors on specified sources.

Sampling of Particulates. Ambient air suspended particulate concentration was traditionally measured gravimetrically over a 24-h period with a "Hi-Vol" sampler. However, in 1987 the EPA changed ambient particulate control to the PM_{10} reference method (24). In the PM_{10} method, a particle size classification head is attached to a Hi-Vol sampler so that only particulates finer than an aerodynamic 10 μm are collected on the filter. Although tape samplers, used for more frequent determination of suspended particulates, have been tied into the EPA Alert Warning System, it is not yet apparent how they will be correlated with PM_{10} monitoring. In the tape method, particulate quantity is measured automatically by light transmittance or β-ray attenuation and converted to an electronic signal for transmission and data processing.

Source sampling of particulates requires isokinetic removal of a composite sample from the stack or vent effluent to determine representative emission rates. Samples are collected either extractively or using an in-stack filter; EPA Method 5 is representative of extractive sampling, EPA Method 17 of in-stack filtration. Other means of source sampling have been used, but they have been largely supplanted by EPA methods. Continuous in-stack monitors of opacity utilize attenuation of radiation across the effluent. Opacity measurements are affected by the particle size, shape, size distribution, refractive index, and the wavelength of the radiation (25,26).

Particle size measurements for particulates extracted by filtration, electrostatic or thermal precipitation, or impaction may be performed using microscopy, sieve analysis, gas or liquid sedimentation, centrifugal classification, or electrical or optical counters (see SIZE MEASUREMENT OF PARTICLES). For aerosol particulate size determination, however, questions arise such as whether the collected particles agglomerate after capture, or whether they are redispersed to the same degree in the measuring media as they were originally. These problems can be avoided mainly by performing particle size measurements on the original aerosol by using devices such as cascade impactors (27), virtual impactors (28), and diffusion batteries and mobility analyzers.

Status of Air Pollution and Control Regulations

There has been considerable improvement, especially in industrial areas, in U.S. air quality since the adoption of the Clean Air Act of 1972. Appreciable reductions in particulate emissions and in SO_2 levels are especially evident. In 1990, however, almost every metropolitan area was in nonattainment status on ozone air quality standards; 50 metropolitan areas exceeded the CO standard and between 50 and 100 exceeded the PM_{10} standard for particulate level (29).

The U.S. Congress adopted a new clean air act in 1990 which has three areas of emphasis: acid rain reduction in the northeastern United States; severe limitation on atmospheric emissions of 189 chemicals on the EPA Hazardous or Toxic Substance list; and tightened regulations on vehicular exhaust, reformulated vehicular fuels, and vehicles capable of using alternative fuels (ozone compliance and smog reduction). Regulations associated with acid rain prevention emphasize reductions in sulfur oxide and NO_x emissions from combustion processes, especially coal-fired power boilers in the midwest. The chemical process industry (CPI) will be increasingly under pressure to eliminate atmospheric releases of VOCs and carcinogenic-suspect compounds.

Control technology requirements vary according to the scale of operation and type of emission problem. For instance, electrostatic precipitator design requirements for fly-ash control from 1000-MW coal-fired power boilers differ from those for a chemical process operation. In the discussion that follows, priority is given to control technology for the CPI as opposed to the somewhat special needs of other industries.

Minimizing Pollution Control Cost

Although the first impulse for emission reduction is often to add a control device, this may not be the environmentally best or least costly approach. Process examination may reveal changes or alternatives that can eliminate or reduce pollutants, decrease the gas quantity to be treated, or render pollutants more amenable to collection. Following are principles to consider for controlling pollutants without the addition of specific treatment devices, ie, the fundamental means of reducing or eliminating pollutant emissions to the atmosphere (30):

Eliminate the source of the pollutant.
    Seal the system to prevent interchanges between system and atmosphere.
        Use pressure vessels.
        Interconnect vents on receiving and discharging containers.
        Provide seals on rotating shafts and other necessary openings.
    Change raw materials, fuels, etc., to eliminate the pollutant from the process.
    Change the manner of process operation to prevent or reduce formation of, or air entrainment of, a pollutant.
    Change the type of process step to eliminate the pollutant.

Use a recycle gas or recycle the pollutants rather than using fresh air or venting.

Reduce the quantity of pollutant released or the quantity of carrier gas to be treated.

Minimize entrainment of pollutants into a gas stream.

Reduce number of points in system in which materials can become airborne.

Recycle a portion of process gas.

Design hoods to exhaust the minimum quantity of air necessary to ensure pollutant capture.

Use equipment for dual purposes, such as a fuel combustion furnace to serve as a pollutant incinerator.

Steps such as the substitution of low sulfur fuels or nonvolatile solvents, change of raw materials, lowering of operation temperatures to reduce NO_x formation or volatilization of process material, and installion of well-designed hoods (31–37) at emission points to effectively reduce the air quantity needed for pollutant capture are illustrations of the above principles.

Selection of Control Equipment

Engineering approaches (38) for the design and selection of pollution control equipment include knowledge of the properties of pollutants: chemical species, physical state, particle size, concentration, quantity of conveying gas, and of effects of pollutant on surrounding environment. The design must consider likely future collection requirements. Advantages of alternative collection techniques must be determined, eg

1. Collection efficiency.
2. Ease of reuse or disposal of recovered material.
3. Ability of collector to handle variations in gas flow and loads at required collection efficiencies.
4. Equipment reliability and freedom from operational and maintenance attention.
5. Initial investment and operating cost.
6. Possibility of recovery or conversion of contaminant into a saleable product.

Known engineering principles must be applied even in areas of extremely dilute concentration.

The physical state of a pollutant is obviously important; a particulate collector cannot remove vapor. Pollutant concentration and carrier gas quantity are necessary to estimate collector size and required efficiency and knowledge of a pollutant's chemistry may suggest alternative approaches to treatment. Emission standards may set collection efficiency, but specific regulations do not exist for many trace emissions. In such cases emission targets must be set by dose–exposure time relationships obtained from effects on vegetation, animals, and humans. With such information, a list of possible treatment methods can be made (see Table 1).

Table 1. Checklist of Applicable Devices for Control of Pollutants

Equipment type	Pollutant classification			
			Particulate	
	Gas	Odor	Liquid	Solid
absorption	•	•		
aqueous solution				
nonaqueous				
adsorption	•	•		
throw-away canisters				
regenerable stationary beds				
regenerable traveling beds				
chromatographic adsorption				
air dispersion (stacks)	•	•	•	•
condensation	•	•		
centrifugal separation (dry)			•	•
chemical reaction	•	•		
coagulation and particle growth			•	•
filtration				
fabric and felt bags				•
granular beds			•	•
fine fibers			•	•
gravitational settling				•
impingement (dry)				•
incineration	•	•	•	•
precipitation, electrical				
dry			•	•
wet	•	•	•	•
precipitation, thermal			•	•
wet collection[a]	•	•	•	•

[a]Includes cyclonic, dynamic, filtration, inertial impaction (wetted targets, packed towers, turbulent targets), spray chambers, and venturi.

Control devices which are too inefficient for a particular pollutant or too expensive can then be stricken from the list. For instance, atmospheric dispersion is not always an acceptable solution; condensation may require costly refrigeration to give adequate collection. Although both absorption and adsorption devices for contaminant gases can be designed for almost any efficiency, economics generally dictate the choice between the two. Grade-efficiency curves should be consulted in evaluating particulate collection devices and the desirability of dry or wet particulate collection, should be considered, especially with respect to material recycle or disposal.

Other factors to be evaluated are capital investment and operating cost, material reuse or alternative disposal economics, relative ruggedness and reliability of alternative control devices, and the ability to retain desired efficiency under all probable operating conditions. Control equipment needs to be both rugged and reliable, in part because managers are often reluctant to shut a process down for control equipment repair. Efficiency of control devices varies with processing conditions, flow rate, temperature, emission concentration, and particle size. Control devices should be designed to handle these variations. Combinations of gaseous and particulate pollutants can be especially troublesome as gaseous removal devices are often unsuitable for heavy loadings of insoluble solids. Concentrations of soluble particulates up to 11 g/m^3 have been handled in gas absorption equipment with some success. However, to ensure rapid particle solution, special consideration must be given to wet–dry interfaces and adequate liquid quantities.

Control of Gaseous Emissions

Five methods are available for controlling gaseous emissions: absorption, adsorption, condensation, chemical reaction, and incineration. Atmospheric dispersion from a tall stack considered as an alternative in the past, is now less viable. Absorption is particularly attractive for pollutants in appreciable concentration; it is also applicable to dilute concentrations of gases having high solvent solubility. Adsorption is desireable for contaminant removal down to extremely low levels (less than 1 ppmv) and for handling large gas volumes that have quite dilute contaminant levels. Condensation is best for substances having rather high vapor pressures. Where refrigeration is needed for the final step, elimination of noncondensible diluents is beneficial. Incineration, suitable only for combustibles, is used to remove organic pollutants and small quantities of H_2S, CO, and NH_3. Specific problem gases such as sulfur and nitrogen oxides require combinations of methods and are discussed separately.

Absorption. Good references for absorption (qv) of gases are 39–42 (see also DISTILLATION). All types of absorption equipment, including plate columns, can be used for air contaminants, but the devices most frequently used are packed columns, open spray chambers and towers, cyclonic spray towers, and combinations of sprayed and packed chambers. For particulate-free gas, the countercurrent packed tower is the usual choice to maximize driving force. Plastic packings with extended surface area and high void space with wetted temperatures below 85°C minimize pressure drop and provide high mass transfer and constant liquid film renewal. Insoluble particulates and heavy loads of soluble ones plug counterflow packing rapidly; concurrent flow tends to reduce plugging. Six months of plug-free operation of a parallel-flow bed absorbing SiF_4 in water has been reported (43).

The cross-flow packed scrubber (Fig. 1) (44) is even more plug-resistant and has been used extensively as a pollutant absorber. Typical design parameters for gas absorption only are gas-flow rate $G = 2.44$ kg/(s·m^2) and liquid flow rate $L = 2.03$ kg/s·m^2. When particulates are present, sprays directed at the bed-retaining grillwork are added upstream. Most of the solids are impacted on the first 150 mm of packing in the gas-flow direction. To remove deposited solids, the liquid rate over the first 300 mm of packing is increased to $L = 13.56$ kg/(s·m^2), maintaining a rate of $L = 2.71$ kg/(s·m^2) over the remainder of the bed. As indicated earlier, solids loading up to 11 g/m^3 has been successfully handled. A single transfer unit has been achieved in a gas-flow depth of 200 mm absorbing HF in water. Because

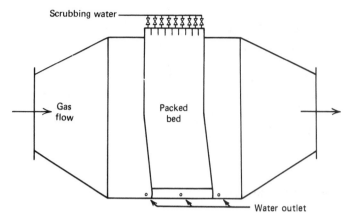

Fig. 1.   Cross-flow packed scrubber.

scale-up can be a problem, a computer program has been developed for the necessary calculations (45).

Open horizontal spray chambers (43) and vertical spray towers have both been used when solids are present. Cyclonic spray towers provide slightly better scrubbing when the optimum spray droplet size is used. Figure 2 illustrates various spray chambers. When a large number of transfer units are required, single tower absorption contacting may be unsatisfactory. Loss of counter-currency resulting from spray entrainment limits the number of transfer units achievable in a single tower. Using vertical spray towers (Fig. 2**b**), 5.8 transfer units have been attained (46). Seven transfer units in a commercial cyclonic spray tower have been reported (47) and 3.5 transfer units have been reported in horizontal spray towers. The venturi scrubber is advantageous for particulate collection, especially submicrometer particles, along with gas absorption, but it has been indicated that these scrubbers are limited to 3 transfer units (46).

Water is the most common absorption liquid. It is used for removing highly soluble acidic gases such as HCl [7647-01-0], HF [7664-39-3], and SiF_4 [7783-61-1], especially if the last contact is with water of alkaline pH; NH_3 [7664-41-7] can also be recovered in water if the final contact is acidic. Problems can arise in the initial absorption stages when contacting high concentration gases and volatile neutralizing agents. Vapor phase reactions can produce a submicrometer smoke which is often difficult to wet and collect. These problems can be avoided if initial contact is made at points in the tower where reactant vapor pressures are low. Gases such as SO_2, Cl_2 [7782-50-5], and H_2S [7783-06-4] can also be absorbed more readily in an alkaline solution. Absorption of SO_2 has been practiced in two coal-fired power plants in England using seawater. Tremendous once-through water quantities have been used; thus the discharge water is still unsaturated with SO_2. Scrubbing of SO_2 using alkaline ammonium salt solutions, as in the Cominco (48) process, has also been practiced. Many absorption processes have been commercialized for removing SO_2 from coal-fired power plant flue gas. Organic liquids such as dialkylaniline, the various ethanolamines, and methyldiethanolamine [105-59-9] can also be used for absorption of particulate-free acidic gases. Low volatility oils and solvents such as kerosene [8008-20-6] can be used to absorb organic vapors as long as the scrub liquid volatility is low enough to prevent

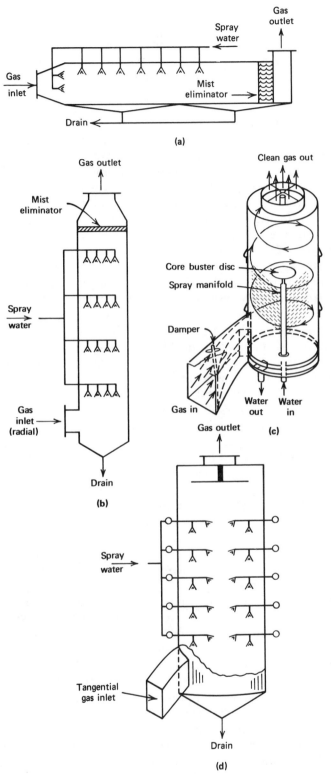

Fig. 2. Types of spray towers: (**a**) horizontal spray chamber; (**b**) simple vertical spray tower; (**c**) cyclonic spray tower, Pease-Anthony type; (**d**) cyclonic spray tower, external sprays.

vapor loss and atmospheric contamination. Control of VOC emissions from small industrial sources by absorption, adsorption, and condensation has been compared (49). Organic compound oxidation and fire are other hazards to be considered.

Disposal of recovered gaseous pollutants can be a problem. Precipitation of the pollutant as an insoluble sludge may be possible through the addition of lime or other reagents. The sludge may be thickened by settling, then dewatered by centrifugation or filtration; however, sludges containing 70% water are not uncommon. Disposal to streams is not feasible and impounding in landfills or tailing ponds is becoming less acceptable. Conversion of the pollutant to a usable form is preferable. Recovered sulfate and sulfite compounds, if ammoniated, may be incorporated into fertilizer or may be used by nearby sulfate–sulfite pulp and paper mills. Recovered halogens can be an even greater problem.

Adsorption.   Adsorption (qv) of gases has been reviewed (40,50) (see also ADSORPTION, GAS SEPARATION). Adsorption, used alone or in combination with other removal methods, is excellent for removing pollutant gases to extremely low concentrations, eg, 1 ppmv. When used in combination, it is typically the final step. Adsorption, always exothermic, is even more attractive when very large gas volumes must be made almost pollutant free. Because granular adsorbent beds are difficult to cool because of poor heat transfer, gas precooling is often practiced to minimize adsorption capacity loss toward the end of the bed. Pretreatment to remove or reduce adsorbable molecules, such as water, competing for adsorption sites should also be considered (41).

Adsorbents may be divided into two categories: general and specific. The most common general adsorbent, activated carbon, does a good job adsorbing organic molecules of all types in the presence of polar molecules such as water vapor. Carbon is one of the few adsorbents that will work in a humid gas stream. However, water vapor will be adsorbed as well and provision for adequate bed capacity must be made. Specific adsorbents are typically simple or complex metal oxides (see MOLECULAR SIEVES). They show greater selectivity than carbon and preferentially adsorb strongly polar molecules. They usually cannot be used for pollution control in the presence of water vapor. Impregnated adsorbents typically contain carbon and a chemically reactive material to remove some particular pollutant. Examples are carbon impregnated with 10–20% bromine to react with olefins, iodine to react with mercury vapor, or lead acetate to remove H_2S.

Desorption.   Pollutants, once adsorbed, may be disposed of by discarding the saturated adsorbent or after adsorbent regeneration. Discarding may be attractive when the quantity of adsorbed material is small, the adsorption process occurs infrequently, or the cost of fresh adsorbent is insignificant compared to the cost or inconvenience of regeneration. In such cases the adsorbent may be contained in a paper carrier or disposable cartridge. If the adsorbed material is not toxic or cannot be removed by landfill leaching, ordinary landfill disposal may be acceptable. However, adsorbate compliance with TSCA (Toxic Substances Control Act) and RCRA (Resource Conservation and Recovery Act) regulations for disposal should be ascertained and utilization of a hazardous waste landfill may be mandated. If both adsorbent and adsorbate are combustible, suitable incineration may be a better disposal method or it may also be possible to return the saturated adsorbent to the manufacturer for regeneration. When economics or toxicity dictate adsorbent regeneration, desorption may be accomplished by heating, evacuating, stripping with an inert gas, displacing the adsorbate with a more readily adsorbed material, or some combination of these methods. The desired form of the recovered pollutant and its subsequent handling may influence the choice. If the bed is heated to the adsorbate's atmospheric boiling point, the adsorbed pollutant will boil off in undiluted form and can then be condensed for process reuse or disposal. Heat-sensitive materials that are apt to decompose and plug the adsorbent are treated at reduced pressure and temperature. Desorption by stripping with an inert gas often produces a more concentrated form of the pollutant than in the original effluent, but separation of the pollutant from the stripping gas is sometimes difficult. This pollutant concentration method may be desireable, however, if the ultimate disposition of the material is incineration. Up to a 40-fold increase in pollutant concentration has been achieved by inert gas stripping (51). Complete separation of pollutant and stripping gas can also be effected by partially condensing the stripped pollutant, reheating, and then recycling. Displacement of adsorbed pollutant with another material is often accomplished through steaming. Once the adsorbent is saturated with water, however, a second regeneration cycle is required to remove the water. In reality, steaming is a combination of methods: displacement, bed heating, and adsorbate stripping.

Regardless of method, desorption is never complete. Adsorbent capacity is always less following regeneration than it is on initial loading of adsorbent. Some adsorbable materials undergo chemisorption: they chemically combine with the adsorbent. An example is the Reinluft process (52) for removing SO_2 from flue gas on activated carbon. The SO_2 is attached to the carbon as sulfuric acid. Desorption occurs only upon heating to 370°C; a mixture of CO_2, evolved from the chemically bound carbon, and SO_2 are driven off.

Adsorption Application.   Figure 3 illustrates pollutant depletion from the carrier gas as it travels through the bed in a typical "adsorption wave." Over time, the adsorbent becomes saturated near the bed inlet and the pollutants must penetrate the bed more deeply before being adsorbed. When the maximum permissible pollutant concentration which can be discharged, C_b, is reached at the outlet, "breakthrough" has occurred, and the adsorbent must be regenerated. The adsorption cycle time is represented by the horizontal displacement between the adsorption curves for the fresh bed and the bed at breakthrough. Cycle times are controlled by the depth of the bed provided and the quantity of gas treated per unit of adsorbent. Stationary beds are used for disposable canisters and for many regenerable applications. For very large gas-flow volumes or high pollutant concentrations, a static bed would be exhausted too rapidly and fluidized-bed and traveling-bed systems are used. In a fluidized-bed adsorbent granules are well-mixed and an adsorption wave does not exist. The bed's capacity may be quite low unless staging is practiced. Fluidization's advantage in single-stage beds is that the adsorbent can be removed continuously from the bed through solids flow into a second regeneration vessel and then continuously returned to the adsorption chamber (see FLUIDIZATION). In traveling-bed systems, spent adsorbent is continuously removed from the bottom, regenerated in another vessel, and continuously recharged at the top of the adsorption bed. Solids flow is downward, and gas flow is upward, ie, countercurrent. A newer development (53), described as chromatographic adsorption, consists of injecting a cloud of adsorbent particles into the

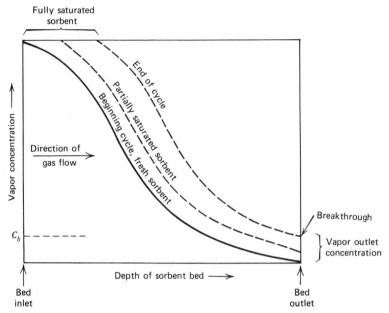

Fig. 3. Passage of the adsorption wave through a stationary bed during the course of an adsorption cycle. The progressing S-shaped curves indicate the nonadsorbed vapor concentration by position in the bed at different time periods. C_b represents the maximum permissible outlet concentration for release to the atmosphere.

gas stream. Adsorption occurs during the concurrent flow of effluent and suspended particles. The adsorbent is then removed in a conventional bag filter. Some final adsorption takes place as the gas flows through the adsorbent layer on the filter bag.

In another new technology, called pressure-swing adsorption (54), the adsorption bed is subjected to relatively short pulses of higher pressure gas containing the species to be adsorbed. The bed is then partially regenerated by reducing the pressure and allowing the adsorbed material to vaporize. By controlling the gas flow and the pressure, the pollutant can be transferred from effluent to another gas stream. This same result can be achieved by temperature fluctuation, but bed temperature changes cannot be as rapidly controlled as bed pressures.

Typical adsorption pollution applications have included odor control in food processing (qv): odor and solvent control in chemical and manufacturing processes such as paints and coating operations, pulp and paper manufacture, and tanneries; odor control from foundries and animal laboratories; and radioactive gas control in the nuclear industry. Additionally, small carbon-filled canisters have been mandated on U.S. automobiles in recent years to reduce evaporative fuel emissions from gasoline engines. Similar adsorption equipment is used to control vapor emissions from fuel-tank filling in states requiring this type of control because of noncompliance with ozone NAAQS regulations. An increasing emphasis on the control of toxic and hazardous organic vapor emissions as presented in the Clean Air Act of 1990, should mean an increase in adsorption applications for air pollution control (49).

Condensation.   Control or reduction of volatile gases and vapors by condensation is most feasible for organic compounds (49,55). Compounds having low vapor pressures at room temperature are treated in water-cooled or air-cooled condensers, but more volatile materials often require two-stage condensation, usually water cooling followed by refrigeration. Minimizing noncondensable gases reduces the need to cool to extremely low dew points. Partial condensation may suffice if the carrier gas can be recycled to the process. Condensation can be especially helpful for primary recovery before another method such as adsorption or gas incineration. Both surface condensers, often of the finned coil type, and direct-contact condensers are used. Direct-contact condensers usually atomize a cooled, recirculated, low vapor pressure liquid such as water into the gas. The recycle liquid is often cooled in an external exchanger.

If condensation requires gas stream cooling of more than 40–50°C, the rate of heat transfer may appreciably exceed the rate of mass transfer and a condensate fog may form. Fog seldom occurs in direct-contact condensers because of the close proximity of the bulk of the gas to the cold-liquid droplets. When fog formation is unavoidable, it may be removed with a high efficiency mist collector designed for 0.5–5-μm droplets. Collectors using Brownian diffusion are usually quite economical. If atmospheric condensation and a visible plume are to be avoided, the condenser must cool the gas sufficiently to preclude further condensation in the atmosphere.

Chemical Reaction.   Reaction of gaseous pollutants can open up new pathways for recovery. Utilization of alkaline scrubbing solutions to collect acidic gases has been discussed. Nitrogen oxides can be decomposed to N_2 and O_2 by reaction with H_2 or CH_4. Many odors can be controlled by scrubbing organic compounds with solutions of strong oxidants such as potassium permanganate [7722-64-7], $KMnO_4$; nitric acid [7697-37-2], HNO_3; hydrogen peroxide [7722-84-1], H_2O_2; hypochlorites; and ozone, O_3. Gas–solid reactions such as the introduction of hydrated lime into a SO_2-containing flue gas stream (56) are also feasible although these schemes usually fall significantly short of 100% pollutant removal. Dry injection of solid sodium bicarbonate [144-55-8], $NaHCO_3$, has been studied for removal of both SO_x and NO_x from flue gas, for HCl removal from waste incineration emission (57), and for other hazardous gases from hazardous waste incineration (58). Flue gas humidification generally aids in achieving a more complete gas–solid reaction. A volatile vapor pollutant can be rendered significantly less volatile by increasing its molecular weight, such as by vapor phase chlorination. An example of a gaseous pollutant control problem that can be changed to a particulate one is the reaction of gaseous HCl with ammonia to produce NH_4Cl smoke.

Incineration.   Gases sufficiently concentrated to support combustion are burned in waste-heat boilers, flares, or used for fuel. Typical pollutants treated by incineration are hydrocarbons, other organic solvents and blowdown gases, H_2S, HCN, CO, H_2, NH_3, and mercaptans. VOC emissions are frequently destroyed by incineration (59–61) as are gases evolving from landfills. Dilute gases, not more concentrated than 25% of their lower explosive limit, are burned in gas incinerators which use either catalytic oxidation or a direct flame contact to provide preheating and ignition. The former require a more steady source of pollutants in regard to both flow and composition; they are more sensitive to overheating and to poisoning. Direct flame incinerators are frequently equipped with heat interchangers to minimize fuel consumption if they are in steady use. (See INCINERATORS.) Flares burning externally are unlikely to destroy hazardous and toxic gases totally because of such variables as atmospheric quenching of flame temperature, variations in wind turbulence, and heat losses to the surroundings. For efficient combustion, the flare should be enclosed on the sides with a lightweight housing containing a high temperature, fiberous insulating material. This reduces heat losses and shields the flame from external turbulence. Temperatures and residence times for the efficient destruction of various hazardous organic vapors are known (62,63). Airborne combustible solids could well be destroyed by properly designed incinerators, but little has been reported in this area. Combustion could be effected following the procedures developed for burning pulverized coal (see COAL CONVERSION PROCESSES) or for burning activated sludge from sewerage treatment.

Specific Problem Gases. Sulfur dioxide, nitrogen oxides, and vehicular exhaust gases are widespread gas pollutants that present specific problems (see AIR POLLUTION; ATMOSPHERIC MODELS). A revised U.S. Clean Air Act has been enacted that requires greater control of the emissions of these gases. Germany and Denmark have already adopted acid rain regulations on sulfur and nitrogen oxide emissions.

Major sources of sulfur dioxide are the combustion of sulfur-containing fossil fuels, the manufacture of sulfuric acid and sludge acid purification, sulfur recovery from petroleum processing, nonferrous smelting, and pulp and paper manufacture. Combustion emissions are controlled by substituting a low sulfur fuel source, by fuel desulfurization and refining (see COAL; PETROLEUM; FUELS, SYNTHETIC), and by sulfur removal either in the combustion process or from the flue gas (see SULFUR REMOVAL AND RECOVERY). Many methods of sulfur removal from flue gas have been developed and voluminous literature is available (64–88). Flue gas desulfurization (FGD) systems can be classified as (1) throwaway vs regenerative and (2) wet vs dry. In throwaway processes, the removed sulfur is discarded, often as calcium sulfate–sulfide sludge; in the regenerative processes, the sulfur is recovered in a useful form. Most of the earlier commercial processes used wet scrubbing and reaction with lime or limestone. Table 2 lists a number of FGD processes. For new boiler installations, fluid bed combustion is often used. The coal is burned in a fluid bed of limestone which reacts during the combustion with the sulfur in the fuel to produce calcium sulfate. A number of newer FGD processes have been investigated or partially commercialized as shown in Table 3. SO_2 emission from sulfuric acid manufacture is controlled by interpass adsorption (91) and, when necessary, by the addition of tail-gas treatment to reduce the effluent below 400–500 ppmv (see SULFURIC ACID and SULFUR TRIOXIDE). Emissions of SO_2 from nonferrous smelters are being controlled by conversion to sulfuric acid even when the acid is subsequently neutralized with limestone (92). Other SO_2 emission control methods for smelters are discussed in references (93,94) and for pulp and paper manufacture in references 95–98 (see PULP; PAPER).

Major sources of nitrogen oxide emission are nitrogen fixation during high temperature combustion, nitric acid manufacture and concentration (see NITRIC ACID), organic nitrations (see NITRATION), and vehicular emissions. During com-

Table 2. Commercialized Flue Gas Desulfurization (FGD) Processes

Process name	Process description	References
	Wet throwaway processes	
dual alkali system	SO_2 absorbed in tower with $NaOH$–Na_2SO_3 recycle solution. $CaOH$ or $CaCO_3$ added externally to precipitate $CaSO_4$, regenerate $NaOH$; make-up $NaOH$ or Na_2CO_3 added. Process attempts to eliminate scaling/plugging problems of limestone slurry scrubbing.	78, 82
limestone slurry scrubbing	Limestone slurry scrubs flue gas. SO_2 absorbed, reacted to $CaSO_3$. Further air-oxidized to $CaSO_4$, settled/ removed as sludge. Lower cost and simpler than other processes. Disadvantages: abrasive/corrosive, plugging and scaling, poor dewatering of $CaSO_4$.	82–84, 87
Dowa process	Similar to dual alkali except $Al_2(SO_4)_3$ solution used in scrubber. Limestone addition regenerates reactant, precipitating $CaSO_4·2H_2O$ crystals which dewater more readily. Reduces plugging/scaling.	
CHIYODA thoroughbred 121 process	Single vessel used to absorb SO_2 with limestone slurry and oxidize product to gypsum.	
forced oxidation	Limestone scrubbing, products air-oxidized to gypsum in separate tank.	
lime spray drying	Wet/dry process. Lime slurry absorbs SO_2 in vertical spray dryer forming $CaSO_3$–CaS, H_2O evaporated before droplets reach bottom or wall. Dry solids collected in baghouse along with flyash.	82
	Dry throwaway processes	
direct injection	Pulverized lime or limestone injected into flue gas (often through burner). SO_2 absorbed on solid particles. High excess alkali required for fairly low SO_2 absorption. Finer grinding, lime preheat, flue gas humidification benefit removal. Particulate collected in baghouse.	79, 80, 86
Trona sorption	Trona (natural Na_2CO_3) or Nacolite (natural $NaHCO_3$) injected into boiler; SO_2 absorbed to higher extent than with dry lime. Product collected in baghouse. Also can capture high quantity of NO_x.	79, 80, 85, 86

Table 2. (*Continued*)

Process name	Process description	References
	Wet regenerative processes	
Wellman-Lord	After flue gas pretreatment, SO_2 absorbed into Na_2SO_3 solution; solids and chloride purged, SO_2 stripped, regenerating Na_2SO_3, and SO_2 processed to S.	81, 82
magnesium oxide process	SO_2 absorbed from gas with $Mg(OH)_2$ slurry, giving $MgSO_3$–$MgSO_4$ solids which are calcined with coke or other reducing agent, regenerating MgO and releasing SO_2.	
citrate-scrubbing	SO_2 absorbed with buffered citric acid solution. SO_2 reduced with H_2S to S. H_2S produced on site by reduction of S with steam and methane.	78
Flakt-Boliden process	SO_2 absorbed with buffered citric acid solution. SO_2 stripped from solution with steam.	
aqueous carbonate	SO_2 absorbed into Na_2CO_3 solution in spray dryer, producing dry Na_2SO_3 particles.	
SULF-X process	FeS slurry absorbs SO_2; product calcined producing S vapors which are condensed.	
Conosox process	K_2CO_3 and K salt solutions absorb SO_2 forming K_2SO_4 which is converted to thiosulfate, $KHSO_3$ which is converted to H_2S, and regenerates K_2CO_3.	
	Dry regenerative processes	
Westvaco	SO_2 adsorbed in activated carbon fluid bed. SO_2, H_2O, and $\frac{1}{2}O_2$ react at 65–150°C forming H_2SO_4. In next vessel, $H_2SO_4 + 3H_2S$ at 150°C gives $4S + 4H_2O$. Bed temperature is increased to vaporize some S. Remaining S reacts with H_2 to H_2S.	
copper oxide adsorption	SO_2 adsorbed on copper oxide bed forming $CuSO_4$. Bed is regenerated with H_2 or H_2–CO mixture giving concentrated SO_2 stream. Bed is reduced to Cu, but reoxidized for SO_2 adsorption.	79

Table 3. Newer FGD Processes and Research

Process type and comments	Paper number in reference 88
direct injection	4A–1, 4A–2, 4A–3, 4A–5
German technology	4A–6
Austrian technology	4A–7
wet calcium-based FGD	
critique of present processes and improvements	5B–2
dual alkali process	5B–10
Austrian experiences	5B–13
German experiences	5B–16
German (Bischoff) process	5B–15
jet bubbling reactor for simultaneous	5B–14
SO_2–particulate removal	
U.S. FGD status	1–1
retrofit economics	2–1, 2–2, 2–3
spray dryer technology	3–1, 3–2, 3–3, 3–4
SO_x–NO_x removal	
activated coke processes	4B–7
dry sorbent injection	4B–8
spray dryer adsorption–oxidation	4B–5
utilizing ferrous chelates	4B–6, also see Ref. 89
gas reburning, sorbent injection	see Ref. 90
European experience	1–3, 1–4, 1–5
Japanese experience	1–6

bustion in the presence of air, N_2 and O_2 react in the high temperature of the flame to produce NO. This reversible reaction favors NO formation as the temperature increases (see Table 4). Because the kinetic rate of the decomposition reaction drops essentially to zero at temperatures below 870°C, reversion of the NO to N_2 and O_2 upon cooling does not occur. Nitrogen present in the fuel tends to produce other nitrogen oxides in addition to NO, which leads to higher exhaust concentrations of NO_x than would exist from the combustion-driven nitrogen fixation alone. NO reacts with O_2 at a slow but steady rate in the atmosphere and thus NO ends up as NO_2. Combustion chemistry and NO_x formation and control have been reviewed (100,101). NO formation in combustion may be reduced by maintaining low excess air (0.5% O_2 or less in flue gas), employing two-stage combustion where the first stage is fuel-rich and reducing (high temperature) and the second stage is oxidizing (1000–1100°C), flue gas recirculation, burner design, combustion chamber modifications, and burner placement. Combustion technology (qv) methods (102–112) are limited in their NO_x reduction capabilities to ranges of 200–300 ppmv NO_x. Target emission levels for acid rain control are expected to be 80–100 ppmv of NO_x. Retro-fitting with low NO_x burners is one of the least expensive ways to achieve NO_x reduction. References 103 and 104 discuss low NO_x burner design. An overview of NO_x control from industrial sources has been presented

Table 4. Time to Form NO in a Gas Containing 75% N_2 and 3% O_2[a]

Temperature, °C	Time to form 500 ppmv NO, s	NO equilibrium concentration, ppmv
1315	1370	550
1538	16.2	1380
1760	1.10	2600
1982	0.117	4150

[a]Ref. 99.

(113); references 114–116 discuss NO_x control retrofit. Other control methods which are being developed for combustion NO_x emissions are selective catalytic reduction (SCR), Thermal Denox, and urea reduction (88,113–122).

SCR is a standard, commercially demonstrated process which has been applied to large coal-fired power boilers in Germany and Japan. Many of the catalysts are preformed, supported catalysts impregnated with vanadium and tungsten oxides (see CATALYSTS, SUPPORTED) that catalyze the reduction of NO_x to N_2 and O_2 by reaction with NH_3 at flue gas temperatures of 250–450°C. A catalyst charge is very expensive; it constitutes as much as 45% of the capital cost of the SCR portion of boiler installation. Moreover, difficulties that have been encountered involving catalyst poisoning point to the need for careful selection of coal sources. Thermal Denox (122,118) destroys NO_x by gas-phase reduction with NH_3. It is limited to temperatures of 900–1100°C and an oxygen content below about 3% by volume. Developed by Exxon, it has been used appreciably in petroleum and process industry combustion gases, as well as in a few coal-fired boilers in Germany. Although less expensive than the SCR process, it has not received significant power industry attention, probably as a result of its limited temperature application window and the need to use an appreciable excess of NH_3 (unless a mixture of both NH_3 and H_2 can be employed). The urea reduction process (119), utilizing a urea–water solution sprayed into the boiler flue gases, has a somewhat wider operating window (800–1200°C, up to 7% O_2) than Thermal Denox. The process is less well developed and has not been commercially demonstrated except for trials on a power boiler in Sweden. Another process undergoing research development destroys NO_x by the reaction of isocyanic acid obtained from cyanuric acid sublimation (120,121). This method appears to have good prospects for diesel engine exhaust NO_x removal.

Control of NO_x emissions from nitric acid and nitration operations is usually achieved by NO_2 reduction to N_2 and water using natural gas in a catalytic decomposer (123–126) (see EXHAUST CONTROL, INDUSTRIAL). NO_x from nitric acid/ nitration operations is also controlled by absorption in water to regenerate nitric acid. Modeling of such absorbers and the complexities of the NO_x–HNO_x–H_2O system have been discussed (127). Other novel control methods have also been investigated (128–129). Vehicular emission control is treated elsewhere (see EXHAUST CONTROL, AUTOMOTIVE).

Control of Particulate Emissions

The removal of particles (liquids, solids, or mixtures) from a gas stream requires deposition and attachment to a surface. The surface may be continuous, such as the wall and cone of a cyclone or the collecting plates of an electrostatic precipitator, or it may be discontinuous such as spray droplets in a scrubbing tower. Once deposited on a surface, the collected particles must be removed at intervals without appreciable reentrainment in the gas stream. One or more of seven physical principles (see Table 5) are frequently employed to move particles from the bulk gas stream to the collecting surface. In some instances, a few other principles such as diffusiophoresis and methods of particle growth and agglomeration have also been used. The magnitude of the force developed to move a particle toward a collecting surface is influenced markedly by the size and shape of the particle. Gravity settling is efficient only for large (D_p > 40–50 μm) particles; flow-line interception and inertial impaction are effective for particles down to 2–3 μm; diffusional deposition and thermal precipitation, increasingly efficient with a decrease in particle size, are highly efficient for particles ≤ 0.5 μm; and electrostatic forces are the strongest forces available to act on fine particles, which are loosely defined as ≤ 2–3 μm. There is a gap in collectability between 0.2 and 2.0 μm. Particles in this range are the most difficult to charge electrically.

Terminal settling velocity can be calculated for spherical particles (eq. 1) in streamline flow (Stokes' law region). Small particles fall faster then predicted as they tend to "slip" between gas molecules, and the Stokes-Cunningham correction factor (eq. 2) must be applied as indicated in Table 5. Figure 4 gives terminal settling velocities of spherical particles in air. For nonspherical particles, multiplication of equation 1 by a sphericity correction constant $K_s = 0.843 \log(\psi/0.065)$ has been recommended (145). The sphericity ψ (146) is defined as the ratio of surface area of a sphere (of volume equal to the particle) to the surface area of the particle. References 147 and 148 give further refinements when sphericity is less than 0.67.

Particles approaching targets such as a baffle, impaction element, fiber, or droplet and having appreciable velocity have sufficient momentum that they may collide with the target. Particles directly in line are collected by flow-line interception and efficiency can be predicted from equation 4. Larger particles outside the streamlines leading to the target may be collected on the sides of the target because of the particle's inertia. Langmuir and Blodgett (149) correlated target efficiency with the inertial separation number of equation 5. Figure 5 gives target efficiencies by impaction number for three target shapes. Many moderate energy collectors capture smaller particles by a combination of direct interception and inertial impaction. It has been suggested (151) that the combined efficiency for the two mechanisms can be obtained from

$$\eta_{\text{total}} = 1 - (1 - \eta_{\text{FL}})(1 - \eta_{\text{IN}}) \tag{13}$$

Efficiency of particle collection is most frequently defined on a mass basis by

$$\eta_{\text{M}} = \left(\frac{\text{mass of particles entering} - \text{mass of particles leaving}}{\text{mass of particles entering}} \right) \tag{14}$$

Table 5. Physical Principles Affecting Particle Movement and Collection

Control principle	Related equations[a]	Conditions and assumptions	References
gravity settling	$$U_t = \frac{D_p^2(\rho_t - \rho_g)g}{18\mu_g} = \left[\frac{4\,D_p(\rho_t - \rho_g)g}{3\rho_g C_D}\right]^{1/2} \quad (1)$$ $$K_m \cong 1 + \frac{A\lambda}{D_p} \quad (2)$$	free falling, rigid sphere; fluid continuum; for $N_{RE} < 0.1$, viscous, streamline flow, no wake formation, $C_D = 24/N_{RE}$; for $N_{RE} > 0.1$, other functions of N_{RE} must be used to calculate C_D Cunningham-Stokes correction factor for small particles when fluid does not behave as continuum; in air, for $D_p = 1\,\mu m$, $K_m = 1.17$; for $D_p = 0.1\,\mu m$, $K_m = 2.7$	130–132
centrifugal deposition	$$\frac{dr}{dt} = U_{t,n} = \frac{D_p^2 \rho_t V_{cT}^2}{18\mu_g r} \quad (3)$$	the tangential velocity, V_{cT}, thus the particle velocity, $U_{t,n}$, is a function of the radius r, usually $V_{cT} \propto 1/r^n$; for free gas rotation and conservation of momentum, $n = 1$; for cyclones, n varies between 0.5 and 0.7	133, 134
flow line interception	$$\eta = \frac{1}{2.00 - \ln N_{RE}}\left[(1 + N_{SF})\ln(1 + N_{SF}) - \frac{N_{SF}}{2}\frac{(2 + N_{SF})}{(1 + N_{SF})}\right] \quad (4)$$	for cylindrical target; N_{SF} is the ratio of particle diameter, D_p, to the diameter of the collector; assumes that particle will be collected if it approaches within $D_p/2$ from collector; none of the particles are reentrained	135, 136
inertial impaction	$$N_{SI} = \frac{U_t U_o}{D_b} = \frac{K_m(\rho_t - \rho_g)D_p^2 U_{og}}{18\mu_g D_b} \quad (5)$$	viscous flow; Stokes law region; physically, N_{SI} is the stopping distance in a quiescent	136, 139

mechanism	equation	description	ref
		fluid of a particle with initial velocity U_0/D_p; Figure 5 relates N_{SI} to collection efficiency	
diffusional deposition	$D_v = \dfrac{K_m kT}{3\pi\mu_g D_g}$ (6)	particle diffusivity from Stokes-Einstein equation; assumes Brownian motion; D_p same order of magnitude or greater than mean free path of gas molecules (0.1 μm at NTP)	140
	$\eta_D = \dfrac{4}{N_{PE}}\left(2 + 0.557\ N_{SC}^{3/8} N_{RE}^{1/2}\right)$ (7)	efficiency for spherical collector; $N_{SC} < 10^5$ or $D_p \leq 0.5$ μm in ambient air	
	$\eta_D = \dfrac{\pi}{N_{PE}}\left(\dfrac{1}{\pi} + 0.55\ N_{SC}^{1/3} N_{RE}^{1/3}\right)$ (8)	efficiency for cylindrical collector; $1 < N_{RE} < 10^4$ and $N_{SC} < 100$	
electrostatic precipitation	$U_e = \dfrac{E_o E_p D_p}{4\ \pi\mu_g K_v}$ (9)	conductive spherical particles; streamline gas flow; gas behaves as continuum; negligible particle acceleration	141
	$\eta = 1 - e^{-(U_e A_e/q_e)}$ (10)	Deutsch-Anderson equation; assumes no reentrainment from collector; well mixed turbulent flow, turbulent eddies small compared to precipitator dimensions	
thermal precipitation	$U_\tau = -\dfrac{1}{5}\dfrac{1}{(1 + \pi a/8)}\dfrac{k_{gTR}}{P}\dfrac{dT}{dX}$ (11)	D_p less than mean free path of gas; free molecular or Knudsen flow regime; a = fraction of inelastic collisions, usually taken as 0.81	142, 143
(thermophoresis)	$U_\tau = -\dfrac{3\ K_m\mu_g}{2\rho_g T\ (2 + k_p/k_g)}\dfrac{dT}{dX}$ (12)	D_p same order of magnitude or greater than mean free path of gas molecules	

^{a}Terms are defined in a nomenclature at the end of this article.

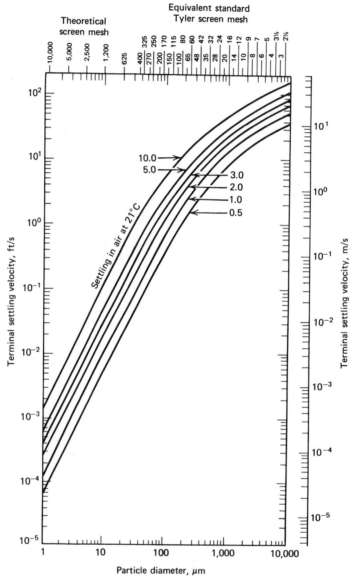

Fig. 4. Terminal velocities in air of spherical particles of different densities settling at 21°C under the action of gravity. Numbers on curves represent true (not bulk or apparent) specific gravity of particles relative to water at 4°C. Stokes-Cunningham correction factor is included for settling of fine particles. The air viscosity is 0.0181 mPa·s (= cP) and density is 1.2 g/L.

Efficiency may also be expressed in terms of the number of particles or area of particles entering and leaving the collector. An alternative terminology, used especially when collection efficiency is high, is that of penetration of particles through a collector. This focuses on the particles lost rather than those caught and penetration is defined as unity minus the fractional efficiency, $P_t = 1 - \eta$. Typical penetration of fine particles through different collectors as a function of

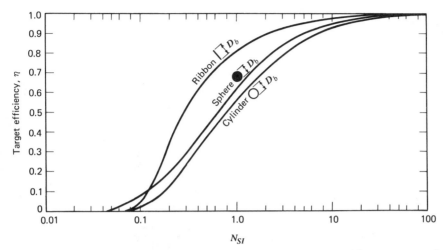

Fig. 5. Target efficiency of spheres, cylinders, and ribbons. The curves apply for conditions where Stokes' law holds for the motion of the particle (see also N_{SI} in Table 5). Langmuir and Blodgett have presented similar relationships for cases where Stokes' law is not valid (149,150). Intercepts for ribbon or cylinder are ¹⁄₁₆; for sphere, ¹⁄₂₄.

particle size is shown in Figure 6. Measurement of the difficulty of particle collection in terms of transfer units has been suggested (153); the number of required collector transfer units is defined as $N_t = \ln[1/(1 - \eta)]$.

Equation 14 applies to the overall efficiency of a collector, but a collector's performance on a specific dust results from the integration of the collector performance for each incremental particle size fraction handled. Typical grade-efficiency curves, graphs of collector efficiency vs particle diameter, for a silica dust (specific gravity 2.3) in many different collectors have been published (154,155). For other particle densities, a correction factor based on the effect of density on the predominating collection mechanism must be applied. The concept of "cutpoint," the particle size collected with 50% efficiency, was introduced to develop generalized grade-efficiency curves (133–134). In a generalized grade-efficiency curve such as that shown in Figure 7, collection efficiency is plotted against the dimensionless ratio D_p/D_{pc}. An improved method for predicting collection equipment performance from particle size distribution has been suggested (156).

Gravity Settling.   The gravity settling chamber is one of the oldest forms of gas–solid separation. It may be nothing more than a large room where the well-distributed gas enters at one end and leaves at the other. Such chambers were used at the turn of the century for collecting products such as lamp-black. Although mechanical conveyors might minimize labor costs, gravity settlers have largely disappeared because of bulky size and low collection efficiency. They are generally impractical for particles smaller than 40–50 μm.

Centrifugal and Cyclonic Collection.   A cyclonic collector is a stationary device utilizing gas in vortex flow, produced either by tangential entry of the gas or by spin baffles with axial gas inlet, to collect particles. Centrifugal force acting on the particles in the gas stream causes them to migrate to the cylindrical

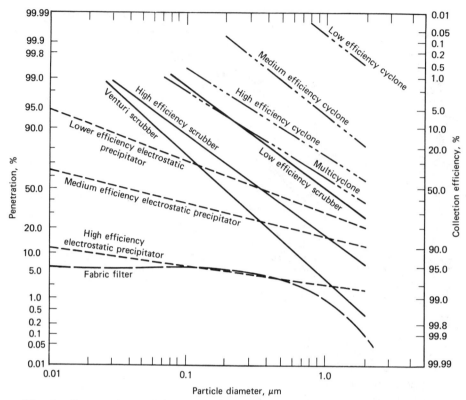

Fig. 6. Penetration and fractional efficiency for fine particles. From ref. 152, courtesy of McGraw-Hill, Inc.

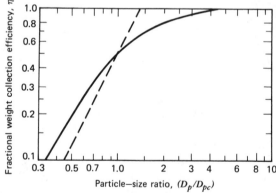

Fig. 7. Cyclone generalized grade-efficiency curves. The solid line is for the Lapple cyclone dimension ratios given in Figure 9. The dotted line is theoretical efficiency based on equations of Rosin and co-workers (157). From ref. 158, courtesy of McGraw-Hill, Inc.

containing wall where they are collected by inertial impaction. Since the centrifugal force developed can be many times that of gravity, very small particles can be collected in a cyclone, especially in a cyclone of small diameter.

Four common types of cyclone design are illustrated in Figure 8. In the

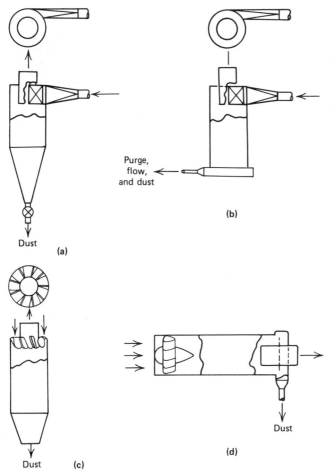

Fig. 8. Cyclone types commonly used (161): (**a**) conventional, large diameter, tangential inlet, axial discharge; (**b**) smaller tube, tangential inlet, peripheral concentrated aerosol discharge; (**c**) small tube axial inlet and discharge; (**d**) smaller tube axial inlet, peripheral concentrated aerosol discharge. Courtesy of American Industrial Hygiene Association.

conventional cyclone (Fig. 8**a**), the gas enters tangentially and spins in a vortex as it proceeds down the cylinder. A conical section causes the vortex diameter to decrease until the gas reverses on itself and spins up the center to the outlet pipe or "vortex finder." The cone causes flow reversal to occur sooner and makes the cyclone more compact than if a cylinder of constant cross section were used. The dust particles collected on the wall flow down in the gas boundary layer to the cone apex and are discharged through an air lock or into a dust hopper serving a number of parallel cyclones. Collection efficiency for a given size particle decreases with increasing diameter. Conventional cyclones are usually 600–915 mm in diameter, although it is possible to build them in other sizes. The cyclones shown in Fig. 8**c** have very similar gas-flow but generally range from 25 to 305 mm in diameter. Because of the small diameter, these have higher collection efficiencies, but lower gas capacity. Large numbers of small cyclones are mounted in a housing with a common tube sheet and dust hopper. Spin vanes in the annular gas inlets produce the vortex flow.

In other cyclone types (see Fig. 8**b** and 8**d**), the dust is not completely removed from the gas stream; instead it is concentrated into about 10% of the total flow. Removal of the dust in aerosol form increases collection efficiency by reducing the dust entrainment losses which always occur at the cone apex in an ordinary cyclone. This purge dust-aerosol flow can increase overall dust collection efficiency by as much as 20–28% of that otherwise lost. Obviously, another separation device, bag filter or another cyclone, is required to complete the separation task on the purge flow. Such a cyclone dust-concentration arrangement may be especially attractive if the increased cyclone efficiency gives adequate particulate removal for a large volume gas stream. The smaller purge stream can then be handled in a more efficient dust separation device. In the type shown in Fig. 8**b**, the gas reverses direction internally as in conventional cyclones; in that shown in Fig. 8**d**, straight-through flow, is convenient for connecting large gas volume breaching without changes in gas direction.

Cyclone Efficiency. Most cyclone manufacturers provide grade-efficiency curves to predict overall collection efficiency of a dust stream in a particular cyclone. Many investigators have attempted to develop a generalized grade-efficiency curve for cyclones, eg, see (159). One problem is that a cyclone's efficiency is affected by its geometric design. Equation 15 was proposed to calculate the smallest particle size collectable in a cyclone with 100% efficiency (157).

$$D_{p\,\min} = \sqrt{\frac{9u_g W_i}{\pi N_e V_c(\rho_p - \rho_g)}} \tag{15}$$

For smaller particles, the theory indicates that efficiency decreases according to the dotted line of Figure 7. Experimental data (134) (solid line of Fig. 7) for a cyclone of Fig. 9 dimensions show that equation 15 tends to overstate collection efficiency for moderately coarse particles and understate efficiency for the finer fraction. The concept of particle cut-size, defined as the size of particle collected with 50% mass efficiency, determined by equation 16 has been proposed (134).

$$D_{pc} = \sqrt{\frac{9u_g W_i}{2\pi N_e V_c(\rho_p - \rho_g)}} \tag{16}$$

This equation is for Figure 9 cyclone dimension ratios. The term N_e, the effective number of spirals the gas makes in the cyclone, was found to be approximately 5 for Lapple's system (134). The solid line grade-efficiency curve of Figure 7 is also used with Lapple's cyclone, which is a somewhat taller, less compact cyclone than many commercial designs.

A more compact cyclone design dimension ratio has been developed for the American Petroleum Institute (API) (162). A third set of cyclone dimension ratios has also been given (163). It is known that cyclone efficiency increases as dust loading increases but most efficiency prediction methods other than that of the API neglect this fact. The generalized grade-efficiency curve of Figure 7 is used

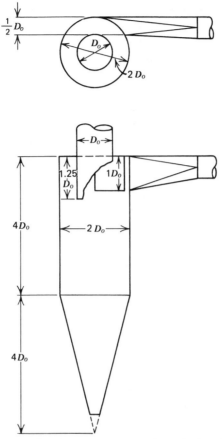

Fig. 9.   Dimension ratios of Lapple (134) cyclone design. From ref. 160, courtesy of Academic Press.

for a Lapple cyclone to predict collection efficiency for particle sizes other than the cut-size. Then an integrated overall collection efficiency is calculated by an interval integration. Needed are a narrow-range particle size analysis of the airborne dust as shown in Table 6 and a grade-efficiency curve for the cyclone for the dust density to be utilized. A narrow-range particle size distribution may be obtained by plotting the actual analysis on log-probability graph paper and interpolating. Each narrow particle interval is represented by its midpoint size and the corresponding collection efficiency shown in Table 6, columns 3 and 4 respectively. The product of columns 2 and 4 is then entered in column 5 as the fraction collected. The summation of column 5 gives overall collection efficiency for the cyclone.

Cyclone Design Parameters.   As indicated earlier, a cyclone's geometric design affects efficiency. The outlet pipe should extend at least differentially below the bottom of the gas inlet to prevent inlet dust from short-circuiting directly to the outlet pipe. The gas inlet should be tangential to the top plate of the cyclone to eliminate eddy flows and turbulence as well as reduce top-plate erosion. Efficiency is increased with a narrower gas inlet since the dust has a

Table 6. Technique for Calculating Cyclone Overall Efficiency for Dust Particles[a]

(1) Size range, μm	(2) % by wt in size range	(3) Median diameter in range, μm	(4) Efficiency of median diameter[b], %	(5) Fraction collected, %
104–150	3	127	100.0	3.0
75–104	7	89.5	98.8	6.9
60–75	10	67.5	97.3	9.7
40–60	15	50	96.3	14.5
30–40	10	35	95.5	9.6
20–30	10	25	95.0	9.5
15–20	7	17.5	93.0	6.5
10–15	8	12.5	90.0	7.2
7.5–10	4	8.75	85.3	3.4
5–7.5	6	6.25	78.5	4.7
2.5–5	8	3.75	65.0	5.2
0–2.5	12	1.25	33.0	4.0
			Total collection	84.2

[a]Columns 1–5 are explained in text.
[b]From a grade efficiency curve.

shorter distance to travel to reach the cyclone wall. However, narrower inlets mean longer, less compact cyclones so that 2:1 rectangular inlets are common. Grade-efficiency is predicated on uniform distribution of gas and dust across the gas inlet. To achieve uniform distribution, duct transitions on the inlet should be gradual (no more than a 15° included angle). Elbows on cyclone inlets should not be used unless they concentrate the dust toward the outside or top of the cyclone.

Traditionally, cyclone dimensions are multiples of outlet pipe diameter D_o. Typical barrel diameters are $2D_o$ but efficiency increases at constant D_o up to a $3D_o$ barrel diameter. Efficiency also improves as barrel and cone length are increased at constant D_o up to the natural length of the vortex. At constant inlet velocity, efficiency increases as outlet diameter (and all ratioed dimensions in a family of cyclones) is decreased. Improved efficiency is attained at the expense of loss in cyclone gas handling capacity. Although cut-size equations indicate that cyclone efficiency can increase without limit (subject only to available pressure drop) as inlet gas velocity is increased, this is not the case. Cyclone collection efficiency goes through a maximum as inlet velocity is increased causing separated dust to be reentrained from the cyclone wall (164). Higher efficiencies can be reached by reducing the gas outlet pipe area in relation to the gas inlet area (164).

The inside wall of a cyclone should be smooth. Bumps and projections create internal turbulence and dust layer reentrainment. Dust discharge from the cyclone cone apex is also important and a smaller apex outlet will get the dust discharge farther from the turbulent area of flow reversal. Theoretically, the apex opening should be less than ¼ of the gas outlet pipe diameter. However, the outlet should not be so small that it bridges easily. Outlets of 75–150 mm are generally desireable. A good gas seal is needed for dust discharge, especially if the cyclone is under vacuum. Atmospheric air sucked in through the apex reentrains collected dust and reduce cyclone efficiency significantly. Dust hoppers are often used and a good head of dust in the hopper helps to seal against leakage through rotary valves and motorized flap gates. It is important to keep the dust level in hoppers pulled down, however, because dust backing up into the cone of the cyclone reduces collection efficiency to zero.

Practically all cyclone performance data have been related to a preset cyclone set of geometric ratios. One model for cyclone grade-efficiency curves has been tested against reported commercial cyclone efficiencies (159). A good fit was obtained.

Cyclone Pressure Drop.   Typical cyclone pressure drops range from 250 to 2000 Pa. Most data are reported for clean air flowing through the cyclone and these data are conservative for design purposes. Many investigators have unsuccessfully attempted to relate pressure drops to inlet and outlet dimension ratios. Manufacturers' calibration curves or experimental measurements on cyclones of similar dimension should be used where possible. If a reliable experimental measurement is available, however, the pressure drop at other conditions can be estimated by first evaluating the constant K_c in equation 17.

$$\Delta P = K_c Q^2 P \rho_g / T \tag{17}$$

Some empirical equations to predict cyclone pressure drop have been proposed (165,166). One (166) reliably predicts pressure drop under clean air flow for a cyclone having the API model dimensions. Somewhat surprisingly, pressure drop decreases with increasing dust loading. One reasonable explanation for this phenomenon is that dust particles approaching the cyclone wall break up the boundary layer film (much like spoiler knobs on an airplane wing) and reduce drag forces.

Cyclone Problems.   Problems may be encountered in cyclone application because of fouling and caking, or from erosion, or when using multiple cyclones. Multiple cyclones are designed so that each cyclone handles a prorated share of gas and dust and the overall efficiency of the system is the same as that calculated for an individual unit. This is the case, however, only when each cyclone receives identical dust fractions (size and loading) and gas flow. Since cyclone efficiency increases with flow and dust loading and is affected by particle-size distribution, the design of the inlet gas distribution system must accomplish the proper distribution. Otherwise, those cyclones with lower gas flow and dust concentration (and perhaps finer dust) will have much poorer efficiency. When multiple cyclones share a common dust hopper, it is important that all cyclones have essentially uniform pressures at the cone apex. Wall caking, unequal gas flow or dust distribution resulting from pressure drop decreases that occur with increases in dust loading, or partial plugging of cone or cyclone inlets can all cause unequal apex pressures. Unequal pressures will cause gas from higher pressure cyclones to flow into the dust hopper and back into the cyclones having lower apex pressure. This short-circuiting can result in heavy dust reentrainment and decreased efficiency.

Fouling of cyclone walls is usually caused by sticky or hydroscopic particulates, or by moisture or other vapor condensation. To prevent particle sticking, a highly polished finish, a graniteware glass coating, or fluorocarbon plastic lining may be used on the walls or alternatively, revolving wall scrapers. Cables or chains are sometimes suspended from the center of the vortex outlet. Vortex flow causes the cable or chain to rotate and thus scrape the wall, sometimes freeing the buildup. Condensation must be prevented by decreasing the dew point of the gas or by heating or insulating the cyclone wall. In extreme situations, the cyclones may be enclosed in a heated chamber.

Erosion can be a severe problem even in well-designed cyclone installations when handling a heavy loading of coarse and abrasive angular particles. One answer is lining the cyclone, using protective materials. The cyclone may be made of wear-resistant plate. Linings may be hard and thick sacrificial castings, wear-resistant applied welded coatings, or of rubber, ceramic shapes, reinforced castable, or brick. Alternatively, cyclones may be used in series, increasing the velocity as dust loading is decreased. In this method, first-stage efficiency will be low, but the bulk of the coarser and more abrasive particles will be collected, permitting higher velocities in second- and even third-stage cyclones. Inlet velocities and dust loadings at which erosion may be excessive have been calculated (Table 7) (167). Dust particles smaller than 5–10 μm do not cause appreciable erosion.

Table 7. Dust Loadings and Cyclone Inlet Velocities above Which Erosion Is Excessive[a]

Dust load, g/m^3	Inlet velocity, m/s
0.7	35
7.0	20
7000	2

[a]Recomputed from ref. 167.

Other problems affecting cyclone efficiency are usually caused by abuse or poor maintenance. Problems may arise from temperature warpage, rough interior surfaces, overlapping plates and rough welds, or misalignment of parts, such as an uncentered (or cocked) vortex outlet in the barrel.

Other Centrifugal Collectors. Cyclones and modified centrifugal collectors are often used to remove entrained liquids from a gas stream. Cyclones for this purpose have been described (167–169). The rotary stream dust separator (170,171), a newer dry centrifugal collector with improved collection efficiency on particles down to 1–2 μm, is considered more expensive and hence has been found less attractive than cyclones unless improved collection in the 2–10-μm particle range is a necessity. A number of inertial centrifugal force devices as well as some others termed dynamic collectors have been described in the literature (170).

Electrostatic Precipitation. An electrical precipitator can collect either solid or liquid particles very efficiently. Using a special design, it can also collect solids and liquids in combination and perform an adequate job of gas absorption. Particles entering a precipitator are charged in an electric field and then move to a surface of opposite polarity where deposition occurs. For particles ≤ 2 μm, electrical forces are stronger than any other collectional force. Thus precipitators have the highest energy utilization efficiency. Advantages are low pressure drops (often 250–500 Pa $=$ 1.9–3.8 mm Hg), low electric power consumption, and low operating costs (mostly capital-related charges). Additionally, particulates are recovered in an agglomerated form, rendering them more easily collectible in case of reentrainment. Unfortunately, precipitators are also the most capital-intensive of all control devices, and mechanical, electrical, and process problems can cause poor on-stream time and reliability.

Precipitators are currently used for high collection efficiency on fine particles. The use of electric discharge to suppress smoke was suggested in 1828. The principle was rediscovered in 1850, and independently in 1886 and attempts were made to apply it commercially at the Dee Bank Lead Works in Great Britain. The installation was not considered a success, probably because of the crude electrostatic generators of the day. No further developments occurred until 1906 when Frederick Gardiner Cottrell at the University of California revived interest (U.S. Pat. 895,729) in 1908. The first practical demonstration of a Cottrell precipitator occurred in a contact sulfuric acid plant at the Du Pont Hercules Works, Pinole, California, about 1907. A second installation was made at Vallejo Junction, California, for the Selby Smelting and Lead Company.

Precipitators can be classed as single- or two-stage. In single-stage units, used for industrial gases, the particles are charged and collected in the same electrical field. Negative discharge (gas-ionizing) polarity is practically always used in single-stage precipitators because higher voltages can be achieved without sparkover. However, negative polarity also produces O_3 from O_2-containing gases. In two-stage precipitation, the particles are charged in an ionizing section and precipitated separately. Two-stage units are used for air purification, air-conditioning (qv), or ventilation. They are operated at lower voltage so there is less electrical hazard, less sparking, and fewer fires. These units have also received some consideration for the collection of condensed hydrocarbon mists. A positive polarity is always used for room air conditioning to avoid ozone formation.

Single-Stage Precipitators. Single-stage precipitators are used for control of fine particulates such as those in mists and smoke, in the 2.5–0.2-μm range. Recently, technology in the development of larger and larger units to treat flue gas from 800–1200-MW coal-fired electric power boilers has been markedly improved. This is a specialized market and these precipitators are well beyond the normal needs of the process industries to which this article is primarily directed.

A complete precipitator consists of (1) discharge electrodes, (2) collecting surfaces (plates or tubes), (3) a suspension and tensioning system for discharge electrodes, (4) a rapping system to remove dust from tubes, (5) dust hoppers and dust-removal system, (6) gas-distribution system and precipitator housing, and (7) power supply and control system. Single-stage precipitators are subclassified as plate or tube type. Typical arrangements are shown in Figure 10. A plate precipitator may have tall (10–15 m), parallel, flat plates with 228–400 mm horizontal spacing and discharge electrodes (wires) that are suspended midway between the plates. Plate precipitators are utilized for collecting flyash, for high gas-flow applications, and for particulates which are comparatively coarser than those caught in the tube-type equipment. Plate-type precipitators are lower in cost because both sides of the plate serve as precipitating surfaces. Tube precipitators are frequently used for liquid mists and sometimes for submicrometer metallurgi-

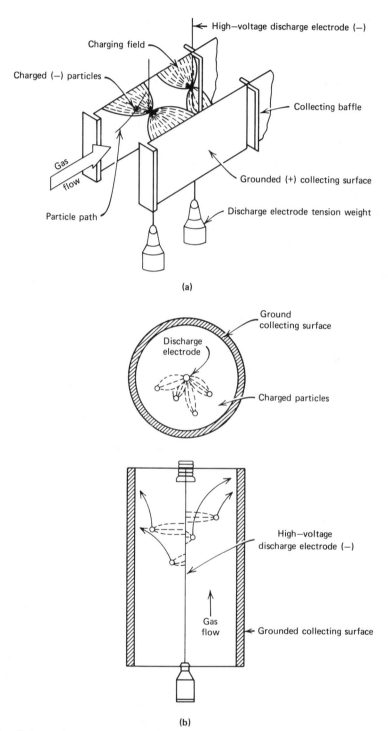

Fig. 10. Schematic arrangement of (**a**) typical plate and (**b**) tube precipitators (172).

cal fumes. Tubes vary in diameter from 150 to 304 mm; a popular size is 273 mm OD by 4.5–5.0 m long. One discharge wire is centered in each tube. Dirty gas generally enters the hopper at the bottom and travels upward through the tubes. A tube sheet may be provided top and bottom; tubes may be rolled or welded into a flat top tube sheet for a mist precipitator. For solids, however, each tube must have a round to square transition at the top end. The tubes are then welded together to form an "egg-crate" sheet. Dust, if deposited on a flat tube sheet, can buildup with a nearly 90° angle of repose and short out the high tension support frame for discharge electrodes.

Figure 11 shows a variety of plate and discharge electrode designs. The most common collecting plate design is flat with vertical, perpendicular baffles at frequent intervals. The plate may be continuous or sectional. The Opzel design is a modification of perpendicular baffles devised to reduce gas-flow resistance. Discharge electrodes may be round wires or small (2.5–4 mm) diameter rods. The smaller the diameter, the higher the intensity of the surrounding electric field. The wire must, however, be sufficiently rugged to withstand both tensioning and thermal stresses. In the event of heavy suspended loads or long spans, wire size may become appreciable. In these cases, a 6 to 9 mm square may be a better design choice because the corners produce a higher intensity field locally. Square bars are often twisted to give one 90° turn in 25–35 mm of length, producing a high intensity field which rotates with the length of the discharge electrode. A 5-pointed-star-shaped discharge electrode yields an even higher field intensity. Discharge electrodes with barbs and punched ribbons may also be used to produce higher intensity electric fields.

Discharge electrodes must be tensioned to hold the wire taut and maintain spacing. Tensioning may be accomplished by attaching a weight (5–10 kg), held in a weight-spacing frame to reduce sway, to the bottom of each wire. In some European designs, now offered in the United States, the wires are stretched between light-weight high-tension pipe frames. The wires may be heavy barbed ribbons that are unlikely to fail from arcing, an important aspect since no provision is made for wire replacement in these rigid frames. The wire supporting frames are hung from high voltage insulators in suitable enclosures on top of the precipitator. Dust, fumes, and mist must be prevented from entering these insulator compartments and coating the insulators with a conductive film. The insulator compartments may be purged with clean air and heated to prevent condensation. In a mist precipitator, liquid drains from the collecting surfaces, but a dry dust precipitator must be rapped at intervals using a weight, hammers, or vibrators. The dust, broken loose from the surface, slides down the plate into the dust hoppers. Dust is removed from the hoppers, usually batch-wise, with suitable conveyors. A precipitator's power supply is usually rectified ac.

Operating Principles. No collection will occur in a precipitator, and no current will flow in the secondary circuit, until gas ionization starts around the discharge electrode. This process is known as corona formation. The starting voltage for corona in air in a tube precipitator is given as

$$V_s = (15.5 \times 10^5) b_r k_\rho D_d \left(1 + \frac{0.0436}{\sqrt{k_\rho D_d}}\right) \ln\left(\frac{D_d}{D_t}\right) \qquad (18)$$

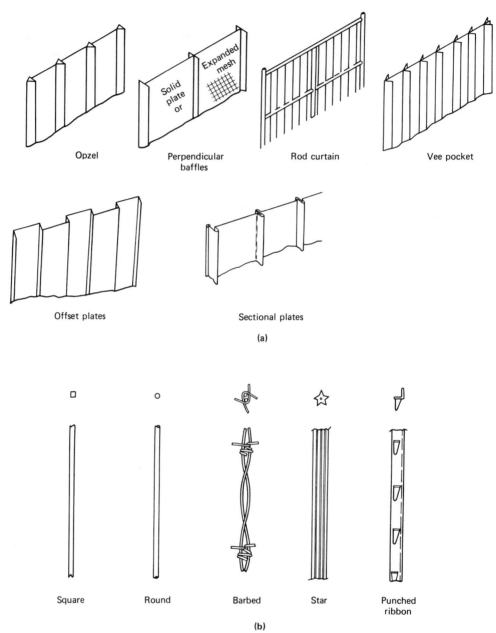

Fig. 11. Types of (**a**) collecting plate and (**b**) discharge electrode designs.

If the effluent is not air-based, it is necessary to multiply by the ratio of the gas electrical breakdown constant to that for air, both taken at 25°C and 101.3 kPa (1 atm).

When corona occurs, current starts to flow in the secondary circuit and some dust particles are precipitated. As potential is increased, current flow and electric field strength increase until, with increasing potential, a spark jumps the gap between the discharge wire and the collecting surface. If this "sparkover" is permitted to occur excessively, destruction of the precipitator's internal parts can result. Precipitator efficiency increases with increase in potential and current flow; the maximum efficiency is achieved at a potential just short of heavy sparking.

Table 8 shows sparking potential in a clean tube precipitator with negative polarity. Sparking potential drops almost directly with decreasing gas density, but increases with moisture content. At elevated temperatures or reduced pressure, the corona starting voltage also drops, but sparkover voltage drops faster. Even low conductivity dust deposits can greatly decrease sparkover voltage until the buildup can be rapped off. For ease of operation and greatest efficiency, the largest possible spread between corona starting voltage and sparkover is desireable. With positive polarity, sparkover occurs at much lower potential and it may become nearly impossible to operate a positive polarity precipitator at elevated temperatures.

Table 8. Sparking Potential for Small Wire Concentric in a Round Tube

Tube diameter, mm	Peak voltage, V[a]	Root mean square voltage, V[a]
102	59,000	45,000
152	76,000	58,000
229	90,000	69,000
305	100,000	77,000

[a]Based on gases at atmospheric pressure, 38°C, containing water vapor, air, CO_2, and mist, using negative polarity electrical discharge. Recalculated from data reported in reference 176.

Dust particles entering the electric field become charged. Both negatively and positively charged ions exist in the small core of ionized gas surrounding the discharge electrode. In a negative polarity precipitator, the positively charged ions are quickly attracted to the discharge electrode and neutralized. Hence only electrons exist outside the corona area and travel through the gas space to the grounded collecting plate. Two mechanisms of particle charging exist: charging by ion bombardment (often called field charging) and charging by ion diffusion. A number of equations to predict particle charging rate have appeared in the literature (141,175). In field charging, electrons collide with dust particles and charge transfer may occur. Eventually, the particle will develop a charge sufficient to repel other electrons, a phenomenon termed charge saturation. Field charging is rapid and most effective on larger particles. The larger the particle, the higher the charge density, and thus the easier is precipitation. Coarse particles (5–30 μm) can be collected almost completely in a precipitator. The lower size range for effective field charging is on the order of 0.5–1.5 μm, depending on particle dielectric constant.

In diffusion charging, particles are too small and mobile for rapid charging by ion bombardment. They become charged by collision caused by motion of the gas molecules. Diffusion charging becomes efficient on particles smaller than 0.2 μm and has been demonstrated to be effective on particles down to 0.05 μm. These fine particles are charged rapidly and have higher charge density at saturation.

Unlike field charging, the rate of charging by diffusion is independent of electric field strength. For particles between 0.2 μm and 1.5–2.0 μm neither charging mechanism is highly efficient, but both operate. Particles in this size range are the most difficult to collect efficiently. Figure 12 compares experimental and calculated particle charge levels for 0.3-μm particles. The field has been reviewed (177) and a particle-charging equation which has been used in computer-modelled precipitator design has been proposed (178,179).

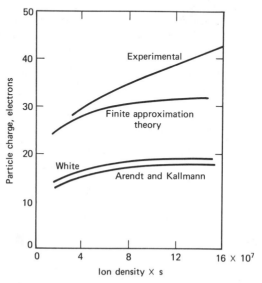

Fig. 12.   Comparison of actual and predicted charging rates for 0.3-μm particles in a corona field of 2.65 kV/cm (141). The finite approximation theory (173) which gives the closest approach to experimental data takes into account both field charging and diffusion charging mechanisms. The curve labeled White (141) predicts charging rate based only on field charging and that marked Arendt and Kallmann (174) shows charging rate based only on diffusion. From ref. 174, courtesy of Addison-Wesley Publishing Co.

It is possible to set up a force balance on a given size particle in a precipitator and hence calculate its trajectory. Three forces tend to move a particle toward the collecting plate: the charge on the particle, the integrated average field strength through which the particle moves, and the electric wind. This last results from the steady flow of ions from the discharge electrodes to the collecting plates; the motion is similar to thermal convective currents. Particle movement is resisted by drag on the particle. In solving a force balance, it is convenient to define the particle migration velocity as the average velocity with which the particle moves toward the collecting surface. This velocity is a function of particle size, field strength, and such particle material properties as electrical conductivity and dielectric constant. The factors affecting migration velocity for conductive particles larger than 1 μm that are in the Stokes' law region can be expressed as (141):

$$U_e = \frac{E_o E_p D_p}{4\pi K_v \mu_g} \tag{19}$$

For conductive particles smaller than 1 μm, equation 19 should be multiplied by the Cunningham correction factor, $(1 + A\lambda/D_p)$. For air at room temperature, the molecular mean free path, λ, is 0.1 μm, and A is 1.72. If the particles to be collected are not electrically conductive, equation 19 should be multiplied by $[1 + 2(\delta - 1/\delta + 2)]$ to correct for the dielectric constant of the material. Equation 19 can be used to predict the relative effect of changes in particle size distribution, gas viscosity, and operating voltage on migration velocity and precipitator efficiency. An interval-integration procedure may be used to calculate average migration velocity for a given particle size distribution, although more often, an average migration velocity is measured experimentally in an existing precipitator for a similar dust. Average migration values are sometimes referred to as drift velocities to distinguish these values from migration velocities which should be related to specific particle sizes. Table 9 lists drift velocities encountered in typical commercial precipitators. Because of the demand for higher efficiency precipitators, drift velocities from less efficient ones must be adjusted for design utilization. To achieve higher efficiency, a greater portion of the difficult to charge particles must be collected; consequently the drift velocity achieved will be lower.

Table 9. Precipitator Applications of Drift Velocities, U_e, cm/s

Precipitator application	Average value[a]	Typical range
pulverized coal fly ash, power boilers	13.1	3.9–20.4
pulp and paper manufacturing particulate	7.6	6.4–9.4
sulfuric acid mist	7.3	6.1–8.5
cement kiln dust (wet process)	10.7	9.1–12.2
nonferrous smelter	1.8	
steel open hearth furnace	4.9	
iron blast furnace	11.0	6.1–14.0
foundry cupola	3.0	

[a]Average values are based on typical efficiencies of precipitators purchased. Drift velocities will drop if higher efficiencies are required. Recomputed from data of reference 175.

The theoretical development of migration velocity involves only field strength, particle charge, and electric wind force; therefore, migration velocity should be independent of forward gas velocity. However, mild turbulent flow can move particles close to the collecting surface, enhancing collection, and giving improved measured migration velocity (180,181). A precipitator developed in Japan to give high migration velocities incorporates a pulse precharger and zigzag collecting electrodes (182). The zigzag collectors probably enhance particle migration through turbulent diffusion.

Collection Efficiency. The classical Deutsch and Anderson equation for predicting particle collection efficiency is

$$\eta = 1 - e^{-(U_e A_e/q_e)} \tag{20}$$

A_e/q_e can be replaced by K_e, which is defined in terms of precipitator geometric parameters and gas passage velocity. For tube precipitators, $K_e = 4L_e/D_t V_e$; for plate precipitators, $K_e = L_e/B_e V_e$. Assumptions involved in applying the Deutsch and Anderson equation to an entire precipitator are that all parallel gas passages have the same gas velocity (this is not necessary if the variation in gas velocity is known because the equation can then be applied passage by passage) and that no reentrainment occurs once the particle is collected. Uniform gas distribution has been improved in some designs with inlet baffling and gas-flow modeling tests (183) and particle reentrainment can be reduced by better dust rapping. Examining K_e indicates that efficiency can be improved by increasing treatment time (longer flow path in the electric field or lower gas velocity) or collection plate area, decreasing the distance from the discharge electrode to the collecting surface, or changing gas or electric field conditions to give higher migration velocity.

Precipitator Operating Problems. Dry dust precipitator operating problems may result from dust reentrainment, dust resistivity, gas distribution, or electrical arcing or lack of voltage control, and some of these problems have been discussed in detail (184). Dust reentrainment can arise from a localized high gas velocity, rapping problems, or dust resistivity. The collected dust becomes agglomerated and is easier to redeposit if entrained. There is an optimum time interval, force intensity, and direction for rapping to minimize reentrainment. A rapping force normal to the plate is most effective. The intensity of a single rap should be just enough to snap the dust cake loose from the plate and allow it to slide en masse down into the dust hopper. Too intense a blow may shatter the cake and project it out into the bulk gas stream as a cloud of small particles; an inadequate blow will require repeated raps to break the cake loose. Electrostatic forces hold the collected dust to the collection surface and the longer the dust layer is in place, the more tightly it is held. When the cake stays in place too long, greater forces are required to dislodge it and chances of reentrainment are greater. One blow of optimum intensity every 1–2 min, continuous intermittent rapping, is better than a burst of high frequency vibration as discussed in rapping parameter experiments (185,186). However, periods of 30–90 min between successive raps may be better (187). Baffles on the collecting plate tend to keep the bulk gas velocity away from the dust layer, providing a quiescent zone through which dust can slide downward during rapping.

Reentrainment can occur at either too high or too low a dust resistivity. With too low a dust resistivity (usually less than 1000 ohm·cm), the dust loses its charge to the collecting plate, randomly tumbles off, and is reentrained. Unburned carbon in flyash is an example. A precipitator is completely unsatisfactory as a total collector for very conductive and fine dusts such as carbon black. However, in such cases precipitators are used as electrostatic agglomerators and final collection occurs in a cyclone. Very high dust resistivity (about $2–5 \times 10^{10}$ ohm·cm) causes a phenomenon known as back corona resulting in poor operation and reentrainment. High resistivity causes a high voltage gradient across the dust layer. The resulting weaker field potential between the discharge electrode and the dust layer surface gives lower drift velocity and collection efficiency. Moreover, the potential drop across the dust layer may be high enough to produce corona discharge, causing the particles to disperse from the collecting surface. These problems have been encountered in some coal-fired power plant precipitators when high sulfur coal is replaced with low sulfur coal. Dust resistivity varies with precipitating temperature and gas moisture content. Resistivity goes through a maximum with temperature increase, but higher moisture levels reduce resistivity at any temperature. Typical remedies are (1) cool the gas (a "cold side" precipitator) (188); (2) treat the gas at a quite high temperature ("hot side" precipitator (189) perhaps requiring alloy construction); and (3) condition the gas by adding steam or moisture. If the dust does not absorb moisture, chemical conditioner may be needed to make the dust hygroscopic. SO_3 is often added for alkaline dusts (190) and NH_3 for acidic ones (191). A correlation for predicting flyash resistivity from composition has been developed (192), and the entire dust resistivity problem has been discussed (193–195).

To achieve maximum efficiency, the precipitator should be operated at the highest voltage possible without excessive sparkover or arcing. Precipitator voltage must be a function of the conditions of the process such as gas temperature, moisture, dust resistivity, and chemical composition. A desirable option for modern precipitators is a voltage-optimizing control circuit which usually contains current-limiting controls and spark-rate sensors. Voltage is automatically varied to maintain light sparking at 50–100 sparks per minute. Strength of sparking can be greatly affected by the pulsation frequency and wave-form of the current. Full-wave, rectified 60–Hz a-c current permits a higher peak voltage than that obtained from a nonpulsing d-c current because the voltage is not at the peak value long enough to produce extensive gas-path electrical breakdown or a heavy arc. Higher frequency current permits the use of higher voltages resulting in higher efficiency without sparkover. The use of a 500-Hz pulse generator and special wave shape has been demonstrated as extremely beneficial (175). Unfortunately, special frequencies or wave shapes have seldom been used commercially because of the cost of power conversion equipment; however, pulse-energization equipment for precipitators (196) has become available in Denmark. Automatic control systems are also available (197). Precipitator on-stream reliability can be enhanced using electrical sectionalization. Fewer discharge electrodes per rectifier result in the loss of a smaller portion of the precipitator if a wire breaks and shorts out. Increasing the number of power rectifiers, however, also increases purchase cost and increases in reliability must be balanced against increases in capital cost.

Power Supply. The preferred power supply for a modern precipitator is a silicon semiconductor power rectifier submerged in oil along a step-up transformer. Typical output voltages are 70–105 kV peak (45–67 root mean square average dc) negative polarity producing output of either two half-waves or one full-wave. Lower voltages are used in some cases of smaller discharge electrode to collecting surface distances. Individual set capacities are typically 15–100 kVA, 250–1500 mA dc. Input power is usually 460 V ±5%, single phase 60 Hz. Power supply capacity must be carefully matched to the needs of the precipitator for maximal operation. The power supply is generally chosen to operate at 70% of capacity or higher. A greatly oversized power supply can result in poor electrical control, high peak currents or excessive sparking, frequent wire burn-out, and overall poor collection efficiency. Too small a power supply can result in its operation being controlled by current limitation and the inability to reach optimum voltage for maximum collection efficiency. Gas temperature and composition, dust loading and resistivity, wire alignment, adequacy of rapping system to keep thin dust layers, collecting surface per electrical section, and geometric design are all factors that can affect power package sizing and selection (198).

Precipitator Application and Cost. There were over 4100 precipitator installations in the United States in 1970, more than 1330 of which were in electric power generating stations for flyash removal (199). About half this number was in each of the next three largest groups of applications: metallurgical, chemical, and fuel-gas detarring. The number of installations may well have doubled by 1990. Design approaches have been discussed (175) and the Air and Solid Waste Management Association (200) has developed an information checklist for precipitator specification and selection. A summary (201) is available as is a review (202) of capital and operating costs for flyash precipitators at TVA steam power plants.

Wet Electric Precipitators. Tube precipitators have long been used for collection of acid mists and for removal of tar from coke oven gas. More recently, plate precipitators employing water sprays have become commercially available for the collection of dry, especially fine particulate, dusts (203). Dust resistivity problems are eliminated because the water-saturated condition renders all particles conductive. All problems with particle reentrainment are also eliminated, because the particles migrate to a water film on the collecting surfaces. Conductive particles have higher migration velocities and therefore there are no problems with high resistivity dust and back corona either. Thus wet precipitators can be smaller for the same collection efficiency. Particle dielectric strength becomes a much more important variable, however, and materials having low dielectric constants, such as hydrocarbon mists, are much more difficult to collect than a water-wettable particle. Relationships for predicting particle charge and relative collection efficiency in a wet precipitator that take the effect of dielectric constant into account have been developed (204). A wet precipitator can also be used to absorb water soluble gases; the pH of the spray liquid can be adjusted to enhance absorption. Comparative tests of wet precipitators with and without the electric field indicate that corona discharge also enhances absorption, but the reason is not known.

There are some disadvantages to wet precipitators. Water can enhance latent corrosion problems and require the utilization of expensive alloy construction instead of the carbon steel often used in a dry precipitator. Some wet precipitators using plastic components have been developed to lower costs in corrosive situations. Additionally, the collection of pollutants in aqueous media may create water treatment and waste handling problems which can equal the cost and complexity of the precipitator installation itself. Spray rate and distribution can be critical in a wet precipitator and must be carefully applied so as not to restrict operating voltage. Recirculated spray water may become supersaturated with low solubility compounds which plate out on surfaces or build up in critical parts of the precipitator. Recirculated suspended solids can erode or plug spray nozzles. Wet precipitators have been most useful in treating mixtures of gaseous and submicrometer particulates such as aluminum pot line and carbon anode baking fumes, fiberglass fume control, coke oven and metallurgical fumes, and phosphate fertilizer emissions.

Two-Stage Precipitators. In a two-stage precipitator, the particles are charged in an initial ionizing stage and then collected in a second section consisting of closely spaced plates as illustrated in Figure 13. Basic modular units are

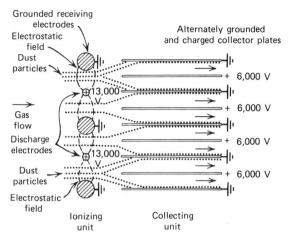

Fig. 13. Operating principle of two-stage electrical precipitator. From ref. 205, courtesy of McGraw-Hill, Inc.

often stacked in banks as needed for the design gas flow. The ionizing electric field is produced by fine tungsten wires, often 0.20 mm in diameter, spaced midway between large diameter (30 mm) grounded rods or flat plates at a 90-mm spacing. Corona discharge and occasional sparking to ground are produced by a 13-kV positive potential. The collecting stage usually consists of a number of parallel 20-gage plates spaced about 6–8 mm apart and separated by insulators. Alternate plates are grounded; the others are charged positively at 5–7.5 kV. Charged particles are attracted to the grounded plates while being repelled by the charged ones. The long, smooth plate surfaces prevent corona or sparking. Thus there is no means in the collector by which entrained particles may be recharged and collected. The plates may be coated with a film of water-soluble adhesive or oil to aid collected particle retention. Collection modules, often 600–1000 mm long, are mass produced. Typical superficial gas velocities are in the range of 1.5–2.8 m/s. The two-stage precipitator has very low (0.042 kW·s/m³) power consumption and a pressure drop in the range of 25–50 Pa (0.2–0.4 mm Hg). Operating voltage is supplied from self-contained power packs that usually have solid state rectifiers and sufficient controls to prevent excessive arcing in the ionizer and to control shorting in the collector.

Because the collecting space for dust is very small, the use of two-stage precipitators is limited to light dust loads or collection of liquid mists. The most frequent use is for air cleaning for ventilation and air conditioning. In many of these units, the collection section must be manually removed for cleaning when dust thickness reaches about 1.5–2.0 mm. Alternatively, they can be cleaned in place with automatically controlled traveling spray cleaning and adhesive coating systems. A major advantage of the two-stage precipitator is its low cost. Collection of liquids is possible, eliminating cleaning needs, but conductive liquids may short the spacing insulators. Two-stage units have also been used for collecting hydrocarbon mists. Single-stage Cottrell precipitators are seldom used on combustible materials because of the danger of fire or explosions from arcs. Since two-stage precipitators operate at much lower voltages and sparking is absent from the collection stage, they are considered to be inherently safer. Sparking can and does occur in the ionizer stage, however, and extreme caution should be exercised if explosive mixtures could be present.

Electrically Augmented Particle Collection.   In the 1970s considerable developmental research was devoted to improved fine particle collection by using a combination of electrostatic forces and other collection mechanisms such as inertial impaction or Brownian diffusion. A number of devices were developed and some successfully applied. Little subsequent progress has been made, presumably because of the reduced availability of governmental funding. As mentioned earlier, the electrostatic attractive force is the strongest collecting force available for particles finer than 2–2.5 μm. It is therefore logical to couple it with other collecting mechanisms in an attempt to build more highly efficient and cheaper fine particle collectors. The forces operating between charged and uncharged bodies decrease in magnitude in the order (206): (*1*) coulomb force, the attraction of a charged particle to an oppositely charged particle or surface; (*2*) charged particle–uncharged conducting collector; (*3*) uncharged particle–charged collector; and (*4*) charged particle in a space-charged repulsive field. These relationships have been quantified (207). The coulomb force, or migration of a particle in a gradient electric field, is, of course, employed in both single-stage and two-stage precipitators. Electrostatics is also operative for charged particles and oppositely charged water droplets. A collection efficiency of 85–95% for industrial gases in such a single-stage vertical spray tower and up to 99% collection efficiency utilizing two towers in series have been demonstrated (208,209). A number of retrofit installations have been made on existing spray towers to enhance collection of submicrometer particles.

An investigation of the collection of charged submicrometer particulates on an oppositely charged 0.8–2.0-mm fluidized bed of sand particles found up to 90% efficiency. The charged particle–uncharged collecting surface principle has been used and marketed in a device where particles are initially charged in a negative polarity ionizer before entering a grounded, irrigated, cross-flow, packed bed of Tellerette packing (211). The charged particles are brought very close to wetted surfaces in the confines of tortuous paths through the bed, inducing a "mirror image" charge of opposite polarity in the water film that causes the particles to impact the water for collection. This device can be used for simultaneous particulate collection and gas absorption. Submicrometer particles have been collected at 85–90% efficiency in single-stage units and at 98% efficiency using two units in series. The principle of charging particles in an ionizer and collecting them dry on uncharged filter media such as cellulose fibers has been practiced for many years in some types of electronic air cleaners. This principle has been applied (212) to improved collection of fine particles in a bag dust filter. Another application of the charged particle–uncharged collector principle was its application to fine particle collection in venturi (wet) scrubbers. Precharging of particles in an ionizer permitted efficient collection of submicrometer particles in the venturi with up to a 50% reduction in venturi pressure drop, a major savings in energy. Unfortunately, a device marketed by Chemical Construction Company disappeared after their bankruptcy, but a similar adaptation is available in France. A charged-droplet scrubber whose commercial design was installed on steel furnace fumes has been described (213). Severe corrosion and electrical loss problems resulted in withdrawal of the scrubber.

Research devoted to optimizing design of various electric augmentation hardware items includes: exploration (214) of parameters for preionizer design; and discussion (215) of detailed mathematical relations for types of charged-droplet scrubbing and means of charging spray particles. Wet scrubbing appears to have a valid place in air pollution control because of its relatively low capital investment compared to other control techniques for fine particles. However, wet scrubbing is a marginal control method because of its poor efficiency–energy relationship on fine particles, and thus electric augmentation should be an attractive means of overcoming the weakness of wet scrubbing at moderate cost.

Particle Filtration.   Filtration devices for particle collection can be divided into three categories: cloth filtration using either woven or felted fabrics in a bag or envelope, paper and mat filters, and in-depth aggregate bed filtration. The first type is used for dry particle removal from gases, but cannot be employed when liquid particulates are present or condensation is imminent. Subclasses of cloth filters are dust filters using cloth in the form of a single envelope (pocket filter) and housings containing rows of stacked cannister filters. The pocket filter has low dust-handling capacity, and when pressure drop becomes too high, the element must be removed and either discarded or manually cleaned. It is used primarily for very low dust loads, occasional emissions, or as a safeguard against broken bags after a normal baghouse filter. Likewise, the multiple canister filters cannot be cleaned and must be replaced once high pressure drop occurs. Filters in the form of fiber pads or pleated paper in frames are used for preparing clean air for process use or ventilation. They have limited dust-holding capacity and are seldom used for air pollution control.

Several types of aggregate-bed filters are available which provide in-depth filtration. Both gravel and particle-bed filters have been developed for removal of dry particulates but have not been used extensively. Filters have also been developed using a porous ceramic or porous metal filter surface. Mesh beds of knitted wire mesh, plastic, or glass fibers are used for the removal of liquid particulates and mist. They will also remove solid particles, but will plug rapidly unless irrigated or flushed with a particle-dissolving solvent.

Dust Filter.   The cloth or bag dust filter is the oldest and often the most reliable of the many methods for removing dusts from an air stream. Among their advantages are high (often 99 + %) collection efficiency, moderate pressure drop and power consumption, recovery of the dust in a dry and often reusable form, and no water to saturate the exhaust gases as when a wet scrubber is used. There are also numerous disadvantages: maintenance for bag replacement can be expensive as well as a sometimes unpleasant task; these filters are suitable only for low to moderate temperature use; they cannot be used where liquid condensation may occur; they may be hazardous with combustible and explosive dusts; and they are bulky, requiring considerable installation space.

Bag filters may be woven or felted, an envelope ("pillow case") supported with an internal wire cage, or a long cylinder or stocking hung freely or containing an internal wire cage, and subject to either shaking or reverse flow for dust removal. Older filter installations employed woven cloth bags and the collected dust was generally removed by shaking. Newer bag cleaning methods such as reverse flow utilize nonwoven felted bags. The availability of fibers is somewhat more limited with felts and one cannot choose a type of weave as with woven fabrics. Three main types of bag filters are illustrated in Figure 14. A fourth type uses backflow of gas, thus essentially providing back-flushing of the cloth. In most baghouse designs, there is some means of slowing the entering gas and deflecting it downward so that coarse dust particles drop out into the hopper. This may be as simple as routing the gas around an inlet baffle. In passing through the filter media, the majority of the dust collects in a dust precoat built up on the bag surface and the cleaned air then flows to the gas outlet. Deposited dust is removed at intervals to prevent excessive pressure drop. In shaker-cleaning filters (envelope and stocking types), automatically controlled dampers shut off the air flow through the bag compartment when the bags are to be shaken. In filters designed for continuous use, a number of parallel compartments are provided to handle flow when one compartment is shut off for cleaning.

In the pulse-jet (reverse-flow) filter (Fig. 14c), air flow is usually not shut off during bag cleaning. Rather, one row or group of bags are cleaned while the rest of the filter remains in service. Dirty gas flows up around the outside of the bag and then through the cloth leaving the dust on the outside of the bag. An internal wire cage keeps the bag expanded and bag and cage are hung from the top tube sheet. Each tube sleeve contains a venturi casting which extends inside the bag and the cleaned air exits through this casting. For bag cleaning, compressed air from an orifice in a manifold pipe above the casting is jetted back down through the venturi. The air pressure, usually at 620–690 kPa (90–100 psi), is released by a solenoid valve actuated by a timer. This jet of air in the venturi induces backward flow of cleaned air from other tubes and an air bubble forms in the top of each bag being cleaned, snaps the cloth away from the cage, and displaces the dust layer. The compressed air pulse must last long enough to allow the bubble to move slowly down the length of the bag to the bottom, cleaning the entire bag surface as it goes. Frequency of cleaning is set to limit the bag pressure drop just before cleaning to the maximum desired level, often 1–1.7 kPa (7.5–13 mm Hg). Degree of cleaning can be adjusted somewhat by changing the pressure of the compressed air and the duration of the pulse. For efficient dust collection, it is desireable to leave a light dust precoat on the bag surface and for long bag life, a minimum number of cleaning cycles is preferred. Since some bags in the compartment are filtering while some are being cleaned, any reentrainment of the dust is followed by redeposition on bags remaining in the filtering operation. Dust redeposition is an appreciable problem even at low filtering velocities; it becomes considerably worse as high filtering velocities are approached (217).

Reverse or backflow bag filters are rather similar in appearance to shaker filters except that the shaking mechanism is eliminated. The stockings are clamped to a tube sheet at the bottom and are closed at the upper end with a metal cap from which they are suspended. Dirty gas enters below the tube sheet and passes upward through the bag. When the bag cleaning cycle begins, the flow of dirty gas is shut off, and a fan forces cleaned gas backward through the bags. A series of rings, sewn into the bags at intervals, prevent the complete collapse of the bag under the reverse-flow conditions. Dust dislodged by the backflow falls down through the bag to a dust hopper below the tube sheet. The quantity of back-

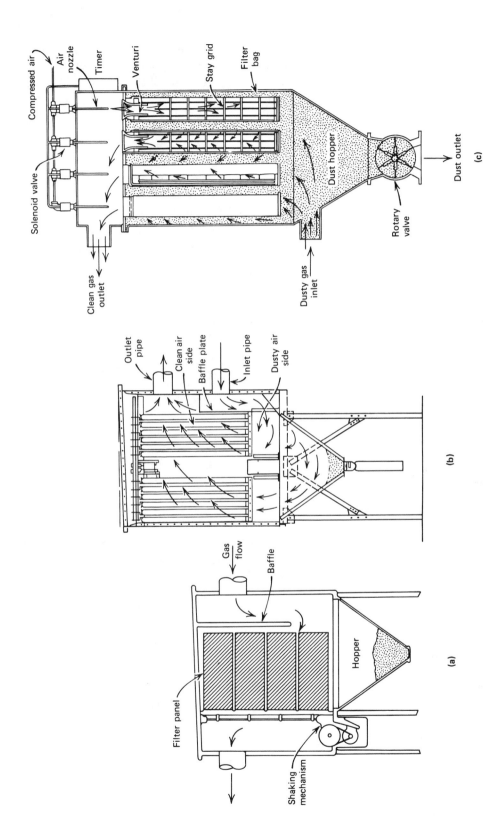

Fig. 14. Types of bag filters: (**a**) panel or envelope filter; (**b**) bag or stocking filter; (**c**) pulse-jet filter (216); (**a**) and (**c**) from ref. 216, courtesy of Academic Press, (**b**) courtesy of Wheelabrator-Frye.

793

flush gas is usually sufficient to produce a reverse-flow superficial velocity of 0.5–0.6 m/min through the bag. Woven fabrics are generally used for reverse-flow bag filters. The small reverse-flow pressures generally used would be insufficient to back-flush felt bags. The principal application for reverse-flow cleaning is in bag-houses using fiberglass bags that handle gas at temperatures above 150°C such as boiler flue gas containing flyash. Bag collapse and reinflation must be sufficiently gentle that excessive stress is not applied to the fiberglass fabric.

Dust Filter Design Considerations. Separation of the dust aerosol from the carrying gas is not a sieving or simple filtration process since filter fabric pore size is much larger than the particles collected. The efficiency of a new bag for fine dust particles is quite low until the bag fibers and interstices are coated with collected dust. A used bag always has higher collection efficiency than a new one because the entrapped dust particles reduce the effective pore size. A precoat of coarser particles serves as the filter media for finer ones and filter efficiency drops momentarily after a bag cleaning that is too thorough. Pressure drop through a cleaned, dust-impregnated bag may be as high as ten times that of a new bag. General particle collection mechanisms, such as direct impingement, inertial impaction, gravity settling, Brownian diffusion, and electrostatic attraction, apply in initial bag coating. An understanding of the importance of these mechanisms and their effect on collector efficiency for fine particles as well as for particle release during cleaning is necessary for the development of improved filtration fabrics (218,219). Direct and inertial impingement are the most important mechanical means for fabric impregnation of larger particles; Brownian diffusion is the most effective for submicrometer particles.

The effect of dust particle electrostatic attraction to fibers has been investigated (220). When air is passed through a fiber bag, it may generate an electrostatic charge on the fiber. The relative magnitude and polarity of the charge depends on fiber composition and surface configuration (eg, filament vs staple) (see FIBERS). Table 10 gives a triboelectric series for various fabrics. If fine aerosol particles develop a charge that is opposite to that of the fabric used in a dust filter, initial collection should be improved. For certain fabrics, collected particles are held so tenaciously that inadequate cleaning occurs and excessive pressure drop develops very quickly. Selection of another fabric having less electrostatic attraction for the dust might well reduce this problem. The effect of charge loss may be a significant factor in cloth cleaning because dust seepage following each cleaning cycle can be a severe problem. Selection of a fabric having greater electrostatic attraction for the dust could be a remedy. Fabric surface characteristics can also affect the retention or release of fine particles. Filament fabrics tend to have slick surfaces and thus easier release than staple fabric. In woven materials, napped fabrics tend to retain fine particles and develop dust precoats better than unnapped materials. Finer and tighter weaves resulting in thicker fabrics (increased fabric weight), less vigorous cleaning, and longer intervals between cleaning cycles all help to reduce leakage of fine particles. In extreme situations, bag precoating with a filter aid after a cleaning cycle may be necessary. Fabric selection and efficiency have been discussed (221).

Pressure drop through a filter bag has two components, the drop through the dust cake and that through the cloth. The flow through both is streamline so the pressure drop varies directly with gas velocity. Pressure drop through the cloth is defined by equation 21. Values of k_c, essentially a drag coefficient for the fabric,

Table 10. Triboelectric Series for Commercial Fabrics[a]

Polarity	Material
+25	
	wool felt
+20	
+15	glass, filament, heat cleaned and silicone treated
	glass, spun, heat cleaned and silicone treated
	wool, woven felt, T-2
+10	nylon-6,6, spun
	nylon-6,6, spun, heat set
	nylon-6, spun
	cotton sateen
+5	Orlon 81, filament
	Orlon 42, needled fabrics
	Arnel, filament
	Dacron, filament
	Dacron, filament, silicone treated
0	Dacron, filament, M-31
	Dacron, combination filament and spun
	Creslan, spun; Azoton, spun
	Verel, regular, spun; Orlon 81, spun
	Dynel, spun
−5	Orlon 81, spun
	Orlon 42, spun
	Dacron, needled
−10	Dacron, spun; Orlon 81, spun
	Dacron, spun and heat set
	polypropylene 01, filament
	Orlon 39B, spun
−15	Fibravyl, spun
	Darvan, needled
	Kodel
−20	polyethylene B, filament and spun

[a]Ref. 220, reprinted with permission of *J. APCA.*

have been tabulated (222) for various types of unused cloth. However, the pressure drop for dust-impregnated fabric can be over ten times as great. Whenever dust cake thicknesses exceed 1.5 mm, the pressure drop across the cloth becomes insignificant. Equation 22 gives the pressure drop across the dust layer at any point in time following cleaning. If it is assumed that dust concentration, C_d and gas flow volume V_f are constant with time, then $m_d = C_d V_f \theta$, where θ is the time since the end of the last cleaning cycle and upon substitution equation 23 results. Despite the squared gas-flow term, flow is streamline, not turbulent.

$$\Delta P_c = k_c \mu_g V_f \tag{21}$$

$$\Delta P_d = k_d \mu_g m_d V_f \tag{22}$$

$$\Delta P_d = k_d \mu_g C_d V_f^2 \theta \tag{23}$$

The term k_d, essentially a drag coefficient for the dust cake particles, should be a function of the median particle size and particle size distribution, the particle shape, and the packing density. Experimental data are the only reliable source for predicting cake resistance to flow. Bag filters are often selected for some desired maximum pressure drop (500–1750 Pa = 3.75–13 mm Hg) and the cleaning interval is then set to limit pressure drop to a chosen maximum value.

Bag filtering area is another variable to be determined. Selection is usually on the basis of the volume of gas to be filtered and a desired superficial gas velocity through the bag surface. The design superficial velocity chosen should depend on the dust concentration in the gas to be filtered, the type of dust, the fineness and abrasiveness, the type of cloth and weave, the cost of the cloth, and the desired frequency of bag cleaning. For woven cloth, the typical range is 0.005–0.04 m/s; although many users like to limit the velocity to 0.020–0.025 m/s. If the dust is very fine or abrasive, or if the dust concentration is unusually high, the velocity drops to 0.015 m/s or less. For shaking collectors, sufficient bag area is necessary that excessive velocities do not exist when one compartment is off-line for shaking. Manufacturers of pulse-jet collectors (felt bags) recommend some-what higher velocities: 0.025–0.075 m/s. Some users, from experience, prefer to use velocities no higher than those employed for cloth bags with shaker cleaning and the experiments (217) involving redeposition would appear to support keeping pulse-jet collector velocities at the lower end of the range.

Bag cleaning and control methods need to be selected. For cloth bags, shaking is well-established, but pulse-jet cleaning is probably the most popular choice today whenever felted bags can be successfully utilized. Pulse-jet cleaning should be more desireable for applications handling heavy dust loads in continuous operation since it is always controlled by an automatic timing device. Automatic timing can also be applied to shaker filters, but for applications of light dust loading or intermittent gas flow, remote manual control may be preferred.

Choice of bag fabric depends upon chemical compatibility with the dust to be collected (see Table 11) and temperature resistance (see Table 12) as well as fabric cost. Additional data on fabric resistance to abrasion, temperature, and specific chemical compounds have been presented (223). There are also other special fabric considerations. Glass fibers do not resist abrasion and flexing well. Surface coatings help, but special cleaning procedures such as reverse flow are required for these fibers, and longer intervals between cleaning are desireable. Nomex nylon has poor resistance to moisture and should be avoided whenever condensation is possible. Teflon, although highly resistant, is extremely expensive; cotton and Dacron are relatively inexpensive.

No bag fabric can withstand truly high temperature, therefore gas cooling is often practiced. The usual methods are indirect cooling, tempering with cold air, direct water spray cooling, or a combination of any of these. Indirect cooling may take place in radiation panels or ducts exposed to the atmosphere, in waste heat boilers, or in heat transfer devices such as finned heat exchangers and heat wheels. Tempering consists of mixing air from the atmosphere and the hot gas in a duct; good mixing must be provided to ensure temperature equilibration. Automatic temperature control can be quite precise and tempering can reduce the dewpoint of a hot, humid gas. The major disadvantage is that tempering increases the gas volume and hence the required size of the dust filter and exhaust fan; it also necessitates additional power to draw the diluted gas through the filter.

Table 11. Chemical Compatibility of Fibers in Dust Collector Bags

Resistance	In acid media	In alkaline media
excellent	polypropylene	polypropylene
	polyethylene	polyethylene
	Saran	Dynel
	Teflon	nylon-6,6
	Orlon	Teflon
good	Dacron	cotton
	Dynel	nylon-6
	glass	Nomex nylon
	wool	Saran
unsuitable	cotton	wool
	nylon-6,6	glass
	nylon-6	
	Nomex nylon	

Table 12. Maximum Desirable Operating Temperature for Filter Bags[a]

Material	Temperature, °C	Material	Temperature, °C
polyethylene	70	Arnel	120
Saran	70	Microtain	125
cotton	80	Kodel	135
Dynel	85	Dacron	140
polypropylene	90	Darvan	150
wool	100	Nomex nylon	230
nylon-6,6 and -6	105	Teflon	260
Orlon	120	glass	290
Acrilan	120		

[a]Longer bag life is obtained if bags are not operated at their maximum temperature.

Direct water spray cooling must be carried out with care. The spray chamber must be designed to ensure complete evaporation of all liquid droplets before the gas enters the baghouse. Spray impinging on the chamber walls can result in a dust mud inside the chamber and any increase in gas dewpoint may result in baghouse problems or atmospheric plume condensation. Spray nozzle wear can result in coarse or distorted spray and wetted bags, and water pressure failure can cause high temperature bag deterioration.

Baghouse safety must also be taken into consideration since the forced ventilation of filtering air can convert a baghouse into a forge in short order, especially if the dust is flammable. Most bags will burn; a few will char but not support combustion. Glass is the only fireproof bag material, but it has a low softening temperature. Large installations need temperature-sensing devices to shut off air flow in the event of fire. In many installations sprinkler heads, CO_2 injection, or dry chemical systems are provided. Because baghouse explosions may also be a problem, large vents should be provided wherever there are susceptible dusts. Checklists of recommendations for safeguarding fabric filters (224) and for specifying and evaluating dust-filter purchase (225) have been prepared.

Explosion venting (226) and optimization of pulse-jet filter design (227) have been discussed. The cost and performance of bag filters has been compared with other particulate control devices (228). Other pertinent reviews (229–231) also exist.

Newer Bag Filter Applications and Developments.   Bag filters have been used extensively for containing dust emissions from low temperature process operations, such as milling and screening, drying, packaging, loading and unloading, and material conveying. Efforts have been made to extend use to the control of fine particle emission from common large-volume, hot-gas streams, such as coal-fired power boilers, and ferrous and nonferrous metallurgical processes. A number of demonstration installations of bag filters on power boiler effluent have been made (232–240). New applications (241) and cost estimation procedures (242) for fabric filters have been discussed. The effects of conventional fabric structure on collection efficiency have been studied (243) and nonconventional fabrics have also been investigated (244). Trilobal fibers gave improved collection efficiency and reduced pressure drop; efficiency also improved when a weave tighter (0.33 tex) than the conventional (0.67 tex = 6 den) weave was used, but at the expense of higher pressure drop. Crimped fibers also improved collection efficiency and reduced pressure drop.

Filtration by Granular Beds.   Fine particle removal by filtering through a bed of granular solids has been known and used for many years, but a basis for predicting efficiency and pressure drop has only been developed in the last decade or two. Granular filters have appeal because of corrosion and temperature resistance, physical robustness, and theoretical ease of cleaning. Both the Lynch filter (245), described in 1936, and the Dorfan Impingo filter (246) developed commercially in 1950, but now unavailable, utilized a downward-traveling gravel bed. A collection efficiency of 98% on 2 to 10-μm particles was claimed for the latter filter, but such efficiency has been found only with a fresh packing charge (247). Another cross-flow granular scrubber, developed by Combustion Power Company (248), used 3–6-mm pea gravel. Gas face velocities were 0.5–0.75 m/s and pressure drops were 0.5–3 kPa (3–23 mm Hg). Collection efficiency on submicrometer particles was low, but subsequently electric augmentation corrected this problem (249); this has been confirmed (250). The process depends on the natural charge on the dust particles passing through the charged (20–30 kV of potential is imposed) field of the gravel bed. Another type of gravel-bed filter, developed in Germany, has been described (251) in which gas, precleaned in a cyclone, is then passed to a stationary gravel bed. When the bed becomes dust saturated, forward flow is stopped and the gravel is back-flushed using air to remove the dust, which is then collected in the cyclone. Tests indicated low collection efficiency on 2-μm particles (252). A cross-flow granular limestone bed has been used to remove both SO_2 and fly ash from power boilers (253). The limestone loses its reactivity because of formation of a gypsum coating. Attempts to remove the fly ash and gypsum using a combination of backflow and bed rapping techniques have not been particularly successful.

In one theoretical equation (254) for collection efficiency in granular beds important parameters are bed thickness, gravel size, and air velocity. A cleanable granular bed filter (Fig. 15) has been commercially tested (255). Filtering granules are contained in annular compartments enclosed in a housing. Dirty gas enters above the static granular bed, travels downward through the bed and emerges

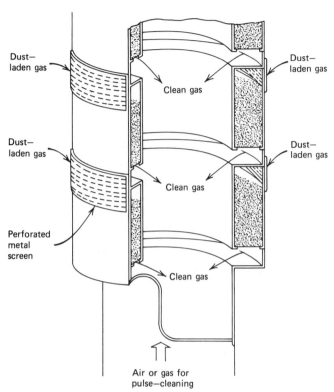

Fig. 15. Zenz-Ducon expandable-bed granular filter. From ref. 256, courtesy of McGraw-Hill, Inc.

clean at the bottom. Collection efficiency is stated to be as high as 99% on 1–3-μm particles. When the beds become dust-laden, forward flow is shut off and a reverse flow of high pressure air in short blasts fluidizes the beds and elutriates the dust. This concentrated dust–air mixture must then be treated in another device for final dust recovery. Granules (590 μm or 30-mesh in size) of various materials have been used. One of the problems appears to be inadequate elutriation of particular dusts so that laboratory tests are conducted using the applicant dust before a unit is recommended for installation.

Particle collection in low fluidization velocity beds has been investigated (254,258); these studies indicate that collection efficiencies higher than 90% would be difficult to obtain. Using shallow fluidized beds of 25-μm particles, 99.7% collection of submicrometer particles through bed staging was obtained (259). Several conditions yielded this efficiency: three stages, each bed 76 mm thick, four stages, each bed 41 mm thick, and five stages, each bed 25 mm thick.

Fiber Bed Mist Filtration. In-depth fiber bed filters are used for the collection of liquid droplets, fogs, and mists. Horizontal pads of knitted metal wire (or plastic fibers), 100–150 mm thick, and gas upflow are used for liquid entrainment removal. Pressure drop is 250–500 Pa (1.9–3.8 mm Hg). Characteristics of the pads vary slightly with mesh density, but void space is typically 97–99% of total volume. Collection is by inertial impaction and direct impingement; thus efficiency will be low at low superficial velocities (usually below 2.3 m/s) and for fine particles. The desireable operating velocity is given as:

$$U_s = K_b \sqrt{\frac{\rho_L - \rho_g}{\rho_g}} \qquad (24)$$

The value for K_b is 0.11 m/s, unless liquid viscosity is high or the liquid loading is very high, when somewhat lower values of K_b are used. The same is true for materials with low liquid surface tension. At desired operating velocities, 99% efficiency will prevail for particles 5 μm and larger. For smaller particles, efficiency drops rapidly; it is about 92% for 3-μm particles. Design factors for wire-mesh mist eliminators have been discussed (260) and the use of two mesh pads in series for collection of finer mist particles at much higher pressure drops have been investigated (261,262).

For collection of fine mist particles, the use of randomly oriented fiber beds is preferable. Three types of such beds are available: chemically resistant glass, polypropylene, or fluorocarbon fibers. Beds 25–50 mm thick held between parallel screens are designated high velocity filters. Installed vertically, with horizontal gas flow and vertical liquid drainage, such beds are sized for 2.5 m/s face velocity and pressure drops of 1–2.5 kPa (7.5–19 mm Hg). These filters collect by inertial impaction. Efficiency improves with increasing velocity and pressure drop, eg, 92–94% efficiency on 1-μm particles at 1–1.5 kPa pressure loss. The performance has been discussed (263). Large cylindrical elements similar to Figure 16, designated high efficiency mist eliminators, are custom designed for collection of submicrometer mist at efficiencies as high as 99.9%. The collection parameters of these units have been described (264). Bed thickness can vary from 75 to 101 mm and face velocities are much lower than those for high velocity units. Pressure drop is in the range of 2.5 to 7.5 kPa (19–56 mm Hg). For particles smaller than 3 μm, the predominant collection mechanism is Brownian diffusion; efficiency increases with reduction in flow and no turndown problems exist. A third type of bed, which is designated spray catcher and is lowest in efficiency, is available for 100% collection of particles 3 μm and larger. It is similar in configuration to the high velocity type, but has a pressure loss of only 138–276 Pa (1–2 mm Hg). Fiber bed mist eliminators, designed primarily for liquid particulates, can also be efficient collectors of water-soluble solid particles if the dust loading is not high and the bed is continuously irrigated with water or another solvent.

Some effluents contain submicrometer oily or tarlike particulates that are difficult to remove from the collection surface. A disposable, dry glass fiber mat, (266), that can handle such submicrometer particulates at 99% efficiency is known. A subsequent modification is the replacement of the disposable mat with a cylindrical drum or endless belt of recticulated foam which is cleaned and recycled. Cleaning procedures, using steaming, detergent washing, solvent cleaning, and ultrasonic cleaning in a variety of fluids (see ULTRASONICS), have been developed.

Wet Scrubbing.   Scrubbers can be highly effective for both particulate collection and gas absorption. Costs can be quite reasonable for the required efficiency, but the addition of water treatment for recycle or for waste disposal

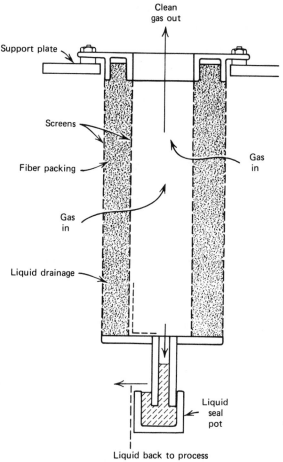

Fig. 16. Brink-Monsanto Envirochem high efficiency mist eliminator element (265).

may make the total cost as great as any other collection method, depending on water treatment complications. Although scrubbers automatically provide cooling of hot gases, the water-saturated effluent may produce offensive plume condensation in cold weather. Many moist effluents become more corrosive than dry ones. Solids accumulation may occur at wet–dry interfaces and icing problems may occur around the stack in winter. For efficient fine-particle collection, energy consumption may also exceed that of dry collectors by an appreciable amount.

Scrubbers make use of a combination of the particulate collection mechanisms listed in Table 5. It is difficult to classify scrubbers predominantly by any one mechanism; but for some systems, inertial impaction and direct interception predominate. Semrau (153,262,268) proposed a contacting power principle for correlation of dust-scrubber efficiency: the efficiency of collection is proportional to power expended and more energy is required to capture finer particles. This principle is applicable only when inertial impaction and direct interception are the mechanisms employed. Furthermore, the correlation is not general because different parameters are obtained for differing emissions collected by different devices. However, in many wet scrubber situations for constant particle-size distribution, Semrau's power law principle, roughly applies:

$$N_t = \alpha P_T^\gamma \tag{25}$$

The constants α and γ depend on the physical and chemical properties of the system, the scrubbing device, and the particle-size distribution in the entering gas stream.

Table 13 can be used as a rough guide for scrubber collection in regard to minimum particle size collected at 85% efficiency. In some cases, a higher collection efficiency can be achieved on finer particles under a higher pressure drop. For many scrubbers the particle penetration can be represented by an exponential equation of the form (271–274)

$$P_t = e^{-A(d_p)^B} \tag{26}$$

where A and B are constants dependent on scrubber design: B is 0.67 for centrifugal scrubbers and essentially 2 for packed towers, sieve plates, and venturi scrubbers. Use of equation 26 permits development of generalized grade-efficiency curves of particle cut-size for a number of different types of scrubbers (see ref. 169). If the entering particles have a log-normal mass distribution, Figure 17 can be used to obtain the integrated particle penetration leaving the scrubber in terms of scrubber cut-size, mass median particle diameter, exponent B, and geometric particle-size distribution. Plotting particle cut-size vs pressure drop for various scrubbers is suitable for developing generalized energy efficiency curves (271). The curves for the various devices outline an imaginary band of pressure drop vs particle size for inertial impaction.

Table 13. Particle Size Collection Capabilities of Various Wet Scrubbers[a]

Type of scrubber	Pressure drop, Pa[b]	Minimum collectible particle dia, μm[c]
gravity spray towers	125–375	10
cyclonic spray towers	500–2,500	2–6
impingement scrubbers	500–4,000	1–5
packed and moving bed scrubbers	500–4,000	1–5
plate and slot scrubbers	1,200–4,000	1–3
fiber bed scrubbers	1,000–4,000	0.8–1
water jet scrubbers		0.8–2
dynamic		1–3
venturi	2,500–18,000	0.5–1

[a]Refs. 155, 269, 270.
[b]To convert Pa to psi, multiply by 1.450×10^{-4}.
[c]Minimum particle size collectible with approximately 85% efficiency.

Wet scrubber collection efficiency may be unexpectedly enhanced by particle growth. Vapor condensation, high turbulence, and thermal forces acting within the confines of narrow passages can all lead to particle growth or agglom-

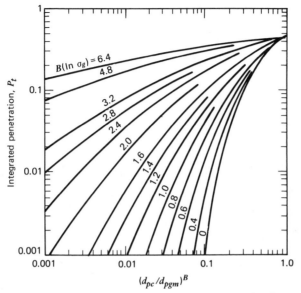

Fig. 17. Overall (integrated) penetration as a function of collector particle cut-size and characteristics and inlet particle parameters for collectors that follow equation 26 (271). Courtesy of *J. APCA*.

eration. Of these, vapor condensation produced by cooling is the most common. Condensation will occur preferentially on existing particles, making them larger, rather than producing new nuclei. Careful experimentation (275) has shown that the addition of small quantities of nonfoaming surfactants to the scrubbing water can enhance the collection of hydrophobic dust particles without further energy expenditure.

Scrubber Performance and Selection. A tremendous number of wet scrubbers have been marketed. References 155, 273, and 276 provide general introductions. An extensive study of wet scrubbers has been done (272), and there are reports on wet scrubber selection and evaluation (277). An open spray tower, such as the countercurrent vertical one of Figure 2**b**, is the simplest type of scrubber. Gas velocities (typically 0.6–1.2 m/s) should be less than the terminal settling velocity of the spray droplets. Optimum spray droplet size for maximum target efficiency has been calculated (278) and can be obtained from Figure 18. Figure 19**a** gives cut-size for vertical spray towers, 19**b** for horizontal spray towers. Tangential inlet of gas at the bottom of vertical spray towers (Fig. 2**b**, **c**) greatly increases the collision of aerosol particles and spray droplets as a result of centrifugal force. The optimum spray droplet size for greatest collection efficiency in centrifugal spray towers is considerably smaller than for gravity towers. Figure 20 shows calculated target efficiencies (140), for spray droplets in a centrifugal field. Cyclonic scrubbers similar to Figure 2**b** have efficiencies of 97% on 1-μm particles (282). A number of commercial scrubbers such as the type N and R Rotoclone and the Microdyne use a combination of water atomization and centrifugal force to capture particles.

Vertical columns with plates or trays are also used for particulate collec-

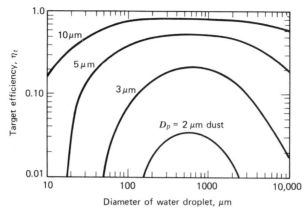

Fig. 18. Target efficiency of a single water droplet in a gravitational spray tower (278,279). From ref. 280, courtesy of McGraw-Hill, Inc.

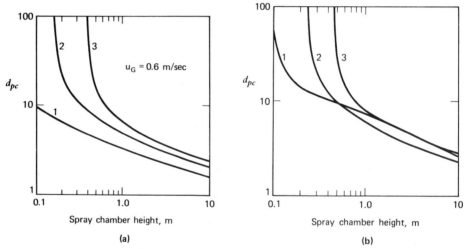

Fig. 19. Predicted performance cut diameter for typical spray towers (271): **(a)** vertical countercurrent spray tower; **(b)** horizontal cross-current spray chamber. Liquid–gas ratio is 1 m³ of liquid/1000 m³ of gas. Drop diameter: curve 1, 200 μm; curve 2, 500 μm; curve 3, 1000 μm. U_G = 0.6 m/s. Courtesy of *J. APCA*.

tion. Almost any type of perforated plate can be used: sieve plates (with or without submerged impingement targets above the holes), slot plates, valve trays, and bubble caps. Collection efficiency is good down to 1-μm particles using pressure drops of 400 Pa (30 mm Hg) per plate. The best efficiency is obtained in all wet scrubbers if the dirty gas is saturated with water vapor before entering the first contact stage. Figure 21 shows predicted cut-size (271) for a simple sieve tray. Froth density must be predicted from complex relationships for sieve plate behavior (272). Packed towers will collect particulates and cut-size is independent of packing type (Fig. 22) (271). Countercurrent towers can be used for mist collection, but plug easily in the presence of solid particulates. Some particulates can be handled in concurrent downflow towers. The ability of a cross-flow packed

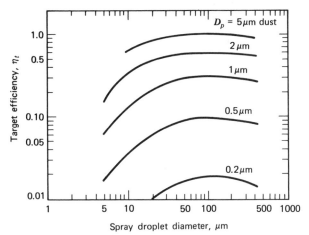

Fig. 20.   Spray droplet target efficiency in a centrifugal spray tower with a centrifugal field of 100 *g* (140,281).

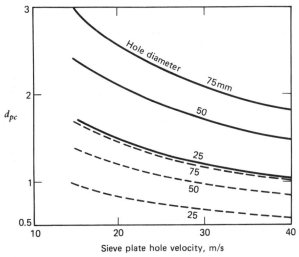

Fig. 21.   Performance cut diameter prediction for typical sieve plate operation on wettable particulates at foam densities *F*: solid line, $F = 0.4$ g/cm^3, dashed line, $F = 0.65$ g/cm^3 (271). Courtesy of *J. APCA*.

tower to handle loads of insoluble dust up to 11 g/m^3 was mentioned earlier under gas absorption. Ninety-five percent collection of 3-μm particles in such a tower has been reported (283). Irrigated fiber pads commercially available are quite efficient on 3-μm particles when operated under a pressure drop of 1 kPa (7.5 mm Hg); they will plug easily on heavy loadings of insoluble dust. Mobile-bed packing of fluidized spheres (turbulent contactors) have both good particulate collection and good mass transfer characteristics. Collection efficiency is good on particles down to 1 μm and the constant movement and rubbing of the spherical targets prevents plugging. For greater efficiency, several stages in series can be used. Typical pressure drop per stage is 0.75–1.5 kPa (5.5–11 mm Hg).

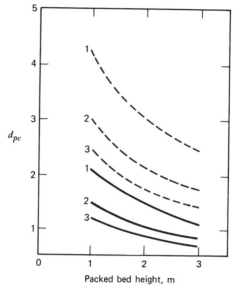

Fig. 22. Performance cut diameter predictions for typical dry packed bed particle collectors as a function of bed height or depth, packing diameter D_c, and packing porosity (void area) ϵ. Bed irrigation increases collection efficiency or decreases cut diameter (271). Solid lines, $D_c = 25$ mm; dashed lines, $D_c = 50$ mm; $\epsilon = 0.75$. Courtesy of *J. APCA*.

In several commercial scrubbers, the dirty gas jets directly into a pool of water where the momentum of the larger particles allows them to penetrate. Some of the water is thus atomized into a spray aiding collision between water droplets and smaller particulate. In one such scrubber careful control of the water level is necessary for efficient particle collection (284). These scrubbers have been used to collect coarser dusts such as from metal grinding; they have good efficiencies on particles down to 3–5 μm. Dynamic collectors employ a motor-driven centrifugal device resembling a water-sprayed centrifugal fan to cause impingement of solid particles in a wetted film. Tests indicate that 1-μm particles can be collected with 50% efficiency in an ordinary water-sprayed centrifugal fan if the wheel tip speed exceeds 90 m/s.

The collection of particles larger than 1–2 μm in liquid ejector venturis has been discussed (285). High pressure water induces the flow of gas, but power costs for liquid pumping can be high because motive efficiency of jet ejectors is usually less than 10%. Improvements (286) to liquid injectors allow capture of submicrometer particles by using a superheated hot (200°C) water jet at pressures of 6,900–27,600 kPa (1000–4000 psi) which flashes as it issues from the nozzle. For 99% collection, hot water rate varies from 0.4 kg/1000 m^3 for 1-μm particles to 0.6 kg/1000 m^3 for 0.3-μm particles.

Venturi and High Energy Scrubbers. The venturi scrubber has been studied more intensively than any other wet scrubber, perhaps because of its ability to efficiently scrub any size particle by changing the pressure drop. The design readily lends itself to mathematical modeling. Gas is accelerated in the throat to velocities of 60–150 m/s where water is added either as a spray, as solid jets, or as a wall-flowing sheet. The water is atomized into very small droplets by the high speed gas and aerosol particles, moving at close to the gas speed, collide with the accelerating liquid droplets and are captured. A cyclonic collector is needed to remove the water mist produced by the venturi. Collection mechanisms have been studied (140,287–292). The liquid drop size produced has been investigated (293); the effect of water-injection method on venturi performance, examined (294); and a generalized pressure-drop prediction method has been developed (295).

Venturi scrubbers can be operated at 2.5 kPa (19 mm Hg) to collect many particles coarser than 1 μm efficiently. Smaller particles often require a pressure drop of 7.5–10 kPa (56–75 mm Hg). When most of the particulates are smaller than 0.5 μm and are hydrophobic, venturis have been operated at pressure drops from 25 to 32.5 kPa (187–244 mm Hg). Water injection rate is typically 0.67–1.4 m^3 of liquid per 1000 m^3 of gas, although rates as high as 2.7 are used. Increasing water rates improves collection efficiency. Many venturis contain louvers to vary throat cross section and pressure drop with changes in system gas flow. Venturi scrubbers can be made in various shapes with reasonably similar characteristics. Any device that causes contact of liquid and gas at high velocity and pressure drop across an accelerating orifice will act much like a venturi scrubber. A flooded-disk scrubber in which the annular orifice created by the disc is equivalent to a venturi throat has been described (296). An irrigated packed fiber bed with performance similar to a venturi scrubber was offered commercially for collection of submicrometer particles at bed pressure drops of 7.5–15 kPa (56–113 mm Hg).

Wet Scrubber Entrainment Separation. Fiber pads and beds to collect fine liquid entrainment have been discussed. Unless fog is present from condensation, scrubber entrainment will be coarse and the high efficiency of fiber beds is not needed. Entrainment separators for scrubbers are usually centrifugal swirl vanes, hook and zigzag eliminators, or momentum separators using a change in direction. Efficiency and pressure drop for zigzag baffles, targets of staggered tube banks, packed beds, and knitted mesh have been compared (297) and these data supplemented with entrainment theory and collection efficiency in centrifugal flow and in sieve plates (298). Entrainment separator design requirements for venturi scrubbers have been discussed (299). Control of recycle water cleanliness is important to good wet scrubber performance and poor scrubber performance has been traced (300) to entrainment of dirty scrubbing liquid and also to temperature flashing (evaporation) of fine particulate contained in recycle spray water.

Developing Particulate Control Technology. Present control methods for particulates are least efficient in the size range from 0.2 to 2.5 μm; this range is the most costly to collect and very energy intensive. Health studies indicate that particles in this size range are also those which penetrate most deeply into, and often become deposited in, the human respiratory system. This is the main reason for the U.S. EPA change in the ambient air quality standard from total suspended particulate (TSP) concentration to a PM$_{10}$ standard (ambient air particles equal to or smaller than 10-μm aerodynamic diameter). The new standard will undoubtedly place even more emphasis on the need to collect particulates in this difficult-to-control fine-particle range, and therefore, collection of this size range is most in need of improvements in technology. Improved collection requires the use of a separating force which is independent of gas velocity or of the growth of particles which can be more readily collected. Particle growth can be accomplished through coagulation (agglomeration), chemical reaction, condensation, and electrostatic attraction. Promising separation forces are the "flux forces" involving diffusiophoresis, thermophoresis, electrophoresis, and Stefan flow. Although particle growth techniques and flux-force collection theoretically can be considered independently, both phenomena are applied in many practical devices.

Thermophoresis may be considered a special form of thermal precipitation. If a hot submicrometer particle is close to a large cold particle or droplet of such relative size that it resembles a wall to the small particle, the kinetic motion of the hot gas molecules opposite the cold particle will bombard and propel the submicrometer particle toward the cold droplet. If cold droplets are introduced into a warmer saturated gas stream, water vapor will condense on the cold droplets reducing the water vapor pressure near its surface and produce a water vapor pressure gradient. The hydrodynamic flow of vapor toward the condensing surface is known as Stefan flow. If the molecular mass of the diffusing vapor is different from the molecular mass of the carrier gas, the motion of small particles is further affected by a density differential. Both of these forces influence the movement of submicrometer particles, and their algebraic sum is known as the force of diffusiophoresis. In condensation, diffusiophoresis tends to move submicrometer particles toward cold droplets; in droplet evaporation, the action is reversed, moving the particles away from the droplet. This is one of the reasons wet scrubbing collection is enhanced if the bulk gas stream is water saturated before its entry to the scrubber, and the scrubbing liquid is cooled below the gas dewpoint. Electrophoresis is the movement of a small particle toward a charged particle. Mathematical description of these forces has been presented (282). Diffusiophoresis has also been discussed (301).

In time, submicrometer particles will coagulate into chains or agglomerates through Brownian motion. Increasing turbulence during coagulation will increase the frequency of collisions and coagulation rate. The addition of fans to stir the gas, or gas flow motion through tortuous passages such as those of a packed bed, will aid coagulation. Sonic energy is also known as an aid to the coagulation process. Production of standing waves in the confines of long narrow tubes can bring about concentration and coagulation of aerosols in band zones in the tube. The addition of water and oil mists to the treated aerosol can improve the effectiveness of sonic agglomeration by improving the tendency of the colliding particles to stick together. Sulfuric acid mist (302) and carbon black (303) have been sufficiently agglomerated so that the coagulated product can be collected in a cyclone. Sonic agglomeration has been tested in the past for many metallurgical fumes, but has generally been found too power intensive for practical consideration. One problem is the low energy efficiency of transforming other energy sources to sonic energy in available sonic generators. The development of more energy efficient sonic generators, coupled with improved knowledge of the phenomenon of sonic coagulation, might justify further investigation. The combination of sonic agglomeration and electrostatic precipitation could result in considerable reduction of precipitator size and perhaps capital cost (304). Coagulation techniques have been discussed in mathematical terms in references 272 and 276.

Particle growth may be brought about (or the charge on particles modified) by the introduction of a gas to react with the particles. Another procedure used is chemical modification of the particle to render it hygroscopic. This can be particularly beneficial for particle growth if the particle continues to absorb moisture to form hydrates. It may also be possible to control the size of aerosols initially formed by chemical means. In studies (305) of chemical reactions producing aerosols it was found that reactions having a large chemical driving force and hence releasing large quantities of energy tended to produce very fine particles (high surface energy); conversely, reactions which occur without release of large amounts of energy tend to produce larger particles (2–6 μm). Therefore, limitation of chemical driving force may often be beneficial to prevent formation of fine aerosols. If steam or other condensable vapors can be condensed while cooling an aerosol, particle growth will occur through vapor condensation on the existing aerosol nuclei. A given mass of smaller particles will present more surface area for condensation than the same mass of larger particles. Thus the smaller particles will selectively grow faster in size than will the larger particles. Also, the addition to the system of particles opposite in electrical charge will result in mutual attraction and aid in particle growth.

Of these various possibilities, those involving electrostatic forces and condensation have received the most interest. Flux-force condensation scrubbing may be desireable for hot gases needing treatment when there is no attractive alternative available to recover the energy. Sizeable amounts of low pressure waste steam are also useful. Reports have appeared on condensation scrubbing in multiple sieve-plate towers (306,307), aspirative condensation (308), and some of the parameters that affect particle growth and collection in a conventional orifice scrubber, with and without condensation (309). Fine particle collection is appreciably improved by scrubbing a hot saturated gas stream with cold water rather than using recirculated hot water. Little collection improvement resulted in cooling below a 50°C dew point, but much better collection was achieved when the hot gas was introduced already saturated close to its initial dew point (rather than admitting it unsaturated with substantial superheat). A decrease in scrubbing efficiency of an evaporative scrubber was also found. Addition of an adiabatic presaturator for hot gases ahead of a scrubber should be quite beneficial.

Odor Control

Odor is a subjective preception of the sense of smell. Its study is still in a developmental stage: information including a patent index has been compiled (310); 124 rules of odor preferences have been listed (311); detection and recognition threshold values have been given (312); and odor technology as of 1975 has been assessed (313). Odor control involves any process that gives a more acceptable perception of smell, whether as a result of dilution, removal of the offending substance, or counteraction or masking (see ODOR MODIFICATION; PERFUMES).

Odor Measurement.   Both static and dynamic measurement techniques exist for odor. The objective is to measure odor intensity by determining the dilution necessary so that the odor is imperceptible or doubtful to a human test panel, ie, to reach the detection threshold, the lowest concentration at which an odor stimulus may be detected. The recognition threshold is a higher value at which the chemical entity is recognized. An odor unit, o.u., has been widely defined in terms of 0.0283 m^3 (1 ft^3) of air at the odor detection threshold. It is a dimensionless unit representing the quantity of odor which when dispersed in 28.3 L (1 ft^3) of odor-free air produces a positive response by 50% of panel members. Odor concentration is the number of cubic meters that one cubic meter of odorous gas will occupy when diluted to the odor threshold. Selection of people to participate in an odor panel should reflect the type of information or measurement required, eg, for evaluation of an alleged neighborhood odor nuisance, the test subjects should be representative of the entire neighborhood. However, threshold determinations may be done with a carefully screened panel of two or three people (314). A general population test panel of 35 people has been described (315).

Static Dilution Methods. A known volume of odorous sample is diluted with a known amount of nonodorous air, mixed, and presented statically or quiescently to the test panel. The ASTM D1391 syringe dilution technique is the best known of these methods and involves preparation of a 100-mL glass syringe of diluted odorous air which is allowed to stand 15 min to assure uniformity. The test panel judge suspends breathing for a few seconds and slowly expels the 100-mL sample into one nostril. The test is made in an odor-free room with a minimum of 15 min between tests to avoid olfactory fatigue. The syringe dilution method is reviewed from time to time by the ASTM E18 Sensory Evaluation Committee, who suggest and evaluate changes. Instead of a syringe, a test chamber may be used which can be as large as a room (316,317). A technique to make threshold determinations for 53 odorant chemicals has been described (316). The test room consisted of two chambers: an antechamber and the actual test chamber. The air in each chamber was circulated through activated carbon to provide a controlled, odor-free background for sample dilution.

Dynamic Dilution Methods. In this method, odor dilution is achieved by continuous flow. Advantages are accurate results, simplicity, reproducibility, and speed. Devices known as dynamic olfactometers control the flow of both odorous and pure diluent air, provide for ratio adjustment to give desired dilutions, and present multiple, continuous samples for test panel observers at ports beneath odor hoods. The Hemeon olfactometer (318,319) uses three ports, each designed as a face piece which surrounds the lower half of the face loosely, and allows three panelists to judge simultaneously. The Hellman odor fountain (320–322) is a similar device. An olfactometer based on forced-choice-triangle statistical design has been constructed (323). To distinguish dynamically obtained group threshold values from ASTM odor units, the ED$_{50}$ (effective dose, 50%) designation may be used. ED$_{50}$ is the concentration at which half the panelists begin to detect odor in a dynamic test. ED$_{50}$ values are 5% higher at the 1000 o.u. level than ASTM odor unit values, 20% higher at 100 o.u. level, and 33% higher at the 20 o.u. level (323). Similar but greater deviations have been obtained when the Hemeon odor meter results have been compared to results of the ASTM static syringe method (323).

Another dynamic instrument, the Scentometer, is the basis for odor regulations in the states of Colorado, Illinois, Kentucky, Missouri, Nevada, and Wyoming, and in the District of Columbia (324). The portable Scentometer (Barneby-Cheney) can produce dilution ratios up to 128:1 in the field. The Scentometer blends two air streams, one of which has been deodorized with activated carbon. The dilution ratio is decreased until the odor becomes detectable (325). Improvements to dynamic methods have been recommended (326).

Odor Control Methods. Absorption, adsorption, and incineration are all typical control methods for gaseous odors; odorous particulates are controlled by the usual particulate control methods. However, carrier gas, odorized by particulates, may require gaseous odor control treatment even after the particulates have been removed. For oxidizable odors, treatment with oxidants such as hydrogen peroxide (qv), ozone (qv), and $KMnO_4$ may sometimes be practiced; catalytic oxidation has also been employed (see EXHAUST CONTROL, INDUSTRIAL). Odor control as used in rendering plants (327), spent grain dryers (328), pharmaceutical plants (329–331), and cellulose pulping (332) has been reviewed (333–335); some reviews are presented in two symposium volumes (336,337) from APCA Specialty Conferences. The odor-control performances of activated carbon and permanganate–alumina for reducing odor level of air streams containing olefins, esters, aldehydes, ketones, amines, sulfide, mercaptan, vapor from decomposed crustacean shells, and stale tobacco smoke have been compared (338). Activated carbon produced faster deodorization in all cases. Activated carbon adsorbers have been used to concentrate odors and organic compounds from emission streams, producing fuels suitable for incineration (339). Both air pollution control and energy recovery were accomplished.

NOMENCLATURE

Symbol	Definition	SI units
A	constant in Stokes-Cunningham correction factor for particle velocity; $A \cong 1.72$, but often considered a function of λ and D_p	dimensionless
A	constant in Calvert correlation for relating penetration and particle size in wet scrubbers, $P_t = \exp(-A\, d_p^B)$	as required by equation 26
A_e	collecting electrode area in an electrical precipitator	m^2
A_p	area of particle projected on plane normal to direction of flow or motion; $A_p = \pi D_p^2/4$ for spherical particles	m^2
a	fraction of inelastic collisions in thermophoresis, equation 11	dimensionless
B	exponential constant in Calvert correlation for relating penetration and particle size in wet scrubbers, $P_t = \exp(-A\, d_p^B)$	as required by equation 26
B_e	distance between the discharge electrode and the collecting plate in a plate type electrical precipitator	m
b_r	roughness factor for discharge electrode in an electrical precipitator; increasing roughness decreases the value of b_r	dimensionless
C_D	overall drag coefficient $= 2\, F_d/\rho_g u^2 A_p$	dimensionless
D_b	representative dimension or diameter of impingement body or target	m
D_c	packing diameter in packed column scrubber	mm
D_d	diameter of the discharge electrode in an electrical precipitator	m
D_o	diameter of outlet pipe on cyclone	mm or m

Symbol	Definition	SI units
D_p	diameter (or equivalent diameter) of a particle	m
D_{pc}	particle cut-size; diameter collected with 50% efficiency in a particulate collector	m
$D_{p\,min}$	diameter of smallest particle collectible with 100% collection efficiency	m
D_t	diameter of a collecting tube in a tubular type electrical precipitator	m
D_v	diffusion coefficient for particle	m^2/s
d_p	diameter of particle (in wet scrubbing and Calvert figures)	μm
d_{pc}	particle cut size in Calvert wet-scrubbing illustrations	μm
d_{pgm}	geometric mass mean particle diameter	μm
E_o	electrostatic charging field strength in an electrical precipitator	V/m or N/C
E_p	electrostatic precipitating field strength in an electrical precipitator	V/m or N/C
e	natural or Naperian logarithm base, 2.718	dimensionless
F	foam density above sieve plates	kg/L
f	Calvert correlation parameter for effect of hydrophobic ($f = 0.25$) or hydrophilic ($f = 0.50$) nature of particles collected in venturi scrubbers	dimensionless
F_d	drag or resistance to motion of a body in a fluid	N
G	superficial gas mass velocity	$kg/(s{\cdot}m^2)$
g	local acceleration due to gravity	m/s^2
K_b	design constant for knitted mesh mist collectors, typically 0.11 m/s	m/s
K_c	cyclone friction factor in equation 17	dimensionless
K_e	geometric design constant for an electrical precipitator; for a plate-type, $K_e = L_e/B_e V_e$, and for a tube-type, $K_e = 4\,L_e/D_t V_e$	s/m
K_m	Stokes-Cunningham correction factor	dimensionless
K_s	sphericity correction constant	dimensionless
K_v	Coulomb's law constant, 8.987×10^9	$(N{\cdot}m^2)/C$
k	Boltzmann constant, 1.380×10^{-23}	J/K
k_c	cloth drag coefficient for a fabric bag type filter	as required by equation 21
k_d	drag coefficient for collected dust cake in a fabric bag type filter	as required by equation 22
k_g	thermal conductivity of gas	$W/(m{\cdot}K)$
k_{gTR}	translational thermal conductivity of gas	$W/(m{\cdot}K)$
k_p	thermal conductivity of particle	$W/(m{\cdot}K)$
k_ρ	density correction factor for corona starting potential in electrical precipitator; ratio of the precipitator gas density to that of air at 25°C and 101.3 kPa (1 atm)	dimensionless
L	superficial liquid mass velocity	$kg/(s{\cdot}m^2)$
L_e	treatment path length in a charged precipitating field in an electrical precipitator	m

Symbol	Definition	SI units
m_d	mass of dust solids collected per unit cloth area in a bag type filter	kg/m^2
N_e	effective number of spirals made by the gas in a cyclonic separator	dimensionless
N_{PE}	Peclet number $= U_o D_b/D_v$	dimensionless
N_{RE}	Reynolds number $= D_p \rho_p u/\mu_g$	dimensionless
N_{SC}	Schmidt number $= \mu_g/\rho_g D_v$	dimensionless
N_{SF}	flow-line separation number $= D_p/D_b$	dimensionless
N_{SI}	inertial separation number $= U_t U_o/D_b$	m/s^2
N_t	number of particulate collection transfer units $= \ln[1/(1-\eta)]$	dimensionless
P	system or collector total pressure	Pa
P_T	total power expended in a contacting device	W
P_t	particle penetration through a collector $1 - \eta$ or $100 - \eta$ (fraction or percentage)	dimensionless
ΔP	pressure drop through a collector or control device	Pa
ΔP_c	pressure drop through the filtering cloth in a fabric bag filter	Pa
ΔP_d	pressure drop through the accumulated dust cake in a fabric bag filter	Pa
Q	volumetric flow rate through collector	m^3/s
q_e	gas volumetric flow rate through an electrical precipitator	m^3/s
r	radius, distance from center line of cyclone separator, or from center line of concentric cylinder electrical precipitator	m
T	system or collector absolute temperature	K
t	time	s
U_e	average-particle migration velocity for a given size particle in an electrical precipitator; also used as the drift velocity which is the average migration velocity for all particles sizes collected	M/s
U_G	superficial gas velocity in wet scrubber	m/s
U_o	linear gas velocity	m/s
U_s	operating superficial velocity through a knitted mesh mist collector	m/s
U_T	thermophoretic velocity of a particle	m/s
U_t	particle terminal settling velocity	m/s
$U_{t,n}$	particle terminal velocity in direction normal to gas velocity	m/s
u	velocity of particle relative to main body of fluid	m/s
V_c	average velocity of gas at cyclone inlet	m/s
V_{cT}	tangential gas velocity component in a cyclone	m/s
V_e	average gas velocity in an individual flow passage in the precipitating field of an electrical precipitator	m/s
V_f	superficial gas velocity through the fabric in a bag filter	m/s

Symbol	Definition	SI units
V_s	minimum potential required for the start of corona discharge in an electrical precipitator	V
W_i	width of gas inlet of cyclonic separator	m
α	intercept constant in Semrau's power law collection principle, equation 25	as required by equation 25
γ	power exponent in Semrau's power law collection principle, equation 25	dimensionless
δ	dielectric constant for nonconductive particles	dimensionless
ϵ	packed bed void fraction in a packed-bed wet scrubber	dimensionless
η	collection efficiency (fraction or percentage)	dimensionless
η_D	diffusional collection efficiency	dimensionless
η_{FL}	flow-line interception efficiency	dimensionless
η_{IN}	inertial deposition efficiency	dimensionless
η_M	collection efficiency by particle mass	dimensionless
η_t	target collection efficiency by inertial impact	dimensionless
λ	molecular mean free path, 0.1 μm for room temperature ambient air	m
μ_g	aerosol carrier gas viscosity	Pa·s
π	geometric constant relating circumference and diameter of a circle, 3.14159	dimensionless
ρ_L	density of liquid or mist droplets	kg/m^3
ρ_g	density of aerosol carrier gas	kg/m^3
ρ_p	apparent or aerodynamic particle density	kg/m^3
ρ_t	true (not bulk) density of solids	kg/m^3
σ_g	geometric standard deviation	dimensionless
θ	time interval at any point since last bag cleaning in a fabric bag filter	s or h, as required, equation 23
Ψ	sphericity, surface area of sphere having the same volume as a particle divided by the actual surface area of the particle	dimensionless

BIBLIOGRAPHY

"Air Pollution Control Methods" in *ECT* 3rd ed., Vol. 1, pp. 649–716, by B. B. Crocker, D. A. Novak, and W. A. Scholle, Monsanto Company.

References have been subdivided by area of interest except that when they cover more than one interest area they are listed only in the area where first cited.

Air pollution, general

1. *Guiding Principles of State Air Pollution Legislation,* U.S. Department of Health, Education, and Welfare, Washington, D.C., 1965.
2. Sect. 1420, Chapt. 111, *General Laws,* Chapt. 836, *Acts of 1969,* the Commonwealth of Massachusetts, Department of Public Health, Division of Environment, Health, Bureau of Air Use Management.
3. D. H. Slade, ed., *Meteorology and Atomic Energy 1968,* U.S. Atomic Energy Commission, July 1968; available as *TID-24190,* Clearinghouse for Federal Scientific and Technical Information National Bureau of Standards, U.S. Department of Commerce, Springfield, Va.
4. D. B. Turner, *Workbook of Atmospheric Dispersion Estimates, US EPA, OAP, Pub.*

AP26, Research Triangle Park, N.C., revised 1970, U.S. Department Printing Office Stock No. 5503-0015.

5. A. D. Busse and J. R. Zimmerman, *User's Guide for the Climatological Dispersion Model, US EPA Pub. No. EPA-R4-73-024,* Research Triangle Park, N.C., Dec. 1973.

6. M. Smith, ed., *Recommended Guide for the Prediction of the Dispersion of Airborne Effluents,* American Society of Mechanical Engineers, New York, 1968.

7. G. A. Briggs, "Plume Rise," *USAEC Critical Review Series TID-25075,* NTIS, Springfield, Va., 1969.

8. *Effective Stack Height: Plume Rise, US EPA Air Pollution Training Institute Pub. SI:406,* with Chapts. D, E, and G by G. A. Briggs and Chapt. H by D. B. Turner, 1974.

9. J. E. Carson and H. Moses, *J. APCA* **19,** 862 (Nov. 1969).

10. H. Moses and M. R. Kraimer, *J. APCA* **22,** 621 (Aug. 1972).

11. G. A. Briggs, *Plume Rise Predictions, Lectures on Air Pollution and Environmental Impact Analyses,* American Meteorological Society, Boston, Mass., 1975.

12. *Guideline on Air Quality Models, OAQPS Guideline Series,* U.S. Environmental Protection Agency, Research Triangle Park, N.C., 1980.

13. G. A. Schmel, *Atmos. Environ.* **14,** 983–1011 (1980).

14. N. E. Bowne, R. J. Londergan, R. J. Minott, D. R. Murray, "Preliminary Results from the EPRI Plume Model Validation Project—Plains Site," Report EPRI EA-1788, Electric Power Research Institute, Palo Alto, Calif., 1981.

15. N. E. Bowne, "Atmospheric Dispersion," in S. Calvert and H. M. Englund, eds., *Handbook of Air Pollution Technology,* John Wiley & Sons, Inc., New York, 1984, pp. 859–891.

16. *Code of Federal Regulations* 40 (CFR 40), *Fed. Reg.,* C-50–99.

17. Ref. 16, part 58.

18. W. W. Lund and R. Starkey, *J. Air Waste Manage Assn.* **40**(6), 896–897 (June 1990).

19. Ref. 16, part 50, appendix A–G.

20. Ref. 16, part 62, appendix B, and subparts C–F, J, M, V.

21. T. A. Gosink, *Environ. Sci. Technol.* **9,** 630–634 (July 1975).

22. S. R. Heller, J. M. McGuire, and W. L. Budde, *Environ. Sci. Technol.* **9,** 210–213 (Mar. 1975).

23. *Fed. Reg.* **40,** 46250 (Oct. 1975).

24. *Fed. Reg.* **52,** 24634–24750 (July 1, 1987); *Fed. Reg.* **54,** 41218–41232 (Oct. 5, 1989).

25. D. S. Ensor and M. J. Pilot, *J. APCA* **21,** 496–501 (Aug. 1971).

26. B. B. Crocker, *Chem. Eng. Prog.* **71,** 83–89 (March 1975).

27. L. E. Sparks, "Particulate Sampling and Analysis," in S. Calvert and H. E. Englund, eds., *Handbook of Air Pollution Technology,* John Wiley & Sons, Inc., New York, 1984, pp. 800–818.

28. B. W. Loo, J. M. Jaklevic, and F. S. Goulding, "Dichotomous Virtual Impactors for Large Scale Monitoring of Airborne Particulate Matter," in B. Y. H. Liu, ed., *Fine Particles, Aerosol Generation, Measurement, Sampling, and Analysis,* Academic Press, Inc., New York, 1976, pp. 311–350.

29. "Hard Realities: Air and Waste Issues of the 90s" (Report of 18th Government Affairs Seminar), *J. Air and Waste Management Assn.* **40**(6), 855–860 (June 1990); also *Proceedings of the 18th Air and Waste Management Assn. Govt. Affairs Seminar,* Air and Waste Management Assn., Pittsburgh, Pa., 1990.

30. R. H. Perry, ed., *Engineering Manual,* 3rd ed., McGraw-Hill, Inc., New York, 1976. Text used with permission.

Control equipment, general

31. *Industrial Ventilation,* 15th ed., Sect. 4, American Conference of Governmental Industrial Hygienists, Committee on Industrial Ventilation, Lansing, Mich., 1978.

32. R. Jorgensen, *Fan Engineering,* 7th ed., Buffalo Forge Company, Buffalo, N.Y., 1970, pp. 471–480.

33. W. E. L. Hemeon, *Plant and Process Ventilation,* Industrial Press Inc., New York, 1954.

34. J. M. Dalla Valle, *Exhaust Hoods,* Industrial Press Inc., New York, 1952.

35. J. L. Alden, *Design of Industrial Exhaust Systems for Dust and Fume Removal,* 3rd ed., Industrial Press Inc., New York, 1959.

36. J. A. Danielson, ed., *Air Pollution Engineering Manual, Pub. No. 999-AP-40,* U.S. Department of Health, Education, and Welfare, Cincinnati, Ohio, 1973, Chapt. 3.

37. B. B. Crocker, "Capture of Hazardous Emissions," in *Proceedings, Control of Specific (Toxic) Pollutants* (Conference, Feb. 1979, Gainsville, Fla.), Air Pollution Control Assn., Pittsburgh, Pa., 1979, pp. 415–433.

38. B. B. Crocker, *Chem. Eng. Prog.* **64,** 79 (Apr. 1968).

Gaseous emission control

39. "Liquid–Gas Systems," in R. H. Perry and D. Green, eds., *Perry's Chemical Engineers Handbook,* 6th ed., McGraw-Hill, Inc., New York, 1984, Chapt. 18.

40. B. B. Crocker and K. B. Schnelle, Jr., "Control of Gases, by Absorption, Adsorption, and Condensation," in S. Calvert and H. M. Englund, eds., *Handbook of Air Pollution Technology,* John Wiley & Sons, Inc., New York, 1984, Chapt. 7, pp. 135–192.

41. T. C. Kenner and D. Zhou, *Environ. Progress,* **9**(1), 40–46 (Feb. 1990).

42. P. N. Cheremisinoff and R. A. Young, eds., *Air Pollution Control and Design Handbook,* Part 2, Marcel Dekker, Inc., New York, 1977.

43. H. O. Grant, *Chem. Eng. Prog.* **60,** 53 (Jan. 1964).

44. A. J. Teller, *Chem. Eng. Prog.* **63,** 75 (Mar. 1967).

45. S. V. Cabibbo and A. J. Teller, "The Crossflow Scrubber—A Digital Model for Absorption," *Paper No. 69-186, APCA 62nd Annual Meeting, New York, N.Y., (June 22–26, 1969).*

46. K. E. Lunde, *Ind. Eng. Chem.* **50,** 293 (Mar. 1958).

47. J. P. Jewell and B. B. Crocker, *Proceedings of the 8th Annual Sanitary and Water Resources Engineering Conference,* Vanderbilt University, Nashville, Tenn., 1969, pp. 211–228.

48. W. W. Lehle in W. W. Duecker and J. R. West, eds., *The Manufacture of Sulfuric Acid,* Reinhold Publishing Corp., New York, 1959, pp. 348–352.

49. J. J. Spivey, *Environ. Progress* **7**(1), 31–40 (Feb. 1988).

50. Ref. 39, Sect. 16.

51. B. Grandjacques, *Pollut. Eng.* **9,** 28–31 (Aug. 1977).

52. D. Bienstock, J. H. Field, S. Katell, and K. D. Plants, *J. APCA* **15,** 459 (Oct. 1965).

53. D. Bienstock, J. H. Field, and J. G. Myers, *J. Eng. Power* **86**(3), 353 (1964).

54. Ref. 39, p. 16-36.

55. S. M. Hall, *J. Air Waste Manage Assn.* **40**(3), 404–407 (Mar. 1990).

56. C. S. Chang and C. Jorgensen, *Environ. Progress* **6**(4), 26–32 (Feb. 1987).

57. K. T. Fellows and M. J. Pilat, *J. Air Waste Manage Assn.* **40**(6), 855–860 (June 1990).

58. P. J. Krall and P. Williamson, *J. APCA,* **36**(11), 1258–1263 (Nov. 1986).

59. V. S. Katari, W. M. Vatavuk, and A. H. Wehe, Part I, *J. APCA* **37**(1), 91 (Jan. 1987); Part II, *J. APCA* **37**(2), 198–201 (Feb. 1987).

60. M. Kosusko and C. M. Nunez, *J. Air Waste Manage Assn.* **40**(2), 254–255 (Feb. 1990).

61. M. A. Palazzolo and B. A. Tichenor, *Environ. Progress* **6,** 172–176 (Aug. 1987).

62. K. C. Lee, H. J. Janes, and D. C. Macauley, *Preprint 78-58.6, APCA 71st Annual Meeting,* (June 25–30, 1978, Houston, Tex.) Air Pollution Control Assn., Pittsburgh, Pa.

63. Ref. 39, p. 26-29.

64. R. W. Coughlin, R. D. Siegel, and C. Rai, eds., *AIChE Symp. Ser.* **70,** (137) (1974).

Contains four papers on flue gas desulfurization, four papers on coal desulfurization, and three papers on petroleum desulfurization.

65. C. Rai and R. D. Siegel, eds., *AIChE Symp. Ser.* **71**, (148) (1975). Contains seven papers on flue gas desulfurization, two on petroleum desulfurization, one on coal desulfurization, and fifteen on NO_x control.

66. J. A. Cavallaro, A. W. Dearbrouck, and A. F. Baher, *AIChE Symp. Ser.* **70**, (137); 114–122 (1974).

67. K. S. Murthy, H. S. Rosenberg, and R. B. Engdahl, *J. APCA* **26**, 851–855 (Sept. 1976).

68. A. V. Slack, *Chem. Eng. Prog.* **72**, 94–97 (Aug. 1976).

69. R. M. Jimeson and R. R. Maddocks, *Chem. Eng. Prog.* **72**, 80–88 (Aug. 1976).

70. *AIChE Symp. Ser.* **68**, (126) (1972). Contains four papers on flue gas desulfurization and two on NO_x control.

71. *Control Technology: Gases and Odors,* APCA Reprint Series, Air Pollution Control Assn., Pittsburgh (Aug. 1973). A reprint of 1970–1973 APCA journal articles: six papers on flue gas desulfurization, three on NO_x control, and two on SO_2 control from pulp and paper mills.

72. *Sulfur Dioxide Processing,* Reprints of 1972–1974 *Chem. Eng. Prog.* articles, AIChE, New York (1975). Contains thirteen papers on flue gas desulfurization, two on SO_2 control in pulp and paper, one on sulfuric acid tail gas, one on SO_2 from ore roasting, and two on NO_x from nitric acid.

73. E. L. Plyler and M. A. Maxwell, eds., *Proceedings, Flue Gas Desulfurization Symposium,* (New Orleans, La., Dec. 1973), *Pub. No. EPA-650/2-73-038,* U.S. EPA, Research Triangle Park, N.C., 1973. Contains 34 papers on flue gas desulfurization.

74. *Proceedings: Symposium on Flue Gas Desulfurization,* (Atlanta, Ga., Dec. 1974), U.S. EPA, Research Triangle Park, N.C.

75. *Proceedings: Symposium on Flue Gas Desulfurization* (New Orleans, Mar. 1976), U.S. EPA, Research Triangle Park, N.C. Contains 36 papers dealing with flue gas desulfurization.

76. *Proceedings of the Third Stationary Source Combustion Symposium, EPA 600/7-79-050a (NTIS PB 292 539)* U.S. EPA, Industrial Environmental Research Laboratory, Research Triangle Park, N.C., Feb. 1979.

77. G. M. Blythe and co-workers, *Survey of Dry SO_2 Control Systems, EPA-600/7-80-030 (NTIS PB 80 166853),* U.S. EPA, Industrial Research Laboratory, Research Triangle Park, N.C., Feb. 1980.

78. *Definitive SO_2 Control Process Evaluation: Limestone Double Alkali and Citrate FGD Process,* EPA Pub. EPA-600/7-79-177, Aug. 1979.

79. J. D. Mobley and K. J. Lim, "Control of Gases by Chemical Reaction," in S. Calvert and H. M. Englund, eds., *Handbook of Air Pollution Technology,* John Wiley & Sons, Inc., New York, 1984, Chapt. 9, pp. 193–213.

80. G. M. Blythe and co-workers, *Survey of Dry SO_2 Control Systems, EPA Pub. EPA-600/7-80-030 (NTIS PB 80-166853),* U.S. EPA, Research Triangle Park, N.C., Feb. 1980.

81. R. Pedroso, *An Update of the Wellman-Lord Flue Gas Desulfurization Process, EPA Pub. EPA-600/2-76-136a,* May 1976.

82. T. W. Devitt, "Fossil Fuel Combustion," in S. Calvert and H. M. Englund, eds., *Handbook of Air Pollution Technology,* John Wiley & Sons, Inc., New York, 1984, Chapt. 15, pp. 375–417.

83. *Flue Gas Desulfurization Systems and SO_2 Control, Pub. GS-6121,* Electric Power Research Institute, Palo Alto, Calif., October 1988. An abstracted bibliography of EPRI Reports and Projects.

84. *Introduction to Limestone Flue Gas Desulfurization, Pub. CS-5849,* Electric Power Research Institute, Palo Alto, Calif., 1988.

85. *SO_2 Removal by Injection of Dry Sodium Compounds, Pub. RP-1682,* Electric Power Research Institute, Palo Alto, Calif., 1987.

86. C. S. Chang and C. Jorgensen, *Environ. Progress* **6**(1), 26 (Feb. 1987).
87. M. T. Melia and co-workers, "Trends in Commercial Applications of FGD," in *Proceedings: Tenth Symposium on FGD, EPRI Report CS-5167,* Electric Power Research Institute, Palo Alto, Calif. (May 1987).
88. *Proceedings of the EPA/EPRI First Combined Flue Gas Desulfurization and Dry SO_2 Control Symposium,* Oct. 25–28, 1988; St. Louis, Mo., Electric Power Research Institute, Palo Alto, Calif.
89. S. S. Tsai and co-workers, *Environ. Progress* **8**(2), 126–129 (May 1989).
90. W. Bartok and co-workers, *Environ. Progress* **9**(1), 18–23 (Feb. 1990).
91. W. Moeller and K. Winkler, *J. APCA* **18**, 324 (May 1968).
92. D. S. Davies, R. A. Jiminez, and P. A. Lemke, *Chem. Eng. Prog.* **70**, 68 (June 1974).
93. R. G. Bierbower and J. H. Sciver, *Chem. Eng. Prog.* **70**, 60 (Aug. 1974).
94. K. T. Semrau, *J. APCA* **21**, 185 (Apr. 1971); its bibliography contains 140 refs.
95. J. E. Walther and H. R. Amberg, *J. APCA* **20**, 9 (Jan. 1970).
96. S. F. Galeano and B. R. Dillard, *J. APCA* **22**, 195 (Mar. 1972).
97. J. E. Walther, H. R. Amberg, and H. Hamby, III, *Chem. Eng. Prog.* **69**, 100 (June 1973).
98. J. D. Rushton, *Chem. Eng. Prog.* **69**, 39 (Dec. 1973).
99. *Control Techniques for Nitrogen Oxide Emissions from Stationary Sources,* Pub. No. AP-67, U.S. Department of Health, Education, and Welfare, Washington, D.C., 1970.
100. "Formation and Control of Nitrogen Oxides," *Catalysis Today* **2,** Elsevier Science Publishers B.V., Amsterdam, 1988, pp. 369–532.
101. J. A. Miller and G. A. Fisk, "Combustion Chemistry," *C&EN,* 22–46 (Aug. 31, 1987).
102. W. Bartok, A. R. Crawford, and A. Skopp, *Chem. Eng. Prog.* **67**, 64 (Feb. 1974).
103. R. K. Srivastava and J. A. Mulholland, *Environ. Progress* **7**(1), 63–70 (Feb. 1988).
104. J. A. Mulholland and R. K. Srivastava, *J. APCA* **38**(9), 1162–1167 (Sept. 1988).
105. W. Bartok, V. S. Engleman, R. Goldstein, and E. G. del Valle, *AIChE Symp. Ser.* **68**(126), 30 (1972).
106. W. Bartok, A. R. Crawford, and G. J. Piegari, *AIChE Symp. Ser.* **68**(126), 66 (1972).
107. H. B. Lange, Jr., *AIChE Symp. Ser.* **68**(126), 17 (1972).
108. D. Thompson, T. D. Brown, and J. M. Beer, *Combust. Flame* **19,** 69 (1972).
109. W. Bartok and co-workers, *Systems Study of Nitrogen Oxide Control Methods for Stationary Sources, Pub. PB-192789,* NTIS, 1969.
110. J. H. Wasser and E. E. Berkau, *AIChE Symp. Ser.* **68**(126), 39 (1972).
111. G. B. Martin and E. E. Berkau, *AIChE Symp. Ser.* **68**(126), 45 (1972).
112. F. A. Bagwell and co-workers, *J. APCA* **21**, 702 (Nov. 1971).
113. T. W. Rhoads and co-workers, *Environ. Progress* **9**(2), 126–130 (May 1990).
114. M. J. Miller, *Environ. Progress* **5**(3), 171–177 (Aug. 1986).
115. J. S. Maulbetsch and co-workers, *J. APCA* **36,** 1294–1298 (Nov. 1986).
116. J. W. Jones, "Estimating Performance and Costs of Retrofit SO_2 and NO_x Controls for Acid Rain Abatement," *ACS Extended Abstract Preprint,* ACS Division of Environmental Chemistry Meeting (June 5–11, 1988, Toronto, Ontario).
117. J. D. Mobley and K. J. Lim, "Control of Gases by Chemical Reaction," in S. Calvert and H. M. Englund, eds., *Handbook of Air Pollution Technology,* John Wiley & Sons, Inc., New York, 1984, Chapt. 9, pp. 203–213.
118. W. Bartok and co-workers, *Chem. Eng. Progress* **84**(3), 54–71 (Mar. 1988).
119. *C&EN* 22 (April 18, 1988).
120. R. A. Perry and D. L. Siebers, *Nature* **324,** 657–658 (1986).
121. U.S. Pat. 4,800,068 (Jan. 24, 1989), R. A. Perry.
122. R. K. Lyon, *Environ. Sci. Technol.* **21**(3), 231–236 (1987).
123. O. J. Adlhart, S. G. Hindin, and R. E. Kenson, *Chem. Eng. Prog.* **67,** 73 (Feb. 1971).
124. D. J. Newman, *Chem. Eng. Prog.* **67,** 79 (Feb. 1971).
125. G. R. Gillespie, A. A. Boyum, and M. F. Collins, *Chem. Eng. Prog.* **68,** 72 (Apr. 1972).

126. R. M. Reed and R. L. Harvin, *Chem. Eng. Prog.* **68,** 78 (Apr. 1972).
127. R. M. Counce and co-workers, *Environ. Progress* **9**(2), 87–92 (May 1990).
128. L. L. Fornoff, *AIChE Symp. Ser.* **68**(126), 111 (1972).
129. B. J. Mayland and R. C. Heinze, *Chem. Eng. Prog.* **69,** 75 (May 1973).

Fine particle dynamics

130. R. Clift and W. H. Gauvin, *Can. J. Chem. Eng.* **49,** 439 (1971).
131. F. A. Zenz and D. F. Othmer, *Fluidization and Fluid Particle Systems,* Reinhold, New York, 1960, pp. 206–207.
132. C. A. Lapple and co-workers, *Fluid and Particle Mechanics,* University of Delaware, Newark, Del., 1956, p. 292.
133. C. B. Shepherd and C. E. Lapple, *Ind. Eng. Chem.* **31,** 972 (1939).
134. C. B. Shepherd and C. E. Lapple, *Ind. Eng. Chem.* **32,** 1246 (1940).
135. K. E. Lunde and C. E. Lapple, *Chem. Eng. Prog.* **53,** 385 (Aug. 1957).
136. C. Y. Shen, *Chem. Rev.* **55,** 595 (1955).
137. P. O. Rouhiainen and J. W. Stachiewicz, *J. Heat Trans.* **92,** 169 (1970).
138. W. E. Ranz, *Penn State Univ. Coll. Eng. Res. Bull.* B-66 (Dec. 1966).
139. W. E. Ranz and J. B. Wong, *Ind. Eng. Chem.* **44,** 1371 (1952).
140. H. F. Johnstone and M. H. Roberts, *Ind. Eng. Chem.* **41,** 2417 (1949).
141. H. J. White, *Industrial Electrostatic Precipitation,* Addison-Wesley Publishing Co., Reading, Mass., 1963.
142. B. Singh and R. L. Byers, *Ind. Eng. Chem. Fund.* **11,** 127 (1972).
143. L. Waldmann, *Z. Naturforsch.* **14A,** 589 (1959).
144. *Control Techniques for Particulate Air Pollutants,* AP-51, U.S. Department of Health, Education, and Welfare, Washington, D.C., 1969, p. 42.
145. E. S. Pettyjohn and E. B. Christiansen, *Chem. Eng. Prog.* **44,** 157 (1948).
146. H. Wadell, *Physics* **5,** 281 (1934).
147. H. A. Becker, *Can. J. Chem. Eng.* **37,** 85 (1959).
148. Ref. 131, pp. 209–216.
149. I. Langmuir and K. B. Blodgett, *U.S. Army Air Forces Technical Report 5418, February 19, 1946,* U.S. Department Commerce, Office Technical Services PB 27565.
150. Ref. 39, p. 20-83, Fig. 20-105.
151. J. H. Seinfeld, *Air Pollution Physical and Chemical Fundamentals,* McGraw-Hill, Inc., New York, 1975, p. 457.
152. A. E. Vandegrift, L. J. Shannon, and G. S. Gorman, *Chem. Eng. Deskbook* **80,** 109 (June 18, 1973).
153. K. T. Semrau, C. W. Margnowski, K. E. Lunde, and C. E. Lapple, *Ind. Eng. Chem.* **50,** 1615 (1958).
154. C. J. Stairmand, *J. Inst. Fuel* **29**(2), 58 (1956).
155. G. D. Sargent, *Chem. Eng.* **76,** 130 (Jan. 27, 1969).
156. C. P. Kerr, *J. APCA* **39**(12), 1585–1587 (Dec. 1989).

Particulate control, gravity, and centrifugal separation

157. P. Rosin, E. Rammler, and W. Intelmann, *Zeit. Ver. Deut. Ind.* **76,** 433–437 (1932) (in German).
158. Ref. 39, p. 20-86, Fig. 20-109.
159. D. Leith and W. Licht, *AIChE Symp. Ser.* **68**(126), 196–206 (1972).
160. A. C. Stern, *Air Pollution,* Vol. III, 2nd ed., Academic Press, Inc., New York, 1968, p. 370.
161. Ref. 160, p. 361; *Air Pollution Manual,* Part II, American Industrial Hygiene Association, Detroit, Mich., 1968, p. 29.

162. *Manual on Disposal of Refinery Wastes, API Pub. 931,* American Petroleum Institute, Washington, D.C., 1975, Chapt. 11.
163. C. J. Stairmand, *Trans. Inst. Chem. Eng.* **29,** 356–383 (1951).
164. B. Kalen and F. A. Zenz, *AIChE Symp. Ser.* **70**(137), 388–396 (1974).
165. R. McK. Alexander, *Proc. Australas. Inst. Min Eng.* **152/3,** 202–228 (1949).
166. C. J. Stairmand, *Engineering* **16B,** 409–412 (Oct. 21, 1949).
167. A. C. Stern, K. J. Caplan, and P. D. Bush, *Cyclone Dust Collectors,* American Petroleum Institute, New York, 1955.
168. H. J. Tengbergen in K. Rietema and C. G. Verner, eds., *Cyclones in Industry,* Elsevier Publishing Co., Amsterdam, 1961 (in English).
169. B. B. Crocker in "Phase Separation" in ref. 39, pp. 18-70–18-88.
170. W. Strauss, *Industrial Gas Cleaning,* 2nd ed., Pergamon Press, Inc., Oxford, 1975.
171. K. J. Caplan in A. C. Stern, ed., *Air Pollution,* Vol. III, 2nd ed., Academic Press, New York, 1968, Chapt. 43.
172. Ref. 144, Figs. 4-43, 4-44.

Particulate control, electrostatic attraction

173. A. T. Murphy, F. T. Adler, and G. W. Penney, *AIIE Trans.* Preprint, Paper 59-102 (1959).
174. Ref. 141, p. 140.
175. S. Oglesby, Jr., and G. B. Nichols, "A Manual of Electrostatic Precipitator Technology," *NTIS Report PB196380,* Southern Research Institute, Birmingham, Ala., 1970.
176. E. Anderson in R. H. Perry, ed., *Chemical Engineers' Handbook,* 2nd ed., McGraw-Hill, Inc., New York, 1941, p. 1873.
177. W. B. Smith and J. R. McDonald, *J. APCA* **25,** 168 (Feb. 1975).
178. J. P. Gooch and N. L. Francis, *J. APCA* **25,** 108 (Feb. 1975).
179. J. P. Gooch and J. R. McDonald, *AIChE Symp. Ser.* **73**(165), 146 (1977).
180. P. Cooperman, "Nondeutschian Phenomena in Electrostatic Precipitation," *Preprint 76-42.2 APCA 69th Annual Meeting, Portland, Oregon, June 27–July 1, 1976.*
181. P. L. Feldman, K. S. Kumar, and G. D. Cooperman, *AIChE Symp. Ser.* **73**(165), 120 (1977).
182. *Chem. Eng.* **83,** 51 (Aug. 30, 1976).
183. C. L. Burton and D. A. Smith, *J. APCA* **25,** 139 (Feb. 1975).
184. A. G. Hein, *J. APCA* **39,** 766–771 (May 1989).
185. W. T. Sproull, *J. APCA* **15,** 50 (Feb. 1965).
186. W. T. Sproull, *J. APCA* **22,** 181 (Mar. 1972).
187. H. W. Spencer, III, and G. B. Nichols, "A Study of Rapping Re-entrainment in a Large Pilot Electrostatic Precipitator," *Preprint 76-42.5, 68th APCA Annual Meeting, Portland, Oregon, June 27–July 1, 1976.*
188. S. Matts, *J. APCA* **25,** 146 (Feb. 1975).
189. A. B. Walker, *J. APCA* **25,** 143 (Feb. 1975).
190. R. E. Cook, *J. APCA* **25,** 156 (Feb. 1975).
191. E. B. Dismukes, *J. APCA* **25,** 152 (Feb. 1975).
192. R. E. Bickelhaupt, *J. APCA* **25,** 148 (Feb. 1975).
193. H. J. White, *J. APCA* **24,** 313 (Apr. 1974).
194. H. J. White, "Control of Particulates by Electrostatic Precipitation," in S. Calvert and H. M. Englund, eds., *Handbook of Air Pollution Technology,* John Wiley & Sons, Inc., New York, 1984, Chapt. 12, pp. 283–329.
195. S. Oglesby, Jr., and G. B. Nichols, *Electrostatic Precipitation,* Marcel Dekker, Inc., New York, 1978, pp. 132–155; resistivity and conditioning discussion.

196. M. R. Schioeth, *Pulse-Energized Electrostatic Precipitators,* paper at the Third Conference on Electrostatic Precipitation, Albano-Padova, Italy, October, 1987, F. L. Smidth Co., Valby, Denmark.

197. V. Reyes, *Comparison between Traditional and Modern Automatic Controllers on Full-scale Precipitators,* Seventh Symposium on Transfer and Utilization of Particulate Control Technology, Nashville, Tenn., March 1988, F. L. Smidth Co., Valby, Denmark.

198. H. J. Hall, *J. APCA* **25,** 132 (Feb. 1975).

199. H. J. White, *J. APCA* **25,** 102 (Feb. 1975).

200. *J. APCA* **25,** 362 (Apr. 1975).

201. T. T. Shen and N. C. Pereira, "Electrostatic Precipitation" in *Handbook of Environmental Engineering,* Vol. 1, The Humana Press, Clifton, N.J., 1979, Chapt. 4, pp. 103–143.

202. J. R. Benson and M. Corn, *J. APCA* **24,** 340 (Apr. 1974).

203. E. Bakke, *J. APCA* **25,** 163 (Feb. 1975).

204. E. Bakke, "The Application of Wet Electrostatic Precipitators for Control of Fine Particulate Matter," *Preprint, Symposium on Control of Fine Particulate Emissions from Industrial Sources, Joint U.S.-USSR Working Group, Stationary Source Air Pollution Control Technology, San Francisco, Calif., Jan. 15–18, 1974.*

205. Ref. 39, p. 20-120, Fig. 20-152.

206. D. W. Cooper, "Fine Particle Control by Electrostatic Augmentation of Existing Methods," *Preprint 75-02.1, 68th APCA Annual Meeting, Boston, Mass., June 15–20, 1975.*

207. K. A. Nielsen and J. C. Hill, *Ind. Eng. Chem. Fundam.* **15,** 149, 157 (1976).

208. M. J. Pilot, *J. APCA* **25,** 176 (Feb. 1975).

209. M. J. Pilot and D. F. Meyer, *University of Washington Electrostatic Spray Scrubber Evaluation, NTIS PB 252653,* Apr. 1976.

210. K. Zahedi and J. R. Melcher, "Electrofluidized Beds in the Filtration of Submicron Particulate," *Preprint 75-57.8, 68th APCA Annual Meeting, Boston, Mass., June 15–20, 1975.*

211. W. L. Klugman and S. V. Sheppard, "The Ceilcote Ionizing Wet Scrubber," *Preprint 75-30.3, 68th APCA Annual Meeting, Boston, Mass., June 15–20, 1975.*

212. Helfritch, *Chem. Eng. Progress* **73,** 54–57 (Aug. 1977).

213. C. W. Lear, W. F. Krieve, and E. Cohen, *J. APCA* **25,** 184 (Feb. 1975).

214. D. H. Pontius, L. G. Felix, and W. B. Smith, "Performance Characteristics of a Pilot Scale Particle Charging Device," *Preprint 76-42.6, 69th APCA Annual Meeting, Portland, Oregon, June 27–July 1, 1976.*

215. J. R. Melcher and K. S. Sachar, "Charged Droplet Scrubbing of Submicron Particulate," *NTIS Pub. PB-241262,* Aug. 1974.

Particulate control filtration

216. Ref. 160, p. 413.

217. D. Leith and M. W. First, *J. APCA* **27,** 534–539 (June 1977); **27,** 636–642 (July 1977).

218. C. E. Billings and J. Wilder, "Handbook of Fabric Filter Technology," *NTIS Pub. PB200648, PB200649, PB200651, PB200650,* 1970.

219. L. Bergmann, *J. APCA* **24,** 1187 (Dec. 1974).

220. E. R. Frederick, *J. APCA* **24,** 1164 (Dec. 1974).

221. R. Dennis, *J. APCA* **24,** 1156 (Dec. 1974).

222. R. L. Lucas in R. H. Perry and C. H. Chilton, eds., *Chemical Engineers Handbook,* 5th ed., McGraw-Hill, Inc., New York, 1973, p. 20–90.

223. V. E. Schoeck in B. E. Kester, ed., *Proceedings, Specialty Conference on Design, Operation, and Maintenance of High Efficiency Particulate Control Equipment,* Air Pollution Control Association, Pittsburgh, Pa., 1973, pp. 103–117.

224. R. A. Gross, *Proceedings of the Specialty Conference on the User and Fabric Filter Equipment II,* Air Pollution Control Association, Pittsburgh, Pa., 1975, pp. 159–163.

225. *J. APCA* **25,** 715 (July 1975). Informative Report No. 7 of the APCA TC-1 Particulate Committee.

226. G. W. Bowerman in S. Calvert and H. M. Englund, eds., *Handbook of Air Pollution Technology,* John Wiley & Sons, Inc., New York, 1984, pp. 153–158.

227. E. Bakke, *J. APCA* **24,** 1150 (Dec. 1974).

228. J. D. Mc Kenna, J. C. Mycock, and W. O. Lipscomb, *J. APCA* **24,** 1144 (Dec. 1974).

229. Ref. 39, pp. 20-97–20-104.

230. J. H. Turner and J. D. McKenna, "Control of Particles by Filters," in S. Calvert and H. M. Englund, eds., *Handbook of Air Pollution Technology,* John Wiley & Sons, Inc., New York, 1984, Chapt. 11.

231. C. Orr, *Filtration, Principles and Practice,* 2 vols., Marcel Dekker, Inc., New York, 1977.

232. M. J. Hobson, *Review of Baghouse Systems for Boiler Plants,* pp. 74–84, ref. 190a.

233. R. A. Gross, *Proceedings of the Specialty Conference on the User and Fabric Filter Equipment II,* Air Pollution Control Association, Pittsburgh, Pa., 1975, pp. 159–163.

234. R. L. Adams, "Fabric Filters for Control of Power Plant Emissions," *Preprint 74-100, 67th APCA Annual Meeting, Denver, Col.,* June 9–13, 1974.

235. R. P. Janoso, "Baghouse Dust Collectors on a Low Sulfur Coal Fired Utility Boiler," *Preprint 74-101, 67th APCA Annual Meeting, Denver, Col., June 9–13, 1974.*

236. D. S. Ensor, R. G. Hooper, R. C. Carr, and R. W. Scheck, "Evaluation of a Fabric Filter on a Spreader Stoker Utility Boiler,"*Preprint 76-27.6, 69th APCA Annual Meeting, Portland, Oregon, June 27–July 1, 1976.*

237. J. D. Mc Kenna, J. C. Mycock, and W. O. Lipscomb, "Applying Fabric Filtration to Coal Fired Industrial Boilers: A Pilot Scale Investigation," *NTIS Pub. PB-245186,* Aug. 1975.

238. R. Dennis and co-workers, "Filtration Model for Coal Fly Ash with Glass Fabrics," EPA Rpt. EPA-600/7-77-095a, *NTIS Pub. PB 276-489/AS,* August 1977.

239. R. L. Chang, T. R. Snyder, and P. Vann Bush, *J. APCA* **39,** Pt. I, 228; Pt. II, 361 (1989).

240. K. M. Kushing, R. L. Merritt, and R. L. Chang, *J. Air Waste Manage Assn.* **40,** 1051–1058 (July 1990).

241. J. H. Turner, *J. APCA* **24,** 1182 (Dec. 1974).

242. R. E. Jenkins and co-workers, *J. APCA* **37,** Pt. I, 749; Pt. II, 1105 (1987).

243. D. C. Drehnel, "Relationship between Fabric Structure and Filtration Performance in Dust Filtration," *NTIS Pub. PB-222237,* 1973.

244. B. Miller, G. E. R. Lamb, and P. Costanza, "Influence of Fiber Characteristics on Particulate Filtration," *NTIS Pub. PB-239997,* Jan. 1975.

245. *Fuel Econ.* **12,** 47 (Oct. 1936).

246. U.S. Pat. 2,604,187, Dorfan.

247. Taub, "Filtration Phenomena in a Packed Bed Filter," Ph.D. Thesis, Carnegie-Mellon University, Pittsburgh, Pa., 1970.

248. Reese, *TAPPI,* **60**(3), 109 (1977).

249. Parquet, *The Electroscrubber Filter: Applications and Particulate Collection Performance,* EPA-600/9-82-005c, p. 363, 1982.

250. Self and co-workers, "Electric Augmentation of Granular Bed Filters," EPA-600/9-80-039c, p. 309, 1980.

251. Schueler, *Rock Prod.,* **76**(7), 66 (1973); **77**(11), 39 (1974).

252. EPA-600/7-78-093, 1978.

253. A. M. Squires and R. Pfeffer, *J. APCA* **20,** 534 (Aug. 1970); L. Paretsky, L. Theodore, R. Pfeffer, and A. M. Squires, *J. APCA* **21,** 204 (Apr. 1971); A. M. Squires and R. A. Graff, *J. APCA* **21,** 272 (May 1971); U.S. Pat. 3,296,775, A. M. Squires.

254. S. Miyamoto and H. L. Bohn, *J. APCA* **24,** 1051 (Nov. 1974); **25,** 40 (Jan. 1975).

255. U.S. Pat. 3,410,055, F. A. Zenz.

256. Ref. 222, p. 20-89.

257. H. P. Meissner and H. S. Mickley, *Ind. Eng. Chem.* **41,** 1238 (1949).

258. D. S. Scott and D. A. Guthrie, *Can. J. Chem. Eng.* **37,** 200 (1959).

259. M. L. Jackson, *AIChE Symp. Ser.* **70**(141), 82 (1974).

260. O. H. York, *Chem. Eng. Prog.* **50,** 421 (1954); O. H. York and E. W. Poppele, *Chem. Eng. Prog.* **59,** 45 (June 1963).

261. O. D. Massey, *Chem. Eng. Prog.* **55,** 114 (May 1959); *Chem. Eng.* **66,** 143 (July 13, 1959).

262. J. W. Coykendall, E. F. Spencer, and O. Y. York, *J. APCA* **18,** 315 (May 1968).

263. J. A. Brink, W. F. Burggrabe, and J. A. Rauscher, *Chem. Eng. Prog.* **60,** 68 (Nov. 1964).

264. J. A. Brink, *Chem. Eng.* **66,** 183 (Nov. 16, 1959); *Can. J. Chem. Eng.* **41,** 134 (1963); J. A. Brink, W. F. Burggrabe, and L. E. Greenwell, *Chem. Eng. Prog.* **64,** 82 (Nov. 1968).

265. Ref. 144, p. 73.

266. J. Goldfield, V. Greco, and K. Gandhi, *J. APCA* **20,** 466 (July 1970).

Particulate control, wet scrubbing

267. K. T. Semrau, *J. APCA* **10,** 200 (1960).

268. K. T. Semrau, *J. APCA* **13,** 587 (1963).

269. Ref. 222, p. 20-98.

270. G. J. Celenza, *Chem. Eng. Prog.* **66,** 31 (Nov. 1970).

271. S. Calvert, *J. APCA* **24,** 929 (1974).

272. S. Calvert, J. Goldshmid, D. Leith, and D. Mehta, "Scrubber Handbook," *NTIS Pub. PB-213016 and PB-213017,* July–Aug. 1972.

273. S. Calvert, *Chem. Eng.* **84,** 54–68 (Aug. 29, 1977).

274. S. Calvert, "Particulate Control by Scrubbing," in S. Calvert and H. M. Englund, eds., *Handbook of Air Pollution Technology,* John Wiley & Sons, Inc., New York, 1984, Chapt. 10, pp. 215–248.

275. H. E. Hesketh, "Atomization and Cloud Behavior in Wet Scrubbers," *U.S.-USSR Symposium on Control of Fine Particulate Emissions,* Jan. 15–18, 1974.

276. W. Strauss, "Particulate Collection by Liquid Scrubbing," in *Industrial Gas Cleaning,* 2nd ed., Pergamon Press, Oxford, 1975, Chapt. 9, pp. 367–408.

277. *J. APCA* **22,** 54 (June 1972).

278. C. J. Stairmand, *Trans. Inst. Chem. Eng.* **28,** 130 (1950).

279. C. J. Stairmand, *J. Inst. Fuel* **29**(2), 58 (1956).

280. Ref. 222, p. 20-97.

281. Ref. 276, p. 374.

282. R. V. Kleinschmidt and A. W. Anthony in L. C. McCabe, ed., *U.S. Technical Conference on Air Pollution,* McGraw-Hill, Inc., New York, 1952, p. 310.

283. S. V. Sheppard, *J. APCA* **22,** 278 (Apr. 1972).

284. H. Doyle and A. F. Brooks, *Ind. Eng. Chem.* **49**(12), 57A (1957).

285. L. S. Harris, *Chem. Eng. Prog.* **42,** 55 (Apr. 1966).

286. H. E. Gardenier, *J. APCA* **24,** 954 (Oct. 1974).

287. H. F. Johnstone, R. B. Field, and M. C. Tassler, *Ind. Eng. Chem.* **46,** 1601 (1954).

288. W. Barth, *Staub* **19,** 175 (1959).

289. S. Calvert and D. Lundgren, *J. APCA* **18,** 677 (Oct. 1968).

290. S. Calvert, D. Lundgren, and D. S. Mehta, *J. APCA* **22,** 529 (July 1972).

291. R. H. Boll, *Ind. Eng. Chem. Fundam.* **12,** 40 (Jan. 1973).

292. H. E. Hesketh, *J. APCA* **24,** 939 (Oct. 1974).

293. R. H. Boll, L. R. Flais, P. W. Maurer, and W. L. Thompson, *J. APCA* **24,** 934 (Oct. 1974).

294. S. W. Behie and J. M. Beeckmans, *J. APCA* **24,** 943 (Oct. 1974).

295. K. G. T. Hollands and K. C. Goel, *Ind. Eng. Chem. Fundam.* **14,** 16 (Jan. 1975).

296. A. B. Walker and R. M. Hall, *J. APCA* **18,** 319 (May 1968).

297. S. Calvert, I. L. Jashnani, and S. Yung, *J. APCA* **24,** 971 (Oct. 1974).

298. S. Calvert, I. L. Jashnani, S. Yung, and S. Stalberg, "Entrainment Separators for Scrubbers—Initial Report," *NTIS Pub PB-241189,* Oct. 1974.

299. E. B. Hanf, *Chem. Eng. Prog.* **67,** 54 (Nov. 1971).

300. T. R. Blackwood, *Environ. Progress* **7**(1), 71–75 (Feb. 1988).

Particulate control, developing technology

These references do not include electrostatic augmentation which are included under "Particulate Control, Electrostatic Attraction."

301. L. E. Sparks and M. J. Pilat, *Atmos. Environ.* **4,** 651 (1970).

302. H. W. Danser, Jr., *Chem. Eng.* **57,** 158 (May 1950).

303. C. A. Stokes, *Chem. Eng. Prog.* **46,** 423 (1950).

304. E. P. Mednikov, *Acoustic Coagulation and Precipitation of Aerosols,* USSR Academy of Science, Moscow, 1963, translated by C. V. Larrick, Consultants Bureau, 1965.

305. G. R. Gillespie and H. F. Johnstone, *Chem. Eng. Prog.* **51,** 74F (Feb. 1955).

306. S. Calvert and N. C. Jhaveri, *J. APCA* **24,** 946 (1974).

307. S. Calvert, J. Goldshmid, D. Leith, and N. Jhaveri, "Feasibility of Flux Force/Condensation Scrubbing for Fine Particle Collection," *NTIS Pub. PB-227307,* Oct. 1973.

308. S. R. Rich and T. G. Pantazelos, *J. APCA* **24,** 952 (Oct. 1974).

309. K. T. Semrau and C. L. Witham, "Condensation and Evaporation Effects in Particulate Scrubbing," *Preprint 75-30.1, 68th APCA Annual Meeting, Boston, Mass., June 15–20, 1975.*

Odor measurement and control

310. J. P. Cox, *Odor Control and Olfaction,* Pollution Sciences Publishing Company, Lynden, Washington, 1975.

311. R. W. Moncrieff, *Odour Preferences,* John Wiley & Sons, Inc., New York, 1966.

312. W. H. Stahl, ed., *Compilation of Odor and Taste Threshold Values Data,* American Society for Testing and Materials Data Series 48, Philadelphia, Pa., 1973.

313. P. N. Cheremisinoff and R. A. Young, eds., *Industrial Odor Technology Assessment,* Ann Arbor Science Publishers, Inc., Ann Arbor, Mich., 1975.

314. J. Wittes and A. Turk, *Correlation of Subjective–Objective Methods in the Study of Odors and Taste,* American Society for Testing and Materials Special Technical Publication 440, Philadelphia, Pa., 1968, pp. 49–70.

315. F. V. Wilby, *J. APCA* **19,** 96 (1969).

316. G. Leonardos, D. Kendall, and N. Barnard, *J. APCA* **19,** 91 (1969).

317. A. Turk, *Basic Principles of Sensory Evaluation,* American Society for Testing and Materials Special Technical Publication #433, 1968, pp. 79–83.

318. W. C. L. Hemeon, *J. APCA* **18,** 166 (1968).

319. W. C. L. Hemeon, *AIChE Symp. Ser.* **73**(165), 260 (1977).

320. T. M. Hellman and F. H. Small, *Chem. Eng. Prog.* **69,** 75 (1973).

321. T. M. Hellman and F. H. Small, *J. APCA* **24,** 979 (1974).

322. T. M. Hellman in Ref. 266, pp. 45–56.

323. A. Dravnieks and W. H. Prokop, *J. APCA* **25,** 28 (1975).

324. G. Leonardos, *J. APCA* **24,** 456 (1974).

325. R. A. Duffee, *J. APCA* **18,** 472 (1968).

326. S. C. Varshney, E. Poostchi, A. W. Gnyp, and C. C. St. Pierre, *Environ. Progress* **5**(4), 240–244 (Nov. 1986).

327. R. M. Bethea and co-workers, *Environ. Sci. Technol.* **7**, 504 (1973).
328. M. W. First and co-workers, *J. APCA* **24**, 653 (1974).
329. G. A. Herr, *Chem. Eng. Prog.* **70**, 65 (1974).
330. D. E. Quane, *Chem. Eng. Prog.* **70**, 51 (1974).
331. D. J. Eisenfelder and J. W. Dolen, *Chem. Eng. Prog.* **70**, 48 (1974).
332. J. E. Paul, *J. APCA* **25**, 158 (1975).
333. J. E. Yocom and R. A. Duffee, *Chem. Eng.* **77**(13), 160 (1970).
334. M. Beltran, *Chem. Eng. Prog.* **70**, 57 (1974).
335. R. M. Bethea, *Engineering Analysis and Odor Control,* Chapt. 13, pp. 203–214, ref. 266.
336. *Proceedings, State of the Art of Odor Control Technology Specialty Conference,* March 1974, Air Pollution Control Association, Pittsburgh, Pa., 1974.
337. *Proceedings, State of the Art of Odor Control Technology Specialty Conference,* March 1977, Air Pollution Control Association, Pittsburgh, Pa., 1977.
338. A. Turk, S. Mehlman, and E. Levine, *Atmos. Environ.* **7**, 1139 (1973).
339. W. D. Lovett and F. T. Cunniff, *Chem. Eng. Prog.* **70**, 43 (May 1974).

General references

A. C. Stern, ed., *Air Pollution,* 3rd ed., Vols. 1–5, Academic Press, Inc., New York (Vols. 1–3, 1976; Vols. 4–5, 1977).
R. G. Bond and C. P. Straub, eds., *Handbook of Environmental Control,* Vol. 1, CRC Press, Cleveland, Ohio, 1972.
K. Wark and C. F. Warner, *Air Pollution: Its Origin and Control,* Dun-Donnelley Publishing Corporation, New York, 1976.
J. H. Seinfeld, *Air Pollution: Physical and Chemical Fundamentals,* McGraw-Hill, Inc., New York, 1975.
H. E. Hesketh, *Air Pollution Control,* Ann Arbor Science Publishers, Inc., Ann Arbor, Mich., 1979.
P. N. Cheremisinoff and R. A. Young, *Air Pollution Control and Design Handbook,* Parts 1 and 2, Marcel Dekker, Inc., New York, 1977.
W. Licht, *Air Pollution Control Engineering: Basic Calculations for Particulate Collection,* Marcel Dekker, Inc., New York, 1980.
H. C. Perkins, *Air Pollution,* McGraw-Hill, Inc., New York, 1974.
C. N. Davies, ed., *Aerosol Science,* Academic Press, Inc., New York, 1966.
W. Ruch, ed., *Chemical Detection of Gaseous Pollutants,* Ann Arbor Science Publishers, Inc., Ann Arbor, Mich., 1966.
B. Y. H. Liu, ed., *Fine Particles—Aerosol Generation, Measurement, Sampling and Analysis,* Academic Press, Inc., New York, 1976.
P. O. Warner, *Analysis of Air Pollutants,* John Wiley & Sons, Inc., New York, 1976.
W. Strauss, *Industrial Gas Cleaning,* 2nd ed., Pergamon Press, Oxford, 1975.
L. Theodore and A. J. Buonicore, *Air Pollution Control Equipment—Selection, Design, Operation and Maintenance,* Prentice-Hall, Englewood Cliffs, N.J., 1982.
S. Calvert and H. M. Englund, eds., *Handbook of Air Pollution Technology,* John Wiley & Sons, Inc., New York, 1984.
L. K. Wang and N. C. Pereira, eds., *Handbook of Environmental Engineering,* Vol. 1—*Air and Noise Pollution Control,* The Humana Press, Clifton, N.J., 1979.

BURTON B. CROCKER
Consultant

AIR SEPARATION.   See CRYOGENICS; MEMBRANE TECHNOLOGY; MOLECULAR SIEVES; NITROGEN.

ALCOHOL FUELS

The use of alcohols as motor fuels gained considerable interest in the 1970s as substitutes for gasoline and diesel fuels, or, in the form of blend additives, as extenders of oil supplies. In the United States, most applications involved the use of low level ethanol (qv) [64-17-5] blends in gasoline [8006-61-9] (see GASOLINE AND OTHER MOTOR FUELS). Brazil, however, launched a major program to substitute ethanol for gasoline in 1976, beginning with a 22% alcohol ethanol–gasoline blend and adding dedicated ethanol cars in the 1980s. By 1985 ethanol cars accounted for 95% of new car sales in Brazil. By 1988 Brazil had 3,000,000 automobiles, about 30% of the total automobile population, dedicated to ethanol. The United States has demonstration vehicles using alcohols (mostly methanol (qv) [67-56-1]), but otherwise has not yet passed beyond the use of limited amounts of alcohols and of ethers produced from alcohols as gasoline components (see OCTANE IMPROVERS). However, proposals continue to be made to implement alcohol programs similar to that of Brazil (1). Other nations such as New Zealand, Germany, and Sweden also investigated the use of alcohols as a transportation fuel.

The benefits of alcohol fuels include increased energy diversification in the transportation sector, accompanied by some energy security and balance of payments benefits, and potential air quality improvements as a result of the reduced emissions of photochemically reactive products (see AIR POLLUTION). The Clean Air Act of 1990 and emission standards set out by the State of California may serve to encourage the substantial use of alcohol fuels, unless gasoline and diesel technologies can be developed that offer comparable advantages.

Properties of Alcohol Fuels

Table 1 summarizes key properties of ethanol and methanol as compared to other fuels. Both alcohols make excellent motor fuels, although the high latent heats of vaporization and the low volatilities can make cold-starting difficult in vehicles having carburetors or fuel injectors in the intake manifold where the fuel must be vaporized prior to being introduced into the combustion chamber. This is not the case for direct injection diesel-type engines using methanol or ethanol. Both methanol and ethanol have high octane values and allow high compression ratios having increased efficiency and improved power output per cylinder. Both have wider combustion envelopes than gasoline and can be run at lean air–fuel ratios with better energy efficiency. However, ethanol and methanol have very low cetane numbers and cannot be used in compression–ignition diesel-type engines unless gas temperatures are high at the time of injection. Manufacturers of heavy-duty engines have developed several types of systems to assist autoignition of directly injected alcohols. These include glow plugs or spark plugs, reduced engine cooling, and increased amounts of exhaust gas recirculation or, in the case of two-stroke engines, reduced scavenging. Additives to improve cetane number have been effective as have dual-fuel approaches, in which a small amount of diesel fuel is used as an ignitor for the alcohol.

There are particular alcohol fuel safety considerations. Unlike gasoline or diesel fuel, the vapor of methanol or ethanol above the liquid fuel in a fuel tank is usually combustible at ambient temperatures. This poses the risk of an explosion should a spark or flame find its way to the tank such as during refueling. Additionally, a neat methanol fire has very little luminosity and, consequently, fire fighting efforts can be difficult in daylight. However, low luminosity also implies low radiative fluxes from the fire. This, combined with the high latent heat of vaporization, means that the heat release of a methanol fire is low relative to one of gasoline or diesel fuel. Because methanol or ethanol are both water-soluble, fires can be successfully controlled by dilution with large amounts of water, a tactic that simply spreads gasoline fires. Nevertheless, fire-extinguishing foams (see FIRE-EXTINGUISHING AGENTS) are the preferred alcohol fire-fighting method.

Some potential problems of alcohol fuels have been addressed by adding small amounts of gasoline or specific hydrocarbons to the fuel, reducing the flammability envelope and providing luminosity in case of fire.

Uses of Alcohol Fuels

Early applications of internal combustion engines featured a variety of fuels, including alcohols and alcohol–hydrocarbon blends. In 1907 the U.S. Department of Agriculture investigated the use of alcohol as a motor fuel. A subsequent study by the U.S. Bureau of Mines concluded that engines could provide up to 10% higher power on alcohol fuels than on gasoline (8). Mixtures of alcohol and gasoline were used on farms in France and in the United States in the early 1900s (9). Moreover, the first Ford Model A automobiles could be run on either gasoline or ethanol using a manually adjustable carburetor (1). However, the development of low cost gasoline pushed other automobile fuels into very minor roles and the diesel engine further solidified the hold of petroleum fuels on the transportation sector. Ethanol was occasionally used, particularly in rural regions, when gasoline supplies were short or when farm prices were low.

Methanol has been used as a motor racing fuel for many decades. Its high latent heat of vaporization cools the incoming charge of air to each cylinder. Increasing the mass of air taken into each cylinder increases the power developed by each stroke, providing a turbocharger effect. Furthermore, methanol has a higher octane value than gasoline allowing higher compression ratios, greater efficiency, and higher output per unit of piston displacement volume. These power increases are advantages in racing as is the simple means by which methanol fuel quality and uniformity can be verified.

The transparency of methanol flames is usually a safety advantage in racing. In the event of fires, drivers have some visibility and the lower heat release rate of methanol provides less danger for drivers, pit crews, and spectators.

Partly for these reasons, methanol has been the required fuel of the Indianapolis 500 since 1965. Methanol is also used in many other professional and amateur races. However, transparency of the methanol flames has also been a disadvantage in some race track fires. The invisibility of the flame has confused pit crews, delayed fire detection, and caused even trained firefighters problems in locating and extinguishing fires.

Table 1. Properties of Fuels[a]

Properties	Methanol	Ethanol	Propane	Methane	Isoctane	Unleaded gasoline	Diesel fuel #2
constituents	CH_3OH	CH_3CH_2OH	C_3H_8	CH_4	C_8H_{18}	$C_nH_{1.87n}$ (C_4 to C_{12})	$C_nH_{1.8n}$ (C_8 to C_{20})
CAS Registry Number	[67-56-1]	[64-17-5]	[74-98-6]	[74-82-8]	[540-84-1]	[8006-6-9]	
molecular weight	32.04	46.07	44.10	16.04	114.23	≈110	≈170
element composition, wt %							
C	37.49	52.14	81.71	74.87	84.12	86.44	86.88
H	12.58	13.13	18.29	25.13	15.88	13.56	13.12
O	49.93	34.73	0	0	0	0	0
density at 16°C and 101.3 kPa[b], kg/m^3	794.6	789.8	505.9	0.6776	684.5	721–785	833–881
boiling point at 101.3 kPa[b], °C	64.5	78.3	6.5	−161.5	99.2	38–204	163–399
freezing point, °C	−97.7	−114.1	−188.7	−182.5	−107.4		< −7
vapor pressure at 38°C, kPa[b]	31.9	16.0	1.297	0.5094	11.8	48–108	negligible
heat of vaporization, ΔH_v, MJ/kg[c]	1.075	0.8399	0.4253	0.5094	0.2712	0.3044	0.270
gross heating value, MJ/kg[c]	22.7	29.7	50.4	55.5	47.9	47.2	44.9
net heating value							
MJ/kg[c]	20.0	27.0	46.2	50.0	44.2	43.9	42.5
MJ/m^3[d]	15,800	21,200	23,400		30,600	32,000	35,600

828

stoichiometric mixture net heating value, MJ/kg[c]	2.68	2.69	2.75	2.72	2.75	2.83	2.74
autoignition temperature, °C	464	363	450	537	418	260–460	257
flame temperature, °C	1,871	1,916	1,988	1,949	1,982	2,027	1,993
flash point, °C	11	13			4	−43 to −39	52–96
flame speed at stoichiometry, m/s	0.43		0.40	0.37	0.31	0.34	
octane ratings							
research	106	107	112	120	100	92–98	
motor	92	89	97	120	100	80–90	
cetane rating	0–5	0–5					>40
flammability limits, vol % in air	6.72–36.5	3.28–18.95	2.1–9.5	5.0–15.0	1.0–6.0	1.4–7.6	1.0–5.0
stoichiometric air–fuel mass ratio	6.46	8.98	15.65	17.21	15.10	14.6	14.5
stoichiometric air–fuel volumetric ratio	7.15	14.29	23.82	9.53	59.55	55	85
water solubility	complete	complete	no	no	no	no	no
sulfur content, wt %	0	0	0	0	0	<0.06	<0.5

[a]Refs. 2–7.
[b]To convert kPa to psi, multiply by 0.145.
[c]To convert MJ/kg to Btu/lb, multiply by 430.3.
[d]To convert MJ/m^3 to Btu/gal, multiply by 3.59.

Low level blends of ethanol and and gasoline enjoyed some popularity in the United States in the 1970s. The interest persists into the 1990s, encouraged by the exemption of low level ethanol–gasoline blends from the Federal excise tax as well as from state excise taxes in many states.

Energy Diversification and Energy Security

The ethanol program in Brazil addressed that country's oil supply problems in the 1970s, at times improved the balance of payments, and served to strengthen the economy of the sugar production portion of the agricultural sector. Although the benefits are difficult to quantify because of the very high inflation rate (10) the Brazilian ethanol program generally met its goals. Ironically the inability of ethanol to substitute successfully for diesel fuel in the heavy-duty truck sector necessitated keeping refinery outputs up to provide sufficient available diesel fuel. Thus gasoline was in surplus in Brazil and oil imports were not as much affected as originally hoped. The Brazilian program demonstrated that petroleum substitution strategies need to address the "whole barrel" product slate of oil refineries.

In the 1990s world events precipitated renewed interest in energy diversification strategies for the U.S. transportation sector. However, few measures are in place to encourage fuel alternatives outside of the exemption from the Federal excise tax on motor fuel granted to ethanol blends. The Alternative Motor Fuels Act of 1988 did extend credits to automobile manufacturers in the calculation of corporate average fuel economy (CAFE) for vehicles that use methanol or ethanol or natural gas; electric vehicles had previously been granted such a credit. Under the Act's provisions, neither ethanol nor methanol is counted as fuel consumed in the calculation of fuel economy. Thus vehicles that use alcohol have very high fuel economy ratings, reflecting the value of these vehicles in reducing oil imports. The credits take effect in model-year 1993. The fuel economy calculation assumes that methanol and ethanol fuels in commerce contain 15% by volume gasoline. Vehicles that can use either alcohol fuels or gasoline receive a reduced CAFE credit. The maximum credit that can be earned by a manufacturer for selling vehicles capable of using petroleum fuels is capped because alcohol fuels usage by these vehicles is not assured.

Alcohol Availability

Methanol.   If methanol is to compete with conventional gasoline and diesel fuel it must be readily available and inexpensively produced. Thus methanol production from a low-cost feed stock such as natural gas [8006-14-2] or coal is essential (see FEEDSTOCKS). There is an abundance of natural gas (see GAS, NATURAL) worldwide and reserves of coal are even greater than those of natural gas.

Natural Gas Reserves.   U.S. natural gas reserves could support a significant methanol fuel program. 1990 proved, ie, well characterized amounts with access to markets and producible at current market conditions U.S. resources are 4.8 trillion cubic meters (168 trillion cubic feet = 168 TCF). U.S. consumption is about one-half trillion m^3 (18 TCF) or 18 MJ equivalents per year, but half of that is imported. Estimates of undiscovered U.S. natural gas reserves range from 14 to

16 trillion m³ (492 to 576 TCF) or roughly a 30-year supply at current U.S. consumption rates. Additional amounts of natural gas may become available from advanced technologies (see FUELS, SYNTHETIC).

If 10% of the U.S. gasoline consumption were replaced by methanol for a twenty year period, the required reserves of natural gas to support that methanol consumption would amount to about one trillion m³ (36 TCF) or twice the 1990 annual consumption. Thus the United States could easily support a substantial methanol program from domestic reserves. However, the value of domestic natural gas is quite high. Almost all of the gas has access through the extensive pipeline distribution system to industrial, commercial, and domestic markets and the value of gas in these markets makes methanol produced from domestic natural gas uncompetitive with gasoline and diesel fuel, unless oil prices are very high.

It is therefore more relevant to examine world resources of natural gas in judging the supply potential for methanol. World proved reserves amount to approximately 1.1×10^{15} m³ (40,000 TCF) (11). As seen in Figure 1, these reserves are distributed more widely than oil reserves.

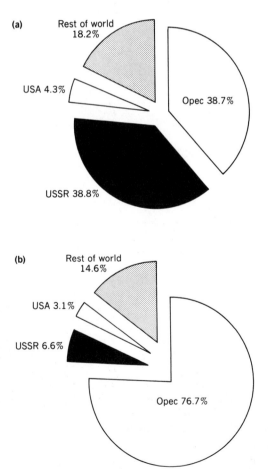

Fig. 1. Distribution of the world proven gas and oil reserves as of January 1, 1988. (**a**) 107.5×10^{12} m³ (3,800 TCF) natural gas; (**b**) 140 m³ (890 barrels) oil.

Using estimates of proven reserves and commitments to energy and chemical uses of gas resources, the net surplus of natural gas in a number of different countries that might be available for major fuel methanol projects has been determined. These are more than adequate to support methanol as a motor fuel.

Coal Reserves.   As indicated in Table 2, coal is more abundant than oil and gas worldwide. Moreover, the U.S. has more coal than other nations: U.S. reserves amount to about 270 billion metric tons, equivalent to about 11×10^{16} MJ (1×10^{20} BTU = 6600 quads), a large number compared to the total transportation energy use of about 3.5×10^{14} MJ (21 quads) per year (11). Methanol produced from U.S. coal would obviously provide better energy security benefits than methanol produced from imported natural gas. At present however, the costs of producing methanol from coal are far higher than the costs of producing methanol from natural gas.

Table 2. World Estimated Recoverable Reserves of Coal in Billions of Metric Tons[a]

Location	Anthracite and bituminous coal[b]	Lignite[c]	Total
North America			
Canada	4.43	2.42	6.85
Mexico	1.91		1.91
United States	231.13	32.71	263.84
total	*237.47*	*35.13*	*272.60*
Central and South America			
total	*5.13*	*0.02*	*5.15*
Western Europe			
total	*32.20*	*58.23*	*90.43*
Eastern Europe and USSR			
USSR	150.19	94.53	244.72
total	*182.82*	*139.17*	*321.99*
Middle East			
total	*0.18*		*0.18*
Africa			
total	*64.67*		*64.67*
Far East and Oceania			
Australia	29.51	36.20	65.71
China	98.79		98.79
total	*130.39*	*38.61*	*169.00*
World total	*652.86*	*271.16*	*924.02*

[a]Ref. 12.
[b]Includes subanthracite and subbituminous.
[c]Includes brown coal.

Biomass.   Methanol can be produced from wood and other types of biomass (see CHEMURGY; FUELS FROM BIOMASS). The prospects for biomass reserves are noted below.

Ethanol.   From the point of view of availability, ethanol is extremely attractive because it can be produced from renewable biomass. Estimates of the amount of ethanol that could be produced from biomass on a sustained basis range from about 3×10^{13} to more than 8×10^{14} MJ/yr (2–50 quad/yr), about half of which would derive from wood crops (11).

Economic Aspects

Alcohol Production. Studies to assess the costs of alcohol fuels and to compare the costs to those of conventional fuels contain significant uncertainties. In general, the low cost estimates indicate that methanol produced on a large scale from low cost natural gas could compete with gasoline when oil prices are around 14¢/L ($27/bbl). This comparison does not give methanol any credits for environmental or energy diversification benefits. Ethanol does not become competitive until petroleum prices are much higher.

Methanol. *Produced from Natural Gas.* Cost assessments of methanol produced from natural gas have been performed (13–18). Projections depend on such factors as the estimated costs of the methanol production facility, the value of the feedstock, and operating, maintenance, and shipping costs. Estimates vary for each of these factors. Costs also depend on the value of oil. Oil price not only affects the value of natural gas, it also affects the costs of plant components, labor, and shipping.

Estimates of the landed costs of methanol (the costs of methanol delivered by ship to a bulk terminal in Los Angeles), vary between 8.9 and 15.6 cents per liter (33.6 and 59.2 cents per gallon). Estimates range from 7.9 to 11.1 cents per liter (30 to 42 cents per gallon) in a large-scale established methanol market and 11.9 to 19.3 cents per liter (45 to 73 cents per gallon) during a small volume transition phase (18).

Estimated pump prices must take terminaling, distribution, and retailing costs into account as well as any differences in vehicle efficiency that methanol might offer. Estimates range from no efficiency advantage to about 30% improvement in fuel efficiency for dedicated and optimized methanol light-duty vehicles. Finally, the cost comparison must take into account the fuel specification for methanol. Most fuel methanol for light-duty vehicles is in the form of 85% methanol, 15% gasoline by volume often termed M85.

The sum of the downstream costs adds roughly 7.9 cents per liter (30¢/gal) and the adjustment of the final cost for an amount of methanol fuel equivalent in distance driven to an equal volume gasoline involves a multiplier ranging from 1.6 to 2.0, depending on fuel specification and the assumed efficiency for methanol light-duty vehicles as compared to gasoline vehicles. The California Advisory Board has undertaken such cost assessment (11).

A cost assessment must also take into account the volume of methanol use ultimately contemplated. A huge methanol program designed to replace most of the U.S. gasoline use would be likely to increase the value of even remotely located natural gas. A big program would also tend to decrease the price of oil, making it more difficult for methanol to compete as a motor fuel. Nevertheless, a balanced assessment of all the studies appears to indicate that a large scale methanol project could provide a motor fuel that competes with gasoline when oil prices are not less than about $0.17/L (1988 U.S. $27/bbl).

In small volume transition phases, methanol cannot compete directly in price with gasoline unless oil prices become very high, with the possible exception of a few scenarios in which low cost methanol is available from expansions to existing methanol plants currently serving the chemical markets for methanol. Energy diversification benefits have not been quantified but the potential air quality benefits have been studied in the work of the California Advisory Board on Air Quality and Fuels. The investment required to introduce methanol might well be justified on air quality grounds, at least in those areas such as California where air pollution programs involve substantial costs. However, the relative advantage of methanol depends on the emissions levels from future vehicles and on the costs of cleaner gasolines that might be able to offer environmental benefits that compete at least to some extent with the environmental benefits of methanol.

Produced from Coal. Estimates of the cost of producing methanol from coal have been made by the U.S. Department of Energy (DOE) (12,17) and they are more uncertain than those using natural gas. Experience in coal-to-methanol facilities of the type and size that would offer the most competitive product is limited. The projected costs of coal-derived methanol are considerably higher than those of methanol produced from natural gas. The cost of the production facility accounts for most of the increase (11). Coal-derived methanol is not expected to compete with gasoline unless oil prices exceed $0.31/L ($50/bbl). Successful development of lower cost entrained gasification technologies could reduce the cost so as to make coal-derived methanol competitive at oil prices as low as $0.25/L ($40/bbl) (17) (see COAL CONVERSION PROCESSES).

These cost comparisons do not assign any credit to methanol for environmental improvements or energy security. Energy security benefits could be large if methanol were produced from domestic coal.

Produced from Biomass. Estimates for methanol produced from biomass indicate (11) that these costs are higher than those of methanol produced from coal. Barring substantial technological improvements, methanol produced from biomass does not appear to be competitive.

Ethanol. Accurate projections of ethanol costs are much more difficult to make than are those for methanol. Large scale ethanol production would impact upon food costs and have important environmental consequences that are rarely cost-analyzed because of the complexity. Furthermore, for corn, the most likely large-scale feedstock, ethanol costs are strongly influenced by the credit assigned to the protein by-product remaining after the starch has been removed and converted to ethanol.

Cost estimates of producing ethanol from corn have many uncertainties (11). Most estimates fall into the range of $0.26 to 0.40 per liter ($1 to 1.50/gal), after taking credits for protein by-products, although some estimates are lower. These estimates do not make ethanol competitive with oil until oil prices are above $0.38/L ($60/bbl) (17).

For these reasons, ethanol is most likely to find use as a motor fuel in the form of a gasoline additive, either as ethanol or ethanol-based ethers. In these blend uses, ethanol can capture the high market value of gasoline components that provide high octane and reduced vapor pressure.

Impact of Incremental Vehicle Costs.   The costs of alcohol fuel usage may include other costs associated with vehicles. Incremental vehicle costs have been estimated by the Ford Motor Company for the fuel-flexible vehicle that can use gasoline, methanol, ethanol, or any mixture of these, to be in the range of $300 per vehicle assuming substantial production (14). This cost may or may not be passed along to the consumer, because of incentives provided to the manufacturer by the Alternative Motor Fuels Act of 1988. There also may be incremental vehicle operating costs resulting from increased lubrication or increased frequency of oil changes.

Infrastructure Requirements.   In general, infrastructure requirements resulting from the expanded use of alcohol fuels are not especially greater than those involved in the production and refining of oil. However, for a corresponding delivery of energy, the capital costs of alcohol infrastructure would presumably be larger, as capital costs of production facilities appear to be larger than corresponding oil refinery costs for an equivalent amount of energy output. Moreover, infrastructure costs for storage and distribution facilities could be higher. Facilities for hydrocarbon fuels are generally not compatible with methanol and ethanol. Thus a program to introduce substantial amounts of alcohol fuel might well require existing infrastructure modifications. These changes can be especially difficult and costly for underground pipelines and tanks, which need to be replaced or modified.

In California, the South Coast Air Quality Management District has implemented a local rule requiring that one new or replacement underground tank at each gasoline retail facility must be suitable for methanol. Replacement of an existing tank can cost $50,000 and perhaps more. But many small retail outlets are being replaced by larger more efficient ones, and many older underground tanks (qv) are being replaced to prevent possible leaks and hence underground contamination. Therefore the compatibility rule allows for the gradual development of a methanol-compatible infrastructure at low costs. The extra costs for methanol-compatible storage ranges from negligible to about $4000 per tank and dispenser, depending on the technical choices that would otherwise be made for gasoline-only facilities.

Vehicle Technology and Vehicle Emissions

One of the reasons that U.S. automobile manufacturers showed more interest in alcohols as alternative fuels in the late 1970s and early 1980s is because alcohol's energy density is closer to gasoline and diesel than other alternatives such as compressed natural gas. They reasoned that consumers would be more comfortable with liquid fuels, envisioning little change in the fuel distribution of alcohols. Most of the research in the 1970s focused on converting light-duty vehicles, to alcohol fuels. Towards the late 1970s, researchers also began to turn their attention to heavy-duty applications. In heavy-duty engines the emissions benefits of alcohols are far clearer than in light-duty vehicles. However, it is also much harder to design heavy-duty engines to use the low cetane number alcohols.

It was not until the early 1980s that the potential air quality benefits of alcohol fuels started to be investigated. It was about five years later that proponents argued that alcohols could provide significant air quality benefits in addition to energy security benefits. Low level blends of ethanol and gasoline were argued to provide lower CO emissions. The exhaust from light-duty methanol vehicles was thought to be less reactive in the formation of ozone. Uses of alcohols fuels in heavy-duty engines showed substantially reduced mass emissions in contrast to light-duty experience which showed about the same mass emissions but a reduced reactivity of the exhaust components.

LIGHT-DUTY VEHICLES

Use of Low Level Blends.   The first significant U.S. use of alcohols as fuels since the 1930s was the low level 10% splash blending of ethanol in gasoline, which started after the oil crisis of the 1970s. This blend, called gasohol, is still sold in commerce although mostly by independent marketers and distributors instead of the major oil companies. EPA provided a waiver for this fuel allowing for a 6.9 kPa (1 psi) increase in the vapor pressure of gasohol over that of gasoline.

In the first years of gasohol use some starting and driveability problems were reported (19). Not all vehicles experienced these problems, however, and better fuel economy was often indicated even though the energy content of the fuel was reduced. Gasohol was exempted from the federal excise tax amounting to a $0.16/L ($0.60/gal) subsidy. Without this subsidy, ethanol would be too expensive for use even as a fuel additive.

Nearly four billion L/yr of ethanol are added to gasoline and sold as gasohol (18). The starting or driveability difficulties have been solved, in part, by the advances in vehicle technology employing fuel feedback controls.

Methanol was also considered as a gasoline additive. Table 3 summarizes some of the oxygenated compounds approved by EPA for use in unleaded gasoline. EPA waivers were granted to: Sun Oil Company in 1979 for 2.75% by volume methanol with an equal volume of tertiary butyl alcohol [75-65-0] (TBA) up to a blend oxygen total of 2% by weight oxygen; ARCO in 1979 for up to 7% by volume TBA; and ARCO in 1981 for the use of the blends containing a maximum of 3.5% by weight oxygen. These last blends are gasoline-grade TBA (GTBA) and OXINOL having up to 1:1 volume ratio of methanol to GTBA. Petrocoal was also granted a waiver to market up to 10% by volume methanol and cosolvents in gasoline but this waiver was revoked in 1986 after automobile manufactures complained of significant material compatibility problems and openly warned consumers against gasolines containing methanol. A waiver was issued to Du Pont in 1985 which allowed addition of up to 5% methanol to gasoline having a mixture of 2.5% cosolvents. None of these additives became very popular (20).

Vehicle Emissions.   Gasohol has some automotive exhaust emissions benefits because adding oxygen to a fuel leans out the fuel mixture, producing less carbon monoxide [630-08-2] (CO). This is true both for carbureted vehicles and for those having electronic fuel injection.

Urban areas such as Denver, Phoenix, and others at high altitudes have problems complying with health-based carbon monoxide standards in part because of automobile emissions. Vehicles calibrated for operation at sea level that

Table 3. EPA Approved Oxygenated Compounds for Use in Unleaded Gasoline[a]

Compound[b]	Broadest EPA waiver	Date	Maximum oxygen, wt %	Maximum oxygenate, vol %
methanol	substantially similar	1981		0.3
propyl alcohols	substantially similar	1981	2.0	(7.1)[c]
butyl alcohols	substantially similar	1981	2.0	(8.7)[c]
methyl tert-butyl ether (MTBE)	substantially similar	1981	2.0	(11.0)[c]
tert-amyl methyl ether (TAME)	substantially similar	1981	2.0	(12.7)[c]
isopropyl ether	substantially similar	1981	2.0	(12.8)[c]
methanol and butyl alcohol or higher mol wt alcohols in equal vol	substantially similar	1981	2.0	5.5
ethanol	gasohol	1979, 1982	(3.5)[c,d]	10.0
gasoline grade tert-butyl alcohol (GTBA)	ARCO	1981	3.5	(15.7)[c]
methanol + GTBA (1:1 max ratio)	ARCO (OXINOL)	1981	3.5	(9.4)[c]
methanol at 5 vol % max + 2.5 vol % min	Du Pont[f]	1985	3.7	[e]
cosolvent	Texas methanol (OCTAMIX)[g]	1988	3.7	[e]

[a]Ref. 21.
[b]All blends of these oxygenated compounds are subject to ASTM D 439 volatility limits except ethanol. Contact the EPA for current waivers and detailed requirements, U.S. Environmental Protection Agency, Field Operations and Support Division (EN-397F), 401 M Street, S.W., Washington, D.C. 20460.
[c]Calculated equivalent for average specific gravity gasoline (0.737 specific gravity at 16°C, NIPER Gasoline Report). Calculated equivalent depends on the specific gravity of the gasoline.
[d]Value shown is for denatured ethanol. Neat ethanol blended at 10.0 vol % produces 3.7 wt % oxygen.
[e]Varies with type of cosolvent.
[f]The cosolvents are any one or a mixture of ethanol, propyl, and butyl alcohols. Corrosion inhibitor is also required.
[g]The cosolvents are a mixture of ethanol, propyl, butyl and higher alcohols up to octyl alcohol. Corrosion inhibitor is also required.

are operated at high altitudes run rich, producing more CO. Blends such as gasohol cause the engine to operate leaner because of the oxygen in the fuel. There are larger CO reductions using oxygenated blends in older, carbureted engines. But even the newer technology vehicles have lower CO emissions using gasohol and other oxygenated fuels because of periods of open loop operation, especially during cold starts.

Blended fuels increase the vapor pressure of the resulting mixture so that more hydrocarbons are evaporated into the atmosphere during operation, refueling, or periods of extended parking. Although these hydrocarbons can react with NO_x emissions in sunlight to form ozone, atmospheric modeling has indicated that ozone is probably not increased as a result of the higher fuel volatility for two reasons (see ATMOSPHERIC MODELS). First, CO is also an ozone precursor so reducing CO reduces ozone. Second, the hydrocarbon species are somewhat less reactive because of the lower reactivity of ethanol. Furthermore, programs for oxygenated fuel use are focused at high CO occurrences during the year. These usually occur in the wintertime, whereas most areas violate ozone standards in the summer months. Therefore, oxygenated fuel programs, as a CO control strategy, do not generally interfere with ozone attainment strategies. However, programs should be individually evaluated (20).

Ethanol blends can also have an effect on NO_x emissions. Scattered data indicate that NO_x may increase as oxygenates are added to the fuel.

Other countries have also investigated the use of low level alcohol blends as an energy substitution strategy as well as to reduce exhaust emissions of lead (qv) (22,23). Brazil implemented low level ethanol–gasoline blends throughout the twentieth century during times of oil shortages or as a hedge against international fluctuations in sugar prices. Blends ranged from 15 to 42%. In 1975 Proalcool, Brazil's ethanol fuel program, was initiated and required the blending of 20% by volume of ethanol in gasoline. This was not totally achieved throughout Brazil until about 1986 when a 22% ethanol–gasoline blend was standardized. Once the fuel was standardized, engine modifications for new vehicles were made, including higher compression and adjustments to the carburetor and timing (22).

Germany also evaluated the gasoline–alcohol blends using methanol. Early programs used 15% methanol added to gasoline (M15). This program required vehicles to be designed for this fuel. Modifications included changes to the fueling system for air–fuel control and vehicle material changes to be compatible with the higher methanol concentrations. The program ended in 1982. M15 was concluded to be feasible if higher vehicle costs could be offset by the possibility of lower fuel costs (24). Lower level blends were also investigated using up to 3% methanol with 2 or more percent of a suitable cosolvent. Unlike M15, gasoline vehicles could use this blend without any modifications (25). Germany has for several years now used low level blends of methanol in their gasoline.

Retrofits. Retrofits are vehicles designed for conventional fuels modified so that the vehicles can operate on alcohol. Generally, because both ethanol and methanol have lower energy densities, the quantity of fuel entering the engine must be increased to get the same power. Also, because the alcohols have slightly different combustion characteristics, engine parameters such as ignition timing need to be adjusted. To optimize performance and fuel economy the compression ratio of the engine should be increased. However, the economics of these conversions are such that the least amount of changes are made and adjustments to engine compression ratio is typically not done.

Retrofits were popular at the beginning of alcohol fuel programs. Kits were introduced that modified only the fuel flow rate into the engine, but material changes were also necessary because both ethanol and methanol are more corrosive to metals than gasoline. Retrofitting allowed maximum market penetration without having to wait for fleet roll over or for manufacturers to market new vehicles. There was some success in converting light-duty vehicles to methanol. Bank of America operated a fleet of 292 converted Ford and Chevrolet vehicles in the late 1970s and into the 1980s before oil prices collapsed. A conversion kit for these vehicles included hardware, material changes, and a fuel additive to help minimize corrosion (26,27).

The California Energy Commission (CEC) evaluated the conversion of 1980 Ford Pintos equipped with 2.3-L four-cylinder engines, feedback-controlled carburetors, and three-way catalysts. Four vehicles were left unmodified, four converted to methanol, and four converted to ethanol. The basic changes required to use the alcohols were: the terneplate coating in the fuel tanks was stripped; fuel level sending units and carburetors were chromated to inhibit corrosion; and the air–fuel ratio, timing, and fuel vaporization mechanisms were recalibrated for proper combustion of alcohol fuels and to comply with emission standards. Two methanol- and two ethanol-fueled engines had special pistons installed to raise compression ratios from 9:1 to 12:1 for better efficiency. These vehicles were operated for 18 months accumulating 272,000 km (169,300 miles). The methanol conversions averaged 25,000 km; ethanol conversions, 21,700 km; and gasoline controls, 22,100 km. Although both the methanol and ethanol vehicles were designed to operate on 100% alcohol, they utilized M94.5 (94.5 vol % methanol and 5.5% isopentane [78-78-4] added to improve cold starts and engine warmup) and CDA-20 (ethanol denatured using 2 to 5% unleaded gasoline) (28,29).

This conversion program indicated that vehicles could be converted to alcohols. Good fuel economy was obtained; methanol vehicles averaged 4.7 km/L (11.0 mpg) or 9.1 km/L (21.3 mpg) on an equivalent energy basis compared to 8.3 km/L (19.5 mpg) for the gasoline control vehicles. No driveability problems were reported and vehicles had no problems starting (lowest temperature was $-1.1°C$). Both the ethanol and methanol Pintos showed increased upper cylinder wear over gasoline engines. Poor lubrication from using alcohols and excess fueling because of carburetor float problems contributed to the higher wear rates. Hydrocarbon, carbon monoxide, and NO_x emissions were less for methanol, 0.14, 3.2, and 0.3 g/km, respectively, than for gasoline, 0.25, 5.6, and 0.6 g/km.

The biggest problems of the CEC Pinto fleet were that vehicle conversions were expensive and alcohol fuels were more expensive than gasoline. Changes to the fuel tank, fuel lines, and the carburetor were too labor-intensive to be done cheaply. However, these changes if designed, could be made during the vehicle manufacturing at little additional cost (30). Brazil priced ethanol at 65% the cost of gasoline (10) so that conversions could be cost-effective because of the savings on the fuel costs.

Other significant disadvantages of retrofits were the quality of the conversion kits and the ability of the conversions to meet emission regulations over the useful vehicle life. The initial phases of the Brazilian ethanol program also suffered because of poor quality vehicle retrofits. The quality was so poor that the program almost failed after a fairly substantial number of vehicles were converted and the ethanol fuel infrastructure was in place. Further incentives and automobile manufacturers introducing new vehicles designed for ethanol stabilized the program (31).

For these reasons, CEC and DOE concluded that the only cost-effective method of getting alcohol fueled vehicles would be from original equipment manufacturers (OEM). Vehicles produced on the assembly line would have lower unit costs. The OEM could design and ensure the success and durability of the emission control equipment.

Dedicated Vehicles.   Only Brazil and California have continued implementing alcohols in the transportation sector. The Brazilian program, the largest alternative fuel program in the world, used about 7.5% of oil equivalent of ethanol in 1987 (equivalent to 150,000 bbl of crude oil per day). In 1987 about 4 million vehicles operated on 100% ethanol and 94% of all new vehicles purchased that year were ethanol-fueled. About 25% of Brazil's light-duty vehicle fleet (10) operate on alcohol. The leading Brazilian OEMs are Autolatina (a joint venture of Volkswagen and Ford), GM, and Fiat. Vehicles are manufactured and marketed in Brazil.

In contrast the California program has some 600 demonstration vehicles (32). Both Ford and Volkswagen participated in the dedicated vehicle phase of the California program. In 1981 Volkswagen provided the first alcohol vehicles produced on an assembly line, forty (19 methanol, 20 ethanol, and 1 gasoline) VW Rabbits and light-duty trucks were manufactured. Design incorporated continuous port fuel injection, 12.5:1 compression ratio, a new ignition system calibration, and a heat exchanger for faster oil warmup. The entire fuel system was designed using materials compatible with methanol and ethanol. These vehicles operated until 1983 and logged 728,000 km of service. This fleet used the same fuel as the ethanol and methanol Pinto retrofits.

In 1981 Ford also provided CEC with 40 Escorts designed to operate on M94.5 and 15 gasoline vehicles to serve as controls. These vehicles had accumulated over 3.4 million km of service as of March 1986. The 1981 Escorts were modified to use methanol. The 1.6-L gasoline engine had a production piston used in European 1.6-L engines to raise the compression ratio to 11.4:1. Other field modifications included spark plug change, a carburetor throttleshaft material change, carburetor float redesign, and the replacement of tin-plated fuel tanks with ones of stainless-steel.

In 1983 the methanol fleet was expanded with the purchase of 506 Ford Escorts . These vehicles are equipped with engines and fuel systems redesigned from Ford's standard 1.6-L gasoline-fueled Escort. Ford also produced five advanced technology vehicles equipped with electronic fuel injection and microprocessor control, with the goal of improving fuel economy and reducing NO_x emissions to 0.25 g/km. The emissions control on these vehicles used the same technology as on the carbureted gasoline versions: standard three-way catalyst, exhaust gas recirculation, and air injection. The 1981 and 1983 Escorts have logged over 48 million km in service and some vehicles have reached gasoline equivalent fuel usage per kilometer over the lifetime of the vehicle.

The methanol fuel specification was changed for the Escorts, at the end of 1983 from M94.5 to a blend of 90% methanol and 10% unleaded gasoline (M90). In the summer of 1984 the fuel was further modified to include 15% gasoline (M85). This change from isopentane was made because gasoline was cheaper. In addition, M85 has a gasoline odor and taste, and in daylight the flame is more visible than either M94.5 or M90. Another safety benefit of the added gasoline is the increased volatility creating a richer air–vapor mixture much less likely to burn or explode in closed containers than neat or 100 percent methanol.

The results of the California fleet demonstrations indicated that fuel economy on an energy basis was equal to or better than gasoline, especially using vehicles having higher compression. None of these engines, however, was fully optimized for methanol so additional improvements are possible. Driveability was also good for the methanol vehicles. An acceleration test of two 1983 methanol-fueled vehicles and a similar 1984 model resulted in: 1983 fuel injection (EFI) methanol, 14.53 s; 1983 carbureted methanol, 15.51 s; and 1984 carbureted gasoline, 19.10 s.

Tests demonstrate that methanol vehicles can meet stringent emission standards for HC, CO, and NO_x as indicated in Figure 2. The primary benefit of

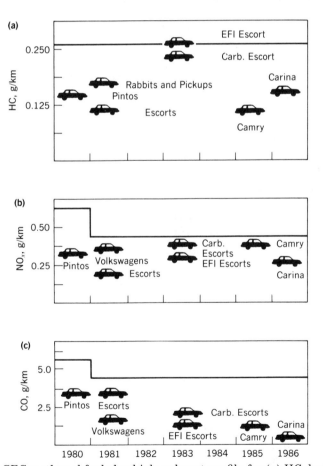

Fig. 2. CEC methanol-fueled vehicle exhaust profile for (**a**) HC, hydrocarbons; (**b**) NO_x; and (**c**) CO. Solid line represents State of California standard maximum emissions. Methanol HC emissions are calculated as $CH_{1.85}$ and not corrected for flame-ionization detector response.

methanol, however, is not the amount of hydrocarbons emitted but rather that methanol-fueled vehicles emit mainly methanol which is less reactive in the formation of ozone than the variety of complex organic molecules in gasoline exhaust. Formaldehyde [50-00-0] emissions from methanol vehicles are increased in comparison to gasoline vehicles. Tests of 1983 Escorts showed tailpipe levels as high as 62 mg/km, well above typical gasoline levels of 2 to 7 mg/km. The 1981 Rabbits ranged from about 6 to 14 mg/km and the 1981 Escorts had levels less than 7 mg/km. All results were obtained on relatively low mileage vehicles. Deterioration of catalyst effectiveness could increase these emissions.

The vehicles investigated were mostly adaptions of gasoline technology. For example, automobile manufacturers recommended that catalytic converters designed for gasoline automobiles be installed on vehicles in Brazil to control acetaldehyde [75-07-0] emissions. Brazil decided against catalysts because gasoline vehicles at that time did not have these systems. Similarly, California adopted M85 to aid in cold starting and to provide some measures of perceived safety. Research to find other additives that would assist in cold starting and provide safety characteristics at a reasonable price have been relatively unsuccessful (33). But another way to overcome the issue of cold starting is to design and optimize engines to operate on 100% methanol (M100). EPA has been pursuing M100 for several years with good success (34). Cold starting is not a problem for direct injected engines where high pressure is used to atomize the fuel and results indicate that a light-duty, direct injection engine can attain very low emissions having good fuel economy and driveability. However, safety concerns of using M100 in general commerce need to be addressed (35).

Fuel Flexible Vehicles. Using dedicated alcohol fuel vehicles pointed to the importance of a wide distribution of fueling stations. Methanol-fueled vehicles require refueling more often than gasoline vehicles.

In 1981, the Dutch company TNO in cooperation with the New Zealand government converted a gasoline engine to a flexible fuel vehicle by adding a fuel sensor. The sensor determined the amount of oxygen in the fuel and then used this information to mechanically adjust the carburetor jets. The initial mechanical system was crude, but the advancement of engine and emission controls, in particular the use of electronics and computers, has brought about substantial refinement (36).

Ford first tried the flexible fuel system on an Escort and called it the "flexible fueled vehicle" or FFV. As seen in Figure 3, the system included building into the electronics any necessary calibrations for gasoline and methanol fuels, adding a sensor to determine the amount of methanol in the fuel, and making necessary material changes to fuel wetted components. The sensor is one of the most critical parts of this system: its output determines parameters such as the amount of fuel to be injected and engine timing. Fuel injectors must also have a wider response range. The engine compression ratio was not changed because the vehicle is designed to operate as well on gasoline as on methanol.

Of course, FFV drivers do not have to use methanol. Emissions benefits are not obtained if methanol is not used, and fuel economy is not optimized for methanol nor are emissions. However the State of California has concluded that advantages offered by the flexibility of the FFV far outweigh the disadvantages (37).

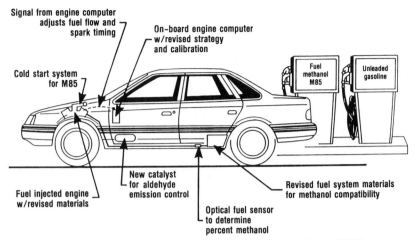

Fig. 3.   Components of a Ford flexible fuel vehicle (FFV).

Many U.S. and foreign automobile manufacturers are developing a fuel flexible vehicle in the 1990s. Ford has developed FFVs for 5-L engines used in Crown Victorias and for 3.0-L engines used in their Taurus car line. GM followed Ford with a variable fueled vehicle (VFV) and applied this technology first to the 2.8-L engine family used in the Corsica, and more recently to the 3.1-L engine family in the Lumina (see Fig. 4). Prototype flexible fuel vehicles are also being developed by Volkswagen, Chrysler, Toyota, Nissan, and Mitsubishi. California is in the process of obtaining an additional 5,000 of these vehicles in the next several model years (MY92 and MY93) to be used by government and private fleets.

The experience using fuel flexible vehicles has been surprisingly successful. California is operating about 200 vehicles and driveability is excellent on whatever fuel is used. Tests performed on a Ford Crown Victoria showed slightly better fuel economy (4%) and better acceleration (6%) on methanol than gasoline (38). The fuel flexible technology is not limited to methanol but with electronic calibration changes also works for ethanol and gasoline combinations. Changes can be made by adjusting the engine maps in the computer.

EPA, the Air Resources Board (ARB), and others are investigating the possible emission benefits of alcohol fueled vehicles. EPA and ARB adopted regulations for hydrocarbon mass emissions which accounted for the oxygen components in the exhaust, so called organic mass hydrocarbon equivalent (OM-HCE). The regulations required methanol vehicles to meet the same OMHCE value as gasoline hydrocarbons, which in California is 0.155 g/km in 1993. The trend is to account also for the total mass and the reactivity of individual hydrocarbon species and the measure being proposed for total mass is non-methane organic gases or NMOG. Reactivity of the individual species that make up NMOG are estimated (39) to give a value of ozone/km.

Vehicle emissions have been monitored over the last several years on fuel flexible vehicles and depending on when the tests were performed, reported in total hydrocarbons, OMHCE, or NMOG. The vehicle testing performed to date has shown that methanol FFVs can provide emissions benefits. Figure 5 shows the

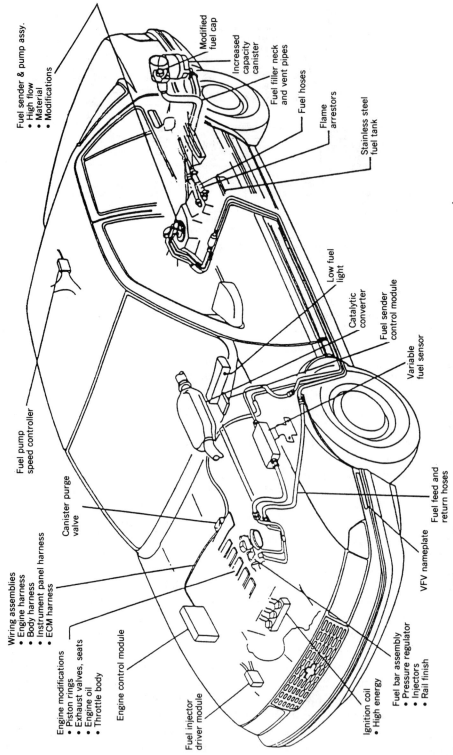

Fig. 4. Components of a Lumina methanol variable fueled vehicle (VFV).

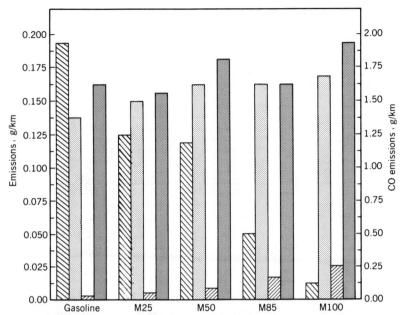

Fig. 5. Emissions from a GM Corsica VFV for gasoline and gasoline–methanol mixtures where ◫ represents total organic material, including hydrocarbons, methanol, and formaldehyde; ▫ represents NO_x; ▨ formaldehyde; and ▥ carbon monoxide.

emissions from a GM Corsica for various gasoline methanol mixtures (40). This vehicle was designed to meet the California standards of 0.155 g/km hydrocarbon, 2.1 g/km CO, and 0.25 g/km NO_x. In addition California requires methanol vehicles to meet a formaldehyde standard of 9.3 mg/km. The total organic emissions decrease with increasing methanol, whereas formaldehyde increases. NO_x and CO vary but appear unaffected by methanol content. The Corsica data were taken on a green catalyst (low vehicle mileage) and some deterioration of these emissions levels can be expected with age.

Figure 6 shows data for four vehicles operated on gasoline and M85, two having an electrically heated catalyst (EHC). The two vehicles equipped with EHC both showed low values of NMOG and estimated ozone production. These data seem to indicate that methanol vehicles result in less ozone than comparable gasoline vehicles. However, the data only include exhaust emissions and not evaporative or running losses. These later sources of emissions should be lower using methanol because of the lower reactivity of the alcohols.

HEAVY-DUTY VEHICLES

The use of alcohols in heavy-duty engines developed more slowly than in light-duty engines primarily because the majority of heavy-duty engines are diesels. Diesels are unthrottled, stratified charge engines which autoignite fuel by heat generated during compression. Engine speed and load are modulated by varying the quantity of fuel injected into the cylinder rather than by throttling the fuel–air mixture as done in the Otto cycle or spark-ignition engine. Unthrottled

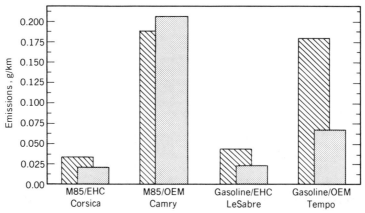

Fig. 6. Comparison of M85 and gasoline emissions (including ozone) corresponding to ▨ estimated ozone (ref. 40) and ▨ NMOG for vehicles having an electrically heated catalyst (EHC) or not (OEM) (ref. 41).

air aspiration reduces pumping losses which in turn increases the engine's thermal efficiency. Diesel engines also are designed for higher compression ratios resulting in further efficiency improvements. The higher efficiency and excellent reliability and durability of these engines make them attractive for heavy-duty applications in trucks, buses, and off road equipment. Unfortunately, their low cetane number limited the compatibility of alcohol fuels with diesel engines without modifications to assist ignition.

Not until the early 1980s were prototype heavy-duty engines developed to operate on methanol and ethanol, having efficiencies equal to or better than corresponding diesel engines. These engines were quieter and had considerably cleaner exhaust emissions. Mass emissions of NO_x and particulates were substantially lower when the cleaner alcohol fuels were used, overcoming the inherent NO_x-particulate tradeoff of diesel engines. The biggest challenge facing engine manufacturers in the 1980s and 1990s was to make the alcohol engines as reliable and durable as heavy-duty diesel engines which operate in the range of 0.3 to 0.5 million kilometers without major engine maintenance and repair.

Technology Options. Because alcohols are not easily ignited in diesel-type engines changes in engine hardware or modifications to the fuel are needed. Engine modifications can include the addition of a spark ignition system or an additional fuel injection system to provide dual fuel capabilities. Fuel modifications involve the addition of cetane improvers. Low level blends do not work because methanol and diesel fuels are not miscible. Some research to emulsify diesel and alcohols was carried out, but was never successful (42). Other investigations involved adding a separate fuel system including two fuel tanks, fuel lines, injection pumps, and injectors. In this approach diesel was used to ignite the fuel mixtures, and at low speeds–low loads diesel was the primary fuel. At high speeds–high loads alcohol was the primary fuel. This dual fuel approach was both cumbersome and expensive (43).

One successful method for using alcohols was fumigation. In this technique alcohol is atomized in the engine's intake air either by carburetion or injection.

Diesel is directly injected into the cylinder and the combined air–alcohol and diesel mixture is autoignited. Diesel consumption is reduced by the energy of the alcohol in the intake air. This approach, although technically feasible, also requires separate fuel systems for the diesel and alcohol fuels. Additionally, the amount of alcohol used is limited by the amount that can be vaporized into the intake air. This approach is more appropriate as a engine retrofit where total energy substitution is not the primary objective (44).

Other possible technologies involve either assisted ignition or cetane improvers. Assisted ignition approaches can be divided between direct injection, stratified charge type engines, and engines converted back to Otto cycles, by throttling and lowering engine compression.

Dedicated Vehicles. As late as 1982, researchers were still arguing the worthiness of alcohols as fuels for heavy-duty engines (45). Pioneer work on multifuel engines led to modifications in diesel engines to burn neat or 100% alcohol. The German manufacturers were the first to provide prototype methanol engines. Daimler-Benz modified their four stroke M 407 series diesel engine to operate on 100% methanol, by converting the diesel version to a spark-ignited, Otto cycle engine. This required lowering the compression, adding a spark ignition system, and carburetion (throttling). To get back some of the efficiency loss caused by going to a throttling and lower compression, the Daimler-Benz design incorporated a heat exchanger to vaporize the methanol using engine cooling water. Vaporized methanol was introduced into the engine using a standard gaseous carburetor–mixing device.

The M 407 hGO methanol engine is a horizontal, water-cooled, inline six-cylinder configuration (46). Basic combustion is similar to the conventional spark-ignition Otto cycle with one significant exception. Lean combustion at part load is possible for two reasons: because of methanol's favorable flammability limits and because methanol is vaporized and introduced as a gas. Equivalence ratios (air–fuel ratio relative to stoichiometric) greater than 2 are possible without misfire, and minimum fuel consumption is obtained at an equivalence ratio of about 1.8. In the higher load range, the engine is controlled by the air–fuel ratio, rather than by intake throttling, so efficiency is increased relative to the conventional spark-ignition engine. Intake throttling is used for control in the lower load range.

The first methanol bus in the world was placed in revenue service in Auckland, New Zealand in June 1981. It was a Mercedes O 305 city bus using the M 407 hGO methanol engine. This vehicle operated in revenue service for several years with mixed results. Fuel economy on an equivalent energy basis ranged from 6 to 17% more than diesel fuel economy. Power and torque matched the diesel engine and drivers could not detect a difference. Reliability and durability of components was a problem. Additional demonstrations took place in Berlin, Germany and in Pretoria, South Africa, both in 1982.

The world's second methanol bus was introduced in Auckland shortly after the first. This was a M.A.N. bus with a M.A.N. FM multifuel combustion system utilizing 100% methanol. The FM system, more similar to a diesel engine, is a direct injection, high compression engine using a spark ignition. Fuel is injected into an open chamber combustion configuration in close proximity to the spark plugs which ignite the air–fuel mixture. Near the spark plugs the air–fuel mixture is rich and combustion proceeds to the lean fuel air mixtures in the rest of the cylinder. The air–fuel charge is thus stratified in the cylinder and these types of engines are often called lean burn, stratified charge. Engine hardware is similar to the diesel version including a high pressure injection pump and a compression ratio comparable to diesel (19:1). This technology was applied to M.A.N.'s 2566 series engines, an inline 6-cylinder engine, and for buses is configured horizontally (47). Like the Mercedes, it is a four stroke engine.

This technology was tested using diesel fuel, gasoline, methanol, and ethanol. A M.A.N. SL 200 bus having the M.A.N. D2566 FMUH methanol engine was also demonstrated in Berlin. Results of these tests were somewhat mixed. Fuel economy was 12% less than a comparable diesel bus, but driveability was very good. Because the methanol fueled bus was not smoke limited at low speeds, higher torque was possible and bus drivers used this advantage to accelerate faster from starts. Emissions results indicated a considerable advantage in using fuels such as methanol. CO and NO_x were reduced compared to diesel engines and particulates were virtually eliminated.

The success of the New Zealand and German programs were instrumental in implementing a similar bus demonstration in California in 1982 (48). The primary objective was to assess the viability of using methanol in heavy-duty engines. The project focused on evaluating engine durability, fuel economy, driveability, and emissions characteristics. CEC also initiated a demonstration project for off-road heavy-duty vehicles using a multifuel tractor capable of operating on either neat methanol or ethanol (30).

M.A.N. and Detroit Diesel Allison, now Detroit Diesel Corporation (DDC), agreed to participate in the California bus program. DDC provided a methanol version of their 6V-92TA engine, which along with the DDC 71 series, is the most commonly used bus engine in the United States. The engine is a compression-ignited, two-stroke design having a displacement of 9.1 liters and power rating of 20,700 W (277 hp). Several design changes were incorporated for operation on methanol, including electronic unit injectors (EUI) for more precise fuel control, an increased compression ratio, a bypass blower, and glow plugs. Compression ignition is achieved by maintaining the cylinder temperatures above the autoignition temperature of methanol. Air is diverted around the blower, reducing the amount of air entering the cylinders. Glow plugs are used as a starting aid and also at low speeds and low loads to maintain the cylinder temperatures necessary for autoignition. The methanol-fueled engine is turbocharged and equipped with a blower (supercharged).

The engine was the first to incorporate compression ignition of alcohols (7). Low cetane fuels can autoignite, however, provided the in-cylinder temperature is high enough and fuel injection correctly timed. This compression ignition works for methanol as well as ethanol and gasoline.

The California bus program was run at Golden Gate Transit District (GGTD) and continued through late 1990 (49). M.A.N. supplied two European SU 240 coaches for this project, one diesel powered and one methanol powered. DDC provided a GM RTS coach powered by methanol. GGTD already had a RTS diesel powered coach. Results indicate that methanol is a viable fuel for heavy-duty engines in general and transit in particular. Driveability including starting, full and partial throttle acceleration, and deceleration was as good or better using the methanol buses as compared to their diesel counterparts. Figure 7 illustrates the comparison of full throttle acceleration. Detailed fuel economy tests were also performed. Figures 8 and 9 compare steady-state and transient fuel consumption tests, respectively. The transient tests were performed using the Fuel Consumption Test Procedure, Type II (50). Methanol is comparable to diesel in steady-state fuel usage tests, but methanol consumption is higher at idle and during accelerations. The idle fuel consumption is higher because methanol can not burn as lean as diesel fuel. Higher transient fuel consumption is a result of poor combustion factors resulting from poor fuel atomization, air control, and over fuelling. The methanol engine should not inherently be worse than the diesel engine during accelerations, if good combustion can be maintained.

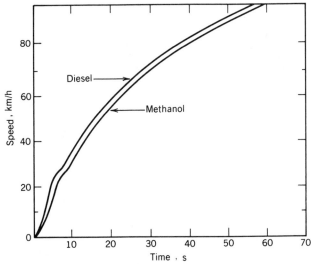

Fig. 7. Full-throttle acceleration for diesel and methanol-powered GM RTS coaches having simulated full-seated passenger loads of 43 passengers.

The biggest problem of the GGTD program was engine and vehicle reliability and durability (see Fig. 10). Components needing the most frequent replacement were electronic unit fuel injectors and glow plugs, followed by the electronic control system (controlled power to the glow plugs), throttle position sensor, fuel pump, and fuel cooler fans. Other problems included increased engine deposits and ring and liner wear (51). The M.A.N. engine had similar but fewer problems; the components having the lowest lifetime were spark-plugs.

California continued the development of methanol powered vehicles primarily because of the substantial emission benefits (52). Then in 1986, the U.S. EPA promulgated technology-forcing standards for on-road, heavy-duty diesel engines which had been basically uncontrolled (53). These standards were also adopted in California by the ARB for buses in 1991 and all heavy-duty engines in 1993. Diesel engine manufacturers have made significant improvements in technology and new diesel engines are projected to meet standards without a particulate trap.

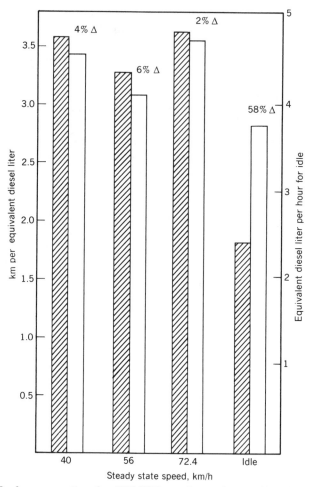

Fig. 8.   Fuel consumption for GM RTS coaches using ▨ diesel and □ methanol as fuel. Differences in fuel consumption are indicated by the symbol Δ. To convert km/L to mpg, multiply by 2.35.

These improvements have made the diesel engine more competitive with alcohol fueled engines.

In 1987 Seattle Metro purchased 10 new American built M.A.N. coaches powered by methanol. Six GM buses powered by DDC methanol engines entered revenue service at Triboro Coach in Jackson Heights, New York, 2 GM buses in Medicine Hat Transit in Medicine Hat, Manitoba, and 2 Flyer coaches in Winnipeg Transit, Winnipeg, Manitoba, Canada. An additional 45 DDC powered methanol buses were introduced in California as indicated by Table 4. Figure 11 shows the distance accumulation of alternate-fueled buses in the four California transit properties.

Many of the development problems identified in the first bus programs were carried into the more recent demonstration projects. Spark-plug life continued to

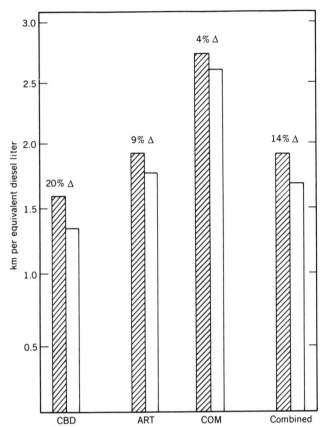

Fig. 9. GM RTS coaches transient fuel consumption for □ diesel and □ methanol. SAE type road test, UMTA ADB duty cycle. Differences in fuel consumption are indicated by the symbol Δ. CBD (Central Business District), 4.3 stops per km, maximum speed 32 km/h, 18 m deceleration, 7 s dwell time, 19.3 km; ART (arterial), 1.24 stops per km, maximum speed 64 km/h, 61 m deceleration, 7 s dwell time, 13 km; COM (commuter), one stop every 6.4 km, maximum speed of 88.5 km/h, 150 m deceleration, 20 s dwell time, 13 km. To convert km/L to mpg, multiply by 2.35.

be an issue at Seattle Metro and the project was terminated in 1990 because of costs of replacement parts. Costs were compounded when M.A.N. decided to discontinue manufacturing buses in the United States. DDC engines also continued to have problems with fuel injectors and glow plugs. Unit injectors were failing for a variety of reasons but the biggest problem was plugging injector tips. Injectors on some buses had to be changed at mileages as low as 1600 to 3200 km compared to diesel injectors which last up to 100 times as long. DDC and Lubrizol have since developed a fuel additive that when added to methanol at 0.06% by volume substantially reduces injector failures (see Fig. 12).

Many improvements have been made in both combustion and emissions control from the first experimental engine operating at GGTD (54). The new DDC preproduction engines have increased compression, 23:1 compared to 19:1, allowing the glow plugs to function only during starting. The rest of the time the

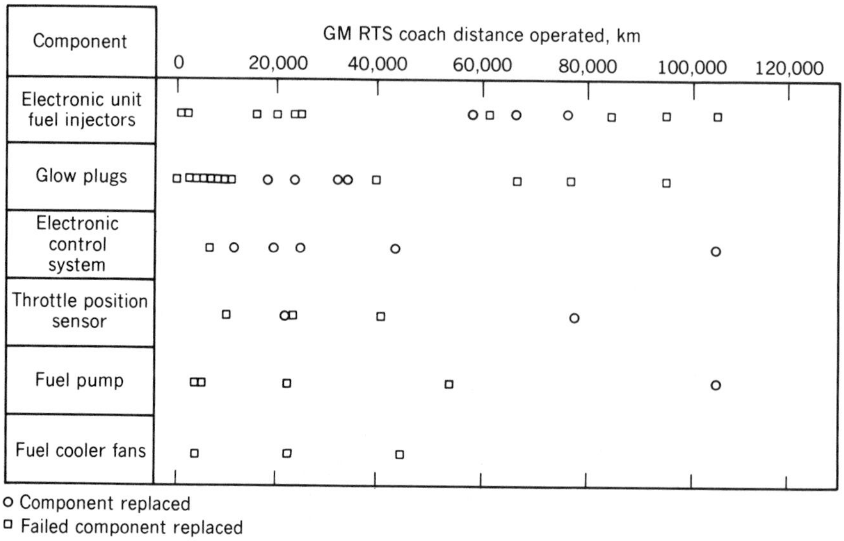

GM RTS coach distance operated, km

Component	0	20,000	40,000	60,000	80,000	100,000	120,000
Electronic unit fuel injectors							
Glow plugs							
Electronic control system							
Throttle position sensor							
Fuel pump							
Fuel cooler fans							

o Component replaced
□ Failed component replaced

Fig. 10. Components of GM RTS methanol-powered coach replaced, ○, and replaced because of component failure, □, as a function of distance operated.

Table 4. Distribution of California's Heavy-Duty Alternative Fuel Demonstration Vehicles

Transit district	No. of vehicles	Fuel	OEM/engine
South Coast Area Transit	1	methanol	DDC 6V-92TA
Riverside Transit District	3	methanol	DDC 6V-92TA
Southern California RTD	30	methanol	DDC 6V-92TA
Southern California RTD	12	methanol/Avocet[a]	DDC 6V-92TA
Southern California RTD	1	methanol	MAN D2566 MUH
Orange County Transit District	2	methanol/Avocet[a]	Cummins L10
Orange County Transit District	2	CNG[b]	Cummins L10
Orange County Transit District	2	LPG[b]	Cummins L10
School bus demo			
various fleets	50	methanol	DDC 6V-92TA
various fleets	10	CNG	Bluebird/Teogen GM 454
Trucking applications			
City of Los Angeles	1	methanol	GMC DDC 6V-92TA
City of Los Angeles	1	methanol/Avocet[a]	Peterbuilt Cummins L10
City of Glendale	1	methanol	Peterbuilt Caterpillar 3306
Golden State Foods	1	methanol	Freightliner DDC 6V-92TA
Federal Express	1	methanol	Freightliner DDC 6-71TA
Arrowhead	1	methanol	Ford/Ford 6.61
SCE	1	methanol	Volvo/DDC 6V-92TA
South Lake Tahoe	1	methanol	International/Navistar DTG-460
Waste Management	2	methanol	Volvo/DDC 6-71TA

[a]Avocet is a cetane improver.
[b]CNG, compressed natural gas, and LPG, liquefied petroleum gas, are also used as alternative fuels.

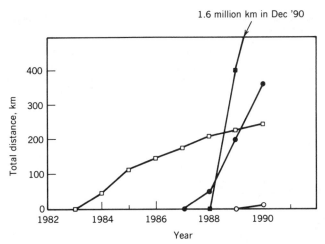

Fig. 11. Distance methanol-fueled transit coaches travelled per year of operation in the □ Golden Gate Transit District (GGTD); ● Riverside Transit Agency (RTA), ■ Southern California Rapid Transit District (SCRTD); and ○ Orange County Transit District (OCTD).

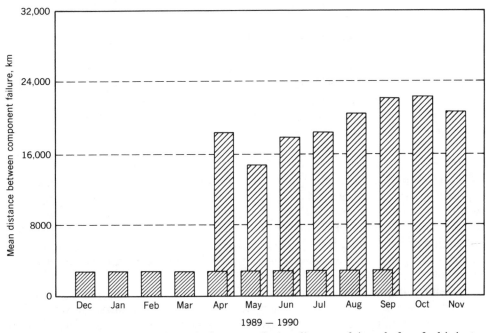

Fig. 12. Methanol and coach fleet cumulative distance driven before fuel injector failure, where ▨ represents operation without, and ▨ represents operation with fuel additive.

cylinder temperatures are high enough to autoignite methanol. This revision increased the life of glow plugs from an average of 11,900 to 22,100 km between failures. Fuel economy has also improved as have exhaust emissions. Tests performed on an engine dynamometer following the Federal test procedure are many times better than California's 1991 bus standard as shown in Table 5.

Table 5. Exhaust Emissions and California 1991 Bus Standards, g/kW·h[a]

Exhaust component	1991 California standards		DC 6V-92TA (catalyst)	
	g/(bhp·h)[a]	g/(kW·h)	g/(bhp·h)[a]	g/(kW·h)
OMHCE	1.3	1.8	0.10	0.14
CO	15.5	21.1	0.22	0.3
NO_x	5.0	6.8	2.3	3.1
particulates	0.10	0.14	0.05	0.07
aldehydes	0.10/0.05[b]	0.14/0.07[b]	0.04	0.05

[a]bhp = brake horsepower; 1 bhp = 0.735 kW.
[b]The 0.10 and 0.14 values apply from 1993 to 1995; the 0.05 and 0.07 values apply beginning in 1996.

Because of the success of these various alcohol fuel programs, heavy-duty demonstrations have been extended to other applications as shown in Table 4. The majority of applications are still either in transit or school buses, but California has also begun a program to utilize methanol engines in heavy-duty trucking applications (55). Domestic engine manufacturers participating in this program include Caterpillar, Cummins, DDC, Ford, and Navistar. The Caterpillar and Navistar engines are four stroke engines having glow plugs to assist in igniting methanol. These engines are very different from the two stroke DDC engine and from each other. The Navistar (56) uses a shield glow plug which neither Caterpillar (57) nor DDC do. Navistar claims these glow plugs provide both good combustion and long life. The combustion chambers and fuel control schemes differ from manufacturer to manufacturer. Ford converted a diesel engine to an Otto cycle and added electronically controlled port fuel injection, throttling, and a distributorless spark-ignition system.

Methanol has been shown to be a viable fuel for a variety of trucking applications in a local yet fairly large geographical area (58). And although the results focus primarily on dedicated methanol engines and vehicles in the U.S., the conclusions are nearly transferrable to ethanol fuels. DDC's 6V-92TA methanol engine operates on ethanol using a different engine calibration optimized for good performance and emissions. Additional hardware modifications are not anticipated.

Cetane Improvers. Compared to dedicated alcohol engines, fewer hardware changes need to be made to heavy-duty engines using cetane improvers. The early research on using cetane improvers and alcohol fuels for heavy-duty diesel engines was performed by the Germans in the late 1970s-early 1980s (59). The possibilities of using cetane improvers with ethanol for heavy-duty vehicles operating in Brazil were investigated (60). The work indicated that nitrates are the most effective cetane improvers for alcohols (61) and the Brazilian program focused on nitrates that could be manufactured from sugar cane, the feedstock for ethanol production. Additives considered included: butyl nitrate, isoamyl nitrates, 2-ethoxyethyl nitrate, and ethylene glycol nitrates. The selection of the improver depends on the method of the additive modifying the ignition delay over the entire speed and load range. Ideally ethanol should match the same ignition delay behavior as diesel.

Four buses converted to ethanol started operation in 1979 and the engine used was an OM 352, 6 cylinder, direct injection engine rated at 96,200 watt (129 hp). The ignition improver was n-hexyl nitrate [20633-11-8] which was later changed to triethylene glycol dinitrate [111-22-8] (TEGDN). TEGDN was mixed with ethanol at 5% or less by volume (60). The test fleet was further expanded to a variety of trucks manufactured and marketed by Mercedes-Benz in Brazil. 1,700 heavy-duty trucks were converted to ethanol for use in the more gruelling sugar cane industry. Engines converted included the OM 352 O (5.7 L, 96,200 watt) and OM 355/5 O (9.7 L, 141,000 watt). Modifications to the engines to use ignition improved ethanol included increasing the fuel delivery capacity of the fuel injection pump, changing the fuel injection nozzles, and making material compatible with the TEGDN/ethanol fuel. The engines provided equal or better power and torque than the unmodified models. Some durability problems, which arose because of lack of lubrication or fuel material incompatibilities, were solved by adding lubricants to the injection pump plungers or by changing materials. Emissions were also generally lower than those from the equivalent diesel engine. Even NO_x was lower because of the lower flame temperatures compared to diesel (62).

The biggest drawback was the cost of ethanol compared to diesel fuel and the cost of the TEGDN and ethanol mixture compared to diesel. Unlike in the United States, diesel fuel in Brazil is considerably less expensive on an energy basis than gasoline, because gasoline is taxed at a higher rate. This is generally the case in Europe as well. So, although technically feasible, the costs were too high for Brazil to convert many heavy-duty vehicles to ethanol.

Additional research for both ethanol and methanol showed that the amount of ignition improver could be reduced by systems increasing engine compression (63). Going from 17:1 to 21:1 reduced the amount of TEGDN required for methanol from 5% by volume to 3%. Ignition-improved methanol exhibited very low exhaust emissions compared to diesels: particulate emissions were eliminated except for small amounts associated with engine oil, NO_x was even lower with increased compression, and CO and hydrocarbons were also below diesel levels.

Auckland Regional Authority converted two M.A.N. buses to use a cetane improver and methanol and South Africa investigated the use of methanol with a proprietary cetane improver. Four Renault buses were converted in Tours, France to operate on ethanol and a cetane improver, Avocet, manufactured by Imperial Chemical Industries (ICI). The results of these demonstrations were also technically successful; slightly better fuel economy was obtained on an energy basis and durability issues were much less than the earlier tests using dedicated engines.

Cetane improvers were first investigated in the U.S. as part of a demonstration project to retrofit DDC 6V-71 two stroke engines to methanol for three transit buses operating in Jacksonville, Florida. This project started out using hardware conversion where the engine was modified to control air flow and a glow plug was added to the system. Although this was basically the same approach as used in the dedicated 6V-92TA methanol engine, the 6V-71 employed mechanical injectors as opposed to electronic injectors. These mechanical injectors were failing even with 1% castor oil [*8001-79-4*] added to the methanol. Jacksonville thus decided to use a nitrate based cetane improver containing a lubrication additive and a corrosion inhibitor and the project proved the viability of cetane-improved methanol.

The first U.S. engine manufacturer to evaluate the use of cetane improvers was Cummins Engine Company. Their methanol designed L10 engines were converted based on the knowledge gained in Brazil with ethanol and 4.5% TEGDN in their 14-L engine (64). They increased the fuel capacity of their injection pumps, for instance, and modified the combustion process to match the diesel start of combustion by increasing the compression ratio from 15.8 to 16.1:1 and adding turbocharging (both increased in-cylinder temperatures). For the methanol L10 development Avocet at 5% by volume mixed with 100% methanol was selected (65). Engine modifications were only made to the fuel system. These included fuel pump changes for increased capacity, a different camshaft to change injection timing, and larger injectors. Changes were also made in various materials to make them compatible with methanol. The resultant L10 matched diesel power and torque, but emissions results were mixed. Data showed low emissions of particulates, but higher HC, CO, and NO_x at low speed, low load operation. These data suggest that additional optimization may be necessary for the L10 methanol engine.

The L10 methanol engines are currently being used in several demonstration fleets in California. The first use was in the City of Los Angeles Peterbilt dump truck (58), a part of the CEC truck demonstration project. Methanol L10s are also being used in the OCTD comparative alternative fuel demonstration project. In this project methanol, CNG, and LPG L10 engines are being compared to each other in a transit bus application. In the Canadian methanol in large engines (MILE) project, two L10 methanol engines were operated for several years in refuse haulers in Vancouver, British Columbia.

The largest retrofit demonstration project in the U.S. is underway at SCRTD where 12 vehicles using DDC 6V-92 with mechanical injectors are being modified to use methanol with Avocet (66). This project has evaluated the changes necessary to optimize the conversions for both fuel economy and emissions. Using the two stroke engine changes had to be made to reduce the air into the engine, increase compression from 17 to 23:1, and use better ring packages. These changes gave both good fuel economy and reduced emissions of NO_x by 50%, considerably lowered particulates, and maintained levels of CO and hydrocarbon. New York has also modified a 8V-71 for use on Avocet-improved methanol and had similar results (67).

The success of these tests indicate that cetane-improved alcohol is technically feasible. Engines can be designed to provide equal diesel power and torque characteristics having lower NO_x and particulate emissions than diesel. However, if it is not necessary to achieve lowest possible emissions, only changes in fuel rate are required, rather than engine changes, and the commercial application of this approach depends mostly on the cost of the cetane improvers. The price of Avocet is about $4/L in small quantities and adding 3.0% in methanol nearly doubles the cost of methanol from $0.13 to $0.22/L. If Avocet were produced in larger quantities its price would drop considerably.

The biggest potential use of the cetane-improver approach may be in vehicle retrofits where for environmental reasons bus and truck fleets may be required to convert to cleaner burning fuels.

Air Quality Benefits of Alcohol Fuels

In the 1970s evaluations of alcohol fuel programs always considered environmental impacts and objectives even though the main thrust of the programs was toward energy security and diversification benefits. Assessments of performance identified these fuels as consistent with environmental goals and by the mid 1980s, the environmental benefits of the alcohol fuels had become the chief driving force for their further consideration. Detailed assessments were made of photochemical smog and air toxics reductions that might be obtained from the wide use of alcohol fuels in light-duty vehicles. Methanol received the most evaluation, because it appeared to be far more cost competitive than ethanol. The potential benefits of alcohols used in heavy-duty diesel-type engines were also studied.

The most comprehensive air quality study, supported by the California ARB and the South Coast Air Quality Management District (68), showed that if gasoline and methanol cars emitted the same amounts of carbon, an assumption that seemed reasonable based on emissions test data taken throughout the 1980s, and if methanol cars had formaldehyde emissions controlled to 9.3 mg/km (equal to the current California formaldehyde emissions standard for methanol automobiles), then substituting M85 for gasoline would produce a 9% reduction in the peak summer-day afternoon ozone level and a 19% reduction in exposure to ozone levels above the Federal standard of 0.12 ppm. These reductions constituted a substantial fraction of the reductions that would be obtained by eliminating all the emissions from vehicles. Additional assumptions were that exhaust carbon emissions were at the level of 0.15 g/km, equal to the planned certification standard for new vehicles in California beginning in 1993 (but not really expected to be characteristic of in-use vehicles) and that the distribution of hydrocarbon species in the exhaust resembled that of cars tested in the 1980s.

The results of the study, conducted at Carnegie Mellon University, are now generally accepted as the best available guides for the smog-reducing benefits of a methanol substitution strategy, at least for the conditions prevailing in the Los Angeles basin. Overall, replacement of a conventional gasoline vehicle by an equivalent M85 vehicle should provide about a 30% reduction in smog-forming potential. A 100% result would be earned by eliminating the vehicle entirely. Vehicles using M100 would provide substantially greater benefits than M85 vehicles.

Benefits depend upon location. There is reason to believe that the ratio of hydrocarbon emissions to NO_x has an influence on the degree of benefit from methanol substitution in reducing the formation of photochemical smog (69). Additionally, continued testing on methanol vehicles, particularly on vehicles which have accumulated a considerable number of miles, may show that some of the assumptions made in the Carnegie Mellon assessment are not valid. Air quality benefits of methanol also depend on good catalyst performance, especially in controlling formaldehyde, over the entire useful life of the vehicle.

Methanol substitution strategies do not appear to cause an increase in exposure to ambient formaldehyde even though the direct emissions of formaldehyde have been somewhat higher than those of comparable gasoline cars. Most ambient formaldehyde is in fact secondary formaldehyde formed by photochemical reactions of hydrocarbons emitted from gasoline vehicles and other sources. The effects of slightly higher direct formaldehyde emissions from methanol cars are offset by reduced hydrocarbon emissions (68).

Methanol use would also reduce public exposure to toxic hydrocarbons associated with gasoline and diesel fuel, including benzene, 1,3-butadiene, diesel particulates, and polynuclear aromatic hydrocarbons. Although public formaldehyde exposures might increase from methanol use in garages and tunnels, methanol use is expected to reduce overall public exposure to toxic air contaminants.

Alcohol Fuels Usage as an Air Quality Strategy

The cost-effectiveness of methanol substitution as an air quality strategy has been studied in some detail. Air quality planners usually rate cost-effectiveness in terms of dollars per ton of reactive hydrocarbons controlled (or removed from the inventory of emissions). Typical costs for controlling reactive hydrocarbon emissions in the U.S. are in the range of several hundreds of dollars per ton. In the Los Angeles area, the average costs of future hydrocarbon control measures average about $500 per metric ton, although some individual measures have cost-effectiveness figures above $10,000 per ton. Methanol substitution appears to be a viable and competitive control strategy. Cost-effectiveness is linked to the price of oil. Methanol appears to have a cost-effectiveness ranging from a few thousand dollars to several tens of thousands of dollars per ton (18,70–73). In heavy-duty engines, the alcohols may offer cost-effective reductions of particulates and NO_x emissions. A recent study indicates that the use of methanol was competitive with the use of cleaner diesel fuels and diesel particulate traps under some circumstances (74).

The potential air quality impacts of ethanol use have not yet been studied in detail.

Global Warming Impacts. Several studies have been made of the global warming impacts of alcohol fuels. The most useful assessments cover the entire life cycle from raw material feedstock production through processing, distribution, and fuel usage. They also consider global gases in addition to carbon dioxide (75,76). Results reflect the influence of assumptions but methanol is expected to provide slight reductions in global warming impacts compared to gasoline. Ethanol evaluations are less certain.

Public Safety Issues

Several investigators have assessed the comparative safety of methanol and conventional hydrocarbon fuels (14,77–79). The ingestion toxicity of methanol has been of some concern because of the number of gasoline ingestions associated with siphoning and in-home accidental ingestions. The use of gasoline in small engines such as those used for lawnmowers, leafblowers, and other small utility applications results in most of the siphoning ingestions and in-home accidental ingestions of gasoline. This potential problem is addressed by discouraging methanol use in small engines, by labelling and public education, and by positive siphoning prevention screens in the fill pipes of vehicles. These screens have been required in recent purchases of methanol vehicles by California agencies.

Skin contact with methanol may present a greater health threat than skin contact with gasoline and diesel fuel and is being evaluated.

The fire hazard of methanol appears to be substantially smaller than the fire hazard of gasoline, although considerably greater than the fire hazard of diesel fuel. The lack of luminosity of a methanol flame is still a concern to some, and M85 (or some other methanol fuel with an additive for flame luminosity) may become the standard fuel for this reason.

In reviewing the full range of health and safety issues associated with all alternative fuels, the California Advisory Board determined that there were no roadblocks that would prevent the near term deployment of either methanol or ethanol, assuming that adequate safety practices were followed appropriate to the specific nature of each fuel (14).

The Future of Alcohol Fuels

In the late 1980s attempts were made in California to shift fuel use to methanol in order to capture the air quality benefits of the reduced photochemical reactivity of the emissions from methanol-fueled vehicles. Proposed legislation would mandate that some fraction of the sales of each vehicle manufacturer be capable of using methanol, and that fuel suppliers ensure that methanol was used in these vehicles. The legislation became a study of the California Advisory Board on Air Quality and Fuels. The report of the study recommended a broader approach to fuel quality and fuel choice that would define environmental objectives and allow the marketplace to determine which vehicle and fuel technologies were adequate to meet environmental objectives at lowest cost and maximum value to consumers. The report directed the California ARB to develop a regulatory approach that would preserve environmental objectives by using emissions standards that reflected the best potential of the cleanest fuels.

The ARB adopted a regulatory package for light-duty vehicles in 1990 that modifies the historically uniform approach to vehicle emissions, in which each and every vehicle in a regulated class must meet the same emissions standard. The new approach adopts emissions standards that apply on the average to the entire sales mix of vehicles sold by each manufacturer in each of several broad weight classes of vehicles. Thus vehicles that use fuels such as methanol and ethanol having air quality benefits in the form of lower levels of photochemical reactivity have the emissions adjusted to reflect the lower smog forming tendency of these fuels. This regulatory approach provides a powerful incentive for vehicle manufacturers to certify at least some of the sales mix of vehicles on fuels such as methanol and ethanol.

The future market response to the new form of emissions regulation is unknown. For the purpose of meeting new vehicle emissions standards, however, it is still not clear whether some combination of new emissions control approaches and reformulated gasolines can provide benefits equal to those of methanol and ethanol. It is possible that the new emissions standards will simply result in improved gasoline technologies, and that, despite the prospective air quality advantages of the alcohol fuels, the market result of the new standards will simply be cleaner gasolines. However, in 1990 the U.S. Alternative Fuels Council agreed on a goal of a 25% share of nonpetroleum transportation fuels by 2005. Although this goal may not become part of a national energy plan, it represents the first official statement of a specific goal to substitute for the use of petroleum in transportation. Alcohol fuels could capture a large part of this 25% share of nonpetroleum fuels, although vehicles powered by natural gas and electric energy will no doubt win some acceptance. For the ordinary passenger car, alcohol fuels may offer the most gasolinelike alternative in terms of range, comparable costs, and compatibility with the current gasoline/diesel storage and distribution infrastructure.

BIBLIOGRAPHY

"Alcohol Fuels," in *ECT* 3rd ed., Suppl. pp. 1–42, by Donald L. Kloss, Institute of Gas Technology.

1. J. W. Shiller, "The Automobile and the Atmosphere" in *Energy: Production, Consumption and Consequences,* National Academy Press, Washington, D.C., 1990, pp. 111–142.
2. *Engineering Data Book,* Gas Processors Suppliers Association, 1972.
3. *Reference Data for Hydrocarbons and Petro-Sulfur Compounds,* Phillips Petroleum Company, 1962.
4. J. B. Heywood, *Internal Combustion Engine Fundamentals,* McGraw-Hill, New York, 1988.
5. *Fire Hazard Properties of Flammable Liquids, Gases, and Volatile Solids,* Pub. # NFPA 325M, National Fire Protection Association, 1984.
6. P. F. Schmidt, *Fuel Oil Manual,* Industrial Press, 1969.
7. R. R. Toepel, J. E. Bennethum, R. E. Heruth, "Development of a Detroit Diesel Allison 6V-92TA Methanol-Fueled Coach Engine," *SAE Paper 831744, SAE Fuels and Lubricants Meeting* (San Francisco, Calif., Oct. 31–Nov. 3, 1983), Society of Automotive Engineers, Warrendale, Pa., 1983.
8. T. Powell, "Racing Experiences with Methanol and Ethanol-based Motor Fuel Blends," *SAE Paper 750124,* Society of Automotive Engineers, Warrendale, Pa., Feb., 1975.
9. Sypher-Mueller International, Inc., *Future Transportation Fuels: Alcohol Fuels,* Energy, Mines and Resources—Canada, Project Mile Report, A Report on the Use of Methanol in Large Engines in Canada, May 1990.
10. S. C. Trindade and A. V. de Carvalho, "Transportation Fuels Policy Issues and Options: The Case of Ethanol Fuels in Brazil" in D. Sperling, ed., *Alternative Transportation Fuels,* Quorum Books, New York, 1989, pp. 163–185.
11. *First Interim Report,* United States Alternative Fuels Council, Washington, D.C., Sept. 30, 1990.
12. Office of Policy, Planning, and Analysis, *Assessment of Costs and Benefits of Flexible and Alternative Fuel Use in the U.S. Transportation Sector,* Technical Report #3 (Methanol Production and Transportation Costs) Pub. # DOE/P/E-0093, U.S. Department of Energy, Washington, D.C., Nov. 1989.
13. Bechtel Corporation, *California Fuel Methanol Cost Study: Chevron Corporation, U.S.,* Vol. 1 (Executive Summary, Jan. 1989); Vol. 2, (Final Report, Dec. 1988), San Francisco, Calif., 1988–1989.
14. California Advisory Board on Air Quality and Fuels, Vol. 1, *Executive Summary;* Vol. 2, *Energy Security Report;* Vol. 3, *Environmental Health and Safety Report;* Vol. 4,

Economics Report; Vol. 5, *Mandates and Incentives Report;* San Francisco, Calif., June 13, 1990.

15. California Energy Commission, *Methanol as a Motor Fuel: Review of the Issues Related to Air Quality, Demand, Supply, Cost, Consumer Acceptance and Health and Safety,* Pub. # P500–89–002, Sacramento, Calif., April 1989.

16. California Energy Commission, *Cost and Availability of Low Emission Motor Vehicles and Fuels,* Sacramento, Calif., Aug. 1989.

17. National Research Council, *Fuels to Drive Our Future,* National Academy Press, Washington, D.C., 1990.

18. Office of Technology Assessment, *Replacing Gasoline: Alternative Fuels for Light-Duty Vehicles,* Pub. # OTA–E–364, U.S. Congress, Washington, D.C., Sept. 1990.

19. J. L. Keller, *Hydrocarbon Process,* May 1979.

20. Office of Mobile Sources, *Analysis of the Economic and Environmental Effects of Methanol as an Automotive Fuel,* U.S. Environmental Protection Agency, Ann Arbor, Mich., Sept. 1989.

21. *Alcohols and Ethers Blended with Gasoline,* Pub. # 4261, American Petroleum Institute, Washington, D.C., Chapt. 4, pp. 23–27.

22. *7th Int. Symp. on Alcohol Fuels,* Institut Francais du Petrole, Editions Technip, Paris, France, 1986.

23. *8th Int. Symp. on Alcohol Fuels,* New Energy and Industrial Technology Development Organization, Sanbi Insatsu Co., Ltd. Tokyo Japan, Nov. 13–16, 1988.

24. H. C. Wolff, "German Field Test Results on Methanol Fuels M100 and M15," *American Petroleum Institute 48th Midyear Refining Meeting* (Session on Oxygenates and Oxygenate-Gasoline Blends as Motor Fuels) May 11, 1983.

25. H. Menrad, "Possibilities to Introduce Methanol as a Fuel, An Example from Germany," *6th Int. Symp. on Alcohol Fuels Technology,* (Ottawa, Canada, May 21–25, 1984), Vol. 2.

26. L. Schieler, M. Fischer, D. Dennler, and R. Nettell, "Bank of America's Methanol Fuel Program: An Insurance Policy that is Now a Viable Fuel," *6th Int. Symp. on Alcohol Fuels Technology,* (Ottawa, Canada, May 21–25, 1984), Vol. 2.

27. R. N. McGill and R. L. Graves, "Results from the Federal Methanol Fleet, A Progress Report," *8th Int. Symp. on Alcohol Fuels* (Tokyo, Japan, Nov. 13–16, 1988).

28. *Alcohol Energy Systems: Alcohol Fleet One Test Report,* Pub. # 500–82–058, California Energy Commission, Sacramento, Calif., Aug. 1983.

29. F. J. Wiens and co-workers, "California's Alcohol Fleet Test Program: Final Results" *6th Int. Symp. on Alcohols Fuels Technology* (Ottawa, Canada, May 21–25, 1984), Vol. 3.

30. Acurex Corporation, *California's Methanol Program: Evaluation Report,* Vol. 2 (Technical Analyses), Pub. # P500–86–012A, California Energy Commission, Sacramento, Calif., June 1987.

31. S. C. Trindade and A. V. de Carvalho, "Utilization of Alcohol Fuels in Brazil: Early Experience, Current Situation, and Future Prospects," *Int. Symp. on Introduction of Methanol-Powered Vehicles* (Tokyo, Japan, Feb. 19, 1987).

32. Acurex Corporation, *California's Methanol Program: Evaluation Report,* Vol. 1 (Executive Summary), Pub. # P500–86–012, California Energy Commission, Sacramento, Calif., Nov. 1986.

33. E. R. Fanick, L. R. Smith, J. A. Russell and W. E. Likos, "Laboratory Evaluation of Safety-Related Additives for Neat Methanol Fuel," SAE Paper 902156, (SP 840), Society of Automotive Engineers, Warrendale, Pa., Oct. 1990.

34. U. Hilger, G. Jain, E. Scheid, and F. Pischinger, "Development of a Direct Injected Neat Methanol Engine for Passenger Car Applications," SAE Paper 901521, *SAE Future Transportation Technology Conf. and Expo.* (San Diego, Calif., Aug. 13–16, 1990).

35. P. A. Machiele, "A Perspective on the Flammability, Toxicity, and Environmental Safety Distinctions Between Methanol and Conventional Fuels," *AIChE 1989 Summer*

National Meeting (Philadelphia, Pa., Aug. 22, 1989), American Institute of Chemical Engineers.

36. J. V. D. Weide and R. J. Wineland, "Vehicle Operation with Variable Methanol/Gasoline Mixtures," *6th Int. Symp. on Alcohol Fuels Technology* (Ottawa, Canada, May 21–25, 1984), Vol. 3.

37. T. B. Blaisdell, M. D. Jackson, and K. D. Smith, "Potential of Light-Duty Methanol Vehicles," SAE Paper 891667, *SAE Future Transportation Technology Conf. and Expo.* (Vancouver, Canada, Aug. 7–10, 1989), Society of Automotive Engineers, Warrendale, Pa.

38. Mobile Source Division, *Alcohol Fueled Vehicle Fleet Test Program: 9th Interim Report,* Pub. ARB/MS–89–09, California Air Resources Board, El Monte, Calif., Nov. 1989.

39. W. Carter, *Ozone Reactivity Analysis of Emissions from Motor Vehicles,* (Draft Report for the Western Liquid Gas Association), Statewide Air Pollution Research Center, University of California at Riverside, July 11, 1989.

40. P. A. Gabele, *J. of Air Waste Management* **40**(3), 296–304 (Mar. 1990).

41. S. Albu, "California's Regulatory Perspective on Alternate Fuels," *13th North American Motor Vehicle Emissions Control Conf.* (Tampa, Fla., Dec. 11–14, 1990), Mobile Source Division, California Air Resources Board, El Monte, Calif.

42. A. Lawson, A. J. Last, A. S. Desphande, and E. W. Simmons, "Heavy-Duty Truck Diesel Engine Operation on Unstabilized Methanol/Diesel Fuel Emulsion," SAE Paper 810346, (SP–480) *Int. Congress and Expo* (Detroit, Mich., Feb. 23–27, 1981) Society of Automotive Engineers, Warrendale, Pa.

43. B. M. Bertilsson, "Regulated and Unregulated Emissions from an Alcohol-Fueled Diesel Engine with Two Separate Fuel Injection Systems," *5th Int. Symp. on Alcohol Fuel Technology* (Auckland, New Zealand, May 13–18, 1982) Vol. 3.

44. R. A. Baranescu, "Fumigation of Alcohols in a Multicylinder Diesel Engine: Evaluation of Potential," SAE Paper 860308, *SAE Int. Congress and Expo.* (Detroit, Mich., Feb. 24–28, 1986) Society of Automotive Engineers, Warrendale, Pa.

45. R. G. Jackson, "Workshop on Diesel Fuel Substitution," *5th Int. Symp. on Alcohol Fuel Technology* (Auckland, New Zealand, May 13–18, 1982), Vol. 4.

46. H. K. Bergmann and K. D. Holloh, "Field Experience with Mercedes-Benz Methanol City Buses," *6th Int. Symp. on Alcohol Fuels Technology* (Ottawa, Canada, May 21–25, 1984), Vol. 1.

47. A. Nietz and F. Chmela, "Results of Further Development in the M.A.N. Methanol Engine," *6th Int. Symp. on Alcohol Fuels Technology* (Ottawa, Canada, May 21–25, 1984), Vol. 1.

48. M. D. Jackson, C. A. Powars, K. D. Smith, and D. W. Fong, "Methanol-Fueled Transit Bus Demonstration," *Paper 83–DGP–2,* American Society of Mechanical Engineers.

49. M. D. Jackson, S. Unnasch, C. Sullivan, and R. A. Renner, "Transit Bus Operation with Methanol Fuel," *SAE Paper 850216, SAE Int. Congress and Expo.* (Detroit, Mich., Feb. 25–Mar. 1, 1986) Society of Automotive Engineers, Warrendale, Pa.

50. *Joint TMC/SAE Fuel Consumption Test Procedure, Type 2,* SAE J1321, SAE Recommended Practice Approved October 1981, Society of Automotive Engineers, Warrendale, Pa., 1981.

51. M. D. Jackson, S. Unnasch, and D. D. Lowell, "Heavy-Duty Methanol Engines: Wear and Emissions," *8th Int. Symp. on Alcohol Fuels* (Tokyo, Japan, Nov. 13–16, 1988).

52. K. D. Smith, "California's Methanol Program," *7th Int. Symp. on Alcohol Fuels,* (Paris, France, Oct. 20–23, 1986).

53. *EPA Emissions Standards, Code of Federal Regulations,* Vol. 40, Chapt. 1, Sect. 86.088–11, 86.091–11, and 86.094–11, U.S. Environmental Protection Agency, U.S.G.P.O., Washington, D.C., 1987.

54. J. Jaye, S. Miller, and J. Bennethum, "Development of the Detroit Diesel Corporation Methanol Engine," *8th Int. Symp. Alcohol Fuels* (Tokyo, Japan, Nov. 13–16, 1988).

55. R. A. Brown, J. A. Nicholson, M. D. Jackson, and C. Sullivan, "Methanol-Fueled Heavy-Duty Truck Engine Applications: The CEC Program." *SAE Paper 890972, SAE 40th Annual Earthmoving Industry Conf.* (Peoria, Ill., April 11–13, 1989) Society of Automotive Engineers, Warrendale, Pa.

56. R. Baranescu and co-workers, "Prototype Development of a Methanol Engine for Heavy-Duty Application-Performance and Emissions," *SAE Paper 891653, SAE Future Transportation Technology Conf.* (Aug. 7–20, 1989), Society of Automotive Engineers, Warrendale, Pa.

57. R. Richards, "Methanol-Fueled Caterpillar 3406 Engine Experience in On-Highway Trucks," *SAE Paper 902160, (SP–840),* Society of Automotive Engineers, Warrendale, Pa., Oct., 1990.

58. M. D. Jackson, C. Sullivan, and P. Wuebben, "California's Demonstration of Heavy-Duty Methanol Engines in Trucking Applications," *9th Int. Symp. on Alcohol Fuels* (Milan, Italy, April 9–12, 1991).

59. W. Bandel, "Problems in the Application of Ethanol as Fuel for Utility Vehicles," *2nd Int. Symp. on Alcohol Fuel Technology* (Wolfsburg, Germany, Nov. 21–23, 1977).

60. W. Bandel and L. M. Ventura, "Problems in Adapting Ethanol Fuels to the Requirements of Diesel Engines," *4th Int. Symp. on Alcohol Fuels* (Guaruja, Brazil, Oct. 1980).

61. A. J. Schaefer and H. O. Hardenburg, "Ignition Improvers for Ethanol Fuels," SAE Paper 810249, SP–840, *SAE Int. Congress. and Expo.* (Detroit, Mich., Feb. 23–27, 1981).

62. E. P. Fontanello, L. M. Ventura, and W. Bandel, "The Use of Ethanol with Ignition Improver as a CI-Engine Fuel in Brazilian Trucks," *7th Int. Symp. on Alcohol Fuels* (Paris, France, Oct. 20–23, 1986).

63. H. O. Hardenburg, "Comparative Study of Heavy-Duty Engine Operation with Diesel Fuel and Ignition-Improved Methanol," *SAE Paper 872093, SAE Int. Fuels and Lubricants Meeting and Expo.* (Toronto, Canada, Nov. 2–5, 1987), Society of Automotive Engineers, Warrendale, Pa.

64. E. J. Lyford-Pike, F. C. Neves, A. C. Zulino, and V. K. Duggal, "Development of a Commercial Cummins NT Series to Burn Ethanol Alcohol," *7th Int. Symp. on Alcohol Fuels* (Paris, France, Oct. 20–23, 1986).

65. A. B. Welch, W. A. Goetz, D. Elliott, J. R. MacDonald, and V. K. Duggal, "Development of the Cummins Methanol L10 Engine with an Ignition Improver," *8th Int. Symp. on Alcohol Fuels* (Tokyo, Japan, Nov. 13–16, 1988).

66. S. Unnasch and co-workers, "Transit Bus Operation with a DDC 6V–92TAC Engine Operating on Ignition-Improved Methanol," *SAE Paper 902161, (SP–840), SAE Int. Fuels and Lubricants Meeting and Expo.* (Tulsa, Oklahoma, Oct. 22–23, 1990).

67. C. M. Urban, T. J. Timbario, and R. L. Bechtold, "Performance and Emissions of a DDC 8V–71 Engine Fueled with Cetane Improved Methanol," *SAE Paper 892064, SAE Int. Fuels and Lubricants Meeting and Expo.* (Baltimore, Md., Sept. 25–28, 1989) Society of Automotive Engineers, Warrendale, Pa.

68. J. N. Harris, A. R. Russell, and J. B. Milford, *Air Quality Implications of Methanol Fuel Utilization, SAE Paper 881198,* Society of Automotive Engineers, Warrendale, Pa., 1998.

69. T. Y. Chang and S. Rudy, Urban Air Quality Impact of Methanol-Fueled Compared to Gasoline-Fueled Vehicles, in W. Kohl, ed., *Methanol as an Alternative Fuel Choice: An Assessment,* John Hopkins Foreign Policy Institute, Washington, D.C., 1990, pp. 97–120.

70. C. B. Moyer, S. Unnasch, and M. D. Jackson, *Air Quality Programs as Driving Forces for Methanol Use,* Transportation Research, Vol. 23A, No. 3, 1989, pp. 209–216.

71. T. Lareau, *The Economics of Alternative Fuel Use: Substituting Methanol for Gasoline,* American Petroleum Institute, Washington, D.C., June 1989.

72. Mobile Source Division, *Alcohol-Fueled Vehicle Fleet Test, 9th Interim Report* ARB/MS–89–09, California Air Resources Board, El Monte, Calif., Nov. 1989.

73. *Proposed Regulations for Low Emission Vehicles and Clean Fuels,* Staff Report, California Air Resources Board, Sacramento, Calif., Aug. 13, 1990.
74. S. Unnasch, C. B. Moyer, M. D. Jackson, and K. D. Smith, *Emissions Control Options for Heavy-Duty Engines, SAE 861111,* Society of Automotive Engineers, Warrendale, Pa., 1986.
75. M. A. DeLuchi, "Emissions of Greenhouse Gases from the Use of Gasoline, Methanol, and Other Alternative Transportation Fuels," in W. Kohl, ed., *Methanol as an Alternative Fuel Choice: An Assessment,* Johns Hopkins Foreign Policy Institute, Washington, D.C., 1990, pp. 167–199.
76. S. C. Unnasch, B. Moyer, D. D. Lowell, and M. D. Jackson, *Comparing the Impacts of Different Transportation Fuels on the Greenhouse Effect, Pub. # 500–89–001,* California Energy Commission, Sacramento, Calif., April, 1989.
77. *Methanol Health and Safety Workshop* (Los Angeles, Calif., Nov. 1–2, 1989) South Coast Air Quality Management District, El Monte, Calif., 1989.
78. P. A. Machiele, "A Perspective on the Flammability, Toxicity, and Environmental Safety Distinctions Between Methanol and Conventional Fuels," *AIChE 1989 Summer National Meeting* (Philadelphia, Pa., Aug. 22, 1989), American Association of Chemical Engineers.
79. P. A. Machiele, "A Health and Safety Assessment of Methanol as an Alternative Fuel," in W. Kohl, ed., *Methanol as an Alternative Fuel: An Assessment,* John Hopkins Foreign Policy Institute, Washington, D.C., 1990, pp. 217–239.

General References

K. Boekhaus and co-workers, "Reformulated Gasoline for Clean Air: An ARCO Assessment," *2nd Biennial U.C. Davis Conf. on Alternative Fuels* (July 12, 1990).
California's Methanol Program, Evaluation Report, Vol. 2 (Technical Analyses), Pub. # P500–86–012A, California Energy Commission, Sacramento, Calif., June 1987.
A. V. de Carvalho, "Future Scenarios of Alcohols as Fuels in Brazil," *3rd Int. Symp. on Alcohol Fuels Technology* (Asilomar, Calif., May 29–31, 1979).
C. L. Gray, Jr. and J. A. Alson, *Moving American to Methanol,* University of Michigan Press, Ann Arbor, Mich., 1985.
P. A. Lorang, Emissions from Gasoline-Fueled and Methanol Vehicles, in W. L. Kohl, ed., *Methanol as an Alternative Fuel Choice: An Assessment* John Hopkins Foreign Policy Institute, Washington, D.C., 1990, pp. 21–48.
J. H. Perry and C. P. Perry, *Methanol: Bridge to a Renewable Energy Future* University Press of American, Lanham, Md., 1990.
D. Sperling, *New Transportation Fuels,* University of California Press, Berkekey, Calif., 1988.
D. Sperling, *Brazil, Ethanol, and the Process of Change, Energy,* Vol. 12, No. 1, pp. 11–23, 1987.
E. Supp, *How to Produce Methanol from Coal,* Springer-Verlag, Berlin, 1990.
J. V. D. Weide and W. A. Ramackers, "Development of Methanol and Petrol Carburation Systems in the Netherlands," *2nd Int. Symp. on Alcohol Fuel Technology,* (Wolfsburg, Germany, Nov. 21–23, 1977).
1987 World Methanol Conf. (San Francisco, Calif., Dec. 1–3, 1987) Crocco and Associates, Houston, Tex.
1989 World Methanol Conf. (Houston, Tex., Dec. 5–7, 1989), Crocco and Associates, Houston, Tex.

MICHAEL D. JACKSON
CARL B. MOYER
Acurex Corporation

ALCOHOLS, HIGHER ALIPHATIC

SURVEY AND NATURAL ALCOHOLS MANUFACTURE

The monohydric aliphatic alcohols of six or more carbon atoms are generally referred to as higher alcohols. Historically, the higher alcohols, particularly those of 12 or more carbon atoms, were derived from natural fats, oils, and waxes and were called fatty alcohols (see FATS AND FATTY OILS); but now similar alcohols are widely available from synthetic processes using petrochemical feedstocks (qv). Although the natural and synthetic alcohols are used interchangeably for many applications, for some applications the distinction still remains. The higher alcohols can be separated into the plasticizer range alcohols, generally 6–11 carbon atoms, and the detergent range alcohols, 12 or more carbon atoms. There is, however, considerable overlap in use. Production of higher alcohols in North America, Europe, and Japan in 1985 was about 2,600,000 tons and United States production was 35% of that total. About three-fourths of the U.S. output was plasticizer range alcohols, which are used primarily as ester derivatives in plasticizers (qv) and lubricants (see LUBRICATION AND LUBRICANTS). The detergent range alcohols are used mainly as sulfate, ethoxy, and ethoxysulfate derivatives in a wide variety of detergent and surfactant applications (see DETERGENCY; SURFACTANTS).

Most higher alcohols of commercial importance are primary alcohols; secondary alcohols have more limited specialty uses. Detergent range alcohols are apt to be straight chain materials and are made either from natural fats and oils or by petrochemical processes. The plasticizer range alcohols are more likely to be branched chain materials and are made primarily by petrochemical processes. Whereas alcohols made from natural fats and oils are always linear, some petrochemical processes produce linear alcohols and others do not. Industrial manufacturing processes are discussed in the SYNTHETIC PROCESSES section, p. 893.

Detergent Range Alcohols. Natural or synthetic detergent range alcohols are usually described as middle cut (12–15 carbon atoms) or heavy cut (16–18 carbon atoms), corresponding to the distillation fractions of coconut alcohol from which these alcohols were first derived. Because middle cut alcohols are preferred for most detergent applications, manufacturers maximize this production through feedstock choice (natural alcohols), or by manipulating processing conditions (synthetic alcohols). The coproduct light cut (6–11 carbon atoms) and heavy cut alcohols are also valuable products. Only a small percentage of detergent range alcohols are sold as pure single carbon chain materials.

The higher alcohols occur in minor quantities primarily as the wax ester (ester of a fatty alcohol and a fatty acid) in many oilseed and marine sources. Free alcohols octacosanol [557-61-9], $C_{28}H_{58}O$, and triacontanol [28351-05-5], $C_{32}H_{66}O$, have been isolated in very small amounts from sugarcane and its products (1). Oil from the sperm whale is rich in wax esters of hexadecanol, octadecenol, and eicosenol; this oil was formerly a major commerical source of these alcohols. The oil of the North Atlantic barracudina fish contains 85% wax esters that consist mainly of hexadecanol and octadecenol (2). Minor amounts of alcohols having 12–26 carbon atoms have been found in both ancient and recent marine sediments, probably having their origin in ocean marine life (3). Wool grease from sheep also contains higher alcohols as wax esters, and is a minor commercial source of alcohol. The seeds of the shrub jojoba which grows in the North American desert give an oil which contains esters of eicosenol and docosenol [629-98-1], and the natural waxes such as carnauba wax [8015-86-9] and candelilla wax [8006-44-8] contain wax esters with alcohols of 26–34 carbon atoms (4). Although higher alcohols could be obtained from any of these plant sources by saponification of the esters, they are not commercially important sources.

   Plasticizer Range Alcohols.   Commercial products from the family of 6–11 carbon alcohols that make up the plasticizer range are available both as commercially pure single carbon chain materials and as complex isomeric mixtures. Commercial descriptions of plasticizer range alcohols are rather confusing, but in general a commercially pure material is called "-anol," and the mixtures are called "-yl alcohol" or "iso. . .yl alcohol." For example, 2-ethylhexanol [104-76-7] and 4-methyl-2-pentanol [108-11-2] are single materials whereas isooctyl alcohol [68526-83-0] is a complex mixture of branched hexanols and heptanols. Another commercial product contains linear alcohols of mixed 6-, 8-, and 10-carbon chains.

Physical Properties

Table 1 provides physical property data for selected pure alcohols (5). The homologous series of primary normal alcohols exhibits definite trends in physical properties: for each additional CH_2 unit the normal boiling point increases by about 20°C, the specific gravity increases by about 0.003 units, and the melting point increases by about 10°C in the lower end of the range and about 4°C in the upper end. The water solubility decreases with increasing molecular weight and the oil solubility increases. In general, the higher alcohols are soluble in lower alcohols such as ethanol and methanol and in diethyl ether and petroleum ether. The solubility of water in 1-hexanol and 1-octanol is appreciable, but drops off rapidly as alcohol molecular weight increases. Enough solubility remains, however, to make even 1-octadecanol slightly hygroscopic. Mixtures of alcohols, such as 1-octadecanol and 1-hexadecanol, are considerably more hygroscopic. Below C_{12} the normal alcohols are colorless, oily liquids with light, rather fruity odors. At room temperature pure 1-dodecanol solidifies to soft, crystalline platelets and the physical form of higher molecular weight alcohols progresses from these soft platelets to crystalline waxes. Although 1-dodecanol has a slight odor, the higher homologues are essentially odorless. The secondary and branched primary alcohols are oily liquids at room temperature and have light, fruity odors. They are soluble in alcohol solvents and diethyl ether, and also show less affinity for water as molecular weights increase. The members of this group do not have well-defined freezing points; they set to a glass at very low temperatures. Physical properties are often ill-defined because of difficulties in obtaining pure samples.

Chemical Properties

The higher alcohols undergo the same chemical reactions as other primary or secondary alcohols. Similar to other chemicals having long carbon chains, however, reactivity decreases as molecular weight or chain branching increase. This lower reactivity and concommitant decreased solubility in water and in other solvents means that more rigorous reaction conditions, or even use of different reaction schemes as compared to shorter chain alcohols, are generally required. Typical reactions of the higher alcohols are as shown.

Esterification

$$ROH + R'COOH \longrightarrow R'COOR + H_2O$$

Sulfation

$$ROH + SO_3 \longrightarrow \quad ROSO_3H$$
$$\text{alkyl sulfuric acid}$$

$$ROSO_3H + NaOH \longrightarrow \quad ROSO_3Na \quad + H_2O$$
$$\text{sodium alkyl sulfate}$$

Etherification

$$ROH + n\ H_2C\!-\!CH_2 \longrightarrow \quad R(OCH_2CH_2)_nOH$$
$$\diagdown\!O\!\diagup$$
$$\text{polyethoxylated alcohol}$$

$$ROH + H_2C\!-\!CH\!-\!CH_2Cl \longrightarrow \quad ROCH_2CH\!-\!CH_2Cl$$
$$\diagdown\!O\!\diagup \qquad\qquad\qquad\qquad\quad |$$
$$\qquad\qquad\qquad\qquad\qquad\qquad OH$$
$$\text{alkyl chlorohydrin ether}$$

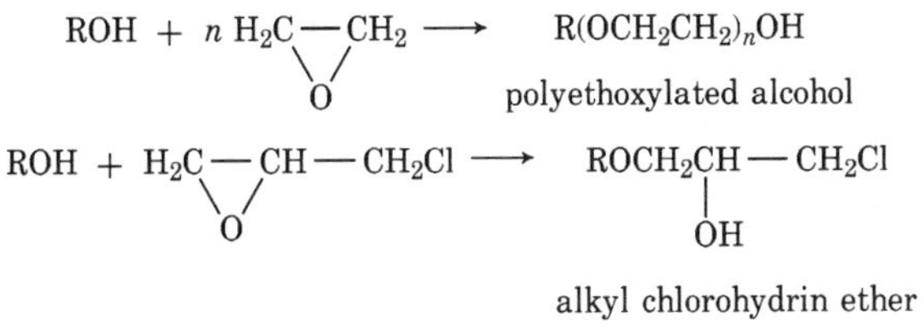

Halogenation

$$3\ ROH + PCl_3 \longrightarrow 3\ RCl + P(OH)_3$$

Dehydration

$$RCH_2CH_2OH \longrightarrow RCH\!=\!CH_2 + H_2O$$

Oxidation

$$RCH_2OH + \tfrac{1}{2}\ O_2 \longrightarrow RCH\!=\!O + H_2O$$

Table 1. Physical Properties of Pure Alcohols

IUPAC name	CAS Registry Number	Molecular formula	Other common names	Specific gravity, 20°C[a]	Refractive index, 20°C[a]	Bp, °C, 101.3 kPa[b]	Mp, °C	Viscosity, mPa·s[a,c]	Solubility, % by wt in water	Solubility, % by wt of water	Solubility in other solvents
				Primary normal aliphatic							
1-hexanol	[111-27-3]	$C_6H_{14}O$	n-hexyl alcohol	0.8212	1.4181	157	−44	5.9	0.59[20]	7.2	petroleum ether, ethanol
1-heptanol	[111-70-6]	$C_7H_{16}O$	n-heptyl alcohol	0.8238	1.4242	176	−35	7.4	0.10[18]		ethanol, petroleum ether
1-octanol	[111-87-5]	$C_8H_{18}O$	n-octyl alcohol	0.8273	1.4296	195	−15.5	8.4	0.06[25]	4.5	
1-nonanol	[143-08-8]	$C_9H_{20}O$	n-nonyl alcohol	0.8295	1.4338	213	−5	11.7			glacial acetic acid, benzene, ethanol, petroleum ether
1-decanol	[112-30-1]	$C_{10}H_{22}O$	n-decyl alcohol	0.8312	1.4371	230	7	13.8		2.8	
1-undecanol	[112-42-5]	$C_{11}H_{24}O$	n-undecyl alcohol	0.8339	1.4402	243	16	17.2	<0.02		petroleum ether, ethanol
1-dodecanol	[112-53-8]	$C_{12}H_{26}O$	n-dodecyl alcohol, lauryl alcohol	0.8306[25]	1.4428	138[1.33]	24	18.8	i	1.3	
1-tridecanol	[112-70-9]	$C_{13}H_{28}O$	n-tridecyl alcohol	0.8238[31]		155[2.0]	30.5				petroleum ether, ethanol
1-tetradecanol	[112-72-1]	$C_{14}H_{30}O$	n-tetradecyl alcohol, myristyl alcohol	0.8165[50]	1.4358[50]	158[1.33]	38		<0.02	nil	
1-pentadecanol	[629-76-5]	$C_{15}H_{32}O$	n-pentadecyl alcohol		1.4408[50]		44				
1-hexadecanol	[36653-82-4]	$C_{16}H_{34}O$	cetyl alcohol, palmityl alcohol	0.8157[60]	1.4392[60]	177[1.33]	49	53[75]	0.06[20]	nil	ethanol, methanol, diethyl ether, benzene
1-heptadecanol	[1454-85-9]	$C_{17}H_{36}O$	margaryl alcohol	0.8167[60]	1.4392[60]		54				
1-octadecanol	[112-92-5]	$C_{18}H_{38}O$	stearyl alcohol, n-octadecyl alcohol	0.8137[60]	1.4388[60]	203[1.33]	58		i	nil	
1-nonadecanol	[1454-84-8]	$C_{19}H_{40}O$	n-nonadecyl alcohol				62				

Name	CAS	Formula	Synonyms								Solvents
1-eicosanol	[629-96-9]	$C_{20}H_{42}O$	eicosyl alcohol, arachidyl alcohol			$251^{1.33}$	66		i	nil	benzene, ethanol, petroleum ether
1-hexacosanol	[506-52-5]	$C_{26}H_{54}O$	ceryl alcohol			$305^{2.67}$	79.5		i		ethanol, ether
1-hentriacontanol	[26444-39-3]	$C_{31}H_{64}O$	melissyl alcohol, myricyl alcohol	0.7784^{95}			87		nil		
9-hexadecen-1-ol	[10378-01-5]	$C_{16}H_{32}O$	palmitoleyl alcohol			$205–210^{2.0}$					
9-octadecen-1-ol	[143-28-2]	$C_{18}H_{36}O$	oleyl alcohol	0.8504^{58}	1.4473^{60}						ethanol, diethyl ether
10-eicosen-1-ol	[28061-39-4]	$C_{20}H_{40}O$	eicosoyl alcohol								
Primary branched aliphatic											
2-methyl-1-pentanol	[105-30-6]	$C_6H_{14}O$	2-methylpentyl alcohol	0.8254	1.4190	148		6.6	0.31	5.4	
2-ethyl-1-butanol	[97-95-0]	$C_6H_{14}O$	2-ethylbutyl alcohol	0.8348	1.4224	146.5	−114				
2-ethyl-1-hexanol	[104-76-7]	$C_8H_{18}O$	2-ethylhexyl alcohol	0.8340	1.4316	184	−70	9.8	0.07	2.6	ethanol, diethyl ether
3,5-dimethyl-1-hexanol	[13501-73-0]	$C_8H_{18}O$		0.8297	1.4250	182.5					
2,2,4-trimethyl-1-pentanol	[123-44-4]	$C_8H_{18}O$		0.839	1.4300	168	−70				ethanol
Secondary aliphatic											
4-methyl-2-pentanol	[108-11-2]	$C_6H_{14}O$	methylamyl alcohol, methylisobutyl-carbinol	0.8083	1.4112	132	−90	5.2	1.7	5.8	ethanol, diethyl ether
2-octanol	[123-96-6]	$C_8H_{18}O$	capryl alcohol	$0.835^{15/4}$	1.4256	178–179	−38	8.2	0.096^{25}		ethanol, petroleum ether
2,6-dimethyl-4-heptanol	[108-82-7]	$C_9H_{20}O$	diisobutylcarbinol	0.8121	1.4231	178	−65	14.3	0.06	0.99	ethanol, diethyl ether
2,6,8-trimethyl-4-nonanol	[123-17-1]	$C_{12}H_{26}O$		0.8193	1.4345	225	−60	21	<0.02	0.60	ethanol, diethyl ether

[a]Temperature, °C, if other than 20°C, is noted as superscript.
[b]Pressure, kPa, if other than 101.3 kPa, is noted as superscript. To convert kPa to mm Hg, multiply by 7.50.
[c]mPa·s = cP.

869

Amination

$$ROH + R'NH_2 \longrightarrow RNHR' + H_2O$$

Oxidation (6,7) and amination (8,9) are discussed in detail elsewhere.

Shipment and Storage

Detergent range alcohols are available in 208-L (55-gal) drums of approximately 160-kg or 23,000-L (6000-gal) tank trucks, in tank cars of 75,000 L (20,000 gal) containing about 60,000 kg, and in marine barges. The tank trucks and cars are usually insulated and equipped with an external heating jacket; the barges have coils for melting and heating the alcohols. High melting alcohols such as hexadecanol and octadecanol are also available as flaked material in three-ply, polyethylene-lined 22.7 kg (50 lb) bags. Detergent range alcohols have a U.S. Dept. of Transportation classification as nonhazardous for shipment. The perfume-grade alcohols, such as specially purified octanol and decanol, are available in bottles and cans; other plasticizer range materials are available in 208-L drums, 23,000-L tank trucks, 75,000-L tank cars, and in marine barges. Because of low melting points, most of these materials do not require transports having heating equipment. Bulk shipments are usually described by the commercial name of the material, such as methylisobutylcarbinol for 4-methyl-2-pentanol. The names hexyl, octyl, or decyl alcohol are used as freight descriptions for the linear or branched alcohols of corresponding carbon number. Linear and branched alcohols of 6–9 carbon atoms, and mixtures containing them, are classified as combustible for shipment by the U.S. DOT because of their low flash points. Alcohols of 10 carbons and above are classified as nonhazardous.

The higher alcohols are not corrosive to carbon steel, and equipment suitable for handling solvents or gasoline is also suitable for the alcohols. However, special storage conditions are often needed to maintain alcohol quality. Lined carbon steel tanks having nitrogen blankets to exclude both moisture and oxygen are recommended for storage of detergent range alcohols (10). Preferred storage temperature is no higher than 10°C above the alcohol melting point and repeated cycles of melting and solidifying must be avoided. Low pressure steam is generally used for heating; for the high melting hexadecanol and octadecanol, hot water can be used in order to reduce exposure to high temperature heating surfaces. Although they are generally considered quite stable, alcohols which are stored either for long periods of time or under improper conditions can undergo such subtle changes as deterioration of color, increase in carbonyl level, or a decrease in acid heat stability. It is sometimes preferable to store high melting alcohols as flakes in bags at ambient temperature rather than melted in a tank at higher temperature.

To prevent rusting and moisture pickup resulting from the hygroscopic nature of plasticizer range alcohols, tanks should be protected from moisture by such devices as a drying tube on the tank or a dry air blanket; nitrogen is usually not needed because ambient storage temperature is adequate for these lower melting materials. In general, plasticizer range alcohols are more storage-stable than the detergent range alcohols. However, to avoid the danger of fire resulting from the low flash points of plasticizer range alcohols, tanks should be grounded, have no interior sources of ignition, be filled from the bottom or have a filling line extending to the bottom to prevent static sparks, and be equipped with flame arrestors.

Economic Aspects

United States production of detergent range alcohol was 354,000 t in 1987, according to the U.S. International Trade Commission, compared to 263,000 t in 1974. About 60% was sold as alcohol on the merchant market; most of the rest was ethoxylated by the producers, then sold as the ethoxylated alcohol or sulfated and sold as the ethoxysulfate surfactant. In the 1960s and early 1970s ethylene-based synthetic alcohols appeared to be the wave of the future. Increases in petroleum prices and stabilization in the price of coconut and palm kernel oils, the primary raw materials for higher alcohols, have led back to natural production. Most alcohol capacity installed in the 1980s uses catalytic hydrogenolysis processes employing natural fats and oils as feedstock to make alcohol. Fatty alcohol capacity is increasingly being built in the coconut and palm oil producing countries. A number of natural alcohol plants have started up or are in various stages of construction in the Philippines, Malaysia, and Indonesia. In the United States however, the lion's share of detergent range alcohol production is by synthetic processes; Shell Chemical is the largest producer in a plant having a 270,000-t capacity. Linear synthetic alcohols can be used interchangeably with natural alcohols except where the presence of minor amounts of chain branching or secondary alcohols preclude use of the synthetics. The more highly branched alcohols are used where branching is not a problem, is desired, or the alcohols are ethoxylated. Ethoxylation reduces the physical and chemical effects of chain branching. Domestic detergent range alcohol producers are shown in Table 2; representative prices are given in Table 3. Manufacturers often adjust coproduct alcohol prices to compensate for shortages or surpluses, keeping the price of the primary material stable.

Table 2. U.S. Manufacturers of Detergent Range Alcohols

Manufacturer	Process	Feedstock	Products
Procter & Gamble	catalytic hydrogenolysis	coconut and palm kernel oils, tallow, palm oil	C_6–C_{18}
Sherex	catalytic hydrogenolysis	tallow	C_{16}, C_{18}, oleyl
Shell Chemical	modified oxo	ethylene/olefins	C_9–C_{15}
Vista	Ziegler	ethylene	C_6–C_{22}
Ethyl	modified Ziegler	ethylene	C_6–C_{22}
Exxon	modified oxo	olefins	C_{13}, C_{15}

Table 3. Prices of Detergent Range Alcohols[a]

Alcohol	Price, U.S.$/kg
lauryl alcohol, fob	1.54
dodecanol/tridecanol, delivered	1.26
hexadecanol, fob	2.01
octadecanol, fob	2.01

[a]November 1989 list prices.

United States production of plasticizer range alcohols was estimated to be 690,000 t in 1988 (11), 44% of which was 2-ethylhexanol. Domestic manufacturers and prices of representative plasticizer range alcohols are given in Table 4. The previous decade has seen a reduction in the number of manufacturers of 2-ethylhexanol and other branched chain alcohols. The volume of most branched alcohols has been static, however, and 2-ethylhexanol volume has doubled; the volume of linear alcohols has also grown. A substantial portion of these materials is used in plasticizers for poly(vinyl chloride) (PVC), so plasticizer range alcohol fortunes are tied to variations in the PVC industry. The plasticizers are mainly diesters of the alcohols and phthalic acid; di(2-ethylhexyl) phthalate [117-81-7] is the highest volume product. Recent price and volume history of 2-ethylhexanol is given in Table 5. Other branched alcohols tend to be priced at, or slightly above, the price of 2-ethylhexanol; the linear alcohols are several cents per kilogram higher. Production costs of plasticizer range alcohols, manufactured either by oxo or Ziegler processes, are strongly dependent on the cost of the ethylene or propylene feedstocks, making them dependent on the cost of crude oil and natural gas.

Table 4. Prices and Manufacturers of Plasticizer Range Alcohols

Material	Price, U.S.$/kg[a]	Manufacturer
hexanol	1.74	Ethyl
		Vista
4-methyl-2-pentanol	1.32	Union Carbide
octanol	2.01	Ethyl
		Vista
octanol, perfumer's grade	3.09	
isooctyl alcohol	0.97	Exxon
2-ethylhexanol	0.93	BASF
		Eastman
		Shell Chemical
		Tenn-USS
		Union Carbide
decanol	1.34	Ethyl
		Vista
decanol, perfumer's grade	1.65	

[a]Delivered price May 1989. The listed price is not necessarily the price listed by the indicated manufacturer.

Table 5. Price and Production Volume of 2-Ethylhexanol[a]

Year	Price, U.S.$/kg	Volume, 10^3 t/yr
1988	0.73	337
1987	0.60	300
1986	0.55	259
1985	0.60	243
1984	0.71	245
1983	0.71	175

[a]Ref. 12.

Most manufacturers sell a portion of their alcohol product on the merchant market, retaining a portion for internal use, typically for the manufacture of plasticizers. Sterling Chemicals' linear alcohol of 7, 9, and 11 carbons is all used captively. Plasticizer range linear alcohols derived from natural fats and oils, for instance, octanol and decanol derived from coconut oil and 2-octanol derived from castor oil, are of only minor importance in the marketplace. The 13–carbon tridecyl alcohol is usually considered to be a plasticizer range alcohol because of its manufacture by the oxo process and its use in making plasticizers. On the other hand, some types of linear 9- and 11-carbon alcohols find major application in detergents.

Analysis

Because the higher alcohols are made by a number of processes and from different raw materials, analytical procedures are designed to yield three kinds of information: the carbon chain length distribution, or combining weight, of the alcohols present; the purity of the material; and the presence of minor impurities and contaminants that would interfere with subsequent use of the product. Analytical methods and characterization of alcohols have been summarized (13).

For the detergent range alcohols, capillary gas chromatography, fast, accurate, and simple to use, is by far the most useful method for determining composition and purity (14). By the proper choice of the capillary stationary phase, carbon chain distribution and the amount of unsaturated, chain branched, or secondary alcohols, as well as the level of minor materials such as esters and hydrocarbons, can be determined. Hydroxyl Value (HV = mg of KOH equivalent to the hydroxyl content of 1 g of alcohol) measures the —OH end group and reflects both the combining weight and the purity of the sample. Saponification Value (SV = mg of KOH required to saponify the esters and acids in 1 g of alcohol), Acid Value (AV = mg of KOH required to neutralize the free fatty acid in 1 g of alcohol), and Ester Value (EV = SV minus AV) are measures of the carboxylic acid impurities present as the free acids or esters. Iodine Value (IV = g of iodine absorbed by 100 g of alcohol) is a measure of carbon–carbon unsaturation present in the alcohol. HV, SV, AV, EV, and IV can all be calculated from the capillary GC analysis. Moisture is also an important criterion of alcohol quality, and the color of the alcohol, usually determined by the APHA (Pt–Co) method, should be as close to water-white as possible. A number of other tests measure attributes important to specific uses. Examples are melting point for the heavy cut alcohols, cloud point of unsaturated alcohols, odor, carbonyl content, peroxide content, and various color stability tests. One of these last is the acid heat stability test. It determines the color change of middle cut alcohol in contact with concentrated sulfuric acid at an elevated temperature as an index of the color of alkyl sulfates that would be made from the alcohol. Test outcome is affected by carbonyl at a level of a few hundred parts per million, and by traces of iron, rust, and dirt particles.

As for detergent range alcohols, extensive use of capillary gas chromatography is also made for composition and purity determination of the plasticizer range materials. For those products that are a broad mixture of various isomers, however, distillation range and Hydroxyl Value are more useful characterizations. From the HV the combining weight can be calculated for subsequent chemical reactions. Carbonyl content is important, especially for those alcohols manufactured from aldehydes by the oxo process. It is often expressed similarly to HV: as the mg of KOH equivalent to the carbonyl oxygen in 1 g of sample. Acidity, expressed in terms of the equivalent weight percent of acetic acid, is used to determine the quality of the alcohol, as are moisture and APHA color. As with the detergent range alcohols, tests which measure color stability in the presence of sulfuric acid are employed to predict the color changes that may occur in subsequent reactions utilizing acid catalysts. Additionally, analytical determinations such as odor, chloride level, hydrocarbon content, and trace metal content, are required for specific uses.

Specifications and Standards

Most of the detergent range alcohols used commercially consist of mixtures of alcohols, and a wide variety of products is available. Table 6 shows the approximate carbon chain length composition of both the commonly used mixtures and single carbon materials; typical properties are given in Table 7. The range of commercially available materials is further described in sales brochures published by the manufacturers (15), who usually can also provide specially tailored blends to meet individual customer needs. Although only even-carbon alcohols are available from natural fats and oils and the Ziegler process, the development of the oxo process for linear alcohols has made odd-carbon alcohols a commercial reality, albeit with some chain branching. Commercial mixtures of these latter alcohols contain both odd and even numbered chain lengths. The major production of detergent range alcohols is in the 12–18 carbon range. Alcohols with 20 carbons and above are available in mixtures such as Vista's Alfol 20+ and Ethyl's Epal 20+. Behenyl alcohol (docosanol) [661-19-8], $C_{22}H_{46}O$, can be made from rapeseed oil. Except for oleyl alcohol, all commercial alcohols are fully saturated.

Both detergent range and plasticizer range alcohols and their derivatives have been accepted by the U.S. government for use in a number of drug and food contact or food additive areas, and plasticizer range alcohols have been accepted as flavoring agents in foods (16). They must meet rigid manufacture, quality

Table 6. Composition of Commercial Detergent Range Alcohols[a]

Alcohol commercial name	Representative trade name	Derived from	C_{12}	C_{13}	C_{14}	C_{16}	C_{18}	C_{20}
lauryl	CO-1214[b] [67762-41-8]	coconut, palm kernel	68		26	6		
	Alfol 1214[c]	ethylene	55		45			
	Epal 1214[d]	ethylene	66		27	7		
	Neodol 23[e]	ethylene	41	57	1			
	Epal 1218 [67762-25-8]	ethylene	49		20	17	14	
	Lauryl Alcohol Special-Type 70[f]	coconut	72		27	1		
	Epal 12	ethylene	99		1			
myristyl	Alfol 14	ethylene	1		99	1		
cetyl	CO-1695 [36653-82-4]	vegetable oil				98	2	
	Epal 16	ethylene			1	98	1	
tallow	TA-1618[b] [67762-30-5]	tallow			2	27	70[h]	1
	Adol 64[g]	fats			4	26	70	
stearyl	CO-1897	vegetable oil				1	98	1
oleyl	HD Oleyl Alcohol D[f]	fats				5	94[i]	1
	Adol 80	fats			4	14	81[i]	1

[a]Approximate composition by wt %, 100% alcohol basis.
[b]Registered trademark for Procter & Gamble alcohols.
[c]Registered trademark for Vista alcohols.
[d]Registered trademark for Ethyl Corporation alcohols.
[e]Registered trademark for Shell alcohols.
[f]Registered trademark for Henkel alcohols.
[g]Registered trademark for Sherex alcohols.
[h]Includes 1% C_{17} alcohol.
[i]Primarily unsaturated.

control, and record keeping requirements. Hexadecanol and octadecanol are used extensively in drug and cosmetic areas which require drug-grade raw materials. For this application they are produced to the specifications of the *National Formulary* (NF) in facilities registered by the U.S. Food and Drug Administration. The NF requirements for hexadecanol are 45–50°C melting point, 2.0 max AV, 5.0 max IV, and 218–238 HV. The NF requirements for octadecanol are 55–60°C melting point, 2.0 max AV, 2.0 max IV, 200–220 HV, and 90% min. octadecanol.

Besides the linear detergent range alcohols, a number of highly branched alcohols of 12 or more carbon atoms made by the oxo process are of commercial importance. Tridecyl alcohol [27458-92-0], $C_{13}H_{28}O$, consisting mainly of tetramethyl-1-nonanols, is one such material; it is generally considered to be a plasticizer range alcohol because of its manufacturing process and use in making plasticizers. Primary alcohols made by the Guerbet process, consisting of alcohols characterized as 2,2-dialkyl-1-ethanols, are available as hexadecyl [68526-87-4], $C_{16}H_{34}O$, octadecyl [27458-93-1], $C_{18}H_{38}O$, eicosyl [52655-10-4], $C_{20}H_{42}O$, and hexacosyl [70693-05-9], $C_{26}H_{54}O$, materials sold by Exxon under the Exxal brand name (17). They should not be confused with linear alcohols having similar names.

Table 7. Properties of Commercial Linear Detergent Range Alcohols

Commercial descriptive name	Hydroxyl Value	Saponification Value	Acid Value	Iodine Value	Melting point, °C	Color, APHA	Moisture, %
lauryl (99% C_{12})	301	0.2	0.02	0.2	23–25	5	0.03
lauryl (68% C_{12})	285	0.2	0.01	0	22	3	0.04
C_{12}–C_{13}[a]	289		0.02		18–22	5	0.02
cetyl	229	0.4	0	0.6	49	6–10	0.04
tallow	208	1.8	0	0.5	53	10–20	0.03
stearyl	206	0.5	0	0.7	58	6–15	0.03
oleyl	206	0.5	0	94	4		0.03

[a]Neodol 23 (registered trademark for Shell alcohols).

Table 8. Typical Properties of Commercial Plasticizer Range Alcohols

Name	Molecular formula	Hydroxyl Value	Acidity, % as acetic	Carbonyl, wt % O	Boiling range, °C	Color, APHA	Moisture, %	Flash point[a], °C
hexyl	($C_6H_{14}O$)		0.001	<0.003	152–160	5	0.05	63
2-ethylhexanol	($C_8H_{18}O$)	431	<0.007	<0.02	182–186	<10	<0.10	84[b]
isooctyl	($C_8H_{18}O$)		0.001	<0.003	184–190	5	0.05	84
isononyl	($C_9H_{20}O$)		0.001	<0.003	202–213	5	0.05	91
hexyl decyl	($C_8H_{18}O$)	408	<0.004	0.003	168–203	5	0.01	81[c]
octanol	($C_8H_{18}O$)	431	<0.005	0.003	184–195	5	0.03	88[c]
decanol	($C_{10}H_{22}O$)	355	<0.01	0.003	226–230	5	0.03	113
tridecyl	($C_{13}H_{28}O$)	283	0.001	<0.003	254–263	5	<0.05	127

[a]Pensky-Martens closed cup unless otherwise noted.
[b]Cleveland open cup.
[c]Tag closed cup.

Isostearyl alcohol [27458-93-1] is a highly branched natural alcohol containing a mixture of C_{18} alcohols derived from isostearic acid.

The sales brochures of the manufacturers describe the plasticizer range alcohols available on the merchant market (18). Typical properties of several commercial plasticizer range alcohols are presented in Table 8. Because in most cases these are mixtures of isomers or alcohols with several carbon chains, the properties of a particular material can vary somewhat from manufacturer to manufacturer. Both odd and even carbon chain alcohols are available, in both linear and highly branched versions. Examples of the composition of several mixtures are given in Table 9.

Table 9. Composition of Commercial Plasticizer Range Alcohols

Material	Component	Composition, wt %
isooctyl	3,4-dimethyl-1-hexanol [19138-79-5] 3,5-dimethyl-1-hexanol [69778-63-8] 4,5-dimethyl-1-hexanol [60564-76-3]	54
	3-methyl-1-heptanol [31367-46-1] 5-methyl-1-heptanol [7212-53-5]	25
	3-ethyl-1-hexanol [41065-95-6]	13
	other primary alcohols	8
hexyl decyl	hexanol	10
(Epal 610)	octanol	44
	decanol	46
octyl decyl	octanol	42
(Alfol 810)	decanol	58

Toxicological Properties

The higher alcohols are among the less toxic of commonly used chemicals and, in general, their toxic effects are reduced as the number of carbon atoms is increased. Table 10 gives data representative of the toxicological properties of the higher alcohols (19–23). Slight differences in material purity, methodology, and grading of results may account for variations in data from different sources, and these data should not be regarded as representing a consistent series. Because the data pertain to animals and not necessarily to humans, they should be used only as a guide. The values for acute oral toxicity may be compared to an LD_{50} of about 3.75 g/kg for sodium chloride ingested by rats. A substance with an LD_{50} of 15 g/kg or above is generally considered to be "practically nontoxic."

Primary human skin irritation of tetradecanol, hexadecanol, and octadecanol is nil; they have been used for many years in cosmetic creams and ointments (24). Based on human testing and industrial experience, the linear, even carbon number alcohols of 6–18 carbon atoms are not human skin sensitizers, nor are the 7-, 9- and 11-carbon alcohols and 2-ethylhexanol. Neither has industrial handling of other branched alcohols led to skin problems. Inhalation

Table 10. Toxicological Properties of Higher Alcohols

Material	Acute oral LD_{50} rats, g/kg[a]	Eye irritation, rabbits[b]	Primary skin irritation, rabbits[c]
hexanol	3.2–4.4	severe	moderate
octanol	18	severe	moderate
decanol	20–26	severe	moderate
dodecanol	>40	moderate	slight
tetradecanol	>8	mild	mild
hexadecanol	>20	mild	mild
octadecanol	>20	mild	mild
4-methyl-2-pentanol	2.6	slight	moderate
2-ethylhexanol	3.7	severe	moderate
mixed isomers			
hexyl	3.7	severe	moderate
isooctyl	>2	severe	moderate
decyl	4.7	severe	moderate
tridecyl	4.7	moderate	moderate

[a]The lethal dose for 50% of the test animals, expressed in terms of g of material per kg of body weight.
[b]Evaluation of the irritation elicited from 0.1 mL of the material applied to the eyes without rinsing.
[c]Evaluation of the irritation elicited from an application of full-strength alcohol left in contact with the skin for 24 h.

hazard, further mitigated by the low vapor pressure of these alcohols, is slight. Sustained breathing of alcohol vapor or mist should be avoided, however, as aspiration hazards have been reported (25).

Manufacture from Fats and Oils

Fats and oils from a number of animal and vegetable sources are the feedstocks for the manufacture of natural higher alcohols. These materials consist of triglycerides: glycerol esterified with three moles of a fatty acid. The alcohol is manufactured by reduction of the fatty acid functional group. A small amount of natural alcohol is also obtained commercially by saponification of natural wax esters of the higher alcohols, such as wool grease.

The carbon chain lengths of the fatty acids available from natural fats and oils range from 6–22 and higher, although a given material has a narrower range. Each triglyceride has a random distribution of fatty acid chain lengths and unsaturation, but the proportion of the various acids is fairly uniform for fats and oils from a common source. Any triglyceride or fatty acid may be utilized as a raw material for the manufacture of alcohols, but the commonly used materials are coconut oil, palm kernel oil, lard, tallow, rapeseed oil, and palm oil, and to a lesser extent soybean oil, corn oil and babassu oil. Coconut and palm kernel oil are the primary sources of dodecanol and tetradecanol; lard, tallow, and palm oil are the primary sources of hexadecanol and octadecanol. Producers of natural fatty alcohols typically make a broad range of alcohol products having various carbon chain lengths. They vary feedstocks to meet market needs for particular alcohols and to take advantage of changes in the relative costs of the various feedstock materials.

The first commercial production of fatty alcohol in the 1930s employed the sodium reduction process using a methyl ester feedstock. The process was used in plants constructed up to about 1950, but it was expensive, hazardous, and complex. By about 1960 most of the sodium reduction plants had been replaced by those employing the catalytic hydrogenolysis process. Catalytic hydrogenation processes were investigated as early as the 1930s by a number of workers; one of these is described in reference 26.

Hydrogenolysis Process. Fatty alcohols are produced by hydrogenolysis of methyl esters or fatty acids in the presence of a heterogeneous catalyst at 20,700–31,000 kPa (3000–4500 psi) and 250–300°C in conversions of 90–98%. A higher conversion can be achieved using more rigorous reaction conditions, but it is accompanied by a significant amount of hydrocarbon production.

$$RCOOCH_3 + 2\ H_2 \xrightarrow[\text{high pressure}]{\text{catalyst}} RCH_2OH + CH_3OH$$

$$RCH_2CH_2OH + H_2 \longrightarrow RCH_2CH_3 + H_2O$$

Fatty esters (wax esters), formed by ester interchange of the product alcohol and the starting material in the hydrogenolysis reactors, are later separated from the product by distillation. Unreacted methyl esters are also converted to fatty esters in the distillation step

$$RCOOCH_3 + R'OH \longrightarrow RCOOR' + CH_3OH$$

so that they too can be separated from the product. Fatty esters are recycled to the hydrogenolysis reactors since they can undergo hydrogenation in a manner similar to methyl esters, in this case yielding two moles of fatty alcohol per mole of ester. Fatty acids can also be used for the higher alcohol production. The fatty acid is pumped into the high pressure reactor and esterified *in situ* using previously made fatty alcohol; the resulting fatty ester then undergoes hydrogenolysis to two moles of fatty alcohol. A recently disclosed process uses the naturally occurring triglyceride ester as the feedstock for hydrogenolysis (27). Although the manufacturing process is simplified by eliminating the production of a methyl ester or fatty acid, degradation of glycerol to 1,2-propanediol also occurs in the high temperature of the reaction and thus degrades a valuable coproduct.

To prepare methyl ester feedstock for making fatty alcohols, any free fatty acid must first be removed from the fat or oil so that the acid does not react with the catalyst used in the subsequent alcoholysis step. Fatty acid removal may be accomplished either by refining or by converting the acid directly to a methyl ester (28). Refining is done either chemically, by removal of a soap formed with sodium hydroxide or sodium carbonate (alkali refining), or physically, by steam distillation of the fatty acids (steam refining) (29). In the case of chemical refining, the by-product soap is acidified to give a fatty acid and these "foots" are used as animal feed or upgraded for industrial fatty acid use. The by-product fatty acid from steam refining is of a higher grade than acidifed foots and is used directly as an industrial fatty acid or as animal feed. In either case, the fatty acid can also be converted to the methyl ester and used as additional alcohol feedstock. Refined oil is dried to prevent the reaction of water with the catalyst during alcoholysis.

Alcoholysis (ester interchange) is performed at atmospheric pressure near the boiling point of methanol in carbon steel equipment. Sodium methoxide [124-41-4], CH_3ONa, the catalyst, can be prepared in the same reactor by reaction of methanol and metallic sodium, or it can be purchased in methanol solution. Usage is approximately 0.3–1.0 wt % of the triglyceride.

$$C_3H_5(OOCR)_3 + 3\ CH_3OH \xrightarrow{\text{NaOCH}_3} 3\ RCOOCH_3 + C_3H_5(OH)_3$$

The alcoholysis reaction may be carried out either batchwise or continuously by treating the triglyceride with an excess of methanol for 30–60 min in a well-agitated reactor. The reactants are then allowed to settle and the glycerol [56-81-5] is recovered in methanol solution in the lower layer. The sodium methoxide and excess methanol are removed from the methyl ester, which then may be fed directly to the hydrogenolysis process. Alternatively, the ester may be distilled to remove unreacted material and other impurities, or fractionated into different cuts. Fractionation of either the methyl ester or of the product following hydrogenolysis provides alcohols that have narrow carbon-chain distributions.

High Pressure Hydrogenolysis.   There are three major hydrogenolysis processes in worldwide use: the methyl ester, slurry catalyst process operated by Procter & Gamble, Henkel, and Kao; the methyl ester, fixed-bed catalyst process operated by Henkel and Oleofina; and the fatty acid, slurry catalyst process developed by Lurgi and operated by several licensees. Each process typically uses a copper chromite or copper–zinc catalyst that is modified to meet the needs of the individual producer. Copper chromite when prepared is nominally a complex mixture of primarily copper(II) oxide and copper(II) chromite. But in use it is believed to be reduced to a mixture of metallic copper, copper(II) oxide, and copper(II) chromite, the metallic copper playing an important, but as yet undefined, role in the catalysis of the reaction. The catalyst is made by reaction of copper nitrate and chromic oxide with ammonia followed by vacuum filtering of the precipitate, water washing, and then roasting in air. The resulting material is a very fine black powder. The roasting operation is continuous, utilizing accurate temperature control to give a catalyst of long life and high activity. Barium, manganese, or other metal ions are sometimes added to improve stability, and silica or other binders may be put in to make a physically strong, fixed-bed catalyst pellet. Hydrogen [1333-74-0] is usually generated on site from methane or propane. The hydrogen should be of high purity to avoid catalyst poisons, such as sulfur and carbon dioxide, and to prevent buildup of inert gases in the system; pressure swing adsorption (PSA) is often used to remove gaseous impurities.

Methyl Ester Hydrogenolysis.   The flow sheet for the continuous methyl ester, catalyst slurry process is shown in Figure 1. The dry methyl ester, hydrogen, and catalyst slurry are fed cocurrently to a series of four vertical reactors operated at 250–300°C and 20,700 kPa (3000 psi). The reactors are unagitated, empty tubes, designed to provide adequate residence time, minimum backmixing, and a reasonable column height. Fresh catalyst powder is slurried with fatty alcohol and recycled catalyst in a weigh tank and metered into the bottom of the

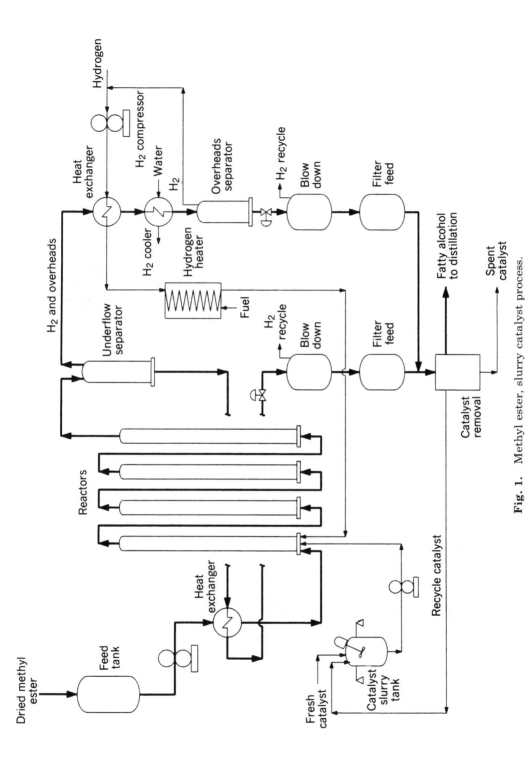

Fig. 1. Methyl ester, slurry catalyst process.

first reactor at approximately 3% of the ester feed rate. The heated hydrogen is fed through a distributor in the bottom of the first reactor. Besides serving as the reducing agent, the hydrogen also provides the principal source of heat and agitation for the reaction, and its flow conveys the mixture of ester, alcohol, and catalyst from one reactor to another. Approximately 30 moles of hydrogen are fed per mole of ester. The product stream from the last reactor, consisting of fatty alcohol, methanol, hydrogen, catalyst, and unreacted ester, enters a gravity separator where the vapor portion, consisting of hydrogen, methanol, and some fatty alcohol, goes overhead. The underflow stream of crude alcohol and catalyst is heat-interchanged with ester feed and depressurized, and the catalyst is removed. Most of the catalyst slurry is recycled but a small amount, to match the amount of fresh catalyst feed, is purged. This keeps a constant catalyst activity. The purged catalyst can be regenerated (30) or sold to a reclaimer to recover copper values. The overhead stream is heat-interchanged with hydrogen feed, cooled, and separated from hydrogen before being depressurized and filtered. An atmospheric stripping column removes methanol from the combined underflow/overhead stream of crude alcohol, and the methanol is recycled to the alcoholysis process. The stripped crude fatty alcohol is distilled in a vacuum column, or fractionated in a series of vacuum columns, to give the finished alcohol. The still bottoms, primarily fatty ester, are mainly recycled, and a small amount of still bottoms is removed from the system as a purge.

The process is controlled by the reaction temperature, feed rate (residence time), catalyst rate, and fresh catalyst usage. It is operated to provide the highest production rate commensurate with high yield and product quality, as well as lowest temperature and fresh catalyst usage. Heat interchange is used wherever possible to minimize energy consumption; low pressure steam is generated from coolers and condensers for use elsewhere in the process. Recycling from the two blowdown tanks recovers the hydrogen dissolved in those streams and reduces the usage of hydrogen feedstock. A fat trap is used to recover minor amounts of fatty alcohol and ester from process water streams and spills to reduce COD (chemical oxygen demand) loadings in the process sewer. The recovered material is then recycled to the process. A minor amount of still bottoms and unusable process remnants is burned as fuel.

The methyl ester, fixed-bed catalyst process is shown in Figure 2. A large excess of hydrogen is mixed with the methyl ester, part of which vaporizes and is carried through one or more fixed beds of catalyst at 200–250°C and a pressure similar to that used in the slurry process (31). After leaving the reactor, the mixture is cooled, then separated into a gaseous phase of mostly hydrogen, which is recycled, and a liquid phase of methanol and fatty alcohol. The liquid phase is depressurized into a blowdown tank, which removes the methanol; the fatty alcohol that remains does not require further purification. The alcohol is fractionated, however, if a product having a narrower carbon chain distribution is desired. The high rate of recirculating hydrogen in this process is claimed to provide fast removal of heat, providing high yields and minimizing side reactions such as hydrocarbon formation.

Fatty Acid Hydrogenolysis. The fatty acid, slurry catalyst process operates at 315°C and a pressure of 31,000 kPa (4500 psi); it is shown in Figure 3 (32,33). This process uses a single large reactor with internal baffles and a

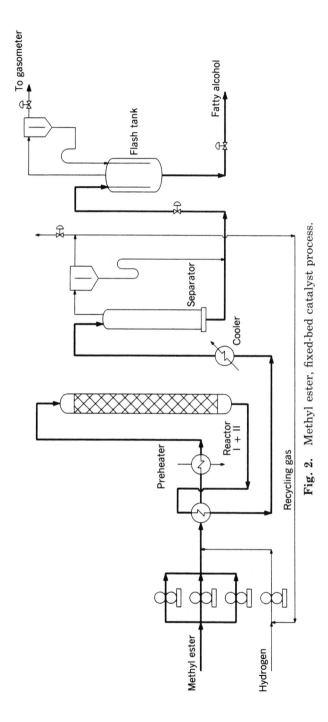

Fig. 2. Methyl ester, fixed-bed catalyst process.

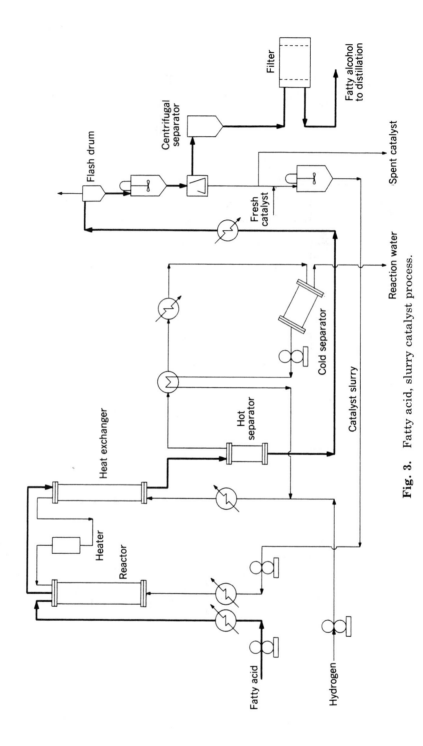

Fig. 3. Fatty acid, slurry catalyst process.

complex flow system. First, previously prepared fatty alcohol reacts with the acid feed to make a fatty ester via the alcoholysis reaction. A mole of water is also released. Then, the fatty ester reacts with hydrogen to give two moles of fatty alcohol per mole of ester. One exits the reactor, the other is recycled to react with the fatty acid feed. In two stages of cooling and separation, the excess hydrogen is separated from the reactor effluent for recycle, the reaction water is separated, and the catalyst containing fatty alcohol is recovered. The catalyst is removed as a slurry in a centrifugal separator for recycle. A small amount of catalyst is continuously purged from the process; an equivalent amount of fresh catalyst is added. After a final polish filtration, the crude fatty alcohol is sent to distillation: single-stage distillation for a broad range of carbon alcohols; fractionation for a narrower range of carbon alcohols.

Production of Unsaturated Alcohols

Unsaturated higher alcohols may be produced by saponification, sodium reduction, or hydrogenolysis of unsaturated fatty acids or esters. Saponification of oil from the sperm whale was a former source, but bans on the slaughter of whales by some nations and a general reduction in whaling have made this method obsolete. Alcohol made by saponification of wool grease (lanolin) is a minor product; sodium reduction of unsaturated esters is no longer an economic process for manufacturing unsaturated alcohols. Hydrogenolysis of unsaturated fatty acids or esters to produce alcohol without loss of the double bond has been a subject of interest for many years. Literature through the mid-1960s has been reviewed (34); and there has also been other work reported (35). In general, the key to double bond retention is a specially designed catalyst to give selectivity coupled with reaction conditions adjusted for the poorer reactivity of this catalyst compared to the copper–chromite catalysts. Cadmium modified catalysts are claimed to be effective, as are zinc chromite and a zinc–lanthanum catalyst (36). A zinc–aluminum catalyst reportedly avoids isomerization of the cis double bond of octadecenoic acid, soybean fatty acid, and linseed fatty acid methyl esters during hydrogenolysis (37). The known commercial hydrogenolysis processes for the production of octadecenol and other unsaturated alcohols are practiced by Sherex Chemical Company in the United States, Henkel K.-G.a.A. in Germany, and the New Japan Chemical Company in Japan. In at least one procedure (38), an unsaturated fatty acid reacts in a continuous process over a fixed catalyst bed at 270–290°C and 19,600 kPa (2800 psi). The catalyst is a complex aluminum–cadmium–chromium oxide that has high activity and exceptionally long life. The process is claimed to give a conversion of ester to alcohol of about 99% retaining essentially all of the original double bonds.

Uses of Detergent Range Alcohols

The detergent range alcohols and their derivatives have a wide variety of uses in consumer and industrial products either because of surface-active properties, or as a means of introducing a long chain moiety into a chemical compound. The major use is as surfactants (qv) in detergents and cleaning products. Only a small amount of the alcohol is used as-is; rather most is used as derivatives such as the poly(oxyethylene) ethers and the sulfated ethers, the alkyl sulfates, and the esters of other acids, eg, phosphoric acid and monocarboxylic and dicarboxylic acids. Major use areas are given in Table 11.

Table 11. Uses of Detergent Range Alcohols

Industry	Use as alcohol	Use as derivative
detergent	emollient, foam control, opacifier, softener	surfactant, softener
petroleum and lubrication	drilling mud	emulsifier, lubricant, dispersant, viscosity index improver, oil field chemical, pour-point depressant, drag reducing agent
agriculture	evaporation suppressant	pesticide, emulsifier, soil conditioner
plastics	mold release agent, antifoam, emulsion polymerization agent, lubricant	plasticizer, emulsion polymerization surfactant, lubricant dispersant, antioxidant, stabilizer, uv absorber
textile	lubricant, foam control, anti-static agent, ink ingredient	emulsifier, finish, softener, lubricant, scouring agent
cosmetics	softener, emollient	emulsifier, biocide, hair conditioner, emollient
pulp and paper	foam control	deresination agent, de-inking agent
food		emulsifier, antioxidant, disinfectant
rubber	plasticizer, dispersant	plasticizer
paint and coatings	foam control	emulsifier
metal working	lubricant, rolling oil	degreaser, lubricant
mineral processing	flotation agent	surfactant

Surfactants.   The detergent range alcohols can be used as building blocks for all of the surfactant types: anionics, cationics, nonionics, and zwitterionics. These alcohols are used for their emulsifying, dispersing, wetting, and cleaning properties and most surfactants (qv) made from them are readily biodegradable. Formulation of nonphosphate heavy duty liquid laundry detergents was made possible by use of these materials as the primary surfactant. The alkyl sulfates derived from C_{12} through C_{15} alcohols are widely used in consumer products such as shampoos, toothpastes, hand dishwashing detergents, and light duty household cleaners. Sodium dodecyl sulfate [151-21-3] is the optimum material for many cleaning compounds because of cleaning ability, mildness, and foaming capability. The alkyl sulfates of C_{16} and C_{18} alcohols are used in powder laundry detergents and other heavy-duty cleaners. Minor amounts of unsulfated alcohol left in the alkyl sulfate detergents serve as foam stabilizers. Surfactants made from polyethoxylated alcohols are in wider use than the alkyl sulfates. They tend to be less irritating to the skin than the alkyl sulfates and perform better in liquid systems such as hand dishwashing detergents and liquid laundry detergents. The ethoxylated materials may be used underivatized as nonionic surfactants. Alternatively, they may be sulfated and then neutralized using a base such as sodium or ammonium hydroxide to give ethoxysulfate anionic surfactants, the largest usage category of detergent range alcohols. Although the amount of ethylene oxide [75-21-8], C_2H_4O, can range from 1 to about 45 moles per mole of alcohol, the degree of ethoxylation of the anionic surfactants is typically 6 to 12, whereas that of the ethoxysulfates typically ranges from 3 to 12. Additionally, ethoxylation yields a broad range of species: for instance, a nominal 3-mole ethoxylate has some alcohol molecules containing up to fourteen units of ethylene oxide, yet it also includes about 15% unreacted alcohol, giving the effect of a mixed surfactant system. Varying the number of parent alcohol carbons, the amount of ethylene oxide used, and to some extent the breadth of the ethylene oxide distribution, gives wide latitude in the hydrophile–lipophile balance (HLB) of the resulting surfactant, which may be used as a nonionic surfactant or sulfated to give an anionic one. This versatility accounts for the broad use of ethoxylates in consumer cleaning products and in industrial applications as wetting agents, cleaning products, dispersing agents, and emulsifiers.

Alkyl glyceryl ether sulfonates are very mild, high foaming surfactants used in bar soaps and shampoos; they are made from the sulfonated alkyl chlorohydrin ether of detergent range alcohols. Alkyldimethyl amines are made from alcohols and then oxidized to give the amine oxide which is used as a mild surfactant in hand dishwashing products, shampoos, and some cosmetic applications. Some specialty cationic quaternary nitrogen surfactants are also made from the alcohols. Specialty phosphate ester surfactants are made from detergent range alcohols and ethoxylated alcohols; these find use mainly as lubricants and wetting agents in the textile industry.

In other surfactant uses, dodecanol–tetradecanol is employed to prepare porous concrete (39), stearyl alcohol is used to make a polymer concrete (40), and lauryl alcohol is utilized for froth flotation of ores (41). A foamed composition of hexadecanol is used for textile printing (42) and a foamed composition of octadecanol is used for coating polymers (43). On the other hand, foam is controlled by detergent range alcohols in applications: by lauryl alcohol in steel cleaning (44), by octadecanol in a detergent composition (45), and by eicosanol–docosanol in various systems (46).

Cosmetics and Pharmaceuticals.   The main use of hexadecanol (cetyl alcohol) is in cosmetics (qv) and pharmaceuticals (qv), where it and octadecanol (stearyl alcohol) are used extensively as emollient additives and as bases for creams, lipsticks, ointments, and suppositories. Octadecenol (oleyl alcohol) is also widely used (47), as are the nonlinear alcohols. The compatibility of heavy cut alcohols and other cosmetic materials or active drug agents, their mildness, skin feel, and low toxicity have made them the preferred materials for these applications. Higher alcohols and their derivatives are used in conditioning shampoos, in other personal care products, and in ingested materials such as vitamins (qv) and sustained release tablets (see CONTROLLED RELEASE TECHNOLOGY).

Lubricants and Petroleum.   Methacrylate esters of detergent range alcohols find use as viscosity index improvers, pour-point depressants, and dispersants (qv) in automobile engine lubricants. The free alcohol, particularly dodecanol (lauryl alcohol), is widely used in aluminum rolling, and also in other metalworking (48). A composition of octadecenol and sodium lauryl sulfate is used for petroleum oil recovery (49). Esters of docosanol are used as drag reducing agents for pipelining of crude petroleum oil, which reduces the power requirements for pumping.

Other Applications.   Alkylbenzyldimethylammonium salts are made from alcohols in the C_{12}–C_{16} range and find use as biocides and disinfectants in a number of areas. Dodecanol, tetradecanol, octadecanol, and tridecyl alcohol esters of thiodipropionic acid are employed as part of the antioxidant system of polyolefin plastics. Higher alcohols are used as antistatic agents (qv), mold release agents, and as additives in olefin polymerization (50); other uses have been reviewed (51). Esters of detergent range alcohols and fatty acids, lactic acid, and maleic acid are used for cosmetics and lubricants. Phosphites and phosphates of detergent range alcohols are also articles of commerce. Triacontanol (C_{32}) has activity as a plant growth regulator, but results have not been consistent enough for commercial use (52). Hexadecanol and octadecanol can be used to retard evaporation of water from reservoirs in arid regions (53). Detergent range alcohols also find application in antifoulant coatings, adhesives, and fabric softeners (54).

Uses of Plasticizer Range Alcohols

The plasticizer range alcohols are utilized primarily in plasticizers, but they also have a wide range of uses in other industrial and consumer products, as shown in Table 12. As in the case of the detergent range alcohols, the plasticizer range materials are little used as is, but rather are employed as the ester derivatives of acids such as phthalic, adipic, and trimellitic.

Plasticizers.   Over 70% of plasticizer range alcohols are ultimately consumed as plasticizers for PVC and other resins. Of this amount, 80% is used as the diester of phthalic acid, for instance di-2-ethylhexyl phthalate (DOP) or diisodecyl phthalate (DIDP) [26761-40-0]. Other plasticizers made from these alcohols are the diesters of adipic acid, azeleic acid, and sebacic acid, plus the triesters of phosphoric acid and trimellitic acid. A small amount of alcohol is used as the terminating agent in specialty polyester plasticizers. The adipates, azelates, and sebacates are employed as specialty materials in some food contact applications and in areas where low temperature flexibility is important, such as automobile interiors; eg, the diadipate ester of hexanol is the plasticizer in poly(vinyl butyral) used for automobile safety glass. The phosphates find application as good low temperature plasticizers and as flame retardant additives, whereas the trimellitates are used for high temperature applications such as the insulation of electrical wiring. The phthalates, however, are the general purpose plasticizers. Phthalate esters of alcohols from 4–13 carbons are available although most are in the C_8 through C_{10} range. All plasticizers are chosen on the basis of performance, cost, and ease of processing; DOP and DIDP are the workhorses of the industry. When

Table 12. Uses of Plasticizer Range Alcohols

Industry	Use as alcohol	Use as derivative
plastics	emulsion polymerization	plasticizer, flame retardant, oxidation and uv stabilizer, heat stabilizer, polymerization initiator
petroleum and lubrication	defoamer	lubricant, grease, lubricant additive, hydraulic fluid, diesel fuel additive
agriculture	stabilizer, tobacco sucker control, herbicide, fungicide	surfactant, insecticide, herbicide
mineral processing	solvent, extractant, antifoam	extractant, surfactant
textile	leveling agent, defoamer	surfactant
coatings	solvent, smoothing agent	surfactant, drying agent, solvent
metal working	solvent, lubricant, protective coating	lubricant, surfactant
chemical processing	antifoam, solvent	solvent
food		flavoring agent
cosmetics	perfume ingredient	

compared to DOP, phthalates of mixed linear alcohols (for instance, mixed heptyl, nonyl, and undecyl alcohols) give improved low temperature properties and resistance to volatile loss whereas those made of higher molecular weight alcohols (for instance, isodecyl or tridecyl alcohols) give improved resistance to extraction and volatile loss but exhibit some loss of plasticizing ability. In general, esters of mixtures of alcohols are favored as plasticizers because they give a broader range of properties than esters of a single alcohol.

Other Plastics Uses. The plasticizer range alcohols have a number of other uses in plastics: hexanol and 2-ethylhexanol are used as part of the catalyst system in the polymerization of acrylates, ethylene, and propylene (55); the peroxydicarbonate of 2-ethylhexanol is utilized as a polymerization initiator for vinyl chloride; various trialkyl phosphites find usage as heat and light stabilizers for plastics; organotin derivatives are used as heat stabilizers for PVC; octanol improves the compatibility of calcium carbonate filler in various plastics; 2-ethylhexanol is used to make expanded polystyrene beads (56); and acrylate esters serve as pressure sensitive adhesives.

Lubricants, Fuels, and Petroleum. The adipate and azelate diesters of C_6 through C_{11} alcohols, as well as those of tridecyl alcohol, are used as synthetic lubricants, hydraulic fluids, and brake fluids. Phosphate esters are utilized as industrial and aviation functional fluids and to a small extent as additives in other lubricants. A number of alcohols, particularly the C_8 materials, are employed to produce zinc dialkyldithiophosphates as lubricant antiwear additives. A small amount is used to make viscosity index improvers for lubricating oils. 2-Ethylhexyl nitrate [27247-96-7] serves as a cetane improver for diesel fuels and hexanol is used as an additive to fuel oil or other fuels (57). Various enhanced oil recovery processes utilize formulations containing hexanol or heptanol to displace oil from underground reservoirs (58); the alcohols and derivatives are also used as defoamers in oil production.

Agricultural Chemicals.   Plasticizer range alcohols are used as intermediates in the manufacture of a number of herbicides (qv) and insecticides, the largest use being that of 2-ethylhexanol and isooctyl alcohol to make the octyl ester of 2,4-dichlorophenoxyacetic acid (2,4–D) [*94-75-7*] for control of broadleaf weeds. Surfactants made from these alcohols are used as emulsifiers and wetting agents for agricultural chemicals. A mixture of octanol and decanol and the proper surfactants is able to kill the young meristemic tissue of some plants without harming more mature tissue. This is the basis for formulations that kill unwanted buds (suckers) in tobacco (59) and other plants and serve as a selective herbicide. Both decanol and 4-methyl-2-pentanol can be used as fungicides (qv) (60).

Surfactants.   A number of surfactants are made from the plasticizer range alcohols, employing processes similar to those for the detergent range materials such as sulfation, ethoxylation, and amination. These surfactants find application primarily in industrial and commercial areas: ether amines and trialkyl amines are used in froth flotation of ores, and the alcohols are also used to dewater mineral concentrates or break emulsions (61). The dialkyl sulfosuccinates of many of the C_8 through C_{13} alcohols also have surfactant applications. Octanol has found an application in a cleaning composition for engine carburetors, and decanol in a detergent for cleaning cotton (62).

Other Applications.   The alcohols through C_8 have applications as specialty solvents, as do derivatives of linear and branched hexanols. Inks, coatings, and dyes for polyester fabrics are other application areas for 2-ethylhexanol (63). Di(2-ethylhexyl) phthalate is used as a dielectric fluid to replace polychlorinated biphenyls. Trialkyl amines of the linear alcohols are used in solder fluxes, and hexanol is employed as a solvent in a soldering flux (64). Quaternary ammonium compounds of the plasticizer range alcohols are used as surfactants and fungicides, similarly to those of the detergent range alcohols.

BIBLIOGRAPHY

"Alcohols, Higher" in *ECT* 1st ed., Vol. 1, pp. 315–321, by H. B. McClure, Carbide and Carbon Chemicals Corporation, Unit of Union Carbide and Carbon Corporation; "Alcohols, Higher, Fatty" in *ECT,* 2nd ed., Vol. 1, pp. 542–559, by K. R. Ericson and H. D. Van Wagenen, The Procter & Gamble Company; "Alcohols, Higher, Synthetic" in *ECT,* 2nd ed., Vol. 1, pp. 560–569, by R. W. Miller, Eastman Chemical Products, Inc. "Alcohols, Higher Aliphatic, Survey and Natural Alcohols Manufacture" in *ECT* 3rd ed., Vol. 1, pp. 716–739, by R. A. Peters, Procter & Gamble Company.

  1. Braz. Pat. Pedido 86 2469A (Jan. 27, 1987), S. Inada and co-workers (to Seitetsu Kagaku Co., Ltd., Shinko Seito Co., Ltd., and Shinko Sugar Production Co., Ltd.); *Chem. Abstr.* **107,** 236087n (1987).
  2. R. G. Ackman, S. N. Hooper, S. Epstein, and M. Kelleher, *J. Am. Oil Chem. Soc.* **49,** 378–382 (1972).
  3. J. Sever and P. L. Parker, *Science* **164,** 1052–1054 (1969).

4. T. K. Miwa, *J. Am. Oil Chem. Soc.* **48,** 259 (1971); A. P. Tulloch, *J. Am. Oil Chem. Soc.* **50,** 367–371 (1973).

5. R. C. Wilhoit and B. J. Zwolinski, *J. Phys. Chem. Ref. Data* **2** (1) (1973).

6. U.S. Pat. 4,097,535 (June 27, 1978), K. Yang, K. L. Motz, and J. D. Reedy (to Continental Oil Co.).

7. D. Landini, F. Montanari, and F. Rolla, *Synthesis* **2,** 134–136 (1979).

8. Eur. Pat. Appl. EP 281,417 (Sept. 14, 1988), P. Y. Fong, K. R. Smith, and J. D. Sauer (to Ethyl Corp.).

9. U.S. Pat. 4,683,336 (July 28, 1987), C. W. Blackhurst (to Sherex Chemical Co.).

10. *Storage and Handling of Shell Neodol Detergent Alcohols, Ethoxylates, and Ethoxysulfates,* SC:133–179, Shell Chemical Company, Houston, Tex., 1979.

11. T. Gibson, *CEH Marketing Research Report: Plasticizer Alcohols,* SRI International, Menlo Park, Calif, 1989.

12. Data from U.S. International Trade Commission.

13. J. A. Monick, *Alcohols, Their Chemistry, Properties and Manufacture,* Reinhold Book Corp., New York, 1968, pp. 519–579.

14. R. E. Oborn and A. H. Ullman, *J. Am. Oil Chem. Soc.* **63,** 95–97 (1986).

15. *Products from the Chemicals Division,* Procter & Gamble Company, Cincinnati, Ohio, 1987; *Adol Fatty Alcohols,* Sherex Chemical Company, Dublin, Ohio, 1986; *Vista Surfactants, Industrial Chemicals, and Plastics,* Vista Chemical Company, Houston, Texas, 1987; *Epal Linear Primary Alcohols,* Ethyl Corporation, Baton Rouge, Louisiana, 1985; *Neodol,* Shell Chemical Company, Houston, Texas, 1987; *Henkel Fat Raw Materials,* Henkel K.-G.a.A., Düsseldorf, Fed. Rep. Germany.

16. *The United States Pharmacopeia,* 21st rev. *The National Formulary,* 16th ed., United States Pharmacopeial Convention, Rockville, Md., 1984; *Food Chemicals Codex,* 3rd ed., National Academy Press, Washington, D.C., 1981.

17. *Exxal Guerbet Alcohols,* Exxon Corporation, Houston, Texas, 1988.

18. *Vista Surfactants, Industrial Chemicals, and Plastics,* Vista Chemical Company, Houston, Texas, 1987; *Epal Linear Primary Alcohols,* Ethyl Corporation, Baton Rouge, Louisiana, 1985; *Exxal Alcohols,* Exxon Chemical Company, Houston, Texas, 1988; *Aristech Alcohols, 2-Ethlyhexanol,* Aristech Chemical Corporation, Pittsburgh, Pa., 1988; *Technical Bulletin, 2-Ethylhexanol,* BASF Corporation, Parsippany, N.J., 1987.

19. D. L. J. Opdyke, ed., *Monographs on Fragrance Raw Materials,* Pergamon Press, Oxford, 1974, pp. 8, 35, 39, 42.

20. R. A. Scala and E. G. Burtis, *J. Am. Ind. Hyg. Assn.* **34,** 493–499 (1973).

21. *Epal Linear Primary Alcohols,* Ethyl Corporation, Baton Rouge, Louisiana, 1985.

22. V. K. Rowe and S. B. McCollister in G. D. Clayton and F. E. Clayton, eds., *Patty's Industrial Hygiene and Toxicology,* Vol. 2C, 3rd ed., John Wiley & Sons, Inc., New York, 1982, pp. 4257–4708.

23. *MSDS for Alfol Alcohols,* Vista Chemical Company, Houston, Texas, 1984, 1985.

24. *J. Am. Coll. Toxicol.* **7,** 359–423 (1988).

25. H. W. Gerarde and D. B. Ahlstrom, *Arch. Environ. Health* **13,** 457–461 (1966).

26. U.S. Pat. 2,091,800 (Aug. 31, 1937), H. Adkins, K. Folkers, and R. Connor (to Rohm & Haas Co.).

27. Ger. Offen. 3,624,812 (Jan. 28, 1988), F.-J. Carduck, J. Falbe, T. Fleckenstein, and J. Pohl (to Henkel K.-G.a.A.).

28. U.S. Pat. 4,608,202 (Aug. 26, 1986), H. Lepper and L. Friesenhagen (to Henkel K.-G.a.A.).

29. F. E. Sullivan, *Chem. Eng. New York* **81,** 56 (April 15, 1974).

30. U.S. Pat. 4,533,648 (Aug. 6, 1985), P. J. Corrigan, R. M. King, and S. A. Van Diest (to The Procter & Gamble Co.).

31. U. R. Kreutzer, *J. Am. Oil Chem. Soc.* **61,** 343–348 (1984).

32. H. Buchold, *Chem. Eng. New York* **90,** 42, 43 (1983).

33. U.S. Pat. 4,259,536 (Mar. 31, 1981), T. Voeste, H. J. Schmidt, and F. Marschner (to Metallgesellschaft A.-G.).

34. H. Bertsch, H. Reinheckel, and K. Haage, *Fette Seifen Anstrichm.* **66,** 763–773 (1964); E. S. Lower, *Spec. Chem.* **2**(1), 30 (1982).

35. U.S. Pat. 3,193,586 (July 6, 1965), W. Rittmeister (to Dehydag, Deutsche Hydrierwerke); J. D. Richter and P. J. Van Den Berg, *J. Am. Oil Chem. Soc.,* **46,** 158–162, 163–166 (1969).

36. Brit. Pat. 1,076,855 (July 26, 1967), A. J. Pantulu, K. T. Achaya, G. S. Sidhu, and S. H. Laheer (to Council of Scientific and Industrial Research, India); Jpn. Kokai 58 210,035 (Dec. 7, 1983) (to Kao Corp.); Ger. Pat. 2,513,377 (Sept. 9, 1976), G. Demmering (to Henkel & Cie.).

37. U.S. Pat. 3,729,520 (Apr. 24, 1973), H. Rutzen and W. Rittmeister (to Henkel & Cie.).

38. Brit. Pat. 1,335,173 (Oct. 24, 1973) (to New Japan Chemical Co.).

39. Eur. Pat. Appl. 296,941 (Dec. 28, 1988), G. Dion Biro and R. De Bona Biro; Ger. Offen. 3,807,250 (Sep. 15, 1988), J. Sulkiewicz (to Anthes Industries, Inc.).

40. Jpn. Kokai 63 176,345 (July 20, 1988), C. Tomizawa and S. Narisawa (to Sumitomo Chemical Co.).

41. Ger. Offen. 3,517,154 (Nov. 13, 1986), W. Von Rybinski and R. Koester (to Henkel K.-G.a.A.).

42. Ger. Offen. 3,535,454 (Apr. 9, 1987), W. Braeuer and P. Diewald (to Bayer A.-G.).

43. Jpn. Kokai 53 101,061 (Sep. 4, 1978), E. Sugawara, S. Shioume, and K. Yorikane (to Dainichi Nippon Cables, Ltd.).

44. Jpn. Kokai 58 221,300 (Dec. 22, 1983) (to Nippon Kokan K.K. and Kao Corp.).

45. Eur. Pat. Appl. 210,721 (Feb. 4, 1987), P. M. Burrill (to Dow Corning Corp.).

46. Ger. Offen. 3,001,387 (July 23, 1981), R. Peppmoeller (to Chemische Fabrik Stockhausen und Cie.).

47. U. Ploog, *Seife. Oele. Fette. Wachse,* **109,** 225–229 (1983).

48. Eur. Pat. Appl. 182,552 (May 28, 1986), M. K. Budd and M. H. Foster (to Alcan International Ltd.); Jpn. Kokai 63 393 (Jan. 5, 1988), K. Nabatake, M. Ogawa, Y. Iwasaki, and T. Mizuta (to Nippon Steel Corp. and Daido Chemical Industry Co., Ltd.); N. P. Korotkova, I. G. Turyanchik, G. I. Cherednichenko, and V. P. Temnenko, *Neftepererab. Neftekhim. (Kiev),* **34,** 16–18 (1988); *Chem. Abstr.* **110,** 98392s (1989).

49. U.S. Pat. 4,213,500 (July 22, 1980), R. L. Cardenas and J. T. Carlin (to Texaco, Inc.).

50. Jpn. Kokai 59 217,782 (Dec. 7, 1984) (to Lion Corp.); U.S. Pat. 4,239,862 (Dec. 16, 1980), D. N. Matthews, W. Nudenberg, and H. A. Petersen (to Uniroyal, Inc.); Jpn. Kokai 61 138,606 (June 26, 1986), T. Tsutsui, M. Kioka, and N. Kashiwa (to Mitsui Petrochemical Industries, Ltd.).

51. E. S. Lower, *Polym. Paint Colour J.* **173,** 506 (1983).

52. S. K. Ries, *CRC Crit. Rev. Plant Sci.* **2,** 239–285 (1985); S. K. Ries and R. Houtz, *HortScience* **18,** 654–662 (1983).

53. U.S. Pat. 3,415,614 (Dec. 10, 1968), R. R. Egan and S. R. Sheeran (to Ashland Oil and Refining Co.).

54. Jpn. Kokai 62 13,471 (Jan. 22, 1987), Y. Yonehara and Y. Nanishi (to Kansai Paint Co., Ltd.); Jpn. Kokai 58 101,182 (June 16, 1983) (to Toshiba Silicone Co., Ltd.); Belg. Pat. 904,142 (July 30, 1986), J. P. Grandmaire and A. Jacques (to Colgate-Palmolive Co.).

55. Eur. Pat. Appl. 190,892 (Aug. 13, 1986), C. J. Chang (to Rhom and Haas Co.); Jpn. Kokai 62 135,501 (June 18, 1987), Y. Kondo, M. Mori, Y. Naito, and T. Chigusa (to Toyo Soda Mfg. Co., Ltd.); Jpn. Kokai 63 89,507 (Apr. 20, 1988), M. Terano, H. Soga, and M. Inoue (to Toho Titanium Co., Ltd.).

56. Jpn. Kokai 58 122,935 (July 21, 1983) (to Sekisui Kaseihin Kogyo K.K. and Eslen Kako

K.K.); Fr. Demande 2,531,971 (Feb. 24, 1984), H. P. Schlumpf, C. Stock, and P. Trouve (to Pluess-Staufer A.-G.).

57. Ger. Offen. 2,910,011 (Sep. 20, 1979), M. J. Rose; Ger. Offen. 3,626,102 (Feb. 11, 1988), M. L. Nelson and O. L. Nelson, Jr. (to Polar Molecular Corp.).

58. U.S. Pat. 4,485,871 (Dec. 4, 1984), B. W. Davis (to Chevron Research Co.); Brit. Pat. 1,542,166 (Mar. 14, 1979), Y.-C. Chiu (to Shell Internationale Research Maatschappij B.V.); U.S. Pat. 4,193,452 (Mar. 18, 1980), P. M. Wilson and J. Pao (to Mobil Oil Corp.).

59. *Off-Shoot-T,* Cochrane Corporation, Memphis, Tenn., 1984.

60. U.S. Pat. 3,778,509 (Dec. 11, 1973), H. L. Lewis (to Cotton, Inc.); Ger. Offen. 2,330,596 (Jan. 10, 1974), E. L. Frick and R. T. Burchill (to National Research Development Corp.).

61. Ger. Offen. 3,018,758 (Dec. 17, 1981), R. Peppmoeller (to Chemische Fabrik Stockhausen und Cie.); U.S. Pat. 4,206,063 (June 3, 1980), C. Dugan, M. E. Lewellyn, and S. S. Wang (to American Cyanamid Co.).

62. Jpn. Kokai 60 155,299 (Aug. 15, 1985), H. Murata and R. Hidaka (to Nitto Chemical Industry Co., Ltd.); U.S. Pat. 4,056,355 (Nov. 1, 1977), J. H. Kolaian, F. C. McCoy, and J. A. Patterson (to Texaco, Inc.).

63. U.S. Pat. 4,711,802 (Dec. 8, 1986), H. P. Tannenbaum (to E. I. du Pont de Nemours & Co., Inc.); Ger. Offen. 3,508,419 (Sep. 11, 1986), G. Neubert, M. Melan, and W. Schultze (to BASF A.-G.); Ger. Offen. 2,413,866 (Oct. 2, 1975), M. Vescia, M. Daeuble, and R. Widder (to BASF A.-G.).

64. Ger. Offen. 3,513,424 (Oct. 23, 1986), W. Kellberg (to Siemens A.-G.).

General References

Fatty Alcohols, Raw Materials, Methods, Uses, Henkel K.-G.a.A., Düsseldorf, 1982. Also published in German as *Fettalkohole.*

J. A. Monick, *Alcohols, Their Chemistry, Properties and Manufacture,* Reinhold Book Corp., New York, 1968.

E. J. Wickson, ed., *Monohydric Alcohols, ACS Symp. Ser. 159,* American Chemical Society, Washington, D.C., 1981.

RICHARD A. PETERS
The Procter & Gamble Company

SYNTHETIC PROCESSES

Higher aliphatic alcohols (C_6–C_{18}) are produced in a number of important industrial processes using petroleum-based raw materials. These processes are summarized in Table 1, as are the principal synthetic products and most important feedstocks (qv). Worldwide capacity for all higher alcohols was approximately 5.3 million metric tons per annum in early 1990, 90% of which was petroleum-derived. Table 2 lists the major higher aliphatic alcohol producers in the world in early 1990.

By far the largest volume synthetic alcohol is 2-ethylhexanol [*104-76-7*], $C_8H_{18}O$, used mainly in production of the poly(vinyl chloride) plasticizer bis(2-ethylhexyl) phthalate [*117-81-7*], $C_{24}H_{38}O_4$, commonly called dioctyl phthalate [*117-81-7*] or DOP (see PLASTICIZERS). A number of other plasticizer primary

Table 1. Synthetic Industrial Processes for Higher Aliphatic Alcohols

Process	Feedstock(s)	Principal products	Worldwide capacity, millions of tons
Ziegler (organoaluminum)	ethylene, triethylaluminum	primary C_6–C_{18} linear alcohols	0.3
oxo (hydroformylation)	olefins based on ethylene, propylene, butylene, or paraffins	primary alcohols	4.2
aldol	n-butyraldehyde	2-ethylhexanol	[a]
paraffin oxidation	paraffin hydrocarbons	secondary alcohols	0.2
Guerbet	lower primary alcohols	branched primary alcohols	[b]
Total			*4.7*

[a]Included in oxo process total.
[b]Less than 0.05.

alcohols in the C_6–C_{11} range are produced, as are large volumes of C_{10}–C_{18} synthetic, mainly primary, alcohols used as intermediates to surfactants (qv) for detergents. Other lower volume synthetic alcohol application areas include solvents and specialty esters.

The Ziegler Process

The Ziegler process, based on reactions discovered in the 1950s, produces predominantly linear, primary alcohols having an even number of carbon atoms. The process was commercialized by Continental Oil Company in the United States in 1962, by Condea Petrochemie in West Germany (a joint venture of Continental Oil Company and Deutsche Erdöl, A.G.) in 1964, by Ethyl Corporation in the United States in 1965, and by the USSR in 1983.

Four chemical reactions are used to synthesize alcohols from aluminum alkyls (see ORGANOMETALLICS) and ethylene (qv).

Triethylaluminum Preparation

$$2\ Al + 3\ H_2 + 6\ C_2H_4 \longrightarrow 2\ (C_2H_5)_3Al$$

Chain Growth

$$(C_2H_5)_3Al + 3x\ C_2H_4 \longrightarrow [C_2H_5(C_2H_4)_x]_3Al$$

Oxidation

$$2\ [C_2H_5(C_2H_4)_x]_3Al + 3\ O_2 \longrightarrow 2\ Al[O(C_2H_4)_xC_2H_5]_3$$

Table 2. Major C$_6$ and Higher Aliphatic Alcohol Producers[a]

Company and location	Capacity 10³ t/yr	Alcohol products	Feedstock
		Ziegler process	
Condea Chemie, Brunsbuettel, Germany	70	n-C$_6$,C$_8$,C$_{10}$,C$_{12}$,C$_{14}$,C$_{16}$,C$_{18}$,C$_{20}$	ethylene
Ethyl Corp, Houston, Tex., U.S.	111	n-C$_6$,C$_8$,C$_{10}$,C$_{12}$,C$_{14}$,C$_{16}$,C$_{18}$,C$_{20}$	ethylene
State, Ufa, USSR	48	n-C$_6$,C$_8$,C$_{10}$,C$_{12}$,C$_{14}$,C$_{16}$,C$_{18}$,C$_{20}$	ethylene
Vista Chemical, Lake Charles, La., U.S.	100	n-C$_6$,C$_8$,C$_{10}$,C$_{12}$,C$_{14}$,C$_{16}$,C$_{18}$,C$_{20}$	ethylene
Ziegler subtotal	329		
		Guerbet process	
Henkel, Duesseldorf, Germany	2	i-C$_{16}$,C$_{18}$,C$_{20}$,C$_{22}$,C$_{24}$,...,C$_{36}$	linear alcohols
Guerbet subtotal	2		
		Caustic fusion process	
Witco Chemical, Dover, Ohio, U.S.	7	2-octanol	castor oil
Caustic fusion subtotal	7		
		Fatty acid hydrogenation processes	
ATOCHEM SA, Lavera, France	7	n-C$_7$	castor oil
Cocochem, Batangas, Philippines	25	n-C$_8$,C$_{10}$,C$_{12}$,C$_{14}$,C$_{16}$	coconut oil
Colgate, Barangay, Philippines	4	n-C$_8$,C$_{10}$,C$_{12}$,C$_{14}$,C$_{16}$	coconut oil
Oleofabrik, Aarhus, Denmark	5	n-C$_{16}$,C$_{18}$	palm oil, tallow
State, Kedzierzyn, Poland	10	n-C$_8$,C$_{10}$,C$_{12}$,C$_{14}$,C$_{16}$	coconut oil
Fatty acid hydrogenation subtotal	51		
		Methyl ester hydrogenation process	
ATUL, India	3	n-C$_8$,C$_{10}$,C$_{12}$,C$_{14}$,C$_{16}$	coconut oil
Aegis, Jalagon, India	5	n-C$_8$,C$_{10}$,C$_{12}$,C$_{14}$,C$_{16}$	coconut oil
Condea Chemie, Brunsbuettel, Germany	30	n-C$_8$,C$_{10}$,C$_{12}$,C$_{14}$,C$_{16}$	coconut oil
Henkel, Duesseldorf, Germany	130	n-C$_8$,C$_{10}$,C$_{12}$,C$_{14}$,C$_{16}$C$_{18}$	coconut oil, tallow

Table 2. (Continued)

Company and location	Capacity 10^3 t/yr	Alcohol products	Feedstock
Henkel, Boussens, France	50	$n\text{-}C_8, C_{10}, C_{12}, C_{14}, C_{16} C_{18}, C_{20}, C_{22}$	coconut oil, other fats
Hüls AG, Marl, Germany	10	$n\text{-}C_8, C_{10}, C_{12}, C_{14}, C_{16}$	coconut oil
Kao Corp, Wakayama, Japan	15	$n\text{-}C_8, C_{10}, C_{12}, C_{14}, C_{16}$	coconut oil
Marchon (Albright & Wilson), Whitehaven, UK	25	$n\text{-}C_8, C_{10}, C_{12}, C_{14}, C_{16}$	coconut oil
New Japan Chemical, Tokushima, Japan	15	$n\text{-}C_8, C_{10}, C_{12}, C_{14}, C_{16}$	coconut oil
Philippinas Kao, Jasaan, Philippines	30	$n\text{-}C_8, C_{10}, C_{12}, C_{14}, C_{16}$	coconut oil
Procter & Gamble, Kansas City, Kan., U.S.	45	$n\text{-}C_8, C_{10}, C_{12}, C_{14}, C_{16}, C_{18}$	coconut oil, tallow
Procter & Gamble, Sacramento, Calif., U.S.	54	$n\text{-}C_8, C_{10}, C_{12}, C_{14}, C_{16}, C_{18}$	coconut oil, palm oil
Sherex, Mapleton, Ill., U.S.	7	oleyl alcohol, $n\text{-}C_{18}$	tallow, soybean oil
Sinopec, Shanghai, China	15	$n\text{-}C_8, C_{10}, C_{12}, C_{14}, C_{16}$	coconut oil
State, Radleben, Germany	10	$n\text{-}C_8, C_{10}, C_{12}, C_{14}, C_{16}$	coconut oil
Synfina-Oleofina, Ertvelde, Belgium	30	$n\text{-}C_8, C_{10}, C_{12}, C_{14}, C_{16}$	coconut oil
Methyl ester hydrogenation process			
subtotal	*474*		
		Oxidation processes	
Japan Catalytic Chemical, Kawasaki, Japan	12	$sec\text{-}C_{11}, C_{12}, C_{13}, C_{14}, C_{15}$	n-paraffins
State, Angarsk, USSR	45	$i\text{-}$ $C_{10}, C_{11}, C_{12}, C_{13}, C_{14}, C_{15}, C_{16}, C_{17}, C_{18}$	n-paraffins
State, Ufa, USSR	90	$sec\text{-}C_{11}, C_{12}, C_{13}, C_{14}, C_{15}, C_{16}$	n-paraffins
State, Volgodonsk, USSR	45	$n\text{-}$ $C_{10}, C_{11}, C_{12}, C_{13}, C_{14}, C_{15}, C_{16}, C_{17}, C_{18}$	n-paraffins

		Oxo process	
Oxidation subtotal	*192*		
Enichem, Augusta, Italy	50	n-C_7 to C_{15}	n-paraffins
Exxon Chemical France, Harnes, France	125	i-C_8,C_9,C_{10};n-C_9,C_{11},C_{13},C_{15}	polygas olefins,alpha olefins
Exxon Chemical Holland, Rozenburg-Europoort, Netherlands	200	i-C_8,C_9,C_{10},C_{13},C_{16}	polygas olefins
Exxon Chemical, Baton Rouge, La., U.S.	295	i-C_6 to C_{10},C_{12},C_{13},C_{16}; n-C_7,C_9,C_{11}	polygas olefins,alpha olefins,butene
Hoechst, Oberhausen-Holten, Germany	40	i-C_{10},C_{13}	propylene
ICI, Teeside, United Kingdom	250	i-C_8,C_9,C_{10};n-C_9 to C_{15}	polygas olefins,alpha olefins
India Nissan Chemical Ind., Baroda, India	13	i-C_7,C_8,C_9,C_{10},C_{11}	polygas olefins
Mitsubishi Kasei, Mizushima, Japan	25	i-C_9	butenes
Mitsubishi Kasei, Mizushima, Japan	30	n-C_7,C_9,C_{11},C_{13},C_{15}	ethylene
Mitsubishi Petrochemical, Yokkaichi, Japan	30	n-C_{12},C_{13},C_{14},C_{15}	n-paraffins
Nippon Oxocol, Ichihara, Japan	85	i-C_7,C_9,C_{10},C_{13}	polygas olefins
Shell Chemical, Stanlow, UK	90	n-C_{10},C_{11},C_{12},C_{13},C_{14},C_{15}	ethylene
Shell Chemical, Geismar, La., U.S.	272	n-C_7,C_8,C_9,C_{10},C_{11},C_{12},C_{13},C_{14},C_{15}	ethylene
Sterling, Texas City, Tex., U.S.	102	n-C_7,C_9,C_{11},C_{13}	alpha olefins
Unipar, Sao Paulo, Brazil	20	i-C_{10},C_{13}	propylene
Oxo process subtotal	*1627*		
		Oxo/aldol processes	
Aristech, Pasadena, Tex., U.S.	86	2-ethylhexanol	propylene
BASF, Ludwigshafen, Germany[b]	100	i-C_9;n-C_9,C_{11},C_{13},C_{15}	butenes,polygas olefins,alpha olefins
BASF, Ludwigshafen, Germany	150	2-ethylhexanol	propylene
BASF, Freeport, Tex., U.S.	30	2-ethylhexanol	propylene
BASF Espanol SA, Tarragona, Spain	30	2-ethylhexanol	propylene

Table 2. *(Continued)*

Company and location	Capacity 10^3 t/yr	Alcohol products	Feedstock
Celanese Mexicana, Celaya, Mexico[c]	70	2-ethylhexanol	acetaldehyde
Chemicke Zavodi, Litwinov, Czechoslovakia	30	2-ethylhexanol	propylene
Chisso, Goi, Japan	50	2-ethylhexanol	propylene
Ciquine, Camacari, Brazil	74	2-ethylhexanol	propylene
Elekieroz do Nordeste, Igarassue, Brazil[c]	15	2-ethylhexanol	acetaldehyde
Hoechst, Oberhausen-Holten, Germany	200	2-ethylhexanol	propylene
Hüls AG, Marl, Germany	200	2-ethylhexanol	propylene
Jilin, Jilin, China	50	2-ethylhexanol	propylene
KII, Koper, Yugoslavia	42	2-ethylhexanol	propylene
Kyowa Yuka, Yokkaichi, Japan	100	2-ethylhexanol	propylene
Lucky, Naju, Korea	120	2-ethylhexanol	propylene
Mitsubishi Kasei, Mizushima, Japan	146	2-ethylhexanol	propylene
National Organic, Bombay, India	8	2-ethylhexanol	propylene
Neste Oxo, Ornskoldsvik, Sweden	10	2-ethylhexanol	n-butyraldehyde
Neste Oxo, Stennungsund, Sweden	126	2-ethylhexanol	propylene
Shell Chemical, Deer Park, Tex., U.S.	27	2-ethylhexanol	propylene
Sinopec, Daqing, China	50	2-ethylhexanol	propylene
Sinopec, Yan Shan, China	20	2-ethylhexanol	propylene
Sinopec, Yueyangshibequ, China	10	2-ethylhexanol	propylene

898

Sinopec, Zibo, China	50	2-ethylhexanol	propylene
Societe Oxo-Chemie, Lavera, France	105	2-ethylhexanol	propylene
State, Burgas, Bulgaria	20	2-ethylhexanol	propylene
State, Beijing, China	10	2-ethylhexanol	propylene
State, Leuna, Germany	40	2-ethylhexanol	propylene
State, Schkopau, Germany	40	2-ethylhexanol	propylene
State, Rimnicu Vilcea, Romania	20	2-ethylhexanol	propylene
State, Timisoara, Romania	60	2-ethylhexanol	propylene
State, Angarsk, USSR	45	2-ethylhexanol	propylene
State, Omsk, USSR	45	2-ethylhexanol	propylene
State, Perm, USSR	90	2-ethylhexanol	propylene
State, Saluwat, USSR	45	2-ethylhexanol	propylene
Texas Eastman, Longview, Tex., U.S.	98	2-ethylhexanol	propylene
Tonen, Kawasaki, Japan	50	2-ethylhexanol	propylene
Union Carbide Corp., Texas City, Tex., U.S.	54	2-ethylhexanol	propylene
Zaklady Azotowe, Kedzierzyn, Poland	100	2-ethylhexanol	propylene
Oxo/aldol subtotal	*2616*		
total world	*5298*		

[a]Data from Refs. 1–6.
[b]Oxo/dimersol process.
[c]Aldol process.

Hydrolysis

$$2\ Al[O(C_2H_4)_xC_2H_5]_3 + 3\ H_2O \longrightarrow 6\ C_2H_5(C_2H_4)_xOH + Al_2O_3$$

This process is currently used by Vista Chemical, successor to Continental Oil Company's chemical business, and by Condea. In the Ethyl Corporation process dilute sulfuric acid is used in place of water in the hydrolysis step; producing alum rather than alumina.

Triethylaluminum Preparation. Triethylaluminum [97-93-8], $C_6H_{15}Al$, can be prepared by a two-step or a one-step process. In the former, aluminum [7429-90-5], Al, powder is added to recycled triethylaluminum and the slurry reacts first with hydrogen [1333-74-0], H_2, to produce diethylaluminum hydride [871-27-2], which in the second step reacts with ethylene [74-85-1], C_2H_4, to produce triethylaluminum. In the one-step process, hydrogen and ethylene are simultaneously fed to the reactor containing the aluminum slurry.

Chain Growth. Triethylaluminum reacts with ethylene in controlled, highly exothermic, successive addition reactions to produce a spectrum of higher molecular weight alkyls of even carbon number. The distribution of chain lengths in the chain growth mixture corresponds closely to the Poisson equation (7). Side reactions lead to small deviations from the Poisson distribution, greater deviations being observed at higher reaction temperatures. Some control of the distribution is obtained by adjustment of triethylaluminum–ethylene ratio as shown in Figure 1. In the Ethyl process, steps are taken to produce a longer chain fraction (predominantly $C_{12}–C_{18}$) that is sent to the oxidation step, and a shorter chain fraction (predominantly $C_2–C_{10}$) that is recycled for additional chain growth. The

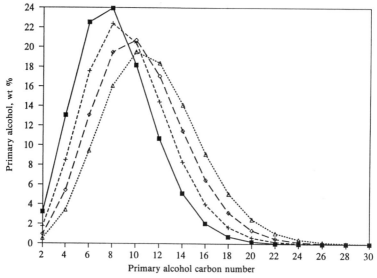

Fig. 1. Ziegler ethylene chain growth. Theoretical (Poisson) distribution of primary alcohols at (–■–) 2.5, (··+··) 3.0, (–◇–) 3.5, and (··△··) 4.0 moles of ethylene per ⅓ mole aluminum. Courtesy of Ethyl Corporation.

final product distribution is about 15–25% C_6–C_{10} and 75–85% C_{12}–C_{18} (8). This approach permits changes in the carbon number distribution of the alcohol product as best fit market demands. A comparison of typical commercial product distributions in the Ethyl and Vista processes is shown in Figure 2.

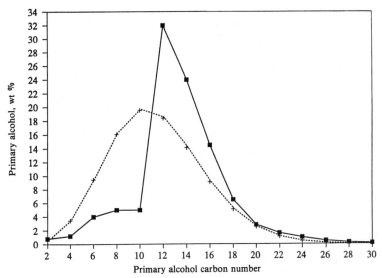

Fig. 2.   Estimated primary alcohol distributions for (■) Ethyl Corporation-modified Ziegler and (+) Vista Corporation Ziegler, at 4.0 moles ethylene per ⅓ mole aluminum. Courtesy of Ethyl Corporation.

There are two important side reactions, particularly above 120°C: (1) aluminum alkyls decompose to form dialkylaluminum hydrides and alpha olefins (the dialkylaluminum hydrides rapidly react with ethylene to regenerate a trialkylaluminum);

$$R_2AlCH_2CH_2R' \longrightarrow R_2AlH + CH_2{=}CHR'$$

and (2) alpha olefins can react with trialkylaluminum to produce branched aluminum alkyls and branched olefins.

$$R_2AlCH_2CH_2R' + CH_2{=}CHR'' \longrightarrow R_2AlCH_2\underset{\underset{R''}{|}}{C}HCH_2CH_2R' \longrightarrow R_2AlH + CH_2{=}\underset{\underset{R''}{|}}{C}{-}CH_2CH_2R'$$

This second reaction leads to the small amount of branching (usually less than 5%) observed in the alcohol product. The alpha olefins produced by the first reaction represent a loss unless recovered (8). Additionally, ethylene polymerization during chain growth creates significant fouling problems which must be addressed in the design and operation of commercial production facilities (9).

Oxidation.   Aluminum alkyls are oxidized to the corresponding alkoxides using dry air above atmospheric pressure in a fast, highly exothermic reaction. In general, a solvent is used to help avoid localized overheating and to decrease the viscosity of the solution. By-products include paraffins, aldehydes, ketones, olefins, esters, and alcohols; accidental introduction of moisture increases paraffin formation. To prevent contamination, solvent and by-products must be removed before hydrolysis. Removal can be effected by high temperature vacuum flashing or by stripping.

Hydrolysis. Aluminum alkoxides are hydrolyzed using either water or sulfuric acid, usually at around 100°C. In addition to the alcohol product, neutral hydrolysis gives high quality alumina (see ALUMINUM COMPOUNDS); the sulfuric acid hydrolysis yields alum. The crude alcohols are washed and then fractionated.

Mild steel is a satisfactory construction material for all equipment in Ziegler chemistry processes except for hydrolysis. If sulfuric acid hydrolysis is employed, materials capable of withstanding sulfuric acid at 100°C are required: lead-lined steel, some alloys, and some plastics. Flow diagrams for the Vista and Ethyl processes are shown in Figures 3 and 4, respectively.

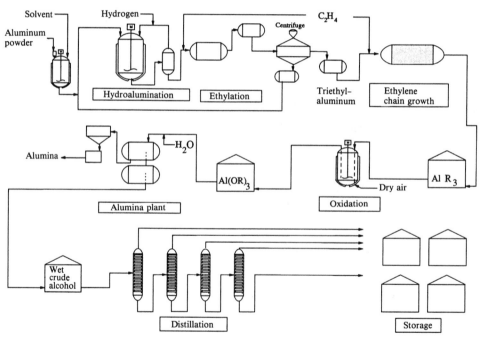

Fig. 3. Flow diagram for the Vista Corporation primary alcohols plant, Lake Charles, Louisiana. Courtesy of Vista Corporation.

Environmental Considerations. Environmental problems in Ziegler chemistry alcohol processes are not severe. A small quantity of aluminum alkyl wastes is usually produced and represents the most significant disposal problem. It can be handled by controlled hydrolysis and separate disposal of the aqueous and organic streams. Organic by-products produced in chain growth and hydroly-

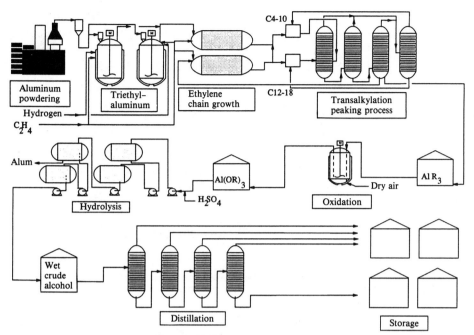

Fig. 4. Flow diagram for the Ethyl Corporation primary alcohols plant, Houston, Texas. Courtesy of Ethyl Corporation.

sis can be cleanly burned. Wastewater streams must be monitored for dissolved carbon, such as short-chain alcohols, and treated conventionally when necessary.

The Oxo Process

The oxo or hydroformylation reaction was discovered in Germany in 1938 (10) and was first used on a commercial scale by the Enjay Chemical Company (now Exxon) in 1948. By 1990 the total world alcohol capacity based on this general technology was over four million metric tons per year (see OXO PROCESS).

The structures and, hence, the properties of the higher oxo alcohols (C_6–C_{18}) are a function of the oxo process and the olefin employed. All the oxo products are primary alcohols and contain one more carbon atom than the feedstock olefin. They differ in two respects from natural alcohols and from Ziegler products, both of which are linear and of even carbon number. First, depending on the feedstock, they contain either even and odd carbon numbers or all odd carbon numbers. Second, the oxo products all have more branching. Branched olefin gives completely branched products; linear olefin gives some 2-methyl branching, the extent of which is dependent on the process. From a conventional cobalt-catalyzed process, the typical product of a linear olefin is 40–50% branched. Modified catalysts reduce branching to 15–25%.

Process Technology. In a typical oxo process, primary alcohols are produced from monoolefins in two steps. In the first stage, the olefin, hydrogen, and carbon monoxide [630-08-0], react in the presence of a cobalt or rhodium catalyst to form aldehydes, which are hydrogenated in the second step to the alcohols.

$$RCH{=}CH_2 + CO + H_2 \xrightarrow{\text{catalyst}} RCH_2CH_2CHO + \underset{\underset{\displaystyle CH_3}{|}}{RCHCHO}$$

$$RCH_2CH_2CHO + \underset{\underset{\displaystyle CH_3}{|}}{RCHCHO} + H_2 \xrightarrow{\text{catalyst}} RCH_2CH_2CH_2OH + \underset{\underset{\displaystyle CH_3}{|}}{RCHCH_2OH}$$

The oxo catalyst may be modified to function as a hydrogenation catalyst as well and, using a 2:1 ratio of hydrogen to carbon monoxide, alcohols are produced directly.

$$RCH{=}CH_2 + CO + 2\,H_2 \xrightarrow{\text{catalyst}} RCH_2CH_2CH_2OH + \underset{\underset{\displaystyle CH_3}{|}}{RCHCH_2OH}$$

These reactions are applicable to most monoolefins and are used to obtain a large number of commercial products.

Cobalt Catalyst, Two-Step, High Pressure Process. The olefin, with re-cycle and makeup cobalt catalyst at 0.1–1.0% concentration, is preheated and fed continuously to the oxo reactor together with the synthesis gas at a 1–1.2:1 H_2 to CO ratio. The reaction is conducted with agitation at 20,300–30,400 kPa (200–300 atm) and 130–190°C. Liquid hourly space velocity (LHSV) in the reactor is 0.5–1.0. The reaction is highly exothermic, 125 kJ/mol (54,000 Btu/lb-mol), and requires cooling. The intermediate aldehyde is hydrogenated to the alcohol at 5,070–20,300 kPa (50–200 atm) and 150–200°C using a catalyst containing copper, zinc, or nickel. The crude product is then fractionated (Fig. 5). The plant may be operated continuously or on a campaign basis with subsequent blending of the alcohols to give the desired product. The reactor and parts exposed to aldehydes or acids are constructed of alloy steel; the remainder is of carbon steel.

The cobalt catalyst can be introduced into the reactor in any convenient form, such as the hydrocarbon-soluble cobalt naphthenate [61789-51-3], as it is converted in the reaction to dicobalt octacarbonyl [15226-74-1], $Co_2(CO)_8$, the precursor to cobalt hydrocarbonyl [16842-03-8], $HCo(CO)_4$, the active catalyst species. Some of the methods used to recover cobalt values for reuse are (11): conversion to an inorganic salt soluble in water; conversion to an organic salt soluble in water or an organic solvent; treatment with aqueous acid or alkali to recover part or all of the $HCo(CO)_4$ in the aqueous phase; and conversion to metallic cobalt by thermal or chemical means.

Modified Cobalt Catalyst, One-Step, Low Pressure Process. The distin-guishing feature of this process, as commercialized by Shell, is catalysis by a cobalt–carbonyl–organophosphine complex such as $[Co(CO)_3P(C_4H_9)_3]_2$ (12). The olefin, using recycle and makeup catalyst at about 0.5% concentration, and synthesis gas at a 2–2.5:1 H_2 to CO ratio, react at 6,080–9,120 kPa (60–90 atm) and 170–210°C for detergent range alcohols. Lower pressures (3,040–7,080 kPa) are employed for *n*-butanol [71-36-3] and 2-ethylhexanol production. LHSV in the reactor is 0.1–0.2. The catalyst is highly selective for hydroformylation of 1-olefins

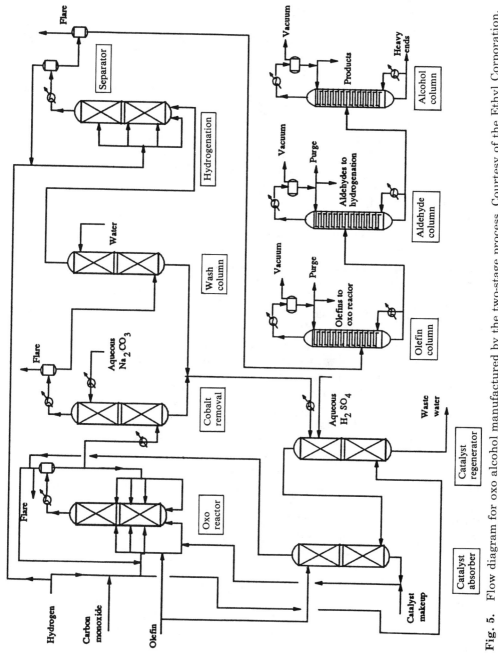

Fig. 5. Flow diagram for oxo alcohol manufactured by the two-stage process. Courtesy of the Ethyl Corporation.

at the terminal carbon atom; this results in a product from a linear feedstock which is up to 75–85% linear, having mainly 2-methyl isomers as branched components. The product is alcohol rather than aldehyde, because the modified catalyst promotes hydrogenation; and, because it is such an effective hydrogenation catalyst, approximately 10% of the olefin feed is also converted to paraffins. Because rapid isomerization of intermediates occurs under the reaction conditions, high primary alcohol selectivity can be obtained from internal olefins as well as from alpha olefins. After degassing and vacuum flashing, the crude alcohols are washed with caustic to convert esters to alcohols, water-washed, and distilled. Purified alcohols are then finished by hydrogenation and filtration (13).

Significant differences in this modified process include use of a lower pressure, slightly higher temperature, lower LHSV, formation of alcohol in one processing step, and a higher hydrogenation of the olefins to paraffins. The process is operated commercially by Shell Chemical U.S.A., Shell Chemical UK, and Mitsubishi Petrochemical exclusively for detergent range alcohols. Detergent range alcohols produced by the Shell process are particularly well-suited for downstream production of ethylene oxide adducts, which are major Shell Chemical products. The process schematic is shown in Figure 6.

Rhodium Catalysts. Rhodium carbonyl catalysts for olefin hydroformylation are more active than cobalt carbonyls and can be applied at lower temperatures and pressures (14). Rhodium hydrocarbonyl [75506-18-2], $HRh(CO)_4$, results in lower n-butyraldehyde [123-72-8] to isobutyraldehyde [78-84-2] ratios from propylene [115-07-1], C_3H_6, than does cobalt hydrocarbonyl, ie, 50/50 vs 80/20. Ligand-modified rhodium catalysts, $HRh(CO)_2L_2$ or $HRh(CO)L_3$, afford n-/iso-ratios as high as 92/8; the ligand is generally a tertiary phosphine. The rhodium catalyst process was developed jointly by Union Carbide Chemicals, Johnson-Matthey, and Davy Powergas and has been licensed to several companies. It is particulary suited to propylene conversion to n-butyraldehyde for 2-ethylhexanol production in that by-product isobutyraldehyde is minimized.

Olefin Sources. The choice of feedstock depends on the alcohol product properties desired, availability of the olefin, and economics. A given producer may either process different olefins for different products or change feedstock for the same application. Feedstocks believed to be currently available are as follows.

Propylene. 2-Ethylhexanol is now produced almost entirely from propylene, with the exception of a minor portion that comes from ethylene-derived acetaldehyde.

Polygas Olefins. Refinery propylene and butenes are polymerized with a phosphoric acid catalyst at 200°C and 3040–6080 kPa (30–60 atm) to give a mixture of branched olefins up to C_{15}, used primarily in producing plasticizer alcohols (isooctyl, isononyl, and isodecyl alcohol). Since the olefins are branched (75% have two or more CH_3 groups) the alcohols are also branched. Exxon, BASF, Ruhrchemie (now Hoechst), ICI, Nissan, Getty Oil, U.S. Steel Chemicals (now Aristech), and others have all used this olefin source.

Other Dimer Olefins. Olefins for plasticizer alcohols are also produced by the dimerization of isobutene [115-11-7], C_4H_8, or the codimerization of isobutene and n-butene [25167-67-3]. These highly branched octenes lead to a highly branched isononyl alcohol [68526-84-1] product. BASF, Ruhrchemie, ICI, Nippon Oxocol, and others have used this source.

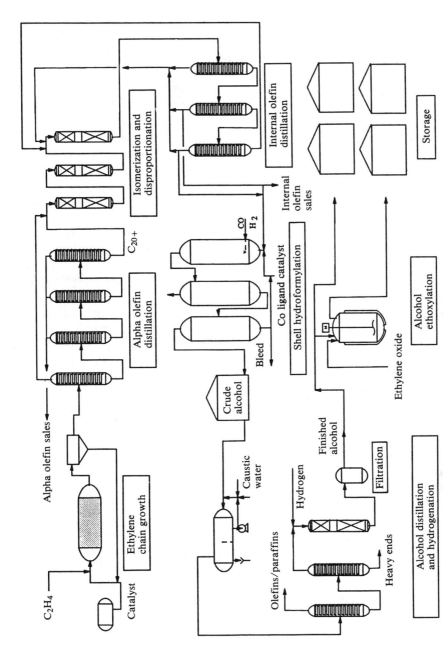

Fig. 6. Flow diagram for the Shell Chemical alcohol-olefin complex, Geismar, Louisiana, and Stanlow, United Kingdom. Courtesy of the Shell Chemical Corporation and the Ethyl Corporation.

The Dimersol process (French Petroleum Institute) produces hexenes, heptenes, and octenes from propylene and linear butylene feedstocks. This process is reported to produce olefin with less branching than the corresponding polygas olefins. BASF practices this process in Europe.

Normal Paraffin-Based Olefins. Detergent range *n*-paraffins are currently isolated from refinery streams by molecular sieve processes (see ADSORPTION, LIQUID SEPARATION) and converted to olefins by two methods. In the process developed by Universal Oil Products and practiced by Enichem and Mitsubishi Petrochemical, a *n*-paraffin of the desired chain length is dehydrogenated using the Pacol process in a catalytic fixed-bed reactor in the presence of excess hydrogen at low pressure and moderately high temperature. The product after adsorptive separation is a linear, random, primarily internal olefin. Shell formerly produced *n*-olefins by chlorination–dehydrochlorination. Typically, C_{11}–C_{14} *n*-paraffins are chlorinated in a fluidized bed at 300°C with low conversion (10–15%) to limit dichloroalkane and trichloroalkane formation. Unreacted paraffin is recycled after distillation and the predominant monochloroalkane is dehydrochlorinated at 300°C over a catalyst such as nickel acetate [373-02-4]. The product is a linear, random, primarily internal olefin.

Ethylene-Based Olefins. *Aluminum Alkyl Chain Growth.* Ethyl, Chevron, and Mitsubishi Chemical manufacture higher, linear alpha olefins from ethylene via chain growth on triethylaluminum (15). The linear products are then used as oxo feedstock for both plasticizer and detergent range alcohols; and because the feedstocks are linear, the linearity of the alcohol product, which has an entirely odd number of carbons, is a function of the oxo process employed. Alcohols are manufactured from this type of olefin by Sterling, Exxon, ICI, BASF, Oxochemie, and Mitsubishi Chemical.

Catalytic Oligomerization. Shell Chemical provides C_{11}–C_{14} linear internal olefin feedstock for C_{12}–C_{15} detergent oxo alcohol production from its SHOP (Shell Higher Olefin Process) plant (16,17). C_9–C_{11} alcohols are also produced by this process. Ethylene is first oligomerized to linear, even carbon–number alpha olefins using a nickel complex catalyst. After separation of portions of the α-olefins for sale, others, particularly C_{18} and higher, are catalytically isomerized to internal olefins, which are then disproportionated over a catalyst to a broad mixture of linear internal olefins. The desired C_{11}–C_{14} fraction is separated; the lighter and heavier fractions are recycled to the isomerization/disproportionation section. The SHOP process has been described in detail in the literature (18) and is shown schematically in Figure 6.

The Aldol Process

The important solvent and plasticizer intermediate, 2-ethylhexanol, is manufactured from *n*-butyraldehyde by aldol addition in an alkaline medium at 80–130°C and 300–1010 kPa (3–10 atm).

$$2\ CH_3CH_2CH_2CHO \xrightarrow{\text{catalyst}} CH_3CH_2CH_2CH{=}CCHO\ +\ H_2O$$
$$\underset{\displaystyle CH_2CH_3}{\phantom{CH_3CH_2CH_2CH{=}CC}|}$$

This step is followed by catalytic hydrogenation at 230°C and 5,070–20,300 kPa (50–200 atm).

$$CH_3CH_2CH_2CH{=}\underset{\underset{CH_2CH_3}{|}}{C}CHO \quad + \; 2\,H_2 \xrightarrow{\text{catalyst}} CH_3CH_2CH_2CH_2\underset{\underset{CH_2CH_3}{|}}{C}HCH_2OH$$

The *n*-butyraldehyde may be obtained from acetaldehyde [75-07-0] by aldol addition followed by hydrogenation, or from propylene by the oxo process. This latter process is predominantly favored (Fig. 7).

The oxo and aldol reactions may be combined if the cobalt catalyst is modified by the addition of organic–soluble compounds of zinc or other metals. Thus, propylene, hydrogen, and carbon monoxide give a mixture of C_4 aldehydes and 2-ethylhexenaldehyde [123-05-7] which, on hydrogenation, yield the corresponding alcohols.

The Paraffin Oxidation Process

Secondary alcohols (C_{10}–C_{14}) for surfactant intermediates are produced by hydrolysis of secondary alkyl borate or boroxine esters formed when paraffin hydrocarbons are air-oxidized in the presence of boric acid [10043-35-3] (19,20). Union Carbide Corporation operated a plant in the United States from 1964 until 1977. A plant built by Nippon Shokubai (Japan Catalytic Chemical) in 1972 in Kawasaki, Japan was expanded to 30,000 t/yr capacity in 1980 (20). The process has been operated industrially in the USSR since 1959 (21). Also, predominantly primary alcohols are produced in large volumes in the USSR by reduction of fatty acids, or their methyl esters, from permanganate-catalyzed air oxidation of paraffin hydrocarbons (22). The paraffin oxidation is carried out in the temperature range 150–180°C at a paraffin conversion generally below 20% to a mixture of trialkyl borate, $(RO)_3B$, and trialkyl boroxine, $(ROBO)_3$. Unconverted paraffin is separated from the product mixture by flash distillation. After hydrolysis of residual borate esters, the boric acid is recovered for recycle and the alcohols are purified by washing and distillation (19,20).

The product secondary alcohols from paraffin oxidation are converted to ethylene oxide adducts (alcohol ethoxylates) which are marketed by Japan Catalytic Chemical and BP Chemicals as SOFTANOL secondary alcohol ethoxylates. Union Carbide Chemical markets ethoxylated derivatives of the materials in the United States under the TERGITOL trademark (23).

The Guerbet Process

Higher molecular weight branched alcohols are produced by condensation of lower alcohols in the Guerbet reaction.

$$2\,RCH_2CH_2OH \longrightarrow RCH_2CH_2\underset{\underset{R}{|}}{C}HCH_2OH + H_2O$$

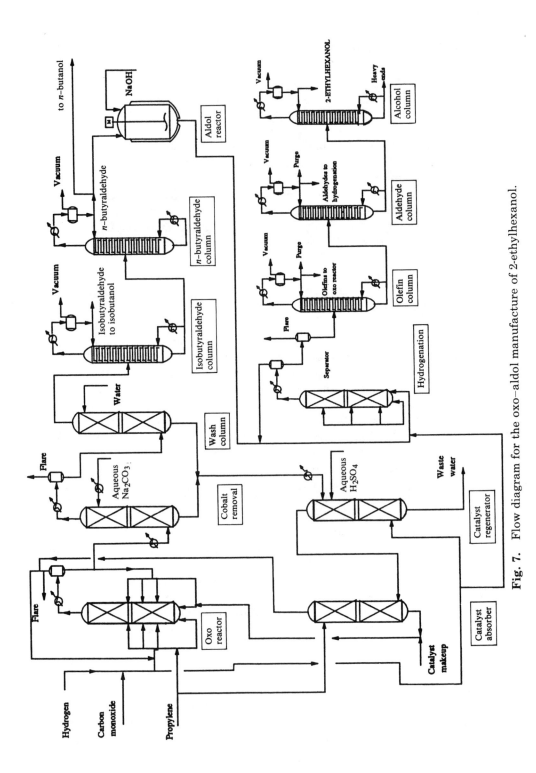

Fig. 7. Flow diagram for the oxo–aldol manufacture of 2-ethylhexanol.

In earlier studies (24), the reaction was carried out at temperatures above 200°C under autogenous pressure conditions using alkali metal hydroxide or alkoxide catalysts; significant amounts of carboxylic acid, RCH_2COOH, were formed as were other by-products. More recent reports describe catalysts which minimize by-products: $MgO-K_2CO_3-CuC_2O_2$ (25), less basic but still requiring high temperatures; Rh, Ir, Pt, or Ru complexes (26); and an alkali metal alkoxide plus Ni or Pd (27), effective at much lower temperatures.

Some 2,000–3,000 t/yr of these specialty alcohols are produced in the United States (Exxon) and in Germany (Henkel) (28). Their high liquidity because of branching permits use of less volatile, higher molecular weight materials, reported to be less irritating than the lower molecular weight linear alcohol materials, in a variety of cosmetic products (29).

BIBLIOGRAPHY

"Alcohols, Higher" in *ECT* 1st ed., Vol. 1, pp. 315–321, by H. B. McClure, Carbide and Carbon Chemicals Corporation; "Alcohols, Higher, Synthetic" in *ECT* 2nd ed., Vol. 1, pp. 560–569, by R. W. Miller, Eastman Chemicals Products, Inc; "Alcohols, Higher Aliphatic, Synthetic in *ECT* 3rd ed., Vol. 1 pp. 740–754 by M. F. Gautreaux, W. T. Davis, and E. D. Travis, Ethyl Corporation.

1. Asociacion Petroquimica Latinomamericana, Anuario Petroquimico Latino Americano 1985, Buenos Aires, 1985.
2. J-P. Davreux, Synfina–Oleofina, 1988.
3. R. F. Modler, "Detergent Alcohols" in *Chemical Economics Handbook,* SRI International, Menlo Park, Calif., 1987.
4. T. Gibson, "Plasticizer Alcohols" in *Chemical Economics Handbook,* SRI International, Menlo Park, Calif., 1985.
5. T. Gibson, "Oxo Chemicals" in ref. 4.
6. G. R. Lappin, J. D. Wagner, Ethyl Corporation, 1989.
7. H. Weslau, *Justus Liebigs Ann. Chem.* **629**, 198 (1960).
8. U.S. Pat. 3,415,861 (Dec. 10, 1968), W. T. Davis and C. L. Kingrea (to Ethyl Corporation).
9. G. R. Lappin in G. R. Lappin and J. D. Sauer, eds., *Alpha Olefins Applications Handbook,* Marcel Dekker, New York, 1989, p. 36.
10. Ger. Pat. 849,548 (Sept. 15, 1952), O. Roelen (to Chemische Verwertungsgesellschaft Oberhausen GmbH).
11. H. Lemke, *Hydrocarbon Process.* **45**(2), 148 (Feb. 1966).
12. U.S. Pats. 3,239,569; 3,239,570; 3,239,571 (Mar. 8, 1966), L. H. Slaugh and R. D. Mullineaux (to Shell Oil Company); Brit. Pats. 988,941; 988,942; 988,943; 988,944 (Apr. 14, 1965) (to Shell Internationale Research Maatschappij NV).
13. E. D. Heerdt, Shell Development Co., personal communication, 1989.
14. Ger. Pat. 953,605 (Dec. 6, 1956), G. Schiller (to Chemische Verwertungsgesellschaft Oberhausen GmbH).
15. Ref. 3, pp. 51–53.
16. Ref. 3, pp. 54–57.
17. U.S. Pat. 3,647,906 (Mar. 7, 1972), F. F. Farley (to Shell Oil Company); U.S. Pat. 3,726,938 (Apr. 10, 1973), A. J. Berger.
18. E. R. Freitas and C. R. Gum, *Chem. Eng. Prog.,* 73 (Jan. 1979).
19. J. Kurata and K. Koshida, *Hydrocarbon Process.* **57**(1), 145 (Jan. 1978); N. J. Steens and J. R. Livingston, Jr., *Chem. Eng. Prog.* **64**(7), 61 (July 1968).

20. N. Kurata, K. Koshida, H. Yokoyama, and T. Goto, in E. J. Wickson, ed., *Monohydric Alcohols, ACS Symp. Ser. 159*, American Chemical Society Washington, D.C., 1981, pp. 113–157.

21. I. M. Towbin and D. M. Boljanskii, *Maslo. Zhir. Prom.* **32**, 29 (1966).

22. H. Stage, *Seifen, Öle, Fette, Wachse* **99** (6/7), 143; (8), 185; (9), 217; (11), 299 (1973).

23. M. Tsuchino, and co-workers, *Paper no. 54 (I&EC Div.)*, 196th National Meeting of the American Chemical Society, Los Angeles, September 25–30, 1988.

24. M. Guerbet, *J. Pharm. Chim.* **6**, 49 (1913); *Chem. Abstr.* **7**, 1494 (1913).

25. M. N. Dvornikoff and M. W. Farrar, *J. Org. Chem.* **22**, 540 (1957).

26. G. Gregorio and G. F. Pregaglia, *J. Organometall. Chem.* **37**, 385 (1972); P. L. Burk, R. L. Pruett, and K. S. Campo, *J. Mol. Catal.* **33**, 1, 15 (1985).

27. J. Sabadie and G. Descotes, *Bull. Soc. Chim. Fr.* **253** (1983).

28. K. Noweck and H. Ridder, *Ullmann's Encyclopedia of Industrial Chemistry*, 5th ed., VCH Verlagsgesellschaft mbh, Weinheim, Germany, 1987, p. 288.

29. K. Klein, P. E. Bator, and S. Hans, *Cosmetics and Toiletries*, **95** 70 (1980); A. J. O'Lenick, Jr., and R. E. Bilbo, *Soap, Cosmet. Chem. Spec.* **52** (April, 1987).

General References

J. A. Monick, *Alcohols, Their Chemistry, Properties, and Manufacture*, Reinhold, New York (1968).

Ziegler chemistry processes; triethylaluminum synthesis

F. Albright, *Chem. Eng.* **74**, 179 (Dec. 4, 1967).
K. Ziegler and co-workers, *Angew. Chem.* **67**, 424 (1955).
K. Ziegler, *Erdöl Kohle* **11**, 766 (1958).
K. Ziegler and co-workers, *Justus Liebigs Ann. Chem.* **629**, 1 (1960).

Ziegler chemistry processes: chain growth

K. Ziegler, *Angew. Chem.* **64**, 323 (1952).
K. Ziegler, *Brennst. Chem.* **35**, 321 (1954).
K. Ziegler, *Angew. Chem.* **68**, 721 (1956).
K. Ziegler and co-workers, *Justus Liebigs Ann. Chem.* **629**, 121, 172 (1960).
K. Ziegler, *Angew. Chem.* **72**, 829 (1960).
K. Ziegler and H. Hoberg, *Chem. Ber.* **93**, 2938 (1960).

Ziegler chemistry processes: displacement reactions

K. Ziegler, H. Martin, and F. Krupp, *Justus Liebigs Ann. Chem.* **629**, 14 (1960).
K. Ziegler, W. R. Kroll, W. Larbig, and O. W. Steudel, *Justus Liebigs Ann. Chem.* **629**, 53 (1960).
K. Ziegler, in H. H. Zeiss, ed., *Organometallic Chemistry*, Reinhold, New York, 1960, p. 218 ff.

Ziegler chemistry processes: oxidation

K. Ziegler, F. Krupp, and K. Zosel, *Justus Liebigs Ann. Chem.* **629**, 241 (1960).
K. Ziegler, F. Krupp, and K. Zosel, *Angew. Chem.* **67**, 425 (1955).

Oxo processes

G. U. Ferguson, *Chem. Ind.* **11**, 451 (1965).
H. Weber and J. Falbe, *Ind. Eng. Chem.* **62**(4), 33 (Apr. 1970).
H. Weber, W. Dimmling, and A. M. Desal, *Hydrocarbon Process.* **55**(4), 127 (1976).

E. J. Wickson and H. P. Dengler, *Hydrocarbon Process.* **51**(11), 69 (1972).

J. Falbe, *Carbon Monoxide in Organic Synthesis,* Springer-Verlag, New York, 1970.

B. Cormels, in J. Falbe, ed., *New Synthesis with Carbon Monoxide,* Springer-Verlag, New York, 1980, p. 1–225.

Plant locations, capacities, feedstocks

Refs. 1 through 6.

JOHN D. WAGNER
GEORGE R. LAPPIN
J. RICHARD ZIETZ
Ethyl Corporation

ALCOHOLS, POLYHYDRIC

Polyhydric alcohols or polyols contain three or more CH_2OH functional groups. The monomeric compounds have the general formula $R(CH_2OH)_n$, where $n = 3$ and R is an alkyl group or CCH_2OH; the dimers and trimers are also commercially significant. Related species where $n = 2$ are discussed elsewhere (see GLYCOLS).

The most important polyhydric alcohols are shown in Figure 1. Each is a white solid, ranging from the crystalline pentaerythritols to the waxy trimethylol alkyls. The trihydric alcohols are very soluble in water, as is ditrimethylol-propane. Pentaerythritol is moderately soluble and dipentaerythritol and tripentaerythritol are less soluble. Table 1 lists the physical properties of these alcohols. Pentaerythritol and trimethylolpropane have no known toxic or irritating effects (1,2). Finely powdered pentaerythritol, however, may form explosive dust clouds at concentrations above 30 g/m^3 in air. The minimum ignition temperature is 450°C (3).

Reactions

Direct acetylation of pentaerythritol using acetic acid in aqueous solution or in toluene produces a mixture of acetates which can be fairly readily separated by chromatographic methods or distillation (8,9). The final product composition can be varied somewhat by altering the amount of water present. Esters of higher homologues and trimethylolpropane can also be synthesized using this procedure. Acrylate and methacrylate monoesters may be produced (10,11) when protecting groups are placed on three of the pentaerythritol hydroxyls. The protected intermediate then reacts with either acryloyl chloride [814-68-6] or methacryloyl chloride [27550-72-7] to give products of the type

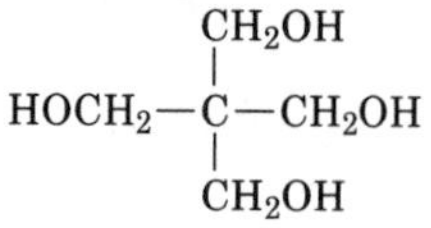

$$\begin{array}{c} CH_2OH \\ | \\ HOCH_2-C-CH_2OH \\ | \\ CH_2OH \end{array}$$

(1)

pentaerythritol, tetramethylolmethane

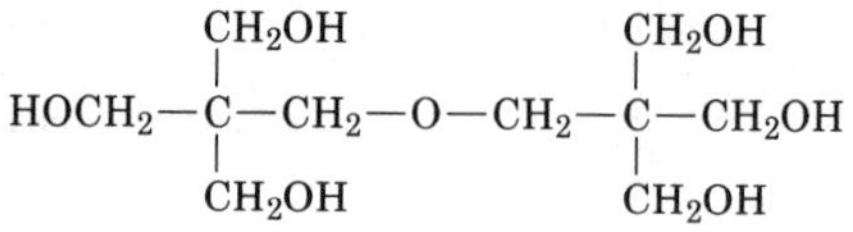

$$\begin{array}{ccc} CH_2OH & & CH_2OH \\ | & & | \\ HOCH_2-C-CH_2-O-CH_2-C-CH_2OH \\ | & & | \\ CH_2OH & & CH_2OH \end{array}$$

(2)

dipentaerythritol

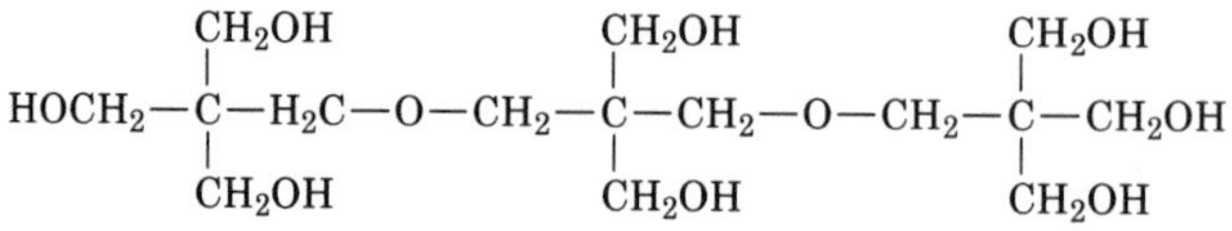

$$\begin{array}{ccc} CH_2OH & CH_2OH & CH_2OH \\ | & | & | \\ HOCH_2-C-H_2C-O-CH_2-C-CH_2-O-CH_2-C-CH_2OH \\ | & | & | \\ CH_2OH & CH_2OH & CH_2OH \end{array}$$

(3)

tripentaerythritol

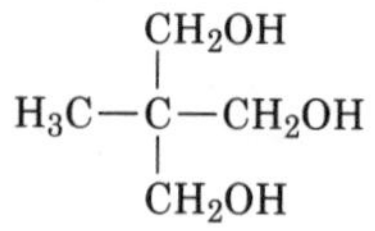

$$\begin{array}{c} CH_2OH \\ | \\ H_3C-C-CH_2OH \\ | \\ CH_2OH \end{array}$$

(4)

trimethylolethane

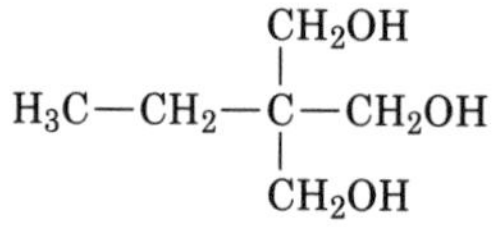

$$\begin{array}{c} CH_2OH \\ | \\ H_3C-CH_2-C-CH_2OH \\ | \\ CH_2OH \end{array}$$

(5)

trimethylolpropane

$$\begin{array}{cc} CH_2OH & CH_2OH \\ | & | \\ H_3C-CH_2-C-CH_2-O-CH_2-C-CH_2-CH_3 \\ | & | \\ CH_2OH & CH_2OH \end{array}$$

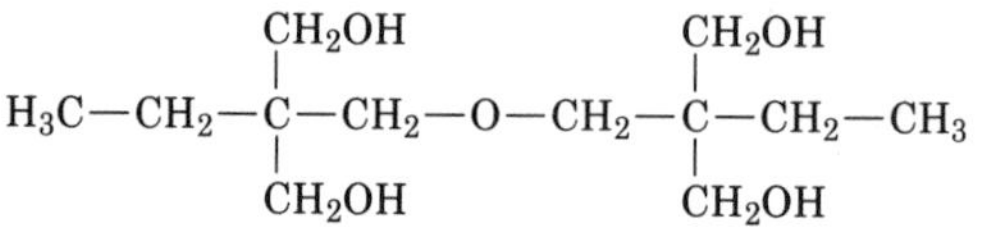

(6)

ditrimethylolpropane

Fig. 1. Polyhydric alcohols. Systematic names are (**1**) 2,2-bis(hydroxymethyl)-1,3-propanediol; (**2**) 2,2-[oxybis(methylene)]-bis[2-hydroxymethyl]-1,3-propanediol; (**3**) 2,2-bis{[3-hydroxy-2,2-bis(hydroxymethyl)propoxy]methyl}-1,3-propanediol; (**4**) 2-hydroxymethyl-2-methyl-1,3-propanediol; (**5**) 2-ethyl-2-hydroxymethyl-1,3-propanediol; and (**6**) 2,2-[oxybis(methylene)]-bis(2-ethyl)-1,3-propanediol.

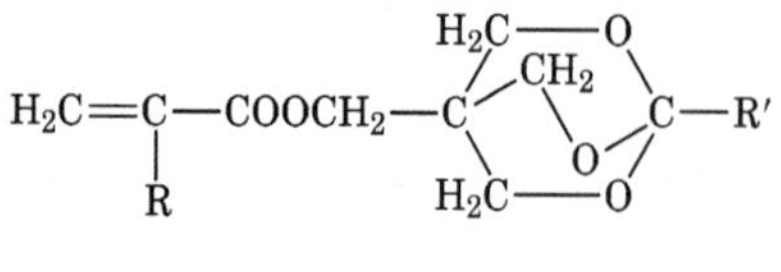

$$H_2C{=}C-COOCH_2-C \begin{array}{c} H_2C-O \\ \diagup CH_2 \diagdown \\ \diagdown \quad O \diagup \\ H_2C-O \end{array} C-R'$$

$$\begin{array}{c} | \\ R \end{array}$$

An alternative synthesis converts monobromopentaerythritol to the ortho ester, followed by reaction with cuprous acrylate.

Long-chain esters of pentaerythritol have been prepared by a variety of methods. The tetranonanoate is made by treatment of methyl nonanoate [*7289-51-2*] and pentaerythritol at elevated temperatures using sodium phenoxide alone, or titanium tetrapropoxide in xylene (12). Phenolic esters having good antioxidant activity have been synthesized by reaction of phenols or long-chain aliphatic acids and pentaerythritol or trimethylolpropane (13). Another ester synthesis employs the reaction of a long-chain ketone and pentaerythritol in xylene or chlorobenzene (14). Mixed esters have been produced using mixed isostearic and cyclohexane carboxylic acids in tribromophosphoric acid, followed by reaction with lauric acid (15).

Polyhydric alcohol mercaptoalkanoate esters are prepared by reaction of the appropriate alcohols and thioester using *p*-toluenesulfonic acid catalyst un-

Table 1. Physical Properties of Polyhydric Alcohols[a]

Property	Pentaerythritol	Dipentaerythritol	Tripentaerythritol	Trimethylolethane	Trimethylolpropane	Ditrimethylolpropane[a]
CAS Registry Number	[115-77-5]	[126-58-9]	[78-24-0]	[77-85-0]	[77-99-6]	[23235-61-2]
molecular formula	$C_5H_{12}O_4$	$C_{10}H_{22}O_7$	$C_{15}H_{32}O_{10}$	$C_5H_{12}O_3$	$C_6H_{14}O_3$	$C_{12}H_{26}O_5$
melting point, °C	261–262[b]	221–222.5[b]	248–250[b]	202[c]	58.8[d]	112–114
boiling point, °C	276 (4 kPa)			283[c]	289[d]	210 (0.12 kPa)
solubility, g/100 g water						
25°C	7.23[b]	0.28[e]	0.018[e]	soluble	soluble	2.6
50°C	16.1[e]	1.1[e]	0.07[e]			8.3
90°C	51.9[e]	6.1[e]	0.51[e]			>200 (completely soluble)
flash point, °C (Cleveland open cup)	260[f]				180[d]	>150
density, g/mL	1.396[f]	1.369[f]	1.30[f]		1.09[d]	1.18
refractive index	1.55 (20°C)[g]				1.472 (70°C)[g]	

[a]Data supplied by Perstorp AB.
[b]Ref. 1.
[c]Refs. 4 and 5.
[d]Ref. 2.
[e]Estimated value.
[f]Ref. 6.
[g]Ref. 7.

915

der nitrogen and subsequent heating (16,17). Organotin mercapto esters are similarly produced by reaction of the esters with dibutyltin oxide (18).

Pentaerythritol can be oxidized to 2,2-bis(hydroxymethyl)hydracrylic acid [2831-90-5], $C_5H_{10}O_5$,

$$
\begin{array}{c}
CH_2OH \\
| \\
HOCH_2-C-COOH \\
| \\
CH_2OH
\end{array}
$$

by direct air oxidation in aqueous solution using a palladium–carbon catalyst (19), or by biological oxidation using corynebacterium or arthrobacter cultures (20).

Bromohydrins can be prepared directly from polyhydric alcohols using hydrobromic acid and acetic acid catalyst, followed by distillation of water and acetic acid (21). Reaction conditions must be carefully controlled to avoid production of simple acetate esters (22). The raw product is usually a mixture of the mono-, di- and tribromohydrins.

Borolane products of mixed composition can be synthesized by direct addition of boric acid to pentaerythritol (23).

Reaction between pentaerythritol and phosphorous trichloride [7719-12-2] yields the spirophosphite, 3,9-dichloro-2,4,8,10-tetraoxa-3,9-diphosphaspiro[5,5]-undecane [3643-70-7], $C_5H_8Cl_2O_4P_2$,

$$
Cl-P
\begin{array}{c}
O-CH_2 \\
O-CH_2
\end{array}
C
\begin{array}{c}
CH_2-O \\
CH_2-O
\end{array}
P-Cl
$$

in the presence of benzene or a methyl acid phosphate catalyst (24,25) followed by removal of hydrogen chloride. Substituents may then replace the remaining chloride on treating the product with an alcohol or phenol (26) in the presence of a hydrogen chloride binding base such as triethylamine. Direct reaction of triethyl phosphite [122-52-1] and pentaerythritol is also possible (27). Pentaerythritol phosphate is similarly prepared by the reaction of pentaerythritol and phosphorus oxychloride [10025-87-3], $POCl_3$, in dioxane (28). Substituted diphosphaspiro compounds are made by reaction of pentaerythritol and either phosphonic anhydrides or trialkyl phosphites and trialkylamine (29,30).

The commercially important explosive pentaerythritol tetranitrate [78-11-5] (PETN), $C_5H_8N_4O_{12}$,

$$
\begin{array}{c}
O_2NO-CH_2 \\
O_2NO-CH_2
\end{array}
C
\begin{array}{c}
CH_2-ONO_2 \\
CH_2-ONO_2
\end{array}
$$

is produced by direct reaction of pentaerythritol in nitric or nitric–sulfuric acid media (31–33).

Aminoalkoxy pentaerythritols are obtained by reduction of the cyanoethoxy species obtained from the reaction between acrylonitrile, pentaerythritol, and lithium hydroxide in aqueous solution. Hydrogen in toluene over a ruthenium catalyst in the presence of ammonia is used (34). The corresponding amino-phenoxyalkyl derivatives of pentaerythritol and trimethylolpropane can also be prepared (35).

Tosylates of pentaerythritol and the higher homologues can be converted to their corresponding tetra-, hexa-, or octaazides by direct reaction of sodium azide (36), and azidobenzoates of trimethylolpropane and dipentaerythritol are prepared by reaction of azidobenzoyl chloride and the alcohols in pyridine medium (37).

Pentaerythritol can be converted to the biscyclic formal, 2,4,8,10-tetra-oxaspiro[5,5]undecane [126-54-5], $C_7H_{12}O_4$,

$$H_2C \underset{O—CH_2}{\overset{O—CH_2}{<}} C \underset{CH_2—O}{\overset{CH_2—O}{>}} CH_2$$

by heating in the presence of formaldehyde or paraformaldehyde and an acid catalyst (38). Alternatively, a cation-exchange resin catalyst may be used and excess water removed by azeotropic distillation (39). Higher aldehydes have also been used to prepare long-chain alkyl and aryl cyclic acetals (40–43).

Simple alkyl and alkenyl ethers of pentaerythritol are produced on direct reaction of the polyol and the required alkyl or alkenyl chloride in the presence of quaternary alkylamine bromide (44). Allyl chloride produces the pentaerythritol tetrallyl ether [1471-18-7],

$$\begin{array}{c} H_2C{=}CH—CH_2—O—CH_2 \quad CH_2—O—CH_2—CH{=}CH_2 \\ C \\ H_2C{=}CH—CH_2—O—CH_2 \quad CH_2—O—CH_2—CH{=}CH_2 \end{array}$$

in high yield by this method (45,46) or by using sodium hydroxide catalyst (47). Polycyclic crown looped and starburst dendrimer ethers are synthesized utilizing blocking–deblocking and high dilution cyclization techniques in reactions of dialcohols and ditosylates (48,49).

Manufacture

Pentaerythritol is produced by reaction of formaldehyde [50-00-0] and acetaldehyde [75-07-0] in the presence of a basic catalyst, generally an alkali or alkaline-earth hydroxide. Reaction proceeds by aldol addition to the carbon adjacent to the hydroxyl on the acetaldehyde. The pentaerythrose [3818-32-4] so produced is converted to pentaerythritol by a crossed Cannizzaro reaction using formaldehyde. All reaction steps are reversible except the last, which allows completion of the reaction and high yield industrial production.

The main intermediates in the pentaerythritol production reaction have been identified and synthesized (50,51) and the intermediate reaction mechanisms deduced. Without adequate reaction control, by-product formation can easily occur (52,53). Generally mild reaction conditions are favored for optimum results (1,54). However, formation of by-products cannot be entirely eliminated, particularly dipentaerythritol and the linear formal of pentaerythritol, 2,2′-[methylenebis(oxymethylene)]bis(2-hydroxymethyl-1,3-propanediol) [6228-26-8]:

$$
\underset{\overset{|}{CH_2OH}}{\overset{\overset{CH_2OH}{|}}{HOCH_2-C-CH_2-O-CH_2-O-CH_2-}} \underset{\overset{|}{CH_2OH}}{\overset{\overset{CH_2OH}{|}}{C-CH_2OH}}
$$

The quantities of formaldehyde and base catalyst required to produce pentaerythritol from 1 mol of acetaldehyde are always in excess of the theoretical amounts of 4 mol and 1 mol, respectively, and mole ratios of formaldehyde to acetaldehyde vary widely. As the mole ratio increases, formation of dipentaerythritol and pentaerythritol linear formal is suppressed. Dipentaerythritol formation may also be reduced by increasing the formaldehyde concentration, although linear formal production increases under those conditions (55,56).

The most common catalysts are sodium hydroxide and calcium hydroxide, generally used at a modest excess over the nominal stoichiometric amount to avoid formaldehyde-only addition reactions. Calcium hydroxide is cheaper than NaOH, but the latter yields a more facile reaction and separation of the product does not require initial precipitation and filtration of the metal formate (57).

A typical flow diagram for pentaerythritol production is shown in Figure 2. The main concern in mixing is to avoid loss of temperature control in this exothermic reaction, which can lead to excessive by-product formation and/or reduced yields of pentaerythritol (55,58,59). The reaction time depends on the reaction temperature and may vary from about 0.5 to 4 h at final temperatures of about 65 and 35°C, respectively. The reactor product, neutralized with acetic or formic acid, is then stripped of excess formaldehyde and water to produce a highly concentrated solution of pentaerythritol reaction products. This is then cooled under carefully controlled crystallization conditions so that the crystals can be readily separated from the liquors by subsequent filtration.

The first stage crystals are rich in pentaerythritol linear formal and may be treated (60,61) to convert this species to pentaerythritol and formaldehyde, which can then be recovered. The concentrated liquors obtained after redissolving are then recrystallized and filtered prior to drying of the final product.

The exact order of the production steps may vary widely; in addition, some parts of the process may also vary. Metal formate removal may occur immediately after the reaction (62) following formaldehyde and water removal, or by separation from the mother liquor of the first-stage crystallization (63). The metal formate may be recovered to hydroxide and/or formic acid by ion exchange or used as is for deicing or other commercial applications. Similarly, crystallization may include sophisticated techniques such as multistage fractional crystallization, which allows a wider choice of composition of the final product(s) (64,65).

Staged reactions, where only part of the initial reactants are added, either to consecutive reactors or with a time lag to the same reactor, may be used to reduce dipentaerythritol content. This technique increases the effective formaldehyde-to-acetaldehyde mole ratio, maintaining the original stoichiometric one. It also permits easier thermal control of the reaction (66,67). Both batch and continuous reaction systems are used. The former have greater flexibility whereas the product of the latter has improved consistency (55,68).

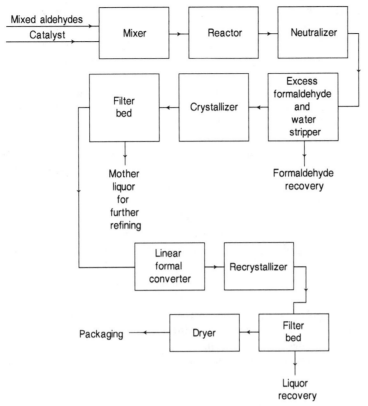

Fig. 2. Flow diagram for pentaerythritol production.

Dipentaerythritol and tripentaerythritol are obtained as by-products of the pentaerythritol process and may be further purified by fractional crystallization or extraction. Trimethylolethane and trimethylolpropane may be prepared by a similar aldol–cross Cannizzaro reaction scheme using propionaldehyde or butyraldehyde, respectively (58) in place of acetaldehyde. Formaldehyde and catalyst requirements are somewhat reduced because of the lower hydrogen content. Ditrimethylolpropane is obtained as a by-product of the trimethylolpropane synthesis (59).

Economic Aspects

Production of pentaerythritol in the United States has been erratic. Demand decreased in 1975 because of an economic recession and grew only moderately to 1980 (69). The range of uses for pentaerythritol has grown rapidly in lubricants (qv), fire-retardant compositions, adhesives, and other formulations where the cross-linking capabilities are of critical importance.

The world's largest producers are Perstorp AB (Sweden, United States, Italy), Hoechst Celanese Corporation (United States, Canada), Degussa (Germany), and Hercules (United States) with estimated 1989 plant capacities of 65,000, 59,000, 30,000, and 22,000 t/yr, respectively. Worldwide capacity for pentaerythritol production was 316,000 t in 1989, about half of which was from the big four companies. Most of the remainder was produced in Asia (Japan, China, India, Korea, and Taiwan), Europe (Italy, Spain), or South America (Brazil, Chile). The estimated rate of production for 1989 was about 253,000 t or about 80% of nameplate capacity.

The world's largest producers of trimethylolpropane are Perstorp AB at 50,000 t/yr, Hoechst Celanese at 23,000 t/yr, and Bayer at 20,000 t/yr. Estimated worldwide capacity is 139,000 t and actual production is on the order of 88,000 t.

Dipentaerythritol is sold by Perstorp AB and by Hercules (United States), ditrimethylolpropane by Perstorp AB both in relatively pure form. Tripentaerythritol is also available; however, the purity is limited. Trimethylolethane is produced commercially by Alcolac (United States) and Mitsubishi Gas Chemicals (Japan).

Pentaerythritol is produced in a variety of grades having differing amounts of dipentaerythritol and small quantities of linear formal. Mono pentaerythritol contains a minimum of 98.0% pentaerythritol with most of the remaining material being dipentaerythritol. Nitration-grade pentaerythritol composition is dependent on the individual customer's demands. The product may be highly pure (>99.5% pentaerythritol) or may contain up to about 1.6% dipentaerythritol. Technical-grade pentaerythritols generally contain at least 8% dipentaerythritol and the normal limit is about 12%, although some specialty products may be even higher. Pure dipentaerythritol may be added to standard technical-grade pentaerythritol. Tripentaerythritol is also present in most technical-grade product.

Analysis

Pentaerythritol may be analyzed by nonspecific wet chemical means such as the hydroxyl method, the results of which include all the usual impurities such as dipentaerythritol, tripentaerythritol, and the formals (70), or the benzal method. A number of gas chromatographic methods allowing facile analysis of the more volatile ethers and esters formed by a number of reagents as well as simultaneous determination of other normal constituents are also available. Acetate esters (71), trimethylsilyl ethers (72), and trifluoroacetate esters (73) give highly satisfactory analyses. ASTM methods D2195 (wet chemical) and D2999 (gas chromatographic) are recognized standards for industrial use.

Health and Safety Factors

Pentaerythritol and trimethylolpropane are classified as nuisance particulate and dust, respectively. They are both nontoxic to animals by ingestion or inhalation and are essentially nonirritating to the skin or eyes (2,74).

Uses

The most important industrial use of pentaerythritol is in a wide variety of paints, coatings, and varnishes, where the cross-linking capability of the four hydroxy groups is critical. Alkyd resins (qv) are produced by reaction of pentaerythritol with organic acids such as phthalic acid or maleic acid and natural oil species.

The resins obtained using pentaerythritol as the only alcohol group supplier are noted for high viscosity and fast drying characteristics. They also exhibit superior water and gasoline resistance, as well as improved film hardness and adhesion. The alkyd resin properties may be modified by replacing all or part of the pentaerythritol by glycols, glycerol, trimethylolethane, or trimethylolpropane, thereby reducing the functionality. Similarly, replacing the organic acid by longer-chain acids such as adipic or altering the quantities of the oil components (linseed, soya, etc) modifies the drying, hardness, and wear characteristics of the final product. The catalyst and the actual cooking procedures also significantly affect overall resin characteristics.

Rosin esters of pentaerythritol prepared from varying amounts of oil yield varnishes of the required oil length for a wide range of outlets such as wood and metal finishing, sealing and jointing formulations, and sand binders for molds. Formulations for some of these complex resins have been described in great detail in the literature, although exact production details are normally guarded jealously by the manufacturers because the order of reaction can also alter the properties of the final product (75–78).

Long-chain esters of pentaerythritol have been used as pour-point depressants for lubricant products, ranging from fuel oils or diesel fuels to the high performance lubricating oils required for demanding outlets such as aviation, power turbines, and automobiles. These materials require superior temperature, viscosity, and aging resistance, and must be compatible with the wide variety of metallic surfaces commonly used in the outlets (79–81).

The explosives and rocket fuels formed by nitration of pentaerythritol to the tetranitrate using concentrated nitric acid (33) are generally used as a filling in detonator fuses. Use of pentaerythritol containing small amounts of dipentaerythritol produces crystallization behavior that results in a high bulk density product having excellent free-flow characteristics, important for fuse burning behavior (82). PETN is also used for medicinal purposes as a vasodilator in the treatment of angina. This product is a dry mixture of pure PETN and an inert carrier, for example, lactose or mannitol, to minimize the usual explosive potential. For the same reason, only small quantities are normally shipped and rigorous packaging is recommended (83).

Pentaerythritol is used in self-extinguishing, nondripping, flame-retardant compositions with a variety of polymers, including olefins, vinyl acetate and alcohols, methyl methacrylate, and urethanes. Phosphorus compounds are added to the formulation of these materials. When exposed to fire, a thick foam is produced, forming a fire-resistant barrier (see FIRE-EXTINGUISHING AGENTS; FLAME RETARDANTS) (84–86).

Polymer compositions containing pentaerythritol are also used as secondary heat-, light-, and weather-resistant stabilizers with calcium, zinc, or barium salts, usually as the stearate, as the prime stabilizer. The polymers may be in plastic or fiber form (87–89).

Pentaerythritol in rosin ester form is used in hot-melt adhesive formulations, especially ethylene–vinyl acetate (EVA) copolymers, as a tackifier. Polyethers of pentaerythritol or trimethylolethane are also used in EVA and polyurethane adhesives, which exhibit excellent bond strength and water resistance. The adhesives may be available as EVA melts or dispersions (90,91) or as thixotropic, one-package, curable polyurethanes (92). Pentaerythritol spiro ortho esters have been used in epoxy resin adhesives (93). The EVA adhesives are especially suitable for cellulose (paper, etc) bonding.

Pentaerythritol and trimethylolpropane acrylic esters are useful in solventless lacquer formulations for radiation curing (qv), providing a cross-linking capability for the main film component, which is usually an acrylic ester of urethane, epoxy, or polyester. Some specialty films utilize dipentaerythritol and ditrimethylolpropane (94,95).

Titanium dioxide pigment coated with pentaerythritol, trimethylolpropane, or trimethylolethane exhibits improved dispersion characteristics when used in paint or plastics formulations. The polyol is generally added at levels of 0.1–0.5% (96).

Photocurable materials for photographic films contain pentaerythritol and dialkylamino and/or nitrile compounds, which have good adhesion and peelability of the layers, and produce clear transfer images (97,98).

Electroless plating on metal substrates can be improved by addition of pentaerythritol, either to a photosensitive composition of a noble metal salt (99), or with glycerine to nickel plating solutions (100). Both resolution and covering power of the electrolyte are improved.

Binary mixtures of pentaerythritol with many other polyol species can produce heat or electrical storage media having excellent retention properties based on the solid–solid, crystal–plastic–crystal phase-transition phenomena. The solid solutions of these mixtures can also be tailored, by adjusting the ratios of the individual components, to exhibit phase transformation and energy release under the required conditions. These media are applicable to domestic hot water, solar heat, and industrial process heat storage and recovery (101–103).

BIBLIOGRAPHY

"Pentaerythritol" in *ECT* 1st ed., Vol. 10, pp. 1–6, by H. Weinberger, Fine Organics, Inc., "Other Polyhydric Alcohols" under "Alcohols, Polyhydric," in *ECT* 2nd ed., Vol. 1, pp. 588–598, by E. Berlow, Heyden Newport Chemical Corp.; in *ECT* 3rd ed., Vol. 1, pp. 778–789, by J. Weber, Celanese Canada, Ltd., and J. Daley, Celanese Chemical Company.

1. E. Berlow, R. H. Barth, and E. J. Snow, *The Pentaerythritols,* Reinhold Publishing Corp., New York, 1958.
2. *Material Safety Data Sheet C41, Trimethylolpropane, Flake,* Celanese Canada Inc., Montreal, 1990.
3. I. Harttmann and J. Nagy, *U.S. Bureau of Mines Report of Investigation No. 3751,* 1944.
4. *Product Bulletin PE-10-55,* Heyden Newport Chemical Corp.
5. *Technical Data Sheets TDS-1 and TB-1,* Commercial Solvents Corp.
6. *Product Bulletin, Pentaerythritol,* Sam Yang Chemical Co., Korea.
7. *Physical Properties Manual,* Celanese Chemical Company, Dallas, Tex., 1981.
8. N. S. Bankar, P. N. Chaudhari, and G. H. Kulkarni, *Indian J. Chem.* **13,** 986–987 (1975).
9. Jpn. Pat. 53/63,306 (June 6, 1978), K. Takeuchi, Y. Matsui, H. Kano, and T. Motohashi (to Ajinomoto Co., Ltd.).
10. A. B. Padias and H. K. Hall, Jr., *Macromolecules* **15,** 217–223 (1982).

11. U.S. Pat. 4,405,798A (Sept. 20, 1983), H. K. Hall and D. R. Wilson (to Celanese Corp.).
12. Brit. Pat. 1,374,263 (Nov. 20, 1974), T. Keating (to Imperial Chemical Industries Ltd.).
13. T. S. Chao, M. Kjonaas, and J. DeJovine, *Preprints Div. Pet. Chem. Am. Chem. Soc.* **24,** 836–846 (1979).
14. U.S. Pat. 3,932,460 (Jan. 13, 1976), D. R. McCoy, T. B. Jordan, and F. K. Ward (to Texaco Inc.).
15. Ger. Pat. 2,758,780 (July 12, 1979), K. H. Hentschel, R. Dhein, H. Rudolph, K. Nuetzel, K. Morche, and W. Krueger (to Bayer A.G.).
16. Jpn. Pat. 57/38,767 A2 (Mar. 3, 1982), (to Asahi Denka Kogyo K.K.).
17. Jpn. Pat. 57/11,959 A2 (Jan. 21, 1982), (to Asahi Denka Kogyo K.K.).
18. Brit. Pat. 1,439,753 (June 16, 1976), H. Coates, J. D. Collins, and I. H. Siddiqui (to Albright and Wilson Ltd.).
19. Jpn. Pat. 52/100,415 (Aug. 23, 1977), T. Kiyoura (to Mitsui Toatsu Chemicals, Inc.).
20. Jpn. Pat. 52/72,882 (June 17, 1977), T. Ooe, T. Nakazato, and R. Yoshikawa (to Mitsubishi Gas Chemical Co., Inc.).
21. Ger. Pat. 2,440,612 (Mar. 6, 1975), Y. Christidis (to Nobel Hoechst Chimie).
22. Ger. Pat. 3,125,338 A1 (Jan. 13, 1983), K. Koenig and M. Schmidt (to Bayer A.G.).
23. A. Kamors, *Tezisy Dokl. Konf. Molodykh Nauchn. Rab. Inst. Neorg. Khim., Akad. Nauk Latv. SSR, 5th,* pp. 17–20.
24. Czech. Pat. 190,732 B (Dec. 15, 1981), J. Holcik, M. Karvas, and J. Masek.
25. Eur. Pat. Appl. EP 113,994 A1 (July 25, 1984), B. E. Johnston and P. R. Napier (to Mobil Oil Corp.).
26. Jpn. Pat. 61,225,191 A2 (Oct. 6, 1986), K. Tajima, M. Takahashi, K. Nishikawa, and T. Takeuchi (to Adeka Argus Chemical Co., Ltd.).
27. Ger. Pat. 2,630,257 (Jan. 12, 1978), H. Habarlein and F. Scheidl (to Hoechst A.G.).
28. U.S. Pat. 4,454,064 A (June 12, 1984), Y. Halpern and R. H. Niswander (to Borg-Warner Corp.).
29. Fr. Pat. 2,489,333 A1 (Mar. 5, 1982), J. Kiefer (to Ciba-Geigy A.G.).
30. U.S. Pat. 4,152,373 (May 1, 1979), M. L. Honig and E. D. Weil (to Stauffer Chemical Co.).
31. L. Desvergnes, *Chim. Ind.* **29,** 1263 (1933).
32. Jpn. Pat. 51/118,708 (Oct. 18, 1976), S. Oinuma and M. Kusakabe (to Agency of Industrial Sciences and Technology).
33. Czech Pat. 221,031 B (Mar. 15, 1986), S. Zeman, M. Dimun, and Z. Cervenka.
34. U.S. Pat. 4,352,919 A (Oct. 5, 1982), E. W. Kluger and C. D. Welch (to Milliken Research Corp.).
35. U.S. Pat. 4,136,044 (Jan. 23, 1979), M. Braid (to Mobil Oil Corp.).
36. W. S. Anderson and H. J. Hyer, *JANNAF Propul. Meet.,* vol. 1, AD-A103 844, CPIA Publ. 340, 1981, pp. 387–398.
37. Jpn. Pat. 56/55,362 (May 15, 1981), (to Teijin Ltd.).
38. U.S. Pat. Appl. 183,707 (Jan. 30, 1981), (to United States National Aeronautics and Space Administration).
39. USSR Pat. 1,035,025 A1 (Aug. 15, 1983), E. B. Smirnova, Z. L. Chilyasova, N. P. Zykova, and L. A. Druganova (to Central Scientific-Research and Design Institute of the Wood Chemical Industry).
40. Ger. Pat. 2,707,875 (Aug. 31, 1978), J. Perner, K. Stork, F. Merger, and K. Oppenlaender (to BASF A.G.).
41. S. Shimizu, Y. Sasaki, and C. Hirai, *Nihon Daigaku Seisankogakubu Hokoku, A,* **15**(1), 47–56 (1982).
42. U.S. Pat. 4,151,211 (Apr. 24, 1979), I. Heckenbleikner and W. P. Enlow (to Borg-Warner Corp.).
43. Ger. Pat. 2,501,285 (July 24, 1975), A. Schmidt (to Ciba-Geigy A.G.).

44. Belg. Pat. 885,670 (Feb. 2, 1981), R. Leger, R. Nouguier, J. C. Fayard, and P. Maldonado (to Elf France).

45. Eur. Pat. Appl. EP 46,731 A1 (Mar. 3, 1982), F. Lohse and C. E. Monnier (to Ciba-Geigy A.G.).

46. Jpn. Pat. 63,162,641 A2 (July 6, 1988), G. Watanabe, N. Nakajima, and Y. Ito (to Yokkaichi Chemical Co., Ltd.).

47. Jpn. Pat. 62,223,141 A2 (Oct. 1, 1987), Y. Fujio, Y. Nishi, and T. Nishimoto (to Osaka Soda Co., Ltd.).

48. E. Weber, *J. Org. Chem.* **47,** 3478–3486 (1982).

49. A. B. Padias, H. K. Hall, Jr., D. A. Tomalia, and J. R. McConnell, *J. Org. Chem.* **52,** 5305–5312 (1987).

50. J. E. Vik, *Acta Chem. Scand.* **27,** 239 (1973).

51. T. G. Bonner, E. J. Bourne, and J. Butler, *J. Chem. Soc.,* 301 (1964).

52. J. E. Vik, *Acta Chem. Scand. B* **28,** 325 (1974).

53. P. Werle, E. Busker, and E. Wolf-Heuss, *Liebigs Ann. Chem.* 1082–1087 (1985).

54. D. I. Belkin, *Zh. Prikl. Khim. (Leningrad)* **52,** 237–239 (1979).

55. L. Kovacic-Beck, I. Beck, and F. Anusic, *Nafta (Zagreb)* **34**(3), 131–135 (1983).

56. M. Lichvar, J. Sabados, V. Macho, and L. Komora, *Chem. Prum.* **36**(2), 57–61 (1986).

57. U.S. Pat. 2,612,526 (Sept. 30, 1952), C. W. Gould (to Hercules Powder Co.).

58. Jpn. Pat. 57/139,028 A2 (Aug. 27, 1982), (to Koei Kagaku Kogyo K.K.).

59. U.S. Pat. 3,962,347 (June 8, 1976), K. Herz (to Perstorp AB).

60. Ger. Pat. 2,930,345 (Feb. 19, 1981), P. Werle and G. Pohl (to Degussa).

61. Czech Pat. 183,405 (May 15, 1980), L. Komora and H. Nitoschneiderova.

62. Span. Pat. 467,714 (Oct. 16, 1978), (to Patentes y Novedades S.A.).

63. Eur. Pat. Appl. EP 242,784 A1 (Oct. 28, 1987), H. V. Holmberg and H. E. Larsson (to Perstorp AB).

64. Czech Pat. 220,256 B (Feb. 15, 1986), J. Ziak.

65. Jpn. Pat. 5,808,028 (Jan. 18, 1983), (to Koei Kagaku Kogyo K.K.).

66. Czech Pat. 181,486 (Jan. 15, 1980), M. Lichvar, J. Sabados, J. Vidovenec, and L. Butkovsky.

67. Jpn. Pat. 57/142,929 A2 (Sept. 3, 1982), (to Koei Kagaku Kogyo K.K.).

68. V. V. Pakulin, A. A. Kruglikov, Y. V. Rogachev, and P. E. Gulevich, *Plast. Massy* **3,** 12–13 (1988).

69. *Synthetic Organic Chemicals,* Publication 776, U.S. International Trade Commission.

70. *Technical Bulletin, Pentaerythritol,* Oxyquim S.A.

71. D. S. Wiersma, R. E. Hoyle, and H. Rempis, *Anal. Chem.* **34,** 1533 (1962).

72. R. R. Suchanec, *Anal. Chem.* **37,** 1361 (1965).

73. *Technical Bulletin, Pentaerythrit,* Degussa.

74. *Material Safety Data Sheet C73, Pentaerythritol,* Celanese Canada Inc., 1990.

75. Belg. Pat. 874,241 (June 18, 1979), (to BASF Farben und Fasern A.G.).

76. Jpn. Pat. 55/82,165 (June 20, 1980), (to Nippon Synthetic Chemical Industry Co., Ltd.).

77. U.S. Pat. 4,690,783 A (Sept. 1, 1987), R. W. Johnston, Jr. (to Union Camp Corp.).

78. Czech Pat. 222,510 B (Mar. 15, 1986), K. Hajek and co-workers.

79. Neth. Pat. 77/11852 (May 2, 1979), (to Hercules Inc.).

80. U.S. Pat. 4,229,310 (Oct. 21, 1980), G. Frangatos (to Mobil Oil Corp.).

81. Ger. Pat. 3,328,739 (Feb. 9, 1984), C. J. Dorer and K. Hayashi (to Lubrizol Corp.).

82. Ger. Pat. 1,901,769 (Oct. 9, 1967), H. Thomas and J. Turbet (to Imperial Chemical Industries Ltd.).

83. *U.S. Pharmacopeia: National Formulary XV,* 20th ed., Mack Publishing Co., Easton, Pa., 1980.

84. Eur. Pat. Appl. 17,609 (Oct. 15, 1980), D. Alt, K. D. Hutschgau, A. Schillmoeller, and B. Fischer (to Siemens A.G.).

85. U.S. Pat. 4,762,746 A (Aug. 9, 1988), L. Wesch and E. Weiss (to Odenwald-Chemie GmbH).

86. J. P. Jain, N. K. Saxena, I. Singh, and D. R. Gupta, *Res. Ind.* **30**(1), 20–24 (1985).

87. Ger. Pat. 2,847,628 (May 31, 1979), B. Sallmen, C. A. Sjoegreen, M. O. Maansson, and K. Ogemark (to Perstorp A.B.).

88. U.S. Pat. 4,162,242 (July 24, 1979), R. House (to Chevron Research Co.).

89. Eur. Pat. 219,427 A1 (Apr. 11, 1987), C. D. G. deMezeyrac, R. Fugier, B. Gicquel, and S. Tetard (to Isover Saint-Gobain).

90. Fr. Pat. 2,378,835 (Aug. 25, 1978), (to E. I. du Pont de Nemours & Co., Inc.).

91. Ger. Pat. 3,410,957 A1 (Sept. 26, 1985), K. J. Gardenier and W. Heimbuerger (to Henkel K.-G.a.A.).

92. Belg. Pat. 890,841 A2 (Feb. 15, 1982), S. B. Labelle and J. A. E. Hagquist (to Fuller H.B. Co.).

93. Jpn. Pat. 61,027,987 A2 (Feb. 7, 1986), A. Matsumaya, H. Ozawa, and S. Hirose (to Mitsui Toatsu Chemicals, Inc.).

94. R. Holman, ed., *UV and EB Curing Formulations for Printing Inks, Coatings and Paints,* SITA Technology, London, 1984.

95. C. G. Roffey, *Photopolymerization of Surface Coatings,* John Wiley & Sons, Inc., New York, 1982.

96. Brit. Pat. 896,067 (Aug. 19, 1960), W. R. Whately and G. M. Sheehan (to American Cyanamid Co.).

97. Jpn. Pat. 54/95,688 (July 28, 1979), S. Kondo, A. Matsufuji, and A. Umehara (to Fuji Photo Film Co., Ltd.).

98. Jpn. Pat. 60,237,444 A2 (Nov. 26, 1985), T. Komamura, M. Iwagaki, and T. Masukawa (to Konishiroku Photo Industry Co., Ltd.).

99. Jpn. Pat. 63,115,395 A2 (May 19, 1988), Y. Takeuchi (to Canon K.K.).

100. USSR Pat. 1,310,460 A1 (May 15, 1987), Y. Y. Lukomski, T. V. Mulina, V. V. Vasiliev, and R. V. Kopteva.

101. Jpn. Pat. 61,204,292 A2 (Sept. 10, 1986), S. Anzai, H. Sakaguchi, H. Yamazaki, K. Shiina, and M. Kurodi (to Hitachi, Ltd.).

102. Jpn. Pat. 61,240,627 A2 (Oct. 25, 1986), Y. Kudo, S. Yoshimura, S. Tsuchiya, and T. Kojima (to Matsushita Electric Industrial Co., Ltd.).

103. U.S. Pat. 4,572,864 A (Feb. 25, 1986), D. K. Benson, R. W. Burrows, and Y. D. Shinton (to United States Dept. of Energy).

WILLIAM N. HUNTER
Celanese Canada Inc.

ALCOHOLS, POLYHYDRIC, SUGAR ALCOHOLS. See SUGAR ALCOHOLS.

ALCOHOLS, UNSATURATED. See ACETYLENE-DERIVED CHEMICALS.

ALDEHYDE RESINS. See ACETAL RESINS.

ALDEHYDES

Aldehydes are carbonyl-containing organic compounds in which the carbonyl function is at a terminal carbon. These compounds are extremely reactive. The carbonyl group is susceptible to both oxidation and reduction, yielding acids and alcohols respectively. Additionally, the carbonyl group is susceptible to nucleophilic addition, providing a means by which to form new chemical bonds. Furthermore, the presence of the carbonyl activates the hydrogens bound to the alpha carbon and thus provides an additional site of reactivity. Ketones are a similar class of compounds where the carbonyl group is nonterminal (see KETONES).

Nomenclature

The common method of naming aldehydes corresponds very closely to that of the related acids (see CARBOXYLIC ACIDS), in that the term *aldehyde* is added to the base name of the acid. For example, formaldehyde (qv) comes from formic acid, acetaldehyde (qv) from acetic acid, and butyraldehyde (qv) from butyric acid. If the compound contains more than two aldehyde groups, or is cyclic, the name is formed using *carbaldehyde* to indicate the functionality. The IUPAC system of aldehyde nomenclature drops the final *e* from the name of the parent acyclic hydrocarbon and adds *al*. If two aldehyde functional groups are present, the suffix *-dial* is used. The prefix *formyl* is used with polyfunctional compounds. Examples of nomenclature types are shown in Table 1.

Table 1. Aldehyde Nomenclature

Structural formula	Name	CAS Registry Number
H—CHO	formaldehyde or methanal	[50-00-0]
CH_3CHO	acetaldehyde or ethanal	[75-07-0]
$(CH_3)_2CHCHO$	isobutyraldehyde or 2-methylpropanal or α-methylpropionaldehyde	[78-84-2]
$CH_3CH{=}CHCHO$	crotonaldehyde or 2-butenal	[4170-30-3]
$CH_3CH_2\overset{\underset{\mid}{CH_3}}{C}HCHO$	2-methylbutyraldehyde or 2-methylbutanal	[96-17-3]
$OHCCH_2CH_2CH_2CHO$	glutaraldehyde or pentandial	[111-30-8]
$OHCCH_2\overset{\underset{\mid}{CHO}}{C}HCH_2CHO$	1,2,3-propanetricarbaldehyde or formylpentandial	[61703-13-7]
$OHCCH_2CH_2CH_2COOH$	4-formylbutanoic acid	[5746-02-1]
$\overline{CH_2CH_2CH_2CH_2CH}CHO$	formylcyclopentane or cyclopentanecarbaldehyde	[872-53-7]

Physical Properties

The C_1 and C_2 carbon aliphatic aldehydes, formaldehyde and acetaldehyde are gases at ambient conditions whereas the C_3 (propanal [123-38-6]) through C_{11} (undecanal [112-44-7]) aldehydes are liquids, and higher aldehydes are solids at room temperature. As can be seen from Table 2, the presence of hydrocarbon branching tends to lower the boiling or melting point, as does unsaturation in the carbon skeleton. Generally, an aldehyde has a boiling point between those of the corresponding alkane and alcohol. Aldehydes are usually soluble in common organic solvents and, except for the C_1 to C_5 aldehydes, are only sparingly soluble in water. The lower, C_1 to C_8, aldehydes have pungent, penetrating, unpleasant odors, some of which may be attributed to the presence of the corresponding acids that form by air oxidation. Above C_8, aldehydes have more pleasant odors and some higher aldehydes are used in the perfume and flavoring industries (see PERFUMES). Interestingly, the C_9 aldehyde, nonanal [124-19-6], is reported to possibly be a human sex pheromone (1). Aldehydes must be kept from contact with air (oxygen) to retain purity.

   Spectroscopic Properties.   Characteristic aldehyde absorptions are summarized in Table 3. A carbonyl stretching frequency between 1720 and 1740 cm^{-1} in saturated compounds is lowered in unsaturated ones where there is conjugation with the carbon–oxygen double bond. The carbonyl group also exhibits a weak ultraviolet absorption near 280 nm as a result of the excitation of one of the unshared electrons on the oxygen atom. Both proton (^{1}H-nmr) and carbon-13 (^{13}C-nmr) nmr spectra are often highly descriptive. The proton adjacent to the carbonyl exhibits a strong downfield (about 10 ppm) resonance relative to the standard tetramethylsilane, TMS, as does the carbonyl carbon which appears at about 200 ppm, again relative to TMS. The low field nmr resonances and the intensity of the stretching frequency are attributable to the polar nature of the carbon–oxygen double bond.

Chemical Properties

Aldehydes are very reactive compounds. Reactions generally fall into two classes: those directly affecting the carbonyl group and those involving the adjacent carbon atom. The polar nature of the carbonyl moiety lends itself to nucleophilic addition and also controls the reactivity of the alpha carbon atom by rendering the hydrogens attached to this carbon relatively acidic. Species having acidic hydrogens are referred to as active methylene compounds and must be protected against inadvertent contact with bases as such contact results in a highly exothermic condensation reaction that may become dangerous. There are many available references offering detailed descriptions of aldehyde reactions (2–6). Many of the reactions involving aldehydes have been named for their discoverers and comprehensive reviews of name reactions are also available (7–9).

   Reduction Reactions.   Aldehydes can be hydrogenated to the corresponding alcohol using a heterogeneous catalyst, for example

$$CH_3CH_2CH_2CHO \xrightarrow{H_2} CH_3CH_2CH_2CH_2OH$$

Table 2. Properties of Aldehydes

Aldehyde	CAS Registry Number	Molecular formula	Molecular weight	Melting point, °C	Boiling point, °C	Solubility, g/100g water
formaldehyde	[50-00-0]	CH_2O	30.03	−92	−21	v sol
acetaldehyde	[75-07-0]	C_2H_4O	40.05	−121	20	v sol
propionaldehyde	[123-38-6]	C_3H_6O	58.08	−81	49	16
butanal (n-butyraldehyde)	[123-72-8]	C_4H_8O	72.1	−99	75	7
2-methylpropanal (isobutyraldehyde)	[78-84-2]	C_4H_8O	72.1	−66	64	11
pentanal (n-valeraldehyde)	[110-62-3]	$C_5H_{10}O$	86.13	−91	103	sl s
3-methylbutanal (isovaleraldehyde)	[590-86-3]	$C_5H_{10}O$	86.13	−51	93	sl s
hexanal (caproaldehyde)	[66-25-1]	$C_6H_{12}O$	100.16	−56	131	sl s
benzaldehyde	[100-52-7]	C_7H_6O	106.13	−26	179	0.3
heptanal (heptaldehyde)	[111-71-7]	$C_7H_{14}O$	114.19	−42	155	0.1
octanal (caprylaldehyde)	[124-13-0]	$C_8H_{16}O$	128.21		171	sl s
phenylacetaldehyde	[122-78-1]	C_8H_8O	120.16	33	194	sl s
o-tolualdehyde	[529-20-4]	C_8H_8O	120.14		202	sl s
m-tolualdehyde	[620-23-5]	C_8H_8O	120.14		199	sl s
p-tolualdehyde	[104-87-0]	C_8H_8O	120.14		205	sl s
salicylaldehyde (o-hydroxybenzaldehyde)	[90-02-8]	$C_7H_6O_2$	122.12	2	197	1.7
p-hydroxybenzaldehyde (4-formylphenol)	[123-08-0]	$C_7H_6O_2$	122.12	116		1.4
p-anisaldehyde (p-methoxybenzaldehyde)	[123-11-5]	$C_8H_8O_2$	136.14	0	248	0.2

Table 3. Spectroscopic Absorptions of Aldehydes

Compound	Ir, cm^{-1a}	Uv, nm	^{1}H-nmr, ppm^b	^{13}C-nmr, ppm^c
acetaldehyde	1730	290	9.80	200
butyraldehyde	1725	283	9.74	202
benzaldehyde	1695	278	10.00	192
2-butenal	1700	301	9.48	193

^{a}Carbonyl stretching frequency.
^{b}Aldehyde proton, relative to TMS.
^{c}Carbonyl carbon, relative to TMS.

Common catalyst compositions contain oxides or ionic forms of platinum, nickel, copper, cobalt, or palladium which are often present as mixtures of more than one metal. Metal hydrides, such as lithium aluminum hydride [16853-85-3] or sodium borohydride [16940-66-2], can also be used to reduce aldehydes. Depending on additional functionalities that may be present in the aldehyde molecule, specialized reducing reagents such as trimethoxyaluminum hydride or alkylboranes (less reactive and more selective) may be used. Other less industrially significant reduction procedures such as the Clemmensen reduction or the modified Wolff-Kishner reduction exist as well.

Oxidation Reactions. In general, the aldehyde function is easily oxidized to form the corresponding carboxylic acid.

$$RCHO \xrightarrow[\text{catalyst}]{O_2} RCO_2H$$

This is the basis for many industrially significant preparations, eg, acetaldehyde to acetic acid [64-19-7] (see ACETIC ACID AND DERIVATIVES), propionaldehyde to propionic acid [79-09-47], furfural [98-01-1] to furoic acid [26447-28-9], and acrolein to acrylic acid [79-10-7] (see ACRYLIC ACID AND DERIVATIVES). Air is the common oxidant and homogeneous metal catalysts are used. Alternatively, a variety of oxidizing reagents are available to perform this transformation. Although both chromium and manganese compounds can be used, an aqueous solution of potassium permanganate [7722-64-7] under either acidic or basic conditions is the more commonly employed reagent. Other reagents that have been used for the oxidation of aldehydes include fuming nitric acid and a suspension of silver oxide in aqueous alkali. The latter provides a very mild and selective method for this type of oxidation.

Aldehydes can undergo an intermolecular oxidation–reduction (Cannizzaro reaction) in the presence of base to produce an alcohol and a carboxylic acid salt. Any aldehyde is capable of participating in such a reaction, however, it is more common for those containing no protons on the alpha carbon, for example

$$2\ C_6H_5CHO \xrightarrow{\text{NaOH}} C_6H_5CH_2OH + C_6H_5CO_2Na$$

The Tischenko reaction is a related transformation where the product is an ester.

$$2\ RCH_2CHO \xrightarrow{\text{Al(OCH}_2\text{CH}_3)_3} RCH_2CH_2O\overset{\displaystyle O}{\overset{\displaystyle \|}{C}}CH_2R$$

Aldol Addition and Related Reactions. Procedures that involve the formation and subsequent reaction of anions derived from active methylene compounds constitute a very important and synthetically useful class of organic reactions. Perhaps the most common are those reactions in which the anion, usually called an enolate, is formed by removal of a proton from the carbon atom alpha to the carbonyl group. Addition of this enolate to another carbonyl of an aldehyde or ketone, followed by protonation, constitutes aldol addition, for example

$$2 \ CH_3CHO \xrightarrow{\text{HCl}} CH_3CH(OH)CH_2CHO$$

The name aldol was introduced by Wurtz in 1872 to describe the product resulting from this acid-catalyzed reaction of acetaldehyde. The addition will occur with base catalysis as well.

The β-hydroxy aldehyde which forms during aldol addition is not usually isolated; it readily dehydrates to form an α,β-unsaturated aldehyde. This overall transformation is commercially employed to produce 2-ethyl-2-hexenal [*26266-68-2*] often referred to as ethylpropylacrolein, from butyraldehyde (qv). The ethylpropylacrolein is then hydrogenated to produce 2-ethylhexanol, a commercially significant plasticizer alcohol (See ALCOHOLS, HIGHER ALIPHATIC). The next higher homologue, a C_{10} alcohol mixture containing 2-propylheptanol, can be produced in an analogous fashion from a valeraldehyde product mixture resulting from hydroformylating a mixed butenes stream (see BUTYLENES). This hydroformylation is accomplished using a new generation of highly active phosphite-promoted rhodium catalysts.

Other reactions similar to the aldol addition include the Claisen and Perkin reactions. The Claisen reaction, carried out by combining an aromatic aldehyde and an ester in the presence of metallic sodium, is useful for obtaining α,β-unsaturated esters.

$$C_6H_5CHO + CH_3CO_2C_2H_5 \xrightarrow[\text{0-5°C}]{\text{Na}} C_6H_5CH{=}CHCO_2C_2H_5$$

The Perkin reaction, utilizing an aromatic aldehyde, an acid anhydride, and a base such as an acid salt or amine, produces the corresponding α,β-unsaturated acid.

$$C_6H_5CHO + (CH_3CO)_2O \xrightarrow{(C_2H_5)_3N} C_6H_5CH{=}CHCO_2H + CH_3CO_2H$$

Analogously, aldehydes react with ammonia [*7664-41-7*] or primary amines to form Schiff bases. Subsequent reduction produces a new amine. The addition of hydrogen cyanide [*74-90-8*], sodium bisulfite [*7631-90-5*], amines, alcohols, or thiols to the carbonyl group usually requires the presence of a catalyst to assist in reaching the desired equilibrium product.

Addition of water or alcohols to aldehydes leads to the formation of a class of compounds known as acetals. This is an acid catalyzed reaction

$$RCHO + 2\ R'OH \xrightarrow{\ H^+\ } RCH\begin{smallmatrix}OR'\\\\OR'\end{smallmatrix} + H_2O$$

and the first addition product, a hemiacetal or gem-diol, is unstable, the equilibrium generally favoring the parent aldehyde. Subsequent steps involve protonation of the —OH which leads to a stabilized carbonium ion, followed by addition of a second alcohol molecule. Acetals have been used in racing car fuels, gasoline additives, and paint and varnish solvents and strippers. Acetals of higher aldehydes have fragrances similar to but not so pungent as the parent aldehydes. Because they are not as sensitive to alkalies or autoxidation, acetals find use as fragrances for alkaline substances such as soaps, shampoos and heavy-duty detergents.

Another very important reaction initially involving nucleophilic attack on an aldehyde carbonyl is the Wittig reaction. An ylid adds to the carbonyl forming a betaine intermediate which then decomposes to produce an olefin and a tertiary phosphine oxide.

$$(C_6H_5)_3P + RR'CHBr \xrightarrow{\ n\text{-}C_4H_9Li\ } (C_6H_5)_3\overset{+}{P}-\overset{-}{C}RR'$$

$$(C_6H_5)_3\overset{+}{P}-\overset{-}{C}RR' + R''CH_2CHO \longrightarrow (C_6H_5)_3PO + RR'C{=}CHCH_2R''$$

Perhaps the most notable example of this chemistry is in the production of vitamin A [*68-26-8*], where the β-ionylidenacetaldehyde is condensed with the ester-ylid to obtain the polyene ester. Reduction then yields vitamin A (see VITAMINS).

The reaction has been extended to include carbanions generated from phosphonates. This is often referred to as the Horner-Wittig or Horner-Emmons reaction. The Horner-Emmons reaction has a number of advantages over the conventional Wittig reaction. It occurs with a wider variety of aldehydes and ketones under relatively mild conditions as a result of the higher nucleophilicity of the phosphonate carbanions. The separation of the olefinic product is easier due to the aqueous solubility of the phosphate by-product, and the phosphonates are readily available from the Arbusov reaction. Furthermore, although the reaction itself is not stereospecific, the majority favor the formation of the trans olefin and many produce the trans isomer as the sole product.

A new class of synthetic aldehyde reactions involves effectively reversing the polarity of the carbonyl group by forming a dithiane intermediate.

Deprotonation yields an anion available for alkylation.

Hydrolysis of the resulting dithiane yields a ketone.

Manufacture

Only a few of the wide variety of synthetic procedures which yield aldehydes as products are used for large-scale commercial preparation. A more complete discussion of syntheses can be found in the literature (10,11).

Hydroformylation of an olefin using synthesis gas, the oxo process (qv), was first commercialized in Germany in 1938 to produce propionaldehyde from ethylene and butyraldehydes from propylene (12).

$$RCH{=}CH_2 + CO + H_2 \xrightarrow{\text{catalyst}} \underset{\underset{CHO}{|}}{RCH}{-}CH_3 + RCH_2CH_2CHO$$

Recent advances in technology involving a new class of highly reactive phosphite-promoted catalysts permit the manufacture of higher (C_7 to C_{15}) aldehydes and the hydroformylation of internal olefins at low temperatures and pressures. Fatty aldehydes, generally produced by dehydrogenation of corresponding alcohols in the presence of a suitable catalyst, can now be manufactured using oxo technology.

The direct oxidation of ethylene is used to produce acetaldehyde (qv) in the Wacker-Hoechst process. The catalyst system is an aqueous solution of palladium chloride and cupric chloride. Under appropriate conditions an olefin can be oxidized to form an unsaturated aldehyde such as the production of acrolein [107-02-8] from propylene (see ACROLEIN AND DERIVATIVES).

Another commercial aldehyde synthesis is the catalytic dehydrogenation of primary alcohols at high temperature in the presence of a copper or a copper-chromite catalyst. Although there are several other synthetic processes employed, these tend to be smaller scale reactions. For example, acyl halides can be reduced to the aldehyde (Rosenmund reaction) using a palladium-on-barium sulfate catalyst. Formylation of aryl compounds, similar to hydroformylation, using HCN and HCl (Gatterman reaction) or carbon monoxide and HCl (Gatterman-Koch reaction) can be used to produce aromatic aldehydes.

$$ArH + CO \xrightarrow[\text{HCl}]{Cu_2Cl_2,\ AlCl_3} ArCHO$$

Additionally, Grignard reagent reacts with an alkyl orthoformate to form an acetal which is then hydrolyzed to the corresponding aldehyde using dilute acid.

$$CH_3CH_2MgX + HC(OR)_3 \longrightarrow CH_3CH_2CH(OR)_2 \longrightarrow CH_3CH_2CHO + ROH$$
$$+$$
$$Mg(OR)X$$

Production and Economic Aspects

As can be seen from Table 4, formaldehyde is clearly the most commerically significant of the aldehydes. Only a small selection of the less commercially significant products are listed, however.

Table 4. Production of Aldehydes

Aldehyde	Production, 10^3 t/yr				1990 prices, \$/kg
	U.S./Canada	Europe	East Asia	*Total*	
formaldehyde	4900	6200	2900	*14000*	0.23
acetaldehyde	340	740	650	*1730*	1.02
propionaldehyde	140	100	100	*340*	0.83
butyraldehyde	1500	1500	1000	*4000*	0.84
phenylacetaldehyde					9.90
salicylaldehyde				*3*	9.24
p-anisaldehyde				*2*	0.20

Table 5 contains a listing of representative aldehyde producers. Only the products identified in Table 4 have been included.

Characterization

Aldehydes can be characterized qualitatively through the use of Tollens' or Fehling's reagents as well as by spectroscopic means. The use of Tollens' reagent, a solution of silver nitrate in dilute, basic, aqueous ammonia, leads to the deposition of a silver mirror in the presence of an aldehyde. Fehling's reagent, an aqueous solution of copper sulfate, sodium potassium tartrate, and sodium hydroxide, produces a reddish-brown precipitate of cuprous oxide in the presence of an aldehyde. Additionally, aldehyde carbonyl groups can be derivatized. Aldehyde oximes, phenylhydrazones, 2,4-dinitrohydrazones, semicarbazones, or sodium bisulfite addition products can be generated, purified, and characterized by their distinctive melting points. Hydrazone derivatives are often useful in isolating the aldehyde as a solid, crystalline material. The carbonyl group may also be oxidized to an acid through the use of hydrogen peroxide or potassium permanganate and the resultant carboxylic acid characterized by its spectroscopic properties and derivatives.

Table 5. Aldehyde Producers

Company	Aldehyde product
	United States and Canada
Aristech Chemical Corp.	butyraldehyde
BASF Corp.	butyraldehyde
Borden Chemical	formaldehyde
E. I. du Pont de Nemours	formaldehyde
Eastman	acetaldehyde, propionaldehyde, butyraldehyde
Georgia-Pacific Corp.	formaldehyde
Givauden Corp.	phenylacetaldehyde, *p*-anisaldehyde
Hoechst Celanese	formaldehyde, acetaldehyde, propionaldehyde, butyraldehyde
Koch Industries	*p*-anisaldehyde
Penta Manufacturing	*p*-anisaldehyde
Reichhold Limited	formaldehyde
Rhône–Polenc Inc.	salicylaldehyde
Union Carbide	propionaldehyde, butyraldehyde
	Europe
BASF	formaldehyde, propionaldehyde, butyraldehyde, *p*-anisaldehyde
Bayer	salicylaldehyde
Borden Chemical	formaldehyde
Chemie Linz	phenylacetaldehyde
Degussa	formaldehyde
Givauden	phenylacetaldehyde, *p*-anisaldehyde
Hoechst AG	acetaldehyde, butyraldehyde
Hüls	acetaldehyde, butyraldehyde
Lonza	acetaldehyde
Neste Oxo AB	butyraldehyde
NORSOLOR	formaldehyde
Oxochimie	butyraldehyde
	East Asia
Chisso Corp.	butyraldehyde
Daicel Chemical	phenylacetaldehyde
Kyowa Yuka	acetaldehyde, butyraldehyde
Lucky Ltd.	butyraldehyde
Midori Kagaki Co.	*p*-anisaldedyde
Mitsubishi Kasei	formaldehyde, acetaldehyde, propionaldehyde, butyraldehyde
Mitsui	formaldehyde, acetaldehyde
Ogawa Co.	*p*-anisaldehyde
Seimi Chemical Co., Ltd.	salicylaldehyde
Showa Denko K. K.	acetaldehyde, phenylacetaldehyde
Showa High Polymer	formaldehyde
Soda Aromatic	*p*-anisaldehyde
Sumitomo	formaldehyde
Tokuyama Petrochemical	acetaldehyde
Tonen Corp.	butyraldehyde

Toxicology

Interest in the toxicity of aldehydes has focused primarily on specific compounds, particularly formaldehyde, acetaldehyde, and acrolein (13). Little evidence exists to suggest that occupational levels of exposure to aldehydes would result in mutations, although some aldehydes are clearly mutagenic in some test systems. There are, however, acute effects of aldehydes.

Irritation and Sensitization.   Low molecular weight aldehydes, the halogenated aliphatic aldehydes, and unsaturated aldehydes are particularly irritating to the eyes, skin, and respiratory tract. The mucous membranes of nasal and oral passages and the upper respiratory tract can be affected, producing a burning sensation, an increased ventilation rate, bronchial constriction, choking, and coughing. If exposures are low, the initial discomfort may abate after 5 to 10 minutes but will recur if exposure is resumed after an interruption. Furfural, the acetals, and aromatic aldehydes are much less irritating than formaldehyde and acrolein. Reports of sensitization reactions to formaldehyde are numerous.

Anesthesia.   Materials that have unquestionable anesthetic properties are chloral hydrate [302-17-0], paraldehyde, dimethoxymethane [109-87-5], and acetaldehyde diethyl acetal. In industrial exposures, however, any action as an anesthesia is overshadowed by effects as a primary irritant, which prevent voluntary inhalation of any significant quantities. The small quantities which can be tolerated by inhalation are usually metabolized so rapidly that no anesthetic symptoms occur.

Organ Pathology.   The principal pathology experimentally produced in animals exposed to aldehyde vapors is that of damage to the respiratory tract and pulmonary edema. In general, the aldehydes are remarkably free of actions that lead to definite cumulative organic damage to tissues. Thus the aldehydes cannot generally be regarded as potent carcinogens. Moreover, the intolerable irritant properties of the compounds preclude substantial worker exposure under normal conditions.

There is a significant difference in the toxicological effects of saturated and unsaturated aliphatic aldehydes. As can be seen in Table 6, the presence of the double bond considerably enhances toxicity. The precautions for handling

Table 6. Effect of Unsaturation on Toxicity of Aldehydes

Compound	Formula	LC_{50}, ppm[a]	LD_{50}, mg/kg[b]	TWA[c]
acetaldehyde	CH_3CHO	20000	1930	200
propionaldehyde	CH_3CH_2CHO	26000	1410	
acrolein	$CH_2{=}CHCHO$	130	25.9	0.1
isobutyraldehyde	$(CH_3)_2CHCHO$	$>8,000^d$	2810	
methacrolein	$CH_2{=}C(CH_3)CHO$	250^d	111	
n-butyraldehyde	$CH_3(CH_2)_2CHO$	60000	2490	
crotonaldehyde (2-butenal)	$CH_3CH{=}CHCHO$	1400	260	2

[a]In rats, an exposure time of 30 min.
[b]In rats, dosage administered orally.
[c]OSHA PEL.
[d]Exposure time of 4 h.

reactive unsaturated aldehydes such as acrolein, methacrolein [78-85-3], and crotonaldehyde (qv) should be the same as those for handling other highly active eye and pulmonary irritants, as, for example, phosgene.

Chemical safety data sheets for individual compounds should be consulted for detailed information. Precautions for the higher aldehydes are essentially those for most other reactive organic compounds, and should include: adequate ventilation in areas where high exposures are expected; fire and explosion precautions; and proper instruction of employees in use of respiratory, eye, and skin protection.

Uses

Aldehydes find the most widespread use as chemical intermediates. The production of acetaldehyde, propionaldehyde, and butyraldehyde as precursors of the corresponding alcohols and acids are examples. The aldehydes of low molecular weight are also condensed in an aldol reaction to form derivatives which are important intermediates for the plasticizer industry (see PLASTICIZERS). As mentioned earlier, 2-ethylhexanol, produced from butyraldehyde, is used in the manufacture of di(2-ethylhexyl) phthalate [117-87-7]. Aldehydes are also used as intermediates for the manufacture of solvents (alcohols and ethers), resins, and dyes. Isobutyraldehyde is used as an intermediate for production of primary solvents and rubber antioxidants (see ANTIOXIDANTS). Fatty aldehydes C_8–C_{13} are used in nearly all perfume types and aromas (see PERFUMES). Polymers and copolymers of aldehydes exist and are of commercial significance.

BIBLIOGRAPHY

"Aldehydes" in *ECT* 1st ed., Vol. 1, pp. 334–342 by E. F. Landau, Celanese Corporation of America, E. I. Becker, Polytechnic Institute of Brooklyn, and O. C. Dermer, Oklahoma Agricultural and Mechanical College; in *ECT* 2nd ed., Vol. 1, pp. 639–648 by L. J. Fleckenstein, Eastman Kodak Co.; in *ECT* 3rd ed., Vol. 1, pp. 790–798, by P. D. Sherman, Union Carbide Corporation.

1. J. Buckingham, *Dictionary of Organic Compounds,* 5th ed., Chapman and Hall, New York, 1982.
2. S. Patai, *The Chemistry of the Carbonyl Group,* Wiley-Interscience, New York, 1966 and 1970.
3. H. O. House, *Modern Synthetic Reactions,* 2nd ed., W. A. Benjamin, Inc., Menlo Park, Calif., 1972.
4. C. D. Gutche, *The Chemistry of Carbonyl Compounds,* Prentice-Hall, Inc., Englewood Cliffs, N.J., 1967.
5. A. T. Nielson and W. J. Houlihan, *Organic Reactions,* Vol. 16, John Wiley & Sons, Inc., New York, 1968.
6. T. Mukaiyama, *Organic Reactions,* Vol. 28, John Wiley & Sons, Inc., New York, 1982.
7. A. R. Surrey, *Name Reactions in Organic Chemistry,* 2nd ed., Academic Press, Inc., New York, 1961.
8. R. C. Denny, *Named Organic Reactions,* Plenum Press, New York, 1969.
9. H. Krauch and W. Kunz, *Organic Name Reactions,* 2nd ed., translated by J. M. Harkin, John Wiley & Sons, Inc., New York, 1964.

10. G. Hilgetag and A. Martini, eds., *Preparative Organic Chemistry,* 4th ed., Wiley-Interscience, New York, 1972, pp. 301–400.

11. *Compendium of Organic Synthetic Methods:* Vol. 1, I. T. Harrison, ed., 1971, pp. 132–176; Vol. 2, I. T. Harrison, ed., 1974, pp. 53–69; Vol. 3, L. S. Hegedus and L. G. Wade, Jr., eds., 1977, pp. 66–87; Vol. 4, L. G. Wade, Jr., ed., 1980, pp. 73–101; Vol. 5, L. G. Wade, Jr., ed., 1984, pp. 92–123; Vol. 6, M. B. Smith, ed., 1988, pp. 51–66, Wiley-Interscience, New York.

12. J. Falbe, *New Syntheses with Carbon Monoxide,* Springer-Verlag, New York, 1980, pp. 1–181.

13. F. A. Patty, *Patty's Industrial Hygiene and Toxicology,* John Wiley & Sons, Inc., New York, 1981, pp. 2629–2669.

DAVID J. MILLER
Union Carbide Chemicals and Plastics Corporation

ALDOL ADDITION. See ALDEHYDES; KETONES.

ALDOSES. See CARBOHYDRATES; SUGAR.

ALE. See BEER.

ALGAL CULTURES. See FOODS, NONCONVENTIONAL.

ALGIN. See GUMS.

ALIZARIN. See DYES AND DYE INTERMEDIATES; DYES, ANTHRAQUINONE.

ALKALI AND CHLORINE PRODUCTS

CHLORINE AND SODIUM HYDROXIDE

PRODUCTION AND MANUFACTURE

Alkali and chlorine products are a group of commodity chemicals which include chlorine [7782-50-5], Cl_2; sodium hydroxide (caustic soda) [1310-73-2], NaOH; sodium carbonate (soda ash) [497-19-8], Na_2CO_3; potassium hydroxide (caustic potash) [1310-58-3], KOH; and hydrochloric acid (qv) (muriatic acid or anhydrous) [7647-01-0], HCl. Chlorine and caustic soda are the two most important products in this group, ranking among the top ten chemicals in the United States. In 1989 chlorine and caustic soda were the eighth and nineth largest volume chemicals, respectively (Table 1). Soda ash, which competes with caustic soda to satisfy the need for alkali (sodium oxide) in many processes, was the eleventh. The applications for chlorine and the alkalies are so varied that there is hardly a consumer product which is not dependent on one or both of them at some manufacturing stage (2).

Table 1. United States 1989 Top 10 Chemicals Production[a]

1989 Rank	1989 Production, 10^6 t	Average annual growth, %		
		1988–1989	1984–1989	1979–1989
1 sulfuric acid	39.4	3.0	0.8	0
2 nitrogen	24.4	2.8	2.3	5.7
3 oxygen	17.1	0.9	3.4	0
4 ethylene	15.9	−6.0	2.2	1.6
5 ammonia	15.3	0.3	0.2	−1.0
6 lime[b]	15.0	3.0	0.6	−2.4
7 phosphoric acid	10.5	−1.0	0.3	1.1
8 chlorine	10.8[c]	−0.9	0.8	−0.9
9 sodium hydroxide	11.3	5.2	0.3	−1.4
10 propylene	9.2	−4.7	5.4	3.6

[a]Ref. 1. Courtesy of American Chemical Society.
[b]Except refractory dolomite.
[c]According to the Chlorine Institute, Inc.

Capacity and Consumption

In 1987 world chlorine capacity totaled 40.7 million tons while the world consumption was 34.5 million tons, so that the apparent world chlorine capacity utilization averaged 85%. As seen in Table 2, the North American continent had the

Table 2. 1987 World Chlorine Capacity and Consumption by Region[a]

	Capacity, 10^3 t	Consumption, 10^3 t	Percentage of world capacity	Percentage of world consumption
North America	12,050	11,710	29.6	34.0
United States	10,240	10,340	25.2	30.0
Canada	1,810	1,370	4.4	4.0
Western Europe	11,540	9,890	28.4	28.7
Eastern Europe	6,840	5,110	16.8	14.8
Japan	3,930	2,940	9.7	8.5
Asia and Pacific	3,390	2,350	8.3	6.8
Latin America	1,790	1,500	4.4	4.3
Middle East	680	590	1.7	1.7
Africa	460	420	1.1	1.2
World total	40,680	34,510	100.0	100.0

[a]Ref. 3. Courtesy of Tecnon Consulting Group.

highest percentage of world consumption (34.0%), followed by Western Europe (28.7%), Eastern Europe (14.8%), Japan (8.5%), Asia (6.8%), and Latin America (4.3%). The two largest world markets for chlorine constitute 39% of its demand. The manufacture of vinyl chloride monomer [75-01-4], C_2H_3Cl, accounts for 26% of world chlorine demand and is the largest chlorine use. The second largest world market for chlorine, at 13% of consumption, is pulp (qv) and paper (qv) manufacture, where it is used primarily to bleach kraft pulp. Other major uses, shown in Table 3, are for the manufacture of ethylene dichloride [107-06-2], $C_2H_2Cl_2$; chlorinated solvents; propylene oxide [75-56-9], C_3H_6O; hypochlorites; and epichlorohydrin [106-89-8], C_3H_5ClO; and for use in water treatment.

Chlorine cannot be stored economically or moved long distances. International movements of bulk chlorine are more or less limited to movements between Canada and the United States. In 1987, chlorine moved in the form of derivatives was 3.3 million metric tons or approximately 10% of total consumption (3). Exports of ethylene dichloride, vinyl chloride monomer, poly(vinyl chloride), propylene oxide, and chlorinated solvents comprise the majority of world chlorine movement. Countries or areas with a chlorine surplus exported in the form of derivatives include Western Europe, Brazil, USA, Saudi Arabia, and Canada. Countries with a chlorine deficit are Taiwan, Korea, Indonesia, Venezuela, South Africa, Thailand and Japan (3).

Caustic soda and chlorine are coproducts and consequently caustic soda production has been limited by chlorine demand. Through 1992 world demand for caustic soda is expected to grow at a higher rate than the demand for chlorine and pressures limiting chlorine growth are expected to tighten world caustic soda supplies (4). This tightening will presumably be addressed by switching to soda ash usage where ever possible. In addition, production of sodium hydroxide by the lime–soda process is again gaining favor (see ALKALI AND CHLORINE PRODUCTS, SODIUM CARBONATE).

United States Chlorine Production. The record high for United States chlorine production occurred in 1979 at 11.2 million tons, followed by a decline

Table 3. 1987 World Chlorine Demand by Consuming Sector[a]

	Quantity consumed	
Chlorine usage	10^3 t	Percentage
vinyl chloride monomer	9,012	26.1
pulp and paper	4,599	13.3
propylene oxide	1,990	5.8
water treatment	1,237	3.6
carbon tetrachloride	934	2.7
perchloroethylene	687	2.0
hypochlorite	681	2.0
epichlorohydrin	600	1.7
1,1,1-trichloroethane	560	1.6
methylene chloride	520	1.5
chloroform	464	1.3
methyl chloride	459	1.3
ethylene dichloride (solvent)	401	1.2
trichloroethylene	368	1.1
chlorobenzene	355	1.0
chloroprene	246	0.7
bromine	129	0.4
ethylene dichloride (trade)	50	0.1
chlorinated polyethylene	16	>0.1
miscellaneous organic	4,988	14.5
miscellaneous inorganic	6,215	18.0
Total	*34,511*	*100.0*

[a]Ref. 3. Courtesy of Tecnon Consulting Group.

resulting in large part from the recessions of 1980 and 1981–1982, which led to a chlorine production low of 8.3 million tons in 1982. Chlorine enjoyed a relatively steady growth throughout the 1980s. Production increased to 10.0 million tons in 1987, then again to 10.5 million in 1988. Furthermore, chlorine production increased to 10.7 million tons in 1989 and remained flat at 10.7 million tons in 1990. Despite the strong economy of recent years, however, United States chlor–alkali production is still below 1979 levels because of the relative maturity of chlorine usage industries and the environmental pressures since the early 1980s aimed at curtailing use. Furthermore, leading edge technologies such as oxygen delignification and an increase in the substitution of chlorine dioxide [10049-04-4], ClO_2, for chlorine are expected to decrease chlorine use in Canadian pulp and paper. This may create a surplus Canadian chlorine supply which could potentially flow into the U.S. market and slow future U.S. chlorine production.

In 1988 diaphragm cells accounted for 76% of all U.S. chlorine production, mercury cells for 17%, membrane cells for 5%, and all other production methods for 2%. Corresponding statistics for Canadian production are diaphragm cells, 81%; mercury cells, 15%; and membrane cells, 4% (5). For a number of reasons, including concerns over mercury pollution, recent trends are away from mercury cell production toward the more environmentally acceptable membrane cells, which also produce higher quality product and have favorable economics.

Chlorine is produced as a gas that is used captively, transferred to customers via pipeline, or liquefied. Liquid chlorine, of higher purity than gaseous chlorine, is either used internally by the producers or marketed. The percentage of U.S. chlorine gas production subsequently liquefied has increased over the past ten years reflecting higher demand for high purity chlorine. This percentage was 60.7% in 1978 and 81.1% in 1987 (5). The majority of this chlorine is consumed captively.

United States Caustic Soda Production. In 1987 U.S. production of caustic soda increased to 10.4 million tons (Fig. 1), more than 10% over that of the previous year. Furthermore, 1988 production was up another 6.7% to 11.1 million tons. The demand for caustic soda has been very strong in recent years as evidenced by both increased U.S. consumption and a strong export demand. In 1987 the United States exported 1.5 million tons, 14.5% of the total caustic soda production (6), representing a 25.5% increase over exports in 1986. Then, in 1988, caustic soda exports grew by another 4.1%. A weak dollar helped boost the 1987 exports. Growth slowed in 1988, however, as a result of an industry (and world) wide caustic soda shortage, which was caused by lower U.S. chlorine consumption and forced allocations. Because industries switched from caustic to soda ash where possible, the lower 1988 export growth was not indicative of caustic soda's export potential.

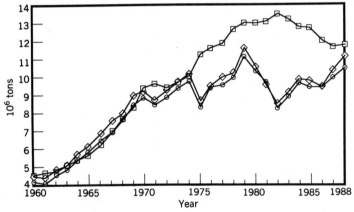

Fig. 1. United States chlorine and caustic soda production: □, annual chlorine capacity; ◇, caustic soda production; ○, chlorine production (6). Courtesy of *Chemical Economics Handbook.*

Because chlorine and caustic soda are electrolysis coproducts and chlorine cannot be stored economically, caustic soda production has been very dependent on both short term and long term chlorine demand and production. Unlike chlorine, however, caustic soda is not under environmental pressure and its growth potential is strong. The molecular weights of caustic soda (39.997) and of chlorine (70.914) result in a theoretical caustic soda to chlorine weight production ratio of 1.128:1 for a 2:1 mole production ratio. However, using Chlorine Institute production data, the calculated 1988 ratio was 1.057:1. This discrepancy exists for a variety of reasons: some chlorine plants do not employ electrolytic methods; some plants that use electrolytic methods produce sodium metal [7440-23-5], Na, as a by-product; and some electrolytic plants use potassium chloride instead of sodium chloride feed to produce potassium hydroxide [1310-58-3], KOH, as the coproduct. FMC presently produces small amounts of caustic soda, used internally as a sodium cyanide feedstock, by causticizing soda ash. Furthermore, Tenneco Inc. recently announced plans to build a plant to produce 68,000 t/yr of 50% caustic soda using a soda ash feed.

Energy Requirements. An electrochemical unit (ECU) in the chlorine industry represents the stoichiometric yield of 1 mole of Cl_2 and 2 of NaOH. The minimum amount of energy required to produce 1 ECU per 1.0 ton of Cl_2 and 1.128 tons of NaOH is 6.05 GJ (5.75×10^6 Btu or 1686 kW·h); to produce 1 ECU per short ton of Cl_2 and 1.128 short tons of NaOH, 5.50 GJ (5.23×10^6 Btu or 1534 kW·h) is needed. The total energy consumed by the chlor–alkali industry in 1988 for U.S. chlorine production (10.5 million tons) was 1% of the total U.S. annual electrical consumption of ~ 2200 billion kW·h.

Chlorine Capacity. In 1982 U.S. chlorine capacity reached a record high of 36,864 t/day (13.2 million tons per year) as shown in Figure 1. After 1982, decreased chlorine demand lowered market prices and some producers discontinued production, so that capacity dropped to 31,888 t/day in 1987 (5). Expansion to meet demand led to an increased capacity in March of 1989 of 32,550 t/day (11.1×10^6 t/yr).

In 1989 chlorine was produced by 25 companies at 52 locations in the United States (Table 4). Approximately half of these plants are located in the Southeast (Fig. 2). Two companies, Dow Chemical USA and Occidental Chemical Corporation, accounted for 54.3% of the total operating capacity; the top five companies

Table 4. Chlorine Plants in the United States[a]

	Annual capacity[b] 10^3 t/yr	Year[c] built	Cell type	Notes[d]
Akzo Chemicals Inc.				
Le Moyne, Ala.	70.8	1965	De Nora 22 × 5 mercury	1
Amax Magnesium Corp.				
Rowley, Utah	13.6	1977	modified IG Farben magnesium	9
American Magnesium Co.				
Snyder, Tex.		1969	VAMI magnesium	8,9
Atochem USA				
Portland, Oreg.	136.1	1947	OxyTech MDC29	1
Tacoma, Wash.	82.6	1929	ICI FM21 membrane '85	1
Cedar Chemical Co.				
Vicksburg, Miss.	32.7	1962	nonelectrolytic	6
Dow Chemical USA				
Freeport, Oyster Creek, Tex.	1,986.8[e]	1940	Dow diaphragm, Dow magnesium	1,9
Pittsburg, Calif.	132.5	1917	Dow diaphragm	1
Plaquemine, La.	1,049.6	1958	Dow diaphragm	1
E. I. du Pont de Nemours & Co.				
Niagara Falls, N.Y.	77.1	1898	Downs sodium	4

Table 4. (*Continued*)

	Annual capacity[b] 10^3 t/yr	Year[c] built	Cell type	Notes[d]
Formosa Plastics Corp., USA				
Baton Rouge, La.	179.6	1937	ICI-DMT diaphragm, '81	1
Fort Howard Corp.				
Green Bay, Wis.	8.2	1968	OxyTech HC3C diaphragm	1,7
Muskogee, Okla.	5.4	1980	ICI FM21 membrane '85	1,7
General Electric Co.				
Mount Vernon, Ind.	49.9	1976	OxyTech H2A diaphragm	1
Burkville, Ala.	23.6	1987	OxyTech H2A diaphragm	1
Georgia Gulf Corp.				
Plaquemine, La.	385.6	1975	OxyTech H4 diaphragm	1
Georgia-Pacific Corp.				
Bellingham, Wash.	81.6	1965	De Nora 18 × 4 mercury	1,7
Brunswick, Ga.	48.1	1967	Ashai Chem membrane '83	1,7
B. F. Goodrich Group				
Calvert City, Ky.	108.9	1966	De Nora 24H5 mercury	1
Hercules, Inc.				
Hopewell, Va.		1939	OxyTech HC3 diaphragm	1,8
La Roche Chemicals				
Gramercy, La.	181.4	1958	OxyTech HC3B, HC3C diaphragm	1
LCP Chemicals				
Acme, N.C.	48.1	1963	Solvay V-200 mercury	1
Ashtabula, Ohio	36.3	1963	Olin E11F mercury	2
Brunswick, Ga.	96.2	1957	Solvay V-100 mercury	1
Moundsville, W.Va.	78.9	1953	Solvay S60 mercury	1
Orrington, Mass.	72.6	1967	De Nora 24H5 mercury	1
Syracuse, N.Y.	[f]	1927	Solvay S60 mercury, '53	1
Mobay Chemical Corp.				
Baytown, Tex.	81.6	1972	Uhde (HCl)	5
Niachlor[g]				
Niagara Falls, N.Y.	199.6	1987	Asahi membrane '87	1
Occidental Chemical Corp.				
Convent, La.	278.5	1981	OxyTech MDC55 diaphragm	1
Corpus Christi, Tex.	417.3	1974	OxyTech MDC55 diaphragm	1
Deer Park, Tex.	347.5	1938	OxyTech MDC21 diaphragm	1
		1938	De Nora 18 × 6 mercury	1
Delaware City, Del.	126.1	1965	De Nora 18 × 4 mercury	3
LaPorte, Tex.	480.0	1974	OxyTech MDC 29 diaphragm	1
Mobile, Ala.	33.6	1964	De Nora 18 × 4 mercury, OxyTech MGC membrane, '91	3
Muscle Shoals, Ala.	132.5	1952	De Nora 12 × 3 mercury	3
Niagara Falls, N.Y.	293.0	1898	OxyTech H4 diaphragm, '74, '78	1
Tacoma, Wash.	202.3	1929	OxyTech H4, diaphragm	1
		1929	OxyTech MGC membrane '88	1
Taft, La.	569.7	1966	OxyTech HC4B, H4 '75 diaphragm	1
		1966	OxyTech MGC membrane '86	1

943

Table 4. (*Continued*)

	Annual capacity[b] 10^3 t/yr	Year[c] built	Cell type	Notes[d]
Olin Corporation				
Augusta, Ga.	101.6	1965	Olin E11F mercury	1
Charleston, Tenn.	230.4	1962	Olin E11F, E812 mercury	1
McIntosh, Ala.	331.1	1952	OxyTech H4 diaphragm, '77–'78	1
Niagara Falls, N.Y.	81.6	1897	Olin E11F mercury '60	3
Oregon Metallurgical Corp.				
Albany, Oreg.	1.8	1971	Alcan magnesium	9
Pioneer Chlor-Alkali Co., Inc.				
Henderson, Nev.	104.3	1942	OxyTech MDC29 diaphragm '76	1
St. Gabriel, La.	159.7	1970	Uhde 30 m^2 mercury	1
PPG Indutries				
Lake Charles, La.	1,041.5	1947	De Nora 48H5 mercury '69	1
		1947	Glanor 1144 diaphragm '77, '80	1
		1947	Bipolar 1161 diaphragm '83	1
New Martinsville, W.Va.	313.0	1943	Columbia N3, N6 diaphragm	1
		1943	OxyTech MDC55 diaphragm '84	1
		1943	Uhde 20-m^2 mercury '58	1
RMI Company				
Ashtabula, Ohio	36.3	1949	Downs sodium	4
Titanium Metals Corp. of America				
Henderson, Nev.		1943	I.G. Farben magnesium	9
Vulcan Chemicals				
Geismar, La.	220.4	1976	OxyTech MDC55 diaphragm	1
Port Edwards, Wis.	65.3	1967	De Nora 24H5 mercury	3
Wichita, Kans.	165.1	1952	OxyTech HC3BT, H4 '75 diaphragm	1
		1952	OxyTech membrane '83	1
Weyerhaeuser Company				
Longview, Wash.	136.1	1957	OxyTech MDC29 diaphragm '75	1,7
Total	*11,135.8[h]*			

[a]Refs. 5, 6. Courtesy of SRI International and The Chlorine Institute, Inc.
[b]Operating capacity as of March 1989. Idled capacity is noted where information is available.
[c]Refers to year chlorine production started at location.
[d]Notes:
  1. Electrolytic plant producing caustic soda, chlorine, and hydrogen from brine.
  2. Electrolytic plant producing caustic potash, chlorine, and hydrogen from brine.
  3. Electrolytic plant producing caustic soda, caustic potash, chlorine, and hydrogen from brine.
  4. Electrolytic plant producing metallic sodium and chlorine from molten sodium chloride.
  5. Electrolytic plant producing chlorine and hydrogen from hydrochloric acid.
  6. Nonelectrolytic plant producing chlorine and potassium nitrate.
  7. Pulp mill.
  8. Not operating.
  9. Electrolytic plant producing magnesium and chlorine from molten magnesium chloride.
[e]Idled capacity = 662.3 t/yr.
[f]Idled capacity = 82.6 t/yr.
[g]A Du Pont–Olin partnership.
[h]Total idled capacity = 744.9 t/yr.

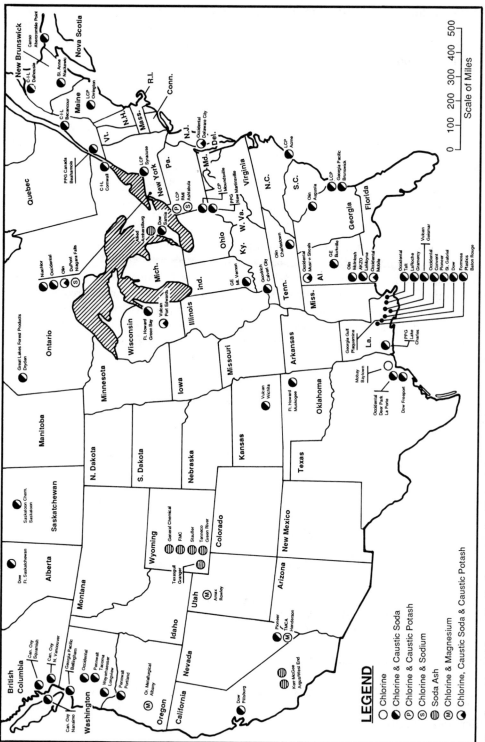

Fig. 2. Operating chlor–alkali plants in the United States and Canada (5). Courtesy of the Chlorine Institute, Inc. (500 miles = 800 km.)

accounted for 77.5%. Although 0.8 million tons per year of chlorine capacity is presently idle (Table 4), announced expansions are expected to bring an additional 660,000 t of new chlorine capacity onstream by 1992, increasing total capacity by about 6%/yr (Table 5).

Table 5. Publicly Announced Chlor–Alkali Expansions

Year	Producer	Location	Chlorine expansions, t/day
1990	Atochem, USA	Portland, Oreg.	100[a]
1992	Formosa Plastics	Point Comfort, Tex.	1720

[a]Conversion to membrane cells.

Chlorine Per Capita Consumption. The U.S. per capita consumption of chlorine increased rapidly from 1955 to 1970, with an average annual growth of 5.8% (Fig. 3). There has been little, if any, growth since 1970 however, and recessionary effects, product maturity, and environmental pressures are evidenced in the cyclic fluctuations. Fluctuations in per capita consumption of chlorine and caustic have more or less tracked each other over the years, although chlorine per capita consumption has consistently exceeded that of caustic soda since 1967.

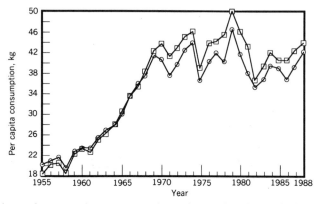

Fig. 3. Annual per capita consumption of caustic (○) and chlorine (□) in the United States (6). Courtesy of *Chemical Economics Handbook.*

Electrolytic Decomposition of Sodium Chloride

Chlorine and caustic soda are coproducts of electrolysis of aqueous solutions of sodium chloride [7647-14-5], NaCl, (commonly called brine) following the overall chemical reaction

$$2\ NaCl + 2\ H_2O \xrightarrow{\text{energy}} 2\ NaOH + Cl_2 + H_2 \qquad (1)$$

This reaction has a positive free energy of 422.2 kJ (100.9 kcal) at 25°C and hence energy has to be supplied in the form of d-c electricity to drive the reaction in a net forward direction. The amount of electrical energy required for the reaction depends on electrolytic cell parameters such as current density, voltage, anode and cathode material, and the cell design.

Conversion of aqueous NaCl to Cl_2 and NaOH is achieved in three types of electrolytic cells: the diaphragm cell, the membrane cell, and the mercury cell. The distinguishing feature of these cells is the manner by which the electrolysis products are prevented from mixing with each other, thus ensuring generation of products having proper purity.

The component electrochemical reactions are the discharge of chloride ions, Cl^-, at the anode,

$$2\ Cl^- \longrightarrow Cl_2 + 2\ e^- \tag{2}$$

and the generation of hydrogen [1333-74-0], H_2, and hydroxide ions, OH^-, at the cathode.

$$2\ H_2O + 2\ e^- \longrightarrow H_2 + 2\ OH^- \tag{3}$$

Chlorine is produced at the anode in each of the three types of electrolytic cells. The cathodic reaction in diaphragm and membrane cells is the electrolysis of water to generate H_2 as indicated, whereas the cathodic reaction in mercury cells is the discharge of sodium ion, Na^+, to form dilute sodium amalgam.

$$Na^+ + Hg + e^- \longrightarrow NaHg \tag{4}$$

This amalgam then reacts separately with H_2O in denuders or decomposers to produce H_2 and NaOH.

$$2\ NaHg + 2\ H_2O \longrightarrow 2\ NaOH + 2\ Hg + H_2 \tag{5}$$

Separation of the anode and cathode products in diaphragm cells is achieved by using asbestos [1332-21-4], or polymer-modified asbestos composite, or Polyramix deposited on a foraminous cathode. In membrane cells, on the other hand, an ion-exchange membrane is used as a separator. Anolyte–catholyte separation is realized in the diaphragm and membrane cells using separators and ion-exchange membranes, respectively. The mercury cells contain no diaphragm; the mercury [7439-97-6] itself acts as a separator.

The catholyte from diaphragm cells typically analyzes as 9–12% NaOH and 14–16% NaCl. This cell liquor is concentrated to 50% NaOH in a series of steps primarily involving three or four evaporators. Membrane cells, on the other hand, produce 30–35% NaOH which is evaporated in a single stage to produce 50% NaOH. Seventy percent caustic containing very little salt is made directly in mercury cell production by reaction of the sodium amalgam from the electrolytic cells with water in denuders.

Efficiency and Energy Consumption. The electrical energy consumed during the electrolysis of brine to produce chlorine gas and sodium amalgam is greater than that used to generate Cl_2 and H_2 in the diaphragm or membrane cell. However, the latter processes also use energy in the form of steam for evaporation of the cell liquor. Table 6 summarizes the major differences in the three technologies. The minimum energy required to convert salt to Cl_2, H_2, and 50% NaOH (6.05 GJ/t of Cl_2) is, of course, the same in all of these processes.

Table 6. Components of Diaphragm, Membrane, and Mercury Cells

Component	Mercury cell	Diaphragm cell	Membrane cell
anode	RuO_2 + TiO_2 coating on Ti substrate	RuO_2-based coating on Ti substrate	RuO_2 based coating on Ti substrate
cathode	mercury on steel	steel or steel coated with activated nickel	steel or Ni based catalytic coating on nickel
diaphragm	none	asbestos, polymer-modified asbestos, or Polyramix (nonasbestos)	ion-exchange membrane
cathode product	sodium amalgam	10–12% NaOH + 15–17% NaCl and H_2	30–33% NaOH + <0.01% NaOH and H_2
decomposer product	50% NaOH and H_2	none	none
evaporator product	none	50% NaOH with ~1.1% salt and solid salt	50% NaOH with ~0.01% salt
steam consumption	none	1500–2300 kg/t NaOH	450–550 kg/t NaOH
cell voltage, V	4–5	3–4	2.8–3.3
current density, kA/m^2	7–10	0.5–3	2–5

Faraday's law states that 96,487 coulombs (1 C = 1 A·s) are required to produce one gram equivalent weight of the electrochemical reaction product. This relationship determines the minimum energy requirement for chlorine and caustic production in terms of kiloampere hours per ton of Cl_2 or NaOH

$$\text{for } Cl_2 \quad \frac{96,487 \times 1000}{60 \times 60 \times 35.45} = 756 \text{ kA·h/t}$$

$$\text{for NaOH} \quad \frac{96,487 \times 1000}{60 \times 60 \times 40} = 670 \text{ kA·h/t}$$

The current efficiency of an electrolytic process ($\eta_{current}$) is the ratio of the amount of material produced to the theoretically expected quantities. Inefficiencies arise from secondary reactions occurring at the anode and cathode and in the bulk.

There are two parasitic reactions offsetting anode efficiency: (*1*) co-generation of oxygen [7782-44-7], O_2, from the anodic discharge of water,

$$2\,H_2O \longrightarrow O_2 + 4\,H^+ + 4\,e^- \tag{6}$$

and (*2*) electrochemical oxidation of hypochlorite ion, OCl^-, to chlorate, ClO_3^-,

$$6\,ClO^- + 3\,H_2O \longrightarrow 2\,ClO_3^- + 4\,Cl^- + 6\,H^+ + {}^3\!/{}_2\,O_2 + 6\,e^- \tag{7}$$

The oxygen contribution from these reactions is dependent on the nature of the anode material and the pH of the medium. The current efficiency for oxygen is generally 1–3% using commercial metal anodes. If graphite anodes are used, another overall reaction leading to inefficiency is the oxidation of carbon to CO_2:

$$C + 2\,H_2O \longrightarrow CO_2 + 2\,H_2 \tag{8}$$

At the cathode, water molecules are discharged yielding H_2 gas and hydroxide ions, OH^-. Some of the caustic generated in the cathode compartment back-migrates to the anode compartment and reacts with dissolved chlorine ($Cl_{2(aq)}$) to form chlorate as follows

$$Cl_{2(aq)} + OH^- \longrightarrow HOCl + Cl^- \tag{9}$$

$$HOCl + OH^- \rightleftharpoons H_2O + OCl^- \tag{10}$$

$$2\,HOCl + OCl^- \longrightarrow ClO_3^- + 2\,H^+ + 2\,Cl^- \tag{11}$$

There are two reactions that influence the cathodic efficiency, namely the reduction of OCl^- and of ClO_3^-:

$$OCl^- + H_2O + 2\,e^- \longrightarrow Cl^- + 2\,OH^- \tag{12}$$

$$ClO_3^- + 3\,H_2O + 6\,e^- \longrightarrow Cl^- + 6\,OH^- \tag{13}$$

Although these reactions are thermodynamically favorable, they are not kinetically significant under normal operating conditions. Hence cathodic efficiency is usually high (>95%) in diaphragm and membrane cells. In mercury cells cathodic inefficiency arises from the discharge of H_2 at the cathode resulting from impurities in the brine. Reactions contributing to anodic inefficiency are the same as in diaphragm or membrane cells.

Current Efficiency. Current efficiency for caustic production in diaphragm and membrane cells can be estimated from collection of a known amount of caustic over a period of time and from a knowledge of the number of coulombs of electricity passed during that time period. An alternative method involves analysis of the gases evolved during electrolysis and determining the anolyte composition. Material balance considerations (7) show the expression for the caustic efficiency for membrane cells to be

$$\eta_{NaOH} = 100 \left\{ \frac{1 - (2F/I)(A - B + Y - D)}{1 + (2\%O_2/\%Cl_2)} \right\} \tag{14}$$

where

$$A = \frac{3q\ C_{\text{NaClO}_3}(d)}{106.45}; \qquad B = \frac{3p\ C_{\text{NaClO}_3}(f)}{106.45};$$

$$Y = \frac{q\ \text{Cl}_2(a)}{70.91}; \qquad D = \frac{p\ X}{70.91};$$

$C_z(y)$ is the concentration in grams per liter of z in medium y; a is anolyte, c is catholyte, f is feed brine, and d is depleted brine; $\text{Cl}_2(a)$ is Cl_2 (soluble) $+$ HOCl $+$ OCl$^-$; F is the faraday constant ($= 96{,}487$ C/mol); I is the load in A; p is the feed brine flow rate in L/s; q is the depleted brine flow rate in L/s; and

$$X = 1.338\ C_{\text{Na}_2\text{CO}_3}(f) + 0.844\ C_{\text{NaHCO}_3}(f) + 1.773\ C_{\text{NaOH}}(f)$$

For estimating Cl_2 efficiency (η_{Cl_2}), the term $\text{Cl}_2(a)$ in equation 14 should be dropped. The corresponding expression for caustic efficiency for diaphragm cells is

$$\eta_{\text{NaOH}} = \left[\frac{100}{1 + (2\%\text{O}_2/\%\text{Cl}_2) + (80/C_{\text{OH}})(E - F)}\right] \qquad (15)$$

where C_{OH} is catholyte caustic concentration in g/L,

$$E = \frac{3C_{\text{NaClO}_3}(a)}{106.45}; \text{ and } F = [\text{Cl}_2(a) - X]/70.91$$

For chlorine efficiency in diaphragm cells

$$\eta_{\text{Cl}_2} = \frac{100}{1 + (2\%\text{O}_2/\%\text{Cl}_2) + (2.2546\ G)/C_{\text{OH}}} \qquad (16)$$

where $G = C_{\text{NaClO}_3}(a) - C_{\text{NaClO}_3}(f)$.

Equation 16 is the correct material balance expression for calculating the chlorine efficiency of diaphragm cells. Whereas many approximate versions are used (8), the one closest to equation 16 is the "six equation":

$$\eta_{\text{Cl}_2} = \left[\frac{100}{1 + (2\%\text{O}_2/\%\text{Cl}_2) + 6\ C_{\text{NaClO}_3}(c)/C_{\text{OH}}}\right] \qquad (17)$$

Current efficiency values based on the six equation are higher by approximately 1.0% than those from equation 16.

Cell Voltage and Its Components. The minimum voltage required for electrolysis to begin for a given set of cell conditions, such as an operational temperature of 95°C, is the sum of the cathodic and anodic reversible potentials and is known as the thermodynamic decomposition voltage, $E°$. $E°$ is related to the standard free energy change, $\Delta G°$, for the overall chemical reaction,

$$\Delta G° = -nFE° \tag{18}$$

where n represents the number of moles of electrons involved in the primary electrode reaction and F is the Faraday constant, expressed in ampere·hour equivalents. $E°$ values for the diaphragm or membrane cells and for mercury cells are presented in Table 7. The $+0.924$ V difference in $E°$ values between membrane cells and mercury cells results from the reaction

$$2\,Na + 2\,H_2O \longrightarrow 2\,NaOH + H_2 \tag{19}$$

which takes place outside the electrolytic cells. Although, in principle, this voltage is recoverable, that has not yet been commercially demonstrated. The $E°$ value for diaphragm or membrane cells at 95°C is 2.23 V for a caustic concentration of 3.5 M. However, these cells operate around 3.0 to 3.2 V at a current density of 2 to 3 kA/m², not at 2.23 V, because in order to drive equation 1 at an acceptable rate, an additional driving force is required to overcome cell resistances and electrode overvoltages.

　　Electrolysis of brine (eq. 1) is endothermic. The overall heat of reaction is 446.68 kJ/mol (106.76 kcal/mol) of chlorine and the thermoneutral voltage, ie, the voltage at which heat is neither required by the system nor lost by the system to the surroundings, would therefore be ~ 2.31 V. In practice, however, chlor–alkali cells operate in the range of 3 to 3.5 V, at an average chlorine efficiency (CE) of $\sim 95\%$, resulting in heat generation (Q) to the extent of 3960 kJ/kg (1710 Btu/lb) of Cl_2, for a voltage of 3.5 V

$$Q = \left(\left[\frac{100}{CE}\right][46.05\ V]\right) - \Delta H \tag{20}$$

Heat produced in these cells operating at voltages of >2.31 V is generally removed by water evaporation and radiation losses.

　　Overvoltage. Overvoltage (η_{ac}) arises from kinetic limitations or from the inherent rate (be it slow or fast) of the electrode reaction on a given substrate. The magnitude of this value can be generally expressed in the form of the Tafel equation

$$\eta_{ac} = k\ \log(i/i_0) \tag{21}$$

where k is the slope of the η_{ac} vs log i curve, i the applied current density, and i_0 the exchange current density of the reaction. The quantity i_0 is a measure of the relative rate of a given reaction, eg, 1 mA/cm² for Cl_2 evolution on dimensionally stable electrodes (DSA). Overvoltage can be lowered by increasing the electrochemically active surface area, which effectively reduces the magnitude of i, or by using catalytic cathodes.

　　Overvoltages for various types of chlor–alkali cells are given in Table 8. A typical example of the overvoltage effect is in the operation of a mercury cell

Table 7. Thermodynamic Decomposition Voltage of Chlor–Alkali Cells at 25°C

	Diaphragm/membrane cell	Mercury cell
anode reaction	$2\,Cl^- \longrightarrow Cl_2 + 2\,e^-$	$2\,Cl^- \longrightarrow Cl_2 + 2\,e^-$
anode potential E^0, V	1.36	1.36
cathode reaction	$2\,H_2O + 2\,e^- \longrightarrow H_2 + 2\,OH^-$	$Na^+ + e^- \longrightarrow Na(amalgam)$
cathode potential E^0, V	-0.828	-1.77
overall reaction	$2\,H_2O + 2\,Cl^- \longrightarrow Cl_2 + 2\,OH^- + H_2$	$2\,Na^+ + 2\,Cl^- \xrightarrow{\text{Hg}} Cl_2 + 2\,Na(amalgam)$
cell potential E^0, V	2.188	3.13

Table 8. Components of Chlor–Alkali Cell Voltages

	Diaphragm cell[a]	Membrane cell[b]	Mercury cell[c]
thermodynamic decomposition voltage			
anode	1.32	1.32	1.32
cathode	0.93	0.93	1.83
overvoltage			
anode	0.03	0.03	0.1
cathode	0.28	0.15	0.4
ohmic drops			
solution	0.12	0.05	0.15
diaphragm	0.38	0.35	
anode and contact to base	0.11		
base	0.06	0.37	0.2
cathode	0.09		
cell voltage	3.32	3.2	4.0

[a] Voltages given are for OxyTech H4 cell operating at 2.3 kA/m^2 (150 kA).
[b] Voltages given are for MGC-26 cell operating at 3.6 kA/m^2 (140 kA).
[c] Voltages given are for a De Nora 24M2 system operating at 10 kA/m^2 (270 kA).

where Hg is used as the cathode material. The overpotential of the H_2 evolution reaction on Hg is high; hence it is possible to form sodium amalgam without H_2 generation, thereby eliminating the need for a separator in the cell.

Ohmic Drops. Another irreversible contribution to the measured cell voltage is the ohmic or *IR* drop across the electrolyte, separator, and cell hardware. The *IR* drop across the hardware can be estimated from Ohm's law and the relationship

$$R = \rho l/A \qquad (22)$$

where R is the resistance (in ohms) of the conductor of length l with a specific resistance of ρ and cross-sectional area A.

The ohmic drop across the electrolyte and the separator can also be calculated from Ohm's law using a modified expression for the resistance. When gas bubbles evolve at the electrodes they get dispersed in and impart a heterogeneous character to the electrolyte. The resulting conductivity characteristics of the medium are different from those of a pure electrolyte. Although there is no exact description of this system, some approximate treatments are available, notably the treatment of Rousar (9), according to which the resistance of the gas–electrolyte mixture, R_{mix}, is related to the resistance of the pure electrolyte, R_{sol}:

$$R_{mix} = R_{sol}(1 + 1.5\epsilon) \qquad (23)$$

where ϵ is the gas void fraction, defined as the ratio of the volume of the gas to the volume of the gas plus the volume of the electrolyte. The *IR* drop in brine solution is generally around 30 to 40 mV/mm at 95°C and a current density of 2.32 kA/m^2. Similarly, for calculating the *IR* drop across the separator, the separator thickness term has to be modified because the distance between the two faces of a separator such as the asbestos diaphragm is not equal to its thickness. The liquid path is tortuous and the area is limited by finite porosity. Thus the IR drop across the separator would be

$$IR_{sep} = xil\rho \tag{24}$$

where x reflects the tortuosity-to-porosity ratio. Typical values of tortuosity for asbestos diaphragms range from 2.2 to 2.8; porosity is generally ~ 0.7–0.8.

The components of the diaphragm, membrane, and mercury cell voltages presented in Table 8 show that, although the major component of the cell voltage is the E^0 term, ohmic drops also contribute to the irreversible energy losses during the operation of the cells.

Direct-Current Electric Power. The operation of a chlor–alkali plant is dependent on the availability of huge quantities of direct-current electric power (8), which is usually obtained from a high voltage source of alternating current. The lower voltage required for an electrolyzer circuit is produced by a series of step-down transformers. Silicon diode rectifiers convert the a-c electricity to d-c for the electrolysis. A set of rectifiers can supply up to 450 kA. Although these diodes can operate at 400 V/diode, a peak a-c voltage of 1500 V corresponding to a d-c output of 1200 V is not exceeded for safety reasons. Rectifier efficiency ($\eta_{rectifier}$) is generally around 97–98%.

The unit cost of the d-c supply decreases with increasing voltage and amperage (see Fig. 4). A chlor–alkali plant is therefore most economical when as many high amperage cells as possible are connected in series.

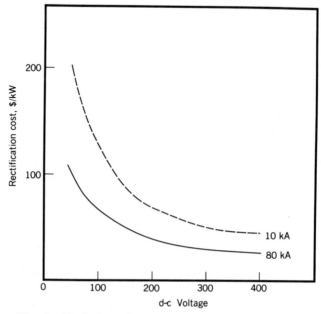

Fig. 4. Variation of rectification costs with voltage.

Overall Energy Consumption.   The voltage efficiency is the ratio of the thermodynamic decomposition voltage to the actual cell voltage (V_{cell}). The energy efficiency (η_{energy}), which is a product of the current and voltage efficiencies, can be expressed for a diaphragm cell as

$$\eta_{energy} = \frac{2.23\ \eta_{current}}{V_{cell}}$$

However, the industry's popular terminology is the energy consumption expressed in terms of kilowatt hours per ton of Cl_2 (E_{Cl_2}) or of NaOH (E_{NaOH}). An estimate of this value requires a knowledge of cell voltage, current efficiency, and the efficiency of the rectifier used to convert a-c power to d-c. The energy consumption for producing a ton of Cl_2 is

$$E_{Cl_2} = \frac{756\ V_{cell}}{\eta_{current}\ \eta_{rectifier}}\quad \left(\frac{\text{a-c kW·h}}{t}\right)$$

and that for a ton of NaOH is

$$E_{NaOH} = \frac{670.1\ V_{cell}}{\eta_{current}\ \eta_{rectifier}}\quad \left(\frac{\text{a-c kW·h}}{t}\right)$$

Manufacturing Processes

The earliest annals of chemistry mention chlorine compounds. In AD 77 Pliny the Elder published a practical collection of chemical reactions including a formula for gold purification that generates hydrogen chloride. Records from more than 800 years later indicate that the Arabic people knew that hydrogen chloride and water produce hydrochloric acid. Sometime around AD 1200 alchemists discovered that a mixture of hydrochloric and nitric acids dissolves gold. There is no record of the heavy, greenish gas generated, however. It was not until 1774 that Swedish apothecary Carl W. Scheele first generated and collected chlorine by the reaction of manganese dioxide and hydrochloric acid. Scheele also discovered chlorine's bleaching action, after placing some leaves and flowers in a bottle containing the gas. Textile producers in France soon heard of the new gas, and in 1789 they bubbled chlorine through a potash solution, producing the first commercial liquid chlorine bleach (10).

Salt was first electrochemically decomposed by Cruickshank in 1800, and in 1808 Davy confirmed chlorine to be an element. In the 1830s Michael Faraday, Davy's laboratory assistant, produced definitive work on both the electrolytic generation of chlorine and its ease of liquefaction. And in 1851 Watt obtained the first English patent for an electrolytic chlorine production cell (11).

Through the 1880s and 1890s producers in Germany, England, Canada, and the United States refined chlorine technology: Siemens developed electric generators; the Griesheim Company (Germany) invented the first practical diaphragm cell; Castner produced the first commercially viable mercury cell; and German producers learned that, although wet liquid chlorine is almost impossible to package, water-free chlorine could be safely shipped in ordinary iron or steel pressure vessels. Thus by the early 1900s chlorine was produced in mercury and diaphragm electrolytic cells and routinely shipped in liquid form. In 1913 Altoona, Pennsylvania became the first of many cities to treat sewage using liquid chlorine. Throughout the early and mid-1900s a myriad of other chlorine uses were discovered, so that today chlorine ranks in the top ten volume chemicals produced in the United States.

Early demand for chlorine centered on textile bleaching, and chlorine generated through the electrolytic decomposition of salt (NaCl) sufficed. Sodium hydroxide was produced by the lime–soda reaction, using sodium carbonate readily available from the Solvay process. Increased demand for chlorine for PVC manufacture led to the production of chlorine and sodium hydroxide as coproducts. Solution mining of salt and the availability of asbestos resulted in the dominance of the diaphragm process in North America, whereas solid salt and mercury availability led to the dominance of the mercury process in Europe. Japan imported its salt in solid form and, until the development of the membrane process, also favored the mercury cell for production.

Anodes.   For over sixty years (1913–1979) graphite was the exclusive anode for chlorine production in spite of the high chlorine overpotential and the dimensional instability caused by the gradual oxidation of carbon to CO_2. Thus a consequence of graphite anode use was an increased electrolyte ohmic drop and a correspondingly high energy consumption. Then, in the late 1960s, the advent of noble metal oxide coatings on titanium substrates by Beer (12) revolutionized the industry. The most widely used anodes were ruthenium oxide [*12036-10-1*] and titanium oxide [*13463-67-7*] activated titanium, which exhibits low chlorine overpotential and excellent dimensional stability. Escalating power costs, triggered by the oil crisis in the mid 1970s, accelerated the transition of the chlor–alkali industry from graphite to metal anodes. All but a very few installations have now converted to the exclusive use of the RuO_2 + TiO_2-coated titanium anodes. These are supplied under the trade name DSA (for dimensionally stable anode) (13). (DSA is a trade mark of Eltech Systems Corp.).

The DSA electrode is coated with a titanium and ruthenium oxide mixture containing up to or over 50 mol % precious metal oxide, the precious metal loading ranging from $\sim$5 to 20 g/m^2. These coatings are formed on titanium mesh substrates by thermal decomposition. Ruthenium and titanium salts are applied to the Ti mesh and fired at $\sim$500°C in air to obtain the mixed oxides. RuO_2 + TiO_2-based coatings are polycrystalline and structurally complex. The oxides exist in the catalytic layer as $Ti_{(1-n)}Ru_nO_2$, where $n < 1$. The coatings exhibit a "mud-cracked" surface morphology having a crystallite size of 10–50 nm and an effective surface area of 200–500 cm^2 per apparent cm^2. The exact value is a function of the heat treatment and the coating composition. Two coatings predominate in the industry: the original Beer coating is believed to contain a Ti:Ru mole ratio of 2:1; a newer three-component coating has a mole ratio of 3Ru:2Sn:11Ti. This coating has a RuO_2 + SnO_2 loading of $\sim$1.6 mg/cm^2. It generates 22 to 25% less oxygen than RuO_2 + TiO_2 coatings. References 14 and 15 give more details about the various patented compositions and methods of preparation.

DSA anodes exhibit long life, very low operating voltage, and high efficiency. They have a reasonable tolerance to a wide range of operating conditions, although performance degradation can occur when the percentage of O_2 generated is high. These conditions include operating at feed brine pH values of more than 11 and at feed brine NaCl concentrations of less than 280 g/L. Exposure to caustic or fluorides results in the dissolution of RuO_2 as RuO_4^- or ruthenium fluoride and of titanium as soluble fluorides. Deposition of foreign matter such as MnO_2 or $BaSO_4$ can also result in blockage of active sites and lead to the failure of the catalytic coatings.

The mechanism of coating failure appears to depend on the type of cell in which the anode is operated. Life in diaphragm cells is at least 12 + yr; in mercury cells, it is considerably shorter, about 3–4 yr. The unavoidable occurrence of minor short circuits, through contact with the mercury cathode, causes gradual physical wear of the anode coating. In the absence of this physical wear, ie, in membrane or diaphragm cells, the limiting factor appears to be passivation of the anode, preceded by a very gradual dissolution of the RuO_2 coating. DSA manufacturers recommend that, to ensure long life, assuming a feed brine concentration of 200 g/L and a pH of <12, the following impurities, in ionic form, be kept at the concentrations indicated in parentheses: Hg (40 ppb); Mn (0.01 ppm); heavy metals (0.3 ppm); total organic content (1 ppm); F^- (1 ppm); Ba (0.4 ppm).

Electrolytic Cell Operating Characteristics.   Currently the greatest volume of chlorine production is by the diaphragm cell process, followed by that of the mercury cell and then the membrane cell. However, because of the ecological and economic advantages of the membrane process over the other systems, membrane cells are currently favored for new production facilities. The basic characteristics of the three cell processes are shown in Figure 5.

A typical energy distribution profile for a diaphragm chlor–alkali operation, illustrated in Figure 6, shows that the electrolytic cells consume about 10.1×10^6 kJ/t of Cl_2, and 9.55×10^6 kJ are required for concentrating the cell liquor containing ~11% caustic and ~15% salt to 50% caustic. These are not optimized energy values since the energy consumption of a diaphragm cell depends on the operating current density, cell voltage, and current efficiency. Similarly, the energy required for evaporation also varies according to the type of evaporator system used. Using a quadruple-effect evaporator, the energy would be ~4.1×10^6 kJ/t of caustic. The cell efficiency of a diaphragm chlor–alkali unit is about 52%. Overall plant efficiency is only ~23%, however, and, as stated, is a function of the evaporator system.

In the case of electrolytic membrane cells, the energy requirements would be ~15 to 20% lower than those of diaphragm cells. The exact value is again a function of the operating current density, cell voltage, and current efficiency. The major energy savings would, however, be in caustic evaporation. The energy needed for concentrating the 33% caustic output from the cells to 50% NaOH would be 2.6×10^6 kJ/t using a single-effect and ~1.6×10^6 kJ/t using a double-effect evaporator. The cell efficiency for membrane cells is ~50%, whereas the overall plant efficiency is ~35–40%.

The Mercury Cell Process.   The mercury cell process (Fig. 7) actually consists of two electrochemical cells. In the electrolyzer, saturated (25.5 wt %) sodium or potassium brine flows through an elongated trough that is inclined

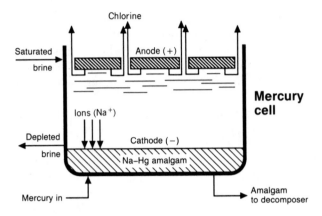

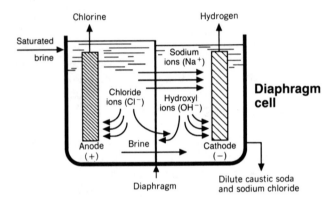

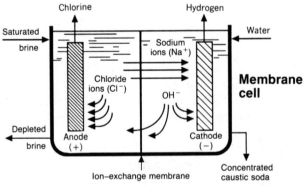

Fig. 5.   Chlorine electrolysis cells. Courtesy of McGraw-Hill, Inc.

approximately 1 to 2.5°. Mercury, which is the cathode, flows concurrently with the brine over a steel base. The sides of the trough are usually rubber lined. Anodes of activated titanium [7440-32-6] are suspended in the brine from above. The anode reaction is represented by equation 2, the cathode reaction by equation 4. The resulting amalgam, containing from 0.25 to 0.5% sodium, flows from the

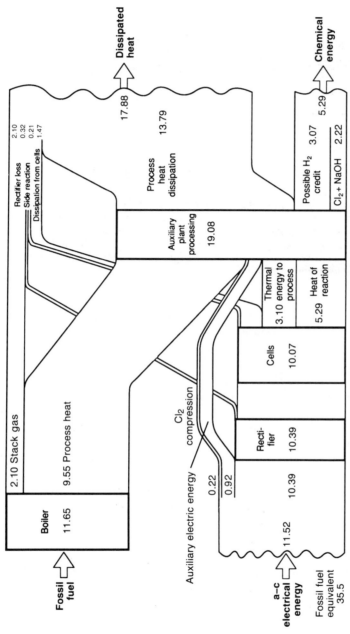

Fig. 6. Energy flow diagram of a typical diaphragm cell operation where numbers represent energy in millions of kilojoules per ton of chlorine. To convert kJ to Btu, multiply by 0.949.

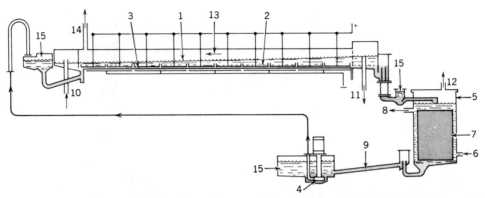

Fig. 7. Mercury cathode electrolyzer and decomposer (11): 1, brine level; 2, metal anodes; 3, mercury cathode, flowing along baseplate; 4, mercury pump; 5, vertical decomposer; 6, water feed to decomposer; 7, graphite packing, promoting decomposition of sodium amalgam; 8, caustic liquor exit; 9, denuded mercury; 10, brine feed; 11, brine exit; 12, hydrogen exit from decomposer; 13, chlorine gas space; 14, chlorine exit; 15, wash water. Courtesy of the Chlorine Institute, Inc.

electrolyzer into a second cell called the decomposer. This is a short-circuited electrical cell in which graphite acts as the cathode and the amalgam as the anode; the reaction is that of equation 5.

The mercury cell operates efficiently because of the higher overpotential of hydrogen on mercury to achieve the preferential formation of sodium amalgam. Certain trace elements, such as vanadium, can lower the hydrogen overpotential, however, resulting in the release of hydrogen in potentially dangerous amounts.

Mercury cells are operated to maintain a 21–22 wt % NaCl concentration in the depleted brine and thus preserve good electrical conductivity. The depleted brine is dechlorinated and then resaturated with solid salt prior to recycling back to the electrolyzer.

Mercury has a high vapor pressure at the normal cell operating conditions; hence it is always found in the reaction products. Although the mercury is almost completely recovered and returned to the process, environmental problems associated with mercury, combined with the less efficient energy utilization compared to the modern membrane cell process, has effectively stopped the building of new mercury cell plants. Furthermore, in the 1990s, membrane cells will most likely replace most of the present mercury cells. For details related to mercury cells, see references 8 and 16 and general references.

The Diaphragm Cell Process. E. A. Le Sueur is credited with the design of the first chlorine cell incorporating a percolating asbestos diaphragm in the 1890s. Brine flows continuously into the anolyte and subsequently through a diaphragm into the catholyte. The diaphragm separates the chlorine liberated at the anode from the sodium hydroxide and hydrogen produced at the cathode. Failure to separate the chlorine and sodium hydroxide leads to the production of sodium hypochlorite [7681-52-9], NaClO, which undergoes further reaction to sodium chlorate [7775-09-9], NaClO$_3$. The commercial process to produce sodium chlorate is, in fact, by electrolysis of brine in a cell without a separator (see CHLORINE OXYGEN ACIDS, SALTS).

The early cells incorporated a horizontal asbestos sheet as the diaphragm. During the 1920s this type of cell was the most widely used in the world and a few are still in operation. Subsequently, three basic types of diaphragm cells have been developed: rectangular vertical electrode monopolar cells, cylindrical vertical electrode monopolar cells, and vertical electrode filter press bipolar cells. These may differ in the choice of diaphragm as well as in electrolyzer design. Electrolyzers are classified as being either monopolar or bipolar depending on the construction or assembly.

Asbestos Diaphragms. The earliest diaphragms were made of asbestos paper sheets. Asbestos (qv) was selected because of its chemical and physical stability and because it is a relatively inexpensive and abundant raw material. The vacuum-deposited asbestos diaphragm, developed in the 1920s, was the diaphragm of choice until 1971, when it was supplanted by the Modified Diaphragm (trademark of OxyTech Systems, Inc.) (17). In its most common form, the Modified Diaphragm contains fibrous polytetrafluoroethylene (PTFE) and a minimum of 75% asbestos. The polymer, following fusion, stabilizes the asbestos, lowers cell voltage, and allows the use of the expandable DSA anode (18) (see Fig. 8), which further lowers the cell voltage (19). The Modified Diaphragm in various formulations is the most common diaphragm in use today.

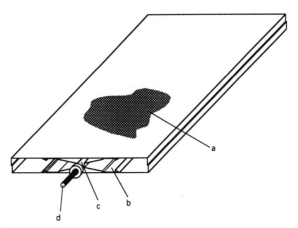

Fig. 8. Anode for monopolar diaphragm cells: a, activated (coated) expanded metal; b, expanding spring; c, titanium-clad copper bar; d, copper thread to fix the anode to the cell base.

The toxicological problems associated with asbestos have been widely published and asbestos has been banned from most uses by the EPA. However, modern diaphragm cell chlorine plants have not had difficulty meeting the required exposure limits for asbestos fibers, and, as of 1990, the chlorine industry had an exemption allowing the continued use of asbestos as a diaphragm material.

Nonasbestos Diaphragms. Many patents relating to the development of nonasbestos replacements for chlorine electrolyzer diaphragms have been issued. One such replacement, the microporous diaphragm (20), incorporates a PTFE sheet containing sodium carbonate as a pore former. Although early difficulties related to fluorocarbon wetting were overcome and the operating results were satisfactory, the microporous diaphragms have not reached commercial status. A mixture of PTFE fibers and zirconia was patented (21), as were synthetic diaphragms having ion-exchange properties similar to the membrane in a membrane cell (22). Neither of these diaphragms has been commercialized.

A vacuum-depositable nonasbestos fiber called Polyramix (trademark of OxyTech Systems, Inc.), having a PTFE backbone and zirconium oxide particles embedded in the structure, has been developed (23). The Polyramix diaphragm has a much longer operating life than an asbestos-based one because of superior chemical stability. In addition, the voltage and current efficiencies have proven equal to or better than modified diaphragms. The first chlorine plants to convert completely from asbestos-based diaphragms to the nonasbestos Polyramix design were G.E. Plastics plants in Mt. Vernon, Indiana and Burkville, Alabama in 1991.

Electrolyzers. Bipolar electrolyzers have unit assemblies where the anode of one cell is directly connected to the cathode of the next cell unit. This assembly minimizes intercell voltage loss because the cells are set up in series like a filter press. The voltage of the electrolyzer is therefore the sum of the individual cell voltages. Bipolar electrolyzers have relatively high voltages and low amperage; thus the cost of electrical rectification is high per unit production capacity (see Fig. 4). Bipolar electrolyzers must either be installed in a large number of electrical circuits or be designed with very large individual cell components. Bipolar cell developers have opted for the latter. In the 1970s, following the development of the DSA anode, the Glanor bipolar electrolyzer was developed.

A monopolar electrolyzer is assembled so that the anodes and cathodes are in parallel. As a result of this setup, the electrical potential of all cells in the electrolyzer is the same. Monopolar electrolyzers operate at a relatively low voltage, 3 to 4 V, and high amperage, allowing circuit construction of up to 200 electrolyzers.

In the United States, 76% of the chlorine produced is from diaphragm cells. Production is equally divided between bipolar and monopolar electrolyzers.

The Dow Bipolar Electrolyzer. Dow has operated bipolar electrolyzers since before 1910, and the design is characterized by simple, rugged construction utilizing relatively inexpensive materials (21,24,25). These cells employ vertical coated titanium anodes, vertical cathodes of mild steel wire mesh bolted to a perforated steel backplate, and a vacuum-deposited asbestos-based diaphragm. A single bipolar element may have 100 m^2 of active area. Copper spring clips attached to the back of the perforated plate of the cathode place the anode of one cell in direct contact with the cathode of the next cell. This electrical connection is immersed in the catholyte during operation. Figure 9 shows these internal cell parts. These electrolyzers are operated at lower current densities than others in the industry, normally using 50 or more cells in one unit or series. One electrical circuit may consist of only two of these series. Figure 10 shows a six-cell series. Treated saturated brine is fed into the anolyte compartment, where it percolates through the diaphragm into the catholyte chamber. The percolation rate is controlled by maintaining the anolyte at a level sufficient to establish a positive, adjustable hydrostatic head. The optimum brine flow rate results in the

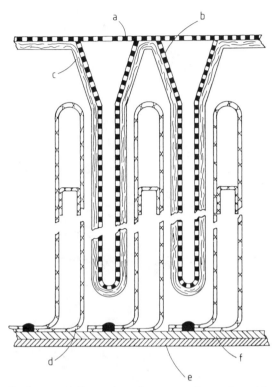

Fig. 9.   Dow diaphragm cell, section view: a, perforated steel back plate; b, cathode pocket; c, asbestos diaphragm; d, DSA anode; e, copper back plate; f, titanium back plate.

decomposition of about 50% of the input NaCl; thus the cell liquor contains 8–12 wt % NaOH and 12–18 wt % NaCl.

These electrolyzers are operated at about 80°C, as opposed to the 95°C common in the industry. This lower temperature allows the use of vinyl ester resins and other plastics for cell construction. Moreover, the Dow diaphragm cell is optimized for low current density (~ 0.5 kA/m^2). It consumes less electrical energy per unit of production than others in the industry. The cell voltage at low current density is only 300–400 mV above the thermodynamic decomposition voltage. Dow does not license its diaphragm cell technology and operating data for these cells are not available in the open literature.

The Glanor Bipolar Electrolyzer.   The Glanor bipolar electrolyzer, jointly developed by PPG Industries and De Nora Permelec, was especially designed for large chlorine plants (26). It consists of a series of bipolar cells clamped between two end electrode assemblies by means of tie rods, forming a filter-press-type electrolyzer. It was designed around and is equipped with DSA anodes. Although each electrolyzer normally consists of 11 cells, up to 12 cells have been operated, and a lesser number could be utilized.

Current is fed into the electrolyzer by means of anodic and cathodic end elements. The anodic compartment of each cell is joined to an independent brine feed tank by means of flanged connections. Chlorine gas leaves each cell from the

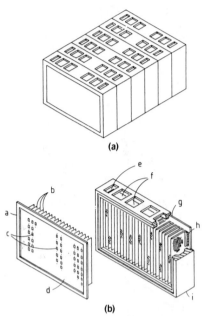

Fig. 10. Dow diaphragm cell: (**a**) Six-cell series. (**b**) Internal cell parts: a, cathode elements; b, cathode pocket elements; c, copper spring clips; d, perforated steel backplate; e, brine inlet; f, chlorine outlet; g, copper backplate; h, titanium backplate; i, anode element.

top, passing through the brine feed tank and then to the cellroom collection system. Hydrogen leaves from the top of the cathodic compartment of each cell; the cell liquor leaves the cathodic compartment from the bottom through an adjustable level connection.

The V-1144 electrolyzer (11 cells of 44 m^2 each) was the first commercial unit, and eight plants utilize this model. The second generation electrolyzer, the V-1161 (11 cells of 61 m^2 each), employs narrower electrode gaps, lower current density, and modified diaphragms, consuming less energy than the V-1144. The operating characteristics of the Glanor electrolyzers are shown in Table 9.

OxyTech Monopolar Electrolyzers. OxyTech Systems (a joint venture company of Occidental Chemical and Eltech Systems) supplies monopolar diaphragm electrolyzers of two designs: the OxyTech "Hooker" H-Type (27,28) shown in Figure 11 and the "Diamond" MDC-Type (28,29) in Figure 12.

The first commercialized vacuum-deposited diaphragm cell was the Hooker type S-1 monopolar cell, introduced in 1929. This design featured vertical graphite plates connected to a copper bus bar and a cathode having woven steel wire cloth fingers between the rows of anodes. An asbestos diaphragm, vacuum-deposited on the cathode, separated the anodic and cathodic compartments. The cathode fingers did not extend completely across the cell, but left a central circulation space. In the following 40 years a family of S-type cells having similar characteristics evolved. Over 12000 were installed in licensed plants.

In 1973 a new H-series of monopolar cells was introduced. These incorpo-

Table 9. Design and Operating Characteristics of Glanor Bipolar Diaphragm Electrolyzers

Item	Model V-1144	Model V-1161
cells per electrolyzer	11	11
active anode area per cell, m^2	35	49
electrode gap, mm	11	6
current load, kA	72	72
current density at 72 kA, kA/m^2	2.05	1.47
cell voltage, V	3.50	3.08
current efficiency, %	95–96	95–96
energy consumption (d-c), $kW \cdot h/tCl_2$	2800	2400
anode gas composition (alkaline brine), %		
$\quad Cl_2$	97.3–98.0	97.0–98.0
$\quad O_2$	1.5–2.2	1.5–2.2
$\quad H_2$	<0.1	<0.1
$\quad CO_2$	0.4	0.4
cell liquor		
$\quad$ NaOH, g/L	135–145	135–145
$\quad NaClO_3$, %	0.03–0.15	0.03–0.15
production per electrolyzer, t/day		
$\quad$ chlorine	24.2	24.2
$\quad$ NaOH	27.0	27.0

rated DSA anodes, operating at higher current densities. The H-series also employs cathode tubes having both ends open and extending across the cell, possible because the circulation space requirement was satisfied by the change from solid graphite to the open DSA anodes (Fig. 8). The anode and cathode row orientation was also rotated 90°, making the design similar to the MDC cells.

The MDC cells feature woven steel cloth cathode screen tubes open at both ends and welded into thick steel tube sheets at each end (28). The tubes, tube sheets, and the outer steel cathode shell form the catholyte chamber (Fig. 13). Copper is bonded to the rectangular cathode shell on the two long sides parallel to the tube sheets and copper connectors attached at the ends of the copper side plates complete the encircling of the cathode to ensure good current distribution. The rectangular design and electrode orientation reduces the electrical path around the circuit as compared to the S-series cells. Modified or Polyramix diaphragms are vacuum-deposited on the cathode tubes (30). Expandable DSA anodes are connected to a copper cell base protected by a TIBAC titanium or a rubber base cover. Orientation of the cathode tubes is parallel to the cell circuit, the opposite of the S-series cells. This arrangement accommodates the thermal expansion of the cell and circuit without changing the anode-to-cathode alignment, allowing for the use of rigid rather than flexible intercell connectors. Over 20,000 t/day of chlorine capacity has been licensed with MDC cells. This technology reduces energy consumption by 15% compared to that of regular asbestos diaphragms and standard box anodes. Table 10 lists the operating characteristics of the four most common OxyTech diaphragm cells.

OxyTech/Uhde HU-Type Designs. The HU-type cells (30) are shown in

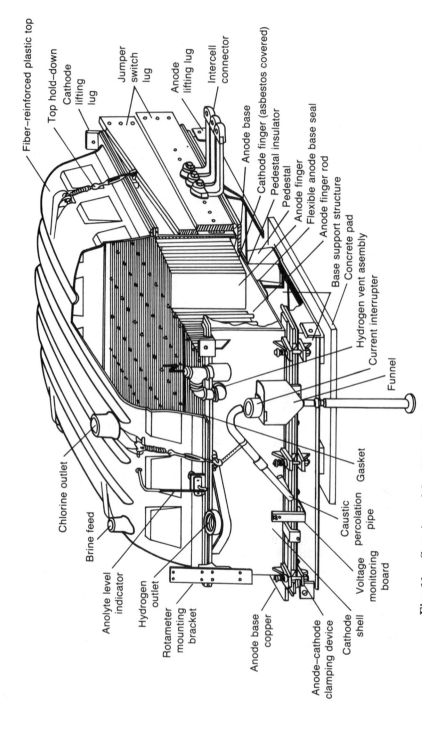

Fiber-reinforced plastic top

Top hold-down

Cathode lifting lug

Jumper switch lug

Anode lifting lug

Intercell connector

Anode base

Cathode finger (asbestos covered)

Pedestal insulator

Pedestal

Anode finger

Flexible anode base seal

Anode finger rod

Base support structure

Concrete pad

Hydrogen vent assembly

Current interrupter

Funnel

Gasket

Caustic percolation pipe

Voltage monitoring board

Cathode shell

Anode–cathode clamping device

Anode base copper

Rotameter mounting bracket

Hydrogen outlet

Anolyte level indicator

Brine feed

Chlorine outlet

Fig. 11. Cut view of OxyTech H-4 diaphragm cell operating at a nominal load of 150 kA.

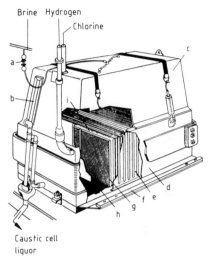

Fig. 12. OxyTech Systems MDC cell: a, brine feed rotometer; b, head sight glass; c, cell head; d, cathode assembly; e, tube sheet; f, grid plate; g, cathode tube; h, grid protector; i, DSA expandable anode.

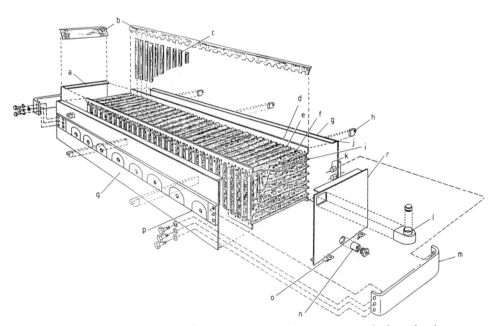

Fig. 13. Exploded view of an OxyTech MDC-55 cell: a, end plate; b, rim screen; c, side screens; d, tube sheet; e, full-cathode tube; f, half-cathode or end tube; g, side plate; h, lifting lug; i, punched and coined stiffener strap; j, bosses; k, end plate, operating aisle end; l, hydrogen outlet; m, connector bar; n, caustic outlet; o, clip angles; p, grid bar, connector side; q, side plate.

Table 10. Operating Capacities and Characteristics of OxyTech Diaphragm Cells

Characteristic	MDC29		MDC55		H2A		H4	
operating range, kA/day	40	80	75	150	50	80	90	150
chlorine capacity, t/day[a]	1.21	2.43	2.26	4.55	1.51	2.42	2.71	4.54
caustic capacity, t/day[a]	1.36	2.74	2.55	5.14	1.70	2.73	3.06	5.13
hydrogen capacity, m^3/day	399	798	748	1,497	498	798	897	1,496
current density, kA/m^2	1.38	2.76	1.37	2.74	1.38	2.21	1.40	2.33
current voltage, V (includes intercell bus)	2.97	3.60	2.97	3.59	2.97	3.35	2.98	3.40
energy consumption, d-c kW·h/t of Cl_2	2363	2847	2363	2839	2363	2655	2371	2694
modified diaphragm life, days	425	200	410	200	410	300	425	275
anode life, yr	10–15	8–10	10–15	8–10	10–15	8–10	10–15	8–10
cathode life, yr	10–15	5–8	10–15	5–8	10–15	5–8	10–15	5–8
distance between cells[b], m	1.60		2.13		2.32		3.05	

[a]To convert t to ton (short), multiply by 1.1.
[b]Distance is measured from center line to center line and side by side positioning with bus connecting.

Table 11. Design and Operating Characteristics of HU Series Diaphragm Cells

Item	Cell type						
	HU 24	HU 30	HU 36	HU 42	HU 48	HU 54	HU 60
number of anodes	24	30	36	42	48	54	60
anode surface area, m^2	20.6	25.8	31.0	36.1	41.3	46.4	51.6
load, kA	30–45	40–60	50–70	55–85	60–95	70–105	80–120
Cl_2 production, t/day	0.90–1.36	1.19–1.82	1.49–2.12	1.64–2.58	1.79–2.88	2.09–3.18	2.39–3.64
NaOH (100%) production, t/day	1.01–1.54	1.35–2.05	1.68–2.39	1.85–2.91	2.02–3.25	2.36–3.59	2.69–4.10
H_2 production, kg/day	25–39	34–52	42–60	47–73	51–82	59–91	68–103
cell length, m	2.1	2.6	3.0	3.5	3.9	4.4	4.8
cell-to-cell distance, m	1.5	1.5	1.5	1.5	1.5	1.5	1.5

Figure 14. The HU-type electrolyzer is more rectangular than even the MDC cell. It is narrow in the direction of current flow, since the anodes are arranged in a single row. Because the cathode is long and narrow, the current density through the thick steel cathode shell is also very low. The cathode design leads itself to closer anode–cathode spacing without the use of the expandable anode. Moreover, this heavy steel construction eliminates the copper around the cathode shell. Another advantage of the long, narrow design is shorter cell circuits, resulting in less piping and savings in other materials. The HU-type cells incorporate a modified diaphragm.

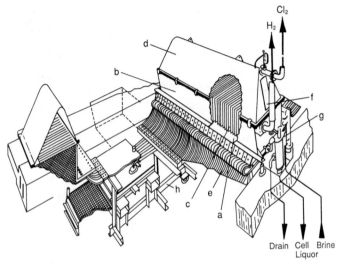

Fig. 14. OxyTech/Uhde HU-type cell: a, cell bottom; b, cathode; c, anode; d, cell cover; e, bus bars; f, brine level gauge; g, brine flow meter; h, bypass switch.

A further novelty of the HU cell system is the design and arrangement of the electrical bypass switch installed underneath, not next to, the circuit of cells. The cells are elevated above the floor, similar to mercury cells, creating a second operating floor; the interconnecting bus bars are flexible and are distributed over the length of the cell. The HU cell design incorporates a bus bar for each individual anode instead of the copper base plate used in the H and MDC series cells. During operation of the bypass switch, connection is made for each individual anode and no additional contact bars are required.

The HU-type cells are offered to cover the 30–150-kA range. All of the different cell types are equipped with cathodes and anodes of identical height and width. The only difference between the various models is the number of anode–cathode elements and consequently the length of the cell. Table 11 lists the characteristics of the various HU cells.

The Membrane Cell Process. In a membrane cell a cation-exchange membrane separates the anolyte and catholyte, as shown in Figure 5. Brine is fed into the anode compartment, where chlorine gas is created and the sodium ions and associated water of hydration migrate through the membrane into the catholyte. Unlike the diaphragm in the diaphragm cell process, the cation-exchange membrane prevents the migration of chloride ions into the catholyte. Depleted brine is discharged from the anolyte to maintain a minimum NaCl concentration. Water is electrolyzed at the cathode and strong caustic (32–35 wt %) is produced either by controlling the water addition rate directly to the catholyte or by recirculating caustic to which water has been added. There is some back-migration of hydroxyl ions into the anolyte, which results in the loss of current efficiency.

Membranes.   The membrane (see MEMBRANE TECHNOLOGY) is the most critical component of this cell technology, and current efficiency and cell voltage, and hence energy consumption, are greatly dependent on its quality. An ideal ion-exchange membrane separator should have (*1*) high selectivity for the transport of sodium or potassium ions, (*2*) negligible transport of chloride, hypochlorite, and chlorate ions, (*3*) zero back migration of hydroxide ion, (*4*) low electrical resistance, and (*5*) good mechanical properties and long term stability for practical use. The membrane in a chlor–alkali electrolyzer is exposed to chlorine on one side and strong caustic on the other. Only perfluoro polymers have been found to withstand these conditions, and, combined with appended groups having ion-exchange properties, they meet the requirements set forth above (31–39).

The first membranes to show significant potential and trigger the development of membrane electrolyzers were made from the perfluorosulfonate polymer called Nafion (40,41).

$$-(CF_2CF_2)_{\overline{n}}-CF-O-CF_2CFO-CF_2CF_2SO_3Na^+ \qquad -(CF_2CF_2)_{\overline{n}}-CF-O-CF_2CF_2CF_2COO^-Na^+$$

$$\begin{array}{ccc} \quad\;\; | & \quad\;\; | & \qquad\qquad\qquad\qquad\;\; | \\ \quad CF_2 & \quad CF_3 & \qquad\qquad\qquad\qquad CF_2 \\ \quad\;\; | & & \qquad\qquad\qquad\qquad\;\; | \end{array}$$

Nafion   Flemion

This first membrane was practical only at low caustic concentrations; it was deficient in limiting hydroxyl ion back migration. The idea of an asymmetric membrane having sulfonic acid groups on the anodic side and converted groups on the cathodic side to overcome the back-migration problem was then developed. These composite membranes operate efficiently at high caustic strengths, individual membranes exhibiting optimal efficiency at a given caustic concentration. Later a perfluorocarboxylate membrane called Flemion (42) which had better resistance to the hydroxyl back migration but also had a much higher electrical resistance was produced. Then the sulfonic acid group on the cathodic side of a Nafion membrane was converted to a carboxylic acid group, achieving the beneficial properties of both membrane types. Today's membranes consist of a film of perfluorosulfonate polymer, a Teflon reinforcing fabric, and a perfluorocarboxylate polymer all bonded together. The performance of such a membrane is shown in Table 12. References 39 and 43 discuss membranes in more detail.

In the design of a membrane electrolyzer utilizing standard membranes, minimizing the voltage drop through the electrolyte gap is accomplished by reducing the gap. When the gap is very small, however, an increase in voltage resulting from the entrapment of hydrogen gas bubbles between the cathode and the hydrophobic membrane is observed. Development membranes, the cathodic

Table 12. Nafion[a] 90209 Performance in Chlor–Alkali Production[b]

current density, kA/m^2	2	3	4
caustic concentration, %	32	32	32
temperature, °C	90	90	90
current efficiency, %	96.0	96.0	96.0
voltage[c], V	2.84	3.07	3.30
energy consumption, kW·h/t caustic soda	1,980	2,140	2,300
water transport (mol H_2O/mol Na^+)	3.9	4.0	4.2

[a]Du Pont registered trademark.
[b]Data, for a 32% caustic concentration at 90°C and a current efficiency of 96.0%, obtained in laboratory cells using a DSA anode and an activated cathode, where the membrane is against the anode at a 3-mm gap.
[c]Electrode to electrode.

surface of which is coated with a thin layer of a porous inorganic material to improve the membrane's hydrophilicity, has solved the bubble effect problem (Fig. 15). These improved, surface-modified membranes have allowed the development of modern electrolyzers called zero-gap or membrane-gap electrolyzers in which there is indeed no gap between the electrodes. Performance data for a membrane-gap electrolyzer are given in Figure 16.

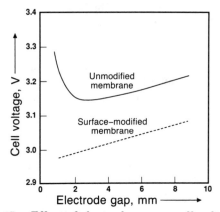

Fig. 15.   Effect of electrode gap on cell voltage.

The electrolysis of potassium chloride [7447-40-7], KCl, to produce chlorine and potassium hydroxide in membrane cells requires similar but unique membranes. Commercial membranes currently employed in high performance membrane electrolyzers include Du Pont's Nafion 900 series and Asahi Glass's Flemion 700 series.

Electrolyzers.   Depending on the manner in which electrical connection is made between cell units, electrolyzers are labeled as monopolar or bipolar. In the monopolar type all the anode and all the cathode elements are arranged in parallel, forming an electrolyzer having high amperage and low voltage. In the bipolar type the cathode of a cell is connected to the anode of the subsequent cell, so that the cells are in series and the resulting electrolyzer has low amperage and

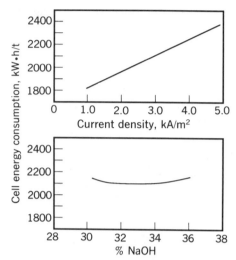

Fig. 16. Performance data obtained in laboratory cells using Nafion NX-961, DSA anode, activated cathode, narrow gap, at 90°C. Energy consumption is expressed as d-c kilowatt hours (electrode to electrode) per ton of NaOH.

high voltage. The bipolar type is advantageous for attaining minimum voltage drop between the cells. But, as the influx and efflux of electrolytes for cells having different electrical potential are gathered into common manifolds, problems of current leakage and corrosion become a concern and must be addressed in the design. In the monopolar cells the voltage loss through the interelectrolyzer bus bars is an inevitable drawback which must be minimized by conservative design. Monopolar electrolyzers are generally designed so that an individual electrolyzer can be short circuited, allowing maintenance cell membrane replacement to be conducted without shutting down the entire circuit.

Asahi Chemical Industry Acilyzer-ML Bipolar Electrolyzers. Asahi Chemical's electrolyzer (Acilyzer-ML), shown in Figure 17, is of the bipolar type, composed of a series of cell frames (42). Each cell consists of a pair of anode and cathode compartments facing each other, having an ion-exchange membrane in between. The anode and cathode compartments are separated by an explosion-bonded titanium–steel or titanium–nickel plate, and vertical support ribs are welded to each side of these partition walls, to which the anode and cathode mesh are spot welded in turn. Each compartment has an electrolyte inlet at the bottom and gas/liquid outlet at the top. These inlets and outlets are connected to the supply and collection headers by PTFE hoses. Both anolyte and catholyte are recirculated back to the electrolyzer from collection tanks. Deionized water is added to the catholyte collection tank to control the caustic concentration, and ultrapurified brine is added to the anolyte collection tank to control the NaCl concentration. A portion of the catholyte is drawn off and is sent to storage or an evaporator for further concentration. A portion of the depleted anolyte is drawn off and is returned to the salt dissolution and primary brine purification system. Supporting arms are attached to both sides of the cell frame; from these the frames

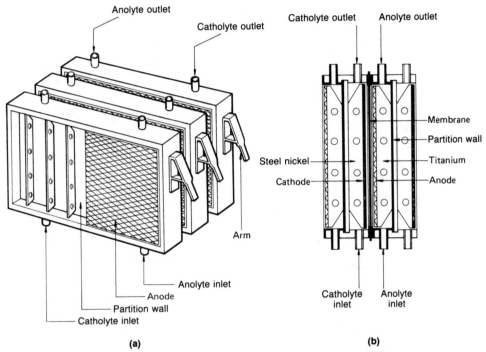

Fig. 17. Schematic of Asahi Chemical Acilyzer-ML bipolar electrolyzer. (**a**) View of cell units; (**b**) structure of cell.

are hung on the side bars of a hydraulic press. The electrolyzers are offered in two sizes: the ML-32, having a production capacity of 10,000 tons of sodium hydroxide per year, is 1.2 by 2.4 m, and the ML-60, having a production capacity of 20,000 tons per year, is 1.5 by 3.6 m. As of January 1989, 23 plants having a combined capacity of 1,711,000 tons of NaOH per year were operating or under construction.

Asahi Glass AZEC Monopolar Electrolyzers. The Asahi Glass Company supplies the four different monopolar electrolyzers shown in Figure 18 (44–46). The AZEC-M electrolyzer has elastomer cell frames having active areas that are 0.2 m wide and 1.0 m tall clamped between fastening plates. The electrical bus connections are from the sides. Natural recirculation is provided for both anolyte and catholyte through gas–liquid separators mounted above the cell frames. The AZEC-MD electrolyzer is a modification. Two M electrolyzers are connected directly side by side, compressed between one set of fastening plates and eliminating one set of intercell connectors. The AZEC-F1 electrolyzer is constructed of metal frames having active areas 2.4 m wide and 1.2 m high. The electrical bus connectors are from the bottom of the cell frames, which are arranged in either single or double stacks. In the latter case, the stacks are electrically in series, but compressed by one set of fastening devices. The anodic and cathodic gas separators are located above the electrolyzers. The AZEC-F2 electrolyzer also has metal frames. The active area is 1.71 m^2 per frame and the electrical bus connections are from the side. The cell frames are supported by the side arms of the supporting

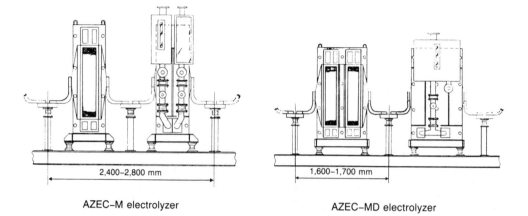

2,400–2,800 mm

AZEC–M electrolyzer

1,600–1,700 mm

AZEC–MD electrolyzer

AZEC–F1 electrolyzer

AZEC–F2 electrolyzer

Fig. 18.   Asahi glass electrolyzers.

unit of the electrolyzer. Raney nickel electrocatalyst is available as an activated cathode for these cells. In 1989 Asahi Glass had 26 plants in operation, producing approximately 1,900,000 tons of sodium hydroxide per year.

Chlorine Engineers CME Monopolar Electrolyzers.   Chlorine Engineers (CEC) (a joint venture company of Mitsui and Company and Mitsui Shipbuilding Company), produces the filter press type of membrane electrolyzer shown in Figure 19 (47). It is of the large element type. Membrane utilization ratio is very high. Uniform electrical current travels into each anode element through titanium-clad, copper-cored conductor rods and current distributors. The current distributor in the electrolyzer serves an additional role as a downcomer, helping the electrolyte self-circulate within the cell to maintain uniformly distributed concentrations as well as good gas release. The internal circulation is intended to eliminate the necessity for an external forced recirculation system. The gasket

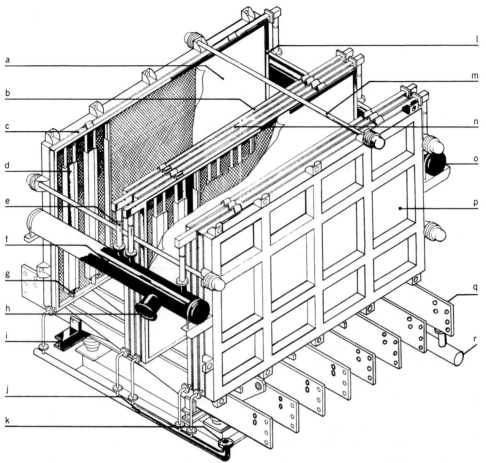

Fig. 19. CME monopolar electrolyzer: a, membrane; b, cathode element; c, half-cathode element; d, current distributor; e, Teflon tube; f, Cl_2 + depleted brine manifold; g, conductor rod; h, Cl_2 + depleted brine outlet nozzle; i, base frame; j, recycled NaOH manifold; k, recycled NaOH inlet nozzle; l, gasket (the gasket-to-element ratio is quite small); m, tie rod; n, anode element; o, H_2 + NaOH manifold; p, end plate; q, under cell bus bar (simplifies piping around the electrolyzers); r, feed brine manifold.

thickness sets the electrode spacing, and either finite- or zero-gap configuration can be accomplished. The anode frame is titanium and the cathode frame is stainless steel. The CME elements are thicker than competing elements for a lower electrolyte gas void fraction. This feature minimizes the drop in the liquid level during shutdowns. CEC offers electroplated activated cathodes. Gas and liquid exit the cell in the stratified overflow mode, as the liquid level is maintained in the upper cell frame. Semitransparent teflon tubes are used to monitor the operation visually. The electrical bus bars are installed underneath and at a right angle to the cell elements, requiring no equalizer between electrolyzers. The bus bar can be used as a short circuiting element by changing the connections. Chlorine Engineers had licensed 15 plants by 1989 with its CME electrolyzers for an annual production capacity of approximately 35,000 t of sodium hydroxide.

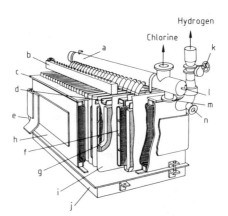

Fig. 20.   Cut view of Chlorine Engineers membrane bag cell: a, manifold; b, frame; c, partition plate; d, sealing plug; e, recirculated NaOH inlet; f, cathode; g, anode; h, cathode can; i, membrane bag; j, base; k, butterfly valve; l, feed brine; m, depleted brine; n, caustic outlet.

Chlorine Engineers MBC Electrolyzer.   Chlorine Engineers retrofits OxyTech MDC monopolar diaphragm cell electrolyzers to convert them into membrane cells (48,49). In retrofitting, CEC installs the membrane in the form of a bag that encloses the anodes (membrane bag cell). In the MBC-29 shown in Figure 20, one bag encloses two anodes. The current conductor bar of the anode passes through a hole in the bottom of the membrane bag for connection to the base. The open end of the bag, facing upward, is fixed to the partition plate by a sealing plug.

ICI FM-21 SP Monopolar Electrolyzers.   ICI's FM-21 SP monopolar electrolyzer incorporates stamped electrodes that are 2 mm thick and of a relatively small (0.2 m^2) size (50). The electrolyte compartments are created by molded gaskets between two of the electrode plates; the electrode spacing is finite and is established by gasket thickness. The electrode frames are supported from rails and are compressed between one fixed and one floating end plate by tie rods. Inlet and outlet streams are handled by internal manifolds. A crosscut view of the electrolyzer is shown in Figure 21. As of 1989, ICI had licensed 20 plants having an annual capacity of 468,250 t of NaOH.

Lurgi Monopolar Electrolyzer.   The Lurgi monopolar electrolyzer, Figures 22 and 23, features louvered plate cathodes (51). Plastic spacers not only set the finite gap between the electrodes but also subdivide the overall cell into a multiplicity of low height cells. The purpose is to force the gases to leave the inner electrode gap and to use the gas lift to create an internal circulation which tends to keep the salt concentration in the gap close to that of the bulk. The anode frame is made of titanium; the cathode frame may be of steel or nickel. An activated nickel cathode is available. The hollow frames of the cells have a multiplicity of orifices and provide ample room for distribution of fluids, reducing the possibility of clogging. The upper horizontal part of the electrodes is designed for gas–liquid disengagement outside the electrical field, avoiding superheating of membranes in the foamy, heat insulating environment.

Oronzio De Nora Technologies DD Monopolar and Bipolar Electrolyzers.   Oronzio De Nora Technologies has the DD-type cell design. Frame

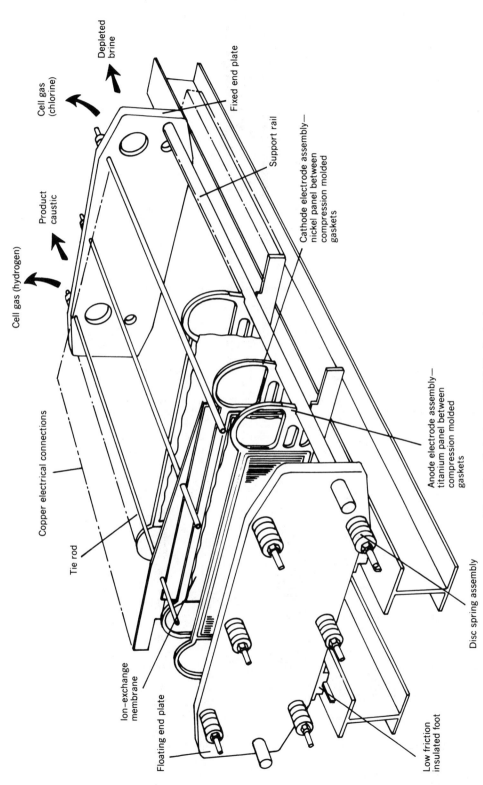

Cell gas (hydrogen)

Product caustic

Cell gas (chlorine)

Depleted brine

Copper electrical connections

Tie rod

Ion–exchange membrane

Floating end plate

Fixed end plate

Support rail

Cathode electrode assembly— nickel panel between compression molded gaskets

Anode electrode assembly— titanium panel between compression molded gaskets

Disc spring assembly

Low friction insulated foot

Fig. 21. Cut view of FM-21 electrolyzer.

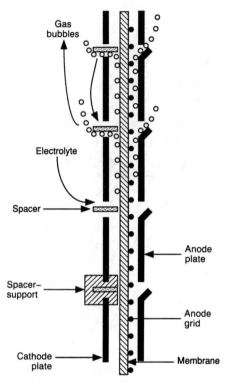

Fig. 22. Schematic of Lurgi monopolar membrane cell.

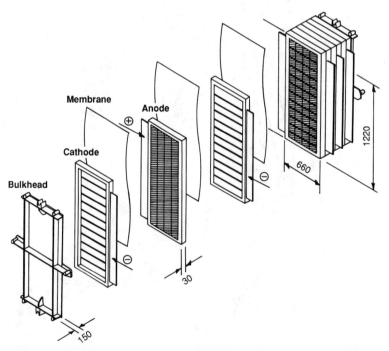

Fig. 23. Lurgi electrolyzer assembly. Numbers are dimensions in mm.

active area ranges from 0.9 to 1.7 m^2 for monopolar electrolyzers and from 0.9 to 5.12 m^2 for the bipolar ones (52,53). The unique design feature of the DD electrolyzers is a thick cast-metal current distributing element, used for collection and distribution of the electrical current. The cast-metal body has a number of protrusions which are used to conduct the current to the electrodes. Protection of the cast-metal bodies against chemical attack is achieved by protective liners of nickel and titanium sheets that are welded to the cast elements. Inlet and outlet nozzles are welded onto the liners. As of 1989, 11 plants representing a 283,700 t/yr NaOH capacity had been licensed.

The monopolar electrolyzer shown in Figure 24 consists of a repetition of anode element–membrane–cathode element series clamped between two end plates, each consisting of a half element. When the electrolyzer is assembled, the nozzles of all inlets, located at the bottom of the electrolyzer, and all outlets, located at the top of the electrolyzer, connect to form an integral manifold external to the cell. Current flows from one electrolyzer to the next in series through flexible copper connectors attached to the cast-metal current distrib-

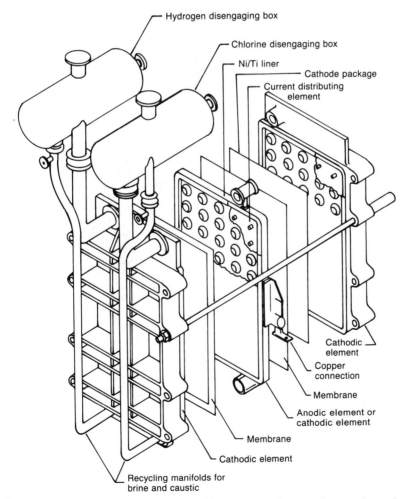

Fig. 24. De Nora Technologies DD-Type monopolar membrane electrolyzer.

uting elements. Each electrolyzer is provided with individual gas separators designed for natural recirculation. The elements are held together by a tie rod clamping system. The cathode includes a removable self-adjusting elastic nickel element and an activated cathode mesh which can be adjusted to the finite-gap or zero-gap configuration. The anode system consists of a removable thin DSA-coated screen against a titanium current conductor.

The De Nora DD-type bipolar electrolyzer is similar in construction to the monopolar electrolyzer except that each cell frame is composed of a pair of anode and cathode compartments opposing each other and having one ion-exchange membrane placed between them. The iron frames are supported by hangers. Each cell is provided with individual flexible PTFE inlet and outlet hoses between the electrolyzer and the feed and collection headers. Recirculation of fluids may be either natural or pressure fed.

OxyTech Systems MGC Monopolar Electrolyzers. The MGC (membrane gap cell) electrolyzer is distinguished by a design in which the electrodes are separated only by the thickness of the membrane itself (54). High surface area DSA anodes and activated cathodes are employed. The anode and cathode sections are alternated in the electrolyzer sandwich and the ion-exchange membrane is placed between them. The anode section consists of a lightweight stamped titanium anode frame having fixed supports welded between the pan and the DSA electrode mesh, a gasket, reticulate nickel interface material, a copper current distribution plate, and a second anode pan electrode assembly. The nickel cathode section is similarly arranged; however, the cathodic electrode is supported by flexible nickel springs, allowing the cathode mesh to press against and conform to the membrane and the opposing anode mesh. The active area of each membrane is $1.5 \, \mathrm{m^2}$ and up to 30 membranes can be assembled into one electrolyzer, providing a production capacity of up to 2810 tons of NaOH per year per electrolyzer.

Connecting spool pieces and gaskets are inserted to create inlet and outlet anolyte and catholyte manifolds external to but integral with the electrolyte compartments, as shown in Figure 25. The electrolyte compartments are sealed with unique O-ring gaskets in a staggered arrangement in which the catholyte gasket is located closer to the fluids than the anolyte O-ring gasket. This design provides a double seal and utilizes the membrane itself to protect the anolyte gasket from the corrosive anolyte. The intercell connections are from the sides, and the electrical bus provides the support for the electrolyzer on a copper current redistribution bus between each electrolyzer. As of the end of 1989, OxyTech Systems had licensed 30 plants having annual production capacity in excess of 750,000 tons of NaOH.

Hoechst-Uhde Bipolar Electrolyzers. The characteristic design feature of the Hoechst-Uhde membrane cell is its single element shown in Figure 26 (55). Each element consists of an anode pan, membrane, gasket, and a cathode pan sealed by means of individually bolted separate flanges instead of the filter press designs used in other electrolyzers. A group of single elements is combined to form an electrolyzer. The current passes from the cathodic back wall of one element to the anodic back wall of the subsequent element by a series of contact strips. Feed and discharge from the individual elements is conducted by flexible PTFE hoses. Both inlet and outlet connections are located at the bottom of the electrolyzers. The electrodes are of the louver type and the electrode spacing can be varied to

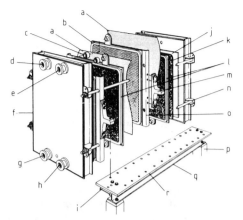

Fig. 25. OxyTech MGC electrolyzer: a, membrane; b, anode assembly; c, manifold spacer; d, anolyte outlet; e, catholyte outlet; f, bulkhead; g, brine inlet; h, NaOH inlet; i, insulating channel; j, bulkhead insulator; k, interface material; l, cathode assembly; m, intercell bus; n, tie rod; o, current distributor; p, electrolyzer support; q, support beam; r, connecting bus bar.

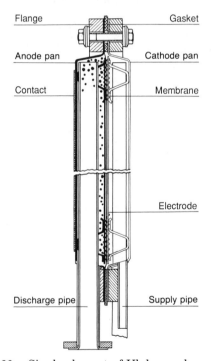

Fig. 26. Single element of Uhde membrane cell.

either the finite-gap or zero-gap configuration. The elements are normally arranged in series (bipolar arrangement) but may be arranged in parallel (monopolar arrangement).

The typical electrode active areas are 1.8 and 2.7 m² and annual electrolyzer production capacity can be up to 16000 tons of NaOH. In 1989 Uhde had 17 plants in operation or under construction having an annual capacity of 800,000 tons of NaOH.

A summary of the current membrane cell technologies is provided in Table 13.

Catalytic Cathodes.   The cathode material generally found in diaphragm cells is low carbon steel in either mesh or perforated form, whereas nickel or stainless steel is used in membrane cell electrolyzers. The hydrogen overvoltage is typically about 350 to 400 mV at 20 A/dm^2 in 2.5 N NaOH at 90°C. The exact value depends on the surface state of the steel or nickel. Energy savings by reducing overpotential as much as 200 to 250 mV are realizable in principle by using catalytic coatings on the cathode substrates. Various surface modification approaches have been adopted since the mid 1970s to decrease the overvoltage of these cathodes. Methods include enhancing the effective area of the cathode by using materials that provide high surface area and better electrocatalytic properties than steel. Composites generally chosen for coating are nickel or noble metal based. They are deposited on the cathode by thermal or plasma and/or electrolytic routes using a second component such as aluminum or zinc, which is leached out in NaOH solutions to achieve a highly electrochemically active surface. Several compositions are mentioned in the literature (8,56–58). However, the coatings that are commercially employed in membrane cells include Ni–S (59), Ni–Al (60), and Ni–NiO mixtures (61), and nickel coatings containing the platinum group metals (62).

Although catalytic cathode technology is practiced in membrane cells, commercialization is still awaited in diaphragm cells. The technical problems confronting catalytic cathode use in diaphragm cells include selection of a coating technique for the complex cathode assembly that would not adversely influence the structural tolerances involved in the fabrication of cathodes, and developing shutdown procedures which would eliminate hypochlorite as quickly as possible to preserve the catalytic activity of the coatings. Such problems do not exist in membrane cells because of the simplicity of the cathode structure and the anion rejection properties of the membrane.

Catalytic cathodes in membrane cell operations exhibit a voltage savings of 100–200 mV and a life of about 2+ yr using ultrapure brine. However, trace impurities such as iron from the caustic recirculation loop can deposit on the cathode and poison the coating, thereby reducing its economic life.

Membrane Cells with Air Depolarized Cathodes.   Substituting an oxygen reduction reaction for the hydrogen evolution reaction at the cathode of a membrane cell reduces the electrical energy required to produce chlorine and sodium hydroxide (63). In the air cathode cell, shown in Figure 27, the anode reaction is the discharge of chloride ions to form chlorine. Brine entering the anolyte chamber is oxidized at the anode to form chlorine gas, and sodium ions migrate through the ion-exchange membrane. Oxygen is reduced at the air cathode to form hydroxide ion which combines with the sodium ion from the anode compartment to form sodium hydroxide. There are three unique features in an air cathode cell. The first is the inclusion of an air chamber behind the electrode to provide oxygen for the cathodic reaction. Second, because no hydrogen is produced, there is no need for hydrogen collection equipment. Third, and most

Table 13. Summary of Current Membrane Cell Technologies

	Bipolar cells			Monopolar cells					
	Asahi Chemical Acilyzer	De Nora Bipolar	Uhde	Asahi Glass AZEC-M	CEC DMC-404x2	De Nora DD	Lurgi	ICI FM-215P	OxyTech MGC
effective membrane area, m^2	2.88–5.4	0.9–5.12	1–3	0.2	3.03	0.9–1.7	0.8	0.21	1.5
cells per electrolyzer	50–100	10–32	2–120	30–540	4	10–32		1–120	2–30
current load, kA	5.8–21.6	2.7–20.5	2–18	18–340	24–48	27–218	1–4	1–100	6–225
current density	2–4	2–4	2–6	3–4	2–4	3–4		1.5–4.1	2–5

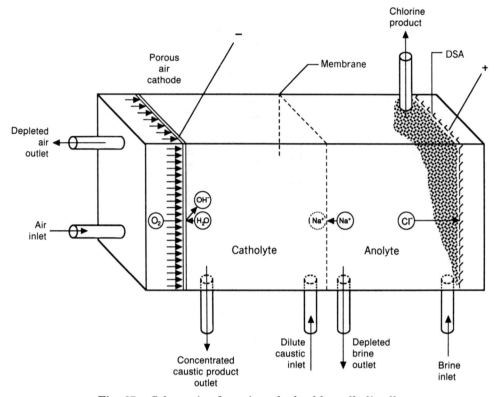

Fig. 27. Schematic of an air-cathode chlor–alkali cell.

important, is that the air cathode cell operates at approximately 1.0 V lower than a conventional hydrogen evolving chlorine cell.

The plant incorporating the air cathode electrolyzer must include a high performance air scrubbing system to eliminate carbon dioxide from the air. Failure to remove CO_2 adequately results in the precipitation of sodium carbonate in the pores of the cathode; this, in turn, affects the transport of oxygen and hydroxide within the electrode. Left unchecked, the accumulation of sodium carbonate will cause premature failure of the cathodes.

Most of the voltage savings in the air cathode electrolyzer results from the change in the cathode reaction and a reduction in the solution ohmic drop as a result of the absence of the hydrogen bubble gas void fraction in the catholyte. The air cathode electrolyzer operates at 2.1 V at 3 kA/m^2 or approximately 1450 d-c kW·h per ton of NaOH. The air cathode technology has been demonstrated in commercial sized equipment at Occidental Chemical's Muscle Shoals, Alabama plant. However, it is not presently being practiced because the technology is too expensive to commercialize at power costs of 20 to 30 mils (1 mil = 0.1 ¢/kW).

Chlorine Plant Auxiliaries. Flow diagrams for the three electrolytic chlor–alkali processes are given in Figures 28 and 29. Although they differ somewhat in operation, auxiliary processes such as brine purification and chlorine recovery are common to each.

Brine Purification. Sodium chloride, whether solution mined or obtained

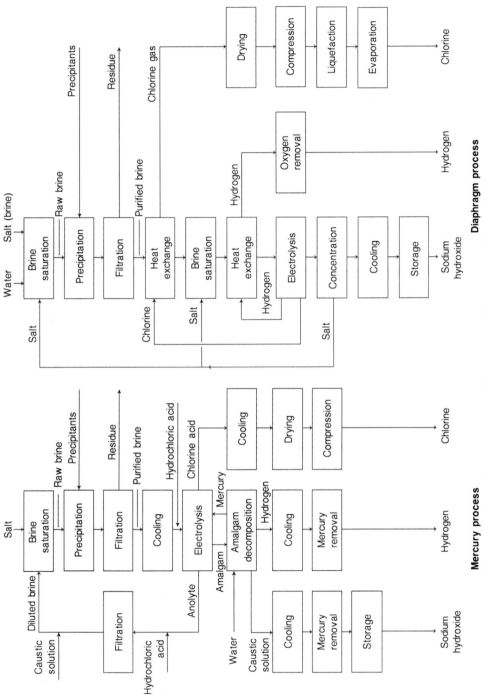

Mercury process

Diaphragm process

Fig. 28. Flow diagrams of the Mercury and Diaphragm chlor–alkali processes.

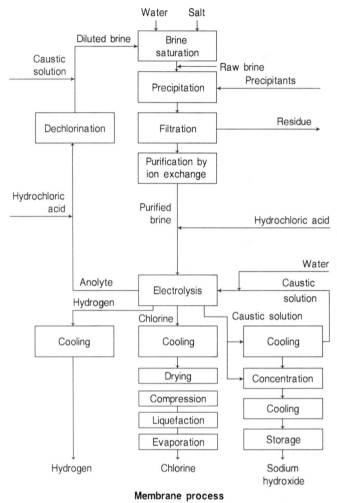

Membrane process

Fig. 29.   Flow diagram of the membrane chlor-alkali process.

as either mined or solar evaporated solid salt, contains contaminants (primarily calcium, magnesium, barium, and sulfates) that are detrimental to the electrolytic process. Trace metals such as iron, titanium, molybdenum, chromium, vanadium, and tungsten can also cause problems in mercury cells by lowering the hydrogen overpotential on mercury. Membranes are especially sensitive to brine impurities, and the membrane cell process requires a higher degree of brine quality than either the diaphragm or mercury system. Impurities that are soluble in the acidic anolyte can become insoluble once they have entered the membrane. In addition to induced flow of cation impurities, neutral or anion impurities can also enter the membrane via diffusion and the considerable water flow from the anolyte to the catholyte. When impurities precipitate, the membrane is physically disrupted; reducing the current efficiency if the disruption occurs in the cathode-side layer of the composite membrane. Even if the impurity is subsequently washed out, the void left behind will still contribute to reduced current efficiency, allowing hydroxyl ions to penetrate further into the membrane from the catholyte. There are synergistic effects, such as the precipitation of complex compounds, usually incorporating calcium and magnesium impurities in the feed brine. Iodine can damage a membrane and reduce current efficiency by precipitating as sodium paraperiodate, $Na_3H_2IO_6$. In addition, some brines contain ammonium ions or organic nitrogen which can be converted to nitrogen trichloride in the electrolyzer and, if concentrated in the downstream process, NCl_3 may explode. Ammonium ions in the brine are removed either by chlorine, producing NH_2Cl which can be removed as gas, or by hypochlorite, which produces nitrogen.

Removal of brine contaminants accounts for a significant portion of overall chlor–alkali production cost, especially for the membrane process. Moreover, part or all of the depleted brine from mercury and membrane cells must first be dechlorinated to recover the dissolved chlorine and to prevent corrosion during further processing. In a typical membrane plant, HCl is added to liberate chlorine, then a vacuum is applied to recover it. A reducing agent such as sodium sulfite is added to remove the final traces because chlorine would adversely react with the ion-exchange resins used later in the process. Dechlorinated brine is then re-saturated with solid salt for further use.

Brines, either from a well in the case of many diaphragm cell plants or from the recirculated, resaturated supply of the mercury or membrane processes, are treated with sodium carbonate to precipitate calcium carbonate, followed by sodium hydroxide to precipitate magnesium hydroxide (64). Most trace metal impurities are also precipitated during this process. The brine is usually heated before treatment to improve the reaction times and the precipitation and settling. Flocculants and other agents are sometimes added, especially if the calcium to magnesium ratio is low, since the calcium carbonate helps to settle the light, fragile magnesium hydroxide flock. In the diaphragm cell process sodium sulfate is controlled by a purge to another process from the caustic evaporators, improving the purity of the salt recovered in the evaporators, which is used to saturate the brine. The dechlorinated brine from the mercury and membrane cell processes is treated with calcium chloride to precipitate calcium sulfate, followed by settling.

Following carbonate and caustic treatment, the precipitates are allowed to settle in a clarifier where most of the solids are removed as a mud. The mud is then washed to recover entrained sodium chloride. The overflow from the clarifiers is filtered, preferably by upflow sand filters followed by precoated polishing filters, to remove all suspended solids. This primary purification step produces brine with less than 4 ppm calcium and 0.5 ppm magnesium, which is satisfactory for both the mercury and diaphragm processes. The membrane process, however, requires brine having less than 20 ppb hardness and this level of purity requires a further purification step. The alkaline brine passes through a series of two or more ion-exchange columns. The quantity of resin in each column must be sufficient to remove all of the impurities (calcium, magnesium, and barium) while the other column is being regenerated with HCl followed by NaOH treatment.

The brine feed to the electrolyzers of all the processes is usually acidified with hydrochloric acid to reduce oxygen and chlorate formation in the anolyte. Table 14 gives the specifications of the feed brines required for the membrane and diaphragm cell process to realize optimal performance.

Table 14. Typical Specifications for Feed Brine to Electrolyzers

	Membrane cell process[a]	Diaphragm cell process[a]
sodium chloride	280–305 g/L	320 g/L[b]
calcium and magnesium	20 ppb	5 ppm
sodium sulfate	7 ppm	5 ppm
silicon dioxide	5 ppm	0.5 ppm
aluminum	50 ppb	0.5 ppm
iron	0.5 ppm	0.3 ppm
mercury	0.04 ppm	1 ppm
heavy metals	0.05 ppm	0.05 ppm
fluoride	1 ppm	1 ppm
iodine	0.4 ppm	
strontium	0.5 ppm	[c]
barium	0.4 ppm	[c]
total organic carbon	1 ppm	1 ppm
pH	2–11	2.5–3.5

[a]All ppb and ppm values represent maximum concentration allowed.
[b]Minimum concentration allowed.
[c]Included with calcium.

Chlorine Cooling, Drying, Liquefaction, and Recovery. Chlorine produced by any of the electrolyzer processes is saturated with water vapor at high temperature. The pressure can range from a slightly negative or positive value to, in the case of some membrane cell plants, up to several hundred kilopascals. In the case of the diaphragm process the chlorine gas stream also carries along droplets of sodium hydroxide and salt. The chlorine is first cooled to usually not less than 10°C to avoid the formation of chlorine hydrate crystals. Cooling is carried out indirectly in titanium tubular heat exchangers in one or two stages using chilled water on the coolest stage. Water and remaining solids are removed in either wet Brinks demisters, which have special filter elements containing glass wool fibers, or electrostatic precipitators.

Chlorine drying is carried out using concentrated (96–98 wt %) sulfuric acid, and, depending on the use intended for the dilute sulfuric acid by-products, the drying can be accomplished in from two to four stages. The sulfuric acid and chlorine flow countercurrently. The drying towers are usually packed towers of brick-lined rubber-lined steel, or PVC-lined glass-fiber-reinforced plastic. After drying, the chlorine gas is passed through demisters to remove sulfuric acid mist. The upper limit for moisture remaining after drying is 50 ppm. Often, the cooled dry gas is scrubbed using liquid chlorine. This precools the chlorine gas prior to compression and provides further purification. Small chlorine plants utilize reciprocating or sulfuric acid sealed liquid-ring compressors. However, turbo compressors are the most economical for larger plants. In all chlorine operations involving compression, great care must be taken to prevent the heat from increasing the temperature of the gas enough to reach the chlorine–steel ignition temperature. It is for this reason that the precooling described above is used. In addition, there are often intercoolers between stages of turbo compressors.

About half of the chlorine produced is used as cooled dry gas which is transported by pipelines to the consuming process; the remainder is liquefied, stored, and shipped. Chlorine can be liquefied over a wide range of pressures and temperatures. Increasing the liquefaction pressure increases the energy required for compression. However, it also reduces the energy required for refrigeration to the extent that the overall energy spent is less. Thus the temperature and pressure at which the chlorine is stored are also factors in selecting the liquefaction pressure.

Any hydrogen contained within the chlorine from the electrolyzer is concentrated in the residual gas from the liquefaction process and must not be allowed to exceed the explosive concentration limit of 5%. Although hydrogen concentration can be controlled by adding dry air to the process, single-stage liquefaction units are not designed to liquefy more than 90–95% of the chlorine. To achieve a higher percentage of liquefaction additional compression and refrigeration are required. This is sometimes followed by the OxyTech chlorine recovery process (65).

Chlorine in the tail gas from liquefaction can be combined with chlorine from plant vessel evacuation systems and that from returned tank cars, trucks, and barges, and be recovered in a chlorine recovery unit (66). The recovered gas mixture and recycle streams are further compressed in a reciprocating compressor. The gas is then cooled, first using water, and then Freon, to $-12°C$. The chilled gas is sent to an absorber while a portion of chlorine which is liquefied is either used as reflux later in the process or sent to storage. The chilled gas enters the bottom of a chlorine absorber flowing upward through two packed sections while cold carbon tetrachloride flows countercurrently. All of the chlorine is absorbed in the carbon tetrachloride and the hydrogen and other inert gases exit the top of the absorber into a solvent recovery unit which removes the residual carbon tetrachloride before venting to the atmosphere. The chlorine-rich carbon tetrachloride leaves the bottom of the absorber and is forced by differential pressure between the towers to the middle of the chlorine stripper. The stripper feed flows downward through two packed sections, releasing chlorine as it is heated. A thermosiphon reboiler provides the heat at the bottom of the stripper. Chlorine stripped from the carbon tetrachloride passes upward through a packed top section of the column where it is cooled and scrubbed of solvent by liquid chlorine reflux. The recovered chlorine in the stripper overhead is sent to the chlorine liquefaction system or recycled to the beginning of the chlorine recovery process to provide an adequate supply of stripper reflux.

Sodium Hydroxide Processing. Sodium hydroxide is usually produced as a 50% water-based solution, although 73% and anhydrous sodium hydroxide are also marketed. High purity sodium hydroxide is available directly from the mercury and membrane cell processes. The mercury cell caustic is produced at 50% strength and only filtration is required to remove mercury droplets. The concentration of sodium hydroxide in the diaphragm cell effluent is only 10–12%. The effluent also contains 13–16% sodium chloride, 0.25% sodium sulfate, and 0.15% sodium chlorate. The solubility of NaCl in 50% caustic is approximately 1–1.5%; thus the salt crystallizes during the evaporation process. From 4.6 to 7.3 tons of water must be evaporated to produce 1 ton of 50% NaOH. Most large diaphragm cell chlor–alkali plants have an associated cogeneration power plant. In these facilities the caustic evaporators, which are triple or quadruple effect, are important for utilizing the by-product steam. The flows of liquor to the different effects varies from plant to plant. The triple-effect system presented in Figure 30 is an example of one of the many different systems. The cell liquor or effluent is fed into the second-effect evaporator. It is partially concentrated and transferred to the third vapor effect through the salt settler system to the first-effect evaporator and finally to the liquor flash tank. Salt is recovered from all three evaporators. Sodium borohydride [16940-66-2] is commonly added to the evaporation system to reduce corrosion of the equipment and to reduce the nickel content in the finished product (67).

Sodium sulfate [7757-82-6], Na_2SO_4, also crystallizes in the latter stages of evaporation. This high sulfate salt can be removed by isolation and not returned to the brine system. Because the diaphragm process produces high quality solid salt and the membrane cell process requires solid salt, it is a good combination to have both diaphragm cells and membrane cells in the same chlor–alkali producing complex. The membrane cell produces 32–35% caustic directly from the electrolyzer. The concentration is normally raised to 50% in a multiple-effect evaporator that is much simpler than the diaphragm caustic evaporator; no solids separation equipment is needed because the salt concentration in this caustic is very low.

Hydrogen Processing. The hydrogen produced in all electrolytic chlor–alkali processes is relatively (>99.9%) pure and requires only cooling to remove water along with entrained salt and caustic. The heat is often recovered into the brine system. Hydrogen from the mercury process must also be scrubbed to remove mercury. The hydrogen is compressed using water-sealed liquid-ring or Roots-type blowers. The hydrogen system, from the electrolyzers to the compressor suction, is always under positive pressure to avoid combination with air, which could form an explosive mixture. Some uses of hydrogen require additional removal of traces of oxygen. This is accomplished by reaction of the hydrogen with oxygen over a platinum catalyst.

The hydrogen can be used for organic hydrogenation, catalytic reductions, and ammonia synthesis. It can also be burned with chlorine to produce high quality HCl and used to provide a reducing atmosphere in some applications. In many cases, however, it is used as a fuel.

Other Chlorine Production Processes. Although electrolytic production of Cl_2 and NaOH from NaCl accounts for most of the chlorine produced, other commercial processes for chlorine are also in operation.

Chlorine from Potassium Hydroxide Manufacture. One of the coproducts during the electrolytic production of potassium hydroxide employing mercury and membrane cells is chlorine. The combined name plate capacity for caustic potash during 1988 totalled 325,000 t/yr and growth of U.S. demand was expected to be steady at 2% through 1990 (68).

Chlorine from HCl. Most organic chlorination reactions consume only half the Cl_2 to produce the desired product; the other half is converted to HCl. Depending on demand and supply of Cl_2 vs HCl, chlorine recovery from hydrochloric acid is sometimes attractive. Two commercial routes are available: electrolysis and oxidation (69).

Electrolysis of HCl. Electrolytic decomposition of aqueous HCl to generate

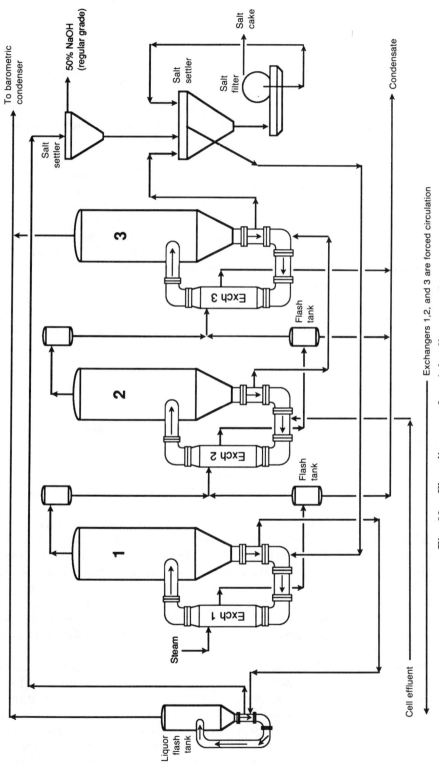

Fig. 30. Flow diagram of a triple-effect caustic evaporator.

Exchangers 1, 2, and 3 are forced circulation

Cl_2 and H_2 follows the overall reaction

$$2 \ HCl \longrightarrow H_2 + Cl_2 \tag{25}$$

A typical electrolytic cell employs graphite electrodes and a PVC diaphragm. Chlorine dissolved in the anolyte diffuses through the diaphragm and is reduced at the cathode resulting in a 2 to 2.5% loss in chlorine efficiency. The cells are operated below 85°C at a HCl concentration of 18.5 wt %. Each cell has an effective surface area of 2.5 m^2 and operates at a current density of 4–5 kA/m^2 corresponding to a cell voltage of ~ 1.90 V and an energy consumption of 1400–1500 kW·h per ton of Cl_2. Spent acid ($\sim 20\%$ HCl) is resaturated after cooling and recycled. The chlorine is dried using sulfuric acid and hydrogen is scrubbed with caustic for removal of Cl_2 and HCl. Hydrochloric acid electrolysis cells are manufactured by Hoechst-Uhde (70). Each electrolyzer consists of 30–36 individual cells connected in series and operates up to a load of 12 kA (69,70). There are a number of these electrolytic operations throughout the world, but only one in the United States, operated by Mobay at Baytown, Texas.

Three other processes that electrolytically convert HCl to Cl_2 using metal chloride catalyst are the Schroeder process, using $NiCl_2$; the Westvaco process, using $CuCl_2$; and the South African process, using $MnCl_2$. None of these processes is commercial. The Schroeder process illustrates the principles involved in these technologies. The Schroeder process is cyclic and involves two steps. The metal chloride is electrolyzed to produce the metal and chlorine:

$$NiCl_2 \longrightarrow Ni + Cl_2 \tag{26}$$

During the second step the metal reacts with HCl to reform the metal chloride, which is then recycled:

$$Ni + 2 \ HCl \longrightarrow NiCl_2 + H_2 \tag{27}$$

The overall reaction is that of equation 25.

Chlorine from the Magnesium Process.   Magnesium is produced by the fused salt electrolysis of $MgCl_2$ (see MAGNESIUM AND MAGNESIUM ALLOYS). The largest magnesium plant in the United States, Dow Chemical at Freeport, Texas, uses calcium–magnesium carbonate as a raw material and the chlorine is recycled within the process. The Rowley, Utah, plant of the AMAX Magnesium Corporation produces magnesium directly from purified magnesium chloride recovered from the Great Salt Lake. Some of the chlorine is recycled in the process, but most of it is sold commercially. SRI International lists AMAX's chlorine capacity at 13,600 t/yr.

Chlorine from the Titanium Process.   Electrolysis of magnesium chloride is a process step in the production of titanium, and one plant produces chlorine by-product for market. Titanium metal is produced by the reaction of titanium tetrachloride with magnesium (see TITANIUM AND TITANIUM ALLOYS). The magnesium chloride from this process step is then electrolyzed as in the production of magnesium. In most titanium plants, the chlorine is recycled in the production of titanium tetrachloride from the titanium oxide ore rutile [1317-80-2]. There is, however, one titanium plant (Oregon Metallurgical Corporation, Albany, Oregon) which purchases titanium tetrachloride [7550-45-0] as its raw material. This plant therefore markets the chlorine generated by the magnesium chloride fused salt cells. SRI International lists Oregon Metallurgical Corporation's capacity at 10,800 t/yr.

Chemical Oxidation of HCl.   Chlorine can be produced from HCl using air or O_2 following the overall reaction

$$4 \text{ HCl} + O_2 \longrightarrow 2 \text{ Cl}_2 + 2 \text{ H}_2O \tag{28}$$

Three routes pursued to achieve this conversion are catalytic oxidation of gaseous HCl, direct oxidation of HCl by an inorganic oxidizing agent, and two-stage processes involving intermediate formation of a metal chloride from either its oxide or oxychloride and subsequent release of the chlorine by air, oxygen, or heat. Only two of these processes have been commercialized.

Shell Chlorine Process.   The Shell process produces Cl_2 from the HCl using air or O_2 in the presence of cupric and other chlorides on a silicate carrier (71). The reaction proceeds at an optimal rate in the temperature range of 430–475°C at an efficiency of 60–70%. A manufacturing unit was built by Shell in the Netherlands (41,000 t/yr) and another in India (27,000 t/yr). Both plants have been closed down.

Kellogg Chlorine Process.   The Kellogg process uses $\sim 1\%$ nitrosylsulfuric acid [7782-78-7] catalyst and a dissimilar material containing a clay desiccant having a reversible water content of ~ 0.5 wt % and a crystalline structure stable to at least 760°C (72,73). Montmorillonite [1318-93-0] is the desired clay desiccant. It absorbs water as it forms, shifting the equilibrium of equation 28 to the right. The basic reaction is carried out on a fluidized bed in which the solids run countercurrent to the gaseous reactants at a temperature of 400–500°C and pressures of 300–1200 kPa (3–12 atm). Nitrosylsulfuric acid catalyst is fed into the top of the stripper column where it reacts with HCl to form nitrosyl chloride which then reacts with O_2 in the oxidizer to produce Cl_2:

$$\text{HCl} + \text{NOHSO}_4 \longrightarrow \text{NOCl} + \text{H}_2\text{SO}_4 \tag{29}$$

$$2 \text{ NOCl} + O_2 \longrightarrow 2 \text{ NO}_2 + \text{Cl}_2 \tag{30}$$

In the absorber–oxidizer HCl is further converted to chlorine

$$\text{NO}_2 + 2 \text{ HCl} \longrightarrow \text{NO} + \text{Cl}_2 + \text{H}_2O \tag{31}$$

and the nitrosyl sulfate formed is recycled to the stripper:

$$\text{NO} + \text{NO}_2 + 2 \text{ H}_2\text{SO}_4 \longrightarrow 2 \text{ NOHSO}_4 + \text{H}_2O \tag{32}$$

$$\text{NOCl} + \text{H}_2\text{SO}_4 \longrightarrow \text{NOHSO}_4 + \text{HCl} \tag{33}$$

The cooled, dried chlorine gas contains $\sim 2\%$ HCl and up to 10% O_2, both of which are removed by liquefaction. A full scale 600-t/day plant was built by Du Pont in 1975. This installation at Corpus Christi, Texas operates at 1.4 MPa (13.8 atm) and 120–180°C and uses tantalum-plated equipment and pipes.

Oxidation of HCl Chloride by Nitric Acid. The nitrosyl chloride [*2696-92-6*] route to chlorine is based on the strongly oxidizing properties of nitric acid (74,75):

$$6 \text{ HCl} + 2 \text{ HNO}_3 \longrightarrow 2 \text{ Cl}_2 + 2 \text{ NOCl} + 4 \text{ H}_2\text{O} \tag{34}$$

$$2 \text{ NOCl} + 2 \text{ H}_2\text{O} + \text{O}_2 \longrightarrow 2 \text{ HCl} + 2 \text{ HNO}_3 \tag{35}$$

The practical problems lie in the separation of the chlorine from the hydrogen chloride and nitrous gases. The dilute nitric acid must be reconcentrated and corrosion problems are severe. Suggested improvements include oxidation of concentrated solutions of chlorides, eg, LiCl, by nitrates, followed by separation

Table 15. Physical Constants of Chlorine

Property	Value
CAS Registry Number	[*7782-50-5*]
atomic number	17
atomic weight	35.453
stable isotope abundance, %	
^{35}Cl	75.53
^{37}Cl	24.47
electronic configuration in the ground state	[Ne]$3s^2 3p^5$
melting point, °C	-100.98
boiling point, °C, at 101.3 kPa^a	-34.05
density relative to air	2.48
critical density, kg/m^3	565.00
critical pressure, MPa^a	7.71083
critical volume, m^3/kg	0.001745
critical temperature, K	417.15
density, kg/m^3 at 0°C and 101.3 kPa^a	3.213
viscosity (gas), Pa·s^b at 20°C	14.0
viscosity (liquid), Pa·s^b at 20°C	340
latent heat of vaporization, J/g^c	287.4
enthalpy of fusion $\triangle H_f$, kJ/kg^c	90.33
enthalpy of vaporization $\triangle H_v$, kJ/kg^c	287.1
standard electrode potential, V	1.359
enthalpy of dissociation, kJ/mol^c	2.3944
electron affinity, eV	3.77
enthalpy of hydration of Cl$^-$, kJ/mol^c	405.7
ionization energies, eV	13.01, 23.80, 39.9, 53.3, 67.8, 96.6, and 114.2
specific heat C_p, kJ/kg·K^c	0.481
specific heat C_v, kJ/kg·K^c	0.357
specific magnetic susceptibility, m^3/kg at 20°C	-7.4×10^{-9}
electrical conductivity of liquid Cl$_2$, $(\Omega \text{ cm})^{-1}$ at -70°C	10^{-16}
dielectric constant for wavelengths > 10 m at 0°C	1.97

^{a}To convert kPa to mm Hg, multiply by 7.5.
^{b}To convert Pa·s to P, multiply by 10.
^{c}To convert J to cal, divide by 4.184.

of chlorine from nitrosyl chloride by distillation at 135°C, or oxidation by a mixture of nitric and sulfuric acids, separating the product chlorine and nitrogen dioxide by liquefaction and fractional distillation.

A survey of nonelectrolytic routes for Cl_2 production was conducted by Argonne National Laboratory; the economics of these processes were examined in detail (76). One route identified as energy efficient and economically attractive is the conversion of waste NH_4Cl to Cl_2.

CHLORINE

Physical Properties

Chlorine, a member of the halogen family, is a greenish-yellow gas having a pungent odor at ambient temperatures and pressures and a density 2.5 times that of air. In liquid form it is clear amber; Solid chlorine forms pale yellow crystals. The principal properties of chlorine are presented in Table 15; additional details are available (77–79). The temperature dependence of the density of gaseous (Fig. 31) and liquid (Fig. 32) chlorine, and vapor pressure (Fig. 33) are illustrated.

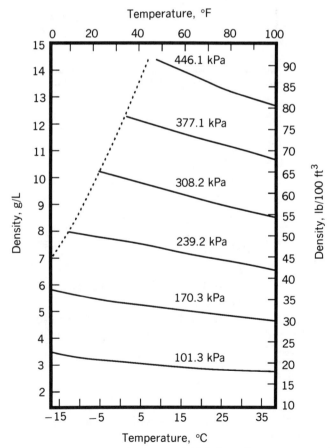

Fig. 31. Density of gaseous chlorine as a function of temperature at various pressures. To convert kPa to psig, multiply by 0.145 and subtract 14.6. (77).

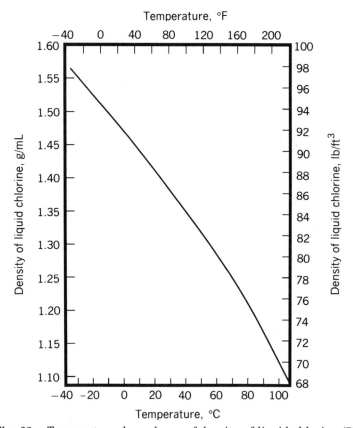

Fig. 32. Temperature dependence of density of liquid chlorine (78).

Enthalpy pressure data can be found in ref. 78. The vapor pressure P can be calculated in the temperature (T) range of 172–417 K from the Martin-Shin-Kapoor equation (80):

$$\ln P = A + \frac{B}{T} + C \ln T + DT + \frac{E(F - T)\ln(F - T)}{FT} \tag{36}$$

where $A = 62.402508$, $B = -4343.5240$, $C = 7.8661534$, $D = 1.0666308 \times 10^{-2}$, $E = 95.248723$, and $F = 424.90$.

Chlorine is soluble in water and in salt solutions, the solubility decreasing with salt strength and temperature (see Fig. 34). It is partially hydrolyzed in aqueous solution as

$$Cl_2 + H_2O \rightleftharpoons HCl + HOCl \tag{37}$$

Below 10°C, chlorine forms hydrates which are greenish-yellow crystals and the Cl_2–H_2O system has a quadruple point at 28.7°C. Solubility data of chlorine in various solvents are given in Table 16.

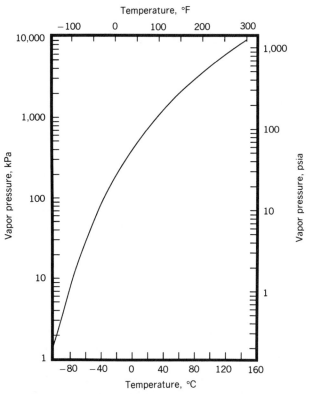

Fig. 33. Vapor pressure of liquid chlorine as a function of temperature (78).

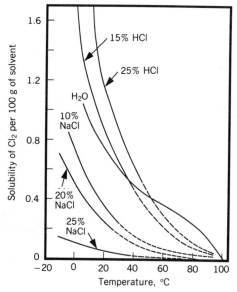

Fig. 34. Solubility of chlorine in water and aqueous HCl and NaCl. Solution concentrations are in wt %.

Table 16. Solubility of Chlorine in Various Solvents

Solvent	Temperature, °C	Solubility[a]
sulfuryl chloride	0	12.0
disulfur chloride	0	58.5
phosphoryl chloride	0	19.0
silicon tetrachloride	0	15.6
titanium tetrachloride	0	11.5
dimethylformamide	0	123[b]
acetic acid, 99.84 wt %	0	11.6[b]
benzene	10	24.7
chloroform	10	20.0
carbon tetrachloride	20	17[c]
hexachlorobutadiene	20	22[c]

[a]Solubility in wt % unless otherwise noted.
[b]Solubility in g/100 mL.
[c]Solubility in mol %.

Chemical Properties

The chemical properties of chlorine have been discussed (79). Chlorine generally exhibits a valence of -1 in compounds, but it also exists in the formal positive oxidation states of $+1$ (NaClO [7681-52-9]), $+3$ (NaClO$_2$ [7758-19-2]), $+5$ (NaClO$_3$ [7775-09-9]), and $+7$ (NaClO$_4$ [7601-89-0]), Molecular chlorine is a strong oxidizer and a chlorinating agent, adding to double bonds in aliphatic compounds or undergoing substitution reactions with both aliphatic and aromatic ones. Significant industrial reaction products are presented in Tables 17 and 18 (81). Chlorine is very reactive under specific conditions but it is not explosive or flammable. Reactions of chlorine with most elements (eg, S, P, I$_2$, Br$_2$, F$_2$) are facile but those with N$_2$, O$_2$, and C are indirect. Chlorine reacts with NH$_3$ to form the explosive NCl$_3$. Chlorine gas does not react with H$_2$ at normal temperatures in the absence of light. However, at temperatures above 250°C, or in the presence of sunlight or artificial light of $\sim$470-nm wavelength, H$_2$ and Cl$_2$ combine explosively to form HCl. Explosive limits of mixtures of pure gases are $\sim$8 vol % H$_2$ and $\sim$12 vol % Cl$_2$ (see Fig. 35). These limits depend on temperature, pressure, and concentration, and can be altered by adding inert gases such as N$_2$ or CO$_2$.

Dry chlorine reacts with most metals combustively depending on temperature: aluminum, arsenic, gold, mercury, selenium, tellerium, and tin react with dry Cl$_2$ in gaseous or liquid form at ordinary temperatures; carbon steel ignites at about 250°C depending on the physical shape; and titanium reacts violently with dry chlorine. Wet chlorine is very reactive because of the hydrochloric acid and hypochlorous acid (see eq. 37). Metals stable to wet chlorine include platinum, silver, tantalum, and titanium. Tantalum is the most stable to both dry and wet chlorine.

Chlorine reacts with alkali and alkaline earth metal hydroxides to form bleaching agents such as NaOCl:

$$Cl_2 + 2\,NaOH \longrightarrow NaCl + NaOCl + H_2O \tag{38}$$

Table 17. Chlorine Derivatives[a]

Chlorine

electrolysis of brine
electrolysis of fused salt
electrolysis of magnesium chloride
electrolysis of hydrochloric acid
catalytic oxidation of hydrochloric acid
chemical reaction of potassium chloride and nitric acid

production of organic chemicals by oxychlorination processes (often interchangeable with oxyhydrochlorination processes)

- ethylene dichloride
- chlorofluoro hydrocarbons
- carbon tetrachloride
- 1,1,1-trichloroethane
- 1,1,2-trichloroethane
- trichloroethylene
- perchloroethylene
- methyl chloride
- phosgene
- methylene chloride
- chloroform
- ethyl chloride
- allyl chloride
- chlorosulfonic acid
- chloroprene
- chloroanthraquinone
- chloroanilines
- dichloropropane (propylene dichloride)
- dichloropropenes—soil fumigants
- methallyl chloride

- amyl chloride
- chlorinated paraffins → high pressure lubricants; fire-proofing agent for textiles (with antimony oxide); plasticizer for poly(vinyl chloride) detergents
- chlorinated waxes → moisture, flame, acid-, and insect-proofing of wood, fabrics, wire, and cable; solvent
- chlorinated naphthalenes
- chloroacetic acid
- 2,4-dichlorophenoxyacetic acid (2,4-D); 2,4,5-trichlorophenoxyacetic acid (2,4,5-T) → herbicides
- chloroacetyl chloride → tear gas (chloroacetophenone)
- chlorobenzene
- dichlorobenzenes
- trichlorobenzene
- tetrachlorobenzene
- benzene hexachloride
- polychlorinated biphenyls → dielectric fluid in transformers and capacitors
- chlorotoluenes
- hexachloroethane
- chlorophenols
- chloral
- hexachlorocyclopentadiene
- perchloromethyl mercaptan → organic synthesis; dye intermediate; fumigant
- tetrachlorophthalic anhydride → flame retardant for plastics

bleach → pulp and paper textiles

production of inorganic chemicals

- sodium chlorate → bleach → textiles; wood pulp
- hypochlorous acid → water purification, antiseptic, epichlorohydrin, lithium hypochlorite; calcium, sodium hypochlorite; chlorinated trisodium phosphate → cleaner and disinfectant
- chlorinated isocyanurates (potassium dichloroisocyanurate, sodium dichloroisocyanurate, trichloroisocyanuric acid) → sanitizers (eg, for swimming pools); household and commercial bleaches; detergents for automatic dishwashers; scouring powders; chlorinated cleaners and sanitizers
- hydrochloric acid
- phosphorus trichloride
- phosphorus pentachloride
- phosphorus oxychloride
- titanium trichloride
- titanium tetrachloride
- ferric chloride
- aluminum chloride, anhydrous
- sulfur monochloride
- sulfur dichloride
- sulfuryl chloride, mercurous chloride
- mercuric chloride
- silicon tetrachloride
- zinc chloride from zinc metal
- antimony pentachloride, antimony trichloride
- stannous chloride
- arsenic trichloride, bismuth trichloride
- chlorine trifluoride
- molybdenum pentachloride
- iodine monochloride → pharmaceutical (antiseptic)
- iodine trichloride → pharmaceuticals

sanitizing and disinfecting agent (eg, for municipal water supplies, swimming pools)

waste and sewage treatment

slimicide

[a]Ref. 81. Courtesy of SRI International.

Table 18. Hydrochloric Acid Derivatives[a]

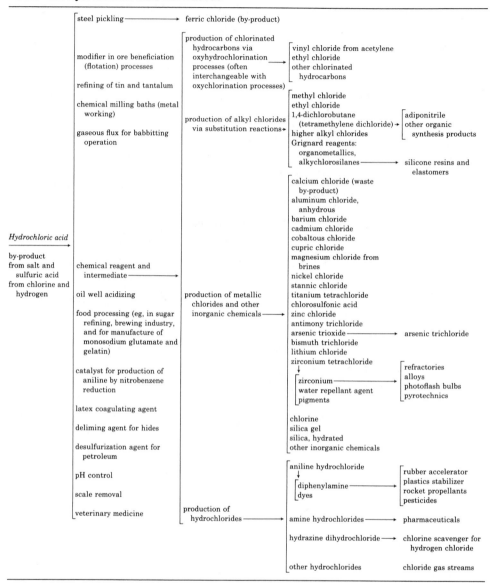

[a]Ref. 81. Courtesy of SRI International.

Reaction with ammonia produces hydrazine.

$$2\,NH_3 + NaOCl \longrightarrow N_2H_4 + NaCl + H_2O \tag{39}$$

TiO_2 reacts with Cl_2 in the presence of carbon during the manufacture of $TiCl_4$, an intermediate step in the production of Ti metal and TiO_2 pigment:

$$TiO_2 + 2\,Cl_2 + C \quad\text{or}\quad 2\,C \longrightarrow TiCl_4 + CO_2 \quad\text{or}\quad 2\,CO \tag{40}$$

$SiCl_4$ is produced by reaction of SiO_2 with Cl_2.

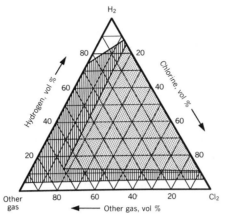

Fig. 35. Explosive limits of Cl_2–H_2–other gas mixtures where ▥ represents the explosive region in the presence of residue gas from Cl_2 liquefaction (O_2, N_2, CO_2) and ▦ in the presence of inert gases (N_2, CO_2).

Chlorine reacts with saturated hydrocarbons either by substitution or by addition to form chlorinated hydrocarbons and HCl. Thus methanol or methane is chlorinated to produce CH_3Cl, which can be further chlorinated to form methylene chloride, chloroform, and carbon tetrachloride. Reaction of Cl_2 with unsaturated hydrocarbons results in the destruction of the double or triple bond. This is a very important reaction during the production of ethylene dichloride, which is an intermediate in the manufacture of vinyl chloride:

$$CH_2{=}CH_2 + Cl_2 \longrightarrow CH_2Cl-CH_2Cl \tag{41}$$

or

$$CH_2{=}CH_2 + \tfrac{1}{2} O_2 + 2\ HCl \longrightarrow CH_2Cl-CH_2Cl + H_2O \tag{42}$$

Direct chlorination of vinyl chloride generates 1,1,2-trichloroethane [79-00-5] from which vinylidene chloride required for vinylidene polymers is produced. Hydrochlorination of vinylidene chloride produces 1,1,1-trichloroethane [71-55-6], which is a commercially important solvent. Trichloroethylene and perchloroethylene are manufactured by chlorination, hydrochlorination, or oxychlorination reactions involving ethylene. Aromatic solvents or pesticides such as monochlorobenzene, dichlorobenzene, and hexachlorobenzene are produced by reaction of chlorine with benzene. Monochlorobenzene is an intermediate in the manufacture of phenol, insecticide DDT, aniline, and dyes (see CHLOROCARBONS AND CHLOROHYDROCARBONS.)

Materials of Construction

The choice of construction material for handling chlorine depends on equipment design and operating conditions. Dry chlorine, with less than 40 ppm by weight of water, can be handled safely below 120°C in equipment made from iron, steel, stainless steels, Monel Metal, nickel, copper, brass, bronze, and lead. Silicone materials, titanium, and high surface area materials such as steel wool should be avoided. Titanium ignites spontaneously in dry chlorine; steel reacts with Cl_2 at an accelerated rate at temperatures greater than 50°C, igniting near 120°C (82). The presence of organic substances or rust increases the risk of steel ignition. For dry chlorine, schedule 80 pipe is used in sizes up to 15-cm diameter; schedule 40 is satisfactory for large diameter pipe.

Liquid chlorine is generally stored in vessels made from nonalloyed carbon steel or cast steel. Fine grain steel with a limited tensile strength is used to ensure proper conditions for welding these containers. Erosion of protective layers on steel is prevented by maintaining the flow velocities to less than 2 m/s. Although organic materials, zinc, tin, aluminum, and titanium are not acceptable for liquid chlorine use, copper, silver, lead, and tantalum are acceptable for some equipment.

Dry chlorine has a great affinity for absorbing moisture, and wet chlorine is extremely corrosive, attacking most common materials except Hastelloy C, titanium, and tantalum. These metals are protected from attack by the acids formed by chlorine hydrolysis because of surface oxide films on the metal. Tantalum is the preferred construction material for service with wet and dry chlorine. Wet chlorine gas is handled under pressure using fiberglass-reinforced plastics. Rubber-lined steel is suitable for wet chlorine gas handling up to 100°C. At low pressures and low temperatures PVC, chlorinated PVC, and reinforced polyester resins are also used. Polytetrafluoroethylene (PTFE), poly(vinylidene fluoride) (PVDF), and poly(tetrafluoroethylene-hexafluoropropylene) (FEP) are resistant at high temperatures. Other materials stable to moist chlorine include graphite, glass, and halogenated plastics.

Gaskets in both dry gas and liquid chlorine systems are made of rubberized compressed asbestos. For wet chlorine gas, rubber or synthetic elastomers are acceptable. PTFE is resistant to both wet and dry chlorine gas and to liquid chlorine up to 200°C. Tantalum, Hastelloy C, PTFE, PVDF, Monel, and nickel are recommended for membranes, rupture disks, and bellows.

Storage and Transportation

Surveys of existing national and international regulations for the handling and transportation of chlorine are available (82). Emergency kits designed to contain leaks during chlorine shipping are available from the Chlorine Institute. Chlorine is liquefied and stored at ambient or low temperature, ensuring that the pressure in the storage system corresponds to the vapor pressure of liquefied chlorine at the temperature in the tank. Whereas permanent storage of chlorine is not recommended, moderate amounts (not exceeding 450 t) are stored in tanks as a liquefied gas. Larger storage capacities employ a low pressure storage system operating at a liquid chlorine temperature of ~ 34°C, which requires a cooling or recompression system to liquefy evaporating chlorine gas that is recycled. A stand-by caustic soda solution absorber absorbs the evaporated chlorine in case of an emergency resulting from failure of the refrigeration unit. The storage system must be degreased, cleaned, and dried to a dew point of -40°C in the purge gas at the outlet of the system. The filling ratio should never exceed 95% of the total volume of the vessel. This corresponds for pressure storage tanks to 1.25 kg of liquid chlorine per liter of vessel capacity at 50°C.

Chlorine is stored and transported as a liquefied gas in cylinders of 45.4-kg or 68-kg capacity that are under pressure and equipped with fusible-plug relief devices. Quantities in the range of 15 to 90 t are transported in tank cars having special angle valves on the manhole cover on top of the vessel. Tank barges of the open-hopper type having several cylindrical uninsulated pressure vessels are used for amounts ranging from 600 to 1200 t. Road tankers are used for capacities of 15 to 20 t.

Chlorine is classified by the U.S. Department of Transportation (DOT) as a nonflammable compressed gas requiring a green label. Chlorine must be packaged in containers complying with DOT and/or Coast Guard regulations related to construction, loading, and labeling, and state, local, and insurance regulations. Transportation of about 70% of liquid chlorine is by rail, 20% by pipe lines, 7% by barges, and the remainder in cylinders.

Analytical Methods

Industrial liquid chlorine is routinely analyzed for moisture, chlorine, other gaseous components, NCl_3, and mercury following established procedures (10,79). Moisture and residue content in liquid chlorine is determined by evaporation at 20°C followed by gravimetric measurement of the residue. Free chlorine levels are estimated quantitatively by thiosulfate titration of iodine liberated from addition of excess acidified potassium iodide to the gas mixture.

Safety

No attempt should be made to handle chlorine for any purpose without a thorough understanding of its properties and the hazards involved. This information is available from the Chlorine Institute and from various handbooks (10,76,79).

Chlorine gas is a respiratory irritant and is readily detectable at concentrations of <1 ppm in air because of its penetrating odor. Chlorine gas, after several

Table 19. Physiological Response to Chlorine

Parameter	Parts of chlorine per million parts of air (volume)
least amount required to produce slight symptoms after several hours of exposure	1
least detectable odor	3.5
maximum amount that can be inhaled for 1 h without serious physiological response	4
least amount required for throat irritation	15.1
least amount required for coughing	30.2
amount causing severe symptoms in 30–60 min	40–60
LD_{50}	
humans, in 30 min	840
rats, in 60 min	290
mice, in 60 min	137
amount expected to affect aquatic life	<0.1 ppm

hours of exposure, causes mild irritation of the eyes and of the mucous membrane of the respiratory tract. At high concentrations and in extreme situations, increased difficulty in breathing can result in death through suffocation. The physiological response to various levels of chlorine gas is given in Table 19.

SODIUM HYDROXIDE (CAUSTIC SODA)

Physical Properties

Sodium hydroxide, NaOH, mol wt 39.998, is a brittle, white, translucent crystalline solid. Because of its corrosive action on all human body tissue, it is also known as caustic soda. Physical properties of the pure material are noted in Table 20. Sodium hydroxide is produced and shipped in the anhydrous state in the form of solid cakes, flakes, or beads, but is used in solution (4). Properties of aqueous sodium hydroxide solutions relevant to industrial operations are available from manufacturers and in handbooks (79). Table 21 lists the significant uses.

Table 20. Physical Constants of Pure Sodium Hydroxide

Property	Value
CAS Registry Number	[1310-73-2]
molecular weight	39.998
specific gravity at 20°C	2.130
melting point, °C	318
boiling point, °C at 101.3 kPa[a]	1388
specific heat, J/g·°C[b] at 20°C	1.48
refractive index at 589.4 nm	
320°C	1.433
420°C	1.421
latent heat of fusion, J/g[b,c]	167.4
lattice energy, kJ/mol[d,e]	737.2
entropy, J/(mol·K)[b] at 25°C and 101.3 kPa[a]	64.45[e]
heat of formation $\triangle H_f$, kJ/mol[a]	
α form	422.46
β form	426.60
heat of transition from α to β form, J/g[b]	103.3
transition temperature, °C	299.6
free energy of formation $\triangle G_f$, kJ/mol[d] at 25°C and 101.3 kPa[a]	−379.5[e]

[a]To convert kPa to mm Hg, multiply by 7.5.
[b]To convert J to cal, divide by 4.184.
[c]To convert J/g to Btu/lb, multiply by 0.4302.
[d]To convert kJ to kcal, divide by 4.184.
[e]Values from ref. 83.

Sodium hydroxide deliquesces on exposure to air, resolidifying because of sodium carbonate formation upon absorbing carbon dioxide. It is very soluble in water and forms hydrates containing 1, 2, 3, 5, and 7 molecules of H_2O, depending on the concentration. Heat is generated during the dilution of concentrated caustic solution, or when solid NaOH is dissolved in water. Figures 36 and 37

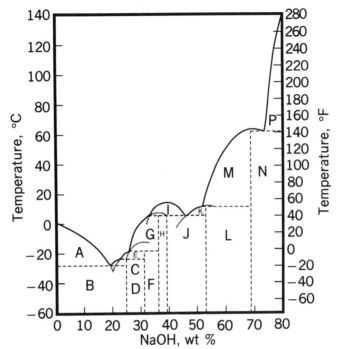

Fig. 36. Freezing points of caustic soda solutions (82): A, ice; B, ice + NaOH·7H$_2$O;
C, NaOH·7H$_2$O; D, NaOH·7H$_2$O + NaOH·5H$_2$O; E, NaOH·5H$_2$O; F, NaOH·5H$_2$O +
NaOH·4H$_2$O; G, NaOH·4H$_2$O; H, NaOH·4H$_2$O + NaOH·3½H$_2$O; I, NaOH·3½H$_2$O; J,
NaOH·3½H$_2$O + NaOH·2H$_2$O; K, NaOH·2H$_2$O; L, NaOH·2H$_2$O + NaOH·H$_2$O; M,
NaOH·H$_2$O; N, NaOH·H$_2$O + NaOH; and P, NaOH. Courtesy of the Chlorine Institute, Inc.

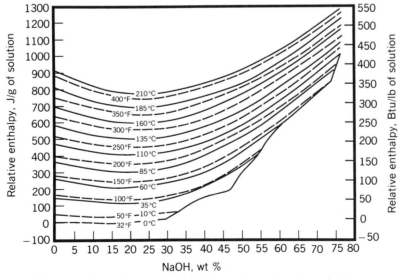

Fig. 37. Enthalpy vs concentration of caustic soda solutions.

Table 21. Sodium Hydroxide Derivatives

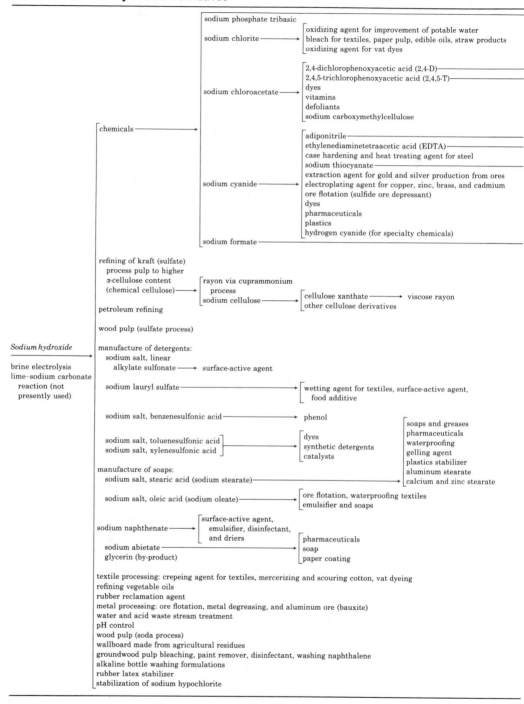

sodium phosphate tribasic

sodium chlorite ⟶
- oxidizing agent for improvement of potable water
- bleach for textiles, paper pulp, edible oils, straw products
- oxidizing agent for vat dyes

sodium chloroacetate ⟶
- 2,4-dichlorophenoxyacetic acid (2,4-D)
- 2,4,5-trichlorophenoxyacetic acid (2,4,5-T)
- dyes
- vitamins
- defoliants
- sodium carboxymethylcellulose

chemicals ⟶

sodium cyanide ⟶
- adiponitrile
- ethylenediaminetetraacetic acid (EDTA)
- case hardening and heat treating agent for steel
- sodium thiocyanate
- extraction agent for gold and silver production from ores
- electroplating agent for copper, zinc, brass, and cadmium
- ore flotation (sulfide ore depressant)
- dyes
- pharmaceuticals
- plastics
- hydrogen cyanide (for specialty chemicals)

sodium formate

refining of kraft (sulfate) process pulp to higher α-cellulose content (chemical cellulose) ⟶
- rayon via cuprammonium process
- sodium cellulose ⟶
  - cellulose xanthate ⟶ viscose rayon
  - other cellulose derivatives

petroleum refining

wood pulp (sulfate process)

Sodium hydroxide

brine electrolysis
lime–sodium carbonate reaction (not presently used)

manufacture of detergents:
sodium salt, linear alkylate sulfonate ⟶ surface-active agent

sodium lauryl sulfate ⟶ wetting agent for textiles, surface-active agent, food additive

sodium salt, benzenesulfonic acid ⟶ phenol

sodium salt, toluenesulfonic acid
sodium salt, xylenesulfonic acid ⟶
- dyes
- synthetic detergents
- catalysts

manufacture of soaps:
sodium salt, stearic acid (sodium stearate) ⟶
- soaps and greases
- pharmaceuticals
- waterproofing
- gelling agent
- plastics stabilizer
- aluminum stearate
- calcium and zinc stearate

sodium salt, oleic acid (sodium oleate) ⟶
- ore flotation, waterproofing textiles
- emulsifier and soaps

sodium naphthenate ⟶
- surface-active agent, emulsifier, disinfectant, and driers

sodium abietate
glycerin (by-product) ⟶
- pharmaceuticals
- soap
- paper coating

textile processing: crepeing agent for textiles, mercerizing and scouring cotton, vat dyeing
refining vegetable oils
rubber reclamation agent
metal processing: ore flotation, metal degreasing, and aluminum ore (bauxite)
water and acid waste stream treatment
pH control
wood pulp (soda process)
wallboard made from agricultural residues
groundwood pulp bleaching, paint remover, disinfectant, washing naphthalene
alkaline bottle washing formulations
rubber latex stabilizer
stabilization of sodium hypochlorite

Table 21. *(Continued)*

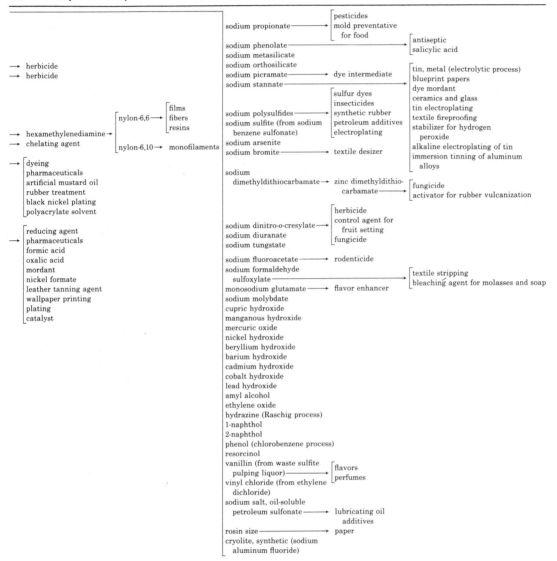

sodium propionate ──→ ⎡ pesticides / mold preventative for food ⎤

sodium phenolate ────────────→ ⎡ antiseptic / salicylic acid ⎤

→ herbicide

→ herbicide

sodium metasilicate
sodium orthosilicate
sodium picramate ──→ dye intermediate
sodium stannate ─────→

sodium polysulfides ──→ ⎡ sulfur dyes / insecticides / synthetic rubber / petroleum additives / electroplating ⎤

sodium sulfite (from sodium benzene sulfonate)
sodium arsenite
sodium bromite ──────→ textile desizer

→ hexamethylenediamine → ⎡ nylon-6,6 → ⎡ films / fibers / resins ⎤ ; nylon-6,10 → monofilaments ⎤

→ chelating agent

→ ⎡ dyeing / pharmaceuticals / artificial mustard oil / rubber treatment / black nickel plating / polyacrylate solvent ⎤

tin, metal (electrolytic process)
blueprint papers
dye mordant
ceramics and glass
tin electroplating
textile fireproofing
stabilizer for hydrogen peroxide
alkaline electroplating of tin
immersion tinning of aluminum alloys

sodium dimethyldithiocarbamate ──→ zinc dimethyldithiocarbamate ──→ ⎡ fungicide / activator for rubber vulcanization ⎤

sodium dinitro-*o*-cresylate ──→ ⎡ herbicide / control agent for fruit setting / fungicide ⎤

sodium diuranate
sodium tungstate

sodium fluoroacetate ──────→ rodenticide

sodium formaldehyde sulfoxylate ──────────→ ⎡ textile stripping / bleaching agent for molasses and soap ⎤

monosodium glutamate ──→ flavor enhancer
sodium molybdate
cupric hydroxide
manganous hydroxide
mercuric oxide
nickel hydroxide
beryllium hydroxide
barium hydroxide
cadmium hydroxide
cobalt hydroxide
lead hydroxide
amyl alcohol
ethylene oxide
hydrazine (Raschig process)
1-naphthol
2-naphthol
phenol (chlorobenzene process)
resorcinol
vanillin (from waste sulfite pulping liquor) ──→ ⎡ flavors / perfumes ⎤
vinyl chloride (from ethylene dichloride)
sodium salt, oil-soluble petroleum sulfonate ──→ lubricating oil additives
rosin size ──────→ paper
cryolite, synthetic (sodium aluminum fluoride)

→ ⎡ reducing agent / pharmaceuticals / formic acid / oxalic acid / mordant / nickel formate / leather tanning agent / wallpaper printing / plating / catalyst ⎤

illustrate solution freezing point and enthalpy data. Because the heat can be excessive to the extent that the temperature of a solution can increase above the boiling point, caution should be exercised during the dilution of caustic from concentrations of >25% by providing proper cooling (84,85).

Chemical Properties

Aqueous solutions of caustic soda are highly alkaline. Hence caustic soda is primarily used in neutralization reactions to form sodium salts (79). Sodium hydroxide reacts with amphotoric metals (Al, Zn, Sn) and their oxides to form complex anions such as AlO_2^-, ZnO_2^{2-}, SnO_3^{2-}, and H_2 (or H_2O with oxides). Reaction of Al_2O_3 with NaOH is the primary step during the extraction of alumina from bauxite (see ALUMINUM COMPOUNDS):

$$Al(OH)_3 + NaOH \longrightarrow NaAlO_2 + 2\,H_2O \tag{43}$$

Caustic soda reacts with weak-acid gases such as H_2S, SO_2, and CO_2.

$$H_2S + 2\,NaOH \longrightarrow Na_2S + 2\,H_2O \tag{44}$$

$$SO_2 + 2\,NaOH \longrightarrow Na_2SO_3 + H_2O \tag{45}$$

$$CO_2 + 2\,NaOH \longrightarrow Na_2CO_3 + H_2O \tag{46}$$

These reactions are used industrially for scrubbing operations and for selective removal of H_2S from natural gas containing CO_2.

Metallic ions are precipitated as their hydroxides from aqueous caustic solutions. The reactions of importance in chlor–alkali operations are removal of magnesium as $Mg(OH)_2$ during primary purification and of other impurities for pollution control. Organic acids react with NaOH to form soluble salts. Saponification of esters to form the organic acid salt and an alcohol and internal coupling reactions involve NaOH, as exemplified by reaction with triglycerides to form soap and glycerol,

$$C_3H_5(COOR)_3 + 3\,NaOH \longrightarrow C_3H_5(OH)_3 + 3\,NaCOOR \tag{47}$$

and with propylene chlorohydrin to form propylene oxide,

$$ClC_3H_6OH + NaOH \longrightarrow C_3H_6O + NaCl + H_2O \tag{48}$$

Reactions of NaOH with natural products are complex. They include solubilization of cotton in rubber reclaiming, starch dextrination, cotton scouring, refining of vegetable oils, and removal of lignin and hemicellulose in the Kraft pulping process. The primary step during the reaction of cellulose, caustic soda, and monochloroacetic acid to form the sodium salt of carboxymethylcellulose is similar to that used in mercerizing cotton and in the preparation of rayon (qv) from cellulose xanthate (see CELLULOSE ETHERS; FIBERS, REGENERATED CELLULOSICS).

Other Processes for NaOH Production

The only caustic soda production process besides electrolysis is the soda–lime process involving the reaction of lime with soda ash:

$$Ca(OH)_2 + Na_2CO_3 \longrightarrow CaCO_3 + 2\,NaOH \tag{49}$$

The lime–soda process is practiced mainly in isolated areas in some process operations, in the Kraft recovery process, and in the production of alumina. It is not as efficient a route as electrolytic production.

In the Kraft recovery process the green liquor, which is an aqueous solution of sodium carbonate, is heated with lime to produce white liquor or caustic soda, which is then returned to the pulp digestion operations. In the production of alumina, lime and soda are fed to bauxite digesters. The $CaCO_3$ produced during the course of the reaction is reburned to lime and is recycled. The main difficulties associated with these processes include the extensive mechanical handling through the use of causticizers, settlers, and repulpers in order to produce caustic that is low in carbonate and the high fuel consumption needed to reconvert $CaCO_3$ to lime.

Special Grades of Caustic Soda

Three forms of caustic soda are produced to meet customer needs: purified diaphragm caustic (50% Rayon grade), 73% caustic, and anhydrous caustic. Regular 50% caustic from the diaphragm cell process is suitable for most applications and accounts for about 85% of the NaOH consumed in the United States. However, it cannot be used in operations such as the manufacture of rayon, the synthesis of alkyl aryl sulfonates, or the production of anhydrous caustic because of the presence of salt, sodium chlorate, and heavy metals. Membrane and mercury cell caustic, on the other hand, is of superior quality and meets the high purity market requirements.

50% Caustic Soda. Diaphragm cell caustic is commercially purified by the DH process or the ammonia extraction method offered by PPG and OxyTech (see Fig. 38), essentially involving liquid–liquid extraction to reduce the salt and sodium chlorate content (86). Thus 50% caustic comes in contact with ammonia in a countercurrent fashion at 60°C and up to 2500 kPa (25 atm) pressure, the liquid NH_3 absorbing salt, chlorate, carbonate, water, and some caustic. The overflow from the reactor is stripped of NH_3, which is then concentrated and returned to the extraction process. The product, about 62% NaOH and devoid of impurities, is stripped free of NH_3, which is concentrated and recirculated. Metallic impurities can be reduced to low concentrations by electrolysis employing porous cathodes. The caustic is then freed of Fe, Ni, Pb, and Cu ions, which are deposited on the cathode.

Purification can also be achieved by cooling the caustic by direct contact with Freon. Pure NaOH·3½H$_2$O crystals are formed and these crystals are centrifuged and separated from the salt containing the mother liquor. However, the purity of this material is not equivalent to that of the mercury grade, although it

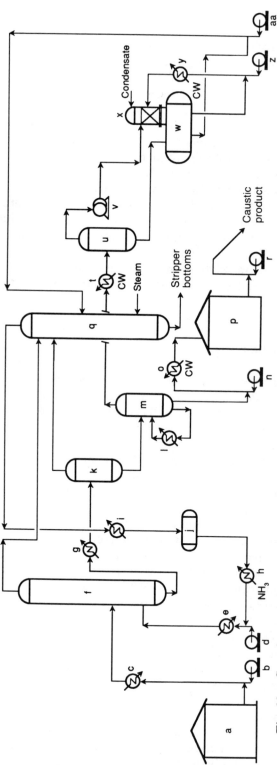

Fig. 38. Caustic purification system: a, 50% caustic feed tank; b, 50% caustic feed pumps; c, caustic feed preheater; d, amonia feed pumps; e, ammonia feed preheater; f, extractor; g, trim heater; h, ammonia subcooler; i, stripper condenser; j, anhydrous ammonia storage tank; k, primary flash tank; l, evaporator reboiler; m, evaporator; n, caustic product transfer pumps; o, purified caustic product cooler; p, purified caustic storage tank; q, ammonia stripper; r, purified caustic transfer pumps; t, overheads condenser; u, evaporator; v, evaporator vacuum pump; w, aqueous storage ammonia tank; x, ammonia scrubber; y, scrubber condenser; z, ammonia recirculating pump; aa, ammonia recycle pump. CW stands for chilled water.

can be used by the rayon industry. This technology, although proven technically, is not practiced commercially in the United States. See reference 87 for a description of the freeze concentration of caustic. Asahi Chemical has reported that pure NaOH can also be made by forming a NaOH–alcohol clathrate compound and distilling it at 100–120°C under reduced pressure (88). This process has been studied only on a laboratory scale and is not yet commercialized.

73% Caustic Soda. About 5% of the caustic soda produced is concentrated to 73% solutions for special usage and for producing anhydrous material. The deciding factor in the production of 73% caustic is the comparative economics of concentrating the 50% solution vs shipping the extra water (the 73% material is made from 50% regular grade caustic). Freezing-point curves of NaOH solutions (see Fig. 36) indicate a relatively flat region from 65 to 75% caustic, where caustic soda freezes at ~62°C, permitting 50% NaOH to be concentrated to 73%. Concentration is carried out in a single-effect forced or natural recirculation mode in steam-heated nickel evaporators. Operation of the evaporators at low temperatures also minimizes nickel corrosion problems and avoids iron contamination in the product. The 73% product is stored in insulated, lined tanks using steam or electrical heating coils at >62°C to prevent freezing.

Anhydrous Caustic. Anhydrous caustic is produced from either 50 or 73% caustic solutions in tubular flash-type nickel or Inconel evaporators at over 400°C. Heat is supplied by direct firing by Dowtherm heat-transfer fluid or by a eutectic mixture of molten potassium nitrate, sodium nitrite, and sodium nitrate. Any sodium chlorate present in 50% caustic feed decomposes at these temperatures and corrodes nickel. Therefore it is removed or destroyed when diaphragm-grade caustic is used by adding reducing agents such as sugar. After the water is removed, the molten caustic is pumped either to a rotary drum flaker or to a drum-loading station and then poured into thin steel drums for shipment. Flaked solid caustic, produced in sizes ranging from ~2 to 20-mm diameters is also shipped in drums.

Materials of Construction

Steel is an acceptable material of construction for handling solutions of up to 50% NaOH below 40°C. Above 40°C the steel corrosion rate increases rapidly and iron is picked up in the solution. Materials for handling 50% NaOH are lined steel for tank cars and lined or unlined steel for tanks and piping.

Nickel is the ideal material for handling caustic at all concentrations and temperatures, including molten anhydrous caustic up to 480°C. Nickel and/or nickel alloys are used for bodies, piping, and heat exchangers of the evaporators and for operations where high solution velocities are encountered, such as centrifugal pumps. Plastics or plastic-lined steel are now commercially available for caustic solutions: fiberglass-reinforced plastic tanks of vinyl ester resins are suitable but prone to mechanical damage; polypropylene is superior from both a cost and performance standpoint. At moderately high temperatures fluorocarbon plastics have been used successfully as pipe linings. Aluminum, zinc, brass, bronze, and copper are readily attacked by caustic and hence are unsuitable for storing caustic solutions.

Storage and Transportation

Caustic soda is classified as a corrosive material by the DOT and DOT regulations and specifications must be followed for handling, labeling, and transportation in containers. Warning labels are recommended for containers of caustic soda solutions and anhydrous caustic soda by the MCA (see ref. 79). The DOT identification number is UN1824 for 50 or 73% liquid, and UN1823 for anhydrous caustic.

Liquid caustic is shipped in tank trucks, tank cars, and barges. Typical tank trucks, constructed of stainless steel without insulation or heating coils, carry 15,000 L or 11.3 t of caustic soda. Tank cars, usually insulated and equipped with heating coils, are made of nickel-clad steel or lined steel. Barges carrying 50% caustic have capacities ranging from 1100 to 2200 t; 25% of the caustic produced in the United States is shipped by barges. Anhydrous caustic marketed as beads and flakes is transported in bulk hopper cars or in steel or fiber drums of 204-kg capacity. They are also packaged in 22.7-kg multiwalled, polyethylene-lined paper bags.

Analytical Methods

Caustic soda solutions are normally tested for general alkalinity and percentages of NaCl, Na_2SO_4, and $NaClO_3$ as well as for Fe and Ni levels. The general methods are outlined in Table 22. Detailed analytical methodologies are available from the major caustic soda suppliers.

Table 22. Analytical Methods for Caustic Soda

Parameter	Method
total alkalinity and NaOH	titration with standard acid
Na_2CO_3	gas-volumetric or gravimetric after decomposing carbonate to CO_2
NaCl	titration with $AgNO_3$ using Volhard method
Na_2SO_4	gravimetric by removing sulfate as $BaSO_4$
$NaClO_3$	volumetric by reaction of the chlorate with $FeSO_4$ in acid solution and titrating the excess $FeSO_4$ with potassium dichromate using diphenylamine sulfonic acid indicator
Fe, Ni	spectrophotometric or atomic absorption spectroscopy

Safety

Caustic soda in liquid or solid form has a marked corrosive action on all body tissue, so that even dilute solutions have a deleterious effect on tissue after prolonged contact. Inhalation of the dust or mist can cause damage to the upper respiratory tract; ingestion causes damage to the mucous membrane or the exposed tissue. It is therefore important that all the properties of caustic and the safety precautions be reviewed before handling. During handling, all persons should wear proper protective clothing, safety goggles (sometimes a full face shield), rubber gloves, boots, and a caustic resistant apron or suit. Safe handling practices are essential at all stages of production, from the laboratory to the manufacturing operations. The safety committee should inspect and advise on processing equipment and be responsible for providing personal protection, eye wash fountains, safety showers, etc.

If caustic soda should come in contact with the eyes, they should be flooded immediately and for at least 15 min, keeping the eyelids apart. If caustic comes in contact with skin or clothing, washing with water must be started immediately to prevent a chemical burn. The reader is advised to consult all the safety and first-aid techniques before handling (79).

Disposal of waste or spilled caustic soda must meet all federal, state, and local regulations and be carried out by properly trained personnel. Accidental spills of dry caustic are shoveled and flushed with water; caustic soda solutions must be diluted and neutralized with acid before discharging into sewers. Dilute acetic acid may be used to neutralize final traces of caustic.

United States Chlorine Market and Future Growth

Chlorine Consumption. In the latter part of the 1980s U.S. chlorine consumption grew an average of 3.2% annually from 9.7×10^6 t in 1986 (6) to 11×10^6 t in 1990. Increased exports of chlorine derivatives, aided by the weaker dollar, have added to chlorine's recent recovery. Because of the varied use pattern, future growth for U.S. chlorine consumption is best determined by compiling a composite based on the forecasted growth for each of its major derivatives. Using this method, average annual growth rates of between 0.9 and 1.3% have been forecasted from 1987 to 1992 (Table 23); other estimates of 0–1% per year are also possible (89). However, because of the unexpected high demand, by 1989 chlorine consumption already equaled the 1992 forecasted high of 11×10^6 t. The high growth rates of the late 1980s are unlikely to continue. Growth in 1990 was flat at 11×10^6 t. A mild decline is expected in 1991, and in 1992 chlorine consumption is expected to recover back to 11×10^6 t. Table 23 provides a good indication of the general trends and the relative importance of individual chlorine derivatives.

Major Chlorine Markets. Forecasted changes in chlorine consumption (Table 23) are addressed below in descending order of the importance of the major derivative, based on 1987 demand.

C_2 *Derivatives.* In 1987, C_2 derivatives accounted for approximately 36% of total U.S. chlorine consumption, and composed chlorine's largest single use category. Chlorine is used primarily to manufacture ethylene dichloride (EDC), the precursor to vinyl chloride monomer (VCM), used in PVC production [see VINYL CHLORIDE AND POLY(VINYL CHLORIDE)]. Approximately 24% of all 1987 U.S. chlorine production was consumed to satisfy growing PVC demand, and through 1992, VCM demand is expected to grow at 4.3% annually (6), due to increasing demand for PVC in the construction, packaging, and other industries. Nearly 85% of all EDC manufactured in the United States is used to produce VCM, and another 11% is exported, mostly for foreign VCM production. VCM exports are expected to grow 3.4% annually through 1992, while annual growth for EDC

Table 23. U.S. Chlorine Consumption[a]

Usage	Quantity, 10^3 t/yr			Average annual growth rate, 1987–1992, %
	Actual		Forecast, 1992	
	1982	1987		
organic chemicals (plus by-product HCl)				
C$_1$ derivatives[b]	696	818	836	0.4
C$_2$ derivatives[c]	2,991	3,696	4,201–4,242	2.6–2.8
fluorocarbons	25	64	54	[d]
C$_3$ derivatives[e]	810	1,154	1,261–1,288	1.8–2.2
C$_4$ derivatives[f]	178	116	109–118	(1.2)–0.3
other hydrocarbon derivatives[g]	226	250	243–254	(0.6)–0.3
chlorinated organics derived from inorganic compounds[h]	557	721	827	2.8
other chlorinated organics	348	272	259	−1
Total organic	*5,831*	*7,091*	*7,790–7,878*	*1.9–2.1*
inorganic chemicals				
titanium tetrachloride[i]	357	409	462–471	2.5–3
hydrogen chloride	210	222	209–227	(1)–1
bromine	86	67	71–74	1–2
phosphorus chlorides	61	73	78–83	1.5–2.5
aluminum and iron chlorides		35	32	−2
Total inorganic	*714*	*806*	*852–887*	*1.3–2.2*
pulp and paper production	1,240	1,608	1,346–1,418	(2.5)–(3.5)
water treatment	400	463	463	
hypochlorites and other uses[j]	305	318	318–349	0–2
Total usage	*8,490*	*10,286[k]*	*10,769–10,995*	*0.9–1.3*

[a]Ref. 6. Courtesy of SRI International.
[b]Chlorinated methanes, silicone.
[c]Ethylene dichloride, vinyl chloride monomer, trichloroethylene, perchloroethylene.
[d]Category includes only direct chlorine consumption; the majority of consumption is included in C$_1$ and C$_2$ derivatives.
[e]Epichlorohydrin, propylene oxide.
[f]Chloroprene.
[g]Monochlorobenzene, dichlorobenzene, linear alkylbenzenes, chlorinated paraffins.
[h]Phosgene (for toluene diisocyanate, diphenylmethane diisocyanate, and polycarbonate resin manufacture), chloroisocyanuric acid, cyanuric chloride.
[i]Titanium dioxide, titanium metal.
[j]Sodium hypochlorite, calcium hypochlorite, chlorinated isocyanurates.
[k]Estimated consumption exceeds apparent consumption by 38×10^3 t.

exports are expected to exceed 10% (6). Chlorinated solvents made from EDC and VCM, such as perchloroethylene (PCE), 1,1,1-trichloroethane, and trichloroethylene (TCE), compose the balance of the C$_2$ chlorine derivatives. Demand for these products is expected to decline through 1992, mainly because of decreased demand for TCE in vapor degreasing and for PCE in dry cleaning (see CHLOROCARBONS AND CHLOROHYROCARBONS, DICHLOROETHYLENE; TRICHLOROETHYLENE; CHLOROETHANES).

Pulp and Paper.   In 1987 the pulp and paper industry accounted for 15.6% of U.S. chlorine consumption. However, concerns over chlorine's potential in forming toxic chlorinated organics are expected to have a negative effect on growth in this industry over the next several years. Growth forecasts for chlorine demand in pulp and paper are -2.5 to -3.5% through 1992. Recent substitutions of oxygen, hydrogen peroxide, and particularly chlorine dioxide for chlorine in pulp bleaching indicate that declines in chlorine consumption for this use may be even greater than forecasted (see CHLOROCARBONS AND CHLOROHYDROCARBONS; TOXIC AROMATICS).

C_3 Derivatives.   The two major C_3 derivatives, propylene oxide (qv) and epichlorohydrin, composed 11% of 1987 chlorine consumption and are expected to grow at between 1.8 and 2.2% through 1992. Propylene oxide is manufactured by two companies, each using a different technology and different raw materials. About 45% is manufactured using the chlorohydrin route by a producer self-sufficient in chlorine (see CHLOROHYDRINS). Average annual growth of propylene oxide at between 1.5 and 2% is forecasted through 1992, based mostly on growing demand for polyether polyol, a propylene oxide derivative used in urethane foam manufacture (6). Chlorine consumption for epichlorohydrin [106-89-8] is expected to grow between 2.0 and 2.5% annually through 1992. Growth will be driven primarily by increased demand for epoxy resins (qv).

Phosgene.   Phosgene [75-44-5] accounted for 5.6% of 1987 chlorine consumption, which is expected to grow 2.8% annually through 1992. Phosgene is used in toluene diisocyanate (TDI), diphenylmethane diisocyanate (MDI), and polycarbonate resin manufacture. The majority of TDI is used to manufacture flexible urethane foams for furniture carpet underlay and bedding; MDI is used predominantly in polymeric applications such as laminate board and spray foam. Polycarbonate resin goes into glazing for sheeting polycarbonate composites, and alloys which are used to replace metal parts for the electronic and automobile industries.

C_1 Derivatives.   The chlorinated methanes, chloroform, methylene chloride, and carbon tetrachloride, consumed approximately 0.8 million tons of chlorine in 1987 and aggregate growth rates from this segment of the industry are expected to remain relatively flat through 1992. Because of its contribution to ozone depletion, carbon tetrachloride use in chlorofluorocarbon manufacture will be phased out in compliance with the recent Montreal Accord. In addition, environmental pressures are expected to continue to impact the use of methylene chloride in aerosol and paint remover applications. Some of the decreases in C_1 derivatives should be offset by positive growth for chloroform in HCFC-22 manufacture, which has not been implicated in ozone depletion (see CHLOROCARBONS AND CHLOROHYDROCARBONS, METHYL CHLORIDE; METHYLENE CHLORIDE; CHLORO-FORM; CARBON TETRACHLORIDE).

Water Treatment.   Chlorine is an excellent bacteriostat, unsurpassed for use in residual water treatment and growth is expected to remain flat through 1992. Attempts by municipal and industrial water treatment facilities to improve economics by increasing chemical efficiency and concerns over chlorine's involvement in the formation of undesirable organic compounds are the reasons for zero growth.

Titanium Tetrachloride.   The major use for titanium tetrachloride [7550-45-0] is in titanium dioxide production, and titanium dioxide [13463-67-7] is enjoying strong growth for use as a filler in pulp and paper manufacture and as a pigment in paint and plastic manufacture. Annual growth for this product is forecasted at between 2.5 and 3.0% through 1992.

Chlorine–Hydrochloric Acid Balance. More than 90% of hydrochloric acid is produced as a by-product during chemical manufacture. The remainder is produced intentionally either by burning chlorine and hydrogen or by reaction of salt and sulfuric acid. EDC manufacture by oxychlorination constitutes the single largest use for by-product HCl, which is also used in chemical manufacture, oil well acidizing, and steel pickling. Intentionally generated HCl is used primarily in food applications and HCl is also used to acidify brine in chlor–alkali production (see HYDROGEN CHLORIDE).

Because HCl is produced as a by-product, supply and demand are out of balance: merchant supply is presently almost double merchant demand. Furthermore, this imbalance is expected to increase over the next several years because of the restrictions imposed on CFC production. Although proposed CFC replacements are expected to require chlorinated precursors for their manufacture, only minor amounts of chlorine will be present in the final product, and the balance will exit the reaction as by-product HCl that will require disposal in an already saturated market. Environmental pressures related to HCl disposal are adding to concerns over the increasing supply. Resistance to HCl deep welling is growing and regulation is increasing. Thus HCl may be forced to find new applications in the chlorine markets. Although the chlor–alkali industry has reacted to environmental concerns by neutralizing many HCl waste streams before disposal, finding appropriate uses for by-product HCl without impacting chlorine production is a challenge facing the industry in the future.

United States Caustic Soda Market and Future Growth

Increases in U.S. demand for caustic soda have been unpredictably high in the last few years. Between 1987 and 1989, the annual increase in demand was about 3% (6). However, the caustic soda market is mature and new areas of significant growth have not surfaced in recent years. The unexpected recent demand is generally related to two factors: the pick-up in the U.S. economy after the slump of 1986 and pulp mills operating at full capacity, leading to less efficient caustic use.

Pulp and paper manufacture accounts for 24% of the total U.S. caustic soda demand. To improve caustic soda efficiency, pulp mills use recovery boilers to reclaim caustic soda from spent pulping liquor for later reuse, but, because recovery boilers are very expensive, few mills have excess boiler capacity. Furthermore, mills having excess boiler capacity buy black liquor from mills that are boiler-limited to reclaim caustic value. In 1987, and especially in 1988, extremely high operating rates in the pulp and paper industry caused maximum loads on recovery boilers, resulting in increased boiler outages and decreased caustic soda recovery. To maintain the maximum production levels needed to meet the high demand for pulp and paper products, bottlenecks at recovery boilers were avoided by placing spent pulping liquor in holding ponds for later reclamation, and make-up caustic soda was purchased externally from chlor–alkali producers. This had the effect of increasing caustic soda demand beyond the relative increase in demand for pulp and paper products and this decreased efficiency has made it difficult to forecast caustic soda consumption.

The latest caustic soda forecast (Table 24) shows a maximum 1992 demand of 10.2 million tons, arrived at by increasing 1987 actual demand by 1.5% annually (6). Because of the unexpectedly high caustic soda demand in 1988, however, the forecasted tonnage for 1992 has already been met. For the reasons outlined above, this strong growth is not expected to continue. In 1989 caustic soda demand grew by 3% over 1988, but between 1989 and 1992 an average annual growth of 1.5% or less is expected.

Table 24. U.S. Sodium Hydroxide Consumption[a]

	Quantity, 10^3 t/yr			Average annual growth rate, 1987–1992, %
	Actual		Forecast, 1992	
Usage	1982	1987		
chemical manufacturing				
inorganic chemicals	1,081	1,363		
organic intermediates, polymers, and end products	2,200	2,390		
other and unidentified uses	619	993		
Total for chemicals	*3,900*	*4,746*	*5,239*	*2.0*
pulp and paper manufacturing	2,070	2,269	2,386	1.0
cleaning products				
soap and other detergents	420	486	524	
household bleaches, polishes, and other cleaning goods	110	118	122	
miscellaneous surface-active agents	40	44	48	
Total cleaning	*570*	*648*	*694*	*1.4*
petroleum and natural gas				
oil and gas production	86	41	32–75	
oil and gas processing	346	387	435	
Total oil and gas	*432*	*428*	*467–510*	*1.8–3.6*
cellulosics				
rayon	111	127	133	
other	62	54	51	
Total fibers	*173*	*181*	*184*	*0.4*
cotton mercerizing and scouring	109	165	127–146	(2.5)–(5)
other[b]	654–794	1,028	1,025	
Total usage	*7,908–8,048*	*9,465*	*10,122–10,184*	*1.4–1.5*

[a]Ref. 6. Courtesy of SRI International.
[b]Essentially for balancing consumption with supply. Includes stock changes.

Caustic Soda Markets. Forecasted changes in caustic soda consumption (Table 24) are addressed below in descending order of the importance of the major industries, based on 1987 demand. Consumption patterns for many derivatives cannot be accurately quantitied and must be estimated.

Chemical Manufacturing. Chemical manufacturing accounts for over 50% of all U.S. caustic soda demand. It is used primarily for pH control, neutralization, off-gas scrubbing, and as a catalyst. About 50% of the total demand in this category, or approximately 25% of overall U.S. consumption, is used in the manufacture of organic intermediates, polymers, and end products. The majority of caustic soda required here is for the production of propylene oxide, polycarbonate resin, epoxies, synthetic fibers, and surface-active agents (6).

Another 29%, 14% of U.S. consumption, is used to manufacture inorganic chemicals, and the remainder is used in the manufacture of miscellaneous chemicals (6). Although significant quantities of caustic soda are used by chlor–alkali producers for brine treatment, alumina production consumes the largest amount of caustic soda in this category. Approximately 0.3 million tons per year of caustic soda is used in alumina production. The U.S. alumina industry is presently running at capacity, but because new alumina capacity is not scheduled to come onstream through 1992, caustic demand for this application will remain relatively flat. Another 166,000 t of caustic soda is used to produce sodium hypochlorite. Zero growth is forecasted in this category also. About 64,000 t of caustic soda goes into the manufacture of sodium cyanide [143-33-9], which is used to mine precious metals. Caustic soda demand for sodium cyanide is expected to grow 12–13%/yr through 1992. Smaller uses for caustic soda and the respective growth rates are: titanium dioxide (2.5–3% growth), silicates (0% growth), zeolites (2–3% growth), and phosphates (0% growth).

Pulp and Paper. Pulp and paper manufacture accounts for 24% of the total U.S. caustic soda demand (6). The caustic soda is used to pulp wood chips, to extract lignin during bleaching, and to neutralize acid waste streams. Despite forecasted growth rates of up to 3%/yr for pulp and paper products, caustic soda demand in this industry is forecasted to grow only 1%/yr (Table 24), and may be even less. Changes in technologies aimed at decreasing chlorine use will also serve to decrease caustic soda requirements. In addition, sodium hypochlorite, which requires caustic soda in its manufacture, is under attack in pulp and paper applications because of potential chloroform formation.

Cleaning Products. Caustic soda is used to produce a wide variety of cleaning products. This segment of the industry comprises 648,000 t or 6.8% of the total U.S. caustic soda consumption and is expected to grow at 1.4% annually through 1992. About 5.1% of the caustic soda consumed goes into the production of soap and other detergent products; another 1.2% is used to produce household bleaches, polishes, and cleaning goods. Growth forecasts for these uses are 1.5 and 1.0%/yr, respectively (6).

Petroleum and Natural Gas. Over 90% of the 428,000 t of caustic soda used in the petroleum and natural gas industry is used to process oil and gas into marketable products, especially by removing acidic contaminants. The remainder is used primarily to decrease corrosion of drilling equipment and to increase the solubility of drilling mud components by maintaining an alkaline pH (6).

Cellulosics. Rayon (qv) and other cellulose products such as cellophane and cellulose ethers (qv) consume 1.9% of U.S. caustic soda demand. Because of competitive products, however, this market has been decreasing since 1965; and forecasted average annual growth through 1992 is less than 0.4% (6) (see COTTON).

Cotton Mercerizing and Scouring. An estimated 1.7% of caustic soda consumption goes into cotton mercerizing and scouring. The majority is used for mercerizing and demand is expected to decline between 2.5 and 5%/yr (6).

Miscellaneous Consumption. An estimated 11% of caustic soda demand is used in a large number of miscellaneous markets: water treatment, food processing, flue-gas scrubbing, mining, glass making, metal degreasing, rubber reclamation, textile dyeing, and adhesive preparations. Composite demand in these markets is expected to remain at zero growth through 1992 (6).

Demand for Caustic Soda Types. Approximately 99% of the sodium hydroxide produced in 1987 was 50% caustic solution (5). Higher concentrations require additional evaporation and therefore increased prices relative to the sodium oxide values. To obtain maximum value, users have learned to adapt manufacturing processes to the 50% caustic soda.

Caustic Soda to Chlorine Balance. In 1988, the ratio of U.S. caustic soda to chlorine consumption was 0.96:1 (see Fig. 39). Since 1968 this ratio has ranged from a low of 0.88:1 (1978 and 1981) to a high of 0.98:1 (1969). No single factor can explain these variations, since caustic soda and chlorine, with few exceptions, have different markets and are therefore not driven by the same economic forces. This ratio is expected to trend upward over the next five years, however, since caustic soda consumption in the United States is forecasted to grow somewhat faster than chlorine consumption. It is expected that this ratio will remain within the range experienced in 1970–1990. Because caustic soda is co-produced with chlorine at a theoretical ratio of 1.1:1, a U.S. consumption ratio below that level results in excess availability of caustic soda. This material is typically shipped

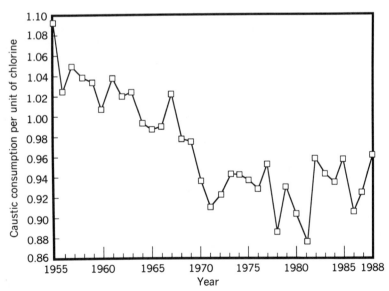

Fig. 39. Variation of caustic consumption from 1955 to 1988 (6). Courtesy of SRI International.

offshore to fill a significant export demand, and in 1988, for example, net U.S. exports of caustic soda amounted to 7.1% of production.

Economic Aspects

The choice of technology, the associated capital, and operating costs for a chlor–alkali plant are strongly dependent on local factors. Especially important are local energy and transportation costs, as are environmental constraints. The primary difference in operating costs between diaphragm, mercury, and membrane cell plants results from variations in electricity requirements for the three processes (Table 25) so that local energy and steam costs are most important.

Table 25. Energy Consumption of Operating Cells, Kilowatt-Hours per Ton of Chlorine

	Cell type		
Energy	Diaphragm	Mercury	Membrane
electricity for electrolysis	2800–3000	3200–3600	2600–2800
steam requirements[a]	600–800	0	200–300
Total	*3,400–3,800*	*3,200–3,600*	*2,800–3,100*

[a]One ton of steam is assumed to be ~ 400 kW·h.

The cost of constructing a grass-roots membrane or diaphragm plant is about $200,000/t·day of chlorine based on 1980 data in Table 26, which is doubled based on 1990 estimates. Environmental considerations have discouraged new mercury cell construction. The cost of converting a mercury cell plant to a membrane cell operation appears to vary between $100,000 to $200,000/t per day of chlorine depending on the process needs, in addition to any disposal and demolition costs required for the conversion. However, the conversion of a diaphragm

Table 26. Plant Capital Costs for 500 Ton per Day Chlorine Production, Millions of Dollars

Parameter	Diaphragm cells	Membrane cells
cells	17.0	34.0
rectifiers	3.4	3.4
brine purification	5.0	10.0
chlorine collection	6.5	6.5
hydrogen collection	2.0	2.0
caustic evaporation	13.5	3.0
caustic storage	5.6	5.2
miscellaneous	7.0	6.3
utilities and offsites	23.0	22.0
overheads	12.0	11.3
engineering	12.0	11.3
Total	*107*	*115*

circuit to a membrane operation may be less, around $80,000 to $150,000/t per day, again depending on the process requirements because the available infrastructure can be readily adapted for the conversion. When converting mercury cell operations to either membrane or diaphragm operations, the cost of cells and the membranes accounts for about 60% of the total investment. Other costs for the conversion to the diaphragm process are less than conversion to membrane cell operation. See references 8, 90, and 91 for more details related to these capital costs.

BIBLIOGRAPHY

"Alkali and Chlorine Industries" in *ECT* 1st ed., Vol. 1, pp. 358–430, by Z. G. Deutsch, Deutsch and Loonam; "Alkali and Chlorine Industries" in *ECT* 2nd ed., Vol. 1, pp. 668–758, by Z. G. Deutsch, Consulting Engineer, and C. C. Brumbaugh and F. H. Rockwell, Diamond Alkali Company; "Alkali and Chlorine Products, Chlorine and Sodium Hydroxide" in *ECT* 3rd ed., Vol. 1, pp. 799–865, by J. J. Leddy, I. C. Jones, Jr., B. S. Lowry, F. W. Spillers, R. E. Wing, and C. D. Binger, Dow Chemical U.S.A.; "Chlorine" in *ECT* 1st ed., Vol. 3, pp. 677–681, by D. G. Nicholson, University of Pittsburgh; "Chlorine" in *ECT* 2nd ed., Vol. 5, pp. 1–6, by D. G. Nicholson, East Tennessee State University; "Chlorine" in *ECT* 2nd ed., Supplement, pp. 167–177, by H. B. Hass, Chemical Consultant.

1. *Chem. Eng. News* 37 (June 18, 1990).
2. R. N. Shreve and J. A. Brink, Jr., *Chemical Process Industries,* 4th ed., McGraw-Hill Book Co., Inc., New York, 1977.
3. *Chlorine and Its Derivatives: A World Survey of Supply, Demand, and Trade to 1992,* Tecnon Consulting Group, London, 1988.
4. *Caustic Soda in the 1990's: A World Survey of Supply, Demand, and Trade,* Tecnon Consulting Group, London, 1989.
5. *North American Chlor–Alkali Industry Plants and Production Data Book,* Pamphlet 10, The Chlorine Institute, Washington, D.C., Jan. 1989.
6. M. Smart, G. Rice, A. Leder, W. Schlegel, and E. Nakamura, "Chlorine/Sodium Hydroxide, CEH Marketing Research Report," *Chemical Economics Handbook,* Menlo Park, Calif., 1989, pp. 733.1000O–733.1000W; 733.1001D–733.1001G; 733-1001K; 733.1001Q–733.1001Z; 733.1002A–733.1003O.
7. B. V. Tilak, S. R. Fitzgerald, and C. L. Hoover, *J. Appl. Electrochem.* **18,** 699 (1988).
8. P. Schmittinger and co-workers, in *Ullmann's Encyclopedia of Industrial Chemistry,* 5th ed., Vol. A6, 1986, p. 399.
9. I. Rousar, V. Cezner, M. Vender, M. Kroutil, and J. Vachuda, *Chem. Prum.* **17/42,** 466 (1967).
10. *Chlorine Handbook,* Diamond Shamrock Chemical Co., Irving, Tex. 1984.
11. D. F. W. Hardie, *Electrolytic Manufacture of Chemicals from Salt,* The Chlorine Institute, Inc., New York, 1975.
12. Brit. Pats. 1,147,442 (May 12, 1965), H. B. Beer (to Chemnor Corp.); 1,195,871 (Feb. 10, 1967), H. B. Beer (to Chemnor AG).
13. V. de Nora and J. W. Kuhn Von Burgsdorff, *Chem. Ing. Tech.* **47,** 125–128 (1975).
14. S. Trasatti, ed., *Electrodes of Conductive Metallic Oxides,* Parts A and B, Elsevier, Amsterdam, 1980, 1981.
15. D. M. Novak, B. V. Tilak, and B. E. Conway in J. O'M. Bockris, B. E. Conway, and R. E. White, eds., *Modern Aspects of Electrochemistry,* Vol. 14, Plenum Press, New York, 1982.
16. J. E. Currey and G. G. Pumplin in J. J. McKetta and W. A. Cunningham, eds.,

Encyclopedia of Chemical Engineering and Design, Vol. 7, Marcel Dekker, Inc., New York, 1978, p. 305.

17. U.S. Pat. 4,410,411 (Oct. 18, 1983), R. W. Fenn III, E. J. Pless, R. L. Harris, and K. J. O'Leary (to OxyTech Systems, Inc., Chardon, Ohio).

18. U.S. Pat. 3,674,676 (July 4, 1972), E. J. Fogelman (to OxyTech Systems, Inc., Chardon, Ohio).

19. U.S. Pat. 3,928,166 (Dec. 23, 1975), K. J. O'Leary, C. P. Tomba, and R. W. Fenn III (to OxyTech Systems, Inc., Chardon, Ohio).

20. E. H. Cook and M. P. Grotheer, *Energy Saving Developments for Diaphragm Cells and Caustic Evaporators,* 23rd Chlorine Plant Manager's Seminar, New Orleans, The Chlorine Institute, Inc., Feb. 6, 1980.

21. U.S. Pats. 4,093,533 (June 6, 1978), 4,142,951 (March 6, 1979), R. N. Beaver and C. W. Becker (to The Dow Chemical Company).

22. U.S. Pat. 4,186,065 (Jan. 29, 1980), C. D. Dilmore, E. V. Hoover, and A. B. Kriss (to PPG Industries).

23. L. C. Curlin, T. F. Florkiewicz, and R. C. Matousek, *Polyramix (TM), A Depositable Replacement for Asbestos Diaphragms,* Paper presented at London International Chlorine Symposium, 1988.

24. R. N. Beaver, *The Dow Diaphragm Cell,* The Dow Chemical Co., Freeport, Tex., 1985.

25. U.S. Pat. 4,497,112 (Feb. 5, 1985), H. D. Dang, R. N. Beaver, F. W. Spillers, and M. J. Hazelrigg, Jr. (to The Dow Chemical Company).

26. *De Nora Glanor Diaphragm Cells,* De Nora Permalec SpA, Milan, Italy.

27. *Diaphragm Cells,* OxyTech Systems, Inc., Chardon, Ohio, 1988.

28. U.S. Pat. 3,591,483 (July 6, 1971), R. E. Loftfield and H. W. Laub (to OxyTech Systems Inc., Chardon, Ohio).

29. U.S. Pat. 4,834,859 (May 30, 1989), L. C. Curlin and R. L. Romine (to OxyTech Systems Inc., Chardon, Ohio).

30. *Alkaline Chloride Electrolysis by the Diaphragm Process; System Hooker,* Uhde GmbH, Dortmund, Germany, 1985.

31. W. G. Grot, *Discovery and Development of Nafion Perfluorinated Membranes,* The Castner Medal Lecture, 3rd London International Chlorine Symposium, June, 1985.

32. J. L. Hurst, *Implementing Membrane Cell Technology Within OxyChem Manufacturing,* International Symposium on Chlor–Alkali Industry, Tokyo, Japan, April, 1988.

33. *Perfluorinated Membranes for the Chlor–Alkali Industry,* E. I. du Pont de Nemours & Co., Inc., Wilmington, Del., 1983.

34. D. L. Peet and J. H. Austin, *Nafion Perfluorinated Membranes, Operation in Chlor–Alkali Plants,* Chlorine Institute Plant Managers Seminar, Tampa, The Chlorine Institute, Feb. 1986.

35. *Nafion 90209, Product Bulletin,* E. I. du Pont de Nemours & Co., Inc., Wilmington, Del., 1988.

36. *Nafion NX-961, Product Bulletin,* E. I. du Pont de Nemours & Co., Inc., Wilmington, Del., 1984.

37. D. L. Peet, *Membrane Durability in Chlor–Alkali Plants,* Electrochemical Society Meeting, Honolulu, Hawaii, Oct. 1987; *Proceedings of the Symposium on Electrochemical Engineering in the Chlor–Alkali and Chlorine Industries,* PV. #88-2, 1988, pp. 329–336.

38. U.S. Pat. 4,025,405 (1972), R. L. Dotson and K. J. O'Leary (to OxyTech Systems, Inc., Chardon, Ohio).

39. H. Ukihashi, M. Yamabe, and H. Miyake, *Prog. Polym. Sci.* **12,** 229 (1986).

40. *Nafion Perfluorinated Membranes for KOH Production, Nafion Product Bulletin,* E. I. du Pont de Nemours & Co., Inc., Wilmington, Del., 1988.

41. *Nafion Perfluorinated Membranes, Introduction,* E. I. du Pont de Nemours & Co., Inc., Wilmington, Del., 1987.

42. *Asahi Chemical Membrane Chlor–Alkali Process,* Asahi Chemical Industry Co., Ltd., Tokyo, Japan, 1987.
43. N. M. Pront and J. S. Moorhouse, *Modern Chlor-Alkali Technology,* Vol. 4, Society of Chemical Industry, Elsevier Applied Sci., London, 1990.
44. Y. Sajima, K. Sato, and H. Ukihashi, *Recent Progress of Asahi Glass Membrane Chlor–Alkali Process, AICHE Symp. Ser.* **82**(248), 108 (1985).
45. T. Yamashita, Y. Sajima, and H. Ukihashi, *The Design and Operating Experiences of Azec Electrolyzers and Recent Development of Flemion Membranes,* London International Chlorine Symposium, London, June 1988.
46. *Azec Electrolyzer Supply Record,* Asahi Glass Co., Ltd., Tokyo, Japan, 1987.
47. *CME Chlorine Engineers Membrane Electrolyzer,* Chlorine Engineers Corp. Ltd., Tokyo, Japan, 1989.
48. M. Esayian and J. H. Austin, *Membrane Technology for Existing Chlor–Alkali Plants,* presented at The Chlorine Institute, Feb. 1984.
49. *Japan's Chlor–Alkali Producers Save Energy by Retrofiting Diaphragm Cells* (Case History), E. I. du Pont de Nemours & Co., Inc., Wilmington, Del.
50. *FM-21 SP Series Membrane Electrolyzer,* ICI Winnington, Northwich, Cheshire, England, 1989.
51. *Lurgi Membrane Electrolyzer,* Lurgi GmbH, Frankfurt am Main, Germany, 1988.
52. *Membrane Chlor–Alkali Process,* Oronzio De Nora Technologies, BV Milan, Italy and Houston, Tex., 1988.
53. G. J. Morris and R. D. Wilson, *Analysis of Oronzio De Nora Technologies Zero Gap Electrode,* Oronzio De Nora Technologies, BV, Milan, Italy and Houston, Tex., 1989; presented at The Electrochemical Society Meeting, Los Angeles, Calif., May 1989.
54. *MGC Monopolar Membrane Electrolyzer,* OxyTech Systems, Inc., Chardon, Ohio, 1988.
55. *Alkaline Chloride Electrolysis by the Membrane Process,* Uhde GmbH, Dortmund, Germany, 1989.
56. B. V. Tilak, A. C. R. Murthy, and B. E. Conway, *Proc. Indian Acad. Sci. (Chem. Sci.)* **97,** 359 (1986).
57. D. L. Caldwell in J. O'M. Bockris, B. E. Conway, E. A. Yeager, and R. E. White, eds., *Comprehensive Treatise of Electrochemistry,* Vol. 2, Plenum Press, New York, 1981, Chapt. 2.
58. F. Hine, B. V. Tilak, and K. Viswanathan in R. E. White, J. O'M. Bockris, and B. E. Conway, eds., *Modern Aspects of Electrochemistry,* Vol. 18, Plenum Press, New York, 1986, Chapt. 5.
59. F. Hine, M. Yasuda, and M. Watanabe, *Denki Kagaku* **47,** 401 (1979).
60. U.S. Pat. 4,024,044 (May 17, 1977), J. R. Brannan and I. Malkin (to OxyTech Systems, Inc., Chardon, Ohio).
61. Jpn. Pat. EP-A 31,948 (Oct. 15, 1986), M. Yoshida and H. Shiroki (to Ashahi Kasei Kogyo).
62. Eur. Pat. Appl. 129,734 (Jan. 2, 1985), N. R. Beaver, L. E. Alexander, and C. E. Byrd (to the Dow Chemical Co.); Eur. Pat. Appl. 129,374 (Dec. 27, 1984), J. F. Cairns, D. A. Fenton, and P. A. Izard (to Imperial Chemical Industry PLC).
63. L. J. Gestaut, T. M. Clere, C. E. Graham, and W. R. Bennett, *Abstract #393;* L. J. Gestaut, T. M. Clere, A. J. Niksa, and C. E. Grahm, *Abstract #124,* Electrochem. Soc. Meeting, Washington, D.C., Oct. 1983.
64. J. T. Keating and K. J. Behling, *Brine, Impurities, and Membrane Chlor–Alkali Cell Performance,* presented at the London International Chlorine Symposium, 1988.
65. *Chlorine Recovery System,* OxyTech Systems, Inc., Chardon, Ohio, 1988.
66. T. A. Liederbach, *Chem. Eng. Prog.* **70,** 64 (1974).
67. U.S. Pat. 4,585,579 (April 29, 1986), T. V. Bommaraju, W. V. Hauck, and V. J. Lloyd (to OxyTech Systems, Inc., Chardon, Ohio).
68. E. J. Rudd and R. F. Savinell, *J. Electrochem. Soc.* **136,** 449c (1989).

69. K. Kerger, *Chem. Eng. Tech.* **43,** 167 (1971).
70. Uhde: Chlor and Wasserstoff aus Salzsauredurch Elektrolyse, 1982.
71. C. W. Arnold and K. A. Kolbe, *Chem. Eng. Prog.* **48,** 167 (1971).
72. The M. W. Kellogg Co., *Hydrogen Chloride to Chlorine, The Kel-Chlor Process,* 20th Chlorine Plant Managers Seminar, New Orleans, La., Feb. 9, 1977.
73. The M. W. Kellogg Co., *Hydrocarbon Process.* **60**(11), 143 (1981).
74. Neth. Appl. 6,407,015 (Dec. 22, 1964) (to Dyamet-Nobel AG). Ger. Pat. DE 1245922-B (1967).
75. Fr. Pat. FR 1294 706 (June 1, 1962) P. Bedague, P. Baumgartner, and J. C. Balaceanu.
76. *A Survey of Potential Chlorine Production Processes,* Contract #31-109-38-2411 by Versar Inc., ANL/OEPM-79-1, Argonne National Laboratory, Argonne, Ill., April, 1979.
77. A. S. Ross and C. Mass, *Can. J. Res. Sect. B* **18,** 55 (1940).
78. Handbooks published by Caustic Soda Suppliers, The Chlorine Institute, Washington, D.C., and The Manufacturing Chemists Assn., 1989.
79. J. J. Martin and D. M. Longpre, *J. Chem. Eng. Data* **29,** 466 (1984).
80. R. Kapoor and J. J. Martin, *Thermodynamic Properties of Chlorine,* University of Michigan, 1957.
81. Gloria M. Lawler and co-workers, *Chemical Origins and Markets, Flow Charts, and Tables,* 5th ed., Chemical Information Services, Menlo Park, Calif., 1977.
82. *Chlorine Manual,* The Chlorine Institute, Washington, D.C., 1986.
83. R. C. Weast, ed., *CRC Handbook of Chemistry and Physics,* 67th ed., CRC Press, Inc., Boca Raton, Fla., 1986–1987.
84. T. P. Hou, *Ind. Eng. Chem.* **46,** 2401 (1954).
85. H. R. Wilson and W. L. McAbe, *Ind. Eng. Chem.* **34,** 565 (1942).
86. *Caustic Purification System,* OxyTech Systems, Inc., Chardon, Ohio, 1988.
87. J. Douglas, *EPRI J.* 21 (Jan./Feb. 1989).
88. Jpn. Pats. J 62132729-A (June 16, 1987), J 62143820-A (June 22, 1987), J 62138322-A (1987), (to Asahi Chemical Industry KK).
89. C. C. Lewis, *CPI Purchasing,* 32 (July 1989).
90. Y. C. Chen, *Process Economics Program Report,* Stanford Research Institute, Menlo Park, Calif., No. 61, 1970; No. 61A, 1974; No. 61B, 1978; No. 61C, 1982.
91. Abam Engineers, *Final Report on Process Engineering and Economic Evaluations of Diaphragm and Membrane Chloride Cell Technologies,* ANL/OEPM-80-9, Argonne National Laboratories, Argonne, Ill., Dec. 1980.

General references

References 8, 16, 43, 57, 58, 89, 90, and 91 are general references.
J. S. Sconce, *Chlorine, Its Manufacture, Properties, and Uses,* Reinhold Publishing Corp., New York, 1962.
D. W. F. Hardie and W. W. Smith, *Electrolytic Manufacture of Chemicals from Salt,* The Chlorine Institute, New York, 1975.
M. O. Coulter, *Modern Chlor–Alkali Technology,* Ellis Horwood, London, 1980.
C. Jackson, *Modern Chlor–Alkali Technology,* Ellis Horwood, Chichester, UK, 1983.
K. Wall, *Modern Chlor–Alkali Technology,* Ellis Horwood, Chichester, UK, 1986.
Diaphragm Cells for Chlorine Production, Society of Chemical Industry, London, 1977.
F. Hine, R. E. White, W. B. Darlington, and R. D. Varjian, eds., *Proceedings of the Symposium on Electrochemical Engineering in the Chlor–Alkali and Chlorate Industries* (Proc. Vol. 88-2), The Electrochemical Society Inc., Pennington, N.J., 1988.
F. Hine, B. V. Tilak, J. M. Fenton, and J. D. Lisius, eds., *Proceedings of the Symposium on Performance of Electrodes for Industrial Electrochemical Processes* (Proc. Vol. 89-10), The Electrochemical Society Inc., Pennington, N.J., 1989.

W. Grot in J. I. Kroschwitz, ed., *Encyclopedia of Polymer Science and Engineering,* 2nd ed., Vol. 16, Wiley-Interscience, New York, 1989, p. 42.

N. M. Prout and J. S. Moorhouse, eds., *Modern Chlor–Alkali Technology,* Vol. 4, Elsevier Applied Science, 1990.

L. Calvert Curlin
OxyTech Systems, Inc.

Tilak V. Bommaraju
Constance B. Hansson
Occidental Chemical Corporation

SODIUM CARBONATE

Sodium carbonate [497-19-8], Na_2CO_3, formula wt 105.99, is a white crystalline solid known as soda ash and less commonly, ash, soda or calcined soda. It is readily soluble in water and is strongly alkaline. It is the eleventh largest world commodity chemical; in 1989 33 million metric tons were produced, including 9 million in the United States. About 75% of world production is synthetic ash made from sodium chloride via the Solvay or similar processes; the remaining 25% is produced from natural sodium carbonate bearing deposits. Over half of the world's production is consumed in the glass industry and another 22% is used in the production of sodium-based chemicals. Sodium carbonate is also used in detergents, pulp (qv) and paper (qv), and environmental control (water treatment and flue gas desulfurization). The normal article of commerce is highly purified (>99%). Differences in bulk density are the only major distinction between the various grades. Minor amounts of sodium carbonate monohydrate [5968-11-6], $Na_2CO_3 \cdot H_2O$, and sodium carbonate decahydrate [6132-02-1], $Na_2CO_3 \cdot 10H_2O$, are sold and used in specialty applications. Aqueous solutions are alkaline. At 25°C the pH of 1, 5, and 10 wt % Na_2CO_3 solutions is 11.37, 11.58, and 11.70, respectively. Physical properties and solubility data are given in Tables 1 and 2, respectively.

Table 1. Physical Properties of Sodium Carbonate[a]

| Property | Value | | | |
	Na_2CO_3	$Na_2CO_3 \cdot H_2O$	$Na_2CO_3 \cdot 7H_2O$	$Na_2CO_3 \cdot 10H_2O$
melting point	825			
bulk density, g/mL	0.59–1.04			
specific gravity	2.533			
heat of formation, ΔH_f, kJ/mol[b] at 0°C	−1131	−1459	−3201	−4082
temperature, °C, stable solid phase	>109	35.4–109.0	32.0–35.4	0–32.0

[a]Refs. 1–4.
[b]To convert kJ/mol to kcal/mol, divide by 4.184.

Table 2. Solubility Data for Sodium Carbonate

Temperature, °C	Na_2CO_3 in saturated solution, wt %
0	6
10	8.5
20	17
30	28
40	32.3
50	32.0
60	31.7
70	31.0
80	30.8
90	30.8
100	30.8

[a]Ref. 3.

Manufacturing Technology

Historically, soda ash was produced by extracting the ashes of certain plants, such as Spanish barilla, and evaporating the resultant liquor. The first large scale, commercial synthetic plant employed the LeBlanc (Nicolas LeBlanc (1742–1806)) process (5). In this process, salt (NaCl) reacts with sulfuric acid to produce sodium sulfate and hydrochloric acid. The sodium sulfate is then roasted with limestone and coal and the resulting sodium carbonate–calcium sulfide mixture (black ash) is leached with water to extract the sodium carbonate. The LeBlanc process was last used in 1916–1917; it was expensive and caused significant pollution.

These disadvantages prompted Ernest Solvay (1838–1922) to develop and commercialize a procedure using ammonia to produce soda ash from salt and limestone. The first plant using the Solvay process was built in 1863; this process or variations are in use in much of the world in the 1990s.

Natural soda ash-containing brines and deposits were found in the United States at Searles Lake, California and Green River, Wyoming in the late 1800s. Sporadic attempts to commercialize these deposits were made in the early 20th century. During the first half of the 1900s the Searles Lake deposit was commercialized and the production facilities improved and expanded. In 1938 large deposits of trona [15243-87-5], $Na_2CO_3 \cdot NaHCO_3 \cdot 2H_2O$, were found in the Green River basin. The first process for producing refined soda ash from this source came onstream in 1953.

Synthetic Processes

SOLVAY-AMMONIA PROCESS

The basic Solvay process remains the dominant production route for soda ash. Its continued success is based on the raw materials, salt and limestone, being more readily available than natural alkali. All soda ash processes are based on the manipulation of saline phase chemistry (6,7) an understanding of which is important both to improving current processes and to the economic development of new alkali resources.

Chemical Reactions.   The overall chemical reaction,

$$2\ NaCl + CaCO_3 \longrightarrow Na_2CO_3 + CaCl_2$$

takes place in a series of process steps. The ammonia formed in step 5 is recycled to step 1; the carbon dioxide from steps 3 and 4 is used in step 2. Overall sodium utilization averages 70%. The remainder is sent to waste. Figure 1 shows a simplified flow diagram (8).

1. $NaCl + NH_3 \longrightarrow$ ammoniated brine
2. ammoniated brine $+ CO_2 \longrightarrow NaHCO_3 + NH_4Cl$
3. $2\ NaHCO_3 \xrightarrow{\Delta} Na_2CO_3 + CO_2 + H_2O$
4. $CaCO_3 \longrightarrow CaO + CO_2$
5. $CaO + 2\ NH_4Cl \longrightarrow 2\ NH_3 + CaCl_2 + H_2O$

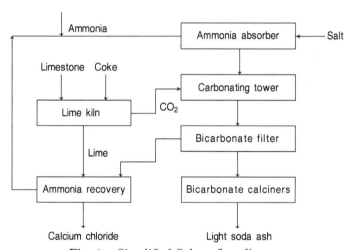

Fig. 1.   Simplified Solvay flow diagram.

Brine Preparation.   Sodium chloride solutions are occasionally available naturally but they are more often obtained by solution mining of salt deposits. Raw, near-saturated brines containing low concentrations of impurities such as magnesium and calcium salts, are purified to prevent scaling of processing equipment and contamination of the product. Some brines also contain significant amounts of sulfates (see CHEMICALS FROM BRINE). Brine is usually purified by a lime–soda treatment where the magnesium is precipitated with milk of lime $(Ca(OH)_2)$ and the calcium precipitated with soda ash. After separation from the precipitated impurities, the brine is sent to the ammonia absorbers.

Ammonia Absorption.   The strong brine is saturated with ammonia gas containing water vapor and carbon dioxide in an absorption tower. The brine descends through the main part of the absorber countercurrent to the rising ammoniacal gases. Most of the ammonia is obtained from various process steps;

however, small amounts are added to make up for losses. During ammoniation, the brine requires cooling (ca 1650 MJ/t = 394 kcal/kg of product soda ash). This operation is generally carried out at slightly less than atmospheric pressure.

Precipitation of Bicarbonate.   The ammoniated brine from the absorber coolers is pumped to the top of one column situated in a group of columns used to precipitate sodium bicarbonate [*144-55-8*], $NaHCO_3$. This column, fouled or partially plugged with bicarbonate after several days of crystallization, is referred to as a cleaning column. Lime kiln gas, compressed to about 414 kPa (60 psi), enters the bottom of the cleaning column and bubbles up through the solution absorbing most of the carbon dioxide. Thus, the carbon dioxide concentration in the liquor leaving the cleaning column is kept below the bicarbonate precipitation level. This liquor is fed in parallel to the top of the remaining columns in the block. These crystallizing or making columns receive a mixture of kiln gas and bicarbonate calciner gas that bubbles up through the solution, precipitating sodium bicarbonate. This process is accompanied by the evolution of considerable heat, 1450 MJ/t (347 kcal/kg), that must be removed to maintain yield. Bicarbonate crystals gradually foul the heat-exchange surfaces and crystallizing columns must be alternately used as the cleaning column. Gases, mostly nitrogen but some carbon dioxide and ammonia, are vented from the crystallizing tower columns and collected before being recycled to the absorber.

Filtration of Bicarbonate.   The slurry from the crystallizing towers is fed to continuous vacuum filters or centrifuges where the crystals are separated from the filter liquor. Air drawn through the vacuum filter (or the vent gas from the centrifuge operation) is returned to the ammonia absorber. The filter cake is carefully washed with fresh water to control residual chloride and to meet customer specifications. It is then conveyed to the calcining operation. Dewatering characteristics of the bicarbonate crystals are dependent on operating conditions in the crystallizing columns. The filter cake, often called crude bicarbonate or ammonia soda, contains 9–10% moisture as residual diluted filter liquor. The cake is made up of sodium bicarbonate and small amounts (5 mol % on a dry basis) of ammonia primarily in the form of ammonium bicarbonate.

Recovery of Ammonia.   The filter liquor contains unreacted sodium chloride and substantially all the ammonia with which the brine was originally saturated. The ammonia may be fixed or free. Fixed ammonia (ammonium chloride [*12125-02-97*]) corresponds stoichiometrically to the precipitated sodium bicarbonate. Free ammonia includes salts such as ammonium hydroxide, bicarbonate, and carbonate, and the several possible carbon–ammonia compounds that decompose at moderate temperatures. A sulfide solution may be added to the filter liquor for corrosion protection. The sulfide is distilled for eventual absorption by the brine in the absorber. As the filter liquor enters the distiller, it is preheated by indirect contact with departing gases. The warmed liquor enters the main coke, tile, or bubble cap-filled sections of the distiller where heat decomposes the free ammonium compounds and steam strips the ammonia and carbon dioxide from the solution.

This carbon dioxide-free solution is usually treated in an external, well-agitated liming tank called a "prelimer." Then the ammonium chloride reacts with milk of lime and the resultant ammonia gas is vented back to the distiller. Hot calcium chloride solution, containing residual ammonia in the form of ammonium hydroxide, flows back to a lower section of the distiller. Low pressure steam sweeps practically all of the ammonia out of the limed solution. The final solution, known as "distiller waste," contains calcium chloride, unreacted sodium chloride, and excess lime. It is diluted by the condensed steam and the water in which the lime was conveyed to the reaction. Distiller waste also contains inert solids brought in with the lime. In some plants, calcium chloride [10043-52-4], $CaCl_2$, is recovered from part of this solution. Close control of the distillation process is required in order to thoroughly strip carbon dioxide, avoid waste of lime, and achieve nearly complete ammonia recovery. The hot (56°C) mixture of wet ammonia and carbon dioxide leaving the top of the distiller is cooled to remove water vapor before being sent back to the ammonia absorber.

Lime Preparation. The most suitable limestone (see LIME AND LIMESTONE), hard and strong with low concentrations of impurities, is graded to a reasonably uniform coarse size. Although other fuels may be used, the limestone is usually mixed with about 7% metallurgical-grade coke or anthracite and then burned in vertical shaft kilns. Air is admitted continuously into the bottom of the kiln; the combustion products are taken off the top. The fuel burns in a zone a little before the middle of the kiln, where the charge is heated to about 950–1100°C, and the stone "burns" to lime. Carbon dioxide is generated both by decomposition of limestone and by combustion of carbon in the fuel. The kiln gases are diluted with nitrogen from the combustion air. This gas usually contains 37–42% CO_2, stone dust, ash particles, and other gaseous impurities. The gas is partially cooled in the kiln by the upper layers of stone. It is further cooled and cleaned before entering the compressors feeding the carbonating columns.

The lime, cooled somewhat by the entering air in the lower parts of the shaft kiln, is discharged intermittently and slaked to calcium hydroxide with an excess of water, usually in rotary slakers. This results in a thick suspension, commonly called milk of lime, stored in agitated tanks. The heat of the reaction produces milk of lime at a temperature of 90–100°C; water addition is controlled to give a free calcium oxide content of 230–300 g/L. In some operations, dry lime is used in place of the milk of lime. The dry lime is pulverized and added continuously to the prelimer in the distillation step, thus reducing the amount of water added and the steam consumed, and producing a more concentrated distiller waste.

Calcining the Bicarbonate to Soda Ash. Crude filtered bicarbonate is continuously calcined by indirect heating. Various techniques are used to heat, 2430 kJ/kg (581 kcal/kg) this material to 175–225°C in the calciners. Carbon dioxide, produced at 95% or higher purity, is compressed and recycled to the carbonating tower in order to enrich the makeup kiln-gas feed. The hot soda ash discharged from the calciner is cooled, screened, and packaged, or shipped in bulk. This product, called light ash has a bulk density of around 590 kg/m³. A certain amount is sold in this form; the majority, however is converted to dense ash.

Dense Ash. Dense soda ash is manufactured by hydrating light ash to produce the larger sodium carbonate monohydrate crystals and then dehydrating them. Hydration may be accomplished in one of two ways: by feeding light ash and water to mixers or blenders, or by adding light ash to a saturated soda ash solution containing a slurry of monohydrate crystals. The monohydrate crystals are then fed to a continuous dryer and the dehydrated product is screened before packing and shipping. Bulk densities typically run between 960–1040 kg/m³.

Waste Disposal. Large volumes of liquid waste containing suspended and dissolved solids are produced in an ammonia–soda plant. The largest quantity comes from the distiller operation where, for every ton of product soda ash, nearly 10 cubic meters of liquid waste are produced. This waste contains about one ton of calcium chloride, one-half ton of sodium chloride, and other soluble and suspended impurities. Traditionally, after settling the suspended solids in large basins, the waste was discharged into local waterways, a practice no longer acceptable in the United States. The costs to comply with environmental regulations and increasing operating costs relative to natural ash production have forced shutdown of all synthetic soda ash plants in the United States.

AMMONIUM CHLORIDE (AC) PROCESS

The ammonium chloride process, developed by Asahi Glass, is a variation of the basic Solvay process (9–11). It requires the use of solid sodium chloride but obtains higher sodium conversions (+90%) than does the Solvay process. This is especially important in Japan, where salt is imported as a solid. The major difference from the Solvay process is that here the ammonium chloride produced is crystallized by cooling and through the addition of solid sodium chloride. The resulting mother liquor is then recycled to dissolve additional sodium chloride. The ammonium chloride is removed for use as rice paddy fertilizer. Ammonia makeup is generally supplied by an associated synthesis plant.

NEW ASAHI (NA) PROCESS

The NA process (9–11) has two variations named the co-production process and the mono-production process. In the co-production process, solid ammonium chloride is produced as a by-product by direct contact cooling crystallization. In the mono-production process, the crystalline ammonium chloride reacts with a lime slurry to release the ammonia for recycle sending the resulting calcium chloride solution to waste. Most Japanese plants have adopted some aspects of the NA process which it is claimed requires less energy than the basic Solvay process and, by using direct contact cooling instead of conventional heat exchangers, reduces fouling.

OTHER SYNTHETIC ROUTES

Two processes taken to some stage of development in the early 1980s have not yet been commercialized.

Hüls Process. Chemische Werke Hüls AG has developed a process to produce soda ash and hydrochloric acid from salt via an amine–solvent system (12). A potential advantage of the Hüls process is that, under some market conditions, hydrochloric acid may be more easily sold than either ammonium or calcium chloride.

Akzo Process. Akzo Zout Chemie has developed a route to vinyl chloride and soda ash from salt using an amine–solvent system catalyzed by a copper–iodide mixture (13). This procedure theoretically requires half the energy of the conventional Solvay processes.

Natural Processes

Sodium carbonate bearing deposits and brines exist around the world. Locations are known in the United States, China, Turkey, Bolivia, Brazil, Venezuela, Mexico, India, Pakistan, USSR, Kenya, Australia, and Botswana (14–20). The overwhelming majority of natural ash production comes from the Green River Basin in southwestern Wyoming. Significant amounts are also produced at Searles Lake in California; lesser amounts at Lake Magadi in Kenya. Minor quantities are reportedly produced in Pakistan, the USSR, and China and small amounts of impure trona come from Owens Lake, California. A plant is currently under construction to recover soda ash from brine at Sua Pan, Botswana. Each deposit has its own distinctive characteristics and each requires different processing techniques.

GREEN RIVER BASIN

Sodium carbonate brines were discovered around Green River, Wyoming before 1900. In 1907 the Western Alkali Corporation processed these brines and produced small amounts of sal soda ($Na_2CO_3 \cdot 10H_2O$) until the end of World War I. The Green River trona ($Na_2CO_3 \cdot NaHCO_3 \cdot 2H_2O$) beds were discovered in 1938 and the first shaft sunk in 1947 by Westvaco Chlorine Products (since acquired by FMC). The first refined soda ash plant became operational in 1953. Prior to that time and for a short period thereafter, small amounts of relatively impure soda ash were produced by hand sorting calcined trona. There are currently five soda ash producers in the Basin; capacities are shown in Table 3.

Table 3. Wyoming Soda Ash Producers

Company	1989 Capacity, 10^6 t/yr
FMC	2.6
General (49% ACI International)	2.0
Rhône-Poulenc	1.7
TGI (Elf Acquitaine)	1.0
Tenneco	1.0
Total	*8.3*

Geology. The Green River Formation was deposited in a large Eocene lake varying in both size and salinity. Trona is found in the Wilkins Peak Member. This represents the stage when the lake covered the smallest area and was the most saline. The bedded trona deposits cover an area of about 3400 km^2 (14,15). At least 25 trona beds have been identified. Beds range in thickness from 1 to 11 m; mining heights are usually 2 to 3 m. The trona is interbedded with fine silt and shale containing associated organic materials (see OIL SHALE). It is estimated that the deposit contains over 100,000 million metric tons of trona, sufficient to satisfy world demand for over 2000 yr. The formation also contains other sodium carbonate salts, Dawsonite [12011-76-6], $Na_3Al(CO_3)_3 \cdot 2Al(OH)_3$, and nahcolite [15752-47-3], $NaHCO_3$, in the Uinta and Piceance Creek areas. Exploitation of these reserves has been evaluated several times over the years with negative results.

Mining.   The trona beds currently mined are between 85 and 95% pure. The balance is predominantly oil shale and associated minerals, but small amounts of sodium chloride and sodium sulfate are also present. Mining is generally confined to one bed at a time. Methods are similar to coal mining (see COAL): room and pillar, longwall, and shortwall mining techniques are used. Equipment, however, is modified to withstand the more abrasive trona.

Room and pillar techniques are used to extract most of the trona. Conventional blasting methods as well as drum (ripper) and borer mining machines are employed. Longwall and shortwall mining techniques which involve hydraulically supporting the overburden while undercutting the trona seam have recently been introduced. In these methods, the hydraulic supports are advanced after the trona is extracted. Eventually the previous cut caves in. These newer techniques allow over 70% recovery of the ore compared to 50% recovery for conventional room and pillar methods.

In 1983 FMC started an experimental program to solution mine trona *in situ* using a proprietary water-based solvent system. This method avoids expensive mine development costs and allows ore recovery in beds that are too deep for conventional mining. Pairs of wells are drilled into the ore bodies and frac-connected using high pressure pumps. The solvent is then circulated through the wells and sent to the processing plant. This well system has been successfully producing commercial quantities of soda ash since 1983.

Another mining process involves the recovery of sodium carbonate deca-hydrate from alkaline ponds. FMC mines this material from its solar evaporation pond using a bucket wheel dredge. The decahydrate slurry is dewatered, melted, and processed to soda ash.

Trona Purification Processes.   Two processes, named the monohydrate and sesquicarbonate according to the crystalline intermediates, are used to produce refined soda ash from trona. Both involve the same unit operations only in different sequences. Most ash is made using the monohydrate process. Figure 2 shows simplified flow diagrams for each.

Monohydrate Process.   In this process (21), trona is ground and calcined (150–300°C) to a crude soda ash using rotary gas-fired and coal grate-fired calciners. The calciner product is then leached with hot water and the clear, hot liquor sent to evaporative recirculating crystallizers (40–100°C) where sodium carbonate monohydrate is produced. Both multiple-effect and mechanical vapor recompression (MVR) crystallizers are used. Some producers send the liquor through activated carbon beds before crystallization to remove trace organics from the solution. These organics, solubilized from the oil shale, can affect crystallizer performance by foaming and changing crystal growth rates and habit. Other trace contaminants are removed from the system by purging small amounts of mother liquor. Weak liquors and insolubles are removed by clarifiers and subsequent filtration. The insolubles are washed again to recover any additional alkali.

Sodium carbonate monohydrate crystals from the crystallizers are concentrated in hydroclones and dewatered on centrifuges to between 2 and 6% free moisture. This centrifuge cake is sent to dryers where the product is calcined

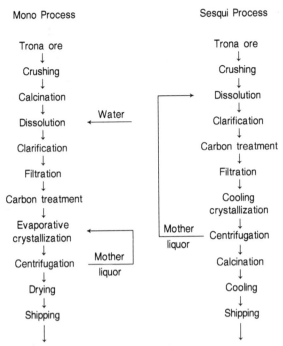

Fig. 2. Simplified flow diagrams for soda ash from trona.

150°C to anhydrous soda ash, screened, and readied for shipment. Soda ash from this process typically has a bulk density between 0.99–1.04 g/mL with an average particle size of about 250 μm.

Sesqui Process. In this process (22), the crushed ore is dissolved in hot (95°C) return liquor, clarified, filtered and sent to cooling crystallizers where sodium sesquicarbonate [6106-20-3], $Na_2CO_3 \cdot NaHCO_3 \cdot 2H_2O$ is formed. Carbon is added to the filters to control any crystal modifying organics. The sesquicarbonate crystals are hydrocloned, centrifuged, and calcined (110–175°C) using either indirect steam heat or gas. This product has a typical bulk density of around 0.89 g/mL. Densities similar to the monohydrate ash may be achieved by subsequently heating the material to about 350°C. Alternatively, the ash can be converted to the monohydrate and calcined as practiced by the Solvay industry.

OTHER NATURAL ASH DEPOSITS

Searles Lake. Searles Lake is a large evaporite deposit about 78 km square and 46 m deep. It contains a complex mixed salt system that includes trona along with potassium, boron, and other salts (23,24). North American Chemical Company recovers soda ash (1.0×10^6 t/yr) from the lake by carbonating and cooling the brine to crystallize sodium bicarbonate (25). The bicarbonate is filtered and calcined to light soda ash which is densified by conversion to the monohydrate followed by calcining. The procedure results in a dense ash with properties equivalent to Wyoming trona derived ash. Other salts, most notably borax, are also recovered from the lake.

Lake Magadi.   Lake Magadi, about 19 km long and 3 km wide, is in the Great Rift Valley of East Africa (26). It is about 6000 m above sea level, has no outlet, and is fed by alkali springs which upon evaporation leave behind trona and other salts. ICI, Ltd. operates a plant at the lake (27,28). Raw trona is dredged up from the lake, crushed, slurried, and sent to an on-shore refining plant where it is washed, screened, and calcined to soda ash. The product is less pure than synthetic soda or the other natural ashes because it contains relatively high amounts of salt, sodium fluoride, and sulfate. About 220,000 t were produced in 1989.

Lake Texcoco.   Lake Texcoco, a few miles northeast of Mexico City, is in the lowest part of the Valley of Mexico. The lake is mostly dry and alkali is recovered from brine wells that have been drilled into the underlying structure. The brine is concentrated first in a spiral flow solar evaporation pond and further in conventional evaporators. This strong brine is carbonated and then cooled to crystallize sodium bicarbonate which is subsequently filtered and calcined to soda ash. Purity of this product is similar to Magadi material (9,29).

Owens Lake.   Owens Lake is a relatively small alkali deposit in Southern California where small amounts of crude trona are recovered and sold (30,31). Several feasibility studies with regard to expanding production at this site have been conducted over the years with negative results.

Sua Pan, Botswana.   A soda ash plant is under construction at Sua Pan in Botswana (32). The plant will recover ash from an alkali brine via a process similar to that at Searles Lake (29).

Economic Aspects

Natural and synthetic soda ash capacity is shown in Table 4. As indicated in Table 5, eight companies represent about 75% of the Western world's soda ash capacity.

Capital and operating costs for soda ash production are extremely site specific (29,10). Key factors include infrastructure development, freight to con-

Table 4. World Distribution of Soda Ash Capacity[a]

Region	Soda ash, 10^6 t/yr	
	Natural	Synthetic
North American	9.5	0.4
Western Europe		6.7
Latin America	0.2	0.8
Africa	0.3	0.1
Asia		4.3
Oceania		0.4
USSR, China, Eastern Europe	0.5	14.5
Total	*10.5*	*27.2*

[a]Refs. 9 and 29.

Table 5. Soda Ash Producers[a]

Company	Capacity, 10^6 t/yr
Solvay & Cie	4.3
FMC	2.6
Rhône-Poulenc	2.4
General Chemical	2.4
ICI	1.3
North American Chemical Co.	1.0
Tenneco	1.0
Elf Acquitaine	1.0
Total	*16.0*

[a]Refs. 9, 29.

sumers, local energy and labor costs, and by-product saleability. 1990 list price of bulk natural soda ash was $108/t, F.O.B. Wyoming.

Although each production process yields ash that is essentially chemically equivalent, the various products differ in physical properties and in contaminants as shown in Table 6. Hopper cars, pneumatic trucks, supersacks, and multiwall kraft bags with polyethylene liners are the usual shipping containers.

Table 6. Typical Properties of Commercial Soda Ash

Property	Synthetic ash		Natural ash (U.S.)	
	Light	Dense	Intermediate	Dense
bulk density, g/mL	0.59	1.04	0.80	1.04
NaCl, ppm	8000–1000	5000–300	300	300
Na_2SO_4, ppm	200	200	300	300
Fe_2O_3, ppm	80–20	80–20	5	3
size distribution, cumulative % on				
sieve no. +40 size, μm 420	3	11	10	11
+100 149	17	90	89	90
+200 74	63	99	99	99
+325 44	86	99	99	99
pore volume, mL/g	0.4	0.1	0.3	0.1
shape[a]	A	M	N	M

[a]Particle shapes are classified as acicular, A, monoclinic and blocky, M, or monoclinic and needle, N.

Dense ash is preferred by the glass industry because its size and bulk density minimize segregation when mixed with other glass batch materials. Light and intermediate grades of soda ash are preferred for some detergent applications where surfactant carrying capacity and dissolution are important. The different physical properties of the various grades are dependent both on the crystalline product calcined (bicarbonate, monohydrate, or sesquicarbonate) and on calcination conditions.

Safety and Handling

Under National Fire Protection Association (NFPA) Designation 704, soda ash is classified as a moderate health hazard. Exposure to soda ash dust may cause severe eye and slight nose and throat irritation. Repeated contact may affect the skin causing redness, dryness, and cracking. First aid procedures are to flush or wash the affected part with water for 15 min, obtaining medical attention if the irritation persists. Simultaneous exposure to the dusts of lime and soda ash should be avoided. The two react to form caustic soda in the presence of moisture or perspiration. Soda ash is strongly alkaline and should not be stored close to acids. It is mildly hygroscopic and generally stored in closed bins or multiwall kraft bags (see PACKAGING MATERIALS, INDUSTRIAL PRODUCTS).

Environmental Aspects

Synthetic Processes. Traditional Solvay plants produce large volumes of aqueous, chloride-containing waste which must be discharged. This fact, in addition to a noncompetitive cost position, is largely responsible for the demise of U.S. synthetic plants. In countries other than the United States, waste is sent to the ocean, rivers, or deep underground wells. The AC and NA coproduct processes produce less aqueous waste than the traditional Solvay and NA mono processes. Related environmental concerns are added whenever a plant complex includes lime quarries and ammonia-producing equipment.

Natural Production Processes. The natural soda ash processes produce no large volumes of associated wastes. The major waste products are the tailings, insoluble shale and minerals associated with the trona and removed during processing. These solids along with purge liquors containing organic and trace impurities are sent to large evaporation ponds where concentration of the aqueous streams is a first step in the eventual recovery of the residual alkali. Because the solids can create dust when dry, several Wyoming producers are experimenting with injecting the tailings into mined out areas underground. Care must also be taken to avoid groundwater contamination by alkali runoff and mine incursion. These risks are not integral to the recovery process, but are affected by plant siting and construction.

By-Products and Coproducts

Calcium Chloride. Distiller waste liquor from synthetic plants can be evaporated in multiple effect evaporators, precipitating residual sodium chloride. The resulting mother liquor is then further evaporated to a molar ratio of $1CaCl_2:2H_2O$ and cooled to produce flakes that are dried in rotary or tray type driers. The product assays at 77–80% calcium chloride [10043-52-4], $CaCl_2$. Most is used as road de-icer; significant amounts are also used as dust control agents and dessicants.

Ammonium Chloride. Most ammonium chloride [12125-02-9], NH_4Cl, is used as rice paddy fertilizer; small amounts are used in dry cells and as fluxing agents (see AMMONIUM COMPOUNDS).

Sodium Bicarbonate. Many soda ash plants convert a portion of their production to sodium bicarbonate [144-55-8], $NaHCO_3$. Soda ash is typically dissolved, carbonated, and cooled to crystallize sodium bicarbonate. The mother liquor is heated and recycled. The solid bicarbonate is dried in flash or tray driers, screened, and separated into various particle size ranges. Bicarbonate markets include food, pharmaceuticals, cattle feed, and fire extinguishers. U.S. demand was approximately 320,000 t in 1989; world demand was estimated at one million metric tons.

FMC makes sodium bicarbonate at the Green River complex by reaction of sesquicarbonate ($Na_2CO_3 \cdot NaHCO_3 \cdot 2H_2O$) with carbon dioxide recovered from a sodium phosphate plant. This fairly recently patented process avoids the energy intensive heating step (33).

Sodium Hydroxide. Before World War I, nearly all sodium hydroxide [1310-93-2], NaOH, was produced by the reaction of soda ash and lime. The subsequent rapid development of electrolytic production processes, resulting from growing demand for chlorine, effectively shut down the old lime–soda plants except in Eastern Europe, the USSR, India, and China. Recent changes in chlorine consumption have reduced demand, putting pressure on the price and availability of caustic soda (NaOH). Because this trend is expected to continue, there is renewed interest in the lime–soda production process. FMC operates a 50,000 t/yr caustic soda plant that uses this technology at Green River; it came onstream in mid-1990. Other U.S. soda ash producers have announced plans to construct similar plants (1,5).

Lime Soda Process. Lime (CaO) reacts with a dilute (10–14%), hot (100°C) soda ash solution in a series of agitated tanks producing caustic and calcium carbonate. Although dilute alkali solutions increase the conversion, the reaction does not go to completion and, in practice, only about 90% of the stoichiometric amount of lime is added. In this manner the lime is all converted to calcium carbonate and about 10% of the feed alkali remains. The resulting slurry is sent to a clarifier where the calcium carbonate is removed, then washed to recover the residual alkali. The clean calcium carbonate is then calcined to lime and recycled while the dilute caustic–soda ash solution is sent to evaporators and concentrated. The concentration process forces precipitation of the residual sodium carbonate from the caustic solution; the ash is then removed by centrifugation and recycled. Caustic soda made by this process is comparable to the current electrolytic diaphragm cell product.

BIBLIOGRAPHY

"Sodium Carbonates" in *ECT* 1st ed., Vol. 12, pp. 601–602, by J. A. Brink, Jr., Purdue University; "Sodium Compounds, Carbonates" in *ECT* 2nd ed., Vol. 18, pp. 458–468, by E. Rau, FMC Corporation. "Alkali and Chlorine Products—Sodium Carbonate" in *ECT* 3rd ed., Vol. 1, pp. 866–883, by A. S. Robertson, Allied Chemical Corporation.

1. J. W. Mellor, *A Comprehensive Treatise on Inorganic and Theoretical Chemistry,* Vol. 2, Suppl. 2, John Wiley & Sons Inc., New York, 1961.
2. F. D. Rossini and co-workers, *Selected Values of Chemical Thermodynamic Properties,* U.S. Government Printing Office, Washington D.C., 1952.
3. A. Seidell, *Solubilities of Inorganic and Metal Organic Compounds,* Van Nostrand Co., Princeton, N.J., 1958.

4. J. Wisniak, *Phase Diagrams,* Elsevier Scientific Publishing Co., New York, 1981.
5. T. P. Hou, *Manufacture of Soda* (ACS Monogr. Ser.) Hafner Publishing Co., New York, 1969.
6. J. Nyvlt, *Solid–Liquid Phase Equilbria,* Elsevier Scientific Publishing Co., New York, 1977.
7. W. C. Blasedale, *Equilibria in Saturated Salt Solutions,* (ACS Monogr. Ser.) Reinhold Publishing Corp., New York, 1927.
8. Z. Rant, *Die Erzeugung von Soda,* F. Enke, Stuttgart, Germany, 1968.
9. *The Economics of Soda Ash,* Roskill Information Services Ltd., London, 1989.
10. K. Tsunashima and K. Nakaya, *The New Asahi (NA) Process for Synthetic Ash Production,* 5th Industrial Mineral International Conference, Madrid, 1982.
11. H. Matsuo, *Kagaku Kogaku* **49**(7), 553–558 (1985).
12. U.S. Pat. 4,320,106 (Mar. 16, 1982), B. Hentschel and co-workers (to Chemische Werke Hüls AG).
13. U.S. Pat. 4,256,719 (Mar. 17, 1981), E. Van Andel (to AKZO N.V. NL).
14. J. J. Fahey, *Saline Minerals of the Green River Formation,* Geological Survey Professional Paper 405, U.S. Government Printing Office, Washington D.C., 1962.
15. R. B. Parker, ed., *Contributions to Geology,* Trona Issue, University of Wyoming, Spring 1971.
16. FMC publication, *The City Below.*
17. P. W. Hynes, *Min. Eng.* 1126 (Nov. 1989).
18. D. R. Delling, *Min. Eng.* 1197 (Oct. 1985).
19. L. N. Post, *Min. Eng.* 1200 (Oct. 1985).
20. R. E. Harris, *Min. Eng.* 1204 (Oct. 1985).
21. U.S. Pats. 2,962,348 (Nov. 29, 1960), 3,655,331 (Apr. 11, 1972), 3,131,996 (May 5, 1964), L. Seglin and H. S. Winnicki (to FMC Corp.); U.S. Pat. 3,244,476 (Apr. 5, 1966), L. K. Smith (to Intermountain Research & Development Corp.); U.S. Pat. 3,260,567 (July 12, 1966).
22. U.S. Pat. 2,639,217 (May 19, 1953), R. D. Pike (to FMC Corp.); U.S. Pat. 2,792,282 (May 14, 1957), R. D. Pike and K. B. Ray (to FMC Corp.); U.S. Pat. 3,028,215 (Apr. 3, 1962), W. R. Frint (to FMC Corp.); U.S. Pat. 3,084,026 (Apr. 2, 1963), W. R. Frint and W. D. Smith (to FMC Corp.); U.S. Pat. 3,309,171 (Mar. 14, 1967), A. B. Gancy (to Intermountain Research & Development Corp.).
23. J. E. Teeple, *The Industrial Development of Searles Lake Brines* (ACS Monogr. Series), University Microfilms Inc., Ann Arbor, Mich., 1979.
24. G. F. Moulton Jr., *Compendium of Searles Lake Operations,* AIME Meeting, Las Vegas, Nev., Feb. 1980.
25. *Trona Soda Ash and the Argus Facility,* Bull. 1400, Kerr-McGee Chemical Corp., 1981.
26. B. H. Baker, *Geology of the Magadi Area, Geological Survey of Kenya,* Report 42.
27. J. J. C. Freeman, *Chem. Eng.* 293 (July 1982).
28. S. C. Hatwell, *Natural Soda Ash from Magadi Glass International,* March 1982, p. 52.
29. A. Russell, *Ind. Min. London* 19 (Jan. 1990).
30. G. D. Dub, *Owens Lake–Source of Sodium Minerals,* Technical Publication No. 2235, American Institute of Mining and Metallurgical Engineers, Sept. 1947.
31. *Soda Ash Industry of Owens Lake,* Vol. 12, No. 10, Mineral Information Service, State of California, 1959.
32. *Chem. Week,* 13 (Nov. 13, 1988).
33. U.S. Pat. 4,654,204 (Mar. 31, 1987), F. Rauh, H. A. Pfeffer, and W. C. Copenhafer (to FMC Corp.).

General References

D. S. Kostick, *Soda Ash and Sodium Sulfate Minerals Yearbook—1988,* U.S. Department of the Interior, Washington D.C., 1988.

R. W. Phelps, "Trona—a Tale of Two Companies," *Eng. Mining J.,* 20 (1990).

M. J. Sagers and T. Shabad, *The Chemical Industry in the USSR,* ACS Professional Reference Book, Westview Press, Boulder, San Francisco and Oxford, 1990.

J. W. Savage and D. Bailey, *Economic Potential of the New Sodium Minerals Found in the Green River Formation,* presented at 61st Annual Meeting of the American Institute of Chemical Engineers, Los Angeles, Calif., Dec. 1–5, 1968.

A. F. Zeller, "Trona," *Eng. Mining J.,* 51 (March 1991).

FRANCIS RAUH
FMC Corporation

ALKALI METALS.   See CESIUM; LITHIUM; POTASSIUM; RUBIDIUM; SODIUM.

ALKALINE EARTH METALS.   See BARIUM; CALCIUM; MAGNESIUM; STRONTIUM.

ALKALOIDS

The actions of the naturally occurring materials now known as alkaloids were probably utilized by the early Egyptians and/or Sumarians (1). However, the beginnings of recorded, reproducible isolation from plants of substances with certain composition first took place in the early nineteenth century. Then in close succession, narcotine [128-62-1] (1, now called noscopine, $C_{22}H_{23}NO_7$) (2) and morphine (2, R = H) (3) (both from the opium poppy, *Papaver somniferum* L.) were obtained.

(1)   (2)

Although their presently accepted structures were unknown, they were characterized with the tools available at the time. Because morphine (2, R = H), $C_{17}H_{19}NO_3$, was shown to have properties similar to the basic soluble salts obtained from the ashes of plants (alkali) it was categorized as a vegetable alkali or alkaloid, and it is generally accepted that it was for this case the word was coined.

However, there is currently no simple definition of what is meant by alkaloid. Most practicing chemists working in the field would agree that most alkaloids, in addition to being products of secondary metabolism, are organic nitrogen-containing bases of complex structure, occurring for the most part in seed-bearing plants and having some physiological activity. An old compendium (4) carefully avoids simple amine bases known to be present in some plants, but does list a variety of compounds such as aristolochic acid I [313-67-7] (3) (from *Aristolochia indica* L., the Dutchman's Pipe) and colchicine (4) (from *Colchicum autumnala* L., the autumn crocus), neither of which is basic, but both of which are physiologically active.

(3)   (4)

In a later reference (5) the list of materials called alkaloids had grown and more structures had been elucidated, but the definition was essentially unchanged. Subsequently, a much more sophisticated definition was proposed (6) which, while meritorious, has apparently been found unworkable, as the most recent catalogue (7), listing nearly ten thousand alkaloids, contains compounds generally fitting within the categories which were used in 1960, but widened still further to include not only nonbasic nitrogen-containing materials from plants, but also substances occurring in animals. Other compounds, the physiological activity of which has not been measured, are also reported (8). Nonetheless, because of their widespread distribution across all forms of life, alkaloids are intimately interwoven into the fabric of existence, and our understanding of some of the roles these substances play is the focus of considerable research as the twenty-first century dawns.

History

From today's perspective, the history of alkaloid chemistry can be divided into four parts. The first part, which doubtlessly developed over aeons prior to the appearance of present-day humanity and about which little is known, deals with the role alkaloids may really play (as divorced from anthropocentric imaginings) in animal and plant defense, reproduction, etc. Second, in the era prior to about 1800, apothecaries' crude mixtures and folk medicinals were administered as palliatives, poisons, and potions. Knowledge of this is based on individual or group records or memory. In the third period, ca 1800–1950, early analytical and isolation technologies were introduced. Good records were kept and techniques honed, so that the wrenching out of specific materials, in truly minute quantities, from the cellular matrices in which they are held could be reproducibly effected.

This time period also saw the beginnings of correlation of the specific structures of those hard-won materials with their properties. Finally, the current era has seen a flowering of structure elucidation as a consequence of the maturation of some analytical techniques, a renaissance in synthetic methods, and the introduction of biosynthetic probes, the last two developments taking advantage of the advances in the first. Most recently, the ability to correlate huge quantities of information at high speed has been developed.

During the first era some insects developed relationships with the plants on which they fed which allowed them to incorporate intact alkaloids for storage and subsequent use. This type of relationship apparently continues to exist. Thus in 1892 there was a report (9) that pharmacophagus swallowtail butterflies (Papilios) obtain and store poisonous substances from their food plants, and some seventy-five years later an investigation (10) showed that the warningly colored and potently odoriferous *Aristolochia*-feeding swallowtail butterfly (*Pachlioptera aristolochiae* Fabr.) is even less acceptable than the unpalatable Danainae to bird predators. Both the plant on which the swallowtail feeds (eg, *Aristolchia indica* L.) and the swallowtail itself contain aristolochic acid I (**3**), $C_{17}H_{11}NO_7$, and related materials. These materials are presumably ingested as larvae feed on the plant, stored during the pupal stage, and carried into the adult butterfly. With regard to the Danainae, the larvae of the butterflies *Danaus plexippus* L. and *Danaus chrysippus* L. feed on *Senecio* spp. which contain, among other compounds, the pyrrolizidine alkaloid senecionine (**5**) (11). (Structure (**5**) appears in Table 4.) Metabolites of this and other related alkaloids apparently serve in courtship and mating, with the more alkaloid-rich individuals having an advantage (12).

There are many other examples of insect use of alkaloids, such as the homotropane alkaloid euphococcinine [*15486-23-4*] (**6**), $C_9H_{15}NO$, which has been noted as a defensive alkaloid in the blood of the Mexican bean beetle (*Epilachna varivestis*) (13); the azaspiroalkene polyzonimine [*55811-47-7*] (**7**), $C_{10}H_{17}N$, an insect repellent produced by the milliped *Polyzonium rosalbum* (14); and the "very fast death factor" (VFDF), anatoxin-a [*64285-06-9*] (**8**), $C_{10}H_{15}NO$, a fish poison, which has been isolated from a toxic strain of microalgae *Anabaena flos-aquae* (15). For (**6**), (**7**), and (**8**), little is yet known about the formation (or genesis) of the alkaloid material.

(**6**)   (**7**)   (**8**)

The period prior to about 1800 includes the history of the crude exudate from unripe poppy pods, which, it is now known, contains narcotine (**1**, noscopine), $C_{22}H_{23}NO_7$, and morphine (**2**, R = H) along with other closely related materials. Also during this time natives of the Upper Amazon basin were making use of crude alkaloid-containing preparations. To help their hunting, some tribes developed the red resinous mixture called tubocurare, containing, among others, the alkaloid tubocurarine [*57-95-4*] (**9**), $C_{37}H_{41}N_2O_6$, obtained primarily from plants of the *Chondrodendron;* others developed Calabash curare, containing, among others, the alkaloid C-toxiferine [*6696-58-8*] (**10**), $C_{40}H_{46}N_4O_2 \cdot 2Cl$, from plants belonging to *Strychnos* spp.

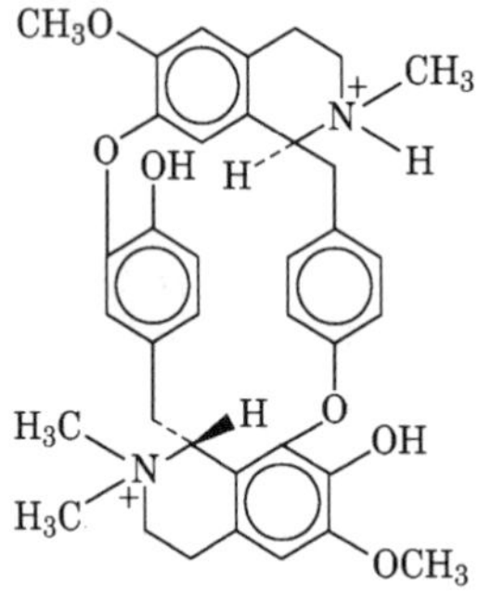

(**9**) 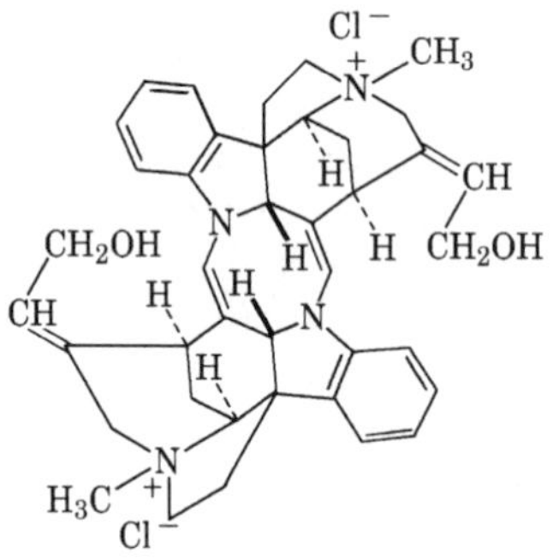   (**10**)

The natives of Peru were learning to ease their physical pains by chewing the leaves of coca shrub (*Erythroxylon truxillence,* Rusby), which contain, among others, the alkaloid cocaine (**11**), and European citizens were recognizing other poisons such as coniine (**12**), from the poison hemlock (*Conium maculatum* L.).

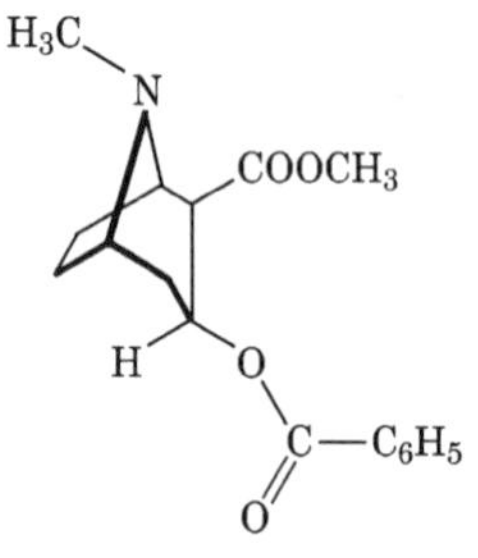

(**11**) 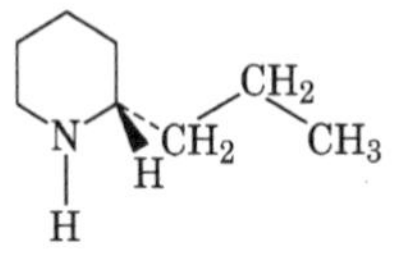   (**12**)

With the introduction of improved analytical techniques, starting around 1817, the evaluation of drugs began and, over a span of about ten years, strychnine (**13**, R = H), emetine (**14**), brucine (**13**, R = OCH$_3$), piperine (**15**), caffeine (**16**), quinine (**17**, R = OCH$_3$), colchicine (**4**), cinchonidine (**17**, R = H), and coniine (**12**) were isolated (16) (Table 1). But, because the science was young and the materials complex, it was not until 1870 that the structure of the relatively simple base coniine (**12**) was established (17) and not until 1886 that the racemic material was synthesized (18). The correct structure for strychnine (**13**, R = H) was not confirmed by x-ray crystallography until 1956 (19) and the synthesis was completed in 1963 (20).

Occurrence, Detection, and Isolation

Given the massive volume of material available, the following discussion is necessarily incomplete and the interested reader is directed to the materials in references 7 and 8, in particular, for more detailed information.

Table 1. Some Alkaloid Drugs

Name	Molecular formula	CAS Registry Number	Structure number	Structure
colchicine	$C_{22}H_{25}NO_6$	[64-86-8]	(4)	[a]
(S)-coniine	$C_8H_{17}N$	[458-58-8]	(12)	[a]
strychnine	$C_{21}H_{22}N_2O_2$	[57-24-9]	(13) (R = H)	
brucine	$C_{23}H_{26}H_2O_4$	[357-57-3]	(13) (R = OCH$_3$)	
emetine	$C_{29}H_{40}N_2O_4$	[483-18-1]	(14)	
piperine	$C_{17}H_{19}NO_3$	[94-62-2]	(15)	
caffeine	$C_8H_{10}N_4O_2$	[58-08-2]	(16)	
cinchonidine	$C_{19}H_{22}N_2O$	[118-10-5]	(17) (R = H)	
quinine	$C_{20}H_{24}N_2O_2$	[130-95-0]	(17) (R = OCH$_3$)	

[a]In text.

The most recent compendium (7) of alkaloids indicates that most alkaloids so far detected occur in flowering plants and it is probably true that the highest concentrations of alkaloids are to be found there. However, as detection methods improve it is almost certain that some concentration of alkaloids will be found almost everywhere. In the higher plant orders, somewhat more than half contain alkaloids in easily detected concentrations. Major alkaloid bearing orders are

Campanulales, Centrospermae, Gentianales, Geraniales, Liliflorae, Ranales, Rhoedales, Rosales, Rubiales, Sapindales, and *Tubiflorae,* and within these orders most alkaloids have been isolated from the families Amaryllidaceae, Apocynaceae, Euphorbiaceae, Lauraceae, Leguminoseae, Liliaceae, Loganiaceae, Menispermaceae, Papveraceae, Ranuculaceae, Rubiaceae, Rutaceae, and Solanaceae. Alkaloids have also been found in butterflies, beetles, millipedes, and algae and are known to be present in fungi, eg, agroclavine (**18**) from the fungus *Claviceps purpurea,* which grows as a parasite on rye and has been implicated, with its congeners, in causing convulsive ergotism (21). They are found in toads, eg, bufotenine (**19**), an established hallucinogen in humans (22); and in the musk deer, muscopyridine [*501-08-6*] (**20**), $C_{16}H_{25}N$. Even in humans morphine (**2**, R = H) is a naturally occurring component of cerebrospinal fluid (23).

(**18**)   (**19**)   (**20**)

The concentration of alkaloids, as well as the specific area of occurrence or localization within the plant or animal, can vary enormously. Thus the amount of nicotine [*54-11-5*] (**21**), $C_{10}H_{14}N_2$, apparently synthesized in the roots of various species of *Nicotiana* and subsequently translocated to the leaves varies with soil conditions, moisture, extent of cultivation, season of harvest, etc and may be as high as 8% of the dry leaf, whereas the amount of morphine (**2**, R = H) in cerebrospinal fluid is of the order of 2 to 339 fmol/mL (23).

(**21**)

Initially, the search for alkaloids in plant material depended largely on reports of specific plant use for definite purposes or observations of the effect specific plants have on indigenous animals among native populations. Historically, tests on plant material have relied on metal-containing reagents such as that of Dragendorff (24), which contains bismuth salts, or Mayer (25), which contains mercury salts. These metal cations readily complex with amines and the halide ions present in their prepared solutions, yielding brightly colored products. Despite false positive and negative responses (26), field testing continues to make use of these solutions. However, it is now clear that newer methods, such as kinetic energy mass spectrometry (MIKE) on whole plant material (27), have the potential to replace these spot tests. After detection of a presumed alkaloid, large quantities of the specific plant material are collected, dried, and defatted by petroleum ether extraction if seed or leaf is investigated. This process usually leaves polar alkaloidal material behind but removes neutrals. The residue, in aqueous alcohol, is extracted with dilute acid and filtered, and the acidic solution is made basic. Crystallization can occasionally be effected by adjustment of the pH. If such relatively simple purification fails, crude mixtures may be used or, more recently, very sophisticated separation techniques have been employed. Once alkaloidal material has been found, taxonomically related plant material is also examined.

Until separation techniques such as chromatography (28,29) and counter-current extraction had advanced sufficiently to be of widespread use, the principal alkaloids were isolated from plant extracts and the minor constituents were either discarded or remained uninvestigated. With the advent of, first, column, then preparative thin layer, and now high pressure liquid chromatography, even very low concentrations of materials of physiological significance can be obtained in commercial quantities. The alkaloid leurocristine (vincristine, **22**, R = CHO), one of the more than 90 alkaloids found in *Catharanthus roseus* G. Don, from which it is isolated and then used in chemotherapy, occurs in concentrations of about 2 mg/100 kg of plant material.

(**22**)

Properties

Most alkaloids are basic and they are thus generally separated from accompanying neutrals and acids by dilute mineral acid extraction. The physical properties of most alkaloids, once purified, are similar. Thus they tend to be colorless, crystalline, with definite melting points, and chiral; only one enantiomer is isolated. However, among nearly ten thousand individual compounds, these descriptions are overgeneralizations and some alkaloids are not basic, some are liquid, some brightly colored, some achiral, and in a few cases both enantiomers have been isolated in equal amounts, ie, the material as derived from the plant is racemic.

Organization

Early investigators grouped alkaloids according to the plant families in which they are found, the structural types based on their carbon framework, or their principal heterocyclic nuclei. However, as it became clear that the alkaloids, as secondary metabolites (30–32), were derived from compounds of primary metabolism (eg, amino acids or carbohydrates), biogenetic hypotheses evolved to link the more elaborate skeletons of alkaloids with their simpler proposed pregenitors (33). These hypotheses continue to serve as valuable organizational tools (7,34,35).

   The building blocks of primary metabolism, from which biosynthetic studies have shown the large majority of alkaloids to be built, are few and include the common amino acids ornithine (23), lysine (24), phenylalanine (25, R = H), tyrosine (25, R = OH), and tryptophan (26). Others are nicotinic acid (27), anthranilic acid (28), and histidine (29), and the nonnitrogenous acetate-derived fragment mevalonic acid (30) (Table 2). Mevalonic acid (30) is the pregenitor of isopentenyl pyrophosphate (31) and its isomer 3,3-dimethylallyl pyrophosphate (32), later referred to as the C_5 fragment. A dimeric C_5 fragment (the C_{10} fragment), ie, geranyl pyrophosphate (33), gives rise to the iridoid loganin [18524-94-2] (34), and the trimer farnesyl pyrophosphate (35). The C_{15} fragment is also considered the precursor to the C_{30} steroid, ie, 2 × C_{15} = C_{30} (see Table 3 and Figure 1).

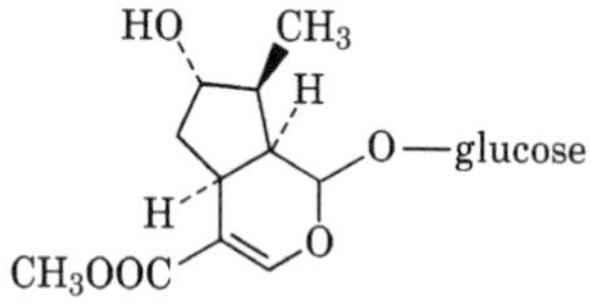

(**34**)

Ornithine-Derived Alkaloids.   Ornithine (23) undergoes biological decarboxylation reductively to generate either putrescine [110-60-1] (36), $C_4H_{12}N_2$, or its biological equivalent, and subsequent oxidation and cyclization gives rise to the pyrroline [5724-81-2], (37), C_4H_7N.

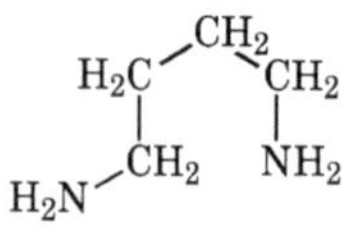

(**36**)   (**37**)

The details have been confirmed by suitable labeling (^{14}C and ^{15}N) and it is fairly certain that either (37) or something very similar to it is available to react with either acetoacetic acid or its biological equivalent to generate the alkaloid hygrine (38) (Table 4). Hygrine is an oily, distillable base found, along with cocaine (11), in the leaves of the Peruvian coca shrub (*Erythroxylon truxillence* Rusby).

   If, instead of an acetoacetate equivalent, a malonyl derivative such as (39) were involved (37), appropriate condensation reactions would lead to the tropane [280-05-7] skeleton (40), $C_7H_{13}N$.

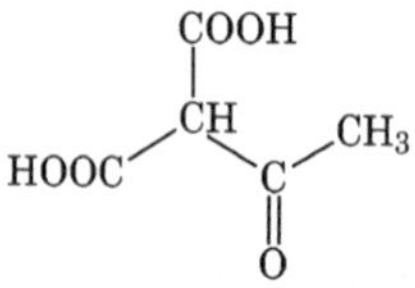

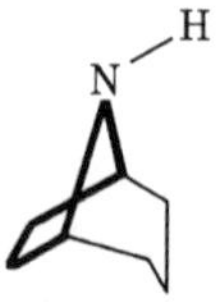

(**39**)   (**40**)

Table 2. Pregenitors of Alkaloids

Name	Molecular formula	CAS Registry Number	Structure number	Structure
ornithine	$C_5H_{12}N_2O_2$	[7006-33-9]	(23)	
lysine	$C_6H_{14}N_2O_2$	[6899-06-5]	(24)	
phenylalanine	$C_9H_{11}NO_2$	[63-91-2]	(25) (R = H)	
tyrosine	$C_9H_{11}NO_3$	[60-18-4]	(25) (R = OH)	
tryptophan	$C_{11}H_{12}N_2O_2$	[73-22-3]	(26)	
nicotinic acid	$C_6H_5NO_2$	[59-67-6]	(27)	
anthranilic acid	$C_7H_7NO_2$	[118-92-3]	(28)	
histidine	$C_6H_9N_3O_2$	[71-00-1]	(29)	
mevalonic acid	$C_6H_{12}O_4$	[150-97-0]	(30)	

The physiologically and commercially important alkaloids of this group of compounds, occurring widely in the Solanaceae and Convolvoulaceae as well as the Erythroxylaceae, include not only cocaine (11) but also atropine (41) and scopolamine (42).

Atropine (41), isolated from the deadly nightshade (*Atropa belladonna* L.) is the racemic form, as isolated, of (−)-hyoscyamine [which is not isolated, of

Table 3. Pyrophosphate Biogenic Precursors[a]

Name	CAS Registry Number	Structure number	Molecular formula
isopentenyl pyrophosphate	[358-71-4]	(31)	$C_5H_9O_7P_2$
3,3-dimethylallyl pyrophosphate	[358-72-5]	(32)	$C_5H_9O_7P_2$
geranyl pyrophosphate	[763-10-0]	(33)	$C_{10}H_{17}O_7P_2$
farnesyl pyrophosphate	[13058-04-3]	(35)	$C_{15}H_{25}O_7P_2$

[a]See Figure 1.

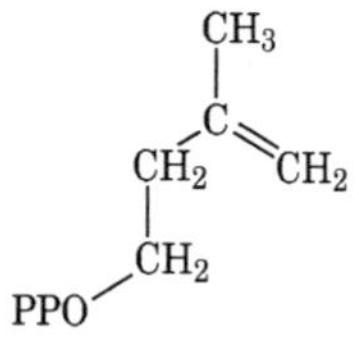

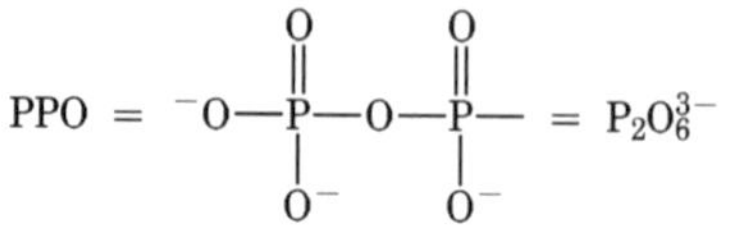

Fig. 1. Pyrophosphate biogenic precursors. See Table 3.

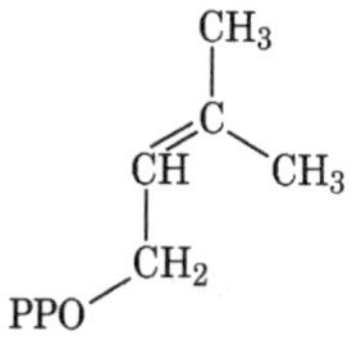

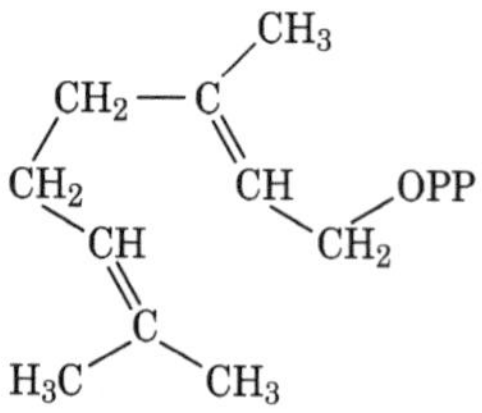

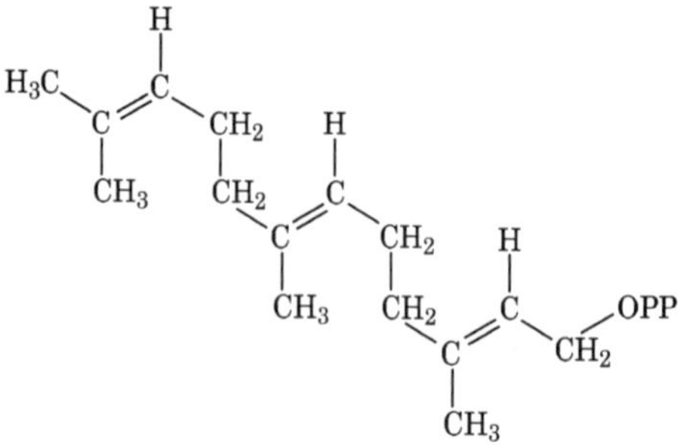

course, from the same plant but is typically found in solanaceous plants such as henbane (*Hyoscyamus niger* L.)]. Atropine is used to dilate the pupil of the eye in ocular inflammations and is available both as a parasympatholytic agent for relaxation of the intestinal tract and to suppress secretions of the salivary, gastric, and respiratory tracts. In conjunction with other agents it is used as part of an antidote mixture for organophosphorus poisons (see CHEMICALS IN WAR).

Scopolamine (**42**), an optically active, viscous liquid, also isolated from Solanaceae, eg, *Datura metel* L., decomposes on standing and is thus usually both used and stored as its hydrobromide salt. The salt is employed as a sedative or, less commonly, as a prophylactic for motion sickness. It also has some history of use in conjunction with narcotics as it appears to enhance their analgesic effects. Biogenetically, scopolamine is clearly an oxidation product of atropine, or, more precisely, because it is optically active, of (−)-hyoscyamine.

Cocaine (**11**) had apparently been used by the natives of Peru prior to the European exploration of South America. Stories provided by early explorers

Table 4. Alkaloids Derived from Ornithine[a,b]

Name	Molecular formula	CAS Registry Number	Structure number	Structure
cocaine	$C_{17}H_{21}NO_4$	[50-36-2]	(11)	
(R)-hygrine	$C_8H_{15}NO$	[496-49-1]	(38)	
atropine	$C_{17}H_{23}NO_3$	[51-55-8]	(41)	
scopolamine	$C_{17}H_{21}NO_4$	[51-34-3]	(42)	
heliotridine	$C_8H_{13}NO_2$	[520-63-8]	(43)	
senecione	$C_{18}H_{25}NO_5$	[130-01-8]	(5)	

[a]Examples in this table exclude those from tobacco.
[b]See references 36 and 37 for additional information.

suggested that the leaves of, for example, *Erythroxylon coca* Lam. were chewed without apparent addiction by the indigenous peoples and with only mild numbing of the lips and tongue in return for increased endurance. Indeed, the recognition of the anesthetic properties possessed by the leaves and the (unwarranted) assumption that addiction was avoidable led to creation of plantations in Bolivia, Brazil, and Java to ensure a continued supply of this valuable material for medicinal purposes. Although it appears that native populations continue the practice of leaf chewing, the purified base obtained by simple extraction of the leaves has become a substance of abuse in the more civilized world. It is now recognized that the alkaloid (**11**) itself is too toxic to be used as an anesthetic by injection.

Condensation of a pyrroline system (**37**) with a second equivalent of ornithine-derived precursor is presumably an alternative to condensation with acetoacetate- or malonate-derived fragments. Indeed, early feeding experiments with, for example, *Senecio istideus* showed that two equivalents of ornithine (**23**) could be accounted for in the structure of the necine, ie, the 1-azabicyclo[3.3.1]heptane or pyrrolizidine portion of the alkaloids, eg, heliotridine (**43**), containing that ring system (**38**). Generally, the pyrrolizidine alkaloids are found esterified with low molecular weight carboxylic acids (or dicarboxylic acids, as in senecionine, (**5**)) at either or both of the hydroxyl groups of the necine. The acids themselves, called necic acids, are generally not found elsewhere in alkaloids, and, although for some time they were believed to arise from acetate or mevalonate, it is now clear that, at least for the few that have been carefully examined, they are themselves derived from simple amino acids.

In addition to the alkaloids in *Senecio* spp. (including asters and ragworts) commented on earlier, the adaptive use of which by butterflies was noted, members of this widely spread group of compounds are found in different genera (*Heliotropium, Trachelanthus,* and *Trichodesma*) within cosmopolitan families (eg, Boraginaceae and Leguminoseae). Most of these alkaloids are toxic, affecting the liver (an organ lacking in moths, butterflies, etc) and their ingestion is manifested in animals with the onset of symptoms associated with names such as horse staggers or walking disease.

Lysine-Derived Alkaloids. Just as putrescine (**36**) derived from ornithine (**23**) is considered the pregenitor of the nucleus found in pyrrolidine-containing alkaloids, so cadaverine [462-94-2] (**44**), $C_5H_{14}N_2$, derived from lysine (**24**) is the idealized pregenitor of the 1-dehydropiperidine [28299-36-7] nucleus (**45**), C_5H_9N, found in the pomegranate, *Sedum, Lobelia, Lupin,* and *Lycopodium* alkaloids (**39**).

(**44**)   (**45**)

As was the case for the alkaloids derived from ornithine (**23**), if either (**45**), its biological equivalent, or something closely resembling it reacts with acetoacetate or its biological equivalent, the pomegranate alkaloid pelletierine (**46**) can arise; note the resemblence to hygrine (**38**). Simple reduction of the carbonyl, with

Table 5. Alkaloids Derived from Lysine[a,b]

Name	Molecular formula	CAS Registry Number	Structure number	Structure
pelletierine	$C_8H_{15}NO$	[4396-01-4]	(46)	
sedridine	$C_8H_{17}NO$	[501-83-7]	(47)	
pseudopelletierine	$C_9H_{15}NO$	[552-70-5]	(48)	
lobeline	$C_{22}H_{27}NO_2$	[90-69-7]	(49)	
(+)-lupanine	$C_{15}H_{24}N_2O$	[550-90-3]	(50) (R = O)	
(−)-sparteine	$C_{15}H_{26}N_2$	[90-39-1]	(50) (R = H,H)	
lycopodine	$C_{16}H_{25}NO$	[466-61-5]	(51)	
annotinine	$C_{16}H_{21}NO_3$	[559-49-9]	(52)	

[a]Examples in this table exclude those from tobacco.
[b]See references 36 and 37 for additional information.

stereospecificity common to enzyme-mediated reactions, can be accommodated and the *Sedum* alkaloid sedridine (**47**) results; cyclization and N-methylation produce pseudopelletierine (**48**) (Table 5). There are somewhat more than 600 annual, biennial, or perennial succulents belonging to the *Sedum* genus of the family Crassulaceae, many of which are characterized by the ability to grow where little else can.

The pomegranate alkaloids, pelletierine (**46**) and pseudopelletierine (**48**) as well as minor accompanying bases, have a long history as salts of tannic acid as an anthelmintic mixture for intestinal pinworms (see ANTIPARASITIC AGENTS, ANTHELMINTICS). The alkaloids themselves (as the tannates) are obtained from pomegranate tree (*Punica granatum* L.) root bark and are among the few bases named after an individual (P. J. Pelletier) rather than a plant.

Isolates from Indian tobacco (*Lobelia inflata* L.), as a crude mixture of bases, have been recognized as expectorants. The same (or similar) fractions were also used both in the treatment of asthma and as emetics. The principal alkaloid in *L. inflata* is lobeline (**49**), an optically active tertiary amine which, unusual among alkaloids, is reported to readily undergo mutarotation, a process normally associated with sugars. Interestingly, it appears that the aryl-bearing side chains in (**49**) are derived from phenylalanine (**25**, R = H) (40).

Feeding experiments utilizing ^{13}C-, ^{15}N-, and ^{2}H-labeled cadaverine (**44**) and lysine (**24**) in *Lupinus augustifolius*, a source of the lupine alkaloids (−)-sparteine (**50**, R = H,H) and (+)-lupanine (**50**, R = O), have been reported which lend dramatic credence to the entire biosynthetic sequence for these and the related compounds discussed above (41). That is, the derivation of these bases is in concert with the expected cyclization from the favored all-trans stereoisomer of the trimer expected on self-condensation of the 1-dehydropiperidine (**45**).

The spores of *Lycopodium clavatum* L. (a club moss), sometimes called vegetable sulfur, have been used medicinally as an absorbent dusting powder; other uses as diverse as additives to gunpowder and suppository coatings have also been recorded. Although for some years the alkaloids common to a number of *Lycopodium* spp., lycopodine (**51**) and annotinine (**52**), were thought to have arisen from suitably folded polyketide chains, it is now accepted that two pelletierine (**46**) or pelletierine-like fragments would suffice. The details of feeding experiments with pelletierine and its precursors appear to indicate, however, that only one pelletierine and, separately, a second acetoacetate and second 1-dehydropiperidine (**45**), which could otherwise be combined to a second pelletierine, are used to generate both of these alkaloids.

Tobacco Alkaloids.   The relatively small number of alkaloids derived from nicotinic acid (**27**) (the tobacco alkaloids) are obtained from plants of significant commercial value and have been extensively studied. They are distinguished from the bases derived from ornithine (**23**) and, in particular, lysine (**24**), since the six-membered aromatic substituted pyridine nucleus common to these bases apparently is not derived from (**24**).

These alkaloids include the substituted pyridone ricinine [*524-40-3*] (**53**), $C_8H_8N_2O_2$, which is easily isolated in high yield as the only alkaloid from the castor bean (*Ricinus communis* L.). The castor bean is also the source of castor oil (qv), which is obtained by pressing the castor bean and, rich in fatty acids, has served as a gentle cathartic.

(27)   (53)   (54)   (55)   (56)

The highly toxic alkaloid (−)-nicotine (21) and related tobacco bases includ-ing such materials as (−)-anabasine [494-52-0] (54), $C_{10}H_{14}N_2$, are obtained from commercially grown tobacco plants (eg, *Nicotiana tabacum* L.). Various tobaccos have differing amounts of these and other bases, as well as different flavoring constituents, some of which are apparently habituating to some individuals. Currently, the assay of the (−)-nicotine (21) content of tobacco, the annual world production of which is in excess of three million tons, is desirable and in some countries mandatory, although the toxicity of the unassayed plant bases may be as high as or higher than that of (−)-nicotine. Interestingly, there appears to be some evidence that cultivation of tobacco increases the alkaloid content, from which it can be argued that increased alkaloid content has insured survival of a particular cultivar.

The pyrrolidine ring of nicotine is derived from ornithine (23), whereas the piperidine ring of anabasine (54) is derived from lysine (24) (42). Also, the carbox-ylic acid functionality of nicotinic acid (27) is lost (along with the C-6 proton) during the biosynthesis in the roots of the tobacco plants from which the bases are subsequently translocated to the leaves. Curiously, whereas nornicotine [494-97-3] (55), $C_9H_{12}N_2$, frequently accompanies nicotine, the former is apparently de-rived by demethylation of the latter rather than the latter undergoing methyl-ation to the former. This is in contrast to what usually seems to occur; that is, methylation at nitrogen and oxygen is usually a late-stage process in alkaloid biosynthesis.

Finally, millions of people in the Far East are apparently addicted to chew-ing ground betel nut, the fruit of the palm tree *Areca catechu* L., which they mix with lime and wrap in betel leaf (*Piper betle* L.) for consumption. They are said to experience a feeling of well-being. Among the alkaloids found in betel nut is arecoline [63-75-2] (56), $C_8H_{13}NO_2$, an optically inactive, steam-volatile base which is used commercially as a vermifuge in dogs and is also a potent muscarinic agent. It is reasonable to assume (evidence lacking) that arecoline may be derived from nicotinic acid (27) by a (rare) reductive mechanism.

Phenylalanine- and Tyrosine-Derived Alkaloids.   Carbohydrate metabo-lism leads via a seven-carbon sugar, ie, a heptulose, derivative to shikimic acid [138-59-0] (57), $C_7H_{10}O_5$, which leads in turn to prephenic acid [126-49-8] (58), $C_{10}H_{10}O_6$ (43).

(57)   (58)

This is the branch-point differentiating phenylalanine (**25**, R = H) from tyrosine (**25**, R = OH). Both phenylalanine and tyrosine contain an aryl ring, a three-carbon side chain (a C_6—C_3 fragment), and a nitrogen. Decarboxylation yields a two-carbon side chain (a C_6—C_2 fragment), eg, 2-phenethylamine (**59**, R = H) from phenylalanine and tyramine (**59**, R = OH) from tyrosine, although it is not certain that in all cases decarboxylation must precede use in alkaloid construction.

(**59**)   (**60**)

After the branching point at prephenic acid (**58**), phenylalanine and tyrosine as well as the amines (**59**) are *not* interconvertible. Finally, deamination and oxidative cleavage of the presumed (and in some circumstances isolated) resulting alkenes yields the equivalent of benzaldehyde (**60**, R = H), C_7H_6O, and *p*-hydroxybenzaldehyde (**60**, R = OH), ie, aromatics with one aliphatic carbon attached (C_6—C_1 fragments).

All of these pieces are used, in conjunction with some of the earlier fragments discussed, as building blocks for alkaloids containing an aromatic ring. In the cases discussed here, a link to either phenylalanine or tyrosine or, in some cases with two aromatic rings to both, has been established by suitable feeding experiments on growing plants.

There is a relatively large number of alkaloids which may be considered as simple phenethylamine [*64-04-0*] (**59**, R = H), $C_8H_{11}N$, or tyramine [*51-67-2*] (**59**, R = OH), $C_8H_{11}NO$, derivatives. These include mescaline (**61**) from the small wooly peyotyl cactus *Lophophora williamsii* (Lemaire) Coult., anhalamine (**62**) and lophocerine (**63**) from other Cactaceae, and the important antamebic alkaloids (−)-protoemetine (**64**), (−)-ipecoside (**65**), and (−)-emetine (**66**) from the South American straggling bush *Cephaelis ipecacuanha* (Brotero) Rich. All of these bases appear to be derived from tyrosine (**25**, R = OH) and not from phenylalanine (**25**, R = H) (Table 6).

Crude preparations of mescaline (**61**) from peyote were first reported by the Spanish as they learned of its use from the natives of Mexico during the Spanish invasion of that country in the sixteenth century. The colorful history (44) of mescaline has drawn attention to its use as a hallucinogen and even today it is in use among natives of North and South America. Although in connection with drug abuse complaints, mescaline is considered dangerous, it has been reported (45) that it is not a narcotic nor is it habituating. It was also suggested that its sacramental use in the Native American Church of the United States be permitted since it appears to provoke only visual hallucination while the subject retains clear consciousness and awareness.

Both of the alkaloids anhalamine (**62**) from *Lophophora williamsii* and lophocerine (**63**) from *Lophocereus schotti* were isolated (after the properties of purified mescaline had been noted) in the search for materials of similar behavior. Interestingly, lophocerine, isolated as its methyl ether, after diazomethane treatment

Table 6. Alkaloids Derived from Tyrosine

Name	Molecular formula	CAS Registry Number	Structure number	Structure
mescaline	$C_{11}H_{17}NO_3$	[54-04-6]	(61)	
anhalamine	$C_{11}H_{15}NO_3$	[643-60-7]	(62)	
lophocerine	$C_{15}H_{23}NO_2$	[19485-63-3]	(63)	
(−)-protoemetine	$C_{19}H_{27}NO_3$	[549-91-7]	(64)	
(−)-ipecoside	$C_{27}H_{35}NO_{12}$	[15401-60-2]	(65)	
(−)-emetine	$C_{29}H_{40}N_2O_4$	[483-18-1]	(66)	

of the alkali-soluble fraction of total plant extract, is racemic. It is not known if the alkaloid in the plant is also racemic or if the isolation procedure causes racemization.

The iridoid loganin (**34**), $C_{17}H_{26}O_{10}$, has been shown to serve as a C_{10} progenitor (see Table 3) and, here, C_{10} fragments are apparent in the alkaloids (−)-protoemetine (**64**), (−)-ipecoside (**65**), and (−)-emetine (**66**). It has been shown that loganin is specifically incorporated into each of these bases in *Cephaelis ipecacuanha* (Brot. A. Rich) and that they are apparently formed sequentially, that is, formation of (−)-ipecoside (**65**) precedes that of (−)-protoemetine (**64**) and (−)-emetine (**66**). The crude dried rhizome and roots from *C. ipecacuanha* which is sometimes known as Rio or Brazilian ipecac, contains all three, as well as other related bases, and has a long history based on native Indian reports of use as an emetic. Purification of the crude extract yields the individual bases, and, because it is relatively more stable and is also present in reasonable quantity, led to the use of emetine (**66**) as its hydrochloride salt in place of the crude plant extract. This use of the pure base rather than crude plant extract has allowed greater certainty in dosage, which is important because, although emetine is quite effective in combating acute amebic hepatitis and is claimed to have some effect against the present scourge of African schistosomiasis, its administration may be accompanied by a rapid drop in blood pressure, irregular heart function, and paralysis of skeletal muscle. The danger of inappropriately large doses, certain to cure the ailment but with the possible death of the patient, is clearly greater with crude extract than with purified alkaloid. Long term, even appropriate, dosage may be accompanied by dermatitis, diarrhea, nausea, etc.

There are only two groups of alkaloids that appear to be derived from tyrosine (**25**, R = OH) utilized as a C_6—C_2 fragment and a C_6—C_1 unit which comes from phenylalanine (**25**, R = H). The first is that small group found only in the Orchidaceae, exemplified by cryptostyline I [*22324-79-4*] (**67**, from *Cryptostylis fulva* Schltr.), $C_{19}H_{21}NO_4$.

(**67**)

The second, a very large group of compounds, is the alkaloids of the Amaryllidaceae. This cosmopolitan family of related compounds includes over one hundred isolated and characterized members of known structure. In every case examined the C_6—C_2 unit is derived from tyrosine and the C_6—C_1 unit comes from phenylalanine, never from tyrosine. For this large number of compounds it is now believed (46) that a single pregenitor derived from the original coupling of the C_6—C_2 unit and a C_6—C_1 unit, ie, norbelladine (**68**, R = H) accounts for all of the compounds isolated. This precursor (**68**, R = H) undergoes a variety of enzyme-catalyzed free-radical intramolecular cyclization reactions, followed by late-stage oxidations, eliminations, rearrangements, and O- and N-alkylations. Working from this generalization as an organizing principle, the majority of known Amaryllidaceae alkaloids can be divided into eight structural classes (47).

These eight classes and alkaloids typical of them are given in Table 7 and Figure 2. The simple base ismine (**75**), isolated from, for example, *Sprekelia formosissima*, along with numerous other alkaloids, has long been considered a degradation product of other bases and is presumably generated in that way from a suitable member of the pyrrolo[*de*]phenanthridine or [5,10*b*]ethanophenanthridine group.

Table 7. Structural Classes of Amaryllidaceae Alkaloids[a,b]

Class	Example	Molecular formula	CAS Registry Number	Structure[b] number
N-benzyl-N-(2-phenylethyl)-amine derivatives	belladine	$C_{19}H_{25}NO_3$	[501-06-4]	(**68**)[c]
pyrrolo[*de*]phenanthridines	lycorine	$C_{16}H_{17}NO_4$	[476-28-8]	(**69**)
[2]-benzopyrano[3,5*g*]indole alkaloids	homolycorine	$C_{18}H_{21}NO_4$	[477-20-3]	(**70**)
dibenzofuran bases	galanthamine	$C_{17}H_{21}NO_3$	[357-70-0]	(**71**)
[5,10*b*]ethanophenanthridine alkaloids	(+)-crinine	$C_{16}H_{17}NO_3$	[510-69-0]	(**72**)
	(−)-crinine	$C_{16}H_{17}NO_3$	[510-67-8]	(**72**)
[2]-benzopyrano[3,4*c*]indole alkaloids[d]	tazettine	$C_{18}H_{21}NO_5$	[507-79-9]	(**73**)
[5,11]methanomorphanthridine bases	manthine[e]	$C_{18}H_{21}NO_4$	[606-51-9]	(**74**)
[f]	ismine	$C_{15}H_{15}NO_3$	[1805-78-3]	(**75**)

[a]A large group derived from tyrosine as a C_6—C_2 fragment and phenylalanine as a C_6—C_1 fragment.
[b]See Figure 2.
[c]R = CH_3.
[d]Can rationally be derived from [5,10*b*]ethanophenanthridine bases.
[e]Can also be derived from suitably substituted [5,10*b*]ethanophenanthridine bases.
[f]Derived from either a pyrrolo[*de*]phenanthridine or a [5,10*b*]ethanophenanthridine.

Lycorine (**69**) was recognized as a potent emetic and a moderately toxic base from the time of its initial isolation from *Narcissus pseudonarcissus* L. (in about 1877) (48). Since that time its isolation from many other Amaryllidaceae, for example, *Lycoris radiate* Herb., has served to establish it as the most cosmopolitan alkaloid of the family. Typically, as much as 1% of the dry weight of daffodil bulbs may consist of lycorine (**69**), which has been reported to crystallize as colorless prisms directly from aqueous acid extract of crude plant material after basification. A high yield synthesis of the racemic base has been reported (49). Galanthamine (**71**) was originally isolated from the Caucasian snowdrop, *Galanthus woronowii* Vel., and as its hydrobromide salt has been proposed for use in regeneration of sciatic nerve. In addition to demonstration of powerful cholinergic activity (50), it is reported to have analgesic activity comparable to morphine (**2**, R = H) (51).

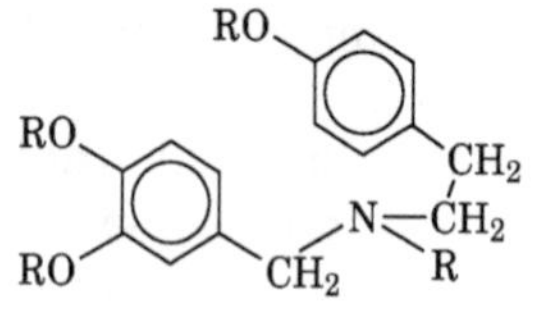

(68) 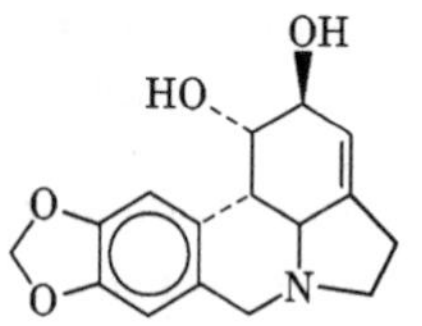   (69) 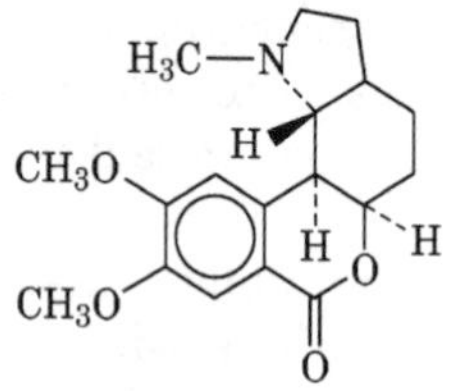   (70)

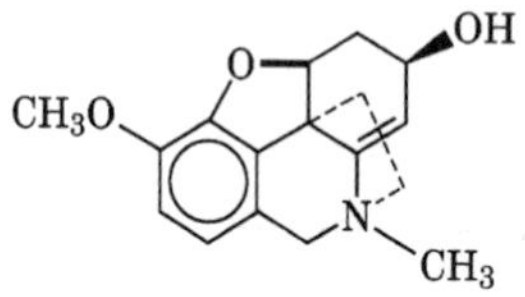

(71) 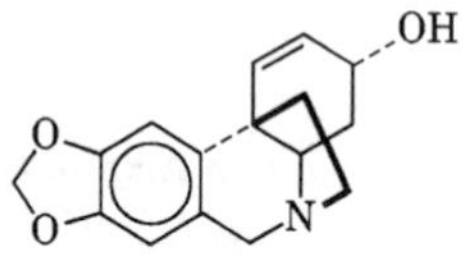   (72) 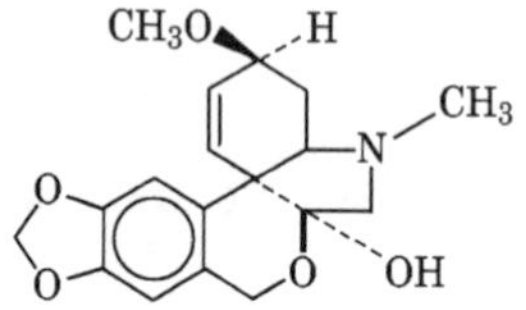   (73)

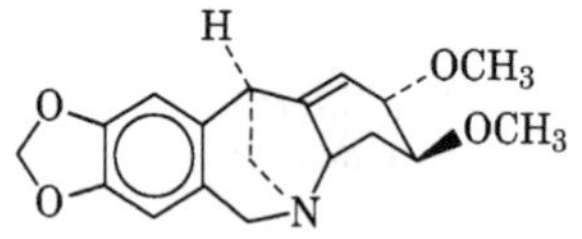

(74) 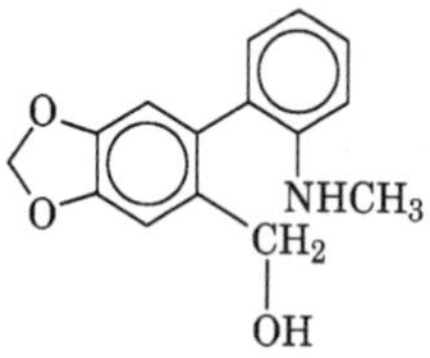   (75)

Fig. 2.   Examples of Amaryllidaceae alkaloids. See Table 7.

(76)

Tazettine (73) has gained notoriety since, subsequent to its isolation from *Sprekelia formosissima* or *Narcisus tazetta* and proof that it was generated *in vivo* from haemanthamine [466-75-1] (76), $C_{17}H_{19}NO_4$,

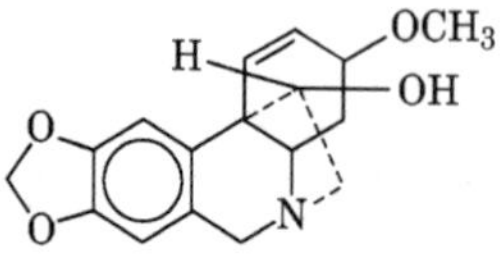

in accord with biosynthetic dogma (52), more careful work (53) in which the strongly basic conditions usually employed in alkaloid isolation were avoided showed that it is an artifact of isolation and that it is readily generated from its precursors during the work-up of the plant material. Manthine (74) occurs, along with several homologues, in South African *Haemanthus* species. Manthine is of interest to the chemical community because it appears that, like tazettine (73), it can be easily generated *in vitro* from a derivative of haemanthamine (76) (54).

Just as norbelladine (68, R = H) can be considered as the precursor of C_6—C_2 + C_6—C_1 alkaloids, norlaudanosoline (77, R = H) seems to be the pregenitor of the vast number of C_6—C_2 + C_6—C_2 alkaloids (Table 8). Laudanosine (77, R = CH₃), isolated from *Papaver somniferum* L. [along with narcotine (1, noscopine), morphine (2, R = H), and numerous other alkaloids], has been shown to have tyrosine (**25**, R = OH) as a specific precursor. Labeling experiments with, for example, [2-^{14}C]tyrosine show that two equivalents of this amino acid are incorporated specifically but not to the same extent, implying that some partitioning has occurred prior to alkaloid formation. Papaverine (**78**), isolated in much greater quantity from *P. somniferum* L. than its tetrahydro derivative laudanosine (**77**, R = CH$_3$), has a long history of use as an antispasmodic for smooth muscle. It is used as its hydrochloride salt, a more stable material than the free base. It is said to be nonhabit-forming, although it is classified as a narcotic by the U.S. Federal Narcotic Laws. Large doses may produce drowsiness, constipation, and increased excitability; if it is given orally, gastric distress may occur.

Phenolic intermolecular coupling (46) of two laudanosoline (**77**, R = H) fragments, which may be preceded or followed by partial O- or N-methylation, gives rise to the dimeric or bisbenzylisoquinoline alkaloids such as oxyacanthine (**79**), obtained along with related materials from the roots of *Berberis vulgaris* L.

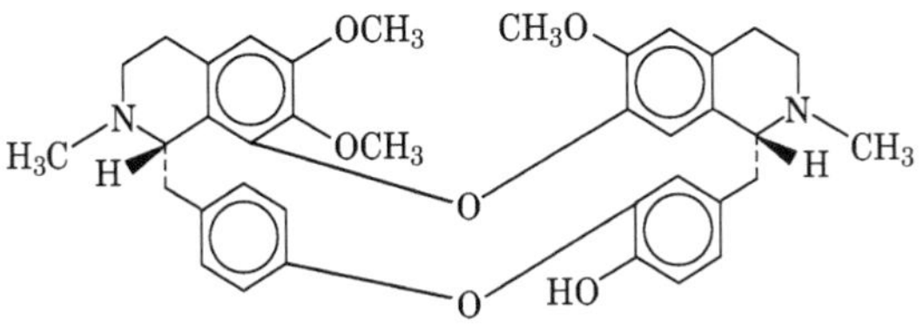

(**79**)

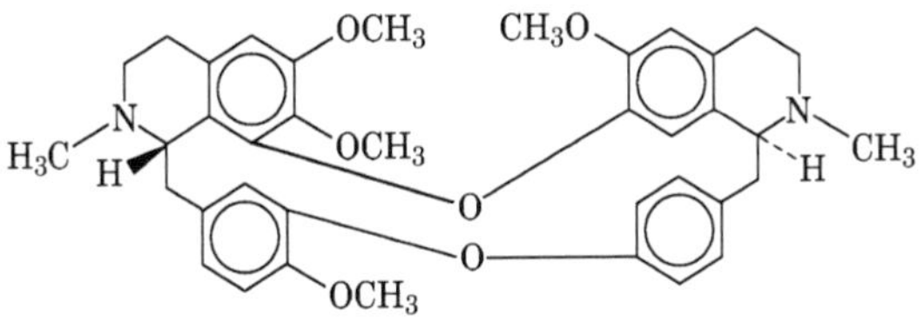

(**80**)

Many other bisbenzylisoquinoline alkaloids, such as tetrandrine (**80**), from *Cyclea peltata* Hook., are also known. Compound (**80**), for example, although it causes hypotension and hepatotoxicity in mammals, in other tests, possessed enough anticancer activity to be considered for preclinical evaluation (55). The arrow poison tubocurare prepared from *Chondrendendron* spp. also contains the bisbenzylisoquinoline alkaloid tubocurarine (**9**).

In an early attempt to understand the genesis of alkaloids from amino acids it was postulated (56) that intramolecular phenolic coupling should lead from benzylisoquinoline bases such as laudanosine (**77**, R = CH$_3$), before it was completely methylated, to aporphine bases such as isothebaine (**81**). For example, between a benzylisoquinoline derived from laudanosoline (**77**, R = H), such as orientaline (**82**), and an aporphine alkaloid such as isothebaine (**81**), there should be a proaporphine alkaloid such as orientalinone (**83**) (56). The isolation of **83** lent credence to the hypothesis. Indeed, the fragile nature of **83** (it readily undergoes the dienone–phenol rearrangement on acid treatment) required unusual skill in obtaining it from total plant extract.

Table 8. Alkaloids from Norlaudanosoline as Pregenitor[a]

Name	Molecular formula	CAS Registry Number	Structure number	Structure
norlaudanosoline laudanosoline	$C_{16}H_{17}NO_4$ $C_{21}H_{27}NO_4$	[4747-99-3] [2688-77-9]	(77) (R = H) (77) (R = CH$_3$)	
papaverine	$C_{20}H_{21}NO_4$	[58-74-2]	(78)	
oxyacanthine tetrandrine	$C_{37}H_{40}N_2O_6$ $C_{38}H_{42}N_2O_6$	[548-40-3] [518-34-3]	(79) (80)	b b
isothebaine	$C_{19}H_{21}NO_3$	[568-21-8]	(81)	
orientaline	$C_{19}H_{23}NO_4$	[27003-74-3]	(82)	

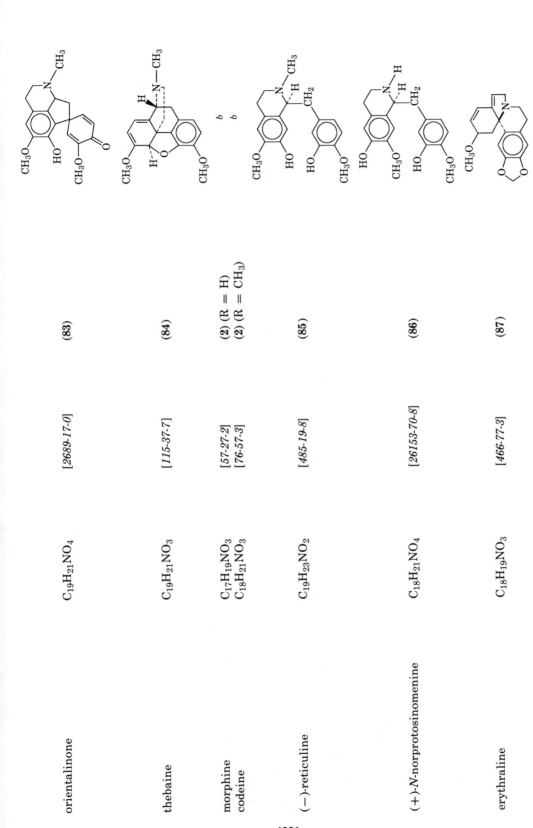

orientalinone   [2689-17-0] $C_{19}H_{21}NO_4$ (83)

thebaine   [115-37-7] $C_{19}H_{21}NO_3$ (84)

morphine   [57-27-2] $C_{17}H_{19}NO_3$ (2) (R = H)
codeine   [76-57-3] $C_{18}H_{21}NO_3$ (2) (R = CH$_3$)

(−)-reticuline   [485-19-8] $C_{19}H_{23}NO_2$ (85)

(+)-N-norprotosinomenine   [26153-70-8] $C_{18}H_{21}NO_4$ (86)

erythraline   [466-77-3] $C_{18}H_{19}NO_3$ (87)

Table 8. (*Continued*)

Name	Molecular formula	Registry Number	Structure number	Structure
argemonine	$C_{21}H_{25}NO_4$	[6901-16-2]	(**88**)	c
berberine	$C_{20}H_{18}NO_4$	[2086-83-1]	(**89**)	c
protopine	$C_{20}H_{19}NO_5$	[130-86-9]	(**90**)	c
chelidonine	$C_{20}H_{19}NO_5$	[476-32-4]	(**91**)	c
rhoeadine	$C_{21}H_{21}NO_6$	[2718-25-4]	(**92**)	c
harringtonine	$C_{28}H_{37}NO_9$	[26833-85-2]	(**93**)	c

[a]This group has tyrosine or phenylalanine joined as C_6—C_2 fragments.
[b]In text.
[c]See Figure 3.

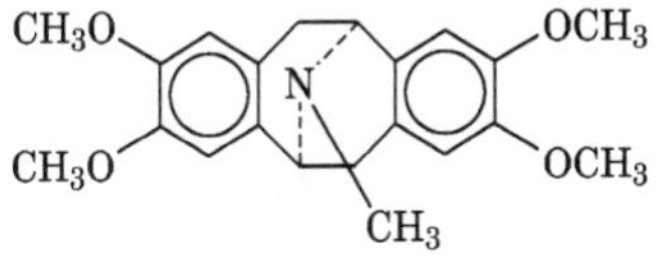

Fig. 3. Oxidative coupling products of methylated derivatives of laudanosoline (77, R = H). See Table 8.

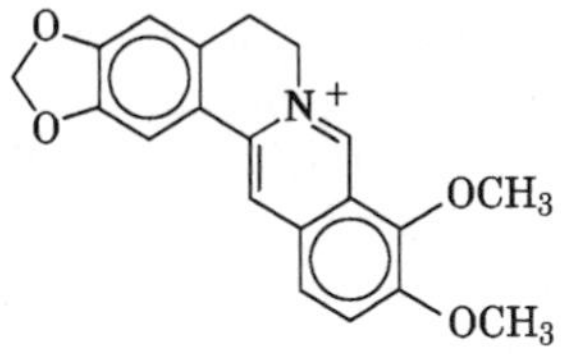

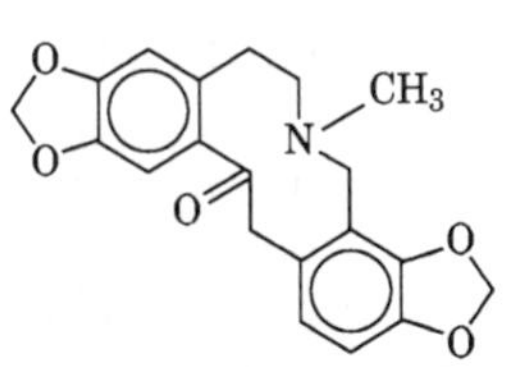

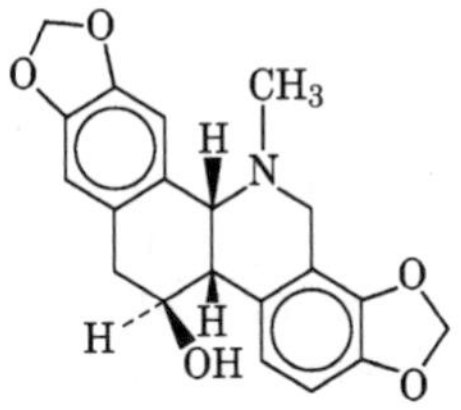

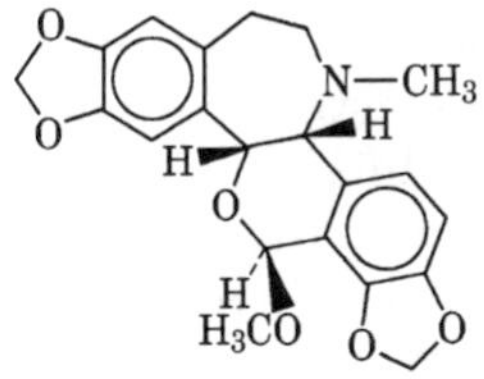

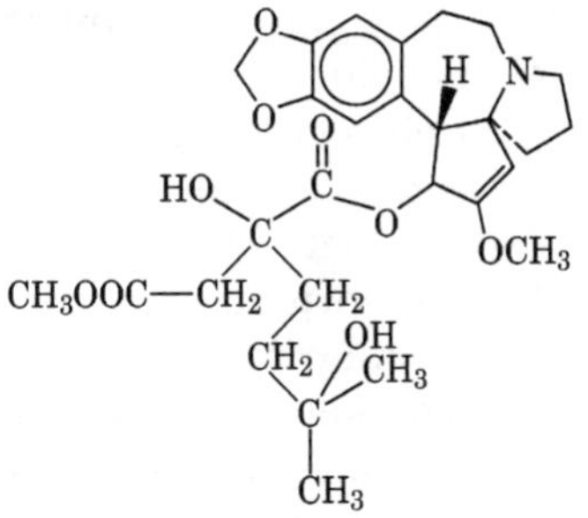

Isothebaine (81), which may be derived from orientalinone (83) in the laboratory, is isolated from the roots of *Papaver orientale* after the period of active growth of the aerial parts and the production of thebaine (84) has ceased. The viscous milky exudate of the unripe seed pods of *P. orientale* as well as the opium poppy *Papaver somniferum* L. is opium. Opium cultivation appears to have spread from Asia Minor to China (via India) and it has been noted that the smoking of opium was common in China and elsewhere in the Far East when trade began in earnest as the eighteenth century closed. Active cultivation was encouraged by the revenues generated from addicts. Today, in the United States, although narcotine (1, noscopine) has some commercial value as a nonaddictive antitussive which occasionally leads to drowsiness, it is morphine (2, R = H), codeine (2, R = CH₃), and thebaine (84), the latter being converted to both of the former, which are of major commercial value.

The importance of morphine (2, R = H) as an analgesic, despite danger of addiction and side effects that include depression of the central nervous system, slowing of respiration, nausea, and constipation, cannot be underestimated, and significant efforts have been expended to improve isolation techniques from crude dried opium extract. Depending on its source, the morphine content of poppy straw or dried exudate may be as high as about 20%. Although the details of current manufacturing processes are closely held secrets, early work (57) has probably not been modified extensively. Usually the crude opium is extracted with water and filtered, and the aqueous extract concentrated, mixed with ethanol, and made strongly basic with ammonium hydroxide. Morphine usually precipitates, while the other bases remain in solution, and is further purified by crystallization as its sulfate.

Codeine (**2**, R = CH$_3$) occurs in the opium poppy along with morphine (**2**, R = H) but usually in much lower concentration. Because it is less toxic than morphine and because its side effects (including depression, etc) are less marked, it has found widespread use in the treatment of minor pain and much of the morphine found in crude opium is converted to codeine. The commercial conversion of morphine to codeine makes use of a variety of methylating agents, among which the most common are trimethylphenylammonium salts. In excess of two hundred tons of codeine are consumed annually from production facilities scattered around the world.

The first synthesis of morphine, and therefore also codeine, was completed in 1956 (58). Although an additional twelve or so syntheses have been reported since then, isolation of morphine remains more important than any synthetic process. However, synthetic endeavors continue to demonstrate new synthetic tools and capabilities and to conduct the search for modified analogues that retain the analgesic properties of morphine but are nonaddicting.

Whereas the particular methylated derivative of laudanasoline (**77**, R = H) called (−)-reticuline (**85**) (59) gives rise to thebaine (**84**), codeine (**2**, R = CH$_3$), and morphine (**2**, R = H), a different derivative of **77** (R = H), ie, (+)-*N*-norprotosinomenine (**86**), serves as the pregenitor of erythraline (**87**), one of the bases found in *Erythrina crista galli* (60). The alkaloids found in all plant parts of *Erythrina* have been intensively studied because many of them produce smooth muscle paralysis, much like tubocurarrine (**9**).

Additional oxidative coupling processes among the various methylated derivatives of laudanosoline yield many other families of bases, including the pavine argemonine (**88**) from *Argemone mexicana* L.; berberine (**89**) from *Hydrastis canadensis* L. which, despite its toxicity, has been used as an antimalarial; protopine (**90**) and chelidonine (**91**) from *Chelidonium majus* L.; rhoeadine (**92**); and the cephalotaxus ester harringtonine (**93**) from Japanese plum yews (*Cephalotaxus* spp.), which is a compound of some significance because it possesses potent antileukemic activity (see Figure 3).

The last group of compounds from tyrosine and phenylalanine is the group derived from utilization of tyrosine (**25**, R = OH) as the C$_6$—C$_2$ fragment and phenylalanine (**25**, R = H) as a C$_6$—C$_3$ fragment (Table 9). They include the 1-phenethyltetrahydroisoquinoline autumnaline (**94**) and the homoproaporphine kreysiginone (**95**), which are typical of their kind and are both isolated from *Colchicum cornigerum* (Schweinf.) and the toxic principle of the autum crocus (*Colchicum autumnale*), colchicine (**4**). In crude form, extracts of *Colchicum* spp. were reportedly known to Dioscorides, a contemporary of Pliny, who served Nero as a physician and was the first to establish systematically the medicinal value of some 600 plants. The use of *Colchicum* spp. extracts in the treatment of gout appears to have begun in the sixteenth century, although it was not until much

Table 9. Alkaloids Derived from Tyrosine as a C_6—C_2 Fragment and Phenylalanine as a C_6—C_3 Fragment

Name	Molecular formula	CAS Registry Number	Structure number	Structure
autumnaline	$C_{21}H_{27}NO_5$	[23068-65-7]	(94)	
kreysiginone	$C_{20}H_{23}NO_4$	[17441-87-1]	(95)	
colchicine	$C_{22}H_{25}NO_6$	[64-86-8]	(4)	a
betanidine	$C_{18}H_{16}N_2O_8$	[2181-76-2]	(96) ($R_1 =$ H, $R_2 =$ COOH)	
isobetanidine	$C_{18}H_{16}N_2O_8$	[4934-32-1]	(96) ($R_1 =$ COOH, $R_2 =$ H)	
betalamic acid	$C_9H_9NO_5$	[18766-66-0]	(97)	

^{a}In text.

later that colchicine (4) was actually isolated (about 1884). Recent interest in colchicine stems from its ability to bring cell division to an abrupt halt at a particular stage.

The structures of the brightly colored (red-violet and yellow) alkaloids found in the order *Centrospermae* (cacti, red beet, etc) remained unknown until the 1960s. In part this was doubtlessly due to the fact that these pigments, called betacyanins or betaxanthins, are relatively unstable and they are water soluble zwitterions. Invariably, in these plants they are found as acetals or ketals of sugars and one of two aglycone fragments called betanidine (**96**, R_1 = H, R_2 = COOH) and isobetanidine (**96**, R_2 = H, R_1 = COOH). Furthermore, it would appear that all betacyanins or betaxanthins may simply be imine derivatives (with the appropriate amino acid) of betalamic acid (**97**).

Tryptophan-Derived Alkaloids.   There are a few simple indole derivatives which are arguably derived, or have actually been shown to be derived, from tryptophan (Table 10) (61). Serotonin (5-hydroxytryptamine [*50-67-9*], (**98**), R = OH) was first isolated (62) as a vasoconstrictor substance from beef serum and shown to be derived from tryptophan. It has also been isolated from bananas and the stinging nettle but its genesis in plants has not yet been established. *N,N*-Dimethylserotonin [*487-93-4*] (bufotenine (**19**)) has been found in such widely diverse sources as the shrub *Piptadenia peregrina*, the seeds of which are said to be the source of a ceremonial narcotic snuff; the parotid gland of the toad (*Bufo vulgaris* Laur.); certain fungi (eg, *Amantia mappa* Batsch.); and human urine. The only slightly more complicated base harmine (**99**) is found widely distributed in the *Leguminoseae* and *Rubiaceae*, extracts of which were at one time used therapeutically against tremors in Parkinson's disease. The seeds of the African rue, *Peganum harmala* L., which are rich in harmine and related alkaloids, have also been used as a tapeworm remedy. Harmine has been shown to be derived from tryptamine (**98**, R = H) by ^{14}C- and ^{15}N-labeling experiments (63).

The C_6 building block, mevalonic acid (**30**), has been shown again and again to lose CO_2 and then serve as the pregenitor of C_5, C_{10}, C_{15}, C_{20}, and C_{30} systems via the isomeric pair 3,3-dimethylallyl pyrophosphate (**32**)–isopentenyl pyrophosphate (**31**) (Table 3). This explains the genesis of a variety of bases containing the pattern defined by the five-carbon branched chain common to them and to the derivatives of mevalonic acid. Included among these are that small group of alkaloids which correspond to a joining of a tryptophan (**26**) and a C_5 unit to produce bases such as the potent uterine stimulant agroclavine (**18**) and its relatives, among which are the peptide amides of lysergic acid (**100**, R = OH).

The pistil of rye and certain other grasses may be infected by the parasitic fungus *Claviceps purpurea* Fries. Unless the infected grain is sieved, the fungus passes into the flour, and bread made from such contaminated flour apparently retains activity from some of the alkaloids elaborated by and present in the fungus. Thus ingestion of the contaminated flour results in the disease called ergotism (St. Anthony's Fire). Convulsive ergotism causes violent muscle spasms which bend the sufferer into otherwise unattainable positions and frequently leaves physical and mental scars; outbreaks of the disease have been recorded into the twentieth century (21). Agroclavine (**18**) and derivatives of lysergic acid (**100**, R = OH) are considered responsible. Nonetheless, extracts of *C. purpurea* have long been used medicinally since they effect smooth muscle contraction and, even today, compounds related to lysergic acid and agroclavine are used for the same purpose. In the early 1940s it was discovered that the diethylamide of lysergic acid [**100**, R = $N(CH_2CH_3)_2$] (LSD-25, as the tartrate salt) could be absorbed through the skin with resulting inebriation. In a bold experiment, it was then demonstrated that oral ingestion resulted in symptoms characteristic of schizophrenia which, although temporary, were quite dramatic (64).

There are currently two medicinally valuable alkaloids of commercial import obtained from ergot. Commercial production involves generation parasitically on rye in the field or production in culture because a commercially useful synthesis is unavailable. The common technique today (65) is to grow the fungus in submerged culture. *Claviceps paspali* (Stevens and Hall) is said to be more productive than *C. purpurea* (Fries). In this way, ergotamine (**100**, R = **101**) and ergonovine (**100**, R = NHCH(CH$_3$)CH$_2$OH) are produced. Ergotamine is obtained from crude extract by formation of an aluminum complex which can then be defatted.

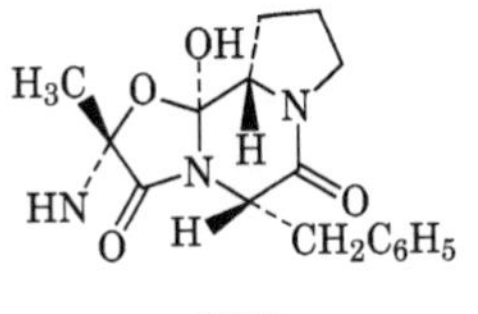

(**101**)

Destruction of the aluminum complex with ammonia then permits hydrocarbon extraction of the alkaloid. The alkaloid is subsequently both isolated and used as its tartrate salt. This nonnarcotic drug, for which tolerance may develop, is frequently used orally with caffeine (**16**) for treatment of migraine; it acts to constrict cerebral blood vessels, thus reducing blood flow to the brain.

Ergonovine (**100**, R = NHCH(CH$_3$)CH$_2$OH) was found to yield lysergic acid (**100**, R = OH) and (+)-2-aminopropanol on alkaline hydrolysis during the early analysis of its structure (66) and these two components can be recombined to regenerate the alkaloid. Salts of ergonovine with, for example, malic acid are apparently the drugs of choice in the control and treatment of postpartum hemorrhage.

Loganin (**34**), the iridoid derived from the dimer (**33**) of the isomeric pair of C$_5$ isoprenoid units, isopentenyl pyrophosphate (**31**) and 3,3-dimethylallyl pyrophosphate (**32**) (Table 3), has been recognized for some years (67) as the C$_9$—C$_{10}$ unit which, along with tryptophan (**26**), makes up the huge group of bases, nearly one thousand well characterized compounds, found in the *Corynanthe-Strychnos, Cinchona, Iboga, Aspidosperma*, and *Eburna*. Loganin is known to undergo oxidative cleavage to secologanin [*19351-63-4*] (**102**),

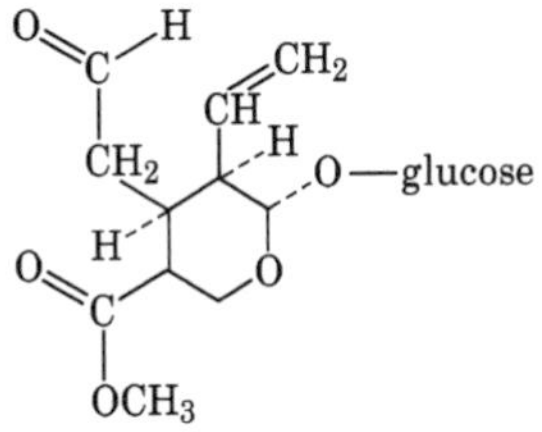

(**102**)

and this fragment, combined with what appears to be tryptamine [*61-54-1*] (**98**, R = H), C$_{10}$H$_{12}$N$_2$, or its biological equivalent, leads to compounds whose permuted structures are often novel and quite complicated. Numerous single examples of rearranged, oxidized, and convoluted structures abound, but the subdivision into the families given above is convenient for description of the majority of structural types of bases (68).

Table 10. Alkaloids from Trytophan

Name	Molecular formula	CAS Registry Number	Structure number	Structure
serotonin	$C_{10}H_{12}N_2O$	[50-67-9]	(98) (R = OH)	
bufotenine	$C_{12}H_{16}N_2O$	[487-93-4]	(19)	
harmine	$C_{13}H_{12}N_2O$	[442-51-3]	(99)	
agroclavine	$C_{16}H_{18}N_2$	[548-42-5]	(18)	
lysergic acid	$C_{16}H_{16}N_2O_2$	[82-58-6]	(100) (R = OH)	
lysergic acid diethylamide	$C_{20}H_{25}N_3O$	[50-37-3]	(100) (R = N(CH$_2$CH$_3$)$_2$)	
ergotamine	$C_{33}H_{35}N_5O_5$	[113-15-5]	(100) (R = **101**)	
ergonovine	$C_{19}H_{23}N_3O_2$	[60-79-7]	(100) (R = NHCHCH$_2$OH)	

1068

ajmalicine	$C_{21}H_{24}N_2O_3$	[483-04-5]	(103)	b
yohimbine	$C_{21}H_{26}N_2O_3$	[146-48-5]	(104)	b
reserpine	$C_{33}H_{40}N_2O_9$	[50-55-5]	(105)	b
ajmaline	$C_{20}H_{26}N_2O_2$	[4360-12-7]	(106)	b
catharanthine	$C_{21}H_{24}N_2O_2$	[2468-21-5]	(107)	b
tabersonine	$C_{21}H_{24}N_2O_2$	[4429-63-4]	(108)	b
vincamine	$C_{21}H_{26}N_2O_3$	[1617-90-9]	(109)	b
strychnine	$C_{21}H_{22}N_2O_2$	[57-24-9]	(13) (R = H)	c
quinine	$C_{20}H_{24}N_2O_2$	[130-95-0]	(17) (R = OCH$_3$)	c
cinchonidine	$C_{19}H_{22}N_2O$	[118-10-5]	(17) (R = H)	c
leurocristine	$C_{46}H_{56}N_4O_{10}$	[57-22-7]	(22) (R = CHO)	a
vincaleukoblastine	$C_{46}H_{58}N_4O_9$	[865-21-4]	(22) (R = CH$_3$)	a
ellipticine	$C_{17}H_{14}N_2$	[519-23-3]	(112) (R$_1$ = CH$_3$, R$_2$ = H)	
oliavacine	$C_{17}H_{14}N_2$	[484-49-1]	(112) (R$_1$ = H, R$_2$ = CH$_3$)	
calycanthine	$C_{22}H_{26}N_4$	[595-05-1]	(113)	a

[a] In text.
[b] See Figure 4.
[c] See Table 1.

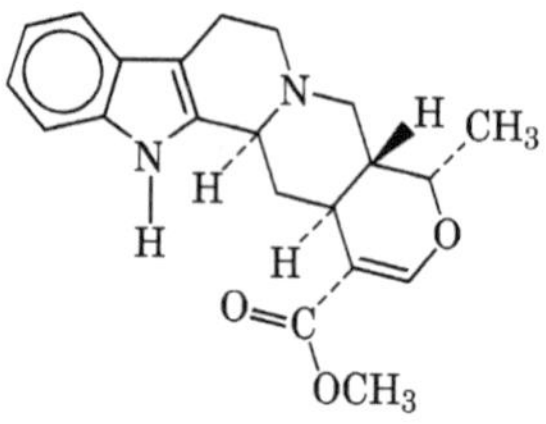

Fig. 4. Alkaloids from secologanin (**102**) and tryptamine (**98**, R = H). (See Table 10).

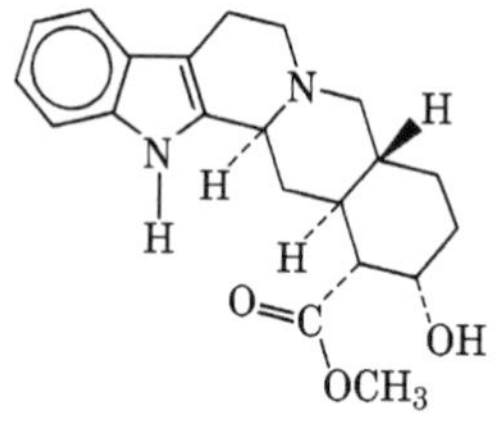

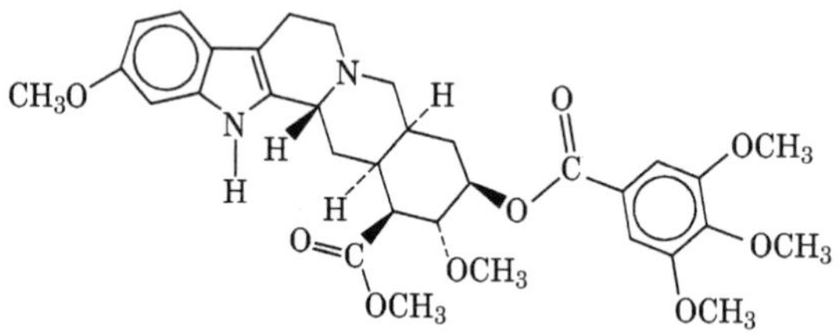

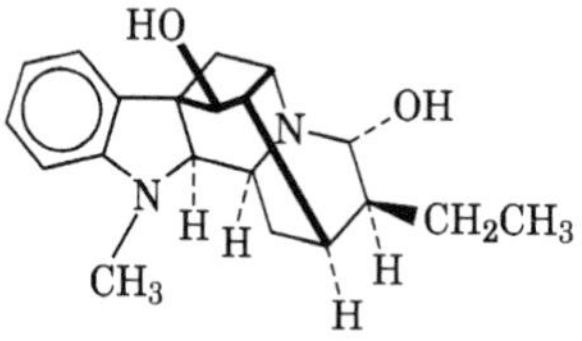

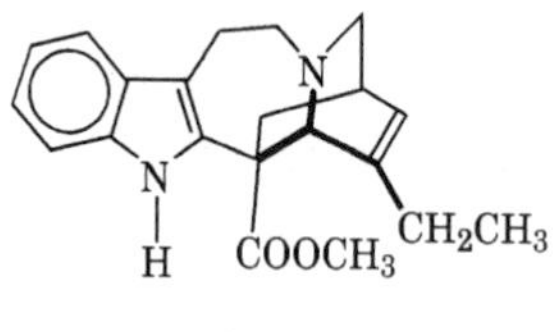

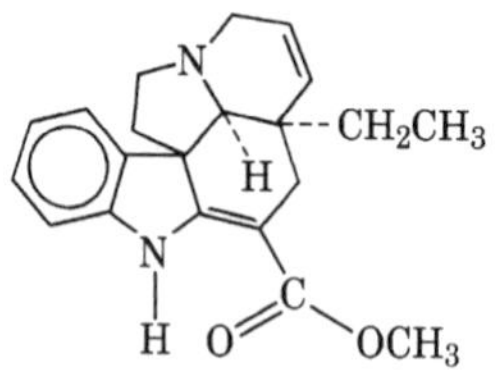

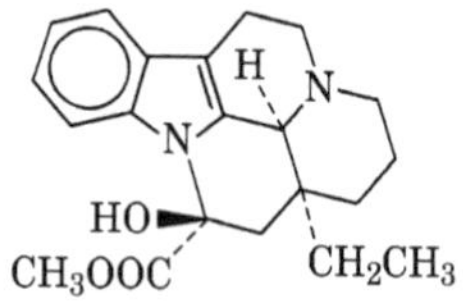

Thus in the *Corynanthe-Strychnos* are found bases such as ajmalicine (**103**), yohimbine (**104**), reserpine (**105**), ajmaline (**106**), and strychnine (**13**); in the *Cinchona*, quinine (**17**, R = OCH₃) and cinchonidine (**17**, R = H); in the *Iboga*, catharanthine (**107**); in the *Aspidosperma*, tabersonine (**108**); and in the *Eburna*, vincamine (**109**).

Ajmalicine (**103**) has been isolated (frequently as the weakest base present) numerous times from a variety of sources, eg, from the bark of *Corynanthe yohimbe* K. Schum. (Rubiaceae), from the roots of *Rauwolfia serpentina* (L.) Benth. (Apocynaceae), and from many other species of the genus *Rauwolfia*, and it is also found in plants of the genus *Catharanthus* (Apocynaceae). It is included here to demonstrate the pattern in secologanin (**102**)–tryptamine (**98**, R = H) coupling, the subsequent elaboration of which will give rise to the other bases to be considered. Thus ajmalicine (**103**) can be visualized as arising from the formation of an imine between the exposed aldehyde in the secologanin (**102**) and the basic nitrogen of tryptamine (**98**, R = H), followed by cyclization to the 2-position of the indole nucleus to form the C ring, opening of the glucose-masked acetal, and·

carbon–carbon bond rotation changing the cis ring junction stereochemistry found in secologanin to the trans stereochemistry of the D/E ring juncture in ajmalicine. The process continues with a second cyclization, now to the freshly exposed aldehyde resulting from opening of the acetal, the latter having swung around so that it is close to the secondary amine of the newly created C ring; a reduction of the imine so created; and a final cyclization of the liberated enol onto the alkene. In short, all ten carbon atoms of secologanin and the entire tryptamine skeleton, as well as the geometry of the product, have been accounted for. This pathway, broadly painted above, is supported in detail by numerous labeling experiments with isotopes of carbon, hydrogen, and nitrogen, as well as the actual isolation of some of the intermediates described. All of the work has been summarized (69), Ajmalicine increases cerebral blood flow and commercial mixtures of ajmalicine with one or more ergot alkaloids have been used in treating vascular disorders and hypertension.

Yohimbine (**104**), also from the bark of *C. yohimbe* K. Schum. and from the roots of *R. serpentina* (L.) Benth., has a folk history (unsubstantiated) of use as an aphrodisiac. Its use has been confirmed experimentally as a local anesthetic, with occasional employment for relief in angina pectoris and arteriosclerosis, but is frequently contraindicated by its undesired renal effects. Yohimbine and some of its derivatives have been reported as hallucinogenic (70). In addition, its pattern of pharmacological activities in a variety of animal models is so broad that its general use is avoided. All ten carbon atoms of secologanin (**102**) as well as the entire skeleton of tryptamine (**98**, R = H) are clearly seen as intact portions of this alkaloid.

Reserpine (**105**), also from the roots of *R. sepentina* (L.) Benth. and other *Rauwolfia* spp., is currently used as a hypotensive. There are reports in the older popular literature showing its use for a wide variety of ailments in the tropics and subtropics where the plants grow. Apparently it was originally used to treat both high blood pressure and insanity. The former use has been replaced by substances of greater value. Even its use as a tranquilizer and sedative, which has shown some apparent successes with neuroses, at lower doses (0.05–1.5 mg/day), is no longer in vogue for treatment of psychoses (at 0.5–5.0 mg/day); better materials have been found. Nonetheless, although no analgesic effect has been noted, reserpine does act as a sedative which reduces aggressiveness. At higher doses, reserpine has been reported to cause depression as well as peptic ulceration. There is some evidence that chronic administration in women results in an increased incidence of breast cancer (71). In other experimental systems it has shown antitumor activity (72). Its interesting structure contains, as expected, a tryptamine unit (this time substituted with a 6-methoxy group), a secologanin (**102**) C_{10} fragment, and a trimethoxybenzoic acid unit. This is presumably derived by methylation of gallic acid [*149-91-7*] (**110**),

(**110**)

typically derived oxidatively from shikimic acid (**57**) and normally associated with tannins in nutgalls, from which it is obtained by hydrolysis. It is not known for *Rauwolfia* spp. if methylation of the gallic acid to produce the trimethoxy-benzoic acid unit found in reserpine (**105**) occurs before the acid is esterified with the remainder of the system or if methylation occurs later. Indeed, the involvement of gallic acid (**110**), $C_7H_6O_5$, itself is, as noted above, presumptive. The total synthesis of reserpine was a landmark synthesis (73).

In ajmaline (**106**), also obtained from the roots of *R. serpentina* (L.) Benth., a more deeply rearranged secologanin (**102**) fragment is embedded in the molecular framework and there has been a decarboxylation from the masked β-keto carboxylic acid to generate a C_9 unit. Current hypotheses argue that the C_9 unit found in ajmaline actually began as the same C_{10} unit already seen in ajmalicine (**103**), yohimbine (**104**), and reserpine (**105**), but the additional cyclization to the C ring and subsequent bonding to the 3-position on the indole nucleus creating a sixth ring is accompanied by decarboxylation. Ajmaline has aroused some interest because it appears to possess antiarrhythmic activity (74), but care is required for this use when there is liver disease.

The synthesis of strychnine was a truly monumental undertaking (20). Strychnine (**13**, R = H), although only moderately toxic when compared to other poisons, both naturally occurring and produced synthetically, probably owes its reputation to its literary use. Obtained from the seeds and leaves of *Strychnos nux vomica* L. and other *Strychnos* spp., it has some history of use as a rodenticide. Poisoning is manifested by convulsions, and death apparently results from asphyxia. As little as 30 to 60 mg has been reported as fatal to humans, although at lower dosage it has received some medical use as an antidote for poisoning by central nervous system depressants, as a circulatory stimulant, and in treatment of delirium tremens. The useful medicinal dosage is normally less than 4 mg. As was the case for ajmaline (**106**), also derived from tryptamine (**98**, R = H) and secologanin (**102**), the pattern for the formation of strychnine (**13**) has been extended to even more deep seated rearrangement as well as the loss of one carbon atom, the same carboxylate as was lost in ajmaline (**106**). In addition, an acetate (C_2) unit has been added (to the indole nitrogen) and another ring created.

Quinine (**17**, R = OCH_3) and cinchonidine (**17**, R = H), which occur together along with other bases, eg, materials epimeric at the one-carbon bridge joining the aromatic nucleus with the 1-azabicyclo[2.2.2]heptane system, are constituents of the root, bark, and dried stems of various *Cinchona* species, but the main source remains *Cinchona officinalis* L. A crude preparation from this source was introduced into use as a palliative for malaria in the seventeenth century but several hundred years then elapsed before the first mixture of crystalline bases was obtained. Until the second World War quinine (**17**, R = OCH_3) and crude *Cinchona* preparations were the only antimalarials available. As supplies became unobtainable, synthetic materials capable of replacing quinine were developed. Recently, however, quinine has again become the treatment of choice for malaria as *Plasmodium falciparum* resistant to other drugs developed. Apparently, resistance to quinine is more difficult for the rapidly changing mosquito population to acquire. Nonetheless, because quinine is not a prophylactic drug but rather a material which suppresses the overt manifestations of malaria, work continues on better treatment. The isolation of quinine from *Cinchona* bark generally involves conversion of the salts of the basic alkaloids to the free bases with, eg, calcium hydroxide (75), and extraction of the alkaloids into benzene or toluene.

Examination of the structures of the alkaloids (17) obtained from *Cinchona* suggests that they probably have a different pattern of formation than those already discussed. However, the differences are more formal than profound. Thus the 1-azabicyclic system is formed from one of the aldehyde equivalent carbons of a secologanin (102) bound fragment with the terminal nitrogen of tryptamine (98, R = H). Cleavage of that nitrogen away from the indole leaves behind the carbon to which it was bound, formally, at the oxidation level of an aldehyde. Then, oxidative opening of the five-membered ring between the indole nitrogen and the adjacent carbon is followed by recyclization from the aryl amine so liberated to the aldehyde function set free in the previous step. Thus the nine expected carbons, one of the original carbons in the C_{10} fragment having been lost by decarboxylation, remain.

An understanding of the chemistry and structure of catharanthine (107), an otherwise minor alkaloid found in *Vinca rosea* Linn. or *Catharanthus roseus* G. Don. which is a potent diuretic in rats, was critical in unraveling the structure of two of the alkaloids co-occuring in *Vinca* which had been shown to be active against leukemia, first in mice and later in humans, ie, leurocristine (vincristine [57-22-7], 22, R = CHO) and vincaleukoblastine (vinblastine [865-21-4], 22, R = CH₃). Interestingly, the genera *Vinca* and *Catharanthus* (Apocynaceae) appear to be used interchangeably, ie, *Vinca rosea* Linn. is frequently called *Catharanthus roseus* G. Don. by some but not by others and vice versa (76). Thus both catharanthine (107) and vincaleukoblastine in concentrated hydrochloric acid, when treated with stannous chloride yielded, among other fragments, the (+)-cleavamine [1674-01-7] (111), so called because it is a broken or cleaved amine.

(111)

After the structure and absolute stereochemistry of cleavamine (111), $C_{19}H_{24}N_2$, was established, its synthesis was shortly completed and impetus to unravel the structure of the dimeric bases (22) was bolstered (77). Again, the C_9 fragment, now only slightly modified from that originally present in secologanin (102), is readily seen in catharanthine (107).

Tabersonine (108), clearly a reduced and simplified version of the second-half of the alkaloids 22, was originally isolated from *Amsonia tabernaemontane* L. and is considered to be a simplified parent of a rather more elaborate subgroup of indole alkaloids.

Among the examples of monoindole bases being discussed, vincamine (109) is the principal alkaloid of *Vinca minor* L. and has received some notoriety because it apparently causes some improvement in the abilities of sufferers of cerebral arteriosclerosis (78). It is believed that this is the result of increasing cerebral blood flow with the accompanying increase in oxygenation of tissue as a result of its action as a vasodilator.

Finally, for this group of tryptophan (**26**)-derived bases there are those in which a tryptamine (**98**, R = H) residue is not obvious. Nonetheless, the pyridocarbazole bases originally isolated from *Ochrosia elliptica* Labill. (Apocynaceae) and subsequently from the genus *Aspidosperma*, among others, which include ellipticine (**112**, R_1 = CH_3, R_2 = H) and olivacine (**112**, R_1 = H, R_2 = CH_3), are derived from tryptophan and the normal C_9—C_{10} fragment expected. These alkaloids are known to inhibit proliferation of cells and continue to be of interest in chemotherapeutic treatments. They appear to inhibit nucleic acid synthesis irreversibly by interacting strongly with DNA.

Bisindole Alkaloids from Tryptophan.   There are two widely different types of alkaloids derived from two tryptophan (**26**) units. The first is a rather small group of compounds based simply on the dimers of tryptophan which includes compounds such as calycanthine (**113**) (Table 10),

(**113**)

isolated from the seeds of the flowering aromatic shrubs Carolina Allspice (*Calycanthus floridus*) and Japanese Allspice (*Chimonanthus fragans*). The second type is that group, reviewed in reference 79, in which the two halves arise in two distinct ways. Both halves may be composed of identical fragments, as in C-toxiferine (**10**), the arrow poison packed in gourds and derived from, eg, *Strychnos froesii* Ducke and *Strychnos toxifera*. The more common and very numerous (nearly one thousand compounds) family is characterized by two halves derived from different fragments, eg, leurocristine (vincristine (**22**), R = CHO) and vincaleukoblastine (vinblastine (**22**), R = CH_3), along with nearly one hundred other compounds, from *Catharanthus roseus* (L.) G. Don, occasionally referred to erroneously as *Vinca rosea* L. (80). This second group has in common the genesis of each half from tryptophan (**26**) and at least one fragment derived from mevalonic acid (**30**) itself (ie, a C_5 fragment) or derived from a monoterpene such as geraniol (**33**, —OH in place of —OPP), ie, loganin (**34**) or secologanin (**102**), a C_{10} fragment.

The search for the bisindole derivatives (**22**) was originally (81) initiated on the basis of folklore. A brew made from Jamacian periwinkle had established itself in local medicine as a treatment for diabetes and it was this material that was investigated and found to contain the cytotoxic compounds (**22**), among others. No materials useful in the treatment of diabetes have been reported from this source.

Although the compounds were isolated in quantities of only a few milligrams per kilogram of crude plant leaves, extensive work on a variety of animal tumor systems led to eventual clinical use of these bases, first alone and later in conjunction with other materials, in the treatment of Hodgkin's disease and acute lymphoblastic leukemia. Their main effect appears to be binding tightly to tubulin, the basic component of microtubules found in eukaryotic cells, thus interfering with its polymerization and hence the formation of microtubules required for tumor proliferation (82).

Initial attempts to synthesize the compounds (**22**) were hampered by the failure to obtain the correct stereochemical configuration about the vindoline–catharanthine linkage, a most difficult problem eventually solved by insight and hard work (83).

Introduction of Nitrogen into a Terpenoid Skeleton. The acetate-derived fragments (**35**) mevalonic acid (**30**), which yields isopentenyl pyrophosphate (**31**) and its isomer, 3,3-dimethylallyl pyrophosphate (**32**); a dimeric C_5-fragment, geranyl pyrophosphate (**33**), which gives rise to the iridoid loganin (**34**); and the trimer farnesyl pyrophosphate (**35**), which is also considered the precursor to C_{30} steroids, have already been mentioned (see Table 3 and Fig. 1). Three of the fragments [(**30**), the pair (**31–32**), and (**34**)] have been invoked as descriptive pregenitors of alkaloids such as emetine (**66**), lysergic acid (**100**), and many other bases already discussed and broadly categorized as monoterpenoid indole alkaloids, eg, ajmalicine (**103**). Although the path that links (**30**) to (**35**) and thence to the steroids is clear (84), the details of the relationships with any of the subfragments on that path and the alkaloids resembling them and included here is less clear. The obscurity arises from two related problems involved in labeling experiments. First, the techniques for feeding suitably labeled precursors such as ^{13}C- and/or ^{14}C-labeled and ^{2}H- and/or ^{3}H-labeled acetate or even larger fragments (those further along on the metabolic pathway to alkaloidal product) such as mevalonic acid (**30**) and loganin (**34**) to many actively growing plants have been worked out, as shown by their incorporation into alkaloids. However, lack of incorporation does not rule out utilization by the plant of the material fed to it, because there is no guarantee that the fed material reached the site of alkaloid synthesis. Thus a negative result may simply mean that the particular technique, stage of plant growth, feeding cycle, photo cycle, etc, was inappropriate for the specific material fed, rather than implying that the material is not capable of incorporation. In this vein, it is generally true that the larger fragments are more difficult to incorporate. Even though they may enter the plant when they might be actively metabolized, the particular form in which they arrive at the cell wall may be wrong for transport across the wall. Smaller fragments are less likely to have transport problems. Thus mevalonic acid (**30**) generated endogenously from exogenous labeled acetate is frequently more easily traced than suitably labeled exogenous mevalonic acid itself. However, the value is correspondingly diminished because everything may be thought of as derived from acetate and whatever the precursor, the label will have been incorporated. Thus a balance must be struck between what can be fed as labeled material and what will be incorporated into the plant. Frequently the largest useful fragment that can be incorporated is mevalonic acid (or its corresponding lactone).

The second experimental problem is that incorporation of a material such as loganin (**34**), or even an amino acid which seems clearly to be a precursor by some biogenetic hypothesis, does not necessarily prove it is a precursor. The material fed may so completely swamp the normal pathways in the plant that the utilization of what was fed generates an aberrant path which nonetheless produces the same product.

These considerations are particularly important when the description of the alkaloids is based on a presumed biosynthesis from terpene fragments because the experimental work linking the smaller pieces with the larger has yet to succeed. That is, the well worked out paths from acetate, through (30) to the steroids, via geranyl pyrophosphate (33), farnesyl pyrophosphate (35), and the universal steroid precursor squalene [111-02-4] (114), $C_{30}H_{50}$, (84) have not been clearly demonstrated to apply in the higher, alkaloid producing, plants. Furthermore, in almost all of the alkaloids whose presumed biosynthesis derives from an insertion of nitrogen into the mevalonic acid-derived fragment, it is not quite clear at what stage the nitrogen insertion occurs. Introduction of the nitrogen at a very late stage might be an artifact of isolation because basification with ammonia of the acidic extract initially employed to isolate the basic materials is common and reaction of water soluble materials with ammonia, followed by cyclizations, etc, might occur.

(114)

When racemic mevalonic acid as the corresponding lactone and labeled at C-2 with ^{14}C is fed to the Chilean shrub *Skytanthus acutus* Meyen., labeled β-skytanthine [24282-31-3] (115) is obtained. *Skytanthus* alkaloids are reputed to be tremorgenic (85).

(115)   (116)   (117)   (118)

Tecomanine [6878-83-7] (116), $C_{11}H_{17}NO$, said to be a potent feline attractant (86) and a material clearly related to β-skythathine (115), $C_{11}H_{21}N$, as well as to nepetalinic acid [485-06-3] (117), $C_{10}H_{16}O_4$, a degradation product of nepetalactone [490-10-8] (118), $C_{10}H_{14}O_2$, which is a major constituent of volatile oil of catnip (*Neteta cataria* L.), is obtained from *Tecoma stans* Juss. Extracts of the latter (87) have some history demonstrating antidiabetic properties. Both of these bases (115 and 116) are derived from geranyl pyrophosphate (33) or loganin (34) or suitable similar precursor(s) before cleavage of the precursor to secologanin (102) or its equivalent.

Alternatively, there are those alkaloids, such as gentianine [439-89-4] (119), $C_{10}H_9NO_2$,

(119)

isolated from *Gentiana tibetica* King, among others, which are presumably derived from secologanin (102) and which exhibit anti-inflammatory action along with being muscle relaxants.

The C_5 trimer farnesyl pyrophosphate (35), in addition to serving as a pregenitor of steroids via squalene (114), is also the pregenitor of the C_{15} compounds known as sesquiterpenes. It has been suggested that farnesyl pyrophosphate (88) similarly serves as the carbon backbone of alkaloids such as deoxynupharidine (120) from *Nuphar japonicum* (Nymphaceae) (water lilies) and dendrobine (121) from *Dendrobium nobile* Lindl. (Orchidaceae) (Table 11). The latter is the source of the Chinese drug Chin-Shih-Hu. Compared to the other families of bases discussed earlier, the numbers of alkaloids supposedly derived from farnesyl pyrophosphate or a close relative is small. However, given the wide variety of plant families containing sesquiterpenes, it is most likely that the numbers of compounds to be found will dramatically increase.

Table 11. Alkaloids from Farnesyl Pyrophosphate

Name	Molecular formula	CAS Registry Number	Structure number	Structure
deoxynupharidine	$C_{15}H_{23}NO$	[1143-54-0]	(120)	
dendrobine	$C_{16}H_{25}NO_2$	[2115-91-5]	(121)	
veatchine	$C_{22}H_{33}NO_2$	[76-53-9]	(123)	a
atisine	$C_{22}H_{33}NO_2$	[466-43-3]	(124)	a
aconitine	$C_{34}H_{47}NO_{11}$	[302-27-2]	(125)	a

a See Figure 5.

Whereas dimerization of two farnesyl pyrophosphates (35) generates squalene (114) on the path to steroids (89), the addition of one more C_5 unit, as isopentenyl pyrophosphate (31) or its isomer, 3,3-dimethylallyl pyrophosphate (32), to the C_{15} compound farnesyl pyrophosphate produces the C_{20} diterpene precursor geranylgeranyl pyrophosphate [6699-20-3] (122).

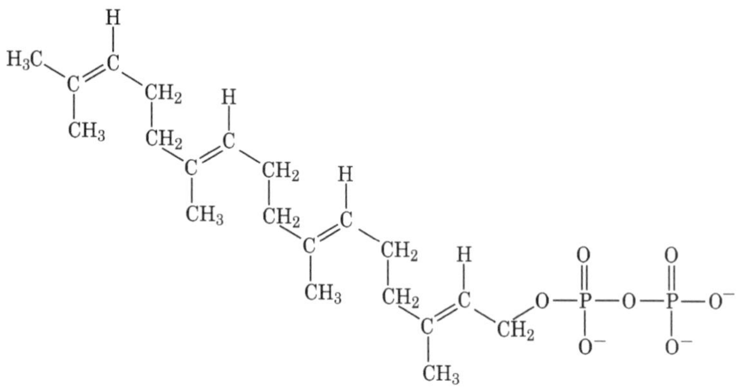

(**122**)

This C_{20} pyrophosphate (**122**), $C_{20}H_{36}O_7P_2$, is thought to provide the carbon framework of the diterpene alkaloids such as veatchine (**123**), atisine (**124**), and aconitine (**125**) (Fig. 5). It is not known at what stage the nitrogen is incorporated into the framework established by the skeleton. The potential for terpene rearrangements and the observation that the alkaloids are frequently found esterified, often by acetic or benzoic acid, as well as free, has led to permutations and combinations producing over 100 such compounds.

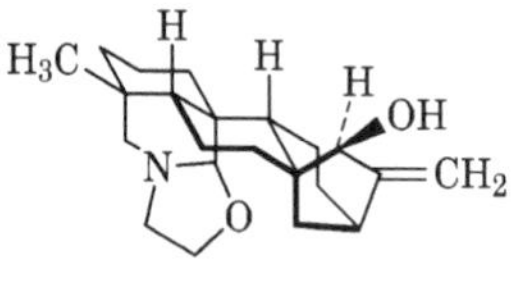

(**123**) 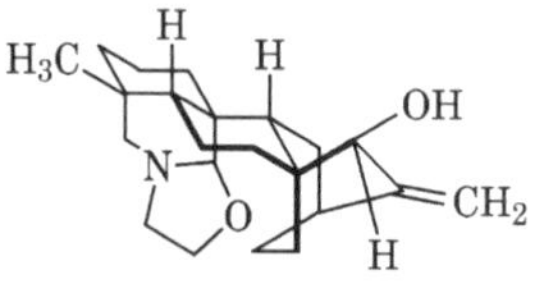   (**124**)

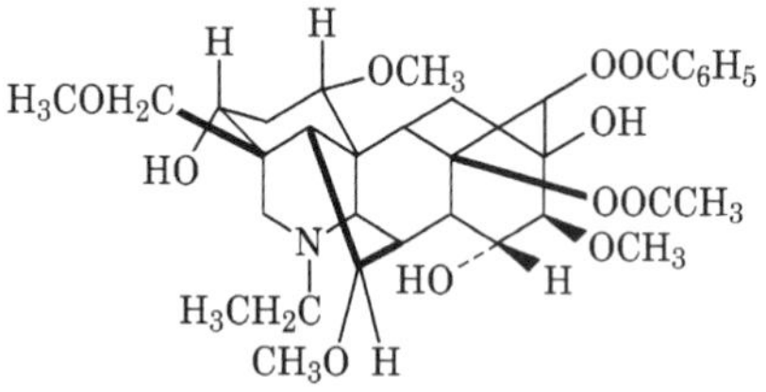

(**125**)

Fig. 5.   Diterpene alkaloids. (See Table 11).

The diterpene alkaloids elaborated by most species of *Aconitum* and *Delphinium* (family Ranunculaceae) are apparently not found in other genera (*Ranunculus, Trollius, Anemone,* etc) in the same family. Similar bases are, however, found in *Garrya* (eg, *Garrya veatchii* Kellog., family Cornaceae). Monkshood (occasionally wolf's bane, friar's cowl, or mouse bane) is obtained from the dried tuberous root of *Aconitum napellus* L. agg., and the plant is said to occur wild (90) in England and Wales as well as in the Swiss and Italian Alps. It is considered among the most dangerous of plants, all parts of it being poisonous, although the bases appear to be most concentrated in the roots. As has usually been the case, this alkaloid-bearing plant, along with the others containing diterpene alkaloids, was initially examined based on folklore and in the hope of finding medicinally valuable palliatives. Thus crude plant material has long been used internally as a febrifuge to lower fever and externally for neuralgia.

The base veatchine (**123**) and related materials are found in the bark of, eg, *G. veatchii* Kellog., and structural elucidation of this complicated and reactive material required massive efforts (91). Its relationship to atisine (**124**) from the roots of the atis plant, *Aconitum heterophyllum* Wall., is clearly seen as that of the well known terpene rearrangement of an *exo*-methylene octa[3.2.1]bicyclic system to that of its [2.2.2] isomer, the remainder of the molecule remaining unchanged. More deep seated rearrangements in the same part of the molecule (ie, a 6–6–5 set of rings with an *exo*-methylene group yielding a 7–5–6 set now incorporating the methylene) generates aconitine (**125**).

The path from squalene (**114**) to the corresponding oxide and thence to lanosterol [*79-63-0*] (**126**), $C_{30}H_{50}O$, cholesterol [*57-88-5*] (**127**), and cycloartenol [*469-38-5*] (**128**) (Fig. 6) has been demonstrated in nonphotosynthetic organisms. It has not yet been demonstrated that there is an obligatory path paralleling the one known for generation of plant sterols despite the obvious structural relationships of, for example, cycloartenol (**128**), $C_{30}H_{50}O$, to cyclobuxine-D (**129**), $C_{25}H_{42}N_2O$. The latter, obtained from the leaves of *Buxus sempervirens* L., has apparently found use medicinally for many disorders, from skin and venereal diseases to treatment of malaria and tuberculosis. In addition to cyclobuxine-D [*2241-90-9*] (**129**) from the Buxaceae, steroidal alkaloids are also found in the Solanaceae, Apocynaceae, and Liliaceae.

(**129**)   (**130**)

The plants of *Solanum* include, among others, the potato (*Solanum tuberosum* L.) and the tomato (*Solanum lycoperisicum* L.). Frequently the plant bases occur as the aglycone portion of a glycoalkaloid bonded to one or more six-carbon sugars. Hydrolysis of the sugar portion and, somewhere along the degradative pathway, excision of the nitrogen (usually via a Hoffmann-type elimination) results in a steroid-like fragment, the analysis of which falls back on the large body of accumulated information about steroids and their degradation products. Solanidine [*80-78-4*] (**130**) is typical of the kind of bases present and has been isolated from a number of *Solanum*.

Interestingly, feeding experiments in *Solanum chacoense* L. (92) demonstrate that cholesterol (**127**), $C_{27}H_{46}O$, can be incorporated into solanidine (**130**),

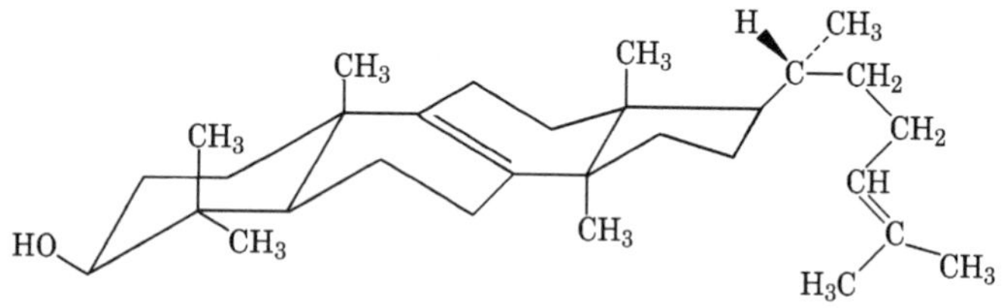

(126)

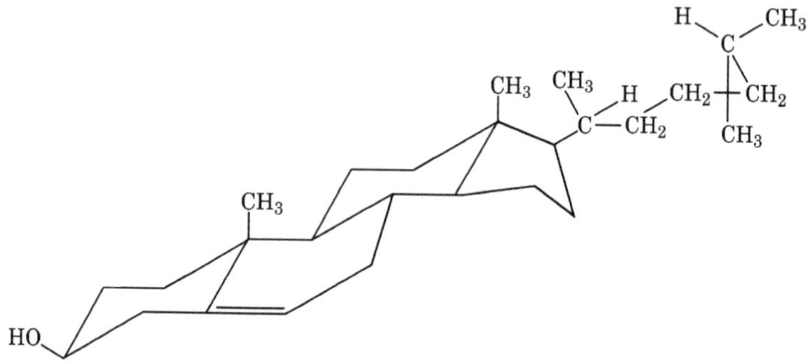

(127)

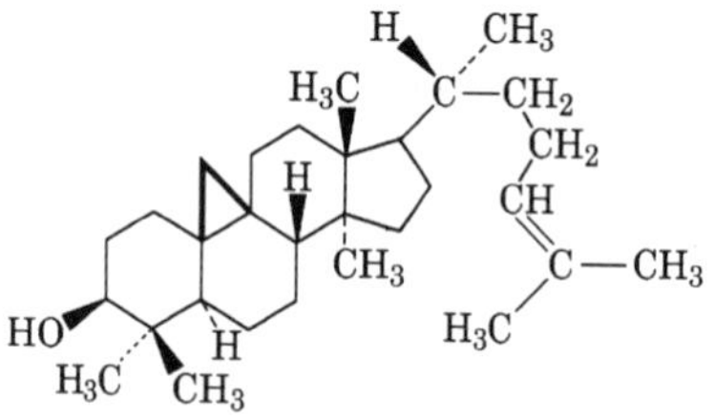

(128)

Fig. 6. Steroids from squalene (114).

$C_{27}H_{43}NO$, but the amount of steroid incorporated is very low and there has been more than one suggestion (93) that the route involves initial degradation of the fed cholesterol to acetate, followed by recreation of the entire skeleton.

In addition to the alkaloids such as cyclobuxine-D (129) and solanidine (130) where the structural similarities to steroids are clear (although it must be remembered that detailed evidence actually linking the compounds is lacking) there are the less obvious (but nonetheless also clearly related) *Veratrum* alkaloids. These compounds, of which protoveratrine A [143-57-7] (131), $C_{41}H_{63}NO_{14}$, obtained from the rhizome of *Veratrum album* L. (Liliaceae), is a typical example, produce dramatic declines in blood pressure on administration and have been received by the medical community as good antihypertensive agents. Generally, however, the dosage must be individualized (slowly) from about 2 mg in 200 mL of saline upward. Because the therapeutically valuable dosage is similar to the toxic dose, and even nonlethal large doses may cause cardiac arrhythmias and peripheral vascular collapse, use of these compounds has frequently been limited to extreme cases where close attention can be accorded the patient.

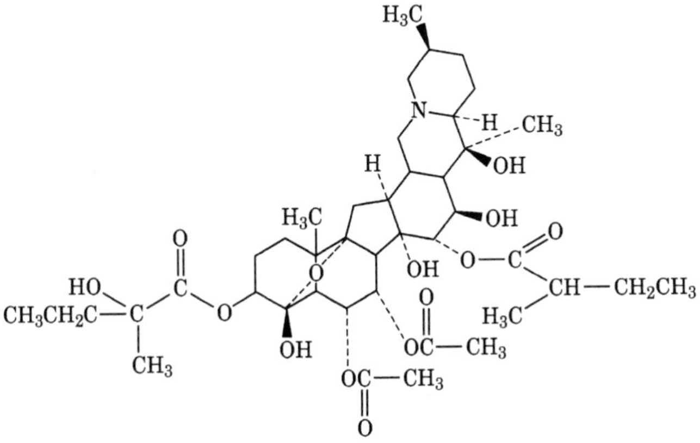

(**131**)

Purine Alkaloids. The purine skeleton is not derived from histidine (**29**), as might be imagined, nor is it derived from any obvious amino acid pregenitor. As has been detailed elsewhere (33,35), the nucleus common to xanthine [*69-89-6*] (**132**), $C_5H_9N_4O_2$, and found in the bases caffeine (**16**), theophylline [*58-55-9*] (**133**, $R_1 = R_2 = CH_3$; $R_3 = H$), and theobromine [*83-67-0*] (**133**, $R_1 = H$; $R_2 = R_3 = CH_3$), $C_7H_8N_4O_2$, is created from small fragments which are attached to a ribosyl unit during synthesis and which can presumably be utilized in a nucleic acid backbone subsequently. All three alkaloids, caffeine, theophylline, and theobromine, occur widely in beverages commonly used worldwide.

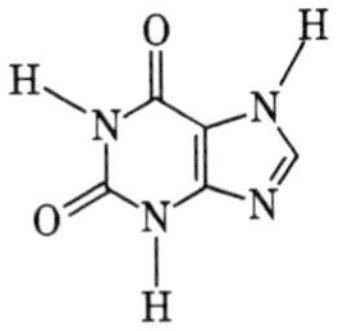

(**132**) 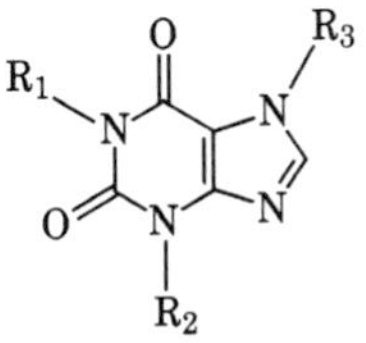   (**133**)

The leaf and leaf buds of *Cammelia sinensis* (L.) O Kuntze and other related plants and most teas contain, depending upon climate, specific variety, time of harvest, etc, somewhat less than 5% caffeine (**16**) and smaller amounts of theophylline (**133**, $R_1 = R_2 = CH_3$; $R_3 = H$) and theobromine (**133**, $R_1 = H$; $R_2 = R_3 = CH_3$). Coffee consists of various members of the genus *Coffea,* although the seeds of *Coffea arabica* L., believed to be indigenous to East Africa, are thought to have been the modern pregenitor of the varieties of coffees currently available and generally cultivated in Indonesia and South America. The seeds contain less than about 3% caffeine which, bound to other agents, is set free during the roasting process. The caffeine may be sublimed from the roast or extracted with a variety of agents, such as methylene chloride, ethyl acetate, or dilute acid (eg, an aqueous solution of carbon dioxide) to generate decaffinated material (94).

Two other commonly found sources of caffeine (**16**) are kola (*Cola*) from the seeds of, for example, *Cola nitida* (Vent.) Schott and Engl., which contains 1–4% of the alkaloid, but little theophylline or theobromine, and cocoa (from the seeds of *Theobroma cacao* L.), which generally contains about 3% theobromine and significantly less caffeine.

All three of these materials are apparently central nervous system (CNS) stimulants. It is believed that for most individuals caffeine causes greater stimulation than does theophylline. Theobromine apparently causes the least stimulation. There is some evidence that caffeine acts on the cortex and reduces drowsiness and fatigue, although habituation can reduce these effects.

Miscellaneous Alkaloids. Shikimic acid (57) is a precursor of anthranilic acid (28) and, in yeasts and *Escherichia coli* (a bacterium), anthranilic acid (*o*-aminobenzoic acid) is known to serve as a precursor of tryptophan (26). A similar but yet unknown path is presumed to operate in higher plants. Nonetheless, anthranilic acid itself is recognized as a precursor to a number of alkaloids. Thus damascenine [*483-64-7*] (134), $C_{10}H_{13}NO_3$, from the seed coats of *Nigella damascena* has been shown (95) to incorporate labeled anthranilic acid when unripe seeds of the plant are incubated with labeled precursor.

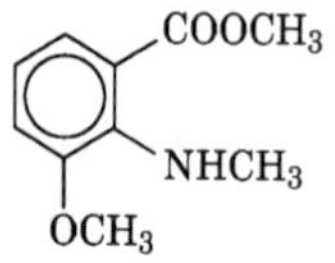

(134)

Similarly, anthranilic acid (28) has been suggested as a reasonable precursor and some early labeling studies have been carried out showing that dictamnine [*484-29-7*] (135), $C_{12}H_9NO_2$, from *Dictamnus albus* and skimminanine [*83-95-4*] (136), $C_{14}H_{13}NO_4$, from *Skimmia japonica* incorporate anthranilic acid in a nonrandom fashion (96,97).

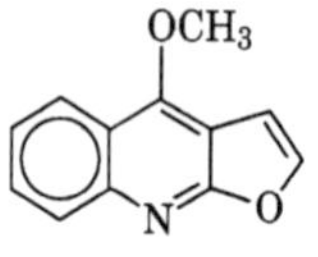

(135) 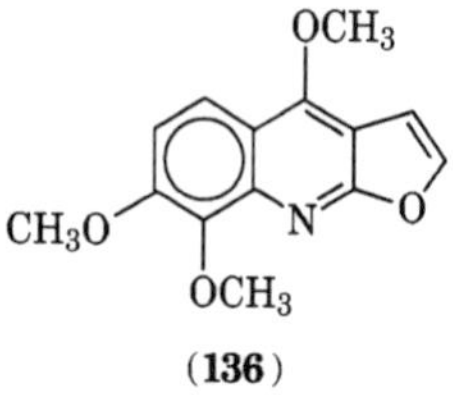   (136)

Securinine [*3610-40-2*] (137), $C_{13}H_{15}NO_2$, is the major alkaloid of *Securinega surroticosa* Rehd. and has been shown to arise from two amino acid fragments, lysine (24) and tyrosine (25).

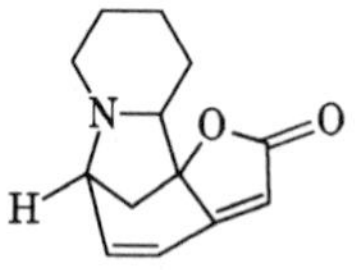

(137)

Reactions at the aromatic nucleus that are quite different from the usual mild condensations and rearrangements which apparently generate the typical alkaloids already discussed must be involved. Securinine (137) is reported to stimulate respiration and increase cardiac output, as do many other alkaloids, but it also appears generally to be less toxic (98).

Coniine (**12**), implicated by Plato in the death of Socrates, is the major toxic constituent of *Conium maculatum* L. (poison hemlock) and, as pointed out earlier, was apparently the first alkaloid to be synthesized. For years it was thought that coniine was derived from lysine (**24**), as were many of its obvious relatives containing reduced piperidine nuclei and a side chain, eg, pelletierine (**46**). However, it is now known (99) that coniine is derived from a polyketooctanoic acid [*7028-40-2*] (**138**), $C_8H_{10}O_5$, or some other similar straight chain analogue.

(**138**)

Economic Aspects

As the twentieth century draws to a close, many alkaloids, such as atropine (**41**) and reserpine (**105**), that have served humanity since early history are being replaced by synthetic materials. Others, such as the *Vinca* bases, eg, vincristine (leurocristine, **22**, R = CHO) remain as powerful medical tools. Replacement of naturally occurring alkaloids is desireable in order to maintain and augment favorable properties while eliminating undesireable properties and effects. Through strides in biochemical research, especially structure–reactivity studies and design of model compounds, synthetic materials have been developed. These new materials, while occasionally related to alkaloids, either because they are derived from alkaloids or because their structures are similar, are not naturally occurring and thus are not alkaloids. However, they are generally much more specific in their action and since they are protected by patents, much more expensive.

There are four broad classes of alkaloids whose general economic aspects are important: (*1*) the opiates such as morphine and codeine (**2**, R = H and R = CH$_3$, respectively); (*2*) cocaine (**11**) (both licit and illicit); (*3*) caffeine (**16**) and related bases in coffee and tea, and (*4*) the tobacco alkaloids such as nicotine (**21**).

The Opiates. The International Narcotics Control Board—Vienna, tracks the licit production of narcotic drugs and annually estimates world requirements for the United Nations. Their most recent publication (100) points out that more than 95% of the opium for licit medical and scientific purposes is produced by India and, in a declining trend, only about 600 t was utilized in 1988. This trend appears to be due to the fact that the United States, the largest user of opium for alkaloid extraction, reduced the amount of opium being imported from about 440 t in 1986 to 249 t in 1987 and 224 t in 1988. The United States used about 48 t of morphine (**2**, R = H) in 1988, most (about 90%) being converted to codeine (**2**, R = CH$_3$) and the remainder being used for oral administration to the terminally ill (about 2 t) and for conversion to other materials of minor commercial import which, while clearly alkaloid-derived, are not naturally occurring.

Cocaine. Production of cocaine [*50-36-2*] (**11**) for licit purposes (100) takes place in Bolivia and Peru. In 1988 the former exported 204 t and the latter 47 t of coca leaves into the United States. The average total licit production of cocaine as a by-product in the extraction of flavoring agents from coca leaves was reported (100) to be only 425 kg in the United States in 1988. This must be weighed against the estimates by the U.S. Drug Enforcement Administration that the value of illegal annual exports of coca from Bolivia (1987) is $2 billion (101). It has been suggested (101) that this dollar sum is three times as much as the earnings from legal exports of tin, coffee, etc and that it is a significant support to the Bolivian economy.

   Caffeine.   About 3% by weight of the roasted coffee bean is caffeine (**16**). The second U.S. Department of Agriculture world coffee crop estimate for 1988–1989 was 4.24×10^9 kg (93.3 million 100-lb bags) (102). World coffee consumption was predicted to rise in the foreseeable future at the rate of 1–2% per year and thus the total amount of caffeine and related alkaloids ingested from this source can also be expected to increase. Caffeine and related bases (eg, theophylline) are also found in various teas but, because most of the major producers (India, China, etc) export relatively little of their crops and keep most for domestic consumption, accurate figures on year-to-year production are more difficult to obtain. Nevertheless, these crops are of significant economic import (103).

   Tobacco.   Tobacco is the principal source of the alkaloid nicotine (**21**), which, it is claimed, is at least partially responsible for the addicting properties of tobacco. A study (104) concerning the world market outlook for this crop claims that in developed countries individual annual consumption of tobacco will decrease from 2.48 to 2.21 kg per adult between 1986 and 2000 while in developing countries, it will increase from 1.59 to 1.75 kg per adult over the same time span. Given expected trends in world population growth, this suggests that total production of leaf tobacco will increase by about 2% annually during the same period. A significant impact on the foreign exchange earnings for developing countries which fail either to produce tobacco for export or to satisfy domestic demand is expected.

BIBLIOGRAPHY

"Alkaloids, Manufacture" in *ECT* 1st ed., Vol. 1, pp. 507–516, by N. Applezweig, Hygrade Laboratories, Inc.; "Alkaloids, History, Preparation, and Use" in *ECT* 2nd ed., Vol. 1, pp. 778–809, by G. H. Svoboda, Eli Lily and Co.; "Alkaloids, Survey" in *ECT* 2nd ed., Vol. 1, pp. 758–778, by W. I. Taylor, Ciba Pharmaceutical Co.; "Alkaloids" in *ECT* 3rd ed., Vol. 1, pp. 883–943, by Geoffrey A. Cordell, College of Pharmacy, University of Illinois.

   1. T. I. Williams, *Drugs from Plants,* Sigma, London, 1947, p. 87; L. S. Goodman and A. Gilman, *The Pharmacological Basis of Therapeutics: A Textbook of Pharmacology,* 3rd ed., Macmillan, New York, 1965, pp. 247–266.
   2. C. Derosne, *Ann. Chim. (Paris)* **45,** 257 (1803).
   3. F. W. Serturner, *Ann. Chim. Phys.* (**2**) 5, 21 (1817).
   4. H.-G. Boit, *Ergebnisse der Alkaloid-Chemie Bis 1960,* Akademie-Verlag, Berlin, 1961.
   5. J. S. Glasby, *Encyclopedia of the Alkaloids,* Vols. 1 and 2, Plenum Press, New York, 1975.
   6. S. W. Pelletier, *Alkaloids. Chemical and Biological Perspectives,* Vol. 1, John Wiley & Sons, Inc., New York, 1983, pp. 25–27.
   7. I. W. Southon and J. Buckingham, eds., *Dictionary of Alkaloids,* Chapman and Hall, New York, 1989.

8. R. H. F. Manske and H. L. Holmes, eds., *The Alkaloids: Chemistry and Physiology,* Vol. 1, Academic Press, Inc., New York, 1950. This series gives a detailed exposition of the chemistry and pharmacology of the alkaloids, by structural class. Vol. 36, A. Brossi, ed., was published in 1989.

9. E. Hasse, *Biblotheca Zoologica* **8,** 1 (1892).

10. J. v. Euw, T. Reichstein, and M. Rothschild, *Israel J. Chem.* **6,** 659 (1968).

11. J. A. Edgar, P. A. Cockrum, and J. L. Frahn, *Experientia* **32,** 1535 (1976).

12. J. Meinwald, *Ann. N.Y. Acad. Sci.* **471,** 197 (1986).

13. T. Eisner, M. Goetz, D. Aneshansley, G. Ferstandig-Arnold, and J. Meinwald, *Experientia* **42,** 204 (1986).

14. J. Smolanoff and co-workers, *Science* **188,** 734 (1975).

15. J. J. Tufariello, H. Meckler, and K. P. A. Senaratne, *J. Am. Chem. Soc.* **106,** 7979 (1984).

16. P. J. Pelleiter and J. B. Caventou, *Ann. Chim. Phys.* **8,** 323 (1818); **10,** 142 (1819).

17. J. Geiger, *Berzelius' Jahresber.* **12,** 220 (1870).

18. A. Ladenburg, *Ber.* **19,** 439 (1886).

19. A. F. Peerdeman, *Acta Crystallogr.* **9,** 824 (1956).

20. R. B. Woodward and co-workers, *Tetrahedron* **19,** 247 (1963).

21. L. R. Caporael, *Science* **192,** 21 (1976).

22. H. Weiland, F. Konz, and K. Mittasch, *Ann.* **513,** 1 (1934).

23. G. J. Cardinale and co-workers, *Life Sci.* **40,** 301 (1987).

24. O. Dragendorff, *Zeitschr. Anal. Chem.* **137** (1866); *The Merck Index,* 5th ed., Merck & Co., Rahway, N.J., 1940, p. 687.

25. H. Mayer, *Am. J. Pharm.* **35,** 20 (1863); *The Merck Index,* 5th ed., Merck & Co., Rahway, N.J., 1940, p. 883.

26. N. R. Farnsworth, *J. Pharm. Sci.* **55,** 225 (1966).

27. R. W. Kondrat, R. G. Cooks, and J. L. McLaughlin, *Science* **199,** 978 (1978).

28. G. Zweig and J. Sherma, eds., *CRC Handbook of Chromatography,* CRC Press, Cleveland, Ohio, 1972.

29. E. Stahl, *Dunnschicht-Chromatographie,* Springer, Berlin, 1969.

30. R. Bentley and I. M. Campbell in M. Florkin and E. H. Stotz, eds., *Comprehensive Biochemistry,* Vol. 20, Elsevier, New York, 1968, pp. 415ff.

31. E. Winterstein and G. Trier, *Die Alkaloide,* Berntrager, Berlin, 1910.

32. Sir Robert Robinson, *The Structural Relations of Natural Products,* Oxford University Press, Oxford, 1955.

33. I. D. Spenser in M. Florkin and E. H. Stotz, eds., *Comprehensive Biochemistry,* Vol. 20, Elsevier, New York, 1968, pp. 231ff.

34. G. A. Cordell, *Introduction to Alkaloids: A Biogenetic Approach,* Wiley-Interscience, New York, 1981.

35. D. R. Dalton, *The Alkaloids—A Biogenetic Approach,* Marcel Dekker, New York, 1979.

36. R. W. Herbert in S. W. Pelletier, ed., *Alkaloids, Chemical and Physiological Perspectives,* Vol. 3, Wiley-Interscience, New York, 1985.

37. E. Leete and S. H. Kim, *J. Am. Chem. Soc.* **110,** 2976 (1988).

38. G. Grue-Sorensen and I. D. Spenser, *J. Am. Chem. Soc.* **105,** 7401 (1983).

39. E. Leistner and I. D. Spenser, *J. Am. Chem. Soc.* **95,** 4715 (1973); T. Hemscheidt and I. D. Spenser, *J. Am. Chem. Soc.* **112,** 6360 (1990).

40. D. G. O'Donovan, D. J. Long, E. Forde, and P. Geary, *J. Chem. Soc. Perkin Trans. 1,* 415 (1975).

41. W. M. Golebiewski and I. D. Spenser, *J. Am. Chem. Soc.* **106,** 7925 (1984).

42. E. Leete and Y.-Y. Liu, *Phytochemistry* **12,** 593 (1973).

43. S. D. Copley and J. R. Knowles, *J. Am. Chem. Soc.* **109,** 5008 (1987); W. J. Guilford, S. D. Copley, and J. R. Knowles, *J. Am. Chem. Soc.* **109,** 5013 (1987).

44. R. E. Schultes and A. Hofmann, *Plants of the Gods,* McGraw-Hill Book Co., Inc., New York, 1979.

45. W. La Barre, D. P. McAllister, J. S. Slotkin, O. C. Stewart, and S. Tax, *Science* **114,** 582 (1952).

46. W. I. Taylor and A. R. Battersby, eds., *Oxidative Coupling of Phenols,* Marcel Dekker, New York, 1967.

47. Ref. 35, pp. 197ff.

48. A. W. Gerrard, *Pharm. J.* **8,** 214 (1877).

49. O. Moller, E. M. Steinberg, and K. Torssell, *Acta Chem. Scand.* **B32,** 98 (1978).

50. J. Bolssier, G. Combes, and J. Pagny, *Ann. Pharm. Fr.* **18,** 888 (1960).

51. T. Kametani and co-workers, *J. Chem. Soc. C,* 1043 (1971).

52. H. M. Fales, J. Mann, and S. H. Mudd, *J. Am. Chem. Soc.* **85,** 2025 (1963).

53. W. C. Wildman and D. T. Bailey, *J. Am. Chem. Soc.* **91,** 150 (1969).

54. W. C. Wildman in R. H. F. Manske, ed., *The Alkaloids,* Vol. 11, Academic Press, New York, 1968, pp. 308ff.

55. E. H. Herman and D. P. Chadwick, *Pharmacology* **12,** 97 (1974); E. J. Gralla, G. L. Coleman, and A. M. Jonas, *Cancer Chemother. Rep. Part 3* **5,** 79 (1974).

56. D. H. R. Barton and T. Cohen, *Festschrift A. Stoll,* Birkhauser Verlag, Basel, 1957, p. 117.

57. M. A. Barbier, *Ann. Pharm.* **5,** 121 (1947).

58. M. Gates and G. Tschudi, *J. Am. Chem. Soc.* **78,** 1380 (1956).

59. H. I. Parker, G. Blaschke, and H. Rapoport, *J. Am. Chem. Soc.* **94,** 1276 (1972).

60. D. H. R. Barton, C. J. Potter, and D. A. Widdowson, *J. Chem. Soc. Perkin Trans. 1,* 346 (1974); D. H. R. Barton, R. D. Bracho, C. J. Potter, and D. A. Widdowson, *J. Chem. Soc. Perkin Trans. 1* 2278 (1974).

61. J. E. Saxton, ed., *Indoles, Part Four, The Monoterpenoid Indole Alkaloids,* Wiley-Interscience, New York, 1983.

62. M. M. Rapport, A. A. Green, and I. H. Page, *J. Biol. Chem.* **176,** 1243 (1948).

63. K. Stolle and D. Groger, *Arch. Pharm. (Weinheim)* **301,** 561 (1968).

64. A. Hofmann, *Botanical Museum Leaflets,* Vol. 20, Harvard University, Cambridge, Mass., 1963, p. 194.

65. Fr. Add. 91,948 (Aug. 30, 1968), J. Rutschmann and H. Kobel (Sandoz Ltd.).

66. W. A. Jacobs and L. C. Craig, *Science* **82,** 16 (1935).

67. A. R. Battersby, A. R. Burnett, and P. G. Parsons, *Chem. Commun.,* 1280 (1968).

68. I. Kompis, M. Hesse, and H. Schmid, *Lloydia* **34,** 269 (1971) (gives a much more elaborate classification scheme).

69. Ref. 35, pp. 443ff.

70. G. Holmberg and S. Gershon, *Psychopharmacologia* **2,** 93 (1961); M. L. Brown, S. Gershon, W. J. Lang, and B. Korol, *Arch. Intern. Pharmacodyn.* **160,** 407 (1966).

71. B. Armstrong, N. Stevens, and R. Doll, *Lancet* **2,** 672 (1974).

72. J. L. Hartwell, *Cancer Treatment Rep.* **60,** 1031 (1976).

73. R. B. Woodward, F. E. Bader, H. Bickel, A. J. Frey, and R. W. Kierstead, *Tetrahedron* **2,** 1 (1958).

74. M. L. Chatterjee and M. S. De, *Bull. Calcutta School Trop. Med.* **5,** 173 (1957); *Chem. Abstr.* **52,** 8356a (1958).

75. J. Schwyzer, *Die Fabrikation Pharmazeutischer and Chemisch, Technischer Produkte,* Springer-Verlag, Berlin, 1931.

76. W. I. Taylor and N. R. Farnsworth, eds., *The Catharanthus Alkaloids, Botany, Chemistry, Pharmacology and Clinical Uses,* Marcel Dekker, New York, 1973; W. I. Taylor and N. R. Farnsworth, eds., *The Vinca Alkaloids, Botany, Chemistry and Pharmacology,* Marcel Dekker, New York, 1973.

77. G. Buchi, P. Kulsa, K. Ogasawara, and R. L. Rosati, *J. Am. Chem. Soc.* **92,** 999 (1970) and references therein.

78. A. Ravina, *Presse Med.* **74,** 525 (1978).
79. G. A. Cordell in ref. 61, pp. 539ff.
80. W. T. Stearn, *Lloydia* **29,** 196 (1966).
81. W. A. Creasey in F. Hahn, ed., *Antibiotics,* Vol. 5, Springer-Verlag, Berlin, 1979, p. 414.
82. E. K. Rowinsky, L. A. Cazenave, and R. C. Donehower, *J. Nat. Cancer Inst.* **82,** 1247 (1990) (deals with antimicrotubule agents, albeit with a nonalkaloidal agent).
83. J. P. Kutney and co-workers, *J. Am. Chem. Soc.* **97,** 5013 (1975); M. E. Kuehne and T. C. Zebovitz, *J. Org. Chem.* **52,** 4331 (1987); M. E. Kuehne, T. C. Zebovitz, W. G. Bornmann, and I. Marko, *J. Org. Chem.* **52,** 4340 (1987).
84. G. Popjak and J. W. Cornforth, *Biochem. J.* **101,** 553 (1966); J. W. Cornforth, R. H. Cornforth, A. Pelter, M. G. Horning, and G. Popjak, *Tetrahedron* **5,** 311 (1959); R. B. Clayton in T. W. Goodwing, ed., *Aspects of Terpenoid Chemistry and Biochemistry,* Academic Press, New York, 1971, pp. 1ff.
85. T. Sakan, *Tampakushitsu Kakusan Koso* **12,** 2 (1967); *Chem. Abstr.* **73,** 42351c (1970).
86. G. L. Gatti and M. Marotta, *Ann. Ist. Super. Sanita* **2,** 29 (1966); *Chem. Abstr.* **65,** 14293e (1966).
87. W. C. Wildman, J. LeMen, and K. Wiesner in W. I. Taylor and A. R. Battersby, eds., *Cyclopentanoid Terpene Derivatives,* Marcel Dekker, New York, 1969, pp. 239ff.
88. O. E. Edwards in W. I. Taylor and A. R. Battersby, eds., *Cyclopentanoid Terpene Derivatives,* Marcel Dekker, New York, 1969, pp. 357ff.
89. T. T. Tchen and K. Block, *J. Am. Chem. Soc.* **77,** 6085 (1955); R. B. Clayton and K. Block, *J. Biol. Chem.* **218,** 319 (1956); L. J. Goad, *Symp. Biochem. Soc.* **29,** 45 (1970).
90. G. A. Swan, *An Introduction to the Alkaloids,* John Wiley & Sons, Inc., New York, 1967, p. 274.
91. S. W. Pelletier, N. V. Mody, and H. K. Desai, *J. Org. Chem.* **46,** 1840 (1981).
92. H. Ripperger, W. Mortiz, and K. Schreiber, *Phytochemistry* **10,** 2699 (1971).
93. S. J. Jadav, D. K. Salunkhe, R. E. Wyse, and R. R. Dalvi, *J. Food Sci.* **38,** 453 (1973).
94. U.S. Pat. 3,108,876 (Oct. 29, 1963), H. H. Turken and T. P. Daley (to Duncan Coffee Co.); U.S. Pat. 3,361,571 (Jan. 2, 1968), L. Nutting and G. S. Chong (to Hills Bros. Coffee, Inc.).
95. E. J. Miller, S. R. Pinnell, G. R. Martin, and E. Schiffman, *Biochem. Biophys. Res. Commun.* **26,** 132 (1967).
96. E. Monkovic, I. D. Spenser, and A. O. Plunkett, *Can. J. Chem.* **45,** 1935 (1967).
97. M. Matsuo and Y. Kasida, *Chem. Pharm. Bull. (Tokyo)* **14,** 1108 (1966).
98. V. A. Snieckus, *Alkaloids (N.Y.)* **14,** 425 (1973).
99. E. Leete, *Accounts Chem. Res.* **4,** 100 (1971).
100. International Narcotics Control Board—Vienna, *Narcotic Drugs, Estimated World Requirements for 1990, Statistics for 1988,* United Nations Publ. E/F/S.89.XI.3, New York, 1989, pp. 33ff.
101. L. Mahnke, *Aachener Geographische Arbeit.* **19,** 137 (1987).
102. USDA, *Foreign Agricultural Service Circular Series* No. FCOF 3-88, Foreign Agricultural Service, U.S. Department of Agriculture, 1988.
103. *Economist (UK),* **296,** 57 (1985).
104. *Tobacco, Supply, Demand and Trade Projections, 1995 and 2000, FAO Economic and Social Development Paper 1990,* No. 86, FAO.

General References

References 7 and 8 are general references.

DAVID R. DALTON
Temple University